Handbook of Solvency for Actuaries and Risk Managers

Theory and Practice

CHAPMAN & HALL/CRC FINANCE SERIES

Series Editor

Michael K. Ong

Stuart School of Business
Illinois Institute of Technology
Chicago, Illinois, U. S. A.

Aims and Scopes

As the vast field of finance continues to rapidly expand, it becomes increasingly important to present the latest research and applications to academics, practitioners, and students in the field.

An active and timely forum for both traditional and modern developments in the financial sector, this finance series aims to promote the whole spectrum of traditional and classic disciplines in banking and money, general finance and investments (economics, econometrics, corporate finance and valuation, treasury management, and asset and liability management), mergers and acquisitions, insurance, tax and accounting, and compliance and regulatory issues. The series also captures new and modern developments in risk management (market risk, credit risk, operational risk, capital attribution, and liquidity risk), behavioral finance, trading and financial markets innovations, financial engineering, alternative investments and the hedge funds industry, and financial crisis management.

The series will consider a broad range of textbooks, reference works, and handbooks that appeal to academics, practitioners, and students. The inclusion of numerical code and concrete real-world case studies is highly encouraged.

Published Titles

Decision Options®: The Art and Science of Making Decisions, **Gill Eapen**

Emerging Markets: Performance, Analysis, and Innovation, **Greg N. Gregoriou**

Introduction to Financial Models for Management and Planning, **James R. Morris and John P. Daley**

Pension Fund Risk Management: Financial and Actuarial Modeling, **Marco Micocci, Greg N. Gregoriou, and Giovanni Batista Masala**

Stock Market Volatility, **Greg N. Gregoriou**

Portfolio Optimization, **Michael J. Best**

Operational Risk Modelling and Management, **Claudio Franzetti**

Handbook of Solvency for Actuaries and Risk Managers: Theory and Pracice, **Arne Sandström**

Proposals for the series should be submitted to the series editor above or directly to:
CRC Press, Taylor & Francis Group
4th, Floor, Albert House
1-4 Singer Street
London EC2A 4BQ
UK

CHAPMAN & HALL/CRC FINANCE SERIES

Handbook of Solvency for Actuaries and Risk Managers

Theory and Practice

Arne Sandström

CRC Press
Taylor & Francis Group
Boca Raton London New York

CRC Press is an imprint of the
Taylor & Francis Group, an **informa** business
A CHAPMAN & HALL BOOK

Chapman & Hall/CRC
Taylor & Francis Group
6000 Broken Sound Parkway NW, Suite 300
Boca Raton, FL 33487-2742

Printed in the United States of America on acid-free paper
10 9 8 7 6 5 4 3 2

International Standard Book Number: 978-1-4398-2130-5 (Hardback)

Library of Congress Cataloging-in-Publication Data

Sandström, Arne.
Handbook of solvency for actuaries and risk managers : theory and practice / Arne Sandström.
p. cm. -- (Chapman & Hall/CRC finance series)
Includes bibliographical references and index.
ISBN 978-1-4398-2130-5 (hbk. : alk. paper)
1. Risk (Insurance) 2. Risk (Insurance)--European Union countries. 3. Asset-liability management. 4. Asset-liability management--European Union countries. 5. Risk management. 6. Risk management--European Union countries. I. Title.

HG8054.5.S26 2011
368.001--dc22 2010042931

Visit the Taylor & Francis Web site at
http://www.taylorandfrancis.com

and the CRC Press Web site at
http://www.crcpress.com

Contents

Preface

WHEN I FINALIZED my first book on solvency in 2005,* I felt that it was not enough. The book, which in many ways was written as an egoistic action—I needed a platform to stand on in the work on solvency and especially on the European Solvency II project—was published during a period when the solvency discussion was intense, both within the European Union and worldwide. In this, my new book, I want to write about the things it lacked. For example, I wanted a stringent and clear definition of solvency. I also thought that the historical review could be made broader and that the theory behind the calculation of the capital requirement could be made more comprehensive.

The goal of this handbook on solvency, and especially on the European Solvency II project, is to be just a *handbook*. The focus has been on the *valuation of assets and liabilities* and *calculation of capital requirement*, and also on the *calculation of the standard formula* within the European Solvency II project.

While I was planning a short course on solvency in Brussels in February 2009 I felt a need to include the development and calibration of the different submodules for the standard formula in my book. This after my friend Professor Jan Dhaene, KU Leuven, had asked me to include that in my solvency course.

The handbook does not include any main discussion on (partial) internal models as that would be the subject for another book. The main audience is professional actuaries and risk managers. I hope that it can be used in the education of actuaries and risk managers, at least as a start and as a handbook.

I am grateful to many friends and colleagues for their valuable comments on different parts of the book. I especially thank Peter ter Berg, Yvette Chrissantonis, Jan Dhaene, Boualem Djehiche, Valérie Kupferman, Mariano Selvaggi, and Mario Wüthrich. I also thank my present and former colleagues Ellen Bramness Arvidsson, Karin Chenon, Lena Friman-Blomgren, Jimmy Hollén, Christian Salmeron, Ellinor Samuelsson, and Patric Thomsson for their valuable comments.

* *Solvency. Models, Assessment and Regulation*. Chapman & Hall, Boca Raton, FL. ISBN: 1-58488-554-8.

Reader's Guide

THE HANDBOOK CAN BE divided into two main sections:

- General ideas about solvency (Parts A, B, and C)
- The European Solvency II project (Parts D, E, and F)

The first section discusses the solvency concept, the historical development, and its place as a part in an enterprise risk management approach. Further, there is a more general discussion on valuation, investment, and own-capital together with modeling and measuring. The last part, Part C, discusses dependence, risk measures, and capital requirements. Subrisks and aggregation are also important parts. The main risks—market, credit, operational, liquidity, and underwriting risks—are discussed in general terms.

The second section, devoted to the European Solvency II project, starts with its general ideas, valuation, investments, and own-funds (Part D). The second part of this section, Part E, is devoted to the standard formula framework. These two parts are based on CEIOPS' final advice. All calibrations done earlier in different quantitative impact studies (QIS), together with the political progress of the project, are given in the appendices (Part F).

The following table shows the structure of the book.

		European Solvency II	
	General Solvency Issues	**CEIOPS' Final Advice**	**Background and Calibration**
General solvency and risk management issues	*Part A* Chapters 1 through 5		
Solvency II—general	Section 2.3	*Part D* Chapter 22	*Part F* Appendices C+D+E
Valuation and investment	*Part B* Chapters 6 through 11	*Part D* Chapters 23 through 25	*Part F* Appendices F+G
Modeling and measuring → Capital requirements	*Part C* Chapters 12 through 21 *Part F* Appendices A+B	*Part E* Chapters 26 through 34	*Part F* Appendices H–P

Web Site Information

For the general part of the book and for general discussions on solvency and accounting I would like to recommend the following Web sites:

IAIS—International Association of Insurance Supervisors:
http://www.iaisweb.org/

IAA—International Actuarial Association:
http://www.actuaries.org

BIS—Bank for International Settlements:
http://www.bis.org

IASB–International Acounting Standards Board:
http://www.iasb.org

For the Solvency II part of the handbook I have mainly used different sources that can be found on the Internet, such as documents from the European Commission, EIOPC, and CEIOPS. Their Web sites are

European Commission & EIOPC—European Insurance and Occupational Pensions Committee:
http://ec.europa.eu/internal_market/insurance/index_en.htm

CEIOPS—Committee of European Insurance and Occupational Pensions Supervisors:
http://www.ceiops.org

There are also other Web sites that the interested reader can visit:

CEA—The European Insurance and Reinsurance Federation:
http://www.cea.eu

CRO Forum—Chief Risk Officers:
http://www.croforum.org

CFO Forum—European Insurance Chief Financial Officers:
http://www.cfoforum.nl

AMICE—Association of Mutual Insurers and Insurance Cooperatives in Europe:
http://www.amice-eu.org

GC—Groupe Consultatif Actuariel Europeen:
http://www.gcactuaries.org

Future Information

My intention was to include CEIOPS' proposal for final advice that was published on mid-November 2009 and at the end of January 2010.

The final advice that will be adopted by the European Commission and accepted by the European Parliament is not included. Level 3 measures, such as standards and guidance, are also not included.

However, I will update any important information regarding the Solvency II development and my book, on my Web site, www.SolvencyII.nu.

Any misprints and corrections of the book will also be published on my Web site.

Arne Sandström
www.SolvencyII.nu

Abbreviations

3L3	Three Level 3 Committees, that is, CEBS, CEIOPS, and CESR
A	Assets
AAA	American Academy of Actuaries, United States
AB	Annualized amount of benefits (Solvency II)
ABI	Association of British Insurers, United Kingdom.
ABS	Asset backed securities
ac	Index for epidemic/accumulation risk (Solvency II)
ACAM	L'Autorité de Contrôle des Assurances et des Mutuelles (the French Supervisory Authority for mutuals)
AccY	Accounting year
ACME	The Association of European Cooperative and Mutual Insurers (now AMICE)
Adj	Adjustment for risk-absorbing effect (Solvency II)
AE	Actual expenses
Af	Average annuity factor (Solvency II)
AFIR	Actuarial Approach for FInancial Risks; AFIR was founded in 1986 as a Section of the International Actuarial Association (IAA)
AGOF	Available group own funds
AIE	Actual inheritance earnings
AISAM	The International Association of Mutual Insurance Companies (now AMICE)
A/L	Asset–Liability (matching)
ALM	Asset–Liability Management
AMA	Advanced Measurement Approaches
AMCR	Absolute minimum capital requirement (Solvency II)
AMICE	The Association of Mutual Insurers and Insurance Cooperatives in Europe
AN	Index for annuities, together with WC (Solvency II)
AO	Index for accident and others insurance risk (Solvency II)
AOF	Ancillary own funds
AP	Published assets (usually book values)

APRA	Australian Prudential Regulation Authority
APT	Arbitrage Pricing Theory
AR	Autoregressive
	Annual replenishment requirement for current annuity table
	Index for epidemic/accumulation risk (Solvency II)
ARCH	Autoregressive Conditional Heteroskedasticity
ARM	Asset risk margin
ARP	The Argentine peso
ART	Alternative risk transfer (considered as a type of reinsurance)
ASB	Actuarial Standards Board, United States
ASM	Available solvency margin (own funds under Solvency II; (regulatory) available capital to cover capital requirements)
ASRF	Asymptotic single-risk factor (model)
ASTIN	Actuarial STudies In Nonlife insurance; ASTIN was created in 1957 as the first section of the International Actuarial Association (IAA)
ATM	Index for Accumulation/trend/modification risks
AUD	The Australian dollar
AY	Accident year
bbl	Barrels
BE	Best estimate
BIA	Basic Indicator Approach (Basel II method for credit risk)
BIPIT	Bivariate probability integral transformation
BIS	Bank for International Settlements, Basel, Switzerland
BK	Black–Karasinski (model)
BOF	Basic own funds
BS	Balance sheet
BSCR	Basic solvency capital requirement (Basic SCR)
C	Capital requirement
	Corporate bond
	Document-classification within EU: Documents relating to official instruments for which the Commission has sole responsibility. Some are transmitted to the Council or Parliament for information
CAPM	Capital–Asset Pricing Model
CaR	Capital at risk
CAS	Casualty of Actuarial Society
CAT/Cat	Catastrophe (risk)
	Index for Cat-risk capital charge (Solvency II)
CBNI	Covered but not incurred
CCI	The comprehensive industrial concentration index
CCR	Current capital-at-risk
cd	Credit derivative
cdf	Cumulative distribution function
CDR	Claims development result
CDS	Credit Default Swaps

CE	Current estimate
	Claims expenditure (Solvency II)
CEA	"Comité Européen des Assurances." The CEA is the European insurance and reinsurance federation
	Current estimate of assets
CEBS	Committee of European Banking Supervisors
CEL	Current estimate of liabilities
CEO	Chief Executive Officer
CEIOPS	Committee of European Insurance and Occupational Pensions Supervisors
CESR	Committee of European Securities Regulators
CET	Central European Time (Greenwich + 1 hour)
CF	Cash flow
CfA	Call for Advice (i.e., calls by the European Commission for CEIOPS' comments on certain aspects of Solvency II. The calls have been grouped in three waves.)
CFO	Chief financial officers
CFT	Cash-flow testing
CHF	The Swiss Franc
CI	Concentration index
CIA	Canadian Institute of Actuaries
CIR	Cox–Ingersoll–Ross (interest rate model)
CLN	Credit-linked notes
CLRM	Complementary loss ratio method
CO	Index for concentration risk (Solvency II)
CoC	Cost-of-capital (rate)
CoCM	Cost-of-capital margin
COM	(European) Commission (Services)
	Document classification within EU: Proposed legislation and other Commission communications to the Council and/or the other institutions, and their preparatory papers. Commission documents for the other institutions (legislative proposals, communications, reports, etc.).
COMP	DG COMP: EC's DG for Competition (European Commission)
	COMP Council (European Council)
COMPASS	Task force of CEIOPS
Contr	Contribution
Coreper	Committee of permanent representatives meeting (Comité des représentants permanents)
Coreper I	Consisting of the deputy head
Coreper II	Consisting of the heads, that is, the "EU Ambassadors"
COSO	Commission of Sponsoring Organizations (of the Treadway Commission)
CP	Consultation paper
	Concentration penalty
CPI	Consumer Prize Index

CR	Capital requirement Capital resources Index for credit risk (Solvency II) Index for currency risk, together with MR (Solvency II) Combined ratio
CRA	Credit Rating Agencies
CRC	Current capital requirement
CRD	Capital Requirements Directive (within EU; Basel II implementation)
CRO	Chief Risk Officer
CS	Credit spread
CSE	Credit spread equivalent (Solvency II)
CSFB	Credit Suisse First Boston (bank)
CSR	Credit spread risk
CSWG	Commission Solvency Working Group (within EU)
CTE	Conditional tail expectation
CTP	Current technical/mathematical provisions
CVaR	Conditional Value-at-Risk
CWG	Council Working Group (European Council)
CY	Current year
D&O	Directors and officers
DB	Deutsche Bundesbank
DCAT	Dynamic capital adequacy testing
DD	Distance to default
DDM	Dividend Discount Model
DFA	Dynamic financial analysis (a simulation technique based on integrated modeling to analyze the overall financial and risk situation of an insurance company over a given time period, thus supporting integrated corporate control taking account of all risk factors)
DFAC	Dynamic financial analysis committee, CAS U.S.
DG	Directorate General (within the European Commission)
DI	*Dagens Industri*, Swedish daily financial paper
DKK	The Danish krona
DP	Discussion Paper
dpf	Discretionary participating features
DPT	Default point
DR	Index for default risk (Solvency II) Index for disability risk together with LR (Solvency II)
DRMC	Dynamic Risk Modeling Committee, CAS U.S.
DSD	Deeply subordinated debt
DSOP	Draft Statement of Principles
DST	Dynamic solvency testing
DT	Index for deferred taxes (Solvency II)

Dur	Duration
DUS	Dödlighetsundersökning, Swedish insurance mortality study
DY	Development year
E&O	Events and occasions
E	Expenses (Solvency II)
EAD	Exposure at default
Earn	Earned premiums (Solvency II)
EBA	European Banking Authority
EC	European Commission
ECB	European Central Bank
ECFIN	DG ECFIN: EC's DG for Economic and Financial Affairs (European Commission)
ECOFIN	Economic and Financial Affairs Council (European Council of Finance Ministers)
ECON	Economic and Monetary Affairs Committee (European Parliament)
ED	Exposure Draft
EDF	Expected default frequency
EDHEC	École De Hautes Études Commerciales du Nord is a French business school founded in 1906
EEA	European Economic Area
EEC	European Economic Community
EEK	The Estonian kroon
EEV	European Embedded Value
EH	Expected health risk capital charge (Solvency II)
EIOPA	European Insurance and Occupational Pensions Authority
EIOPC	European Insurance and Occupational Pensions Committee
EL	Expected loss
EMPL	Employment and Social Affairs Committee (European Parliament)
	DG EMPL: EC's DG for Employment, Social Affairs and Equal Opportunities (European Commission)
	EMPL Council (European Council)
ENTR	DG ENTR: EC's DG for Enterprise and Industry (European Commission)
EP	European Parliament
	Earned premiums (Solvency II)
EPD	Expected policyholder deficit
EQ	Equity
EQU	Market value of overall equity and UCITS exposure (Solvency II)
ER	Index for equity risk, together with MR (Solvency II)
	Index for endowment risk
	Index for expense risk, together with LR or HR (Solvency II)
ERM	Enterprise risk management
ERM-II	Enterprise Risk Management Institute International, United States

ERM II Exchange Rate Mechanism II
ES Expected shortfall
ESA European Standard Approach
ESFS European System of Financial Supervisors
ESMA European Securities and Markets Authority
ESRB European Systemic Risk Board
ESRC European Systemic Risk Council, now changed for ESRB
EU European Union
EVC Extreme-Value Copulas
EVT Extreme-Value Theory
EWS Early-Warning System
Ex Index for expense risk (Solvency II)
exp Index for health expense risk (Solvency II)
Exp Expense risk capital charge (Solvency II)
F Failure (index)
FA Final Advice
FASB Financial Accounting Standard Board, United States
FAST Financial Analysis Solvency Tools, used by NAIC, United States (old FAST)
Financial Analysis and Surveillance Tracking, used by NAIC, United States
FC Fixed cost risk
FCD Financial Conglomerate Directive
FD Framework Directive
FDB Future discretionary benefits
FfC Framework for Consultation
FFSA Féderation Francaise des Sociétés d'Assurance (the French Insurers Association)
FI Financial instruments
Market value of fixed income assets (Solvency II)
FIN-USE Forum of User Experts in the Area of Financial Services (set up by the European Commission in 2004)
FinReq Financial Requirements Expert Group, CEIOPS
FQR Financial quota share reinsurance
FR France
FSA Financial Services Authority, United Kingdom
FTK Financial Assessment Framework, the Netherlands
FX Index for Foreign Exchange (risk)
g gross, is used as gCR, gKC, and so on (Solvency II)
GA Granularity adjustment
GAAP Generally Accepted Accounting Principles
GARCH Generalized Autoregressive Conditional Heteroskedasticity
GBP The British sterling
GC Groupe Consultatif Actuariel Européen (association of European actuaries)
General Collateral
GCM Grand correlation matrix (Grand dependence matrix)

GD	Generalized duration
GDV	Gesamtverband der Deutschen Versicherungswirtschaft e. V. (German Insurance Association)
GE	Index for General premium and reserve risk, together with WC (Solvency II)
gen	Generalized
GER	Germany
GISG	General Insurance Study Group, United Kingdom
GMAD	Generalized mean-absolute deviation
GNAIE	Group of North American Insurance Enterprises
GOF	Group (total) own funds
GP	Gross premium
GPD	Generalized Pareto distribution
gr	Gross
GR	Gross reserves
GSCR	Group SCR
GSV	Guaranteed Surrender Value
GWP	Gross written premiums
H	Also used for HHI Health risk capital charge (Solvency II)
H5N1	Virus ("the flu virus")
HGP	Total health gross premium (Solvency II)
HHI	The Herfindahl–Hirschman index
HJM	Heath–Jarrow–Morton (interest rate model)
HKI	The Hannah-Key index
h.o.t.	Higher-order terms
HR	Index for health risk capital charge (Solvency II)
HRT	Heavy Right Tail
HTI	Hall–Tideman index
IA	Impact Assessment
IAA	International Actuarial Association
IAD	Insurance Accounting Directive (with EU)
IAIS	International Association of Insurance Supervisors
IAS	International Accounting Standards
IASB	International Accounting Standards Board
IASC	International Accounting Standards Committee
IBNER	Incurred but not enough reported
IBNR	Incurred but not reported
IBNS	Incurred but not settled
IC	Insurance Committee
ICAS	Individual Capital Adequacy Standards, United Kingdom
ICMS	Individual company market share (GDV model)
IF	Income factor (GDV model)
IFRS	International Financial Reporting Standards
IGD	Insurance Group Directive

IGT	Intragroup Transactions
i.i.d.	Independent and identically distributed
IL	Illiquid liabilities
	Incurred losses (Solvency II)
ILS	Insurance-linked securities
ILW	Industry Loss Warranties
IM	Internal Model
IORP	Institutions for Occupational Retirement Provisions
IPD	Investment Property Databank
IR	Index for interest rate risk (Solvency II)
IRB	Internal rating-based (Basel II method for credit risk)
IRIS	Insurance Regulation Information System (used by NAIC, United States)
IRTS	Interest rate term structure (risk-free)
ISG	Indicators Subgroup of the Social Protection Committee, Employment and Social Affairs DG (within European Commission)
	Interservice Steering Group
IT	Income bearer (GDV model)
	Individuated t-copula
	Information Technology
IV	Intrinsic Value
IWG	Insurance Working Group (European Council)
JLS	DG JLS: EC's DG for Justice, Freedom, and Security (European Commission)
JPY	The Japanese yen
JTF	Joint Task Force (for European supervisory committees)
JURI	Legal Affairs Committee (European Parliament)
KB	Kickback risk
KC	The KC-factor gives risk mitigating effect due to future profit sharing in the capital charge (Solvency II)
KMV	Kealhofer, McQuown, and Vasicek (founded the KMV model)
KR	Kickback risk
L	Liabilities
LA	Liquid assets
	Index for lapse risk (Solvency II)
La	Index for lapse risk (Solvency II)
LCR	Liquidity coverage ratio
LDI	Liability-driven investments
LG	Liquidity gap
LGD	Loss-given default
Liab	Liability
LIBOR	The London Interbank Offered Rate
LL	Liquid liabilities
LO	Index for longevity risk (Solvency II)
LOB/LoB	Line of business
Long	Longevity

LP	Published liabilities
LPM	Lower partial moments
LPT	Loss portfolio transfer
LR	Loss ratio
	Index for life risk capital charge (Solvency II)
LRM	Liability risk margin
LSAD	Lower semi-absolute deviation
LT	Index for long-term health risk (Solvency II)
LTL	The Lithuanian litas
LUR	Life underwriting risk
LWG	London Working Group (also Sharma WG)
MARKT	DG MARKT: EC's DG for Internal market and services (European Commission)
MA	Hidden reserves on the asset side of the balance sheet
MAD	Mean-absolute deviation
MAT	Marine–Aviation-Transport
MCCSR	Minimum Continuing Capital and Surplus Requirement (Canada)
MCR	Minimum capital requirement
MEL	Mean excess loss
MEP	Member of the European Parliament
MGF	Minimum guarantee fund (Solvency I and II)
ML	Implicit margin of the published liabilities
	Market loss
	Maximum likelihood
MMP	Multiyear/multiline products
MO	Index for morbidity risk capital charge (Solvency II)
mod	Modified (duration)
MoU	Memorandum of Understanding
Morb	Morbidity
Mort	Mortality
MP	Mathematical provisions
MR	Index for market risk (Solvency II)
	Index for mortality risk, together with LR (Solvency II)
	Index for excessive loss/mortality/cancellation risk together with HR (Solvency II)
	Index for claims/mortality risk, together with LT (Solvency II)
MS	Market share
MSCI	Formerly Morgan Stanley Capital International Inc.
MTP	Multitrigger products
MSE	Mean square error
MV	Market value
MVA	Market value of assets
MVL	Market value of liabilities
MVM	Market value margin

n	Net is used as nCR, nKC, and so on (Solvency II)
NAIC	National Association of Insurance Commissioners, United States
NatCat	Natural Catastrophe
NCE	Net claims expenditure
ncp	Index for capital requirement for noncontrolled participations in a group (Solvency II)
nd	Nondiversifiable (Solvency II)
NIBR	Not incurred but reported (not used)
NL_PL	Nonlife expected profit or loss
NL	Index for nonlife risk capital charge (Solvency II)
	Netherlands
NOI	Net operating income
NP	Normal power
	Net earned premiums
	Nonproportional reinsurance
NSAR	Net sum at risk
NSFR	Net Stable Funding Ratio
NTP	Net technical provisions
NAV	Net value of assets minus liabilities (Solvency II)
obs	Index for observed
OE	Operating expenses
OECD	Organization for Economic Cooperation and Development
OEEC	Organization for European Economic Cooperation
OF	Own funds
OFL	Own funds liability
ofs	Index for capital requirement from other financial sectors within a group (Solvency II)
OJ	*Official Journal* (within EU)
OR	Index for other risk
	Index for operational risk (Solvency II)
ORIC	Operational Risk Insurance Consortium
ORSA	Own risk and solvency assessment
ORX	Operational Risk eXchange Association
OSFI	Office of the Superintendent of Financial Institutions, Canada
ot	Index for capital requirement for other business not solo-adjusted in groups (Solvency II)
OTC	Over-the-counter
P	Premium (net earned)
	Property: P1, P2, and so on (risk measures)
PC	Probability of concordance
PCO	Net provision for claims outstanding
PCR	Prescribed Capital Requirement, *cf.* SCR

PD	Probability of default
	Probability of discordance
PDI	Projected disability increase
pdf	Probability density function
PES	European Parliament's Socialist Party
pgf	Probability-generating function
PIM	Partial Internal Model
PMI	Projected mortality increase
PML	Probable maximum loss
PR	Index for premium risk, together with NL (Solvency II)
	Index for property risk, together with MR (Solvency II)
prem	Index for premium
PSM	Published solvency margin
PV	Present value
PXL	Prospective XL cover
QBRM	Quantile-based risk measures
QIS	Quantitative Impact Study
QS	Quota share
R	Reserve
RB	Risk bearer
	Total amount of claims against policyholders and insurance agents (Solvency II)
RBC	Risk-based capital
RBNS	Reported but not settled
RC	Risk charge
	Rating category
	Risk concentration
RD	The effective duration of credit risk exposure (Solvency II)
Re	Index for reinsurance default risk (Solvency II)
RE	Market value of property exposure (Solvency II)
red	Index for redemption (Solvency II)
ref	Index for reference (Solvency II)
res	Index for reserve (Solvency II)
Rev	Revised (used as RevEQ) (Solvency II)
REX	German Interest rate index (Renteindex)
RF	Risk factor
RI	Rosenbluth index
RKI	Robert Koch Institute
RL	Released liabilities
RM	Risk Management
	Risk margin
RO	Index for run-off (Solvency II)
RoEC	Return on Economic Capital

RORAC	Return on risk-adjusted capital approach
ROZ-IPD	Cooperation between ROZ) Raad Voor Onroerende Zaken in the Netherlands and IPD
RP	Recurrence period
	Capital charge for reserve and premium risk (Solvency II)
	Risk premium
RPo	Replicating portfolio
RPS	Reduction for profit sharing
RR	Index for reserve risk, together with NL (Solvency II)
	Index for revision risk, together with LR (Solvency II)
RSAS	Royal Swedish Academy of Science
RT	Risk bearer (GDV model)
RTS	Report to Supervisors
RXL	Retrospective XL cover
SA	Standard Approach
	Standardized Approach (Basel II)
	Sum assured (Solvency II)
SANCO	DG SANCO: EC's DG for Health and Consumers (European Commission)
SAS	SAS Institute
SEC	Securities and Exchange Commission, United States
	Document classification with EU: Documents that cannot be classified in any of the other series
SecGen	Secretary General
SCR	Solvency capital requirement
SEK	The Swedish krona
SFCR	Solvency and Financial Condition Report
SKK	The Slovak koruna
SME	Small- and medium-sized enterprises
SOA	Society of Actuaries, United States
SPV	Special purpose vehicle
SR	Strengthened interest rate bases
	Shortfall risk
	Desirable solvency capital requirement
	Index for spread risk (Solvency II)
SRP	Supervisory Review Process
SST	Swiss Solvency Test
ST	Index for (accident and health) short-term risk (Solvency II)
struct	Structured
sup	Index for supplementary
SWE	Sweden
T-VaR	See Tail-VaR
T1	Tier 1
T2	Tier 2

T3	Tier 3
Tail-VaR	Tail-Value-at-Risk (the risk measure that describes the expected value of the loss in the event of a loss subject to a defined confidence interval)
TB3MS	U.S. 3-month Treasury Bill
TBSA	Total Balance Sheet Approach
TCR	Target Capital Requirement (earlier version of SCR) Trend catastrophe risk
TE	Timing error (Solvency II)
ton	Index for tontine
TP	Technical provision
TPE	Technical provisions and earned premiums (Solvency II)
TRS	Total return swaps
TS	Technical Specification
TV	Tail-VaR
UAOF	Unavailable own funds
UCITS	Undertaking for collective investment in transferable securities
UE	Univariate elliptical (distributions)
U.K./UK	United Kingdom
UL	Unit-linked contracts (or more generally contacts where the policyholder bears the investment risk)
ULAE	Unallocated loss adjustment expenses
UPM	Upper partial moments
UR	Unstrengthened interest rate bases Underwriting risk (Solvency II)
U.S./US	United States
USD	United States dollar
USP	Undertaking-specific parameters
UW	Index for underwriting risk (Solvency II)
V	VaR
VaPo	Valuation Portfolio
VaR	Value-at-Risk (the risk measure which describes the loss subject to a defined confidence interval)
VDO	Value of the default option
Vol	Volume
VR	Volatility risk
W	Weighted
WC	Index for workers compensation risk (Solvency II)
WG	Working Group(s)
WHO	World Health Organization, United Nations
WLR	Whole life risk
wp	Index for with-profits (Solvency II)
WPFS	Working Party of Financial Services, within the European Council
WPI	Working Party on Insurance, within the European Council

WT	Wang Transform
WWII	World War II
wwcons	Index for consolidated capital requirement in a group (Solvency II)
WXL	Working XL reinsurance
XL	Excess of Loss reinsurance
xs	Index for excessive loss/mortality/cancellation risk (Solvency II)

PART A

Solvency Introduction

Companies make money by taking risks and lose money by not managing risks. ERM is a partner in business—not a support function.

Fulvio Conti

CEO Enel, Italy at the AFIR Colloquium in Rome 2 October, 2008

As a background we first define the concepts of solvency and to be solvent in Chapter 1. The historical development of solvency assessments is discussed in Chapter 2 and the managing of risks in Chapter 3. A brief summary is tabulated in Chapter 4.

The solvency assessment framework is briefly introduced and discussed in Chapter 5. The valuation of assets and liabilities, elements of a solvency assessment system, and the modeling of capital requirements—all these issues will be discussed in more detail in the subsequent chapters of Part B.

CHAPTER 1

Solvency

IN GENERAL TERMS, *solvency* is the state of having more assets than liabilities. To be *solvent* means that one is in the state of solvency.

In general, you cannot answer the question "How solvent is the company?" because solvency–insolvency is a two-state space. However, in a decision process you may need a more flexible view. Theoretically, this can be done using fuzzy set theory. Fuzzy set theory, see, for example, Derrig and Ostaszeewski (2004), erases the distinction between the two-state space and replaces it with a probability distribution that gives the likelihood that an element is a member of a particular set. In practice, this depends on the jurisdiction the company is in, and how the regulator defines the theoretical solvency levels. If the company has assets larger than some of these theoretical levels, then the company is in a "regulatory solvency state," satisfying the statutory financial requirements.

To be consistent with the framework laid down below, we need to state that if the regulator defines a target capital level and a minimum capital level of requirements, then the regulatory solvency state exists if the company has assets above the minimum level—you cannot be "insolvent in two different states or stages." This is true even if we use a fuzzy set environment. This is also in line with the definition given by Bennet (2004): *An insurer is solvent for regulatory purposes when its assets exceed its liabilities by the regulatory minimum margin.*

To be solvent means that the company is in the state of solvency. This could also be interpreted as being able to pay future claims as they fall due (a going concern situation), see, for example, Owen and Law (2005), or to be able to pay all liabilities on an immediate liquidation (a breakup situation) or transfer all liabilities to a willing partner (a run-off situation).

In general terms, a *solvency margin* is a buffer in a company's assets that covers one or both of the theoretical solvency levels required by the regulator. For the supervisor it is important that the policyholders are protected. But it is also important for the supervisor to ensure the stability on the financial market.

The general Available Solvency Margin (ASM), is the difference between the assets (A) and the liabilities (L), meaning that the company is in the general solvency state if the ASM is larger than zero. The state of insolvency appears if ASM is less than zero. This definition, in terms of solvency margin, was given in, for example, Pentikäinen (1952).

The theoretical or target capital requirement can in some jurisdictions be the minimum amount required by the regulator so that the insurers can continue its business in some form. In other jurisdictions, the theoretical (target) capital requirement is just a target or an early warning signal. This means that if the insurer has an ASM above the theoretical requirement, it can continue as a going concern, that is, being in the regulatory solvency state. Otherwise it has to be a dialogue between the insurer and the supervisor on what steps to take to be sure that the ASM is above the theoretical capital requirement. Some systems, for example, the risk-based capital system in the United States (introduced by the National Association of Insurance Commissioners, NAIC) and the upcoming European Solvency II system, have an intervention ladder between the upper target level and the absolute minimum level. Introducing intervention ladders in this way is a practical application of the fuzzy set theory. This is also outlined as the third key feature in the IAIS' (*International Association of Insurance Supervisors*) paper on regulatory capital requirements, IAIS (2007e): "The solvency regime should include a range of solvency control levels which trigger different degrees of intervention by the supervisor in a timely manner. The solvency regime should have due regard to the coherence of the solvency control levels established and any associated corrective action that may be at the disposal of the insurer, and of the supervisor, including options to reduce the risks being taken by the insurer as well as to raise more capital."

In the sequel we will assume a jurisdiction with two regulatory capital requirements[*]; the theoretical or target requirement will be called the solvency capital requirement (SCR) and the absolute minimum requirement will be called the minimum capital requirement (MCR). In terms of the IAIS (2007d), the SCR is called the prescribed capital requirement (PCR). Thus, ideally we have

$$\text{MCR} < \text{SCR} \leq \text{ASM}, \tag{1.1}$$

where the MCR should be interpreted as a "hard level" of intervention. This could also be interpreted such that if the company is in the state of having assets lower than the MCR level, then the company is not in the regulatory state of solvency.

These definitions are illustrated in Figure 1.1.

Beard (1964) considered an insurance undertaking "as a collection of liabilities against which a group of assets are held. If the assets are in the aggregate sufficient to meet the liabilities then in a certain sense it can be said that the undertaking is solvent."

The definition of solvency given in Benjamin (1977) gives rise to the two concepts of solvency as discussed above, that is, the two extremes of a range of possibilities:

- The liabilities are those paid on an immediate liquidation of the company (a breakup or a runoff of the company) or if its liabilities could be transferred to a willing partner, or
- The company is regarded as solvent if it pays all its debts as they mature (the going concern approach)

[*] If we only have one level of requirement, then SCR = MCR.

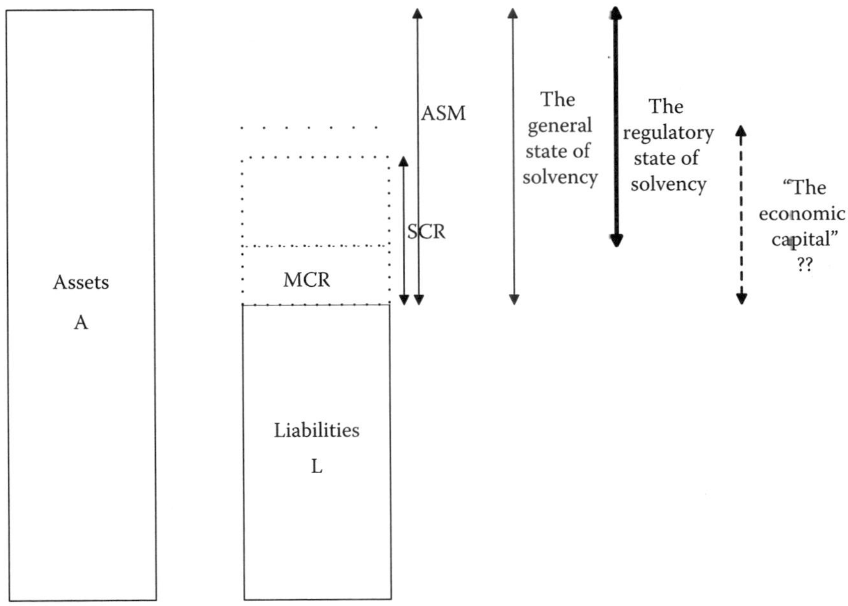

FIGURE 1.1 The state of solvency. This illustrates the different states of solvency. MCR: minimum capital requirement, SCR: solvency capital requirement, ASM: available solvency margin. The exact position of the economic capital is above the liabilities, but it depends on the preferences of the insurance company.

The first position may be obtained when ASM $\leq$ MCR, that is, when the insurer breaks the minimum floor, the supervisor will intervene and decide if the company must break up (all businesses are closed) or if the business should be put in runoff (no new business is allowed, but all old contracts are to be fulfilled), that is, being in the regulatory insolvency state. The second position may be obtained when ASM $\geq$ SCR. In Campagne (1961a, 1961b), the term "dynamic solvency" was used for the going-concern approach and "static solvency" for the breakup situation; see also Kastelijn and Remmerswaal (1986).

Note that the liability concept refers to the obligations set out in the insurance contracts. The technical provision is the value of the insurance obligations set aside in the balance sheet. Traditionally, the technical provision includes implicit margins of prudence.

Traditionally, the strength of solvency has been evaluated by the use of risk theory techniques (*cf.* also Pentikäinen, 2004). The ruin probability is the probability that the insurer, having an initial ASM, ASM $\geq$ SCR, will become insolvent during a chosen time horizon $(0,T]$. By "insolvent" in its legal meaning, we mean the break of the MCR within the time interval $(0,T]$. T is, usually, at least one year and is usually chosen according to the accounting period.

We have mainly talked about solvency in terms of regulation. Pentikäinen (1967) discussed two different ways of looking at the solvency concept:

1. *From the point of view of the supervisory authorities.* The benefits of the claimants and policyholders must be secured. This is the regulatory view discussed above.

2. *From the point of view of the management of the company.* The continuation of the function and existence of the company must be secured. This may be seen as the economic capital view, described below.

Definition (1) is narrow as it does not demand continuity of the company but allows it to be wound up. It could be approved as a basis of the legal system. "… the supervising authorities and the legal security measures shall be restricted to the minimum i.e. to secure the insured benefits only, but otherwise each company shall have freedom to develop its function as it itself desires." As stated in Pentikäinen (1984), the first case indicates the maintenance of the insurer's ability to meet his obligations for a short period of time, say one year, that is, the concept of solvency discussed above.

In IAIS (2007d), the first key feature describes the purpose and definition of regulatory capital requirements and is in line with the first definition above: "Regulatory capital requirements should be established at a level such that the amount of capital that an insurer is required to hold should be sufficient to ensure that, in adversity, an insurer's obligations to policyholders will continue to be met as they fall due. These requirements should be defined such that assets will exceed technical provisions and other liabilities with a specified level of safety over a defined time horizon."

In the second case, the objective is to guarantee the continued existence of the insurer. This is the economic capital view, which is a more complex situation than the latter, including the first case as well. This second view may be seen as an approach taken by the company itself to set its own solvency level for its internal control. The capital calculated by the company to meet its own needs and internal requirements is usually called "economic capital". The objective is to get a maximum value creation. A more general discussion of economic capital is found in Lelyveld (2006) and an introduction in Bhatia (2009). This economic capital view indicates the use of an own risk and solvency assessment (ORSA) approach as proposed by the IAIS (2007b, 2007c).

If we take the regulatory view as the basis, then the company's existence can be left to the management. This could be done by means of adequate reserves, loadings of premiums, and reinsurance.

In an environment, as discussed in Chapter 5, with a three-pillar approach, the second pillar with supervisory qualitative measures will build a bridge between the two approaches. The two ways of looking at solvency can be combined through internal models and the ORSA, but also through the pillar system *per se*.

Another way of looking at insolvency (i.e., not being in a solvency state) is to consider the various rules in company law regarding insolvency and bankruptcy. This type of "judicial insolvency" is not discussed in the sequel.

In the literature, see Chapter 2, there are a number of different formulas for the assessment of the capital requirement, especially for the MCR. In the beginning of the 1990s, there was a tendency to seek rules that took into account all risks that the insurance companies are facing. Systems of this kind are usually called risk based and hence a risk-based capital requirement. A thorough discussion of the insurers' risks is given in IAA (2004), see also Sandström (2005). The steps toward a risk-based capital requirement are one of many steps that have been taken to reach the nowadays common management umbrella called the enterprise risk management (ERM) see Chapter 3, where the economic capital requirement and the regulatory capital requirement are two of the issues.

In constructing an SCR, we need to discuss at least the following fundamental issues:

- *Valuation* of assets and liabilities
- Risk margins for *uncertainty* in assets and liabilities
- Risk *measures* for the volatility in assets and liabilities
- The *modeling* (risk categories, risk mitigation, diversification, etc.)

but also, as stated by Brender (2004), the insurance company or its supervisor "must first choose a degree of probability of continued survival and future time horizon over which the insurer will continue to operate." All these elements together constitute the solvency assessment framework.

This is discussed briefly in Chapters 5 through 6 and more deeply in Chapters 7 through 21.

Before doing so, we will briefly summarize the historical development of assessing the capital requirements and managing risks.

CHAPTER 2

A Historical Review

LOOKING AT SOLVENCY from a historical perspective we can divide the development into two main eras, the classical approach period and the economic approach period, where assets and liabilities are looked upon at the same time and where their interaction is a major issue.

By the "classical approach period," we mean the time between World War II and approximately 1980. The time period after that we call the "economic approach period." Before the mid-1940s, a company that wanted to be in the state of solvency (without explicitly talking about it) had prudent margins in its reserves or in the mortality or interest rate assumptions. Classical theories such as the collective risk theory and the theory around "ruin probabilities" were developed. Before the term "solvency" was introduced, concepts such as statutory reserves and stabilization reserves were often used, "which have been formed in the course of years and which serve as an extra guarantee for fulfilling the obligations undertaken" (Campagne et al., 1948, p. 338).

One important year is 1986. This was the first time "traditional actuaries," mainly from Europe and using the classical ruin theory, met with "modern financial actuaries/economics," mainly from the United States. The latter used modern financial theory and the pricing of insurance undertakings. The meeting was the First International Conference on Insurance Solvency, held in Philadelphia, 18–20 June; see Cummins and Derrig (1988, 1989). This first conference on solvency was met with skepticism from, for example, the UK General Insurance Study Group (GISG). In its 1987 report (GISG, 1987), it is stated that "One of the most challenging aspects of the Philadelphia conference for the UK participants was trying to understand the entirely different approaches being adopted by American academics and others, from the standpoint of financial economics. Whilst a first reaction might have been to reject much of it as high flow academic work out of touch with the reality of managing insurance companies, it is apparent that some of the research work *is* finding practical application in the United States and it seems worthwhile and important that we should understand what these American researchers are saying." The 2nd conference was held in Brighton, United Kingdom, in 1988; see Cummins and Derrig (1991).

The influence of "financial economists/actuaries" as demonstrated at the International Conferences on Insurance Solvency in 1986 and 1988 has been important in developing new tools. D'Arcy (1988) referred to four main differences in the approach of financial economists and actuaries.

i. The definition of risk

   Actuaries: are concentrating on uncertainty, which is important for reinsurance and solvency purposes, but which could be ignored in pricing and reserving.

   Financial economists: see some risks as diversifiable, by combining risks together, while others are systematic or undiversifiable.

ii. The return on investment

   Actuaries: have a one-dimensional view, a single rate of interest for discounting.

   Financial economists: are concerned about volatility in interest rates, market interest rates, and interrelationships.

iii. The profitability

   Actuaries: is based on profits (premiums + investment income − claim costs − expenses).

   Financial economists: is based on return on capital or equity; profit is related to investment required.

iv. The valuation of assets

   Actuaries: use historic book values

   Financial economists: use market values

The First International Conference on Insurance Solvency was held during a "transition period" between the old traditional actuarial view on insurance and a more "modern" financial view. In 1987 Professor Hans Bühlmann introduced the actuary of the third kind (Bühlmann, 1987). The state of the art in actuarial science and financial economics at that time is summarized in D'Arcy (1989, 2009). In op. cit., a comparison between actuarial science and financial economics is given together with indications on current issues in financial economics and some major applications of financial economics to insurance. This and other forces gave birth to a new section within the International Association of Actuaries (IAA): the Actuarial Approach for FInancial Risks (AFIR) section—the study of financial-related subjects in the insurance industry.

2.1 CLASSICAL APPROACH

For the nonlife insurance business, the methods were either based on fluctuation in aggregate claims or based on combined ratios (mainly on the loss ratio part). These methods are presented first. We then look at the classical methods for the life insurance business and on some multivariate approaches.

2.1.1 Fluctuation in Aggregate Claims and Ruin Theory

The pioneering work on nonlife insurance was done by Teivo Pentikäinen in Finland in the beginning of the 1950s. The Finnish system, introduced in 1953, was based on the calculation of equalization reserves and enabled to obtain bounds on the reserves such that the effect of the stochastic fluctuations in the annual claims amount was taken care of. In the system, both minimum and maximum bounds on the equalization reserves were calculated. The total claims amount was assumed to follow a compound Poisson distribution. The minimum requirement was that it should, in combination with the free assets (own funds) and a loaded premium, be sufficient to pay an assumed total claim amount with a probability of $1-\varepsilon$. An adjustment for an assumed interest rate was also made. The maximum amount was calculated for tax purposes as the equalization reserve, up to an upper limit, was exempted from tax. The final upper and lower limits of the equalization reserve were obtained using a normal power approximation. For details, see, for example, Kastelijn and Remmerswaal (1986), Pesonen (1967), Pörn (1968), and Hovinen (1969). In Pentikäinen (1952) the minimum solvency margin was discussed in terms of risk theory and reinsurance/net retention using the Esscher transform (Esscher, 1932). As pointed out by Jukka Rantala at one of the plenary sessions at the 39th ASTIN Colloquium in Helsinki in 2009, the Finnish system has many similarities with Solvency II. The Finnish system included standard methods for the calculation of technical provisions (if the standard method was not suitable for the company it could use its own assumptions), two capital requirements (an absolute minimum requirement for a run-off situation and one risk-based target level for a going-concern situation). The solvency test, with detailed instructions given by the regulator, was based on a one year time horizon with 1% ruin probability. In the going-concern perspective, the equalization reserve had an upper limit set for tax-free capital and reserves. Implementing studies, such as the Solvency II Quantitative Impact Studies, were accomplished. If the standard formula was not applicable, the companies could use own data and calculations (internal models and ORSA).

An extension of the calculation of stabilization reserves, required as an adjustment of deviations from risk, from life insurance to fire insurance, is given in Campagne et al. (1947); see also Willemse and Wolthuis (2005).

In 1980, the Belgian insurance organization UPEA (since 2004, Assuralia) also proposed to study the fluctuation of the aggregate claims amount following a compound Poisson distribution. This was done using a normal power approximation of the initial reserve (or solvency margin at time zero), and writing the initial reserve as a linear function of the square root of the risk premium; see, for example, Kastelijn and Remmerswaal (1986).

A Norwegian system, developed in the beginning of the 1980s, was also based on claims amount. It was based on rules for determining the technical provisions and statistical data needed for the calculations. The system has been described in Norberg and Sundt (1985), Kastelijn and Remmerswaal (1986), and Sandström (2005); see also Norberg (1986).

Norberg (1993) also looked at a solvency system for life insurance companies.

Classical ruin theory seeks the probability that a company, with an initial reserve (= initial solvency margin) and some premium loading, will break a predefined ruin barrier within a specific time horizon (could be infinity). For a review of the literature, see, for example, Kastelijn and Remmerswaal (1986). In terms of the definition of solvency in Chapter 1, the barrier should be in terms of the MCR level.

Assuming that the aggregate claims amount followed a compound Poisson distribution, Pentikäinen (1962) obtained bounds on the probability of ruin in terms of the net retention. Bohman (1974, 1977) used a random walk approach to get some rule of thumb to approximate the probability of ruin. A similar result was obtained by de Vylder (1978) using a "practical solution."

2.1.2 Combined Ratios and Other Ratios

During the 1950s, Professor Cornelis Campagne in the Netherlands made a research study for the OEEC* Insurance Committee. As his research was recognized as pioneering in the approach for assessing an extra minimum reserve for both nonlife and life insurance companies, he was asked to present a report ("Minimum Standards of Solvency for Insurance Firms") in 1957 to the OEEC Insurance Committee (Campagne, 1957). His work was further developed and a final report was presented in 1961 (Campagne, 1961a, 1961b).

In nonlife, Campagne's model was simple but elegant. He studied the combined ratio, that is, the sum of the expense ratio and the loss ratio. Let the net retained premium be 100%. From this, a constant fraction equal to the average expense ratio is deducted; from the European data of 1952–1957, it was fixed at 42%. The remaining part is what remains for claims payment. Campagne assumed that the net loss ratio followed a beta distribution. In his first report (Campagne, 1957), the Poisson distribution was used. With data from different European countries he estimated the value-at-risk (VaR) of the loss ratio at the 99.97 percentile as 83%. Thus the combined ratio will be 42% + 83% = 125%. In other words, the company will need an extra 25% of the premium during 1 year to meet the requirements. After further works during the 1960s and political negotiations, this framework became the base for the first nonlife directive in Europe in 1973 (see below).

Campagne died in 1963 and the Italian Professor Bruno de Mori continued his work for the OEEC. His working group proposed that the loss ratio should follow a normal distribution; see Mori (1965). The loss ratio will be within the range of 3σ around the mean with a probability of 0.997. The solvency margin, taken as a fraction of a specific quantity Q, was defined as 3σ times the ratio between the claims reserve and the total claims amount divided by the ratio between Q and the total earned premium. The quantity Q could be the total premium income, the averaged total claim amount over 3 years or the technical reserves. The approach is discussed in Kastelijn and Remmerswaal (1986).

de Wit and Kastelijn (1980) recreated the Campagne model and updated it with Dutch data from the period 1976 to 1978; see also Kastelijn and Remmerswaal (1986). Ramlau-Hansen (1982), commenting on the de Wit and Kastelijn discussion, assumed that the loss

* Organization for European Economic Cooperation, now called OECD, the Organization for Economic Cooperation and Development.

ratio should follow a lognormal distribution. The Ramlau–Hansen approach was that the fluctuation in the lognormal variable stemmed from the difference in the expected value for different companies, difference in the circumstances from year to year and a stochastic fluctuation in the results. The model was estimated using a credibility approach.

Bouro et al. (1980) suggested a development of the methods proposed by Campagne and de Mori and suggested to use the combined ratio, which was assumed to follow a normal distribution.

The leverage* and other related ratios as measures of insolvency are discussed in Kahane (1979).

In the prediscussion of the first nonlife directive within the European Union (EU), there was a proposal to have a solvency margin based on a combination of 24% of gross premiums written, 34% of incurred claims, and 19% of technical provisions. The last part was removed, as there was no possibility to reach an agreement on how to value the technical provisions.

Some countries believed that the proposed percentages for solvency were too high and some others that they were only sufficient; a compromise was eventually achieved; see Schlude (1979). The (required) solvency margin should be the higher of two indices, namely the *premium index* and the *claims index.*

The premium index:

- 18% of gross premiums up to 10 million units†
- 16% of gross premiums in excess of 10 million units

The claims index:

- 26% of gross average incurred claims up to 7 million units
- 23% of gross average incurred claims in excess of 7 million units

The average in the claims index is usually taken over the last 3 years (7 years for certain risks such as storm and hail). The result is reduced for reinsurance by the ratio (Net paid claims)/(Gross paid claims) with a maximum reduction of 50%. The shift from the premium index to the claims index will normally take place when the loss ratio is approximately 69% ($18/26 \approx 16/23 \approx 69\%$). As stated by Schlude (1979, p. 28), one objection against this system is that it does not take account of the structure of a company's losses. One could argue that the solvency requirement for a company with 10 claims each amounting to 100,000 units should be higher than the requirement for a company with 1000 claims each amounting to 1000 units.

2.1.3 Change in the Capital Position

Because the risk on investments is the most important factor for life insurance companies and because the technical provisions are the most important invested amount, Professor

* The ratio between liabilities and assets. In United Kingdom, the term "gearing" is usually used.

† Article 5 (a) of the First nonlife directive, EEC (1973), defines the units of account: "means that unit which is defined in Article 4 of the Statute of the European Investment Bank."

Campagne considered a percentage of the technical provisions as a minimum solvency margin. Initially, Campagne called this type of reserve for life insurance business a "stabilization reserve"; see Campagne et al. (1948). In the study conducted after the World War II, Campagne used Dutch data from 10 of the largest companies between 1926 and 1945.

When Campagne was asked to do the study for the OEEC (see above), he adopted the same approach as in the 1940s; see Campagne (1957, 1961a, 1961b), Kastelijn and Remmerswaal (1986), and Willemse and Wolthuis (2005). Campagne asked "how great has the extra reserve to be, so that with a probability smaller than 1/100 respectively 1/1000 this can be expressed to be insufficient for the financing of investment losses and deviations of foundations; in which case furthermore distinctions have to be made between cases in which the stabilization reserve has to be sufficient for one year or more years" (Campagne et al., 1948, pp. 342–343). A Pearson type IV distribution seemed to fit the data best. Campagne concluded that an extra reserve of 6% of the technical provision would be adequate with a probability of 99%. With a probability of 95%, the percentage of the extra reserve became 4% and this was the extra reserve proposed by Campagne. It was implemented as a part of the first life directive within the EU* in 1979.

The European Commission had to decide whether a common European solvency margin should be based on explicit items, implicit items, or a mixture of both; see Pool (1990). *Explicit items* means the same type of items making the solvency margin for nonlife insurance undertakings and *implicit items* should reflect the future performance of the life assurance undertakings. This was not a new discussion as an OECD report in 1971 favored the implicit item approach. This study was chaired by Mr Buol from Switzerland (the Buol report†); see Pool (1990, p. 36) and Kastelijn and Remmerswaal (1986, p. 31).

In the Buol report, the margin is calculated on the basis of the difference between reserves based on *strengthened* (SR) and *unstrengthened* (UR) interest rate bases. The unstrengthened interest rate, UR, is based on the interest rates of the past 20 years: UR $= 2/3 *$ (minimum interest rate over the past 20years) $+ 1/3 *$ (90% of the most recent rate). The strengthened interest rate, SR, is defined as SR $= 80\% *$ UR. The working group also considers 85% as a possibility.

The difference between the UR-based reserve and the SR-based reserve can for different types of insurance be expressed as a percentage of the UR-based reserve in combination with a percentage of the capital sum at risk. The Buol report suggested that for an average portfolio the result could be expressed as

$$9\% * (\text{UR-reserve}) + 6\% * (\text{sum assured at risk}).$$

Short-term contracts, which lead to little or no reserving, were considered separately by a subgroup to Buol's working group.

The final EU Directive for life insurance companies was probably inspired by the combined work of Campagne and the Buol report. The basic formula for the (required) solvency

* At that time the EEC, the European Economic Community.

† Buol et al. (1971): Les garanties financieres requises des enterprises d'assurance vie, OECD.

margin for life assurance companies was set up in the first life directive, EEC (1979), mainly as the sum of

First result: 4% of the mathematical reserves (gross of reinsurance)

+

Second result: 0.3% of the capital sum at risk.

The European solvency systems, which were seen as early warning systems, are described in, for example, Daykin (1984) and Sandström (2005).

2.1.4 Multidimensional Systems

All the classical systems discussed above were based on either one or two measures, for example, the European nonlife system that had premium and claims amount as critical levels. The NAIC in the United States introduced a multidimensional early warning system in 1972. It was called the Insurance Regulation Information System (IRIS), and it is indeed an information system built up in two phases. It is primarily intended to assist state insurance departments in executing their statutory mandates to oversee the financial condition of insurance companies operating in their respective states.

The system is based on a "total balance sheet approach"—a balance sheet based on conservative valuations. In the first phase, the statistical phase, the companies should calculate some 11–12 different ratios depending on the type of insurance business conducted. The ratios are like "premium to surplus" and "reserves to surplus" leverage ratios, and so on. It is then decided how many of the ratios are outside some predetermined ranges. Companies with the highest number outside the ranges are recorded as being of the highest priority for the second phase, the analytical phase. In this second phase, regulatory experts review the results of the statistical phase. The main objective of this was to determine which companies required regulatory attention.

One presentation on IRIS can be found in Kastelijn and Remmerswaal (1986), but more up-to-date information can be found at NAIC's Web site.* IRIS has been in use since 1972 and is a tool used in conjunction with other tools, such as the risk-based capital (RBC) system introduced in the 1990s (see below). IRIS is also a part of NAIC's audit system, Financial Analysis Solvency Tools (FAST).

The IRIS system is based on the calculation of the numbers of ratios outside given ranges. Another way to handle this "ranking problem" is to use a multivariate approach, such as Discriminant analysis. One drawback is that you need a fairly large number of distressed companies. Five papers on the application of discriminant analysis are mentioned in Kastelijn and Remmerswaal (1986)—they all use the same set of data in their analysis!

In 1989, the NAIC adopted a solvency policing agenda, which resulted in a number of changes in the solvency regulation in the United States. A new system, FAST (Financial

* www.naic.org

Analysis and Surveillance Tracking), was adopted and implemented in 1993. FAST could be considered as an information set that encompasses nearly all of the information in the IRIS and consists of approximately 30 ratios and corresponding scores for each ratio. The overall FAST score, which equals the sum of the individual insurer's audit ratios multiplied by the corresponding scores, is the ultimate output from FAST.

Hershbarger and BarNiv (1991) also used discriminant analysis, together with nonparametric analysis and logit analysis, to analyze the financial distress in the US life insurance industry. They used IRIS' 12 ratios together with an additional eight financial ratios from the rating agency A.M. Best. Data from 1975 to 1985 were used. The analyses correctly classified between 82% and 91% of the life insurance companies to be insolvent 1 or 2 years prior to the entrance into the insolvency state. For other works in this area, see the brief review of the literature in Hershbarger and BarNiv (1991).

2.2 ECONOMIC APPROACH

In the full economic approach, all the risks faced by an insurance company should be taken into account when calculating the capital requirement, and both sides of the balance sheet should be subjected to a market-consistent valuation. Several steps in this direction have been made during the last 30 years. From the beginning, there were two main parallel trends: one in Europe mainly based on the classical risk and ruin theory but also based on risk management techniques such as immunization (see Chapter 3) and one in North America mainly based on new financial mathematics.

Kahane (1979) used econometric modeling in establishing the relationships between solvency margins, underwriting results and return on investments. He examined three models. "The analysis is based on the proposition that capital requirement must be related to the overall performance of the insurance company. The overall performance is a function of both underwriting and investment incomes and their risks"; see Kahane (1979, p. 3). In one of the models the effect of diversification was taken into account. A review of the solvency literature with an emphasis on microeconomic and financial aspects is given in Kahane et al. (1989).

Research on solvency assessment was initiated in Europe as many countries got the nonlife and life directives during the 1970s implicating minimum solvency margins. The initial work done in Finland during the 1980s influenced very much the work done in United Kingdom by both the Scottish Faculty of Actuaries and the English Institute of Actuaries. The Scottish work was on life insurance business and the Finnish and the English work on nonlife business.

For the Finnish work, see Pentikäinen (1982), Rantala (1982), and in a more general setting Pentikäinen (1984), and Beard et al. (1984). A more complete simulation model was published in 1989; see Pentikäinen et al. (1989). The Finnish study was a motivation to use simulations and internal models and this influenced the work done in the United Kingdom. Both the Finnish and the UK systems made efforts to relate the capital requirement of an insurance company with the risks to which it is exposed.

The UK work is described in, for example, Daykin (1984) and Daykin et al. (1984, 1987), and as a risk management tool in Daykin and Hey (1990). The work done by the Faculty, see Hardie et al. (1984), introduced the Wilkie model for investment simulations.

The Finnish solvency study on nonlife insurance presented two objectives for a solvency capital requirement, namely to be a safeguard for the policyholder and a guarantee for the long-term continuation of the insurance company. They defined the available solvency margin as the sum of the free reserves (equity capital), the hidden reserves (underestimation of assets), and the equalization reserve (overestimation of liabilities). They used simulation as a tool to gain knowledge about the behavior of some of the risks. The study resulted in a risk theory model and the calculation of a solvency margin, new bounds on the equalization reserves, reserve for catastrophic risks (CAT risks), and an approximation of the ruin probability. The initial study did not include the claims reserve risk, the asset risk, and exchange rate risk.

The GISG was influenced by the Finnish work, and presented biannual reports on solvency, for example, in 1981, 1983, 1985, and 1987. In the first report from 1981, different aspects on the calculation of capital requirement were discussed, namely technical risk, investment risk, and other risks, such as reinsurance risk. The 1983 report was published by Daykin et al. (1984). The study is interesting as it considered and discussed the bases for valuation of assets and liabilities. Using the notations used in this book, let CEL be the current estimate of the liabilities ("best estimate") and CEA the corresponding estimate of the assets. LP is the "published liabilities" including implicit margins and AP is the published assets (usually book values). Then $ML = LP - CEL$ is the implicit margins in the published liabilities and $MA = CEA - AP$ is the hidden reserves in the asset side. The ASM is defined as $ASM = CEA - CEL = PSM + MA + ML$, where PSM is the published solvency margin. Here we have an embryo of the economic approach to solvency, discussing market-consistent valuation of both assets and liabilities. The report was criticized for the idea that the technical provisions should contain prudent margins for solvency purposes instead of using the solvency margin as "risk buffer"; see Daykin et al. (1987).

The working group also proposed to use a five-component risk model, using today's terminology:

- Risk 1: Market risk (asset depreciation)
- Risk 2: Reserve risk (extreme fluctuation in claims run-off)
- Risk 3: Premium risk (underwriting risk)
- Risk 4: Credit risk (reinsurance risk)
- Risk 5: Operational risk ("other risk")

This is a structure similar to the one proposed within Solvency II in Europe! Risk charge from Risk 1, RC1, is calculated using stress tests, the other risk charges (RC2–RC5) are calculated using factor-based approaches and for RC5 an additional fixed amount. The solvency margin, as expressed as a solvency capital requirement, SCR, is calculated as if all risks are fully correlated: $SCR = RC1 + RC2 + RC3 + RC4 + RC5$.

Skerman (1984) discussed "a realistic actuarial valuation in essence consists of determining the discounted value of a flow of income and outgo in the future." Slee (1984) stated that the "best estimate" should not include any implicit margins "for caution." Instead this margin should be part of the available capital ("free reserves").

The asset risk and the claims reserve risk were not covered in the traditional risk theory literature. But later, as a result of the Anglo-Finnish discussions that were held, these risks were examined; see, for example, Pentikäinen (1984), Pentikäinen and Rantala (1986), and Daykin et al. (1987, 1989). The work done by the working party of the GISG was presented at the first International Conference on Insurance Solvency in 1986 and, at the ASTIN (Actuarial STudies In Nonlife insurance) meeting in 1986, showed a change in emphasis as compared with earlier work (Daykin et al., 1984). The new proposal included a cash-flow model, incorporating the Wilkie model, based on Coutts and Devitt (1989) and a new approach regarding uncertainty. Uncertainty, in all forms, is taken care of at the level of the totality of the company. This means that it is less important if prudence is taken care of in the technical reserves or in the solvency margin. Therefore, the working party used the accounting term "technical provisions" instead of "technical reserves." In today's terms, the approach would be nontransparent.

The proposed model can be viewed in the light of three components: (1) liabilities arising from existing business, (2) future premiums and liabilities resulting from the risks underwritten, and (3) asset returns and asset value movements.

In 1982 the Faculty of Actuaries of Scotland initiated a research on the nature and assessment of the solvency of life insurance undertakings. This, together with the work done in both Finland and the United Kingdom, inspired a large quantity of research using simulation techniques in understanding and assessing capital requirement. Some of these researches were done in direct coordination with the Finnish Study Group, and the Working Parties of the Faculty and of the GISG. The Scottish Faculty of Actuaries presented their report on solvency of life insurance companies in 1984; see Hardie et al. (1984). The investment risks were investigated using the simulation model developed by Wilkie, the Wilkie model; see Wilkie (1984, 1986a, 1986b). The Wilkie model was later on also used by the GISG's working party for nonlife insurance (see above). The Wilkie investment model was later on updated and extended; see Wilkie (1995).

Ryan (1984) described a stochastic simulation approach for nonlife insurance. For a presentation of other early simulation studies on solvency, see Kastelijn and Remmerswaal (1986).

The GISG model was later developed into a tool for risk management and ALM; see Daykin and Hey (1990).

The Anglo-Finnish simulation models used an emerging costs approach to solvency, meaning a quantification of the likelihood that the insurer will be unable to meet its obligations. The emerging costs approach differentiates between winding up, runoff and going-concern situations. All elements affecting an insurer's financial strength, both on the asset side and the liability side, are considered; see also Daykin et al. (1994). Feldblum (1992) compared the Anglo-Finnish approaches with the "total balance sheet approach" used by NAIC in the United States; *cf.* Section 2.1.4.

The Committee on Valuation and Related Problems of the Society of Actuaries, appointed in 1977, published a report in 1979, the Trowbridge Report (1979), proposing an RBC approach. The basic principles of the Trowbridge report were that the values of the assets and liabilities should be the expected present value of future cash flows generated by the ownership of the assets and liabilities. It is stated that "Traditionally actuaries have spent a great deal of time on the determination of the liabilities. Now it appears necessary for actuaries to be concerned not only with the liabilities but also with the assets, and the degree of consistency with which the two values are obtained."

It was suggested that risk could be categorized into three parts, and that capital charges could be calculated for each risk category. The solvency capital was termed "contingency reserves." The three risk categories were, using the terminology used today,

- C1: Credit risk (asset depreciation)
- C2: Insurance risk (pricing inadequacy)
- C3: Asset/liability mismatch risk (interest rate fluctuation)

The exact terms used in the report are given in parentheses. The treatment of fluctuation in interest rates is discussed in Chapter 3 on the development of risk management.

As formal provisions for interest rate risk had not been an explicit element in asset and liability valuation, the Trowbridge Committee concentrated much of its attention to interest rate risk and immunization; see Chapter 3.

Regarding the valuation of assets and liabilities, the Committee thought that

1. The assets of an insurance enterprise can be viewed as the present value of the income stream arising from investments already owned. The only theoretical problem is the determination of the discount rate. The traditional methods of asset valuation are special cases.
2. The liabilities of an insurance enterprise are the present value of a disbursement stream arising from insurance and annuity contracts already on the books. Again, the theoretical problem is the determination of the valuation interest rate.
3. Clearly, there should be consistency, if not full equity, between the valuation rates for the assets and liabilities.

The Committee suggested a conceptual framework for a balance sheet of an insurance company.

Assets	**Liabilities**
Investments	Liabilities reserves
Cash	C1: Contingency reserve 1
	C2: Contingency reserve 2
	C3: Contingency reserve 3
	Surplus

As the focus was on the C3-risk, the Society of Actuaries formed a task force under the Trowbridge Committee to perform the research needed to identify and measure the C3-risk. It was named C-3 Risk Task Force.

Based on the proposal in the Trowbridge report, work on an RBC system began in Canada in 1984 and a first test calculation was established in 1989. The Canadian Office of the Superintendent of Financial Institutions (OSFI) did the work on the RBC system. After field tests in the life insurance industry the RBC system for life companies took effect in 1992. This system is known as the minimum continuing capital and surplus requirement (MCCSR). In the earliest version of the MCCSR calculation, it was implicitly assumed that all risks were fully correlated; thus it was the sum of the individual risk charges. The system was, in a way, static and was later amended and made more dynamic.

Similar systems for nonlife business took effect in 2002; see, for example, Sandström (2005). In parallel, the NAIC, in the United States, began to work on a similar model. Their formula is named after the approach, RBC, and includes a fourth risk category:

- C4: Operational risk (business risk),
  which is, as stated by Brender (2004), a "catch-all" risk category covering everything from bad management to bad luck. The capital requirement for the life insurance business is calculated as

$$\mathrm{SCR} = [(\mathrm{C1} + \mathrm{C2})^2 + \mathrm{C3}^2]^{1/2} + \mathrm{C4}.$$

The NAIC's RBC system was introduced for life and health insurance in 1992 and nonlife in 1993; see, for example, Brender (2004) and Sandström (2005). According to NAIC, this was a first step toward scenario testing and dynamic modeling, which were to follow. These steps have not been taken in the RBC system, but it became a stepping stone to dynamic financial analysis (DFA), which was introduced in the mid-1990s by the work done by the Casualty Actuary Society, CAS.

A comparison of NAIC's RBC system and its FAST audit ratio system with a cash flow simulation model is given by Cummins et al. (1999). They found that the FAST system dominated the RBC as a static method for predicting insurer insolvency.

Risk-based systems were also discussed and introduced in Australia, Singapore, and Japan and within the EU. At the same time, waiting for a new European system, different solvency assessment systems were introduced in the United Kingdom, Switzerland, and the Netherlands. Traffic light systems based on stress tests were introduced in Denmark, and as a supervisory tool in Sweden; see, for example, Sandström (2005).

Historically, there have been problems in comparing the available and required solvency margins between companies (and especially between companies in different countries). Assets have either been defined as historical book values or as market values. But the main problem has been the technical provisions as these have included implicit margins to protect policyholders. These implicit margins have been set by the actuaries and have been reflecting the prudence of the company. Even in the European Union, with its life and nonlife insurance directives, the incomparability of the technical provisions has been recognized and discussed.

2.3 EUROPEAN SOLVENCY II PROJECT

The works done by Campagne were the base for the solvency directives within the EU. The first solvency directives from the 1970s have been amended in the second and third directives from the 1980s and 1990s; see, for example, Sandström (2005). Based on the discussions in the Müller report (1997), the EU Parliament in 2002 adopted revised directives, Solvency I, and at the same time worked on a future risk-based system.* This new solvency system, called Solvency II, will be implemented in Europe on October 31, 2012.† The basic ideas of Solvency II are given in EU Commission (2006) and the proposals for a Framework Directive was published on July 10, 2007; see EU Commission (2007).

2.3.1 Basic Architecture of Solvency II

The European Commission has published on its Web site the basic architecture of Solvency II (http://ec.europa.eu/internal_market/insurance/solvency/architecture_en.htm).

Solvency II is based on a three-pillar approach, which is similar to the banking sector (Basle 2) but has been adapted for insurance.

The first pillar contains the quantitative requirements. There are two capital requirements, the SCR and the MCR, which represent different levels of supervisory intervention. The SCR is a risk-based requirement and the key solvency control level. Solvency II sets out two methods for the calculation of the SCR: the European Standard Formula or firms' own internal models. The SCR will cover all the quantifiable risks an insurer or reinsurer faces and takes into account any risk mitigation techniques. The MCR is a lower requirement and its breach triggers the ultimate supervisory intervention: the withdrawal of authorization.

The second pillar contains qualitative requirements on undertakings such as risk management as well as supervisory activities.

The third pillar covers supervisory reporting and disclosure. Firms will need to disclose certain information publicly, which will bring in market discipline and will help ensure the stability of insurers and reinsurers (disclosure). In addition, firms will be required to report a greater amount of information to their supervisors (supervisory reporting).

Solvency II will also streamline the way insurance groups are supervised and recognizes the economic reality of how groups operate. The new regime will strengthen the powers of the group supervisor, ensuring that group-wide risks are not overlooked, and demand greater cooperation between supervisors. Groups will be able to use group-wide models and take advantage of group diversification benefits.

As a solvency directive would be technically advanced and any changes that need to be done, for example, changed dependence structures or changed levels of different factors, could take 1–3 years to implement, it was decided to use the four-level Lamfalussy procedure in adopting the directive. The Lamfalussy procedure, which is discussed in Appendix D, Section D.1.1, means that at the first level we have a framework directive given the frame to the *work of art* that constitutes the details. The details are decided in the second level

* For more information on the Solvency II project, see http://ec.europa.eu/internal_market/insurance.

† The implementation will be postponed until December 31, 2012.

as a regulation that could be adopted during a 3-month period. On the third level in the procedure, we have standards and guidance published by CEIOPS.

The general ideas of the Solvency II framework directive are discussed in Chapter 22, but they are also given in the preamble to the directive text; see Appendix D, Section D.8.

The Solvency II project can be divided into three phases:

i. 1999/2000–2003: The learning phase; see Appendix D, Section D.1

ii. 2003–2009: The framework directive phase; see Appendix D, Sections D.2 through D.5

iii. 2008/2009–2012/2013: The implementing phase; see Appendix D, Sections D.6 through D.7

The European Parliament adopted the framework directive on April 22, 2009 and the ECOFIN Council made a note of it on May 5, 2009. The formal adoption, before publication, was made by the ECOFIN Council on November 10, 2009. The directive text was published in *Official Journal* L 335 on December 17, 2009 (OJ, 2009). The Framework Directive (FD) entered into force on January 5, 2010 with a deadline for transposition into national law set at October 31, 2012.*

The important implementation and calibration phase is discussed in Appendix D, Sections D.6 through D.7.

The general structure of the framework directive is given in Table 2.1. A more detailed list of contents is given in Appendix D, Section D.9.

2.3.1.1 General Ideas

The general ideas behind Solvency II and its final implementation are discussed in Chapter 22 with some background notes given in Appendix E. Here we also briefly discuss Pillar II and Pillar III issues as well as Group issues. Internal modeling is also mentioned.

2.3.1.2 Valuation and Investment

Asset valuation is discussed in Chapter 23 and Appendix F.

Liability valuation is discussed in Chapter 24 and Appendix G.

Investments are discussed in Chapter 25.

2.3.1.3 Standard Formula of SCR and MCR

The overall SCR standard formula is discussed in Chapter 26 and the background calibration in Appendix H, Section H.1.

The market risk is discussed in Chapter 27 and the background calibration in Appendix H, Section H.2.

The credit risk is discussed in Chapter 28 and the background calibration in Appendix H, Section H.3.

* As the supervisory committee CEIOPS will be transformed into a new authority from January 1, 2011, parts of the FD text have to be changed. These changes were proposed early in 2010. At the same time the European Commission proposed to extend the deadline for transposition to December 31, 2012.

TABLE 2.1 General Structure of the Framework Directive

	Titles	The Directive Includes	Article No.
I	General rules on the taking up and pursuit of direct insurance and reinsurance	Pillar I–III issues—standard formula, internal models, ERM, transparency	1–177
II	Specific provisions for insurance and reinsurance	Insurance contracts, information to policyholders, provisions for nonlife (coinsurance, assistance, legal expenses, health and accident at work), provisions for life, rules for reinsurance (finite, special purpose vehicles)	178–211
III	Supervision of insurance and reinsurance undertakings in a group	Group issues (excl. group support)	212–266
IV	Reorganization and winding-up of insurance undertakings	Winding-up proceedings; special register	267–296
V	Other provisions	Court, euro, EIOPC	297–304
VI	Transitional and final provisions	Derogation and abolition of measures, transitional period, entry into force	305–312
Annex I	Classes of nonlife insurance		
Annex II	Classes of life insurance		
Annex III	Legal forms of undertakings		
Annex IV	SCR standard formula		
Annex V	Groups of nonlife insurance classes for the purposes of Article 157		
Annex VI	Part A: Repealed Directives, Part B: List of time limits		
Annex VII	Correlation table (between articles)		

Reading notes on Solvency II.

The operational risk is discussed in Chapter 29 and the background calibration in Appendix H, Section H.4.

The liquidity risk is discussed in Chapter 30.

The nonlife underwriting risk is discussed in Chapter 31 and the background calibration in Appendix H, Section H.5.

The life underwriting risk is discussed in Chapter 32 and the background calibration in Appendix H, Section H.6.

The health underwriting risk is discussed in Chapter 33 and the background calibration in Appendix H, Section H.7.

The MCR is discussed in Chapter 34 and the background calibration in Appendix H, Section H.8.

A structured reading instruction for the whole Handbook is given in the Preface.

CHAPTER 3

Managing Risks and the Enterprise

ONE OF THE stepping stones to the development of asset–liability management (ALM) approaches for the insurance industry was the high inflation during the 1970s. We start this chapter with a brief summary of the inflation during this period.

3.1 STEPPING STONES TO MANAGING ASSETS AND LIABILITIES

The inflation during the twentieth century can be illustrated by the United States inflation, as measured by its consumer price index, CPI. The U.S. CPI between 1913 and 2007 is depicted in Figure 3.1. What we see is that the inflation was growing slowly during the years up to the mid-1960s and then started to grow rapidly. The relative change in the CPI, measured by $[CPI(t)/CPI(t-1) - 1]$, is given in the lower part of Figure 3.1. We see that even though the growth was modest, as compared with the latest, say, 40 years, the relative changes were very volatile. Note also the volatile period during the 1970s.

The 1970s period is usually called the "great inflation period." The causes of this inflation period are still debated; see, for example, Collard and Dellas (2004) and De Long (1995). "Loose monetary policy" played an important part, but it is not agreed which factors were responsible for such a policy.

Different measures show three cycles of inflation in the late 1960s and during the 1970s, each larger than that before (De Long, 1995). The first peak was in 1969, the second in 1973–1974 and the third and last peak, but also at the highest level, in 1980–1981. This is illustrated by the U.S. 3-month Treasury Bill in Figure 3.2.

As stated by De Long (1995), the inflation during this period was "broad-based." It was not a shift in relative prices, but a shift in absolute price levels. This shift in absolute price levels is illustrated by the U.S. oil prices in Figure 3.3. One explanation to what set off the "great inflation" is the Arab oil embargo that resulted from Egypt's attack on Israel in October 1973. Even if the embargo was lifted less than half a year later, the oil price did not go back to its earlier level. For a discussion of the effect of oil prices on inflation, see Trehan (2005), where it is clearly shown that there has been a shift in the impact of oil prices on inflation.

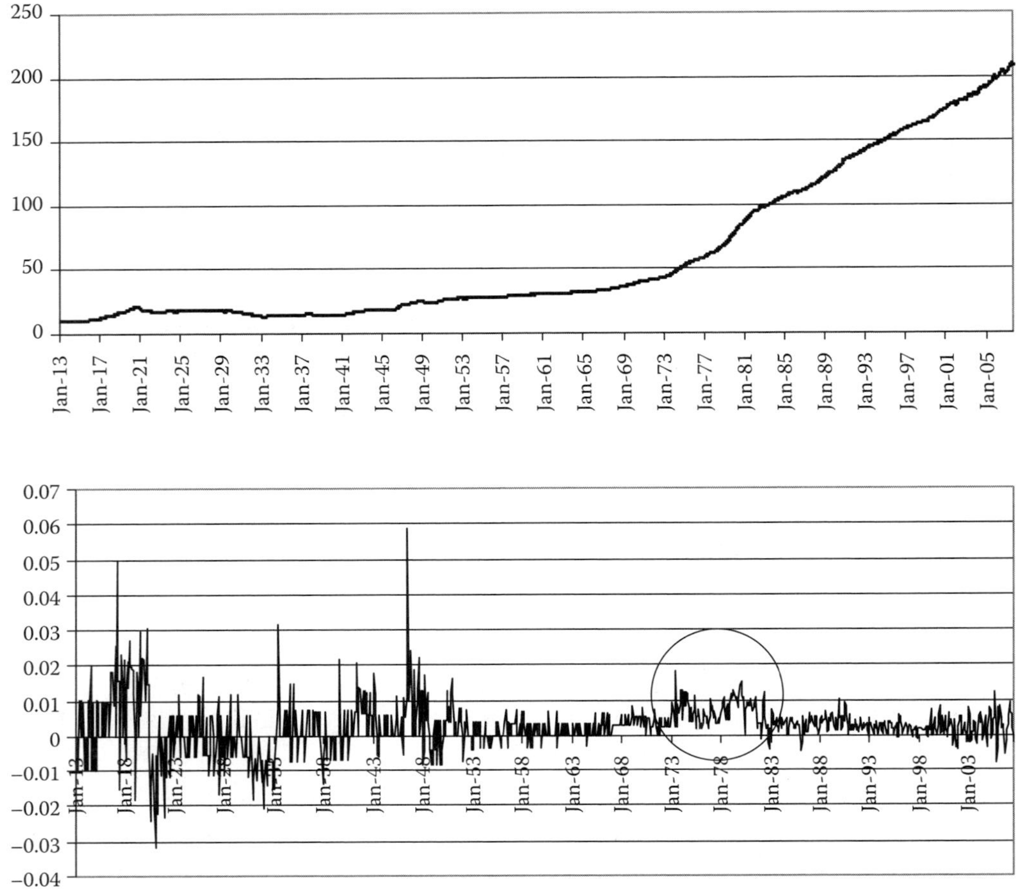

FIGURE 3.1 The development of the CPI in the United States from January 1, 1913 to August 1, 2007. Monthly data. The upper diagram shows the CPI and the lower one shows the relative changes during the same period. (Data from IPM Informed Portfolio Management AB, Sweden.)

Research shows that oil price shocks fail to predict inflation after the "great inflation period." Trehan (2005) also argues that the oil shocks during the 1970s were assigned too large a role in the inflation during the "great inflation period" and ignoring inflation expectations and monetary policy.

The two oil price peaks are shown in Figure 3.3.

Up to the mid-1960s, the yield on U.S. long-term government securities was in the range between 2% and 6%. This changed dramatically during the 1970s where we saw an accelerating inflation with both an increased level and volatility of interest rates. This is illustrated in Figure 3.4.

Applying a GARCH(1,1) (Generalized Autoregressive Conditional Heteroskedasticity) model to the relative changes in the yield curve gives the conditional variance $h_t = 0.000183 + 0.700228\varepsilon_{t-1}^2 + 0.175651h_{t-1}$. This means that 18% of the variance shock in the relative changes remains during the next month. A presentation and discussion of

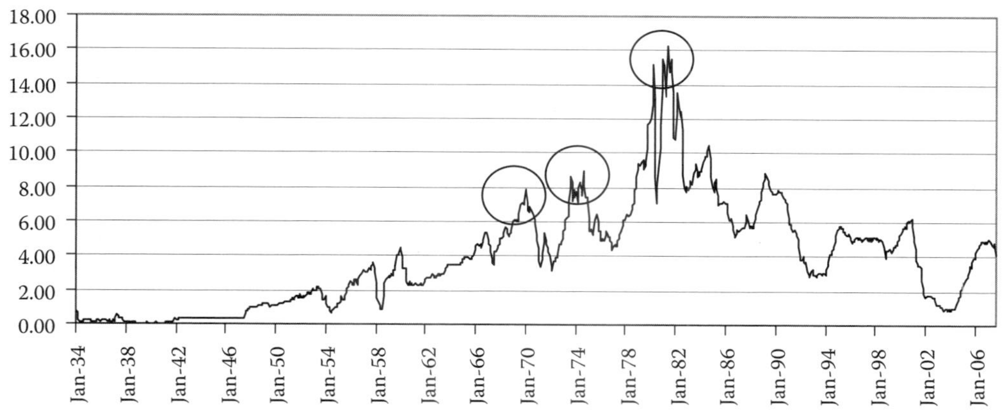

FIGURE 3.2 The U.S. 3-month Treasury Bill: Secondary Market Rate. Monthly data. Period: 1934-01-01–2007-08-01. The three inflation peaks are seen around 1969, 1973–1974 and 1980–1981. (Data from Economic Research. Federal Reserve Bank of St. Louis. Web site: research.stlouisfed.org/fred2/series/TB3MS?cid=116.)

ARCH (Autoregressive Conditional Heteroskedasticity) and GARCH modeling is given in, for example, RSAS (2003) and Engel (2001).

In 1971, the American president announced that the US dollar no longer should be based on the gold standard. The floating exchange rates promoted new theoretical approaches to bonds and their yields. Instead of discretionary focus on each maturity, the analyzers started to draw a curve through all yields. This was also the start for researchers to develop models on bond prices as a function of interest rates. The first was the Vasicek model presented in 1977. The ARCH and GARCH techniques to explain the clustering of changes in financial time series were developed in the first part of the 1980s; see, for example, RSAS (2003).

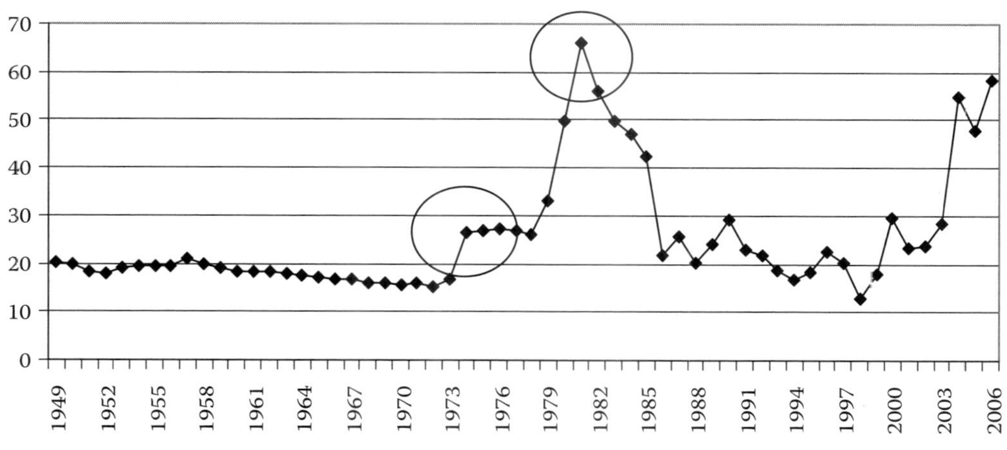

FIGURE 3.3 U.S. oil prices; average in US dollar/bbl. Inflation adjusted 2006. Note the two peaks around 1973–1974 and in the beginning of the 1980s. (Data from www.inflationdata.com/inflation/inflation_rate/Historical_Oil_Prices_Table.asp.)

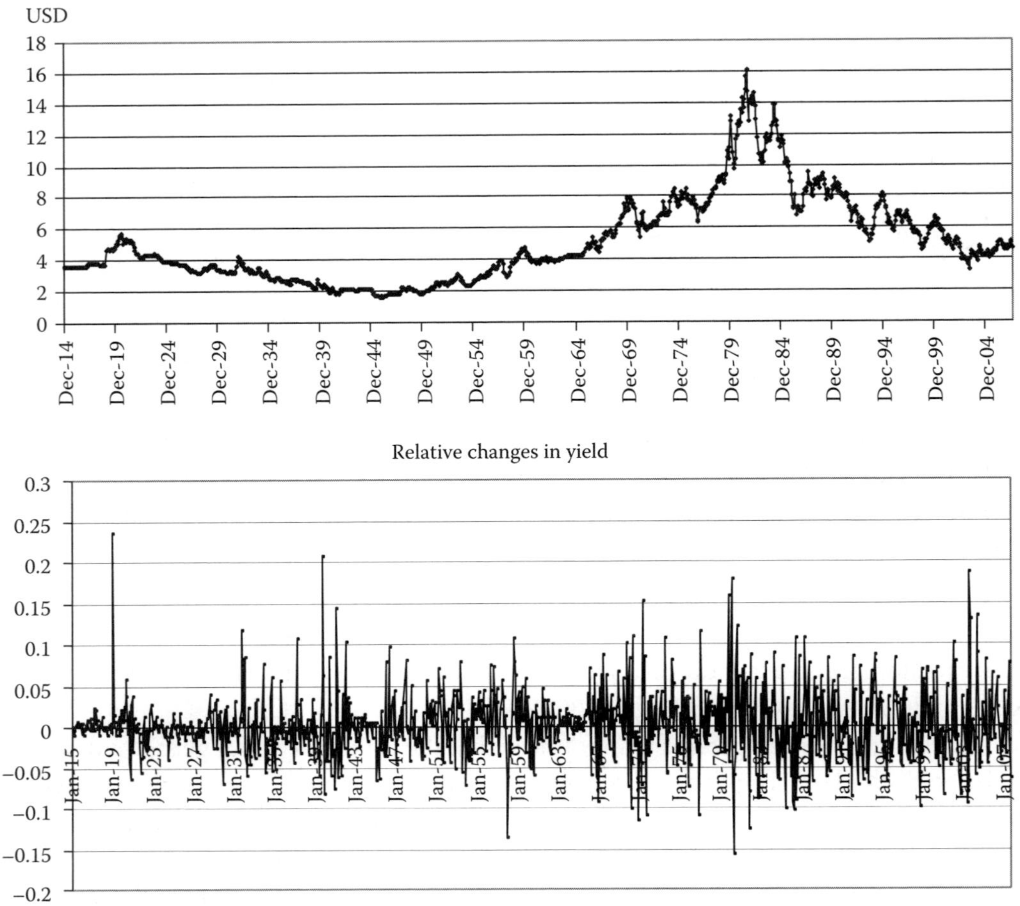

FIGURE 3.4 The 10-year yield curve for U.S. Government Benchmarks, Bid, Average USD. Monthly data. The relative changes, as measured by $[Y(t)/Y(t-1)-1]$, where $Y(t)$ is the 10-year yield, is shown in the lower part of the figure. (Data from IPM Informed Portfolio Management AB, Sweden.)

All these changes during the 1970s were also a stepping stone to the development of ALM approaches for the insurance business; see, for example, Henderson (2004, p. 176). At the beginning of the development of ALM, only the interest rate was focused on; later on, the ALM was broadened to include other risks too.

3.2 RISK MANAGEMENT AND ALM

One fundamental aspect of the insurance business is uncertainty, both in number of claims and claim severity. Even for a certain event, such as death, the time is uncertain. The core activity within the insurance business is to manage these risks. From a historical view, risk management (RM) was mainly on the life insurance side to begin with, but during the last decades it is also within nonlife businesses.

In general terms, the purpose of managing risks is to limit the undertaking's risk exposure to a level that is acceptable for the undertaking's managing board. One way to handle this is to use some kind of risk mitigation technique, such as reinsurance. A system of RM is an arrangement involving one or more processes such as risk identification, risk control, risk mitigation, and financing.

"Risk appetite" is a new term including both negative and positive elements as compared to, for example, risk tolerance, which has only negative aspects and there is a certain amount of risk that can be borne. An introduction to risk appetite is given by D'Arcy (2009).

Doff (2007) gives five reasons why the managing of risks in the insurance industry has grown in importance during the last decades.

1. The industry has undergone deregulation
2. Privatization for reducing government costs, and so on has emerged
3. The industry is acting more and more across borders—internationalization
4. The capital market has become more volatile—globalization
5. Growing risks and insolvency

Due to these changes the industry has changed its way of handling risks. Step by step the industry has moved from a silo approach of managing risks, that is, concentration on one risk at a time and seeking a way to minimize this risk, to a more holistic view, where all risks and their interdependence are taken care of at the same time as the focus is on the managing board, its being fit and proper, and so on. The latter case is called the ERM; see Section 3.3. Adding a corporate finance dimension to the RM could create value for the enterprise; see Shimpi and Lowe (2006).

The steps in managing risks leading to the ERM can be divided into four main parts:

1. Model office tools
2. Tools to manage asset–liability interactions
3. Simulation tools and testing
4. DFA tools

Below, we will summarize the above steps leading to the full ERM approach.

3.2.1 Model Office Tools

The model office is used to get an understanding of the evolvement of insurance business in the future under changing conditions, such as the effect of new businesses, changing mortality intensities, and assets returns. An introduction to the model office approach is given in Macdonald (2004), where it is stated that the first construction of a model office was probably done by Manley (1869). Manley, not using the term "model office," stated that when a business has been established, "there is no part of its management so important as that relating to the periodical valuation of the liabilities under its polices." He also argued that the periodical valuation should be twofold:

1. To determine the exact financial positions of the business at stated periods
2. To ascertain, if any, what surplus may be available and disposable as realized profits

Manley's approach was for projecting the policies of a business using different mortality tables and methods of valuation (different interest rates, premium loadings, etc.). His first "hypothetical tables" were for annuities taken at the age of 30 and valued at 10, 20, 30, 40, 50, and 60 years using three different interest rates (3, 3.5, and 4) and different mortality tables.

The term "model office" was introduced in Valentine (1875), where Valentine referred to Manley's constructed tables as representing "a model office; showing assumed amounts of policies taken out at the end of stated periods, and that, having done so, he formed a second table giving reserves, in the case of such model office, at different stages in its career, according to various data." To simplify the definition of a model office is to introduce the pocket model office concept, that is, using a few policies to replace the many and simplifying the calculation; *cf.* Ward (1968).

Following these pioneering papers, different model offices were constructed; see, for example, Macdonald (2004). One milestone in the use of a model office was the work done by Haynes and Kirton (1952), see Section 3.2.2, on cash-flow matching. Later on, these model offices became representative policy models, what is nowadays called *model points* and needed in modern tools such as simulations and DFA. A model point is thus a representative policy for the model calculation.

In the early days without computers, the model offices were disused as they failed to allow for changes in time in the portfolio in an easy way. These early models were in this sense static. With computers it became easy to change the parameters in the model office to get outputs of a changing environment. The steady-state approach as given by, for example, Manley and Valentine is called a *static model* approach, as compared to *dynamic models* showing a changing environment; *cf.* Ward (1968), where Ward discusses the evolution of model offices.

A discussion on "models of model offices" is given in, for example, Macdonald (1994, 1997).

3.2.2 Tools to Manage Asset–Liability Interactions

In the early stages of actuarial science, the managing of risks was one-sided, mainly on the liability side and less on the assets side, but not on both sides at the same time. Fluctuations in interest rates will have an influence on both sides. It is generally understood that there are three main approaches to this problem: cash-flow testing, immunization, and cash-flow matching. At the end of this section, we will discuss ALM. A good introduction to asset management is given in Sherris (2004), where the historical development and principles, strategies, and models are discussed. A brief history of ALM is also given by Martellini (2006).

3.2.2.1 Cash-Flow Testing

Cash-Flow Testing, CFT, has become an important aspect of the insurance industry. The CFT is defined by ASB (2002) as "A form of cash flow analysis involving the projecting and comparison of the timing and amount of cash flows resulting from economic and other assumptions." A cash-flow analysis is any evaluation of the risks associated with the timing or amount of cash flow. A first actuarial standard of practice was published in 1988; see ASB (1988).

A New York State regulation from 1986 required insurance companies underwriting annuities and guaranteed income contracts to perform an asset–liability analysis, namely a cash-flow test. In 1993, the U.S. NAIC adopted a Standard Valuation Law, which requires insurers to perform a CFT to verify that they have sufficient reserves. The testing should follow the ASB (2002). The scenarios, that is, the set of economic and other assumptions that the CFT is based on, are the key elements of the test. The number of scenarios should be consistent with the purpose of the analysis (ASB, 2002). The scenarios are either deterministic or stochastic.

The NAIC testing involved assessing the impact of seven interest rate scenarios, for example, no deviation, increasing, decreasing, and so on. For an illustration of these tests, see Sigma (2000). In NAIC's RBC system, see Section 2.2, the C3-risk is charged using a CFT procedure with 12 or 50 scenarios; see NAIC (2006). Cash-flow testing as a management tool is discussed in Behan (1993). The CFT is a useful ALM diagnostic tool, but it suffers from some basic shortcomings, for example, it can be manipulated to produce desired results and it is focusing only on the interest rate risk.

3.2.2.2 Immunization

In the pioneering work by Haynes and Kirton (1952), the interaction and relationship between the assets and liabilities of a life insurance business were considered using a model office. The guarantees in life insurance contracts are based on the results of the investments done by the insurance company. Haynes and Kirton (1952) asked the question "How can a life office offer such guarantees?" They concluded that the guarantees must be backed either by *matched assets* (and providing an equivalent guarantee of capital and interest) or by sufficient surplus to cover possible adverse effects of departure from the "matched assets" position.

Of course, the accepted principle in the life insurance business was that bearing in mind that in distributing its assets in investments, the outstanding term of existing liabilities and its date of redemption should be taken care of; see Haynes and Kirton (1952, p. 142). They also maintained that "it is difficult in practice to determine where the optimum date distribution of assets lies—the distribution which, so far as possible, will insulate the fund from the effect of fluctuations in the market rate of interest." They used simple model offices to match assets and liabilities.

A couple of months later, another pioneering work was published, namely that on *immunizations* by Redington (1952). In 1938, Macaulay introduced a summary statistic of the effective maturity of bonds termed "duration" or "Macaulay duration" (Macaulay, 1938). The Macaulay duration is a direct measure of a bond's volatility and could be seen as a weighted average term to maturity where the years are weighted by the present value of the related cash flow. The combination of size and timing of coupon payments with the time to maturity has been developed and discussed by Ferguson (1983) and Tilley (1988).

The idea of Redington was to immunize the portfolio value against interest rate changes using the Macaulay duration. Let, as before, A be the value of assets and L the value of the liabilities. We introduce the value of the cash flow at time t, generated by the assets from dividends, coupons, and so on, by $A(t)$ and we denote the value of the cash flow at time t from outstanding insurance policies by $L(t)$. Let V denote the interest rate function, $P_A(t)$ the probability that the asset cash flow will occur, and $P_L(t)$ the probability that the liability cash flow will occur. The fundamental principle behind the present value of the assets V_A and the liabilities V_L is exactly the same, namely

$$V_A = \sum_t v^t P_A(t)A(t) \quad \text{and} \quad V_L = \sum_t v^t P_L(t)L(t).$$

Redington introduced the idea that if $V_A = V_L$, then they must also be equally responsive to changes in the rate of interest, that is, if the interest rate changes from δ to $(\delta + \varepsilon)$ and a change of V_A and V_L to V'_A and V'_L, then by using a Taylor expansion, we obtain

$$V'_A - V'_L = (V_A - V_L) + \varepsilon \frac{\mathrm{d}(V_A - V_L)}{\mathrm{d}\delta} + \frac{\varepsilon^2}{2!} \frac{\mathrm{d}^2(V_A - V_L)}{\mathrm{d}\delta^2} + \cdots. \tag{3.1}$$

As $V_A = V_L$, the first term will vanish. Redington proposed as a satisfactory immunization policy that

$$\frac{\mathrm{d}(V_A - V_L)}{\mathrm{d}\delta} = 0, \tag{3.2}$$

and

$$\frac{\mathrm{d}^2(V_A - V_L)}{\mathrm{d}\delta^2} > 0. \tag{3.3}$$

Equation 3.2 is called a first-order immunization. If the second derivative is positive, Equation 3.3, then because ϵ^2 is positive irrespective of whether the change is positive or negative, any change in the interest rate will result in a profit. This result holds if the change is not so large that the higher terms in the Taylor expansion will have an influence on the result.

	Immunization
$\sum_t v^t\,[P_A(t)A(t) - P_L(t)L(t)] = 0$	There is no surplus or deficit initially
$\sum_t tv^t\,[P_A(t)A(t) - P_L(t)L(t)] = 0$	Duration matching is achieved, that is, the Macaulay durations are the same for assets and liabilities (*volatility matching*)
$\sum_t t^2v^t\,[P_A(t)A(t) - P_L(t)L(t)] > 0$	The portfolio of assets has a greater degree of convexity than the liabilities. If equality, we have *convexity matching*

Cairns (2004b) refers to immunization as a means of managing the risks associated with parallel shifts in the yield curve. This theory suggests the possibility of arbitrage; *cf.* Cairns (2004b) for a discussion of some problems arising from the classical theory of immunization. To make the matching more accurate we could set Equation 3.3 equal to zero. This is usually called convexity matching or a second-order immunization.

Vanderhoof (1972) brought the immunization theory to North America. In the Trowbridge report (1979), resulting in the proposal for a risk-based capital requirement (*cf.* Section 2.2), Mr Samuel Turner summarized the discussion on immunization as follows: "I believe we are equally convinced that recognition of immunization will also result in increased awareness of interest risks, and therefore is a sounder managed company." Vanderhoof (1973) introduced two new concepts to the problem of matching assets and liabilities. The first was the use of reinvestment rates, that is, the rates anticipated for new investments in future years, and the second was the recognition that investment cash flows are a function of the economical environment (i.e., of future reinvestment rates). Tilley (1980) recognized that insurance cash flows are also a function of the economic environment. He also showed that it was possible to use asset–liability matching (A/L matching) to shape the business investment strategy.

For a historical discussion on duration and immunization, see Leibowitz (1983). Immunization is one application of a hedging approach.

In the model proposed by Redington there was no distinction between short-term and long-term interest rates. All yield curves were assumed to be flat. The interest rate shocks were also assumed to be small. Redington's theory was generalized to the case where the interest rate is a function of time and the shocks are of arbitrary magnitude in Shiu (1986).

Interest rate modeling has become an important part in stochastic simulations. Two good references are Cairns (2004a, 2004c). Yield elasticity as a measure of interest rate risk is discussed in Fisher (2005).

In Simester (1982) there was a discussion on immunization versus dedication, that is, cash-flow matching. A matched portfolio is a special case of the immunized portfolio, where the portfolio is structured so that each payout is exactly matched by coupons and maturity proceeds of assets. A dedicated portfolio is one where the exact matching is the goal, but proceeds of investments are allowed to come a little early where exact precision is not possible or practical (Simester, 1982, p. 1280).

"In 1951 the sessional meetings committee of the Institute of Actuaries asked me to write the paper on life office valuations for delivery in Spring 1952 that now appears in J.I.A. 78 (1952), p. 286. Although the paper was, in this sense, commandeered, it was a task which I undertook with enthusiasm because valuation was both my job and my hobby. Although I am an inveterate re-polisher the task was completed to my satisfaction some fortnight before the date for delivery to the scrutineers. At that stage the paper was in its final form except that instead of the passages on immunization in the earlier part of the paper, there was a conventional passage on the necessity of matching assets with liabilities on such matters as currency, degree of risk and spread over time. I have no doubt that this passage was as plausible as my predecessors had always been on these subjects but attitudes were changing, a spirit of enquiry was abroad, and these merely ritual genuflexions were ceasing to satisfy. Matching had not been one of the foremost issues which had been speeding my pen, but as I lay in bed one Saturday morning about a fortnight before the paper was due, relaxed because it was a weekend and my paper was finished, the little cloud of doubt about matching found an empty mind in which to grow.

.....

These equations, which are very nearly self-evident, were clearly established in my mind that Saturday morning before I raised my head from the pillow, and it is the principles which they embody to which I gave the rather impetuous title of immunisation.

......

Perceptive readers of my paper must have been puzzled by an odd shift of attitude between the early passages on immunization and the later sections on valuation which, although it is not apparent, were, of course, written earlier."

Source: Data from Frank M. Redington, The Actuary, The Newsletter of the Society of Actuaries, Volume 16, No. 1, January, 1982.

Insurers are advised to consider duration matching a useful rule of thumb to be supplemented with more sophisticated ALM techniques (Sigma, 2000, p. 15).

3.2.2.3 Cash-Flow Matching

Cash-flow matching was suggested in a formal way by Koopmans (1942) as a way of managing assets and liabilities; *cf.* Shiu (2004). The main challenge is to find the cheapest portfolio of fixed income securities such that the accumulated net cash flows are nonnegative at all planning time periods. References to the literature on modeling cash-flow matching can be found in Shiu (2004). The practice of matching assets to the liabilities is known as cash-flow matching or dedication.

3.2.2.4 Asset–Liability Managing

ALM was originally developed to address interest rate risk. The interest rate risk became a concern during the 1970s when the interest rates increased and became more volatile than in the past; *cf.* the beginning of this chapter. The insurance industry developed a full arsenal of techniques to deal with the interest rate risk. As the techniques evolved they incorporated other risks, and became an important tool for managing both product-specific risks and company-specific risks. As discussed in Sigma (2000), ALM mainly consists of four techniques: not only the three mentioned earlier, namely CFT, immunization, and cash-flow matching, but also DFA. The latter is discussed in Section 3.2.4.

ALM and duration analysis were introduced to the nonlife insurance business (at least in North America) by Ferguson (1983). One ALM technique that has evolved mainly within the nonlife industry is DFA, discussed in Section 3.2.4.

The ALM provides a framework for assessing and managing a company's risk exposure. The Society of Actuaries, SOA, has defined ALM (SOA, 1998).

ALM is the practice of managing a business so that decisions on assets and liabilities are coordinated; it can be defined as the ongoing process of formulating, implementing, monitoring, and revising strategies related to assets and liabilities in an attempt to achieve financial objectives for a given set of risk tolerances and constraints.

Financial objectives and acceptable risk levels are defined by the organization. ALM is relevant to, and critical for, the sound management of the finances of any institution that invests to meet liabilities.

SOA (1998) includes references to the literature that the SOA is recommending. Basics of financial economics that are relevant to ALM can be listed under the following topics:

- Markowitz portfolio selection
- Capital–Asset Pricing Model, CAPM, and other models for evaluating a security's risk and return
    - Index models
    - Arbitrage Pricing Theory, APT
- Derivatives
- Behavioral finance

The pricing of nonlife insurance contracts is discussed in the pioneering work by D'Arcy and Doherty (1988). Financial economic theory such as pricing an insurance contract is discussed in D'Arcy (1989).

Bank ALM and VaR is discussed in SOA (1998). In general, banks manage risks by matching assets and liabilities (asset–liability mismatch risk), by matching duration, and by hedging and securitization. Much of the techniques stems from the Black–Scholes model and related works. A very good source is Field (2003). The book consists of a collection of 55 papers on general financial RM.

Securitization is a financial technique that transforms assets into tradable securities. It is used as a tool to reduce the capital liquidity by creating securities. It has evolved on the bank side since the 1970s in parallel with the techniques of derivatives. It became a new technique for the insurance industry in the beginning of the 1990s.

"To hedge" means offsetting a risk inherent in any market position by taking an equal but opposite position in the market. Thus, any loss on the original investment will be hedged, or offset, by a corresponding profit from the hedging instrument. A risk is said to be hedgeable if it can be avoided or mitigated by an offsetting transaction, for example, by the use of financial derivative instruments, such as options or futures. A summary of hedging and RM is given in Cairns (2004d).

The GISG simulation model, see Section 2.2, was later developed into a tool for RM and ALM; see Daykin and Hey (1990).

General ALM techniques, including immunization, were introduced for pension funds in the beginning of the 2000s under the name Liability-Driven Investments, LDI. The main goal of LDI, as an investment strategy, is to gain sufficient assets for the pension fund to meet all its liabilities. Nothing is new under the sun! For a discussion on LDI see, for example, AON (2005) and JPMorgan (2006); see also Martellini (2006).

3.2.3 Simulation Tools and Testing

The more general ALM models are based on simulations. As discussed in Section 2.2, the pioneering work on solvency and simulations was done by a working party within the Finnish Society of Actuaries, and working parties within the Institute and Faculty of Actuaries in the United Kingdom. Two basic sources are Pentikäinen et al. (1989) and Daykin and Hey (1990).

Inflation and asset movements were modeled by using autoregressive stochastic processes—the Wilkie model (Wilkie, 1984, 1986a, 1986b, 1995).

Many sources on simulation models can be found in the literature, but the two main sources given above are among the best.

The main goal of a simulation approach to ALM is not only to capture all risks that insurance undertakings are facing, and the interaction between assets and liabilities, but also capturing diversification effects.

In the "early days," the simulation models were usually deterministic, meaning a "what if" analysis. The values of the variables in the model were predefined. Nowadays, they are usually stochastic in nature, meaning that the variables are selected from assumed probability distributions. But it is important to remember that a deterministic model may help us to interpret the results from a stochastic model.

Scenario testing, or deterministic scenario testing, projects the trends of a company's financial conditions under various scenarios. Stress testing involves the worst-case and unlikely scenarios and is as such an extreme case of the scenario testing. Using a stochastic simulation approach means that numerous scenarios are randomly generated. This approach is usually termed stochastic scenario testing. The trend has been from deterministic modeling techniques to stochastic modeling techniques and from static to dynamic. Much focus has also been on the interrelationship between the variables, leading to dynamic financial analyses; see Section 3.2.4.

Valuating derivatives and other financial securities is best done by using Monte Carlo simulation methods. An application-based survey and textbook is found in McLeish (2005). In this book, McLeish considers the simulation techniques as "the mathematical equivalent of the invention of the printing press."

3.2.4 DFA Tools

There has been a significant difference between life insurance and nonlife insurance industries in using and practicing ALM. As was seen above in Section 3.2.2, the first ALM models were focusing on a single risk, namely the interest rate risk. Typically, these models are considering a single line of business and assets supporting that line. The Finnish and British simulation approaches were the first attempts toward a more holistic view of the business and its risks. Typically, the DFA models the reaction of all quantifiable risks and the assets covering those risks. The DFA originated from the pure ALM approaches. The models used by the nonlife insurance industry include the underwriting risk and others. As such it is also a part of the ERM; see the next chapter. Early variants of DFA have been dynamic solvency testing (DST) for the life insurance business and dynamic capital adequacy testing (DCAT).

In CAS' handbook, three definitions of DFA are given:

- A systematic approach to financial modeling in which financial results are projected under a variety of possible scenarios, showing how outcomes might be affected by changing internal and/or external conditions.
- A way of studying the behavior of industrial systems to show how policies, decisions, structure, and delays are interrelated to influence growth and stability. DFA integrates the separate functional areas of management—marketing, investment, research, personnel, production, and accounting. Each of these functions is reduced to a common basis by recognizing that any economic or corporate activity consists of flows of money, orders, materials, personnel, and capital equipment.
- An analysis that considers all the material components of a given system (e.g., an insurance enterprise) and their interrelationships in one integrated model, which projects income and cash flows over a period of time.

With its RBC system introduced in 1992 and 1993, the U.S. NAIC proposed that scenario testing and dynamic modeling were to follow in order to better reflect the solvency status of the insurance industry. This made the American and Canadian actuarial organizations to release different handbooks on dynamic modeling, for example, *the Casualty Actuarial Society's DFA Handbook,* which was released in September 1995, see CAS (1995); *the Society of Actuaries' Handbook*, see SOA (1996); and *the Canadian Institute of Actuaries' educational note on Dynamic Capital Adequacy Testing*, see CIA (1996). The CAS' release of its *Handbook on DFA* in 1995 and its research committee, Dynamic Financial Analysis Committee (DFAC) have been the driving forces behind the development of DFA for the nonlife business. Due to the evolving of DFA techniques, the DFAC changed its name to Dynamic Risk Modelling Committee (DRMC) in 2006. Accordingly, the name of the handbook has also changed to *Dynamic Risk Modelling Handbook.* For a description of DFA and its development, see Blum and Dacorogna (2004), Erling and Parnitzke (2007), Hardy (2004), and Shiu (2006). A model framework and an application of DFA are given in Lowe and Stanard (1997). The latter paper is also a step toward an ERM application.

In setting up a DFA model, one needs to go through a couple of steps in the conducting process; see, for example, Shiu (2006).

- Investigate the risks faced by the insurer and the factors affecting its performance
    - Make up a list of risk categories—both on the liability and the asset sides
    - Conduct a statistical/economic analysis to determine which risks to incorporate into the model

- Choose reward and risk measures
    - Select measures in accordance with the purpose of the analysis
    - Typical reward measures: expected surplus, policyholder surplus, and shareholder's equity
    - Typical risk measures: deviation of the reward measures, VaR, Tail VaR
- Determine projection time
    - The projection period depends on the characteristics of the risks
    - It should be long enough to capture the effects of all risks (e.g., ultimo)
- Build a DFA model
    - Usually, a model office is assigned model points; the model office could either be deterministic or stochastic
    - A scenario generator comprising stochastic models for the risk factors (RFs) is set up
    - The scenario generator is calibrated—suitable parameters are defined
- Project cash flow
    - Future cash flow is projected using the scenario generator
- Interpret the results and provide feedback
    - Discussion and interpretation of results
- Prepare a written report and present it to the Board
    - Communicate the results

With a deterministic model office, scenarios are prespecified to be used for projecting the company's assets and liabilities. One way is to use some stress tests. In a stochastic model office, a stochastic model is used to generate the projecting. The parameters used are usually taken from historical data.

3.3 ENTERPRISE RISK MANAGEMENT

ERM is an umbrella approach of different, but coherent, processes in the identifying, assessing, planning, organizing, leading, controlling, and acting activities, such as measuring, monitoring, and mitigating risks, within an organization in order to minimize the effects of risks of potential events that may affect the organization.

One part of the ERM is the solvency assessment. During the 1990s, we got RBC systems implemented in various countries and the development of DFA approaches; *cf.* Chapter 2 and Section 3.2. The IAIS and the IAA in their framework on solvency assessment have adopted the Basel II three-pillar approach. Other projects in different countries, such as the Solvency II project in Europe, have implemented a "total risk view," meaning that all risks that the organizations face should be part of the solvency assessment, either measured quantitatively or qualitatively.

External factors in the globalization, such as the corporate accounting scandals (such as the Enron, WorldCom, and Anderson Consulting affairs), have had profound implications worldwide and led, for example, to adoption of the U.S. Sarbanes–Oxley Act of 2002. These new American laws included RM processes aimed at keeping an up-to-date internal control process.

Bennet (2004) defines ERM as "The process by which organizations assess, control, exploit, finance and monitor risks from all sources for the purpose of increasing the organizations short- and long-term value to shareholder' (from Casualty Actuarial Society; CAS). It combines a whole range of financial risks with insurance risks and seeks to optimize the manner of its risk taking"; see also CAS' Web site on ERM.[*] The SOA perceives ERM as an organized, systematic way of managing risks throughout an organization on an ongoing process; *cf.* SOA (2006). It is a process that should become an integral part of how an organization operates. Wang (2004) explains, "ERM will not replace existing specializations such as asset–risk modeling, credit–risk modeling, and so on. Instead ERM is a new specialization that coordinates the risk-taking activities of various business units, reconciles diverse perspectives and harmonizes different economic interests and incentives for the ultimate benefit of the enterprise." IAIS (2007b) identifies that the ERM involves the self-assessment of all reasonably foreseeable and relevant risks that an insurer faces and their interrelationships. The European Solvency II is an example of an ERM approach.

Some synonyms to ERM are, for example, business risk management, holistic risk management, and strategic risk management.

Traditional RM focuses on the risk associated with individual investments and insurance portfolios. The ERM has a holistic view of the enterprise that forms the difference to RM, as it is the risk environment of the entire enterprise that is the focus in ERM. This means that also the people involved and the way they act are of major importance as a part of the ERM! The traditional RM approach and the ERM approach are illustrated in Figures 3.5a and 3.5b, respectively.

As the ERM is an ongoing process, it is not just compliance. It has to be a part of the culture of the whole organization and, as such, has to be embedded into the organization's strategic goals. One part of the process is Corporate Governance, meaning the relationship among various participants in determining the direction and performance of corporations (SOA, 2006). The people involved have to be "fit and proper" and the ERM must start from the top level of the management team and the board. The enterprise must have access to expertise, as the ERM is multidisciplinary (e.g., the board, actuaries, risk officers,[†] investors, accountants, etc.). However, the drivers behind ERM are not only corporate governance, management controls, financial transparency, and governmental regulations, but also international regulatory authorities (Basel II, Solvency II, and IAIS), rating agencies, and other international organizations (e.g., IASB and IAA).

[*] http://www.casact.org/research/erm

[†] For example, the Chief Risk Officer; see CRO Forum at http://www.croforum.org

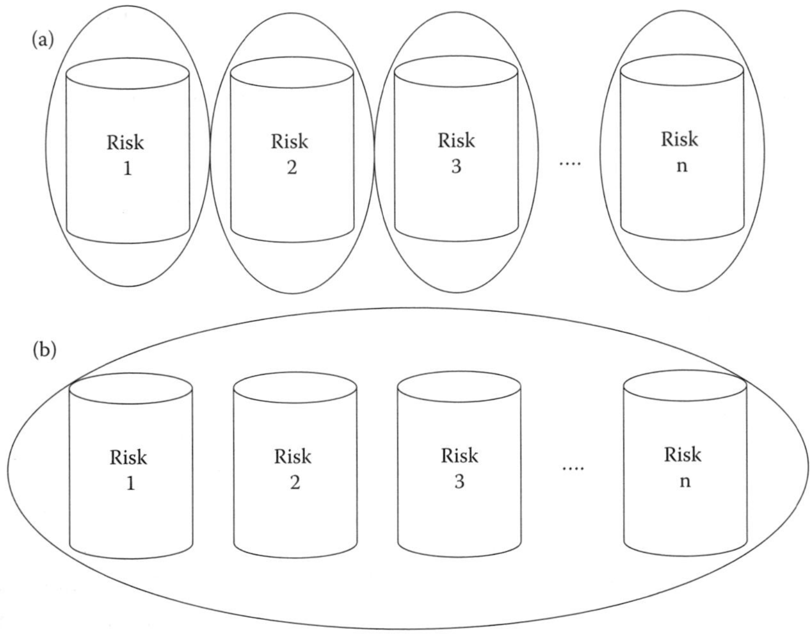

FIGURE 3.5 (a) The traditional managing of risks is based on the silo approach: each risk is identified and managed using different techniques; and (b) ERM is based on a holistic approach: all risks of the enterprise are considered all together and managed using a risk process including the managing staff and so on.

3.3.1 COSO's Definition of ERM

In 2004, the Commission of Sponsoring Organizations of the Treadway Commission, COSO,* published guidance on the implementation of a consistent ERM framework (COSO, 2004a). The COSO approach is very much qualitative and focuses on the people involved in the process: "… a process, affected by an entity's board of directors, management and other personnel, applied in a strategy setting and across the enterprise, designed to identify potential events that may affect the entity, and manage risk to be within its risk appetite, to provide reasonable assurance regarding the achievement of entity goals."

In COSO's report Enterprise Risk Management—Integrated Framework (COSO, 2004a), an ERM architecture in terms of four categories of objectives and within each of them eight components are to be considered. The eight interrelated components for ERM apply to each level of an organization and should follow the following four categories of *objectives*:

- *Strategic*—high-level goals, aligned with and supporting its mission
- *Operations*—effective and efficient use of its resources
- *Reporting*—reliability of reporting
- *Compliance*—compliance with applicable laws and regulations

* See http://www.coso.org

The eight interrelated components of ERM are an integrated part of the management process and are derived from the way management runs the enterprise. As the ERM is a multidirectional and iterative process, any one component may influence any other component. The *components* are (*cf.* COSO, 2004b)

- *Internal environment*—The internal environment encompasses the tone of an organization, and sets the basis for how risk is viewed and addressed by an entity's people, including RM philosophy and risk appetite, integrity and ethical values, and the environment in which they operate.
- *Objective setting*—Objectives must exist before management can identify potential events affecting their achievement. ERM ensures that management has in place a process to set objectives and that the chosen objectives support and align with the entity's mission and are consistent with its risk appetite.
- *Event identification*—Internal and external events affecting achievement of an entity's objectives must be identified, distinguishing between risks and opportunities. Opportunities are channeled back to management's strategy or objective-setting processes.
- *Risk assessment*—Risks are analyzed, considering likelihood and impact, as a basis for determining how they should be managed. Risks are assessed on an inherent and a residual basis.
- *Risk response*—Management selects risk responses—avoiding, accepting, reducing, or sharing risk—developing a set of actions to align risks with the entity's risk tolerances and risk appetite.
- *Control activities*—Policies and procedures are established and implemented to help ensure that risk responses are effectively carried out.
- *Information and communication*—Relevant information is identified, captured, and communicated in a form and a timeframe that enable people to carry out their responsibilities. Effective communication also occurs in a broader sense, flowing down, across, and up the entity.
- *Monitoring*—The entirety of ERM is monitored and modifications are made as necessary. Monitoring is accomplished through ongoing management activities, separate evaluations, or both.

As these objectives and components apply to each level of the organization, it can be illustrated as a three-dimensional cube, where the third dimension is the entity's organizational level; see COSO (2004b).

3.3.2 SOA's Definition of ERM

The SOA felt that the strategic objective could be further developed and that greater emphasis should be on the following topics (SOA, 2006):

- Risks external to the entity and outside of management's control
- Interdependent risks and cross-functional issues
- Coordination of RM within the entity
- Transparency of the RM system
- Reputation of the firm
- Quantification of risk, including consideration of risks for which not much data exist—such as future or infrequent risks, or correlated risks
- Long-term evaluation of risk, including scenario planning and stress testing

The SOA proposes the activities of ERM to be organized into four main themes or processes. They are

- *Risk Control*—the process of identifying, monitoring, limiting, avoiding, offsetting, and transferring risks.
- *Strategic Risk Management*—the process of reflecting risk and risk capital in the strategic choices that a company makes.
- *Catastrophic Risk Management*—the process of envisioning and preparing for extreme events that could threaten the viability of the enterprise.
- *Risk Management Culture*—the general approach of the firm for dealing with its risks. A positive Risk Management Culture will incorporate ERM thinking automatically into all management decision making.

The SOA discusses these themes in detail in SOA (2006).

3.3.3 ERM-II–CAS–SOA Definition

In contrast to the qualitative approach taken by the COSO, the ERM Institute International,* ERM-II, has taken a more quantitative approach. The ERM-II, the CAS, and the CAS/SOA Risk Management Section have published a joint report on ERM for property–casualty undertakings (Wang and Faber, 2006). The general conclusion in the report also holds for life insurance undertakings. The report defines ERM as *the discipline of studying the risk dynamics of the enterprise, the interactions of internal/external players and forces, and how players' actions (including the risk management practices) influence the behaviours of the risk dynamics, with the ultimate goal of improving the performance and resiliency of the system*. The definition takes an engineering-like approach and paves the way for a "scientific" approach; see Wang (2006). To understand the above definition the *risk dynamics* has to be defined. With risk dynamics they mean the interactions of forces and players within and outside

* http://www.ermii.org

the enterprise. As there are usually multiple risk dynamics at work, some may dominate others in influencing the behaviors of the enterprise. The multiple risk dynamics and their interaction define a system of risk dynamics on the enterprise (Wang and Faber, 2006). In the report, it is also stated that ERM for the insurance industry must be considered fundamentally different from other industries, as insurance is an option to pay someone else's risk for a fee, whereas other industries primarily manage risks that are relevant to business processes.

The collective theoretical basis for ERM is based on five principles according to the ERM-II–CAS–SOA.

ERM-principle #1: Risk dynamics exist as objective states of nature, of which we can gain more knowledge through experience, insights, and modelling of internal forces within the system. We need to understand the various forces within risk dynamics to be able to measure and predict the strength and the direction of the forces. We also need to have a form of approach to react to them.

ERM-principle #2: An enterprise has multiple risk dynamics at multiple levels with multiple forces. To gain an overall picture we need to understand the interactions of risk dynamics at different levels and to reconcile the multiple perspectives.

There are both local risk dynamics (inherent to the business operations) and macro dynamics (segment and company levels) that are impacting the whole enterprise. They could be harder to identify on a "individual risk level."

ERM-principle #3: Market valuations and internal valuations are among the major forces impacting the dynamics of the enterprise. A big part of ERM is to study the dynamics of external market valuations and their impact on the assets and liabilities of the enterprise, and to use internal valuations to influence the behaviors of the enterprise.

One of the main objectives of an enterprise is to maximize the value for the stakeholders. Therefore, the changes in value, creates a manifestation of the risk dynamics and most of the risks are measured by their economic values. The market valuation will become an even more important factor in the valuation of assets and liabilities as the enterprises move toward a more volatile valuation as the "fair valuation" of IASB.

ERM-principle #4: Properly constructed risk metrics and valuation models can shed light on the behavior of risk dynamics, and are powerful forces and essential tools for taking a structured and disciplined approach that aligns business strategy with the process, people, technology, and knowledge in the organization.

The companies need to develop risk metrics and capital models to show the business activities. Risk valuation models should assess the company's risk profile and capital needs, and to direct the allocation of resources.

ERM-principle #5: Action taken by key participants within risk dynamics (e.g., insurance company executives, underwriters and actuaries, rating agencies and regulators) can exert great influence on the behaviour of these risk dynamics.

The companies need to understand the psychological and behavioural characteristics and their interaction of different players.

In the report, the ERM is considered as the brain and nerve system for the enterprise. The comparison is striking, as the company should collect information in the same way as a human being is using all the senses and instincts to collect information about the external environment. It should be able to interact and respond to feedbacks, for example, by tools for correction of actions. An early warning system (EWS) is one approach to get early information and to use it correctly to give early feedback. It is also stated in the report that ERM is the science of balancing different coexisting forces within risk dynamics.

As this approach is a step away from the qualitative approach of COSO, see above, the quantitative risk valuation models are seen as an essential tool for implementing ERM. The enterprise's sensitivity to risk factors is captured by the risk assessment, which requires knowledge of the enterprise's total risk exposure accumulations and how identified risk factors affect it.

A risk dynamic system may behave in a very curious way under a stressed situation. That is why linear correlation models break down (Wang and Faber, 2006). As stated, risk valuation models can and should be developed. They should be developed for different levels of the business operations, such as the company, business/segment, and segment level. The model could also be built on different risk types such as those proposed by IAA (2004). To develop and implement risk valuation models, Wang and Faber (2006) propose some high-level principles, such as "integrated valuation," "forward-looking," "model robustness and benchmarking," "understanding," "commitment," and "monitoring of behaviors."

Rating agencies are now including an ERM evaluation as a part of their rating review. Standard and Poor's will evaluate the ERM quality of an insurance firm according to five areas (see Ingram, 2006): (1) risk management culture, (2) risk controls, (3) emerging risk management, (4) risk and economic capital models, and (5) strategic risk management.

Practical, quantitative solutions are presented in, for example, Brehm et al. (2007) and Wang and Faber (2006). For a historical overview of the development of the ERM, see, for example, SOA (2006) and Brehm et al. (2007).

3.3.4 IAIS ERM Standards and Guidance

In IAIS' framework paper on insurer solvency (IAIS, 2005), not only the governance process and controls in areas such as the Board, directors, and so on, but also the fit and proper testing of directors and management are referred to. Guidance on the establishment and operations of an ERM framework and its importance from a supervisory perspective in underpinning robust solvency assessment is given in IAIS (2007b).

The IAIS has developed a number of principles, standards, and guidance papers to help promote the development of global well-regulated insurance markets. In October 2008, IAIS approved and published its standards and guidance papers on ERM (IAIS, 2008a, 2008b). IAA has developed a Note on this guidance; see Section 3.3.6 and IAA (2009a).

IAIS recognizes that there are a number of commonly used terms to describe the process of identifying, assessing, measuring, monitoring, controlling, and mitigating risks. These activities are a description of the ERM activities.

The IAIS best practice and its requirements for ERM are given below and described in Figure 3.6. The IAIS standards (key features) on ERM are given below.

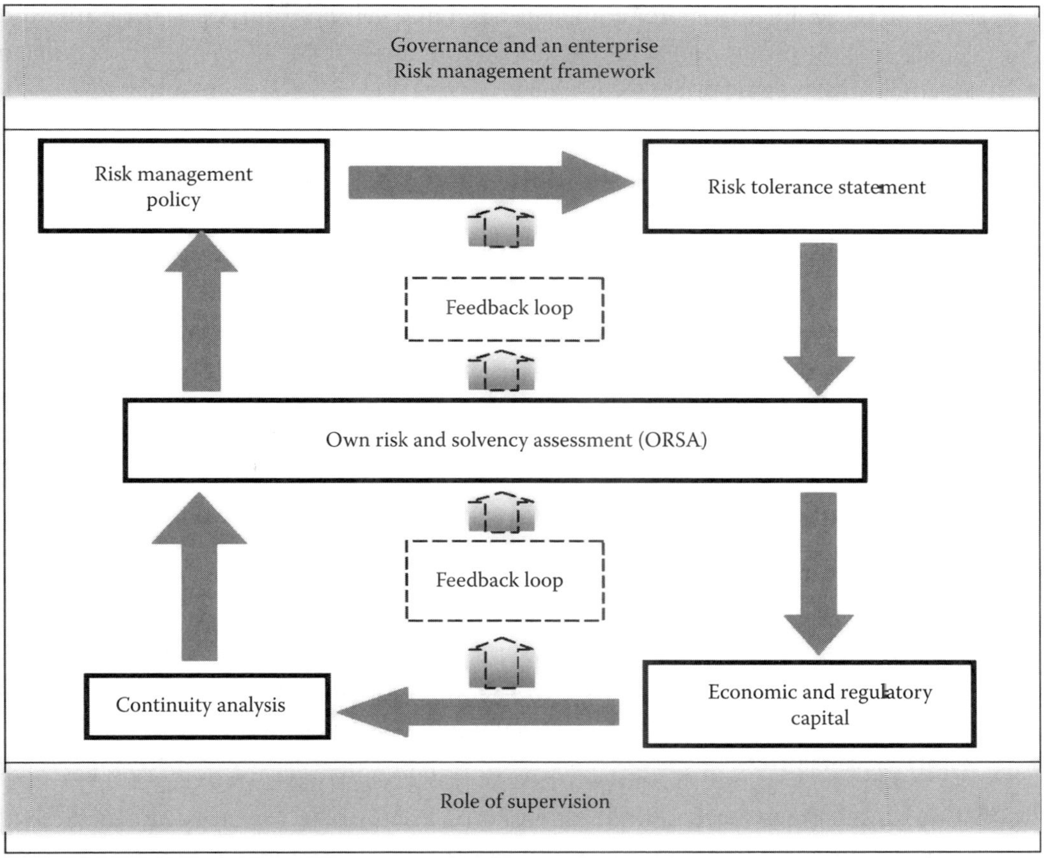

FIGURE 3.6 The best practice of ERM as proposed by IAIS. The features are references to the Key Features below. (Adapted from IAIS. 2008b. Guidance Paper on Enterprise Risk Management for Capital Adequacy and Solvency Purposes. International Association of Insurance Supervisors. Approved October. Guidance Paper No. 2.2.5.)

The eight key features of the IAIS Standard on ERM are as follows—a combination of IAIS (2007b) and IAIS (2008b):

Key Feature 1—Governance and an ERM Framework

1. As part of its overall governance structure, an insurer should establish, and operate within, a sound ERM framework that is appropriate to the nature, scale, and complexity of its business and risks.
2. The ERM framework should be integrated with the insurer's business operations and culture, and address all reasonably foreseeable and relevant material risks faced by the insurer in accordance with a properly constructed RM policy.
3. The establishment and operation of the ERM framework should be led by and overseen by the insurer's board and senior management.

4. For it to be adequate for capital management and solvency purposes, the framework should include provision for the quantification of risk for a sufficiently wide range of outcomes using appropriate techniques.

5. Measurement of risk should be supported by accurate documentation providing appropriately detailed descriptions and explanations of risks.

Key Feature 2—Risk Management Policy

6. An insurer should have a RM policy that outlines the way in which the insurer manages each relevant and material category of risk, both strategically and operationally.

7. The policy should describe the linkage with the insurer's tolerance limits, regulatory capital requirements, economic capital, and the processes and methods for monitoring risk.

Key Feature 3—Risk Tolerance Statement

8. An insurer should establish and maintain a risk tolerance statement that sets out its overall quantitative and qualitative tolerance levels and defines tolerance limits for each relevant and material category of risk, taking into account the relationships between these risk categories.

9. The risk tolerance levels should be based on the insurer's strategy and be actively applied within its ERM framework and RM policy.

10. The defined risk tolerance limits should be embedded in the insurer's ongoing operations via its RM policies and procedures.

Key Feature 4—Risk Responsiveness and Feedback Loop

11. The insurer's ERM framework should be responsive to change.

12. The ERM framework should incorporate a feedback loop, based on appropriate and good quality information, management processes, and objective assessment, which enables the insurer to take the necessary action in a timely manner in response to changes in its risk profile.

Key Feature 5—Own Risk and Solvency Assessment (ORSA)

13. An insurer should regularly perform its ORSA to provide the board and senior management with an assessment of the adequacy of its RM and current, and likely future, solvency position.

14. The ORSA should encompass all reasonably foreseeable and relevant material risks including, as a minimum, underwriting, credit, market, operational, and liquidity

risks. The assessment should identify the relationship between RM and the level and quality of financial resources needed and available.

Key Feature 6—Economic and Supervisory Capital

15. As part of its ORSA an insurer should determine the overall financial resources it needs to manage its business given its own risk tolerance and business plans, and to demonstrate that supervisory requirements are met.
16. The insurer's RM actions should be based on consideration of its economic capital, regulatory capital requirements, and financial resources.

Key Feature 7—Continuity Analysis

17. As part of its ORSA, an insurer should analyze its ability to continue in business, and the RM and financial resources required to do so over a longer time horizon than typically used to determine regulatory capital requirements.
18. Such a continuity analysis should address a combination of quantitative and qualitative elements in the medium- and longer-term business strategy of the insurer and include projections of the insurer's future financial position and analysis of the insurer's ability to meet future regulatory capital requirements.

Key Feature 8—Role of supervision in RM

19. The supervisor should undertake reviews of an insurer's RM processes and its financial condition. The supervisor should use its powers to require strengthening of the insurer's RM, including solvency assessment and capital management processes, where necessary.

3.3.5 IAA's Note on IAIS' Key Features

In order to support the standards and guidance developed by IAIS, see Section 3.3.4 and IAIS (2008a, 2008b), the International Actuarial Association, IAA, has developed a Note on ERM (IAA, 2009a). This Note is based on industry experience, supervisory practice, and other published frameworks. The IAA Note "unpacks" each of IAIS' eight key features and explains them in detail. The explanations are presented as issues to consider and information about solutions that others have given. The Note is based on a combination of IAIS (2007b) and IAIS (2008b).

Furthermore, the IAA Note discusses the definition of ERM, what it is, and the stages of its maturity in implementation. There are also case studies and examples of a risk committee charter, the key roles and responsibilities of the chief risk officer (CRO), and topics and structure of an RM policy.

3.3.6 EU Solvency II ERM Approach

The European Solvency II project is an ERM application. One of the main features of the project is the internal RM process (the corporate governance, fit, and proper) and the companies' ORSA having control on all risks. The European supervisory body, CEIOPS,* has issued a paper on RM and corporate issues (CEIOPS, 2007b). An ERM framework is discussed in Cruz (2009).

3.3.7 The Financial Crisis and ERM

The financial crisis (mainly) of 2007–2009 has shown that there is a need for treating RM as a strategic indispensability. As pointed out by Shimpi (2009) the subprime crisis represents a failure to manage risks effectively and not a failure in RM as a critical business activity. Shimpi also suggested that three aspects of ERM implementation need to be strengthened:

1. RM is a strategic imperative and should be treated as such
   a. Reinforce the role of the CRO
   b. Increase board engagement on risk
   c. Align incentives to reflect risk
2. Financial managers should urgently reassess the adequacy of their current RM capabilities
   a. Recognize operational risk as substantial
   b. Fungibility should be stress tested
3. The greatest shortcoming is cultural: Management should improve the engagement of employees, as well as the board members and senior executives responsible for RM
   a. Establish clear guidance on accountability
   b. Assess your risk culture regularly

In JRMS (2008), a series of essays on the financial crisis and lessons learned, the core lesson is that the crisis is not a result of a failure of the ERM process *per se*, but more a failure to implement ERM processes at all. The authors have pointed out that key to ERM is the realization of a corporate culture aligning desired performance with incentives and the matching of the authority to make decisions with accountability for the decisions that are made. The points given by Shimpi above are also discussed in JRMS (2008). Berliet pointed out in JMRS (2008) that companies that have withstood the turmoil best have been disciplined about

- Managing strategic risks
- Holding sufficient capital

* Web site: www.ceiops.org

- Aligning interests of shareholders and managers

He also pointed out that their discipline has demonstrated that they have been taking risk governance and RM seriously!

CEIOPS (2009a) also discussed the lessons learned from the crisis and the effect it will have on their final advice on the level 2 implementing measures on the European Solvency II project. This will include

- Further refinement of the existing solvency calibrations
- Strengthened governance, RM, and internal controls (in the insurance sector)
- A strong emphasis on own RM
- A rethinking on the scope of regulation and supervision, focusing more on consolidated entities (groups) than on solo entities only

CHAPTER 4

Summary of the Development of ERM and Solvency

A SUMMARY OF THE DEVELOPMENT is given in the following tables. Only the main steps are outlined. Table 4.1 show the classical steps until the mid-1980s and Tables 4.2 and 4.3 show the "modern" steps mainly using an economic approach. The steps shown here focus not only on solvency assessments, but also on the development of RM and financial economics.

TABLE 4.1 The Classical Steps in the Development of Managing Insurance Business During the "Classical Approach Period"

Time Period	Approach	Comments
	"Pre-Classical Time": 1850–1950	
1850–1900	Model office: 1869: Manly 1875: Valentine	(3.2.1)
1900–1950	1900: *Bachelier, diffusion process for investment*	(LH), (PF)
	1909: *Lundberg, diffusion process for collective risk theory*	
	1938: Macaulay duration	
	1942: Cash-flow matching, Koopmans	(3.2.1)
	1947: Campagne's life solvency approach	(3.2.2.1)
	1952: *Markowitz portfolio theory*	(2), (PF)
	Classical Time: 1950–1980	
1950–1970	1952: Immunization Haynes and Kirton, Redington	(3.2.2.2)
	1953: Finnish equalization reserves	(2.1.1)
	1957/61: Campagne's solvency approaches	(2.1.2–2.1.3)
	1957–1962: *de Finetti and Borch: maximize the value of the firm, utility theory and risk exchanges*	
	1959: Copulas—Sklar's theorem	(13.1)

continued

TABLE 4.1 (continued) The Classical Steps in the Development of Managing Insurance Business During the "Classical Approach Period"

Time Period	Approach	Comments
	Classical Time: 1950–1980	
	1961–66: *Capital Asset Pricing Model (CAPM)*	(LH), (PF)
1970–1980	*Accelerating inflation → different research and modeling developments*	(3.1)
	1972:	
	NAIC's IRIS	(2.1.4)
	Immunization to North America	(3.2.2.2)
	1973:	
	EU: 1st nonlife directive	(2.1.2)
	Option Pricing Model: Black–Scholes, Merton	(LH), (VH)
	1976: *Arbitrage Pricing Model (APM): Ross*	(PF)
	1977: *Interest Rate Modeling: Vasicek model*	(PF)
	1979:	
	EU: 1st life directive	(2.1.3)
	SOA's RBC approach, the Trowbridge report	(2.2)

Note: For references to financial economic developments not discussed in the preceding chapters, see, for example, Field (2003). References within parentheses refer to chapters in this book. LH: Hughston (1999), VH: Henderson (2004), PF: Field (2003).

TABLE 4.2 The Modern Steps in Managing an Insurance Business up to the Dynamical Financial Analysis, DFA, During the "Economic Approach Period"

Time Period	Approach	Comments
	"Pre-Modern Time": 1980–1990	
1980–1990	1982:	
	Finnish first simulation model	(2.2)
	ARCH models	(RSAS)
	1982–1987:	(LH), (PF)
	Diffusion Processes in Finance	
	1984:	
	Faculty's life simulation model (Wilkie model)	(2.2)
	Institute's first nonlife simulation model	(2.2)
	Start of development of RBC system in Canada	(2.2)
	1984: different stochastic simulation models	(2.2)
	1984–1986: The Wilkie model	(2)
	1985–1986: *Interest rate Modeling: Cox–Ingersoll–Ross (CIR), Ho-Lee*	(PF)
	1986:	
	Cash-Flow Testing	(3.2.2.1)
	GARCH models	(RSAS)
	Generalized immunization	(3.2.2.2)
	1st Int. Conf. on Insurance Solvency, Philadelphia, United States	(2.2)
	1987:	
	Institute's 2nd nonlife simulation model	(AS)

TABLE 4.2 (continued) The Modern Steps in Managing an Insurance Business up to the Dynamical Financial Analysis, DFA, During the "Economic Approach Period"

Time Period	Approach	Comments
	"Pre-Modern Time": 1980–1990	
	1988:	
	Basel I Accord	(AS)
	EU: 2nd nonlife directive	(2.2)
	2nd Int. Conf. on Insurance Solvency, Brighton, United Kingdom	(2.2)
	1989:	
	2nd Finnish simulation model	(2.2)
	Canadian tests of RBC	
	D'Arcy's paper on Actuaries of the third kind	
	1990:	
	Institute of Actuaries' model as a RM tool	
	KMV's Credit risk model	(16.2)
	EU: 2nd life directive	(AS)
	Modern Time I: 1990–2000	
1990–2000	1992:	
	RBC system in Canada (Life)	(2.2) and (AS)
	RBC system in United States (Life and Health)	(2.2) and (AS)
	Advanced CAT-models (following hurricane Andrew)	
	EU: 3rd life and nonlife directives	(AS)
	Interest Rate Modeling: Heath-Jarrow–Morton (HJM)	(PF)
	1993:	
	RBC system in the United States (nonlife)	(2.2) and (AS)
	RiskMetric's VaR	(14.2 to 14.3)
	1995:	
	Extended Wilkie model	(2.2)
	CAS-definition: Dynamic Financial Analysis, DFA	(3.2.4)
	1997:	
	IASB starts its work on Insurance Accounting	(AS)
	J.P. Morgan's CreditMetric model	(16.2)
	CSFB's CreditRisk + model	(16.2)
	Tail-VaR	(14.2 to 14.3)
	EU: The Müller report	(AS)
	1999:	
	The European Union sketches what became the Solvency II project	(AS)
	Coherent risk measures and Tail-Var	(14.2 to 14.3)

Note: For references to financial economic developments not discussed in the preceding chapters, see, for example, Field (2003). References within parentheses refer to chapters in this book. AS: Sandström (2005), LH: Hughston (1999), PF: Field (2003), RSAS: RSAS (2003).

TABLE 4.3 The Modern Steps in Managing an Insurance Business up to the Enterprise Risk Management, ERM, during the "Economic Approach Period"

Time Period	Approach	Comments
	Modern Time II: 2000–2012	
2000–2005	2000–2003:	
	Solvency II—the learning phase	(AS)
	2002:	
	Australia: new solvency system	(AS)
	Germany: first GDV solvency model	(AS)
	EU: Solvency I	(AS)
	EU: the Sharma report	(2.3.1.4)
	2003:	
	Singapore: new RBC system	(AS)
	Denmark: Traffic Light System	(AS)
	IASB Exposure Draft on insurance contracts	(AS)
	2003–2008/2009:	
	EU: Solvency II—the framework directive phase	(AS)
	2004:	
	COSO-definition: ERM	(3.3)
	United Kingdom: new Solvency System with Internal models	(AS)
	2006:	
	Basel II Accord for banks	(AS)
	Switzerland: new Solvency System with Internal models	(AS)
	Dutch solvency system for pension institutes	(AS)
	Sweden: Traffic Light System as a supervisory tool	(AS)
	SOA-definition of ERM	(3.3)
	ERMII-definition of ERM	(3.3)
	2007:	
	IASB Discussion Paper on Insurance Contracts	
	Two books on Market Valuation of insurance liabilities are published	(W), (MS)
	EU Solvency II: draft framework directive	(2.3.2)
	2008:	
	Regulator's condition for the solvency capital requirement	(14.1)
	2008/2009–2012:	
	Solvency II—the implementing phase	
	2009:	
	The Solvency II Framework Directive is adopted	(Appendix D)

Note: References within parentheses refer to chapters in this book. AS: Sandström (2005), W: Wüthrich et al. (2007), MS: Møller and Steffensen (2007).

CHAPTER 5

Elements of Solvency Assessment Systems

IN THE BANKING supervisory system, Basel II, a three-pillar approach was introduced. The IAA (IAA, 2004) and the European Union Solvency II project have adopted this approach. The main difference is that the solvency system is much more advanced. Nevertheless, it consists of the following three pillars:

Pillar I: The quantitative requirements
Pillar II: The qualitative requirements—the supervisory review process
Pillar III: Statutory and market reporting

The first pillar includes the calculation of the capital requirements according to a standard model (e.g., factor based) or the introduction of partial or full internal models. It also includes rules on provisioning and eligible capital. The second pillar focuses on the supervisors and their review process, for example, a company's internal control and RM, the approval of using partial or full internal models in Pillar I and its validation. The supervisor can also impose a company to increase its SCR, so-called capital add-ons, if it believes that the capital is not adequate or that the management is insufficient. The third pillar includes the reporting to both the supervisor and the market. The latter case will promote market discipline and greater transparency, including harmonization of accounting rules.

The IAIS has illustrated the framework and common structure of the assessment of the insurer solvency as outlined in Figure 5.1. The three pillars are formulated as financial (Pillar I), governance (Pillar II) and market conduct (Pillar III).

A solvency regime should be risk sensitive. To achieve this it could use some or all of the following:

- *Financial requirements*
- *Quantitative limits*

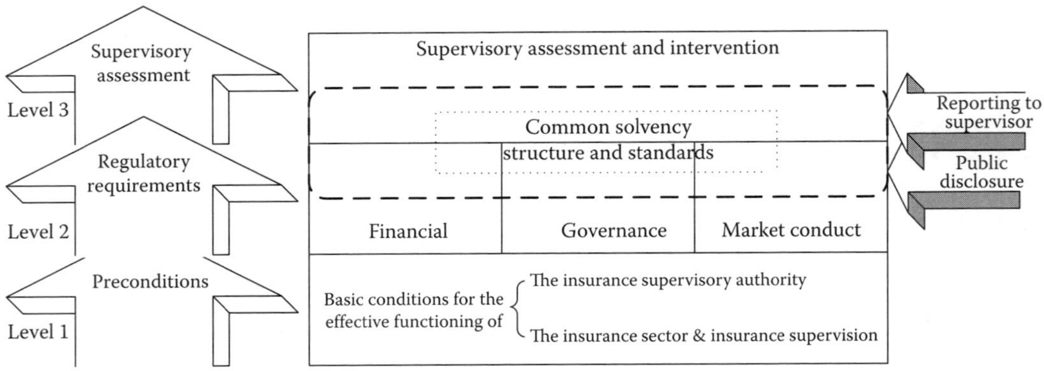

FIGURE 5.1 The IAIS' levels and blocks for a common structure of insurance solvency assessment. (Adapted from IAIS. 2007e. The IAIS Common Structure for the Assessment of Insurer Solvency. International Association of Insurance Supervisors. 14 February; IAIS. 2007f. Summary of IAIS Positions on The Valuation of Technical Provisions. International Association of Insurance Supervisors, October.)

- *Qualitative requirements*
- *Additional capital requirements or safety measures arising from supervisory assessment.* (IAIS, 2007e)

The first two bullets correspond to the first pillar and the last two to the second pillar. Parts of Pillar III issues are set out in the following sentence.

> *Public disclosure of information enhances market discipline, imposing strong incentives on insurers to conduct their business in a safe, sound and efficient manner. Insurer solvency and solvency assessment thus benefit from appropriate public disclosure.* (IAIS, 2007e)
>
> *Regulatory financial requirements need to be firmly rooted in economic valuation.* (IAIS, 2007e)

The IAIS has suggested that risk factors and their distinct components should be represented in the technical provisions, that is, in the liability risk margin (LRM) within the market value of liabilities (MVL), see Figure 5.4, and in the capital requirements as follows:

- Risk that is to be reflected both in the LRM and the capital requirement:
  - Uncertainty and residual market volatility in underwriting risk
  - Unhedgeable mismatch risk
- Risk that is reflected only in the capital requirement and not in the LRM:
  - Volatility other than residual market volatility in underwriting risk
  - Hedgeable mismatch risk

This is outlined in Structure Elements 9–11 in IAIS (2007e). On way of distinguishing between risks taken care of in the risk margin with a time horizon to the ultimate and the solvency capital requirement with a short time horizon, say $T = 1$ year is to use the mean square function.

We illustrate the different elements of the supervisory system and requirements in the following way; see Appendix A, Section A.1 for more details.

Let $\mathbb{F} = \sigma\{\mathbb{M}, \Theta\}$ be an σ-algebra, where $\mathbb{M}$ is a finite set of models (including trends) and Θ is a finite set of parameters in the models. Let $\hat{\theta} = \hat{\theta}(X_1, X_2, \ldots, X_n)$ be an estimator of the true current estimate of liabilities (CEL), θ and let $X_1, X_2, \ldots, X_n$ be a random sample of size n from a probability distribution function with parameter θ, $f_\theta(\cdot)$, and $\theta \in \Theta$, the parameter space. The mean square error (MSE) is the expectation of the squared error loss in estimating θ by $\hat{\theta}$:

$$E\left[(\hat{\theta} - \theta)^2\right] = E\left[\left(E(\hat{\theta}) - \theta\right) + \left(\hat{\theta} - E(\hat{\theta})\right)\right]^2 = (\text{bias})^2 + V(\hat{\theta}). \tag{5.1}$$

The last term in Equation 5.1 can be rewritten in terms of the sigma algebra $\mathbb{F}$ as

$$V(\hat{\theta}) = E_{\mathbb{F}}\left(V(\hat{\theta}|\mathbb{F})\right) + V_{\mathbb{F}}\left(E(\hat{\theta}|\mathbb{F})\right). \tag{5.2}$$

Combining Equations 5.1 and 5.2, we obtain

$$\text{MSE}(\hat{\theta}) = (\text{bias})^2 + E_{\mathbb{F}}\left(V(\hat{\theta}|\mathbb{F})\right) + V_{\mathbb{F}}\left(E(\hat{\theta}|\mathbb{F})\right). \tag{5.3}$$

In Equation 5.3, the first term, squared bias, is an issue for the supervisor, that is, in the three-pillar approach outlined above, it is a Pillar II issue. The second term is the expected volatility, which is taken care of as a part of the SCR, but with a shorter time horizon. The third term represents the uncertainty in models and parameters and is the volatility of the level of the CEL. This term is the one that constitute the LRM. Note that in nonlife insurance this term is mainly a function of the liabilities, but in life insurance the liabilities could also be a function of the assets.

Assume that the time horizon $(0, T)$ is split into T uncorrelated time buckets, $1, 2, \ldots, T$, each representing a financial year (or accounting year), $[0, 1), [1, 2), \ldots, [T - 1, T)$. The uncertainty term in Equation 5.3 could be split up into two parts, one representing the time bucket [0,1) and the second the time bucket $[1, T)$. The first part will be included as a part of the SCR and the second will constitute the base for the risk margin.

5.1 IAIS CAPITAL REQUIREMENTS STANDARDS AND GUIDANCE

The IAIS has developed a number of principles, standards, and guidance papers to help promote the development of global well-regulated insurance markets. In October 2008, IAIS approved and published its standards and guidance papers on the structure of regulatory capital requirements—IAIS (2008c, 2008d). An earlier version was published in 2007—IAIS (2007d).

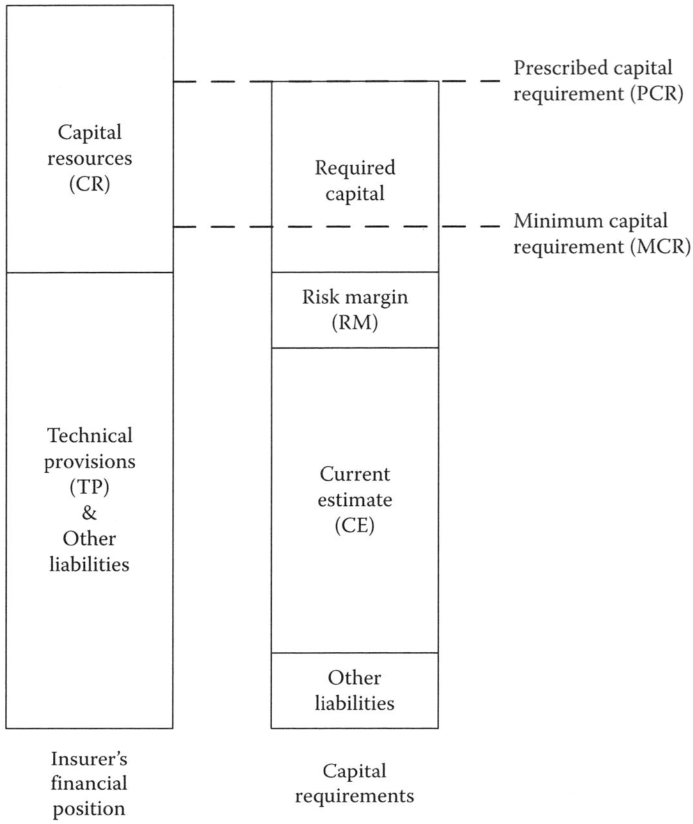

FIGURE 5.2 Solvency control levels, capital requirement, and the building blocks of the IAIS structure. (Adapted from IAIS. 2008d. Guidance Paper on the Structure of Regulatory Capital Requirements. International Association of Insurance Supervisors. Approved October. Guidance Paper No. 2.2.1.)

The IAIS best practice and the structure of regulatory capital requirements are given below. It is illustrated in Figure 5.2. The IAIS standard on the structure of regulatory capital requirements is given in Section 5.1.1.

The guidance paper provides guidance on 15 principles-based requirements for a solvency regime in relation to regulatory capital requirements as set out in the Standard on the structure of regulatory capital requirements; see Section 5.1.1. The purpose of the guidance paper is to support the enhancement, improved transparency and comparability and convergence of the assessment of insurer solvency internationally. The preconditions in a particular supervisory regime, among other factors, will determine the specifics of effective supervision within that regime, including the specific requirements of the solvency regime in relation to regulatory capital requirements.

The guidance paper addresses the structure of regulatory capital requirements in a supervisory regime for solvency assessment. While the broader issues in relation to capital resources are identified to establish the context of the solvency assessment process, the requirements regarding the nature and quality of capital resources are not covered in depth in the guidance paper but will be the focus of a separate standard and guidance paper.

The target capital is named prescribed capital requirement (PCR) by IAIS.

REGULATORY CAPITAL REQUIREMENTS

Requirement 1

A total balance sheet approach should be used in the assessment of solvency to recognise the interdependence between assets, liabilities, regulatory capital requirements and capital resources and to ensure that risks are appropriately recognised.

ESTABLISHING REGULATORY CAPITAL REQUIREMENTS

Requirement 2

Regulatory capital requirements should be established at a level such that the amount of capital that an insurer is required to hold should be sufficient to ensure that, in adversity, an insurer's obligations to policyholders will continue to be met as they fall due.

SOLVENCY CONTROL LEVELS

Requirement 3

The solvency regime should include a range of solvency control levels which trigger different degrees of intervention by the supervisor with an appropriate degree of urgency.

Requirement 4

The solvency regime should ensure coherence between the solvency control levels established and the associated corrective action that may be at the disposal of the insurer and/or the supervisor. Corrective action may include options to reduce the risks being taken by the insurer as well as to raise more capital.

REGULATORY CAPITAL REQUIREMENTS AS TRIGGERS FOR SUPERVISORY INTERVENTION

Requirement 5

The regulatory capital requirements in a solvency regime should establish a solvency control level which defines the level above which the supervisor would not require action to increase the capital resources held or reduce the risks undertaken by the insurer. This is referred to as the PCR.

Requirement 6

The PCR should be defined such that assets will exceed technical provisions and other liabilities with a specified level of safety over a defined time horizon.

Requirement 7

The regulatory capital requirements in a solvency regime should establish a solvency control level which defines the supervisory intervention point at which the supervisor would invoke its strongest actions, if further capital is not made available. This is referred to as the Minimum Capital Requirement (MCR).

Requirement 8

The solvency regime should establish a minimum bound on the MCR below which no insurer is regarded to be viable to operate effectively.

APPROACHES TO DETERMINING REGULATORY CAPITAL

Requirement 9

The solvency regime should be open and transparent as to the regulatory capital requirements that apply. It should be explicit about the objectives of the regulatory capital requirements and the bases on which they are determined.

Requirement 10

In determining regulatory capital requirements, the solvency regime should allow a set of standardised and, if appropriate, other approved more tailored approaches such as the use of (partial or full) internal models.

RISKS TO BE ADDRESSED

Requirement 11

The solvency regime should be explicit as to where risks are addressed, whether solely in technical provisions, solely in regulatory capital requirements or if split between the two, the extent to which the risks are addressed in each. The regime should also be explicit as to how risks and their aggregation are reflected in regulatory capital requirements.

CALIBRATION OF REGULATORY CAPITAL REQUIREMENTS

Requirement 12

The supervisor should set out appropriate target criteria for the calculation of regulatory capital requirements, which should underlie the calibration of a standardised approach.

Requirement 13

Where the supervisory regime allows the use of approved more tailored approaches such as internal models for the purpose of determining regulatory capital requirements, the target criteria should also be used by those approaches for that purpose to ensure broad consistency among all insurers within the regime.

SUPERVISORY REVIEW

Requirement 14

The solvency regime should be designed so that any variations to the regulatory capital requirement imposed by the supervisor are made within a transparent framework, are proportionate according to the target criteria and are only expected to be required in limited circumstances.

SUPERVISORY REPORTING AND PUBLIC DISCLOSURE

Requirement 15

The solvency regime should be supported by appropriate public disclosure and additional confidential reporting to the supervisor.

The different capital requirement blocks and the insurer's financial position are illustrated in Figure 5.2.

5.2 MODELING CAPITAL REQUIREMENT

5.2.1 General Structure

The MVL and the capital requirement, in terms of SCR, have somewhat different role in a solvency regime (IAIS, 2007a). The LRM is a safeguard for the CEL.

The SCR provides further safeguards for the policyholders by protecting both the MVL and the market value of assets (MVA), and their interaction. The SCR should be calibrated such that it could withstand current year claims experience in excess of current estimate (CE) and that assets still exceed the MVL at the end of a defined time horizon, say 1 year, with a certain degree of confidence, say 99.5%, or that the available capital (ASM) can withstand a range of predefined shocks or stress scenarios over the defined time horizon, *cf.* IAIS (2006b).

Let $X = \text{MVL}$ and Y be two stochastic variables, where Y is the assets covering the MVL; see Figure 5.3. The SCR is now defined as a function of $Z = h(X, Y)$. We can assume an unknown and probably skewed distribution function of Z with $\mu_Z = \text{MVL}$.

The solvency capital requirement is now defined by the use of a stochastic variable $\text{SCR} = f(Z) - \mu_Z$, where $f(Z)$ usually is taken as the VaR or Tail VaR; see Chapter 14. The MSE of an estimate of SCR could be evaluated in a similar way as in Equation 5.3 for a time horizon of $T = 1$ year. The function $h(X, Y)$ includes both the risks inherent in the liabilities (X) and the assets covering them (Y) as well as the interaction between the assets and liabilities; see Sandström (2007b, 2008).

Risk neutral probability measures have been introduced in financial economics for pricing different financial instruments. But when discussing probability distributions of financial results, such as the SCR, real-world probability measures are used. Hence, in the context of the calculation of the capital requirements, we use real-world probabilities.

In many solvency assessment models the capital requirement is defined in terms of changes in value between two dates. But as risk and capital requirements are related to the variability of future values of a given position, *cf.*, for example, Artzner et al. (1999), which are due to insurance events and market conditions, it is better to only consider future values. This means that we do not need to think of whether the egg or the hen was first.

Modeling is discussed in detail in Chapter 12 (general modeling) and Chapter 15 (capital requirement modeling).

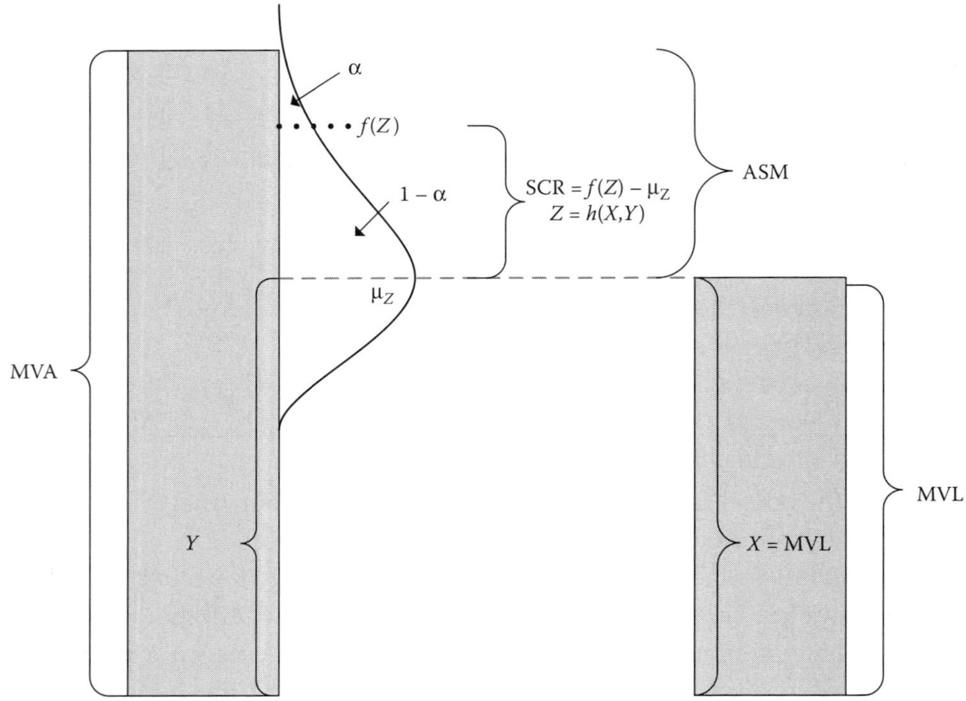

FIGURE 5.3 The SCR is calculated as the difference between a function of $h(X,Y)$ and the mean of the distribution. The distribution is a function of $X = \text{MVL}$ and the corresponding assets covering the liabilities (Y).

To model the SCR, the IAA (2004) has proposed five main *risk categories*, mainly based on the Basel II Accord and insurance characteristics:

1. Insurance risk (or underwriting risk)
2. Credit risk
3. Market risk
4. Operational risk
5. Liquidity risk

The insurance risk is associated both with the peril covered by the specific line of insurance business (fire, motor, liability, death, etc.) and with the specific processes associated with the conduct of the insurance business.

Credit risk is the risk of default and change in the credit quality of issuers of securities, counterparties, and intermediaries, to which the company has an exposure (e.g., reinsurers). Market risks come from the level of volatility of market prices of assets. They involve the exposure to movements in the level of financial variables (e.g., stock prices, interest rates, exchange rates, etc.), but also the mismatch between assets and liabilities.

Operational risks can be defined as the risks of loss resulting from inadequate or failed internal processes, people, and systems or from external events.

Liquidity risk is the exposure to loss due to insufficient liquid assets being available. If a risk category is not possible to model and be treated as a Pillar I risk, it should be treated as a Pillar II assessment, that is, under the supervisory reviews process. These five risks are used not only for solo entities but also for insurance groups or financial conglomerates. There is also a sixth main risk category that can be introduced for groups or conglomerates: group risk or participating risk. Examples of the latter type are an internal reinsurance program within an insurance group or the possibility that a bank is insuring its credit risk to an insurance company in the conglomerate.

The modeling is made in a top-down fashion. Each of the main risk categories is, in the next step of modeling, split up into subrisks, which in turn could be split up into sub-sub-risks, and so on. On the other hand, the calculation of the capital requirement is made by a bottom-up approach, see Groupe Consultatif (2005), starting from the lowest level.

The risks and subrisks are discussed from a general viewpoint in Chapters 16 through 21. From a European Solvency II perspective, they are discussed in Chapters 26 through 33. The minimum capital requirement proposed in Solvency II is discussed in Chapter 34.

Dependence is discussed in Chapter 13 and general risk measures in Chapter 14.

To model the capital requirements, as described by SCR in Figure 5.3, we have two main problems to deal with: the nonnormality and the nonlinearity. These issues are discussed in Section 15.3.

5.2.2 Diversification and Mitigation

Diversification and mitigation are generic terms as they could be distinguished by its members. We start with the term "risk diversification" and end up with the term "risk mitigation." There is a clear connection between the two generic terms. The definition of the generic term "diversification" follows the proposal given by IAA in its answer and comments to IAIS' paper IAIS (2006); see IAA (2006).

Insurers are *pooling* risks in order to benefit from the "law of large numbers." By "pooling" it is meant to aggregate similar risks that are similarly managed.

Diversification involves accepting risks that are not similar in order to benefit from the lessened correlation of contingent events.

Hedging, or offsetting risks, involves accepting risks with a strong negative correlation as compared to diversification, which merely requires the absence of a strong positive correlation.

The most well-known *risk mitigation* technique in the insurance context is reinsurance. But pooling, diversification, and hedging also give risk mitigation benefits to the insurers. Risk mitigation techniques are discussed in Section 9.1.

5.3 VALUATION OF ASSETS AND LIABILITIES

The total balance sheet approach was introduced by IAA, see IAA (2004), and should recognize not only the asset and liability sides of the balance sheet, but also the interdependence between them and the impact of the SCR, MCR, and the eligibility of capital covering the requirements. The technical provisions are the reserves set aside to cover the liabilities the company faces according to the insurance contracts. It will usually also include a risk margin, cf. below. The valuation of assets and liabilities is discussed in more detail in Part B, Chapters 6 through 11.

A total balance sheet approach is based on common valuation methodologies. It should make optimal use of information provided by the financial markets in getting market values where they exist or getting market-consistent values where market values do not exist. This holds for both the assets and the liabilities and is called the economic value.*

In a traditional actuarial valuation of the present value, a deterministic interest rate function is used. In market-consistent valuation, the deterministic interest rate is changed for a stochastic function, a deflator, reflecting the market price.

The valuation should be prospective and all cash flows related to assets and liabilities should be discounted and valued at the CE. The expected present value of the future cash flows should use a relevant risk-free yield curve and should be based on current, credible information and realistic assumptions.

A risk margin covering the uncertainty linked to future cash flows over their whole time horizon is added to the CE.

The basic concepts of the valuation of liabilities, that is, estimated cash flows, discounting and a risk margin, are usually seen as independent of each other. If the cash flow is based on deterministic discounting, it makes no difference in which order the estimation of cash flow or discounting is made. However, if we introduce stochastic discounting using deflators, this is not true. Also, see IAA (2009b, Section 3.4.6), if the market directly affects cash flows, such as if lapse rates are related to market conditions, the expected cash flows and discount factors are dependent. Here we should calculate expected discounted cash flows instead of discounted expected cash flows.

The risk-adjusted CE of the liabilities is called "economical" technical provisions, or MVL. The corresponding risk-adjusted assets are called the MVA. The risk margins should be determined in a way that enables the insurance obligations to be transferred to a third party or to be put in runoff.

In the valuation procedure, hedgeable assets and liabilities should be valuated by a mark-to-market approach, as any risk margins are implicit in observed market prices. Non-hedgeable assets and liabilities, on the other hand, should be valued by a mark-to-model approach, that is, a CE requires the calculation of an explicit risk margin. This is illustrated in Figure 5.4.

In terms of a total balance sheet approach, see Chapter 6, the available capital ("available solvency margin") should be written as

$$\text{ASM} = \text{MVA} - \text{MVL}. \tag{5.4}$$

* Economic value: The value of assets or liability cash flows, derived in such a way as to be consistent with current market prices where they are available or using market consistent principles, methodologies, and parameters; see CEA-Groupe Consultatif (2006).

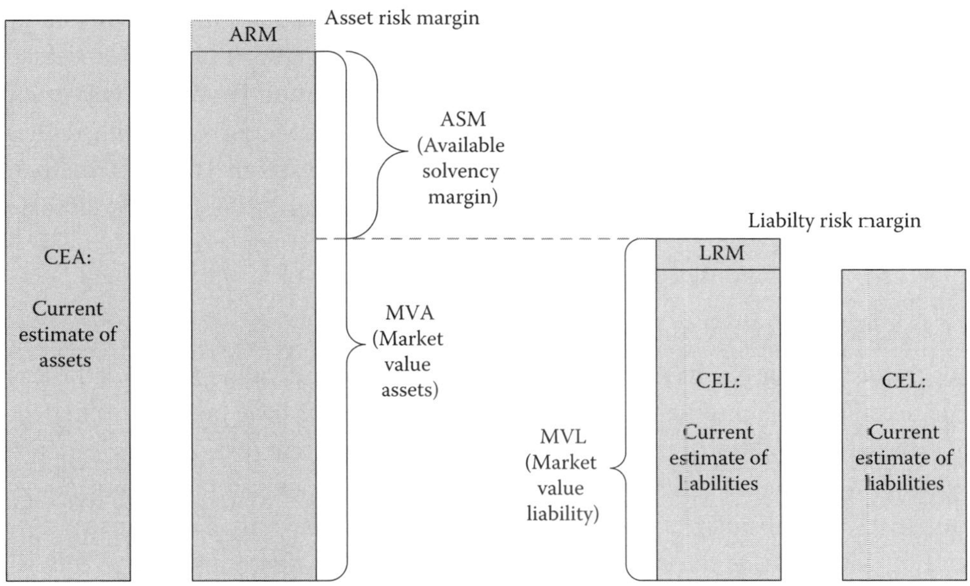

FIGURE 5.4 The total balance sheet approach as described in the text.

Asset risk margin (ARM): ARM is a risk margin deducted from the CE of assets due to uncertainty in models and parameters. Valuations of derivative instruments may include uncertainty in the models used. Valuation of property depends very much on how frequently it is made. Yearly valuation may give accurate calculation of the property value, but if you do daily valuations you need some model behind the calculation; this gives rise to an uncertainty in the valuation due to models and parameters. Valuation of assets is discussed in Chapter 7.

Liability risk margin (LRM): As there is usually no liquid market for insurance risks, we need to use a mark-to-model approach to determine the MVL, that is, to model a risk margin on top of the CEL. It should take account of the uncertainty of models and parameters and be such that the insurance contracts could be sold to a "willing buyer" or put in runoff. In economic terms, the risk margin is often called a market value margin (MVM).

Using an economic approach, a proxy of the LRM (or MVM) can be given by a cost-of-capital, CoC, approach. "The cost of capital approach bases the risk margin on the theoretical cost to a third party to supply capital to the company in order to protect against risks to which it could be exposed" (CEA, 2006a); see also CEA-CRO Forum (2006). CEA is the European insurance and reinsurance federation.

The CoC approach was first introduced in the solvency context in the Swiss Solvency Test, see SST (2004) and Sandström (2005), where the risk margin is defined as the hypothetical cost of regulatory capital necessary to runoff all liabilities, following financial distress of the company. Let SCR_t be the capital requirement for the year $t, t = 1, \ldots, T$, for the liabilities in runoff. Then the risk margin is calculated as* $\mathrm{LRM} \approx \mathrm{CoC}\% \sum_{t=1}^{T} \mathrm{SCR}_t$.

* If you consider the transformation of the business to a "willing buyer" that is not putting the business in run-off, but considering it as a "going concern," then the risk margin could be calculated as $\mathrm{LRM} \approx \mathrm{CoC}\% \cdot \mathrm{SCR}_1$.

One way in market-consistent valuation is to use a replicating portfolio in valuing the liabilities, that is, a portfolio of assets that replicates the cash flow of liabilities most closely. The liability cash flow can be obtained by the use of a Valuation Portfolio (VaPo); see, for example, Wüthrich et al. (2007). The VaPo can be used to catch the corresponding replication portfolio on the asset side. Valuation of life insurance liabilities from a financial mathematic view is discussed in Møller and Steffensen (2007). Valuation of liabilities is discussed in Chapter 8.

5.4 CONSERVATIVE VALUATION REGIMES

During the last decades we have seen an increased globalization, influencing not the least the financial sector. Large international (re)insurance companies have been consolidated, and the number of multinational companies that operate in different jurisdictions is increasing. The globalization makes it necessary to have similar and transparent disclosure regimes. A company with subsidiaries in different countries may have to use as many different accounting standards as the number of countries they are operating in and this is very burdensome.

The work done by IASB* on insurance contracts and a global accounting system for the insurance industry is therefore vital. As part of that, the collaboration between the IASB and the FASB† in the United States is very important in reaching a global view on accounting and measuring insurance contracts. In August 2006, the Committee of European Securities Regulators (CESR), and the U.S. Securities and Exchange Commission (SEC), launched a work plan for a common financial reporting. During the spring of 2007, the U.S. SEC announced that they would issue a proposal allowing foreign private issuers to file financial reports using either the International Financial Reporting Standards (IFRS), of the IASB, or the U.S. GAAP.‡ For non-U.S. companies the requirement to use U.S. GAAP in financial reporting would be removed from 2009.

In August 2008, the U.S. SEC announced a road map for shifting from the U.S. GAAP to the new global accounting system set by IASB. This means that U.S. companies have to switch to the IFRS beginning from 2014.

There is a wide diversity in approaches for valuing assets and liabilities across the world. For assets it is usually the local accounting system that governs the adopted valuation approach.

For the liabilities, on the other hand, both the accounting systems and the general conservative approaches adopted worldwide by the regulators govern the valuation approach. The different methodologies adopted by the insurance companies also give rise to mishmash. This makes it more or less impossible to compare the financial strength of companies.

Due to the increased globalization and changing supervisory and legal frameworks, deregulations of financial markets have taken place. As a consequence, the need to compare the

* International Accounting Standard Board (www.iasb.co.uk).

† Financial Accounting Standard Board (www.fasb.org).

‡ Generally Accepted Accounting Principles.

financial strength of insurance undertakings has increased. The insurance industry, by tradition and statutory rules, has used conservative approaches in the valuation of assets and liabilities and thus implicitly has put "risk margins" on the valuations. This could be done by using conservative mortality rates or by adjustments to the discount rates. Hence, the conservative valuation has created hidden surpluses and deficits that ought to be recognized for at least solvency assessment purposes. Any such hidden surpluses or deficits would undermine useful comparisons between undertakings.

Having a conservative capital requirement that might be well beyond economic capital levels would have a negative impact on the development of the insurer's capital. A failure could thus be hidden behind a conservative regime (IAA, 2004). As accounting has been largely based on book valuation any deficit due to changes in market prices may not be recognized. This problem was one of the major motivations behind the adoption of *asset–liability management* techniques during the 1980s.

As the insurer's capital is determined from its (statutory) balance sheet as the difference between the value of the assets and the liabilities, it will be impossible to compare two companies that bear the same risk profiles if the underlying valuation regimes differ.

If not the largest, but at least one of the largest items on the balance sheet is the technical provisions, as calculated by the actuary. In a conservative regime, the financial strength of a company will thus depend not only on the quality of the work done, but also on the methodology used, by the actuary.

Within the EU, a precursor of the first nonlife directive that was set up in 1973 presupposed that the technical provisions were set up in a uniform manner in the member states; *cf.* Section 2.1.2. As this was not the case, the approach of using the technical provisions as a part of the solvency assessment was not adopted. During the 1970s the European Commission set up two working groups to study the harmonization of technical reserves. No decision was taken; see, for example, Sandström (2005). In the European Solvency II project, the European Commission has stated that "an increased level of harmonization for technical provisions is a cornerstone of the new solvency system" (EC Commission, 2006).

Earlier we mentioned an implicit "risk margin" due to a prudential view of the technical provisions. As a matter of fact, there are also more modern explicit risk margins put on top of a "best or current estimate" of the liabilities that are conservative in one sense. The quantile method used by the Australian regulator APRA* is of recent origin (APRA, 2000). The use of a 75% confidence level was used as a proxy for what market participants would consider as representing a "reasonable value" or as commensurate with prudently set technical provisions under the old "prudential regime"; see Section 8.4 and IAA (2008a).

We can summarize the problems of the old conservative regulation. Assets are usually valuated as either book value or market value. This depends on the statutory regime or the local accounting system. The liabilities as valued by the technical reserves (in accounting terms: *technical provisions*) set aside depend, on the one hand, on the regulatory framework and, on the other, on the valuation approach adopted by the undertakings and their actuaries. The regulation may state that the assumptions behind, for example, the mortality rate used

* APRA: Australian Prudential Regulation Authority.

to calculate the technical reserves should be prudent and conservative. Thus, in that case, it would be up to the actuary to formulate the approach adopted by the company. In many countries, the insurers may use conservative interest rates in their valuations, only limited to, for example, a maximum rate defined by the regulator.

The Solvency Working Party[†] of the International Association of Actuarial, IAA, believed "that a proper assessment of an insurer's true financial strength for solvency purposes requires appraisal of its total balance sheet on an integrated basis under a system that depends upon realistic values, consistent treatment of both assets and liabilities and does not generate hidden surplus or deficit" (IAA, 2004, p. 4). "Regulatory financial requirements (..) need to be firmly rooted in economic valuation" (p. 5 in IAIS, 2007a).

Economic value, as measured by economic capital, is defined as the company's own amount of capital needed to meet future obligations arising from the existing business with a high degree of certainty over a defined time horizon and to maintain its external credit rating.

We adopt this approach and outline the total balance sheet approach in Chapter 6.

[†] This WP was set up by IAA to give input on solvency assessment to the International Association of Insurance Supervisors, IAIS. In 2004 the WP was reformed as a solvency subcommittee under the IAA Regulatory Committee.

PART B

Valuation, Investments, and Capital

All money is a matter of belief.

Adam Smith, 1721–1790
Scottish Economist

IN CHAPTER 6 we look at the basic cornerstone of the valuation of assets and liabilities, that is, the *economic total balance sheet approach* (TBSA). The TBSA means that assets and liabilities should be valued in a market-consistent way and their interactions should be a part of the solvency assessment. In Section 6.2, we discuss the time value of money (the present value and discounting rates). Hedging is discussed in Section 6.3 and risk margins (RMs) in Section 6.4.

In Chapter 7, we look at asset valuation and the asset risk margin, and in the subsequent chapter, we look at liability valuation: Section 8.1 discusses the current estimate of liabilities and Section 8.2 gives a brief discussion of discretionary participating features. In Section 8.3, the replicating portfolio technique is discussed. The cash flow (CF) of the liabilities is constructed using the Valuation Portfolio technique (VaPo). The VaPo can be used to catch a corresponding replication portfolio on the asset side. The liability RM is discussed in Section 8.4 and the CoC approach as a measure of the RM (market-value-margin) is discussed in Section 8.5. Other issues, such as risk mitigation techniques and segmentations, are discussed in Chapter 9.

Chapter 10 gives a brief discussion of investments and the eligibility of capital. We also give a brief discussion on accounting valuation in Chapter 11.

CHAPTER 6

Total Balance Sheet Approach

> **Total balance sheet approach**
>
> Principle which states that the determination of an insurer's capital that is available and needed for solvency purposes should be based on all assets and liabilities, as measured in the regulatory balance sheet of the insurer, and the way they interact.
>
> *Source:* Data from CEA-GC, 2007. Solvency II Glossary, Comité Européen des Assurances and Groupe Consultatif Actuariel Européen. March 2007.

Under the IAA framework, "the capital requirements and risk oversight process in two jurisdictions with similar business, legal, economic and demographic environments and supervisory philosophy and controls should be comparable" (IAA, 2004, p. 9). One cornerstone in the IAA framework, as stated above, is the TBSA. It is also the second key feature in the IAIS' guidance on regulatory capital requirements (IAIS, 2007d): "A total balance sheet approach should be used to recognise the interdependence between assets, liabilities, regulatory capital requirements and capital resources and to ensure that risks are appropriately recognised."

The CEA* states that as the main focus of the capital requirement is policyholder protection, an *economic balance sheet* should be based on the policyholder perspective (CEA, 2007). By "an economic balance sheet" we mean a balance sheet (BS) statement based on an accounting approach using market-consistent values for all current assets and current obligations relating to in-force business, including off-BS items (see CEA-GC, 2007). One definition of the term *market consistent valuation* is given by Smith and Sheldon (2004): "A valuation algorithm is a method for converting projected CFs into a present value.

*Comité Européen des Assurances—Insurers of Europe; http://www.cea.assur.org

A valuation algorithm may be specified with reference to a set of calibration assets. We say a valuation is market consistent if it replicates the market prices of the calibration assets to within an acceptable tolerance."

In this approach, the risks associated with all the BS items and their potential impact on these items should be taken into account. According to IAIS, the allowance for quality of the items that comprise the capital resources of an insurer can be made in the requirements of the required capital or by applying limits or "prudential filters" to adjust them when determining available capital for solvency purposes; see IAIS (2007e), *cf.* also Section 10.2.

In an economic approach there should not be any artificial limits on the available capital. In the Solvency II project the European Commission has adopted the banking terminology for available capital, that is, *own funds*. The building blocks of a TBSA are

- Market consistent valuation of assets and liabilities (see Section 6.1)
- The interdependence between assets, liabilities, capital requirement, and resources
- The time value of money (see Section 6.2)

A TBSA is based on common valuation methodologies. It should make optimal use of information provided by the financial markets (EU Commission, 2006), in getting market values where they exist or getting market-consistent values where market values do not exist, of both the assets and the liabilities. The valuing of assets and liabilities based on market values where available is called a mark-to-market approach, and where market values are not available, the valuation is based on market-consistent valuation techniques, which is called a mark-to-model approach (see CEA-GC, 2007). This is called an economic value technique. We discuss these topics in more detail in the following chapters.

The IAIS (2007e) uses the term capital resources for own funds and defines it in a broad sense as the amount of assets in excess of the amount of liabilities. In this context, they define liabilities as including technical provisions, but only other liabilities ("senior liabilities") that are not treated as capital resources. Liabilities that can absorb losses and that are not ranked as senior liabilities are here called own funds liabilities and are treated as part of the own funds; see Figure 6.1 and Section 8.6.

In a traditional actuarial valuation of the present value, a deterministic interest rate function is used. In a market-consistent valuation the deterministic interest rate is changed for a stochastic function, a deflator, reflecting the market price. The valuation should be prospective and all CFs related to assets and liabilities should be discounted and valued at the current estimate (CE). In Australia the term "central estimate" has been used for the current estimate and in the European Solvency II project the term "best estimate" is used. The IAA and IAIS have adopted the concept of current estimate. The expected present value of the future CFs should use a relevant risk-free yield curve and should be based on current, credible information and realistic assumptions. A RM covering the uncertainty linked to future CFs over their whole time horizon is deducted or added to the current estimate depending on whether it is the current estimate of the assets or of the liabilities; see Sections 7.1 and 8.4.

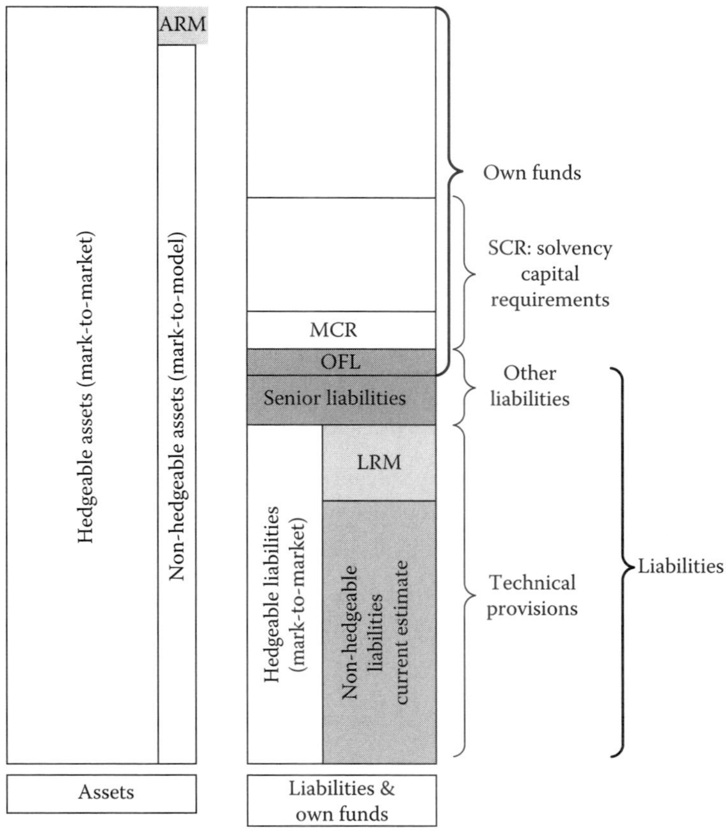

FIGURE 6.1 The building blocks of the Total Balance Sheet Approach. ARM: asset risk margin (*deducted*), LRM: liability risk margin (*added*), MCR: minimum capital requirement, and OFL: own funds liabilities. Own funds that are part of the assets, are illustrated in the opposite side.

The basic components of the valuation of liabilities, that is, estimated CFs, discounting, and a RM, are usually seen as independent of each other. If the CF is based on deterministic discounting it makes no difference in which order the estimation of CF or discounting is made. However, if we introduce stochastic discounting using deflators, this is not true. Also, see IAA (2009b, Section 3.4.6), if the market directly affects CFs, such as if lapse rates are related to market conditions, the expected CFs and discount factors are dependent. Here we should calculate expected discounted CFs instead of discounted expected CFs.

A general discussion of RMs on the liabilities is given in Mourik (2005). Using an explicit, unbiased, market-consistent, probability-weighted estimate of the contractual CFs and using a market discount rate adjusting for the time value of money gives us a value of the liabilities that is called a current exit value. If assets or liabilities could be market valuated these are called hedgeable. Otherwise they are called nonhedgeable; see Section 6.3.

The interaction between assets and liabilities does not mean that there should be a full matching of assets to the liabilities, but that, for example, a changed interest rate should be visualized by changing assets and liabilities.

In Malamud et al. (2007) a comprehensive unified equilibrium theory of asset and liability pricing is given. The mathematical framework generates pricing algorithms for nonhedgeable (*cf.* Section 6.3) insurance risks.

The International Actuarial Association (IAA) and the International Association of Insurance Supervisors (IAIS) have issued standards and guidance regarding the assessment of insurer solvency. In IAIS (2007a) the main concepts of the solvency assessment are summarized.

- *Structure Element 4 (of IAIS, 2007a):* A total balance sheet approach should be used to recognise the interdependence between assets, liabilities, capital requirements and capital resources and to ensure that risks are fully and appropriately recognised.
- *Cornerstone IV (of IAIS, 2007a):* the solvency regime requires a valuation methodology which makes optimal use of and is consistent with information provided by the financial markets and generally available data on insurance technical risks.
- *Structure Element 7 (of IAIS, 2007a):* Given the intrinsic uncertainty of insurance obligations, the technical provisions need to include a risk margin over the current estimate of the cost of meeting the policy obligations. The risk margin should be calibrated such that the value of the technical provisions is equivalent to the value that an insurer would be expected to require in order to take over the obligations.

We are talking about regulatory capital, as for example used for solvency purposes. This capital is usually not the same as the working capital of a company. They may look the same on a BS, but their purposes are somewhat different. In the first case the capital should have the main focus on policyholders' protection, and in the second case the main focus is on company protection.

The different building blocks of the TBSA from a policyholder's perspective are illustrated in Figure 6.1 and are discussed in Chapters 7 and 8.

The asset valuation is discussed in Chapter 7 and the asset risk margin, ARM, is discussed in Section 7.1. The liability side of the total BS is discussed in Chapter 8. The current estimation is discussed in Section 8.1 and the discretionary participating features, in Section 8.2.

Valuating nonhedgeable liabilities using a replicating portfolio is discussed and illustrated in Section 8.3. The VaPo is a portfolio of liabilities. It is introduced in Section 8.3.2 and could be used as a tool to get the asset mirror of the liabilities, that is, a replicating portfolio. For nonhedgeable liabilities, the LRM is introduced in Section 8.4.

The CoC approach for assessing the RM is discussed and illustrated in Section 8.5. Other liabilities, subdivided into senior liabilities and own fund liabilities, are discussed in Section 8.6.

6.1 MARKET–CONSISTENT VALUATION

The concept of *market-consistent valuation* has been discussed frequently in both accounting and regulatory reporting fields during the last years. The topic is not simple and, as stated in

Ernst & Young (2008), there are as many different opinions about what the concept actually means as there are proponents and opponents.

We will mainly follow op. cit. in discussing market-consistent valuation, that is, we start with the concept of a deep and liquid market. A liquid market as defined by the Bank for International Settlements, BIS, is defined as a market where participants can rapidly execute large-volume transactions with a small impact on prices and a deep market denotes either the volume of trades possible without affecting prevailing market prices or the amount of orders on the order-books of market-makers at a given time.

In such a deep and liquid market, agents (buyers and sellers) constantly trade and hence define prices for the securities. It is only in the interplay of many buyers and sellers that the price will be defined. Hence deep and liquid markets supply market prices that have some attractive properties, for example, they react fast to changes in relevant information and the values are additive, do not depend on the specifics of the buyer and seller and are unique at a given point of time. Because of irregularities of the market, the price of a security can be different from its value and the depth and liquidity of the market can change. Under a financial distress or turmoil, markets that earlier were deep and liquid can turn into a freeze-up. Hence, in this case, the observed prices may not be good proxies to the values. Any reference markets will become smaller. This in turn means that the nonhedgeable part will be larger.

The CF from deeply traded securities does not depend on the owner. The CF from an insurance liability, specified by claims, expenses, changed environment, and other factors also depends on the insurer holding the liability in its portfolio.

In Ernst & Young (2008) it is established that the underlying goal of a market-consistent valuation of insurance liabilities is to transfer the problem into a setting where observable market prices are reliable and useful. This is done by determining the CF of the insurance liability, for example, using the VaPo technique in Section 8.3.2 and then replicating it by using CFs of deeply traded financial instruments ("replicating portfolio").

A fixed CF has two components. The first is the part that can be replicated exactly by the use of deeply traded securities or financial instruments (from a given reference market). This is the hedgeable part. The second is the part that cannot be replicated, that is, the nonhedgeable part.

An important consequence is that if a market is not deep and liquid, then there is no unique market-consistent value of assets or liabilities.

A market-consistent valuation framework depends on its purpose. Three main variants of market-consistent valuation approaches are usually discussed. They all differ in the choice of CFs that are associated with them. They are

- *Current exit value*: the value is determined by market values in such a way that the liabilities could be transferred to another insurer: "market specific"
- *Fulfillment value/production cost value*: the value of the liabilities is determined with reference to the insurer itself: "entity specific"
- *Distress valuation*: the value is defined with reference to the entity in the case of runoff in a financial distress

In the *current exit value approach*, the purpose is to determine a price that could be charged to transfer them to a knowledgeable third party. It defines a transfer value based on market assumptions. The price would depend on the portfolio that is transferred and the portfolio of the third party. The market-consistent value is independent of the entity holding it but depends on the hypothetical buyer.

If you assume a large number of buyers, then you could model that hypothetical deep and liquid market. As stated in Ernst & Young (2008) this is in the realm of theory. On the other hand, if you assume one single buyer, then the current exit value could be based on this single buyer, which is in contrast to a real market. It is, on the other hand, an example of a "reference entity" approach.

In the *fulfillment or production cost value approach* the purpose is to determine the cost for an insurer to hold the insurance liabilities in its portfolio ultimate. The company's experience is behind both the expected claims and expense costs and the expected costs for capital to support nonhedgeable risks. There is no need for hypothetical market values or buyer-specific values.

In the *distress valuation approach* new business is not allowed and the portfolio is assumed to be in runoff or transferred to a third party.

In a deep and liquid market, the current exit value and production cost valuation would be equal. In a real-world situation, that is a "mixed situation," the position is not that the current exit value and the settlement value approach opposing frameworks. They are complementary frameworks.

In theory you can have the current exit value ("market specific") approach versus the fulfilment/settlement value approach ("entity specific"). But, *in practice*, you will have a mixed market-consistent approach that consists of both market values (e.g., interest rates) and entity-specific values (if possible at portfolio level: mortality rates, and at entity level: expenses; if not possible: market mortality rates).

6.2 TIME VALUE OF MONEY

The time value of money is obtained by applying a discount rate to the future CFs. The method used and the measurement obtained differ depending on the context and the objective.

General principles of choosing discount rates are given in IAA (2009b, ch. 5). The discussion in op. cit. is for regulatory and general purpose financial reporting and not explicitly for capital requirements. However, the discussion on discount rates is general and also applies to capital requirements. In the following chapter we briefly discuss the present value concept. Then we discuss risk-free discount rates.

6.2.1 Present Value

The current estimate of the insurance liabilities is discussed in Section 8.1. However, it is the present value, PV{.}, of the unbiased expected CFs arising from the insurance contracts. Thus it reflects the time value of money.

Assume that $\mathrm{X} = (X_1, X_2, X_3, \ldots)$ is a random vector variable whose distribution is the outcome of the valuation of the insurance liabilities and takes into account every factor influencing the corresponding CFs. We also assume that $r = (r_1, r_2, r_3, \ldots)$ is a curve of risk-free interest rates. Thus the current estimate, CE, is defined as

$$\text{CE} = PV\,\{E[X_1], E[X_2], E[X_3], \ldots\} = \frac{E[X_1]}{1+r_1} + \frac{E[X_2]}{(1+r_2)^2} + \frac{E[X_3]}{(1+r_3)^3} + \cdots.$$

The expected CFs should be based on *realistic* underwriting principles used by the company: realistic mortality rates, realistic claims frequencies, realistic surrender rates, and so on.

Instead of using a deterministic discount function $v_t = (1 + r_t)^{-t}$, we could use a more general estimation procedure of the current estimate using a state price density process or, in actuarial terms, a stochastic deflator function ρ_t. The value at t would then be $E(\rho_t X_t)$ instead of $v^t E(X_t)$, as above. Stochastic deflators are discussed in Jarvis et al. (2001). Deflators (ρ_t) and span-deflators $(Y_t = \rho_t/\rho_{t-1})$ are thoroughly discussed in Wüthrich et al. (2007). A span-deflator Y_t "transports" cash amount at time t to value at time $t-1$. Note that $\rho_t = \prod_{k=1}^{t} Y_k$.

The main differences between a deterministic discount factor and a deflator is that, in the former case, the factor is known at the beginning of the period $(0,t]$, whereas ρ_t is only known at the end of the period. If we deal with a deterministic CF we can work with state-independent deterministic discount factors. By contrast, the state-dependent discount factors (deflators) incorporate the dependence structure also between the deflators and the CFs. Deflators allow for modeling of options and guarantees in the policies; *cf.* Wüthrich et al. (2007).

Let $\sum_{i=1}^{n} X_i$ be the aggregate claims and $v = (1 + r)^{-1}$ be the deterministic discount factor. The present value at time 0 to be paid at time 1 is given by $S = v \cdot \sum_{i=1}^{n} X_i$. Assume that the $X_i, i = 1, \ldots, n$, are independent and identical distributed, then the average payment per policy has mean and variance $E[S/n] = v \cdot E[X]$ and $\text{Var}[S/n] = v^2 \cdot (\text{Var}[X]/n)$.

If we assume that the discount factor is stochastic (Y) and replacing the deterministic discount factor ν by the random variable Y, which we assume is independent of the payment X_i, $i = 1, \ldots, n$ (see also Dhaene et al., 2002a), then $S = Y \cdot \sum_{i=1}^{n} X_i$ and the average payment per policy has mean and variance $E[S/n] = E[Y] \cdot E[X]$ and $\text{Var}[S/n] = E[Y^2] \cdot \text{Var}[X]/n) + (E[X])^2 \cdot \text{Var}[Y]$. Hence, the law of large numbers will not eliminate the risk involved.

6.2.2 Discounting Rates

There are different types of discount rates that could be used to measure the present value of the CFs of future liabilities. In the environment of capital requirements, a risk-free, or unlinked, discount rate is usually preferred. But, of course, other types of discount rates could be used; see IAA (2009b, ch. 5). As insurance obligations, especially for life business, have long maturities, interest rates with long maturity have to be investigated further. Techniques for interpolation and extrapolation are discussed in Section 6.2.3.

By the "yield" we mean the annual rate of return on an investment, expressed as a percentage, and a yield curve is the term structure of interest rates; the yield curve is a graph that plots the yields of similar-quality bonds against their maturities and ranges from the shortest to the longest. A yield curve that shows the same yield for short-term maturity bonds as for long-term maturity is called a flat yield curve.

The IAA (2009b, ch. 5.3) discusses three possible sources of risk-free discount rates. These are, with or without any possible adjustments,

- Government bond rates
- Corporate bond rates
- Swap rates

Depending on the circumstances, a single average discount rate may be used. But normally, if there are no relevant observable investment return rates, the most similar available yield curve or interest rates should be used. For more details, see IAA (2009b, ch. 5.3).

Government bonds rates: these are usually the most prudent basis for risk-free yields. The disadvantages with these are that usually there are a limited number of long-dated bonds so you only have a few points to base the yield curve on. The prices could be distorted because of high demand from insurance undertakings, especially if there are regulatory requirements favoring government bonds.

Government bond rates + adjustment: these are government bonds adjusted for market distortions not relevant in estimating the CFs of insurance liabilities. One such distortion, which is usually difficult to quantify, is the short supply of bonds at the long end of the yield curve. In an efficient market with bond sale and repurchase ("government bond repo") the possibility to earn an extra premium is reflected in a corresponding lower bond yield. According to IAA (2009b, ch. 5.3.2), the repo rates exceed the bond yields by 5–10 basis points.

Corporate bond rates − adjustment: From a high-quality corporate bond rate, a deduction for a margin for the default risk is made. This would give us a proxy for a risk-free rate. The default risk is discussed at length in IAA (2009b, ch. 5.3.3). This measure is seen as the least robust choice of a risk-free rate.

Swaps − adjustment: The advantage of the swap market is that its liquidity is often extended a lot further than the government bond market. Thus it may be a more robust and reliable basis for long-term discounting.

A swap can be seen as an agreement between two counterparties to exchange CFs or streams of payments linked to two different indices on one or more dates in the future. The two sets of payments being swapped initially are set to have equal market value, so ignoring the costs there is no money exchange.

Swaps have been used in conjunction with indices relating to interest and exchange rates as well as commodity and equity prices. With *interest rate swaps*, the CFs that are exchanged consist of interest payments having different characteristics but based on a common underlying or notional principal amount, which in general is not exchanged. The most common ("plain vanilla") interest rate swap consists of one party undertaking payments linked to a short-term floating interest rate index such as LIBOR* and receiving a stream of fixed interest payments; the other counterparty undertakes the opposite set of transactions.

With *currency swaps* and *commodity swaps* the CFs that are exchanged consist of payments indexed to interest rates (a fixed leg or a floating leg) in different currencies (and typically also include the exchange of the underlying principal amounts at maturity) and to prices of commodities, respectively; see Alworth (1993). An interest rate swap consists of two legs: fixed rate and floating rate. The fixed rate leg payer pays a fixed rate, the swap rate, and receives a stream of floating payments. The floating rate leg payments are calculated at the prevailing market interest rate on that payment date.

The transaction of swaps does not, *per se*, reflect the credit risk of the parties transacting them. Instead the credit risk of swaps refer to the risk inherent in achieving the floating leg of the swap, meaning that the company has to deposit the underlying nominal of the swap with another financial institution. Thus this nominal would be subject to credit risk, which should be eliminated in a risk-free measure. Swap rates are discussed at length in IAA (2009b, ch. 5.3.4). Because of the credit risk and an adjustment to the spread between the LIBOR rate and a "General Collateral" (GC), repo rate unadjusted swap spreads are not justified as risk-free.

To illustrate the various methods we compare them in Figure 6.2, which is given in IAA (2009b, Section 5.3.5).

6.2.3 Long Maturities and Liquidity Considerations

Any appropriate risk-free interest rate term structure has to be constructed from a finite number of data points of sufficient liquidity. Hence, we need to *interpolate* between these data points and *extrapolate* beyond the last available data point of sufficient liquidity. The discount factor often increases with the time to maturity. This means that any extrapolation of the risk-free curve beyond the last available data point significantly impacts the present value of long-term insurance liabilities. Therefore, the technique of extrapolation needs to adhere to the desired risk-free criteria set out in this advice, with the exception of liquidity. Different extrapolation techniques have been discussed by CEIOPS (CEIOPS, 2009d, p. 2); see also Section 24.3.3.

In highly developed markets we are probably able to observe liquid bonds or interest swaps rate with long maturity, perhaps up to 40–50 years. In less developed markets, liquid bonds might be limited to a few years. Even in many developed markets the observable liquid interest rates may have a maturity of only 10–15 years. In the latter case there may be less liquid instruments with higher maturity, say 25 years, but these are usually few and exposed to thin trading. This could cause a problem for some markets as life liabilities tend to be longer than 20, 30, or more years.

There are different approaches to deal with this "illiquidity problem." In general terms, this could be dealt with by introducing some cutoff points on the interest rate duration timescale:

* LIBOR: The London Interbank Offered Rate is a daily reference rate based on the interest rates at which banks offer to lend unsecured funds to other banks in the London wholesale interbank market.

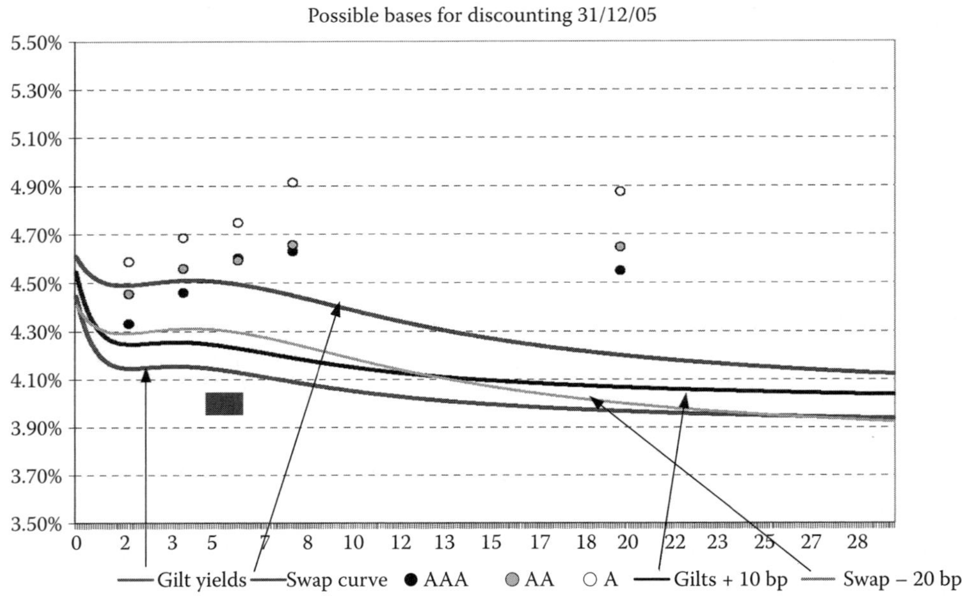

FIGURE 6.2 Possible bases for discounting as risk-free rates. The gilts refer to Government bonds. [From IAA, 2009b. Measurement of Liabilities for Insurance Contracts: Current Estimates and Risk Margins. Prepared by the ad hoc Risk Margin Working Group, International Actuarial Association, 15 April. An International Actuarial Research Paper. ISBN: 978-0-9812787-0-4 (Figure 5.10). With permission.]

$0 < T_1 < T_2 < T_U$, where T_1 is the cutoff point up to where we have an interest rate in a deep and liquid market, T_2 is a point where an equilibrium interest rate will be determined, and T_U is some ultimate duration, say 120 years.

Duration $[0,T_1]$ is the liquidity area where we use observed interest rates. If the market data are discontinuous it is possible to interpolate between the discontinuous points and even to smooth the observed curve; see, for example, Barrie and Hibbert (2008) for such techniques. They are also discussed in Antonio et al. (2009).

In Barrie and Hibbert (2008) and Antonio et al. (2009), the equilibrium point T_2 coincides with the ultimo point T_U to an unconditional forward rate. They extrapolate between T_1 and $T_U = 120$ years. In op. cit., the yield curve is represented by continuously compounded (log) forward interest rates.

Another approach, discussed in Faltinsen and Samuelsson (2009a), is to determine the equilibrium point T_2 and to interpolate between T_1 and T_2, and to make an extrapolation from T_2 onwards. The extrapolation is done by fixing a constant long-term equilibrium level of interest rate; see Figure 6.3.

Duration $[T_2,T_U]$ is the long-term interest rate that will fluctuate around a mean, that is, the long-term equilibrium level. This is the point where the interest rate is assumed to be equal to the long-term real interest rate in an economy, calibrated from market data, and the future inflation target. In the Faltinsen and Samuelsson (2009a) approach, a term "premium" is added to the real rate and the inflation target. The term "premium" is the premium for

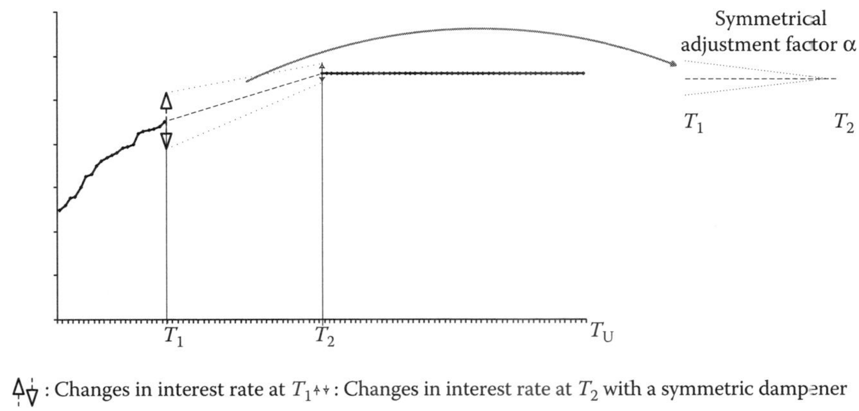

FIGURE 6.3 A description of the Faltinsen and Samuelsson (2009a) model. *Left part*: The duration interval $[0,T_1]$ is the liquidity area where we use observed interest rates. The duration interval $[T_2,T_U]$ is the long-term interest rate, that is, the long-term equilibrium level. A linear interpolation is made between T_1 and T_2. A change in the interest rate at T_1 is reflected in the equilibrium level by a symmetric adjustment, which depends on the duration length between T_1 and T_2. *Right part*: The longer T_1 is from T_2, the higher is the adjustment α.

holding long maturities and is derived from the market as the difference between the quoted interest rate at the liquidity cutoff area $[0,T_1]$ and the sum of the real rate and the future inflation target.

The long-term interest rate might also include a tuning component, designed to absorb observed market movements and to mitigate distortions depending on imbalances between supply and demand especially in times of market distress. The adjustment is continuously made via a factor, α, which is basically the ratio between the available interest risk in the bond market and the interest sensitivity of the aggregated pension liabilities. The factor is applied to a change in the observed market rate at T_1 and added to the long-term interest rate with some restrictions. It will thus have the effect of a symmetrical dampener, limiting procyclicality effects.

Let the dampener effect be δ, which will be dependent on the length between T_1 and T_2. The dampener effect will be highest when we are close to T_1 and zero when T_1 is close to T_2, say at $T_2 - t_\delta$, where, for example, $t_\delta = 1$ year. Hence, the dampener effect δ is a decreasing function from 1 to 0.

If the volume of supplied duration increases, that is, with a fixed T_2, the duration gap $[T_1, T_2]$ will decrease. The market movement adjustment factor will approach one as the liquidity increases, that is, when $T_1 \rightarrow T_2$, and the method will ensure that the interest rate curve used for valuation of liabilities will converge with a mark-to-market model. The adjustment factor can be defined as $\alpha = \left(1 - \delta_{[T_2 - t_\delta - T_1]}\right)$, where T_1 is variable and $T_2 - t_\delta > T_1$. As an example, if the duration gap $[T_1, T_2]$ is long enough, we may set $\delta = 0.75$, hence having an adjustment of 25%. When the gap is only 1 year, that is, $t_\delta = 1$, then $\delta = 0$ and hence the adjustment would be 1. Thus, there is no contradiction between the interest rate curve

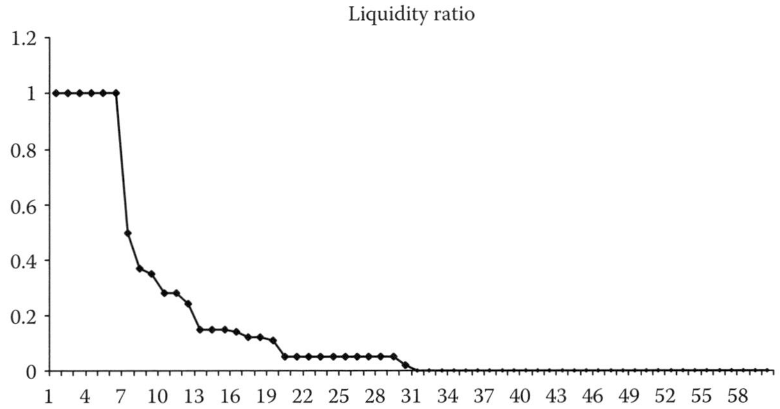

FIGURE 6.4 The liquidity ratio. Supply and demand are reasonably matched up to 7 years in this example, as indicated by the figure. After 13 years the liquidity is lower, and after 30 years where there are no bonds, there is no liquidity whatsoever.

derived by the normal interest rate method and the interest rate observed in the market when the market is deep and liquid.

In the Faltinsen and Samuelsson (2009a) approach, the interest rate in the duration gap $[T_1, T_2]$ will be determined by interpolation, for example, by a linear approach.

In a later paper, Faltinsen and Samuelsson (2009b), an interpolation method was introduced. It is based on the interpolation within the liquid cutoff area, determined by a liquidity ratio. The macroeconomic model, called the Kaplan model, combines this liquidity ratio and a macro interest rate model.

To develop an appropriate liquidity measure would be to look at the total size of liabilities and compare this with the free-floating part of the bond market.

Liquidity ratio: the ratio between the cumulative sum of bonds and the cumulative sum of liabilities. The ratio is capped at one; see Figure 6.4. If the ratio is one, this indicates that the market should be liquid. However, a ratio closer to zero indicates an illiquid market. It indicates whether a market is liquid today, but also when one could expect the market to be liquid in a distressed scenario.

The Kaplan model: in this macroeconomic model the interest rate consists of both market rates and the long-term equilibrium rate; the liquidity ratio decides how much of each. For any point in time the liquidity ratio is used to calculate a forward rate that consists of a share, equal to the liquidity ratio, of the forward market rate and another share, equal to one minus the liquidity ratio, of the long-term equilibrium rate. If the liquidity ratio equals one, the model will use the market forward rate. On the other hand, if the liquidity ratio equals zero, the model will use the long-term equilibrium rate. And in between if the liquidity ratio is for example 60%, the model will use 60% of the market forward rate and 40% of the equilibrium rate. This model was suggested by Peter Kaplan at the Stockholm office of Royal Bank of Scotland; see Faltinsen and

Samuelsson (2009b). Once a model forward rate curve is constructed, this is used to construct a spot rate curve.

6.2.4 Liquidity Premium

With liquidity we refer to the extent to which an asset or a liability can be transformed to cash (or something equivalent) without a substantial price discount at a given point in time; *cf.* IAA (2009b; Section 5.4). Two assets (or liabilities) with the same future CFs may have different characteristics. One can be costly to buy or sell and the other can have more or less zero dealing costs. Assets that are costly to buy or sell, that is, are having trading illiquidity, will have a lower price as compared to similar assets that are cheaper to trade (trading liquidity).

If the valuation is expressed in terms of yields, then the illiquid asset will give a higher yield than the liquid asset. A yield may be decomposed into its risk-free part, the compensations for expected default losses and unexpected default losses (credit risk premium) plus a residual spread compensation for illiquidity and the costs (risks) of selling prior to maturity (the (il)liquidity premium). The terms "liquidity premium" and "illiquidity premium" may be somewhat confusing, but in the sense described they are interchangeable; see, for example, Hibbert (2009).

The above discussion of the risk-free interest rate is based on rates earned from highly liquid securities; see Section 6.2.2. These rates could also be based on the characteristics of liquidity. Introducing liquidity characteristics would thus introduce interest rates that are somewhat larger than those based on highly liquid securities.

If liabilities are replicated with assets incorporating a liquidity premium, such a premium would directly be included in the value of the liability. This implies that the risk-free interest rate used for discounting should include an addition, namely an (il)liquidity premium.

For many insurance obligations the liability is reasonably well predictable and hence the insurance company needs less liquid assets to back these liabilities. Buying illiquid assets to cover these liabilities would hence be cheaper: for discounting, the interest rate would then include a liquidity premium. But for most insurance obligations the liabilities can be characterized by a certain level of "illiquidity," for example, because of different characteristics such as longevity risk, lapse risk, and surrender options. This is also reflected in the nature of the company's investments, that is, the assets invested will match the nature of liabilities. Also this will require a liquidity premium. This could be explained in the following way. If we use illiquid assets to match similar illiquid liabilities, it would be appropriate to look at the liquidity premium of the assets as a proxy for the liquidity premium included in the market-consistent valuation of the liabilities.

Estimating the liquidity premium is discussed in IAA (2009b) with reference to different research reports. A review of the literature on both theoretical and empirical evidence for the existence of a liquidity premium is given in Hibbert et al. (2009). CRO Forum has set up seven principles for the recognition of liquidity premium (CRO, 2009b).

Both the macroeconomic approach, as discussed in Section 6.2.3, and the illiquidity premium are discussed in GDV (2009).

6.3 HEDGING

Hedging is an important and accepted risk management tool for risk mitigation. One way of thinking of hedging is to think in terms of insurance. If you hedge, you are insuring yourself against a negative event. Hedging is a technique by which you will not make money, but by which you can reduce a potential loss. If you buy property insurance, you hedge yourself against fire, theft or other unforeseen events.

To *hedge* means offsetting a risk inherent in any market position by taking an equal but opposite position in the market. Thus, any loss on the original investment will be hedged, or offset, by a corresponding profit from the hedging instrument. A risk is said to be hedgeable if it can be avoided or mitigated by an offsetting transaction, for example, by use of financial derivative instruments, such as options or futures.

A *long hedge* means buying futures contracts to protect against possible increasing prices of an asset or a commodity, or to reduce the impact of price fluctuations. On the other hand, a *short hedge* involves selling futures contracts to protect against possible declining prices of an asset or a commodity.

A traditional technique is to design a portfolio of (financial) assets with CFs that offset (replicate) other CFs in certain scenarios; see IAIS (2007a). One example of a hedgeable risk in an insurance contract is an interest rate guarantee.

In more technical words, a "hedge" means an investment in two securities with negative correlation. If you reduce risk you will also reduce a potential profit. With a hedging strategy you will reduce the volatility in prices and make it easier to plan for the future.

In the valuation procedure, hedgeable assets and liabilities should be valuated by a mark-to-market approach, as any RMs are implicit in observed market prices. Nonhedgeable assets and liabilities should be valued by a mark-to-model approach, that is, a current estimate requires the calculation of an explicit RM, *cf.* Figure 6.1. Traditional nonhedgeable risks could be transformed to hedgeable risks using, for example, replicating portfolio techniques, see Section 8.3, only leaving uncertainty risk in models and parameters.

Financial risks that are nonhedgeable are, for example, different embedded financial options and guarantees in life insurance contracts that are not traded on the financial market. But also risks where the duration exceeds a reasonable extrapolation from durations traded on the financial market and traded financial instruments that are not available in sufficient quantities; see CEIOPS (2007a).

6.4 RISK MARGINS

The IAIS (2006) describes the role of a RM as

> The nature of the margin over CE {current estimate} is frequently described differently depending on the viewpoint. In an accounting sense it is often thought of as the amount that would be required to compensate a transferee for the risk inherent in a transfer of the liabilities. It is also sometimes thought of as a shock absorber. In solvency terms, this margin in the technical provision tends to be thought of in terms of prudence or a confidence level, which, together with the capital requirement in addition to the technical provision, contributes to the overall sufficiency of

> the solvency assessment regime. In both cases the IAIS believes that one of the key characteristics of the margin is to reflect the level of uncertainty in the calculation of the CE.
>
> At this stage, we do not see any reason why conceptual differences should arise in the methodologies for calculating the margin over CE within the context of insurance liabilities for both accounting and solvency purposes (IAIS, 2006, para 37+38).

The idea that the overall sufficiency of the solvency assessment regime could be split into the uncertainty in the current estimate and the capital requirement is formalized below.

The risk adjusted current estimate of the nonhedgeable liabilities together with the hedgeable liabilities is called "economical" technical provisions, or the market value of liabilities, MVL. The corresponding risk adjusted assets are called the market value of assets (MVA). The RMs should be determined in a way that enables the insurance obligations to be transferred to a third party or to be put in runoff; *cf.* also EU Commission (2006).

In terms of a TBSA, the own funds (OF), that is, the available capital or the "available solvency margin," should be written as

$$\text{OF} = \text{MVA} - \text{MVL}.$$

The RM should reflect only the *uncertainty* in the current estimate, not the inherent *volatility* in the process underlying the estimate (process risk). This latter risk is reflected in the capital requirement. This could be illustrated by the use of the mean square error (MSE), *cf.* Chapter 5 for more details,

$$\text{MSE}\left(\hat{\theta}\right) = (\text{bias})^2 + E_{\mathbb{F}}\left(V\left(\hat{\theta}|\mathbb{F}\right)\right) + V_{\mathbb{F}}\left(E\left(\hat{\theta}|\mathbb{F}\right)\right). \tag{6.1}$$

In Equation 6.1, the first term, squared bias, is an issue for the supervisor, that is, in the three-pillar approach outlined in Chapter 5, it is a Pillar II issue. The second term is the expected volatility, which is taken care of as part of the capital requirement, but with a shorter time horizon than the uncertainty. The third term represents the uncertainty in models and parameters and is the volatility of the level of the current estimate. This term is the one that constitutes the liability RM, LRM. Note that in nonlife insurance, this term is mainly a function of the liabilities, but in life insurance the liabilities could also be a function of the assets.

As stated by Ruygt (2006), "as long as risk absorption capacity of product features is taken into account, the MVL should not include arbitrary liability floors such as surrender value floors and should not be based on the market values of minimum guaranteed liability cash flows, as this may not appropriately reflect the value of the liabilities. The discussion on arbitrary floors related to individual contract-level valuations, where the imposition of minimum values/floors without regard to portfolio effects would significantly overstate the portfolio-level value of the liabilities."

From a financial reporting and economic capital perspective, the RM is discussed at length in IAA (2009b); see also Sections 8.4 and 11.4.

6.5 ACCOUNTING

We have not yet discussed the important concept of accounting. Not to burden the insurance undertaking with both a regulatory requirement and a statutory accounting requirement for solvency purposes, it would be desirable to have one reporting system that could be used for both purposes.

The International Accounting Standard Board (IASB*) has been working towards a consistent approach to insurance accounting. The main difference in its new approach is its functional view (on insurance contracts) as compared to earlier approaches that have been an institutional view (on insurance companies). The work done by IASB has been in coordination with the U.S. Financial Accounting Standard Board (FASB†). A presentation of IASB's work is given in Wright (2006).

A discussion on the accounting valuation is given in Chapter 11.

* For more information see its Web site: www.iasb.org

† For more information see its Web site: www.fasb.org

CHAPTER 7

Asset Valuation

THE VALUATION OF an asset may be used for a number of purposes. It could be used to generate information needed for internal control and reporting, for resource allocation and for company performance assessment. It could also be used for getting the basis for realization of assets, to determine insurance cover and for risk exposure, but also for external reporting. There could be a choice of valuation method and this depends mainly on the purpose of the valuation. However, it could also depend on the nature of the assets involved.

Asset valuation is the process of estimating the market value of a financial asset. Valuations can be done on assets invested in marketable securities such as stocks, options, business enterprises, or intangible assets such as patents and trademarks. Valuations are required in many contexts including investment analysis, capital budgeting, merger and acquisition transactions, financial reporting, taxable events to determine the proper tax liability, and in litigation.

The first and initial valuation is applied at the time of acquisition, and this corresponds to the cost of acquisition. Revaluations are undertaken at periodic intervals with a frequency chosen to reflect the nature of the class of assets concerned.

A market value of assets is more reflective of the economic reality as compared to the cost of acquisition. One shortcoming of market value accounting could be that it is more subjective than book value accounting. Real estate, intellectual property, and artwork are assets that are not having market value every time. Thus they have to be estimated.

Valuation of financial assets is usually done using one or more of different types of models, such as

Relative value models: used to determine the value based on the market prices of similar assets.

Absolute value models: used to determine the value by estimating the expected future earnings from owning the asset, discounted to the present value.

Option pricing models: used for certain types of financial assets (e.g., warrants, put options, call options, employee stock options and investments with embedded options such as a callable bond) and are a complex present value model.

Local accounting standards usually require a company to classify its investments into one of three categories when they are purchased. This classification is based on the company's intended use of that security and the classification dictates the accounting treatment.

1. *Held-to-maturity:* assets that the company has the positive intent and ability to hold to maturity are classified as such and are reported at *amortized cost.* The impact of temporary fluctuations in fair value of the asset is not reflected in the company's financial statements. Since equity securities have no maturity date, they cannot be classified as held-to-maturity.

2. *Trading:* assets that are bought and sold principally for the purpose of selling them in the near term are classified as trading assets and are reported in the financial statements at *fair value.* Changes in the fair value from period to period are reported as a component of net income.

3. *Available-for-sale:* assets not classified either as held-to-maturity or trading are considered available-for-sale and are reported at *market value* or *fair value.* Changes in the fair value from period to period are not reported as a component of net income but are charged or credited directly to equity. This process is called "mark-to-market."

Intangible assets such as patents, copyrights, software, trade secrets, and customer relationships can be valued by the use of a valuation model. Since few sales of benchmark intangible assets can be observed, one often values these sorts of assets estimating the costs to recreate it or using a present value model. Valuations of intangible assets are often necessary for financial reporting and intellectual property transactions.

Hedgeable assets should be valued at market value. If there is a reliable and observable market price the asset value should be equal to that price (this presupposes a deep and liquid market). As stated above, this is usually known as the mark-to-market approach. The appropriate market price for assets with long positions is the bid price (C^{bid}) at the valuation date. A *long position* means "a position held by a dealer in securities, commodities, currencies, and so on, in which his holding exceeds his sales, because he expects prices to rise enabling him to sell his longs at a profit"; see Pallister and Isaacs (2003).

On the other hand, for assets with short positions we use the ask price (C^{ask}) at the date of valuation. A short position means a position in which sales exceeds holdings as the dealer is expecting the prices to fall, cf. Pallister and Isaacs (2003). The ask price, or offer price, is the price a seller of a commodity is willing to accept for it and the bid price is the price offered by a buyer (bidder) when he buys the commodity or the price a buyer is willing to close a deal. The difference between these prices is referred to as the "bid-ask spread" or just "spread."

If the market price is *not reliable,* for example, due to illiquidity or nontradability, approximations of the market value should be used. This is usually done by using some model consistent with any relevant market information (mark-to-model). Because of the uncertainty in the model used, a risk margin (RM) should be deducted from the current estimate of these assets. The deduction, also referred to as the ARM, is also due to the credit and liquidity risks attached with these instruments.

7.1 ASSET RISK MARGIN: A DEDUCTION OF NONHEDGEABLE ASSETS

A RM should be deducted from the current estimate of the nonhedgeable assets, owing to uncertainty in models and parameters. Valuations of derivative instruments may include uncertainty in the models used; see, for example, Cont (2006). Valuation of property depends very much on how frequently it is made. Annual valuation may give an accurate calculation of the property value, but if you do daily valuations you need some model behind the calculation; this gives rise to an uncertainty in the valuation due to models and parameters.

One way of estimating the ARM is to take a proportion λ, $0 < \lambda < 1$, of the spread, that is, the difference between the ask price, C^{ask}, and the bid price, C^{bid}, that is, ARM $\approx \lambda|C^{\text{ask}} - C^{\text{bid}}|$. A coherent measure of model uncertainty is given in Cont (2006); see also Figures 5.1 and 6.1 in Chapters 5 and 6.

CHAPTER 8

Liability Valuation

THE IAIS (2007a) has suggested that RFs and their distinct components should be represented within the MVL, in the LRM, and in the SCR, as follows:

- Risk that is to be reflected both in the LRM and in the capital requirement:
  - Uncertainty and residual market volatility in underwriting risk (UR)
  - Unhedgeable mismatch risk
- Risk that is reflected only in the capital requirement and not in the LRM:
  - Volatility other than residual market volatility in UR
  - Hedgeable mismatch risk

One way of distinguishing between risks taken care of in the risk margin, LRM, with a time horizon to the ultimate, and the SCR, with a short time horizon, say $T = 1$ year, is to use the mean square function; *cf.* Chapter 5 and Section 6.4.

The MVL can be derived as the expected value of future CFs [current estimate (CE), see Section 8.1] and the cost of managing the risks underlying the business on an ongoing basis, a market value margin (MVM) or LRM. Thus, MVL represents the market-consistent value at which the liabilities can be transferred to "a willing, rational, diversified counterparty in an arms' length transaction under normal business condition," *cf.* CRO (2006).

As insurance liabilities are usually not traded on a deep and liquid market the MVL cannot be determined from the capital market directly. However, the use of market-consistent valuation techniques, such as replication portfolios, *cf.* Section 8.3, would be useful. The components of MVL are shown in Figure 8.1; *cf.* Ruygt (2006) and CRO (2006).

Market valuation, from a financial mathematical viewpoint, of life and pension liabilities is discussed in Møller and Steffensen (2007). The book itself, together with the references, constitutes a basis for market valuation in life and pension insurance and includes a discussion on interest rate theory in insurance and the use of the "binomial method" and the Black–Scholes model. Market valuation of the mortality risk is discussed in van Broekhoven (2002).

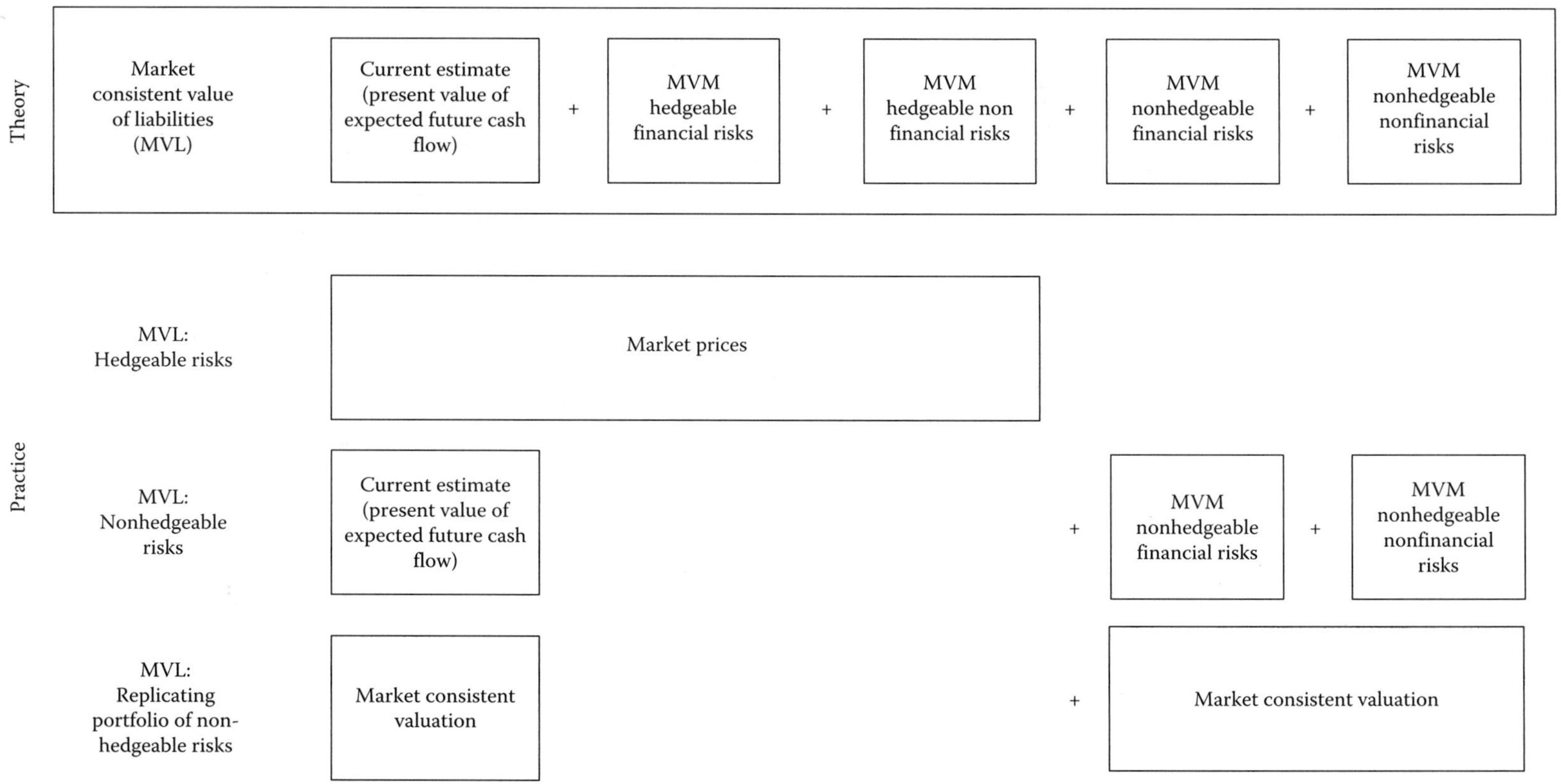

FIGURE 8.1 The decomposition of the market-consistent valuation of liabilities (MVL) into hedgeable risks, nonhedgeable risks, and risks that could be valuated using a replication portfolio technique; see Section 8.3. MVM stands for the market value margin. The IAIS' position on the valuation of TP is summarized in IAIS (2007f).

The interaction between investment and insurance risk is discussed in Parker (1997), where it is shown that the riskiness of a life portfolio can be divided into an insurance risk and an investment risk. This is done in a stochastic mortality and interest rate environment.

8.1 CE OF LIABILITIES

The time value of money is discussed in Section 6.2. In general, the risk-free discount rate can be used in the current estimation of the liabilities. However, if the CFs of, for example, profit-sharing contracts, see Section 8.2, are considered, then the expected investment return on the specified assets hold for these contracts and their expected reinvestment rates should be taken into consideration.

The International Accounting Standards Board (IASB) gave the following overall principles for estimating future CFs (IASB, 2007b, Appendix E). These principles were proposed for accounting, but they are, nevertheless, also valid for solvency purposes.

In estimating the current value of insurance liabilities, the insurer should calculate CFs that are explicit and as consistent as possible with observable market prices. The CFs should incorporate all available unbiased information about the amount, timing, and uncertainty arising from the contractual obligations. They should correspond to conditions at the end of the reporting period, that is, current CF, and exclude entity-specific CF, that is, any CF that would not arise for another entity holding an identical obligation. The accounting notation of exit value is discussed in Chapter 11.

As a starting point for calculating, the CE is a range of scenarios reflecting all possible, or at least the full range of, outcomes. Each scenario should not only specify the CFs' amount and timing, but also the probability of the outcome. After taking the time value of money, that is, discounting the CF, the scenarios should be weighted with the probability of the outcome. This would give us a probability-weighted expected present value (PV).

The market consistency of the prices means that the variables should be as consistent as possible with observable market prices. A market price, see IASB (2007b, E9), is a blend of wide-ranging views about possible future outcomes reflecting risk preferences of the market participants. This is a mark-to-market approach in valuation. If there is no market value (MV) we need to use a mark-to-model approach. One such approach, the replicating portfolio (Rpo) approach, is discussed in Section 8.3.

In IAA (2009b), IAIS discusses its proposal that "similar obligations with similar risk profiles should result in similar liabilities," even when the obligations are in different entities. IAIS state that the individual entity experience cannot serve as the sole basis of measurement of risk margins. If that is the case, the risk margin, and the liability that includes that risk margin, would be larger in a small entity than in a large entity, both identical except that the small entity has fewer such risks. In addition, to do so would mean that the two entities, one with risks from the LoB X and the other with the same number of identical risks of type X but also having risks of type Y (where the risks X and Y are not perfectly correlated), would record different liabilities; the entity with risks X and Y would measure a smaller risk margin (RM) per unit of risk X than would the entity with only risk X.

One way to achieve the IAIS objective would be to measure the reporting entity RM by considering how the RM in the reporting entity portfolio would be valued by a potential standardized entity, notionally representing a transferor. This is called a *reference entity* or equivalently we talk about measuring the risk margins for the reporting entity liabilities as part of a *reference portfolio*. The use of a "reference" entity or reference portfolio in this context constitutes a new approach. One definition of a *reference entity* could be a large, multiline, diversified insurer with business similar in nature to the portfolios subject to the valuation. As the use of a reference entity that cannot be observed is a relatively new concept, IAA sees that it may be difficult to obtain agreement with respect to specific assumptions without further guidance or research. The reference entity would likely not be a particular entity in the industry.

The CE has been discussed in detail by IAA (2005, 2009b) in a financial reporting context. CEs for nonlife reserves are discussed in, for example, Barnett and Zehnwirth (2000).

The IAA's key characteristics of CEs in the financial reporting context are discussed in IAA (2009b, ch. 4). A CE reflects the current expectation based on all the currently available information about the relevant CFs associated with the financial item being measured. In the context of capital requirements they could be seen as follows:

- All relevant CFs should be included
  - All relevant contractual rights and obligations, such as contractual options and guarantees, should be included in a prospective way
  - Future expected catastrophic and calamity risks should be included
- The CE must be consistent with the scope and context of capital requirements
  - Before a CE is determined it is important to define the scope of the estimation, that is, to make it compatible with the regulated jurisdiction.
  - Can income taxes be excluded in the calculation?
- The use of market and nonmarket inputs
  - For most insurance contracts, there is usually no market-based input.
  - The regulation might require the use of a specific risk-free interest rate from an active market.
  - Models or other valuation techniques may be used when observed values are non-reliable. As an example, mortality rates from a portfolio may not be based on a sufficiently large dataset. A modeled mortality rate based on a larger set of data may be used instead.
  - An RPo might be used to measure a portfolio of liabilities that are nonmarket based (see Section 8.3).
- Nonmarket assumptions should be determined on a portfolio-specific basis
  - A preferred source of assumptions is experienced from the portfolio measured. Thus, the unit of account would be the portfolio under consideration.

    - If there is no fully credible portfolio-specific data available, then industry or population data can be used.
- Consistency of assumptions
    - If two or more assumptions are related, for example, correlated, the application of these assumptions should be reflected in a consistent manner
    - Assumptions should be consistent over time
- Determination of valuation techniques and its input
    - When the valuation techniques have been determined, mark-to-market or mark-to-model, the input parameters have to be derived
    - Using multiple valuation techniques may enhance the credibility of the CE
    - The valuation technique used may not be unique and, in some cases, are portfolio specific
- Asymmetry of ELs or benefits should be reflected
    - Nonsymmetric probability distributions apply as a result of a fat tail or catastrophic tail or a one-sided limit on possible values
    - Guarantees and reinsurance retention limits are other examples of asymmetry
- Approximations
    - Approximations to individual assumptions or to aggregate estimates may be used if they give reasonable estimates
    - A mid-year assumption for CFs may be used
- The quality of data is important
    - If limited or unrealistic data is all that is available, other relevant data sources or experience sources may be used

8.2 DISCRETIONARY PARTICIPATING FEATURES (dpf)

Many life insurance contracts provide the policyholder a known guaranteed benefit, for example, a 3% return on investments, and the possibility to get an additional discretionary increase to payouts. These types of contracts are known as profit-sharing or with-profit business, see, for example, CEA (2007), and are said to have dpf.

The term guaranteed benefits means "payments or other benefits to which a particular policyholder or investor has an unconditional right that is not subject to the contractual discretion of the issuer" (CEA-GC, 2007). The guarantee is irrespective of how any extra benefits are described, for example, as vested, declared, or allotted.

The discretionary feature depends on the contract and the decision by the company (the board). The benefits of the dpf contracts are typically in the form of an annual increase of the guaranteed benefits or as a bonus, either on an annual basis or at maturity. The form of the dpf shows varying flexibility, depending on the jurisdiction and its profit-sharing regime, its legal and statutory restrictions, the degree of the policyholder expectation on the "profit sharing," and on the loss absorption capacity in adverse situations. The forms may be differentiated by vested, declared, or allotted contracts.

A contract with a *vested dpf* is expected to generate a future profit sharing, which, according to the legal system, should be paid out to policyholders, and thus should be treated as a liability and accounted for in the CE of the *technical provisions* (policyholder's reasonable expectation).

A *declared or allotted dpf*, which depends on the legal form and the construction of the contract, should be treated as other liabilities. Depending on the legal form it should be treated as *senior liabilities* or *own funds liabilities (OFLs)*, *cf.* Section 8.6. Contracts with declared dpf are treated as senior liabilities, as they usually can absorb losses only under certain limited circumstances. On the other hand, allotted dpf contracts should be treated as OFLs, that is, they can be fully used to cover any losses. In this case, the management has the discretion to reduce future discretionary bonuses to compensate for losses under adverse circumstances. CFs arising from profit reserves appearing in a balance sheet (BS) where they may be used to cover any losses that may arise and where they have not been made available for distribution to policyholders (i.e., surplus funds) should be treated as OFLs.

Example In Sweden the discretionary participation rights have the legal format of and are accounted for as equity (own funds). A recognition of these as risk-absorbing liabilities would therefore require change to the relevant company code.
It is important to note that there are jurisdictions where discretionary benefits are truly risk absorbing. In Sweden it is perfectly possible for discretionary benefits that have been declared, but not paid out, to be used to fund losses. Indeed, this was recently the subject of legal proceedings, where policyholders complained about the reduction in pension payouts. The finding was in favor of the insurance companies' position. In other words it is a well-established fact that the accumulated discretionary benefits can be treated as risk capital.

The valuation of contracts with dpf should be done in a market-consistent way.

If the companies are confirming the policyholders that they will have an extra guaranteed benefit of, say X Euro, then this amount must be seen as vested. This extra benefit must then be unbundled from the total future "possible" and discretionary benefits and set aside as a liability.

It should be noted that the IAIS do not agree on the classification made above: "amounts relating to future policyholder distributions in respect of both the guaranteed and discretionary elements of participating contracts should be treated as liabilities based upon the expected future cash flows. To treat them as equity would misrepresent the financial position of the company" (IAIS, 2006).

8.3 VALUATION TECHNIQUES

One way of valuating liabilities in a market-consistent way is to construct an RPo, or hedge portfolio, of assets. It is defined as a portfolio of assets that closely matches the corresponding CFs of the liabilities and is discussed in Section 8.3.1.

Another tool that could be used to valuate the liabilities is to use a VaPo, which is a portfolio of liabilities. This technique is discussed in Section 8.3.2.

In Section 8.3.3 we illustrate the construction of an RPo from the CFs of a VaPo.

8.3.1 Replicating Portfolios

If we could match the future CFs of the liabilities exactly, we would have a market-consistent value of the liabilities. The closer we could match the portfolio of liabilities by a portfolio of assets, the better we would understand and communicate with managers; especially the communication between actuaries and investors would improve. Using the RPo as a proxy would help in performing more quickly and easily the revaluation and projection of the liabilities; see, for example, Ogrodzki (2007). For an illustration of a replication portfolio approach, see CRO (2006), Figure 4.

As pointed out by Ogrodzki (2007), there is an understandable skepticism about the possibility to replicate a complex with-profit insurance fund with different options. However, as pointed out by Finkelstein (2005) and Schrager (2008), different insurance contracts could be compared and replicated by different exotic options. Exotic options are more complex than commonly traded ones (vanilla options). A portfolio replicating an (exotic) option could be constructed by using certain amounts of asset underlying the option and bonds. This is usually referred to as synthetic assets. The following diagram illustrates this:

$$\text{insurance contracts} \quad \longleftrightarrow \quad \text{exotic options}$$
$$\longleftrightarrow \quad \text{synthetic assets(replicating portfolio)}$$

Profit-sharing contracts, where the profit sharing is obtained when returns are high but not obtained when returns are low, could be replicated by the use of a call option on a stock or payer swaptions (an interest rate derivative that pays out when the interest rate is above a strike level). Another example given by Schrager (2008) is a guarantee in a unit-linked (UL) contract. This is nothing else than a put option on the underlying investment funds.

An RPo could be used for different purposes. Examples are

- In sensitivity analysis and in the calculation of economic capital
- As a BS estimator (proxy effects of market movements on the liabilities are quickly obtained)
- As a tool for scenario testing; see Ogrodzki (2007)

One technique that is used by insurance firms to get a market-consistent value of their liabilities is based on a set of risk neutral scenarios. The scenarios are usually many and widely different, say, 1000 with different risk parameters. This would be very time consuming. If we could reduce a liability portfolio to standard financial instruments this would save time and make the calculation more or less instantaneous; see Schrager (2008).

The use of an RPo approach within a portfolio management and an enterprise risk management framework is discussed in Dembo and Rosen (2000). They pointed out that it is a powerful tool to financial problems such as hedging in complete and incomplete markets, asset and capital allocation, benchmark tracking, design of synthetic products, and so on.

But the most important application of an RPo is to get a proxy of the future CFs of the insurance liabilities and the calculation of a market-consistent value of the insurance liabilities. Schrager (2008) defines an RPo as *a portfolio of standard financial instruments which matches the cash flows generated by the liabilities as good as possible.*

Ogrodzki (2007) distinguished between RPos constructed to fit the MVL under current conditions and a one-year stress scenario (MV fitting) and those constructed to fit the liability CFs in a range of usually stochastically scenarios (CF fitting).

IAA (2009b, pp. 204–205) defines a replicating portfolio as:

Replicating portfolio to an insurance liability (also referred to as a *minimum risk portfolio*). A portfolio of assets providing cash flows that exactly match the cash flows from the liability in all scenarios. Its aim is to appropriately reflect the value of the options and guarantees in the contract in establishing a discount rate. In practice, in the context of insurance, it is a portfolio of assets that minimizes the variation between the asset and liability cash flows across all scenarios. Replicating assets can include all types of traded instruments, including financial options. Hence, in this context, a replicating portfolio reflects the effects of contractual options and guarantees on the cash flows, but may not fully replicate the impact of other risks. Unreplicated liability risks are allowed for outside the discount rate.

The replicating portfolio concept is closely related to other concepts, including those of a minimum risk portfolio and a matching formula, as can be seen in the following: A *minimum risk portfolio* is a portfolio of assets that minimizes the variation between asset and liability cash flows, across all scenarios. Its aim is, as far as possible, to reflect the uncertainties in the contract outcome, in establishing a discount rate. This is a slightly different concept from replicating portfolio, since it explicitly accepts the possibility of approximate replication of duration, option and guarantee effects. In practice, in the context of insurance, the two terms are synonymous. Other liability risks are allowed for outside the discount rate.

A *matching portfolio* is a notional portfolio of "risk-free" fixed interest assets providing cash flows that exactly match the expected cash flows from the liability. Such a portfolio eliminates any gain or loss from interest rate movements and can be used to establish risk-free discount rates for the portfolio, but does not respond to features, such as options, guarantees and insurance risk, that give rise to potential departures from the expected cash flows. These other liability risks are allowed for outside the discount rate. The term is also used for an investment strategy that seeks to minimize interest rate risk with assets that approximate the expected liability cash flows.

The general procedure is to find candidate assets and to use some kind of goodness-of-fit tests on the assets and liabilities. To generate the optimal asset portfolio, which depends on the "nature of the liabilities," several steps have to be followed. First, decide if the aim is to fit CFs in each year or to group the CFs into buckets. Second, decide if different components in the contracts may be unbundled to ease the modeling, that is, to replicate the different components separately and then to sum them up. Third, chose an asset portfolio.

Fourth, make up a procedure for goodness-of-fit testing the chosen asset portfolio. The most common approach is to use regression techniques. These steps are discussed with illustrating examples in Ogrodzki (2007), and in these examples, the candidate assets used were equities, equity put options, receiver swaps, and receiver swaptions.

In Algo Risk (2007) three steps to build an RPo are presented by the use of an optimization program. The first is to make a preoptimization (create interest rate scenarios, equity returns, generate the CFs for each liability, and under each scenario, define the assets eligible for the RPo, and finally generate the CF for all combinations). The second step is the optimization (the goodness-of-fit test, specifying constraints on the optimal portfolio and running the program). The third step is to analyze the results.

A simple example of a liability CF could be the annual payout for eight years: 1, 1, 1, 50, 1, 1, 1, and 101. An RPo could be based on an 8-year bond with a 1% dividend, and a 4-year zero-coupon bond with a maturity value of 49. This is illustrated in the following scheme.

		Replicating Portfolios		
Year	Liability CF	8-year Bond with 1% Dividend	4-year Zero-Coupon Bond with Maturity Value of 49	CF of the Replicating Portfolio
1	1	1	0	1
2	1	1	0	1
3	1	1	0	1
4	50	1	49	50
5	1	1	0	1
6	1	1	0	1
7	1	1	0	1
8	101	101	0	101

In the next chapter we will use an approach that has been developed by Swiss actuaries, and with Hans Bühlmann as the leading one. The approach is called the VaPo technique. It is not a replication portfolio as the VaPo is a portfolio of insurance liabilities. But it could be used as a tool to construct an RPo of corresponding assets (or candidate assets).

8.3.2 Valuation Portfolio

The VaPo is not an RPo. It is a portfolio of liabilities derived from the insurance obligations as opposed to an investment portfolio of assets available for covering these insurance obligations.

As pointed out by Bühlmann and Merz (2007), it is important to distinguish between a *portfolio*, which is a list of financial instruments showing how many of each instrument the company holds, and the *value of the portfolio*, that is, the value which in our sense would be a MV (e.g., in euros or US$).

The VaPo will be used as a tool to get an RPo. It is a technique that can be used especially in internal modeling.

The general structure of the VaPo technique is discussed and developed in Bühlmann (2002, 2003, 2004), and Bühlmann and Merz (2007). The VaPo technique applied to life insurance is outlined in Baumgartner et al. (2004) and for nonlife insurance in Buchwalder et al. (2007). The techniques are summarized and discussed in detail in Wüthrich et al. (2007).

Bühlmann's basic point (Bühlmann, 2002) was that the classical actuarial technique for dealing with the time value of money was far from economic reality and had to be fundamentally revisited. *The trick is to understand all financial obligations of an insurance company, which involve the time value of money, as financial instruments.* This approach had been in use in some countries for more than 10 years before the Bühlmann 2002 provoke; see Aase and Persson (2003). A presentation on the Italian profit-sharing system is given in De Felice and Moriconi (2004, 2005).

In illustrating the VaPo technique, we follow Bühlmann (2004), Bühlmann and Merz (2007), and also Wüthrich et al. (2007).

Let $X = (X_0, X_1, \ldots, X_k, \ldots, X_N)$ be a random vector of stochastic flow of payments, where the random variable X_k is the stochastic payment at time k. The insurance contract starts at $k = 0$ and ends at $k = N$. Here N is either deterministic or stochastic. The unit of time could be days, weeks, months, or years. X_k is understood as the claim payments in $[k-1, k], k = 1, \ldots, N$, *minus* the premiums paid during the same period. In life insurance, N is usually defined by the contract. The general valuation principle for a deterministic technical risk, see Wüthrich et al. (2007, ch. 3.3), is made in two main steps:

1. For every portfolio of policies with CF $\boldsymbol{X}$, we construct the VaPo($\boldsymbol{X}$) in first defining the *units* (in the examples we have used zero-coupon bonds) and then determining the *number or amount* of each unit. Summing up gives us the VaPo($\boldsymbol{X}$).

2. Apply an accounting principle A on the VaPo to obtain a monetary value.

Let ρ_t be a stochastic deflator; see Section 6.2.1. The PV of the ultimate loss at time $k = 0$ is thus defined as $U(\rho) = E\left[\sum_{k=0}^{N} \rho_k X_k\right]$. In this case, we have a modern actuarial discounting approach giving an economic value. This is the principle we use, *cf.* Wüthrich et al. (2007, p. 33), that is, by using the zero-coupon bonds, as mentioned in Step 1 above.

If we buy VaPo($\boldsymbol{X}$) as assets, then the resulting portfolio is called an RPo.

In Section 8.3.2.1 we consider an illustrative and simple example for life insurance but also discuss some aspects of more general applications, such as securitization and lapses. In Section 8.3.2.2 we discuss nonlife insurance examples and in Section 8.3.2.3 we discuss the VaPo technique in general.

8.3.2.1 VaPo for Life Insurance

In life insurance it is convenient to construct the VaPo for each policy and then aggregate the VaPos to the complete portfolio.

We use as the basis the example in Bühlmann (2004), but modify it a bit, to illustrate the technique; *cf.* also Wüthrich et al. (2007). We have a nonparticipating Endowment Policy for the amount of 100 currency units (e.g., dollars or euros) written at age $x = 50$ for a period of $N = 5$ years against an annual premium of P. A death benefit of 100 currency units is paid to the family if a person dies during the period. The deterministic model, with no mortality uncertainty, is calculated using a mortality table with survival function values $l_x, l_{x+1}, \ldots, l_{x+5}$. Let $l(x), l(x+1), l(x+2), l(x+3), l(x+4)$, and $l(x+5)$ be the number of persons alive at $x+t, t=0,1,\ldots,5$, that is, the portfolio of the insured at time t. With no new policies sold, this means that $l(x) \geqslant l(x+1) \geqslant \ldots \geqslant l(x+5)$. The relation between l_x and $l(x)$ is given in the example below.

The number of persons of age x that die in the age interval $[x+t, x+t+1]$ is defined by $d(x+t) = l(x+t) - l(x+t+1)$.

We assume that the premiums are paid at the beginning of each year, deaths occur in the middle of the year, and the death benefits are paid at the end of the year and in time before the closing of the book. During the year $t=0$, a premium amount of $P * l(50)$ is paid to the company and $100 * d(50)$ is paid out before the end of the year $t=0$.

The financial instruments needed are zero-coupon bonds $Z^{(t)}$ maturing after $t = 0, 1, 2, 3, 4$, and 5 years. The zero-coupon bonds are taken at PV. The CF presented in Table 8.1 is without any securitization, which will be introduced when we valuate the CF below. For solvency purposes we want to valuate the portfolio on December 31, year $t=0$, and we denote this VaPo by VaPo_0.

Summing the CF in Table 8.1 and setting it equal to zero gives us the risk premium (RP) of the portfolio, *without any securitization.*

VaPo_1 at time $t=1$, or more precisely on December 31, year $t=1$, is a part of VaPo_0. It is easily seen that going from VaPo_0 to VaPo_1 is just to add the cash stream from and to the $l(50)$ policies compensating for the "disappearing" number of units; see, for example, Wüthrich et al. (2007, ch. 3.4). The step from VaPo_t to VaPo_{t+1} is financed from the CF generated by the policies to which the VaPo is refereeing. This is called the *self-financing property.*

The next step is to valuate the CF, which is summarized in Table 8.2. To be sure that we can pay out the 100 units at the time of death (to securitize) we buy a put option. The "put" is a financial contract allowing the buyer (or holder, here the insurer) the right, but not the obligation, to sell a commodity or financial instrument (the underlying assets) to the seller (the writer) of the option at a certain time for a certain price, the strike price. The seller

TABLE 8.1 CF for a Nonparticipating Endowment Life Insurance Contract Written at Age $x = 50$ for a Period of $N = 5$ years

Number of Units			
Premium	**Survival Benefit**	**Death Benefit**	**CF**
$-\text{P}\cdot l(50)$		$100\cdot d(50)$	$X_{50} = -Z^{(0)}\text{P}\cdot l(50) + 100\cdot d(50)$
$-\text{P}\cdot l(51)$		$100\cdot d(51)$	$X_{51} = -Z^{(1)}\text{P}\cdot l(51) + 100\cdot d(51)$
$-\text{P}\cdot l(52)$		$100\cdot d(52)$	$X_{52} = -Z^{(2)}\text{P}\cdot l(52) + 100\cdot d(52)$
$-\text{P}\cdot l(53)$		$100\cdot d(53)$	$X_{53} = -Z^{(3)}\text{P}\cdot l(53) + 100\cdot d(53)$
$-\text{P}\cdot l(54)$		$100\cdot d(54)$	$X_{54} = -Z^{(4)}\text{P}\cdot l(54) + 100\cdot d(54)$
	$100\cdot l(55)$		$X_{55} = Z^{(5)}100\cdot l(55)$

Note: The number of insurance contracts with persons aged 50 is $l(50)$. A deterministic mortality table is used.

TABLE 8.2 Valuation Scheme for $l(50)$ Nonparticipating Endowment Life Insurance Contracts Written at Age $x = 50$ for a Period of $N = 5$ years

	Number of Units			
Unit	**Premium**	**Survival Benefit**	**Death Benefit**	**CF**
$Z^{(0)}$	$-\mathrm{P} \cdot l(50)$		$100 \cdot d(50)$	$-Z^{(0)}\mathrm{P} \cdot l(50) + 100 \cdot d(50)$
$Z^{(1)}$	$-\mathrm{P} \cdot l(51)$		$[100 + \mathrm{Put}^{(1)}(100)] \cdot d(51)$	$-Z^{(1)}\mathrm{P} \cdot l(51) + [100 + \mathrm{Put}^{(1)}(100)] \cdot d(51)$
$Z^{(2)}$	$-\mathrm{P} \cdot l(52)$		$[100 + \mathrm{Put}^{(2)}(100)] \cdot d(52)$	$-Z^{(2)}\mathrm{P} \cdot l(52) + [100 + \mathrm{Put}^{(2)}(100)] \cdot d(52)$
$Z^{(3)}$	$-\mathrm{P} \cdot l(53)$		$[100 + \mathrm{Put}^{(3)}(100)] \cdot d(53)$	$-Z^{(3)}\mathrm{P} \cdot l(53) + [100 + \mathrm{Put}^{(3)}(100)] \cdot d(53)$
$Z^{(4)}$	$-\mathrm{P} \cdot l(54)$		$[100 + \mathrm{Put}^{(4)}(100)] \cdot d(54)$	$-Z^{(4)}\mathrm{P} \cdot l(54) + [100 + \mathrm{Put}^{(4)}(100)] \cdot d(54)$
$Z^{(5)}$		$[100 + \mathrm{Put}^{(5)}(100)] \cdot l(55)$		$Z^{(5)}[100 + \mathrm{Put}^{(5)}(100)] \cdot l(55)$

has the obligation to purchase the assets at the strike price (here, 100 units) if the buyer needs to exercise the option. An American put option allows the buyer to exercise the option at any time during the life of the option, while a European put option allows the buyer to exercise it just for a short period before expiration. The greater the time period before the expiration date of the option, the more value it has, as there is a greater chance that the underlying assets have *decreased in value*. Therefore, there is a time premium to be paid by the buyer. The buyer (or holder) will only exercise the put option if the price on the assets has decreased below the strike price 100, otherwise the holder will earn on selling them on the open market.

A *portfolio insurance* is a strategy for an investor who needs downside protection and desires upside potential. We assume that the premium payment is invested in some asset, for example, in equities, and therefore we need portfolio insurance. An alternative in this simple case would be to buy zero-coupon bonds that would not need any insurance. In Wüthrich et al. (2007, Example 3.4), the authors illustrated the theory with a UL payout with a minimal interest rate guarantee, that is, with a payout that was the maximum between the asset and the minimal guarantee, which required a put option.

As a portfolio insurance we introduce a European put option in the modeling, assuring that the company would not lose anything if they have to sell assets during the payout periods, that is, to have an option to exchange the asset portfolio for an amount that equals the value of the liabilities, see, for example, Panjer (2001, Section 9.5). As stated in op. cit., portfolio insurance using a put option is the term to be used for dynamic strategies that are replicating the payoff on a portfolio. To get the PV we have to change the numéraire.

For payments during the first year, $t = 0$, we hold cash, so we only need to buy European put options for expiration dates $t = 1, \ldots, 5$. We denote these by $\mathrm{Put}^{(t)}(100)$, $t = 1, \ldots, 5$. At the expiration date, or exercise date, $t = 1, \ldots, 5$, the value of the option for the holder is $\max(100 - S_t, 0)$, where S_t denotes the price of the underlying asset at time t.

Summing the CF in Table 8.2 and setting it equal to zero gives us the RP for the portfolio, *with securitization.*

TABLE 8.3 Yield Curve of Zero-Coupon Bonds as Approximated by Swedish Swap Rates, and the Approximated Zero-Coupon Bonds $Z^{(t)}$

	Swap Rates as Proxies for the Zero-Coupon Yield Curve, $S(s,t)$									
s	$t-s=1$	$t-s=2$	$t-s=3$	$t-s=4$	$t-s=5$	$t-s=6$	$t-s=7$	$t-s=8$	$t-s=9$	$t-s=10$
2007: $S(s,t)$	0.0473	0.0471	0.0473	0.0477	0.0481	0.0484	0.0487	0.0489	0.0491	0.0493
	$t=1$	$t=2$	$t=3$	$t=4$	$t=5$	$t=6$	$t=7$	$t=8$	$t=9$	$t=10$
2007: $Z_s^{(t)}$	0.9538	0.9101	0.8678	0.8263	0.7862	0.7480	0.7111	0.6762	0.6428	0.6108

The VaPo calculated on December 31, year $t = 0$, gives us the CE of the insurance portfolio in consideration.

The construction of a VaPo could in principle be made for every type of insurance contract. All that we need is financial instruments corresponding to the insurance contracts. A discussion on insurance policies and the use of, for example, exotic options is given in Finkelstein (2005), presented at the XV AFIR colloquium held in Zürich in 2005.

Before we look at a concrete example, we have to price the European put option that is used. In order to use financial mathematics we switch from a discrete to a continuous time model. We assume a nonconstant interest rate. Let $Z_s^{(t)}, s \leqslant t$, be the price process of the zero-coupon bonds paying 1 at time t for different times s. The zero-coupon bonds are approximated by the use of swap rates* as the zero-coupon bond yields taken from a Swedish financial newspaper.† The price process of the zero-coupon bonds is defined by

$$Z_s^{(t)} = \exp\left(-(t-s) \cdot R(s,t)\right),$$

where $R(s,t)$ is approximated by the swap rates $S(s,t)$ in Table 8.3. The strike price is 100 at $t = 1, 2, \ldots, 5$. If we take the PV of the strike price, we obtain $K_s^{(t)} = 100 \cdot Z_s^{(t)}$. Let S_0 be the current asset price of 100 units.

The following price process gives the general option pricing formula for European vanilla put options, see Panjer (2001, Section 9.5), Elliott and Kopp (2005, Chapter 9.4) and Wüthrich et al. (2007, Chapter 3.8),

$$\text{Put}^{(t)}(100) = K_s^{(t)} \cdot N(-d_2(s,t)) - S_s \cdot N(-d_1(s,t)), \tag{8.1}$$

where $N(*)$ is the cumulative probability distribution function for a standard normal variable, and

$$d_1(s,t) = \frac{\log\left(S_s/K_s^{(t)}\right) + \sigma^2(t-s)/2}{\sigma\sqrt{t-s}},$$

$$d_2(s,t) = d_1(s,t) - \sigma\sqrt{t-s}.$$

*See, for example, Riksbank (2006). The Riksbank is Sweden's central bank.

†Dagens Industri (DI), 13 December 2007. DI is a Swedish Financial Newspaper published 5 days a week.

If the zero-coupon-bonds yield curves $R(s,t)$ had been constant $(= r)$, then Equation 8.1 would have been the well-known Black–Scholes formula. For the Black–Scholes formula and option theory, see Henderson (2004) for a brief summary, but also Baxter and Rennie (1996), Björk (1998), Kwok (1998), and Elliott and Kopp (2005). All the classical papers by Samuelson, Black, Scholes, Merton, and so on are reprinted in Hughston (1999).

Example The above example is applied to Swedish data for both men and women. We assume that there are $l_{\mathrm{M}}(x) = 10,000$ men and $l_{\mathrm{W}}(x) = 10,000$ women. The zero-coupon bonds are approximated as above by the use of swap rates taken from a Swedish financial newspaper. The Makeham-adjusted mortalities for Swedish insured men and women, who are deterministic, are taken from mortality studies done in Sweden (DUS, 2007). The Makeham-adjusted mortalities for the individual insured are (x is age):

$$\text{Men}: \mu_{\text{Men}}(x) = 0.0004 + 0.000002342 \cdot \mathrm{e}^{0.124x}$$

$$\text{Women}: \mu_{\text{Women}}(x) = 0.0003 + 0.000001271 \cdot \mathrm{e}^{0.127x}.$$

To calculate the $l(x)$'s we use the relation $l_x = l_{[0,x]} = \exp\left(-\int_0^x \mu(s)\mathrm{d}s\right)$ and divide it by l_{50} and multiply with 10,000. If we write l_x as $l_{[0,x]}$, as above, then $(l_{x+t}/l_x) = \left(l_{[0,x+t]}/l_{[0,x]}\right) = l_{[x,x+t]} = \exp\left(-\int_x^{x+t} \mu(s)\mathrm{d}s\right)$. Multiplying $l_{[x,x+t]}$ with 10,000 we can define $l(x+t) = 10,000 \cdot l_{[x,x+t]}$ with $l_{[x,x]} = 1$.

Before calculating the CF and the VaPo, we summarize the calculation of the Put(100) option assuming $S_0 = 102$ and $\sigma = 0.15$ (Table 8.4).

The monetary value of the CF on December 31, year $t = 0$, is just the sum of the CFs excluding the first year. Note that the VaPo is an 11-dimensional vector: Five dimensions from the premium in-flow, five from death benefits, and one from survival benefits; *cf.* Wüthrich et al. (2007).

To get the full CF from Table 8.5, each CF unit has to be multiplied by the units (zero-coupon bonds). Doing so and setting the sum equal to zero gives us the pure RP. The annual premium (RP) would be 17.99 units for men and 17.95 for women. Summing up the VaPo_0 in Table 8.6 would give us zero. VaPo_1 is calculated on December 31, time $t = 0$.

The CE of the liabilities at time $t = 1$, as calculated by the VaPo on December 31, time $t = 0$, is thus, for the portfolio of men, 178,190.07 units and is 178,404.94 units for women. Using the zero-coupon bonds as an accounting principle has given us a market-consistent valuation.

TABLE 8.4 Calculation of the Put Option Assuming $S_0 = 102$ and $\sigma = 0.15$

$s = 0$	$t = 1$	$t = 2$	$t = 3$	$t = 4$	$t = 5$
$K_0^{(t)}$	95.38	91.01	86.78	82.63	78.62
$d_1(0,t)$	0.522359	0.643483	0.751890	0.852000	0.943909
$N(-d_1(0,t))$	0.300710	0.259955	0.226059	0.197107	0.172608
$d_2(0,t)$	0.372359	0.431351	0.492082	0.552000	0.608499
$N(-d_2(0,t)$	0.354813	0.333107	0.311331	0.290474	0.271428
$\text{Put}^{(t)}(100)$	3.1696	3.8006	3.9593	3.8970	3.7337

TABLE 8.5 CF Units for an Endowment Insurance Policy Applied to Swedish Data

Unit Zero-Coupon Bonds	Price at $t = 0$	Swedish Data		CF Units				
					Men		Women	
		Men	Women	Premium	Survival Benefits	Death Benefits	Survival Benefits	Death Benefits
$Z^{(0)}$	1	$l(50) = 10,000$	$l(50) = 10,000$	$-Pl(50)$		1600		1100
$Z^{(1)}$	0.9538	$l(51) = 9984$	$l(51) = 9989$	$-Pl(51)$		1857		1238
$Z^{(2)}$	0.9101	$l(52) = 9966$	$l(52) = 9977$	$-Pl(52)$		2076		1349
$Z^{(3)}$	0.8678	$l(53) = 9946$	$l(53) = 9964$	$-Pl(53)$		2183		1455
$Z^{(4)}$	0.8263	$l(54) = 9925$	$l(54) = 9950$	$-Pl(54)$		2494		1662
$Z^{(5)}$	0.7862	$l(55) = 9901$	$L(55) = 9934$		1,027,067		1,030,490	

Note: The zero-coupon bonds are approximated by the use of Swedish swap rates. The number of insured alive, $l(x)$, are calculated by the Swedish Insurance Federation (DUS, 2007).

TABLE 8.6 VaPo for an Endowment Insurance Policy Applied to Swedish Data

Unit Zero-Coupon Bonds	Price at $t = 0$	Swedish Data		CF: VaPo	
		Men	Women	Men	Women
$Z^{(0)}$	1	$l(50) = 10,000$	$l(50) = 10,000$	$-178,190.07$	$-178,404.94$
$Z^{(1)}$	0.9538	$l(51) = 9984$	$l(51) = 9989$	$-169,438.14$	$-169,842.64$
$Z^{(2)}$	0.9101	$l(52) = 9966$	$l(52) = 9977$	$-161,181.23$	$-161,763.60$
$Z^{(3)}$	0.8678	$l(53) = 9946$	$l(53) = 9964$	$-153,284.77$	$-153,950.58$
$Z^{(4)}$	0.8263	$l(54) = 9925$	$l(54) = 9950$	$-145,385.93$	$-146,209.70$
$Z^{(5)}$	0.7862	$l(55) = 9901$	$l(55) = 9934$	807,480.13	810,171.46
			$\text{VaPo}_1 =$	178,190.07	178,404.94

Note: The zero-coupon bonds are approximated by the use of Swedish swap rates. The number of insured alive, $l(x)$, are calculated by the Swedish Insurance Federation (DUS, 2007).

One example of an endowment portfolio, without securitization, is given in Bühlmann (2004) and another example of a unit or equity-linked endowment portfolio is discussed in Bühlmann and Merz (2007) and in more detail in Wüthrich et al. (2007). Profit-sharing systems are discussed in De Felice and Moriconi (2004, 2005). In Møller and Steffensen (2007), both the UL and with-profit systems are discussed in detail.

Death benefits during the contracts period, here between $0 < t < 5$, may be paid using the premium income and hence no securitization would be needed except for $t = 5$.

We can also introduce corrections for lapses. Assume that during the years $0 < t < 5$, the proportion of nonlapses are $0 \leqslant s_{x+t} \leqslant 1, 0 < t < 5$. Let the number of persons alive and still in the portfolio be defined by $l^*_{x+t} = s_{x+t} \cdot l_{x+t}, t = 1, 2, 3, 4,$ and 5. The nonlapsed intensity could, in a more general setting, be written as $S(x, x+t) = \mathrm{e}^{-\int_x^{x+t} u(v)\mathrm{d}v}$, where $t \geqslant x$. Assuming a constant nonlapsed function u, we have $S(x, x+t) = \mathrm{e}^{-u(x,x+t)\cdot(t+x-x)} = \mathrm{e}^{-u(x,x+t)\cdot t}$. The lapses are not affecting the size of the portfolio at start $l(x)$ at $t = 0$. Assuming an annual lapse rate of 1%, the VaPo on December, time $t = 0$, would be 172,755 units for men and 172,955 units for women. The RP would be 17.44 units for men and 17.41 for women, taking the annual lapses into consideration.

8.3.2.2 VaPo for Nonlife Insurance

As in Bühlmann and Merz (2007) and in Wüthrich et al. (2007), we look at pooled data of claims experience from one branch of nonlife insurance. The claims experience is usually summarized in a claims development triangle as in Table 8.7.

Let $X_{ik,j}$ denote the incremental payments for accident year i, $i = 1, 2, I$, in development year j, $j = 0, 1, \ldots, J$. We assume that there are no further payments after development year J.

By using the well-known chain ladder method, we can predict the payments made in the lower triangle in Table 8.7. We use a deterministic approach here. This is illustrated in two examples below. For each accident year $i > I$, we then calculate the CF: $X_{I+1}, X_{I+2}, \ldots$. The expected future value of the CF each development year is calculated using conditional expectation. The condition is made both on the observed realizations of the development

TABLE 8.7 A General Claims Development Triangle with $I = J - 1$

	Development Year j								
Accident Year i	**0**	**1**	**2**	**3**	**4**	**…**	j	**…**	J
1	$X_{1,0}$	$X_{1,1}$							$X_{1,J}$
2									
	Realizations for random variate $X_{i,j}$ are observed								
							$X_{i-1,j}$		
i	$X_{i,0}$								
	Predicted payments $X_{i,j}$								
I	$X_{I,0}$								

as shown in the upper part of Table 8.7, but also on known changes to the insurance environment (e.g., changes in the payment structure and inflation).

Example Once again we use Swedish data to illustrate the calculations. In the first example, we look at a short-tailed business with $J = 2$ development years, and in the second example, we have a long-tailed business with $J = 8$ development years. The Swedish Financial Supervisory Authority has produced the examples to be used as proxies for smaller companies. This is done in its work together with the European supervisors, CEIOPS, within the European Solvency II project. The data can be seen as "average" data from the Swedish market. All numbers of units are in 1000s.

In the first example, we have Home Property, a short-tailed business, where all claims are assumed to be settled after two development years; see Table 8.8.

We start in Table 8.8 with the increments in the left part and then go (black arrow) to the right part of the table with cumulative observed data. The deterministic development factors are estimated to be 1.622 and 1.093. This gives us the cumulative estimates in the lower gray part of the right table. The gray arrow illustrates the step back to the observed increments where the estimated increments are shown in the gray lower part. For the accident years 2008 and 2009 the estimated payments are shown in Table 8.9.

Table 8.10 illustrates the CF. As units or basis elements we use the zero-coupon bonds $Z^{(t)}$ and as the number of units we choose the conditional expected liabilities at time $t = I$: $\lambda_{I+k} = E\left[X_{I+k}|F_I\right]$, for $k \geqslant 1$, and $F_I = \{X_{i,j}; i + j \leqslant I\}$.

Hence from Table 8.10 we obtain the pure valuation scheme as illustrated in Table 8.11.

TABLE 8.8 Home Property, Short-Tailed Business

Observed Increments, in Currency Units					Observed Cumulative Data, in Currency Units			
AY/DY	**0**	**1**	**2**		**AY/DY**	**0**	**1**	**2**
1999	1689	904	459		1999	1689	2593	3052
2000	1871	1397	219		2000	1871	3268	3487
2001	2069	1183	199		2001	2069	3252	3451
2002	2218	1273	303	→	2002	2218	3492	3795
2003	2119	1425	245		2003	2119	3545	3790
2004	1964	1254	172		2004	1964	3218	3390
2005	1902	1131	479		2005	1902	3033	3512
2006	2138	1361	324		2006	2138	3500	3824
2007	2139	1330	322	←	2007	2139	3469	3791

Source: Swedish data—Data from Swedish Financial Supervisory Authority.
Note: The figures in gray-shaded cells are estimated using an ordinary development approach. AY: Accident year; DY: Development year. Currency in 1000 units.

TABLE 8.9 Home Property, a Short-Tailed Business

AY/AccY	$I+1=2008$	$I+2=2009$
$I-1=2006$	324	
$I=2007$	1330	322
Payments	1654	322

Note: Swedish data—Estimate of the payments the next 2 years. AY: accident year; AccY: accounting year. Currency in 1000 units.

TABLE 8.10 Cash Flow

AccY	CF	Units	Price at $t=0$	No. of Units
$I+1=2008$	X_{I+1}	$Z^{(1)}$	0.9538	$\lambda_{(I+1)}=1654$
$I+2=2009$	X_{I+2}	$Z^{(2)}$	0.9101	$\lambda_{(I+2)}=322$

Note: Home Property, a short-tailed business. Swedish data—As units, or basis elements, we have used $Z^{(t)}$, for $t=1,2$, see Table 11.3. The number of units is the payments according to Table 8.9. Currency in 1000 units.

TABLE 8.11 Valuation Portfolio

AccY	Price at $t=0$	No. of Units = CF	Discounted Value at $t=0$
$I+1=2008$	0.9538	1654	1578
$I+2=2009$	0.9101	322	293
	Monetary value of VaPo_0:		1871

Note: Home Property, a short-tailed business. Swedish data—Data from: Table 8.10. Currency in 1000 units.

The PV of the future CF, that is, the VaPo at time I, is $1578+293=1871$, that is, the CE of the future payments based on the reserve.

As the next example, we have Motor Third Party, a long-tailed business, where all claims are assumed to be settled after eight development years; see Table 8.12. The procedure is

TABLE 8.12 Motor Third Party, a Long-Tailed Business

Observed Increments, in Currency Units									
AY/D Y	**0**	**1**	**2**	**3**	**4**	**5**	**6**	**7**	**8**
1999	1217	1041	653	3	186	6	150	174	237
2000	1233	1374	134	235	24	167	166	280	249
2001	1264	1229	322	147	167	141	152	236	253
2002	1301	1538	270	214	158	185	180	265	284
2003	1527	1732	296	198	264	158	205	302	323
2004	1655	1683	261	160	189	155	201	297	318
2005	1753	1414	246	174	180	148	192	283	303
2006	1765	1237	328	170	176	144	187	276	296
2007	1900	1825	407	211	218	179	233	343	367
Observed Cumulative Data, in Currency Units									
AY/DY	**0**	**1**	**2**	**3**	**4**	**5**	**6**	**7**	**8**
1999	1217	2258	2911	2915	3101	3107	3257	3431	3668
2000	1233	2607	2741	2977	3000	3167	3333	3613	3863
2001	1264	2493	2815	2962	3129	3270	3422	3658	3910
2002	1301	2839	3109	3322	3481	3666	3846	4111	4395
2003	1527	3259	3555	3753	4017	4175	4380	4681	5005
2004	1655	3338	3599	3759	3948	4103	4304	4601	4918
2005	1753	3166	3413	3587	3767	3915	4107	4390	4693
2006	1765	3001	3329	3499	3675	3819	4007	4283	4578
2007	1900	3725	4133	4344	4562	4741	4973	5316	5683
Development Factors									
	1.9603	1.1093	1.0511	1.0502	1.0393	1.0491	1.0689	1.0690	1

Source: Swedish data—Data from: Swedish Financial Supervisory Authority.

Note: The figures in gray are estimated using an ordinary development approach. AY: Accident year; DY: Development year. Currency in 1000 units.

exactly the same as for the short-tailed business above. Therefore we drop some of the explanations below.

The PV of the future CF, that is, the VaPo at time I, is 8860, that is, the CE of the future payments based on the reserve.

The chain–ladder approach is also discussed in Buchwalder et al. (2006), Merz and Wüthrich (2007), and Wüthrich and Merz (2008).

8.3.2.3 *Some General Aspects on the VaPo Technique*

In the preceding chapters we have obtained what we were seeking, that is, the CEs of liabilities (the PV of the future CFs of existing contracts). In Wüthrich et al. (2007), the VaPo approach is extended from a deterministic life model, where the table is given and only the financial instruments describe a stochastic process, to a stochastic one (with stochastic mortality, meaning, e.g., the use of a Cox process for modeling the random evolution of the mortality of an individual) and to include a protection against technical risks. This type of risk protection could be obtained from reinsurance, risk loadings (LRM; *cf.* Section 8.4) or the SCRs. A VaPo with this additional protection is called VaPo protected against technical risks; see Wüthrich et al. (2007, p. 35).

Financial risks derive from the fact that the existing asset portfolio A differs from the VaPo RPo. In controlling financial risks, the use of Margrabe options is made to get the right to exchange one asset portfolio for another; see Bühlmann (2004) and Wüthrich et al. (2007, Chapter 4) and their references. The original paper by Magrabe from 1978 is reprinted in Hughston (1999).

These parts of the VaPo technique can be used in the determination of the LRM and the SCR and are thus not discussed here any further.

For information about useful financial methods to use in the VaPo approach for general life insurance products such as with-profit business and UL business, see, for example, Møller and Steffensen (2007). There is also a discussion in De Felice and Moriconi (2004, 2005).

The deterministic nonlife approach used above can be extended to a stochastic version using, for example, the Mack approach to protect against technical risks; see, for example, Bühlmann and Merz (2007), Buchwalder et al. (2006, 2007), Wüthrich et al. (2007), and Wüthrich and Merz (2008).

In the case of the nonlife VaPo, the choice of the units (above we used zero-coupon bonds) is less obvious than for a life VaPo. This is one of several open questions that remain to be solved; see Wüthrich et al. (2007, pp. 104–105). The premium reserve risk (RR) is not discussed above. One way of treating this could be to set $\lambda_I = -P_I$. The problem of using such an approach is discussed in op. cit, p. 105.

8.3.3 Valuating the Liabilities

To illustrate the construction of an RPo from the CFs of a VaPo, we use the second of the nonlife examples with data in Table 8.14, that is, the PV of the CF from Table 8.13.

The "hypothetical replicating portfolio" in Table 8.15 is a very simple illustrative example. For more complex CFs, more complex solutions can be made, for example, using exotic options.

8.4 LRM: AN ADDITION TO NONHEDGEABLE LIABILITIES

For insurance risks that are nonhedgeable, that is, where there is no general liquid market for them, we need to use a mark-to-model approach to determine the MVL, that is, to model an RM on top of the CE of liabilities. It should take account of the uncertainty of models and parameters and be such that the insurance contracts could be sold to a "willing buyer" or put in runoff.

Any such RMs should reflect the market perception of risk, and as defined by the market we can call it an MVM. This means that the investor would require reward for taking the risk and that defines the risk margin. As the MVM, by definition, is the reward or compensation that the market requires or demands to take over the risk in the liabilities, it defines an

TABLE 8.13 Motor Third Party, a Long-Tailed Business

AY/AccY	2008	2009	2010	2011	2012	2013	2014	2015
1999	0							
2000	249	0						
2001	236	253	0					
2002	180	265	284	0				
2003	158	205	302	323	0			
2004	189	155	201	297	318	0		
2005	174	180	148	192	283	303	0	
2006	328	170	176	144	187	276	296	0
2007	1825	407	211	218	179	233	343	367
Payments:	3339	1635	1322	1174	967	812	639	367

Note: Swedish data—Estimate of the CF the next 8 years. AY: accident year; AccY: accounting year. Currency in 1000 units.

TABLE 8.14 Valuation Portfolio

AccY	Price at $t = 0$	No. of Units = CF	Discounted Value at $t = 0$
$I + 1 = 2008$	0.9538	3339	3185
$I + 2 = 2009$	0.9101	1635	1488
$I + 3 = 2010$	0.8678	1322	1147
$I + 4 = 2011$	0.8263	1174	970
$I + 5 = 2012$	0.7862	967	760
$I + 6 = 2013$	0.7480	812	607
$I + 7 = 2014$	0.7111	639	454
$I + 8 = 2015$	0.6762	367	248
Monetary value of VaPo_0:			8860

Note: Motor Third Party, a long-tailed business. Swedish data—As units or basis elements we have used $Z^{(t)}$, for $t = 1, 2, \ldots, 8$; see Table 8.3. Currency in 1000 units.

exit value. Such a current exit value is defined by IASB (2007a, 2007b) as "The amount the insurer would expect to pay at the reporting date to transfer its remaining contractual rights and obligations immediately to another entity." This is an economic RM and it has the following features (*cf.* IASB, 2007a, p. 44).

- The RM as viewed from the current exit model perspective:
  - RM reduces as the insurer is released from risk
  - Profit is affected due to both adverse and favorable changes in estimation of CFs when they occur
  - RM reflects the increases/decreases in the amount of risk at the end of the period
  - RM reflects the amount of risk remaining at the end of the period

TABLE 8.15 The CF of a "Hypothetical Replication Portfolio" Compared with the CF of the VaPo of the Example from Table 8.14

		RPos									
Year	CF from Table 2.14	#: 2455 8-year Bonds	#: 4470 7-year Bonds	#: 5940 6-year Bonds	#: 7396 5-year Bonds	#: 9402 4-year Bonds	#: 11061 3-year Bonds	#: 14328 2-year Bonds	1-year Zero-Coupon Bond with Value 3,129,948	2–8 Year Zero-Coupon Bonds with Value	Discounted Value of the CF
2008	3,339,000	2455	4470	5940	7396	9402	11,061	14,328	3,129,948		3,185,000
2009	1,635,000	2455	4470	5940	7396	9402	11,061	1,447,128		148	1,488,000
2010	1,322,000	2455	4470	5940	7396	9402	1,117,161			176	1,147,000
2011	1,174,000	2455	4470	5940	7396	949,602				137	970,000
2012	967,000	2455	4470	5940	746,996					139	760,000
2013	812,000	2455	4470	599,940						135	607,000
2014	639,000	2455	451,470							75	454,000
2015	367,000	247,955								45	248,000

Note: All 8- to 2-year bonds are assumed to have a 1% dividend. The 1- to 8-year zero-coupon bonds are assumed to have maturity values in accordance with the columns in the table.

- The exit model causes the insurer to recognize additional expenses at the end of the period, followed by income in a later period, when he recognizes an increase in the amount of risk or an increase in the price of risk

A comparison between the above-stated exit value and another concept, namely the entry value as discussed by IASB, is given in Chapter 11.

The MVM, as defined by the exit model, is defined as the cost of risk, that is, a margin in addition to the expected PV of future CFs required to manage the business on an ongoing basis; *cf.* CRO (2006). The economic approach in measuring this margin is to use a cost-of-capital (CoC) approach; see Section 8.5.

As stated above, using an economic approach, the LRM, or the MVM, can be given by a CoC approach. "The cost of capital approach bases the risk margin on the theoretical cost to a third party to supply capital to the company in order to protect against risks to which it could be exposed" (CEA, 2006a); see also CEA-CRO Forum (2006). A comparison between different RM approaches has been given in GC (2006), but the main reference to the LRM is the work done by a "Risk Margin Working Group" within IAA (IAA, 2009b, Ch. 6).

The key criteria for a good RM are (*cf.* GC, 2006)

- Ease of calculation
- Stability of calculation between classes and years
- Consistency between different companies
- Consistency with the overall solvency system
- Consistency with IASB's Insurance Contract project
- As close as possible to market consistency

In addition to this, an RM should

- Sit on top of the CE
- Capture uncertainty in parameters, models, and trends to the ultimate
- Be harmonized between countries
- Provide a sufficient level of policyholder protection together with the minimum capital requirements and SCRs

Note that these additions to the RM have been slightly reformulated as originally they were focused on the European Solvency II project.

In Australia, the RM for nonlife insurance is calculated as the 75th percentile of the distribution function where the unbiased mean equals the CE of liabilities. "In APRA's*

*APRA: Australian Prudential Regulation Authority.

view, a 75 per cent probability of sufficiency would be commensurate with prudently set technical provisions under the current regime" (APRA, 2000, p. 18). This illustrates that the use of the percentile approach was not motivated by economic terms, but just to have technical provisions set at the "old level" with embedded prudence.

The discussion made by IAA (IAA, 2009b) is summarized in Chapter 11. The CoC approach is noted in op. cit. as a generally accepted method for setting profit margins in premium rates and, in a simplified form, was also used for reporting embedded values. As such, it is a method that market participants could reasonably be expected to consider in setting terms for the cost of bearing risk. To apply the CoC method, the applicable capital and CoC are needed at the reporting date and at each period of development of the runoff of the obligations. To estimate the required capital amounts, we need the expected CFs measured at each future reporting period until the claim/contract obligations are settled.

8.5 CoC APPROACH TO ESTIMATE THE RM

The CoC approach has a theoretical basis in determining the market-consistent value of the liabilities (MVL). Shareholders of an insurance undertaking need to provide capital to support the acquisition of a portfolio, and would therefore require compensation for the capital that is supplied; *cf.* CEA (2006a). The CoC approach distinguishes between hedgeable and nonhedgeable liabilities and, as financial theory shows that shareholders would require compensation only for nonhedgeable liabilities, we apply it to nonhedgeable risks.

It has been recognized that the CoC varies across industries due to heterogeneity of the risks facing the firms; see Cummins and Phillips (2005) where the CoC models are developed for a going concern to reflect the LoB characteristics of undertakings in property liability insurance industry. It is also observed that the CoC is inversely related to the firm's size and that long-tail commercial lines of property-liability insurance tend to have higher CoC than short-tail lines.

Different methods to estimate CoC are used in Cummins and Phillips (2005), for example, full information CAPM and Fama-French 3-Factor estimates. In the latter case,* the CoC is 18.1% for property-liability insurance, 18.8% for life insurance, 16.9% for health insurance, and 21.1% for finance excluding insurance.

In Strommen (2006), the mathematical connection between the margin for uncertainty (in valuating liabilities) and the market price of risk as measured by the CoC approach is discussed. MVMs and the CoC approach are also discussed in Kriele and Wolf (2007).

The CoC approach was first introduced in the solvency context in the Swiss Solvency Test; see SST (2004) and Sandström (2005), where the RM is defined as the hypothetical cost of regulatory capital necessary to runoff all liabilities, following financial distress of the company.

* Full Information Fama-French 3-Factor Estimates with sum beta adjustment, market value weighted, see Cummins and Phillips l (2005).

To calculate the capital needed to hold the RM, we can follow the CRO's three steps (CRO, 2006):

1. Project the SCR net of diversification benefits, for nonhedgeable risks from time $t = 1$ until the runoff of the portfolio ($t = T$), $t = 1, 2, \ldots, T$
2. Calculate the capital charge at each projection year ($t = 1, 2, \ldots, T$) as the SCR_t times the CoC charge (c_t) in order to arrive at an LRM_t, $t = 1, 2, \ldots, T$
3. Discount the projected capital charges to determine the overall LRM

Let PV[·] be the PV operator, as before. Hence $\text{PV}[c_t \cdot \text{SCR}_t] = c_t \cdot \text{PV}[\text{SCR}_t]$ and

$$\text{LRM} = \sum_{t=1}^{T} c_t \text{PV}[\text{SCR}_t],$$

where SCR_t is the projected SCR. If we let the annual risk-free interest rate be r_t, $t = 1, 2, \ldots, T$, then

$$\text{LRM} = \sum_{t=1}^{T} c_t \frac{\text{SCR}_t}{(1 + r_t)^t}. \tag{8.2}$$

The projected $\text{SCR}_t, t = 1, 2, \ldots, T$, should be calculated for the nonhedgeable risks net of diversification effects. For some risks, a stochastic projection may be needed. As most risks requiring stochastic projections are hedgeable financial risks, the MV of those liabilities is used.

In the Swiss Solvency Test, SST, see, for example, CEA (2006a), there are two methods for calculating the LRM:

- The full and "sophisticated approach" as the one outlined above. This means that capital charge (SCR) is calculated for each year, projecting assets, liabilities, and risks.
- A simplified approach, meaning that the relation between the capital charge at $t = 0$ and the CE of the technical provisions (CE) remains the same throughout the runoff period.

For the projected SCR, the SST approach is to assume that composing an optimal RPo would reduce all hedgeable risks. As this could normally not be done immediately, there will be some market risks for the first year(s) in the runoff period. The CoC is set to 6% above the risk-free interest rate for each projection year t, $t = 1, 2, \ldots, T$. In SST (2004), the calculation of the RM is discussed. Reviews of the methodology for quantifying the RM, within the SST, is discussed by Furrer (2004); see also Luder (2005) and SST (2006a).

Here we assume that hedgeable risks are swapped instantaneously as in CEA (2006a); the use of the simplified approach from the SST system is also proposed by CEA (2006a). Let

$p_0 = \mathrm{SCR}_0/\mathrm{CE}_0$ be the ratio between the calculated SCR at $t = 0$ and the CE at the same time. We now assume that $\mathrm{SCR}_t = p_0 \cdot \mathrm{CE}_t$ and hence we obtain

$$\mathrm{LRM} = p_0 \sum_{t=1}^{T} c_t \frac{\mathrm{CE}_t}{(1 + r_t)^t} \tag{8.3a}$$

and if $c_t = c$ for all t, then

$$\mathrm{LRM} = p_0 \cdot c \cdot \sum_{t=1}^{T} \frac{\mathrm{CE}_t}{(1 + r_t)^t}. \tag{8.3b}$$

As the capital charge for the solvency requirement is calculated as a function of the CE of the liabilities plus the RM, we would have circularity in the calculation. However, the RM is a reflection of the uncertainty in the CE and is a capital charge for that, and hence it should not be included in the calculation of the capital charge for solvency, that is, it should not be included in the SCR. This also means that the assumption made in Equation 8.3a and 8.3b is motivated. Ohlsson and Lauzeningks (2008) show, by using a backward approach, that there is no circularity in letting the RM enter into the SCR calculation. The noncircularity is also discussed in Kriele and Wolf (2007).

If you consider the transformation of the business to a "willing buyer," that is, not putting the business in runoff, but considering it as a "going concern," then the RM could be calculated as $\mathrm{LRM} \approx c_1 \cdot \mathrm{SCR}_1$; see Sandström (2005).

Salzmann and Wüthrich (2009) developed a rigorous multiperiod CoC approach for a general insurance liability runoff. They considered four different approaches:

- *The Regulatory Solvency Approach:* This approach is risk based with respect to the next accounting year, but not for all successive accounting years (it uses a proxy for later accounting years).
- *The Split of the Total Uncertainty Approach:* This approach is risk based for all accounting years. The risk measures quantify the risk in all accounting years with respect to information available at the beginning.
- *The Expected Stand-alone Risk Measure Approach:* This approach incorporates risk measures for each accounting year that are risk based, that is, measurable with respect to the previous accounting year. These risk-adjusted claims reserves are self-financing on the average, but lack protection against possible shortfalls in the CoC CF.
- *The Multiperiod Risk Measure Approach:* This approach gives a complete, methodologically consistent view via multiperiod risk measures. Hence it is much more technical and complex than the other three.

Based on a case study, it is shown that approaches 2 and 3 serve as good approximations to approach 4. Due to its simplicity, approach 2 is shown as preferable from a practitioner's point of view. The example also shows that the CoC margins have a substantial implication on premiums that have to be accounted for in premium calculation. The discussion was based on nominal values.

In the following example we make an assumption that simplifies the calculation. The runoff pattern is assumed to have a linear structure over 10 years.

8.5.1 Example from CEA (2006a)

For details, we refer to CEA (2006a). As discussed above the projection of the SCR is based on the CE of the liabilities. The runoff period is assumed to be 10 years and the runoff pattern will be linear: $\mathrm{CE}_{t-1} - (\mathrm{CE}_0/10), t = 1, 2, \ldots, 10$, that is, we deduct one-tenth of the original current estimate (CE_0) from each runoff year. The market risk is assumed to be hedged immediately, and thus its contribution to the nonhedgeable part of SCR equals zero. A simplified BS for this life undertaking is given in Table 8.16.

The total SCR equals 588.8 and the nonhedgeable part equals 219.0 units. The proportion of the CE to the SCR at time zero is $p_0 = (219.0/17,057.0) = 1.3\%$. Thus, each runoff year we deduct 21.9 units from the nonhedgeable SCR. This is given in Table 8.17.

This gives us the new "corrected" BS, which is given in Table 8.18.

TABLE 8.16 A Simplified BS in Units

Assets	Liabilities
MVAs:	OFs: 1223
19,205	Other liabilities: 925
	Current estimate of liabilities: 17,057

Source: Data from CEA. 2006a. *CEA Document on Cost of Capital*, CEA, Brussels, April 21. Available at http://www.cea.eu.

Note: The LRM is here set to zero.

TABLE 8.17 The CoC Charge and the Calculation of the LRM (MVM)

Runoff Period	Nonhedged SCR	CoC (%)	Capital Charge	Discount at Risk-Free Interest Rate (%)	Present Value of Capital Charge
0	219.0				
1	197.1	4	7.9	97	7.7
2	175.2	4	7.0	94	6.6
3	153.3	4	6.1	92	5.6
4	131.4	4	5.3	89	4.7
5	109.5	4	4.4	86	3.8
6	87.6	4	3.5	84	2.9
7	65.7	4	2.6	81	2.1
8	43.8	4	1.8	79	1.4
9	21.9	4	0.9	77	0.7
				LRM:	35.0

Source: Data from CEA. 2006a. *CEA Document on Cost of Capital*, CEA, Brussels, April 21. Available at http://www.cea.eu.

TABLE 8.18 Corrected BS

Assets	Liabilities
MVAs:	OFs: 1188
19,205	Other liabilities: 925
	LRM: 35
	Current estimate of liabilities: 17,057

Source: Data from CEA. 2006a. *CEA Document on Cost of Capital*, CEA, Brussels, April 21. Available at http://www.cea.eu.

Note: The market-consistent value of the technical provisions is 17,092 units $(17,057 + 35)$.

Simplifications for the calculation of the RM using the CoC approach in QIS4 were given by CEA (CEA, 2008a).

8.6 OTHER LIABILITIES

The *other liabilities* are subdivided into *senior liabilities* and *OFLs*. The senior liabilities are those liabilities that rank equal to or higher than the technical provisions ("policyholders liabilities"). OFLs are those that could be used as available capital. The definition of other liabilities depends very much on the jurisdiction and local requirements.

Examples of senior liabilities could be not only prepaid premiums, tax, and outstanding salaries to the employees but also deposits and debt with first call secured future CFs and collateral. If the premium period starts on January 1st and the premiums are prepaid on, say, December 27th, the accumulated premiums paid during these last days in December would constitute a senior liability. For solvency purposes, reinsurance liabilities should be ranked equal to direct policyholder liabilities.

Examples of OFLs could be subordinated debt, general creditors, and deferred tax; see CEA (2007a). In a situation of winding-up they would be able to absorb losses and thus, in the solvency regime, be a part of the capital requirement. See also Section 8.2 on dpf.

CHAPTER 9

Other Valuation Issues

In Section 9.1, *risk mitigation* techniques are discussed, such as pooling, diversification, hedging, reinsurance, and alternative risk transfer (ART); caution for *risk enhancing* is discussed in Section 9.2 and *segmentation* in Section 9.3.

9.1 RISK MITIGATION

We have already discussed one risk mitigation technique, namely the hedging approach in Section 6.2. Hedging is a type of insurance, where assets, for example, are invested in two different securities with negative correlation, meaning that any loss in one of the investments will be offset (or hedged) by a profit in the other hedging instrument. Risk mitigation techniques are discussed in IAA (2009b, ch. 7).

Diversification and mitigation are generic terms, as stated in Section 5.2.2. We start with the term "risk diversification" and end up with the term "risk mitigation." There is a clear connection between the two generic terms. The definition of the generic term "diversification" follows the proposal given by IAA in its answer and comments to IAIS' paper IAIS (2006); see IAA (2006).

Diversification benefits are a fundamental cornerstone in the insurance business. The benefits arise not only from the aggregation of risks within and between lines of business (LOBs), but also between risk modules, and the risks within and between legal entities and countries.

The most well-known *risk mitigation* technique in the insurance context is reinsurance. But pooling, diversification, and hedging also give risk mitigation benefits to the insurers.

We distinguish between in-house and external risk mitigation techniques.

9.1.1 In-House Risk Mitigation

By "in-house risk mitigation," we mean techniques that are naturally used by insurance firms to minimize risks. Some of them are in fact building blocks of insurance: for example, pooling and diversification.

9.1.1.1 Pooling

Insurers are *pooling* risks in order to benefit from the "law of large numbers." By "pooling," it is meant to aggregate similar risks that are similarly managed. "The statistical concept is that mutually independent risks, when aggregated, will have experience that reflects a well behaved and measurable probability distribution function about the statistical mean. Note that aggregation of risks of significantly disparate size does not "(ensure) that volatility of future cash flows is at an economically sustainable level" if the largest risks accepted are too large relative to the size of the total financial resources of the insurer" (IAA, 2006).

In IAA (2009b), the terms "pool" and "portfolio" of obligations are used as synonyms meaning that pooling is the same as putting together a portfolio of insurance contracts with similar risks and risk exposures, that is, the *action* that two portfolios are merged. If the insurance firm is built up on different pools or portfolios, the pooling is very much similar to the segmentation; see Section 9.2.

The pooling is a way of balancing the portfolio, so it consists of homogeneous risks, making it possible to estimate the behavior of the portfolio as a whole; see IASB (2007a). Examples of pools are a portfolio of private cars or a portfolio of policyholders that are assumed to have similar mortality characteristics, such as age, gender, and so on.

9.1.1.2 Diversification

"*Diversification* involves accepting risks that are not similar in order to benefit from the lessened correlation of contingent events"; see IAA (2006).

A pool or portfolio of risks is diversifiable if it is sufficiently large and the risks are uncorrelated, such that the variability of the total portfolio is less than the sum of the variability of each component. In one way, the diversification can be seen as the result of a pooling of systematic risk combinations and the independent effects of the law of large numbers.

A multiline insurer diversifies the risks by selling many different types of insurance contracts, namely different pools or portfolios of contracts; see IASB (2007a).

The exposure to risk can be classified at three main levels:

- Risks arising at the entity level: *Diversifiable* or *idiosyncratic*
- Risks faced by the insurance industry: *Systematic* and usually *nondiversifiable*. The risks affect the entire insurance market or some market segments. Some diversification and asset allocation can protect against systematic risks, as different portions of the market tend to perform differently at different times.
- Risks faced by the whole economy and the whole society: *Systemic* and *nondiversifiable*. The risks affect the entire financial market or the whole system, and not just some specific participants.

Everyone will die and therefore it is not possible to take away the mortality risk from the society *per se* (the systemic part) but an insurance company can, by a health declaration, make a selection that diminishes the risk of dying earlier than expected according to actuarial assumptions (a systematic part).

By *diversifiable* we mean that if a risk category can be segmented; see Section 9.2, into risk classes and the risk charge of the total risk is not higher than the sum of the risk charges

TABLE 9.1 In-House Techniques: Risks and Dependence

	Risks
Technique (chapter)	Application/dependence
Pooling (9.1.1.1)	Law of large numbers (independent risks)
Diversification (9.1.1.2)	Not strong positive correlations ("zero," negative)
Systematic risks (9.1.1.2)	Usually nondiversifiable
Systemic risks (9.1.1.2)	Nondiversifiable
Hedging (9.1.1.3)	Strong negative correlation

of each subrisk, then we have the effect of diversification ("subadditivity," see below). This effect can be measured as the difference between the sum of several capital charges and the total capital charge when dependency between them is taken into account. If the risk charge of risk X is $C(X)$ and Y is another risk with charge $C(Y)$, then the effect of diversification can be written as

$$C(X+Y) \leq C(X) + C(Y).$$

Diversification is discussed in Groupe Consultatif (2005).

Increasing dependence between losses has a decreasing effect on diversification benefits as well as an increasing effect on the shortfall. This and related issues are discussed in Dhaene et al. (2009).

9.1.1.3 Hedging/Offsetting Risks

Hedging, or offsetting risks, involves accepting risks with a strong negative correlation as compared to diversification, which merely requires the absence of a strong positive correlation; see IAA (2006). Hedging is a technique by which you will not make money, but by which you can reduce a potential loss. If you buy house insurance, you hedge yourself against fire, theft, or other unforeseen events.

The negative correlation of an uncertainty associated with a set of obligations is used to offset risks. One example is the offsetting between whole life insurance contracts and whole life payout annuity contracts; see IAA (2009b). As the level of mortality for life insurance and for life annuities is different, they will compensate for each other. On the other hand, the trends in the mortalities are highly correlated.

Hedging is also discussed in Section 6.3.

The in-house risk mitigation techniques discussed above are summarized in Table 9.1.

9.1.2 External Risk Mitigation

External risk mitigation comprises techniques using an external source to hedge or offset the risks, such as reinsurance and ART, but also the hedging of assets.

9.1.2.1 Reinsurance

To protect itself against risks such as losses, an insurance company can buy insurance (cede) from another insurer, a reinsurance company. In other words, reinsurance companies sell insurance to insurance companies. The insurance company buying reinsurance is called the cedent. A reinsurance company may also buy reinsurance to protect itself, known as retrocession.

There are many reasons why an insurance company buys reinsurance. The main reason is of course that it wants to hedge the risks it is facing. This is done by the transfer of risks to the reinsurance company. Usually a company has a limit on the risk that it can cover. This may be defined in its articles of incorporation/association as the retention, that is, the highest amount of a single risk that the company can stand for itself. Covering risks above the retention level entails the company to buy reinsurance or in another way minimize the risk exposure. The reinsurance is a way of getting a surplus relief.

To buy reinsurance is a way to not only protect the policyholders, but also defend the existence of the company as a going concern. It is also a way of making the results of an insurer more predictable by letting the reinsurer absorb the largest losses. This will also reduce the amount of capital that the company needs to provide coverage.

Nonlife reinsurance is dominating over life reinsurance; the proportions are approximately 4:1. This is mainly due to the fact that many of the life insurance products are savings products, which are generally not reinsured. For a life insurer it is important to not only reduce the negative impact of high sums insured for individual risks, but also reduce the accumulation of mortality risks, especially for group covered. Nonlife reinsurance contracts are usually renewed every year. Life reinsurance contracts have, on the other hand, usually the same time frame as the insurance contract period.

There are two main types of reinsurance: proportional and nonproportional; see, for example, SwissRe (1998). Reinsurance contracts cover usually more than one policy (treaty), a portfolio of policies. It is also possible to buy reinsurance on a per policy basis known as facultative reinsurance.

Proportional reinsurance means that the reinsurer takes a predefined proportion of the business that the insurer writes. The reinsurer will get the predefined share of each premium and will, on the other hand, pay the same share of the losses, meaning that the insurer reduces any risk by the predefined percentage. The reinsurer will pay the cedent company a commission in compensation for the costs of writing and administrating the portfolio reinsured. This main type of proportional reinsurance is called *absolute quota share reinsurance.* There is also a quota share limit. Risks above this limit may be covered by another type of arrangement.

A *variable quota share reinsurance*, or *surplus reinsurance* (surplus *line* treaty), is another proportional reinsurance type. It allows different percentages for the retention that the insurance company holds and for the business ceded to the reinsurer. A retained line is defined as the insurance company's retention. In a 5-line surplus treaty, the reinsurer would accept up to five times the retention, meaning that the insurance company has a *capacity* of maximum six times the retention.

Example The insurance company's retention is 100,000 units and they buy a surplus reinsurance of five lines, that is, coverage above 100,000 units on 500,000. The total coverage, or the capacity, is at maximum 600,000 units. The insurer (cedent company) will keep the premiums corresponding to the retention. Premiums corresponding to the five lines will be transferred to the reinsurer.

Surplus reinsurance is, of course, more complex and the administration costs more expensive than for pure quota share reinsurance. Proportional reinsurance can be complemented with facultative reinsurance for the largest risks. A facultative risk would then take care of risks above the lines defined in a surplus reinsurance.

A proportional reinsurance business is quite simple to deal with in solvency assessment. This is not the case for nonproportional reinsurance. A nonproportional reinsurance will give the cedent company payment if the loss exceeds the retention or priority (the deductible). The two main forms of nonproportional reinsurance are *excess of loss*, XL, and *stop loss*. The limit that a nonproportional reinsurance covers is called a *layer*.

Example The insurance company's retention (or deductible) is 100,000 units and they buy reinsurance on the risk above this, but up to, say, 200,000 units. The insurance company will retain any loss up to 100,000, get paid from the reinsurer up to 200,000 units, and pay any losses above 300,000 units.

Excess of loss, XL, reinsurance can have three forms:

- *Working XL, per risk XL (WXL/R):* The limit of the insurance policy is greater than the retention: The reinsurer will pay the cedent for losses above the retention up to the limit.
- *Working XL, per occurrence/event XL (WXL/E):* The insurer will recover from reinsurance in the event of multiple policy losses in one event, an event cover like a hurricane. The reinsurance is irrespective of the number of possible risks affected by the loss.
- *The accumulation loss, Cat XL:* The Cat XL provides special protection against accumulation losses. They are not affected by one loss on one risk.

Another nonproportional reinsurance form is the *stop-loss reinsurance*. Typical for this type of reinsurance is a definition of an event, for example, a storm affecting forests, extending to all such events during a year. This means that a stop-loss reinsurance covering storms takes into consideration all such losses during a year. The retention and the stop-loss cover are fixed each year. This aggregate excess of loss (stop loss) gives a "frequency protection" to the insurer. This is the most comprehensive reinsurance protection an insurance company can have. As such, it is very difficult to predict its outcome in any solvency assessment model.

9.1.2.2 Alternative Risk Transfer

ART is an umbrella, aiming at increasing the efficiency of risk transfer and covers different risk mitigation or risk transfer techniques. ART provides risk-bearing entities, such as insurance companies protection, and is an approach that covers the insurance and financial markets. A description of ART is found in, for example, Sigma (1999) and Neuhaus (2004).

During the period from the 1970s to the 1990s, a series of insurance capacity crises drove the buyers of traditional reinsurance to look for other ways to buy risk mitigation coverage. Large enterprises started to buy insurance cover from their own captive insurance companies, leading to less costs, but usually also leading to an increased risk consciousness. This is a type of *funding the risk transfer* and should be seen as a part of an enterprise risk management approach.

Key features of ART solutions, according to Sigma (1999), are

- Tailored to specific client solutions.
- Multiyear, multiline cover.
- Spread of risks over time and within the policyholder's portfolio. This is what makes the assumption of traditionally uninsurable risks possible.
- Risk assumption by non(re)insurers.

The ART concept is including the transformation of risks from the capital market into (re)insurance forms. This is either made through the policy itself or through the use of a transformer reinsurer, important for the credit risk markets and weather markets.

The ART can be classified into three alternative types: the channels, the solutions, and the risk carriers (Sigma, 1999). Some types of ART are listed below.

- Distribution channels:
  - *Captive insurance:* A captive is a normal insurance or reinsurance company that belongs to a company not active in the insurance business itself and that mainly insures the risk of its parent company. This results in the parent company retaining control of funds that normally would have been paid in insurance premiums.
- Solutions:
  - *Finite Risk Reinsurance:* Finite risk reinsurance comprises alternative solutions that are financing instead of transferring risks. It is based on spreading individual risks over time. To finite risk reinsurance contracts, the so-called experience accounts will be attached. At the end of the contract period, at the latest, the experience account balance will be repaid to the cedant (if > 0), or repaid to the reinsurer (if < 0). As the risk transfer is finite it makes it possible for a reinsurer to cover risks that normally would be uninsurable, usually named *Insuratization*.

    Two main types of finite reinsurance can be made.
  - *Retrospective contracts:* Losses already incurred but not yet settled
    - Loss portfolio transfer (LPT)
    - Retrospective excess of loss cover (RXL)
  - *Prospective contracts:* Claims not yet incurred but anticipated in the future
    - Financial quota share reinsurance (FQR)
    - Prospective excess of loss cover (PXL)

- *Multiyear/multiline products (MMP):* MMP combines different classes of insurance and it bundles insurance and finance risks into multiyear contracts with aggregate retention (Integrated Contracts). An MMP can also include a finite risk transfer to allow for uninsurable risks.
- *Multitrigger products (MTP)*: An MTP stipulates that at least two events must occur within a specific time frame to trigger the reinsurance activity. Usually insurance and noninsurance loss events occur simultaneously.
- *Contingent capital:* Contingent capital solutions are a way of financing an insurance loss by means of equity or loan capital at terms agreed upon. It is a kind of option solution to raise capital in the event of a severe insurance loss.

- Risk carriers
  Risk securitization is one major sector of ART. It includes, for example, insurance bonds (CAT bonds), derivative transactions, and reinsurance sidecars. Essentially, it involves the sale of assets or rights to future cash flows to a Special Purpose Vehicle (SPV) or reinsurance sidecar, in return for an immediate cash advance. The SPV, in turn, funds the purchase through the issue of bonds backed by the income from these assets.
  - *Insurance bonds:* Catastrophe bonds, CAT bonds: A way of securitize catastrophe risk portfolios. See, for example, Cairns et al. (2008) for a discussion on mortality CAT bonds.
    - *SPV or reinsurance sidecar:* An SPV is a specialized reinsurance company that issues conventional reinsurance policy to policyholders. The policyholders do not run any credit risk as the capital is made available before any losses and is invested in safe, short-term securities. The SPV is usually financed by a sponsoring reinsurer and an investor such as a hedge fund or from private equity. Mortgages and other assets such as auto loans and credit cards receivables can be securitized by an SPV. A security issued on fixed-income assets is called a collateralized debt obligation (CDO); see, for example, Donnelly and Embrechts (2009) for a brief introduction and a discussion on the financial crises in 2007–2009.
    - *Longevity bonds:* Transfers longevity risk to the financial market; see, for example, Cairns et al. (2008).
    - *Mortality swaps:* Counterparties swap fixed series of payments in return for series of payments linked to the number of survivors in a given cohort; see, for example, Cairns et al. (2008).
    - *Excess mortality bonds:* Transfers an "excess mortality" risk to the financial market.
    - *Life embedded value:* Monetizes an expected future surplus on a portfolio of life policies.

- *Insurance derivatives:* Futures and options for natural CAT risks traded on different markets or "over the counter.[*]" An option based on a company-specific loss event corresponds to a conventional stop-loss treaty.
  - *Weather derivatives:* These are used to reduce risk associated with adverse or unexpected weather conditions (e.g., "Global warming index").
  - *Climate change derivatives:* A derivative index allowing investors to bet on the combined impact of carbon emissions and rising global temperatures (e.g., "Greenhouse index," launched in early 2008).
  - *Industry loss warranties, ILW:* A derivative contract through which the purchaser's protection is based on the total loss arising from an event to the entire insurance industry.

An alternative division of ART instruments is given in CEIOPS (2009c):

- Insurance derivatives
- Equity-like instruments
- Insurance-linked securities (ILS)

ILS, which is the main topic of CEIOPS' report, is defined as an asset class and characterized as a pooling of insurance-related cash flows transformed into tradable securities. The underlying assets and associated income streams serve as collateral. In many cases, an SPV serves as an intermediary between the cedent (also called sponsor) and the investors. Longevity-linked securities are discussed in, for example, Cairns et al. (2008).

Two main classes of ILS are CAT bonds and life bonds. Advantages and disadvantages of ILS are discussed in Cairns et al.(2008).

9.1.2.3 Hedging

To *hedge* means offsetting a risk inherent in any market position by taking an equal but opposite position in the market. Thus, any loss on the original investment will be hedged, or offset, by a corresponding profit from the hedging instrument. A risk is said to be hedgeable if it can be avoided or mitigated by an offsetting transaction, for example, by the use of financial derivative instruments, such as options or futures.

9.2 RISK ENHANCING

Risk concentration is a major risk-enhancing issue. It has been observed when implementing different practical solvency capital assessment approaches such as the European Solvency II.

[*] Over-the-counter, OTC, means that the contract details are not standardised, but individually negotiated between the client and the dealer. For definitions on different derivative securities, see Cuthbertson (2004).

Risk concentration was also observed by IAA (2009b, ch. 7.9). Risks based on the geographical location of the risk, for many insurance coverages, are an important risk consideration for insurance entities, especially where there is potential for weather-related losses such as hurricanes, earthquakes or tsunamis, or a potential for large losses resulting from terrorism or from changes in a judicial ruling that impacts only one geographical area.

By writing policies over larger geographical areas, the risk of one single loss event impacting all of the policies would be lower. The impact is that the current estimate of losses will be more certain and the entity's risk margin as a proportion of current estimates would be lower. Risk concentration is not restricted to location, as it can relate a relatively large exposure to any characteristic of insured objects.

Risk diversification and risk concentration should be treated in a consistent manner in the measurement of insurance contract liabilities.

9.3 SEGMENTATION

Segmentation is a way to reflect the pooling of insurance contracts, that is, grouping contracts with similar risks. This could be done in order to facilitate the valuation of the insurance liabilities and to determine the technical provisions. Having a segment (or pool) of a similar contract would make it possible to valuate the group instead of doing a contract-by-contract valuation. In different jurisdictions, different segmentations may be used in order to consider the diversification effects between different risks when calculating the capital requirement.

A pool of risks could be segmented in a granular way. For instance, take a pool of motor third-party liability insurance contracts. This pool could be segmented into private cars and other cars (e.g., commercial cars), where the latter could consist of industrial cars, trucks, and so on.

One way of segmenting insurance contracts is to split them up according to "lines of business," LOBs. LOBs, which should be seen as homogeneous risk groups, could either be defined by the company and its company structure or by the reporting requirements made by, for example, the supervisory authority. A third way to define LOB segmentation would be to look at the lines or classes used by the authority to license a company to do business.

Segmentation could also be a package policy that covers both property, for example, fire, burglary, water damage, and so on—and liability, such as "homeowners insurance."

An insurance company could specialize in different markets or in different sectors. For instance, Allied Specialty Insurance in Florida, specializes in the entertainment and amusement industry. Their segmentation in LOBs consists of Amusement Facility Insurance, Fair and Festival Insurance Program, Motor Sports, Outdoor Amusement Insurance, and Special Insurance.

In the United States, the different states have their own terminology for the LOBs.* For example, the state of Arizona has the following 17 LOBs:

- Casualty with workers' compensation
- Casualty without workers' compensation

* See the Web site of NAIC: http://www.naic.org/industry_ucaa.htm.

- Disability
- Life (includes annuities)
- Variable annuity
- Variable life
- Marine and transportation
- Mortgage guaranty
- Prepaid legal
- Property
- Surety
- Title
- Vehicle
- Life and disability reinsurer
- Health care services organization
- Health, medical, dental, optometric service corporations
- Prepaid dental plan organization

On the other hand, the state of California has the following 23 LOBs:

- Life
- Fire
- Marine
- Title
- Surety
- Disability
- Plate Glass
- Liability
- Workers' compensation
- Common carrier liability
- Boiler and machinery
- Burglary

- Credit
- Sprinkler
- Team and vehicle
- Automobile
- Mortgage
- Aircraft
- Mortgage guaranty
- Insolvency
- Legal
- Miscellaneous
- Financial guaranty

Within the European Union there are, according to the insurance directives, 18 nonlife insurance classes and nine life insurance classes; see, for example, Sandström (2005).

CHAPTER 10

Investments and Own Funds

10.1 INVESTMENTS

Insurance companies, like other companies, use assets to invest and to gain capital. Another viewpoint of investment is to tailor the flows of expenditures and receipts over a period of time; see, for example, Luenberger (1998). For more details of investment issues, see op. cit.

The investments made by insurance companies need to be done in the best interest of the policyholders. Traditionally this has meant that the investments have been limited by regulatory requirements, and it has usually not been enough with the use of ALM techniques such as immunization; see Section 3.2.2.2. The limits could have been both on the type of permitted assets and on the maximum share of that asset, for example, 25% of the total assets invested.

Under the economic TBSA, where the capital requirement should capture all quantifiable risks, any such investment limits could be relaxed. Investment in a "bad" asset should instead be reflected in a higher capital requirement, reflecting the credit and liquidity characteristics of assets. Instead of any arbitrary quantitative limits one could set up qualitative rules, such as the prudent person rule.

10.1.1 Prudent Person Rule

All investments held by insurance undertakings, that is, assets covering technical provisions, plus assets covering the capital requirement and the free assets (OFs), should be invested, managed, and monitored in accordance with the prudent person principle. The prudent person principle requires insurance undertakings to invest assets in the best interest of policyholders, adequately match investments and liabilities, and pay due attention to financial risks, such as liquidity and concentration risk.

The prudent person principle rule was stated by Judge Samuel Putnum in 1830: Those with responsibility to invest money for others should act with prudence, discretion, intelligence, and regard for the safety of capital as well as money. A prudent man is a person that does not place all of his risks into one investment, nor does he choose to unwisely pay too much. With this principle, an investment decision should be made on the basis of a standardized code of conduct.

An Enterprise Risk Management approach, as outlined in Section 3.3, includes the principles of investment from the holistic view and not only from individual investments.

10.2 AVAILABLE CAPITAL: ELIGIBLE OWN FUNDS

Traditionally, regulatory capital refers to specific categories of capital used to meet regulation requirements on the available capital (OFs, see Figure 6.1). Regulatory defined *tiers of capital* attempt to reflect different liquidity characteristics and different levels of capacity to absorb losses. The potential for loss is defined partly by how regulators assign assets to risk categories. Regulatory capital requires usually government approval and is subject to and enforced by compliance with strict and detailed laws.

Here we have developed the view of an *economic TBSA*, meaning that the determination of the regulatory (available) capital needed by a company for solvency purposes should be based not only on all assets and liabilities, but also on the interaction between them. The valuation should be market consistent and include both on-balance sheet items and off-balance sheet items. The capital requirement should capture all quantifiable risks, and as such, there need not be any derogation in the capital used, as any depreciation of capital should be reflected in the capital requirement. The less risk-absorbing an item is, the higher the capital requirements will be.

Nevertheless the Basel Accord (Basel II) applied to the bank industry introduced the eligibility of capital and eligible OFs ("available capital") to cover the capital requirement. To have an arbitrage-free solution between the bank and insurance industries, many jurisdictions will probably have a set of rules and regulations for the insurance industry similar to Basel II; see BIS (2006a). Because of that, we will look at one solution about the eligibility of capital, namely the European Solvency II. In recital number 29 in the proposal of a Framework Directive, COM (2007a), the following explanation for introducing a classification of OF items is given:

> Since all financial resources do not provide full absorption of losses in the case of winding-up and on a going-concern basis, own fund items should be classified in accordance with quality criteria, and the eligible amount of own funds to cover capital requirements should be limited accordingly. The limits applicable to own fund items should only apply to determine the solvency standing of insurance and reinsurance undertakings, and should not further restrict the freedom of those undertakings with respect to their internal capital management.

The regulatory view on capital is that its main purpose is to act as a buffer against unexpected losses and to be enabled to absorb unforeseen losses (over a finite time horizon and a predefined confidence level). The solvency capital requirement, SCR, see Figure 6.1, reflects these unforeseen losses. The main purpose of eligible capital would be to cover the SCR.

The basic OFs, defined as the excess of assets over liability, are an important part of the solvency assessment. The qualitative difference between the SCR, acting as a soft level of intervention, and the MCR, acting as a hard level of intervention, is that the MCR is related to a stressed situation where the company may be closed and that the SCR is related to a going-concern situation. This qualitative difference is also reflected in that the MCR needs to be covered with higher quality elements than the SCR.

The OFs are the sum of [Article 85 in COM (2007a)]

- *Basic OFs:* On-balance sheet items, and
- *Ancillary OFs:* Off-balance sheet items.

The basic OFs are the sum of "assets in excess of liabilities" and "subordinated liabilities." The first part is pure capital defined under a TBSA and is therefore available as a loss-absorbing buffer. From the same argumentation there is neither any argument to require an assessment of quality nor a classification of the items included in that buffer. Even if there may be arguments for "subordinated liabilities" to be of different quality (dated subordinated liabilities may be seen as of lower quality than undated), these quality aspects should be taken care of in the capital requirements. This fact is ignored in the following discussion.

By *eligibility of capital* is meant a categorization of various capital elements, depending on its capacity for loss absorbing and its availability.

The following criteria of eligibility capital have been set up in the proposed framework directive, COM (2007a) Article 92, Characteristics used to classify OFs into tiers:

> Own fund items shall be classified into three tiers on the basis of the following characteristics:
>
> 1. In the case of winding-up, the repayment of the item is refused to its holder until all other obligations, including insurance and reinsurance obligations towards policyholders and beneficiaries of insurance and reinsurance contracts, have been met (*subordination*);
> 2. The total amount of the item, rather than only part of it, is available to absorb losses in the case of winding-up (*loss-absorbency*);
> 3. The item is available, or can be called up on demand, to absorb losses on a going-concern basis, as well as in the case of winding-up (*permanence*);
> 4. The item is not dated, or has a duration which is sufficient taking into account the duration of the insurance and reinsurance obligations of the undertaking (*perpetuality*);
> 5. The item is free from mandatory fixed charges and requirements or incentives to redeem the nominal sum and is clear of any encumbrances (*absence of mandatory servicing costs*).

In the Basel Accord (and Basel II), a bank's capital is split up into three tiers reflecting different liquidity characteristics and different levels of capacity to absorb losses.

- *Tire 1:* Highest quality capital (with subtiers: Core Tier 1, Noninnovative Tier 1, and Innovative Tier 1).

- *Tier 2:* Supplementary elements that are eligible up to specified limits (with subtiers: Upper Quality Tier 2, Lower Quality Tier 2).
- *Tier 3:* Elements at the discretion of the national authority, to support market risks only.

The classification of capital into tiers, as proposed in the European solvency framework directive, can be described referring to Table 10.1.

Capital elements that could be included in the three tiers are as given in Table 10.2. This is only a proposal taken from the European Insurance Supervisory Committee, CEIOPS (CEIOPS, 2006a, 2006b), and should not be seen as a definitive classification. CEIOPS has also proposed a subdivision of Tier 1 and Tier 2 into subtiers as in the Basel II environment. In the framework directive, COM (2007a), it is decided that there should be subtiers, but these are to be defined in implementing the measures set down by the European Insurance and Occupational Pension Committee, EIOPC. There may also be adjustments to capital, for example, a deduction of certain intangible assets such as goodwill.

The MCR should only be covered by Tier 1 and Tier 2 eligible basic OFs capital. The difference between SCR and MCR, cf. Figure 10.1 below, could also, apart from Tier 1 and Tier 2 on BS items, include Tier 3 capital and Tier 2 eligible ancillary OFs.

Let T1 be Tier 1, T2 be Tier 2, and T3 be Tier 3. The indices BOF and AOF stand for basic own funds and ancillary own funds, respectively. The index OF stands for OFs. Hence from Table 2.20 we have

$$\begin{aligned} \mathrm{T1_{OF}} &= \mathrm{T1_{BOF}}, \\ \mathrm{T2_{OF}} &= \mathrm{T2_{BOF}} + \mathrm{T2_{AOF}}, \\ \mathrm{T3_{OF}} &= \mathrm{T3_{BOF}} + \mathrm{T3_{AOF}}. \end{aligned}$$

TABLE 10.1 Classification of OF into Tiers Depending on the Quality of Capital Items and whether If they are in On- or Off-BSs

Quality of Items		BS Items			
Good	Not good	On-BS BOF	Off-BS AOF	→ Tier	Main limits
√		√		1	Tier 1 > 1/3 OF → T1 > 1/2 (T2 + T3)
√			√	2	
	√	√			
	√		√	3	Tier 3 < 1/3 OFs → T3 < 1/2 (T1 + T2)
	√ (Poor)	√			

Source: The main limits are from Article 97, COM. 2007a. Proposal for a Directive of the European Parliament and of the Council on the taking-up and pursuit of the business of Insurance and Reinsurance. SOLVENCY II.COM (2007a) 361 final, 2007/0143 (COD), Brussels, July 19.

Note: T1: Tier 1, T2: Tier 2, and T3: Tier 3.

TABLE 10.2 A First Proposal for the Classification of Capital into Tiers and Subtiers

Tiers	Subtiers	Criteria	Capital Elements
Tier 1	Core tier 1	Fully loss absorbent	Paid-up voting shares or initial fund
Highest quality capital		Permanent Fully subordinated	Disclosed reserves Retained earnings Parts of participation fund surplus and future benefits that may be used to cover losses Any difference in the valuation of technical provisions
	Noninnovative tier 1	Fully loss absorbent Permanent Fully subordinated	Perpetual noncumulative preference shares Subordinated members' accounts Budgeted members' calls
	Innovative tier 1	Hybrid capital instruments	Deeply subordinated debt
Tier 2 Supplementary elements that are eligible up to specified limits	Upper quality tier 2	Partly loss absorbent Subordinated to the rights of policyholders Perpetual	Perpetual cumulative preference shares Perpetual subordinated debt and tier 2 hybrid capital
	Lower quality tier 2	Partly loss absorbent Subordinated to the rights of policyholders Perpetual	Dated preference shares Dated subordinated debt
Tier 3 Subject to the prior approval of the supervisory authority		Contingent capital that provides loss absorption under particular circumstances Capital elements that are specific to the insurance industry	Potential members' call Unpaid element of partly paid equity Letters of credit Tier 2 elements substituted to tier 3, subject to certain limits

Source: Adapted from CEIOPS. 2006a. Answers to the European Commission on the third wave of Calls for Advice in the framework of the Solvency II project. CEIOPS-DOC-03/06, May. Available at www.ceiops.org; CEIOPS. 2006b. Draft Advice to the European Commission in the Framework of the Solvency II project on Pillar I issues—further advice. Consultation Paper 20, CEIOPS-CP-09/06, 10 November. Available at www.ceiops.org.

from above, this means that the

MCR should be covered by $T1_{BOF} + T2_{BOF}$-capital, subject to $T1_{BOF} > T2_{BOF}$.
SCR should be covered by $T1_{OF} + T2_{OF} + T3_{OF}$-capital, subject to the main limits given in Table 10.1.

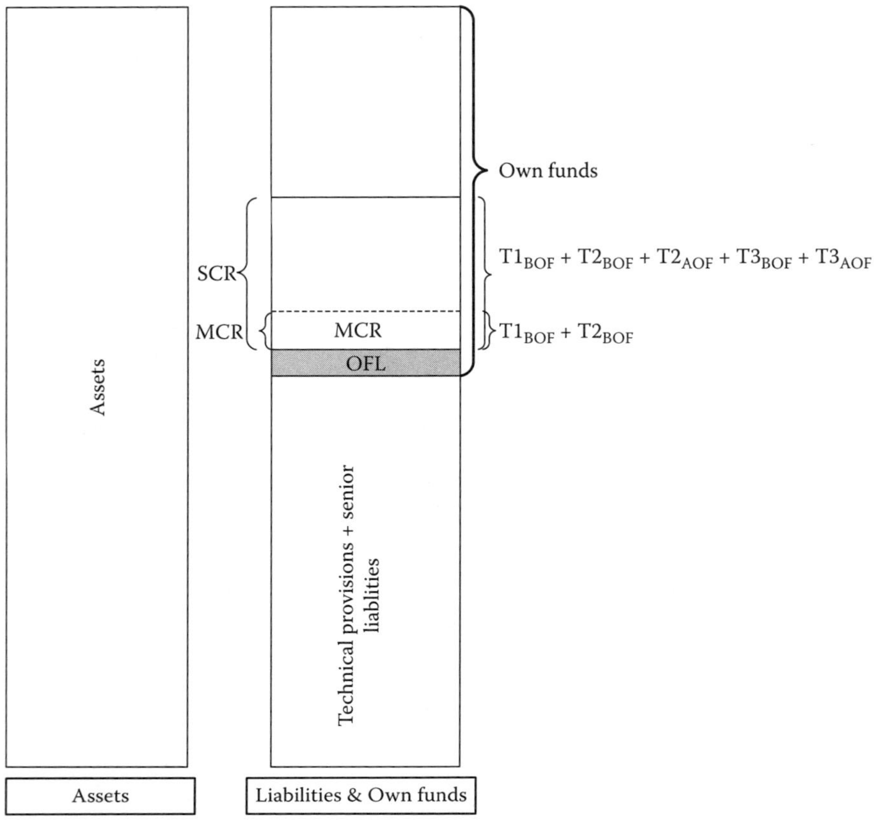

FIGURE 10.1 The MCR is restricted to Tier 1 and Tier 2 BOF capital. The difference between SCR and MCR may include not only Tier 2 and Tier 3 AOF capital, but also Tier 3 BOF. Own funds that are part of the assets, are illustrated in the opposite side.

A comparison between capital instruments eligible in the European banking, insurance, and securities regulation is described in JTF (2006).

CHAPTER 11

Accounting Valuation

11.1 BACKGROUND

In 1997, the International Accounting Standards Committee (IASC) started a project with the objective of developing an International Accounting Standard (IAS) for the insurance industry. In March 2001, IASC was transformed into a board, International Accounting Standards Board (IASB). New International Accounting Standards are now called International Financial Reporting Standards (IFRS).

At the start of the project, there was a wide range of accounting standards used by insurers, and these standards often differed from accounting standards for other enterprises in the same country. The objective of the project was "To produce a single set of high quality, understandable and enforceable global accounting standards that require high quality, transparent and comparable information in financial statements."

IASB had earlier started to develop a new standard for financial instruments and this influenced the work on insurance standards. It was expected that the valuation of financial instruments should be based on the *fair value* concept. The fair value concept used for insurance contracts should be consistent with accounting principles for other financial sectors such as banking and the securities industry, but since these were not in a position to move to fair value, other valuation concepts were proposed. In Leigh (2004), fair value is defined as the sum of the discounted best estimate and a market-value margin.

At its meeting on July 17, 2000, the European Council of Finance Ministers endorsed the Commission's proposal, from June the same year, that all listed EU companies, including insurance undertakings, should prepare their consolidated accounts in accordance with the IAS by 2005. The IAS-based regulation came into force on July 19, 2002; see COM (2002a). *The introduction of the new accounting system for the insurance industry is much more than a technical issue and will result in fundamental changes to the way the industry reports and do businesses* (COM, 2002a).

11.2 INTERNATIONAL DEVELOPMENTS

A Steering Committee, established by the IASC, finalized a report to the new IASB based on the comments on the issue papers: a Draft Statement of Principles (DSOP), for an IAS for insurance (IASB, 2001). Not less than 138 comments were sent to IASB on the DSOP. The

Steering Committee observed at an early stage that insurance contracts are not traded in a deep and liquid market as other assets (or liabilities). Hence, the determination of fair value of insurance liabilities gives rise to difficult conceptual and practical issues. The view of the Steering Committee was that a fair value approach should be based on assumptions that an independent marketplace participant would make in determining *the charge* it would make to acquire the liability. The Steering Committee also discussed a *nonfair valuation* based on a company's own assumptions and expectations. This approach is referred to as the entity-specific valuation.

It was required that an asset and liability reporting approach, a *full (total) balance sheet approach*, be used, instead of a deferral and matching reporting approach; see Abbink and Saker (2002). The reasons for this were that an asset and liability approach will

- Provide greater transparency
- Produce accounts that are more understandable
- Make it easier for users to make comparisons between different sets of accounts

The new accounting standard should be applied to *insurance contracts* (a functional view) and not, as earlier, to insurance companies (the institutional view).

In the early phase of the work it was understood that an insurance contract should be unbundled into its insurance and noninsurance (financial or service) part and the accounting would be according to three different standards:

- IAS 18 for service contracts
- IAS 39 for financial instruments
- IFRS 4 for insurance contracts

Hence, the definition of insurance contracts is crucial.

In May 2002, IASB decided to split the work on standards for insurance accounting into two phases. Phase I was an interim measure and one of the main objectives for this first phase was the definition of insurance contracts, as opposed to financial instruments. Full implementation would occur with its Phase II.

In 2003, IASB produced an Exposure Draft on Insurance Contracts, ED 5, and with the publication of *IFRS 4 Insurance Contracts* in March 2004, IASB ended phase I (IASB, 2004).

From mid-2004, the IASB started to intensify its work on insurance reporting. It set up an Insurance Working Group (IWG), made up of analysts, actuaries, auditors, and regulators. The IASB and its IWG had meetings with different organizations, for example, with the U.S. Financial Accounting Standards Board (FASB) and the International Association of Insurance Supervisors (IAIS).

During 2005, the IASB added a Fair Value Measurement project to its agenda. This resulted in a discussion paper (DP), namely IASB (2006). The paper consists of the IASB's preliminary views of the provisions of the FASB's Statement on Financial Accounting Standards No. 157

Fair Value Measurement. The FASB's definition of fair value was "the price that would be received to sell an asset or paid to transfer a liability in an orderly transaction between market participants at the measurement date."

In May 2007, the IASB published a DP on its preliminary views on insurance contracts (IASB, 2007a, 2007b).

There was also a joint project "Memorandum of Understanding," MoU, between IASB and the U.S. FASB. The MoU pledged to use their best efforts to make their existing financial reporting standards fully compatible as soon as is practicable and to coordinate their future work programs to ensure that once achieved, compatibility is maintained. It also set out a road map of convergence between IFRSs and U.S. GAAP during the coming years. Among the specific projects were one on discount rates and one on revenue recognition.

In August 2008, the U.S. Securities and Exchange Commission (SEC), proposed a road map toward global accounting standards. The road map sets out that in 2011 the SEC will make a decision whether an adoption of IFRS would be in the public interest and would benefit the investors. It could lead to an implementation of the IFRS in the United States in 2014.

Under the U.S. FASB's fair valuation accounting, it had been possible for banks to shift their holdings to a cost basis under some circumstances. During the financial turmoil in October 2008, the IASB decided to adjust its fair valuation rules, so it became easier to reclassify some of the holdings to cost rather than the fair value ("market value") during specific circumstances. During this period, banks and insurers wrote down billions of the value of their holdings. With this change, a step from the market valuation was taken.

11.3 INSURANCE CONTRACTS, REVENUE RECOGNITION, AND FINANCIAL INSTRUMENTS

The IFRS 4 defined an *insurance contract* as a contract under which one party (the insurer) accepts significant insurance risk from another party (the policyholder) by agreeing to compensate the policyholder if a specified uncertain future event (the insured event) adversely affects the policyholder (IASB, 2004). The definition was open, so it could be redefined in the further work. In discussing the concept of insurance contracts, the IASB considered the following issues.

- For insurance contracts, the revenue (the premiums) is generally known (and received) in advance and the costs (claims and benefits) are not known until later. In other industries, the costs are known before any revenue.
- There are insurance contracts that expose the insurer to risks that will not be resolved for many years.
- Pooling of risks arising from similar contracts, see Section 9.1.1.1, will give an insurer a statistical basis for estimating the amounts, timing, and uncertainty of the cash flows (CFs) rising from insurance contracts.

- The insurer may be exposed to a moral hazard, that is, the existence of insurance contracts will increase the number and level of losses.
- Many insurance contracts contain implicit or explicit investment or deposit components.
- Long-term insurance contracts often give the policyholder the option to continue the contract at fixed or constrained prices even if risks have changed or to cancel the policy.
- An option is often used if it is favorable to the policyholder.
- The acquisition costs (costs for originating the contracts) could be high.
- The administrative costs could be high over the life of some contracts.
- Generally, there is no liquid and active secondary market in liabilities and assets arising from insurance contracts. By an "insurance liability" is meant an insurer's obligation under an insurance contract and by an "insurance asset" is meant an insurer's rights under an insurance contract.
- With profit, contracts give policyholders the right to share profits that is made. A discretionary participating feature gives the right to share profits made depending on the insurance company's board's discretion.
- A policyholder may suffer large losses if the insurance firm is unable to pay the claims. That is the reason why the insurance industry is regulated.

11.3.1 Service Contracts

Regarding the first issue on revenue, the IASB and the U.S. FASB have undertaken a joint project to develop a new standard for *revenue recognition*. In IFRSs, the new standard will replace the existing standards on revenue recognition, IAS 11 *Construction Contracts* and IAS 18 *Revenue*. The project was part of the MoU. In December 2008, the IASB published a DP on Preliminary Views on Revenue Recognition in Contracts with Customers. See, for example, IASB's update published on its Web site for more information about the project.

11.3.2 Financial Instruments

In July 2009, the IASB published an Exposure Draft, ED, on *financial instruments* (IASB, 2009d). The classification and measurement approach developed will measure almost all financial assets and liabilities at either amortized cost or fair value. If a financial instrument has only "basic loan features" and is managed on a contractual yield basis, it would be measured at amortized cost. In other cases, the measurement will be at fair value, except for some specific items. The replacement of IAS 39 on financial instruments is divided into three parts:

- *Phase 1—classification and measurement*. A near final draft, *IFRS 9 Financial Instruments*, was published in November 2009 (IASB, 2009h).

- *Phase 2—amortized cost and impairment.* An Exposure draft was published in November 2009 (IASB, 2009i). Comments to be received by June 30, 2010.
- *Phase 3—hedge accounting.* An exposure draft is planned for the first quarter of 2010.

Full replacement of IAS 39 will be done during 2010 and the new requirements will be mandatory from January 2013.

11.3.3 Insurance Contracts

For an insurance contract, the three main building blocks of a liability are defined by IASB (2007a) as

a. An estimate of the future CFs (IASB, 2007a, p. 28)
   - The estimates are explicit.
   - The estimates of the CFs are as consistent as possible with observable market prices.
   - The estimates incorporate, in an unbiased way, all available information about the amount, timing, and uncertainty of all CFs arising from the contractual obligations.
   - The estimates of the CFs are current; in other words, they correspond to conditions at the end of the reporting period.
   - Entity-specific CFs are excluded. CFs are entity specific if they could not arise for other entities holding an identical obligation.

b. The effect of the time value of money

   The discounting is discussed in Section 6.2.

c. A margin
   - If the financial reporting should represent faithfully the difference between a liability with fixed CFs and a liability with uncertain CFs, the measuring of the liabilities need to include a margin reflecting the uncertainty.

The three building blocks are discussed in more detail below—the first two in Section 11.4 and the last one in Sections 11.5 and 11.6. IASB has been discussing the measurement attributes in detail. The attributes have also been discussed in, for example, IAA (2009b, Section 3.4.1) in principle-based terms. For financial reporting purposes, several possible objectives could be applied. They include

- *Historical cost:* Reflects a past price or inputs that reflect original assumptions
- *An exit value:* Represents the amount that another party, that is, a market participant, would agree to pay to transfer the financial item or the group of contracts
- *A settlement value:* Would be the amount that can be obtained to immediately settle the obligation

- *A fulfillment value:* Is a measure of the amount to the obligation as it becomes due, that is, the CFs associated with the settlement of the obligations over the course of their lifetime as they come due

As a "current estimate" is based on the current set of rights and obligations, the first model based on historical costs does not fit.

Many stakeholders, more than 160, commented on the DP from IASB (IASB, 2007a, 2007b). There was a significant amount of opposition to the current exit value concept. The opposition was focusing on three factors.

- There is a lack of relevance of current exit value if an entity could not transfer the liability
- Current exit value excludes entity-specific CFs
- Current exit value of a liability reflects its credit characteristics

Based on the comments, the IASB discussed a *settlement value* at its meeting in February 2008. In general, this concept is quite similar to current exit value, but is modified to (a) have a slightly more entity-specific flavor and (b) exclude the credit characteristics of the liabilities. In its September 2008 meeting, the IASB suggested to use the term *current fulfillment value* as a better description of the notion as compared to settlement value. This issue is discussed in more detail in Section 11.4, where also the final proposals are presented.

The Board aimed to publish an Exposure Draft on Insurance Contracts in April 2010; see their Web site for more information (www.iasb.org). A final standard will be published in 2011. The effective date for this requirement will probably be January 1, 2014.

SOA in the United States has conducted a project on assessment of the extent to which practical models can address some of the key issues involved in applying the IASB's proposal and to compare it with U.S. GAAP values. They presented a report, SOA (2008), for life, health, and annuity products. Income expected to be reported under IASB's DP proposal differed significantly from that resulting from the U.S. GAAP, particularly at the time of contract issuance. The overall patterns of resulting liabilities after the first year were broadly the same.

The direction and extent of the initial profit or loss can differ significantly between the compared methods, depending on type of contract, product design, and underlying profitability of the product.

11.4 AN ESTIMATE OF THE FUTURE CASH FLOWS

Current estimation of the liabilities has been discussed at length by IAA (2005, 2009b); see also Section 8.1.

The estimation of future CFs for liabilities should be based on all the current information. There are two main approaches for estimating the CFs during the period before a claim, as discussed in the DP on Insurance Contracts (IASB, 2007a).

- Entry Value: "Premiums Received"
  Estimates are done at inception and then "locked in at inception," meaning that they are used throughout the life of the contract and by ignoring information that will start becoming available later.
  IASB's definitions: (1) The amount that the insurer would charge a policyholder today for entering into a contract with the same remaining rights and obligations as the existing contract. (2) The amount that a rational insurer would charge a policyholder today for entering into a contract with the same remaining rights and obligations. Of these two definitions, the IASB preferred the second (IASB, 2007a, pp. 60–61).

- Current Exit Value: "Cost of Fulfilling Obligations"
  The current exit value uses all currently available information in making estimates.
  IASB's definition: The amount the insurer would expect to pay at the reporting date to transfer its remaining contractual rights and obligations immediately to another entity (IASB, 2007b, p. 77).

The IASB also considered other measurement methods and their attributes. Among these were fair value and embedded value. At the time of publishing the DP on the preliminary views on insurance contracts, the fair value project was not finalized. The IASB was not in a position to determine whether the fair value as defined within the project and the current exit value were the same. They stated, however, that they had not identified any significant differences between them.

Embedded value is defined as the present value of estimated profit that will flow to an insurer from its existing contracts. CFO Forum has defined another version of embedded value called the European Embedded Value (EEV) (CFO, 2004). The EEV is defined as the present value of the shareholders' interests in the earnings distributable from assets allocated to the covered business after sufficient allowance for aggregate risks in the covered business. It consists of the following components: (1) free surplus allocated to the covered business, (2) required capital, less the cost of holding capital, and (3) the present value of future shareholder CFs from in-force covered business. The value of a future new business is excluded from the embedded value calculation.

The IASB's preliminary views on the measurement of the liabilities are that they should be explicit, unbiased, market-consistent, probability-weighted, and current estimates of the contractual CFs. The IASB was initially in favor of the current exit value definition. As this is a valuation approach, it does not mean that an insurer can, will or should transfer its insurance liabilities to a third party.

Important parts of all the currently available information used in defining the exit value are, for example, customer behavior, such as lapses and surrender, customer relationship, and acquisition costs. An insurer should recognize acquisition or transaction costs as an expense when the insurer incurs them. In general, a surrender value of an insurance contract is not a lower limit for the current exit value.

Some insurance and investment contracts give the policyholder a right to participate in a favorable performance of the contracts, but the insurer has usually some discretion over the amount or the timing of the distribution or payout to the policyholders. Here there are also often some constraints over the discretion. Such contracts are usually called profit insurance contracts; *cf.* also Section 8.2. In IASB (2007b), a liability is defined as a present obligation of the entity arising from past events, the settlement of which is expected to result in an outflow from the entity of resources embodying economic benefits. IAS 37, Nonfinancial Liabilities, identifies two categories of obligations: a legal obligation and a constructive obligation.

- *A legal obligation:* An obligation that derives from a contract (through its explicit or implicit terms), legislation, or other operations of law (IAS 37)
- *A constructive obligation:* A present obligation that arises from an entity's past actions when:
    a. By an established pattern of past practice, published policies or a sufficiently specific current statement, the entity has indicated to other parties that it will accept particular responsibilities.
    b. As a result, the entity has created a valid expectation in those parties that they can reasonably rely on it to discharge those responsibilities.

  From IAS 37 exposure draft, June 2005.

The IASB's preliminary view was that the CFs used in measuring a with-profit business should satisfy a legal or a constructive obligation that exists at the reporting data.

IASB had also a project to amend the nonfinancial liability standard IAS 37, Provisions, Contingent Liabilities and Contingent Assets (the updated IAS 37 model), and to analyze how the measurement approach developed therein could be applied to insurance contracts.

Other measurement issues discussed were

a. *Assets backing insurance contracts:* Accounting mismatches must be eliminated; with accounting, mismatch arises if changes in economic conditions affect assets and liabilities to the same extent, but they do not respond equally to the economic changes.
b. *Unit of account:* There was a discussion on whether the unit of account should be an individual contract or some aggregation of contracts (portfolios).
c. *Reinsurance:* Reinsurance liabilities should be measured at the current exit value, that is, the same measurement method applies to both direct insurers and reinsurers.
d. *Unbundling:* As stated earlier, the IASB had the initial opinion, see IFRS 4, that an insurance contract should be unbundled into its insurance and noninsurance (financial or service) parts and the accounting would be according to three different standards: IAS 18 for service contracts, IAS 39 for financial instruments and IFRS 4 for insurance contracts.

   In the DP, IASB (2007a), the IASB's preliminary view was that if an insurance contract includes both an insurance component and a deposit component, then

- If the two components are so interdependent that they can only be measured on an arbitrary basis, they should not be unbundled
- If the two components are not interdependent, then the insurance contract should be unbundled
- If the two components are interdependent, but can be measured separately in a nonarbitrary way, then the insurance contract should be unbundled

e. *Credit characteristics of insurance liabilities:* The IASB's preliminary view was that if an insurer measures its liabilities at current exit value, then the measurement should reflect the liability's credit characteristics.

f. *Investment contracts:* No preliminary suggestions on investment contracts were given by the IASB.

The reaction to the proposal to use the current exit value in the DP was mainly that it excluded the use of entity-specific values and reflected the credit characteristics of the entity. Subsequent discussions in the IASB's Insurance Working Group first identified a settlement value approach. This concept was quite similar to current exit value, but modified to (a) have a slightly more entity-specific flavor and (b) exclude the credit characteristics of the liabilities.

In its September 2008 meeting, the IASB suggested the use of the term *current fulfillment value* as a better way of describing the notion as compared to settlement value. A definition of a fulfillment value was proposed: "The expected present value of the cost of fulfilling the obligation to the policyholder over time."

There were mixed views on the margin and on day one profits. The IASB identified two approaches on how to deal with the risk margin (RM) and the day one profit issue (IASB, 2008a).

Approach A: Treated the cost of bearing risk as one of the costs of fulfilling the obligation. The cost of bearing risk was seen as a form of RM. Current fulfillment on day one should not include the day one difference, where we have defined "day one difference" as the difference between the premium and the expected present value of the CFs.

Approach B: Did not treat the cost of bearing risk as a cost. An additional rule was required to stipulate that the margin was included and to prescribe how to determine it. The current fulfillment value of the liability included the day one difference.

To support the discussion in the Board meeting, a comparison was made between the current exit value (as defined by the DP), the current fulfillment value (as proposed from the comment), and the value in use (as defined in IAS 36, Impairment of Assets).

At its October 2008 and February 2009 meetings, the IASB identified and discussed five viable different candidate measurement approaches (IASB, 2008c):

1. Current exit value as proposed by the DP (IASB, 2007a, 2007b)
2. Current fulfillment value, including a RM based on the cost of bearing risk

3. Current fulfillment value, including a RM based on the cost of bearing risk and, separate from the RM, an additional margin as the difference between the premium and the expected value of the CFs plus the margin for bearing risk
4. Current fulfillment value including a single margin calibrated at inception to the premium
5. Unearned premium for the preclaims liability of short-duration contracts

Candidates 1–4 all use a building block approach for measuring the insurance liabilities and could be applied to all types of insurance contracts. The similarities and differences between the candidates were discussed in IASB (2008c). A tabular comparison was discussed both at the October 2009 meeting and at the February 2009 meeting (IASB, 2009a).

At its June 2009 meeting the IASB continued its discussion on candidate measurement approaches for insurance contracts. They decided to tentatively include a new candidate based on the updated IAS 37 model (modified to exclude day one gains). The updated IAS 37 model will be used for other types of uncertain liabilities and therefore it was considered natural for IASB to consider this approach as a candidate for insurance contracts liabilities. After removing some of the earlier candidates, the following three remained.

- *Measurement approach based on updated IAS 37 model* (for all insurance liabilities): The amount the entity would rationally pay at the end of the reporting period to be relieved of the present obligation, plus a "residual margin," based on day one difference.
- *Current fulfillment value* (for all insurance liabilities): The expected present value of the cost of fulfilling the obligation to the policyholder over time, excluding the cost of bearing risk; plus a "composite margin" based on the day one difference. (Previously candidate 4.)
- *Unearned premium* (only for preclaim short duration insurance liabilities): An implicit building block approach that includes (1) expected cash flows, (2) time value of money, and (3) a margin. All locked-in at inception. (Previously candidate 5.)

For more details of the three candidates, see IASB (2009c, 2009f). The first two candidates are built up by the three building blocks discussed in IASB (2007a); see above. They were discussed and compared in IASB (2009e).

Both the measurement approach and the current fulfillment approach measure the insurance liability from the perspective of the insurer—not from a market participant. They would also measure the liability by looking at fulfillment of an insurance obligation over time; hence IASB did not expect any difference between the two approaches in relation to financial market variables such as the discount rates. They also use the expected present value and measure these cash flows from the insurer's perspective. Hence the IASB did not exclude a specific insurer's cash flow. One difference in the cash flow could be that in the measurement approach, the IAS 37 model requires the insurer to use a subcontractor's charge as a service margin, which is not necessarily included in the current fulfillment approach. The difference is also discussed in Section 11.5.

No conclusion was made at the July 2009 meeting, but the IASB tentatively decided that the third candidate, the unearned premium approach, would provide decision-useful information about preclaims liabilities of short-duration insurance contracts and to require the use of an unearned premium approach for those liabilities. IASB also discussed the insurance contract project with the FASB at the end of July 2009.

At the September 2009 meeting, there were only two candidates left: the first two, that is, the approach based on the updated IAS 37 model and the current fulfillment value with composite margin. The Board tentatively decided to go for the updated IAS 37 model. But as there was a "significant minority" supporting the current fulfillment value approach, the exposure draft should include both approaches. If the updated IAS 37 model was the main proposal of IASB at this time, the current fulfillment value approach was the main candidate of FASB.

At the meeting, the Board discussed discount rates for insurance liabilities. They decided not to publish any detailed guidance on how to determine the discount rate. However, the discount rate for an insurance liability should conceptually adjust estimated future CFs for the time value of money in a way that captures the characteristics of that liability rather than using a discount rate based on expected returns on actual assets backing those liabilities (IASB, 2009g).

At the December 2009 meeting, the staff provided the IASB and FASB boards with a summary of reasons why the insurance contract project has been pursuing an approach that measures insurance liabilities by reference to future CFs, rather than an approach that applies the principles being developed in the project on revenue recognition. The boards also tentatively decided that the risk adjustment should measure the insurer's view of the uncertainty associated with the future CFs and that the measurement of an insurance liability should not be updated for changes in the risk of nonperformance by the insurer.

An exposure draft was expected by April 2010.

11.4.1 Effect of the Time Value of Money

The IASB's preliminary view was that the current market discount rates that adjust the estimated future CFs for the time value of money should be used.

The discounting has been addressed by IAA (2009b) and is discussed in Chapter 6.

11.5 A MARGIN: IASB'S DISCUSSION

The IASB's preliminary view on the margin was that it should be an explicit and unbiased estimate of the margin that market participants require for bearing risk (a RM) and for providing other services, if any (a service margin). The two margins should both be explicit.

The RM should be determined for a portfolio of insurance contracts that are subject to similar risks and are managed as a single portfolio (the unit of account), but should not reflect any diversification benefits between portfolios.

The objective of the RM was to give users useful information about the uncertainty associated with the liabilities incorporating model and parameter risks (IASB, 2007b). This is fully in line with the discussions in Section 6.4. The IASB saw that the RM should reflect all risks associated with the liability. Hence the RM should not reflect risks that do not arise from the liability such as investment risks, asset–liability mismatch risk, or operational risks related to future transactions except when these risks affect the amount of payouts to the insurance policyholder. The approach of measuring the RM should incorporate tail risks from skewed distributions (payoffs from embedded options or low-frequency high-severity risks).

Approaches to determining RMs are briefly discussed in IASB (2007b), but are discussed at length in IAA (2009b, ch. 6). A prudent and traditional regulatory view is that the RM is expected to cover adverse deviation expected under normal circumstances. This could be put in another way. Both RM and capital requirement are available to finance the cost of adverse events—the RM provides a first level of protection and the capital requirement a second level.

At its April 2009 meeting, the IASB discussed the margins included in the five-candidate measurement approaches discussed in Section 11.2. Candidate 2 was seen as inconsistent with the Boards' decision on day one difference and therefore was not discussed. Below, the four candidates, their margin components and the objective of the margin are summarized; see IASB (2009b).

Candidate	Margin Components	Objective of the Margin
1 Current exit value	Risk margin	Compensation for bearing risk required by a market participant
	Service margin	Margin required by market participant for services other than the service of bearing risk
	Deferred day one difference	Avoid any positive day one difference from being recognized in profit or loss
3 Current fulfillment value—margin for cost of bearing risk + an additional margin	Risk margin	The cost of bearing risk, measured from the particular insurer's perspective
	Additional margin	Avoid any positive day one difference from being recognized in profit or loss (consistency with revenue recognition)
4 Current fulfillment value—composite margin	Composite margin	Capture the margin as implied by the actual transaction with the policyholder (premium) (consistency with revenue recognition)
5 Unearned premium	Implicit margin	N/A

The components of the remaining candidates after the June 2009 and July 2009 meetings where these were discussed as candidates 4 and 5 above plus the new measurement approach are based on the updated IAS 37 model. The latter consisted of three margin components:

- *Risk margin:* The amount the entity would pay to be relieved of risk

- *Service margin:* The amount required by the contractor for other services
- *Residual margin (calibrated to premiums):* Day one difference (between the actual margin and the required margin)

The risk and service margins reflect the fact that an insurer would rationally pay different amounts to be relieved of the two liabilities that differ in riskiness; see IASB (2009e). The IASB staff could not find any definition for margins (both risk and service margins) that comes directly from the definition of the current fulfillment approach.

11.6 A MARGIN: DISCUSSIONS BY IAIS AND IAA

An insurance contract contains an agreement by the insurer to provide, in exchange for a premium, benefits upon the occurrence of future events. This is a way of hedging risks. It may thus be appropriate to discuss the nature of an insurance contract before discussing the objective of a RM; IAA (2009b, Chapter 6.1):

a. *The policyholder view:* Policyholders are subject to certain risks (e.g., frequency, timing, and severity) of certain contingent events, which they usually do not wish to bear themselves. They see the advantage in transferring those risks to an insurer.

b. *The insurer view:* The insurer can manage the risks by using different risk management techniques such as pooling homogeneous risks, diversifying multiple pools risks, reinsurance, securitization of risks, and so on; *cf.* Section 9.1.

A general view is that insurance liabilities should be valuated by an estimate of the expected value of future CFs plus a RM reflecting the *remaining* uncertainty. A RM for solvency purposes could differ from one for accounting purposes, but the aim must be to have the one and the same margin for uncertainty.

The IASB's three-block model assumes that a rational transferee requires something more than the current estimate. Otherwise he would not expect to receive anything for taking on a risk that could work out unexpectedly. This amount, the RM, could therefore be regarded as an additional amount associated with the uncertainty inherent in the future financial return from the contract. This RM would reflect the compensation to the transferee for the risk of taking on an obligation to pay uncertain CFs.

The IAA (2009b) defines a RM as "(also referred to as *margin over current estimate*). The portion of a liability associated with the risk and uncertainty associated with insurance risk. An amount or *margin* reflecting an assessment of the uncertainty inherent in an insurance risk with certain attributes based on a specific measurement approach."

This is also in line with the ideas of IASB that only nonhedgeable risks, that is, credit risks and market risks that cannot be replicated, should be reflected in the liabilities.

In the solvency assessment discussed in this book, *all* risks should be reflected in the capital requirements—only the uncertainty in nonhedgeable risks should be reflected in a RM.

The IAA report discussed two perspectives on assessing a RM:

- The provision for cost of bearing risks approach (exit value approach) and
- The policyholder protection approach (prudent approach)

The first one is in line with the general approach adopted in this book. The latter one is in line with a conservative valuation approach (prudent regulatory approach) discussed above and in Section 5.4. This approach is not discussed further.

The IAA report suggested that a rational methodology to assess the margin over the current estimate was to put oneself in the position of the transferee and put a question like "What thought processes might the transferee go through in order to work out what extra amount it might require over the current estimate?" One answer would probably be "as much as possible," but, as stated in IAA (2009b), in a market in equilibrium, the margin would be based on a reasonable return reflecting the risk of uncertainty. It could also reflect the returns of a likely diversified portfolio of investments.

If we take the view of a jurisdiction with a regulatory requirement, then one answer could be that the transferee would like to have an amount corresponding to the capital requirement, or at least an amount corresponding to the CoC for that. This is in line with the approach taken in Sections 8.4 and 8.5.

A conceptual framework for the role of the RM is given in Strommen (2008). The conceptual basis for describing valuation of the liabilities is the cost of capital approach. Other methods that are discussed in op. cit. are percentile methods, market-consistent (risk neutral) methods, explicit assumption methods, and undiscounted estimated future payments methods.

The IAIS recognized that both the policyholder protection and the cost of risk bearing objectives must be satisfied (IAIS, 2006, p. 5). "The IAIS stresses that any transfer would need to be made to an entity capable of accepting the transfer which, in the case of a regulated industry like insurance, implies that the transferee would also need to be regulated and capable of settling the obligation to the claimant/beneficiary. Accordingly, the IAIS believes that any transfer notion should be strongly influenced by the settlement obligations that the transferee would undertake." Therefore, the transferee would need to provide capital or at least to prove its ability to cover losses from its resources. "The risk margin is the cost for providing that capital or equivalent guarantee" (IAA, 2009b).

According to the IAIS (2006), acceptable methods for calculating a RM should have the following characteristics:

IAIS-1. The less that is known about the current estimate and its trend; the higher the RMs should be

IAIS-2. Risks with low frequency and high severity will have higher RMs than risks with high frequency and low severity

IAIS-3. For similar risks, contracts that persist over a longer timeframe will have higher RMs than those of shorter duration

IAIS-4. Risks with a wide probability distribution will have higher RMs than those risks with a narrower distribution

IAIS-5. To the extent that emerging experience reduces uncertainty, RMs will decrease, and vice versa

The same properties were also identified by IASB (IASB, 2007b). The IAA noted that the third characteristic has two interpretations:

Interpretation 1: Liabilities that persist over a longer time frame have increased exposure to risks and hence will have higher RMs than shorter tail liabilities that are otherwise exposed to similar risks. We call this IAIS-3.1.

Interpretation 2: For two sets of liabilities with the same riskiness in their distribution of ultimate settlement values (i.e., having similar risks), the RM should be higher for the liabilities that settle over a longer time period. We call this IAIS-3.2.

A RM should be market consistent, be easy to calculate, be consistent between classes of business, provide information about earnings from current business and the difference between actual and expected results from obligations not yet settled, and show earnings only when they are sufficiently reliable to provide users of financial statements with guidance useful for making decisions (IAA, 2009b, ch. 6).

The IAA believed that a desirable RM, in addition, should have the following characteristics; see IAA (2009b, ch. 6.2), partly from Group Consultatif, GC (2006).

IAA-1. Apply a consistent methodology for the entire lifetime of the contract

IAA-2. Use assumptions consistent with those used in the determination of the corresponding current estimates

IAA-3. Be determined in a manner consistent with sound insurance pricing practices

IAA-4. Vary by product (class of business) based on risk differences between the products

IAA-5. Be easy to calculate

IAA-6. Be consistently determined between reporting periods for each entity; that is, the RM varies from period to period only to the extent that there are real changes in risk

IAA-7. Be consistently determined between entities at each reporting date; that is, two entities with similar business should produce similar RMs using the methodology

IAA-8. Facilitate disclosure of information useful to stakeholders

IAA-9. Provide information that is useful to users of financial statements

IAA-10. Be consistent with regulatory solvency and other objectives

IAA-11. Be consistent with IASB objectives

Characteristics IAA-9–11 were not discussed further in IAA (2009b).

According to an IAA classification, IAA (2009b), four different basic groups of methods of calculating RMs can be found. They are

1. "Quantile methods": including percentile or confidence levels and the related methods that are often considered with quantile methods, specifically conditional tail expectation (also called Tail Value-at-Risk or Tail VaR) and multiples of the second and higher moments of the risk distribution, for example, VaR
2. The CoC method
3. Discount-related methods, including deflator-adjusted CFs and risk-adjusted discount rates
4. Explicit assumptions

IASB (2007b) also suggested a fifth method for assessing the RM. It was a CAPM* approach usually related to assets and was used in capital allocation. IASB reject the use of implicit margins, irrespective of how these are set. Other methods that could be used are based on utility theory and hazard transforms.

The IAA discussed and compared the different approaches in detail. The methods are ranked. The following discussion on the IAIS characteristics is taken from IAA (2009b, ch. 6.11.1).

1. *Quantile methods:* All of the quantile methods fail to comply with IAIS-3.2. Consider two products that have the same risk distribution for unsettled contract obligations at the reporting date, but have obligations that involve settlement over two different time periods. To comply with IAIS-3.2 the margins should be different. However, the RMs for the two products determined based on confidence level, CTE, number of standard deviations, or any method that relies only on characteristics of the risk distribution would not be different.

   In addition, the confidence level method does not necessarily satisfy IAIS-1, 2, 4, and 5. IAA showed that very skewed distributions can result in negative RMs, as increase in skewness is accompanied by a decrease in the rate of increase of RMs. More generally, the examples also show that as distributions become more dispersed and more skewed, the RMs implied by a fixed confidence level include fewer standard deviations. This violates the spirit of IAIS-1, 2, 4, and 5 throughout and the latter of those in the extreme.

   Tail-VaR and methods based on multiples of standard deviation generally satisfy IAIS-1, 2, 4, and 5 better than those based on the confidence level method. However, while Tail-VaR is more sophisticated, in that it can provide a better insight into the tail amounts, its general approach is similar to that of confidence levels.

2. *Cost-of-capital method:* Only the CoC method would generally satisfies all six IAIS characteristics. IAIS-1–5 require that the RM increases as the risk distribution becomes "wider" and/or more skewed. This holds for the CoC method. IAIS-3.2 requires that for two products with the same risk distribution, the product with a longer settlement

*CAPM: Capital Asset Pricing Model.

period will have a larger RM. This is also true for the CoC method because the CoC RM will be the sum of risk contributions over a longer period.

3. *Discount methods:* Discount methods, other than methods that use a zero discount, satisfy guidelines IAIS-3.1 and -3.2. The methods satisfy the other characteristics only to the extent that interest rate risk adjustments vary by product and settlement duration. Discount methods have not been applied in that way in the past.

   The use of a zero discount does not routinely satisfy guidelines IAIS-3.1 or IAIS-3.2 because changes in risk-free rate, with no change in uncertainty, would cause a change in the RM implied by the use of a zero discount.

4. *Explicit assumptions:* Explicit (or implicit) assumptions could be constructed in a manner to address the characteristics, but do not necessarily satisfy any of the characteristics. Each product would need its own set of assumptions. As an implementation approach, explicit assumptions, selected by product, might be used to approximate the percentile, CoC, or discount methods. If the approximation were sufficiently close, the explicit assumption approach would satisfy the guidelines to the same extent as the method it approximates.

IAA also discussed the compliance of the four main methods to its own desirable characteristics. The following summary is also from IAA (2009b, chs. 6.11.2 through 6.11.5):

IAA-1–2: All four methods can be applied based on a consistent methodology for the entire lifetime of the contract. Moreover, to the extent that each of the methods utilizes assumptions relevant to current estimates, they would be implemented in a manner that is consistent with emerging experience as the experience affects the current estimates.

IAA-3: The cost-of-capital method is a common actuarial pricing methodology. Some quantile methods are also used in insurance pricing, although less frequently.
Risk adjusted discount rates have also been used in pricing insurance products that have features connected with financial markets, for example, financial guarantees or deposits. These approaches are less relevant to nonhedgeable aspects of other insurance risks.

IAA-4: The confidence level method would need to vary between classes in order to reflect the targeted degree of consistency. However, both the Tail VaR and multiples of standard deviation methods can provide consistency between classes. Risk adjusted discount rates would be consistent between classes only if the risk adjustment was selected appropriately on a current basis. Although explicit assumptions could be designed to achieve consistency, this attribute is not automatically achieved through this approach.

IAA-5: Regarding the ease of use of a calculation benchmark, IAA described the mechanical application of formulas or the use of models that requires no judgmental inputs as "easier" than methods that require judgment in addition to calculations. Methods that require less simulation of future results were also characterized as "easier" than methods that require more extensive simulation of future results.

Cost-of-capital: At each reporting date it is needed to determine the required capital levels both at the reporting date and at each subsequent reporting date. For the first year, we need "n" estimates of capital at various projected dates. For the second year we need "$n-1$" estimates of capital, and so on. In total, over the course of the runoff we need $(n^2+n)/2$ distributions. If it takes 10 years to settle all the obligations, we need to determine 10 risk distributions in the quantile method and 110/2 or 55 risk distributions in the CoC method.

Moreover, the methods of determining the "cost" in the CoC method have not yet been well established. It would be easy if the cost, for regulatory financial reporting purposes, were determined by regulation, as has been done for the Swiss Solvency Test, or if the cost did not require routine adjustments. On the other hand, it might involve extensive calculations and application of judgment. In addition, if the "cost" in CoC for regulatory financial reporting is specified by the regulator in a way that is not consistent with what the market would require, then the value might not be suitable for general purpose financial reporting.

Quantile: In the family of quantile methods, at each reporting date we need to estimate the quantile or moment information only at that reporting date. The release of risk over time is not considered in the quantile method. If it takes n years for obligations to settle, we will need "n" (one for each year for year-end reporting purposes) estimates over the course of the runoff. Although it is simply mechanical to run a model $n$times, it may take significant time to develop n number of confidence levels to reflect the risk in each period adequately.

Discount: Risk-adjusted discount rates could be easy or more difficult, depending on details involved in the risk adjustment.

Explicit assumptions: Although explicit assumptions could be very simple to apply, relatively complex models might be applied to, say, the individual assumptions. If explicit assumptions were used as an implementation approach, then periodically the assumptions must be tested to confirm that the approximation remains valid. The testing would be particularly important if there were environmental changes (e.g., interest, inflation, equity markets, and court decisions) that are likely to affect the validity of the approximation. This testing requirement, particularly during times of environmental change, reduces the "ease of calculation" of explicit assumptions.

IAA-6–7: These characteristics depend on whether the RM method is properly sensitive to risk, the six extended IAIS characteristics and IAA-4, and also sufficiently easy to implement, IAA-5. Since the evaluation overlaps with the characteristics already discussed, we have not separately rated these two characteristics.

IAA-8: The minimum level of likely disclosure would be the amount of RM and the basis for deriving that amount. Any approach other than implicit assumptions would allow for the minimum level of disclosure.

TABLE 11.1 A Comparison of Four Methods of Assessing the Risk Margin Made by IAA

Issue	Cost-of-Capital	Quantile	Discount	Explicit Assumptions
Source of risk used to measure risk margin	Time and shape	Shape	Time	Varies depending on selected assumptions
Market consistency—in theory	1	2 for Tail-VaR and standard deviation 3 for quantiles	4	4
Market consistency—in practice	Unknown	Unknown	Unknown	Unknown
Complies with IAIS-1, 2, 3.1, 4, and 5	1	1 for Tail-VaR and standard deviation 2 for quantiles	3	4
Complies with IAIS-3.2	1	3	2	4
Complies with IAA-1–IAA-2	1	1	1	1
Complies with IAA-3	1	2	3	4
Complies with IAA-4: Consistency among classes of business	1	2 for Tail-VaR and standard deviation 3 for quantiles	3	4
Complies with IAA-5: Ease of calculation	4	3	2	1
Complies with IAA-8	1	1	1	1

Source: From IAA. 2009b. Measurement of Liabilities for Insurance Contracts: Current Estimates and Risk Margins. An International Actuarial Research Paper, ISBN: 978-0-9812787-0-4. With permission.

Note: 1: Best meets the criteria, 4: Least meets the criteria.
Time: The rate at which risk is released over time.
Shape: The risk distribution of possible outcomes around the mean value at the reporting date over a specific time horizon.

For the other methods, the methodology chosen and the key parameters in the calculations should be disclosed.

Market consistency: The CoC approach measures risk, expressed as required capital, at the reporting date, measures how that risk declines over time, and applies a capital charge for the cost of holding that capital. This is a framework that is familiar to banking and to major investment decision making in all industries.

- Some quantile methods and, less frequently, risk-adjusted discount rates, are used in risk assessment, but not typically for pricing, outside of insurance.
- The CoC method attempts to utilize market information, for example, required returns, while the other methods have no specific connection with market information.

There is no available information that allows us to determine whether any particular calibration actually produces liabilities close to transfer values. We therefore split our assessment of these characteristics into theory and practice.

The analysis is summarized in Table 11.1 in accordance with the different key characteristics of a suitable RM.

The result of the IAA analysis shows that the CoC best meets the criteria, except for the ease of calculation. If we take the regulatory perspective and take the CoC approach as outlined in Section 8.5, the rank would definitively be between 1 and 2 for ease of calculation.

PART C

Modeling and Measuring

Everything should be made as simple as possible, but not simpler

Albert Einstein (1879–1955)
Swiss-German-US physicist

Models are imperfect approximations of reality. They are valuable, but incomplete, abstractions and only as good as their underlying assumptions.

Michael C. Schmitz and Susan J. Forray
Consulting actuaries. From JRMS (2008)

A new model could be more relevant, but is at the same time more uncertain

Philipp Keller
Swiss actuary and friend (2008)

MODELING AND MEASURING risks are discussed in Chapters 12 through 21.

General modeling and approximations using Taylor expansion is discussed in Chapter 12. The important concept of dependence using copulas is discussed in Chapter 13. Risk measures and different properties of risk measures are discussed in Chapter 14. Especially VaR and Tail-VaR are discussed. The modeling and measuring capital requirement is discussed in Chapter 15. A pragmatic proposal is also given for the problem of nonnormality and skewness.

In Chapters 16 through 21 we discuss different risks and subrisks. The main risks discussed are market, credit, operational, liquidity, and underwriting/insurance risks. Different proposals for assessing the capital charge from these risks are discussed.

For a general discussion on risk analysis, see, for example, Aven (2003).

CHAPTER 12

Developing a Model

IN THIS CHAPTER, we will start with an *exact* relationship assuming that a variable y can be exactly determined by a function of another variable x, that is, $f(x)$. We use the Taylor expansion, dropping some higher-order terms to get approximations. This is discussed in Section 12.1. The Taylor expansion of this function, dropping some higher-order terms, gives us either a linear approximation or a nonlinear approximation.

A linear approximation, developed into a linear regression model, is discussed in Section 12.2. The special case of homogeneous functions, which are important in explaining the use of factor-based models, is introduced in Section 12.3.

Nonlinear approximations are discussed in Section 12.4. This chapter includes the use of the Taylor expansion up to the second-order partial derivatives: "second-order approximations." However, the interesting part is the quadratic approximation in Section 12.4.2.1.

The chapter ends (Section 12.5) with an explanation of how to use these approximations. This is based on the discussion presented in IAA (2004, Appendix H).

12.1 ANALYTIC APPROXIMATION OF AN EXACT MODEL

A basic assumption in modeling is that all phenomena can be completely determined by causal relationships. Let y be a variable that is exactly determined by a function of another variable x. To stress the causal character of the relation between y and x we write it as $y = f(x)$ meaning that y will satisfy the equation $y = f(x)$. We assume f to be continuous and infinitely differentiable.

In reality, we will probably not be able to measure y by the relation $f(x)$ exactly but only in a neighborhood ξ of x. We can expand $f(x)$ and y around ξ and $y_0 = f(\xi)$ by a Taylor series:

$$y = y_0 + \frac{\partial f}{\partial x}(x - \xi) + \frac{1}{2!}\frac{\partial^2 f}{\partial^2 x}(x - \xi)^2 + (h.o.t.), \tag{12.1}$$

where $h.o.t.$ stands for higher-order terms and the derivatives $\partial f/\partial x = f'(\xi)$ and $\partial^2 f/\partial x^2 = f''(\xi)$ are evaluated at $x = \xi$ and are therefore simply constants.

If we use only the first part of the Taylor expansion, we can write Equation 12.1 as

$$y = y_0 + \frac{\partial f}{\partial x}(x - \xi) + u, \tag{12.2}$$

where u is a disturbance term, including the higher-order terms from Equation 12.1, reflecting the uncertainty in the simplification made by using only the first part of the Taylor expansion.

Going from an exact model $y = f(x)$, which you cannot use as it is, to an approximation by using the first part of a Taylor expansion around some point ξ close to x as given by Equation 12.2 gives the essentiality of modeling.

In more general terms, assume that y is a function of x_i, $i = 1, 2, \ldots, D$, that is, we have an exact model $y = f(x_1, x_2, \ldots, x_D)$.

12.1.1 First Approximation of the Exact Model

We may assume that a large part of the variables x_i is more or less redundant in the modeling as they may not contribute significantly to the exactness in the modeling. We can assume that we only need the first d variables, $d < D$. Thus the first approximation we do is to write

$$y = f(x_1, x_2, \ldots, x_D) \approx f(x_1, x_2, \ldots, x_d), \quad d < D.$$

12.1.2 Second Approximation of the Exact Model

As in Equation 12.1, we can expand $y = f(x_1, x_2, \ldots, x_d)$, where we have written an identity between y and the first approximation f made above, around the point $y_0, \xi_1, \xi_2, \ldots, \xi_d$, where $y_0 = f(\xi_1, \xi_2, \ldots, \xi_d)$. Hence we obtain

$$y = y_0 + \sum_{i=1}^{d} \frac{\partial f}{\partial x_i}(x_i - \xi_i) + \frac{1}{2!}\sum_{i=1}^{d}\sum_{j=1}^{d} \frac{\partial^2 f}{\partial x_i \partial x_j}(x_i - \xi_i)(x_j - \xi_j) + (h.o.t.). \tag{12.3}$$

12.2 LINEARIZATION

If we use only the first part of the Taylor expansion, that is, the first-order partial derivatives, we can write Equation 12.3 as

$$y = y_0 + \sum_{i=1}^{d} \frac{\partial f}{\partial x_i}(x_i - \xi_i) + u, \tag{12.4}$$

where u, as above, is a disturbance term. On using only the first part of the model, we have linearized it. As the first-order partial derivatives are understood to be evaluated at the point $y_0, \xi_1, \xi_2, \ldots, \xi_d$ we can replace those partial derivatives by constant coefficients $\beta_i = \partial f / \partial x_i$, $i = 1, 2, \ldots, d$.

If we also introduce $\beta_0 = y_0 - \sum_{i=1}^{d} \beta_i \xi_i$, we obtain $y = \beta_0 + \sum_{i=1}^{d} \beta_i x_i + u$.

This is the *linear regression model* and it depends on that we dropped higher-order terms in Equation 12.4 and instead introduced a disturbance term u.

If we instead evaluate the function $y = f(x_1, x_2, \ldots, x_d)$ around the stochastic point $(X_1, X_2, \ldots, X_d)$ with $\xi_i = \mu_i = E(X_i)$, that is, the expectation of X_i, $i = 1, 2, \ldots, d$, then we obtain $Y = Y_0 + \sum_{i=1}^{d} \partial f/\partial\mu_i(X_i - \mu_i) + U$. Hence, $E(Y - Y_0) = E(U)$ and a usual assumption is that the latter is defined as equal to zero.

12.3 HOMOGENEOUS FUNCTIONS

We start with a definition of a homogeneous function.

Definition: A function $f(x_1, x_2, \ldots, x_d)$ is homogeneous of degree n if and only if for every positive value of λ

$$f(\lambda x_1, \lambda x_2, \ldots, \lambda x_d) = \lambda^n f(x_1, x_2, \ldots, x_d). \tag{12.5}$$

For a more exact definition, see, for example, Widder (1965).

12.3.1 Euler's Theorem

It can be proved that (see, e.g., Widder, 1965)

$$\sum_{i=1}^{d} \frac{\partial f}{\partial x_i} x_i = n \cdot f(x_1, x_2, \ldots, x_d). \tag{12.6a}$$

If the function is homogeneous of degree one, then we have from the theorem that

$$\sum_{i=1}^{d} \frac{\partial f}{\partial x_i} x_i = f(x_1, x_2, \ldots, x_d) = y. \tag{12.6b}$$

A sketch of a proof for the degree one case is given in Appendix A, Section A.3.1. The theorem states that under homogeneity of degree one, a function $f(x)$ could be reduced to the sum of the arguments (x_i, $i = 1, 2, \ldots, d$) multiplied by their first partial derivatives.

Thus, if we assume f to be homogeneous of degree one, then an equation similar to Equation 12.4, with $\tau_i = \partial f/\partial x_i$, can be written as

$$\begin{aligned} y &= y_0 + \sum_{i=1}^{d} \frac{\partial f}{\partial x_i}(x_i - \xi_i) + u = y_0 - \sum_{i=1}^{d} \frac{\partial f}{\partial x_i}\xi_i + \sum_{i=1}^{d} \frac{\partial f}{\partial x_i} x_i + u \\ &= y_0 - \sum_{i=1}^{d} \frac{\partial f}{\partial x_i}\xi_i + y + u, \end{aligned} \tag{12.7}$$

with the use of Equation 12.6b.

Hence, ignoring the rest term u and changing ξ_i for x_i, we obtain with $y = f(x_1, x_2, \ldots, x_d)$ and $f(0, 0, \ldots, 0) = 0$

$$f(x_1, x_2, \ldots, x_d) = \sum_{i=1}^{d} \tau_i x_i. \tag{12.8}$$

Hence, under a linear approximation and homogeneity, we can approximate the function with a *factor-based formula*; see also IAA (2004).

12.3.2 Corollary to Euler's Theorem

A corollary to Euler's theorem is that if the function $f(x_1, x_2, \ldots, x_d)$ is homogeneous of degree one, then

$$\sum_{i=1}^{d} \frac{\partial^2 f}{\partial x_i \partial x_j} x_i = 0, \quad \text{for any } j. \tag{12.9}$$

A sketch of a proof is given in Appendix A, Section A.3.2.

12.4 NONLINEAR APPROXIMATIONS

If the linear approximations outlined above are not good enough, it is possible to develop higher-order approximations to y, y^2, or some other function of y. We start with the second-order approximations from the Taylor expansion (12.3). Then we look at a quadratic approximation and end up with some comments on higher-order approximations.

12.4.1 Second-Order Approximation

Using the Taylor expansion up to the second part, that is, including the second-order partial derivatives, we can write the general Equation 12.3 as

$$\begin{aligned} y &= y_0 + \sum_{i=1}^{d} \frac{\partial f}{\partial x_i}(x_i - \xi_i) + \frac{1}{2!}\sum_{i=1}^{d}\sum_{j=1}^{d} \frac{\partial^2 f}{\partial x_i \partial x_j}(x_i - \xi_i)(x_j - \xi_j) + u \\ &= y_0 + \sum_{i=1}^{d} \beta_i (x_i - \xi_i) + \frac{1}{2}\sum_{i=1}^{d} \gamma_{ii}(x_i - \xi_i)^2 + \frac{1}{2}\sum_{i \neq j=1}^{d}\sum^{d} \gamma_{ij}(x_i - \xi_i)(x_j - \xi_j) + u, \end{aligned} \tag{12.10}$$

where we replaced the partial derivatives by constant coefficients $\beta_i = \partial f/\partial x_i$ and $\gamma_{ij} = \partial^2 f/\partial x_i \partial x_j$, $i, j = 1, 2, \ldots, d$ as the partial derivatives are understood to be evaluated at the point $y_0, \xi_1, \xi_2, \ldots, \xi_d$.

If we, as above in Equation 12.10, instead evaluate the stochastic function $Y = f(X_1, X_2, \ldots, X_d)$ around the stochastic point $(X_1, X_2, \ldots, X_d)$ with $\xi_i = \mu_i = E(X_i)$, that is, the expectation of X_i, $i = 1, 2, \ldots, d$, we obtain

$$Y = Y_0 + \sum_{i=1}^{d} \frac{\partial f}{\partial \mu_i}(X_i - \mu_i) + \frac{1}{2!}\sum_{i=1}^{d}\sum_{j=1}^{d} \frac{\partial^2 f}{\partial \mu_i \partial \mu_j}(X_i - \mu_i)(X_j - \mu_j) + U. \tag{12.11}$$

Taking the expectation of $Y - Y_0$, we obtain

$$E(Y - Y_0) = \frac{1}{2!}\sum_{i=1}^{d}\sum_{j=1}^{d}\frac{\partial^2 f}{\partial\mu_i\partial\mu_j}\mathrm{Cov}(X_i, X_j)$$

$$= \sum_{i=1}^{d} r_{ii}V(X_i) + \sum_{i\neq}\sum_{j=1}^{d} r_{ij}\mathrm{Cov}(X_i, X_j), \tag{12.12}$$

where $r_{ij} = (1/2)(\partial^2 f/\partial\mu_i\partial\mu_j)$, $V(X_i) = E[(X_i - \mu_i)^2]$, $\mathrm{Cov}(X_i, X_j) = E[(X_i - \mu_i)(X_j - \mu_j)]$, and $E(U) = 0$, that is, V and Cov are the variance and covariance, respectively.

Using the notation $\sigma_i^2 = V(X_i)$ and for the correlation coefficient $\rho_{ij} = \mathrm{Cov}(X_i, X_i)/\sigma_i\sigma_j$, we can rewrite Equation 12.12 as

$$E(Y - Y_0) = \sum_{i=1}^{d} r_{ii}\sigma_i^2 + \sum_{i\neq}\sum_{j=1}^{d} r_{ij}\sigma_i\sigma_j\rho_{ij}. \tag{12.13}$$

12.4.2 Higher-Order Functions

If we believe that the linear approximation as outlined in Section 12.2 is not good enough, we can develop higher-order approximations to a function $y^m = f^m(x_1, x_2, \ldots, x_d)$, that is, a function f of power $m > 1$.

We have special interest in the quadratic approximations to $y^2 = f^2(x_1, x_2, \ldots, x_d)$, but we will also mention higher-order approximations. *We assume that f is homogeneous of degree one.*

12.4.2.1 Quadratic Approximation

We use the Taylor expansion up to the second part, as in Equation 12.3, and drop the higher-order terms. For the quadratic function $f^2(x_1, x_2, \ldots, x_d)$, we obtain

$$f^2(x_1, x_2, \ldots, x_d) = f^2(\xi_1, \xi_2, \ldots, \xi_d) + \sum_{i=1}^{d}\frac{\partial f^2}{\partial x_i}(x_i - \xi_i)$$

$$+ \frac{1}{2}\sum_{i=1}^{d}\sum_{j=1}^{d}\frac{\partial^2 f^2}{\partial x_i\partial x_j}(x_i - \xi_i)(x_j - \xi_j), \tag{12.14}$$

ignoring higher-order terms.

By using Equations 12.6b and 12.9 for functions that are homogeneous of degree one, see Appendix A, Section A.3.3, we obtain

$$f^2(x_1, x_2, \ldots, x_d) = \sum_{i=1}^{d}\sum_{j=1}^{d} r_{ij}x_ix_j, \tag{12.15}$$

because $f^2(0, 0, \ldots, 0) = 0$, where

$$r_{ij} = \frac{1}{2}\frac{\partial^2 f^2}{\partial \xi_i \partial \xi_j}.$$

12.4.2.2 Higher-Order Approximation

In a similar way as for the quadratic functions in Section 12.4.2.1, Equation 12.14 can be extended to arbitrary power m if we use the Taylor expansion of f^m:

$$f^m(x_1, x_2, \ldots x_d) = \frac{1}{m!}\sum_{i_1 i_2 \ldots i_m}^{d} \frac{\partial^m f^m}{\partial \xi_{i_1} \partial \xi_{i_2} \ldots \partial \xi_{i_m}} x_{i_1} x_{i_2} \ldots x_{i_m},$$

where we have ignored all higher-order terms; see also IAA (2004, Appendix H.6).

12.5 RISK MODELS

The models discussed in the previous sections are independent of any distributional assumptions underlying the variables. This is important as most risk variables in modeling are nonnormal.

We use C as the capital requirement function. We let $x_j, j = 1, 2, \ldots, d$, be an exposure measure for the jth risk component among n. An exposure measure could be the amount of equities hold, the amount of bonds hold with duration of 5 years, premiums, technical provisions, and so on.

12.5.1 Homogeneous Models of Degree One

The homogeneity property (of degree one) is important. Consider the change of currency. The homogeneity implies that the risk measure is independent of the currency:

$$C(\lambda x_1, \lambda x_2, \ldots, \lambda x_d) = \lambda C(x_1, x_2, \ldots, x_d). \tag{12.16}$$

Assume that the supervisory authority defines some baseline exposures $x_1^0, x_2^0, \ldots, x_d^0$, which is referred by the IAA (2004, Appendix H) as the *target point* or *target mix*. These exposures represent some "ideally representative company." From Euler's theorem, see Equation 12.8, it follows that we could write the capital requirement as a factor model:

$$C(x_1, x_2, \ldots x_d) = \sum_{i=1}^{d} \tau_i x_i, \tag{12.17}$$

where $\tau_i = \partial C/\partial x_i^0$ is the *factor*, based on the target mix, that the *risk volume measure* x_i should be multiplied with to give the capital charge.

In the simplest case with only one risk ($n = 1$), we obtain

$$C(x) = \tau x. \tag{12.18}$$

Here τ is the factor that the risk volume measure should be multiplied with to get the capital requirement.

If this model is not as accurate as you believe in, it is possible to get a quadratic approximation to C^2. From Equation 12.15, we obtain

$$C^2(x_1, x_2, \ldots, x_d) = \sum_{i=1}^{d}\sum_{j=1}^{d} r_{ij}x_i x_j, \tag{12.19}$$

where r_{ij} is defined at the target risk, cf. Equation 12.17. Hence, the approximation $C(x_1, x_2, \ldots, x_d) = \sqrt{\sum_{i=1}^{d}\sum_{j=1}^{d} r_{ij}x_i x_j}$ is valid in a neighborhood of the target mix.

Panjer (2002) has shown that this approximation is exact for a standard deviation or a TVaR risk measure if the underlying distribution is multivariate normal.

As stated by IAA (2004, Appendix H), if the actual mix of risks is close to the target mix, as defined by the supervisory authority, then the capital requirement could be approximated by a *factor-based model*. The factors are derived from the derivatives of the capital requirement function at the target risk mix.

12.5.2 Second-Order Risk Models

The second-order approximation given in Section 12.4.1 shows that, irrespective of the underlying distributions, you could model the *expected capital requirement*, with $Y_0 = 0$, as

$$E[C(x_1, x_2, \ldots, x_d)] = \sum_{i=1}^{d}\sum_{j=1}^{d} \rho_{ij}C_{ij}^*, \tag{12.20}$$

where $C_{ij}^* = \sigma_i\sigma_j r_{ij}$ and $r_{ij} = (1/2)\partial^2 C/\partial\mu_i\partial\mu_j$. The capital requirement is here expressed as the standard deviations between two risks and a factor based on the derivatives of the capital requirement function at a mean value of the target mix.

CHAPTER 13

Dependence

In the baseline approach proposed by IAA, which is introduced in Chapter 14, the total squared capital requirement is the sum of the cross-products of different capital requirements from the risks incorporated into the model. They are added using a correlation matrix. As this approach is based on a multivariate normal distribution serving as a first-order approximation, the correlations used are the linear correlation coefficients; see, for example, Equation 14.3b.

In Chapter 12 we tried to model the capital charge from a sum of different risks. The expected capital requirement, as expressed in Equation 12.20, is also shown as the product of correlation time risk charges.

The correlation, or co-relation, is a concept that refers to the departure of two variables from independence. The *linear correlation coefficient* measures in some sense the strength and direction of a *linear relationship* between two stochastic variables X and Y such as $Y = a + bX$, and a and b are real numbers with $b \neq 0$. In the world of multivariate normal distributions (generally in the world of spherical and elliptical distributions), the linear correlation measures the dependency between pairs of stochastic variables.

In insurance and finance the normal distribution assumption is usually not true. Loss distributions are usually skewed and heavily tailed and nonlinear derivative products contradict the use of linear correlation. As noted by Embrechts et al. (2003), "market data returns tend to be uncorrelated, but dependent, they are heavy tailed, extremes appear in clusters and volatility is random." Discussions on the upper tail are also given in, for example, Venter (2002) and Charpentier (2003).

As a tool for understanding the dependence and relations among multivariate variables, copulas were introduced for actuaries by Frees and Valdez (1998); see also the discussion following that publication. For a general introduction to the theory of copulas, see Nelsen (1999). For a general discussion on copulas and current research, see Embrechts (2007), and for a practical discussion on inference on copula models, see Genest and Favre (2007). Other very good references on copulas are, for example, Cech (2006) and Strassburger (2006). A brief discussion on copulas and the subprime mortgage crisis 2007–2009 is given in Donnelly and Embrechts (2009).

13.1 DEPENDENCE STRUCTURE

It is important to distinguish between *dependence structure* and *dependence strength*. The dependence between stochastic variables $X_1, \ldots, X_d$ is completely described, that is, both its structure and strength, by their joint distribution function

$$F(x_1, \ldots, x_d) = P\left[X_1 \leq x_1, \ldots, X_d \leq x_d\right].$$

The main idea in distinguishing between the structure and the strength, see Embrechts et al. (2002), is to separate F into one part describing the dependence structure and parts describing the marginal behavior. This leads us to the concept of *copulas*. An n-dimensional copula is an n-dimensional distribution function with uniform marginal distributions.

The dependence structure is entirely determined by the copula, independent of scale and shape; see below. On the other hand, the scaling and shape (mean, standard deviation, skewness, and kurtosis) are fully determined by the marginal probability density functions (pdfs). This division is shown by Equation 13.2.

One of the most important results regarding copula is Sklar's representation theorem; see, for example, Nelsen (1999) and Denuit et al. (2005). It states that the dependence structure between $X_1, \ldots, X_d$ is described by the copula C if the joint distribution function F is given by

$$F(x_1, \ldots, x_d) = C\left[F_1(x_1), \ldots, F_d(x_d)\right], \tag{13.1}$$

where F_i denotes the marginal distribution function of x_i. This means that the joint distribution function of the quantiles of $x_1, \ldots, x_d$ is given by the copula function C. We assume here, and throughout the text, that the $F_i, i = 1, \ldots, d$, are continuous, which implies that the copula is unique. The pros and cons of modeling joint distributions with copulas of random variables with discrete margins are discussed in Genest and Nešlehová (2007). One definition of a cupola is that it is the distribution function of a random vector in $\mathbb{R}^d$ with standard uniform marginals, that is, $U_i = F_i(x_i), i = 1, \ldots, d$, are uniformly distributed random variables on [0,1].

This means that the joint distribution function F is separated into the dependence structure given by C and into uniform marginals.

One definition of a copula is that it is a function that associates the quantiles of one random variable with the quantiles of another random variable.

Equation 13.1 could be rewritten as $C(u_1, \ldots u_d) = F(F_1^{-1}(u_1), \ldots, F_d^{-1}(u_d))$ for $(u_1, \ldots, u_d)^t \in [0,1]^d$. For independent random variables, the copula takes the form $C_i(u_1, \ldots u_d) = u_1 \ldots u_d$. When we wish to fit copulas to data and estimating parameters and rank correlations, see below, it is useful to define the copula density. If the copula C is absolutely continuous and of an absolutely continuous joint distribution function F with strictly increasing, continuous marginals $F_1, \ldots, F_d$, we may differentiate $C(u_1, \ldots, u_d)$. The copula density c is given by

$$c(u_1, \ldots, c_d) = \frac{\delta C(u_1, \ldots, c_d)}{\delta u_1 \ldots \delta u_d} = \frac{f(F_1^{-1}(u_1), \ldots, F_d^{-1}(u_d))}{f_1(F_1^{-1}(u_1)) \ldots f_d(F_d^{-1}(u_d))},$$

where f is the joint pdf of F; $f_1, \ldots, f_d$ are the marginal densities; and $F_1^{-1}, \ldots, F_d^{-1}$ are the inverses of the marginal distribution functions.

One interesting result is given in Property 4.2.14 in Denuit et al. (2005). It states that if the marginal distribution functions F_1 and F_2 are continuous with pdfs f_1 and f_2, respectively, then the joint pdf of $(X_1, X_2)^t$ can be written as

$$f(x_1, x_2) = f_1(x_1) \cdot f_2(x_2) \cdot c(F_1(x_1), F_2(x_2)), \tag{13.2}$$

where c is the copula density function defined by

$$c(u_1, u_2) = \frac{\delta^2}{\delta u_1 \delta u_2} C(u_1, u_2), u_1, u_2 \in [0, 1].$$

Equation 13.2 has a nice interpretation. The first part of the right-hand side, $f_1 f_2$, is the joint pdf that corresponds to independence. The second part, c, that distorts independence is therefore interpreted as the local dependence measure. If X_1 and X_2 are independent, then $c = 1$.

In Wang's discussion on Frees and Valdez (1998), he shows that copulas can be applied to probability-generating functions and characteristic functions.

We discuss estimation in Section 13.5.

One important fact, which relies on the fact that copulas describe dependence between variables on the level of quantiles, is that the dependence structure as summarized by a copula is invariant under increasing and continuous transformations of the marginals. Assume that $T_1, \ldots, T_d$ are increasing continuous functions, then both $(X_1, \ldots, X_d)^t$ and $(T(X_1), \ldots, T(X_d))^t$ have the same copula C. For a proof, see, for example, McNeil et al. (2005). As stated in Embrechts et al. (2002) this means, for example, that one could change our model from percentage returns to a model with logarithmic returns. Other practical applications are given in IAA (2004, p. 135):

- Sensitive claims data could, after being transformed by an increasing function, be made publicly available, as data are not back traceable. It is still possible to use the data for estimating copulas.
- Copula for gross losses can be assumed to be the same as for net losses!

The following two figures show what is meant with dependence structure (Figure 13.1).

Copulas can be used for combining risks when the marginal distributions are estimated individually; see Chapter 15. Thus the joint density will preserve the characteristics of the marginals. As described by Rosenberg and Schuermann (2004), a normal dependence relation can be preserved using normal copula at the same time when the marginals are entirely general, for example, with Weibull and lognormal marginals.

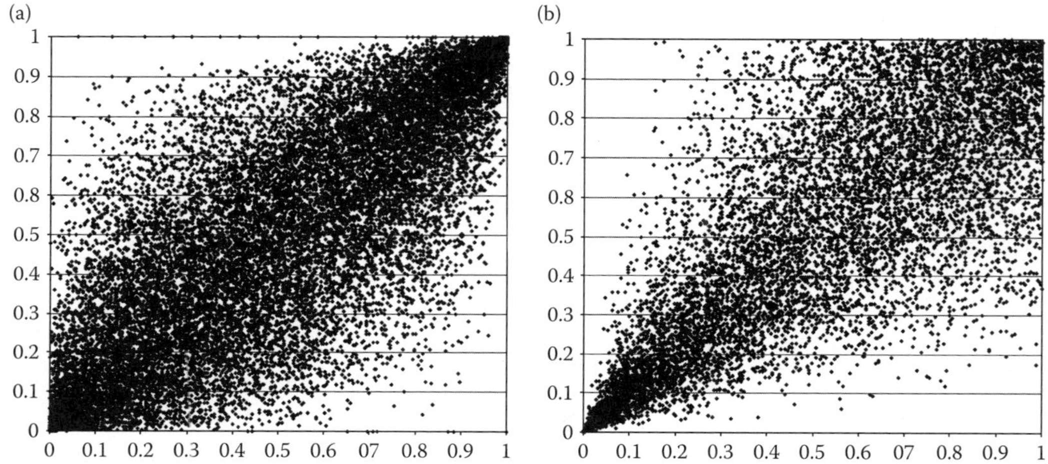

FIGURE 13.1 Two distributions with identical linear correlations $\rho = 0.8$. (a) Gaussian copula and (b) Clayton copula. As we can see, they differ in their dependence structure. 10,000 simulations.

13.1.1 More on Copulas and Dependence

For every copula $C(u_1, \ldots, u_d)$, the lower and upper Fréchet–Hoeffding bounds hold:

$$C_l(u_1, \ldots, u_d) = \max\left(\sum_{i=1}^{d} u_i + 1 - d; 0\right) \leq C(u_1, \ldots, u_d)$$

$$\leq \min(u_1, \ldots, u_d) = C_u(u_1, \ldots, u_c).$$

In general, it holds that $-1 \leq C_l \leq C(u_1, \ldots, u_d) \leq C_u \leq 1$. If $d = 2$, the bounds C_l and C_u are themselves copulas and $C_l = \max(u_1 + u_2 - 1, 0)$ is the bivariate distribution function of the vector $(U, 1 - U)^t$ and C_u is the bivariate distribution function of the vector $(U, U)^t$. C_u and the independent copula C_i are d-copulas for $d \geq 2$. C_l is not a copula for $d \geq 3$, but in some sense it is the best lower bound; see Embrechts et al. (2003).

The Fréchet–Hoeffding lower bound $C_l(u_1, u_2)$ is smaller than every 2-copula and every copula is smaller than the upper bound $C_u(u_1, u_2)$. If C_1 and C_2 are copulas, we say that C_1 is smaller than C_2 and we write that as $C_1 \prec C_2$ if $C_1(u, v) \leq C_2(u, v)$ for all u and v. This partial ordering is called a *concordance ordering*.

In the case where $d = 2$, we say that C_l describes perfect negative dependence, in the sense that u_2 is almost surely a strictly decreasing function of $\boldsymbol{u}_1$, and C_u describes perfect positive dependence, in the sense that they are almost surely strictly increasing functions of each other. C_l and C_u for $d = 2$ are described in Figure 13.2.

The perfect positive dependence and perfect negative dependence, as described above, are both monotonic-dependent structures; see Figure 13.2. We therefore introduce the two dependence concepts: comonotonicity and countermonotonicity; see, for example, Wang and Dhaene (1998) and Denuit and Dhaene (2003). Comonotonicity and aggregating non-independent risks are discussed in, for example, Dhaene et al. (2002a, 2002b) and (2007); see

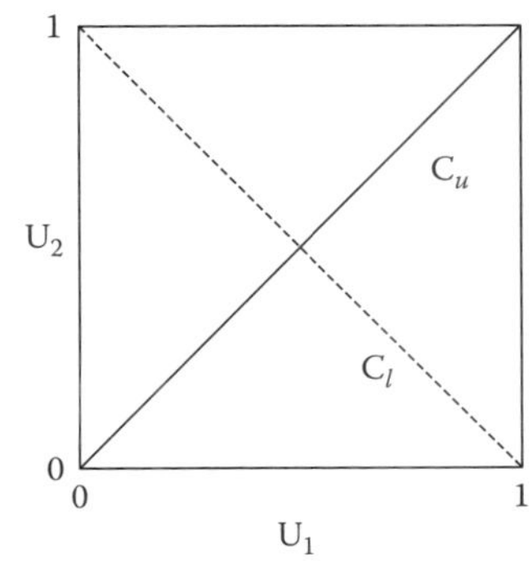

FIGURE 13.2 Perfect negative dependence (C_l) and perfect positive dependence (C_u) for $d = 2$ as described by the Fréchet–Hoeffding bounds.

also Denuit et al. (2005). When comparing two random variables, the "most favorable" way of assuming their interrelation would be to assume mutual independence between them. The "least favorable" way would be to assume a comonotonic dependence structure between them. Assume a sum of dependent random variables, whose marginal distributions are known but with unknown (or complicated) joint distribution. Dhaene et al. (2002a, 2002b) have shown that a conservative approach would be to replace the original sum by one with a simpler dependence structure, which is considered to be less favorable by risk-averse decision makers. This conservative sum involves the components of the comonotonic version of the original random vector.

Comonotonic random variables: If $(X, Y)^t$ has copula C_u, then X and Y are said to be comonotonic. The comonotonic copula ensures that risks always move in the same direction, which is the worst-case scenario in the insurance business (perfect positive dependence). The comonotonicity concept can be seen as an extension of the concept of positive perfect correlation.

Countermonotonic random variables: If $(X, Y)^t$ has copula C_l, then X and Y are said to be countermonotonic. The countermonotonic copula ensures that risks always move in the opposite direction (perfect negative dependence).

What we would like to have is a *scalar valued measure of dependence* between two random variables X and Y in a similar manner as the linear correlation coefficient that takes on the value -1 for the case of countermonotonicity and the value $+1$ for the case of comonotonicity. As a matter of fact, *rank correlation* is the concept that is defined at copula level and will handle dependence in the way we wish.

Let $(x_1, y_1)^t$ and $(x_2, y_2)^t$ be two observations from a vector $(X, Y)^t$ of continuous random variables. We now define concordant and discordant variables.

Concordant random variables: $(x_1, y_1)^t$ and $(x_2, y_2)^t$ are said to be concordant if $(x_1 - x_2)(y_1 - y_2) > 0$.

Discordant random variables: $(x_1, y_1)^t$ and $(x_2, y_2)^t$ are said to be disconcordant if

$$(x_1 - x_2)(y_1 - y_2) < 0.$$

There is a theorem, see Nelsen (1999) and Embrechts et al. (2003), that states that if Q denotes the difference between the probability of concordance and discordance of $(x_1, y_1)^t$ and $(x_2, y_2)^t$, that is,

$$Q = P[(X_1 - X_2)(Y_1 - Y_2) > 0] - P[(X_1 - X_2)(Y_1 - Y_2) < 0] = \mathrm{PC} - \mathrm{PD},$$

where PC is the probability of concordance and PD is the probability of discordance, then

$$Q = Q(C_1, C_2) = 4 \int_0^1 \int_0^1 C_2(u, v)\, \mathrm{d}C_1(u, v) - 1, \tag{13.3}$$

where $C_1(u, v)$ denotes the copula of $(x_1, y_1)^t$ and $C_2(u, v)$ denotes the copula of $(x_2, y_2)^t$.

A real-valued measure r of dependence between two continuous random variables X and Y, with copula C, is a measure of concordance if it satisfies certain properties; see, for example, Embrechts et al. (2002) and Denuit et al. (2005).

1. Symmetry: $r(X, Y) = r(Y, X)$
2. Normalization: $-1 \le r(X, Y) \le 1$
3. Comonotonicity: $r(X, Y) = 1$ if and only if X and Y are comonotonic
4. Countermonotonicity: $r(X, Y) = -1$ if and only if X and Y are countermonotonic
5. For $t : \mathbb{R} \to \mathbb{R}$ strictly monotonic:

$$r(t(X), Y) = \begin{cases} r(X, Y) & \text{if } t \text{ is increasing} \\ -r(X, Y) & \text{if } t \text{ is decreasing} \end{cases}.$$

A consequence of the definition of concordance is that if Y is almost surely an increasing function of X, then $r(X, Y) = C_u = 1$, and if Y is almost surely a decreasing function of X, then $r(X, Y) = C_l = -1$. A property like $r(X, Y) = 0 \Leftrightarrow X, Y$ are independent contradicts property 5. It is stated in Embrechts et al. (2002) that there is no dependence measure that satisfies property 5 and this latter property.

Positive dependence is discussed in, for example, Denuit et al. (1999) and its relation to the effect on aggregate claims in Denuit et al. (2001).

13.2 DEPENDENCE STRENGTH: RANK CORRELATION

Two classical rank correlation coefficients, the Spearman's rho and Kendall's tau rank correlations, are nonparametric measures at copula level and can be considered as measuring the degree of monotonic dependence between two random variables. We will define the two rank correlations between two random variables. This can easily be extended to $d > 2$ dimensions in the same way as for linear correlations, that is, introducing pairwise correlations in a $d \times d$-matrix.

To illustrate rank correlation, assume that X and Y are two random variables and $R_{(X)i}$ and $R_{(Y)i}$ are the ranks of the outcomes of X and Y. Using the indicator function $I\{\cdot\}$ we can define the empirical rank functions for the marginal data as

$$R_{(X)i} = \sum_{k=1}^{d} I\{X_k \le X_i\}$$

and

$$R_{(Y)i} = \sum_{k=1}^{d} I\{Y_k \le Y_i\}.$$

The dependence between ranks would be perfect if and only if $R_{(X)i} = R_{(Y)i}$, for all $i = 1, 2, \ldots, d$. It is therefore logical to use the difference $d_i = R_{(X)i} - R_{(Y)i}$, which indicates the disparity between the two sets of rankings. The larger the d_i values, the less perfect the association or dependence between the two variables. For more "classical discussions" on the two rank correlation measures, we refer to Siegel (1956).

If X and Y are continuous random variables with copula C, then Kendall's tau and Spearman's rho are measures of concordance; see Nelsen (1999). Note that the linear correlation coefficient is not invariant under strictly increasing transformations of X and Y and is therefore not a measure of concordance; see, for example, Embrechts et al. (2003). Kendall's tau and Spearman's rho depend only on the dependence structure—you could shrink and stretch the axis without changing the dependence structure. On the other hand, Pearson's linear correlation coefficient depends not only on the dependence structure but also on the marginal distributions.

The idea behind the dependence concept of concordance is that two random variables X and Y are concordant when "large" values of X go with "large" values of Y and "small" values of X go with "small" values of Y. Concordance measures are reviewed and discussed in Denuit and Dhaene (2003).

13.2.1 Spearman's Rho

Definition: Let X and Y be random variables with distribution functions F_1 and F_2 and joint distribution function F and copula C. Then Spearman's rank correlation is defined by

$$\begin{aligned}\rho_S(X,Y) &= 3Q(C, C_i) = 12\int_0^1\int_0^1 uv\,\mathrm{d}C(u,v) - 3\\ &= 12\int_0^1\int_0^1 C(u,v)\,\mathrm{d}u\,\mathrm{d}v - 3,\end{aligned} \tag{13.4a}$$

with Q defined in Equation 13.3 and C_i is the independent copula. Spearman's rho can also be written as $12\mathrm{E}(UV) - 3$.

It can be shown that

$$\rho_S(X,Y) = \rho(F_1(X), F_2(Y)), \tag{13.4b}$$

where ρ is the usual Pearson's linear correlation coefficient.

This nonparametric measure is named after Charles Spearman who published the measure in 1904. It is just an ordinary linear correlation coefficient based on ranks.

13.2.2 Kendall's Tau

Definition: Let (X_1, Y_1) and (X_2, Y_2) be two independent pairs of random variables from F. Then Kendall's tau (rank correlation coefficient) is given by

$$\rho_\tau(X,Y) = P[(X_1 - X_2)(Y_1 - Y_2) > 0] - P[(X_1 - X_2)(Y_1 - Y_2) < 0]. \qquad (13.5a)$$

Let X and Y be continuous random variables with copula C. Then we can write Kendall's tau for $(X,Y)^t$ as Equation 13.3

$$\rho_\tau(X,Y) = Q(C,C) = 4\int_0^1\int_0^1 C(u,v)\,\mathrm{d}C(u,v) - 1, \qquad (13.5b)$$

or as 4E[C(U,V)] − 1.

This nonparametric measure is named after Maurice Kendall who published the measure in 1938. It measures the degree of correspondence between rankings.

The rank correlation coefficient is based on "natural orderings" of ranks. In this way, it is usually more cumbersome to calculate by hand. One advantage, as compared to Spearman's rank correlation coefficient, is that it could be generalized to a partial correlation coefficient.

13.3 TAIL DEPENDENCE

Tail dependence relates to the amount of dependence in the upper-right-quadrant tail or the lower-left-quadrant tail of a bivariate distribution. Tail dependence between two continuous random variables X_1 and X_2, with marginal distribution functions F_1 and F_2, is a *copula property* and hence the amount of tail dependence is invariant under strictly increasing transformations of X_1 and X_2.

The dependence between two variables in the tails can be defined as, see IAA (2004) and McNeil et al. (2005),

Upper tail dependence:

$$\lambda_U := \lambda_U(x_1,x_2) = \limsup_{u\to 1} P[X_1 > F_1^{-1}(u)|X_2 > F_2^{-1}(u)], \qquad (13.6a)$$

provided that $\lambda_U \in [0,1]$ exists, and

Lower tail dependence:

$$\lambda_L := \lambda_L(x_1,x_2) = \limsup_{u\to 0} P[X_1 \le F_1^{-1}(u)|X_2 \le F_2^{-1}(u)], \qquad (13.6b)$$

provided that $\lambda_L \in [0,1]$ exists.

These tail dependencies can be derived directly from the copula for x_1 and x_2. If the linear correlation between two variables is $\rho \neq 1$, then the tail dependency for x_1 and x_2 is zero.

If $\lambda_U \in (0, 1]$, then X_1 and X_2 are said to be asymptotically dependent in the upper tail and if $\lambda_U = 0$, X_1 and X_2 are said to be asymptotically independent in the upper tail.

If the bivariate copula is such that

$$\lim_{u \to 1} \frac{1 - 2u + C(u, v)}{1 - u} = \lambda_U$$

exists, then the C has upper tail dependence if $\lambda_U \in [0, 1]$ and upper tail independence if $\lambda_U = 0$; see Embrechts et al. (2003). Similar results can be found for lower tail dependence; see Embrechts et al. (2003). If C is an exchangeable copula, that is, $C(u, v) = C(v, u)$, then $\lambda_U = 2 \lim_{u \to 1} P[V > u | U = u)$.

As we have assumed that F_1 and F_2 are continuous distribution functions, we can get simple expressions for λ_L and λ_U in terms of the unique copula C of the bivariate distribution; see McNeil et al. (2005):

$$\lambda_L = \lim_{u \to 0^+} \frac{C(u, u)}{u},$$

and

$$\lambda_U = \lim_{u \to 1^-} \frac{\widehat{C}(1 - u, 1 - u)}{1 - u} = \lim_{u \to 0^+} \frac{\widehat{C}(u, u)}{u},$$

where $\widehat{C}$ is the survival copula of C defined as $\widehat{C}(1 - u, 1 - v) = 1 - u - v + C(u, v)$.

Hence,

$$\lambda_U = \lim_{u \to 1^-} \frac{1 - 2u + C(u, u)}{1 - u}.$$

If the copula has a simple closed form, the calculation of these lower and upper coefficients is straightforward.

Conditional copulas of (U, V), given, for example, $U \leq u, V \leq v$, are discussed in Charpentier (2003).

13.4 COPULA CLASSES AND FAMILIES

There are many different copulas and methods to construct them presented in the literature; see, for example, Nelsen (1999). It is not always convenient to identify a copula. One method to construct a copula is the method of compounding, which is described and illustrated in, for example, Frees and Valdez (1998); see also Nelsen (1999) for methods of constructing copulas. Copulas are usually classified in classes of copulas with different copula families; for different copula classifications, see, for example, Nelsen (1999), McNeil et al. (2005), Embrechts et al. (2003), and Panjer (2006), and the references therein.

Gaussian and t-copulas, see below, are copulas implied by known multivariate distribution functions and do not have simple closed forms—they are therefore called implicit copulas. On the other hand, there are explicit copulas having simple closed forms, such as the Clayton family $C_\theta(u, v) = \max[(u^{-\theta} + v^{-\theta} - 1)^{-1/\theta}, 0]$; see below. As we are focusing on standard formulas, we will mainly look at bivariate copulas, but as multivariate copulas are of interest in internal modeling, we will also mention some extensions to higher dimensions ($n > 2$).

In some copula families, such as the Gaussian family, the upper tail dependence is zero ($\lambda_U = 0$). As we are most interested in copula families with tail dependence, that is, with $\lambda_U > 0$, we will mainly discuss such copula families (and classes).

13.4.1 Copula Class: Archimedean

The Archimedean class of copulas, see, for example, Nelsen (1999), includes many interesting copula families, such as the Clayton, Frank, and Gumbel families, and they allow for a great variation of different dependent structures. All commonly discussed Archimedean copulas have closed form expressions. They are not derived from multivariate distribution functions using Sklar's representation theorem. This means that the extensions of bivariate copulas to n-copulas need to be asserted; see, for example, Embrechts et al. (2003) and McNeil et al. (2005). Generating Archimedean copula families is discussed in Genest, Ghoudi, and Rivest in the discussion on Frees and Valdez (1998).

The Archimedean d-copula can be expressed in the form

$$C(u_1, \ldots, u_d) = \varphi^{-1}\{\varphi(u_1), \ldots, \varphi(u_d)\},$$

where $\varphi : (0, 1] \to [0, \infty)$, such that $\varphi(1) = 0$ and $((-1)^i \mathrm{d}^i/\mathrm{d}x^i)\varphi^{-1}(x) > 0, i \in (1, \ldots, d)$. The function φ is called a generator of the copula. If $\varphi(0) = \infty$, then it is called a strict generator. We refer to the above references for more details; see also Nelsen (1999) and Genest et al. (2006). Genest and Rivest (1993) were the first to propose a procedure for identifying a generator in empirical applications.

A bivariate Archimedean copula is defined by (see, e.g., Genest and Rivest, 1993; Denuit et al., 2005)

$$C(u, v) = \begin{cases} \varphi^{-1}[\varphi(u) + \varphi(v)] & \text{if } \varphi(u) + \varphi(v) \neq \varphi(0) \\ 0 & \text{otherwise} \end{cases} \tag{13.7}$$

for $0 \leq u, v \leq 1$, where the generator φ is a convex function, that is, $\varphi' < 0$ and $\varphi'' > 0$, and $\phi(1) = 0$.

If (U, V) has distribution function C, where C is an Archimedean copula generated by φ, then the function $K_C(t) = t - (\varphi(t)/\varphi'(t)), t \in [0, 1]$ is the distribution function of the random variable $C(U, V)$; see, for example, Genest and Rivest (1993) and Embrechts et al. (2003) and Genest and Favre (2007). For random variables having an Archimedean copula, Kendall's tau can be expressed in terms of the generator, see op. cit., as

$$\rho_\tau = 1 + 4 \int_0^1 \frac{\varphi(t)}{\varphi'(t)} \, \mathrm{d}t. \tag{13.8}$$

This is just a special case for dimension $d = 2$ from a multivariate version of Kendall's tau; see Embrechts et al. (2003).

In Embrechts et al. (2003, Theorem 6.6), it is shown that if the generator φ is a strict generator, then, if $\varphi'^{-1}(0)$ is finite, then $C(u,v) = \varphi^{-1}[\varphi(u) + \varphi(v)]$ does not have upper tail dependence. If C has upper tail dependence, then $\varphi'^{-1}(0) = -\infty$ and the upper tail dependence is given by

$$\lambda_U = 2 - 2 \lim_{s \to 0^+} \left(\frac{\varphi'^{-1}(2s)}{\varphi^{-1}(s)} \right).$$

In Nelsen (1999), 22 different one-parameter families of Archimedean copulas are listed. Some of them are considered here.

In Charpentier (2003), it is shown that the conditional Archimedean copula of (U, V), given, for example, $U \leq u, V \leq v$, is still an Archimedean copula.

The *Gumbel family* has generator $\varphi(t) = (-\ln t)^\theta$ and is defined by

$$C_\theta^{\mathrm{Gu}}(u,v) = \exp\left\{ -\left[(-\ln u)^\theta + (-\ln v)^\theta \right]^{1/\theta} \right\}, \quad 1 \leq \theta \leq \infty. \tag{13.9}$$

The Gumbel family of copulas has a strict generator and has upper tail dependence; see Table 13.1, at the end of Chapter 13, for details. It can be shown that Kendall's tau is $\rho_{\tau\theta} = 1 - (1/\theta)$, which could be used for estimating θ; see Section 13.5.3. The upper tail dependence is $\lambda_U = 2 - 2^{1/\theta}$, when $\theta > 1$. The tail dependence is zero when $\theta = 1$ and tends toward 2 as θ becomes large.

In the d-dimensional case, $d > 2$, this one-parameter copula family is known as the *Gumbel–Hougaard copula*; see Panjer (2006).

$$C_\theta^{\mathrm{GH}}(u_1, \ldots, u_d) = \exp\left\{ -\left[\sum_{i=1}^{d} (-\ln u_i)^\theta \right]^{1/\theta} \right\}.$$

A multivariate extension of Archimedean copulas is also discussed in Embrechts et al. (2003). The theory in op. cit. is illustrated with a discussion of a d-dimensional Gumbel family, which, in the 3-dimensional case with two parameters, is given by

$$C_{\theta_1,\theta_2}(u_1,u_2,u_3) = \exp\left\{ -\left(\left[(-\ln u_1)^{\theta_2} + (-\ln u_2)^{\theta_2} \right]^{\theta_1/\theta_2} + (-\ln u_3)^{\theta_1} \right)^{1/\theta_1} \right\}, \theta_1 \leq \theta_2.$$

The Gumbel copula (Figure 13.3.) is also an extreme value copula; see Section 13.4.3.

The *Clayton family* (Figure 13.4) with generator $\varphi(t) = (1/\theta)(t^{-\theta} - 1)$ is defined by

$$C_\theta^{\mathrm{Cl}}(u,v) = (u^{-\theta} + v^{-\theta} - 1)^{-1/\theta}, \quad \theta \geq -1. \tag{13.10}$$

The Clayton family of copulas has a strict generator if $\theta \geq 0$ and has lower tail dependence. The Kendall's tau is $\rho_{\tau\theta} = (\theta/\theta + 2)$ and the lower tail dependence is $\lambda_L = 2^{-1/\theta}$.

TABLE 13.1 The Spearman's Rho, Kendall's Tau, and Lower and Upper Tail Dependences for Some Copula Families

Copula	Spearman's Rho $\rho_S =$	Kendall's Tau $\rho_\tau =$	Lower tail Dependence, $\lambda_L =$	Upper Tail Dependence, $\lambda_U =$
Archimedean				
Gumbel	No closed form	$1-\frac{1}{\theta}$	0	$2-2^{1/\theta}$
Clayton	Complicated form	$\frac{\theta}{(\theta+2)}$	$2^{-1/\theta}$	0
Frank	$1-\frac{12}{\theta}[D_1(\theta)-D_2(\theta)]$	$1-4[1-D_1(\theta)]/\theta$	0	0
Elliptical				
Gaussian	$\frac{6}{\pi}\arcsin\left(\frac{1}{2}\rho\right)$	$\frac{2}{\pi}\arcsin(\rho)$	0	0
Student's t	$\frac{6}{\pi}\arcsin\left(\frac{1}{2}\rho\right)$	$\frac{2}{\pi}\arcsin(\rho)$	$2t_{\upsilon+1}\left(-\sqrt{\frac{(\upsilon+1)(1-\rho)}{1+\rho}}\right)$	
Other				
Marshall–Olkin	$\frac{3\alpha_1\alpha_2}{2\alpha_1+2\alpha_1-\alpha_1\alpha_2}$	$\frac{\alpha_1\alpha_2}{\alpha_1+\alpha_1-\alpha_1\alpha_2}$	0	$\min(\alpha_1\alpha_2)$
Fréchet		$\theta(\theta+2)/3$	θ	θ

Note: D_1 and D_2 are the first Debye functions; see Section 13.4.1.

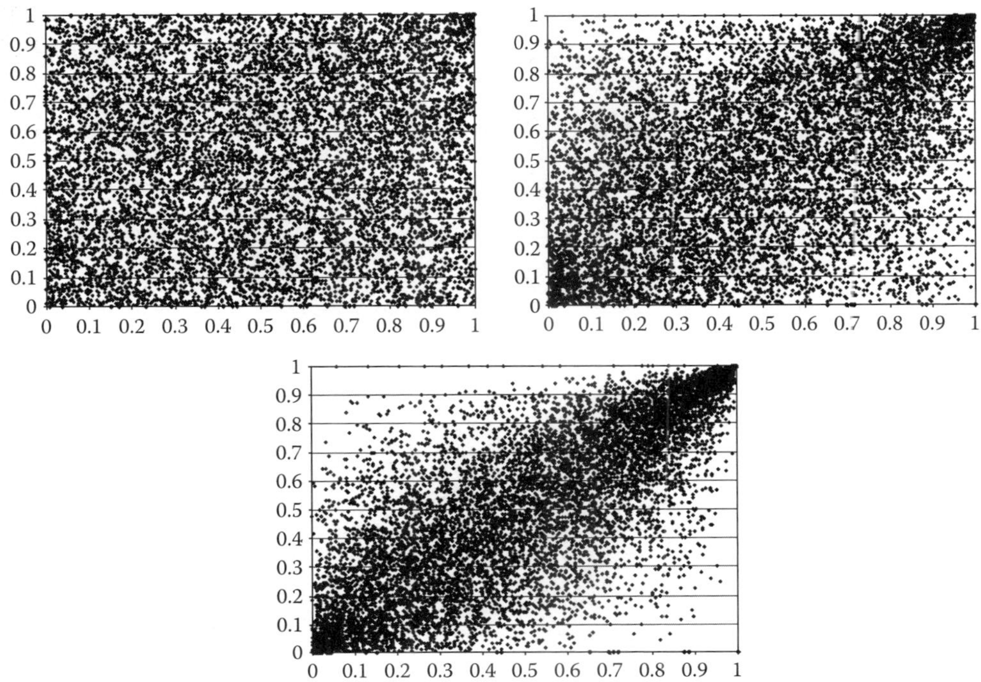

FIGURE 13.3 The Gumbel copula. The linear correlation is $\rho = 0.1, 0.5$, and 0.8, respectively. 10,000 simulations.

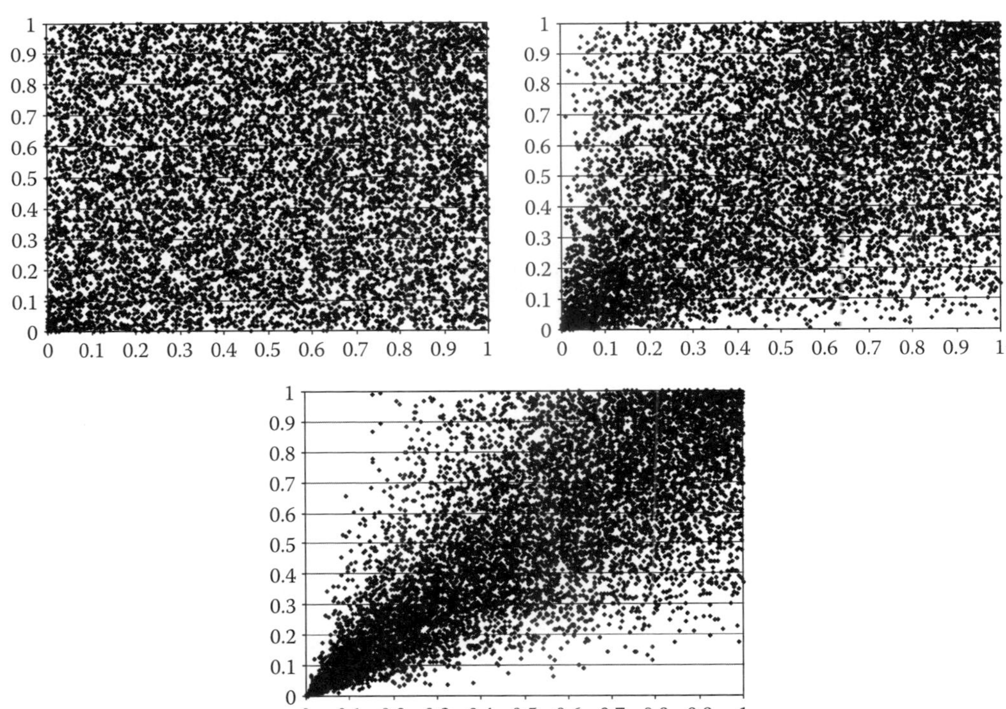

FIGURE 13.4 The Clayton copula. The linear correlation is $\rho = 0.1, 0.5$, and 0.8, respectively. 10,000 simulations.

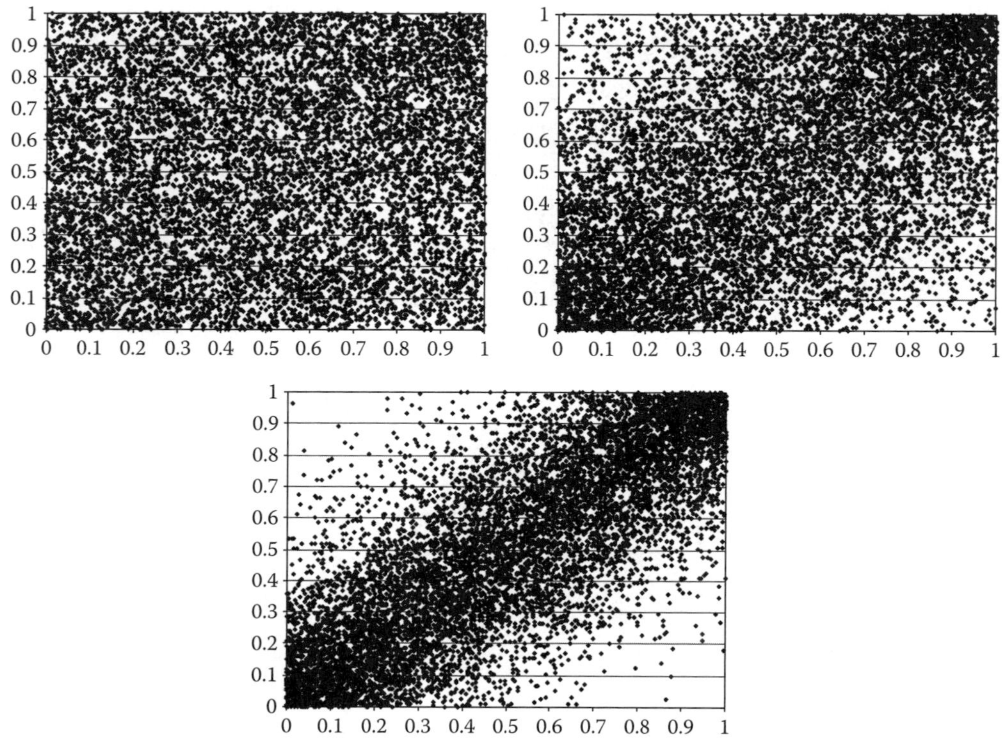

FIGURE 13.5 The Frank copula. The linear correlation is $\rho = 0.1, 0.5$, and 0.8, respectively. 10,000 simulations.

In the d-dimensional case, $d > 2$, this one-parameter copula family is known as the *Cook–Johnson copula*; see Panjer (2006).

$$C_\theta^{CJ}(u_1, \ldots, u_d) = \left[\sum_{i=1}^{d} u_i^{-\theta} - d + 1\right]^{-1/\theta}, \quad \theta > 0.$$

The *Frank family* (Figure 13.5) of copulas has a generator $\varphi(t) = -\ln((e^{-\theta t} - 1)/(e^{-\theta} - 1))$ and is defined, in the bivariate case, by

$$C_\theta^{Fr}(u, v) = -\frac{1}{\theta}\ln\left[1 + \frac{(e^{-\theta u} - 1)(e^{-\theta v} - 1)}{(e^{-\theta} - 1)}\right], \quad \theta \in R\backslash\{0\}. \tag{13.11}$$

The Frank family of copulas has a strict generator, but does not have either upper or lower tail dependency. The Kendall's tau is $\rho_{\tau\theta} = 1 - 4(1 - D_1(\theta)/\theta)$ and Spearman's rho $\rho_{S\theta} = 1 - (12/\theta)[D_1(\theta) - D_2(\theta)]$, where $D_1(\theta)$ and $D_2(\theta)$ are the first and second Debye functions given by

$$D_k(x) = \frac{k}{x}\int_0^x \frac{t^k}{e^t - 1}\,dt, \quad k = 1, 2, \ldots$$

For the multivariate version, see, for example, Panjer (2006).

Other Archimedean copulas are, for example, the *Ali-Mikhail-Haq family*, with upper tail dependence equal zero, and the *Joe family* with the same upper tail dependence as the Gumbel family (there is no simple closed form for Kendall's tau). Joe's BB1, BB3, BB6, and BB7 copula families have all upper tail dependencies; see Panjer (2006).

The independent copula C_i, see Denuit et al. (2005) and Panjer (2006), is an Archimedean copula. It can be shown that the Fréchet upper bound C_u is not a member.

13.4.2 Copula Class: Elliptical

The class of elliptical distributions includes many well-known distribution functions, such as normal and Student's t. A good introduction to elliptical distributions is Landsman and Valdez (2003). Elliptical copulas are the copulas of elliptical distributions. When the elliptical distribution is 1-dimensional it coincides with the class of 1-dimensional symmetric distributions. The copulas have no simple closed forms.

It is worth mentioning that the multivariate normal distribution is the only one in the class of elliptical distributions where uncorrelation implies independence.

An elliptical distribution is uniquely determined by μ, Σ, and ϕ, where $\mu \in \mathbb{R}^d$ is some location parameter, Σ is a nonnegative definite, symmetric matrix, and ϕ is a function of the quadratic form $z^t \Sigma z$ such that the characteristic function $\varphi_{X-\mu}(z) = \phi(z^t \Sigma z)$. The copula of a nondegenerated elliptically distributed random vector is uniquely determined by R and ϕ, where R is the linear correlation matrix of the random vector.

For bivariate normal and t-distributions it can be shown that Kendall's tau and Spearman's rho can be written in terms of the linear correlation coefficient ρ; see, for example, McNeil et al. (2005). They are

$$\rho_\tau = \frac{2}{\pi} \arcsin(\rho) \tag{13.12a}$$

and

$$\rho_S = \frac{6}{\pi} \arcsin\left(\frac{1}{2}\rho\right), \tag{13.12b}$$

and hence a nonparametric estimator of ρ is either, by Kendall's tau, $\hat{\rho} = \sin(\pi \hat{\rho}_\tau / 2)$, or by Spearman's rho, $\hat{\rho} = 2\sin(\pi \hat{\rho}_S / 6)$.

In the bivariate case, the *Gaussian copula* family is defined by

$$C_\rho^{\text{Ga}} = \int_{-\infty}^{\Phi^{-1}(u)} \int_{-\infty}^{\Phi^{-1}(v)} \frac{1}{2\pi(1-\rho^2)^{1/2}} \exp\left\{-\frac{s^2 - 2\rho st + t^2}{2(1-\rho^2)}\right\} ds\, dt, \tag{13.13}$$

where ρ is the usual linear correlation coefficient of the corresponding bivariate normal distribution.

It is shown in Embrechts et al. (2003, Example 3.4) that $\lambda_U = 0$ for $\rho < 1$, which means that the Gaussian copula does not have upper tail dependence. The same is true for lower tail dependence (Figure 13.6).

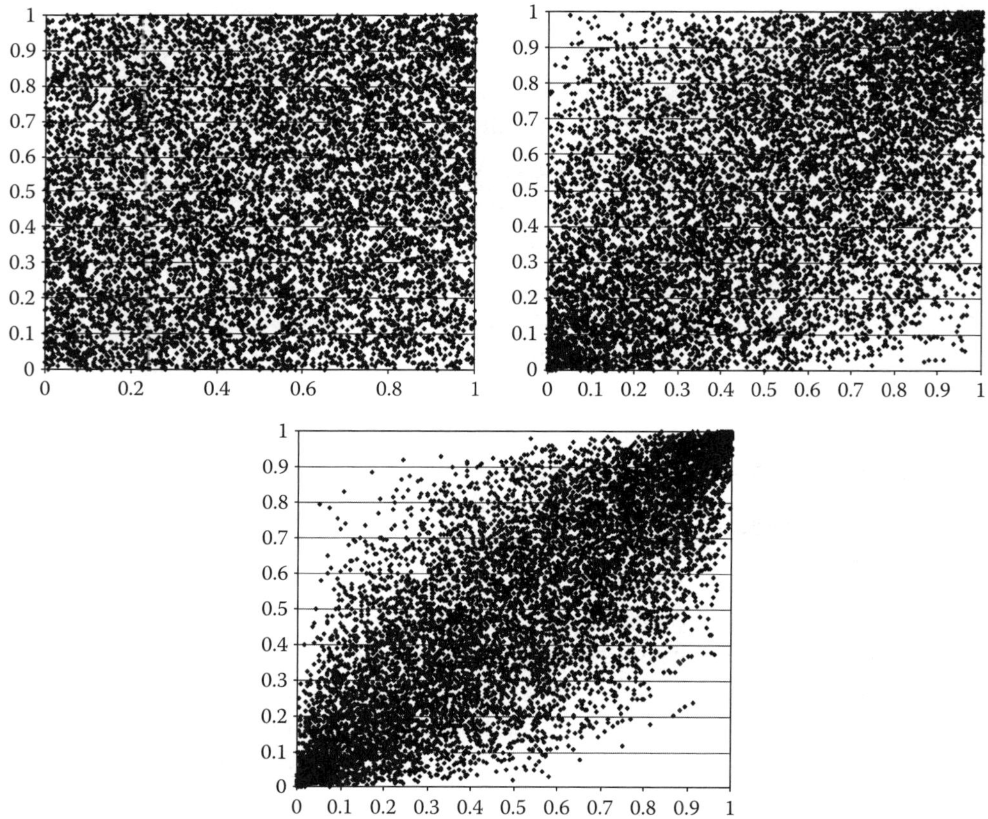

FIGURE 13.6 The Normal copula. The linear correlation is $\rho = 0.1, 0.5$, and 0.8, respectively. 10,000 simulations.

The Student's *t-copula family* is in its bivariate case defined by

$$C^t_{\upsilon,\rho}(u,v) = \int_{-\infty}^{t_\upsilon^{-1}(u)} \int_{-\infty}^{t_\upsilon^{-1}(v)} \frac{1}{2\pi(1-\rho^2)^{1/2}} \times \left\{1 + \frac{s^2 - 2\rho st + t^2}{\upsilon(1-\rho^2)}\right\}^{-(\upsilon+2)/2} \mathrm{d}s\,\mathrm{d}t, \tag{13.14}$$

where ρ is the usual linear correlation coefficient of the corresponding bivariate t_υ distribution if $\upsilon > 2$, and υ is the degree of freedom, df.

The lower and upper tail dependences are identical.

$$\lambda_{\mathrm{L}}(\upsilon,\rho) = \lambda_{\mathrm{U}}(\upsilon,\rho) = 2t_{\upsilon+1}\left(-\sqrt{\frac{(\upsilon+1)(1-\rho)}{1+\rho}}\right), \tag{13.15}$$

provided that $\rho > -1$. The upper (lower) tail dependence increases with ρ and decreases with υ. As the dfs $\upsilon \to \infty$, that is, to the standard normal distribution, the upper and lower tail dependences tend to zero.

In Figure 13.7, we illustrate the t-copula with 2, 10, and 30 dfs.

Spearman's rho and Kendall's tao are the same as in the normal case above.

An algorithm for random variate generation from t-copula $C^t_{\upsilon,\rho}$ is given by Algorithm 5.2 in Embrechts et al. (2003).

A discussion on the d-dimensional t-copula is given in, for example, Embrechts et al. (2003), McNeil et al. (2005), and Panjer (2006).

In practice, if the model has more than two variables, the choice of copula family often comes down to the Gaussian or t-copula; see Venter et al. (2007). As the t distribution tends to the Gaussian as df tends to infinity, the choice is mainly the decision on the value of df, which could be interpreted as a common shock effect in the model. The t-model takes a correlation parameter for each pair of variables $(X_i, X_j), i, j = 1, \ldots d$, and any correlation matrix can be used.

A main problem with t-copulas is the symmetry between lower and upper tail dependences. Another problem is that it has only one df parameter. We are mainly interested in the upper tail dependence, which corresponds to the insurance situation where the strongest tail dependence happens in the upper tail. T-copulas are discussed in Venter (2003), where its use is discussed with the common shock effect of the strength of storms on three LOBs.

One alternative is to use the *grouped t-copula*; see Daul et al. (2003). It uses different df parameters υ for different subgroups of variables, and was motivated by describing the risk of credit portfolios driven by a large set of risk factors, which are grouped into eight groups. Each grouped variables are t-distributed with different dfs. The df parameters, $\upsilon_i, i = 1, \ldots, 8$, were estimated using maximum-likelihood estimation. If the number of groups reduces to one, we have the so-called meta-t distribution, that is, distributions with t-copula and arbitrary marginal distributions.

Venter et al. (2007) introduced the *individuated t-copula*, IT-copula, by letting each variable have its own df parameter. The IT-copula has the form

$$C^{\mathrm{IT}}(u_1, \ldots, u_n) = \int_0^1 \frac{\prod_{i=1}^n \left\{\sqrt{h_i(y)}\Gamma(\upsilon_i/2)\left[1 + t_i^2/\upsilon_i\right]^{(1+\upsilon_i)/2} / \Gamma((1+\upsilon_i)/2)\right\}}{\sqrt{\det(R)(2\pi)^n} \cdot \exp\left\{\sum_{i,j}^n (J_{i,j} t_i t_j/2)\sqrt{h_i(y)h_j(y)/\upsilon_i\upsilon_j}\right\}}\, \mathrm{d}y,$$

where $h_i(y)$ is the inverse chi-square distribution quantile with probability y and df υ_i, while Γ represents the gamma density with parameters $\alpha = \upsilon_i/2$ and $\beta = 2/\upsilon_i$; $\det(R)$ is the determinant of the correlation matrix R and J is the matrix inverse of R.

Spearman's rho and Kendall's tau are not exactly the same as for the t-copula, but they are quite close. The tail dependence index for the IT-copula for X_i and $X_j, i, j = 1, \ldots, n$, is

$$\lambda_{i,j} = \int_0^\infty \bar{F}\left(c_i y^{1/\upsilon_i}, c_j y^{1/\upsilon_j}\right) \mathrm{d}y,$$

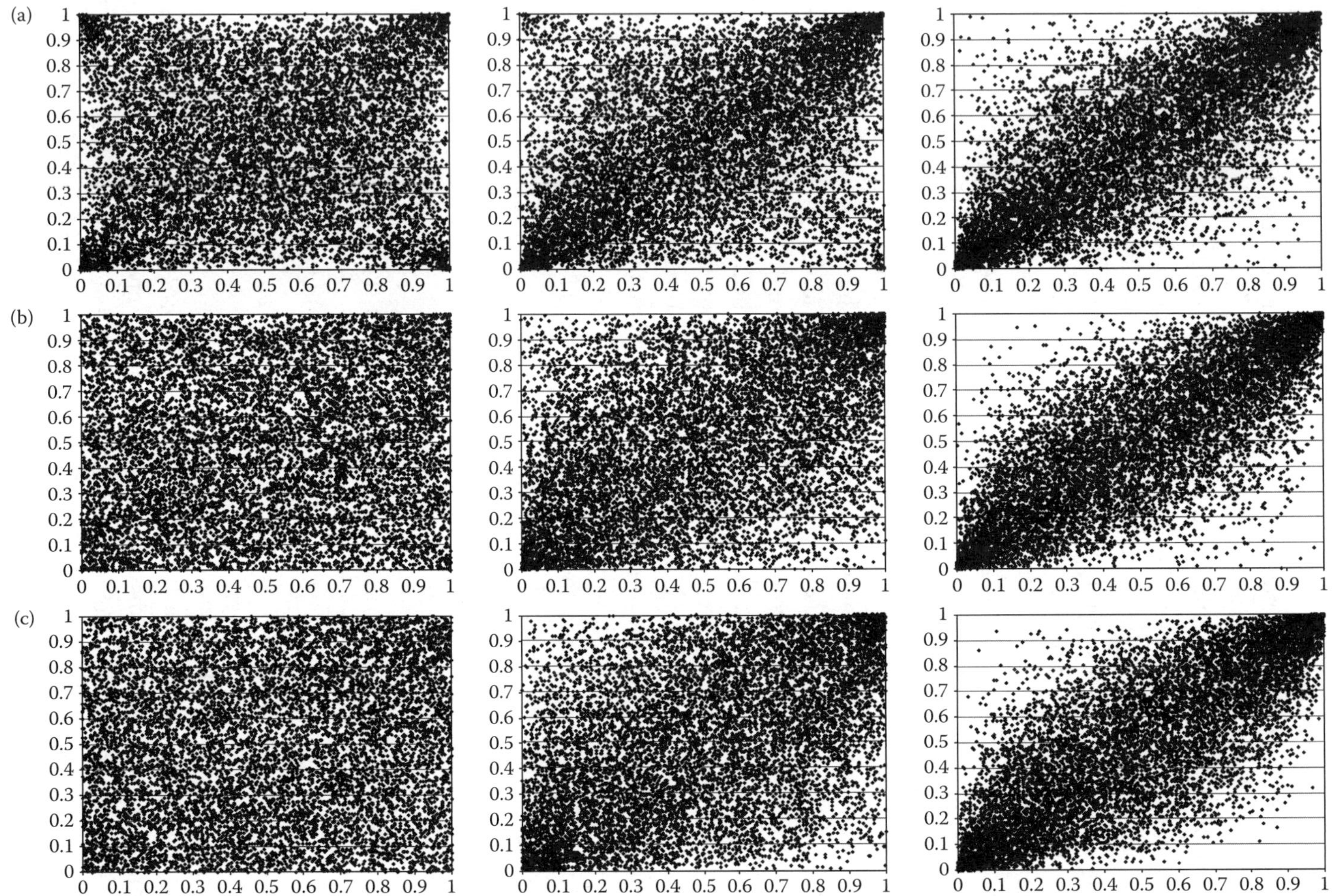

FIGURE 13.7 The t-copulas with $\upsilon = 2$(a), 10(b), and 30(c) in respective rows. The linear correlation is 0.1 in the left column, 0.5 in the middle column, and 0.8 in the right column. 10,000 simulations each.

where $\bar{F}$ denotes the standard normal survival function $\bar{F}(x,y) = \Pr(X > x, Y > y)$ and

$$c_i = \sqrt{2}\left[\Gamma\left(\frac{1+\upsilon_i}{2}\right)/\sqrt{4\pi}\right]^{1/\upsilon i}.$$

The IT-copula may be fitted using maximum likelihood; see Venter et al. (2007).

Demarta and McNeil (2005) discuss t-copula and related copulas as skewed t-copula, grouped t-copula, and also t extreme-value copulas.

13.4.3 Other Classes of Copulas

13.4.3.1 Extreme-Value Copulas

Extreme-value copulas, EVC, are discussed in, for example, McNeil et al. (2005), Panjer (2006), and Genest and Favre (2007). EVC is also discussed in Genest, Ghoudi, and Rivest in the discussion by Frees and Valdez (1998). In the bivariate case, it can be seen that the EVC can be represented as

$$C^{\text{EVC}}(u,v) = \exp\left\{\ln(uv)\cdot A\left(\frac{\ln u}{\ln(uv)}\right)\right\},$$

where A is a dependence function, usually called Pickand's dependence function, satisfying

$$A(t) = \int_0^1 \max\left[x(1-t), t(1-x)\right]\, \mathrm{d}H(x),$$

for $t \in [0,1]$ and where H is a distribution function on [0,1]; see Panjer (2006).

Any differentiable, convex function, satisfying the inequality $\max[t, 1-t] \le A(t) \le 1, 0 < t < 1$, can be used to construct a copula. If $\mathrm{A}(t) = 1$, we obtain the independence copula, and taking the other extreme, that is, $\mathrm{A}(t) = \max(t, 1-t)$, having perfect correlation, and hence perfect dependence with $C(u,u) = u$.

The upper tail dependence can be represented in terms of the dependence function as

$$\lambda_{\mathrm{U}} = 2 - 2A\left(\frac{1}{2}\right);$$

see Panjer (2006). The Gumbel copula, see Section 13.4.1, is also an EVC with dependence function $A(t) = \left[t^{\theta} + (1-t)^{\theta}\right]^{1/\theta}, \theta \ge 0$, and hence by setting $t = 1/2$, we obtain the upper tail dependence $\lambda_{\mathrm{U}} = 2 - 2^{1/\theta}$.

An asymmetric version of the bivariate Gumbel copula, also known as Tawn copula, is defined as

$$C^{\text{AGu}}_{\theta,\alpha,\beta}(u,v) = u^{1-\alpha}v^{1-\beta}\exp\left\{-\left[(-\alpha\ln u)^{\theta} + (-\beta\ln v)^{\theta}\right]^{1/\theta}\right\}, \tag{13.16}$$

and is a family in the EVC class; see McNeil et al. (2005) and Panjer (2006).

Another EVC is the Galambos copula, with dependence function $A(t) = 1 - \left[t^{-\theta} + (1-t)^{-\theta}\right]^{-1/\theta}, \theta > 0$. The Galambos copula, which is not an Archimedean copula, has the bivariate form

$$C_\theta^{\mathrm{Gal}}(u,v) = uv \exp\left\{\left[(-\ln u)^{-\theta} + (-\ln v)^{-\theta}\right]^{-1/\theta}\right\},$$

with upper tail dependence of $2^{-1/\theta}$.

The Heavy Right Tail (HRT) copula is discussed in, for example, Venter (2002). The HRT is constructed as a "flipped" Clayton copula, where the flipped copula is defined by the relation $C_F(u,v) = u + v - 1 + C(1-u, 1-v)$. Other extreme value copulas are the Generalised Marshall–Olkin, Hüsler–Reiss, and Joe's BB5; see, for example, Panjer (2006).

As discussed in Panjer (2006), Archimedean copulas and EVC can be combined into a single family of copulas—the *Archimax family*. They can be represented by the form

$$C^{\mathrm{AM}}(u,v) = \varphi^{-1}\left[\{\varphi(u) + \varphi(v)\} \cdot A\left\{\frac{\varphi(u)}{\varphi(u) + \varphi(v)}\right\}\right],$$

where ρ is an Archimedean generator and A is a dependence function.

Joe's BB4 copula is an Archimax copula with the Clayton copula generator $\varphi(u) = u^{-\theta} - 1, \theta \geq 0$, and dependence function $A(t) = 1 - \left\{t^{-\delta} + (1-t)^{-\delta}\right\}^{-1/\delta}, \theta > 0, \delta > 0$. This gives the BB4 copula

$$C_{\theta,\delta}^{\mathrm{BB4}}(u,v) = \left\{u^{-\theta} + v^{-\theta} - 1 - \left[(u^{-\theta} - 1)^{-\delta} + (v^{-\theta} - 1)^{-\delta}\right]^{-1/\delta}\right\}^{1/\theta}.$$

13.4.3.2 The Marshall–Olkin Copulas

The *Marshall–Olkin copulas* are discussed in Embrechts et al. (2003). We will consider the bivariate case. For a multivariate extension, see op. cit. The Marshall–Olkin copula is defined as

$$C_{\alpha_1,\alpha_2}^{\mathrm{MO}}(u,v) = \min\left\{u^{1-\alpha_1}v, uv^{1-\alpha_2}\right\} = \begin{cases} u^{1-\alpha_1}v & \text{if } u^{\alpha_1} \geq v^{\alpha_2} \\ uv^{1-\alpha_2} & \text{if } u^{\alpha_1} \leq v^{\alpha_2} \end{cases}. \tag{13.17}$$

It can be shown that Spearman's rho is

$$\rho_S(\alpha_1,\alpha_2) = \frac{3\alpha_1\alpha_2}{2\alpha_1 + 2\alpha_2 - \alpha_1\alpha_2},$$

and that Kendall's tau is

$$\rho_\tau(\alpha_1,\alpha_2) = \frac{\alpha_1\alpha_2}{\alpha_1 + \alpha_2 - \alpha_1\alpha_2}.$$

The copula has upper tail dependence given by $\lambda_U = \min(\alpha_1, \alpha_2)$.

Marshall–Olkin described a bivariate exponential distribution playing an important role in a 2-dimensional Poisson process. The Marshall–Olkin copula is stemming from such a process; see Nelsen (1999) and Embrechts et al. (2003).

13.4.3.3 Fréchet Copulas

Fréchet copulas are discussed in Genest et al. (2006). They are a mixture of the independence copula C_i and the upper Fréchet–Hoeffding bound, C_u, that is,

$$C_\theta^{\mathrm{FH}}(u,v) = (1-\theta)uv + \theta \min(u,v), \quad \theta \in [0,1]. \tag{13.18}$$

Kendall's tau is $\rho_\tau = \theta(\theta+2)/3$, and hence $\theta = \sqrt{3\rho_\tau + 1} - 1$, and the lower and upper tail dependences are equal: $\lambda_{\mathrm{L}} = \lambda_{\mathrm{U}} = \theta$.

13.4.3.4 Farlie–Gumbel–Morgenstern

The Farlie–Gumbel–Morgenstern copula is defined as

$$C_\theta^{\mathrm{FGM}}(u,v) = uv[1 + \theta(1-u)(1-v)], \quad \vartheta \in [-1,1]. \tag{13.19}$$

Kendall's tau is $\rho_\tau = (2/9)\theta \in [-(2/9),(2/9)]$ and Spearman's rho is $\rho_{\mathrm{S}} = (1/4) + (\theta/36)$. The copula has neither lower nor upper tail dependence.

13.4.4 A Summary

For some of the copulas discussed above, we summarize the corresponding rank correlation (Spearman's rho and Kendall's tau) and the lower and upper tail dependences in Table 13.2.

In our work, we are mainly interested in situations with non-negative dependence and especially situations with the upper tail dependence $\lambda_{\mathrm{U}} > 0$, as this, in some sense, may be seen as the dangerous dependence.

Using Kendall's tau and/or Spearman's rho, we can use their estimates to estimate the copula parameters and hence at least one of the parameters in the upper tail dependence index. If there are other parameters in the upper tail index, we may use other methods to determine them. As an example, using the t-copula for the dependence structure, we may estimate ρ from data using one of the rank correlation measures. The closer the dfs are to 2, the higher the tail dependence will be. Expert knowledge may help us here.

TABLE 13.2 A Brief Summary of Some of the Estimates that Can be Made from Kendall's Tau (r_τ) and Spearman's Rho (r_{S}): the Copula Parameter and the Lower and Upper Tail Dependences

Copula	**Parameter, $\hat{\theta}$**	$\hat{\lambda}_{\mathrm{L}}$	$\hat{\lambda}_{\mathrm{U}}$
Archimedean			
Gumbel	$\frac{1}{1-r_\tau}$	0	$2 - 2^{1-r_\tau}$
Clayton	$\frac{2r_\tau}{1-r_\tau}$	$2^{-(1-r_\tau)/2r_\tau}$	0
Elliptical			
Student's t	—	$2t_{\upsilon+1}\left(-\sqrt{\frac{(\upsilon+1)(1-r)}{1+r}}\right); r = \sin\left(\frac{\pi r_\tau}{2}\right)$ or $r = 2\sin\left(\frac{\pi r_{\mathrm{S}}}{6}\right)$	

13.5 ESTIMATION AND TESTING

Consider a random pair $(X_i, Y_i), i = 1, \ldots, n$, and its ranks $(R_{X(i)}, R_{Y(i)})$ taking values from 1 to n, see Section 13.2. Here n is defined as the size of the sample of n from (X, Y). $R_{X(i)}$ stands for the ranks of X_i among $X_1, \ldots, X_n$, and, in a similar way, $R_{Y(i)}$ stands for the ranks of Y_i among $Y_1, \ldots, Y_n$. As we have assumed continuous random variables, there will be no ties; that is, no two values x_i and x_j will be equal. As we have seen in Section 13.4, for some families of copulas, Spearman's rho and/or Kendall's tau can be expressed as functions of the copula parameters. This can be used to estimate the copula parameters from empirical data. Fitting copulas to data is discussed in, for example, Frees and Valdez (1998), McNeil et al. (2005), and Genest and Favre (2007).

13.5.1 Empirical Copula

The domain of the empirical copula, see, for example, Genest and Favre (2007), is obtained by rescaling the rank axis by $1/(n+1)$. This gives us a set of points in the unit square $[0, 1] \times [0, 1]$.

The empirical copula is defined by

$$C_n(u, v) = \frac{1}{n} \sum_{i=1}^{n} 1\left\{ \frac{R_{X(i)}}{n+1} \leq u, \frac{R_{Y(i)}}{n+1} \leq v \right\}, \tag{13.20}$$

where $1\{\cdot\}$ denotes the indicator function. This empirical copula C_n is a rank-based estimator of C. Its large sample distribution is normal and is centered at C.

13.5.2 Estimating Dependence

The natural way to estimate the rank correlation coefficients discussed in Section 13.2 is to use the rank pairs $(R_{X(i)}, R_{Y(i)})$, or equivalently, the points $(R_{X(i)}/(n+1), R_{Y(i)}/(n+1))$.

Using Pearson's linear correlation coefficient on ranks gives us Spearman's rho:

$$r_S = \frac{\sum_{i=1}^{n} \left(R_{X(i)} - \bar{R}_X\right)\left(R_{Y(i)} - \bar{R}_Y\right)}{\sqrt{\sum_{i=1}^{n} \left(R_{X(i)} - \bar{R}_X\right)^2 \sum_{i=1}^{n} \left(R_{Y(i)} - \bar{R}_Y\right)^2}} \in [0, 1], \tag{13.21}$$

where

$$\bar{R}_X = \bar{R}_Y = \frac{1}{n} \sum_{i=1}^{n} i = \frac{n+1}{2}.$$

Another way to express Spearman's rho is

$$r_S = \frac{12}{n(n+1)(n-1)} \sum_{i=1}^{n} R_{X(i)} R_{Y(i)} - 3\frac{n+1}{n-1}.$$

This is an unbiased estimator of ρ_S defined in Equation 13.4a. Take this latter equation and change C to C_n:

$$12\int_0^1\int_0^1 uv\,\mathrm{d}C_n(u,v) - 3 = \frac{12}{n}\sum_{i=1}^{n}\frac{R_{X(i)}}{n+1}\frac{R_{Y(i)}}{n+1} - 3 = \frac{n-1}{n+1}r_S.$$

As $C_n \to C$ as $n \to \infty$, it can be shown that r_S is an asymptotically unbiased estimator of ρ_S.

Under the null hypothesis $H_0 : C = C_i$ of independence between X and Y, the distribution of r_S is approximately normal $N(0, 1/n+1)$. Hence H_0 may be rejected at level α if

$$\sqrt{n-1}\,|r_S| > k_{1-\alpha/2},$$

where $k_{1-\alpha/2}$ is the $(1-\alpha/2)$-quantile of the standard normal distribution. For $\alpha = 0.05$, we obtain $\sqrt{n-1}|r_S| > 1.96$.

In practice, calculating Spearman's rank correlation is easiest if one first calculates the differences as elucidated above. With n pairs of ranks, we obtain the following way to calculate the rank correlation:

$$r_S = 1 - \frac{6\cdot\sum_{i=1}^{n} d_i^2}{n^3 - n}.$$

As stated in Genest and Favre (2007), r_S is superior to Pearson's linear correlation coefficient $\hat{\rho}$ as $E(r_S) = \pm 1$ if and only if X and Y are *functionally dependent* (where the underlying copula is either C_l or C_u). $E(\hat{\rho}) = \pm 1$ is obtained if and only if X and Y are *linear functions* of one another. Spearman's r_S is always well defined, even for heavy-tailed distributions where the theoretical value of the linear correlation coefficient may not exist.

Let PC_n and PD_n be the number of concordant and discordant pairs, respectively; see Section 13.1.1. The empirical version of Kendall's tau is defined by

$$r_\tau = \frac{PC_n - PD_n}{\binom{n}{2}} = \frac{4}{n(n-1)}PC_n - 1. \tag{13.22}$$

As $(X_i - X_j)(Y_i - Y_j) > 0$ if and only if $(R_{X(i)} - R_{X(j)})(R_{Y(i)} - R_{Y(j)}) > 0$, we see that r_τ is a function of the ranks of the observations. r_τ is a function of C_n, as shown in Genest and Favre (2007).

Let $W_i = (1/n)\#\{j : X_j \le X_i, Y_j \le Y_i\}$, where # stands for the cardinality of a set, and $\bar{W} = (1/n)\sum_{i=1}^{n} W_i$. Then $PC_n = -n + n^2\bar{W}$ and

$$r_\tau = 4\frac{n}{n-1}\bar{W} - \frac{n+3}{n-1}.$$

This is an asymptotically unbiased estimator of ρ_τ given by Equation 13.5b.

Under the null hypothesis $H_0 : C = C_i$ of independence between X and Y, the distribution of r_τ is approximately normal $N(0, 2(2n+5)/9n(n-1))$. Thus, H_0 may be rejected at level α if

$$\sqrt{\frac{9n(n-1)}{2(2n+5)}}\,|r_\tau| > k_{1-\alpha/2}.$$

Other measures and tests of dependence have been proposed in the literature; see, for example, Genest and Favre (2007). Another tool for detecting dependencies is to use scatter plots or other graphical tools. Two such tools, the Chi-plots and K-plots, are discussed and illustrated in Genest and Favre (2007). The K-plot is a graphical representation of a nonparametric estimation ($K_{\theta n}$) of the distribution function $K_C(t)$ of the random variable $C(U, V)$ (see Section 13.4.1) and the corresponding distribution function of a given copula family.

13.5.3 Estimating Copula Families

Assume that we have decided to use a specific parametric family of copulas, C_θ, as a model for dependence between random variables X and Y. The question is how to estimate θ from a given random sample $(X_1, Y_1), \ldots, (X_n, Y_n)$ from (X, Y). The discussion given here is based on Genest and Favre (2007). Statistical inference for bivariate Archimedean copulas is discussed by Genest and Rivest (1993).

As the ranks of the observations $(R_{X(i)}, R_{Y(i)})$ are the best summary of the joint behavior of the pairs of random variables, we will only consider rank-based estimators.

With the parameter θ of the copula family C_θ, we can start the estimation procedure by using the estimates of ρ_S or ρ_τ and letting

$$\theta_S = h_1(\rho_S) \quad \text{or} \quad \theta_\tau = h_1(\rho_\tau),$$

where h_1 and h_2 are two smooth functions representing the relation between the parameters and the population value of Spearman's rho and Kendall's tau, respectively.

The two estimators $\widehat{\theta}_{Sn} = h_1(r_S)$ and $\widehat{\theta}_{\tau n} = h_1(r_\tau)$ may be referred to as the Spearman-based estimator and the Kendall-based estimator of θ, respectively.

For the Gumbel family of copulas we have $\rho_\tau = 1 - (1/\theta)$ and hence we can estimate the parameter by using an estimate r_τ of $\rho_\tau : \hat{\theta} = 1/(1 - r_\tau)$. The upper tail dependence is thus estimated by $\hat{\lambda}_U = 2 - 2^{(1-r_\tau)}$. For the Clayton family of copulas, we obtain, in a similar way, $\hat{\theta} = 2r_\tau/(1 - r_\tau)$ and lower tail dependence $\hat{\lambda}_L = 2^{-(1-r_\tau)/2r_\tau}$.

As is shown in Genest and Favre (2007), an approximately $100(1-\alpha)\%$ confidence interval for θ, by using Kendall's tau, is given by

$$\widehat{\theta}_{\tau n} \pm k_{1-\alpha/2} \cdot \frac{1}{\sqrt{n}} \cdot 4 \cdot S \cdot \left|h_2'(r_\tau)\right|, \tag{13.23}$$

where

$$S = \sqrt{\sum_{i=1}^{n} \left(W_i + W_i^* - 2\bar{W}\right)^2},$$

$$W_i = \frac{1}{n}\# \left\{j : X_j \leq X_i, Y_j \leq Y_i\right\},$$

$$W_i^* = \frac{1}{n}\# \left\{j : X_i \leq X_j, Y_i \leq Y_j\right\},$$

$$\bar{W} = \frac{1}{n}\sum_{i=1}^{n} W_i.$$

It can be found that the estimator r_S of Spearman's rho is asymptotically normal distributed. Hence, an approximately $100(1-\alpha)\%$ confidence interval for θ, by use of Spearman's rho, is given by

$$\widehat{\theta}_{Sn} \pm k_{1-\alpha/2} \cdot \frac{1}{\sqrt{n}} \cdot \sigma_n \cdot \left|h_1'(r_S)\right|, \tag{13.24}$$

where σ_n^2 is a suitable estimator of the asymptotic variance σ^2, which depends on the underlying copula C. For details, see Genest and Favre (2007) and references therein.

A consistent estimator of σ^2 is given by, see Genest and Favre (2007),

$$\sigma_n^2 = 144\left[-9A_n^2 + B_n + 2C_n + 2D_n + 2E_n\right],$$

where

$$A_n = \frac{1}{n}\sum_{i=1}^{n} \frac{R_{X(i)}}{n+1}\frac{R_{Y(i)}}{n+1},$$

$$B_n = \frac{1}{n}\sum_{i=1}^{n} \left(\frac{R_{X(i)}}{n+1}\right)^2 \left(\frac{R_{Y(i)}}{n+1}\right)^2,$$

$$C_n = \frac{1}{n^3}\sum_{i=1}^{n}\sum_{j=1}^{n}\sum_{k=1}^{n} \frac{R_{X(i)}}{n+1}\frac{R_{Y(i)}}{n+1} 1\left\{R_{X(k)} \leq R_{X(i)}, R_{Y(k)} \leq R_{Y(j)}\right\} + \frac{1}{4} - A_n,$$

$$D_n = \frac{1}{n^2}\sum_{i=1}^{n}\sum_{j=1}^{n} \frac{R_{Y(i)}}{n+1}\frac{R_{Y(j)}}{n+1} \max\left[\frac{R_{X(i)}}{n+1}, \frac{R_{X(j)}}{n+1}\right],$$

$$E_n = \frac{1}{n^2}\sum_{i=1}^{n}\sum_{j=1}^{n} \frac{R_{X(i)}}{n+1}\frac{R_{X(j)}}{n+1} \max\left[\frac{R_{Y(i)}}{n+1}, \frac{R_{Y(j)}}{n+1}\right].$$

As an illustration we take the Farlie–Gumbel–Morgenstern family of copulas $C_\theta(u,v) = uv - \theta uv(1-u)(1-v),\ u, v \in [0,1]$ defined for $\theta \in [-1,1]$; see Genest and Favre (2007).

As Kendall's tau can be written as $\rho_\tau = (2/9)\theta$, we have an estimator $\widehat{\theta}_{\tau n} = (9/2)r_\tau$. In a similar way, Spearman's rho can be written as $\rho_S = \theta/3$ and hence an alternative nonparametric estimator would be $\widehat{\theta}_{Sn} = 3r_S$.

The estimation procedures used above are constructed as a nonparametric adaptation of the method of moments. Other estimation methods, such as the maximum pseudolikelihood method, are discussed in Genest and Favre (2007).

Klugman and Patie (1999) fitted bivariate insurance data with copulas and used maximum likelihood estimation. They also fitted a median regression line to Frank's copula.

13.5.4 Goodness-of-Fit Tests

Before deciding to use some specific family of copulas, we have a choice between different families of copulas. Suppose that we have two parametric copula families $C_{1\theta}$ and $C_{2\xi}$ to choose among and that the two copulas $C_{1\theta n}$ and $C_{2\xi n}$ were fitted by some method. One natural question would be "Which of the two models gives us the best fit to the observed data?" Having bivariate data, one natural way to answer the question is to compare the scatter plot of the pairs $(R_{X(i)}/(n+1), R_{Y(i)}/(n+1))$, that is, the support of the empirical copula C_n, with data generated from a large sample from $C_{\theta n}$. Simulation algorithms are available; see, for example, Genest and Favre (2007) and reference therein. A review of Goodness-of-Fit Tests is given in Genest et al. (2009).

Another approach, which is related to the aforementioned K-plot, consists of comparing the empirical distribution K_n of the variables $W_1, \ldots, W_n$, where

$$W_i = \frac{1}{n}\#\left\{j : X_j \le X_i, Y_j \le Y_i\right\} = C_n\left(\frac{R_{X(i)}}{n+1}, \frac{R_{Y(i)}}{n+1}\right),$$

with $K_{\theta n}(t) = P[C_{\theta n}(U,V) \le t]$, the theoretical distribution of $W = C_{\theta n}(U,V)$. K_n and $K_{\theta n}$ can be plotted on the same graph.

One goodness-of-fit test statistic that has been proposed is of the Cramér–von Mises form and was introduced for Archimedean models with generator ρ.

$$S_{n\xi} = n\int_{\xi}^{1} [K_n(t) - K_{\theta n}(t)]^2\, dt,$$

where $\xi \in [0,1]$, and $K(t) = t - \rho(t)/\rho'(t), t \in [0,1]$ is the distribution function of $C(U,V)$; cf. Section 13.4.1. $K(t)$ is called the bivariate probability integral transformation, BIPIT. This is discussed in Genest and Favre (2007).

Genest et al. (2006) have proposed two alternative, and more general, test statistics. The first one is a Cramer–von Mises statistic and the second one a Kolmogorov–Smirnov statistic:

$$S_n = n\int_0^1 [K_n(t) - K_{\theta n}(t)]^2\, k_{\theta n}(t)\, dt,$$

and

$$T_n = \sup_{0\le t\le 1} \left|\sqrt{n}\,[K_n(t) - K_{\theta n}(t)]\right|,$$

where $k_{\theta n}(t) = \mathrm{d}K_{\theta n}(t)/\mathrm{d}t$. These test statistics could be calculated by

$$\widehat{S}_n = \frac{n}{3} + n\sum_{j=1}^{n-1} K_n^2\left(\frac{j}{n}\right)\left\{K_{\theta n}\left(\frac{j+1}{n}\right) - K_{\theta n}\left(\frac{j}{n}\right)\right\} - n\sum_{j=1}^{n-1} K_n\left(\frac{j}{n}\right)$$
$$\times\left\{K_{\theta n}^2\left(\frac{j+1}{n}\right) - K_{\theta n}^2\left(\frac{j}{n}\right)\right\},$$

and

$$\widehat{T}_{\mathrm{n}} = \sqrt{n} \max_{i=0,1;0\le j\le n-1}\left\{\left|K_n\left(\frac{j}{n}\right) - K_{\theta n}\left(\frac{j+1}{n}\right)\right|\right\}.$$

Pros and cons regarding these test statistics are discussed in Genest et al. (2006) and in Genest and Favre (2007).

One other way to construct a goodness-of-fit test is to consider the distance between C_n and $C_{\theta n}$. The limiting distribution of the process $\sqrt{n}\,\{C_n - C_{\theta n}\}$ is complex and it could probably only be implemented through the use of bootstrap methods. Of course, kernel estimation methods are also available. One bootstrap-based goodness-of-fit test based on the Cramér–von Mises statistic,

$$\mathrm{CM}_n = n\sum_{i=1}^{n}\left\{C_n\left(\frac{R_{X(i)}}{n+1}, \frac{R_{Y(i)}}{n+1}\right) - C_{\theta n}\left(\frac{R_{X(i)}}{n+1}, \frac{R_{Y(i)}}{n+1}\right)\right\}^2$$
$$= n\sum_{i=1}^{n}\left\{W_i - C_{\theta n}\left(\frac{R_{X(i)}}{n+1}, \frac{R_{Y(i)}}{n+1}\right)\right\}^2,$$

is used in Genest and Favre (2007).

13.5.5 Estimating Regression Functions

Copulas can help us to understand a full joint multivariate distribution. Frees and Valdez (1998) discussed how copulas could be used to estimate regression functions. Even if it is possible to evaluate general regression functions in terms of specific copula families, the copula theory is more suited for the alternative *quantile regression*.

Define the pth quantile, or the VaR, to be the solution of the equation $p = F_k(x_p|x_1, \ldots, x_{d-1})$. In the case of $d = 2$, we have

$$p = F_2(x_p|X_1 = x_1) = c_1[F_1(x_1), F_2(x_2)] = c_1[u_1, u_2],$$

where $c_1(\cdot,\cdot)$ is the first partial derivative with respect to F_1.

For the Gumbel copula, the first partial derivative is

$$c_1^{\mathrm{Gu}}(u,v) = \frac{\partial C(u,v)}{\partial u} = \left[\frac{\ln u}{\ln C(u,v)}\right]^{\theta-1} \frac{C(u,v)}{u}, \tag{13.25}$$

and for the Frank's copula, we have

$$c_1^{\mathrm{Fr}}(u,v) = \frac{\mathrm{e}^{\theta u}(\mathrm{e}^{\theta v}-1)}{\mathrm{e}^{\theta u}-1+(\mathrm{e}^{\theta u}-1)(\mathrm{e}^{\theta v}-1)};$$

see Frees and Valdez (1998).

To get the pth percentile of the Gumbel copula we use Equation 13.25 and $x_{\mathrm{p}} = F_2^{-1}(v_p)$, where v_p is the solution to

$$\left[\frac{\ln F_1(x_1)}{\ln C_\theta^{\mathrm{Gu}}(F_1(x_1), v_{\mathrm{p}})}\right]^{\theta-1} \frac{C_\theta^{\mathrm{Gu}}(F_1(x_1), v_{\mathrm{p}})}{F_1(x_1)} = p.$$

An illustration of this is given in Frees and Valdez (1998).

CHAPTER 14

Risk Measures

FINANCIAL RISKS can be measured as to their strength or the financial consequences of a specific economic activity. In solvency considerations, we would put a capital requirement on the risks: the higher the risk is, the higher the capital requirement should be. The interpretation of a risk measure's outcome depends very much on the context where it is used. As shown in Tsanakas (2007), there have been three main areas of application of risk measures.

- As representations of risk aversion in asset pricing models, with as a leading paradigm the use of variance as a risk measure in Markowitz portfolio theory.
- As tools for the calculation of the insurance price corresponding to a risk. Under this interpretation, risk measures are called premium calculation principles in the classical actuarial literature.
- As quantifiers of the economic capital that the holder of a particular portfolio or risks should safely invest in.

We are mainly interested in the third interpretation of risk measures, as economic capital is an insurer's internal capital requirement and as such is identical in nature to the regulator's solvency capital requirement.

If the risk is measured by $C(X)$, where X is a random variable quantifying the economic variate, then the capital requirement could be a factor times the measured risk $C(X)$, that is, $aC(X)$, $a > 0$. For an introduction to risk measures, see, for example, Albrecht (2004a). Denuit et al. (2005) defines a risk measure as (Definition 2.2.1 in op. cit.):

A risk measure is a functional C that maps a risk X to a nonnegative real number $C(X)$, possibly infinite, representing the extra cash that has to be added to X to make it acceptable.

This definition is equal to the implicit definition above if $a = 1$. Risk measures are discussed in, for example, Panjer (2006), McNeil et al. (2005), and Brehm et al. (2007). A review of the development of measuring financial risks is given in Embrechts et al. (2008). As in Albrecht (2004a), we will mainly discuss risks capturing financial events that could have an affect on the solvency requirement.

Albrecht (2004a) distinguish between two categories of risk measures.

1. Risk as the magnitude of deviation from a target, $T(X)$: $C_1(X) = T(X) + aR(X)$, where $aR(X)$ is a risk loading in terms of premium principles or a capital requirement in terms of solvency
2. Risk as a capital (or premium) requirement: $C_2(X) = aR(X)$

As seen above, there is an intuitive correspondence between the two categories. As an example and letting the target be the expected value $E(X)$, the second category could be related to the first by

$$C_2(X) = C_1(X) - E(X) = aR(X). \tag{14.1}$$

The main focus here is on capital requirement for solvency purposes, and as such we will mainly consider measures of the second category, that is, by Equation 14.1. This category, as expressed by Equation 14.1, defines the economic capital or the solvency capital requirement. The $E(X)$ part covers the expected losses, and technical provisions, and $C_1(x)$ can thus be interpreted as a cushion against unexpected losses.

To be formal, following Embrechts et al. (2008), the risk is represented by a random variable X or $X(t)$, defined on a filtered probability space $(\Omega, F, (F_t)_{t\in[0,T]}, \mathbb{P})$. The filtration $(F_t)_t$ is assumed to be (a) right-continuous and (b) F_0 contains all null sets, such as: if $B \subset A \in F_0$ with $\mathbb{P}[A] = 0$, then $B \in F_0$. Risks are modeled as nonnegative random variables and measuring risk is the same as establishing a relation C between the set of random variables and real numbers $\mathbb{R}$.

How should a risk measure be constructed? There is, of course, no single answer to this question. One way is to set up some desirable properties for a risk measure and then to test commonly used measures to see if they fulfill the desired properties. Another way is to define a measure for the purpose it should be used for.

In Section 14.1, we will look at different properties for risk measures that have been proposed in the literature, and in Section 14.2, we will look closer at different families of risk measures that have been proposed. Two of the most commonly used risk measures, the VaR, and the Tail VaR (TVaR) will be discussed in Section 14.3. It is important to distinguish between measures of dispersion and measures of concentration. The latter type, which is important in quantifying capital requirements, will be discussed in Section 14.4.

For the reading, we start with a definition of the quantile concept.

Quantiles: The $(1-\alpha)$-quantile (or percentile) of a random variable X is any value $x_{1-\alpha}$ such that $F(x_{1-\alpha}^-) \leqslant 1-\alpha \leqslant F(x_{1-\alpha})$. In our discussions we will usually have small α values, for example, 0.005 or 0.001.

14.1 PROPERTIES OF RISK MEASURES

In setting up desirable properties for a risk measure, we mainly follow Albrecht (2004a). A desirable risk measure may or may not satisfy these properties. The first property is from Denuit et al. (2006) and implies that the risk measure should be based on a distribution function. The property, called law invariance or objectivity, ensures that F_X contains all information to measure the riskiness (P = property).

P1—objectivity: The risk measure C does not depend on the risk itself but only on its underlying distribution, that is, $C(X) = C(F_X)$, where F_X is the distribution function of X.

Property one can be reformulated as $X =_d Y \Rightarrow C(X) = C(Y)$, where $=_d$ denotes equality in distribution.

Pedersen and Satchell (1998) saw risk as a deviation from a location parameter. They introduced the following four desirable properties.

P2—nonnegativity: $C(X) > 0$

P3—positive homogeneity: $C(aX) = aC(X), a \geqslant 0$

P4—subadditivity: $C(X + Y) \leqslant C(X) + C(Y)$

P5—shift invariance: $C(X + b) \leqslant C(X), \forall b \in \mathbb{R}$

The second property is understood from the introduction and the third property means that the risk of a certain multiple of the financial position (financial risk) equals the multiple of the risk of the financial position. It is important to note that positive homogeneity does not imply "currency independence"; see, for example, Denuit et al. (2006), and also Section 12.3.

The fourth property means that the risk of combined financial positions is less than the sum of the risks of the separate positions. In an investment context, this means that we allow for the effect of diversification, and in an insurance context, that we allow for the effect of pooling risks. The subadditivity property can be interpreted as "a merger does not create extra risk" (Denuit et al., 2006). Both P3 (positive homogeneity) and P4 (subadditivity) have been a subject for discussion. It has been demonstrated that these properties make risk measures insensitive to liquidity risk. Subadditivity also neglects the notion of "residual risk."

The diversification effect is usually defined as

$$\text{Div} = \sum_{i=1}^{n} C(X_i) - C\left[\sum_{i=1}^{n} X_i\right]. \tag{14.2}$$

Diversification is always positive for subadditive risk measures.

The fifth property means that the risk measure is location-free, that is, it is invariant to the addition of a constant to the financial position.

The two properties P3 and P4 together imply that a constant random variable will have zero risk. P3 and P5 together imply that the risk measure is convex. Aggregating them, we get the convex property,

P6—convexity: $C[\lambda X + (1 - \lambda)Y] \leqslant \lambda C(X) + (1 - \lambda)C(Y), \lambda \in [0, 1]$, whatever be the dependency structure of X and Y; see, for example, Denuit et al. (2006).

As risks are understood to be location-free from P2–P5, these properties are well suited for the first category of risk measures.

One of the most important and influential systems of properties for financial risk measures is due to Artzner et al. (1999)—*coherent risk measures.*

P7—translation invariant: $C(X + b) = C(X) + b, \ \forall b \in \mathbb{R}$

P8—monotonicity: $X \leqslant Y \Rightarrow C(X) \leqslant C(Y)$.

A risk measure satisfying P3, P4, P7, and P8 is said to be coherent.

Property seven (P7) means that adding a fixed amount to the initial financial position should increase the risk by the same amount. The property implies that by adding $-C(X)$ to the initial financial position, we obtain a neutral position as $C[X - C(X)] = 0$; see Denuit et al. (2006). This could be seen as a desirable property for many risk measures, but, as stated in Denuit et al. (2006), this is not the case for a risk measure used to calculate solvency capital requirement (C_{SCR}) or economic capital (C_{EC}). These risk measures constitute the requirements, or amounts, of money for safety in addition to the provisions. A risk measure of this kind should be of the form $C_{\text{SCR}} = C'[(X - C(X))_+]$, for some risk measures C' and C. Here we have used the definition of the shortfall of a portfolio with loss X: $\max(X - C(X), 0) \equiv [X - C(X)]_+$.

Changing C for a general threshold d, the shortfall $(X - d)_+$ also defines a left-censored and shifted random variable; see Panjer (2006).

A requirement $C(b) = b$, for all real-valued b, could be defined in insurance pricing, but the definition of a risk measure $C(b)$ could be equal to the utility u, that is, $C(b) = u(b)$, where u is a utility function. This property of coherence is thus closer to premium calculation principles; see Denuit et al. (2006).

The eighth property (P8) means that if $P(X \leqslant Y) = 1$, then $C(X) \leqslant C(Y)$, meaning that the capital required as a cushion against loss X is always smaller than the corresponding amount for Y, when Y exceeds X (*weak monotonicity*). The stronger concept of *strong monotonicity*, using stochastic dominance, is discussed in, for example, Denuit et al. (2006).

Coherent risk measures can be characterized as the class of convex risk measures satisfying positive homogeneity (P3). The latter class, which is an extension of the class of coherent risk measures, is also called the class of weakly coherent risk measures.

A coherent risk measure is well suited for the second category of risk measures. Note that the definition of coherent risk measures is made with respect to the four properties P3, P4, P7, and P8, but there is no set of properties that are universally accepted. This means that a measure not fulfilling these properties may be as adequate as any "coherent measure."

Combining P3 (positive homogeneity) and P7 (translation invariant) guarantees that C is a linear risk measure, that is, $C(aX + b) = aC(X) + b, a > 0, b \in \mathbb{R}$.

Rockafellar et al. (2003) introduced a system of properties for risks of the second category from properties P3, P4, P7, and P9; see also Rockafellar et al. (2006a, 2006b):

P9—expectation boundedness: $C(X) > E(-X)$ for all nonconstant X, and $C(X) = E(-X)$ for all constant X.

Risk measures following P3, P4, P7, and P9 are called expectation-bounded. Adding P8, we get an expectation-bounded coherent risk measure.

The following two properties are based on P4—subadditivity, but with equality, that is, additivity. In this case, the dependence structure is specified, as we see in P10 and P11.

COHERENT RISK MEASURES AND GENERALIZED SCENARIOS

As discussed by Artzner et al. (1999), see also Dowd and Blake (2006), the outcomes of scenario analyses, or stress tests, can be interpreted as coherent risk measures. This surprising implication provides a risk-theoretical justification for using stress tests!

In Artzner et al. (1999), see also Artzner (1999) and Meyers (2000), the representation of coherent risk measures give as an by-product the concept of generalized scenarios. A generalized scenario is a probability measure on the state of nature. A coherent risk measure $C[X]$ is defined as

$$C[X] = \sup\{E_P[X] | P \in \mathbb{P}\}$$

Where P in E_P indicates that the expectation is computed under the probability distribution P and $\mathbb{P}$ is a nonempty set of probability measures. The elements of $\mathbb{P}$ are called generalized scenarios.

We can think of a simple example having five possible outcomes:

Scenarios	P_1	Outcomes, X:	P_2	Outcomes, X:
asset: up 3%	0.15	0	0.25	8
asset: up 1%	0.15	2	0.40	6
asset: no change	0.05	2	0.05	2
asset: down 1%	0.40	6	0.15	2
asset: down 3%	0.25	7	0.15	0
"Worst case":		7		8
	$E_{P1} = 4.55$		$E_{P2} = 5.60$	

A coherent risk measure is defined through an "expected value after scenarios" and the supremum operation guarantees that if several scenarios are given, risk is measured by the worst case obtained. In the above example the supremum, that is, the maximum expected value taken over $\mathbb{P} = \{P_1, P_2\}$ is 5.60.

Classical scenario analysis usually refers to the assignment of a set of scenarios as a set of possible outcomes, chosen either from historical data or from the regulator's subjective view. This is also known as stress testing. In this classical approach we can think of a set A of scenarios and taking the worst case. The five scenarios above can be seen as a set and the worst case would be the worst outcome. In the generalized scenario approach we can think of a set $\mathbb{P}$ of probability measures and think of each member $P \in \mathbb{P}$ as a generalized scenario. If the set of probability measures are chosen as a set of pointmass scenarios, the concept of generalized scenarios is reduced to the concept of classical scenarios. These concepts are also discussed and given mathematical interpretations in McNeil et al. (2005).

A tenth property, comonotonic additivity, was introduced by Wang et al. (1997).

P10—comonotonic additivity: $C(X+Y) = C(X) + C(Y)$ for all comonotonic random variables X and Y.

Adding comonotonic risks never decreases the riskiness of the situation, see, for example, Denuit et al. (2005, 2006): "Comonotonic risks are bets on the same event and neither of them is a hedge against the other" (op. cit.).

An eleventh property, additivity for independent risks, ensures that a policyholder has no interest in splitting a risk and asking for coverage from several insurers. This makes this property a reasonable requirement for a premium calculation principle (Denuit et al., 2006).

P11—independent additivity: $C(X+Y) = C(X) + C(Y)$ for all independent risks X and Y.

Dhaene et al. (2008a) introduced a complement to the subadditivity property (P4) by a regulator's condition. The reason for this is that the subadditivity property is usually imposed on measures on solvency capital requirements, which could lead to situations where the shortfall risk increases by a merger. A shortfall risk, or residual risk, is a downside risk measured relative to a target variable.

In op. cit. it is shown that for a specific confidence level, the VaR (see the next chapter) satisfies the regulator's condition and is the most efficient capital requirement in the sense that it minimizes some reasonable cost function. It is also shown that for risk measures fulfilling the subadditivity property (within a class of concave distortion risk measures), also for a specific confidence level, the TVaR (see the next chapter) is the optimal solvency capital requirement satisfying the regulator's condition.

P12—regulator's condition: For any random couple (X,Y) and a given number $0 < \varepsilon < 1$, the solvency capital requirement C_{SCR} has to satisfy the condition

$$\begin{aligned} &E[(X+Y-C_{\mathrm{SCR}}[X,Y])_+] + \varepsilon\cdot C_{\mathrm{SCR}}(X+Y) \\ &\quad \leqslant E[(X-C_{\mathrm{SCR}}[X])_+] + E[(Y-C_{\mathrm{SCR}}[Y])_+] + \varepsilon\cdot C_{\mathrm{SCR}}(X) + \varepsilon\cdot C_{\mathrm{SCR}}(Y). \end{aligned} \tag{14.3}$$

The condition in P12 can be seen as a compromise between the requirement of "subadditivity" and the requirement of "not too subadditive." The number ε can be equal to the required CoC rate, but it could also be smaller.

Let us explain the idea behind this property. We consider a portfolio with future loss X. As discussed in Dhaene et al. (2008a), the regulator wants the solvency capital requirement that is related to X to be "sufficiently" large to ensure that the shortfall is "sufficiently" small. To get this, we let the regulator introduce a risk measure for the shortfall risk $C^{\mathrm{SR}}[X]$:

$$C^{\mathrm{SR}}[(X-C_{\mathrm{SCR}}[X])_+]. \tag{14.4}$$

Hence two different "risk measures" are involved. The $C_{\mathrm{SCR}}[X]$ determines the SCR and $C^{\mathrm{SR}}[X]$ determines the shortfall risk. It is natural to assume that C_{SCR} satisfies

$$C_{\mathrm{SCR},1}[X] \leqslant C_{\mathrm{SCR},2}[X] \;\Rightarrow\; C^{\mathrm{SR}}[(X-C_{\mathrm{SCR},1}[X])_+] \geqslant C^{\mathrm{SR}}[(X-C_{\mathrm{SCR},2}[X])_+]. \tag{14.5}$$

Equation 14.5 says that an increase of the capital requirements (left-hand side) implies a reduction of the shortfall risk (right-hand side). Hence, the larger the capital requirement $C_{SCR}[X]$ is the smaller will $C^{SR}[(X - C_{SCR}[X])_+]$ be. From the regulator's point of view, the shortfall risk, as expressed by Equation 14.4, should be "sufficiently" small.

The insurer has to hold the required capital $C_{SCR}[X]$, which involves a capital cost, say $\varepsilon C_{SCR}[X]$, where ε, $0 < \varepsilon < 1$, is the excess return on capital or CoC rate, usually defined by the regulator, *cf.* Sections 8.4 and 8.5 on risk margins. To take this into account, Dhaene et al. (2008a) consider the cost function

$$c[X, C_{SCR}[X]] = C^{SR}[(X - C_{SCR}[X])_+] + \varepsilon \cdot C_{SCR}[X], \tag{14.6}$$

which takes into account the capital requirement, the shortfall risk, and the capital cost. The optimal capital requirement $C_{SCR}[X]$ can be determined as the smallest amount d that minimizes the cost function $c[X, d]$.

What we now have is a minimizing problem, which is a compromise between minimizing the cost of insolvency (high capital) and minimizing the CoC (low capital).

Dhaene et al. (2008a) use the expectation to measure the shortfall risk, that is, $C^{SR}[X] = E[X]$. Condition (14.5) is satisfied by this choice of shortfall risk measure. It also satisfies properties P1, P3, P4, P6, P7, P8, and P10. It is shown in op. cit. (Theorem 1) that the smallest element in the set of minimizers to the cost function $c[X, d]$ as defined by Equation 14.6 with $C_{SCR}[X] = d$, $0 < \varepsilon < 1$, is given by

$$C_{SCR}[X] = \mathrm{VaR}_{1-\varepsilon}[X]. \tag{14.7}$$

The minimization of the cost function $c[X, d]$ above is taken over all possible values d.

Dhaene et al. (2008a) also consider the case where the cost function is taken over a restricted set of possible values for d. Theorem 2 in op. cit. gives the results for minimizing $\min_{d \in A} c(X, d)$, where the cost function is defined as above and the set A is defined as

$$A = \{C_g[X] | g \text{ is a concave distribution function and } C_g[X] \geqslant \mathrm{VaR}_{1-\varepsilon}[X]\}.$$

Then

$$C_{SCR}[X] = \mathrm{TVaR}_{1-\varepsilon}[X], \tag{14.8}$$

at level $1 - \varepsilon$, $0 < \varepsilon < 1$.

The theorem states that if you want to set the SCR such that it belongs to the class of concave distortion risk measures (which fulfill P4—subadditivity) that is the smallest minimizer of Equation 14.6 with $C_{SCR}[X] = d$, and such that it is not smaller than the smallest minimizer of the unconstrained problem (i.e., $\geqslant \mathrm{VaR}[X]$), then the optimal SCR is given by Equation 14.8.

In property P12 above, the cost function given by Equation 14.6, with $C^{SR}[X] = E[X]$, is also considering the subadditivity requirements; see Dhaene et al. (2008a). By Theorem 6 in op. cit., it is shown that the SCR $C_{SCR}[X] = \text{VaR}_{1-\varepsilon}[X]$ fulfills the regulator's condition given by Equation 14.3. Any subadditivity SCR $C_{SCR}[X] \geqslant \text{VaR}_{1-\varepsilon}[X]$ fulfills P12.

Property P12 says that the expected shortfall of the merger between two portfolios will be less than the sum of the expected shortfalls of each portfolio, that is, "a merger decreases the shortfall"; (Dhaene et al., 2008a). To sum up, minimizing the cost function gives the VaR as a capital requirement measure, but minimizing the total cost gives us the TVaR.

In a study by Desmedt and Walhin (2008) it is shown that if TVaR is used as a risk measure, merging risks could be of interest both of policyholders (regulators defending them) and shareholders. It is also shown that TVaR is not too subadditive under a wide range of dependence structures (using copulas).

The difference between *decision making under risk* and *decision making under uncertainty* is that in the former case the probabilities of the possible events are given, while they are not given in the latter case. In the former case, the probability measure may be unknown but it is nevertheless fixed and unique. As stated by Denuit et al. (2006), a coherent risk measure is not a measure of risk but a measure of uncertainty.

The standard theory for decision making under risk is the expected utility theory. In Section 14.2, we will briefly look at measures derived from expected utility theory. Risk measurement and utility theory are discussed in, for example, Albrecht (2004a) and Denuit et al. (2005, 2006).

14.2 FAMILIES OF RISK MEASURES

The classification of risk measures is not unique. One member of a family of risk measures may be a member of several other families. This classification, which is mainly based on Albrecht (2004a) and Brehm et al. (2007), is just aimed to assist in finding suitable risk measures. Risk measures are discussed in several books and papers. One example of papers discussing risk measures, including inequality measures and measures from operational research, management science and psychology, is Pedersen and Satchell (1998); see also, for example, Dowd and Blake (2006). Income inequality measures are also discussed in Piesch (1975) and Nygård and Sandström (1981). Two specific risk measures are further discussed in Section 14.3: the VaR and TVaR.

14.2.1 Stone's Three-Parameter Family of Risk Measures

Stone's family of risk measures has three parameters: z, k, and c, and is defined as

$$C^S(X) = \left[\int_{-\infty}^{z} (|x - c|)^k f(x)\, \mathrm{d}x \right]^{1/k}. \tag{14.9}$$

This family contains, for example, the standard deviation, the semistandard deviation, the mean-absolute deviation (MAD) and the more general measure $E[|X - E(X)|^k]^{1/k}$; see, for example, Albrecht (2004a) and below.

14.2.2 Pedersen and Satchell's Five-Parameter Family of Risk Measures

Pedersen and Satchell's family of risk measures, see Pedersen and Satchell (1998), has five parameters: z, c, a, b, and a weight function w. It is defined as

$$C^{\mathrm{PS}}(X) = \left[\int_{-\infty}^{z} (|x - c|)^{a} w[F(x)] f(x)\,\mathrm{d}x \right]^{b}. \tag{14.10}$$

Letting $a = k$, $b = 1/k$, and $w(\cdot) = 1$ gives us Stone's family. The family also includes the variance, the semivariance, the lower partial moments (LPM), as well as others; see Albrecht (2004a) and below.

14.2.3 Expected Utility Theory–Based Risk Measures

The expected utility theory is the standard theory for decision under risk. Let u be the utility function, specific to each decision maker. As the preference function is $\Psi(X) = E(u(X))$, it implies that there is no separate measurement of risk or value in the utility theory. They are considered simultaneously. However, it is possible to derive an explicit risk measure from a specific utility function. This is given by the *Jia and Dyer's standard measure of risk*, defined as

$$C^{\mathrm{JD}}(X) = -E[u(X - E(X))],$$

that is, risk is measured as the negative expected utility of the transformed variable $X - E(X)$, making it location-free. Different risk measures are given in the following table.

Utility Function, $u(x) =$	**Risk Measure, $C(x)$**	**Comments**
$ax - bx^2$	$\mathrm{Var}(X) = E[(X - E(X))^2]$	The variance
$ax - bx^2 + cx^3$	$\mathrm{Var}(X) - cE[(X - E(X))^3]$	Corrected variance by the magnitude of the third central moment
$ax - \lvert x\rvert$	$\mathrm{MAD}(X) = E[\lvert X - E(X)\rvert]$	The mean-absolute deviation

Risk aversion, which is an important topic in the economics of insurance, is an attitude toward risks and uncertainty, that is, to buy insurance. A person is said to be risk averse if he always prefer a certain return $E[X]$ to a risky prospect X, irrespective of the distribution of X. Hence, from Jensen's inequality, his utility function will satisfy $E[u(X)] \le u(E[X])$, for all X, if u is concave. We usually assume the utility function $u(x)$ to be strictly increasing and concave, that is, $u'(x) > 0$ and $u''(x) \leqslant 0$. A person with a concave utility function is risk averse.

Risk aversion and expected utility theory are two theoretical approaches that are usually used for premium principles; see, for example, Denuit et al. (2005, 2006). Different premium principles, such as the Esscher premium principle, could be used as risk measures. An elementary premium principle property is that $E(S) < \Pi(S) < \max(S)$, where S is a claim variable and Π is the premium principle. Several of the properties of the risk measures given above have been imposed as desirable properties from premium principles, such as P3—positive homogeneity, P7—translation invariance, and P4—subadditivity for independent risks.

A closed system of properties for premiums was introduced by Wang et al. (1997). They required P8—monotonicity, certain continuity properties, and P10—comonotonic additivity.

Expected utility theory is discussed in, for example, Denuit et al. (2005, 2006). Subject utility theory for decision under uncertainty differs from the expected utility theory only by its interpretation of the underlying probability distribution.

14.2.4 Distorted Risk Measures

From the discussion of premium principles, $\Pi(X)$, Wang et al. (1997) proved, under certain additional conditions, see Section 14.2.3, that the following equality holds:

$$\Pi(X) = \int_0^\infty g[1 - F(x)]\,\mathrm{d}x, \tag{14.11}$$

where F is the distribution function of the nonnegative X, and g is a distortion function that is nondecreasing with $g(0) = 0$ and $g(1) = 1$. If g is concave, then the resulting premium principle will be a coherent risk measure. This gives us an explicit method to construct coherent risk measures. Equation 14.11 is usually called Wang's risk measure ($C_g^W[X]$). This risk measure can be written as a mixture of VaRs; see Denuit et al. (2005). Mixtures of VaRs are called spectral measures of risk.

Consider the fact that, with $\bar{F}_X(x) = 1 - F_X(x) = P(X > x)$, the expectation of X can be written as

$$E[X] = -\int_{-\infty}^0 (1 - \bar{F}_X(x))\,\mathrm{d}x + \int_0^\infty \bar{F}_X(x)\,\mathrm{d}x.$$

A "distorted expectation" is defined for a nondecreasing function g: $[0, 1] \to [0, 1]$ with $g(0) = 0$ and $g(1) = 1$ as

$$E_g[X] = -\int_{-\infty}^0 (1 - g[\bar{F}_X(x)])\,\mathrm{d}x + \int_0^\infty g[\bar{F}_X(x)]\,\mathrm{d}x. \tag{14.12}$$

Note that the right-hand integral in Equation 14.12 equals that in Equation 14.11. The distortion function g distorts the probabilities $\bar{F}_X(x)$ before the calculation of the "expected value" is done.

In general, E_g is not an expectation of a transformed random variable, but with additional assumptions on g it will be. Hence, there exist distortion functions g such that Y is preferred to X if and only if $E_g[X] \leqslant E_g[Y]$.

The distortion expectation theory can be seen as a dual theory of choice under risk in the sense that it uses the concept of distortion functions instead of utility functions. As seen by Denuit et al. (2005), the distortion expectation theory corresponds to Yaari's dual theory. A comparison between expected utility theory and expected distortion theory is given in op. cit.

Let $\bar{g}(x) = 1 - g(1-x), x \in [0,1]$ be the *dual distortion function* to the distortion function g. The dual distortion function is also a distortion function such that $\bar{\bar{g}} \equiv g$ and g convex $\Leftrightarrow$ $\bar{g}$ concave.

As an extension of Equation 14.12, we have

$$C_g(X) = -\int_{-\infty}^{0} [1 - \bar{g}[\bar{F}_X(x)]]\,\mathrm{d}x + \int_{0}^{\infty} \bar{g}[\bar{F}_X(x)]\,\mathrm{d}x. \tag{14.13}$$

VaR (see Denuit et al., 2006):

For any α in (0,1) the (1 − α)-quantile risk measure for a random variable X is defined by $F_X^{-1}(1-\alpha) = \inf\{x \in \mathbb{R} | F_X(x) \geqslant 1-\alpha\}, \alpha \in [0,1]$. Here $F_X^{-1}(1-\alpha)$ corresponds to the distorted function

$$\bar{g}(u) = \begin{cases} 1 & \text{if } u > 1-\alpha \\ 0 & \text{otherwise} \end{cases} \quad \text{for } 0 \leqslant u \leqslant 1.$$

TVaR (see Denuit et al., 2006):

The TVaR, at level $1-\alpha$, is defined by $\mathrm{TVaR}_{1-\alpha}(X) = (1/\alpha)\int_{1-\alpha}^{1} F_X^{-1}(1-\alpha)\mathrm{d}u$, $\alpha \in [0,1]$, corresponding to the distortion function

$$\bar{g}(x) = \min\left(\frac{x}{\alpha}, 1\right), \quad 0 \leqslant x \leqslant 1.$$

Distortion risk measures with concave $\bar{g}$ are subadditive. They follow the properties P1, P3, P4, P7, P8, and P10.

A person is said to be risk-averse, under distortion expectation theory, if his distortion function is convex, which is seen from the fact that a convex distortion function satisfies $g(p) \leqslant p$ for all $p \Rightarrow g[\bar{F}_X(x)] \leqslant \bar{F}_X(x), x \in \mathbb{R}$. As shown in Denuit et al. (2005), we have for a convex distortion function g that $E_g[X] \leqslant E[X] = E_g[E[X]]$.

This means that a risk-averse person will always prefer certain fortune to a random fortune with the same expected value. A person is risk-neutral if $g(p) = p$. In general, distorted risk measures follow properties P1, P3, P7, P8, and P10. If the distorted function is continuous, then the distorted risk measure is coherent; see Dowd and Blake (2006). The VaR's distortion function, see above, is not continuous and hence not, in general, coherent.

A distortion function leading to risk measures giving different weights to upside and downside risks is the Wang transformation; see Wang (2000).

The Wang transformation is usually called the normal transform risk measure. For any $0 < \alpha < 1$, the Wang transform is defined by the distortion function

$$g_\alpha(\beta) = \Phi[\Phi^{-1}(\beta) + \Phi^{-1}(\alpha)], \qquad 0 < \beta < 1, \quad 0 < \alpha < 1, \tag{14.14}$$

where Φ is the standard normal distribution function. If X is normally distributed $N(\mu, \sigma^2)$, then the Wang transform gives a new normal distribution with $\mu^* = \mu + \lambda\sigma$ and $\sigma^* = \sigma$, where $\lambda = \Phi^{-1}(1-\alpha)$. Hence, for normal distributions, the Wang transform, $\mathrm{WT}(1-\alpha)$, is identical to $\mathrm{VaR}_{1-\alpha}[X]$; see also Wang (2002b, 2002c). For lognormal distributions, the Wang transform is $\mathrm{WT}(1-\alpha) = e^{\mu+\lambda\sigma+\sigma^2/2}$ with $\lambda = \Phi^{-1}(1-\alpha)$. A variation of the Wang transformation with volatility multiplier is discussed in Wang (2002a). For a review of distortion risk measures, see, for example, Dhaene et al. (2006).

Coherent distortion risk measures, which are tail-preserving, and economic capital are discussed in Hürlimann (2004); see also Bellini and Caperdoni (2007).

One subfamily of distorted risk measures is those where the probabilities are shifted toward the unfavorable outcomes and the risk measures are computed with the transformed probabilities. In Brehm et al. (2007), these risk measures are called *probability transforms*. The main example of a transformed measure is the expected loss under the transformed probabilities. The usual asset pricing formulas, such as the capital asset pricing model, CAPM, and the Black–Scholes options pricing formula, can be expressed as transformed means.

Another subfamily of distorted risk measures is *distorted-exponential* risk measures defined as $C^{\mathrm{DE}}(X) = (1/\alpha)\ln[C_g(e^{\alpha X})]$, where C_g is the distorted risk measure. The measure combines the properties of the exponential premium principle (see Section 14.2.5.1) and those of distorted risk measures; see Tsanakas (2007). The measure follows properties P1, P6, P7, and P8.

Transformed means are promising types of risk measures as they provide the market value of the risk measure. The mean under the Wang transform has been found to approximate market prices, both in the case of bonds and CAT-bonds. The minimum martingale transform and the minimum entropy martingale transform are two transforms for pricing in incomplete markets.

The tail-based measures discussed in the next chapter can be used with transformed probabilities. Usually, a W as for "weighted" is added to the abbreviations of these risk measures, such as WVaR, WTVaR, and so on.

14.2.5 Other Types of Risk Measure Classifications

In this chapter, we will look at different risk measures from a different angle than in the previous chapters. We will classify them according to whether the risk measures are two- or one-sided. The two-sided measures measure the total variability, but the one-sided measures measure either only the favorable deviations from expectations (upside results) or only the adverse deviation from expectations (downside results); see Albrecht (2004a) and Witcraft (2004). For solvency purposes, we will be more interested in the one-sided risk measures, as we are mainly interested in category two measures; see the introduction of this chapter.

If the expected value is the target, we have a two-sided risk measure that measures the magnitude of the distance from the realizations of X to $E(X)$, such as $C_1(X) = E(X) \pm C_2(X)$. If we have quadratic deviations, measuring the volatility, we get the variance or, by taking the square root, we get the risk measure standard deviation $\sigma = \mathrm{Var}(X)^{1/2}$. These two volatility measures have been the main risk measures in economics and finance since Markowitz's pioneering work on portfolio theory in the 1950s. As it is the downside risk that we are mainly interested in, we see that this view is contradicted by using a two-sided volatility measure. The variance does not account for fat tails in the underlying distribution and hence does not account for the tail risk. There have been proposals to extend the volatility measures to higher central moments, including skewness and kurtosis. One such measure is the generalized mean-absolute deviation (GMAD)

$$\mathrm{GMAD}_{k,l}(X) = E\left[|X - E(X)|^k\right]^{1/l}, \quad k, l > 0. \tag{14.15}$$

Equation 14.15 could be further generalized as $E\left[f(X - E(X))^k\right]^{1/l}$, a family of risk measures studied by Rockafellar et al. (2003). The MAD is obtained by letting $k = l = 1$.

These risk measures could also be used for one-sided risk measures and, usually, vice versa. In the following, we will focus on one-sided risk measures, both of the type $C_1(X) = E(X) + C_2(X)$ and the type $C_2(X)$ according to the definitions made in the introduction of this chapter.

14.2.5.1 Moment-Based Risk Measures

We have already mentioned two of the most popular volatility risk measures, which are members of the moment-based measures:

$$\sigma^2 = \mathrm{Var}(X) = E[(X - E(X))^2] \tag{14.16a}$$

and

$$\sigma = +\sqrt{\mathrm{Var}(X)} = +\sqrt{E[(X - E(X))^2]}, \tag{14.16b}$$

that is, the variance and the standard deviation, respectively. Measuring the economic volatility, the variance will be in units of (currency)^2 and the standard deviation in units of currency. As stated in Brehm et al. (2007), either measure has the disadvantage that they measure both the favorable deviations and the unfavorable deviations. It follows properties P1, P3, P4, and P7. The tail-based semistandard deviation, see below, measures only the unfavorable deviation.

The expected value principle is usually used in insurance pricing. The risk measure is defined by $C^{\mathrm{EV}}(X) = \lambda \cdot E(X), \lambda > 1$. It follows properties P1, P3, P8, P10, and P11.

As for the two-sided risk measures, risk measures such as (normalized) skewness, kurtosis, and MAD have been proposed. These could be summarized by the GMAD in Equation 14.15.

One approach, to overcome the negative view of the variance/standard deviation, is to use *exponential measures*, which in some sense encapsulate all moments; see Brehm et al. (2007):

$$C^{\mathrm{EM}}(X) = E[X \cdot \mathrm{e}^{cX/E(X)}]. \tag{14.17}$$

It is scaled to currency units and captures the effect of large losses exponentially. It exists under some limits; for example, there must be a maximum possible loss.

The exponential premium principle has been used as a risk measure. It is defined as $C^{\mathrm{EPP}}(X) = (1/\alpha) \ln E[\mathrm{e}^{\alpha X}], \alpha > 0$. In the classical ruin problem, it gives the required level of premium associated with the Cramér–Lundberg bounds for ruin probabilities; see, for example, Tsanakas (2007). In the financial literature, it has also the name entropic risk measures. It follows properties P1, P6, P7, P8, and P11.

14.2.5.2 Tail-Based Measures: Measures of Shortfall Risks

The shortfall risk is, as stated before, a downside risk measured relative to a target value, for example, the expected value $\mu = E(X)$ or any arbitrary deterministic or stochastic target $T(X)$; see, for example, Albrecht (2004a). The shortfall can be interpreted as the part of the loss that cannot be covered by the insurer if the target is put to the capital requirements, for example, by VaR or TVaR. In the literature, it is also named the residual risk, the insolvency risk, or policyholder's deficit; *cf.* the expected policyholder's deficit below.

One family of risk measures is the LPM, of degree $k = 0, 1, 2, \ldots$, given by

$$\mathrm{LPM}_k(T(X); X) = E[\max\{T(X) - X, 0\}^k], \tag{14.18a}$$

or, in its normalized form $k > 1$,

$$C_k^{\mathrm{LPM}}(X) = \mathrm{LPM}_k\,(T(X); X)^{1/k}. \tag{14.18b}$$

Special cases that are important in applications are, letting the target be deterministic $z = T(X)$,

$$k = 0 - \text{the shortfall probability} : \mathrm{SP}_z(X) = \mathrm{LPM}_0(z; X) = P(X \leqslant z) = F_X(z) \tag{14.19a}$$

$$k = 1 - \text{the expected shortfall} : \mathrm{SE}_z(X) = \mathrm{LPM}_1(z; X) = E[\max(z - X, 0)] \tag{14.19b}$$

$$k = 2 - \text{the shortfall variance} : \mathrm{SV}_z(X) = \mathrm{LPM}_2(z; X) = E[\max(z - X, 0)^2]. \tag{14.19c}$$

The normalized version of Equation 14.19c is the shortfall standard deviation

$$C_z^{\mathrm{SSD}}(X) = \mathrm{LPM}_2(z; X)^{1/2} = E\left[\max(z - X, 0)^2\right]^{1/2}. \tag{14.19d}$$

If we let $z = E(X)$, the expected value, we get from Equation 14.19b the *lower semiabsolute deviation* (LSAD). The *semivariance* is obtained from Equation 14.19c, and the *semistandard deviation*, mentioned above, equals the shortfall standard deviation.

We can also consider *conditional shortfall risk measures*, such as the mean excess loss or the conditional shortfall expectation, defined by

$$\mathrm{MEL}_z(X) = E[z - X | X \leqslant z] = \frac{\mathrm{SE}_z(X)}{\mathrm{SP}_z(X)}. \tag{14.20}$$

This is the "worst-case" risk measure! In insurance consideration, Equation 14.20 is usually reformulated to the expected shortfall risk measure of the right tail risk $\mathrm{ES}_z(X) = E[X - z | X \geqslant z]$. In a similar way, we may reformulate all LPM measures to *upper partial moments* (UPM).

VaR: Given some confidence or probability level $1-\alpha, \alpha \in (0,1)$, and a given risk X with cumulative distribution function $F_X(x)$, the VaR, denoted by $\mathrm{VaR}_{1-\alpha}(X)$, is defined as

$$\mathrm{VaR}_{1-\alpha}(X) = F_X^{-1}(1-\alpha) = \inf\{x \in \mathbb{R} | F_X(x) \geqslant 1-\alpha\}. \tag{14.21a}$$

VaR is a quantile function that always exists and is expressed in units of currency—"the lost money"; see, for example, Denuit et al. (2005). As VaR is a risk measure of category one, we obtain the category two counterpart version by using

$$C_{1-\alpha}^{\mathrm{VaR}}(X) = \mathrm{VaR}_{1-\alpha}(X) - E(X). \tag{14.21b}$$

VaR is discussed in Section 14.3. It follows properties P1, P3, P7, P8, P10, and P12, and, for elliptically distributed risks, also P4. A shortcoming of VaR is that it does not consider any information above the quantile $1-\alpha$ and hence does not take the upper right tail into consideration. The next measures do that.

TVaR and CTE: Given some confidence or probability level $1-\alpha, \alpha \in (0,1)$, and a given risk X with cumulative distribution function $F_X(x)$, the TVaR, denoted by $\mathrm{TVaR}_{1-\alpha}(X)$, and the conditional tail expectations CTE are defined as follows.

$$\mathrm{TVaR}_{1-\alpha}(X) = \frac{1}{\alpha}\int_{1-\alpha}^{1} \mathrm{VaR}_{\xi}(X)\,\mathrm{d}\xi, \quad 0 < \alpha < 1. \tag{14.22a}$$

The TVaR could be interpreted as the arithmetic average or mixture of the VaRs of X from $1-\alpha$ to 1. The alternative definition is the CTE, that is, the conditional expected loss given that the loss exceeds its VaR-loss:

$$\mathrm{CTE}_{1-\alpha}(X) = E[X | X > \mathrm{VaR}_{1-\alpha}(X)]. \tag{14.22b}$$

The CTE is the average loss in the worst $100\alpha\%$ cases.

If the distribution function $F_X(x)$ is continuous, then $\mathrm{TVaR}_{1-\alpha}(X) = \mathrm{CTE}_{1-\alpha}(X)$. For a discussion of the discontinuous case, see, for example, Denuit et al. (2005).

Both TVaR and CTE are category one risk measures. Other category one risk measures on the upper tail are the conditional VaR (CVaR) and the expected shortfall.

Conditional VaR, CVaR: The CVaR is defined as the expected value of the losses exceeding VaR and is defined as

$$\mathrm{CVaR}_{1-\alpha}(X) = E[X - \mathrm{VaR}_{1-\alpha}(X) | X > \mathrm{VaR}_{1-\alpha}(X)] = \mathrm{CTE}_{1-\alpha}(X) - \mathrm{VaR}_{1-\alpha}(X). \tag{14.23}$$

If we change $\mathrm{VaR}_{1-\alpha}(X)$ to a general threshold d with $P(X > d) > 0$, Equation 14.23 is the mean excess loss function denoted by $e_X(d)$; see, for example, Panjer (2006).

Expected shortfall, ES: the expected shortfall, see above, is defined as

$$\mathrm{ES}_{1-\alpha}(X) = E[(X - \mathrm{VaR}_{1-\alpha}(X))_+]. \tag{14.24}$$

As shown by Denuit et al. (2005), the following relationships between these tail risk measures hold (Property 2.4.2 in op. cit.).

1. $\text{TVaR}_{1-\alpha}(X) = \text{VaR}_{1-\alpha}(X) + \dfrac{1}{\alpha}\text{ES}_{1-\alpha}(X)$
2. $\text{CTE}_{1-\alpha}(X) = \text{VaR}_{1-\alpha}(X) + \dfrac{1}{P[X > \text{VaR}_{1-\alpha}(X)]} \cdot \text{ES}_{1-\alpha}(X)$
3. $\text{CVaR}_{1-\alpha}(X) = \dfrac{1}{P[X > \text{VaR}_{1-\alpha}(X)]} \cdot \text{ES}_{1-\alpha}(X).$

If $F_X(x)$ is continuous, then $\alpha = P[X > \text{Var}_{1-\alpha}(X)]$, and we obtain $\text{TVaR}_{1-\alpha}(X) = \text{CTE}_{1-\alpha}(X)$ and $\text{CVaR}_{1-\alpha}(X) = (1/\alpha) \cdot \text{ES}_{1-\alpha}(X)$.

These risk measures are discussed in Section 14.3. In some publications in the literature, the name "expected shortfall" is given for CVaR; so the reader should be aware of the definitions made in the literature. For continuous distributions, the TVaR follows properties P1, P3, P4, P6, P7, P8, P10, and P12.

Other versions of these measures are the tail-risk measures minus the expected value, that is, category two risk measures. The corresponding risk measure, or economic capital, to the TVaR is the excess TVaR, denoted by XTVaR, defined as

$$C_{1-\alpha}^{\text{TVaR}}[X] = \text{XTVaR}_{1-\alpha}(X) = \text{TVaR}_{1-\alpha}(X) - E(X) = \text{VaR}_{1-\alpha}(X) + \frac{1}{\alpha}\text{ES}_{1-\alpha}(X) - E(X). \tag{14.22c}$$

As shown in Brehm et al. (2007), if the mean is financed by funds like the technical provisions, the capital need above the mean has to be funded by other sources. This need, or capital requirement, is captured by XTVaR or other similar category two risk measures, such as

$$C_{1-\alpha}^{\text{VaR}}[X] = \text{XVaR}_{1-\alpha}(X) = \text{VaR}_{1-\alpha}(X) - E(X); \textit{cf.}\ \text{Equation 14.21b}.$$

The *expected policyholder deficit* (*EPD*), is calculated by multiplying (TVaR − VaR) by the probability level α, that is,

$$\text{EPD}_{1-\alpha}(X) = \alpha[\text{TVaR}_{1-\alpha}(X) - \text{VaR}_{1-\alpha}(X)] = \frac{1-\alpha}{\alpha}\text{ES}_{1-\alpha}(X).$$

In general, the EPD could be defined for any threshold τ defined as

$$\text{EPD}(X;\tau) = P[X > \tau] \cdot [\text{TVaR} - \tau].$$

The EPD can be interpreted as the expected value of losses that cannot be funded by the surplus (the threshold surplus τ).

As shown in Brehm et al. (2007), the insurer would unlikely be able to reinsure all default possibility away by paying for the EPD. The market value of that is called the *value of the default option* (VDO). Usually it is estimated by options pricing methods.

Other tail-based risk measures, under the name of quantile-based risk measures (QBRM) are discussed in Dowd and Blake (2006). They also include convex risk measures (P6), dynamic risk measures, comonotonicity approaches, Markov bounds approaches, and "best practice" risk measures.

14.2.5.3 Generalized Moments

The family of generalized moments is more general than those discussed in Section 14.2.5.1. They also include some of the tail-based measures of Section 14.2.5.2. These measures are discussed in Brehm et al. (2007). One family of generalized moment risk measures has the form of

$$C^{\mathrm{GM}}(X) = E[X \cdot \eta(F_X(x))], \tag{14.25}$$

for nonnegative scalar functions η. These risk measures are called spectral risk measures.

If η is the indicator function $\eta(F_X(x)) = 1[X \geqslant F_X(1-\alpha)]$, we obtain, in the continuous case, the TVaR and the CTE.

Another spectral risk measure is given by the weight function $\eta(\xi) = \mathrm{e}^{-\theta(\xi-(1-\alpha))^2}$, that is, the weight function is the squared distance from the target percentile $1-\alpha$. This is called a "blurred VaR."

14.3 VAR AND TVAR

In this section, we will closely look at those two risk measures that have been the focus of attention in the financial and actuarial literature for about 15 years. In Section 14.3.2 we will look at the VaR, and in Section 14.3.3 at the TVaR. We conclude with a summary of the merits of the VaR and TVaR in Section 14.3.4, but present an application of the Cornish–Fisher expansion that we will use for VaR and TVaR.

14.3.1 Assuming Nonnormality

Many used models are based on a normal distribution explicitly or even implicitly without any discussion. Here we will relax this assumption and assume that the distribution is positively skewed, but is still unknown. One way to tackle the problem with skewed distributions is to use a Cornish–Fisher expansion to transform the quantiles and the tail expectations of the skewed distributions into a standard normal distribution; see Appendix B. In the actuarial literature this is usually known as the "normal power approximation" or NP approximation. This is discussed, for example, in Beard et al. (1984), Daykin et al. (1994), and Sandström (2005). The TVaR transformation is discussed in, for example, Giamouridis (2006). If the skewness is considered not to be sufficient, then also the kurtosis could be used. In Giamouridis (2006), a second-order Cornish–Fisher expansion is given for both the VaR and TVaR; see also Appendix B. We illustrate this in Figure 14.1.

If we had any idea of the skewness of our unknown distribution, that is, at least a guess of the magnitude of the skewness parameter, then we could use the NP approximation in the risk charges discussed in Chapter 15 to arrive at the total capital requirement. The NP approximation can thus be used for risk measures such as VaR and TVaR in terms of the "standard deviation principle."

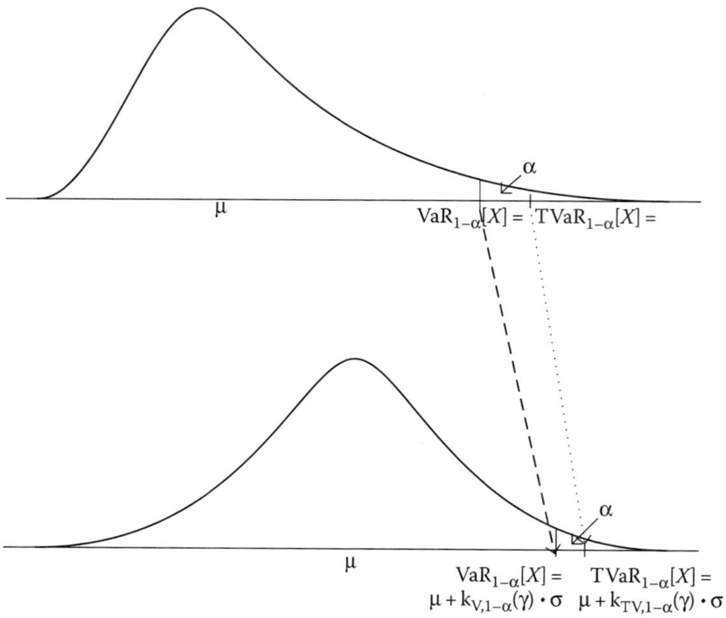

FIGURE 14.1 The percentiles of a complicated and skew distribution can be defined in terms of the percentiles of the standard normal distribution by the NP approximation. The factor k depends on the skewness in the original distribution!

We assume that all risk charges are distributed either by a symmetric distribution (standard normal) or by a positively skewed distribution. Assume that the skewness is measured by

$$\gamma_i = \frac{E[(Y_i - \mu_i)^3]}{\sigma_i^3}, \quad i = 1, \ldots, r, \text{ and } \gamma_i \geqslant 0.$$

If the risk charges are having distribution functions with $\gamma_i \geqslant 0$, $i = 1, \ldots, d$, then the skewness of the total risk charge distribution will be positively skewed: $\gamma_{SCR} \geqslant 0$. We use a Cornish–Fisher expansion to redefine a quantile or the tail expectation of a general skewed distribution in terms of the standard normal distribution; see Cornish and Fisher (1937), Fisher and Cornish (1960), Johnson and Kotz (1970), and McCune and Gray (1982).

In the traditional approximation, the $1 - \alpha$ quantile of a skew distribution is written in terms of the $1 - \alpha$ quantile of the standard normal distribution with a correction for the skewness: $\mu + k_{V,1-\alpha}(\gamma) \cdot \sigma$, where $k_{V,1-\alpha}(\gamma)$ is a new quantile of the standard normal distribution and a function of the skewness in the original distribution. V stands for VaR.

The approach can be generalized so that the tail expectation of a skew distribution can also be written in terms of the tail expectation of the standard normal distribution with a correction for the skewness: $\mu + k_{TV,1-\alpha}(\gamma) \cdot \sigma$, where $k_{TV,1-\alpha}(\gamma)$ is a new quantile of the standard normal distribution and a function of the skewness in the original distribution. TV stands for TVaR.

14.3.2 Value-at-Risk

VaR is just a percentile or quantile of a given distribution of aggregate risks. The VaR on a portfolio is the maximum potential loss we expect over a given period, say 1 year, at a given level of confidence or under a certain probability of ruin (α%).

The concept of VaR is not new, but the name stems from J.P. Morgan's introduction to RiskMetrics, which was published in 1993. Nowadays, VaR is the benchmark concept for determining market risk capital charges in the bank sector. It is also the key concept in both Basel II and the European Solvency II system.

There has been a lot of discussions about pros and cons of the VaR as a risk measure. The VaR does not consider what happens during the holding period, only the result at the end of the period. Another deficit is that it does not measure the potential size of a loss, given that the loss exceeds the VaR. However, the main criticism of the VaR is that it, in general, lacks the property P4—subadditivity. But as has been seen by McNeil et al. (2005) if the distribution is "nice," that is, not too skewed, not too fat-tailed, then VaR will follow the subadditivity property. It has been shown that in most practical situations VaR is subadditive; see, for example, Denuit et al. (2006). It has been shown that VaR will not fulfill P4 if the distributions are "superfat," that is, the tails are so fat that the first moment (mean) is not defined. VaR is subadditive for elliptical risks; see McNeil et al. (2005).

VaR is defined in Section 14.2.5.2, Equation 14.16a, and is illustrated in Figure 14.2.

VaR as a risk measure is discussed in, for example, Denuit et al. (2005), Dhaene et al. (2006), Dowd and Blake (2006), McNeil et al. (2005), Panjer (2006), and Embrechts et al. (2008). Discussions on estimating VaR are given in, for example, Dowd (2004), Dowd and Blake (2006), and Manganelli and Engle (2001).

VaR follows the following properties:

P1—objectivity

P3—positive homogeneity

P4—subadditivity for elliptical risks

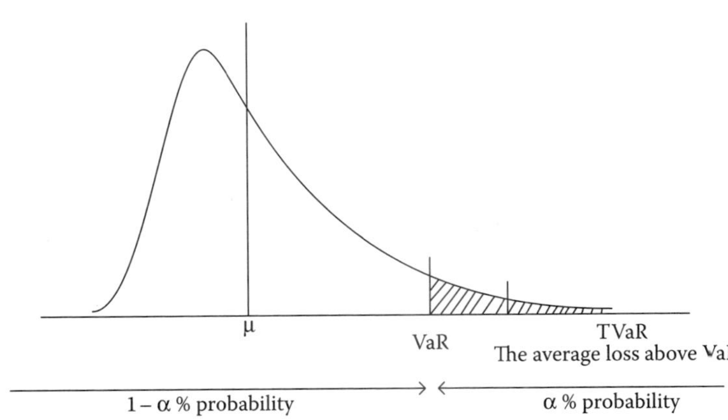

FIGURE 14.2 A description of VaR and TVaR.

P7—translation invariant

P8—monotonicity

P10—comonotonic additivity

P12—the regulator's condition

In McNeil et al. (2005), it is shown that P10 implies that if $\alpha \in (0,1)$ and $X_1, \ldots, X_d$ are comonotonic random variables with cumulative distribution functions $F_1, \ldots, F_d$ that are continuous and strictly increasing, then

$$\mathrm{VaR}_{1-\alpha}\left[\sum_{i=1}^{d} X_i\right] = \sum_{i=1}^{d} \mathrm{VaR}_{1-\alpha}[X_i].$$

A more general form is also given. Let $\Psi : \mathbb{R}^d \to \mathbb{R}$ be an increasing and left-continuous function in each argument and let $X_1, \ldots, X_d$ be comonotonic random variables, then $\mathrm{VaR}_{1-\alpha}[\Psi(X_1, \ldots, X_d)] = \Psi[\mathrm{VaR}_{1-\alpha}[X_1], \ldots, \mathrm{VaR}_{1-\alpha}[X_d]]$.

P12 is discussed in Section 14.1 and we will therefore discuss it only briefly. In Dhaene et al. (2008a) it is shown that coherent risk measures used for solvency capital requirements can be "too subadditive" in the sense that the capital requirements may lead to an increase of the shortfall risk in case of merger. This must be seen as a deficit from both the perspective of the insurer and the perspective of the regulator. Theorem 1 in op. cit. (see also Section 14.1) justifies the use of VaR as a measure of the solvency capital requirement. The risk measure we have used and want to keep "sufficiently" small is the shortfall risk measure $C^{SR}[X]$ measured by $E[(X - C_{\mathrm{SCR}}[X])_+]$.

As was shown in Section 14.1, the VaR was the solution to the problem of finding an SRC that fulfills the regulator's condition (P12) and also making the cost function $c[X, C_{\mathrm{SCR}}[X]]$, given by Equation 14.6, minimal for every X.

VaR is a distorted risk measure, but the distortion function is neither concave nor continuous; see Section 14.2.4.

As noted by Dowd and Blake (2006), seeing VaR as a spectral risk measure implies that the user is risk-loving, that is, has negative risk aversion, in the tail region. Looking at VaR as a distorted risk measure implies that it is a poor risk measure because it is based on a "badly behavioral" distortion function (not continuous).

The European Insurance and Occupational Pensions Committee (EIOPC) has summarized the merits of VaR and TVaR that have been discussed by the body of the European insurance industry, CEA, and the European supervisors, CEIOPS; see EIOPC (2006). The summary table is given in Section 14.4.

Let $\phi(x)$ and $\Phi(x)$ denote the pdf and the cdf respectively, of the standard normal distribution $N(0,1)$, that is, $\phi(x) = \Phi'(x)$. We use the notation $k_{1-\alpha} = \Phi^{-1}(1-\alpha)$ in the following paragraphs. The variable $k_{1-\alpha}$, which can be seen as a distance factor, is tabulated in most standard statistical textbooks. Some values of $k_{1-\alpha}$ are given in Table 14.3.

TABLE 14.1 VaR for Different Distribution Functions

Distribution Function		$\text{VaR}_{1-\alpha}[X] =$	Comments
Normal	$X \sim N(\mu, \sigma^2)$	$\mu + k_{1-\alpha} \cdot \sigma$	D(97), P(52)
Lognormal	$\ln X \sim N(\mu, \sigma^2)$	$e^{\mu + k_{1-\alpha} \cdot \sigma}$	D(97)
Student t	$X \sim t_v(\mu, \sigma^2)$	$\mu + t_{v,1-\alpha} \cdot \sigma$	$v > 2$ df M(40), P(52)
Exponential	$X \sim \exp(1/\theta)$	$\theta \cdot \ln \alpha$	D(135), P(52)
General skewed pdf	Cornish–Fisher expansion	$\mu + k_{V,1-\alpha} \cdot \sigma$, where $k_{V,1-\alpha} = k_{1-\alpha} + \frac{\gamma}{3}[k_{1-\alpha}^2 - 1]$	Sandström (2007a)

Note: The number within parentheses is the page number, D(p): Denuit et al. (2005), M(p): McNeil et al. (2005), and P(p): Panjer (2006).

For Student t distribution, we use the notations t and T for the pdf and cdf, respectively. In a similar way as above, we let $t_{v,1-\alpha} = T_v^{-1}(1 - \alpha)$, which is tabulated in most standard statistical textbooks.

A comparison between the Cornish–Fisher expansion ("normal power approximation") and the lognormal distribution is given in IAA (2009b, Appendix C4).

Table 14.1 gives explicit expressions for VaR for different distribution functions.

14.3.2.1 Variance of a VaR Estimator

VaR has become the most popular and used risk measure. As such, it is important to have a tool to quantify the statistical precision of an estimated VaR, $V\hat{a}R$, that is, the sampling error in the estimate. This problem is studied in a paper by Manistre and Hancock (2005) for a TVaR estimator. As a "by-product" they also arrive at a variance estimator for VaR; see also Manistre and Hancock (2008). For a brief discussion on the variance of an estimator of TVaR and the covariance between VaR and TVaR estimators, see Section 14.3.3.1.

The asymptotic expression for the variance of an estimator $V\hat{a}R$ based on a sample of n is given by

$$\sigma^2(\text{V\^aR}_n) = \text{VAR}[\text{V\^aR}_n] \approx \frac{\alpha(1-\alpha)}{n \cdot [f(\text{V\^aR}_n)]^2}. \tag{14.26}$$

Hence, the variance of the quantile estimator depends on the value of the probability density of the underlying distribution at the quantile point $f(\text{VaR})$. The estimation of this quantity from data can be problematic. In Manistre and Hancock (2005), the following estimator of $f(\text{VaR})$ is used:

$$\hat{f}(\text{VaR}) = \frac{\xi}{\hat{F}_n^{-1}(1-\alpha) - \hat{F}_n^{-1}(1-\alpha-\xi)},$$

where $\xi = 1/100$. This empirical density function is sensitive to the choice of ξ.

14.3.3 TVaR

The fact that VaR, in general, does not fulfill property P4—subadditivity—was the starting point for looking at other risk measures fulfilling the property. Another deficit with VaR as a risk measure is that it does not inform us about loss size beyond this quantile.

Artzner et al. (1999) have listed the properties that they believe a risk measure should fulfill. The concept of coherent risk measures was born. TVaR fulfills these properties. As a measure of risk, it was introduced by Wirch (1997); see also Wirch and Hardy (1999).

The TVaR is defined by Equation 14.22a. In the continuous case, TVaR is identical with the conditional tail expectation given by Equation 14.22b. If we define the expected shortfall as $\mathrm{ES}_{1-\alpha}[X] = E[(X - \mathrm{VaR}_{1-\alpha}[X])_+]$, as given by Equation 14.24, then we have for continuous distribution functions (see Denuit et al., 2005 and Section 14.2.5.2)

$$\mathrm{TVaR}_{1-\alpha}[X] = \mathrm{CTE}_{1-\alpha}[X] = \mathrm{VaR}_{1-\alpha}[X] + \frac{1}{\alpha}\mathrm{ES}_{1-\alpha}[X], \tag{14.27}$$

where the last term is the CVaR: $\mathrm{CVaR}_{1-\alpha}[X] = (1/\alpha)\mathrm{ES}_{1-\alpha}[X]$. The CVaR is sometimes termed the mean excess loss function. There is some confusion in the literature. Sometime TVaR and CTE are called expected shortfall. We have used the terminology of Denuit et al. (2005).

For continuous distributions, the TVaR follows properties P1, P3, P4, P6, P7, P8, P10, and P12.

As shown by McNeil et al. (2005), the asymptotic of the "shortfall-to-quantile ratio," $\mathrm{TVaR}_{1-\alpha}(X)/\mathrm{VaR}_{1-\alpha}(X)$, as $\alpha \to 0$, is interesting. For the normal distribution the ratio tends to 1, but for the t distribution with $v > 1$, we have $\lim_{\alpha \to 0}(\mathrm{TVaR}_{1-\alpha}(X))/(\mathrm{VaR}_{1-\alpha}(X)) = v/(v-1) > 1$. This is an indication of the difference between TVaR and VaR for heavy-tailed distributions.

The TVaR is illustrated in Figure 14.2. TVaR is a distorted risk measure with a continuous distortion function given in Section 14.2.4.

The EIOPC has summarized the merits of VaR and TVaR that have been discussed by the body of the European insurance industry, CEA, and the European supervisors, CEIOPS; see EIOPC (2006). The summary table is given in Section 14.3.

We use the same notations as for VaR in Section 14.3.2.

Elliptical distributions can be seen as extensions of multivariate normal distributions having elliptical contours. Univariate elliptical (UE) distributions are the corresponding marginal distributions. Normal and Student t are examples of UE distributions. The reader is referred to Denuit et al. (2005), McNeil et al. (2005), Dhaene et al. (2006) and Panjer (2006) for further reading. Landsman and Valdez (2003) showed that all UE distributions with finite mean and variance could be written as

$$f(x) = \frac{c}{\sigma} g\left[\frac{1}{2}\left(\frac{x-\mu}{\sigma}\right)^2\right],$$

where $g(x)$ is a function of $[0, \infty)$ with $\int_0^\infty g(x)\,\mathrm{d}x < \infty$. Let $G(x) = c\int_0^x g(y)\,\mathrm{d}y$, $\bar{G}(x) = G(\infty) - G(x)$, $F(x) = \int_{-\infty}^x f(y)\,\mathrm{d}y$, and $\bar{F}(x) = 1 - F(x)$. Then $\mathrm{TVaR}_{1-\alpha}(X)$ can be written as in Table 14.2a with $x_{1-\alpha} = \mathrm{VaR}_{1-\alpha}(X)$, $\alpha < 0.5$. This result is a generalization of the TVaR for normal distributed random variables that were shown by Panjer (2002); see also Panjer (2006) for a proof.

TVaR in multivariate elliptical distributions and for sums of elliptical risks is discussed in Landsman and Valdez (2003).

The conditional density function of X given that $x_{1-\alpha} \leqslant X < \infty$ is said to be a normal distribution singly truncated from below with density

$$g(y) = \begin{cases} \frac{1}{\sigma}\phi\left(\frac{y-\mu}{\sigma}\right)/[1-\Phi(x_{1-\alpha})] & x_{1-\alpha} \leqslant y < \infty \\ 0 & \text{otherwise} \end{cases}$$

The mean and variance of the singly truncated normal distribution are given by (see, e.g., Johnson and Kotz, 1970)

$$\mu_{CT}(\alpha) = \mu + \sigma \cdot h(k_{1-\alpha})$$

and

$$\sigma_t^2 = \sigma\left\{1 + z_{1-\alpha}h(k_{1-\alpha}) - h^2(k_{1-\alpha})\right\},$$

where $h(k_{1-\alpha}) = (\phi(k_{1-\alpha}))/\alpha$ is the hazard function; see, for example, Odeh et al. (1977). The hazard function can be written as $h(x) = 1/R(x)$, where $R(x)$ is *Mills' ratio* (see, e.g., Kotz et al., 1985). Tables of Mills' ratio are usually published in most books on statistical tables. Approximations for Mills' ratio are usually derived from expansions and inequality bounds.

Let $k_{1-\alpha}^* = h(k_{1-\alpha})$ be the distance factor; then

$$\text{TVaR}_\alpha(X) = \mu + k_{1-\alpha}^*\sigma$$

and

$$\sigma_t^2 = \sigma^2\{1 + k_{1-\alpha}^* z_{1-\alpha} - k_{1-\alpha}^{*2}\}.$$

For Student t distribution, we use the notations t and T for the pdf and cdf, respectively. In a similar way as above, we let $t_{v,1-\alpha} = T_v^{-1}(1-\alpha)$, which is tabulated in most standard statistical textbooks.

Table 14.2a gives explicit expressions for TVaR for different distribution functions.

We will also give the expected shortfall, ES, for the normal and lognormal distributions in Table 14.2b.

14.3.3.1 Variance of a TVaR Estimator

The TVaR has become an important measure of risk; hence, it is important to have a tool to quantify the statistical precision of an estimated TVaR, $TV\hat{a}R$. This problem has been studied by Manistre and Hancock (2005, 2008).

Assume that we have a random sample of size n and we rank them in reversed order, that is, $x_{[1]} \geqslant x_{[2]} \geqslant \cdots \geqslant x_{[n]}$. A TVaR estimator at the $1-\alpha = k/n$ level is given by the average of the k highest order statistic, that is,

$$\text{TV\hat{a}R} = \frac{1}{k}\sum_{j=1}^{k} x_{[j]}.$$

TABLE 14.2a The TVaR for Different Distribution Functions

Distribution Function		$\mathrm{TVaR}_{1-\alpha}[X] =$	Comments
Univariate elliptical	$X \sim UE(\mu, \sigma^2)$	$\mu + \lambda \cdot \sigma^2$, with $\lambda = \dfrac{(1/\sigma)\bar{G}[(1/2)((x_{1-\alpha} - \mu)/\sigma)^2]}{\bar{F}(x_{1-\alpha})}$	$x_{1-\alpha} = \mathrm{VaR}_{1-\alpha}(X)$ L(61), P(95)
Normal	$X \sim N(\mu, \sigma^2)$	$\mu + k^*_{1-\alpha} \cdot \sigma$ or $\mu + \lambda \cdot \sigma^2$ with $\lambda = \dfrac{(1/\sigma)\phi(k_{1-\alpha})}{\alpha} = \dfrac{1}{\sigma}k^*_{1-\alpha}$	L(62), P(52)
Student t	$X \sim t_v(\mu, \sigma^2)$	$\mu + \dfrac{t[t_{v,1-\alpha}]}{\alpha} \times \left\{\dfrac{v + [t_{v,1-\alpha}]^2}{v-1}\right\} \cdot \sigma$	$v > 2$ df M(46), P(52), see also L(63–64)
Logistic	$X \sim \mathrm{Log}(\mu, \sigma^2)$	$\mu + \lambda \cdot \sigma^2$, with $\lambda = \dfrac{1}{2}\left[\dfrac{(1/\sigma)\phi(z_{1-\alpha})}{(1/\sqrt{2\pi}) + \phi(z_{1-\alpha})} \times \dfrac{(1/\sigma)\phi(z_{1-\alpha})}{\bar{F}(z_{1-\alpha})}\right]$	$z_{1-\alpha} = \dfrac{x_{1-\alpha} - \mu}{\sigma}$ L(65), P(96)
Lognormal	$\ln X \sim N(\mu, \sigma^2)$	$\dfrac{\Phi[\sigma - k_{1-\alpha}]}{\alpha} \cdot e^{\mu+\sigma^2/2}$	D(98)
Exponential	$X \sim \mathrm{Exp}(1/\theta)$	$\theta \cdot [1 + \ln \alpha]$	P(52)
General skewed pdf	Cornish–Fisher expansion	$\mu + k_{\mathrm{TV},1-\alpha} \cdot \sigma$, where $k_{\mathrm{TV},1-\alpha} = k^*_{1-\alpha} \cdot \left[1 + \dfrac{\gamma}{6}k^3_{1-\alpha}\right]$	Sandström (2007a)

Note: The number within parentheses is the page number, D(p): Denuit et al. (2005), L(p): Landsman and Valdez (2003), M(p): McNeil et al. (2005), and P(p): Panjer (2006). For exponential power distribution, see Landsman and Valdez (2003, p. 65).

TABLE 14.2b The TVaR for Different Distribution Functions

Distribution Function		$ES_{1-\alpha}[X] =$	Comments
Normal	$X \sim N(\mu, \sigma^2)$	$\{\phi(k_{1-\alpha}) - \alpha k_{1-\alpha}\} \cdot \sigma$	D(99)
Lognormal	$\ln X \sim N(\mu, \sigma^2)$	$\Phi[\sigma - k_{1-\alpha}]e^{\mu + k_{1-\alpha}\cdot\sigma^2/2} - \alpha e^{\mu + k_{1-\alpha}\cdot\sigma}$	D(99)

Note: The number within parentheses is the page number, D(p): Denuit et al. (2005).

To study the standard deviation of $TV\hat{a}R$ we actually estimate two quantities. These are

- The $1 - \alpha = k/n$ quantile: we use the kth reversed sample order statistic $x_{[k]}$ for this
- The conditional expectation of X, given that it exceeds the $(1 - \alpha)$-quantile

As the uncertainty in both quantities is contributing to the uncertainty in the TVaR estimate, Manistre and Hancock (2005) conditioned the variance on the observation of the kth reversed order statistic; *cf.* Appendix A.

$$\sigma^2(\text{TV}\hat{\text{a}}\text{R}) = \text{VAR}[\text{TV}\hat{\text{a}}\text{R}] = E\{\text{VAR}[\text{TV}\hat{\text{a}}\text{R}|X_{[k]}]\} + \text{VAR}\{E[\text{TV}\hat{\text{a}}\text{R}|X_{[k]}]\}. \quad (14.28)$$

An unbiased estimator of the first term on the right-hand side of Equation 14.28 is given by $(1/k) \cdot \text{VAR}\left(x_{[1]}, x_{[2]}, \ldots, x_{[k]}\right)$. The second term can be estimated in the large sample limit, that is, when $n \to \infty$.

The asymptotic expression for the variance of the TVaR-estimator is given by

$$\sigma^2(\text{TV}\hat{\text{a}}\text{R}) = \text{VAR}[\text{TV}\hat{\text{a}}\text{R}] \approx \frac{\text{VAR}[X|X > \text{VaR}] + (1-\alpha)\cdot[\text{TVaR} - \text{VaR}]^2}{n \cdot \alpha}.$$

The corresponding variance for the VaR-estimator is given by Equation 14.26 and the covariance is given by

$$\text{Cov}\left[\text{TV}\hat{\text{a}}\text{R}, \text{V}\hat{\text{a}}\text{R}\right] \approx \frac{(1-\alpha)\cdot[\text{TVaR} - \text{VaR}]}{n \cdot f(\text{VaR})}.$$

As we have that TVaR $\geqslant$ VaR, it is shown that the variables are positively correlated. If $\alpha = 1$, then $\text{TV}\hat{\text{a}}\text{R} = \hat{\mu}$ and obviously the variance equals $\sigma^2(\hat{\mu}) = \text{VAR}[X|X > -\infty]$.

The behaviour of theses estimators is discussed in Manistre and Hancock (2005, 2008). Similar results related to other risk measures are discussed in Jones and Zitikis (2003).

14.3.4 Tabled Distance Functions

We end this chapter with a table with different values of the distance functions for the normal distribution and t-values for the t-distribution (Table 14.3a). In Table 14.3b, we have given the corresponding distance functions for VaR and TVaR from a general skewed distribution using the first terms in a Cornish–Fisher expansion.

TABLE 14.3a Distance Functions $k_{1-\alpha}$ and $k^*_{1-\alpha}$ for Different Values of α for $\mathrm{VaR}_{1-\alpha}$ and $\mathrm{TVaR}_{1-\alpha}$ in Case of Normality

	Normal Distribution		$t_{v,1-\alpha}, v =$						
$1-\alpha$	$k_{1-\alpha}$	$k^*_{1-\alpha}$	3	4	5	8	10	20	30
0.990	2.33	2.67	4.54	3.75	3.37	2.90	2.76	2.53	2.46
0.995	2.58	2.90	5.84	4.60	4.03	3.36	3.17	2.85	2.75
0.999	3.09	3.37	10.21	7.17	5.89	4.50	4.14	3.55	3.39
0.9995	3.29	3.55	12.92	8.61	6.87	5.04	4.59	3.85	3.65
0.9999	3.72	3.96	22.20	13.03	9.68	6.44	5.64	4.54	4.23

Note: $t_{v,1-\alpha}$ is also given for Student t for different values of α and v dfs.

TABLE 14.3b Distance Functions $k_{1-\alpha}$ and $k^*_{1-\alpha}$ for Different Values of α for $\mathrm{VaR}_{1-\alpha}$ and $\mathrm{TVaR}_{1-\alpha}$ in Case of Normality

	$\mathrm{VaR}_{1-\alpha}[X] =, \alpha \in (0,1)$		$\mathrm{TVaR}_{1-\alpha}[X] =, \alpha \in (0,1)$	
$1-\alpha$	$k_{1-\alpha}$	$k_{V,1-\alpha}$	$k^*_{1-\alpha}$	$k_{TV,1-\alpha}$
0.990	2.33	$2.33 + 0.74\gamma$	2.67	$2.67 + 5.63\gamma$
0.995	2.58	$2.58 + 0.94\gamma$	2.90	$2.90 + 8.30\gamma$
0.999	3.09	$3.09 + 1.42\gamma$	3.37	$3.37 + 16.57\gamma$
0.9995	3.29	$3.29 + 1.64\gamma$	3.55	$3.55 + 21.07\gamma$
0.9999	3.72	$3.72 + 2.14\gamma$	3.96	$3.96 + 33.98\gamma$

Note: $k_{V,1-\alpha}$ and $k_{TV,1-\alpha}$ are the corresponding distance functions where a general skewed distribution has been approximated by using the Cornish–Fisher expansion. The skewness in the underlying distribution is denoted by γ.

14.3.5 Summary of the Merits of VaR and TVaR

The EIOPC has summarized the merits of the VaR and TVaR that have been discussed by the body of the European insurance industry, CEA, and the European supervisors, CEIOPS; see EIOPC (2006) (Table 14.4).

14.4 CONCENTRATION MEASURES

It is important to distinguish between dispersion measures and concentration measures. When concentration is high, the dispersion is low. In this sense, dispersion measures are similar in character to disparity measures. In financial economics, dispersion is usually measured by the standard deviation. A review of the literature on concentration measures (focusing on banks) is given in Bikker and Haaf (2002). In the following, we will mainly follow the work of Piesch (1975).

There are two dimensions of concentration measures. In one, we have the market, that is, the size of a firm within a comparative market. The other is the geographical dimension, considering the concentration in different geographical markets.

Assume that we have n ordered observations such that $x_{(1)} \leqslant x_{(2)} \leqslant \cdots \leqslant x_{(n)}$ and its relative severities or sizes $p_{(1)} \leqslant p_{(2)} \leqslant \cdots \leqslant p_{(n)}$, where $p_{(i)} = x_{(i)}/n\bar{x}$ and $\bar{x} = (1/n)\sum_{i=1}^{n} x_i$, $\sum_{i=1}^{n} p_i = 1$, and $p_{(i)}$ is the ith ordered relative severity/size. We use the usual notation of x_i and p_i to denote the unordered values.

TABLE 14.4 Merits of VaR and TVaR Summarized by EIOPC (2006)

Perspective	VaR	TVaR
Theoretical qualities	VaR is not a "coherent" measure, which means that it does not respect all actuarial qualities for such a measure. In particular, it is not "subadditive": VaR (risk A + risk B) might be higher than the sum of VaR (risk A) and VaR (risk B), which is not logical since there should be risk mitigation between A and B. Under certain circumstances, VaR may underestimate the level of exposure On the other hand, from a practical perspective, VaR in the tail region of the distribution (e.g., 99.5%) would be roughly subadditive	TVaR is a coherent risk measure; it is therefore recommended by the International Actuary Association. In particular, it is subadditive. Therefore, unlike VaR, TVaR never underestimates the level of exposure
Day-to-day risk management	VaR is easy to explain to top management and other stakeholders. It is therefore easy to implement throughout the company and to embed in the company's risk culture (so as to improve risk management). 99.5% VaR focuses on the worst 9950th loss out of 10,000 simulations: it is the worst scenario under "normal" circumstances. It does not focus on extreme events	TVaR requires more mathematical background to be understood and implemented. It is more difficult to embed in the company's culture. 99.5% TVaR focuses on what happens in the 50 worst scenarios out of 10,000 simulations: it tackles large risks and extreme events. Such risks often trigger bankruptcy: studying them should improve extreme events management and limit the probability of insolvency
Implementation	VaR refers to "normal" circumstances (see the previous box): it is easier to collect data and make realistic assumptions. Since VaR refers to one worst-case scenario (the 99.5% confidence level), it is easy to design a proxy stress-test to calculate VaR. You just have to define the 99.5% scenario	Companies often lack the necessary data to simulate extreme events (e.g., an event like Katrina happens only once in 35 years; so it is impossible to conduct proper statistical analysis). Companies sometimes have to make haphazard assumptions to assess TVaR; so results are potentially marred with a significant modeling error
Consistency with the other financial sectors	VaR is the most commonly used risk measure in all financial sectors. For instance, banks use VaR to assess market risk	99.5% TVaR is similar to an "expected loss under default," where default corresponds to breaching the 99.5% percentile. Such a concept is used under the Basel II most advanced approach to credit risk ("loss given default" or LGD)
In place within companies/ Administrative burden	VaR is already often used by direct insurers (77% of respondents to a worldwide SOA survey)	TVaR is the preferred risk measure of reinsurance companies, which often have already developed internal models

Source: From EIOPC. 2006. Choice of a risk measure for supervisory purposes: Possible amendments to the Framework for Consultation. European Insurance and Occupational Pensions Committee. MARKT/2534/06-EN, November 2006. The European Commission, Available at http://ec.europa.eu/internal_market/. With permission.

In measuring disparity, two extreme distributions, that is, the "equality distribution" and the "inequality distribution," play a special role. The most extreme equality distribution is obtained from $p_i = 1/n, i = 1, 2, \ldots, n$, and we say that there is no disparity. Hence, such a measure should be zero when there is no disparity. On the other hand, the most extreme inequality distribution is at hand when $p_{(n)} = 1$, and $p_{(1)} = p_{(2)} = \cdots = p_{(n-1)} = 0$. A disparity measure should, in this case, equal 1 or should be normalized such that it equals $1 - (1/n)$.

For concentration, there are also two extreme distributions: the total concentration that equals the most extreme inequality distribution, that is, when $p_{(n)} = 1$, and $p_{(1)} = p_{(2)} = \cdots = p_{(n-1)} = 0$, and the "total equality distribution." The latter distribution does not need to be equal to the "extreme equality distribution" with $p_i = (1/n), i = 1, 2, \ldots, n$. Usually one introduces a reference value n_0; see, for example, Münzner (1963) and Piesch (1975). In the following, we assume that $n = n_0$.

Bikker and Haaf (2002) discuss concentration indices in their general form, that is, in the form $\text{CI} = \sum_{i=1}^{n} w_i p_i$, where w_i is a weight for the market share $p_i, i = 1, 2, \ldots, n$. Having "self-weights," that is, $w_i = p_i, \forall i$, gives us the well-known Herfindahl–Hirschman index, HHI. If the weights are a function of the rankings, we get, for example, the Hall–Tideman index and the Rosenbluth index. As a third weight class, we can take a logarithm function of the market shares. This class includes the entropy measure.

These three classes of concentration measures include the following indices:

$$\text{(1) The Herfindahl-Hirschman index: HHI} = \sum_{i=1}^{n} p_i^2 \tag{14.29a}$$

$$\text{(2a) The Hall-Tideman index: HTI} = \frac{1}{2\sum_{i=1}^{n} R_{x(i)} p_{(i)} - 1} \tag{14.30a}$$

$$\text{(2b) The Rosenbluth index: RI} = \frac{1}{2\sum_{i=1}^{n} (1 + n - R_{x(i)}) p_{(i)} - 1}, \tag{14.30b}$$

where $R_{x(i)}$ denotes the rank order of $x_{(i)}$, that is, $1 \leqslant R_{x(i)} \leqslant n$

$$\text{(3) The Entropy measure: } E(p) = \sum_{i=1}^{n} p_i \log \frac{1}{p_i} = -\sum_{i=1}^{n} p_i \log p_i. \tag{14.31}$$

14.4.1 Herfindahl–Hirschman Index

The Herfindahl–Hirschman index, HHI, also denoted the Herfindahl index, was in its origin used to measure the size of firms in relation to the industry, but also as an indication of the amount of competition. It was first used by Hirschman in the 1940s for market shares (as a fact, he used the square root of H, that is, $\sqrt{H}$). Herfindahl used the measure in his thesis in 1950 to describe the copper industry. It became known as the Herfindahl index due to work done by Rosenbluth in the 1950s. Nowadays, the abbreviation HHI, instead of only HI, is usually used. The index is the most used summary measure of concentration in the literature. It is also used as a benchmark when comparing other measures. The Antitrust Division of the Department of Justice in the United States uses the HHI to determine mergers. They consider an HHI between 0.10 and 0.18 to indicate a "moderately concentration." An index below 0.10 would be seen as indicating "no concentration" and an index above 0.18 as indicating "concentration."

Let x_i be the size of firm I, $i = 1, 2, \ldots, n$, and $n\bar{x}$ the total market size. Then the index is readily seen as the sum of squares of the market shares ($p_i = x_i/n\bar{x}$). If H decreases it is interpreted as a loss of market power and at the same time an increase in competition. If all firms have an equal share, the reciprocal $1/H$ shows the number of firms in the industry ($1/H = n$). When the firms have unequal shares, the reciprocal indicates the equivalent number of firms in the industry. A value of, say, 1.25 indicates a market structure of 1.25 firms of the same size.

From Equation 14.29a, we get $(1/n) \leqslant H \leqslant 1$. A normalized version of Equation 14.29a is

$$H^* = \frac{H - 1/n}{1 - 1/n} = \frac{1}{1 - 1/n}\left(\sum_{i=1}^{n} p_i^2 - \frac{1}{n}\right), \tag{14.29b}$$

where $0 \leqslant H^* \leqslant 1$.

Defining $\bar{p} = 1/n$, we can write Equation 14.29a as

$$H = \frac{1}{n} + n\text{Var}(p), \tag{14.29c}$$

where $\text{Var}(p)$ is the variance of the relative severities/sizes, that is, $\text{Var}(p) = (1/n)\sum_{i=1}^{n}(p_i - (1/n))^2$. Combining Equations 14.29b and 14.29c gives us

$$H^* = \frac{n^2}{n-1}\text{Var}(p).$$

14.4.2 Hall–Tideman and Rosenbluth Indices

The concentration indices developed by Rosenbluth and Hall and Tideman resemble one another in form and in character. The basic difference between these measures and the HHI is that they include the number of firms in the calculation of the concentration measures. This makes the measures reflect the conditions of a particular industry.

In the Hall–Tideman index, the ranking of the ith firm is number I, that is, the smallest firm gets rank 1 and the largest firm gets rank n. In the Rosenbluth index, the ranking is reversed, that is, the largest gets rank 1 and the smallest gets rank n.

For the index given by Equation 14.30b, we have that $1/n \leqslant \text{RI} \leqslant 1$. On the other hand, the HTI has the following boundaries, the first one when all firms have the same share and the last if all the market is concentrated in one firm: $(1/n) \geqslant \text{HTI} \geqslant 1/(2n-1)$.

The Rosenbluth index has the form $\text{RI} = 1/2C$, where C can be interpreted as an area above the concentration curve; see, for example, Piesch (1975) and Bikker and Haaf (2002). Hence, RI can be seen to be linked to the well-known Gini coefficient, G. It can be seen that $C = \sum_{i=1}^{n}(1 + n - R_{x(i)})p_{(i)} - 1/2$.

Note that $\text{RI} = 1/(n(1 - G))$, where G is the Gini coefficient. In a similar way, we can write $\text{HTI} = 1/(n(1 + G))$.

14.4.3 Entropy Measure

The entropy measure has its origin in information theory and it is measured with log-base 2. This could be relaxed by changing the scale: $\log_2 p_i = \log p_i / \log 2$. The measure ranges from 0 to $\log_2 n$ and it varies inversely to the degree of concentration. However, an alternative entropy measure has been suggested in the literature: $\varepsilon(p) = -H(p)$ with $-\log n \leqslant \varepsilon(p) \leqslant 0$. The normalized versions to Equation 14.28 and the above-mentioned alternative are $H^*(p) = H(p)/\log n$ and $\varepsilon^*(p) = \varepsilon(p)/\log n$.

14.4.4 Other Measures of Concentration

The comprehensive industrial concentration index (CCI), also known as the Horvath index, is related to the HHI and has the form $\text{CCI} = p_{(n)} + \sum_{i=1}^{n-1} p_{(i)}^2[1 + (1 - p_{(i)})]$, where $p_{(n)}$ is the relative size, market share, of the leading firm. The CCI gives more weight to the $n-1$ smallest firms as compared to the HHI ($2 - p_{(i)}$ as compared to 1); see, for example, Piesch (1975).

Another measure is the Hannah–Kay index, which has the form $\text{HKI} = (\sum_{i=1}^{n} p_i^{\alpha})^{1/(1-\alpha)}$, $\alpha > 0, \neq 1$. Renyi's α-order entropy index is defined as log HKI; see Piesch (1975).

These two measures and others such as the U index and the Hause index are discussed in Bikker and Haaf (2002).

14.4.5 Concentration in Solvency Assessment

The measures of concentration discussed above have been focusing on a firm's relative size on a specific market. In solvency consideration, we do not look at a firm and its market, but we focus on an insurance firm and how its assets and/or risks are concentrated. Then its "asset concentration" can be measured by some of the above-mentioned concentration measures.

The insurance company buys reinsurance as a protection. From a credit perspective, it is better for the company to buy reinsurance from many different reinsurance companies than from only one. This type of credit risk, or default risk, could be measured by using a concentration index.

For a specific line of business it could be better, in the sense of spreading the risks, to have the risks insured in different geographical regions, for example, within a country or between countries. Letting n be the number of different geographical regions where the company sells its insurance policies, a measure of concentration could be used to measure this type of risk.

CHAPTER 15

Capital Requirement

Modeling and Measuring

In this chapter we will discuss different approaches to capital requirement (CR) assessment. In Section 15.1 we briefly look at different approaches for risk modeling that have been used to aggregate risks. They are especially suitable for internal modeling. As we are focusing on a simple standard formula, we cannot use copulas directly, but we should always have the copula thinking in mind. For applications using copulas to aggregate risks and arriving at an economic capital, see, for example, Cech (2006) and Tang and Valdez (2006).

Wang (1998) considered a variety of models combining different aggregate loss distributions that could be used to derive the required capital. The list of methods was extended by Dhaene et al. (2005) using the concept of comonotonicity.

The first approach that we discuss in more detail is a top-down approach usually called capital allocation (Section 15.2). The second approach is a bottom-up approach. For the latter case, we will use the IAA's so-called baseline model discussed in Section 15.3.

In many solvency assessment models the CR is defined in terms of changes in value between two dates. But as risk and CR are related to the variability of future values of a given position, *cf.*, for example, Artzner et al. (1999), which are due to insurance events and market conditions, it is better to only consider future values.

15.1 GENERAL CONSIDERATIONS

One of the main technical problems, or difficulties, in risk management for financial firms (insurance, banking, and securities) is the aggregation of a diverse set of risks. One such diverse set of risks is illustrated in Figure 15.1. These risks are discussed in Chapters 16 through 21.

Aggregating these risks and the CR implied by them can be technically challenging. A discussion on financial conglomerates* is given by Kuritzkes et al. (2003). They use a

* A financial conglomerate is defined by Joint Forum (2001) as "any group of companies under common control whose exclusive or predominant activities consists of providing significant services in at least two different sectors (banking, securities, insurance)".

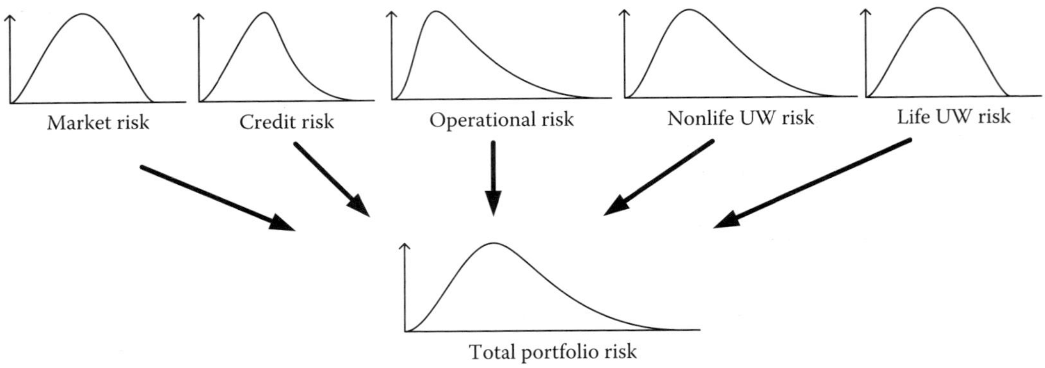

FIGURE 15.1 One example of a risk manager's view of the marginal distributions of different risk types.

"building block" approach to aggregate risks at three successive levels at which risk is usually managed.

Level 1: Aggregation of the stand-alone risks within a single risk type (aggregation of subrisks to a risk type, e.g., market risk).
Level 2: Aggregation across different risk factors within a single business line (e.g., within an insurance company or a bank).
Level 3: Aggregation across different business lines, that is, at the conglomerate level.

Examples of Level 1 aggregations are aggregating different market subrisks (equities, interest rates, and foreign exchange) to market risk and aggregating different underwriting subrisks (mortality, longevity, and disability) to life underwriting risk. At the second level, we aggregate all main risk types to the business level, for example, to an insurance company level. The general characteristics of the three levels of aggregation are stated by Kuritzkes et al. (2003) to be

Level 1: Has many risk factors that are neither strongly concentrated nor highly correlated. As a result, there should be a high diversification benefit at Level 1. The achievable diversification benefit will be driven by the scope of activity.
Level 2: Has fewer risk factors, and they are likely to be more concentrated and more correlated. Consequently, there should be lower diversification benefit at Level 2 than at Level 1.
Level 3: There are only a few factors (business units); some are likely to be much more highly concentrated than others, and correlations will tend to be high. Level 3 should therefore yield the smallest diversification effects of the three levels.

These relationships are examined by Kuritzkes et al. (2003). They also summarize different sources of data (linear correlation coefficients).

In Rosenberg and Schuermann (2004), it is pointed out that all financial firms, no matter which sector they operate in, are influenced by three core risk types: market, credit, and operational risks; see Chapters 17 through 19 for more details. They also note that market risk is usually symmetric and often approximated as normal. On the other hand, both credit and operational risks generate more skewed distributions because of occasional, extreme losses; *cf*. Figure 15.1.

As a loss variable, Rosenberg and Schuermann (2004) use a k-period ahead portfolio return $r_{t+k} = \ln(Y_{t+k}) - \ln(Y_t)$ and compare VaR and TVaR as risk measures. For bank i, return from risk type j (market, credit, or operational) at time t, the CR is written as $r_{ijt} = \alpha_{ij} + \sum_{i=1}^{n} \beta_{ji} x_{jti} + \varepsilon_{ijt}$, where the second term is a factor model as in Equation 12.17 for a homogeneous model of degree one and where x_{jti}, $i = 1, \ldots, n$, are a set of observable risk factors.

The total returns of the portfolio are defined as a weighted sum of risk returns. As an example they use three returns and total return $r_{\mathrm{p}} = w_1 r_1 + w_2 r_2 + w_3 r_3$, $w_1 + w_2 + w_3 = 1$. Mean and variance in an elliptical setting, *cf.* Chapter 13, are defined as $\mu_{\mathrm{p}} = \sum_{j=1}^{3} w_j \mu_j$ and $\sigma_{\mathrm{p}}^2 = \sum_{j=1}^{3} w_j^2 \sigma_j^2 + 2 \sum_{i\neq}^{3} \sum_{j=1}^{3} w_i w_j \sigma_{ij}$. The portfolio VaR is then written as $\mathrm{VaR}_{1-\alpha}(r_{\mathrm{p}}) = \mu_{\mathrm{p}} + \sigma_{\mathrm{p}} F_{\mathrm{p}}^{-1}(1-\alpha)$. Hence, the VaR for the portfolio return is

$$\mathrm{VaR}_{1-\alpha}(r_{\mathrm{p}}) = \mu_{\mathrm{p}} + \sqrt{\sum_{j=1}^{3} w_j^2 \sigma_j^2 \left[F_{\mathrm{p}}^{-1}(1-\alpha)\right]^2 + \sum_{i\neq}^{3} \sum_{j=1}^{3} w_i w_j \sigma_{ij} \left[F_{\mathrm{p}}^{-1}(1-\alpha)\right]^2}. \quad (15.1)$$

The VaR of the portfolio can be written in terms of the second moments of the marginal returns and the inverse cumulative distribution function of the portfolio return density. The quantiles $F_{\mathrm{p}}^{-1}(1-\alpha)$ can be seen as a scaling factor for each volatility.

For the family of elliptical distributions, including the normal, the quantiles of the individual standardized returns are the same as for the portfolio return, that is, $F_{\mathrm{p}}^{-1}(1-\alpha) = F_{\mathrm{i}}^{-1}(1-\alpha)$, $i = 1, 2, 3$.

Changing the quantile $F_{\mathrm{p}}^{-1}(1-\alpha)$ for $F_{\mathrm{i}}^{-1}(1-\alpha)$, $i = 1, 2, 3$, in Equation 15.1, we get what Rosenberg and Schuermann (2004) calls a hybrid VaR:

$$H\text{-}\mathrm{VaR}_{1-\alpha}(X_{\mathrm{p}}) = \mu_{\mathrm{p}} + \sqrt{\sum_{j=1}^{3} w_j^2 \sigma_j^2 \left[F_i^{-1}(1-\alpha)\right]^2 + \sum_{i\neq}^{3} \sum_{j=1}^{3} w_i w_j \sigma_{ij} F_i^{-1}(1-\alpha) F_j^{-1}(1-\alpha)},$$

when all risks are perfectly correlated, that is, $\rho_{ij} = 1$, $\forall i, j$, we have an additive VaR:

$$\text{Add-}\mathrm{VaR}_{1-\alpha}(X_{\mathrm{p}}) = \mu_{\mathrm{p}} + \sqrt{\sum_{j=1}^{3} w_j^2 \sigma_j^2 \left[F_i^{-1}(1-\alpha)\right]^2} = \sum_{j=1}^{3} w_i \mathrm{VaR}_{1-\alpha}(X_i).$$

They also assume that the joint distribution is multivariate normal. Hence each marginal is normal and we get the normal VaR, N-VaR, by setting $F_i^{-1}(1-\alpha) = \Phi^{-1}(1-\alpha)$, that is, equal to the inverse standard normal cumulative distribution function.

Rosenberg and Schuermann (2004) compare the H-VaR, Add-VaR, and N-VaR together with a fourth VaR based on a copula approach. In this approach, they fit market-, credit-, and operational risks-related returns, and use them to generate the inverse marginal cumulative distribution functions as inputs to the copula to generate the total risk distribution. The time horizon over which the risks are measured and modeled was harmonized. For example, banks usually measure market risk at a daily frequency. The daily to annual conversion is made by using the "root-t rule": $\text{VaR}_{1\text{ year}} = \sqrt{252} \cdot \text{VaR}_{1\text{ day}}$, where 252 is the number of trading days ($\sqrt{252} \approx 15.9$).

To aggregate the three risk types, you need to know their marginal distributions, specify a copula, assign correlations, and have a view on the relative weight or contribution each marginal would have on the joint distribution (using different business risks).

In the comparison, both normal and Student t copulas were used. The general result was that Add-VaR < H-VaR < Normal Copula VaR < N-VaR and that H-VaR and the Normal Copula VaR have a similar shape and are close to each other. They state that "it is striking how close H-VaR is to Copula-VaR, though always more conservative."

Rosenberg and Schuermann (2004) also compared the VaR results with results obtained using expected shortfall (ES) as a risk measure. Their general conclusions are that there are no differences in the shape in ES as compared to VaR. The ES is, by definition, always greater than VaR and their sensitivities to changes in business mix and correlation are very similar.

To measure the VaR sensitivity to the choices of copula, Rosenberg and Schuermann (2004) also used a Student t-copula with 5 and 10 degrees of freedom. As we know from Chapter 13, lower degrees of freedom increases tail dependence and tail risk. The Copula-VaR is typically larger by approximately 0.03% for a Student $t(10)$ copula than for the normal copula. For a Student $t(5)$ copula, the difference is approximately 0.06% as compared with the normal. They also found that for a strong tail dependence, Student $t(5)$ copula, the H-VaR approach gives a fairly accurate estimate of the copula VaR. It should be noted that both Student t and the normal distributions are symmetric distributions.

The approaches described above are called top-level aggregation approaches by Aas et al. (2005) and are used by them. Other similar works in this area were done by Ward and Lee (2002) and the IAA's baseline model discussed in Section 15.3. We call this approach a "bottom-up approach" as one develops marginal models for the annual loss distribution of each risk type independently and then "adds them up" or merges them to a joint distribution using a correlation structure or a copula function.

Another approach was used by Alexander and Pézier (2003), called a base-level aggregation approach by Aas et al. (2005). This approach is based on a common risk factor/type model used for all risk types. At the end the marginal distributions are added together. The common risk factor model identifies the economic risk factors that have most influence on different risk types and use a simultaneous risk model for these risk factors. The model includes a description of the dependence structure among risk factors through a correlation structure or a copula function.

The models we have briefly discussed above are possible approaches for internal modeling. In developing a standard formula for the CR we cannot use the copula theory directly, but we can have the copula thinking in mind!

15.2 TOP-DOWN APPROACH: CAPITAL ALLOCATION

Assume that we have captured the overall solvency CR SCR and want to allocate this overall SCR to different main risk types, such as the underwriting risk, the market risk, and the credit risk. The reason for allocating the SCR to subrisks could be due to regulation or internal risk management requirements in setting risk-adjusted profitability.

The allocation of the overall SCR can also show the contribution of each business unit to the total company risk; see, for example, Brehm et al. (2007). It can also be used for premium setting and getting limits on each LoB. The allocation of capital is also discussed by, for example, Meyers (2003), Tsanakas (2007), and Dhaene et al. (2008c, 2009).

The capital allocation problem is not only focusing on CRs but is also an approach that could be used by investors. Following McNeil et al. (2005), we assume an investor who can invest in d different investment areas with losses represented by the random variables $X_i, i = 1, 2, \ldots, d$. Depending on the area of application, three different economic interpretations are given in McNeil et al. (2005):

- *Performance measurement:* The investor is a financial institution and the X_i's represent the negative profit and loss of the d different LoBs.
- *Loan pricing:* The investor is a loan book manager responsible for a portfolio of d loans.
- *General investment:* The investor is either an individual or institutional investor, and the X_i's represent the negative profits and losses corresponding to a set of investments in various assets.

Usually this type of performance is measured by some return on risk-adjusted capital approach (RORAC), such as the ratio expected profit/risk capital.

In practice, a two-step procedure is used; see McNeil et al. (2005). We formulate this in terms of the overall CR:

1. Compute the overall solvency capital requirement (SCR) as $C(X)$, where $X = \sum_{i=1}^{d} X_i$ and C is a particular risk measure as discussed in Chapter 14, for example, the VaR or TVaR.
2. Allocate the overall CR $C(X)$ to the individual risk types according to some mathematical capital allocating principle such as, if $R(X_i)$ denotes the CR allocated to the risk type X_i,

$$C(X) = \sum_{i=1}^{d} R(X_i). \tag{15.2}$$

In this chapter we will focus on the second step, that is, the procedure to allocate the overall solvency CR to risk types. This is called a *capital allocation principle*. For a more formal setup, the reader is referred to Chapter 6.3 in McNeil et al. (2005).

One approach for allocating $C(X)$ to the d risk types is to calculate the risk measure separately for each risk type, for example, for each business unit, and then spreading the overall risk measure proportionally. This usually assumes that $\sum_{i=1}^{d} C(X_i) \neq C(X)$, depending on diversification/aggregation effects; see, for example, Brehm et al. (2007).

Capital allocation is also discussed in Albrecht (2004b). Dhaene et al. (2009) reviews different allocation principles and propose an optimal capital allocation approach. The approach is based on a decision criteria saying that *capital should be allocated such that for each business unit the allocated capital and the loss are sufficient close to each other.* The appropriate allocation is based on an optimization problem involving a deviation function, weight factors to the different possible outcomes of the deviation function, and a measure of exposure or business volume. By choosing different measures, weights, and deviation functions, different classes of allocation principles are derived. These classes are

- Business-unit-driven allocations
- Aggregate-portfolio-driven allocations
- Market-driven allocations
- Allocation with respect to the default option.

CTE-based allocation principles are also discussed in Dhaene et al. (2008c).

15.2.1 Marginal Decomposition and the Euler Capital Allocation Principle

The marginal, or incremental, impact of a business segment (business unit or a risk type) on a firm's total risk measure is the decrease in the overall risk measure that would result from an incremental proportional decrease to that risk type or business segment; see, for example, Venter et al. (2006), Venter (2007), and Brehm et al. (2007). If all these incremental impacts sum up to the firm's overall risk, then we call this allocation a marginal decomposition.

In Equation 15.2, we define the marginal decomposition as

$$R(X_i) = \lim_{\varepsilon \to 0} \frac{C[X + \varepsilon X_i] - C[X]}{\varepsilon}. \tag{15.3}$$

A measure has to be positive homogeneous, or scaleable, to have a marginal decomposition; so we restrict our discussion to risk measures that follow property P3 of Section 14.1, that is, are positive homogeneous: $C(aX) = aC(X),\ a \geqslant 0$.

Assume that we have portfolio weights $\lambda_i \neq 0, i = 1, 2, \ldots, d$, such as $X(\lambda) = \sum_{i=1}^{d} \lambda_i X_i$, and a total risk measure defined by $C[X(\lambda)] = \sum_{i=1}^{d} \lambda_i R_i(\lambda)$, where the mapping R_i is called a per-unit capital allocation principle associated with the risk measure C. We also assume that C is differentiable and then, from Euler's theorem, see Section 12.3.1, we obtain that $C[X(\lambda)] = \sum_{i=1}^{d} \lambda_i (\partial C[X(\lambda)]/\partial \lambda_i)$.

Then the mapping $R_i(\lambda) = \partial C[X(\lambda)]/\partial \lambda_i$ will be called the *Euler capital allocation principle*; see definition 6.25 in McNeil et al. (2005). The following results are given in op. cit.

$$\text{The covariance principle: } \mathrm{Var}(X) = \sum_{i=1}^{d} \mathrm{Cov}(X_i, X). \tag{15.4}$$

From Equation 15.4 we also obtain the standard deviation principle, that is, $\sqrt{\mathrm{Var}(X)} = \sqrt{\sum_{i=1}^{d} \mathrm{Cov}(X_i, X)}$.

$$\text{The VaR principle: } \mathrm{VaR}_{1-\alpha}[X] = \sum_{i=1}^{d} E[X_i | X = \mathrm{VaR}_{1-\alpha}[X]], \tag{15.5}$$

where $\lambda_i = 1, \ \forall i$.

$$\text{The Tail-VaR principle: } \mathrm{TVaR}_{1-\alpha}[X] = \sum_{i=1}^{d} E[X_i | X \geqslant \mathrm{VaR}_{1-\alpha}[X]]. \tag{15.6}$$

These measures are also discussed in Venter et al. (2006), Venter (2007), and Brehm et al. (2007).

The Euler allocation principle is also discussed for elliptical distributions in McNeil et al. (2005).

15.2.2 Co-Measures

One approach to allocate risk measures is the *method of co-measures*; see, for example, Venter et al. (2006), Venter (2007), and Brehm et al. (2007). The co-measures generalize the idea of covariance.

Definition: If the total CRs $C(X)$ can be written as

$$C(X) = E[h(X)L(X)|g(X)], \tag{15.7a}$$

where g is some condition about X, for example, $X > b$, L is some function for which the conditional expectation exists, and h is an additive function, that is, $h(X+Y) = h(X) + h(Y)$. Then the co-measure is defined by

$$R(X_i) = E[h(X_i)L(X)|g(X)], \tag{15.7b}$$

such that $C(X) = \sum_{i=1}^{d} R(X_i) = \sum_{i=1}^{d} E[h(X_i)L(X)|g(X)]$.

Many risk measures can be written in this form. Usually only h and L are needed. By the additivity of h the co-measure given by Equation 15.7b can be decomposed further:

$$R(X_i) = E\left[\sum_{j=1}^{d} h_j(X_i)L(X)|g(X)\right].$$

Covariance: Let $h(X) = L(X) = X - E(X)$, then $R(X_i) = E[(X_i - E(X_i))(X - E(X))] = \text{Cov}(X_i, X)$ and we have the decomposition given by Equation 15.4. We have used the fact that "unconditional" is a special case of conditional here.

Co-VaR: Let $h(X) = X$, and $L(X) = 1$ and $g(X) : X = \text{VaR}_{1-\alpha}[X]$, then the co-VaR is defined by Equation 15.5, that is, $R(X_i) = E[X_i | X = \text{VaR}_{1-\alpha}[X]]$.

Co-TVaR: Let $h(X) = X$, and $L(X) = 1$ and $g(X) : X \geqslant \text{VaR}_{1-\alpha}[X]$; then the co-TVaR is defined by Equation 15.6, that is, $R(X_i) = E[X_i | X \geqslant \text{VaR}_{1-\alpha}[X]]$.

The co-TVaR is also discussed in Landsman and Valdez (2003). Other risk measures and their co-measures are discussed in Venter et al. (2006) and Brehm et al. (2007). For example, the exponential risk measure defined by Equation 14.16 has both a simple co-measure, which is not a marginal decomposition, and a more complicated version having a marginal decomposition; see Venter et al. (2006). Spectral measures and cospectral measures, including the TVaR, are also discussed in op. cit. Distortion risk measures are discussed in, for example, Tsanakas (2007).

15.3 BOTTOM-UP APPROACH: IAA'S BASELINE MODEL

15.3.1 First-Order Approximation

Let X be an $N(\mu, \sigma^2)$-random variable. Let $\phi(\cdot)$ denote the probability density function of the standard normal distribution and $\Phi(\cdot)$ its cumulative distribution function.

The *cumulant generating function* is unique for each distribution. For a general continuous variable, it can be written as a series expansion

$$\varphi_Y(t) = \mu t + \sigma^2 \frac{t^2}{2} + \kappa_3 \frac{t^3}{3!} + \kappa_4 \frac{t^4}{4!} + \cdots,$$

where μ and σ^2 are the expectation and variance, respectively. κ_3 and κ_4 are the higher cumulants of the distribution. The normal distribution has the following cumulant-generating function:

$$\varphi_Y(t) = \mu t + \sigma^2 \frac{t^2}{2},$$

with all higher cumulants equal to zero. *Thus the normal distribution can be viewed as a first-order approximation of the true unknown distribution.* As in IAA (2004, Appendices H1 and H2), if we apply this to all risk components, then the aggregate risk will result in a multivariate normal distribution that serves as a first-order approximation or a *baseline model.*

15.3.2 Standard Deviation Principle as a Baseline Risk Measure

If we use the standard deviation as the baseline risk measure, then the CR could be defined as the VaR minus the mean ("best estimate")

$$C = \mu + k_{1-\alpha}\sigma - \mu = k_{1-\alpha}\sigma, \tag{15.8}$$

that is, the solvency capital level minus the mean, that is, a factor k times the standard deviation, where, for example, $k_{0.99} = 2.33$ and $k_{0.995} = 2.58$ depending on the level of confidence (99% or 99.5%, respectively). The choice of the "correct" level of confidence is a regulatory decision rather than an actuarial. Risk measures in general are discussed in Chapter 14.

When the standard deviation is used as a risk measure and the CR is a multiple of the standard deviation*

$$C_j = k_{1-\alpha}\sigma_j, \tag{15.9}$$

cf. Equation 15.8 above, then the CR of the aggregate risk can be written as (summing over r risks)

$$C = \sqrt{\sum_{i=1}^{d}\sum_{j=1}^{d} \rho_{ij}C_iC_j} = k\sqrt{\sum_{i=1}^{d}\sum_{j=1}^{d} \rho_{ij}\sigma_i\sigma_j} = k\sqrt{\sum_{i=1}^{d} \sigma_i^2 + \sum_{i\neq}^{d}\sum_{j}^{d} \rho_{ij}\sigma_i\sigma_j}, \tag{15.10a}$$

see, for example, IAA, Appendix H (IAA, 2004). The last term of Equation 15.10a can be also written as

$$C = \sqrt{\sum_{i=1}^{d} C_i^2 + \sum_{i\neq}^{d}\sum_{j}^{d} \rho_{ij}C_iC_j}. \tag{15.10b}$$

If the distribution is skewed, that is, has a heavier tail to the right; the factor $k_{1-\alpha}$ should be larger than that for the normal distribution. The $k_{1-\alpha}$ might be different for different risk categories, depending on the form of the specific distribution function.

The baseline approach is also called the linear correlation approach in some jurisdictions. This approach is correct if the risks are assumed to be multivariate normally distributed. The approach also holds for a more general framework shown in Dhaene et al. (2005). The framework in op. cit. can be seen as an extension of the work done by Wang (1998). In Dhaene et al. (2005), a general framework within a given class of translation-scale invariant distributions was given, together with a situation when both normal and lognormal random variables were involved.

As we will relax the linear correlation assumptions, we will not, in general, use this terminology. We will also change the wording from "correlation" to "dependency" and hence be able to use a dependence matrix that could consist of rank correlations, tail correlations, and so on.

15.3.3 Assume Nonnormality

Our unknown and hypothetical distribution was assumed to follow a normal distribution in the previous section. Following the discussion in Section 14.3.1, we will relax this assumption

* Another way of writing the capital charge would be to rewrite it as $C_j = k_{1-\alpha}\sigma_j = \mu_j k_{1-\alpha} \upsilon_j$, where μ_j is the expected value and $\upsilon_j = \sigma_j/\mu_j$ is the coefficient of variation.

and assume that the distribution is positively skewed, but still unknown. Similar results hold for negatively skewness.

We will use a Cornish–Fisher expansion to transform the quantiles and the tail expectations of the skewed distribution into a standard normal distribution.

In the traditional approximation, the $1-\alpha$ quantile of a skew distribution is written in terms of the $1-\alpha$ quantile of the standard normal distribution with a correction for the skewness: $\mu + k_{V,1-\alpha}(\gamma)\sigma$, where $k_{V,1-\alpha}(\gamma)$ is a new quantile of the standard normal distribution and a function of the skewness in the original distribution. V stands for VaR.

The approach can be generalized so that the tail expectation of a skew distribution also can be written in terms of the tail expectation of the standard normal distribution with a correction for the skewness: $\mu + k_{TV,1-\alpha}(\gamma)\,\sigma$, where $k_{TV,1-\alpha}(\gamma)$ is a new quantile of the standard normal distribution and a function of the skewness in the original distribution. TV stands for TVaR.

These results are summarized in Tables 14.1, 14.2a, 14.3a, and 14.3b.

This illustrates that we can use the Cornish–Fisher approximation for a skewed distribution to arrive in a new "baseline model" such as

$$C^* = \sqrt{\sum_{i=1}^{r}\sum_{j=1}^{r} \rho_{ij} C_i^* C_j^*} = \sqrt{\sum_{i=1}^{r}\sum_{j=1}^{r} \rho_{ij} k_{1-\alpha}(\gamma_i)\sigma_i k_{1-\alpha}(\gamma_j)\sigma_j}. \tag{15.11}$$

15.3.4 A Pragmatic Solution

To model the CRs we have two main problems to deal with:

- *The nonnormality:* Was discussed above in Section 14.3.1
- *The nonlinearity:* Could be solved using a benchmark approach and tail correlations (Taking into consideration dependence structure and copulas)

We use the IAA's baseline as a start (IAA, 2004). The general structure of the model is taken *from* a linear correlation structure. We use four risk categories and define the general baseline structure letting $C_i = k\sigma_i$, where k is a quantile function, which is clearly defined for a standard normal distribution. Then the total risk is

$$C = \sqrt{\begin{array}{c} C_1^2 + C_2^2 + C_3^2 + C_4^2 + 2\rho_{12}C_1C_2 + 2\rho_{13}C_1C_3 + 2\rho_{14}C_1C_4 \\ + 2\rho_{23}C_2C_3 + 2\rho_{24}C_2C_4 + 2\rho_{34}C_3C_4 \end{array}}.$$

Assume a situation where we have the left-hand matrix with unknown dependence structure. Assume that we believe in that the fourth risk category is fully correlated with the other three, as described in the middle matrix. If the linear correlation were exact, then we would have the right-hand matrix, that is, if the risks are fully correlated, then $C = C_1 + C_2 + C_3 + C_4$. But this is not always what we believe in! In the Müller report (1997), the early version of the NAIC risk-based system was described: once all RBC values

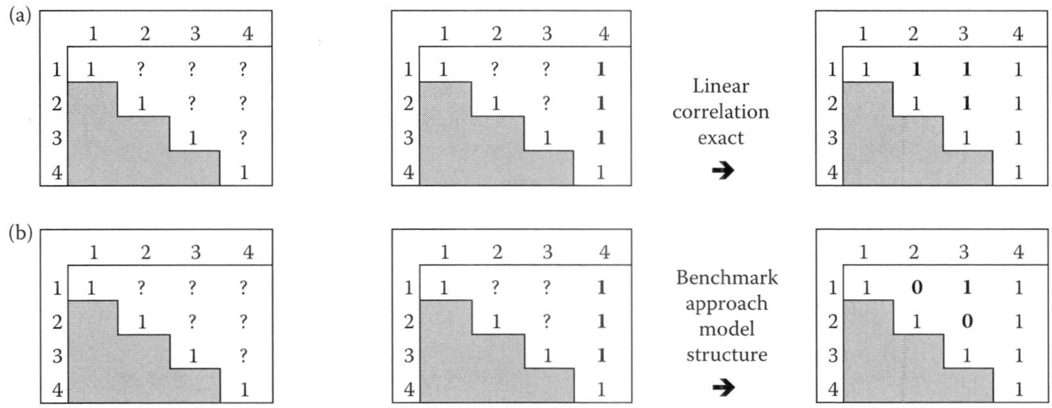

FIGURE 15.2 Illustrations of how the final correlation structures will be using (a) the linear correlation approach and (b) the benchmark approach.

of the individual categories have been calculated, they are combined into the total RBC. For this the individual values are, however, not simply added up but compensation is made because not all risks will cause losses simultaneously. If it is assumed that both asset risk and interest rate risk ($C1$ and $C3$) are completely correlated and the technical risk ($C2$) is not related to either of them and in addition that the business risk ($C4$) is completely correlated with the other three risks, this will result in a total RBC in life insurance ($\mathrm{RBC_{LV}}$) as follows:

$$\mathrm{RBC_{LV}} := C_4 + \sqrt{C_2^2 + (C_1 + C_3)^2}.$$

The "Benchmark approach" is a pragmatic solution for nonlinear relationships. In terms of the above matrices, the additional assumption that first and third risks are fully correlated and that the second risk is uncorrelated with the first and third risks could be described as given in Figure 15.2. Hence the matrix will not usually be positive definite.

We proceed as follows. The fourth risk is fully correlated with all the other three, that is, $\rho_{4,(123)} = 1$, giving us the following pragmatic structure: $C = C_4 + C_{(123)}$. Consider now $C_{(123)}$. The second risk is uncorrelated with the other two risks, that is, $C_2^2 + C_{(13)}^2$. The last risk is $C_{(13)}^2 = (C_1 + C_3)^2$, since they are fully correlated. This gives us the RBC structure above! Note that each main risk category can be thought of as consisting of different subrisks; for example, we can have the following structure for risk C_2: $C_2^2 = C_{21}^2 + (C_{22} + C_{23} + C_{24})^2 + 2\rho_{21,25}C_{21}C_{25} + C_{25}^2$.

15.3.5 Calibration for Skewness*

In the European Solvency II project, the target measure is the SCR. This is a theoretical capital level that an insurance undertaking shall fulfill to be on the safe side.

The European Commission (COM) has stated that "The parameters in the SCR should be calibrated in such a way that the quantifiable risks to which an institution with a diversified portfolio of risks is exposed are taken into account and based on the amount of

*This chapter is based on Sandström (2007b).

economic capital corresponding to a ruin probability of 0.5% (Value at Risk of 99.5%) and a one year time horizon .... The methods used to check that this level is effective must be defined. The SCR should be based on a going-concern basis. These principles shall apply regardless of whether a standard formula or an internal model is used." COM (2006, para 19).

The standard formula for the overall capital requirement, SCR, will be calculated using a *modular approach*. This means that each risk charge, for example, from the underwriting risk, market risk, credit risk, and operational risk, should be calculated in a similar way, but with different methods, such as a factor-based approach or a scenario-based approach, could be used. It is also easier for the supervisory authority to compare the influence of some of the modules to the overall CR, and also to compare the development over time for a single company.

The risk charges, as calculated by the modular approach, should be aggregated using the baseline approach as proposed by IAA (2004) and COM (2006). Tail dependency should be taken care of in the dependence structure used. Dependence and tail dependence were discussed in Chapter 13.

Each main risk module may be split up into subrisk modules; for example, market risk could be split up into interest rate, equity, property, spread, currency, and concentration risk modules. These modules will be aggregated using the baseline approach.

Using the baseline approach, we obtain, with d risk modules,

$$k_{1-\alpha}^2\sigma_{\text{SCR}}^2 = \sum_{i=1}^{d} k_{1-\alpha}^2\sigma_i^2 + 2\sum_{i\neq}^{d}\sum_{j}^{d} k_{1-\alpha}^2\rho_{ij}\sigma_i\sigma_j, \qquad (15.12a)$$

where k is the $1-\alpha$ percentile (one-sided) of the standard normal distribution, σ is the standard deviation of the ith normal distribution, and ρ is the dependence between the risk variables. Rewriting Equation 15.12a in general risk-charge terms, we obtain

$$C_{\text{SCR}}^2 = \sum_{i=1}^{d} C_i^2 + 2\sum_{i\neq}^{d}\sum_{j}^{d} \rho_{ij}C_jC_j. \qquad (15.12b)$$

In the standard formula for the SCR, each risk charge $C_i = k_{1-\alpha}\sigma_i$ is calculated either using a factor-based model or a scenario model. But underlying these approaches is the idea that they should be calibrated so as to fulfill the assumption made for all risk charges, that is, the same risk measure, the same confidence level, and time horizon.

In CEIOPS'* consultation paper no. 20, CEIOPS (2006b), the calibration is developed further:

- The calibration of the standard formula should be consistent with the key aspects of the SCR's design.

*CEIOPS: Committee of European Insurance and Occupational Pensions Supervisors, for more information see www.ceiops.org

- A first step toward achieving this aim is to ensure that the different modules of the standard formula are calibrated in a consistent manner. The most intuitive approach seems to be to apply the calibration objectives for the overall SCR (confidence level, time horizon, etc.) to each individual risk module. So, for example, the choice of parameters, factors, and scenarios for assessing interest rate risk should be consistent with a 99.5% probability of survival over 1 year, while also taking account of any model error arising from the particular technique chosen to assess that risk.
- This approach would have a number of advantages:
  - It provides an unambiguous reference for the calibration of each module, without the need for arbitrary decisions on the relative weight each risk should contribute to the overall SCR
  - It would give supervisors a clearer picture of the composition of the overall SCR and of the influence of individual risk drivers on the overall risk profile of each insurer, informing the choice of supervisory action
  - It would facilitate the use of partial internal models

  But, clearly, the aggregation of the modules needs to reflect cross-risk diversification effects to avoid overstating CRs.
- In principle, the approach to aggregation should also be consistent with the calibration objectives for the SCR. If linear correlation techniques are used, correlation coefficients between risk pairs should be chosen for consistency with the SCR objectives expressed in Section 2 of CEIOPS (2006b). But, as noted above, such techniques are problematic because they are only mathematically correct in the case of multivariate normally distributed risks. The practical consequence of this is that the resulting CRs would be too low in the case of risks that follow a heavily skewed distribution or for risks where the dependency relationship is nonlinear.

This means that all risk charges C_i should be calculated in a consistent way with the C_{SCR}. If all the underlying probability distributions are normal (with $\gamma = 0$) or if all underlying distributions have the same skewness, then the approach is straightforward. The problem is that most of the underlying probability distributions are differently skewed (to the right); *cf*., for example, Figure 15.1. A pragmatic solution is used to illustrate the effect of skewed distributions to the overall capital requirement, SCR. The solution is also used to calibrate the standard formula.

15.3.5.1 General Calibration Problems

The general problem in using the baseline approach for aggregating the different risk modules as proposed for the European SCR, as discussed above, is discussed in Pfeifer and Strassburger (2008). They show that it is not sufficient for general purposes with a calibration of the standard SCR for the skewness. For the class of risk distributions they consider, the square root formula tends to underestimate, and in some cases also overestimate, the true SCR. This is a general drawback of the baseline formula outside the world of elliptical risk distributions.

In their view, the only reasonable "all purpose" calibration is to apply the maximum value of one to all correlations, that is, having $C_{SCR} = \sum_{i=1}^{d} C_i$.

In the next chapter we will introduce another way of calibrating the CR formula using the above-mentioned Cornish–Fisher expansion.

15.3.5.2 Calibration

In Equation 15.12b, all the risk charges C_i, $i = 1, \ldots, d$, will be calculated with the same risk measure (e.g., VaR or TVaR), the same confidence level, and time horizon as the overall risk charge C_{SCR}.

We assume that the skewness of each probability distribution is known, especially that of the overall risk requirement SCR. When we discuss the examples in the next section we will assume that the n skewness γ_i, $i = 1, 2, \ldots, d$, are known and use them to estimate the skewness of the unknown distribution for the capital requirement SCR. Using the Cornish–Fisher expansion, see Appendix B, we can develop the capital charge for SCR in terms of the transformed risk measure (VaR or the TVaR) in terms of the d risk modules:

$$
\begin{aligned}
C_{SCR}^2 &= k_{\cdot,1-\alpha}^2(\gamma_{SCR})\sigma_{SCR}^2 \\
&= k_{\cdot,1-\alpha}^2(\gamma_{SCR})\sum_{i=1}^{d}\sigma_i^2 + 2k_{\cdot,1-\alpha}^2(\gamma_{SCR}) \times \sum_{i\neq}^{d}\sum_{j}^{d}\rho_{ij}\sigma_i\sigma_j,
\end{aligned}
$$

where the dot in '$k_{\cdot}$' indicates either a V (VaR) or a TV (TVaR). The functional forms of '$k_{\cdot}$' are given in Tables 14.1, 14.2a, and 14.3b.

This model is in conflict with the "consistency approach" adopted by CEIOPS as the $k_{\cdot}$ to each d risk standard deviation is not giving a correct risk charge. This is achieved by adding the "correct" quantiles or hazard functions:

$$
\begin{aligned}
C_{SCR}^2 &= \sum_{i=1}^{d}\left[\frac{k_{\cdot,1-\alpha}^2(\gamma_{SCR})}{k_{\cdot,1-\alpha}^2(\gamma_i)}\right]k_{\cdot,1-\alpha}^2(\gamma_i)\sigma_i^2 \\
&\quad + 2\sum_{i\neq}^{d}\sum_{j}^{d}\rho_{ij}\left[\frac{k_{\cdot,1-\alpha}^2(\gamma_{SCR})}{k_{\cdot,1-\alpha}(\gamma_i)k_{\cdot,1-\alpha}(\gamma_j)}\right]k_{\cdot,1-\alpha}(\gamma_i)\sigma_i k_{\cdot,1-\alpha}(\gamma_j)\sigma_j \\
&= \sum_{i=1}^{d}\frac{k_{\cdot,1-\alpha}^2(\gamma_{SCR})}{k_{\cdot,1-\alpha}^2(\gamma_i)}C_i^2 + 2\sum_{i\neq}^{d}\sum_{j}^{d}\rho_{ij}\frac{k_{\cdot,1-\alpha}^2(\gamma_{SCR})}{k_{\cdot,1-\alpha}(\gamma_i)k_{\cdot,1-\alpha}(\gamma_j)}C_iC_j \\
&= \sum_{i=1}^{d}f_{\cdot,i}^2C_i^2 + 2\sum_{i\neq}^{d}\sum_{j}^{d}\rho_{ij}f_{\cdot,i}C_if_{\cdot,j}C_j, \qquad (15.13)
\end{aligned}
$$

where f_i denotes the calibration factor for the ith risk charge and the dot is for either VaR or TVaR.

Comparing Equation 15.13 with the starting Equation 15.12b shows that each risk charge has to be calibrated through the multiplication with a calibration factor f.

These calibration factors are

$$\text{VaR risk measure: } f_{V,i} = \frac{k_{V,1-\alpha}(\gamma_{SCR})}{k_{V,1-\alpha}(\gamma_i)} = \frac{z_{1-\alpha} + \dfrac{\gamma_{SCR}}{6}\left(z_{1-\alpha}^2 - 1\right)}{z_{1-\alpha} + \dfrac{\gamma_i}{6}\left(z_{1-\alpha}^2 - 1\right)}$$

$$= \frac{6z_{1-\alpha} + \gamma_{SCR}\left(z_{1-\alpha}^2 - 1\right)}{6z_{1-\alpha} + \gamma_i\left(z_{1-\alpha}^2 - 1\right)}, \tag{15.14a}$$

$$\text{TVaR risk measure: } f_{TV,i} = \frac{k_{TV,1-\alpha}(\gamma_{SCR})}{k_{TV,1-\alpha}(\gamma_i)} = \frac{1 + (\gamma_{SCR}/6)z_{1-\alpha}^3}{1 + (\gamma_i/6)z_{1-\alpha}^3}$$

$$= \frac{6 + \gamma_{SCR} z_{1-\alpha}^3}{6 + \gamma_i z_{1-\alpha}^3}. \tag{15.14b}$$

As there will always be experts having knowledge about the different distribution functions and the dependence between the variables, we assume that all skewnesses are known, except for the overall CR distribution. One way of estimating the γ_{SCR} is to calculate it as a weighted mean of the n skewness parameters $\gamma_i, i = 1, \ldots, d$. As weights we can use the capital charges: $\hat{\gamma}_{SCR} = \sum_{i=1}^{d} w_i\gamma_i$, where $w_i = C_i / \sum_{j=1}^{d} C_j$.

Some simple examples, illustrating the difference between no calibration and the case where a calibration has been done is given in Sandström (2007a). These illustrations are summarized in Table 15.1, showing the effect with and without calibration.

The calibration approach discussed above was discussed and compared to other aggregation methods in Savelli and Clemente (2009). The other methods compared were a

TABLE 15.1 Summary of the Calculations Made in Sandström (2007a), First Assuming Normality, that is, Ignoring the Assumed Skewness, and then Incorporating the Skewness, without and with Calibration

	SCR	
	VaR, $1 - \alpha = 0.995$	**TVaR, $1 - \alpha = 0.99$**
Assuming normality	100	100
Assuming skewness	119.1	214.1
Without calibration		
With calibration	121.3	260.0

Note: For comparison we have put the normality case to 100. The assumptions made in these examples are $\sigma_1 = \sigma_2 = 10,\ \rho_{12} = 0.5,\ \gamma_1 = 0,\ \gamma_2 = 1.0$.

simulation approach as per an Internal Model, aggregation according to QIS3, and a formula based on an empirical multiplier.

15.3.5.3 More on Skewness

Rosenberg and Schuermann (2004) simulated marginal risk distributions for market, credit, and operational risks; cf. also Section 15.1. The skewness they found from 200,000 simulations was $\gamma_{MR} = 0.0$ for market risk, $\gamma_{CR} = 1.1$ for credit risk, and $\gamma_{OR} = 4.5$ for operational risk. For the latter two risks we have changed the sign on the skewness to illustrate the fact that we talk of losses > 0 and not in negative terms. The total portfolio risk after simulating the weighted average of the marginal returns from a joint density using a normal copula gave a skewness of $\gamma_P = 1.0$. In the same study, the kurtosis is also calculated. For market risk the kurtosis was 3.9, for credit risk 5.2, and for operational risk 35.3. This shows that the operational risk has the fattest tail and the symmetric market risk has the thinnest. It is, of course, possible to include the kurtosis in the Cornish–Fisher expansions, but we have not done it here.

The k-values for these three risk types are given in Table 15.2 for both VaR and TVaR.

Alexander and Pézier (2003) discussed different subrisks constituting part of the market risk. They calculated skewness from historical data and obtained the following values given in Table 15.3.

15.3.6 A Suggestion

We propose to use IAA's baseline model as a standard formula as described above, but we make some adjustments, at least to some extent, to capture the two main problems

TABLE 15.2 Distance Functions for $1 - \alpha = 0.995$ and VaR and TVaR

$k_{\cdot,0.995}$	**Skewness, γ =**	**VaR: $k_{V,0.995}$ =**	**TVaR: $k_{TV,0.995}$ =**
Market risk	0.0	2.58	2.90
Credit risk	1.1	3.61	12.03
Operational risk	4.5	6.81	40.25

Note: Skewnesses for market, credit, and operational risks are from Rosenberg and Schuermann (2004).

TABLE 15.3 Skewness for Different Market Subrisks

Market Risk: Subrisks	**Subrisk Factor**	**Skewness, γ =**
Interest rate	Parallel shift	−1.103
	Slope	0.652
	Volatility	0.955
Equities	Overall level	0.171
Spread	Volatility	0.288
	Credit spread	−0.363

Source: Adapted from Alexander, Carol and Jacques Pézier. 2003. *On the Aggregation of Market and Credit Risks*. ISMA Centre Discussion Papers in Finance 2003-13. ISMA Centre, University of Reading, UK.

encountered earlier, namely

- nonnormality
- nonlinearity

As risk measures we use either VaR or TVaR; see Chapter 14 for more details.

We start with the main baseline model given by Equation 15.10b and use the Cornish–Fisher expansion to change the distance functions '$k_\cdot$' to capture the skewness in the underlying distribution.

The nonlinearity is captured by first using a pragmatic solution given in Section 15.3.4. But as we have stated earlier, we shall always have dependence structure and copulas in mind. We cannot use this directly in a standard formula but as another pragmatic solution we can calculate Kendall's tau assuming an elliptical distribution and "change" the linear correlation coefficients with $\hat{\rho}_{ij} = \sin((\pi/2)\hat{\rho}_{\tau,ij})$ for each pair i and j; *cf*. Table 13.1. Some values for Kendall's tau and the "estimated" linear correlation coefficients are given in Table 15.4a.

Another way would be to use a Gumbel copula approach. If we have any information about the upper tail dependence, we could estimate Kendall's tau by the upper tail dependence; see Table 13.1. Values of the upper tail dependence, Kendall's tau and the corresponding linear correlation coefficient are given in Table 15.4b. As an example, if we "know" that the upper tail dependence must be, say, 0.7, then the corresponding linear correlation coefficient to use would be 0.8284.

The capital charges C_i, $i = 1, 2, \ldots, d$, in (15.10b) can be calculated in different ways. If we think that one of the models for capital charges is homogeneous of degree one, we can use a factor model as described in Section 12.5.1.

If we use a coherent risk measure, such as TVaR, we can use a stress test to capture the risks capital charge; see Section 14.1. But in a pragmatic world we may also use stress tests for noncoherent risk measures if we could calculate the capital charge for the stress test in a

TABLE 15.4a Values of Kendall's Tau and Corresponding Values of the Linear Correlation Coefficient Using the Relation $\hat{\rho}_{ij} = \sin((\pi/2)\hat{\rho}_{\tau,ij})$; *cf*. also Table 13.1

$\hat{\rho}_{\tau,ij} =$	$\hat{\rho}_{ij} =$	$\hat{\rho}_{\tau,ij} =$	$\hat{\rho}_{ij} =$
−1	−1	0	0
−0.9	−0.9877	0.1	0.1564
−0.8	−0.9511	0.2	0.3090
−0.7	−0.8910	0.3	0.4540
−0.6	−0.8090	0.4	0.5878
−0.5	−0.7071	0.5	0.7071
−0.4	−0.5878	0.6	0.8090
−0.3	−0.4540	0.7	0.8910
−0.2	−0.3090	0.8	0.9511
−0.1	−0.1564	0.9	0.9877
		1	1

TABLE 15.4b Upper Tail Dependence (Gumbel Copula), the Corresponding Kendall's Tau and Linear Correlation Coefficient

$\hat{\lambda}_U =$	$\hat{\rho}_{\tau,ij} =$	$\hat{\rho}_{ij} =$
0	0	0
0.1	0.0740	0.1160
0.2	0.1520	0.2365
0.3	0.2345	0.3600
0.4	0.3219	0.4844
0.5	0.4150	0.6067
0.6	0.5146	0.7231
0.7	0.6215	0.8284
0.8	0.7370	0.9159
0.9	0.8625	0.9768
0.91	0.8757	0.9810
0.92	0.8890	0.9848
0.93	0.9024	0.9883
0.94	0.9159	0.9913
0.95	0.9296	0.9939
0.96	0.9434	0.9961
0.97	0.9574	0.9978
0.98	0.9714	0.9990
0.99	0.9856	0.9997

way that is consistent with, for example, a VaR calculation. This means a calculation at the same confidence level and the same time horizon.

Calibration, as discussed above, is also very important to consider when building a standard formula.

CHAPTER 16

Risks and Subrisks

IN THE EARLIER CHAPTERS, we discussed risks, risk types, and main risks. These are the building blocks in the assessment of the SCR. This holds both for a standard formula or the use of internal models. These risk types, which we will define in accordance with IAA (2004), are composed of underlying subrisks.

In the previous chapter, we discussed the aggregation of these risks or, to be more specific, the aggregation of the risk charges put on each risk. The main objective of this chapter is to first discuss the main risks and their subrisks and then to discuss different approaches to capture the risk charges, that is, to look at different risk models.

Risk is inherent in all actions carried out by an insurance undertaking: from designing, pricing, and marketing their products, the underwriting procedure, the calculation of liabilities and technical provisions, and the selection of assets backing these provisions, to the overall claims and risk management. A thorough discussion of the insurer risks is given in, for example, IAA (2004).

Some of the risks are faced by the entire economic environment and some just by the insurance environment. The latter risks are difficult to protect against. For other risks there are different ways of protection or at least actions that can be made to diminish the effects of a realization of a "risky event."

The exposure to risk can be classified into three main levels:

- Risks arising at the entity level (diversifiable).
- Risks faced by the insurance industry (systematic and usually nondiversifiable): Risk affects the entire insurance market or some market segments. Some diversification and asset allocation can protect against systematic risk, as different portions of the market tend to perform differently at different times.
- Risks faced by the whole economy and the whole society (systemic and nondiversifiable): Risk affects the entire financial market or the whole system and not just some specific participants.

Everyone will die and therefore it is not possible to take away the mortality risk from the society *per se* (the systemic part) but an insurance company can, by a health declaration, make a selection that diminishes the risk of dying earlier than expected according to actuarial assumptions (a systematic part).

Risk mitigation techniques, such as pooling and diversification, are discussed in Section 9.1.

The five main categories proposed by IAA are underwriting, credit, market, operational, and liquidity risks. Credit, one of the main risk categories, could in turn be divided into, for example, counterparty default risk, migration risk, spread risk, sovereign risk, and concentration risk.

In Rosenberg and Schuermann (2004), it is pointed out that all financial firms, no matter which sector they operate in, are influenced by three core risk types: market, credit, and operational risks. I would add a fourth and a fifth core risk type, namely the liquidity risk and the legal risk. These are the risks discussed in Basel II project; cf. Sandström (2005). The legal risk is not discussed further.

The five main categories proposed by IAA (2004) are the underwriting risk and the four core risks mentioned above (excluding the legal risk).

- *Underwriting risk* (or the insurance risk) This is the risk an insurance company faces by underwriting the insurance contracts. They include both the perils covered by the specific LoBs and the specific processes associated with the management and conduct of the business. There are a lot of specific risks. Some of them are the subrisks underwriting process risk, pricing risk, claims risk, net retention risk, and the reserving risk.

- *Credit risk* This is the risk of default and change in credit quality of issuers of securities in the investment portfolio, the reinsurance counterparty risk. Other risks are, for example, the direct counterparty default risk, spread risk, sovereign risk, and concentration risk.

- *Market risk* This risk arises from the level or volatility of market prices on assets and involves different movements in the level of various financial variables. Some subrisks are, for example, the asset/liability mismatch risk, reinvestment risk, currency risk, interest-rate risk, and risks in equity and property.

- *Operational risk* This risk is usually seen as "residual risks" not measured by the other categories. It is often defined in terms of risk of losses due to inadequate or failing processes, people, and systems such as fraud, and also risks from external events.

- *Liquidity risk* This risk is the exposure to loss due to the fact that insufficient liquid assets are available to meet CF requirements. As there has been a concentration within the financial sector to a small number of large institutions operating globally, the risk of contagion increases. "The fact that many banks adopt similar modelling techniques, assumptions, and risk management standards, and place a greater reliance on dynamic hedging might also exacerbate such market movements. Common sense suggests it would be prudent to hold liquidity cushions against such a possible outcome, particularly given the opacity of the concentrations" (Large, 2004).

There is also a risk of being a member of a group or a financial conglomerate. We will call this the *participating risk*. This sixth risk type includes partly the *reinsurance counterparty risk* that we have if one of the group members cedes parts of its business to one of the other members (the "reinsurer"). This type of business is usually taken care of in the insurance risk. It also includes *a credit risk type*. Assume a financial conglomerate between an insurer and a bank. The bank insures its credit risk at the group's insurance company. A default in the bank will affect both the assets and the liabilities sides of the insurer. This means that we do not only have a "positive" effect of being a member of a group (in terms of diversification) but also a possible "negative" effect from the participating risk!

When we discuss the different risks and the capital charges put on them, it is important to be aware of the time horizon that is chosen for the specific consideration. The main consideration is to capture a picture of the firm's current financial position. As stated by IAA (2004), a reasonable period for the solvency assessment time horizon is about 1 year. A time horizon of 1 year for the solvency assessment should not be confused with the time horizon needed for valuation of assets and liabilities. When a firm is calculating its own solvency assessments for economic capital, the time horizon will probably be set to a longer period, say, 3–5 years.

The amount of required capital for the solvency assessment, given a fixed time horizon, must be sufficient with a high level of confidence, such as 99% or 99.5%, to meet all obligations. So the time horizon and the level of confidence are two interacting parts in a solvency assessment framework.

Risks and subrisks together with aggregation are also discussed in Bhatia (2009).

The main risks and its subrisks are discussed in Chapters 17 through 21. We start with the four core risk types: market risk (Chapter 17), credit risk (Chapter 18), operational risk (Chapter 19), and liquidity risk (Chapter 20). In Chapter 21 we discuss the underwriting risk (insurance risk).

16.1 AGGREGATION OF (SUB)RISKS

The simplest way to aggregate risk charges is just to add them together. Taking a linear correlation approach or the IAA's baseline approach, see Section 15.3, into consideration, this simply means that just summing up the charges would be the same as assuming that all risks, or risk charges, are fully linearly correlated or dependent. This would be the simplest way of assuming the dependence structure between the risk charges.

Risk charges can be "correctly" consolidated by risk aggregation. Any risk aggregation needs to quantify the dependencies between the risks. Even if the dependency is zero, the diversification effect can be sizeable.

Aggregation of capital charge distributions has traditionally been done for independent variables using convolutions or compound distributions; see, for example, Daykin et al. (1994) and Sundt and Vernic (2009). Useful tools in this context are Panjer and de Pril recursions. As an example, in the Swiss Solvency Test, the aggregation of the nonlife insurance risk is made by convolutions: first between normal and major claims and then with the provision risk; see SST (2006b).

One drawback with these approaches is that the assumption of independence between the capital charges is usually violated. Dependence between risk charges is the rule in practice—independence is an exception. The "least favorable way" to approximate the dependence structure would be to assume the comonotonic dependence structure; see Section 13.1, Dhaene et al. (2002a, 2002b) and Jonk (2008); see also Denuit et al. (2005).

If the company uses an internal model, the aggregation could be built into the model. If, for example, the marginal distributions of the risks are used, then copula aggregation could be used. This is discussed in, for example, Cech (2006) and Tang and Valdez (2006). Nonlinear dependencies–copulas are built into a DFA model by Eling and Toplek (2009) using copula aggregation.

Other, perhaps more classical, statistical methods could also be used. We leave the discussion of aggregating (dependent) variables and internal modeling to the statistical literature—but one important reference is Denuit et al. (2005); see also Jonk (2008).

Different hierarchical trees using Archimedean copulas are discussed in Savelli and Clemente (2009). Different aggregation approaches, such as sum of squares, correlations, copulas, and quadrant correlations ("eggbox approach"), are discussed by Smith (2009).

We are mainly looking at a standard formula approach and the aggregation of risk charges calculated in different risk modules. Both the GDV model and the model proposed by CEA are based on IAA's baseline approach. The calculation of the different main risk modules in the GDV and CEA models is discussed in Chapters 17 through 21. In GDV (2005), it is explicitly stated that their approach is in conformance with the IAA's approach. In the CEA model (CEA 2006b), the conformity with the IAA's model is indirect as they talk about aggregation of the risk charges with diversification and loss absorbance (by liabilities).

To aggregate different risk charges based on risk modules, we propose to use the pragmatic approach using a combination of IAA's baseline model, rank correlations, and tail dependencies, and also the Cornish–Fisher expansion. These approaches are discussed in Section 15.3.

CHAPTER 17

Market Risk

MARKET RISK is introduced into an insurer's operations through variations in financial markets that cause changes in asset values, products, and portfolio valuation (IAIS, 2004).

Market risks relate to the volatility of the market values of assets and liabilities due to future changes of asset prices (/yields/returns). In this respect, the following should be taken into account:

- Market risk applies to all assets and liabilities
- Market risk must recognize the profit-sharing linkages between the asset CFs and the liability CFs (e.g., liability CFs are based on asset performance)
- Market risk includes the effect of changed policyholder behavior on the liability CFs due to changes in market yields and conditions (IAA, 2004)

The market risk arises from the level or volatility of market prices on assets and liabilities and involves the exposure to movements in financial variables such as interest rates, bond prices, equities (stocks) and property (real estate) prices, and exchange rates. It should include all risks that result from the volatilities in capital market prices.

An isolated analysis on the impact of changes in market prices, as in banks, is not sufficient. Indeed, for an insurer, the impact on assets and liabilities must be examined simultaneously.

Hence, we will mainly look at the *asset/liability mismatch risk* (A/L risk), which results from the simultaneous volatility and uncertainty risk inherent in market values of future CFs from the insurers' asset and liabilities. We will usually consider the retained business only, as risks related to reinsurance are taken into account by the default credit risk.

When assessing the A/L risk, only the part of the business for which the company is bearing a risk should be taken into consideration, for example, UL business where the policyholder bears the risk is excluded. It is important to remember that a company bears the risk that the CFs generated by the assets do not match the payouts due to the market volatility, credit default, or changes in the interest rate curve.

As stated above, it is essential to simultaneously consider the volatilities of assets and liabilities when there are common background factors affecting both of them. The company can manage these risks by a suitable choice of assets, by a dynamic investment policy or by product design.

This interplay between assets and liabilities should be reflected in the capital requirements (CRs).

The assessment of the A/L risk should be based on best estimates and a simple starting point for quantifying the asset/liability risk could be a multiple of the volatility of asset return. Purchasing hedging instruments, such as caps and floors, can significantly decrease the interest rate risk, and the volatility risk of equities can be reduced by the purchase of puts and calls. Those hedging strategies have a major impact when setting up an asset liability policy and are an essential tool to prevent insolvency. Using hedging instruments reduces the asset volatilities, and this should be taken into account.

The principal risk sources of market risk for insurers are, according to IAA (2004),

- *Interest Rate Risk:* risk of exposure to losses resulting from fluctuations in interest rates; see Section 17.2
- *Equity and Property Risk:* risk of exposure to losses resulting from fluctuation of market values of equities and other assets; see Sections 17.3 and 17.4, respectively
- *Currency Risk:* risk that relative changes in currency values decrease the values of foreign assets or increase the value of obligations denominated in foreign currencies; see Section 17.5
- *Basis or Spread Risk:* risks that yield instruments of varying credit quality, liquidity, and maturity do not move together, thus exposing the company to market value variation independent of liability values; see Section 18.4
- *Concentration Risk:* risk of increased exposure to losses due to concentration of investments in a geographical area or other economic sectors; see Section 18.5

Other risks to consider are

- *A/L Risk:* to the extent that the timing or amount of the CFs from the assets supporting the liabilities and the liability CFs are different (or can drift apart), the insurer is subject to asset/liability mismatch risk
- *Reinvestment Risk:* risk that the returns on funds to be reinvested will fall below anticipated levels
- *Off-Balance Sheet Risk:* risk of changes in values of contingent assets and liabilities such as swaps that are not otherwise reflected in the BS

The resulting asset volatilities should be matched to the corresponding volatilities in liabilities: a change in interest rates may cause simultaneous changes in asset values and in liabilities, and the company should be only required to hold sufficient capital to cover the net effect of these changes. This principle can be applied relatively straightforwardly for *bonds* covering technical provisions, as it is relatively easy to compare the corresponding CFs.

However, this matching principle should not be limited to bonds only. Some insurance liabilities, typically deferred annuities, have very long duration. The market may not offer bonds of sufficient duration to back those liabilities. So, risk managers invest more widely in *equities* and *property* to provide an adequate assets/liability policy.

Market risk can only be measured appropriately if the market value of assets and liabilities are measured adequately, *cf.* the valuation chapters, Chapters 10 and 11.

As discussed above, the A/L mismatch risk, as defined above, would be an inherent part of the other risks.

IAA (2004) differentiates between Type A and Type B risks.

Type A risks: Shorter-term insurance contracts without complex embedded options or guarantees are subject to Type A risk. The essential ingredients to Type A market risk are (IAA, 2004)

- Projected future asset and liability CFs
- Nature of embedded options
- Time horizon
- Confidence level
- Current economic scenario
- Series of possible future economic scenarios

The determination of Type A risks could be simplified by different approximations due to these ingredients and would result in a range of standardized approaches.

One such approximation might require the grouping of future CFs into various term "buckets." In Basel II, the term "maturity method" is used. Another term used is "duration bands." The sum of the CFs in these buckets would be multiplied by factors to produce the CR.

Another approximation might use option-adjusted durations to represent the price sensitivity of CFs, the current market value of future CFs, and a set of investment return shocks. The shocks would need to be designed to reflect the time horizon and confidence level desired as well as the possible pattern of adverse scenarios. In this regard, it may be desirable to recognize the more active investment management conducted on closely managed blocks of business (i.e., when the active management holding period is less than the standard 1-year time horizon) (IAA, 2004, p. 57).

The simplest approximation is to multiply the BS value of insurer assets and liabilities by a table of factors reflecting the presumed presence and size of Type A risk.

Type B risks: The long duration of both life and nonlife insurance liabilities requires the consideration of long-term rates of reinvestment, since replicating portfolio assets of sufficiently long duration may not be currently offered in the market. Measuring market risk for these liabilities entails considerable uncertainty about the composition of the replicating portfolio and the manner of its reinvestment to mature the underlying CFs. Life insurance contracts may contain various complex, long-term options and/or guarantees for which replicating market positions may not currently exist. These types of market risk will be called Type B risk.

As stated by IAA, the development of standardized approaches for capturing Type B risks is difficult. Where these risks are material in an insurance company, the supervisor should encourage or even require the insurer to perform appropriate advanced approaches for modeling their Type B market risk. Standardized approaches might include

- For long-term interest guarantees in life insurance and annuity products, determining the present value of future liability CFs on the presumption that long-term reinvestment returns revert to a conservative view of historical long-term averages
- For complex options, deriving appropriately conservative factors based on rigorous stochastic modeling of industry-wide data to adequately capture the tail of the loss distribution for the confidence level required

In a standard formula, the number of asset categories should not be large, but it should be possible to enlarge the number of categories if the firm is planning for an internal model.

In the European Standard Approach (ESA), proposed by the European insurance industry body CEA (2006b), the market and ALM risks have the main principle: "Market risk should be considered at the company level. The capital requirement for market risk can be assessed by analysing the impact on available capital of shocks in the underlying market risk factors." The subrisks proposed in the model are interest rate risk, equity risk, real estate (property) risk, currency risk, and credit spread risk.

Stress tests for interest rate risk, equity risk, and property risk and also for credit spread risk and lapse risk are given in a study by Watson Wyatt for the British FSA (Watson Wyatt, 2004a).

The standard model for market risks in the SST is described in SST (2006b, ch. 4.1) and the references given therein. The field test made in 2006 examined 74 different market risk factors.

17.1 DIFFERENT MARKET RISK ISSUES

In the next three chapters, we will discuss three approaches that can be used for market risks, and especially for interest rate risks. The first one is the square-root-of-time scaling that could be applied to all types of assets. The second is duration and the third is different interest rate models.

17.1.1 Scaling: The Square-Root-of-Time Scaling

In contrast to market risks for banks having a time horizon in terms of days or weeks, the time horizon of an insurer market risk is more appropriately determined in terms of years.

For the CR we have adopted a 1-year time approach. For a discussion about time horizons, see IAA (2004). Existing bank models allow for a good measurement of market risks of trading books over small time intervals such as days up to two weeks.

Assume that we decide to use a bank model that is calibrated for a time horizon $h < 1$ year, say some days. To transform the model up to a time horizon of 1 year we need a scaling rule for the gap between h and 1 year. For an insurance firm, assets and liabilities can only be priced occasionally, and risk exposures must be measured and managed over these long spans, for example, 1/2 year or 1 year.

The simplest scaling rule is given by

$$\sigma_{1\text{ year}} \approx \sqrt{k} \cdot \sigma_{h\text{ days}}, \tag{17.1}$$

where $\sigma_{1\text{ year}}$ is the upscaled volatility to the 1 year time horizon, $\sigma_{h\text{ days}}$ is the h days volatility, and k is a scale factor. As the number of trading days per year is approximately 261, the scaling factor would be $k = 261/h$.

McNeil et al. (2005) discuss the time scaling. Kaufmann and Patie (2003) have studied methodologies that are proposed to model the evolution of risk factors over a long time horizon, for example, random walk, AR(p), GARCH(1,1) and heavy-tailed distributions. It is well known that daily returns are dependent on volatility clustering. On the other hand, returns over a longer time horizon (e.g., months) are close to being independent. Because of the weaker dependence for lower frequency data, some types of time aggregation rules may perform well as compared to the short-term case.

The square-root-of-time rule given by Equation 17.1 can be motivated by normal assumptions. Assume that the expected values of the log-returns are equal to zero and have a normal distribution with volatility σ over 1 day. Also assume that the log-returns are i.i.d. This logarithmic asset price follows a random walk with zero drift. Aggregating over k days gives us a normal distributed variable with zero mean and volatility $\sqrt{k} \cdot \sigma$.

Under normality, both VaR and TVaR could be scaled by this rule, that is, if $\text{VaR}^{(1)}$ and $\text{TVaR}^{(1)}$ are the VaR and TVaR measured over a 1 day time horizon, then the k day measures are $\text{VaR}^{(k)} = \sqrt{k}\,\text{VaR}^{(1)}$ and $\text{TVaR}^{(k)} = \sqrt{k}\,\text{TVaR}^{(1)}$. For a 1 year time horizon, we would get the scale factor equal to $\sqrt{261} \approx 16.2$.

17.1.2 Duration

A general discounted CF model can be defined as

$$W_0 = \sum_{t=1}^{n} \frac{\text{CF}_t}{(1 + r_t)^t}, \tag{17.2}$$

where W_0 = the current, or present, value of the asset; CF_t = the expected cash flow at period t; r_t = the required rate of return for each period's CF (the discount rate); and n = the time to maturity of the asset.

The CF is usually defined per year. If it is paid out during the year, both the CF and the rate of return are divided by the number of periods per year; for example, for half-year payment with two periods per year, Equation 17.2 becomes

$$W_0 = \sum_{t=1}^{n} \frac{\mathrm{CF}_t/2}{(1 + r_t/2)^t}.$$

The *duration* is defined as the weighted average life of the asset

$$D_n = \sum_{t=1}^{n} t w_t, \tag{17.3}$$

where

$$w_t = \frac{1}{W_0} \frac{\mathrm{CF}_t}{(1 + r_t)^t}.$$

Let us now assume $r_t = r, \forall t$, that is, we assume a flat or level interest rate curve. The *modified duration* is defined as $D_{n,m} = D_n/(1 + r)$, and $D_{n,m} W_0$ is the modified dollar duration. The price *sensitivity* of the asset is defined as

$$\frac{\mathrm{d}W_0}{\mathrm{d}r} = -\sum_{t=1}^{n} \frac{tCF_t}{(1 + r)^{t+1}} = -\frac{1}{(1 + r)} D_n W_0 = -D_{n,m} W_0. \tag{17.4}$$

Using the approximation $\mathrm{d}W_0/\mathrm{d}r \approx \Delta W_0/\Delta r$, we obtain

$$\Delta W_0 \approx -D_{n,m} W_0 \Delta r, \tag{17.5}$$

or the relative change as $\Delta W_0/W_0 \approx -D_{n,m}\Delta r$. Paying attention to the units, if $D_{n,m}$ is measured in years, then Δr refers to change per year.

Note that $D_{n,m}$ measures the relative slope of the price–interest rate curve at a given point. This gives a straight-line (first order) approximation to the price–interest rate curve and is a useful measure as a means of assessing risk and as a procedure for controlling it.

Djehiche and Hörfelt (2005) used stochastic processes to model the equity, interest rate, and property risks.

17.1.3 Interest Rate Models

During the high inflation period in the 1970s, different models of interest rates were developed. For a reference, see, for example, Cairns (2004a, 2004c). Following Cairns (2004c, ch. 4.1) we consider one-factor models for the risk-free interest rate within a continuous time framework. The risk-free rate, $r(t)$, is an Itô process with a general stochastic differential equation

$$\mathrm{d}r(t) = a(t)\,\mathrm{d}t + b(t)\,\mathrm{d}W(t).$$

Here, $W(t)$ is a standard Brownian motion under a real-world measure P, and $a(t) = a(r(t))$ and $b(t) = b(r(t))$ are processes such that the process $r(t)$ is Markovian and time

TABLE 17.1 One-Factor Time-Homogeneous Models for the Interest Rate $r(t)$ and Some Characteristics

Model	$a(r)$	$b(r)$	$r(t) > 0$	**AR**	**Simple**
Merton, M (from 1973)	μ	σ	N	N	Y
Dothan, D (from 1978)	μr	σr	Y	N	N
Vasicek, V (from 1979)	$\alpha(\mu - r)$	σ	N	Y	Y
Cox–Ingersoll–Ross, CIR (from 1985)	$\alpha(\mu - r)$	$\sigma\sqrt{r}$	Y	Y	Y
Pearson–Sun, PS (from 1994)	$\alpha(\mu - r)$	$\sigma\sqrt{r - \beta}$	Y if $\beta > 0$	Y	N
Brennan–Schwartz, BS (from 1979)	$\alpha(\mu - r)$	σr	Y	Y	N
Black–Karasinski, BK (from 1991)	$\alpha r - \gamma r \log r$	σr	Y	Y if $\gamma > 0$	N

Source: Adapted from Cairns, Andrew J.G. 2004c. *Interest Rate Models: An Introduction*. Princeton University Press, Princeton and Oxford, ISBN 0-691-11893-0.

Note: N: no, Y: yes, AR: autoregressive, and Simple: simple formulae.

homogeneous. For Itô processes, stochastic differential equations, and Brownian motions, see, for example, Cairns (2004c). Some basic characteristics that are desirable are

- Interest rates should be positive
- The interest rate $r(t)$ should be autoregressive, meaning that it cannot drift away to infinity or down to zero, but will go back to some long-term interest rate target

Even if a model is simple and elegant, we need to demonstrate that the proposed model gives a good approximation to what has been observed (Cairns, 2004c, p. 53). Table 17.1 gives seven different one-factor, time homogeneous, models for $r(t)$.

The Vasicek model is discussed in detail in Cairns (2004c, ch. 4.5), the CIR model in op. cit. ch. 4.6, and the two models are compared in op. cit. ch. 4.7. The generalized lognormal model proposed by Black and Karasinski is discussed in Cairns (2004c, ch. 4.9).

17.2 INTEREST RATE RISK

This initial discussion is based on the duration; see Section 17.1.2. Assume that $n = T$ is the time to maturity for the bond. The market value of the bond is defined by

$$B_0 = \sum_{t=1}^{T} \frac{c_t^*}{(1+r)^t} + \frac{M}{(1+r)^T} = B_{0,\text{cp}} + B_{0,\text{M}}. \tag{17.6a}$$

Here we have used the following notations:

B_0: the bond price, that is, the total discounted value of the bond accounted for by the future payments.

c_t^*: the (annual) coupon payment, that is, the CF at time t.

r: the yield.

T: time to maturity (in years).

M: the face value, that is, the amount to be repaid as a lump sum at maturity.

$B_{0,\text{cp}}$: the bond price due to the coupon payment.

$B_{0,\text{M}}$: the bond price due to the payment of maturity.

If we define $c_T = c_T^* + M$ and $c_t = c_t^*, t = 1, 2, \ldots, T-1$, we obtain

$$B_0 = \sum_{t=1}^{T} \frac{c_t}{(1+r)^t}. \tag{17.6b}$$

A summary statistic of the effective maturity of the bond is the *Macaulay duration*. It was independently proposed by different authors, but first by Macaulay in 1938. From Equation 17.3, we obtain

$$D_{\text{B}} = \frac{1}{B_0} \sum_{t=1}^{T} \frac{tc_t}{(1+r)^t}. \tag{17.7}$$

The sensitivity of the bond price can be expressed in terms of the Macaulay duration as, *cf.* Equation 17.4 above,

$$\frac{\text{d}B_0}{\text{d}r} = -\sum_{t=1}^{T} \frac{tc_t}{(1+r)^{t+1}} = -\frac{1}{(1+r)} D_B B_0. \tag{17.8a}$$

It can also be written as

$$-\frac{1}{(1+r)} \left[D_{\text{B,cp}} + T \frac{M}{B_0(1+r)^{\text{T}}} \right],$$

where $D_{\text{B,cp}}$ is the Macaulay duration for the coupon payment period.

The first-order approximation (from the Taylor expansion) gives us, *cf.* Equation 17.5 above,

$$\Delta B_0 \approx D_{B,m} B_0 \Delta r, \tag{17.8b}$$

where $D_{B,m}$ is a modified Macaulay duration $(= -D_B/(1+r))$.

The approximation Equation 17.8 is only good for small changes in the yield. This is the usual argument for improving Equation 17.8 by introducing a second term, the so-called *convexity*. This is the suggestion made by IAA (2004, Appendix D, para 7.15).

Let $B_0(r) = \sum c_t(1+r)^{-t}$, $B_0'(r) = -\sum tc_t(1+r)^{-t-1}$, and $B_0''(r) = \sum t(t+1)c_t (1+r)^{-t-2}$. Then, from Taylor expansion, we obtain

$$B_0(r + \Delta r) - B_0(r) \approx D_{B,m} B_0 \Delta r + \frac{1}{2} C(\Delta r)^2, \tag{17.9}$$

where the first term is due to duration and the second one is due to convexity.

$$C = \frac{1}{B_0(1+r)^2} \sum_{t=1}^{T} t(t+1) \frac{c_t}{(1+r)^t}.$$

Note that Δr is the change in yield per year (or per period), depending on the units of $D_{B,m}$ and C. The second term approximates the percent price change due to convexity, that is, the part of the price change due to the curvature of the price–yield relationship.

In the above discussion, it was assumed that we had a single bond. In reality, we have a bond portfolio consisting of several different bonds with different yields and durations. The duration of the bond portfolio is equal to the weighted mean of the different durations, with weights equal to the relative bond price:

$$D = \sum_{i=1}^{m} \omega_i D_{Bi},$$

where $\omega_i = B_{0i}/B_0$ and $B_0 = \sum_{i=1}^{m} B_{0i}$; see, for example, Luenberger (1998, Section 3.5, p. 62).

If we split the liabilities and the corresponding assets into duration bands, or buckets, we could look at each duration band as if they constitute its own portfolio. As an example, as proposed by IAA (2004, p. 143), we have the following buckets:

d_1 : [0−1] year (median duration: $md_1 = 0.5$)

d_2 : [1−2] years (median duration: $md_2 = 1.5$)

d_3 : [2−5] years (median duration: $md_3 = 3.5$)

d_4 : [5−8] years (median duration: $md_4 = 6.5$)

d_5 : [8−12] years (median duration: $md_5 = 10.0$)

d_6 : [12−16] years (median duration: $md_6 = 14.0$)

d_7 : [16−24] years (median duration: $md_7 = 20.0$)

d_8 : [24−] years (median duration: say $md_8 = 28$)

The assets corresponding to each duration band will be matched to the liabilities of similar duration.

IAA (2004) discuss the classical standardized approach in calculating a mismatch position using a Macaulay duration analysis. The approach has a number of drawbacks and IAA mentions three of them:

- The duration approach is based on a first-order Taylor approximation of the interest sensitivity of the present value. This approximation is not very good for larger interest changes. A better approximation is possible by including the second-order term, that is, the so-called convexity.

- More importantly, the duration approach assumes a parallel shift of the spot yield curve, while nonparallel shifts are equally possible, and possibly even more "dangerous" for the company. Nonparallel shifts can be taken into account by applying the approach for some duration bands individually and summing the results. Such an alternative approach can also be considered as an approach that allows for correlations between the changes of the "average" spot yields per duration band that are less than one.
- Still requires a fair degree of complex modeling by the company.

17.2.1 IAA Model

The IAA discusses different approaches, more or less simple, to capture the interest rate risk (IAA, 2004). The risk of fixed-income investments depends on some properties of these investments. The relevant properties are

- *Duration:* sensitivity if an interest rate increases
- *Rating:* which matters for assessing the credit risk

Apart from ordinary bonds, there are mortgage-backed and asset-backed securities, which behave in a similar way, except for prepayments in a period of falling interest rates. Bonds denominated in a foreign currency are affected by the currency risk; see Section 17.5.

The market risk for fixed-income investments is dominated by the risk of increasing interest rates. When the relevant interest rates or the whole yield curve increases by 1%, the value of a bond (portfolio) decreases by the amount of duration times 1%. The duration can either be exactly computed from the CFs of the bond and the current zero-coupon yield curve, or it can be approximately assessed as 80% of the mean time to maturity.

IAA example: a bond with a time to maturity of 10 years will lose about 8% of its value when the interest rate level increases by 1%.

To capture the interest rate risk, we need the probability distribution of interest rate changes over a time horizon. This can be done through a statistical analysis of empirical economic data. The variance of interest rate changes is slightly higher for short maturities than for long ones. According to IAA (2004, p. 139), a very rough estimate, for major currencies, of the standard deviation of yearly interest rate changes is of about 1.25%. Thus we obtain an approximate standard deviation of the annual value change (IR: interest rate):

$$\sigma_{\text{IR}} \approx \frac{\text{“Mean time to maturity”}}{100} \times \text{“Asset Value”}.$$

Using this approximation instead of a detailed statistical calculation, IAA proposes that it should be loaded with a prudent factor. The formula can be applied to whole portfolios, not only individual bonds, as the risk reduction due to diversification between bonds is rather small.

17.2.2 GDV Model

In the standard formula proposed by the German GDV* in 2005, GDV (2005), it is clearly stated, as we have done above, that the impact of changes in market interest rates on assets is not sufficient. Instead, the impact on assets and liabilities must be examined simultaneously.

An increase or decrease in the capital market interest rate means that the market value of both fixed-interest securities in the company's investment portfolio and that of the company's liabilities will change. According to the GDV model, the company has an interest change risk if the modified durations or market values of assets and liabilities are different ("not hedged").

If the duration on the asset side is longer than that for liabilities, the company suffers a net loss in market value in the event of a rise in interest rates; this risk must be measured accordingly. For nonlife insurers, this generally concerns all lines of business (LoB) except liability insurance, including motor liability insurance, and accident insurance with premium refunds. For life and health insurers, asset duration is generally shorter than liability duration (in terms of guaranteed benefits). Nevertheless, a rise in market interest rates may pose a risk in the event of cancellations and guaranteed surrender values. Therefore, the risk of interest rate changes for life insurers is also known as the *ALM risk due to cancellation.*

By the same token, a decline in capital market interest rates represents a risk if asset duration is shorter than the duration on the liability side. This is generally the case for life insurers and, in the case of nonlife insurers, for liability insurance and accident insurance with premium refunds.

Additional security measures such as swaptions are not sufficiently captured by the duration-based method and consequently by the standard method. These should be taken into account within the supervisory ladder or through internal models.

Therefore the following procedure is used in the GDV model (GDV, 2005, Section 5.1.1.3.1):

The various LoB are combined into two categories: nonlife insurers (not including accident insurance with premium refunds) and life insurers, as well as accident insurance with premium refunds. Accident insurance with premium refunds is to be calculated in a manner identical to the life model below.

- For each category, a distinction is made between a rise and decline in interest rates
- Since the risks are entirely opposite, the higher of the two net risks is taken

The change in market values as a result of a change in interest rates Δ, that is, a negative Δ in the event of a decrease, and a positive Δ in the event of an increase in interest rates, can generally be quantified in the following way:

Market value change for fixed-income titles on the asset side, using Equation 17.8b, gives us $\Delta \mathrm{MV_A} = \mathrm{MV_A} \cdot D_\mathrm{A} \cdot \Delta$. In a similar way, the market value change for the insurance liabilities is defined as $\Delta \mathrm{MV_L} = \mathrm{MV_L} \cdot D_\mathrm{L} \cdot \Delta'$. Here MV stands for the market value of

*GDV: Gesamtverband der Deutschen Versicherungswirtschaft e.V; that is, the German Insurance Association.

assets (bonds, fixed-income portfolios) and liabilities, respectively, prior to a shock or stress D for the modified duration of the bond portfolio (fixed-income portfolio) and the liabilities. Δ and Δ' are the market interest rate changes on the asset and liability sides, respectively. The negative sign in Equation 17.8a is implicit in Equation 17.8b.

Both formulas for market value changes mentioned above can be considered as a product of a *risk factor* (i.e., market interest rate changes Δ, Δ' on the assets or liabilities side) and a *risk bearer* (i.e., the product from market values and modified durations). Looking at these two formulas, it is seen that the modified durations are the measure for the changes in market values in the event of an interest rate stress.

The basis for determining the interest rate stress, upward or downward, is the Black–Karasinski (BK) model; see Section 17.1.3. The BK model is simplified by assuming that the time parameter t is set to 1 year. Hence the BK model becomes, for the interest rate after 1 year,

$$\log(r_1) = \alpha \cdot \mu + (1 - \alpha) \cdot \log(r_0) + \sigma \cdot \varepsilon,$$

where α represents the drift (mean reversion factor), $\exp(\mu)$ the long-term mean (mean revision level), and σ the momentary volatility of the process. r_0 represents the initial interest rate and r_1 the interest rate after 1 year. The variable ε has a standard normal variable and the $\log(r)$ has a normal distribution.

In the modeling of the BK model, GDV used 10-year euro swap rates for the risk-free interest rates for life insurance. The REX index, based on the most liquid segments of the German bond markets, was used to estimate the parameters using ordinary least squares. From data for 15 years (01.01.1991–01.01.2005), the following parameter values were obtained: $\alpha = 0.20$, $\mu = -3.02$, that is, $\exp(\mu) = 0.0488$, and $\sigma = 0.19$.

The GDV model defines the capital charge for the interest rate downward stress as

$$\mathrm{CR}_{\mathrm{MR,IR}}^{\mathrm{DOWN}} = \max\left\{\mathrm{MV_A} \cdot D_\mathrm{A} \cdot \Delta_{\mathrm{DOWN}} - \mathrm{MV_L} \cdot D_\mathrm{L} \cdot \Delta'_{\mathrm{DOWN}}; 0\right\} - E_{\mathrm{Basic}},$$

where E_{Basic} is the income from some basic requirements defined in the model; see GDV (2005, Chapter 5.1.1.3.1.3). Δ_{DOWN} is the change in interest rates on the assets side, calculated on the basis of the 99.5% confidence level of the 10-year euro swap rate (in this case, negative), and Δ'_{DOWN} is the change in interest rates on the liabilities side. The capital charge for the interest rate upward stress is defined as

$$\begin{aligned}\mathrm{CR}_{\mathrm{MR,IR}}^{\mathrm{UP}} = \max\{&\mathrm{MV_A} \cdot D_\mathrm{A} \cdot \Delta_{\mathrm{UP}} - \mathrm{MV_L} \cdot D_\mathrm{L} \cdot \Delta_{\mathrm{UP}} \\ &+ 1_{\mathrm{L\&A}} \cdot \mathrm{CR}_{\mathrm{Lapse}}; 0\} - E_{\mathrm{Basic}},\end{aligned}$$

where Δ_{UP} is the interest change based on a 99.5% confidence level (in this case, positive), $1_{\mathrm{L\&A}}$ is an indicator function that equals 1 for life and accident insurance with premium return and equals 0 otherwise; and $\mathrm{CR}_{\mathrm{Lapse}}$ is a capital charge for the insurance contracts that is expected to be cancelled after the stress. The $\mathrm{CR}_{\mathrm{Lapse}}$ is given in GDV (2005, ch. 5.1.1.3.1.2).

17.2.3 CEA Model

CEA proposed in 2006 an ESA for solvency assessment as another contribution to the debate on a European standard formula; see CEA (2006b).

In the CEA's ESA, the modified duration is used as a proxy of the sensitivity of both assets and liabilities to parallel changes in the level of interest rates. The interest rate movement can either have a positive or negative effect on the company's solvency position depending on the matching situation of assets and liabilities. Under the ESA, the CR should be calculated both for an upward and a downward movement of interest rates. The CR is then calculated considering all risks, diversification, and risk absorption by future profit sharing. The worse outcome for the company is then chosen for the final capital charge. Under the proposed factor-based ESA implementation, only simple hedging strategies are modeled. The approach taken considers only the modification in the modified duration of the asset being covered provided by the hedge. Using scenarios or partial internal models under the framework could assess the impact of more complex strategies. Under the ESA, template bonds and mortgages are considered as asset classes. The model could automatically be extended to any other interest rate-sensitive assets (CEA, 2006b).

The capital charge is based on Equation 17.8b and is given by

$$\mathrm{CR}_{\mathrm{MR,IR}} = \max\left\{\mathrm{CR}_{\mathrm{MR,IR}}^{\mathrm{UP}}; \mathrm{CR}_{\mathrm{MR,IR}}^{\mathrm{DOWN}}\right\},$$

where the upwards interest rate stress is defined by

$$\mathrm{CR}_{\mathrm{MR,IR}}^{\mathrm{UP}} = \mathrm{MV}_{\mathrm{A}} \cdot D_{\mathrm{A}} \cdot \Delta_{\mathrm{UP}} - \mathrm{MV}_{\mathrm{L}} \cdot D_{\mathrm{L}} \cdot \Delta_{\mathrm{UP}},$$

and the downward interest rate stress is defined by

$$\mathrm{CR}_{\mathrm{MR,IR}}^{\mathrm{DOWN}} = \mathrm{MV}_{\mathrm{A}} \cdot D_{\mathrm{A}} \cdot \Delta_{\mathrm{DOWN}} - \mathrm{MV}_{\mathrm{L}} \cdot D_{\mathrm{L}} \cdot \Delta_{\mathrm{DOWN}}.$$

MV is the market value of Assets and Liabilities, respectively.

D is the modified duration of the bond portfolio (including cash) and the net liability portfolio.

For a life insurance company, the asset impact for both the upward and downward stresses, that is, $\mathrm{MV}_{\mathrm{A}} \cdot D_{\mathrm{A}} \cdot \Delta_{*}$, is modified by adding a risk mitigation effect defined as $\mathrm{CR}_{\mathrm{Mitigation}}^{\mathrm{Life}} = \mathrm{MV}_{\mathrm{hedged\ A}} \cdot \Delta D \cdot \Delta_{*}$, where $\mathrm{MV}_{\mathrm{hedged\ A}}$ is the market value of hedged assets and ΔD is the change in duration of hedged assets provided by strategy. Adding a mass lapse risk component defined as $\mathrm{CR}_{\mathrm{Mass\ Lapse}}^{\mathrm{Life}} = (\text{Mass Lapse Rate}) \cdot [\mathrm{GSV} - \mathrm{MV}_{\mathrm{L}} \cdot (1 - D_{\mathrm{L}} \cdot \Delta_{\mathrm{UP}})]$ also modifies the upward stress. GSV stands for the guaranteed surrender value.

17.3 EQUITY RISK

This is the risk of exposure to losses resulting from fluctuation of market values of equities.

17.3.1 IAA Model

The IAA discusses different approaches, more or less simple, to capture the equity risk (IAA, 2004, pp. 140–141). Equity positions are subject to Type A market risk when these assets are used to fund similarly performing policyholder liabilities, for example, UL funds with no material guarantees, or represent free surplus. In these situations, market risk results from short-term volatility in the market value of the underlying assets. The longest time horizon to be considered in this case (as discussed earlier) is 1 year.

The variance of equity returns has been analyzed in numerous studies. The volatility, or the annual standard deviation, is higher than for bonds. Even for the best-diversified portfolios as represented by index-tracking portfolios, the standard deviation of annual returns may easily be 20% of the asset value. For individual equities of reasonable quality, it may be about 30%.

Some individual equity titles may have distinctly higher risks. These risks have to be quantified, based on empirical data.

If equity is denominated in foreign currency (FX), the standard deviation is $\sigma_{\text{Foreign Equity}} = \sqrt{\sigma^2_{\text{Equity}} + \sigma^2_{\text{FX}}}$. This is a conservative formula. Foreign equity investments often have a standard deviation of returns lower than this, mainly in the long run. When aggregating equity investments of different currency zones, we should add their standard deviations, assuming total dependence, rather than adding the squares (assuming independence). This is a conservative assumption, which may be refined by a detailed analysis of correlations between equity indices of different countries.

17.3.2 GDV Model

In the standard formula proposed by the German GDV (2005), the performance of equities is assumed to follow a lognormal distribution. The risk factor is defined as the 0.5% quantile of that spread. The expected value for equities is defined as the current 10-year risk-free return on annuities plus a risk mark-up (currently 3.5%). The GDV model is presented in GDV (2005, Annex 11). The model is the same for equity and property risks. We summarize it here.

17.3.2.1 Normal Distribution of Returns and Lognormal Distribution of Prices

If S represents the current market value of a stock (or property), in an analysis of the market risk for price ST, T time units later, $S_T = Se^{rT}$ will be assumed, with $r = \ln(ST/S)/T$ representing the "continuous return," since for a small time span T the approximation $S_T - S \approx r \cdot T \cdot S$ applies, that is, the appreciation in time span T equals the return r on the price, S.

Normal distribution is assumed for the continuous return r. If one examines the case of $T = 1$, that is, a 1-year period, then S_1 represents the equities price at the end of the year. Due to $S_1 = Se^r$, that is, $r = \log(S_1/S)$, the relative change S_1/S is lognormally distributed.

Since S_1/S is lognormally distributed, the price at the end of the year S_1 is lognormally distributed. Therefore, the investment model assumes lognormal distribution of equities (and property) prices for the analysis of market risk.

It should be noted that the continuous return r differs from the return R applicable to the year under observation, which is defined by $R = (S_1 - S)/S$ or $S_1 = S \cdot (1 + R)$. For the connection between r and R, the following applies: $r = \log(1+R)$ and

$$R = e^r - 1 = r + \frac{r^2}{2} + \frac{r^3}{6} + \cdots, \quad \text{and} \quad R > r.$$

In the model, μ represents the expected value of the return R, while σ represents the standard deviations in the continuous return r, that is, standard deviation is estimated, as usual, as the average annual continuous return. From the expected value μ of the variable R, one arrives at the expected value of variable r as $\log(1+\mu) - \sigma^2/2$. The term $-\sigma^2/2$ may be interpreted as a convexity adjustment for the log transformation. The derivation is described below.

17.3.2.2 Expected Value of the Continuous Return

In financial models, one often postulates for the equity prices a geometric Brownian motion $dS = \mu S\,dt + \sigma S\,dW$, with dW being a Wiener process. Upon discrete examination of this process, this means that the return R on the equity price is normally distributed, $R = (S_T - S)/S \sim N(\mu T, \sigma\sqrt{T})$.

After application of Ito's Lemma, one can derive the process for $\log(S)$ from the process S. One sees then that S_T is lognormally distributed and that $\log(S_T)$ is normally distributed:

$$\log(S_T) \sim N\left(\log S + \left(\mu - \frac{\sigma^2}{2}\right) T, \sigma\sqrt{T}\right).$$

For the continuous return r, therefore $r = \log(S_T/S) \sim N((\mu - \sigma^2 2)T, \sigma\sqrt{T})$, and the expected value of the continuous return T in a one-year view is thus $\mu - \sigma^2/2$.

17.3.2.3 Risk Factors for Equities (and Property/Real Estate)

Unlike for normally distributed risks, income is included in these risk factors. The reason is that the mean of the lognormal distribution, $e^{\mu+\sigma^2/2}$, depends on the standard deviation and therefore cannot be examined separately, as can be done for normally distributed risks.

From $\log(S_T) \sim N(\log S + (\mu - \sigma^2/2)T, \sigma\sqrt{T})$, the following follows for $S = 1$ (per Euro 1 price value) and $T = 1$, the one-year time horizon: $S_1 \sim \text{LN}(\mu - \sigma^2/2, \sigma)$ with LN representing the lognormal distribution function.

The RF is derived as RF = (S − Quantile) and, with $S = 1$, as RF = 1 − Quantile so that

$$\text{RF} = 1 - \text{LN}^{-1}\left(1 - \alpha; \mu - \sigma^2/2, \sigma\right),$$

with $\text{LN}^{-1}(1-\alpha)$ representing the $1-\alpha$ quantile of the lognormal distribution function.

For equities, the following variables were used: μ = 3.75% risk-free interest rate +3.5% risk premium for equities = 7.25% and σ = 17%. Data are given in GDV (2005, Annex 12). The stress test should be based on an equity fall of 31.8%.

17.3.3 CEA Model

CEA proposed in 2006 an ESA for solvency assessment as another contribution to the debate on a European standard formula; see CEA (2006b).

The equities are classified into volatility categories with different shock factors being applied to each of the categories. The CR for equity risk is calculated by applying a factor to the market value of equity holdings within the company. Concentration in individual equities is taken into account in the proposed ESA by applying an additional capital load to those equities for which the company has holdings above a certain threshold. The CR for equity risk includes the combinational impact of a drop in equity markets together with an increase in lapse rates for products with a guaranteed surrender value (for life insurance).

Under the proposed ESA factor implementation, only simple hedging strategies are modeled. Using scenarios or partial internal models under the framework could assess the impact of more complex strategies. In the ESA model, a placeholder for equity categories has been included. These categories consider different equity markets: Europe, North America, Asia and developing countries, and others. These categories are only a placeholder.

The CR is defined as

$$\mathrm{CR_{MR,ER}} = \sum_{\text{Equity}} \mathrm{MV_E} \cdot f_\mathrm{E} - 1_{\text{Life}} \cdot (\text{Mitigation-Conc1-Lapse}) + (1 - 1_{\text{Life}}) \cdot \text{Conc2},$$

where the sum is taken over the different types of equities, MV is the market price of equity E, f_E is a factor (Europe: 0.15, North America: 0.20, Asia: 0.25, and others: 0.30), and 1_{Life} is an indicator function taking the value 1 if life insurance and 0 otherwise.

$$\begin{aligned}
&\text{Mitigation} = (\text{market value of hedged Equity}_\mathrm{E}) * \min\{f_\mathrm{E} - \text{MaxLoss}_\mathrm{E}; f_\mathrm{E}\},\\
&\text{Conc1} = \text{Conc2} * (\Sigma \mathrm{MV_E} * f_\mathrm{E} - \text{Mitigation}),\\
&\text{Conc2} = (\text{Concentration penalty factor}) * (\%\text{ equities above threshold}),\\
&\text{Lapse} = (\text{Mass Lapse Rate}) * (\text{Guaranteed Surrender Value}\\
&\qquad - \mathrm{MV}_{(\text{L,with guaranteed surrender value})}),
\end{aligned}$$

17.3.4 Equity Duration

In FFSA* (2003), a working paper from the French insurance association FFSA, it was proposed that the CEA initial interest rate mismatch model, of the form $D_\mathrm{B}\mathrm{MV_B} - D_\mathrm{L}\mathrm{MV_L}$, where D is the duration and MV the market values, should be extended to also include a duration analysis for equities and property.

* FFSA: Féderation Francaise des Sociétés d'Assurance.

Here we will consider the "duration" of equities and property as the "holding period," that is, the *mean time* a company is holding its equities and property. The duration d will also be, in accordance with IAA (2004), split up into duration bands.

Define a band of *mean holding periods* for a company's equities. This can be done by historical company data: Duration band, say d_4 : [5–8] years (median duration: $md_4 = 6.5$). This model is discussed in Sandström (2005, Appendix A, Section A.2.1.3 for both equities and properties).

The CF of equities (or stocks) is termed *dividends* (d_t). For property (or real estate), the term is *net operating income* (NOI_t). The discussion that follows applies to equities (or stock) but the same approach will be used for property, as this investment class resembles equities in many ways.

Djehiche and Hörfelt (2005) used a stochastic process to model the equity risk together with both interest rate and property risks.

17.3.4.1 Dividend Discount Model

John Burr Williams introduced the dividend discount model (DDM) in a thesis in 1937. The idea was to forecast the future dividends a company would pay to its stockholders and then discount them using a suitable interest rate.

From Equation 17.2, we obtain

$$S_0 = \sum_{t=1}^{n} \frac{d_t}{(1+r)^t}, \tag{17.10}$$

where S_0 is the present value of the stock (the intrinsic or theoretical value), d_t is the dividend paid at t, and r is the required return (discount rate) for each year t.

Note that the maturity is perpetual (that means perhaps 20 years or so in reality). If a company buys equity after time H (the holding period for the seller), he assumes to get a profit from it in the future. S_0 can be written as $d_1/(1+r) + S_1$, where S_1 is the sales price at the end of $t = 1$ equals Equation 17.10 with $t = 2, 3, \ldots$.

The Williams basic model Equation 17.10 has been modified many times. Gordon and Shapiro introduced one of the most common modifications in 1956. In this model, it is assumed that the dividends grow with a constant rate (g).

17.3.4.2 Constant Growth DDM (the Gordon–Shapiro Model or Simply the Gordon Model)

Assume that $d_t = d_0(1+g)^t$, where g is the constant growth rate and d_0 is the most recent per-share dividend (and a value that is known for certainty). It is also assumed that $r > g$ and n is perpetual. From Equation 17.10 we now obtain

$$S_0 = \frac{d_0(1+g)}{(r-g)} = \frac{d_1}{r-g}. \tag{17.11}$$

17.3.4.3 Zero Growth DDM

This is the simplest form of Equations 17.10 and 17.11 assuming constant dividends, that is, zero growth. Assuming $g = 0$ gives us*

$$S_0 = \frac{d_0}{r}. \tag{17.12}$$

Before looking at other modifications of the DDM, we will shortly look at the duration and sensitivity. The duration can be defined as

$$D_{S,n} = \sum_{t=1}^{n} tw_t = \frac{1}{S_0}\sum_{t=1}^{n}\frac{td_t}{(1+r)^t}.$$

We can now define the sensitivity of equities in terms of this general "perpetual duration" as

$$\text{Sensitivity: } \frac{1}{S_0}\frac{\mathrm{d}S_0}{\mathrm{d}r} = -(1+r)^{-1}D_{S,n}.$$

17.3.4.4 Two-Part Dividend Stream

Assume that a company holds equities for a period of H years (holding period) and then sells them. For the holding period ($t = 1, 2, \ldots, H$) we can assume the DDM approach (Equation 17.10) and for the future period ($t = H+1, H+2, \ldots$) we assume the constant growth model (the Gordon–Shapiro model). This is also a common model for new technology equities, which are expected to grow rapidly for a few years and then to slow down as they approach "maturity." For this model, we have

$$S_0 = \sum_{t=1}^{H}\frac{d_t}{(1+r)^t} + \frac{d_H(1+g)}{(r-g)}\left(\frac{1}{1+r}\right)^H. \tag{17.13a}$$

This valuation model is similar to the bond valuation model Equation 17.6a. Assume $d_H^* = d_H + (d_H(1+g)/r - g)$ and $d_t^* = d_t$ for $t = 1, 2, \ldots, H-1$. Then we have

$$S_{0,H} = \sum_{t=1}^{H}\frac{d_t^*}{(1+r)^t}. \tag{17.13b}$$

We can now define the duration by

$$D_{S,H} = \sum_{t=1}^{H} tw_t = \frac{1}{S_{0,H}}\sum_{t=1}^{H}\frac{td_t}{(1+r)^t}. \tag{17.14}$$

* The simplest way is to use Equation 17.11. Using Equation 17.10 we obtain $S_0 = d_0\sum_{t=1}^{\infty} 1/(1+r)^t$. Multiplying both sides with $(1+r)$ gives us $(1+r)S_0 = d_0 + d_0\sum_{t=1}^{\infty} 1/(1+r)^t$ and subtracting both sides with S_0 gives $rS_0 = d_0$ and hence Equation 17.12.

A first-order approximation (Taylor expansion) gives us the change in equity value as measured in terms of change in interest rate r, *cf.* Equation 17.8:

$$\Delta S_{0,H} \approx D_{S,m} S_{0,H} \Delta r, \tag{17.15}$$

where

$$D_{S,m} = -\frac{1}{1+r} D_{S,H}.$$

The change in equity value is thus approximated by modified Macaulay duration of the equity's "duration" *times* the intrinsic equity value (present value) *times* the change in rate of return during the year. The estimation (Equation 17.15) can be improved by using the second-order term as in Equation 17.9. Note that for the Gordon–Shapiro model (Equation 17.11), we obtain the duration as $(1+r)/(r-g)$ and therefore we have

$$\Delta S_{0,H} \approx -\Delta r d_0 \frac{(1+g)}{(r-g)^2}.$$

17.3.4.5 Some Other Models

Estep and Hanson extended the Gordon–Shapiro model in 1980 to incorporate the effect of inflation by defining $(1+r) = (1+rr)(1+ei)$ and $(1+g) = (1+rg)(1+ftc)$, where rr is the real rate of return, ei is the expected rate of inflation, rg is the real growth, and ftc is a flow through coefficient. Other extensions include, for example, property market cycles.

17.4 PROPERTY RISK

This is the risk of exposure to losses resulting from fluctuation of market values of equities.

17.4.1 IAA Model

The IAA discusses different approaches, more or less simple, to capture the property (real estate) risk (IAA, 2004). Real estate investments can be treated as equity. Real estate indices take the role of equity indices. The diversification between different countries may be slightly stronger than the analogous diversification effect for equity. Real estate prices tend to increase when mortgages are becoming cheap, that is, when interest rates fall.

17.4.2 GDV Model

In the standard formula proposed by the German GDV (2005), the performance of properties is assumed to follow a lognormal distribution as for equities. The model is described above in Sections 17.3.2.1 through 17.3.2.3.

For real estate, the following variables are used: $\mu = 5.5\%$ and $\sigma = 7.4\%$. Data are given in GDV (2005, Annex 13). The stress test should be based on a property fall of 13.0%.

17.4.3 CEA Model

CEA proposed in 2006 an ESA for solvency assessment as another contribution to the debate on a European standard formula; see CEA (2006b).

The property assets are classified into volatility categories with different shock factors being applied to each of the categories. The CR for real estate risk is calculated by applying a factor to the market value of property holdings within the company.

In the ESA model, a placeholder for real estate categories has been included. These categories consider different types of property: residential, commercial, and others. These categories are only a placeholder.

The CR is defined as

$$\mathrm{CR}_{\mathrm{MR,PR}} = \sum_{\text{Property}} \mathrm{MV}_{\mathrm{P}} \cdot f_{\mathrm{P}} - 1_{\mathrm{Life}} \cdot (\text{Mitigation-Concentration-Lapse}), x$$

where the sum is taken over the different types of properties, MV is the market price of property P, f_{P} is a factor (residential: 0.05, commercial: 0.10, and others: 0.15), 1_{Life} is an indicator function taking the value 1 if life insurance and 0 otherwise.

$$\text{Mitigation} = (\text{market value of hedged Equity}_{\mathrm{P}}) * \min\{f_{\mathrm{P}} - \text{MaxLoss}_{\mathrm{P}}; f_{\mathrm{P}}),$$

$$\text{Concentration} = (\text{Concentration penalty factor}) * (\Sigma\mathrm{MV}_{\mathrm{P}} * f_{\mathrm{P}} - \text{Mitigation}) * (\%\text{ equities above threshold}),$$

$$\text{Lapse} = (\text{Mass Lapse Rate}) * (\text{Guaranteed Surrender Value} - \mathrm{MV}_{(\text{L, with guaranteed surrender value})}).$$

17.4.4 Property Duration

As for equity, this was a proposal from the French insurance industry; see Section 17.3.4.

Define a band of *mean holding periods* for a company's property. This can be done by historical company data: Duration band, say d_5 : [8−12] years (median duration: $md_5 = 10.0$). The model is exactly the same as the one discussed in Section 17.3.4.

17.5 CURRENCY RISK

Currency risk is, to a large extent, taken into account in the A/L risk if the calculations are made by currencies. If there is a need for a specific term for the currency mismatch risk, it could simply be defined as *coefficient times the mismatch.* A discussion of this risk can be found in IAA (2004, pp. 139–140).

17.5.1 IAA Model

In the IAA model (IAA, 2004), the solvency requirement for currency risk can be defined in a similar way as for equity and property risks (i.e., by setting it equal to the actual market value of the assets denominated in foreign currency times a conservative estimate of the potential change of value within the next year). The "potential change" factor can include the effects of both the potential change of the yields (/prices) and the potential change of the currency (IAA, 2004, p. 142).

17.5.2 GDV Model

In the standard formula proposed by the German GDV (2005) the foreign currency risk takes into account potential losses from exchange rate fluctuations.

Such fluctuations pose a risk if the claims by insured persons are denominated in a currency other than that of the investments covering those claims (nonhedged). If there is no currency hedging, a risk arises if the asset currency declined in value relative to the obligation currency. The *risk bearer* for the foreign currency risk is the unhedged market exposure in fixed-interest titles and real estate. The equities investments of insurance companies are increasingly diversified on an international scale, so that currency-related volatility in the equities portfolio is declining due to this effect.

The model is based on the principle that the overall risk of a foreign-currency equities portfolio plus fluctuation in exchange rates may not exceed the equities risk in the local currency. Based on this principle, the foreign currency risk of equities is set at zero.

Normal distribution is assumed for the currency risk and the calculation is outlined in GDV (2005, Annex 14). The expected earnings are set at zero, like the correlations, since the latter are not stable over time (GDV, 2005, ch. 5.1.1.3.3). Based on these definitions, the following risk factors currently result, relative to other currencies, from the 0.5% quantile of the corresponding normal distribution:

	Volatility	**Risk Factor (RF)**	**Risk Bearer (RB)**
US dollar (USD)	10.2%	26.3%	Unhedged market exposure of fixed-income titles and real estate in the relevant currency
Pound (GBP)	7.5%	19.3%	
Yen	11.0%	28.3%	
Other European currencies	10.0%	25.8%	

17.5.3 CEA Model

CEA proposed in 2006 an ESA for solvency assessment as another contribution to the debate on a European standard formula; see CEA (2006b).

The currency risk calculation is based on the net currency position of the company:

- Both asset and liability foreign currency holdings are considered
- Foreign currency positions are considered the net of any hedges

Net positions are classified by the currency in which they are denominated and different factors are applied for each currency. The capital charge is denied as

$$\mathrm{CR}_{\mathrm{MR,CR}} = \sum_{\text{Currency}} \mathrm{NP}_{\mathrm{C}} \cdot f_{\mathrm{C}},$$

where NP is the net position of currency C, defined as $\mathrm{NP}_{\mathrm{C}} = |\mathrm{MV}_{\mathrm{A,C}} - \mathrm{MV}_{\mathrm{L,C}}|$, where MV is the market value in currency C of assets and liabilities. Only unhedged currency positions are considered in the model. f_C is a currency factor (US dollar: 0.20, GB pound: 0.20, yen: 0.20, Swiss franc: 0.20, others: 0.20—for illustration only).

17.6 OTHER CATEGORIES

Cash (and others)

Duration band: say $d_1 : [0-1]$ year (median duration: $md_1 = 0.5$).

Cash is usually modeled as bonds; see Section 17.1.

In the CEA model, the credit spread risk is part of the market risk. This model is discussed in Section 18.4.1.

CHAPTER 18

Credit Risk

CREDIT RISK is the risk of financial loss resulting from default or movement in the credit quality of issuers of securities (in the company's investment portfolio), debtors (e.g., mortgagors), or counterparties (e.g., on reinsurance contracts, derivative contracts, or deposits given) and intermediaries, to whom the company has an exposure (IAIS, 2004).

For a general financial firm, the credit risk is "the risk of not receiving promised repayments on outstanding investments such as loans and bonds, because of the 'default' of the borrower" (McNeil et al., 2005). Especially for an insurer, we have the default of a reinsurer as a main credit risk.

Gielens (2004) describes the credit risk as follows: "Credit risk captures the risk on a financial loss that an institution may incur when it lends money to another institution or person. This financial loss materializes whenever the borrower does not meet all of its obligations specified under its borrowing contract. These obligations have to be interpreted in a large sense. ... Moreover, credit risk also plays an important role in pricing financial assets and hence influences the interest rate borrowers have to pay on their loans."

There are two main determinants of credit risk. First we have the *probability of default* (PD), that is, the probability that the debtor does not pay, and second we have the *loss given default* (LGD). The LGD is usually smaller than the amount of loan. The PD is usually determined by firm-specific structural aspects, the sector the borrower belongs to and the general economic environment; see, for example, Gielens (2004). There are two kinds of models for determining the PD addressed in the literature: accounting-based models and market-based models; *cf.* Aas (2005). The latter models are based on the value of the firm as set by the market. Moody's KMV model is one example; see Section 18.2. For a general discussion on credit risks, see, for example, Lütkebohmert (2008), Martin (2004), and McNeil et al. (2005).

The subrisks highlighted below follow from the discussion in IAA (2004).

The *default credit risk* is the risk that an undertaking will not receive fully or partially (or receives delayed) cash flows or assets to which it is entitled because a party with which it has a (bilateral) contract defaults in one or another way.

In this risk, we also include downgrade/migration risk and indirect credit or spread risk. By *downgrade/migration risk*, we mean a future risk that changes the PD by an obligor that will affect the present value of the contract with him. This could be measured by changed rating. By *indirect credit risk*, we mean the risk due to market perception of increased risk on a micro or macro basis; see IAIS (2004). The reason for this is that we want to minimize the possibility for arbitrage of the risk from banking to the insurance sector, and vice versa.

The *settlement risk* arises from the lag between the value date and the settlement date of securities transactions. By *sovereign risk*, we mean the risk of exposure to losses due to the decreased value of foreign assets or the increased value of obligations denominated in foreign currencies. This is dealt with under the currency risk in Section 17.5.

By *concentration risk*, we mean various types of concentrations or exposures, for example, in investments (asset concentration) and catastrophic events (concentration of liabilities). For example, investment in a high proportion of specific equities can be considered as risky (concentration). This risk is also due to concentration in geographical areas, economic sectors, and connected parties.

The *counterparty risk*, including reinsurance counterpart risk, is the risk of values of reinsurance, contingent assets, and liabilities.

Following IAA (2004), some of the key drivers of credit risk include

- *Credit quality:* The credit quality of an investment or an enterprise refers to the probability that the issuer will meet all contractual obligations. This also implies to reinsurance firms. One of the common measurements used in assessing the credit quality is to use the rating of the issuer.
- *Maturity:* The longer the term to maturity of an investment is, the higher is the probability of an issuer's default.
- *Concentration by industry:* Conditions that trigger credit events have a tendency to impact the entire economy at the same time.
- *Concentration by geography:* Credit risk has been shown to carry a high degree of contagion. Periods with few credit events are followed by periods with high default experience. In the same way, economically depressed regions tend to produce high default experience. These regions change over time.
- *Size of expected loss (EL):* The size of losses due to credit events varies widely. It could be an effect of delay in timing of a scheduled payment.

Crosbie and Bohn (2003) group the elements of credit risks into standalone and portfolio risks:

Standalone Risk

- *Default probability:* the probability that the counterparty or borrower fails to service obligations
- *LGD:* the extent of the loss incurred in the event that the borrower or counterparty defaults
- *Migration risk:* the probability and value impact of changes in default probability

Portfolio Risk

- *Default correlation:* the degree to which the default risks of the borrowers and counterparties in the portfolio are related
- *Exposure:* the size, or proportion, of the portfolio exposed to the default risk of each counterparty and borrower

To overcome some of the credit risks, the risk manager has to have sound underwriting practices in place and lending limits within the firm. Some hedging strategies, as defined by IAA (2004), include

- Letters of credit
- Contingency deposits
- Securitization of mortgages ("Mortgage-Backed Securities")
- Securitization of other assets ("Assets-Backed Securities")
- Credit derivatives (such as credit default swaps, collateralized debt obligation, total return swaps, credit spread options, etc.)

Usually VaR is used as a measure of credit risk. To develop a framework for this, you need a distribution function of losses, and to derive a loss distribution for default, two main paths have been used in credit risk modeling. Either you use historical data to make *simulations* to get an "empirical" loss distribution or you make theoretical assumptions so that the loss distribution can be presented in an *analytical form.*

As we have mentioned above, there are several issues to consider before the modeling work starts. As an example, the spread risk is related to both the market risk and to the credit risk, but the downgrade risk is a pure credit spread risk. Hence, adding the capital charges for spread risk and downgrade risk may lead to double counting.

An appropriate credit risk model should address both the migration risk (credit spread risk) and the default risk in an integrated framework. Changes in market and economic conditions affect the profitability of firms and hence also the exposure to the counterparts, the PD, and also the migration between two credit ratings.

The PD of a firm is mainly based on three elements (Crosbie and Bohn, 2003):

- *Value of assets:* the market value of assets, measuring the present value of the future free cash flows produced by the firm's assets discounted back at an appropriate discount rate.
- *Asset risk:* the uncertainty or risk of the asset value, measuring the firm's business and industry risk. As the value of the firm's asset is an estimate, it is uncertain.
- *Leverage:* the extent of the firm's contractual liabilities, measured as the difference between the assets' market value and the book value of the liabilities.

We will look closer at some credit risk models used. Before looking at the bank model from Basel II, we briefly look at three models developed during the 1980s and 1990s, the KMV, the CreditMetrics, and the CreditRisk+ models. These three models are discussed in Section 18.1 and the Basel II approach in Section 18.2. In Section 18.3, we look closer at a model especially constructed for reinsurance counterparty default risk. In Section 18.4, we discuss credit spread risk models, and in Section 18.4, models for the concentration risk. A mathematical framework for credit risk modeling is discussed in Chen (2003).

The standard model for credit risks in the SST is described in SST (2006b, ch. 4.2) and references given therein. It is based on the Basel II requirements for banks.

18.1 DIFFERENT PUBLIC CREDIT RISK MODELS

We will briefly present three public credit risk models that had their inception during the 1990s. They are Moody's KMV model, J.P. Morgan's CreditMetrics model, and Credit Suisse First Boston's (CSFB's) CreditRisk+ model. In Gordy (2000), the two latter models are compared, and in Crouhy et al. (2000), all three models plus a fourth one, McKinsey's CreditPortfolioView, are compared. See also Lütkebohmert (2008) for a discussion on these models.

The following variables for each firm i are used in the following discussion.

EAD_i: exposure at default; gives an estimate of the amount outstanding (drawn amounts plus likely future drawdown of yet undrawn lines) in case of borrower defaults.

PD_i: unconditional probability of default; calculated per rating grade. This gives the average percentage of obligors that default in this rating grade in the course of 1 year. In Basel II the PDs are supposed to be obtained from the internal rating system of the bank. Usually data from different rating institutes are used to estimate the PDs.

LGD_i: loss given default; gives the percentage of exposure the bank might lose in case the borrower defaults. These losses are usually shown as a percentage of EAD, and depend, among others, on the type and amount of collateral as well as the type of borrower and the expected proceeds from the workout of these assets. The Basel Committee has not proposed any specific rules for estimating the LGDs. Banks are instead required to provide their own estimates. One minus the LGD, $1 - \text{LGD}$, is called *recovery*. As there is a possibility to have a recovery >1, we might have negative LGDs. A discussion on LGDs is given in Schuermann (2004).

For a given maturity, these three parameters are used to estimate two types of EL; see Schuermann (2004).

- EL as an amount: $\text{EL} = \text{PD} \times \text{LGD} - \text{EAD}$
- EL as a percentage of the EAD: $\text{EL\%} = \text{PD} \times \text{LGD}$

The KMV model uses an extensive database to get default probabilities and to simulate loss distributions for both default probabilities and migration risks. Their methodology relies upon the default probability concept of Expected Default Frequency (EDF) for each issuer. The methodology is based on Merton's firm value model (Merton, 1974).

The CreditMetrics model is based on a credit migration analysis, that is, to estimate the probabilities of moving between different credit standings, including the default state, within a given time horizon. The time horizon is usually set to 1 year. The model values any bond or loan portfolio 1 year forward and the changes in values are only related to credit migration. The model, which uses simulations, is also based on Merton's firm value model (Merton, 1974), but differs from the KMV's methodology in how the simplifying assumptions are made.

The CreditRisk+ model, using theoretical assumptions, makes it possible to arrive, in an analytic form, at the loss distribution and focuses on modeling the default risk. It is assumed that the default of bonds or loans follows a Poisson process. Credit migration risk is not explicitly modeled, but it allows for stochastic default rates, which could be seen as accounting for migration risk.

The three models, KMV, CreditMetrics, and CreditRisk+, are all suited for use in internal modeling and to be a part of capturing economic and regulatory capital for banks. It is important to be aware of the fact that the models, including the Basel II IRB model, do not measure typically the causes of bank failures: the excessive concentration of capital in a small number of exposures with related risks; *cf.* Dwyer and Qu (2007).

Both the KMV credit risk model and the CreditMetrics model, see below, are based on simulations, while the CreditRisk+ and the Basel II Credit Risk IRB have analytical forms of the loss distributions.

18.1.1 Merton and Vasicek Models

The prototype of all firm-value models is Merton's single asset model, where loans are modeled as a claim on the value of a firm using the Black–Scholes option pricing model. Both Merton and Black–Scholes proposed a simple model of the firm that allows us to relate the credit risk to the capital structure of the firm. In this simple model, the value of the firm's assets, $V_{\mathrm{A}t}$, is assumed to follow a geometric Brownian motion with constant mean. The firm is assumed to have issued two classes of securities, namely equity and debt. The debt is a pure discount bond where the payment of K is promised at time t. No dividends are paid out. If the firm's asset value is exceeding the promised payment K at time t, the lenders are paid the promised amount and the shareholders are receiving the residual asset value. If the asset value is less than K, then the firm defaults. This means that the lenders receive a payment equal to the asset value (= the recovery). The shareholders do not get anything.

The Merton model, Merton (1974), is discussed in, for example, Hull et al. (2004), Lütkebohmert (2008), Martin (2004), and McNeil et al. (2005).

Let a firm's assets value be $V_{\mathrm{A}t}$ and its equity $V_{\mathrm{E}t}$ at some time $t > 0$. For $t = 0$, we have today the value of the assets and the equity today: $V_{\mathrm{A}0}$ and $V_{\mathrm{E}0}$. The payment to the shareholders at time t is in the Merton framework equal to

$$V_{\mathrm{E}t} = \max[V_{\mathrm{A}t} - K; 0] = [V_{\mathrm{A}t} - K]_+.$$

The equity is thus a call option on the assets of the firm with strike price equal to the promised debt payment K. The current price is therefore equal to (the Black–Scholes formula)

$$V_{E0} = V_{A0}\Phi(d_1) - Ke^{rt}\Phi(d_2), \tag{18.1}$$

where

$$d_1 = \frac{\log\left(\frac{V_{A0}\,e^{rt}}{K}\right) + \left(\frac{\sigma_A^2}{2}\right)t}{\sigma_A\sqrt{t}} \quad \text{and} \quad d_2 = d_1 - \sigma_A\sqrt{t},$$

and σ_A is the volatility of the asset value and r is the risk-free interest rate. Both are assumed to be constant. We can rewrite d_1 as

$$d_1 = \frac{\log\left(\frac{V_{A0}}{K}\right) + \left(\frac{r + \sigma_A^2}{2}\right)t}{\sigma_A\sqrt{t}}.$$

The parameter d_2 is the distance to default (DD) in the KMV model below.

An informal description and interpretation of the Black–Scholes formula is given by Stone (2007).

A firm's assets value is assumed to follow a standard geometric Brownian motion, that is,

$$V_{At} = V_{A0} \cdot \exp\left\{\left(r - \frac{\sigma_A^2}{2}\right)t + \sigma_A\sqrt{t}Z_t\right\},$$

where $Z_t \sim N(0,1)$, $r = \mu_A$ and σ_A^2 are the mean and variance of the future rate of return on the assets of the firm, dV_{At}/V_{At}. V_{At} is lognormally distributed with the expected value at time t equal to $E(V_{At}) = V_{A0}\,e^{rt}$. The firm defaults if $V_{At} < K$, where K is the firm's liability or threshold; see, for example, Crosbie and Bohn (2003).

Under the Black–Scholes model, the Merton model gives the PD by

$$\text{PD} = \Phi(-d_2) = \Phi\left(\frac{\ln\left(\frac{K}{V_{A0}}\right) - \left(r - \left(\frac{1}{2}\right)\sigma_A^2\right)t}{\sigma_A\sqrt{t}}\right), \tag{18.2}$$

where K is the firms liability set as the threshold, V_{A0} is the firms asset value at $t = 0$, and r, σ_A^2 are constants and t the maturity. Eom et al. (2002) compare the performance of the Merton model with alternatives using bond spreads.

The model states that the counterparty defaults as it cannot meet its obligations as the value of its assets is lower than its due amount. The area under the normal distribution of the asset value below the debt level represents the PD.

Models that descend from Merton's firm-value model are based on a latent variable approach. They underlie some of the most important models used, such as KMV and CreditMetrics, but also the Basel II Credit Risk IRB model. For a discussion on this model, see Gordy (2000). The unobserved latent variables y_i are taken to be linear functions of RFs and diversifiable (idiosyncratic) effects ε_i:

$$z_i = zw_i + \eta_i\varepsilon_i. \tag{18.3a}$$

The vector of factor loadings w_i determines the relative sensitivity of obligor (debtor) i to the RFs. The weight η_i determines the relative importance of idiosyncratic risk for the obligor. The default of an obligor occurs if a latent variable, usually interpreted as the value of the obligor's assets, falls below some threshold, usually interpreted as the value of the obligor's liability. Dependence between default events is caused by dependence between latent variables. In the KMV and CreditMetrics models, the latent variables are assumed to be multivariate normal. The aggregate portfolio loss distribution is sensitive to the choice of the multivariate distribution of the latent variables; see Frey et al. (2001).

The ε in model (Equation 18.3a) is assumed to be multivariate Gaussian distributed with mean zero and variance–covariance matrix Ω. You can also assume that the diagonals in the matrix are ones, so all marginal distributions are $N(0,1)$. To each start-of-period, rating grade G is a threshold value K_G. When the latent variable z_i falls below the threshold K_{Gi} the obligor defaults, that is, $zw_i + \eta_i\varepsilon_i < K_{Gi}$. The thresholds are set so that the unconditional PD for rating grade G is PD$_G$, that is, $PD_G = \Phi(K_G)$, where Φ is the standard normal cumulative distribution function.

The simplest credit portfolio is one consisting of N loans of equal size that are worth zero at default and $1/N$ otherwise. The capital requirements of assets are determined from risk-weighted formulas, developed from this simple credit portfolio model that is called the Asymptotic Single-Risk Factor (ASRF) model. The model is discussed in, for example, Finger (1999, 2001) and Gordy (2003). The ASRF is derived from an adaptation of Merton's single asset model mentioned above.

In the ASRF model, the credit risk in a portfolio is divided into two categories: the systematic risk and the diversifiable or idiosyncratic risk; *cf.* Section 9.1.1.2.

The systematic risk is facing the whole industry and is nondiversifiable. It also represents the effect of unexpected changes in the macroeconomic and financial market conditions on the performance of the borrowers.

The diversifiable or idiosyncratic risk affects the individual firms. The main idea behind the ASRF model is that as the portfolio becomes more and more fine-granulated, the largest individual exposures account for a smaller and smaller share of the portfolio exposure. The model assumes that the bank credit portfolio consists of a large number of relatively small exposures and the systematic risk is modeled with only one risk factor.

The value of the firm's assets is measured by the price at which the total of the firm's liabilities can be purchased. Hence, the total value of the firm's assets is the sum of its stock and the value of its debt. We follow Aas (2005) in describing this model.

The normalized assets return Z_i of firm i in the credit portfolio is driven by a single systematic risk factor Z and an idiosyncratic noise component ε_i. This is a special case of the model described in Equation 18.3a above.

$$Z_i = \sqrt{\rho_i} \cdot Z + \sqrt{1 - \rho_i} \cdot \varepsilon_i, \tag{18.3b}$$

where Z and ε_i are independent and identically distributed (i.i.d.) standard normal distributed variables. The $\sqrt{\rho_i}$ is the linear correlation coefficient between the asset return Z_i and the common factor Z. The $\sqrt{\rho_i}$ is interpreted as the sensitivity to systematic risk. The variable Z_i has a standard normal distribution. Z is a common risk to all firms (systematic) representing the macroeconomy. This factor is usually referred to as the *economy*. The noise component ε_i represents the risks specific to the institutions i. From this model it is clear that the assets of all firms are multivariate normal and the assets of two firms i and j have a correlation coefficient equal to $E(Z_i Z_j) = \sqrt{\rho_i \rho_j}$. In accordance with Merton (1974), where he proposed a model for assessing the credit risk of a company by characterizing the company's equity as a call option on its assets, we define a binary random variable for each firm and subportfolio k: $I_{ik} = 1$ if $Z_{ik} \leqslant \Phi^{-1}(\text{PD}_k)$ and $I_{ik} = 0$ if $Z_{ik} > \Phi^{-1}(\text{PD}_k)$, where $\Phi(\cdot)$ indicates the cumulative standard normal distribution. The parameter PD_k is the unconditional PD of subportfolio k.

The (conditional) PD^c, given the outcome of the systematic risk factor $Z = z$ and suppressing the subportfolio index k, is then

$$\begin{aligned}
\text{PD}^c &= P\left(I_i = 1 | Z = z\right) = P\left(Z_i < \Phi^{-1}(\text{PD}_i) | Z = z\right) \\
&= P\left(\sqrt{\rho_i} \cdot Z + \sqrt{1 - \rho_i} \cdot \varepsilon_i < \Phi^{-1}(\text{PD}_i) | Z = z\right) \\
&= P\left(\varepsilon_i < \frac{\Phi^{-1}(\text{PD}_i) - \sqrt{\rho_i} \cdot Z}{\sqrt{1 - \rho_i}} | Z = z\right) = \Phi\left(\frac{\Phi^{-1}(\text{PD}_i) - \sqrt{\rho_i} \cdot z}{\sqrt{1 - \rho_i}}\right) \qquad (18.4)
\end{aligned}$$

When $\rho_i = 0$, the default rate as measured by Equation 18.4 equals PD_i irrespective of z, and when ρ_i approaches 1, the conditional default probability only takes two values, that is, we have the Bernoulli distribution taking the value 0 when $Z > \Phi^{-1}(\text{PD}_i)$ and the value 1 when $Z < \Phi^{-1}(\text{PD}_i)$.

Merton's single asset model can be extended to a model for the whole portfolio. This was shown by Vasicek (2002, 1987, 1991). We call Equation 18.4 the Vasicek model or the Vasicek distribution. This model is sufficiently accurate if the different numbers of loans/firms in the portfolio are large and exact for a portfolio of infinitely many loans/firms (full granularity).

In Vasicek (2002), the model (Equation 18.4) was adjusted for granularity. Consider a portfolio of uniform credits with single exposure weights $w_1, w_2, \ldots, w_n$, and let $\delta = \sum_{i=1}^{n} w_i$. Vasicek showed in op. cit. that we could take into account the fact that a portfolio is composed of a discrete number of relatively large exposures (and not by a very large number of identical small exposures) by changing the correlation coefficient ρ in Equation 18.4 to $\rho + \delta \cdot (1 - \rho)$.

In the Merton model, we take the default threshold and then infer the PD. This logic approach is reversed in the Vasicek model as in this case the PD is taken as input and the default threshold is inferred using the standard normal distribution. Hence the model measures by how many standard deviations the current asset value is higher than the current default threshold. What we get is the distance-to-default, DD; see Section 18.1.2. The DD is an average DD in an "average situation." The DD is then transformed into a new DD in an economic downturn. This is done by using a single factor model, that is, the *economy*, using Equation 18.4, where $\Phi^{-1}(\text{PD}_i)$ is the DD and z is the distance-from-*economy*.

18.1.2 Moody's KMV Model

The abbreviation KMV stands for the three creators Stephen Kealhofer, John McQuown, and Oldrich Vasicek who started the firm KMV Corporation in 1989. In 1990, they released its first version of the Expected Default Frequency, EDF, credit measure model. In 2002, the firm was bought by Moody's Corporation and was renamed "Moody's KMV" and the credit measure was renamed "Moody's KMV EDF." The model itself is called the Vasicek–Kealhofer model. In 1997, the model incorporated a credit migration model. Version 8 of the EDF model was released in 2008; *cf.* Dwyer and Qu (2007). A brief history of the KMV model is given in Dvorak (2008).

Distribution Type: Simulation Based

In the KMV model, the actual PD, that is, the EDF, is derived for each obligor from Merton's firm-value model, based on an option pricing model. The credit risk is mainly driven by the dynamics of the asset value of the issuer. The PD is thus a function of the firm's capital structure, the volatility of the asset returns, and the current asset value. Crouhy et al. (2000) view the EDF as "cardinal ranking" of obligors relative to default risk, as compared with "ordinal ranking" proposed by rating agencies. When the stochastic process for the asset value has been specified, the real PD, for any time horizon, can be derived.

The EDF measure, which does not contain any information on the LGD, is an absolute measure of default risk. The EDF measure is a mapping converting a DD, see below, into an actual PD. We take Merton's model (Equation 18.1) and the default inequality $V_{At} < K$ and introduce an index i for firm i. We can now rewrite Equation 18.2 using the inequality. This gives us

$$\frac{\ln\left[\dfrac{K_i}{V_{A0,i}}\right] - \left(\mu_{Ai} - \left(\dfrac{\sigma_{Ai}^2}{2}\right)\right)t}{\sigma_{Ai}\sqrt{t}} > Z_i. \tag{18.5a}$$

Let the left-hand side of Equation 18.5a be $-\mathrm{DD}_i$, that is, minus the distance to default. Hence, DD is equal to

$$\mathrm{DD} = \frac{\ln\left[\dfrac{V_{A0,i}}{K_i}\right] + \left(\mu_{Ai} - \left(\dfrac{\sigma_{Ai}^2}{2}\right)\right)t}{\sigma_{Ai}\sqrt{t}}. \tag{18.5b}$$

The DD index is the number of standard deviations between the mean of the distribution of the asset value (lognormal) and a critical threshold, the default point. The default point is set at the par value of current liabilities including short-term debt (over the time horizon) plus half of the long-term debt. The DD is thus the distance between the expected asset value in 1 year, $E(V_1)$, and the default point, DPT, expressed in terms of standard deviation of future asset returns; see Crouhy et al. (2000) and Crosbie and Bohn (2003):

$$\mathrm{DD} = \frac{E(V_1) - \mathrm{DPT}}{\sigma_A}. \tag{18.5c}$$

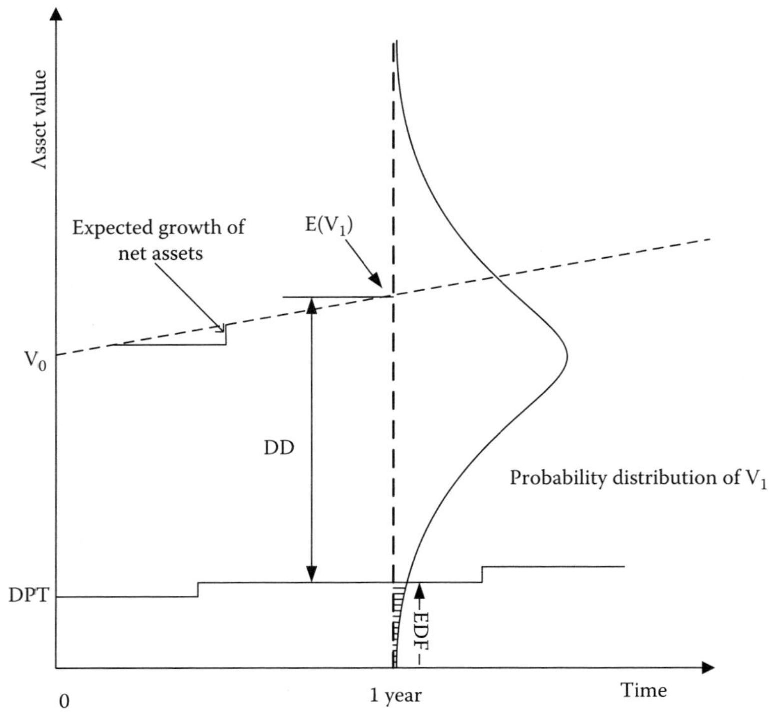

FIGURE 18.1 DD and EDF.

To calculate the DD, we need to know $V_{A0,i}$, μ_{Ai}, and σ_{Ai}. The variable $V_{A0,i}$ cannot be observed, but could be calculated by using the Black–Scholes option pricing model, where the firm's equity $V_{E,i}$ is observed in the market. This gives us the possibility of obtaining $V_{A0,i}$ and σ_{Ai}; see, for example, Hull et al. (2004). The DD is illustrated in Figure 18.1.

The Vasicek–Kealhofer model does not use the probabilities $P(Z_i < -\text{DD}_i) = \Phi(-\text{DD}_i)$ to estimate the EDF_i. Instead they use a historic database including DDs and search for firms that at the same time have a similar DD and then observe the default frequency for these.

To calculate the EDF for the portfolio, you need to consider that there is dependence between the firm's assets.

The EDF credit risk measure can also be used to derive credit categories relative to the measure of risk. This mapping converts the median of different EDF levels of Aaa, Aa, Baa, Ba, B, Caa, Ca, and C credits into EDF ratings.

18.1.3 CreditMetrics Model

CreditMetrics was introduced by the company J.P. Morgan as a tool for credit risk measurement during the spring of 1997. This model, which was presented in Gupton et al. (1997), is also discussed in, for example, Gordy (2000) and Crouhy et al. (2000). RiskMetrics was founded in 1994 as an internal function within J.P. Morgan that developed a VaR model. In 1998, RiskMetrics was spun out of J.P. Morgan as a separate company.

TABLE 18.1 One-Year Transition Matrix (%)

	Rating at Year-End (%)							
Initial Rating	**AAA**	**AA**	**A**	**BBB**	**BB**	**B**	**CCC**	**Default**
AAA	93.00	6.18	0.66	0.07	0.08	0.01	0.00	0.000
AA	0.61	91.03	7.53	0.64	0.09	0.08	0.01	0.005
A	0.08	1.99	91.69	5.55	0.49	0.18	0.01	0.008
BBB	0.03	0.26	4.05	89.70	5.05	0.76	0.07	0.083
BB	0.04	0.11	0.56	5.26	83.80	8.95	0.73	0.548
B	0.00	0.07	0.23	0.50	4.67	84.36	5.71	4.448
CCC	0.06	0.01	0.34	0.56	1.10	7.99	47.02	42.896

Source: Data from Standard and Poor's rating histories 1981–2004.

Distribution Type: Simulation Based

To estimate the loss distribution, CreditMetrics use Merton's model, and the "value" of the assets, depending on the rating, is assumed to follow the geometrical Brownian motion given by Equation 17.1.

The model described by Equation 18.3a is the base model of CreditMetrics, which in its simplest case, as described by Finger (1999, 2001), is given by Equation 18.3b. The model includes not only estimates of default probabilities, but also estimates of the transition probabilities of moving between different crating ratings. A 1-year transition matrix is given in Table 18.1; see also Schuermann (2008).

The model can be summarized as a two-step building block: step 1 is a VaR due to credit for single counterparties and step 2 is a VaR at the portfolio level.

In the first step, we specify a rating system with rating categories, together with migration probabilities from one credit quality to another; *cf.* Table 18.1. This matrix is the key building block of the credit-VaR model. It is assumed that all issuers within the same rating class are credit-homogeneous with the same transition probabilities and the same default probabilities. The risk horizon is specified, usually 1 year, and the forward discount curve at the risk horizon for each credit category is specified. In the case of default, the recovery rate is set to face value or par value. In a final step, all this is transformed into the forward distribution of the changes in portfolio value conservative to credit migration; see, for example, Crouhy et al. (2000).

The simple model is estimated by simulation, by first taking a single draw from the portfolio drawing a z from an $N(0,1)$ and a set of i.i.d. $N(0,1)$ idiosyncratic ε. The latent Z_i is formed for each counterpart and compared to the threshold, K_G, to determine the default status I_i. Given the value of $Z = z$, everything else is independent. This conditional independence is the crucial part, as this gives us the Vasicek model as described by Equation 17.4. The PD is given by $PD_F = \Phi(-DD_F)$, where DD_F is given by Equation 17.5b and F stands for "failure." If the counterparty is rated CCC at time t, that is, if the value of the assets is $V_t : V_F < V_t < V_{CCC}$, then the probability of being in that class is given by $PD_{CCC} = \Phi(-DD_{CCC}) - \Phi(-DD_F)$. If the portfolio is simulated 100,000 times, then the estimated VaR at 99.5 percentile of the loss distribution is given by the 99,500th element of the sorted loss outcomes.

The conditional probabilities of default are dependent on the "market weights" $\sqrt{\rho_i}$. If the weight is close to 1, meaning that the asset correlations are high, it is most affected by the market and if the weight is lower, the more the counterpart's randomness is due to the idiosyncratic term. A factor model can be used to estimate the correlations.

18.1.4 CreditRisk+ Model

In the autumn of 1997, the CSFB introduced its credit risk measurement tool called CreditRisk+. The model, presented in Wilde (1997), is also discussed in, for example, Gordy (2000) and Crouhy et al. (2000).

Distribution Type: Analytic Form

The CreditRisk+ is a model of default risk and the counterparty has only two possible end-of-period states: default or nondefault. If the counterparty defaults, the lender suffers a fixed loss, that is, the LGD. The distributional assumptions of the total portfolio losses allow us to get a convenient analytic form. The framework consists of three building blocks. The first contains the frequency of default events, the second contains the severity of the losses, and the third is the derivation of the distribution of default losses.

The default correlations in the model are assumed to be driven by a vector of K RFs $x = (x_1, x_2, \ldots, x_K)^t$ and, conditional on x, the defaults of individual counterparties are assumed to be independently Bernoulli distributed draws, that is,

$$I_i = \begin{cases} 1 & \text{if } i \text{ default} \\ 0 & \text{otherwise} \end{cases}.$$

The conditional PD, "drawing a default," is a function of the rating grade G of the counterparty, the realization of the risk factors x and a vector of factor loadings $w = (w_{i1}, w_{i2}, \ldots, w_{iK})^t$ measuring the sensitivity of counterparty i to each of the RFs. It is also assumed that the factor loadings sum up to 1.

The conditional PD of CreditRisk+ is specified, for each counterpart, as $\text{PD}_i^c(x) = \text{PD}_{Gi} \cdot \left[\sum_{k=1}^{K} w_{ik} x_k\right]$, where PD_G is the unconditional PD given credit-standing G. The RFs x serve as a scaling up or down function. Hence we obtain $E\left[\text{PD}_i^c\right] = \text{PD}_{Gi}$. The distribution of default losses is not calculated directly. Instead the probability-generating function (pgf) of defaults is calculated. The pgf $\eta_Y(t)$ of a discrete random variable Y is a function of an auxiliary variable t, defined by $\eta_Y(t) = E(t^Y) = \sum_{k=0}^{\infty} t^k P(Y = k)$. The pgf has two nice properties: the sum of two independent random variables has the product of the two pgfs and if $\eta_Y(t|X)$ is the conditional pgf, given the value X having a distribution function $H(x)$, then the unconditional pgf $\eta_Y(t)$ is simply $\eta_Y(t) = \int_x \eta_Y(t|x)\, \mathrm{d}H(x)$.

For a single counterparty, we obtain $\eta_Y(t|x) = t^0 P(I = 0) + t^1 P(I = 1) = 1 - \text{PD}^c(x) + \text{PD}^c(x) \cdot t = (1 + \text{PD}^c(x))\,(t - 1)$, where we have dropped the index i. We can use the approximation $\log(1 + x) \approx x$ for $x \approx 0$ and hence write $\eta_Y(t|x) = \exp\left\{\log\left(1 + \text{PD}^c(x)\right)(t - 1)\right\} \approx \exp\left\{\text{PD}^c(x)(t - 1)\right\}$ for each counterpart. Gordy (2000) refers this to the "Poisson approximation." Conditional on the risk factors x, default events are independent among the counterparties. Hence, the pgf of the sum of the counterparties' defaults is the product of the individual pgfs:

$$\begin{aligned}\eta(t|x) &= \prod_i \eta_i(t|x) \approx \prod_i \exp\left\{\mathrm{PD}^c(x)(t-1)\right\} \\ &= \exp\left\{\mu(x)(t-1)\right\}, \quad \text{where } \mu(x) = \sum_i \mathrm{PD}_i^c.\end{aligned}$$

Expanding the pgf in its Taylor series shows that the probability of n defaults is Poisson distributed; see Wilde (1997).

To get the unconditional pgf we have to integrate out the RFs x. In CreditRisk+ the RFs are assumed to be independent gamma-distributed random variables with mean 1 and variance σ_k^2, $k = 1, 2, \ldots, K$. It can be shown that the unconditional pgf is

$$\eta(t) = \prod_{k=1}^{K} \left[\frac{1-\delta_k}{1-\delta_k t}\right]^{1/\sigma_k^2}, \quad \text{where } \delta_k = \frac{\sigma_k^2 \mu_k}{1+\sigma_k^2 \mu_k} \text{ and } \mu_k = \sum_i w_{ik} \mathrm{PD}_{Gi}.$$

This pgf shows that the total number of defaults in the portfolio is a sum of K-independent negative binomial variables.

The last step in the model is to find the pgf $\Psi(t)$ for losses. LGD is assumed to be a constant fraction λ of the loan size L_i for the counterpart i. The loss exposures λL_i are assumed to be integers and expressed in integer multiples of a fixed unit of loss. The base unit of loss is denoted by v_0 and the integer multiples are called standardized exposure levels, defined for counterparty i as $v_i = \lambda L_i / v_0$, and rounded to the nearest integer. The round-off figures build up exposure bands or buckets. Each band is viewed as an independent portfolio of loans. The pgf is first calculated for each band and then for the entire portfolio. See, for example, Wilde (1997) for a discussion on exposure bands.

The $\Psi_i(t)$ denotes the pgf for counterpart i. The probability of loss of v_i units on a portfolio of only one counterpart i equals i defaults. Hence we obtain $\Psi_i(t|x) = \eta_i\left(t^{v_i}|x\right)$ for one counterpart, and for all counterparties, $\Psi(t) = \prod_i \Psi_i(t) = \exp\left\{\sum_{k=1}^{K} x_k \sum_i \mathrm{PD}_{Gi} w_{ik}(t^{v_i}-1)\right\}$. Integrating out the RFs we obtain, see Gordy (2000),

$$\Psi(t) = \prod_{k=1}^{K} \left[\frac{1-\delta_k}{1-\delta_k P_k(t)}\right]^{1/\sigma_k^2}, \quad \text{where } P_k(t) = \frac{1}{\mu_k} \sum_i w_{ik} \mathrm{PD}_{Gi} t^{v_i},$$

and δ and μ as above.

In CreditRisk+, the unconditional probability that there will be n units of loss v_0 in the total portfolio is given by the coefficient on z^n in the Taylor expansion of $\Psi(t)$. Different methods, such as Panjer recursion or saddle-point approximation, have been used.

Several extensions of the basic one-period model are proposed within the CreditRisk+ framework, such as the multiperiod model; see Wilde (1997). The calculation of "fair" VaR and TVaR from the loss distribution is discussed in, for example, Haaf and Tasche (2002).

18.2 BASEL II: CREDIT RISK IRB

The Basel II capital accord, as proposed by the Basel Committee on Banking Supervision, includes three of the core risks discussed above. One of them is the credit risk. For the background to Basel II and the credit risk mitigation, see, for example, Sandström (2005). The banks are allowed to use either a simple approach, similar to the one in the Basel I Accord (1988 Accord), or a comprehensive approach. The simpler one is discussed in op. cit. We will look at the comprehensive approach.

Distribution Type: Analytic Form

In Gordy (2003), a rigorous presentation of the model used to generate the capital requirements, against credit risk, is presented. Two quick introductions to the model are given by, for example, Finger (2001) and Resti (2002). The internal ratings-based approach is also discussed in Gordy (2004) and Aas (2005).

Aas (2005) states that the Basel II Credit Risk internal rating-based (IRB) approach is a hybrid between a very simple statistical model of capital needs for credit risk and a negotiated settlement. The capital requirement is calculated in a bottom-up approach. At the first level, the capital requirements are calculated for different assets and then at the second level the capital requirements are added up to the total.

Let w_i be an exposure weight defined as $w_i = E_i / \sum_{i=1}^{N} E_i$.

The EL per dollar of exposure for the portfolio is given by $E(L) = \sum_{i=1}^{N} w_i \cdot \mathrm{LGD}_i \cdot \mathrm{PD}_i$.

The portfolio loss per dollar of exposure is defined by $L = \sum_{i=1}^{N} w_i \cdot \mathrm{LGD}_i \cdot Z_i, L < 0$. Let $y = \Phi^{-1}(\alpha)$; then we obtain the model presented by Gordy (2003):

$$q_\alpha(L) = \sum_{i=1}^{N} w_i \cdot \mathrm{LGD}_i \cdot \Phi\left(\frac{\Phi^{-1}(\mathrm{PD}_i) - \sqrt{\rho_i} \cdot \Phi^{-1}(\alpha)}{\sqrt{1-\rho_i}}\right), \tag{18.6}$$

which gives the capital charge per dollar of exposure for large N.

If we have K homogeneous subportfolios with the loss of subportfolio k given by $L_k = E_k \cdot \mathrm{PD}_k \cdot \sum_{i=1}^{n_k} I_{ik}$ and $L = \sum_{k=1}^{K} L_k$, then Equation 18.6 can be written as

$$q_\alpha(L) = \sum_{k=1}^{K} E_k \cdot \mathrm{LGD}_k \cdot \Phi\left(\frac{\Phi^{-1}(\mathrm{PD}_k) - \sqrt{\rho_k} \cdot \Phi^{-1}(\alpha)}{\sqrt{1-\rho_k}}\right); \tag{18.7}$$

see, for example, Aas (2005).

The Basel II Credit Risk IRB capital charge per dollar of exposure is given by

$$C_\alpha(L) = q_\alpha(L) - E(L) = \sum_{k=1}^{K} w_k \cdot \mathrm{LGD}_k \cdot \left[\Phi\left(\frac{\Phi^{-1}(\mathrm{PD}_k) - \sqrt{\rho_k} \cdot \Phi^{-1}(\alpha)}{\sqrt{1-\rho_k}}\right) - \mathrm{PD}_i\right]. \tag{18.8}$$

The Basel Committee has proposed a maturity adjustment to be multiplied with each term in Equation 18.8. For firm i, this adjustment is given by $m_i = 1 + (M - 2.5)b(\mathrm{PD}_i)/$

$[1 - 1.5b(\text{PD}_i)]$, where $b(\text{PD}_i) = \left[0.11852 - 0.05478 \cdot \log(\text{PD}_i)\right]^2$. Here, M is the maturity and is set to 2.5 years for all rating grades. This maturity adjustment is only done for the corporate borrower portfolio; see Aas (2005).

An extension of the one-factor model is given in, for example, Hamerle et al. (2003). For the asset correlations used in the IRB model mentioned above, the Basel Committee has provided different formulae for different business segments (BIS, 2005).

18.3 REINSURANCE COUNTERPARTY DEFAULT RISK

The general Vasicek model is valid when we have a large homogeneous population. This is the main drawback when considering the reinsurance counterparty default risk, as the reinsurance market is limited and insurance companies usually buy reinsurance from a small number of heterogeneous reinsurers, perhaps only one single reinsurer. With a small number we mean that the number is finite, say, at maximum 50–100.

As is seen in Appendix J, during two of the quantitative impact studies conducted for the European Solvency II project, QIS3 and QIS4, the Vasicek model has been tested. The approaches in QIS3 and QIS4 have been discussed and criticized by Hürlimann (2008b). The correlation between the reinsurer has been approximated by a function of the concentration in the reinsurance counterparty portfolio using the Herfindahl–Hirshman concentration index; see Section 18.4.1 and Appendix J. This approach has also been criticized, for example, by ter Berg (2008). A general discussion on the modeling of reinsurance credit rating is given in Shaw (2007).

An alternative approach is proposed by ter Berg (2008). He assumes that the PD is a function of a common shock size, which is a latent random variable. This unobserved random variable generates correlations between the reinsurance companies reflecting its rating.

To construct the *ter Berg model*, we perform the following steps:

1. Define a model for the common shock variable S.
2. Define a model for the PD, given the common shock $S = s$: PD(s), for each counterparty.
3. Define a baseline PD, p_b, independent of S, depending on which rating class the counterparty belongs to.
4. Let the expected default probability equal the PD given by a rating agency, PD_r; then the baseline PD can be written as a function of the PD given by a rating agency.
5. We introduce the set of reinsurance counterparties that an insurance company buys reinsurance from. In terms of ter Berg (2008), we have a "bouquet of reinsurers," and we define a random variable for reinsurance default risk by Z. Taking the VaR of Z gives us the capital charge.

18.3.1 Common Shock Model

Assume that S is an annual common shock variable, $S \in [0, 1]$. If S is close to zero it will have a minor impact on the industry, but if it is close to one, the shock will affect the whole industry, that is, we will have a worldwide CAT scenario. Hence, a default event will be more probable for large values of S.

S is assumed to have a beta distribution with parameters α and $\beta = 1$, that is, the beta probability density function (pdf) will be, given α,

$$f(s|\alpha) = \alpha \cdot s^{\alpha-1}, \quad 0 \leqslant s \leqslant 1, \quad 0 < \alpha < 1.$$

The function shows a monotone decreasing probability for the shock size and it will be important in deriving the first moments of default probabilities as functions of the random shock size. In ter Berg (2008), the global parameter α was set as 0.1 as an illustration. This pdf is shown in Figure 18.2.

PD as a Function of the Shock Size

To define the PD as a function of the shock size we introduce a baseline for this probability, p_b. This probability is specified for each of the reinsurance counterparties $i = 1, \ldots, k$. The shock-modified default probability is defined in the *ter Berg model* as

$$\text{PD}(s) = p_b + (1 - p_b) \cdot s^{\tau/p_b}, \quad 0 < p_b < 1, \quad \frac{\tau}{p_b} > 0.$$

This shock-modified PD has two parameters: the baseline PD, p_b, and a power parameter $\tau > 0$. If the power parameter $\tau > p_b$, then we have a convex curve and the baseline default

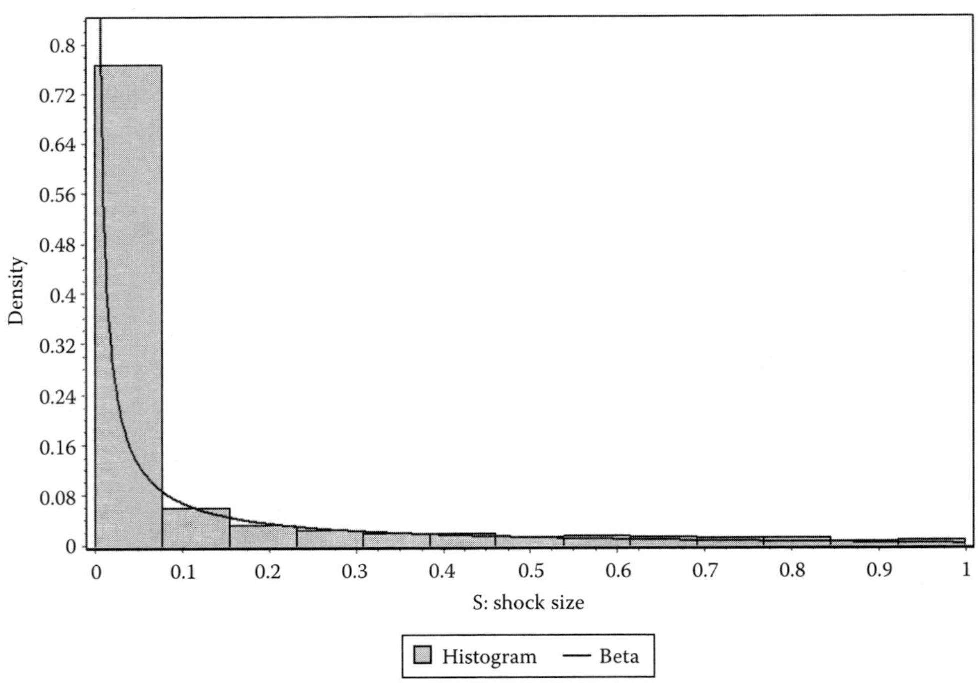

FIGURE 18.2 The probability density function beta (0.1,1) illustrating the common shock variable S and its density function. A small shock would be likely to occur.

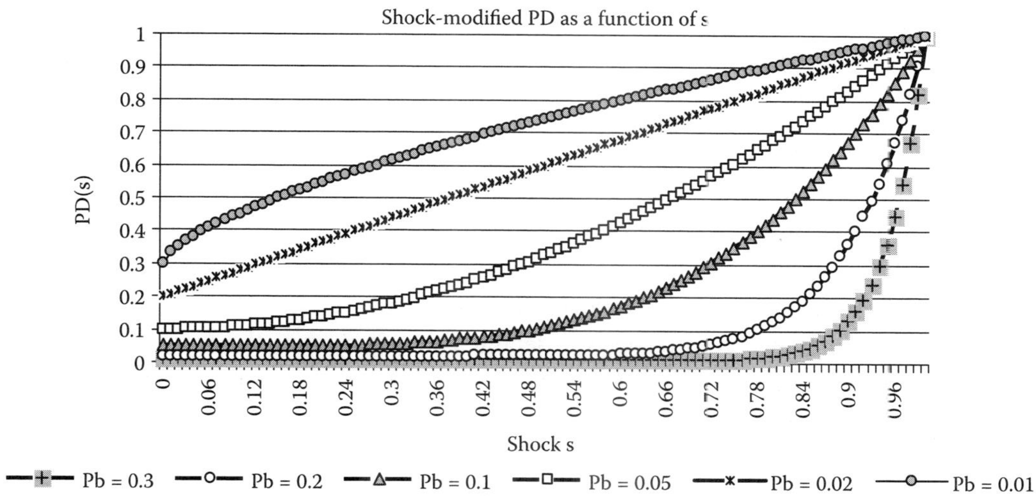

FIGURE 18.3 The shock-modified default probability PD(s) as a function of the shock size s. The global parameter τ is set to 0.2. The baseline default probabilities p_b are 0.01, 0.02, 0, 05, 0.1, 0.2, and 0.3.

probabilities are viewed as low (if they are equal, we will have a straight line), and if $\tau < p_b$, we will have a concave curve, where the baseline default probabilities are viewed as high.

The parameters α and τ are global, but p_b is dependent on each counterparty i, $i = 1, \ldots, k$, and in fact on which rating class it belongs to. In ter Berg (2008), the global parameter τ was set to 0.2 for illustration (Figure 18.3).

18.3.2 Baseline Default Probabilities

The expected value of the PD is given by the use of the probability density function $f(s|\alpha)$ and the shock-modified default probability function PD(s), where we have put $\delta = \tau/p_b$ for each counterparty (the index is suppressed):

$$\begin{aligned} \text{PD} = E\left[\text{PD}(S)\right] &= \int_0^1 \text{PD}(s) \cdot f(s|\alpha)\,\mathrm{d}s = \int_0^1 \left[p_b + (1 - p_b)s^{\delta}\right] \cdot \alpha \cdot s^{\alpha - 1}\,\mathrm{d}s \\ &= \frac{\delta \cdot p_b + \alpha}{\delta + \alpha} = \frac{(\tau + \alpha) \cdot p_b}{\tau + \alpha \cdot p_b} \end{aligned} \tag{18.9}$$

Now, use the PD as given by a rating agency as the value of PD:PD$_r$. This means that we can solve for each rated reinsurer its baseline default probability. This would be

$$p_b = \frac{\text{PD}_r \cdot \tau}{\alpha \cdot (1 - \text{PD}_r) + \tau}. \tag{18.10}$$

It is important to note that we have suppressed the index for the counterparty $i, i = 1, \ldots, k$, in the formulas given above. As α and τ are global, the PD_r for reinsurance counterparty i will define the baseline $p_{bi}, i = 1, \ldots, k$.

18.3.3 Total Default Loss

Assume that we have a "bouquet of reinsurance counterparties"; $i = 1, \ldots, k$. The default process is indicated through a set of Bernoulli random variables (indicator functions), $I_1, \ldots, I_k$, where $I_i = 1$ if counterparty i defaults and 0 otherwise, $i = 1, \ldots, k$. For $i \neq j$, the indicator functions I_i and I_j are independent random variables.

Given that reinsurance counterparty i defaults, that is, $I_i = 1$, the loss, LGD_i, will be denoted by y_i. We assume that this LGD is nonrandom, but it is possible to incorporate an additional variance for its randomness.

Let Z be the random variable for the reinsurance default risk and define it by

$$Z = \sum_{i=1}^{k} I_i \cdot y_i.$$

Given an observed shock $S = s$, we can easily get the expectation, variance, and probability of no default. First note that

$$\Pr\left[I_i = 1 | S = s\right] = E\left[I_i | S = s\right] = E\left[I_i^2 | S = s\right] = E\left[I_i^3 | S = s\right] = \text{PD}_i(s).$$

Now we have

$$E\left[Z | S = s\right] = \sum_{i=1}^{k} \text{PD}_i(s) \cdot y_i,$$

$$V\left[Z | S = s\right] = \sum_{i=1}^{k} \text{PD}_i(s) \cdot y_i^2 - \left(\sum_{i=1}^{k} \text{PD}_i(s) \cdot y_i\right)^2,$$

$$\Pr\left[Z = 0 | S = s\right] = \prod_{i=1}^{k} (1 - \text{PD}_i(s)).$$

What is interesting for us are the unconditional moments, that is, $E(Z)$ and $E(Z^2)$, giving us the variance. Assuming a normal distributed random variable Z, we could use the moments to get the capital charge using VaR or TVaR.

The unconditional expectation of Z is, using Equation 18.9,

$$\begin{aligned}\mu_Z = E(Z) &= \int_0^1 \sum_{i=1}^{k} y_i \cdot \text{PD}_i(s) \cdot f(s)\,\text{d}s \\ &= \sum_{i=1}^{k} y_i \cdot \int_0^1 \text{PD}_i(s) \cdot f(s)\,\text{d}s = \sum_{i=1}^{k} y_i \cdot \frac{(\tau + \alpha) \cdot p_{bi}}{\tau + \alpha \cdot p_{bi}}. \qquad (18.11a)\end{aligned}$$

It is shown in ter Berg (2008) that the variance is given by

$$\sigma_Z^2 = \mathrm{Var}(Z) = \sum_{i=1}^{k}\sum_{j=1}^{k} \omega_{ij} y_i y_j, \tag{18.11b}$$

where

$$\omega_i = \mathrm{PD}_i \cdot (1 - \mathrm{PD}_i), \quad i = 1, \ldots, k,$$

and

$$\omega_{ij} = \frac{\alpha \cdot (1 - p_{\mathrm{b}i}) \cdot (1 - p_{\mathrm{b}j})}{\alpha + \dfrac{\tau}{p_{\mathrm{b}i}} + \dfrac{\tau}{p_{\mathrm{b}j}}} - (\mathrm{PD}_i - p_{\mathrm{b}i}) \cdot (\mathrm{PD}_j - p_{\mathrm{b}j}),$$

$$i \neq j, \quad i, j = 1, \ldots, k. \tag{18.12a}$$

The covariance function (Equation 18.12a) can be reformulated as the correlation between two arbitrary counterparties as $\rho_{ij} = \omega_{ij}/\sqrt{\omega_i \omega_j}$. It has been shown* by ter Berg that the covariance function (Equation 18.12a) can be written as

$$\omega_{ij} = \frac{\mathrm{PD}_i \cdot (1 - \mathrm{PD}_i) \cdot \mathrm{PD}_j \cdot (1 - \mathrm{PD}_j)}{(1 + \gamma) \cdot \left[\mathrm{PD}_i + \mathrm{PD}_j\right] - \mathrm{PD}_i \cdot \mathrm{PD}_j}, \tag{18.12b}$$

where $\gamma = \tau/\alpha$. As $\omega_i = \mathrm{PD}_i \cdot (1 - \mathrm{PD}_i)$, the correlation between two counterparties is

$$\rho_{ij} = \frac{\sqrt{\mathrm{PD}_i \cdot (1 - \mathrm{PD}_i) \cdot \mathrm{PD}_j \cdot (1 - \mathrm{PD}_j)}}{(1 + \gamma) \cdot \left[\mathrm{PD}_i + \mathrm{PD}_j\right] - \mathrm{PD}_i \cdot \mathrm{PD}_j}. \tag{18.13a}$$

If we assume that all counterparties belong to one and the same rating class and have the same "equicorrelation" ρ, then from Equation 18.13a, it can be seen that

$$\rho = \frac{1 - \mathrm{PD}}{2 + 2\gamma - \mathrm{PD}}. \tag{18.13b}$$

If we assume $\tau = 0.2$ and $\alpha = 0.1$, then $\gamma = 2$ and hence $\rho = (1 - \mathrm{PD})/(6 - \mathrm{PD})$. For $\mathrm{PD} = -1$, we get $\rho = 2/7$, for $\mathrm{PD} = 0$, we get $\rho = 1/6$, and for $\mathrm{PD} = 1$, we get $\rho = 0$.

18.3.4 Rating Classes

Assume that we have C rating classes R_c, $c = 1, \ldots, C$. C is typically seven or eight. A counterparty i, $i = 1, \ldots, k$, will belong to one of the rating classes such as $i \in R_c$.

Assume that α and τ are given by the regulator. From Equation 18.10, we see that the baseline PD is a function of the PD given by the rating agency and as ω_{ij}, $i, j = 1, \ldots, k$, only depends on the PD given by the rating agency we see that the ω_{ij}, $i, j = 1, \ldots, k$, only

*Private communication.

depends on the rating classes. This means that we can rewrite Equations 18.11a and 18.11b as

$$\mu_Z = E(Z) = \sum_{i=1}^{k} y_i \cdot \frac{(\tau+\alpha)\cdot p_{bi}}{\tau+\alpha\cdot p_{bi}} = \sum_{c=1}^{C} Y_c \cdot \frac{(\tau+\alpha)\cdot p_{bc}}{\tau+\alpha\cdot p_{bc}}, \tag{18.14a}$$

and

$$\begin{aligned}\sigma_Z^2 = \mathrm{Var}(Z) &= \sum_{i=1}^{k}\sum_{j=1}^{k} \omega_{ij} y_i y_j = \sum_{i\neq j} \omega_{ij} y_i y_j + \sum_{i=1}^{k} \omega_i y_i^2 \\ &= \sum_{i=1}^{k}\sum_{j=1}^{k} \omega_{ij} y_i y_j + \sum_{i=1}^{k} [\omega_i - \omega_{ii}]\, y_i^2 \\ &= \sum_{c=1}^{C}\sum_{d=1}^{C} \omega_{cd} Y_c Y_d + \sum_{c=1}^{C} [\omega_c - \omega_{cc}] \cdot Y_{2c},\end{aligned} \tag{18.14b}$$

where $Y_c = \sum_{i\in R_c} y_i$, $Y_{2c} = \sum_{i\in R_c} y_i^2$, $\omega_c = \mathrm{PD}_c \cdot (1-\mathrm{PD}_c)$, $\omega_{cc} = \alpha\cdot(1-p_{bc})^2/[\alpha + (2\tau/p_{bc})] - (\mathrm{PD}_c - p_{bc})^2$, and $\omega_{cd} = \alpha\cdot(1-p_{bc})\cdot(1-p_{bd})/[\alpha+\tau/p_{bc}+\tau/p_{bd}] - (\mathrm{PD}_c - p_{bc})\cdot(\mathrm{PD}_d - p_{bd})$, $c \neq d$, $c, d = 1,\ldots,C$. For each rating class c, $c = 1,\ldots,C$ we also have from Equation 18.10 that $p_{bc} = \mathrm{PD}_c \cdot \tau/[\alpha\cdot(1-\mathrm{PD}_c)+\tau]$.

Let

$$\begin{aligned} w_c &= \mathrm{PD}_c - p_{bc}, \\ u_{cd} &= \frac{\alpha\cdot(1-p_{pc})\cdot(1-p_{pd})}{\alpha+\tau/p_{bc}+\tau/p_{bd}}, \end{aligned}$$

and $v_c = \omega_c - \omega_{cc}$. Then we have $\omega_{cd} = u_{cd} - w_c \cdot w_d$ and hence Equation 18.14b could be written as

$$\begin{aligned}\sigma_Z^2 = \mathrm{Var}(Z) &= \sum_{c=1}^{C}\sum_{d=1}^{C} (u_{cd} - w_c\cdot w_d)\cdot Y_c \cdot Y_d + \sum_{c=1}^{C} v_c \cdot Y_{2c} \\ &= \sum_{c=1}^{C}\sum_{d=1}^{C} u_{cd}\cdot Y_c\cdot Y_d + \sum_{c=1}^{C} v_c\cdot Y_{2c} - \left(\sum_{c=1}^{C} w_c \cdot Y_c\right)^2.\end{aligned} \tag{18.14c}$$

18.3.5 Capital Charge

Assuming a standard normal distribution, the capital charge or requirement will, using the VaR as a measure, be (α here is not the parameter used in this model, but the general probability of "ruin")

$$\mathrm{CR}_{1-\alpha}(Z) = k_{1-\alpha}\cdot\sigma_Z. \tag{18.15a}$$

Let $c_Z = \sigma_Z/\mu_Z$ be the coefficient of variation. Hence the capital charge given by Equation 18.13a could be written as a volume measure times a factor:

$$\mathrm{CR}_{1-\alpha}(Z) = \mu_Z \cdot [k_{1-\alpha} \cdot c_Z]\,. \tag{18.15b}$$

Here the volume is the expected LGD. A pure volume should be obtained from the total LGD, taken over all reinsurance counterparties. Assume that we have k reinsurance counterparties. Then we could take $Y_{(k)} = k \cdot \mu_Z$ as the volume measure and the factor as $[k_{1-\alpha} \cdot c_Z/k]$. This could, of course, be simplified by assuming that the volume measure is $Y_{\mathrm{Sum}} = \sum_{i=1}^{k} y_i$, that is, the total sum of the LGDs. As a factor we could take $[k_{1-\alpha} \cdot \sigma_Z/Y_{\mathrm{Sum}}] = \left[k_{1-\alpha} \cdot \hat{c}_Z/k\right]$, where $k_{1-\alpha} = 2.58$ for $\alpha = 0.005$ and $\hat{c}_Z = \sigma_Z/\bar{Y}_{\mathrm{Sum}}$.

Example 18.1: One reinsurance counterparty If the insurer has only one reinsurance counterparty, then the variance of Z could be written as

$$V(Z) = y^2 \cdot \mathrm{PD} \cdot (1 - \mathrm{PD}),$$

and hence the capital charge would be

$$\mathrm{CR}_{1-\alpha}(Z) = y \cdot k_{1-\alpha} \cdot \sqrt{\mathrm{PD} \cdot (1 - \mathrm{PD})} = \mathrm{LGD} \cdot \text{“factor,”}$$

that is, the volume (LGD) times a factor.

Example 18.2: One pool of reinsurance counterparties Assume that the insurance company has a pool of reinsurance counterparties all having the same rating $c = r$. From Equation 18.12b, we obtain

$$\begin{aligned}\sigma_Z^2 &= \omega_{rr} Y_r^2 + [\omega_r - \omega_{rr}]\, Y_{2r} = Y_r^2 \cdot \left[\omega_{rr} + (\omega_r - \omega_{rr}) \cdot \frac{Y_{2r}}{Y_r^2}\right] \\ &= Y_r^2 \cdot \left[\omega_r \cdot \frac{Y_{2r}}{Y_r^2} + \omega_{rr} \cdot \left(1 - \frac{Y_{2r}}{Y_r^2}\right)\right].\end{aligned}$$

Then the capital charge would be

$$\begin{aligned}\mathrm{CR}_{1-\alpha}(Z) &= Y_r \cdot k_{1-\alpha} \cdot \sqrt{\left[\omega_r \cdot \frac{Y_{2r}}{Y_r^2} + \omega_{rr} \cdot \left(1 - \frac{Y_{2r}}{Y_r^2}\right)\right]} \\ &= \mathrm{LGD}_{\mathrm{pool}\ r} \cdot \text{“factor,”}\end{aligned}$$

where Y_r is the sum of the LGDs of the pool, Y_{2r} is the sum of squared LGDs of the pool, and PD_r is the probability-of-default for the equal rated pool.

With equicorrelation within the pool, the "factor" can be written as

$$\sqrt{\left[\mathrm{PD}_r \cdot [1-\mathrm{PD}_r] \cdot \left\{\frac{Y_{2r}}{Y_r^2} + \frac{1-\mathrm{PD}_r}{2+2\gamma-\mathrm{PD}_r} \cdot \left(1-\frac{Y_{2r}}{Y_r^2}\right)\right\}\right]}.$$

18.4 CREDIT SPREAD RISK MODELS

One component of spreads, or credit spreads, is the EL on corporate bonds due to default. This could be measured in percentage as EL% = PD × LGD. But, as has been seen in the literature, the EL accounts for only a small fraction of spreads irrespective of rating and maturity; see Amato and Remolona (2003) and Table 18.2. In op. cit., it is argued that the spreads are so wide because they are pricing undiversified credit risk. Other explanations are the role of taxes, risk premia, and liquidity risk.

As the credit spread could be seen as the excess return demanded by the market for assuming a certain credit exposure, it fits more naturally within a market risk model. Therefore it is not a part of the CreditRisk+ model; see, for example, Wilde (1997).

The credit spread is the difference between the yield of a corporate bond and a government bond at each point of maturity, and the migration (downgrade) risk is a part of the credit spread risk. The last part has been modeled in both the KMV and CreditMetrics models discussed above.

The Merton model, see Section 18.1, can be used to explain risk debt yields. Let B_0 be the market price of the debt at time $t = 0$, that is, $B_0 = V_{\mathrm{A}0} - V_{\mathrm{E}0}$, and using Equation 18.1, we obtain

$$B_0 = V_{\mathrm{A}0}\left[\Phi(-d_1) + L \cdot \Phi(d_2)\right], \tag{18.16a}$$

where $L = K\,\mathrm{e}^{-rt}/V_{\mathrm{A}0} = K^*/V_{\mathrm{A}0}$ is a measure of leverage and the numerator is the present value of the threshold (or promised debt). The yield to maturity on the debt is implicitly

TABLE 18.2 Spreads and ELs, EL%, in Basis Points

	Maturity							
	1–3 Years		3–5 Years		5–7 Years		7–10 Years	
Rating	**Spread**	**EL%**	**Spread**	**EL%**	**Spread**	**EL%**	**Spread**	**EL%**
AAA	49.50	0.06	63.86	0.18	70.47	0.33	73.95	0.61
AA	58.97	1.24	71.22	1.44	82.36	1.86	88.57	2.70
A	88.82	1.12	102.91	2.78	110.71	4.71	117.52	7.32
BBB	168.99	12.48	170.89	20.12	185.34	27.17	179.63	34.56
BB	421.20	103.09	364.55	126.74	345.37	140.52	322.32	148.05
B	760.84	426.16	691.81	400.52	571.94	368.38	512.43	329.40

Source: From Amato, Jeffery D. and Eli M. Remolona. 2003. The credit spread puzzle. *BIS Quarterly Review*. December. http://www.bis.org/forum/research.htm (free of charge). With permission.

Note: In basis points, that is, 1 bp = 0.0001. Spreads are averages over the period January 1997–August 2003 of the Merrill Lynch option-adjusted spread indices for U.S. corporate bonds. See text for details of computation of EL.

defined as, *cf.* Hull et al. (2004) and Martin (2004),

$$B_0 = K\,e^{-yt} = K^*\,e^{(r-y)t}. \tag{18.16b}$$

Substituting $V_{A0} = K^*/L$ into Equation 18.16a and combining the two Equations 18.16a and 18.16b, we obtain the credit spread, that is, the difference between the yield and the risk-free rate, implied by the Merton model as

$$CS_M = y - r = \frac{-\ln\left[\Phi(d_2) + \Phi(-d_1)/L\right]}{t}. \tag{18.17}$$

As for the risk neutral PD in Equation 18.2, this implied that credit spread only depends on the leverage L, the asset volatility σ_A and the time to maturity t.

To model the credit spread from the KMV model, Crouhy et al. (2000) uses a single cash flow approach and shows that the implicit discount rate y, which accounts for default risk, is $y = r + \text{CS}$ and is the solution to

$$\frac{(1-\text{LGD})}{1+r} + \frac{\text{LGD}(1-\text{PD})}{1+r} = \frac{1}{1+r+\text{CS}},$$

where PD is the risk neutral PD. The solution is

$$CS_C = \frac{\text{LGD}\cdot\text{PD}\cdot(1+r)}{1-\text{LGD}\cdot\text{PD}} = \frac{\text{EL}\%}{1-\text{EL}\%}\cdot(1+r). \tag{18.18}$$

This model can be generalized to the valuation of a stream of cash flows.

18.4.1 CEA Model

The body of the European insurance industry, CEA, proposed in 2006 a European Standard Approach (ESA) for solvency assessment as another contribution to the debate on a European standard formula; see CEA (2006b).

The credit risk Credit spread risk is included as part of market risk. A credit spread widening is a market event that affects all credit risky assets. In many cases, credit ratings downgrade and default risk is limited for well-diversified portfolios over a 1-year period. These effects are assumed to be implicitly captured in the concentration penalty factor included in the model. Assets should be classified attending to their credit quality since credit spread volatility differs depending on the quality of the assets.

Capital requirement for the credit spread risk is assessed by applying a simple factor to the market-consistent value of credit risk assets. Assets are classified by their credit rating, and different factors are applied to each rating.

The following rating classes are considered in the model:

- Government
- AAA

- AA
- A
- BBB
- BB
- B
- CCC or worse
- Unrated

The capital charge is defined as

$$\text{CR}_{\text{MR,CS}} = \text{CSR} + \text{CP},$$

where CSR: credit spread risk, defined as $\sum_{\text{Rating}} \text{MV}_{\text{A,R}} \cdot \text{MD}_{\text{R}} \cdot f_{\text{R}}$, where MV is the market value of assets with rating R, MD is the modified duration for rating R, and f_R is a rating factor and CP: concentration penalty, given by CP = (concentration penalty factor) $*$ CSR $*$ (% bond above threshold).

18.5 CONCENTRATION RISK

The term "concentration risk" is the general notion of the risk arising from an uneven distribution of counterparties in credit or other business relationships, such as reinsurance, or from a concentration in various business sectors or geographical regions. Hence, we have the following categories of concentration risk, *cf.* DB (2006):

- *Single-name concentration—Granularity:* concentration of loans to individual borrowers or those buying reinsurance
- *Sectoral concentration:* an uneven concentration across sectors of industry or geographical regions

Also a third category, one can distinguish

- *Bilateral concentration:* a concentration of exposures to enterprises connected with one another through bilateral business relations

Different methods for measuring the single-name concentration or granularity have been discussed in the literature. A study of the credit risk concentration was presented by BIS (2006b). In DB (2006), a distinction is made between model-free (heuristic) and model-based methods. The first one includes, for example, simple ratios such as the ratio of the sum of exposures to the 20 or 30 largest single borrowers to the total exposure. A general discussion on concentration risk in credit portfolios is given by Lütkebohmert (2008).

Other measures of single-name concentration are different concentration measures such as the Gini coefficient or the Herfindahl–Hirschman index as discussed in Section 18.4.1; see also Lütkebohmert (2008).

A granularity adjustment (GA) for the ASRF forms the basis of the IRB model of the Basel II Credit Risk. The GA can be calculated as the difference between unexpected loss in the real portfolio and unexpected loss in an infinitely granular portfolio with the same risk characteristics. The GA is calculated as $\text{GA} = \text{VaR}_{1-\alpha}(L) - \text{VaR}_{1-\alpha}\left[E\left(L|X\right)\right]$ and taking the second-order Taylor expansion to the quantile of the portfolio loss L; X is the systematic factor.

A simplified model for the sectoral concentration shows the basic structure of multifactor models used in the banking sector. The simplified model equals that given by Equation 18.3b and the more general model equals that given by Equation 18.3a. This is discussed in DB (2006). In Equation 18.3b, we have incorporated a subscript for sector $s = 1, \ldots, S$:

$$Z_{si} = \sqrt{\rho_{si}} \cdot Z_s + \sqrt{1 - \rho_{si}} \cdot \varepsilon_{si}. \tag{18.3b$'$}$$

The $S \times S$ correlation matrix, Ω, gives the correlation between the sectoral factors. The asset correlation between two firms i and j in sectors s and t, respectively, is then given by

$$\rho(Z_{si}, Z_{tj}) = \sqrt{\rho_{si}} \cdot \sqrt{\rho_{tj}} \cdot \Omega_{s,t}.$$

The percentage portfolio loss L at the end of the risk horizon can be determined as

$$L = \sum_{s=1}^{S} \sum_{i=1}^{M_s} w_{si} \cdot \Lambda_{si} \cdot I\left\{Z_{si} \leqslant \Phi^{-1}(\text{PD}_{si})\right\}, \tag{18.19}$$

where w_{si} is the share of credit exposure of the ith firm in sector s in the overall portfolio, Λ_{si} is the relative LGD. It is assumed that this loss ratio is independent of the default event and can be replaced by its expected value $E[\Lambda_{si}]$ for risk measurement purposes, and $I\{*\}$ is the indicator function, with PD being the PD of the ith firm in sector s.

Hence we can derive the economic capital (capital charge) by deducting the EL

$$\text{EL} = \sum_{s=1}^{S} \sum_{i=1}^{M_s} w_{si} \cdot E[\Lambda_{si}] \cdot \text{PD}_{si},$$

from the 99.5% quantile of the distribution of L; this can be made by simulations. This model and simplifications are discussed in Düllmann and Masschelein (2006).

CHAPTER 19

Operational Risk

OPERATIONAL RISK is the risk of loss resulting from inadequate or failed internal processes, people, systems, or from external events (IAA, 2004).

This concept has primarily emerged from the banking industry and the Basel II project. It can be seen as consisting of more or less all risks other than insurance, market, credit, and liquidity risks. There is no unique definition of what is to be understood under operational risk although many existing systems equate it, for instance, to a collection of exposures to

- Failure in control and management
- Failure in IT processes
- Human errors
- Fraud
- Jurisdictional and legal risk

As such, rather than being a truly isolated risk component, operational risk is often associated and overlapping with other risk factors. The nature of operational risk in insurers differs from that in banks because of the different nature of the businesses.

In many models, such as the U.S. NAIC RBC system, it is assumed that the operational risk is truly correlated to all other risks, meaning that the correlation is ρ (operational risk, other risks) $= 1$.

Operational risk is thoroughly discussed in Brehm et al. (2007) and a textbook on modeling losses, using frequency and severity, modeling potential losses, using statistical tools, and calculating economic capital to support operational risk is Panjer (2006). Operational risk is also discussed in McNeil et al. (2005). Stochastic models for operational risk such as Extreme Value Theory (EVT), heavy-tailed ruin-type estimates, and others are discussed in Embrechts et al. (2004). Other techniques to capture the operational risk are given in, for example, Chavez-Demoulin et al. (2006) and Degen et al. (2007). Aggregation of risk capital with applications to operational risk is discussed in, for example, Chavez-Demoulin et al.

(2006) and Embrechts and Puccetti (2006). A Bayesian approach is discussed in Lambrigger et al. (2007, 2008).

In CRO (2009a), CRO Forum has published the guiding principles underlying the establishment of best practice for the management of operational risk within insurance and reinsurance companies.

19.1 DATA GATHERING FOR THE INSURANCE INDUSTRY

Because of the current general lack of sufficient insurer quantitative data, there can only be a very simple capital requirement for insurers (and banks). In the future, when data gathering becomes a natural element in the risk management environment, there will be a possibility to estimate the loss distribution of operational risk. One way to do this is to combine different database sources. We will distinguish between three types of databases.

Type 1 data: Consists of internal company data of mainly small- and medium-sized losses. It is important that the companies start to collect data for this purpose.

Type 2 data: Data from a specific insurance database. Several companies need to join such a database.

Type 3 data: Data from a global operational risk database. This type of data includes very large losses from different sources and markets. The data could be used for the upper tail of the loss distribution.

Internal loss data are often limited and biased. Therefore an external source of loss data provides a benchmark against which a company's loss experience can be compared. In 2002, a Type 2 database was introduced for banks under Basel II. The database, Operational Risk eXchange Association (ORX, www.orx.org) is administrated from Zürich, Switzerland. Three years later, the Association of British Insurers, ABI, started a similar consortium for the insurance industry, the Operational Risk Insurance Consortium (ORIC, www.abioric.com). In both cases the databases are designed to supplement internal loss data (Type 1 data) for economic and regulatory capital modeling purposes. In ORIC only losses larger than GBP 10,000 are included. SAS Institute has a global, general operational risk database with data for approximately 25 years. This database of Type 3 is called OpRisk Global data.

There are other databases for the bank industry, for example, Algo OpData and the British BBA's database.

There has been a tendency for companies to undertake scenario analysis to get additional data points (Type 1). These data points are often merged with the actual historic loss data points for modeling purposes.

Both the ORX and ORIC are based on the Basel II main operational risks. The British ABI has adjusted the Basel II subcategories to the insurance industry. The main operational risks in Basel II, ORX, and ORIC are

- Internal fraud
- External fraud

- Employment practices and workplace safety
- Clients, products, and business practice
- Damage to physical assets
- Business disruption and system failures
- Execution, delivery, and process management

The members of ORIC are insurance companies from Europe and Australia. Based on ORIC's database, Selvaggi (2009) studied methods for scaling the size (severity) and the number of external losses (frequency) to make them equivalent to a company's internal loss events. This is an important issue when a company is going to merge its internal loss data with data from an external database, such as ORIC.

Severity analysis: Selvaggi (2009) used an econometric approach based on regression techniques. It was assumed that the size of operational losses depends on general (systemic and systematic) factors and idiosyncratic (diversifiable) factors. The latter factors are also called exposure metrics. Let L_i be the loss amount for operational loss i. The functional form in the econometric model is $L = A\,\mathrm{e}^{\sum \beta X + \delta Z}$, where A captures the general aspects and X and Z refer to company-specific and event-specific measures, respectively. Taking logarithms on both sides, we obtain an equation for the log-losses that can be estimated by using ordinary least squares techniques. The following explanatory variables were used in the ORIC linear regression models that were tested:

- Size of the insurer

  GWP: Average premium income

  FTE: Average number of full-time employees

- Company-specific effects (indicator variables)

  Identifies the five top firms in terms of number of operational loss events

- Recovery

  reco: Recovery amounts

- Business line (indicator variable)

  life: 1 for loss events in the life business and 0 otherwise

- Business function (indicator variables)

  CS: Customer Service/Policy Administration

  SD: Sales and Distribution

  claims: Claims

- Loss event type (indicator variables at ORIC Level 2 event types)

  adv: Advisory Activities

  custom: Customer Account Management

  syst: Systems

  tran: Transaction Capture, Execution, and Maintenance

With this econometric model it is possible to *scale down or up* the ORIC loss amounts for a specific insurance company C from data based on company A:

$$\text{Loss}_\text{C} = \frac{\hat{h}_\text{C}}{\hat{h}_\text{A}} \text{Loss}_\text{A},$$

where the function $\hat{h}_i$ is the stochastically significant regressor of the baseline model:

$$\hat{h}_i = b_1 \cdot \text{GWP}_i + b_2 \cdot \text{FTE}_i + d_1 \cdot \text{reco}_i + d_2 \cdot \text{CS}_i + d_3 \cdot \text{SD}_i + d_4 \cdot \text{claims}_i + d_5 \cdot \text{adv}_i + d_6 \cdot \text{custom}_i.$$

In summary, Selvaggi (2009) found that

- The size of the insurer is positively correlated with the size of the operational losses. Both the premium income [gross written premiums (GWP)] and the number of FTE have positive impacts on the size of the loss.
- The severity of operational losses is more sensitive to headcounts than to premiums.
- Customer- and claims-related business functions are negatively associated with the size of the operational losses (other things being equal). Advisory activities are associated with bigger losses.

Scaled data were also used to fit a distribution function to data. Four families of distribution functions for nonnegative values were used in the study:

- Generalized beta distribution
- Exponential distribution
- Lognormal distribution
- Pareto distribution

Using goodness-of-fit tests suggested that the exponential distribution was most closely aligned with loss data. For both internal company data and scaled "external" data, the lognormal distribution provided the best fit for the distribution of operational losses.

Frequency analysis: Selvaggi (2009) used an econometric analysis based on models for count outcomes. The dependent variable is the number of losses experienced by 18 insurance companies per quarter in 2005–2008 (288 observations). The explanatory variables (exposure metrics) used were

- GWP: Average gross premium income
- FTE: Average number of full-time employees
- life: Proportion of losses corresponding to the life business unit

In the study, two regression models, specifically designed for discrete outcomes of the dependent variable, were used:

- Poisson regression
- Negative binomial regression

The regression results were highly aligned between the two models. The Poisson regression model has one serious shortcoming in that it assumes equality of the conditional mean and variance functions.

In summary it was found that

- The size of the insurer (GWP) was strongly and positively correlated with the number of losses.
- The number of FTEs did not appear to be strongly correlated with the frequency.
- Life business units were strongly correlated with higher frequency.

A discussion on using internal and external data and expert opinion is given in Lambrigger et al. (2007, 2008).

Based on operational risk modeling experience from the Basel II AMA approach (see Section 19.2.3), Folpmers (2008a, 2008b) illustrates how an approach taking interdependencies into account can be put into practice. Data used are from the IT domain and consist of 11 operational IT-risk events with recorded frequency and severity data (minimum, mode, and maximum losses). For the frequency, a Poisson distribution is assumed, and for the severity, a triangle distribution is assumed. The operational expected loss for the 11 risk events was 1.5 million euros. To calculate the VaR at 99.9% for operational risk, a simulation study was set up. Without taking the interdependence into account, that is, assuming independence between the events, the VaR gave a capital charge of 5.6 million euros. To also include the interdependence, a copula approach was used to model the VaR calculation. This was done by mapping a quantile from a marginal normal distribution to a marginal Poisson distribution. If the dependences between the 11 risk events are all set equal to 0.3, the loss distribution will be more skewed than the one illustrated in Figure 19.1. The VaR

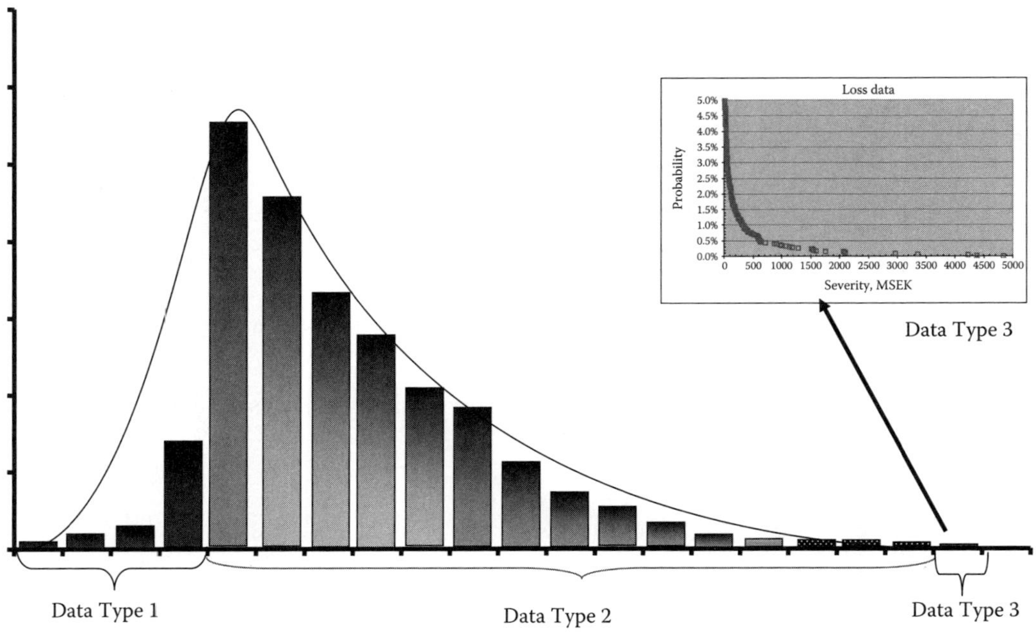

FIGURE 19.1 Construction of an empiric operational loss distribution using three different data sources: (1) Type 1 data: internal company data—mainly small and medium losses, (2) Type 2 data: a specific insurance database, and (3) Type 3 data: data from a global database. Note that Type 1 data are also reflected in the Type 2 region (dark black).

is now equal to 6.5 million euros. Hence, taking the interdependence into account increases the capital charge by 16%.

19.2 BASEL II CAPITAL CHARGE

In a standard formula for the insurance capital charge, we believe that for operational risk it should be similar to the standard approach of Basel II; see also Section 5.1 in Sandström (2005). The main reason for this is any possibility for arbitrage from the banking to the insurance sector, and vice versa. This means that the operational risk formula follows as closely as possible one of the models used by the banking sector.

The Basel II framework presents three methods for calculating operational risk capital charges in a continuum of increasing sophistication and risk sensitivity (see BIS, 2004):

- The Basic Indicator Approach (BIA)
- The Standardized Approach (SA)
- Advanced Measurement Approaches (AMA)

The banks are encouraged to move along the spectrum of available approaches as they develop more sophisticated operational risk measurement systems and practices. Basel II also defines qualifying criteria for the SA and the AMA. A bank will not be allowed to choose to revert to a simpler approach once it has been approved for a more advanced approach without supervisory approval.

19.2.1 Basic Indicator Approach

Banks using the BIA must hold capital for operational risk equal to a fixed percentage (denoted α) of average annual gross income over the previous three years. The charge may be expressed as follows (see BIS, 2004, p. 137):

$$K_{\mathrm{BIA}} = \frac{\alpha}{n^+} \sum_{i=1}^{3} \mathrm{GI}_i^+,$$

where K_{BIA} is the capital charge under the BIA and; GI_i^+ is the annual, positive, gross income over the previous three years; n^+ is the number of the previous years for which the gross income is positive; and $\alpha = 15\%$, which is set by the Committee.

Gross income is defined as net interest income plus net noninterest income. It is intended that this measure should

- Be gross of any provisions (e.g., for unpaid interest)
- Be gross of operating expenses, including fees paid to outsourcing service providers
- Exclude realized profits/losses from the sale of securities in the banking book
- Exclude extraordinary or irregular items as well as income derived from insurance

19.2.2 Standardized Approach

The banks' activities are divided into eight business lines in the standardized approach (SA): corporate finance, trading and sales, retail banking, commercial banking, payment and settlement, agency services, asset management, and retail brokerage. The business lines are defined in detail in BIS (2004).

The gross income is an indicator that serves as a proxy for the scale of business operations (within each line of business).

The capital charge for each business line is calculated by multiplying gross income by a factor β assigned to that business line.

The β serves as a proxy for the industry-wide relationship between the operational risk loss experience for a given business line and the aggregate level of gross income for that business line.

The total capital charge is calculated as the 3-year average of the regulatory capital charges across each of the business lines (the positive part of each line). The total capital charge may be expressed as

$$K_{\mathrm{SA}} = \frac{1}{3} \sum_{i=1}^{3} \max \left[\sum_{j=1}^{8} (\mathrm{GI}_j \times \beta_j), 0 \right],$$

where K_{SA} is the capital charge under the SA and; GI_j is the annual gross income in a given year, as defined above in the BIA, for each of the eight business lines ($j = 1, \ldots, 8$); see below; β_j is a fixed percentage, set by the Committee, relating the level of required capital to the level of gross income for each of the eight business lines ($j = 1, \ldots, 8$); see below.

Business Lines	**Beta Factor**
1: Corporate finance	0.18
2: Trading and sales	0.18
3: Retail banking	0.12
4: Commercial banking	0.15
5: Payment and settlement	0.18
6: Agency services	0.15
7: Asset management	0.12
8: Retail brokerage	0.12

There is also an Alternative Standardized Approach (ASA) that can be used if the bank can prove to the supervisory authority that this approach is superior to the ordinary SA; see footnote 97 on page 139 in BIS (2004).

19.2.3 Advanced Measurement Approaches

Under the AMA, the regulatory capital requirement will be equal to the risk measure generated by the bank's internal operational risk measurement system using the quantitative and qualitative criteria for the AMA discussed in the Basel Accord; see BIS (2004, p. 140).

19.3 GDV MODEL

In the standard formula proposed by the German GDV (2005), the operational risk is defined in a similar way as for Basel II, that is, as the risk of loss either as a result of the inadequacy or failure of internal procedures, persons, and systems or as a result of external events. This definition includes legal risks.

Risks from general business activity can be divided into four groups:

- *Legal risks:* Risks from the outcome of trials, organizational risks, and IT risks
- *Personnel risk:* Mismanagement, poor advice, fraud, and the risk of default by insurance agents, unless represented separately
- *External risks:* Market and legal risks (e.g., changes in legal conditions such as in tax laws and legal rulings)
- *Catastrophe risk:* Natural disasters, but only insofar as they relate to points 1 through 3 above. The underwriting risk of accumulation risks due to natural events, major risks, and epidemics is included in the relevant underwriting risk category of life, nonlife, and health insurers.

Since this risk category is difficult to quantify and since there is no database available in this regard, the capital requirement for the operational risk is computed as follows:

$$\mathrm{CR_{OpR}} = \max\{x\% * \text{earned gross premiums}; y\% * \text{provisions}\}.$$

The risk factors are estimated for each sector using internal company calculations in accordance with the risk catalogue pursuant to the German Corporate Control and Transparency Act.

For life insurers, the following risk factors are used: 6% related to earned gross premiums and 0.6% related to technical provisions. For nonlife insurers, the following risk factors are used: 3% related to earned gross premiums and 3% related to technical provisions. For health insurers, the following risk factors are used: 3% related to earned gross premiums and 0.3% related to technical provisions.

19.4 CEA MODEL

The body of the European insurance industry, CEA, proposed in 2006 a European Standard Approach (ESA) for solvency assessment as another contribution to the debate on a European standard formula; see CEA (2006b).

Under the simple factor ESA approach, operational risks should be considered at the company level by applying a simple risk factor to relevant volume measures.

For life insurance companies, the placeholder volume measures included in the proposed ESA are technical provisions and premiums:

$$CR_{\mathrm{OpR}} = f_{\mathrm{TP}}\, TP + f_{\mathrm{P}}\, P,$$

where f is a risk factor for the technical provisions (TPs) and for the premiums (P), respectively.

For nonlife insurance companies, the placeholder volume measures included in the proposed ESA are premium and reserve indicators.

- For insurance undertakings in runoff (and with limited premium volume), the gross reserves can be used as a volume measure.
- For new insurance undertakings (with limited reserves and assets), the GWP can be used as a volume measure.

The capital charge is defined as

$$\mathrm{CR}_{\mathrm{OpR}} = \max\left\{f_{\mathrm{GWP}}\,\mathrm{GWP}; f_{\mathrm{GR}}\,\mathrm{GR}\right\},$$

where f is a risk factor for the GWP and for the gross reserves (GR), respectively.

CHAPTER 20

Liquidity Risk

THE LIQUIDITY RISK is the risk that an insurer, although solvent, has insufficient liquid assets to meet his obligations (such as claims payments and policy redemptions) when they fall due. The liquidity profile of an insurer is a function of both its assets and liabilities; see IAIS (2004).

An insurer needs cash to pay the policyholder benefits. At the same time, at least in a going concern perspective, it receives liquid assets in the form of premiums. These two CFs are not identical; see Doff (2007). If an insurer unexpectedly needs to pay out more benefits than expected, due to redemptions or mass claims, then he could be illiquid in the sense that he has insufficient liquid assets. If the beneficiaries do not receive the expected compensation for the claims because of liquidity problems, the confidence in the insurer would be in danger, irrespective of whether the company is solvent or not. This in turn could lead to cancellations or surrender of the insurance contracts with disastrous consequences.

CRO Forum (CRO, 2008) gives a generic definition of liquidity risk as the risk that cash sources are insufficient to meet cash needs under either current conditions or possible future environments. At the heart of effective liquidity measurement is a clear understanding of a company's cash sources and cash needs.

Liquidity is the *oxygen for a healthy market.* Its risk is an inherent part of the financial industry, but it is more important for banks than for insurers.

20.1 MANAGING LIQUIDITY RISK

IAA distinguishes between different levels of liquidity management (IAA, 2004, p. 33):

- Day-to-day cash management, which is commonly a Treasury function within the company.
- Ongoing CF management, which typically monitors cash needs for the next 6 or 24 months.
- Stress liquidity risk, which is focused on CAT-risk.

IAA also points out triggers for an unexpected demand for liquidity: cash calls following major loss events, a credit rating downgrade, negative publicity, whether justified or not, deterioration of the economy, reports of problems of other companies in the same line of business (LOB) or similar LOBs, extent of reliance on and performance of secured sources of funding and their terms, for example, the line of credit capacity and conditions, and breadth of funding and accessibility/liquidity of capital markets (e.g., through catastrophe bonds).

The IAIS defines different subrisks of the liquidity risk, including (see IAIS, 2004)

- *Liquidation value risk:* The risk that unexpected timing or amounts of needed cash may require the liquidation of assets in such a way that this could result in loss of realized value.
- *Capital funding risk:* The risk that the insurer will not be able to obtain sufficient outside funding as its assets are illiquid at the same time it needs it to meet large claims.

A third risk discussed by IAIS is

- *Affiliated investment risk:* The risk that an investment in a member company of the conglomerate or group may be difficult to sell, or the affiliates may create a drain on the financial or operating resources from the insurer. This latter risk is part of the *participating risk* as discussed in the introduction to Chapter 16.

In September 2008, the Basel Committee on Banking Supervision published its *Principles for Sound Liquidity Risk Management and Supervision* (which was a revised version of a set of guidelines from 2000) (BIS, 2008). They set up 17 principles for a sound liquidity risk management and supervision. These principles underscore the importance of establishing a robust liquidity risk management framework that is well integrated into the bank-wide risk management process. During the period before the financial crisis, the liquidity risk and its management did not receive the same level of scrutiny and priority as other risk areas. The crisis illustrated not only how quickly and severely liquidity risks can crystallize, but also the inaccurate and ineffective management of liquidity risk. The primary objective of the guidance is to raise banks' resilience on liquidity stress. Among other things, the principles seek to raise the standards in the following areas:

- Governance and the articulation of a firm-wide liquidity risk tolerance
- Liquidity risk measurement, including the capture of off-balance sheet exposures, securitization activities, and other contingent liquidity risks that were not well managed during the financial market turmoil
- Aligning the risk-taking incentives of individual business units with the liquidity risk exposures that their activities create for the bank

- Stress tests that cover a variety of institution-specific and market-wide scenarios, with a link to the development of effective contingency funding plans
- Strong management of intraday liquidity risks and collateral positions
- Maintenance of a robust cushion of unencumbered, high-quality liquid assets to be in a position to survive protracted periods of liquidity stress
- Regular public disclosures, both quantitative and qualitative, of a bank's liquidity risk profile and management

According to this proposal, banks should conduct stress tests on a regular basis for a variety of stress scenarios to identify sources of potential strain and to ensure that current exposures remain within its established liquidity risk tolerance (from Principle 10).

In August 2009 the Basel Committee decided to publish a set of *quantitative liquidity requirements* for public consultation at its meeting at the end of 2009. The document, BIS (2009), was published in December.

It is proposed that the requirements will consist of two quantitative measures:

- *Liquidity coverage ratio (LCR)* would use a risk-sensitive approach, based on a short period of acute liquidity stress, to calculate a requirement for a pool of high-quality assets that banks would be mandated to hold in a liquidity buffer. Stress period: probably 30 days.
- *Net stable funding ratio, NSFR* (a supplementary funding requirement) would encourage banks to put greater reliance on stable long-term funding for long-term assets, with quantitative constraints on banks' structural funding to be calculated in due course. Stress period: probably up to 1 year. Two measures are discussed: core funding ratio and NSFR.

These measures are discussed in Section 20.2.

It is likely that such quantitative rules on the liquidity risk will be a complement to the qualitative rules set out in the Capital Requirements Directive (CRD) for banks. It is also very likely that any such rules set up by CEBS for future amendments of the CRD will influence the work by CEIOPS on capital requirements for insurance undertakings.

The liquidity risk is usually assumed to be an issue for the supervisory review process; see IAA (2004, p. 34). The management of the liquidity risk in financial groups has been discussed by the Joint Forum (2006).

The CRO Forum (CRO, 2008) has also set up principles of liquidity risk management. They believe that adequate liquidity, whether from internal or external sources, must be maintained at all times to manage through even extreme liquidity risk events and that it is inappropriate to expect any amount of required capital to protect against insolvency arising from this risk. The best line of defense is a strong liquidity policy and management framework where liquidity risk is robustly measured, monitored, and managed. This framework should include an operational plan to help the company manage through liquidity stress conditions.

CRO Forum set up the following eight principles:

- *Principle 1:* Liquidity risk is an asset/liability concern; it is neither solely an asset risk nor a liability risk. Effective liquidity risk management starts with a careful assessment of the liquidity characteristics of a company's assets and liabilities and a realistic assessment of any potential liquidity mismatches.
- *Principle 2:* Management should set its tolerance for liquidity risk by using qualitative and quantitative tools, including consideration for its tolerance of other risks. Management must understand where the company is vulnerable to liquidity risk in order to set this tolerance.
- *Principle 3:* The cost of securing adequate liquidity should be reflected in product design and valuation. Necessary liquidity can be provided through liability product design as well as by the investment portfolio backing the products and/or through external or contingent lines.
- *Principle 4:* A company's strategic asset allocation and contingent liquidity planning should directly reflect the expected and contingent liquidity needs of its liabilities and potential sudden extreme shifts of liquidity in the financial markets.
- *Principle 5:* A company should manage its access to the financial markets and have an ongoing presence in its chosen funding channels. It should regularly gauge its capacity to use these channels in order to better monitor the liquidity value of its assets and to safeguard its reputation during times of stress.
- *Principle 6:* Management should require that a written liquidity risk policy be maintained. The policy should be approved by senior management and reviewed regularly to ensure that it remains current and operational.
- *Principle 7:* A company should maintain a written liquidity stress management plan that is approved by senior management. The plan should reflect the company's advance planning for times of liquidity stress, and will guide the company's management actions during a liquidity crisis.
- *Principle 8:* Requiring capital to provide for liquidity risk is an ineffective means of managing the risk. Liquidity risk is a risk to be managed at all times—before, during, and after any stress event—and no amount of capital can replace comprehensive liquidity risk management.

20.2 MODELING LIQUIDITY RISK

A market for a security is termed liquid if investors can buy and sell large amounts of the security in a short time without affecting its price very much (McNeil et al., 2005, p. 41).

The literature on liquidity has been focusing on the determinants of bid–ask spreads and transaction costs; see, for example, Bangia et al. (1999) and Le Saout (2002). Yet there is no agreement on how to properly measure illiquidity and the capital charge for it. In BIS (1999), the Bank for International Settlements, BIS, defines asset liquidity according to at least one of three dimensions:

- *Depth* defines the maximum number of shares that can be traded without affecting prevailing quoted market prices.
- *Tightness* is measured by bid–ask spread and indicates how far transaction price diverges from the mid-price.
- *Resiliency* defines the speed at which price fluctuations resulting from trades are dissipated or how quickly markets clear order imbalance.

Tightness is discussed in Bangia et al. (1999). They split the uncertainty in market value into the uncertainty in asset returns and due to liquidity risk. The latter risk is divided into

- Endogenous liquidity, which is a result of market characteristics and affects all market players, and
- Exogenous liquidity, which is specific to the firm's position in the market.

In op. cit., it is proposed to adjust the VaR model for the market risk into one asset returns component and one component based on the liquidity risk. The VaR is modeled from the mid-price. A correction is made for the fat-tailed returns distribution.

One bank model for withdrawals is proposed by Tobin and Brown (2004). It is based on weekly withdrawals data and uses a Cornish–Fisher expansion including both the skewness and the kurtosis.

A simple liquidity model would be to estimate the distribution of the "liquidity gap" (LG), for example, as the ratio between "illiquid liabilities" (IL) and the difference between total assets (A) and liquid assets (LA). The nominator would be the difference between possible released liabilities (RL) due to redemptions or mass claims and "liquid liabilities" (LL). The denominator is thus the asset that is seen as illiquid. Hence the liquidity gap could be defined as

$$\mathrm{LG} = \frac{\mathrm{IL}}{\mathrm{A} - \mathrm{LA}} = \frac{\mathrm{RL} - \mathrm{LL}}{\mathrm{A} - \mathrm{LA}}.$$

Das and Hanouna (2009) looked at the transmission of illiquidity from equity markets to credit default swaps (CDS). They compared different measures of liquidity, and their theoretical and empirical results showed that credit spreads and illiquidity are positively correlated.

The Basel Committee on Banking Supervision has published proposals for two measures of liquidity risk exposures (BIS, 2009). The standards establish minimum levels of liquidity for internationally active *banks*. Banks are expected to meet these standards as well as adhere to all the principles set out in BIS (2008). The proposed measures were sent out for consultation with a deadline of April 2010.

The proposed measures were developed to achieve two separate, but complementary, objectives (BIS, 2009):

- *First objective:* To promote the short-term resiliency of the liquidity risk profile of institutions by ensuring that they have sufficient high-quality liquid resources to survive an acute stress scenario lasting for 1 month. The Liquidity Coverage Ratio (LCR) was developed to achieve this objective.
- *Second objective:* To promote resiliency over longer-term time horizons by creating additional incentives for banks to fund their activities with more stable sources of funding on an ongoing structural basis. The NSFR was developed to capture structural issues related to funding choices.

A survey by the Basel Committee conducted in 2009 showed that more than 25 different measures and concepts were used globally by supervisors. For consistency and harmonization, the two measures proposed are considered as the minimum types of information the supervisors should use in monitoring the liquidity risk profiles. Additional metrics may be used by the bank supervisors. In addition to the metrics outlined below to be used as standards, the Basel Committee also proposed that supervisors utilize other metrics as consistent monitoring tools. These metrics, discussed in BIS (2009), capture specific information related to a bank's CFs, balance sheet structure, available unencumbered collateral and certain market indicators.

20.2.1 Liquidity Coverage Ratio

LCR is defined as

$$\mathrm{LCR} = \frac{\mathrm{HQLA}}{\mathrm{NCO}_{30}} \geqslant 1,$$

where HQLA and NCO_{30} are defined as follows:

HQLA: Stock of high-quality liquid assets. In order to qualify as a HQLA, assets should be liquid in markets during a time of stress and, ideally, be central bank eligible.

NCO_{30}: Net cash outflows over a 30-day time period.

The LCR builds on traditional liquidity coverage ratio methodologies used by banks to assess the exposure to contingent liquidity events. Net cumulative cash outflows for the scenario are to be calculated for 30 calendar days into the future. The standard requires that the value of the ratio should not be lower than 100%, that is, the stock of liquid assets should at least equal the estimated net cash outflows. Banks are expected to meet this requirement continuously and hold a stock of unencumbered, high-quality assets as a defense against the potential onset of severe liquidity stress. Banks and supervisors are also expected to be aware of any potential mismatches within the 30-day period and ensure that sufficient liquid assets are available to meet any CF gaps throughout the month (BIS, 2009).

The definitions of the numerator HQLA and the denominator NCO are discussed in detail in BIS (2009). The ratio identifies the amount of unencumbered, high-quality liquid assets an institution holds that can be used to offset the net cash outflows it would encounter under an *acute short-term stress scenario* specified by supervisors. The specified scenario entails both institution-specific and systemic shocks built upon actual circumstances experienced in the global financial crisis. The scenario entails

- A significant downgrade of the institution's public credit rating
- A partial loss of deposits
- A loss of unsecured wholesale funding
- A significant increase in secured funding haircuts
- Increases in derivative collateral calls and substantial calls on contractual and non-contractual off-balance sheet exposures, including committed credit and liquidity facilities

As part of the LCR, banks are also required to provide a list of contingent liabilities, both contractual and noncontractual, and their related triggers.

20.2.2 Net Stable Funding Ratio

NSFR is defined as

$$\text{NSFR} = \frac{\text{ASF}}{\text{RSF}} > 1,$$

where ASF and RSF are defined as follows:

ASF: The available amount of stable founding; see Table 1 in BIS (2009, pp. 21–22).
RSF: The required amount of stable funding; see Tables 2 and 3 in BIS (2009, pp. 23–24).

The NSFR builds on the traditional net liquid asset and cash capital methodologies used by internationally active banking organizations, bank analysts, and rating agencies. However, the proposed measure expands general industry conventions of these concepts to account for the potential liquidity risk of off-balance sheet exposures and various types of maturity mismatches involved in short-term secured funding of long-dated assets that traditional forms of these measures may ignore. The NSFR must be greater than 100%.

Stable funding is defined as those types and amounts of equity and liability financing expected to be reliable sources of funds over a 1-year time horizon under conditions of extended stress. The amount of such funding required of a specific institution is a function of the liquidity characteristics of various types of assets held, off-balance sheet contingent exposures incurred, and/or the activities pursued by the institution (BIS, 2009). Definitions of available and required stable funds are given in BIS (2009).

CHAPTER 21

Underwriting/Insurance Risk

INSURANCE COMPANIES assume risk through the insurance contracts they underwrite; see IAA (2004, p. 29).

The *underwriting process risk* refers to the risk related to the businesses that will be written during the following year. In general, we should consider it *net of reinsurance*, as the reinsurance will be dealt with in the default credit risk category. The underwriting process risk will thus be highly correlated with the credit risk.

This is the main nonlife insurance risk, but is also important for annuities and can be measured by the volatility of the underwriting result and by a factor measuring the expected profitability of the business. The volatility includes a separate term for the *nonsystematic underwriting risk*, that is, the risk that is due to the variation in the frequency and severity of the claims, which can be diversified through a greater portfolio and an adequate reinsurance policy.

Another term is needed to describe the variance of *systematic underwriting risk*, that is, the risk that is common to the whole insurance market and that cannot be reduced by pooling.

The calculations should be made according to some kind of segmentation, for example, by homogeneous risk groups, by LOBs, by the insurance classes introduced by the European Commission (EC), see Sandström (2005, Appendix B), or according to the examples given in Section 9.3. Also the effect of having business in different geographical regions should be captured. Geographical regions could be either different regions within a country or different countries.

The company's own loss experience should be used to measure the risk exposure and expected profitability.

IAA (2004) did not list all specific hazards, but looked at different generic risks that apply to most LOBs:

- *Underwriting Process Risk*—risk from exposure to financial losses related to the selection and approval of risks to be insured.
- *Pricing Risk*—risk that the prices charged by the company for insurance contracts will be ultimately inadequate to support the future obligations arising from those contracts.

- *Product Design Risk*—risk that the company faces risk exposure under its insurance contracts that were unanticipated in the design and pricing of the insurance contract.
- *Claims Risk* (for each peril)—risk that many more claims occur than expected or that some claims that occur are much larger than expected claims resulting in unexpected losses. This includes both the risk that a claim may occur and the risk that the claim might develop adversely after it occurs.
- *Economic Environment Risk*—risk that social conditions will change in a manner that has an adverse effect on the company.
- *Net Retention Risk*—risk that higher retention of insurance loss exposures results in losses due to catastrophic or concentrated claims experience.
- *Policyholder Behaviour Risk*—risk that the insurance company's policyholders will act in ways that are unanticipated and have an adverse effect on the company.
- *Reserving Risk*—risk that the provisions held in the insurer's financial statements for its policyholder obligations (also "claim liabilities," "loss reserves," or "technical provisions") will prove to be inadequate.

Ensuring the adequacy of *claim provisions* is mainly a pillar II issue. However, the pure random *run-off risks* always remain, for example, in the estimation of the IBNR and IBNER, and the change in the claims provision, during the time horizon of one year, should be captured. We call this the *reserve risk.*

We will deal with the diversifiable underwriting risk in the business to be written in the following year and call this the *premium risk.*

The capital requirement for *surrender and lapse risks* will be determined based on the assumed trend parameters and lapse ratios (given by the EC) multiplied by the guaranteed surrender value minus best estimate liability (given by the company) but never below zero. These risks are mainly related to life assurance but will also be introduced for nonlife business. As an increase in interest rates lowers the market value of bonds and probably also increases surrender rates and lapse rates, it is assumed that it is highly correlated with the market risk (asset/liability risk).

The *expense risk* consists of, for example, acquisition, administrative, and claims settlement costs. Some of the costs cannot be allocated to any particular LOB and they are called overhead expenses. The risk will consist of two parts: the first is the overhead expenses part and the second is the part depending on the LOB.

For unit-linked contracts the principal risks are concentrated in expenses and lapses. The major insurance (underwriting) risk is borne by the insured. In this case the expenses are administrative costs and the risk that these will exceed what can be earned on the policies. We will only consider the retained business and the risk will consist of weighted net technical provisions (NTP).

The *biometric risk* consists of two parts: one part measuring volatility and trends in mortality and another measuring volatility and trends in sickness. We assume that they are uncorrelated with each other.

To sum up, the main risk category "insurance risk" consists of the following subrisks:

- Underwriting risk
- Biometric risks
  - Mortality risk
  - Sickness risk
- Surrender and lapse risk
- Cost/expenses risk

In Section 21.1, we discuss the nonlife underwriting risk, and in Section 21.2, the life underwriting risk. In Section 21.3, we will briefly discuss health-underwriting risk. In some jurisdictions, this subrisk is part of either the nonlife underwriting risk or the life underwriting risk.

21.1 NONLIFE UNDERWRITING RISK

In developing a standard formula, there are some key characteristics of nonlife insurance that require special considerations (IAA, 2004):

- Heterogeneity of risk (even within established "classes" of insurance business)
- Substantial effects of correlation between different underwriting risks
- Difference between outstanding claims liabilities and liabilities because of unexpired risk inherent in unearned premiums
- Annual renewal basis for the vast majority of the business
- Significant role played by reinsurance (especially in relation to concentration of risk)
- Difficulty in estimating separate claim incidence and severity in projecting experience for a minority of the business

Any standardized approach for nonlife insurance will need to take account of these characteristics and will require the classification of all nonlife insurance business in each regulatory jurisdiction into defined LOBs—the level of detail in the definition effectively being under the control of the supervisor in the jurisdiction under consideration.

When considering the nonlife underwriting risk, it is common to divide the claims into a few large claims (CAT risks) and into the majority of small- or medium-sized claims. The latter case could be divided into a

- *Reserve risk*: relates to claims from two sources: (1) the absolute level of the claims provisions may be misestimated. (2) Because of the stochastic nature of future claims payouts, the actual claims will fluctuate around their statistical mean value.

- *Premium risk*: relates to future claims arising during and after the period up to the time horizon for the solvency assessment.

Different models have been proposed in the literature to capture the nonlife underwriting risk and determine a capital charge. Some of them are briefly introduced and referenced in the subsequent chapters. A one-year time horizon model is discussed in Section 21.1.4.

21.1.1 Different Factor-Based Models Used

In the IAA model (IAA, 2004, Appendix B3), it is assumed that the aggregate loss distribution is lognormal and a Tail VaR (TVaR) at 99% is taken as a capital charge. The model of the claims is the classical mixed-compound Poisson model; see, for example, Beard et al. (1984) and Sandström (2005, Appendix A.1.1).

The Finnish model is also based on the classical mixed-compound model (Sandström, 2005, Appendix A.1.1). No assumption on the loss distribution is made in this model. The model captures mainly the premium risk.

As these models and most of the models discussed so far in the literature do not explicitly take the solvency time horizon into consideration, they will not be dealt with in detail.

The standard model for nonlife insurance risks in the Swiss Solvency Test (SST) is described in SST (2006b, ch. 4.4) and references given therein. A comparison between the SST model and the Solvency II insurance risk model used in QIS4 is discussed in Gisler (2009).

We will, in the sequel, look closer at two models proposed in connection with the European Solvency II work: first the German GDV model from 2005 and then the model proposed by the European industry, the CEA model. We will end up this part with a discussion on a one-year time horizon approach.

21.1.2 GDV Model

A first GDV[*] nonlife model, which was published in 2002, was based on Standard and Poor's RBC model. The model included the premium risk and the loss reserve risk.

In 2005, the GDV launched a new standard formula for the capital requirement; see GDV (2005). This was a contribution to the debate on a European standard formula within the Solvency II project. As CEIOPS in the spring of 2006 published its first proposal for a standard formula, CEIOPS (2006d), the GDV model was not developed any further.

The nonlife underwriting risk included premium and reserve risks, as well as the natural catastrophe (NatCat) risk storm. The charge for premium and reserve risks was calculated using company net combined ratios per LOB taking into account correlations between LOBs determined on market data. The model did not distinguish between the premium risk and the reserve risk; however, it valued the entire underwriting risk in a single factor with premium as the risk bearer ("capital requirement = factor risk bearer"). The approach was justified with respect to the reserve risk when the ratio of reserves to premiums for individual LOBs did not fluctuate too much.

[*] GDV: Gesamtverband der Deutschen Versicherungswirtschaft e.V.; that is, the German Insurance Association.

Let the combined ratio of LOB k be

$$\mathrm{CR}_k = \frac{\mathrm{NCE}_k + \mathrm{OE}_k}{\mathrm{NP}_k},$$

where NCE_k is the net claims expenditure, that is, the run-off results for claims related to the insurance period plus the claims expenditure for the accounting year; OE_k is the operating expenses; and NP_k is the net earned premiums.

The combined ratio is taken from the income statement, that is, it includes the ultimo run-off of the reserve risk, and before changes in the German equalization provision and similar provisions.

The reserve risk was not explicitly divided into one part for the volatility in the settlement (reserve risk) and one for the volatility in claims ratio for the accounting year (premium risk), as is suggested in Section 21.1.4. For an argumentation for not doing this split, see GDV (2005, Section 5.3.1.2).

To develop the capital charge, that is, the solvency capital requirement, we consider the combined ratios CR_k for different LOBs k as random variables with expectations $\mu_k = E(\mathrm{CR}_k)$ and variance $\sigma_k^2 = \mathrm{Var}(\mathrm{CR}_k)$. For the whole business, we write $\mathrm{CR} = \sum_k w_k \cdot \mathrm{CR}_k$, where $w_k = \mathrm{NP}_k / \sum_k \mathrm{NP}_k$. We also have $\mu = \sum_k w_k \cdot \mu_k$ and $\sigma^2 = \sum_j \sum_k w_j w_k \sigma_j \sigma_k r_{jk}$, where r_{jk} is the correlation between the combined ratios of LOBs j and k.

Let SR denote the desirable solvency capital of the company expressed as a ratio of the net earned premium risk in a similar way to the combined ratio. The main idea of the GDV model in calculating the SR was that a critical underwriting situation is characterized by expenses exceeding revenues, that is, $\mathrm{SR} > 1$. This model is similar to the model proposed for the OECD in Campagne (1961),* where Campagne proposed that the loss ratio part followed a beta distribution; see also Section 2.1.2 and Sandström (2005).

Thus, the SR should only be exceeded, with only a low probability α, in the case of poor underwriting development CR-1:

$$P(\mathrm{CR} - 1 > \mathrm{SR}) \leqslant \mathrm{a} \iff P(\mathrm{CR} \leqslant \mathrm{SR} + 1) \geqslant 1 - \alpha \iff \mathrm{SR} + 1 \geqslant q_{1-\alpha},$$

where $q_{1-\alpha}$ denotes the $(1-\alpha)$ quantile of the distribution of CR.

The next step in the GDV model is to represent the "risk measure" as a multiple of the standard deviation, given the confidence level. Each quantile can be represented by a factor a; a "score value" of the quantile in the form $q_{1-\alpha} = \mu + a \cdot \sigma$, that is, the score value is $a = (q_{1-\alpha} - \mu)/\sigma$. Hence, the desirable solvency capital in % is given by

$$\mathrm{SR} \geqslant \mu + a \cdot \sigma - 1. \tag{21.1}$$

For LOB k, we let $\mathrm{SR}_k = a \cdot w_k \cdot \sigma_k$ and hence the standard deviation in Equation 21.1 can be written as

$$a^2 \cdot \sigma^2 = \sum_j \sum_k r_{jk} \mathrm{SR}_j \mathrm{SR}_k.$$

*This proposal, with some changes, was used as a part of the Solvency I capital charge by the European Union.

The solvency capital charge in euros would be $\text{NP} \cdot (\mu + a \cdot \sigma - 1)$.

To make the estimates of the mean and variance of the combined ratios, the companies were requested to have data for 15 years. For companies having less than 15 years of data for some LOBs or for new insurance companies, a simplified method is used. The SR for LOB k is calculated as

$$\text{SR}_k = a \cdot w_k \cdot sf \cdot f_k,$$

where sf is a size factor and f_k a variability factor depending on LOB. In order to ease the transition from using the "lump sum" $sf \cdot f_k$ instead of the company-specific standard deviation σ_k, the company can use a weighted sum of the two parts as soon as it has at least 11 years of data. From the year the company has 11 years of data, the SR is calculated as

$$\text{SR}_k = a \cdot w_k \cdot \left[\frac{15 - y}{5} \cdot sf \cdot f_k + \frac{y - 10}{5} \cdot \sigma_k \right], \quad y = 11, 12, \ldots, 15.$$

The values of sf and f_k were given by GDV.

The initial version of the GDV model only included storm risks for Germany as a NatCat risk. They were integrated into the model as a separate LOB. The NatCat risk is denoted by NC and is modeled using a *market loss spread*. It was assumed that the market loss spread will give probabilities of possible losses indicated as a percentage of the insured sum.

The market loss spread is based on a collective model with expected value E, standard deviation σ, and fixed quantiles q_1 and q_2. The quantile q_1 corresponds to a risk tolerance selected for the model with significance level α. The other quantile, q_2, reflects the fact that NatCat events are already reflected in the combined ratios, and double counting is to be avoided. As the combined ratios are based on a 15-year time horizon, the q_2-quantile is to reflect events with a recurrence period (RP) of at least 15 years. Let $\hat{q} \equiv q_1 - q_2$ and assume that it can be defined as a factor times the standard deviation, that is, $\hat{q} = \hat{\alpha} \cdot \sigma$.

Let V_M and V_U be the market-wide and company-specific insured sums, respectively. The corresponding premiums are P_M and P_U. The capital requirement SR_NC is defined in the GDV model in the following way. If the market loss spread is accurate for the company and the expression $\hat{q} \equiv q_1 - q_2$ corresponds to the company's risk tolerance, then $\text{SR}_\text{NC} = \hat{q} \cdot V_\text{U} = \hat{\alpha} \cdot \sigma \cdot V_\text{U}$. This defines the absolute value of the gross capital requirement.

For simplicity, assume that $(V_\text{U}/V_\text{M}) \approx (P_\text{U}/P_\text{M})$. Then we have the approximation $\text{SR}_\text{NC} = \hat{q} \cdot V_\text{M} \cdot (P_\text{U}/P_\text{M})$. If all market-specific parameters are combined into a constant $\alpha \equiv \hat{\alpha} \cdot (V_\text{M}/P_\text{M})$, then $\text{SR}_\text{NC} \equiv \alpha \cdot \sigma \cdot P_\text{U}$.

German data for 2003 showed that $P_\text{M} = 4854$ billion euro and $V_\text{M} = 10,843$ bn euro.

The *probable maximum loss*, PML, for a specific RP is defined as quantile x of the loss spread distribution of a collective model. That model was assumed to follow a Poisson distribution with parameter λ (= average number of claims per year) and a loss spread distribution F.

The PML for a specific RP is defined as the quantile x of F fulfilling the following equation:

$$\text{RP}(x) = \frac{1}{\lambda \cdot (1 - F(x))}.$$

Excluding reinsurance, the difference $\hat{q} = q_1 - q_2 = 0.076\%$ is multiplied by the insured sum. The total gross market loss, with the German data from 2003, would then be expected, with an annuity of 200, as $q \cdot V_M = 0.076\% \cdot 10,843$ bn euro $= 8.2$ bn euro.

Each direct insurer has a gross capital share in this amount, that is, $\text{SR}_{\text{NC}} = (P_{\text{U}}/P_{\text{M}}) \cdot$ 8.2 bn euro.

The capital charge is reduced depending on the reinsurance coverage:

- *Proportional reinsurance:* a proportional reduction of c is made, that is, $\text{SR}_{\text{NC}} = (1 - c) \cdot \hat{q} \cdot V_{\text{M}} \cdot (P_{\text{U}}/P_{\text{M}})$
- *CatXL/stop-loss reinsurance:* a CatXL/stop-loss liability X beneath the 200 year PML may be deducted from (not below zero) $\text{SR}_{\text{NC}} = \max\left[(1 - c) \cdot \hat{q} \cdot V_{\text{M}} \cdot (P_{\text{U}}/P_{\text{M}}); 0\right]$. If the CatXL coverage is structured in the opposite way, that is, quota cover on the CatXL deductible, the steps in the calculation must be reversed.

21.1.3 CEA Model

CEA proposed in 2006 a European Standard Approach, ESA, for solvency assessment as another contribution to the debate on a European standard formula; see CEA (2006b). The model was based on a single framework where standard factors were applied to all risk types to calculate the SCR. The framework was also made such that a specific company could calculate the capital requirement using company-specific data to calibrate the factors.

Premium, reserve, and catastrophe risks were included in the underwriting risk. All calculations were made for separate LOBs and aggregated using correlations.

Reserve risk, or reserving risk, was defined as the risk that the technical provisions held to cover incurred claims, for coverage already provided, may be inadequate. It is calculated as

$$\text{CR}_{\text{RR},k} = \text{CP}_k \cdot f_{\text{RR},k},$$

where CP_k is the net claims provisions for LOB k, at market value and $f_{\text{RR},k}$ is a reserve risk factor for LOB k; it can be seen as a multiple of the standard deviation of the distribution of the technical provision. The multiple is determined such that the required capital and the market value of the technical provisions are 99.5% sufficient to cover incurred claims for coverage already provided.

Premium risk is defined as the risk that the volume of ultimate losses for future claims that have occurred or are still to occur at the valuation date, comprising both losses paid during the time horizon of the cover and provisions made at its end, plus expenses, is higher than the premiums received for the cover period.

New business and renewals arising over the next year create an additional source of risk. The exposure measure should take these risks into account.

As a proxy for the business written of the upcoming year the written premiums over the previous year can be used.

The capital charge is calculated as

$$\text{CR}_{\text{PR},k} = [\text{WP}_k + \text{UP}_k] \cdot f_{\text{PR},k},$$

where WP_k: the net written premium for LOB k; UP_k: the unearned premium for LOB k; and $f_{PR,k}$: a premium risk factor for LOB k; it is a multiple of the standard deviation of the distribution of the loss ratio. The multiple is determined such that the required capital plus premium written in the coming year plus the net unearned premium reserve are 99.5% sufficient to cover expenses over the cover period and also to cover both losses paid and provisions made that are related to the premium written in the coming year plus the net unearned premium reserve.

CAT risk: It is assumed that certain CAT scenarios would be specified for the market. The exposure would be measured by the gross market share of the company.

Let $CR^{netQS} = CL_{gross} \cdot (1 - QS\%)$. Then the capital charge is defined according to the ESA as

$$CR_{CAT} = \begin{cases} CR^{netQS} & \text{if no CAT-XL reinsurance} \\ \max\left[CR^{netQS} - \text{reinsurance limit}; 0\right] & \\ \quad + \min\left[\text{retention}; CR^{netQS}\right] & \text{if CAT-XL reinsurance} \end{cases}$$

where CL_{gross}: the company's gross CAT loss $= ML \cdot WP_{comp}/WP_{Market}$; QS%: the percentage of losses that are ceded to the quota share reinsurer; and ML: the market loss for one severe CAT event in an LOB; specified by the local supervisor.

21.1.4 A One-Year Time Horizon Approach

The classical claims reserving methods have all an ultimo time horizon, meaning that they look at the total runoff; see, for example, Wüthrich and Merz (2008). The ultimo time horizon is used when valuing the liabilities, see, for example, Section 8.3.2.2.

For solvency purposes, we will consider the one-year time horizon. This is illustrated in Table 21.1. The calculation below could be done for either gross or net of reinsurance.

To cope with the definitions of the reserve risk and the premium risk above, we need the following notations given in Table 21.2.

To simplify the presentation, we look at Table 21.1b. It is a stylized version of Table 21.1a.

In a study on the nonlife long tail business, AISAM-ACME* (2007) declared that the calibration of the reserve risk needs to reflect strictly the one-year time horizon rather than the classical full runoff approach.

In the discussion given below, we partly follow Ohlsson and Lauzeningks (2008) and Hürlimann (2008a).

The *reserve risk* is taken over the one-year time horizon and could be defined as the risk in the one-year runoff result over that period. Let R^0 be the opening reserve at the beginning of the "current year," and R^{01} be a best estimate of the closing reserve at the end of the "current year." The "01" indicates that it includes the changes made in the reserve during the one-year period. It does not include "new claims." The C^{01} is an estimate of the amount

* AISAM and ACME merged into AMICE, Association of Mutual Insurers and Insurance Cooperatives in Europe. Web site: www.insurance-mutuals.org

TABLE 21.1a The General Claims Development Triangle for a Time Horizon of One Year

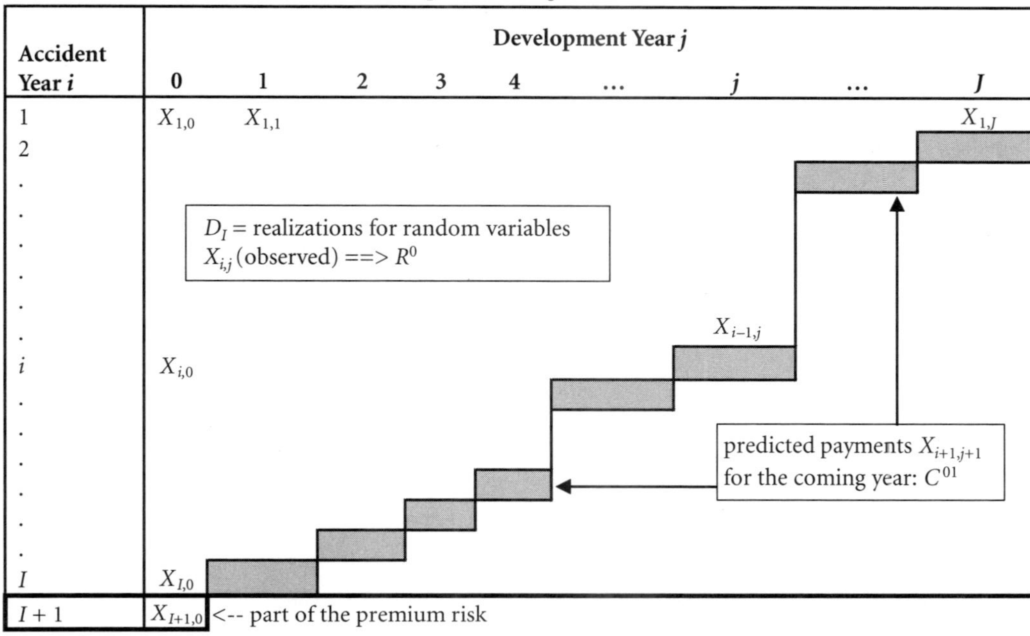

Note: The lower right part of the matrix is not used for solvency assessment purposes. Row $I + 1$, corresponding to the current year, is used in the calculation of the premium risk. See also Table 21.1b below.

TABLE 21.1b A Stylized Version of Table 21.1a

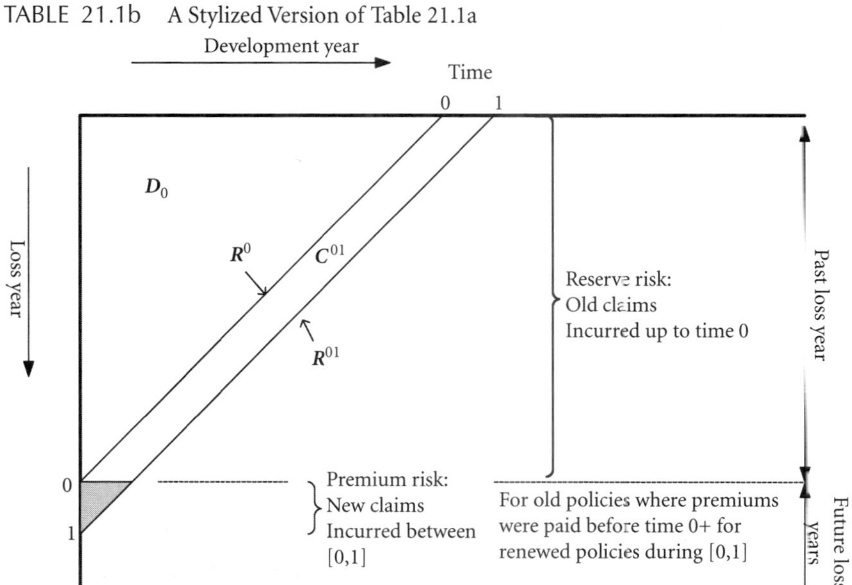

Note: The *reserve risk* is based on old claims incurred up to time 0. The *premium risk* is based on claims incurred between 0 and 1 based on old contracts (premiums paid before time 0) and renewed contracts (premiums paid during 0 and 1).

TABLE 21.2 Notations Used to Explain the Two Concepts of Reserve Risk and Premium Risk

t=0		*t=1*	
Reserve risk ← only claims up to $t = 0$ →			
Opening reserve R^0		R^{01}	Closing reserve
	Amount paid during 0–1:	C^{01}	
	Premium risk ← only claims incurred during 0–1 →		
Opening premium reserve $\tilde{U}^0$		$\tilde{U}^{01}$	Closing premium reserve
	Written premiums:	P^{01}	
	Earned premiums:	$\tilde{P}^{01}$	
	Payment for claims incurred during 0–1:	$\tilde{C}^{01}$	
		$\tilde{R}^{01}$	Closing claims reserve for claims incurred during 0–1
	Expenses for claims incurred during 0–1:	E	

paid out during the year. The technical run-off result (loss) is then

$$L_R^{01} = R^{01} + C^{01} - R^0. \tag{21.2}$$

A negative loss is a profit. The probability distribution of L_R^{01} captures the one-year reserve risk, given the observations made at time $t = 0$. Only R^0 is known at time $t = 0$. The result given by Equation 21.2 is called *claims development result* (CDR) in, for example, Wüthrich et al. (2009). The CDR for Mack's distribution-free chain ladder model is analyzed in op. cit. and in Merz and Wüthrich (2008). Bühlmann et al. (2009) analyzed the CDR within a credibility chain ladder model. In Dahms et al. (2008), the CDR is analyzed for the complementary loss ratio method (CLRM). The CLRM considers simultaneously claims paid and claims incurred data and was presented in Dahms (2008).

Let the reserve risk volume measure be $V_R = R^0$. Then the "reserve loss ratio" could be defined as

$$LR_R = \frac{L_R^{01}}{V_R} = \frac{R^{01} + C^{01}}{R^0} - 1 = X_R - 1. \tag{21.3a}$$

Historical volatility could be calculated by using historic data for the underwriting years $(n, n+1]$, where $n+1$ are before our considerations at (0,1], that is, based on

$$LR_R^{n,n+1} = \frac{R^{n,n+1} + C^{n,n+1}}{R^n} - 1 = X_R^{n,n+1} - 1. \tag{21.3b}$$

In a similar manner, we can define the *premium risk* taken over the one-year time horizon. As above, we have the opening premium reserve denoted by $\tilde{U}^0$. During the year, we assume written premiums of P^{01}, and at the end of the year, a closing premium reserve of $\tilde{U}^{01}$. The earned premium during the year is thus $\tilde{P}^{01} = \tilde{U}^0 + P^{01} - \tilde{U}^{01}$. The last term in the earned premiums includes, as stated in CEIOPS's definition of the premium reserve, future claims arising during and *after the period* up to the time horizon for the solvency assessment; see above.

With $\tilde{C}^{01}$ we define the first-year payments for the claims made during this year. Let E be the operating expenses during the year. The technical result (loss) could thus be written as

$$\begin{aligned} L_P^{01} &= \tilde{R}^{01} + \tilde{C}^{01} + E - \left(\tilde{U}^0 + P - \tilde{U}^{01}\right) \\ &= \tilde{R}^{01} + \tilde{C}^{01} + E + \tilde{U}^{01} - \tilde{U}^0 - P^{01}. \end{aligned} \tag{21.4}$$

If we define $\tilde{P}^{01}$, the earned premium, as the volume measure for the premium risk, V_P, then the premium loss ratio could be defined as

$$LR_P = \frac{L_P^{01}}{V_P} = \frac{\tilde{R}^{01} + \tilde{C}^{01} + E}{\tilde{P}^{01}} - 1 = X_{\tilde{P}} - 1. \tag{21.5a}$$

Historical volatility could be calculated by using historical data for the underwriting years $(n, n+1]$, where $n+1$ are before our considerations at $(0,1]$, that is, based on a similar expression as in Equation 21.3b:

$$\mathrm{LR}_P^{n,n+1} = X_P^{n,n+1} - 1. \tag{21.5b}$$

Combining the reserve risk from Equation 21.2 and the premium risk from Equation 21.2 gives us the *combined risk*

$$\begin{aligned} L_C^{01} &= (R^{01} + C^{01} - R^0) + \tilde{R}^{01} + \tilde{C}^{01} + E - \left(\tilde{U}^0 + P - \tilde{U}^{01}\right) \\ &= (R^{01} + C^{01} - R^0) + \tilde{R}^{01} + \tilde{C}^{01} + E + \tilde{U}^{01} - \tilde{U}^0 - P^{01}. \end{aligned} \tag{21.6}$$

We could also take the sum of the opening reserve and the earned premium as a grand volume measure. Thus we get

$$LR_C = \frac{L_C^{01}}{V_C} = \frac{(R^{01} + C^{01}) + \tilde{R}^{01} + \tilde{C}^{01} + E}{R^0 + \tilde{P}^{01}} - 1 = X_C - 1. \tag{21.7}$$

Following Hürlimann (2008a), the random loss vanishes in the average due to the actuarial equivalence principle, that is, $E(L_x^{01}) = 0$, where x denotes the reserve risk, premium risk, or the combined risk. This means that the expected target of the three risks has the expectation of one. Dropping the index x for R, P, or C, we get $E(X) = 1$.

As we see in Appendix M, we could assume X to be lognormally distributed, that is,

$$X \sim \mathrm{LN}(\mu_X, \sigma_X^2).$$

The *portfolio volatility parameter* is $\sigma = \sqrt{\mathrm{Var}(X)}$ and hence $\sigma^2 = \mathrm{Var}(L^{01}/V)$. As X is lognormal with mean one, we get $E(X) = \mathrm{e}^{\mu_X + (1/2)\sigma_X^2} = 1$ and $\sigma^2 = \mathrm{Var}(X) = \mathrm{e}^{\sigma_X^2 - 1}$. This, on the other hand, implies that

$$\mu_X = -\frac{1}{2}\sigma_X^2 \quad \text{and} \quad \sigma_X^2 = \ln(1 + \sigma^2). \tag{21.8}$$

The VaR and the TVaR both capture the capital requirement of the portfolio at a confidence level of $1-\alpha$.

Combining the results from Equation 21.8 above and the lognormal expression for VaR from Table 14.1, where $k_{1-\alpha} = \Phi^{-1}(1-\alpha)$, we obtain for the VaR

$$\mathrm{VaR}_{1-\alpha}(L) = \mathrm{VaR}_{1-\alpha}\left(\frac{L}{V}\right)V = [\mathrm{VaR}_{1-\alpha}(X) - 1]\cdot V = \rho_{1-\alpha}(\sigma)_{\mathrm{VaR}}\cdot V, \qquad (21.9\mathrm{a})$$

where the volatility-dependent function is

$$\rho_{1-\alpha}(\sigma)_{\mathrm{VaR}} = \frac{\exp\left[k_{1-\alpha}\cdot\sqrt{\ln(1+\sigma^2)}\right]}{\sqrt{1+\sigma^2}} - 1. \qquad (21.9\mathrm{b})$$

On the other hand, if the capital requirement is measured by TVaR, we obtain, by using the expression in Table 14.2a and the expressions from Equation 21.8, the TVaR capital charge and a volatility expression similar to Equations 21.9a and 21.9b, given by

$$\mathrm{TVaR}_{1-\alpha}(L) = \mathrm{TVaR}_{1-\alpha}\left(\frac{L}{V}\right)V = [\mathrm{TVaR}_{1-\alpha}(X) - 1]\cdot V = \rho_{1-\alpha}(\sigma)_{\mathrm{TVaR}}\cdot V \qquad (21.10\mathrm{a})$$

and

$$\rho_{1-\alpha}(\sigma)_{\mathrm{TVaR}} = \frac{1-\Phi\left[k_{1-\alpha}-\sqrt{\ln(1+\sigma^2)}\right]}{\alpha}, \qquad (21.10\mathrm{b})$$

where $\Phi(\cdot)$ is the cumulative standard normal distribution.

We have used the fact that both VaR and TVaR are translation invariants, that is, follow property P7 of Section 14.1.

In Wüthrich and Bühlmann (2009), the volatility of discounted claims reserves for the one-year claims development is studied. The nominal liability CF was modeled using a Bayes chain ladder model and the discount rates were modeled using the Vasicek model for zero-coupon bond prices.

21.2 LIFE UNDERWRITING RISK

In developing a standard formula, there are some key characteristics of life insurance that require special considerations (IAA, 2004):

- Heterogeneity of risk (even within established "classes" of insurance business)
- Importance of mortality/morbidity, lapse, and expense (underwriting) risks
- Substantial effects of correlation between different underwriting risks
- Long-term nature of the majority of the business

- Significant role played by reinsurance (especially in relation to concentration of risk)
- Difficulty in modeling policyholder behavior for some products
- Importance of adjustable product features in some products (e.g., participating or with-profits policies, etc.)

Any standardized approach for life insurance will need to take account of these characteristics and will require the classification of all life insurance business in each regulatory jurisdiction into defined product types—the level of detail in the definition effectively being in the control of the supervisor in the jurisdiction under consideration.

When considering the life underwriting risk, IAA proposed to consider three different subrisks, namely

- *Mortality risk:* This is the key risk component for a life company and its risk components are
  - volatility
  - catastrophe
  - trend uncertainty
  - level uncertainty.
- *Lapse risk:* The risks posed to an insurer by an unanticipated rate of policy lapses, terminations or surrenders (collectively referred to here as 'lapse risk') are varied and complex. The treatment of lapse risk within a capital requirement will also vary from jurisdiction to jurisdiction.
- *Expense risk:* Operating expenses of an insurance company represent a considerable portion of an insurer's annual costs. The other major elements of annual costs include the change in policy liabilities, that is, reserves or technical provisions, and policy benefits/claims.

21.2.1 IAA Model

21.2.1.1 Mortality Risk

We consider first the mortality risk. The IAA assumes that the expected mortality claim level, or risk premium RP, is defined by

$$\mathrm{RP} = \sum_i q_i \cdot X_i,$$

where q_i is the mortality rate and X_i the amount of insurance for the ith insured person. The number of deaths is assumed to be Poisson distributed and the total claim level follows a Compound Poisson distribution. This means that the standard deviation and the skewness of the distribution are given by $\sigma = \sqrt{\sum_i q_i \cdot X_i^2}$ and $\gamma = \left(\sum_i q_i \cdot X_i^3\right)/\sigma^3$, respectively.

Mortality Risk: Volatility (IAA, 2004, p. 44)
Traditional volatility risk (VR) is often calculated using a simulation model with many scenarios generated based on parameter input(s) into a Monte Carlo process. An alternative is to use the Cornish–Fisher expansion, *cf.* Section 14.3, using the first three moments of the Compound Poisson distribution. This approach will be less time-consuming than simulation models.

The VaR at a 99.5% confidence level is

$$\mathrm{CR}_{\mathrm{Volatility}} = \sigma \cdot [2.58 + 0.94 \cdot \gamma].$$

Under typical circumstances this approach can be further simplified. Assuming that # is the number of insured risks and the average q_i is around 0.0025, the capital can be calculated as follows:

$$\mathrm{CR}_{\mathrm{Volatility}} = \left[\frac{77.4}{\sqrt{\#}} + \frac{942.7}{\#}\right] \cdot \mathrm{RP}.$$

Mortality Risk: Catastrophe (IAA, 2004, p. 45)
Beyond "normal" random fluctuations (volatility) in mortality experience from one period of time to the next, extra capital is needed for extreme events that result in high positive deviations in the claim level. These events can be caused by severe epidemic (e.g., Spanish Flu in 1918), NatCat (e.g., earthquake), and terrorist attack (e.g., events of 9/11).

Due to the lack of data it is difficult to model this kind of risk and a very simple approach may be the most useful and appropriate. The capital for catastrophe risk can, for example, be based on portion of the expected number of deaths during one year. Based on the experience of the Spanish Flu epidemic, a doubling of one year's expected deaths may be appropriate.

Mortality Risk: Trend Uncertainty (IAA, 2004, p. 46)
Another mortality risk component is trend uncertainty, the difficulty in accurately assessing the future direction (e.g., improvement) of the mortality assumption in future years. With many product terms extending for the lifetime of the insured, this can be a considerable risk, especially for payout annuities. It is difficult to model mortality trend uncertainty in a simple way. The result depends on product, duration, and interest rate. A simplified approach to provide for trend uncertainty could be to apply a factor multiplied by the present value amount of the liabilities.

In IAA (2009, Appendix D3.1), a method is proposed and applied to Dutch data.* The mortality assumption for the current estimate is based on a projection of Dutch population mortality, adjusted for use as insured mortality by using a factor of 0.80 (times q_x).

The average age of the portfolio of contracts is assumed to be 12 years and annual mortality data from 1950 through 1998 were available. In developing the current estimate mortality rates, the current estimate trend is based on the average trend experienced between 1988 and 1995. In 1988, there was a significant change in trend observed. Within the 48 years of observations, nine separate trends are observed: there is an average trend between 1950 and 1960 ($i = 1$), 1955 and 1965 ($i = 2$), and so on.

* The work was done by Henk van Broekhoven.

Based on each generation table i, a corresponding liability can be calculated. This results in nine different liabilities L_i. For these nine liabilities, a standard deviation can be calculated as

$$s_{\text{trend}} = \sqrt{\frac{9}{8}\left\{\left(\frac{1}{9}\sum_i L_i^2\right) - \left(\frac{1}{9}\sum_i L_i\right)^2\right\}}.$$

The trend uncertainty calculated this way follows a Student's t distribution with eight degrees of freedom (dfs). Hence the capital requirement can be calculated as

$$\text{CR}_{\text{trend}} = t_{8,1-\alpha} \cdot s_{\text{trend}}.$$

Mortality Risk: Level Uncertainty (IAA, 2004, p. 45)
Level uncertainty is caused by the volatility observed in the past. This can make it difficult to estimate the "real" or "true" current average mortality. The same kind of model as in the VR can be used to calculate this risk. However, the potential impact on the liability must be determined because level uncertainty involves the misestimating the mortality assumption for all future years. This makes it difficult to find a simple factor approach.

One approach would be to "shock" the present value amount of the policy liabilities using best estimate mortality rates. To find this shock, the same kind of approach can be used as for volatility.

In IAA (2009, Appendix D3.4), a method is proposed. It is based on the Cornish–Fisher approximation of the compound Poisson distribution in terms of the standard normal distribution.

Define the ratio between the expected mortality rate for insured persons and the whole population by dividing the observed deaths over a certain period by the expected deaths over the same period, based on the population mortality or an industry reference table:

$$f_{\text{be}} = \frac{\mu_{\text{obs}}}{\mu_{\text{ref}}}.$$

In the mortality level uncertainty we reproduce the uncertainty in the observations μ_{obs} by means of an adjustment factor $f_{\text{ec}} = (\mu_{\text{obs}} + (-) \cdot \sigma \cdot [s_{ci} + t_{ci} \cdot \gamma]) / \mu_{\text{ref}}$, where the factors s and t depend on the time horizon and the confidence level. The standard deviation (σ) and skewness (γ) are defined as in the beginning of this section.

Mortality Risk: Total
The final capital requirement for mortality risks should provide for each of the components described above. To the extent that the mortality experience is shared with the policyholders, the corresponding credit should be granted in the capital requirements.

21.2.1.2 Lapse Risk
According to IAA, there are two primary effects of unanticipated lapse rates (IAA, 2004, p. 47).

Type one: The first one involves the payment of surrender or termination values. The relationship of the amount of a surrender payment to the value of the liability being held in respect of a particular policy is of great importance. When a policy lapses, the company pays the surrender value and "receives" the actuarial reserve that is released by the policy's termination. If surrender values are lower than policy reserves, the company is at risk from lapse rates that are lower than expected, particularly if high lapse rates were anticipated in the pricing of a product. The case that surrender values exceed policy reserves results in higher lapse rates being unfavorable to the insurer. In some jurisdictions these risks are mitigated by regulations. A requirement that a company holds policy liabilities at least as large as surrender values provides partial protection against overly high lapse rates while minimum required surrender values reduce the likelihood that insurers will price their products using an assumption of high lapse rates. It is important to recognize that the relationship between the surrender value and the actuarial reserve is not fixed; it will generally vary with the duration of a particular policy.

A capital requirement with respect to this type of lapse risk requires the division of an insurance company's policies into two classes:

1. Those policies for which actuarial liabilities L are greater than surrender values S and
2. Those policies for which $S > L$. The capital requirements would then be of the form $j(L–S)$ or $k(S–L)$, respectively, for appropriately chosen factors j and k.

Type two: The second primary effect of unanticipated lapse rates is that the insurer may not realize the expected recovery from future premiums of initial policy acquisition expenses. These acquisition expenses may be recognized implicitly in financial statements through the use of modified net level premium valuation methods. These implicit methods generally do not include any provision for unfavorable variations in lapse rates. Recovery of acquisition expenses may also be recognized explicitly through a reduction in policy liabilities or through the introduction of a receivable asset. In this latter case, the adjustment to financial values is made subject to a form of recoverability test. Under the second primary effect, the risk to insurers is generated by lapse rates that are greater than expected.

A capital requirement in respect of this type of lapse risk could be of the form mU where m is an appropriately chosen factor and U is the present expected value of acquisition expenses recoverable from future premiums.

In the case that lapses are recognized explicitly in the valuation of actuarial liabilities, another approach to capital requirements in respect of the first type of lapse risk is available. This requires the division of policies into two classes:

1. Those for which an increase in lapse rates results in an increase in policy liabilities and
2. Those for which policy liabilities increase when assumed lapses decrease

The capital requirement is of the form of the difference between a special valuation of policy liabilities and the normal valuation. For the special valuation, the lapse assumption is multiplied by a specified factor greater than one for policies in the first class and by a factor less than one for policies in the second class.

21.2.1.3 Expense Risk

It is important for an insurer to understand its expenses and their component parts for the purposes of proper product pricing, provisioning, solvency assessment, and so on (IAA, 2004, p. 48).

Most important in any analysis of insurer expenses is to obtain the split of expenses between acquisition and maintenance and also between fixed and variable. A table like this should be developed:

The fixed expenses are those expenses that do not vary in proportion to the volume of the total new and existing business at least over the short term. Especially important for assessing the adequacy of provisions and for solvency assessment is a proper determination of the split of expenses between acquisition and maintenance. This split is based on insurer judgement. If too many expenses are allocated to the acquisition category, then a forward-looking view of the company's ongoing maintenance expenses will be understated. This may result in the underprovisioning of such expenses in the liabilities and an overly optimistic view of the company's future financial condition.

The capital charge could be calculated as *factor* ∗ *(Expense source from Table 21.3)*. This is discussed in IAA (2004, pp. 49–50).

21.2.2 GDV Model

In 2005 the GDV launched a new standard formula for the capital requirement; see GDV (2005). The life underwriting risk is here called the calculation risk and comprises the costs risk, the policyholder and insurance agent default risk (DR) as a form of lapse risk, and biometric risks. Biometric risks can be classified into fluctuation, accumulation, trend, and modification risks, each of which is calculated for the following segmentations: life, endowment, disability (BU), and other insurance (OI).

The risk categories examined are consistent with those of the IAA model (IAA, 2004).

The risk factors for the fluctuation risk are taken from a distribution assumption; for the other risks, the relevant factors are derived from supervisory returns and/or heuristic methods. The calculation should be based on the historical values of the individual company. In order to account for volatility, the maximum is ascertained for all individual company data in the last three years. If a partial risk is negative, that is, amounts to income, the risk is set at zero. As a result, the overall life underwriting risk cannot be negative.

TABLE 21.3 Expense Table

Expenses	**Fixed Costs**	**Variable Costs**	
Acquisition	#	#	Total acquisition
Maintenance	#	#	Total maintenance
	Total fixed	Total variable	Total expenses

Source: From IAA. 2004. *A Global Framework for Insurer Solvency Assessment.* International Actuarial Association, Ontario, ISBN: 0-9733449-0-3. With permission.

The calculation risk represents the only risk in the model that is specific to life insurance. The annuity option is implicitly taken into account in the calculation of duration through assumptions regarding the probability of a lump-sum settlement. Other guarantees and options are not explicitly represented in the model, but are adequately considered through conservative assumptions.

Proportional reinsurance is taken into account in relation to the fluctuation risk, and accumulation reinsurance in relation to the accumulation risk. On the other hand, surplus, stop loss, and other reinsurance contracts are not included in the standard method.

21.2.2.1 Cost Risk

A distinction is made between the fixed costs risk and the kickback (KB) risk (if borne by the insurer) in the case of investments for the account and at the risk of policyholders (unit-linked life/annuity insurance).

The fixed costs risk (risk bearer: FC) takes into account the fact that, unlike variable costs, fixed costs cannot be reduced directly. The capital requirement for this risk equals the total fixed costs for the year, that is, fixed acquisition costs, maintenance and settlement costs, as well as the net total of other income and expenses. Administrative cost revenue from the portfolio and amortization charges may be subtracted from that total. The result is expressed as a percentage of fixed costs, with documentations of supervisory returns serving as the basis, maximized over the last three years.

The KB risk (risk bearer: KR) considers the reduced reimbursement of administrative expenses of the funds based on changes in market value. For this risk, the factor 0.175% (= 35% slump in equities value of a 50% KB of a 1% management fee) is applied to the market value of investments.

The capital charge is defined as

$$\mathrm{CR}_{\mathrm{LUR,CR}} = f_{\mathrm{FC}} \cdot \mathrm{FC} + f_{\mathrm{KR}} \cdot \mathrm{KR},$$

where f is a risk factor for fixed cost (FC) risk and KB risk, respectively.

21.2.2.2 Insurance Agent+ Policyholder DR

This risk includes the potential loss of claims against insurance agents and policyholders in the event of (multiple) terminations by policyholders without the option of asserting claims for the return of commissions in an equal amount.

The capital requirement for this risk is therefore equal to the reduction in Zillmer claims based on the company's supervisory returns, minus reversed commissions from early disposal plus depreciation and allowances on accounts receivable from agents for reversed commissions. The result is then multiplied by a factor of 2, in order to arrive at a double cancellation rate, as in the IAA model. The result is expressed in the percentage of Zillmer claims and claims against agents and policyholders and maximized over the last three years. Separate from this risk is the risk of cancellation due to a rise in interest rates. The risk is assigned to investment risk.

The capital charge is defined as

$$CR_{LUR,DR} = f_{DR} \cdot DR,$$

where f is a risk factor for the DR of policyholders.

21.2.2.3 Biometric Risk

All biometric risks are divided into fluctuation, and accumulation, trend, or modification risks.

The fluctuation risk considers the deviation of benefits from the expected value. The accumulation risk refers to the simultaneous occurrence of multiple claims caused by a single event (e.g., an earthquake). The trend or modification risk considers the possibility of a misassessment (e.g., of the longevity trend or the risk that expected mortalities do not conform to the portfolio). The accumulation, trend, or modification risks are combined into a single risk category.

Since for conservative reasons the standard method takes account of the expected income from biometric risks only in relation to the trend or accumulation risk, comparison with the 99% security level on which mortality tables are based is only possible to a limited extent.

21.2.2.4 Fluctuation Risk

As is usual in the differentiation of mortality tables, the standard deviations in the loss spread are estimated using the following formulas.

$$\text{Life insurance}: \quad \sigma_{\text{life}} = \sqrt{\frac{(CR - AE) \cdot AE}{n}},$$

where CR is the capital at risk in portfolio, AE is the actual expenses for claims in the financial year, and n is the number of insurance contracts.

$$\text{Endowment insurance}: \quad \sigma_{\text{endowment}} = \sqrt{\frac{(MP - AIE) \cdot AIE}{n}},$$

where MP is the mathematical provisions minus insured death benefits, AIE is the actual inheritance earnings, and n is the number of insurance contracts.

For disability and OI, the same formula as for life insurance applies. The capital charge for the fluctuation risk is taken as the sum of 2.58 times the volatility of the different products. The capital at risk is estimated as 2.58 times the standard deviation; 2.58 represents precisely the 0.5% quantile of the standard normal distribution. The risk bearer is already included in the formulas. The specific risk mixture of the individual insurance company is also taken into account through the number of insurance contracts, capital at risk, and actual expenses.

The fluctuation risks are assumed to be fully correlated and added together. On account of the assumption of homogeneous portfolios, the fluctuation estimate is increased by a 100% security surcharge. If a proportional reinsurance contract exists, the reinsurer's share in the capital charge is calculated without reinsurance.

21.2.2.5 Accumulation, Trend, and Modification Risk

As in the IAA model, the model's calculation of the accumulation and modification risk for whole life insurance is based on a one-year doubling of mortality. Thus, the capital requirement is defined as net risk minus actual expenses for claims based on the company's supervisory returns. The same procedure is followed regarding the accumulation and modification risk for disability insurance.

If a reinsurance contract exists for accumulation of risks, it may be used in the model to reduce the capital requirement. The maximum scope of coverage for an accumulation of risks may be used to reduce the capital requirement once. However, that coverage is then added back to the capital requirement, weighted with the probabilities of reinsurer default (specified in the model).

The capital requirement for the trend risk is estimated, for example, for endowment insurance as in the IAA method for the mortality accumulation risk based on the average potential annual replenishment requirement. The selected factor is equal to an average 0.35% of mathematical provisions for annuities. The factor "annual replenishment for current annuity table" must take account of both the replenishment amount for traditional annuity insurance and the replenishment amount for unit-linked annuity insurance with guaranteed or partly guaranteed annuity factors. With respect to OI, for example, capital redemption products, tontines or dread disease, the risk is estimated at 0.2% of the capital at risk.

The capital charge is defined as

$$\mathrm{CR}_{\mathrm{LUR,ATM}} = f_{\mathrm{WLR}} \cdot \mathrm{CR}_{\mathrm{WLR}} + f_{\mathrm{ER}} \cdot \mathrm{AR} + f_{\mathrm{DR}} \cdot \mathrm{CR}_{\mathrm{DR}} + f_{\mathrm{OI}} \cdot \mathrm{CR}_{\mathrm{OI}},$$

where f is a risk factor for whole life risk (WLR), endowment risk (ER), disability risk (DR), and OI. The risk bearers are capital at risk (CR), and annual replenishment requirement for current annuity table (AR).

Depending on the type of risk, accumulation/trend/modification risks are fully correlated with the fluctuation risk; as to the relationship between the types of risk themselves, they are assumed as being not correlated. The capital requirements for all subrisks of the biometric risk are maximized at zero.

21.2.3 CEA Model

CEA, the body of the European insurance industry, proposed in 2006 a "European Standard Approach" (ESA) for solvency assessment as another contribution to the debate on a European standard formula; see CEA (2006b).

The main principle for the life underwriting risk model is that simple factors are applied to exposure volume measures for each of the sun risk types. The following risk subtypes are considered under life underwriting risk:

- Mortality
- Longevity
- Morbidity

- Lapse
- Expense

Reinsurance is considered as the main type of risk mitigation. It is considered under the framework by using volume-measures net of reinsurance.

With the one-year time horizon approach, the risks and profits arising from new business sold over the next year should be considered. However, for most life companies, new policies incorporated over one year are relatively unimportant compared to the in-force portfolio of the company. For this reason, the ESA does not consider new business over the next year.

The potential risks associated with the companies, where there are substantial changes in premium income (increase or decrease), are better addressed in Pillar II.

21.2.3.1 Mortality Risk

Mortality risk is defined in the ESA model as unexpected deviation on the mortality experience for products providing death coverage. Risk components as classified by the IAA are volatility, uncertainty (level risk and trend risk), and extreme event (catastrophe).

Increasing the size of the portfolio can reduce VR and this should be taken into account when calculating the required capital. Uncertainty and catastrophe risk are not diversifiable by increasing the size of the portfolio and for simplicity both risks are treated together under the factor approach.

The required capital calculation should take into account the term of products and the flexibility of the insurer to adjust premiums. Products with shorter terms will face less trend or level uncertainty risks than products with longer ones.

The required capital for mortality risk uses simple factors applied to the sum-at-risk net of reinsurance. Sum at risk is defined as the sum assured less technical provisions. Different term groupings are considered and products for which the insurer has the ability of changing premiums over their term are grouped apart (e.g., annual renewable term products). So the placeholder groups under the proposed model are (1) annual renewable term products, (2) 0–5 years, (3) 5–10 years, and (4) >10 years. Size of the portfolio is taken into account for VR by introducing a factor that is sensitive to the number of insured heads.

The capital requirement is defined as

$$\mathrm{CR}_{\mathrm{LUR,MR}} = \mathrm{VR} + \mathrm{TCR},$$

where the VR is defined as $\mathrm{VR} = f_V \cdot \sum(\mathrm{NSAR}_{\mathrm{term}})$, f is a volatility factor as defined below, NSAR is the net sum at risk, and trend catastrophe risk (TCR) is defined as $\mathrm{TCR} = \sum(\mathrm{NSAR}_{\mathrm{term}}) \cdot f_{\mathrm{term}}$. The volatility factor f_V is defined by

$$f_V = \alpha \cdot \sqrt{\frac{q_x(1-q_x)}{n}},$$

where α is a parameter representing the number of standard deviations to be covered; for example, 2.58 for the normal distribution with 0.5% quantile. q_x is the death parameter and n is the number of insured persons.

21.2.3.2 Longevity Risk

Longevity risk is defined as the unexpected deviation on the mortality experience for products providing coverage in case of life. The required capital calculation should take into account the term of products. Products with shorter terms will face less trend or level uncertainty risks than products with longer ones.

The required capital for longevity risk uses simple factors applied to the technical provision net of reinsurance coverage. Different term groupings are considered. The placeholder groups under the proposed model are (1) 0–5 years, (2) 5–10 years, and (3) >10 years. No distinction is made between volatility and uncertainty risks for longevity.

The capital requirement is defined as

$$\mathrm{CR}_{\mathrm{LUR,LR}} = \sum (\mathrm{NTP}_{\mathrm{term}}) \cdot f_{\mathrm{term}},$$

where NTP is the net technical provisions and f is a risk factor.

21.2.3.3 Morbidity Risk

In many jurisdictions, products providing long-term life companies write disability or critical illness products. Therefore the ESA considers morbidity risk as a component of the capital requirement for life insurance.

Morbidity risk is defined here as unexpected deviation in the morbidity experience for products providing death coverage. Risk components are the same as those considered for mortality risk: volatility, uncertainty (level risk and trend risk), and extreme event (catastrophe).

The treatment of these components is the same as for mortality risks: the volatility factor is sensitive to the size of the portfolio; uncertainty and catastrophe risk are treated together as one factor. In the volatility factor given above, we change q_x for i_x, that is, the morbidity probability.

21.2.3.4 Lapse Risk

In the ESA model, the lapse risk is defined as unexpected deviation in the expected lapse rate. The required capital for lapse risk uses simple factors applied to the technical provision.

Different term groupings are considered. The placeholder groups under the proposed model are (1) 0–5 years, (2) 5–10 years, and (3) >10 years.

The capital requirement is defined as

$$\mathrm{CR}_{\mathrm{LUR,La}} = \sum (\mathrm{TP}_{\mathrm{term}}) \cdot f_{\mathrm{La,term}},$$

where TP is the technical provisions and f is a lapse risk factor.

21.2.3.5 Expense Risk

The capital allowance for expense risk under the proposed ESA covers the risk that the company faces due to its inability to reduce fixed expenses under certain circumstances, for example, lower production than expected, and mass lapses. Fixed expenses over the last year are considered as the volume measure.

The capital requirement is defined as

$$CR_{LUR,Ex} = \text{fixed expenses} \cdot \text{factor}.$$

21.3 HEALTH UNDERWRITING RISK

In the standard formula proposed in GDV (2005) by the German GDV, health underwriting risk is a separate subrisk category. We will therefore briefly present this model.

The standard model for health insurance risks in the SST is described in SST (2006b, ch. 4.5) and references given therein.

21.3.1 Risk from Net Costs

The cost risk arises if the cost surcharges factored into the gross premiums are not enough to cover the actual costs accruing in the accounting year. The possible causes of such a shortfall are many. Therefore, all cost items of private health insurers are taken into account. The following can be identified: (1) direct acquisition expenses, (2) indirect acquisition expenses, (3) claims settlement expenses, (4) net administrative expenses, (5) other income, and (6) other expenses.

The annual net costs are determined by adding together or subtracting from each other these six components and are related to gross premiums earned in order to ensure comparability with other periods.

In order to quantify fluctuations in net costs, the mean value for the 10 year time period under examination (currently: 1995–2004) is determined, and the standard deviation σ is defined. This standard deviation is then used to define the risk factor (RF) for the fluctuation in net costs, based on a significance level of 99.5%.

The risk bearer (RT) is the gross premium earned for the accounting year (at present: 2004).

The mean value is used for measuring fluctuations in net costs in the last three financial years, at present: 2002–2004. This income factor (IF) takes into account expected losses (IF < 0) and expected income (IF > 0) from net costs. Income bearers (IT) are the gross premiums earned for the accounting year (at present: 2004).

A risk from fluctuations in net costs arises if the per capita loss and formula for measuring reduction in the group of insured persons (mortality and cancellation) factored into the calculations do not correspond to the actual per capita loss, mortality, and cancellations. The following three components of this risk can be distinguished:

- The *excessive loss risk* or per capita loss risk arises when the actual per capita loss is greater than the loss factored into the calculation.
- The *mortality risk* exists if the actual funds from provisions for increasing age becoming available due to death are lower than those factored into the calculation.
- The *cancellation risk* exists if the actual funds from provisions for increasing age becoming available due to cancellations are lower than those factored into the calculation.

The specific calculation is made using the appropriate net risk amounts from internal documentations. In order to ascertain fluctuations in net risks, the model uses data from supervisory returns (documentations) of the last 10 years (at present: 1995–2004).

The net risks result from the difference between the actual and the calculated inheritances from cancellation (= cancellation risk) and death (= mortality risk) and the actual and calculated per capita claims of insured (= excess claim risk).

Within net risks these three components of the underwriting risk are fully and precisely reflected by private health insurance companies, that is, the correlations between those components are considered in a risk-appropriate fashion.

Net annual risks are normalized using gross premiums earned in order to guarantee the comparability of net risks with other periods.

In order to quantify fluctuations in net risks, the mean value in the 10 year time period under examination (at present: 1995–2004) is represented, and standard deviation σ is calculated.

This standard deviation is then used to calculate the RF for fluctuations in net risks, based on a significance level of 99.557%. The risk bearers (RT) are the gross premiums earned for the accounting year (at present: 2004).

The mean value is used for measuring fluctuations in net costs in the last three financial years, at present: 2002–2004. This IF takes into account expected losses (IF < 0) and expected income (IF > 0) from net risks. Income bearers (IT) are the gross premiums earned for the accounting year (at present: 2004).

21.3.2 Epidemic/Accumulation Risk

Owing to effective risk mitigation measures, such as airport closures, the distribution of respirators and the distribution of medicine, it is considered that a severe outbreak, for example, of influenza is out of the question. Therefore, the RF is set at 1% with sufficient security. Risk bearers (RT) are divided among the individual companies in accordance with their market shares (individual company market share = ICMS).

21.3.3 Security Surcharge

A security surcharge of at least 5% of the gross premium must be incorporated into the insurance premium. This surcharge must not be already incorporated in other technical bases. It is an important underwriting income source. In the standard model, the individual company security surcharge of the accounting year (at present: 2004) is taken into account in the available solvency margin.

PART D

European Solvency II General Ideas, Valuation and Investment: Final Advice

Solvency II will be the crown jewel of the European Union.

Allessandro Iuppa
President, NAIC, Chair, IAIS, at the EU Commission's public hearing in June 2006

The art of progress is to preserve order amid change and to preserve change amid order.

Alfred North Whitehead (1861–1947)
English "mathematician, philosopher"

Chapters 22 through 25 discuss the European Solvency II project: its general ideas, the valuation of assets and liabilities, and investment. The presentation is mainly based on official documents published by the European Commission, EIOPC and CEIOPS. Also documents from the European Parliament are used.

The general ideas are discussed in Chapter 22. Asset valuation according to Solvency II is discussed in Chapter 23 and Liability valuation in Chapter 24. Part D ends with a discussion on investments and EOFs in Chapter 25. Background information is available in the appendices.

The presentation here is based on Level 1 texts, that is, the Framework Directive (FD), and Level 2 implementing measures, that is, detailed measures complementing the FD, proposed by CEIOPS in its draft proposals and its final advice as published up to January 2010. The last final advices given by CEIOPS were published in March 2010. Based on all these advices the European Commission developed Level 2 implementing measures, mainly as EU regulation. These will be published in late 2011.

When we talk about an *undertaking* in both Part D and Part E chapters, we mean both an insurance undertaking and a reinsurance undertaking, unless otherwise is explicitly mentioned.

CHAPTER 22

European Solvency II

General Ideas

THE GENERAL DEVELOPMENT of the European Solvency II system and the basic architecture has been discussed in Section 2.3 and in Appendix D. The background to the general ideas is also discussed in Appendix E.

In the following sections mentioned, we will discuss the following specific general issues:

- The general structure: Section 22.1
- An ERM approach—Governance: Section 22.2
- The proportionality principle: Section 22.3 (see also Appendix E, Section E.1)
- Internal and partial internal models: Section 22.4 (see also Appendix E, Section E.2)
- Group issues: Section 22.5 (see also Appendix E, Section E.3)
- Actions to be taken: Section 22.6
    - Capital add-on: Section 22.6.1
    - Extension of recovery period: Section 22.6.2
- Reporting and disclosure: Section 22.7

22.1 GENERAL STRUCTURE

The background of the FD is discussed in Appendix D. The basic architecture of Solvency II, explained in Section 2.3.1, is also discussed there. It has also been explained in different publications such as CEA-Tillinghast (2006), CEA (2007b, 2007c) and in the rapporteur Peter Skinner's report, EP (2008).

As with Basel II for the banking industry, Solvency II is building a new regulatory framework for the European insurance sector. With the new solvency framework, as outlined in the FD the Commission has introduced a risk-sensitive approach with incentives for risk-increased management. One goal of Solvency II is to compel companies to better measure, monitor, and manage their risks, but also to create incentives to measure and monitor the risks correctly. Companies that do not measure or manage their risks adequately will be penalized through larger capital requirements.

Solvency II and the valuation of assets and liabilities and the calculation of the capital requirements are part of an Enterprise Risk Management (ERM) approach. This is reflected in the FD text regarding, for example, the governance, fit and proper requirements, different functions, but last but not least the own risk and solvency assessment (ORSA).

An economic approach will lead to a better and optimal allocation of capital, taking into account *market-consistent valuation of assets and liabilities*. It also aligns solvency rules with a realistic valuation according to the present market assessment of assets and liabilities of a company.

The FD is principle based and a Lamfalussy compliant approach; see Appendix D, Section D.1.1 with the basis for adoption of implementing measures at Level 2 and with instructions for supervisory work at Level 3 of the Lamfalussy process. Its architecture is structured in the form of *three pillars*. The main objective of pillar I requirements is to ensure that insurance and reinsurance companies are able to meet their obligations when due, with a 0.5% probability of ruin on a one-year time horizon. Therefore the Solvency II framework is structured on a *total balance sheet approach* (TBSA); balancing all the liabilities with all assets the company needs to hold and not valuating BS items in isolation.

The three pillars developed for Basel II were the models for Solvency II, but the similarities are limited. The business model of the insurance industry is very different to that of banks. The three pillars are the following:

Pillar I: It defines the capital resources that a company needs to hold in order to be considered solvent, that is, measure the assets, liabilities, and capital. The quantitative approach is holistic, as it should measure all quantifiable risks. Two thresholds are defined: the SCR and the MCR. The capital requirement can be based on the use of a standard formula or an (partial) internal model.

Companies need to hold enough assets to meet the following:

- Technical provisions (best estimate and a risk margin),
- MCR,
- SCR.

The *valuation of assets and liabilities* is built on a TBSA, leading to a better and optimal allocation of capital, taking into account *market-consistent valuation of assets and liabilities*. It also aligns solvency rules with a realistic valuation according to the present market assessment of assets and liabilities of a company. The valuation of assets and liabilities are discussed in Chapters 23 and 24 with background notes provided in Appendices F and G, respectively. *Investments* and *own funds* are discussed in Chapter 25 and Appendix E, Section E.4.

FD texts:
Valuation: Article 75
Technical provisions: Articles 76–86
Own funds: Articles 87–99
Investments: Articles 132–135

The *SCR* can be calculated by using a standard formula or an (partial) internal model. The *standard formula* is discussed in Chapters 26 through 33; the calibration and development is discussed in Appendices H to O.

An increasing number of companies have developed *internal models* to better measure the risks they take. Partial or full internal models are customised to the company's risk profile and a central tool for the company's risk management. The standard formula is intended to be a simple and cost-efficient alternative that provides some of the benefits of risk weighting of internal models for those firms who will not, or decide not to, develop their own internal model. The standard formula aims at capturing the risk profile of an average company. This lack of customisation may in certain cases, in particular if companies are writing high or complex risk business, induce a slightly more conservative risk assessment, which in turn may represent a *cost* in terms of needing more capital.

FD texts:
SCR—Standard formula: Articles 100–102 and 103–111
SCR—Full and partial internal models: Articles 100–102 and 112–127

The calculation of the minimum or ultimate, capital requirement (MCR) is discussed in Chapter 34 and Appendix P.

FD texts:
MCR: Articles 128–131

Pillar II: Defines the qualitative requirements as a supplement to Pillar I. It defines the framework of the supervisory review process (SRP) and the ORSA.

Pillar II defines a framework of supervisory control focusing on internal processes, such as the ORSA, governance, aspects of operational risk including an effective internal control system, risk management systems, actuarial function, internal audit, rules on outsourcing, and so on, that is, a full ERM approach. Liquidity risk is initially a part of the Pillar II supervisory process. Pillar II also defines the possibility of imposing capital add-on if the supervisor finds that the capital requirement is too low depending on, for example, the risk profile. Early-warning indicators, investment policies, asset–liability matching, and risk-mitigating programs are all parts of the Pillar II process.

The FD enhances tools for supervisory activities, including definition of supervisory powers and provisions for cooperation between supervisors as well as for supervisory convergence. It is important for these provisions to be in line with the provisions in securities and banking sectors and hence to achieve the cross-sectoral consistency and convergence.

The ORSA shall be conducted on a regularly basis and be an integral part of the business process and be a part of the strategic decisions that the company makes. As a minimum, it should include

- The overall solvency needs, including nonquantifiable risks.
- Compliance with the requirements related to the technical provisions and capital.
- Any deviations between the company's own risk profile and the assumptions underlying the SCR calculation, resulting in recalibration of (partial) internal models.

FD texts:
Pillar II issues: Articles 27–34, 36–38, 40–50, and 192–194

Pillar III: Defines the disclosure requirements and transparency and imposes greater discipline on the industry.

The main objective of Pillar III was the disclosure of information to underpin market discipline, supervisory reporting, and transparency requirements. On the supervision side, it encourages supervisory cooperation and convergence, enhances the role of CEIOPS, introduces an early-warning mechanism and outlines a framework for a more effective Group supervision. There was a need to converge the rules on supervisory reporting in order to deliver a comparable format and content, especially important when talking of Group supervision as well as reporting obligations via national authorities to CEIOPS.

FD texts:
Pillar III issues: Articles 35, 51–56

Third country equivalence is discussed in CEIOPS (2009g). Three main issues are discussed:

- Reinsurance supervision—Article 172 of the FD,
- Group solvency calculation—Article 227 of the FD, and
- Group supervision—Article 260 of the FD.

The approach used by CEIOIPS is to identify the key supervisory principles in the Solvency II FD and the objectives each supervisory principle seeks to achieve. In order to be considered equivalent, a third country regime will have to meet each of the principles and objectives laid out in CEIOIPS (2009g). For each principle and objective, the "indicators" of equivalence are also outlined—namely, those factors that provide guidance in determining whether the relevant principles and objectives are achieved.

22.1.1 Extracts ("Recitals") from the FD Preamble

The recitals to the FD give us the ideas behind the FD. In each of Chapters 22 through 26 we are quoting the recitals that give the background to the topics discussed. Recitals not quoted in these chapters will be discussed in Appendix D, Section D.8. The number to the left before the recital is the numbering in the preamble. The headings are given for "easy reading."

Harmonization

(2) In order to facilitate the taking-up and pursuit of the activities of insurance and reinsurance, it is necessary to eliminate the most serious differences between the laws of the Member States as regards the rules to which insurance and reinsurance undertakings are subject. A legal framework should therefore be provided for insurance and reinsurance undertakings to conduct insurance business throughout the internal market thus making it easier for insurance and reinsurance undertakings with head offices in the Community to cover risks and commitments situated therein.

(3) It is in the interests of the proper functioning of the internal market that coordinated rules be established relating to the supervision of insurance groups and, with a view to the protection of creditors, to the reorganisation and winding-up proceedings in respect of insurance undertakings.

(11) Since this Directive constitutes an essential instrument for the achievement of the internal market, insurance and reinsurance undertakings authorised in their home Member States should be allowed to pursue, throughout the Community, any or all of their activities by establishing branches or by providing services. It is therefore appropriate to bring about such harmonisation as is necessary and sufficient to achieve the mutual recognition of authorisations and supervisory systems, and thus a single authorisation which is valid throughout the Community and which allows the supervision of an undertaking to be carried out by the home Member State.

General Features

(14) The protection of policyholders presupposes that insurance and reinsurance undertakings are subject to effective solvency requirements that result in an efficient allocation of capital across the European Union. In light of market developments the current system is no longer adequate. It is therefore necessary to introduce a new regulatory framework.

(16) The main objective of insurance and reinsurance regulation and supervision is the adequate protection of policy holders and beneficiaries. The term beneficiary is intended to cover any natural or legal person who is entitled to a right under an insurance contract. Financial stability and fair and stable markets are other objectives of insurance and reinsurance regulation and supervision which should also be taken into account but should not undermine the main objective.

(17) The solvency regime laid down in this Directive is expected to result in even better protection for policy holders. It will require Member States to provide supervisory authorities with the resources to fulfill their obligations as set out in this Directive. This encompasses all necessary capacities, including financial and human resources.

(18) The supervisory authorities of the Member States should therefore have at their disposal all means necessary to ensure the orderly pursuit of business by insurance and

reinsurance undertakings throughout the Community whether pursued under the right of establishment or the freedom to provide services. In order to ensure the effectiveness of the supervision, all actions taken by the supervisory authorities should be proportionate to the nature, scale, and complexity of the risks inherent in the business of an insurance or reinsurance undertaking, regardless of the importance of the undertaking concerned for the overall financial stability of the market.

(22) The supervision of reinsurance activity should take account of the special characteristics of reinsurance business, notably its global nature and the fact that the policyholders are themselves insurance or reinsurance undertakings.

(44) Insurance undertakings pursuing both life and nonlife activities should manage those activities separately, in order to protect the interests of life policy holders. In particular, those undertakings should be subject to the same capital requirements as those applicable to an equivalent insurance group, made up of a life insurance undertaking and a non-life undertaking, taking into account the increased transferability of capital in the case of composite insurance undertakings.

Supervision

(23) Supervisory authorities should be able to obtain from insurance and reinsurance undertakings the information which is necessary for the purposes of supervision, including, where appropriate, information publicly disclosed by an insurance or reinsurance undertaking under financial reporting, listing, and other legal or regulatory requirements.

(24) The supervisory authorities of the home Member State should be responsible for monitoring the financial health of insurance and reinsurance undertakings. To that end, they should carry out regular reviews and evaluations.

(25) Supervisory authorities should be able to take account of the effects on risk and asset management of voluntary codes of conduct and transparency complied with by the relevant institutions dealing in unregulated or alternative investment instruments.

(37) In order to ensure effective supervision of outsourced functions or activities, it is essential that the supervisory authorities of the outsourcing insurance or reinsurance undertaking have access to all relevant data held by the outsourcing service provider, regardless of whether the latter is a regulated or unregulated entity, as well as the right to conduct on-site inspections. In order to take account of market developments and to ensure that the conditions for outsourcing continue to be complied with, the supervisory authorities should be informed prior to the outsourcing of critical or important functions or activities. Those requirements should take into account the work of the Joint Forum and are consistent with the current rules and practices in the banking sector and Directive 2004/39/EC and its application to credit institutions.

(40) It is necessary to promote supervisory convergence not only in respect of supervisory tools but also in respect of supervisory practices. The Committee of European Insurance and Occupational Pensions Supervisors (CEIOPS) established by Commission Decision 2009/79/EC1 should play an important role in this respect and report regularly to the European Parliament and the Commission on the progress made.

22.2 AN ERM APPROACH

The Solvency II FD gives us the ground for an "ERM" system. The holistic ERM view is not only looking at the valuation of assets and liabilities and the capital requirements, but also on the SRP and the disclosure. However, the main building blocks in the ERM system is the governance and the requirements on the people and the functions they have to obey.

As this is not the main objective of this book we will briefly mention only some of the final advice that CEIOPS has given on the governance and the functionality of different functions. The reader is referred to CEIOPS work CEIOPS (2009b08) and the final advice in CEIOPS (2009f09).

22.2.1 Extracts ("Recitals") from the FD Preamble

The recitals to the FD give us the ideas behind the FD. The number to the left before the recital is the numbering in the preamble.

(29) Some risks may only be properly addressed through governance requirements rather than through the quantitative requirements reflected in the Solvency Capital Requirement. An effective system of governance is therefore essential for the adequate management of the insurance undertaking and for the regulatory system.

(30) The system of governance includes the risk-management function, the compliance function, the internal audit function and the actuarial function.

(31) A function is an administrative capacity to undertake particular governance tasks. The identification of a particular function does not prevent the undertaking from freely deciding how to organise that function in practice save where otherwise specified in this Directive. This should not lead to unduly burdensome requirements because account should be taken of the nature, scale and complexity of the operations of the undertaking. It should therefore be possible for those functions to be staffed by own staff, to rely on advice from outside experts or to be outsourced to experts within the limits set by this Directive.

(32) Furthermore, save as regards the internal audit function, in smaller and less complex undertakings it should be possible for more than one function to be carried out by a single person or organisational unit.

(33) The functions included in the system of governance are considered to be key functions and consequently also important and critical functions.

(34) All persons that perform key functions should be fit and proper. However, only the key function holders should be subject to notification requirements to the supervisory authority.

(35) For the purpose of assessing the required level of competence, professional qualifications and experience of those who effectively run the undertaking or have other key functions should be taken into consideration as additional factors.

(36) All insurance and reinsurance undertakings should have, as an integrated part of their business strategy, a regular practice of assessing their overall solvency needs with a view to their specific risk profile (own-risk and solvency assessment). That assessment neither requires the development of an internal model nor serves to calculate a capital requirement different from the Solvency Capital Requirement or the Minimum Capital Requirement.

The results of each assessment should be reported to the supervisory authority as part of the information to be provided for supervisory purposes.

22.2.2 Reference to the FD

The main reference in the FD is found in Articles 38 and 40–50.

22.2.3 Main Building Blocks

The *four-eyes principle*, that is, the principle that prior to implementing any significant decision concerning the undertaking at least two persons should review any such decision, should be complied with by all insurance undertakings. The administrative, management, or supervisory body is the focal point of the governance system. CEIOPS has proposed that an undertaking's system of *governance* should; see CEIOPS (2009f09):

(a) Establish, implement, and maintain effective cooperation, internal reporting, and communication of information at all relevant levels within the undertaking;

(b) Be robust with a clear and well-defined organizational structure that has well-defined, clear, consistent, and documented lines of responsibility across the organization;

(c) Ensure that the members of the administrative, management, or supervisory body possess sufficient professional qualifications, knowledge, and experience in the relevant areas of the business to give adequate assurance that they are collectively able to provide a sound and prudent management of the undertaking;

(d) Ensure that it employs personnel with the skills, knowledge, and expertise necessary to discharge properly the responsibilities allocated to them;

(e) Ensure that all personnel are aware of the procedures for the proper discharge of their responsibilities;

(f) Establish, implement, and maintain decision-making procedures;

(g) Ensure that any performance of multiple tasks by individuals does not and is not likely to prevent the persons concerned from discharging any particular function soundly, honestly, and professionally;

(h) Establish information systems that produce sufficient, reliable, consistent, timely, and relevant information concerning all business activities, the commitments assumed, and the risks to which the undertaking is exposed;

(i) Maintain adequate and orderly records of its business and internal organization;

(j) Safeguard the security, integrity, and confidentiality of information, taking into account the nature of the information in question;

(k) Introduce clear reporting lines that ensure the prompt transfer of information to all persons who need it in a way that enables them to recognize its importance; and

(l) Establish and maintain adequate risk management, compliance, internal audit, and actuarial functions, the characteristics of which are set out below.

The undertakings have to ensure that potential source of conflicts of interest is identified and procedures have to be established so that those involved with the implementation of the strategies and policies understand where conflicts of interest could arise and how these should be addressed, for example, by establishing additional controls. The undertakings have to regularly identify the risks for which contingency plans have to be in place taking account of the areas where they are especially vulnerable. All plans should be regularly tested and updated to ensure that they are and will remain effective, and communicated to all relevant management and staff. The governance is outlined in Article 41.

The financial crisis has triggered a reflection on some common principles related to internal governance; CEIOPS (2009f31). One issue, perhaps the most "prominent," is related to *remuneration practices* applied to the members of the administrative, management, or supervisory body and senior management of financial entities, as well as to personnel undertaking activities that involve risk-taking. In the footsteps of the financial crisis, several documents have been published on the need of revising remuneration policies and schemes. For example, CEIOPS believes that high-level principles developed by CEBS for the bank industry are also, in general, applicable to the insurance sector.

Therefore, an overall remuneration policy and practice shall be adopted that is in line with the undertaking's business and risk strategy, risk profile, objectives, values, risk management practices, and long-term entity-wide interests and performance. There will be a clear, transparent, and effective governance structure around remuneration, including the definition of the remuneration policy and its oversight.

In this sequel we briefly highlight only different important issues.

Fit and proper requirements, Article 42: Undertakings shall notify the supervisory authority, which persons effectively run the undertaking, and which, if any, other key function holders are identified for the undertaking. Key functions are those considered important and critical in the system of governance and include risk management, compliance, internal audit, and actuarial functions. Other functions may be considered as key functions in accordance with the proportionality principle.

Risk management system, Article 44: An effective risk management system covers all material risks and requires at least the following:

a. A clearly defined and well-documented risk management strategy that includes the objectives, key principles, risk appetite, and assignment of responsibilities across all the activities of the undertaking and is consistent with the undertaking's overall business strategy;

b. Adequate written policies that include a definition and categorization of the risks faced by the undertaking, by type, and the levels of acceptable risk limits for each risk type, implement the undertaking's risk strategy, facilitate control mechanisms, and take into account the nature, scope, and time horizon of the business;

c. Appropriate processes and procedures that enable the undertaking to identify, assess, manage, monitor, and report the risks it is or might be exposed to;

d. Appropriate reporting procedures and feedback loops that ensure that information on the risk management system, which is coordinated and challenged by the risk management function, is continuously monitored and managed by all relevant staff and the administrative or management body;

e. Reports that are submitted to the administrative or management body by the risk management function on the material risks faced by the undertaking and on the effectiveness of the risk management function;

f. An appropriate ORSA process.

Areas to be covered by the risk management system:

- Underwriting and reserving:
  - Suitable processes and procedures should be in place to ensure the reliability, sufficiency, and adequacy of both the statistical and accounting data to be considered both in the underwriting and reserving processes.
- Asset—liability management:
  - The undertaking shall develop written ALM policies that especially take into account the interrelation with different types of risks, such as market risks, credit risks, liquidity risks, and underwriting risks, and establish ways to manage the possible effect of options embedded in the insurance products. Hence, the ALM policy shall provide for
    - A structuring of the assets that ensures that the undertaking holds sufficient cash and diversified marketable securities of an appropriate nature, term, and liquidity to meet its obligations, including obligations to pay bonuses to policyholders, as they fall due;
    - A plan to deal with unexpected cash outflows, or changes in expected cash inflows and out-flows and
    - The identification of mitigation techniques and their impact on embedded options, and the assessment of the possible effects these can have throughout the life of the insurance policies and/or reinsurance contracts.
- Investment, including derivatives and similar commitments:
  - The investment policy should be defined in line with what a competent, prudent, and expert manager would apply in order to pursue the investment strategy.
  - Investments are subject to market risk. Market risk is the risk of loss, or of adverse change in the financial situation, resulting directly or indirectly from fluctuations in the level and in the volatility of market prices of assets, liabilities, and financial instruments.

- It is required that assets and investments are identified that fulfill the prudent person principle.
- Special management, monitoring, and control procedures for the undertaking's investment assets is important, in particular in relation to investments that are not quoted in a market and to complex structured products.

- Liquidity risk management:
  - Liquidity risk refers to the risk that undertakings are unable to realize investments and other assets in order to settle their financial obligations when they fall due.
  - It is the undertaking's responsibility to have sound liquidity management practices that cover both short- and long-term considerations. Short-term liquidity, or cash management, covers the day-to-day cash requirements under normally expected or likely business conditions.
- Concentration risk management:
  - Concentration risk means all risk exposures with a loss potential that is large enough to threaten the solvency or the financial position of undertakings.
  - Undertakings need to have in place adequate procedures and processes for the active monitoring and management of concentration risk to ensure that it stays within established policies and limits and mitigating actions can be taken as necessary. The monitoring of concentration risk shall include an analysis of possible contagion lines.
- Operational risk management:
  - The administrative or management body should be aware of the major categories and exposures of the undertaking's operational risks as a distinct risk category that should be managed, and should approve, implement, and periodically review the undertaking's operational risk management framework. The undertaking shall implement an effective process to regularly identify, document, and monitor exposure to operational risk and track-relevant operational risk data, including near misses.
- Reinsurance and other risk mitigation techniques:
  - Reinsurance and similar techniques such as alternative risk transfer (ART) techniques may enable the undertaking to prudently manage and mitigate, in particular the insurance specific risk. However, they also carry new potential risks, such as the risk of counterparty default.
  - As part of their reinsurance management strategy, undertakings should have adequate procedures and processes for the selection of suitable reinsurance programs. The level of sophistication for these processes and procedures should be proportionate to the nature, scale, and complexity of the undertaking's risks and

to the capabilities of the undertaking to manage and control the risk mitigation technique used.

- When undertakings use SPVs, the following principles shall be considered taking into account the requirements and guidelines set out in Chapter 26 on SPVs:
    - The fully funded requirement shall be actively monitored by the undertaking through its system of governance;
    - Any remaining risk from the SPV shall be fully taken into account in the undertaking through its risk management system and also taken into account within the calculation of its regulatory capital requirements.
- Financial risk mitigation techniques should only be used where it is appropriate to do so as part of an overall risk management policy and reinsurance management strategy, where both qualitative and quantitative features shall be appropriately considered.

- Credit risk management:
    - The process of risk management should be capable of identifying and mitigating any credit risk in relation to internally defined limits. The undertaking should be alert to changes in individual credit ratings as well as credit portfolio risk through regular appropriate and proportionate monitoring processes, and capable of evaluating relevant parameters such as probabilities of default even where exposures are unrated. Exposure to speculative grade assets should be prudent and undertakings facing larger credit risk exposures should be capable of hedging credit risk, for example, via derivatives to protect against a protracted fall in credit quality or turn in the credit cycle.
- Strategic risk:
    - Strategic risk is defined as the risk of the current and prospective impact on earnings or capital arising from adverse business decisions, improper implementation of decisions, or lack of responsiveness to industry changes. The overall strategy of the undertaking should incorporate its risk management practices. In this sense, the undertaking should have a process for setting strategic high-level objectives and translating these into detailed shorter-term business and operation plans.
- Reputational risk:
    - Reputational risk is defined as the risk of potential loss to an undertaking through deterioration of its reputation or standing due to a negative perception of the undertaking's image among customers, counterparties, shareholders, and/or supervisory authorities. To that extent it may be regarded as less of a separate risk, than one consequent on the overall conduct of an undertaking. The administrative or management body of the undertaking should be aware of potential reputational risks it is exposed to and the correlation with all other material risks.

- Risk management function:
  - The undertaking needs to establish a risk management function within its organizational structure that is proportionate to the scale, nature, and complexity of risks inherent within the business. The embedding of the risk management function in the organizational structure of the undertaking and the associated reporting lines shall ensure that the function is objective and free from influence from other functions and from the administrative, management, or supervisory body.

ORSA, Article 45: The ORSA is outlined in Article 45 and CEIOPS has published an issue paper, see CEIOPS (2008f). ORSA is one of the linchpins of the ERM system.

The FD describes the ORSA as a tool of the risk management system that requires the undertakings to properly assess their own short- and long-term risks and the amount of OFs necessary to cover them. At the same time, the ORSA represents an important source of information for the supervisory authorities, and the undertakings are obliged to describe the process they have undertaken to satisfy the ORSA requirements through the regulatory reporting requirements. Hence, the ORSA can be defined as the entirety of the processes and procedures employed to *identify, assess, monitor, manage, and report* the short- and long-term risks an undertaking faces or may face and to determine the OFs necessary to ensure that the undertaking's overall solvency needs are met at all times.

One goal of ORSA is enhancing the awareness of the interrelationships between the risks an undertaking is currently exposed to, or may be facing in the long term, and the internal capital needs that follow from this risk exposure, whether it uses the standard formula or an internal model to calculate the SCR: *First of all the ORSA represents the undertaking's opinion and understanding of its risks, overall solvency needs and own funds held.* As such, it is important to perform the ORSA as it helps the undertaking to obtain a real and practical understanding of its risks it is exposed to.

The ORSA does not require an undertaking to develop an internal model, is not a capital requirement different from the SCR and the MCR, and should not be too burdensome.

Other main issues that have been discussed by CEIOPS (2009b08, 2009f09) are briefly mentioned below.

Internal control, Article 46: The internal control system should ensure that an undertaking's systems, whether manual or based on information technology, are appropriate to the undertaking's strategies and data needs and consistent with the nature and complexity of its activities. One important issue in the internal control system is the compliance function. Compliance risk is defined as the risk of legal or regulatory sanctions, material financial loss, or loss to reputation an undertaking may suffer as a result of not complying with laws, regulations and administrative provisions as applicable to its activities.

Internal Audit, Article 47: The internal audit function needs to be independent of the organizational activities audited and carry out its assignments with impartiality. The principle of independence entails that the internal audit function should only operate under the oversight of the administrative, management or supervisory body, reporting to this body or an audit committee. At the same time, it has to be ensured that the internal audit function is not subject to instructions of the administrative, management or supervisory body when performing the audit and when evaluating and reporting the audit results. Only the right to change and approve audit plans remains unaffected by the required abstention from issuing instructions.

Actuarial Function, Article 48: The process of developing *European actuarial guidelines*, issued by CEIOPS, shall involve participants with appropriate knowledge and experience of actuarial issues and shall represent in balanced proportions the insurance industry, the actuarial profession and the academic community. The actuarial function shall have access to the appropriate resources and information systems that provide all necessary information, relevant for the discharge of its responsibilities.

The actuarial function should coordinate the calculation of the technical provisions. In order to accomplish this task it should, at least,

a. Apply methodologies and procedures to assess the sufficiency of technical provisions ensuring that their calculation is consistent with the underlying principles;

b. Assess the uncertainty associated with the estimates;

c. Produce judgment whenever this is needed, making use of appropriate information and experience of the persons that are in charge of the function;

d. Ensure that problems related to the calculation of technical provisions arising from insufficient data quality are dealt with appropriately and that, where it is impracticable to apply common methods of calculating technical provisions because of insufficient data quality, the most appropriate alternatives to common methods applied are found, taking into consideration the principle of proportionality;

e. Ensure that homogeneous risk groups for an appropriate assessment of the underlying risks are identified;

f. Consult any relevant market information and ensure that it is integrated into the assessment of technical provisions;

g. Compare and justify any material differences among the estimates for different years; and

h. Ensure that an appropriate assessment of options and guarantees embedded in liabilities is provided.

In order to be able to provide its opinions free from influence from other functions and the administrative, management, or supervisory body, the actuarial function shall be constituted by persons who have a sufficient level of independency.

Outsourcing, Article 49: The undertaking's outsourcing policy should include considerations of the impact of outsourcing on its business and the reporting and monitoring arrangements to be implemented when an outsourcing contract has been agreed. The policy should be regularly assessed and updated, with any necessary changes implemented.

Remuneration: Remuneration is a governance issue. It is discussed in, for example, CEIOPS (2009f31).

22.3 PROPORTIONALITY PRINCIPLE

Owing to the level of sophistication of the new Solvency II regime, one of the key questions has been how to gear it to the *nature, scale, and complexity of the risks* to which an insurance undertaking is exposed, particularly, but not only, with regard to small- and medium-sized undertakings. This gives a brief sum up of the proportionality principle, which is discussed in, for example, CEIOPS (2008b).

In the FD, particular care has been taken to ensure that the new solvency regime is not too burdensome for low risk profile undertakings, which are often small- and medium-sized undertakings. Appropriate treatment of these undertakings has to be achieved through the application of the *principle of proportionality*.

The FD established the proportionality principle as a general principle that applies throughout the Directive, highlighting it in several provisions and leaving its concrete implementation to Level 2 measures and Level 3 guidance.

22.3.1 Extracts ("Recitals") from the FD Preamble

The proportionality principle is quoted in recitals (19), (20), and (21) to the FD. Earlier thoughts, discussions, and calibrations are discussed in Appendix E, Section E.1. Draft final advice on the assessment of the principle of proportionality is given by CEIOPS (2009d07) and their final advice was a part of the new draft final advice, CEIOPS (2009e14).

(19) This Directive should not be too burdensome for small- and medium-sized insurance undertakings. One of the tools by which to achieve that objective is the proper application of the proportionality principle. That principle should apply both to the requirements imposed on the insurance and reinsurance undertakings and to the exercise of supervisory powers.

(20) In particular, this Directive should not be too burdensome for insurance undertakings that specialize in providing specific types of insurance or services to specific customer segments, and it should recognize that specializing in this way can be a valuable tool for efficiently and effectively managing risk. In order to achieve that objective, as well as the proper application of the proportionality principle, provision should also be made specifically to allow undertakings to use their own data to calibrate the parameters in the underwriting risk modules of the standard formula of the SCR.

(21) This Directive should also take account of the specific nature of captive insurance and captive reinsurance undertakings. As those undertakings only cover risks associated with the industrial or commercial group to which they belong, appropriate approaches should thus be provided in line with the principle of proportionality to reflect the nature, scale, and complexity of their business.

22.3.2 Reference to the FD

Even if the proportionality principle is only mentioned explicitly in Article 29 (3a), General principles of supervision, it should be applied to all Articles in the FD. It is also indirectly mentioned when the directive text talks about the nature, scale, and complexity of different activities and calculations.

22.3.3 Interpretation of the Proportionality Principle

The proportionality principle is a generally acknowledged principle of the due course of law and is not therefore comprehensively defined in the FD. It applies throughout the Directive. It has two aspects:

(a) Proportionality has to be taken into account when implementing the requirements laid down in the FD; and

(b) supervision has to be carried out in a proportionate manner.

Proportionality does not mean the introduction of automatic and systematic simplifications for certain undertakings. The principle will be applied where it would be disproportionate to

- The nature,
- The scale, and
- The complexity

of undertakings' business to apply general quantitative and qualitative rules without relief. The individual risk profile should be the primary guide in assessing the need to apply the proportionality principle.

The principle of proportionality applies to all the provisions in the directive and, as a consequence, to all future implementing measures. The proportionality principle should be applied in a coherent way across the three pillars as well as the group provisions.

The proportionality principle should be applied regardless of whether it is explicitly mentioned in a provision or not. The mention of the principle in certain Articles should not lead to the conclusion *per se* that it does not apply or applies less where it is not explicitly mentioned.

Proportionality works two-ways: It justifies simpler and less burdensome ways of meeting requirements for low risk-profile portfolios, but also increases the likelihood that undertakings in fulfilling requirements will need to apply more sophisticated methods and techniques for more complex risk portfolios.

In considering the *nature* of the risks, supervisors will take into account the underlying risk profiles of the classes of business an undertaking is writing, for example, whether it is long- or short-tailed business, or whether it is a low-frequency and high-severity business or consists of high-frequency and low-severity risks. The specific nature of risks inherent to the reinsurance business and to the captives business should also be taken into account.

Complexity is linked to the nature of the business as certain kinds of business may dictate the use of more demanding methods or an advanced system of governance. In particular, a more sophisticated risk management system may be regarded in order to deal properly with all risks that the undertaking faces. However, it may also be introduced via the investment strategy of the undertaking or because the insurer chooses to employ challenging methods or processes in some areas that require a commensurate degree of complexity in other areas of the undertaking. It is also linked to the complexity in the evaluation of the commitments, for example, unlimited motor liability, or investment in a complex option, or annuities (as opposed to a lump sum), or nonproportional reinsurance (as opposed to a straightforward direct insurance business).

CEIOPS (2009d07, 2009e14) considers the fact that nature and complexity of risks are closely related, and for the purposes of an assessment of proportionality could best be characterized together. Complexity could be seen as an integral part of the nature of risks, which is a broader concept. In mathematical terms, the nature of the risks underlying the insurance contracts could be described by the probability distribution of the future CFs arising from the contracts. This encompasses the following characteristics:

- The degree of homogeneity of the risks;
- The variety of different subrisks or risk components of which the risk is comprised;
- The way in which these subrisks are interrelated with one another;
- The level of certainty, that is, the extent to which future CFs can be predicted;
- The nature of the occurrence or crystallization of the risk in terms of frequency and severity;
- The type of the development of claims payments over time; or
- The extent of potential policyholder loss, especially in the tail of the claims distribution.

Through *scale* a size criterion is introduced making a distinction between *small and large*. Relating to the valuation of assets, liabilities or risks, this criterion resembles a materiality requirement and the approach applied should ensure an appropriate relative or absolute approximation of the theoretically correct value. The undertaking could use a measurement of scale to identify risks where the use of simplified methods would likely be appropriate, provided this is also commensurate with the nature and complexity of the risks. The measurement of scale could also be used to introduce a distinction between *material and nonmaterial* risks. There are different interpretations to scale. It could be seen as the degree of vulnerability to risk, where the scale would increase as either the likelihood or the impact of the risk increases (scale = likelihood × impact). It could also be related to the SCR as the vulnerability to risk under a worst-case scenario. A third interpretation would be in terms of the best estimate of the underlying obligations. Based on these considerations, CEIOPS suggests the following relative assessments of scale:

Scale = (relative) size of the best estimate
Scale = likelihood × (relative) impact
Scale = SCR/(volume measure).

In assessing what is proportionate, the focus must be on the combination of all three criteria to arrive at a solution that is adequate to the risk an undertaking is exposed to. For instance, a business may well be small-scale but could still include complex risk profiles or on the contrary, it may be large-scale with a simple risk profile. In the first case, it should not be allowed to use simplified methods, while the possibility may be considered in the second case under very specific circumstances. By the interrelation between the three indicators the focus should be on the combination of all three factors. CEIOPS (2009d07) concludes that this overall assessment of proportionality would ideally be more qualitative than quantitative, and cannot be reduced to a simple formulaic aggregation of isolated assessments of each of the indicators.

In terms of nature and complexity, the assessment should seek to identify the main qualities and characteristics of the risks, and should lead to an evaluation of the degree of their complexity and predictability. In combination with the "scale" criterion, the undertaking may use such an assessment as a filter to decide whether the use of simplified methods would be likely to be appropriate. For this purpose, it may be helpful to broadly categorize the risks according to a two-dimensional "scale" and "complexity/predictability" diagram; see CEIOPS (2009d07, p. 20).

As it will probably tend to be the small- and medium-sized undertakings that will find relief in the application of the proportionality principle through simpler ways of meeting supervisory requirements, it is actually imprecise to talk of proportionality as size-based. Consequently, CEIOPS will not find a definition for small and medium-sized undertakings, SMEs, or to develop a set of simplified requirements to be applied only to these SMEs.

Proxies have been developed in case there is insufficient company-specific data of appropriate quality to apply a reliable statistical actuarial method. Therefore, proxies can be regarded as special types of simplified methods that are positioned at the "lower end" of continuum of methods that could be applied. In the FD text, the term *approximations* is used.

In CEIOPS (2009d07, 2009e14), the assessment of the proportionality of the valuation process of the underlying risks is described as a three-step process:

Step 1: Assess nature, scale, and complexity of underlying risks.

Step 2: Check whether valuation methodology is proportionate to risks as assessed in step 1, with regard to the degree of model error resulting from its application.

Step 3: Back test and validate the assessment carried out in steps 1 and 2.

This process is intended to set out expectations on undertakings and the supervisor in their application of the proportionality when selecting a valuation methodology. CEIOPS is of the opinion that it is important that a flexible and principle-based framework is maintained.

The first step is discussed above. The second step concerns the assessment whether a specific valuation methodology can be regarded as proportionate to the nature, scale, and complexity of the risks as analyzed in the first step. To carry out this assessment, CEIOPS wants the undertaking to analyze whether the valuation methodology in question takes into account the properties and characteristics of risks identified in the first step in a proportionate way, and also has due regard to the scale of the risks.

For the best estimate, see Chapter 24, this means that a given valuation technique should be seen as proportionate if the resulting estimate is not expected to diverge materially from the "true" best estimate which is given by the mean of the underlying risk distribution, that is, if the model error implied by the measurement is immaterial.

In the third step and as part of the actuarial control cycle, it should be checked whether the best estimates calculated in past years turn out to be appropriate in subsequent years. Such back testing is considered to be part of the validation process the undertakings are expected to carry out when calculating technical provisions. Back testing should always be done when an undertaking's risk profile has significantly changed over time.

Simplifications and proxies for the best estimate, risk margin, and reinsurance recoverables are discussed in Chapter 24. Simplifications for the capital charges in the SCR calculation are discussed in Chapters 26 through 33.

22.4 INTERNAL AND PARTIAL INTERNAL MODELS

By approval from the local supervisor the company may use (partial) internal models. For a given subrisk or risk module, a company can exchange the standard formula for its own methodology to create a model for the capital charge. If only one or more subrisks are changed, they are called partial internal models. They combine elements of internal models and enable companies to incrementally improve their risk management. With an internal model (or full internal model), we mean a model that captures the total capital requirement. A partial internal model captures one or more risks in the total model. Rest of it is captured by the standard formula.

The use of partial or fully internal models for the risk measurement has the intention of giving the company incentives to evaluate and control their risks more accurate.

22.4.1 Extracts ("Recitals") from the FD Preamble

The recitals to the FD give us the ideas behind the FD. The number to the left before the recital is the numbering in the preamble.

The use of partial and full internal models is quoted in recital (68) to the FD:

(68) In accordance with the risk-oriented approach to the SCR, it should be possible, in specific circumstances, to use partial or full internal models for the calculation of that

requirement rather than the standard formula. In order to provide policyholders and beneficiaries with an equivalent level of protection, such internal models should be subject to prior supervisory approval on the basis of harmonized processes and standards.

22.4.2 Reference to the FD

The main reference for internal models (and related issues) in the FD is found in Articles 100–102, 112–127, and for groups in Articles 231 and 234.

22.4.3 Approval of Internal Models

The approval process of (partial) internal models is given in CEIOPS' (final) draft advice, CEIOPS (2009b12, 2009e03, 2009f08, 2010a03) and for group internal models in CEIOPS (2009b13). Tests and standards for the approval of internal models are discussed in the (draft) final advice by CEIOPS (2009d18). Partial internal models are discussed in Section 22.4.4. One key issue is that a full or partial internal model should be seen as an integral part of the risk management organization of the undertaking or the group.

The approval process consists of several steps to follow.

Step 1—preapplication: at this stage the undertaking has the possibility to have initial talks with the supervisor. This initial discussion should at least include

- An indication that the undertaking intends to apply for approval to use its internal model to calculate the SCR and when it plans to apply;
- The scope of the internal model application and for which risks/entities/LOBs/major business units are covered by the model;
- An initial view from the undertaking on how the internal model meets the requirements for approval in the Level 1 text (i.e., a self-assessment of internal model readiness); The self assessment of internal model readiness shall not be a substitute for the internal model requirements in the Level 1 text. The format of the self-assessment might develop over time, based on the supervisory authorities experience of internal model approvals, or could be based on experience in other countries/industries. Supervisory authorities will need to consider to what extent a standard format could be developed for all undertakings, and how the format should vary according to the nature, scale, and complexity of risks borne by the undertaking and how the undertaking is managed.
- The undertaking shall also be able to explain their concrete project plan for meeting the internal model requirements by the date of the application.
- Any information the (re)insurance undertaking deems necessary and relevant to understand the model at the provisional stage of preapplication (e.g., a draft of the information to be submitted later for the internal model approval application). An undertaking indicating that they intend to apply for internal model approval should be expected to be well on the way to preparing the documentation for the application.

- Access to any draft documentation of the internal model as set out by Article 125 of the Level 1 text.

Step 1—Groups: In addition to the solo requirements, CEIOPS expects for groups that

- The preapplication will assess the scope of the group internal model and its consistency with the scope of group supervision according to Article 214 or deviates, as well as the architecture of the model, for example, how responsibilities are split within the group.
- All supervisory authorities establish jointly a cooperative and consultative framework.
- For each individual preapplication, supervisory authorities develop together with the group an overall supervisory plan of action that covers each step of the approval process including priority issues and a timetable.

Step 2—Application: The required minimum documentation for an application shall be divided into

a. *Cover letter* requesting approval approved and signed by the administrative or management body of the insurance undertaking.
   - A written declaration from the administrative or management bodies of the insurance undertaking to confirm that all clarifications and supporting documents have been provided and no material facts and/or details relevant to the approval have been knowingly concealed.
   - A copy of the application approval of the companies administrative or management bodies as set out in Article 116 (e.g., extracts from the relevant minutes)
   - The results of the latest ORSA and details of the undertaking's business and risk strategies as set out in Article 45.

b. *Scope of application* for full and partial internal models and model coverage. The undertaking shall explain what elements, including material quantifiable risks, LOBs, and entities, are proposed to be covered by the internal model.

c. Risk management process and risk profile.

d. Self-assessment.

e. Technical characteristics.

f. External models and data.

g. Model governance, systems and controls, including documentation on organization charts.

h. Up-to-date independent review/Validation report.

i. *Policy for changing* the full and partial internal models and other policies for internal model governance.

j. Plan for future model improvement.

k. Capital requirement.

Step 2—Groups: In addition to the solo requirements, CEIOPS expects for groups that An application form including a list of the undertakings and major business units included the group, the list of their relevant supervisory authorities, and the method used to derive the consolidated accounts.

- A list of the undertakings that will use an internal model to calculate their solo capital requirements.
- The capital requirements for the group and each of the related undertakings intending to use the group internal model to assess its SCR.
- A description of the structure of the group and of the intragroup transactions (IGTs).
- Where applicable, the list of related undertakings or business units excluded from the scope of the internal model, the rationale for this choice, and the method used to assess risks in these undertakings or business units.
- Where applicable, the transitional plan to include undertakings not yet taken into account in the internal model.
- Where applicable, the regulatory capital requirements for related undertakings included in the scope of the group internal model, but subject on a solo basis to other solvency requirements than the one described in the Level 1 text.
- The cover letter requesting approval shall be approved and signed by all the administrative, management, or supervisory bodies of the undertakings applying for permission to calculate the solo SCR with the group internal model, and of the undertaking submitting the application. The application pack shall be sent to the group supervisor in an official language of its Member State, unless the group supervisor agrees that information is provided in another language or a combination of both an official language and another language.

Step 3—Policy for changing the full and partial internal models: The undertaking has the full responsibility for creating the policy for changing the full and partial internal model. Changes are classified as major or minor.

Step 3—Groups: Should be consistent with the solo requirements.

Step 4—Assessment: The supervisory authorities shall decide on the application within six months of receipt of the complete application. The approval cannot be delegated from the supervisory authority to any third party. The supervisor analyzes the information submitted. The assessment procedure is seen as an iterative process. The assessment shall at least comprise a technical review of the model (i.e., its scope, design, build, integrity, and applications), its coverage and ability to calculate the SCR for the undertaking, documentation, the risk management process, senior management role, and their understanding of the model and shall be partly carried out by a set of on-site examinations.

Step 4—Groups: The supervisory authorities concerned shall be the group supervisor and supervisory authorities of all the Member States in which the head office of all subsidiary undertakings is situated. Nevertheless the other supervisory authorities involved in the group supervision may be consulted during the approval process. The purpose of this is to ensure that the group does not derive any benefit from excluding parts of the business from the group model scope.

Step 5—Decision making process: The output from the assessment of the model is used by the supervisor to reach a decision. The decision alternatives are

- Approval
- Rejection
- Limited approval

Step 5—Groups: The decision about the application is sent by the group supervisor to the group, and to all the concerned supervisory authorities. It shall be written in an official language of the Member State of the group supervisor, and if needed, in a language commonly understandable by all the other supervisory authorities.

The specific issues for approval of group internal models, such as information about the structure of the group, the scope and model coverage, and language and signatory, are discussed in CEIOPS (2009b13, 2009f08).

The following tests and standards are discussed in CEIOPS (2009d18, 2009f28):

Use test: The undertaking's use of the internal model shall be sufficiently material to result in pressure to improve the quality of the internal model. Seven principles are set out for the use test. This is outlined in Article 129 of the FD. The following elements may be taken into account when assessing the use of an internal model:

- Impact on policyholders;
- Impact on RM and use of policies, especially on risk mitigation, ALM, risk appetite, risk strategy, reinsurance program design, limit system, and analysis of new products;
- Impact on capital management, capital measurement, and allocation;
- Whether the scale of use of the internal model reflects the nature, scale, and complexity of the risks inherent in the business of the undertaking;
- Impact on the level playing field between undertakings;
- Consistency between supervisory authorities' decisions.

Internal model governance: The model governance should encourage the organization of a dialogue between every user of the model, likely to be the business units, and the risk management function about the characteristics of the internal model in order to increase understanding of the model and its outputs. This should lead to proposals for improvements to the model, enabling it to better reflect the risk profile of the undertaking.

Statistical quality standards: Includes statistical methodology, calculation methodology, and assumptions (adequate actuarial and statistical techniques), data quality, risk rankings, diversification, aggregation, risk mitigation, guarantees and options, future management actions, and payments.

Calibration standards: Different time periods and different risk measures are main issues here.

Profit and loss attribution: The undertaking shall regularly review the sources and causes of profit and loss for each major business unit. This exercise provides information used for the system of governance.

Validation: Validation is a set of tools and processes used by the undertaking to gain confidence over the results, design, workings, and other processes within the internal model. These tools and processes used for validation will be quantitative as well as qualitative.

Documentation standards: The documentation of an internal model shall be thorough, sufficiently detailed, and sufficiently complete to satisfy the criterion that an independent knowledgeable third party could form a sound judgment as to the reliability of the internal model and the compliance with Articles 120–126 and could understand the reasoning and the underlying design and operational details of the internal model.

External models and data: External models and data must be consistent with the standards and requirements set out for the use of an internal model to calculate the SCR.

Internal modeling is not the scope of this book and is therefore not discussed explicitly any further.

22.4.4 Partial Internal Models

Partial internal models are discussed in CEIOPS (2009e03, 2010a03). According to the FD, Article 112, undertakings may use partial internal models for the calculation of

- One or more risk modules,
- Or one or more submodules of the Basic SCR,
- The capital requirement for operational risk, and
- The adjustment for the loss-absorbing capacity of technical provisions and deferred taxes.

However, partial modeling may also be applied to the whole business of undertakings, or only to one or more major business units. An undertaking employing partial internal models may also use different risk categorizations than those in the standard formula or jointly model two or more different subrisks, not necessarily belonging to the same risk modules, as one. Annex A in CEIOPS (2009e03, 2010a03) gives a set of examples of different forms that partial internal models may assume.

For partial internal models some adaptations to the approval steps and the tests and standards discussed for internal models has to be made. These are discussed in CEIOPS (2009f08, 2009e03, 2010a03).

For the approval of partial internal models, the undertaking must comply with the provisions set out in Article 112 for internal models of the FD; see Section 22.4.3. Additional three specific provisions for partial internal models are set out in Article 113 of the FD.

Specific Provision 1: Justification of the Limited Scope of the Model

The reason for the limited scope of application of the model has to be properly justified by the undertaking and it should be set out in the application process and agreed by the supervisory authority. The undertaking should demonstrate that the limiting scope is properly justified.

For two undertakings within a group, with similar risk profile, the group is expected to calculate the capital requirement for the two entities in the same manner (standard formula, partial, or full internal models). If a group has acquired a new legal entity that is using an approved (partial) internal model, then this would not be seen as "cherry-picking."

Specific Provision 2: Better Reflection of the Risk Profile

Undertakings shall demonstrate that partial internal models reflect their risk profile more appropriately than the standard formula and that the resulting capital requirement meets the principles set out in Articles 100–102 of the FD; that is, going-concern assumptions, coverage of risks, and risk mitigation effects.

Specific Provision 3: Integration of Partial Internal Model Results into the Standard Formula

The design of the partial internal model should be consistent with the principles set out in the above-mentioned Articles of the FD allowing the partial internal model to be fully integrated into the SCR standard formula.

The supervisory authority has the option to require a transitional plan for the approval of a partial internal model.

There are different ways of integrating the results of a partial internal model with the results from the standard formula. For all these options there are two tests:

- *Feasibility test:* Is it possible to use the standard formula correlation matrix to integrate the partial internal model results with the results of the standard formula?
  - The feasibility test for an integration method is to determine whether it is possible to integrate the partial internal model with the standard formula using the chosen integration method.
- *Appropriateness test:* Is there any strong evidence that it is inappropriate to integrate the partial internal model's results into the standard formula's results?

- The appropriateness test for the integration mechanism looks at whether it is appropriate to use an integration method to integrate the partial internal model and the standard formula to produce the SCR for the undertaking. CEIOPS requires that undertakings provide "strong evidence" to the relevant supervisory authority that this integration method is inappropriate to be allowed to move to the next stage of selecting an integration mechanism.

  For showing "strong evidence," the undertaking has to analyze at least some or all of the following elements:

  - Equivalence of the SCR
  - Risk profile
  - Data
  - Use test

If the standard formula dependence matrix is neither feasible nor appropriate, then different options are available:

- *Option 1:* Integration of partial internal models using only coefficients *prescribed by supervisory authorities*;
- *Option 2:* Integration of partial internal models using techniques provided by supervisory authorities or—if these are not possible or there is strong evidence that these are inappropriate—dependency structures and parameters *provided by the undertaking.*
- *Option 3:* Integration of partial internal models using dependency structures and parameters provided by the undertaking—if these are not approved by the supervisory authority—techniques *provided by supervisory authorities.*

At Level 3 Standards or Guidance, CEIOPS will provide the companies specific aggregation techniques to use. This list has to be up to date and hence refined, based on methodology developments, etc.

A major business unit as regards to partial internal models is defined as a functional unit in an undertaking (either a solo entity or a group) which is managed with independence and with dedicated governance processes and for which it makes sense to calculate profit and losses, and capital charge for one or more risk modules (or the adjustment for loss-absorbing capacity of technical provisions and deferred taxes). The classification and justification of the major businesses units are a part of the internal model governance. Its definition should be consistent and stable. For example, ring-fenced funds, branches, and geographical regions may be considered as major business units.

Specific risks that may arise either at the solo or at the group level, and are not explicitly considered in the SCR standard formula, may be included. Such risks may, for instance, be underwriting cycle's risk, liquidity risk, and commodity or contagion risks; see CEIOPS

(2009e03, 2010a03). When dealing with partial internal models, undertakings and/or groups have several options to consider these specific risks. For example,

- Assume that these specific risks are linked to existing risks of the standard formula. This kind of approach may be particularly suitable for integrated stochastic models based on economic scenario generators;
- Establish a new risk module to take into account these risks;
- Assume that the risk is linked to a specific business unit and build a full model with respect to this business unit, which takes into account these specific risks.

22.5 GROUP ISSUES

The assessment of group solvency is discussed in CEIOPS (draft) final advice; CEIOPS (2009d21, 2009f32). Supervision of risk concentration (RC) and IGTs are discussed by CEIOPS in CEIOPS (2009d22, 2009f33) and cooperation and colleges of supervision in CEIOPS (2009d23, 2009f34). In CEIOPS (2009e04, 2010a04), the differentiation between centralized RM and group-wide RM is discussed. The centralized RM as discussed in op. cit. is a "relic" from the old draft FD and was a condition for the use of group support, which was carved out before the FD was adopted; see Appendix E, Section E.3.

22.5.1 Extracts ("Recitals") from the FD Preamble

The recitals to the Framework Directive (FD) give us the ideas behind the FD. The number to the left before the recital is the numbering in the preamble.

Group issues are quoted in recitals (95) through (116) to the Framework Directive:

(95) Measures concerning the supervision of insurance and reinsurance undertakings in a group should enable the authorities supervising an insurance or reinsurance undertaking to form a more soundly based judgment of its financial situation.

(96) Such group supervision should take into account insurance holding companies and mixed-activity insurance holding companies to the extent necessary. However, this Directive should not in any way imply that Member States are required to apply supervision to those undertakings considered individually.

(97) Whilst the supervision of individual insurance and reinsurance undertakings remains the essential principle of insurance supervision it is necessary to determine which undertakings fall under the scope of supervision at group level.

(98) Subject to Community and national law, undertakings, in particular mutual and mutual-type associations, should be able to form concentrations or groups, not through capital ties but through formalised strong and sustainable relationships, based on contractual or other material recognition that guarantees a financial solidarity between those undertakings. Where a dominant influence is exercised through a centralised coordination, those undertakings should be supervised in accordance with the same rules as those provided for groups constituted through capital ties in order to achieve an adequate level of protection for policy holders and a level playing field between groups.

(99) Group supervision should apply in any case at the level of the ultimate parent undertaking which has its head office in the Community. Member States should however be able to allow their supervisory authorities to apply group supervision at a limited number of lower levels, where they deem it necessary.

(100) It is necessary to calculate solvency at group level for insurance and reinsurance undertakings forming part of a group.

(101) The consolidated Solvency Capital Requirement for a group should take into account the global diversification of risks that exist across all the insurance and reinsurance undertakings in that group in order to reflect properly the risk exposures of that group.

(102) Insurance and reinsurance undertakings belonging to a group should be able to apply for the approval of an internal model to be used for the solvency calculation at both group and individual levels.

(103) Some provisions of this Directive expressly provide for a mediatory or a consultative role for CEIOPS, but this should not preclude CEIOPS from also playing a mediatory or a consultative role with regard to other provisions.

(104) This Directive reflects an innovative supervisory model where a key role is assigned to a group supervisor, whilst recognising and maintaining an important role for the solo supervisor. The powers and responsibilities of supervisors are linked with their accountability.

(105) All policy holders and beneficiaries should receive equal treatment regardless of their nationality or place of residence. For this purpose, each Member State should ensure that all measures taken by a supervisory authority on the basis of that supervisory authority's national mandate are not regarded as contrary to the interests of that Member State or of policy holders and beneficiaries in that Member State. In all situations of settling of claims and winding-up, assets should be distributed on an equitable basis to all relevant policy holders, regardless of their nationality or place of residence.

(106) It is necessary to ensure that own funds are appropriately distributed within the group and are available to protect policyholders and beneficiaries where needed. To that end, insurance and reinsurance undertakings within a group should have sufficient own funds to cover their solvency capital requirements.

(107) All supervisors involved in group supervision should be able to understand the decisions made, in particular where those decisions are made by the group supervisor. As soon as it becomes available to one of the supervisors, the relevant information should therefore be shared with the other supervisors, in order for all supervisors to be able to establish an opinion based on the same relevant information. In the event that the supervisors concerned cannot reach an agreement, qualified advice from CEIOPS should be sought to resolve the matter.

(108) The solvency of a subsidiary insurance or reinsurance undertaking of an insurance holding company, third-country insurance, or reinsurance undertaking may be affected by the financial resources of the group of which it is part and by the distribution of financial resources within that group. The supervisory authorities should therefore be provided with

the means of exercising group supervision and of taking appropriate measures at the level of the insurance or reinsurance undertaking where its solvency is being or may be jeopardised.

(109) Risk concentrations and intra-group transactions could affect the financial position of insurance or reinsurance undertakings. The supervisory authorities should therefore be able to exercise supervision over such risk concentrations and intra-group transactions, taking into account the nature of relationships between regulated entities as well as non-regulated entities, including insurance holding companies and mixed activity insurance holding companies, and take appropriate measures at the level of the insurance or reinsurance undertaking where its solvency is being or may be jeopardised.

(110) Insurance and reinsurance undertakings within a group should have appropriate systems of governance, which should be subject to supervisory review.

(111) All insurance and reinsurance groups subject to group supervision should have a group supervisor appointed from among the supervisory authorities involved. The rights and duties of the group supervisor should comprise appropriate coordination and decision-making powers. The authorities involved in the supervision of insurance and reinsurance undertakings belonging to the same group should establish coordination arrangements.

(112) In light of the increasing competences of group supervisors, the prevention of arbitrary circumvention of the criteria for choosing the group supervisor should be ensured. In particular, in cases where the group supervisor will be designated taking into account the structure of the group and the relative importance of the insurance and reinsurance activities in different markets, internal group transactions as well as group reinsurance should not be double counted when assessing their relative importance within a market.

(113) Supervisors from all Member States in which undertakings of the group are established should be involved in group supervision through a college of supervisors (the College). They should all have access to information available with other supervisory authorities within the College and they should be involved in decision-making actively and on an on-going basis. Cooperation between the authorities responsible for the supervision of insurance and reinsurance undertakings as well as between those authorities and the authorities responsible for the supervision of undertakings active in other financial sectors should be established.

(114) The activities of the College should be proportionate to the nature, scale, and complexity of the risks inherent in the business of all undertakings that are part of the group and to the cross-border dimension. The College should be set up to ensure that cooperation, exchange of information, and consultation processes among the supervisory authorities of the College are effectively applied in accordance with this Directive. Supervisory authorities should use the College to promote convergence of their respective decisions and to cooperate closely to carry out their supervisory activities across the group under harmonised criteria.

(115) This Directive should provide a consultative role for CEIOPS. Advice by CEIOPS to the relevant supervisor should not be binding on that supervisor when taking its decision. When taking a decision, the relevant supervisor should, however, take full account of that advice and explain any significant deviation therefrom.

(116) Insurance and reinsurance undertakings which are part of a group, the head of which is outside the Community should be subject to equivalent and appropriate group supervisory arrangements. It is therefore necessary to provide for transparency of rules and exchange of information with third-country authorities in all relevant circumstances. In order to ensure a harmonised approach to the determination and assessment of equivalence of third-country insurance and reinsurance supervision, provision should be made for the Commission to make a binding decision regarding the equivalence of third-country solvency regimes. For third countries regarding which no decision has been made by the Commission, the assessment of equivalence should be made by the group supervisor after consulting the other relevant supervisory authorities.

22.5.2 Reference to the FD

Articles 221, 225–229 in the FD give access to group solvency, and Articles 219, 220, 222–224, 227, and 230–233 give the frame for group solvency calculation. The insurance group directive was one of the insurance directives that were consolidated in the new "Solvency II directive".

22.5.3 Assessment of Group Solvency

We first have to look at the concept of a group. This is briefly done in the next section. Calculations of the capital requirement are discussed in Section 22.5.3.2. For more details, the reader is referred to Appendix E, Section E.3 where more details, based on QIS 4, is given. In the last section, we introduce RC, IGTs, and college of supervisors. Specifics related to group partial/full internal models are discussed above in Sections 22.3 and 22.4; see also CEIOPS (2009f08).

22.5.3.1 Groups

With an *insurance group,* we mean two or more insurance undertakings (direct insurer, reinsurer, etc.) that act as a *group.* A member of a group could be a participating undertaking (at least 20% or more of the voting rights or capital are held), its subsidiaries, but also an entity in which an undertaking holds a participation. Undertakings linked to each other by a relationship can be included in a group.

Mutuals and mutual-type associations (subject to Community and national law) can come together by constituting concentrations or groups. Those groups are usually not constituted with capital ties but through formalized strong and sustainable relationships, based on contractual or other material recognition that guarantees a financial solidarity between the mutuals or mutual-type associations.

Two important concepts are significant and have dominant influence:

- *Significant influence:* Where an undertaking has 20% or more of the shareholders' or members' voting rights in another undertaking. This means that the group has the power to participate in financial and operational policy decisions, but not control them.

- *Dominant influence:* Where an undertaking has more than 50% of the shareholders' or members' voting rights in another undertaking (a subsidiary undertaking); CEIOPS (2009d21). A dominant influence may exist for the purpose of assessing group solvency, but not for the establishment of the statutory consolidated accounts.

Regulated financial entities with capital requirement should be included in group calculation using the deduction aggregation method, see below, when it is not appropriate to include them through the default method.

Participation in an insurer or reinsurer with a dominant influence will imply a full integration of the participation in the accounts or a proportional integration. If significant influence is exercised the contribution to the group SCR with respect to the participation should be calculated as the group's share in the participation multiplied by the solo SCR of this participation. For other types of institutions, the treatment of participations should be as follows:

Other financial regulated entities: Consistent with the Financial Conglomerate Directive (FCD).

Related credit institutions, investment firms, and financial institutions: Shall be included in the group calculation using sectoral requirements and not allowing for diversification, in accordance with the FCD.

Financial nonregulated undertakings: A notional capital requirement shall be calculated.

Institutions for occupational retirement provision, IORP: Shall be included in the group calculation using sectoral requirements and not allow for diversification.

Participations in entities outside the financial sector, both dominant and significant influence: Should be consolidated through the equity method. Treatment in the group SCR should then be consistent with the treatment in the solo SCR.

The FD text provides for the inclusion of third-country entities in the group calculations. The recognition of diversification from those entities may be challenging. Issues such as professional secrecy, access to information, and the fungibility or transferability of OFs may restrict the recognition of diversification. These restrictions may lead the group supervisor to require the application of the accounting consolidation method pursuant to Article 220 or the deduction method pursuant to Article 229.

The equivalence of the third-country prudential and solvency regimes was discussed by CEIOPS (2009g).

22.5.3.2 Calculation Methods

Two methods for calculating the *group capital requirement* are given in the FD:

- *Default method—Accounting consolidation:* This method recognizes diversification benefits between different group entities, including between EEA and non-EEA insurance entities and with-profit businesses. The calculation also takes into account any participation in insurance entities. No diversification benefits are recognized for non-insurance participations or participations with no control relationship. All worldwide insurance undertakings of the group, including any non-EEA insurance undertakings, should be taken into account in the calculations. Groups may take into account geographical diversification benefits where these are permitted in the Standard Formula for the solo entity. The group solvency margin is the difference between OFs eligible to cover the SCR and the group SCR calculated on the basis of consolidated data (the consolidated group SCR).

- The group SCR calculated with the consolidated-accounting method is
- The SCR calculated on the (re)insurance part of the group composed by all (re)insurance undertakings for which diversification is recognized, named group SCR*;
  - The sum of $SCRs_j$ calculated on each other regulated undertaking j for which no diversification is recognized:
  - Insurance undertakings on which significant influence is exercised,
  - Credit institutions, investment firms, and financial institutions,
  - Any other regulated entities for which a specific capital requirement should be assessed, and
  - Where supervisors decide not to recognize diversification with an insurance undertaking.
- *Alternative method—Deduction and aggregation method:* It calculates the group solvency as the difference between the sum of the aggregated OFs in the group and the aggregated SCRs in the group. Diversification effects are already recognized in solo calculations, but the deduction and aggregation method does not allow for additional diversification effects at the group level as the group SCR represents the sum of the solo SCRs. When calculating the aggregated group SCR the capital charge on IGTs should be eliminated in order to avoid double charging for the same risk.

The accounting consolidation method is the default and preferred method. However, if the exclusive application of the default method is not appropriate, then the alternative method, or a combination thereof, may be used. The decision is made by the group supervisor in consultation with other supervisors concerned.

For the calculation of group OFs, it is necessary to analyze the group OFs in order to identify the eligible parts and to assess these within each entity that are not able to absorb losses in other entities within the group.

There may be restrictions on the fungibility and transferability of OFs. CEIOPS (2009f32) has defined these two important concepts as

- *Fungibility:* An element of OFs that can fully absorb any kind of losses within the group, regardless of the undertaking within which those OFs are held or where the commitments arise (in compliance with the local prudential and legal rules). Fungible OFs in this sense are thus not dedicated to a certain purpose. Fungibility of OFs at solo level does not automatically imply fungibility at the group level.

- *Transferability:* The ability to transfer OFs from one undertaking to another within the group. Transferability leads to increase/decrease of OFs in a solo entity without increasing/decreasing the group OFs, except the likely cost of the transfer. The time and the costs of the transfer have to be taken into account.

OFs at the group level are assessed following a five-step approach for the default method:

Step 1: The solo BSs are consolidated according to the accounting consolidation rules.

Step 2: The regulatory consolidated BS is calculated by applying prudential filters:

i. Adjustment of the scope of supervision.

ii. Treatment of related credit institutions, investment firms, and financial institutions.

iii. Treatment of related undertakings for which the necessary information is unavailable.

iv. Any other necessary adjustments or deductions.

Step 3: Assessment of the contribution of each undertaking to the group SCR:

The contribution to group OFs of an undertaking's unavailable own funds (UAOFs) is then limited by its contribution to the group SCR. For each entity included in the group calculation, the excess UAOFs is the difference, if positive, between its UAOFs and its contribution to the group SCR; see below.

Step 4: The available group own funds (AGOFs) to cover the group SCR is calculated by deducting from the regulatory group OFs the sum of excess UAOFs (determined for all entities of the group);

Step 5: In order to be eligible to cover the group SCR, the AGOF must comply with the tiers' limits that are laid out in the Directive.

Example: Calculation of the AGOFs.

Step 1—calculate the group total own funds (GOFs), including capital unavailable at the group level.

Step 2—calculate *j:s* contribution to Group SCR (GSCR): $\mathrm{GSCR}_j = \mathrm{GSCR} \cdot \mathrm{SCR}_j / \sum_i \mathrm{SCR}_i$. The GSCR_j gives member company *j:s* contribution to the AGOF. GSCR is the group SCR.

Step 3—from the GOFs, which includes *j:s* total own funds OF_j, we deduct the UAOFs from company j and add its contribution GSCR_j, that is,

$$\mathrm{AGOF} = \mathrm{GOF} - \mathrm{UAOF}_j + \mathrm{GSCR}_j.$$

In this example we have assumed only one company where there have been any restrictions. If there are restrictions within several companies we just add their UAOFs) and their contributions according to Step 2.

The group may use a group-specific assessment of the contributions of the related insurance undertakings where a group internal model has been used for the assessment of the group SCR. The sum of all individual contributions shall be equal to the group SCR and then the allocated diversification effects, which are not affecting the solo SCR, should be assessed for any restrictions on the availability of solo OFs. The sum of the excess of OFs identified as unavailable should then be deducted from group OFs.

It is important to take notice of with-profit business, ring-fenced funds, and transferability, but also hybrid capital, subordinated liabilities, and minority interests. These issues and group-specific risks such as reputational risk, contagion risk, impact of IGTs, and operational risk are discussed in CEIOPS (2009d21, 2009f32).

22.5.3.3 *Other Group Issues*

RC and IGTs are two intragroup relationships that can influence the risk profile of an undertaking; Articles 244–245. IGT includes transactions between an undertaking and a related, participating or a related undertaking of a participating undertaking of the undertaking. It also involves transactions between an undertaking and a natural person that holds participation in one of the combinations above. The scope of IGT should also include third-country undertakings, branches, other regulated entities, and unregulated entities. In CEIOPS (2009d22, 2009f33), there is a nonexhaustive list of specific transactions that should be subject to supervision.

There is a difference between RC and concentration risk. This is explained by CEIOPS as "Risk concentration" is a broad term that covers single risk exposures and combinations of risk exposures that may arise in a group. RC within a specific risk category is usually referred to as "concentration risk." For example, concentrated exposures may arise with reference to specific counterparties or specific industry sectors or countries; CEIOPS (2009f33).

The College of Supervisors shall be a permanent platform for cooperation and coordination dedicated to enhance the exchange of information among supervisory authorities involved; Article 248 and CEIOPS (2009d23, 2009f34). It will aim at facilitating exchange of information, views, and assessments among supervisors in order to allow for a more efficient and effective group and solo supervision and timely action. The College of Supervisors will enable supervisors to develop a common understanding of the risk profile of the group as the starting point for a risk-based supervision at both the group level and the solo level.

22.6 ACTIONS TO BE TAKEN

There is a possibility for the supervisor to require capital add-on if they consider the adequacy of the capital requirement insufficient. This is discussed in Section 22.6.1. In the event of exceptional fall in financial markets, there is a possibility to extend the recovery period as discussed in Section 22.6.2 (Pillar II dampener).

22.6.1 Capital Add-On

The main basis for the (draft) final advice is discussed in CEIOPS (2009d19, 2009f29).

22.6.1.1 Extracts ("Recitals") from the FD Preamble

The recitals give us an idea behind the capital add-on text.

(26) The starting point for the adequacy of the quantitative requirements in the insurance sector is the SCR. Supervisory authorities should therefore have the power to impose a capital add-on to the SCR only under exceptional circumstances, in the cases listed in this Directive, following the SRP. The SCR standard formula is intended to reflect the risk profile of most insurance and reinsurance undertakings. However, there may be some cases where the standardized approach does not adequately reflect the very specific risk profile of an undertaking.

(27) The imposition of a capital add-on is exceptional in the sense that it should be used only as a measure of last resort, when other supervisory measures are ineffective or inappropriate. Furthermore, the term exceptional should be understood in the context of the specific situation of each undertaking rather than in relation to the number of capital add-ons imposed in a specific market.

(28) The capital add-on should be retained for as long as the circumstances under which it was imposed are not remedied. In the event of significant deficiencies in the full or partial internal model or significant governance failures, the supervisory authorities should ensure that the undertaking concerned makes every effort to remedy the deficiencies that led to the imposition of the capital add-on. However, where the standardised approach does not adequately reflect the very specific risk profile of an undertaking the capital add-on may remain over consecutive years.

Recital 41 is also of interest for capital add-ons.

22.6.1.2 Reference to the FD

The main reference in the FD is found in Articles 37, 51–52, 119, and 231–232.

22.6.1.3 CEIOPS' Principles for Solo Capital Add-Ons

CEIOPS has set up five principles regarding the capital add-on issues. We briefly mention these principles below. The convergence of supervisory practices and particularly in the case of capital add-ons will be crucial to guarantee a level playing field for undertakings across the EU. Article 51 also prescribes specific disclosure requirements for capital add-ons. The interested reader is referred to CEIOPS (2009d19, 2009f29) for more details.

Principle 1—Objective: Setting a capital add-on is a supervisory power aimed at ensuring an adequate level of SCR, thereby protecting policyholders' interests and preserving a level playing field.

CEIOPS has classified a capital add-on into two types: Capital add-on triggered by a significant deviation from the risk profile embedded in the SCR calculation, either calculated by the standard formula or by an internal model, referred to as a Risk Profile Capital Add-On, and Capital add-on triggered by a significant governance deficiency, which will be referred to as a Governance Capital Add-On.

A *Risk Profile Capital Add-On* aims to ensure that the SCR corresponds to the level defined in the Level 1 text, that is, a confidence level of 99.5% over a one-year period. Setting such a capital add-on is not a substitute for either the development of an adequate (partial) internal model, the improvement of an inadequate (partial) internal model, or any other measure such as using undertaking specific parameters.

A *Governance Capital Add-On* aims to protect policyholders' interests in situations where a significant flaw in an undertaking's system of governance prevents it from being able to properly identify, measure, monitor, manage, or report its risks and where this cannot be remedied within an appropriate time frame. Setting a Governance Capital Add-On is a measure taken to ensure that the SCR is high enough to cover the increased risks arising from the significant governance deficiency. The capital add-on does, however, not absolve the undertaking from complying with governance requirements as specified in the Level 1 text. The undertaking still has to remedy the deficiencies identified.

Principle 2—Due process for setting capital add-ons: The setting of a capital add-on should follow a due process. The supervisory authority should give proper consideration to whether a capital add-on is an adequate supervisory measure, taking into account the position of the undertaking concerned.

Principle 3—Identification: Any deviation by an undertaking from the requirements set out in the Level 1 text and further specified in Level 2 implementing measures should be addressed by the undertaking in order to establish compliance subject to the principle of proportionality. Only significant risk profiles or governance deviations are relevant for the purpose of setting a capital add-on.

Principle 4—Follow-up: The setting and the amount of a capital add-on should be reviewed more frequently than annually if there are indications that the situation that led to the setting of the capital add-on has changed based on valid evidence.

Principle 5—Communication: The disclosure made by the undertaking on the amount and justification of the capital add-on should be compliant with the supervisory decision.

22.6.1.4 CEIOPS' Advice for Group Capital Add-Ons

With necessary adaptation, all five principles for solo capital add-ons also holds for group capital add-ons. At the group level, a risk profile capital add-on relates to a deviation from the risk profile of the group, that is, the group SCR. A governance capital add-on relates to the deficiencies of the group's overall system of governance.

One of the main issues in the Level 2 advice on group solvency is how to capture group-specific risks in the group SCR. Unique risks arise at the level of the group, for example, contagion risk, and those risks are generally not adequately captured by the standard formula, as it is principally designed for a solo undertaking.

If the deduction and aggregation method is applied, any capital add-on applied at the solo level would flow directly through into the aggregated group SCR.

When applying the accounting-consolidated-based method, the group SCR will be calculated based on the consolidated accounts using either the standard formula or an internal model. Then, the group supervisor, in cooperation with the College, will assess whether there is a need for a capital add-on at the group level. In doing that, add-ons existing at the solo level will have to be considered. It implies that the group supervisor will have to check whether the rationale for the solo capital add-ons remains at the group level.

The interested reader is referred to CEIOPS (2009d19, 2009f29) for more details.

22.6.2 Extension of the Recovery Period

The main basis for the (draft) final advice is given by CEIOPS (2009e02, 2010a02). This approach is also called the Pillar 2 dampener.

22.6.2.1 Recital and Reference to the FD

The main recital is (61) given in Section 26.1 and the main reference in the FD is found in Articles 138 and 143.

22.6.2.2 CEIOPS' Proposals

According to the FD the supervisory authority has the possibility to extend the recovery period, in the case of an exceptional fall in financial markets, when an undertaking is no longer complied with the SCR.

The supervisor *shall* require the undertaking to take necessary measures to reestablish the level of EOFs within 6 months from the observed noncompliance with the SCR (or a reduction of its risk profile to ensure the compliance with SCR). The period of six months *may* be extended by 3 months by the supervisor. The periods of six and three months *may* in cases of exceptional fall be extended. This latter flexibility for the supervisor has the goal of not implying measures that could have procyclical effects. Procyclicality refers to the effect on the economic, financial, or insurance cycle as a consequence of the actions of regulation and supervision. It specifically describes a situation where the overall impact of actions caused by regulation increases the severity of these cycles.

CEIOPS (2009e02, 2010a02) proposes that the maximum extension shall be set at 21 months in the event of exceptional falls in financial markets. Hence, the maximum period for recovery will be 30 months. Any such extension will only be granted following an explicit request by the undertaking concerned.

Any decision to permit an extension as well as the duration is at the discretion of the supervisor. All relevant factors have to be considered. The following external and internal factors should be taken into account.

External factors:

- Detrimental impact on policyholders;
- Financial market stability (including systemic risk), in particular the procyclical impact of distressed sales of assets on the financial markets;
- Ability of financial markets to provide extra capital at a reasonable price;
- Availability of an active market and liquidity of the market;
- Availability in financial markets of adequate financial mitigation instruments (e.g., hedges) at a reasonable price;

- Capacity of the reinsurance market to provide reinsurance cover at a reasonable price;
- Anticipated policyholders' behavior.

Internal factors:

- The causes leading to the noncompliance with the SCR;
- Degree of noncompliance with the SCR;
- The composition of OFs held by the undertaking;
- The composition of the undertaking's assets;
- Nature and duration of technical provisions and other liabilities;
- Solutions effectively available to the undertaking;
- Potential availability of support from other group entities (if applicable)
- The size or significance of the undertaking relative to the market, that is, the impact on the market and on the policyholders if the undertaking were to experience severe financial problems; and
- Steps taken by the undertaking to limit the outflow of capital and the deterioration of its solvency situation.
    - These two factors should not be taken into account in the decision procedure.
- The point in time when normal conditions are expected to be reestablished.
- How much time the undertaking needs to resolve the breach of the SCR without negative economic effects on its standing.

The undertaking should report to the supervisor every 3 months on its progress toward reestablishment of compliance with the SCR. According to the FD, the report has to include information about the measures taken by the undertaking as well as the progress made toward reestablishing compliance with the SCR. In case such a report shows that no significant progress has been achieved in reaching the aim of reestablishing compliance, the extension is to be withdrawn; CEIOPS (2009e02, 2010a02).

22.7 REPORTING, DISCLOSURE, AND EXCHANGE OF INFORMATION

The main basis for the (draft) final advice is given by CEIOPS (2009d20). Transparency and accountability, which are important issues for disclosure, is discussed in CEIOPS (2009b09, 2009f10).

22.7.1 Extracts ("Recitals") from the FD Preamble

The recitals give us an idea behind the reporting, disclosure, and exchange of information. Also recital (23) relates to this area.

(39) Provision should be made for exchanges of information between the supervisory authorities and authorities or bodies which, by virtue of their function, help to strengthen the stability of the financial system. It is therefore necessary to specify the conditions under which those exchanges of information should be possible. Moreover, where information may be disclosed only with the express agreement of the supervisory authorities, those authorities should be able, where appropriate, to make their agreement subject to compliance with strict conditions.

(41) The objective of the information and report to be presented in relation to capital add-ons by CEIOPS is not to inhibit their use as permitted under this Directive but to contribute to an ever higher degree of supervisory convergence in the use of capital add-ons between supervisory authorities in the different Member States.

(42) In order to limit the administrative burden and avoid duplication of tasks, supervisory authorities and national statistical authorities should cooperate and exchange information.

(43) For the purposes of strengthening the supervision of insurance and reinsurance undertakings and the protection of policyholders, the statutory auditors within the meaning of Directive 2006/43/EC of the European Parliament and of the Council of 17 May 2006 on statutory audits of annual accounts and consolidated accounts should have a duty to report promptly any facts which are likely to have a serious effect on the financial situation or the administrative organisation of an insurance or a reinsurance undertaking.

Disclosure

(38) In order to guarantee transparency, insurance and reinsurance undertakings should publicly disclose—that is to say, make it available to the public either in printed or electronic form free of charge—at least annually, essential information on their solvency and financial condition. Undertakings should be allowed to disclose publicly additional information on a voluntary basis.

22.7.2 Reference to the FD

The main reference in the FD is found in Articles 35–36, 51, 53–56, 112, and 254, 256. Other relevant articles for the public disclosure of internal models are Articles 120–124 and 126.

Article 35(1) sets out the requirements for the Report to Supervisors, RTS, and Article 51 sets out the requirements for the Solvency and Financial Condition Report, SFCR.

Supervisory disclosure is found in Article 31.

22.7.3 CEIOPS' Proposals

The undertakings and groups shall report to the supervisory authority the RTS and the SFCR on a regular basis.

They should both be stand-alone reports and contain a qualitative report, including quantitative data, where necessary, and quantitative reporting templates; CEIOPS (2009f30).

- The *RTS* is the private report through which undertakings submit regular information to the supervisory authority. The RTS shall enable supervisors to carry out the SRP, and also forms the basis for the supervisory dialogue between undertakings and their supervisory authority. The supervisor shall review the RTS to ensure that the RTS fulfills the established requirements for this document and is consistent with the SFCR.
- The *SFCR* is the public report through which undertakings disclose to the public, including supervisory authorities, information to enable the public to analyze their solvency, and financial condition. The undertaking has responsibility to compile and publish this report. The supervisor shall review the SFCR, using a risk-based approach, to ensure that the SFCR fulfills the established requirements for this document that the information presented in the SFCR is appropriate and consistent with the information provided under the RTS, so that it allows a proper understanding of the solvency and financial condition of the undertaking.

The structures of the SFCR report and the RTS report are given in CEIOPS (2009d20, 2009f30). The reporting requirements being developed will be applicable to *all* undertakings. However, the proportionality principle introduces the following differentiation in the supervisory reporting and public disclosure requirements:

- The detail of information to be received by supervisors will be commensurate with the nature, scale, and complexity of the risks inherent in the business of the undertaking concerned. Undertakings with complicated risk profiles are likely to have more to report and disclose and explain to fulfill supervisory reporting and public disclosure requirements than undertakings with less complex risk profiles;
- Undertakings will not be required to fulfill reporting or disclosure requirements that are not applicable to them. In such cases it will suffice to state that the requirements are not applicable to them; so there is a degree of proportionality inherent in the supervisory reporting and public disclosure requirements; and
- The frequency with which an undertaking has to provide the full qualitative information through the RTS will be linked to the intensity of the SRP. Those undertakings that are not required to submit a full qualitative RTS on an annual basis would be required to submit details of only material changes to the full qualitative information or report that no material changes have occurred.

The undertaking's systems of governance are described in the SFCR. It includes, for example, details of the ORSA, and also on the valuation of assets and liabilities, the structure and amount of OFs (and their quality), and the amount of the SCR and MCR.

The RTS will contain all the regularly reported information necessary for the purposes of supervision, within a private document sent to the supervisory authority.

CEIOPS' discussion also included draft templates for reporting. The process of reporting is summarized in the following table; see CEIOPS (2009f30).

	SFCR	RTS	Quantitative Reporting Templates (to be Included in the SFCR and RTS as CEIOPS' Work Develops)
Frequency	Annual	Full RTS annual for undertakings subject to annual detailed assessment as part of SRP Annual RTS on material changes to the full requirements for undertakings not subject to annual detailed assessment as part of SRP.	Quarterly, and private to the supervisor, and annual.
Submission date	Within 14 weeks of an undertaking's financial year end. Groups: Up to additional 4 weeks.	Within 14 weeks of an undertaking's financial year end. Groups: Up to additional 4 weeks.	Within 4 weeks for quarterly quantitative reporting templates after the quarter end. Within 14 weeks for the full quantitative reporting templates after undertaking's financial year end.
Format	Electronically, following a common structure as developed by CEIOPS.	Electronically, following a common structure as developed by CEIOPS.	Electronically, following a common standardized template format as developed by CEIOPS.
Internal approval by administrative or management body	Yes	Yes	Yes

22.7.3.1 Supervisory Disclosure

Supervisory disclosure aims to make information related to supervision available in a timely manner to all interested parties, including undertakings, brokers and intermediaries, other market participants, other supervisory authorities, and potential policyholders. It has two main objectives:

a. Enhancing the effectiveness of supervision and
b. Helping to foster convergence of supervisory practices and thus promoting a level playing field throughout Europe.

This has been discussed and elaborated in CEIOPS (2009b09, 2009f10).

CHAPTER 23

European Solvency II

Asset Valuation

23.1 EXTRACTS ("RECITALS") FROM THE FRAMEWORK DIRECTIVE PREAMBLE

The recitals from the Framework Directive (FD) enlighten us about the ideas behind the FD. In each of Chapters 22 through 26, we have quoted the recitals that give a background to the topics discussed. Recitals not quoted in these chapters are left to Appendix D.8. The number to the left before the recital is the numbering in the preamble.

The recitals relate to general valuation of both assets and liabilities.

(15) In line with the latest developments in risk management, in the context of the International Association of Insurance Supervisors, the International Accounting Standards Board, and the International Actuarial Association and with recent developments in other financial sectors, an economic risk-based approach should be adopted that provides incentives for insurance and reinsurance undertakings to properly measure and manage their risks. Harmonization should be increased by providing specific rules for the valuation of assets and liabilities, including technical provisions.

(45) The assessment of the financial position of insurance and reinsurance undertakings should rely on sound economic principles and make optimal use of the information provided by financial markets, as well as generally available data on insurance technical risks. In particular, solvency requirements should be based on an economic valuation of the whole balance sheet.

(46) Valuation standards for supervisory purposes should be compatible with international accounting developments, to the extent possible, so as to limit the administrative burden on insurance or reinsurance undertakings.

23.2 REFERENCE TO THE FRAMEWORK DIRECTIVE

The main basis of the (draft) final advice on asset valuation given by CEIOPS (CEIOPS, 2009b10) can be primarily found in Article 75 in the FD. The provisions given in the FD should be read in relation to the recitals given in Section 23.1. The main building block is the market-consistent approach given by the Level 1 text. This advocates an economic balance sheet.

23.3 VALUATION PRINCIPLES

Earlier thoughts, discussions, and calibrations are discussed in Appendix F. CEIOPS has discussed the valuation of assets in CEIOPS (2009b10, 2009f11).

It is assumed that the undertaking will carry on its business as a going concern and not on a "stress scenario" assumption. As IASB has provided principles and guidance for calculating fair value for almost all assets and liabilities, CEIOPS' view is that it is more efficient to refer to IFRS as a proxy for the determination of an economic valuation than developing its own valuation principles. The Level 1 texts largely coincide with the current definition of fair value under IFRS. For greater flexibility, CEIOPS is of the opinion that it is important that the Level 2 implementing measures should only include a reference to the general IFRS framework avoiding references to individual IAS/IFRS. To make it easier for the reader, references are made to different IAS/IFRS in the text.

CEIOPS recommends adopting IFRS, as endorsed within the EU, as a reference framework for building a coherent economic balance sheet reflecting the valuation principles laid down within Solvency II. This implies that additional specifications only need to be provided where the IFRS are noncompatible with Article 75 of the FD. As an example, some items under IFRS should be measured at historical costs. This is typically not compatible with the economic balance sheet approach.

The Level 2 implementing measures include mainly high-level principles. More details will be provided by Level 3 guidance.

For building a Solvency II economic balance sheet using IFRS as a reference, the companies should determine if the figures provide for an economic valuation. If not, they have to adjust the IFRS accounting figures, unless they are of minor importance. The proportionality principle should be taken into account.

If possible, the fair value of assets must be based on a mark-to-market approach. Where it is not possible, mark-to-model procedures should be used. In the latter case the companies need to have an "appropriate degree of qualification" on the selection of the model and parameters used. They also need to have adequate systems and controls that would be sufficient to give both the management and the supervisor confidence in their valuation estimates. CEIOPS also wants the companies to have external value verification, at least every 3 years.

23.4 VALUATION OF CERTAIN ASSETS

23.4.1 Intangible Assets (Including Goodwill)

Goodwill on acquisitions is defined by IFRS 3 and IFRS 4 and refers to an extended presentation for insurance contracts acquired in a business combination or transfer. CEIOPS considers the economic value for solvency purposes to be nil.

The present value of future profits that are not incorporated into the valuation of the technical provisions should be valued at nil.

Intangible assets, other than goodwill, are defined in IAS 38, which is seen as a good proxy if and only if intangible assets can be fair valued according to this IAS. If it is not possible to make a fair valuation, intangible assets should be valued at nil for solvency purposes.

23.4.2 Properties

Property, plant, and equipment are dealt within IAS 16. They have tangible items if they are held for use in the production or supply of goods or services and are expected to be used during more than one period. All these items are accounted for in accordance with IAS 16, except when another accounting standard requires or permits a different treatment: as an example, if they are held for sale according to IFRS 5.

They are recognized as assets, according to IAS 16, if and only if it is probable that future economic benefits associated with the item will flow to the entity and the cost of the item can be measured reliably. Hence spare parts and servicing equipment have to be recognized immediately at profit or loss. Other aspects to be included in the valuation would be, for example, renovations and extensions. According to IAS 16, property, plant, and equipment are initially measured at cost. For subsequent measurements, the undertakings could choose between a *cost model* (cost, less any depreciation and impairment loss) and a *revaluation model* (fair value at the date of revaluation, less any depreciation or impairment; the model can be used if the fair value could be measured reliably).

For solvency purposes CEIOPS considers the revaluation model as a reasonable proxy, provided that the IAS 16 is applied such that the revaluations are made with sufficient regularity.

If they are not measured at economic value, they have to be remeasured at fair value for solvency purposes (applying a revaluation approach).

Investment property is defined in IAS 40 as property held to earn rentals, or for capital appreciation, or for both, rather than for use in the production or supply of goods or services or for administrative purposes or sale in the ordinary course of business. After recognition, the IAS 40 admits cost (initial cost, less depreciation and impairments) or fair value measurement. CEIOPS sees the fair value model of IAS 40 as in line with Article 74 of the FD. Hence any investment properties measured at cost should be revaluated at fair value for solvency purposes.

CEIOPS wants the companies to have independent external valuation or confirmation of revaluation of property as soon as significant changes occur, but at least every 3 years.

23.4.3 Participations

Definitions of elements of participations/associates, subsidiaries and joint ventures, and special purpose vehicles (SPV) are provided in IAS 27, IAS 28, and IAS 31. The valuation and treatment for own funds and groups purposes are not dealt with here.

IAS 27 states that a consolidated financial statement shall include all subsidiaries of the parent. According to IAS 28 and IAS 31, except when classified as held for sale (in accordance with IFRS 5), investments in an associate in a consolidated financial statement shall be accounted as equity. An interest in a jointly controlled entity shall be accounted using the proportionate consolidation approach or the equity approach. Investments in subsidiaries, jointly controlled entities and associates, not classified as held for sale in accordance with IFRS 5, has to be accounted for as cost in accordance with IAS 39.

Under Solvency II, the definition of participations includes investments in associates, subsidiaries, and joint ventures. This definition is for the economic valuation process.

Valuation for Solvency II purposes could be done in different ways: it could depend on whether the companies are listed or unlisted or it could be based on whether the undertaking exercises control or only a significant influence. A third way to valuate participations is the use of a market-consistent methodology.

Investments in participations that are associates and joint ventures shall be valued based on the advice for financial instruments.

If the investments retained interest in an SPV or other investments in SPVs, the valuation for Solvency II purposes should be based on either a mark-to-market or a mark-to-model approach.

23.4.4 Financial Assets

Financial assets are defined according to IAS 39, where it is stated that a company has to recognize a financial asset on its balance sheet when it becomes a party to the contractual provisions of the instrument. According to IAS 39, financial instruments can be split into four categories: financial assets or financial liabilities at fair value (through profit and loss), held-to-maturity investments, loans and receivables, and available-for-sale financial assets.

Valuation for Solvency II purposes should be done at fair value, in accordance with IAS 39.

23.4.5 Other Assets

Contingent assets are defined according to IAS 37 as a possible asset that arises from past events and whose existence will be confirmed only by the occurrence or nonoccurrence of one or more uncertain future events not wholly within the control of the entity. According to IAS 37, contingent assets shall not be recognized in a company's accounts and should not be recognized in its financial statements (as this could result in the recognition of assets that may never be realized). When its realization is virtually certain, it is not a contingent asset anymore and would be recognized.

For Solvency II purposes, it is assumed that the principles established in IAS 37 are reasonably compatible with Article 75 of the FD.

Useful definitions of *deferred tax assets* are given in IAS 12, where it is stated that they are the amounts of income taxes recoverable in future periods in respect of deductible temporary differences, the carryforward of unused tax losses, and the carryforward of unused tax credits. Temporary differences are those between the carrying amount of an asset in the statement of financial position and its tax base.

For Solvency II purposes, the valuation of deferred tax assets shall only be taken into account when they are linked to specific and identifiable assets on the solvency balance sheet. Unused tax losses and tax credits should be valued at nil if there is no link to any specific identifiable asset on the Solvency II balance sheet. Deferred tax assets should not be discounted and should be measured at the tax rates expected to apply when the asset is realized.

CHAPTER 24

European Solvency II Project

Liability Valuation

24.1 EXTRACTS ("RECITALS") FROM THE FRAMEWORK DIRECTIVE PREAMBLE

The recitals to the FD give us the ideas behind the FD. Chapters 22 through 26 quote the recitals that provide the background to the topics discussed. Recitals not quoted in these chapters will be discussed in Appendix D, Section D.8. The number to the left before the recital is the numbering in the preamble.

(53) In order to allow insurance and reinsurance undertakings to meet their commitments towards policyholders and beneficiaries, Member States should require those undertakings to establish adequate technical provisions. The principles and actuarial and statistical methodologies underlying the calculation of those technical provisions should be harmonised throughout the Community in order to achieve better comparability and transparency.

(54) The calculation of technical provisions should be consistent with the valuation of assets and other liabilities, market consistent, and in line with international developments in accounting and supervision.

(55) The value of technical provisions should therefore correspond to the amount an insurance or reinsurance undertaking would have to pay if it transferred its contractual rights and obligations immediately to another undertaking. Consequently, the value of technical provisions should correspond to the amount which another insurance or reinsurance undertaking (the reference undertaking) would be expected to require to take over and fulfill the underlying insurance and reinsurance obligations. The amount of technical provisions should reflect the characteristics of the underlying insurance portfolio. Undertaking-specific information, such as that regarding claims management and expenses, should therefore be used in their calculation only insofar as that information enables insurance and reinsurance undertakings better to reflect the characteristics of the underlying insurance portfolio.

(56) The assumptions made about the reference undertaking assumed to take over and meet the underlying insurance and reinsurance obligations should be harmonised throughout the Community. In particular, the assumptions made about the reference undertaking

that determine whether or not, and if so to what extent, diversification effects should be taken into account in the calculation of the risk margin should be analysed as part of the impact assessment of implementing measures and should then be harmonised at Community level.

(57) For the purpose of calculating technical provisions, it should be possible to apply reasonable interpolations and extrapolations from directly observable market values.

(58) It is necessary that the expected present value of insurance liabilities is calculated on the basis of current and credible information and realistic assumptions, taking account of financial guarantees and options in insurance or reinsurance contracts, to deliver an economic valuation of insurance or reinsurance obligations. The use of effective and harmonised actuarial methodologies should be required.

(59) In order to reflect the specific situation of small and medium-sized undertakings, simplified approaches to the calculation of technical provisions should be provided for.

24.2 REFERENCE TO THE FD

The main legal basis for the (draft) final advice given by CEIOPS on the liability valuation is primarily found in Articles 75 and 76–86 in the FD. The provisions given in the FD should be read in connection with the Recitals given in Section 24.1.

24.3 VALUATION PRINCIPLES

Earlier thoughts, discussions, and calibrations are discussed in Appendix G.

CEIOPS (draft) final advice on the valuation of TPs is given in different consultation papers:

- CEIOPS (2009b01, 2009f01): Methods and statistical techniques for calculating the BE
- CEIOPS (2009b02, 2009f02): Segmentation
- CEIOPS (2009b05, 2009f05): Treatment of future premiums
- CEIOPS (2009b07, 2009f07): Assumptions about future management actions
- CEIOPS (2009d01, 2009f13): Actuarial and statistical methodologies to calculate the BE
- CEIOPS (2009d02, 2009f14): Risk-free interest rate term structure (IRTS)
- CEIOPS (2009d03, 2009f15): Circumstances in which TPs shall be calculated as a whole
- CEIOPS (2009d04, 2009f16): Calculation of the risk margin
- CEIOPS (2009d05, 2009f17): Standards for data quality
- CEIOPS (2009d06, 2009f18): Counterparty default adjustment to recoverables from reinsurance contracts and SPVs
- CEIOPS (2009d07, 2009e14, 2010a11): Simplified methods and techniques to calculate TPs

In general, the liabilities consist of TPs and other liabilities. Other liabilities are discussed in Section 24.6. TPs are built up of a BE and a RM. The FD states that the BE should be equal to the probability weighted average of future CFs taking account of the time value of money, using the relevant risk-free IRTS. This in effect acknowledges that the BE shall allow for uncertainty in the future CFs.

In CEIOPS (2009d03, 2009f15), it is considered that the general rule set out in the FD text is that the TPs should be calculated as the sum of two explicit components that are the BEs plus an appropriate RM. Both components should be valued separately.

The FD concludes that there should be implementing measures laying down the circumstances when the TPs should be calculated as the sum of the BE and a risk margin (RM) or as a whole. The calculation of TPs *as a whole* is only admissible under three indispensable circumstances, which should be assessed in a strict manner:

- The future CFs associated with insurance or reinsurance obligations can be replicated reliably
- This replication shall be provided by financial instruments
- Those financial instruments shall have reliable market values, which are observable

A main conclusion made by CEIOPS after the financial crisis is the necessity of limiting the scope of mark-to-model practices and nonactively traded assets; CEIOPS (2009d03, 2009f15). CEIOPS uses the lessons drawn from the crisis to interpret the expression "financial instruments for which a reliable market value is observable," which is taken from the FD text. It should be understood as financial instruments quoted in "deep, liquid and transparent market," which requires meeting all the following requirements:

- Market participants can rapidly execute large-volume transactions with little impact on the prices of the financial instruments used in the replications
- Current trade and quote information of those prices is readily available to the public
- The properties specified above are expected to be permanent

CEIOPS considers that "future CFs associated with an undertaking's (insurance) obligations" should not be considered as replicable when one or several features of the future CF depend

- *On any type* of biometric development or on the behavior of the policyholder (its expected value, its volatility, or any other feature)
- *On the behaviour* of the policyholder (unless such behavior does not affect the value of the obligation) (its expected value, its volatility, or any other feature)
- *To any extent* on the development of magnitudes internal to the undertakings, such as expenses or acquisition costs
- *To any extent* on the development of magnitudes external to the undertaking for which there are no financial instruments for which reliable market values are observable

CEIOPS illustrates the main case when the possibility to calculate the TP as a whole is met as the case when the (insurance) obligation, according to the clauses of the contract, consists in delivering a portfolio of financial instruments for which a reliable market value is observable or in the portfolio's price at the moment of the payment of the benefit. Residually there could be limited cases of CFs that can be replicated reliably, such as a future fixed benefit in an insurance contract where the policyholder cannot lapse the contract.

In the FD, it is inferred that the *quality of data* used to valuate the TPs should be assessed by scrutinising three criteria:

- *Appropriateness:* Data are considered to be appropriate if it is suitable for the intended purpose and relevant to the portfolio of risks being analyzed (i.e., directly relates to the underlying risk drivers).
- *Completeness:* Data are considered to be complete if it allows for the recognition of all the main homogeneous risk groups within the insurance or reinsurance portfolio.
- *Accuracy:* Data are considered to be accurate if it is free from material mistakes, errors, and omissions. Most of these will be caused by human error or IT failures; thus a particular link exists with operational risk (OR), in particular the systems and processes employed by the company.

Hence, data used to valuate the TPs should be based on data that are considered to be complete, accurate, and appropriate for that purpose. This is discussed in CEIOPS (2009d05, 2009f17). The concept data refer to all information that are directly or indirectly needed in order to carry out a valuation of TPs, in particular enabling the use of appropriate actuarial and statistical methodologies, in line with the underlying obligations, undertaking's specificities, and with the principle of proportionality. Moreover, data comprise numerical, census, or classification information but not qualitative information.

According to CEIOPS, the assessment of the quality of data should in principle be done at the portfolio level, more specifically taking into account the set of available data that are necessary and relevant to carry out the intended analysis. This includes both internal and external information used by the undertaking. The assessment of the accuracy criteria should be carried out at a more granular level, relating to the individual items.

This applies, in particular, when a set of data is used to set a particular assumption. The set of data used for this purpose should be checked for verification of three criteria, to ensure that the assumptions used in the valuation of TPs are as much as possible adequate, up-to-date, prospective, realistic, and credible.

CEIOPS considers that the role of both the auditors and the actuarial function requires that some degree of analysis is performed with regard to the quality of the data, although the focus, the objectives, and the techniques employed for such an assessment will be different. The actuarial function should judge how much credibility should be assigned to historical data and to prospective assumptions. This judgment has to be based on a careful analysis of the underlying liabilities, the company and portfolio's experience, and relevant qualitative information.

24.3.1 Best Estimate

In order to capture the uncertainty in the CFs of the BE, the undertaking should ideally, CEIOPS (2009b01)

- Consider all possible future scenarios
- Estimate the likelihood/probability of each of those scenarios
- Calculate the CFs receivable/payable by the insurer in each of those scenarios
- Discount the projected CFs to reflect the time value of money in each of those scenarios
- Take the probability weighted average of the discounted CFs from each of those scenarios

The standard mentioned above is unlikely to be practical and the undertaking should consider how far the assumptions, underlying the valuation approach, are likely to differ from this ideal. In choosing an appropriate actuarial and statistical method to calculate the BE, the undertaking should consider whether the assumptions underlying the valuation technique appropriately reflect the nature of their obligations and the element of uncertainty inherent in the CFs.

The causes of uncertainty in the CFs that should be allowed for in the application of the valuation technique, may, according to CEIOPS (2009b01, 2009f01), include the following:

- Fluctuation in the timing, frequency, and severity of claim events
- Fluctuation in the period taken to settle claims and/or expenses
- Fluctuation in the amount of expenses
- Changes in the value of an index/market values used to determine claim amounts
- Changes in both entity- and portfolio-specific factors such as legal, social, or economic environmental factors, where relevant. This may include changes as a result of legislation
- Uncertainty in policyholder behavior
- The exercise of discretionary future management actions by the undertaking (which may depend on the above-mentioned causes of uncertainty and also on entity-specific factors)
- Path dependency This is where the CFs depend not only on economic conditions on the CF date but also on economic conditions at previous dates. A CF that has no path dependency can be valued by, for example, using an assumed value of the equity market at a future point in time. However, a CF with path dependency would need additional assumptions as to how the level of the equity market evolved (the equity market's path) over time in order to be valued.

- Interdependency between two or more causes of uncertainty.

Some risk drivers may be largely independent of the other factors that determine the Cfs. Other risk drivers may be heavily influenced by or even determined by several other risk drivers (interdependence). For example, a fall in market values may influence the undertaking's exercise of discretion in future participation, which in turn affects policyholder behavior. Another example would be a change in the legal environment or the onset of a recession that could increase the frequency or severity of nonlife claims.

The valuation technique chosen shall meet the following requirements; CEIOPS (2009f01):

- The insurance undertaking shall be able to demonstrate the appropriateness, including the robustness of the techniques and assumptions used, with regard to the nature, scale, and complexity of risks. In order to meet this requirement, an insurance undertaking shall be able to provide sound rationale for the choice of one technique over other relevant techniques. This also applies to simplified techniques, approximations.
- The insurance undertaking shall assess the degree of judgement required in each method and to what extent the undertaking is able to carry out such a judgement in an objective and verifiable manner according the requirements set out in CEIOPS' advice on actuarial and statistical methodologies to calculate the BE.
- The insurance undertaking shall be able to demonstrate that the valuation technique and the underlying assumptions are realistic and reflect the uncertain nature of the CFs.
- The valuation technique shall be chosen on the basis of the nature of the liability being valued and from the identification of risks that materially affect the underlying CFs.
- The assumptions underlying the valuation technique shall be validated and reviewed by the insurance undertaking.
- The valuation technique and its results shall be capable of being audited.
- If policy data are grouped, the insurance undertaking shall be able to demonstrate that the grouping process appropriately creates homogeneous risk groups that allow for the risk characteristics of the individual policies. This applies to either claims or policy data.
- With regard to the previous points, (i.e., having ensured that the valuation technique is appropriate and robust given the nature, scale, and complexity of the risk), insurance undertakings shall ensure that their capabilities (e.g., actuarial expertise, IT systems) are commensurate with the actuarial and statistical techniques used.

For many types of uncertainty, there are a large variety of possible future scenarios. Actuarial and statistical techniques have been developed to form a practical approach of estimating the value of liabilities, including stochastic simulation, and deterministic and analytical techniques; see CEIOPS (2009b01, 2009f01).

When selecting the valuation technique, insurance undertakings shall consider the following factors and the material impact on the value of the liability, subject to proportionality:

- Whether the CFs are materially path dependent
- Existence of material nonlinear interdependencies between several drivers of uncertainty
- Whether the liability CFs are materially affected by the potential future management actions
- Presence of risks that have a material asymmetric impact on the value of the CFs, in particular if contracts include material-embedded options and guarantees or if there are complex reinsurance contracts in place
- Whether the value of options and guarantees is materially affected by the policyholder behavior assumed in the model
- The availability of relevant data taking into account the requirements on data quality set out in CEIOPS' advice on standards for data quality

Where simulation techniques are used, economic scenario files are a main building block. Market-consistent asset models could produce such scenario files. This calibration relies both on expert judgement and the availability of market data. The application of more sophisticated techniques is limited to cases where sufficiently robust knowledge/data are available. The following areas should be taken into account when considering the advice on the use of simulation techniques:

- *Management actions:* The undertaking shall apply management actions, which are objective, reasonable, and verifiable.
- *Setting assumptions:* The model may require a large number of parameters that a more limited number of (external) people have the experience to calibrate. Although assumptions are based on past experience and current conditions as far as possible, judgement shall be used for some assumptions.
- *Validation:* Due to the additional dimension in the assumption set, it is insufficient to check the result obtained is accurate through a combination of summary statistics, spot checks, and rough estimates. The use of simulation approaches therefore means that the results require different techniques/tools to audit.
- *Interpretation:* With all approaches, interpretation of the results may require a clear understanding of the assumptions underlying the technique where this materially

affects the overall results. With a simulation approach, particular attention shall be paid to the behavior of the asset–liability model in extreme scenarios.

- *Model points:* The undertaking shall measure the potential for additional error and review the grouping accordingly to ensure that important risk characteristics of the portfolio are not neglected.

According to CEIOPS (2009d01, 2009f13), valuation of the TPs requires the analysis of the underlying liabilities and the collection of qualitative and quantitative information. It is a process that requires expert judgement about the credibility to assign to historical data, to what extent one should rely on prospective modeling given the knowledge about experience, and needs to consider the estimation uncertainty. Sometimes there are situations where an assessment needs be done on inadequate and scarce data and on information that could not be treated as reliable. Therefore, the value of TPs should not rely solely on models. It should rely on a variety of techniques including the application of judgement based on sound reasoning and business logic. To be able to produce such judgements, based on sound reasoning and business logic, the valuation process of TPs could not be performed by anyone but require the person with sufficient knowledge how to use actuarial and financial mathematics, understand the nature and complexity of the insurance risk and have adequate experience in performing valuation assessment. Further considerations on the actuarial function are included in CEIOPS (2009b08).

24.3.1.1 CF Projection

In calculating the BE of the TPs, all the cash in-flows and out-flows required to settle the obligations over their lifetime should be used; CEIOPS (2009f13). It should be calculated gross, without deduction of the amounts recoverable from reinsurance contracts and SPVs, and should reflect realistic future demographic, legal, medical, technological, social, and economical developments (including future inflation). Recoverables from reinsurance and SPV contracts are discussed in Section 24.3.1.6. In the case of coinsurance, the CFs of each coinsurer should be calculated as their proportion of the expected CFs without deduction of the amounts recoverable from reinsurance and SPVs.

The time horizon should be set ultimate, that is, the full lifetime of the portfolio of contracts on the date of valuation. The BE should be calculated separately for obligations in different currency. All lists given in this chapter is nonexhaustive.

To determine the BE, the following list of cash in-flows and out-flows should be included.

- Gross cash in-flows
  - Benefits
    - Claims payments
    - Maturity benefits
    - Death benefits
    - Disability benefits

  - Surrender benefits
  - Annuity benefits
- Expenses—Take account of expenses for servicing the obligations over the lifetime thereof:
  - Administrative expenses
  - Investment expenses
  - Claims management expenses and handling expenses
  - Acquisition expenses including commissions that are expected to be incurred in the future
  - The expenses should include both allocated and unallocated expenses
- Other gross CF items
- For nonlife obligations, the undertaking has to allocate expenses between premium provisions and claims provisions (where appropriate)

- Gross cash out-flows
  - Taxation payments that are charged to policyholders

Other different CF features to take into account in the calculation of the BE and its CF are

- Options and guarantees; see Section 24.3.1.2
- Policyholders behavior; see Section 24.3.1.3
- Management actions; see Section 24.3.1.4
- Distribution of extra benefits; see Section 24.3.1.5

Future premiums are related to the settlement of existing obligations and which future premiums relate to the settlement of future obligations. As an obligation always arises from an insurance contract the definition of an existing obligation should be seen as equivalent to the definition of an existing contract. Only CFs that relates to existing obligations should be recognized in the solvency balance sheet. The best estimate calculation should only include future CFs with existing insurance contracts. Recognition and boundaries of existing contracts are discussed in CEIOPS (2009b05, 2009f05).

There are situations where a contract can include a policyholder option to increase the future premiums by means of an option to renew the contract, to extend the insurance coverage to another person, to extend the insurance period, to increase the insurance coverage, or to establish new insurance cover, without the possibility for the insurance undertaking to react.

In CEIOPS view, see CEIOPS (2009f05), if the undertaking *expects a loss* from the additional future premiums, then this loss relates to the existing contract because the obligation to pay the benefits that lead to the loss is already part of the existing contract. On the other hand, if the undertaking *expects a profit* from the additional future premiums, then this profit does not relate to the existing contract as the existing contract does not include a policyholder obligation to pay the future premium.

For the calculation of the BE, the boundaries of an existing insurance contract should be defined as, CEIOPS (2009f05)

a. Where the insurance or reinsurance undertaking has a unilateral right to cancel the contract, a unilateral right to reject the premium, or an unlimited ability to amend the premium or the benefits at some point in the future, any premiums received beyond that point (and any resulting benefit payments to policyholders, expenses, etc.) do not belong to the existing contract.

b. Where the undertaking's right to cancel the contract or to reject the premium or the ability to amend the premium or the benefits relates only to a part of the contract, the same principle as defined in (a) should be applied to the part in question.

c. Future premiums and any resulting benefit payments to policyholders, expenses, and so on which relate to an option or guarantee that provides rights under which the policyholder can renew the contract, or a part of the contract, belonging to the existing contract if, and only if, the inclusion of the renewals *increase* the BE. *The Commission has pointed out an inconsistency of this treatment of future premiums and the economic approach of Solvency II.*

d. The same principle as defined in (c) is applied to options or guarantees that allow the policyholder to extend the insurance coverage to another person, to extend the insurance period, to increase the insurance coverage, or to establish new insurance cover.

e. All other CFs relating to the contract should be included in the calculation of the BE. In particular, future premiums, and any other resulting benefit payments to policyholders, expenses, and so on should be included if their payment by the policyholder is legally enforceable.

Life insurance obligations: As a starting point, CEIOPS (2009d01, 2009f13) assumes that a policy-by-policy approach is used for calculating the BE. If a best estimate element of TPs is negative it should not be set to zero. Reasonable actuarial methods and approximations, to group the policies, for the calculation of the BE may be used. Therefore the segmentation is important; see Section 24.4. No implicit or explicit surrender value floor should be assumed for the valuation of the contracts.

Nonlife insurance obligations: The valuation of BE should be carried out separately for

- *Provisions for claims outstanding:* They relate to the claims events that have occurred before or at the valuation date—whether the claims arising from those events have been reported or not, that is, including IBNR claims. The CFs projected should comprise all future claims payments as well as claims management and claims administration expenses arising from these events.

- *Premium provisions:* They relate to claims events occurring after the valuation date and during the remaining in-force period of policies held by the undertaking. The CF projections should comprise all future claims payments and claims management expenses arising from those events CFs arising from ongoing administration of the in-force policies and expected future premiums stemming from existing policies.

  The BE of premium provisions should be calculated as the expected present value of future in- and out-going CFs, being a combination of, for example,

  - Future premium payments; regarding the boundaries of the future premiums, see CEIOPS (2009b05, 2009f05)
  - CFs resulting from future claims events
  - CFs arising from allocated and unallocated claims management expenses
  - CFs arising from ongoing administration of the in-force policies

If nonlife insurance policies give rise to payment of annuities, whose estimate requires the use of appropriate life actuarial techniques, the provisions for claims outstanding should be carried out separately for annuities and other claims. For premium provisions, a separate calculation of annuity obligations should be performed if a substantial amount of incurred claims will give rise to the payment of annuities.

The principle of substance over form: When discussing valuation techniques for calculating TPs, it is common to refer to a distinction between a valuation based on life techniques and a valuation based on nonlife techniques. The distinctions between life and nonlife techniques are aimed toward the nature of the liabilities (*substance*), which may not necessarily match the legal form (*form*) of the contract that originated the liability. The choice between life or nonlife actuarial methodologies should be based on the nature of the liabilities being valued and from the identification of risks, which materially affect the underlying CFs.

24.3.1.2 Options and Guarantees

Embedded options and guarantees are important components of TPs that need to be continuously monitored by the insurer. The potential for nonlinear behavior, existence of path dependencies, and inherent complexity and uncertainty requires the use of relatively sophisticated valuation methodologies to deliver accurate results; CEIOPS (2009d01, 2009f13).

Contractual options: Defined as the right to change benefits, to be taken at the choice of its policyholder, on terms that are established in advance. Thus, in order to trigger an option, a deliberate decision of its holder is necessary.

Examples of contractual options:

- *Surrender value option:* The policyholder has the right to fully or partially surrender the policy and receive a predefined lump sum amount

- *Paid-up policy option:* The policyholder has the right to stop paying premiums and change the policy to a paid-up status
- *Annuity conversion option:* The policyholder has the right to convert a lump survival benefit into an annuity at a predefined minimum rate of conversion
- *Policy conversion option:* The policyholder has the right to convert from one policy to another at prespecific terms and conditions
- *Extended coverage option:* The policyholder has the right to extend the coverage period at the expiry of the original contract without producing further evidence of health

Financial guarantees: These are present when there is a possibility to pass losses to the insurer or to receive additional benefits as a result of the evolution of financial variables, solely or in conjunction with nonfinancial variables (e.g., investment return of the underlying asset portfolio, performance of indices, etc.). In the case of guarantees, the trigger is generally automatic, the mechanism would be set in the policy's terms and conditions, and thus not dependent of a deliberate decision of the policyholder/beneficiary. In financial terms, a guarantee is linked to option valuation.

The following is a nonexhaustive list of examples of common financial guarantees embedded in life insurance contracts:

- Guaranteed invested capital
- Guaranteed minimum investment return
- Profit sharing (i.e., future discretionary benefits)

Nonfinancial guarantees: A situation where the benefits provided would be driven by the evolution of nonfinancial variables, such as reinstatement premiums in reinsurance, experience adjustments to future premiums following a favorable underwriting history (e.g., guaranteed no claims discount).

The BE of contractual options and financial guarantees

- Must capture the uncertainty of CFs, taking into account the likelihood and severity of outcomes from multiple scenarios combining the relevant risk drivers
- Should reflect both the intrinsic value and the time value

When valuing the BE of contractual options and financial guarantees, the segmentation considered should not inappropriately distort the underlying risks by, for example, forming groups containing policies that are "in the money" and policies that are "out of the money."

24.3.1.3 Policyholders' Behavior

When valuing future CFs, future policyholders' behavior should be taken into account. Policyholders' behavior is taken into account in the valuation of the future CF by, for example, making assumptions about contractual option exercise rates including surrender rates and paid-up rates; CEIOPS (2009d01, 2009f13).

Policyholders' option to

- *Surrender* is often dependent on financial markets and undertaking-specific information, in particular the financial position of the undertaking
- *Lapse* and also in certain cases to surrender are mainly dependent on the change of policyholders' status such as the ability to further pay the premium, employment, divorce, and so on
- *Exercise* other contractual options that are based on the risk drivers that have the potential to materially affect the level of moneyness

24.3.1.4 Management Actions

Future management action may be reflected in the projected CF; CEIOPS (2009d01, 2009f13, 2009b07, 2009f07). For the calculation of the best estimate liability, the assumed future management actions in the future CFs should have the following qualities: objectivity, realism, and verifiability.

24.3.1.5 Distribution of Extra Benefits

Future CFs need to be split into guaranteed and discretionary benefits (bonuses) because the loss-absorbing capacity of TPs is limited by the TPs relating to the future discretionary benefits. The risk mitigation effect provided by future discretionary benefits shall be no higher than the sum of TPs and deferred taxes relating to those future discretionary benefits.

The guaranteed and discretionary benefits/bonuses should take the loss absorbing capacity of TPs into account. It should also be in line with IFRS. CEIOPS (2009d01, 2009f13) defines it as

- *Guaranteed benefit:* This represents the value of future CFs that does not take into account any future declaration of future discretionary bonuses. The CFs take into account only those liabilities to policyholders or beneficiaries to which they are entitled at the valuation date. Guaranteed benefits at the valuation date are those benefits that cannot be reduced whatever the future state of the of the world.
- *Conditional discretionary benefit:* This is a liability based on declaration of future benefits influenced by legal or contractual declarations and performance of the undertaking/fund. It could be linked with IFRS definition of "discretionary participation features" as additional benefits that are contractually based on
  - The performance of a specified pool of contracts or a specified type of contract
  - Realized and/or unrealized investment return on a specified pool of assets held by the issuer; or
  - The profit or loss of the company, fund or other entity that issues the contract

- *Pure discretionary benefit:* This represents the liability based on the declaration of future benefits that are at the discretion of the management. It could be linked to IFRS definition of "discretionary participation features" as additional benefits whose amount or timing is contractually at the discretion of the issuer.

Discretionary benefits: Correspond to the sum of the *conditional discretionary benefit* and *pure discretionary benefit* items. The definitions of conditional discretionary benefit and pure discretionary benefit should not be understood as requirement that they should be valued separately. Only a distinction between guaranteed benefits and discretionary benefits is required.

CEIOPS has observed that Member States differently define guaranteed and discretionary benefits. As an example, some define it as guaranteed what the policyholders are already entitled to at the valuation date, and what they will be entitled to due to contractual or legislative obligations.

Guaranteed benefits at the valuation date are those benefits that cannot be reduced whatever the future state of the world.

24.3.1.6 Recoverables from Reinsurance and SPV Contracts

The BE that corresponds to the probability-weighted average of future CFs, taking account of the time value of money using the relevant risk-free interest rate (see Section 24.3.3), should be calculated gross, without deduction of amounts recoverable from reinsurance contracts, and SPVs; CEIOPS (2009d01, 2009f13).

The amounts recoverable from reinsurance and SPVs contracts should be shown separately, on the asset side of undertakings' balance sheet as "recoverable from reinsurance contracts and SPVs." The calculation of recoverables from SPVs and finite reinsurance should also be done separately. Separate figures for the premium and claims provisions contained within amounts recoverable shall be calculated. In the claims provisions, part of the recoverable should comprise the compensation payments for the claims accounted for the gross claims provisions. All other payments should be considered in the premium provisions part of the recoverable.

If there is a marked divergence between the timing of recoveries and of direct payments is observed, this should be taken into account in the projection of CFs. Where such timing is sufficiently similar to that for direct payments, the undertaking shall have the possibility of using the timing of direct payments.

The amount of recoverables from reinsurance and SPVs contracts should be adjusted in order to take account of expected losses due to counterparty default, whether this arises from insolvency, dispute, or another reason; CEIOPS (2009d01, 2009f13, 2009d06, 2009f18).

In CEIOPS (2009d06, 2009f18), it is stated that the *adjustment for counterparty default* should approximate the expected present value of the LGD of the counterparty, weighted with the PD of the counterparty. Possible default events during the whole run-off period of the recoverables should be taken account of. The LGD is the expected present value of the change in CFs underlying the recoverables, resulting from a default of the counterparty at a certain point in time.

PDs are usually assumed to be nonconstant over time. In this regard it is possible to distinguish between *point-in-time estimates*, which try to determine the current default probability, and *through-the-cycle estimates*, which try to determine a longtime average of the default probability. CEIOPS suggests that point-in-time estimates should be the default approach, but it should be possible to use through-the-cycle estimates if point-in-time estimates cannot be derived in a reliable, objective, and prudent manner or their application would not be in line with the proportionality principle. If point-in-time estimates are used in the assessment, the time dependence of the probabilities should be taken into account. If through-the-cycle estimates are applied, it can usually be assumed that the probability of default does not change during the run-off of the recoverables.

The assessment of the PD should be based on the fact that the probability increases with the time horizon of the assessment, that is, the probability that the counterparty defaults during the next 2 years is higher than the probability of default during the next year.

Usually the probability of default during the following year is known (as an estimate). For example, if this probability is expected to be constant over time, then the probability that the counterparty defaults during year t can be estimated as $\text{PD}_t = \text{PD} \cdot (1 - \text{PD})^{t-1}$.

The *recovery rate* (i.e., the share of the debts that the counterparty will be able to pay) of the counterparty is not easy to determine. CEIOPS has decided that, due to the financial crisis, the highest rate to use if no reliable estimate of the recovery rate is available, should be 50%; CEIOPS (2009f18).

Calculation of the Adjustments

Assume that the recoverables are denoted by R_i in year i, $i = 1, \ldots, n$. The probability of default during year i is denoted by PD_i. If it is assumed that the recovery rate is equal to 45% in case of default during the coming years, and ignoring the time-value of money, then the adjustment for counterparty default could be written as

$$\text{Adj}_{\text{CD}} = -\sum_{i=1}^{n}(1 - 0.45)\text{PD}_i \cdot \left[\sum_{j=i}^{n} R_j\right] = -\sum_{i=1}^{n} 0.55 \cdot \text{PD}_i \cdot \left[\sum_{j=i}^{n} R_j\right],$$

where 55% of the recoveries are lost.

CEIOPS considers that the adjustment for counterparty default should be calculated separately, at least for each LoB and each counterparty, in order to be able to allocate the credit risk to the segments and be able to identify risk concentrations. For the same reason, the adjustment should be calculated separately for nonlife premium provision and nonlife claims provisions.

24.3.1.6.1 Simplifications Simplifications of reinsurance and SPV recoverables are discussed in CEIOPS (2009d07, 2009e14, 2010a11).

The calculation of the amount of recoverable from reinsurance contract of life insurance business should be based on a policy-by-policy approach. For the calculation of the probability-weighted average CF of the recoverables or net payments to the policyholder, the same simplifications as for the calculation of BE of life insurance policies could be applied; see Section 24.5. This is also true in general.

For reinsurance recoverables (RRs) a simplification may be calculated in an indirect manner as the difference between the TPs gross of reinsurance and the TPs net of reinsurance, given that the latter has been adjusted for the expected losses due to the counterparty defaults.

The determination of nonlife RRs and TPs net of reinsurance should be based on gross-to-net techniques, provided that the criteria set out below are met and that further refinements have been made where this would be necessary to ensure the gross-to-net technique applied is proportionate to the risks. gross-to-net techniques, developed in CEIOPS-GC (2008) are briefly discussed below.

Criteria for using gross-to-net techniques:

- For a given homogeneous risk group, the techniques should be applied separately to each of the best estimate components gross of reinsurance, leading to the same best estimate components net of reinsurance as gross of reinsurance.
- The RRs per homogenous risk groups could then be calculated as the sum of the differences between the best estimate TPs gross and net of reinsurance for the premium provision and the claims provisions, respectively.
- When applying the techniques, it should be assessed whether an allowance for the expected counterparty defaults is reflected in a satisfactory manner in the best estimate net of reinsurance.
- The calculation of the best estimate net of reinsurance and of the RRs should be carried out at a level being sufficiently granular with respect to the impact of reinsurance programs within the various homogenous risk groups and with respect to the impact of changes in the reinsurance program over time.
- Where gross-to-net techniques are applied, the following conditions should be met:
  - With respect to premium provisions, the applied techniques should as a minimum distinguish between LoB
  - With respect to claims provisions, the applied techniques should as a minimum distinguish between LoB and—for a given LoB—between the accident years not finally developed

Gross-to-net techniques are developed in CEIOPS-GC (2008). For both the premium provisions and the provisions for claims outstanding, it is assumed at the outset that the Gross-to-net methods should be stipulated for the individual LoB.

The calculation of the gross and net values should be based on the same RM; hence the calculation of the RRs could be further simplified as $\text{RR} = \text{BE}_G - \text{BE}_N$, where BE_G and BE_N are the BEs, gross and net respectively.

Gross-to-Net techniques for premium provisions: For simplicity, it is assumed that the gross-to-net techniques can be represented by a multiplicative factor to the gross provision.

The relationship between the provisions on a gross basis ($PP_{Gross,k}$), the provisions on a net basis ($PP_{Net,k}$), and the gross-to-net factor ($GN_k(ck)$)—for LoB k—can be represented in a somewhat simplified manner as follows

$$PP_{Net,k} = GN_k(c_k) \cdot PP_{Gross,k},$$

where c_k is a parameter-vector representing the relevant characteristics of the reinsurance program covering the CBNI claims related to LoB no. k at the balance sheet day.

Gross-to-net techniques for provisions for outstanding claims: Separate techniques should be stipulated for each accident year not finally developed (for a given LoB). Accordingly, the relationship between the provisions on a gross basis ($PCO_{Gross,k,t}$), the provisions on a net basis ($PCO_{Net,k,t}$), and the Gross-to-Net factor ($GN_{k,t}(c_{k,t})$) for LoB k and accident year t can be represented in a somewhat simplified manner as follows:

$$PCO_{Net,k,t} = GN_{k,t}(c_{k,t}) \cdot PCO_{Gross,k,t}$$

where $c_{k,t}$ is a parameter-vector representing the relevant characteristics of the reinsurance program for this combination of LoB and accident year.

In many cases *the adjustment* for counterparty default will be rather small as compared to the RRs. In order to reduce the burden of the calculation of the adjustment on the undertaking, CEIOPS (2009e14, 2010a11) has provided a simplification for the calculation of the adjustment. The simplification should only be applied if the adjustment can be expected to be small and there are no indications that the simplification formula leads to a significant underestimation. The simplified calculation is defined as

$$Adj_{CD} = -\max\left\{(1 - RR) \cdot BE_{Rec} \cdot Dur_{mod} \cdot \frac{PD}{1 - PD}; 0\right\},$$

where Adj_{CD}: adjustment for counterparty default; RR: recovery rate of the counterparty; BE_{Rec}: best estimate of recoverables not taking account of expected loss due to default of the counterparty; Dur_{mod}: modified duration of the recoverables; and PD: probability of default of the counterparty for the time horizon of 1 year.

24.3.2 Risk Margin

The RM is discussed in general terms in Section 6.4, Sections 8.4 and 8.5, and from an accounting perspective in Section 11.5. Earlier thoughts, discussions, and calibrations are discussed in Appendix G, Section G.6.

According to Article 77, in the FD, the RM should "be such as to ensure that the value of the TPs is equivalent to the amount insurance and reinsurance undertakings would be expected to require taking over and meeting the insurance and reinsurance obligations." The RM should "be calculated by determining the cost of providing an amount of eligible own funds (EOFs) equal to the SCR necessary to support the insurance and reinsurance obligations over the lifetime thereof. The rate used in the determination of the cost of providing that amount of EOFs (CoC rate) shall be the same for all insurance and reinsurance undertakings and shall be reviewed periodically."

According to CEIOPS (draft) final advice, CEIOPS (2009d04, 2009f16), the implementing measure regarding the risk calculations should focus on the following aspects:

- The definition of the reference undertaking; see Section 24.3.2.1
- The calibration of the CoC rate; see Section 24.3.2.2
- The general/overarching methodology for calculating the RM in accordance with the CoC approach; see Section 24.3.2.3
- Simplified methods including the criteria to be fulfilled in order to apply these simplifications; see Section 24.3.2.4

24.3.2.1 Reference Undertaking

The definition of a reference undertaking is a key issue for CEIOPS in the valuation of the RM; CEIOPS (2009d04, 2009f16). Hence they set up 10 assumptions that a reference undertaking must fulfill if the objectives are to be achieved.

Assumption 1: The reference undertaking is not the undertaking itself (i.e., the original undertaking), but another undertaking.

Assumption 2: The reference undertaking is an empty undertaking in the sense that it does not have any insurance or reinsurance obligations and any OFs before the transfer takes place.

Assumption 3: After the transfer, the reference undertaking has eligible OFs corresponding exactly to the amount of SCR that is necessary to support the transferred insurance and reinsurance obligations.

Assumption 4: After the transfer of insurance and reinsurance obligations, the reference undertaking has assets to cover the vest estimate net of reinsurance and SPVs, the RM and the SCR. These assets should be considered to minimize the market risk of the undertaking (for the purposes of calculating the RM). The reference undertaking should only be subject to market risk that is unavoidable in practice.

The unavoidable market risk could be seen as the remaining market risk that is left if the CF is taking over, say 40 years, and it is only possible to match this CF for, say 25 years, with risk-free instruments. Hence the unavoidable market risk is the illiquidity in the remaining part of the CF.

Assumption 5: The SCR of the reference undertaking consists of

a. Underwriting risk with respect to the transferred insurance and reinsurance obligations

b. Counterparty default risk with respect to ceded reinsurance and SPVs

c. Operational risk

d. Unavoidable market risk

It is assumed that the reference undertaking does not carry any counterparty default risk to financial derivatives contracts.

Assumption 6: The loss-absorbing capacity of TPs in the reference undertaking corresponds to those of the original undertaking.
Assumption 7: There is no loss-absorbing capacity of deferred taxes related to the reference undertaking.
Assumption 8: The insurance and reinsurance obligations of each LoB are transferred to the empty reference undertaking in isolation. Hence, no diversification benefit between LoB arises.
Assumption 9: The internal model of the original undertaking (partial or full) can be used to measure the SCR of the reference undertaking to the extent that these models cover at least the risks referred to in Assumption 5 as defined by the standard formula.
Assumption 10: The CoC RM is defined as net of reinsurance and SPVs.

24.3.2.2 CoC Rate

CEIOPS explains the CoC rate as an annual rate applied to a capital requirement in each period. Because the assets covering the capital requirement themselves are assumed to be held in marketable securities, this rate does not account for the total return but merely for the spread over and above the risk free rate.

The RM shall guarantee that sufficient TPs for a transfer are available even in a stressed scenario. Hence, the CoC rate has to be a long-term average rate, reflecting both periods of stability and periods of stress.

CEIOPS suggestion is, based on "available information," that a

Cost-of-Capital rate of at least 6%

reflects the cost of holding an amount of EOFs for an undertaking being capitalized corresponding to a confidence level of 99.5% VaR over a one-year time horizon.

24.3.2.3 General Methodology for Calculation of the Risk Margin

In CEIOPS (2009d04, 2009f16), the general methodology for calculating the RM is laid out. Let CoC_M denote this CoC-based RM.

From *Assumption 8*, regarding the reference undertaking that the RM should be calculated per LoB and that no diversification effects should be taken into account, we obtain

$$\text{CoC}_M = \sum_k \text{CoC}_{M;k}, \tag{24.1}$$

where k denotes the different LoBs, for example, $k = 1, \ldots, L$, and $\text{CoC}_{M;k}$ is the RM, based on the CoC approach, for LoB k.

In line with *Assumptions 2 and 3*, laid down for the reference undertaking, it is empty before the transfer of the obligations takes place, whereas it, after the transfer, has EOFs corresponding exactly to the SCR that is necessary to support the transferred obligations. This means that the reference undertaking at time $t = 0$ (when the transfer takes place) will capitalize itself to the required level of EOFs, that is, for the reference undertaking we obtain

$$\text{EOF}(0) = \sum_k \text{SCR}_k(0), \tag{24.2}$$

where EOF(0) is the EOFs raised by the reference undertaking at time $t = 0$, when the transfer takes place, and $\text{SCR}_k(0)$ is the capital requirement for a given LoB k at time $t = 0$ as calculated for the reference undertaking.

The cost of providing this amount of EOFs equals the CoC rate times the amount.

Assume that the risk-free rate for maturity t is r_t. Let the CoC rate be c. Hence we can write the RM, based on the CoC approach, as

$$\begin{aligned}\text{CoC}_M &= c \cdot \sum_{t=0}^{r} \frac{\text{EOF}(t)}{(1+r_t)^{t+1}} = c \cdot \sum_{t=0}^{r} \sum_k \frac{\text{SCR}_k(t)}{(1+r_t)^{t+1}} = \sum_k \left[c \cdot \sum_{t=0}^{r} \frac{\text{SCR}_k(t)}{(1+r_t)^{t+1}} \right] \\ &= \sum_k \text{CoC}_{M;k}.\end{aligned} \tag{24.3}$$

From Equation 24.3 we see that for each LoB the RM is calculated as the discounted value of providing the future capital requirements. The calculation of the EOFs equal to the SCR is done for all future years (up to r).

If the standard formula is used to calculate the capital requirement for all SCRs, $\text{SCR}_k(t)$, $t = 0, 1, \ldots, r$, for a given LoB k should be calculated as

$$\text{SCR}_k(t) = \text{BSCR}_k(t) + C_{\text{OR},k}(t) - \text{Adj}_k(t),$$

where BSCR is the Basic SCR for the given LoB k and at year t as calculated for the reference undertaking, $C_{\text{OR},k}(t)$ is the capital charge for the OR for LoB k at year t, and $\text{Adj}_k(t)$ is the adjustment for the loss-absorbing capacity of TPs for the given LoB k at year t as calculated for the reference undertaking.

The calculation of the BSCRs should be based on the correlation assumptions laid down in FD; see Chapter 26, and although only the unavoidable market risk and the counterparty default risk with respect to ceded reinsurance is taken into consideration.

With respect to the LoB within nonlife insurance, the RM, as calculated per line of business, should be attached to the overall BE, that is, no split between RMs for premiums provisions and for provisions for claims outstanding should be done.

24.3.2.4 Simplifications

In general, simplifications should be based on the principles laid down in Section 24.5. In light of the considerations made there, this indicates that it may be appropriate to include in Level 2 at least some level of detail concerning the calculation of the RM using simplified techniques. This is discussed in CEIOPS (2009d07, 2009e14, 2010a11).

Based on the general principles, a possible hierarchy regarding the methods to be used for projecting future SCRsper LoB can be summarized as follows:

1. Make a full calculation of all future SCRs without using any simplifications
2. Use simplifications to some or all modules and submodules to be used for future SCR calculations
3. Use simplifications to derive the whole SCR for each future year, for example, by using the proportionality principle
4. Estimate all future SCRs at once, for example, by using the duration approach
5. Approximate the RM by calculating it as a percentage of the BE

This structure implies that the simplifications are getting simpler step by step. To assist undertakings in deciding which simplified methods would be appropriate to determine the RM, each step in this hierarchy should be accompanied with appropriate eligibility criteria based on quality and materiality considerations. When deciding which level of the hierarchy is most appropriate could be structured along the following lines:

1. Start from the bottom
2. If level no. n is appropriate, then use it
3. Otherwise, go upward in the hierarchy to level no. $n-1$

CEIOPS (2009e14, 2010a11) has proposed different simplifications for the SCR calculation.

Simplification 1: Overall SCR for Each Future Year

This simplification takes into account the maturity and the run-off pattern of the obligations net of reinsurance. However, it is based on an assumption that the risk profile linked to the obligations remains unchanged over the years. This could mean that for all years, for example,

- The composition of the subrisks in underwriting risk is the same (all underwriting risks)
- The average credit standing of reinsurers and SPVs is the same (counterparty default risk)
- The unavoidable market risk in relation to the net best estimate is the same (market risk)

- The proportion of reinsurers' and SPVs' share of the obligations is the same (OR)
- The loss absorbing capacity of the TPs in relation to the net best estimate is the same (adjustment)

By using a representative example of a proportional method, the reference undertaking's SCR for a given LoB and year t could be fixed in the following manner:

$$\mathrm{SCR}_{\mathrm{RU},k}(t) = \frac{\mathrm{BE}_{\mathrm{Net},k}(t)}{\mathrm{BE}_{\mathrm{Net},k}(0)} \cdot \mathrm{SCR}_{\mathrm{RU},k}(0), \quad t = 1, 2, 3, \ldots$$

where $\mathrm{SCR}_{\mathrm{RU},k}(0)$: the SCR as calculated at time $t = 0$ for the reference undertaking's portfolio of (re)insurance obligations in an individual LoB k; $\mathrm{BE}_{\mathrm{Net},k}(0)$: the best estimate TPs net of reinsurance as assessed at time $t = 0$ for the undertaking's portfolio of insurance obligations in an individual LoB k; and $\mathrm{BE}_{\mathrm{Net},k}(t)$: the best estimate TPs net of reinsurance as assessed at time t for the undertaking's portfolio of insurance obligations in an individual LoB k.

Simplification 2: Estimation of at Future SCR "at once"

This simplification takes into account the maturity and the run-off pattern of the obligations net of reinsurance. However, it is based on the following simplified assumptions:

- The length of the contracts is one year at the most, that is, there is no premium and catastrophe risk after year 0 (non life underwriting risks)
- The average credit standing of reinsurers and SPVs remains the same over the years (counterparty default risk)
- The modified duration is the same for obligations net and gross of reinsurance (OR and counterparty default risk)

Simplification 2 for Nonlife Insurance

With respect to nonlife insurance, excluding nonlife annuities, the duration approach implies that the RM for an individual LoB ($\mathrm{CoC}_{\mathrm{M},k}$) can be calculated in the following manner:

$$\begin{aligned}\mathrm{CoC}_{M,k} &= \mathrm{CoC} \cdot \left[\frac{\mathrm{SCR}_{\mathrm{RU},k}(0)}{1 + r_1} + \sum_{t>0} \frac{\mathrm{SCR}_{\mathrm{RU},k}(t)}{(1 + r_{t+1})^{t+1}}\right] \\ &\approx \left[\frac{\mathrm{CoC}}{1 + r_1}\right] \cdot \left\{\mathrm{SCR}_{\mathrm{RU},k}(0) + \mathrm{UW}_{\mathrm{RU},k,>0} + \mathrm{OP}_{\mathrm{RU},k,>0} + \mathrm{CD}_{\mathrm{RU},k,>0}\right\},\end{aligned}$$

where $\mathrm{SCR}_{\mathrm{RU},k}(0)$: the SCR as calculated at time $t = 0$ for the reference undertaking's portfolio of (re)insurance obligations; $\mathrm{UW}_{\mathrm{RU},k,>0}$: an approximation of the sum of all future SCRs covering the underwriting risk related to the reference undertaking (as discounted to

$t = 1$); $OP_{RU,k,>0}$: an approximation of the sum of all future SCRs covering the OR related to the reference undertaking (as discounted to $t = 1$); $CD_{RU,k,>0}$: an approximation of the sum of all future SCRs covering the counterparty default risk related to ceded reinsurance and SPVs related to the reference undertaking (as discounted to $t = 1$); and CoC: the CoC rate.

Within this setup, the approximated sums of future SCRs related to each of the three main kinds of risks to be covered by the RM calculations are estimated as follows for the given LoB k

$$\begin{aligned} UW_{RU,k,>0} &= Dur_{Mod,k}(1) \cdot 3 \cdot \sigma_{NL,RR,k} \cdot PCO_{Net,k}(1) \\ OP_{RU,k,>0} &= Dur_{Mod,k}(1) \cdot \lambda \cdot PCO_{Gross,k}(1) \\ CD_{RU,k,>0} &= Dur_{Mod,k}(1) \cdot SCR_{RU,CD,k}(0) \cdot \frac{PCO_{Re,k}(1)}{PCO_{Re,k}(0)}, \end{aligned}$$

where $PCO_{Net,k}(1)$: the best estimate provision for claims outstanding net of reinsurance as calculated at $t = 1$; $PCO_{Gross,k}(1)$: the best estimate provision for claims outstanding gross of reinsurance as calculated at $t = 1$; $PCO_{Re,k}(t)$: reinsurers' share of the best estimate provision for claims outstanding as calculated at $t = 0, 1$; $SCR_{RU,CD,k}(0)$: the capital charge for the counterparty default risk related to ceded reinsurance and SPVs as allocated to the given LoB at $t = 0$; $Dur_{Mod,k}(1)$: the modified duration of reference undertaking's insurance obligations net of reinsurance at $t = 1$; $\sigma_{NL,RR,k}$: the standard deviation for reserve risk as defined in the premiums and reserve risk module of the SCR standard formula; and λ: the percentage to be applied on the best estimate TPs gross of reinsurance as defined in the OR module of the SCR standard formula.

Simplification 2 for Life Insurance

This simplification takes into account the maturity and the run-off pattern of the obligations net of reinsurance. However, it is based on the following simplified assumptions:

- The composition and the proportions of the risks and subrisks do not change over the years (basic SCR)
- The average credit standing of reinsurers and SPVs remains the same over the years (counterparty default risk)
- The modified duration is the same for obligations net and gross of reinsurance (OR, counterparty default risk)
- The unavoidable market risk in relation to the net BE remains the same over the years (market risk)
- The loss-absorbing capacity of the TPs in relation to the net best estimate remains the same over the years (adjustment)

The RM for a given LoB within life insurance ($CoC_{M,k}$) could be calculated according to the following formula:

$$CoC_{M,k} = \left[\frac{CoC}{1+r_1}\right] \cdot Dur_{Mod,k} \cdot SCR_{RU,k}(0),$$

and the variables are defined for nonlife insurance above.

Simplification 2 for a Combination of Nonlife and Life Insurance & Health Insurance
If the LoB comprises both traditional nonlife obligations and obligations in the form of annuities, the RM is calculated by combining the results of a nonlife calculation and a life calculation.

Simplification 3: A simple Method Based on Percentages of the Best Estimate
This is a proxy method. The RM for a given LoB ($CoC_{M,k}$) should be calculated as a percentage of the best estimate TPs net of reinsurance at $t = 0$.

$$CoC_{M,k} = a_k \cdot BE_{Net,k}(0),$$

where $BE_{Net,k}(0)$: the best estimate TPs net of reinsurance as assessed at time $t = 0$ for the undertaking's portfolio of (re)insurance obligations within the given LoB; and α_k: a fixed percentage for the given LoB.

Simplifications for Individual Modules and Submodules
A more sophisticated approach to the simplifications proposed earlier would be to focus on the individual modules or submodules in order to approximate the individual risks and/or subrisks covered by the relevant modules. These are discussed in CEIOPS (2009e14, 2010a11).

For the market margin, the main case of risk is an unavoidable mismatch between the CFs of the insurance liabilities and the financial instruments available to cover these liabilities. Such a mismatch is unavoidable if the maturity of the available financial instruments is lower than the maturity of the insurance liabilities. If such a mismatch exists it usually leads to a capital requirement for interest rate risk under the downward scenario. The focus of the simplification is on this particular kind of market risk. The contribution of this risk to the RM could be approximated for a given LoB as $CoC_{M,k,MR} = CoC \cdot UM_{RU,k,\geqslant 0}$, where the approximated sum of the present and future SCRs covering the unavoidable market risk is calculated as

$$UM_{RU,k,\geqslant 0} = \max\left\{0.5 \cdot BE_{Net,k}(0) \cdot \left[Dur_{Mod,k} - n\right] \cdot \left[Dur_{Mod,k} - n + 1\right] \cdot \Delta r_n; 0\right\},$$

where $BE_{Net,k}(0)$: the best estimate net of reinsurance as assessed at time $t = 0$ for the undertaking's portfolio of insurance liabilities in the given LoB; $Dur_{Mod,k}$: the modified duration of the undertaking's insurance liabilities net of reinsurance in the given LoB at $t = 0$; n: the longest duration of available risk-free financial instruments (or composition of instruments) to cover the insurance liabilities in the given line of business; and Δr_n: the absolute decrease of the risk-free interest rate for maturity n under the downward stress scenario of the interest rate risk submodule.

Quarterly Calculations

Quarterly calculations of the TPs, that is both the best estimate and the RM, are required due to the MCR-calculations. It may also be circumstances where an insurance undertaking will have to calculate its TPs even more frequently.

It seems reasonable to base the simplified calculations of the RM to be carried out during the year on the RM calculated at the beginning of the year. Since no full calculations of the SCR are carried out during the year a likely candidate for these simplifications may be to fix the RM for an individual line of business at a given point in time during the forthcoming year [i.e., $\mathrm{CoC}_{M,k}(t)$] as follows:

$$\mathrm{CoC}_{M,k}(t) = \frac{\mathrm{BE}_{\mathrm{Net},k}(t)}{\mathrm{BE}_{\mathrm{Net},k}(0)}\mathrm{CoC}_{M,k}(0), \quad 0 < t < 1,$$

and $\mathrm{CoC}_{M,k}(0)$ is the RM as calculated at time $t = 0$ for the reference undertaking's portfolio of (re)insurance obligations in an individual line of business, $\mathrm{BE}_{\mathrm{Net},k}(0)$ is the best estimate TPs net of reinsurance as assessed at time $t = 0$ for the reference undertaking's portfolio of (re)insurance obligations in an individual line of business; and $\mathrm{BE}_{\mathrm{Net},k}(t)$ is the best estimate TPs net of reinsurance as assessed at time t for the reference undertaking's portfolio of (re)insurance obligations in an individual line of business.

It is important that this formula is not applied in cases where the BEs are expected to decrease, in relative terms to the business, for example, in cases where significant new business may generate even negative BEs or BEs close to zero. Other simplifications could be possible; see the discussion in CEIOPS (2009e14, 2010a11).

24.3.3 Risk-Free IRTS

One of the main building blocks within Solvency II is the choice of a risk-free IRTS. The methodology and choice have very important consequences for the insurance industry under the Solvency II environment.

For each currency, a relevant risk-free IRTS should be defined using a uniform methodology. It will be used to measure the time value of money of the CFs payable in that currency.

CEIOPS concludes, CEIOPS (2009d02, 2009f14), that for each valuation date, the IRTS should be determined on the basis of market data relevant for that date. For a given currency and valuation date each undertaking should use the same relevant IRTS, which should consist of rates for all relevant maturities.

CEIOPS believes that a relevant IRTS should ideally meet the following *risk-free rate criteria:*

- *No credit risk:* The rates should be free of credit risk
- *Realism:* It should be possible to earn the rates in practice

- *Reliability:* The determination of the rates should be reliable and robust
- *Highly liquid for all maturities:* The rates should be based on financial instruments from deep, liquid, and transparent markets
- *No technical biases:* The rates should have no technical bias; Technical bias for government bonds is illustrated in IAA (2009)

The criteria are also enlarged by the following quality criteria:

- *Available for all relevant currencies:* There could be a negative impact on the level playing field.
- *Proportionality:* The choice of risk-free interest rate should not depend on the proportionality principle, that is, on the nature, scale, and complexity of the risks.

CEIOPS considers that the main options available, *cf*. Section 6.2.2, for the derivation of the risk-free interest rates are

- Government bond rates
- Government bond rates plus an adjustment for technical bias
- Swap rates
- Swap rates minus an adjustment for credit risk

For each currency, CEIOPS proposes a three-stage approach to determine the IRTS:

Stage 1: If government bonds are available that meet the IRTS criteria above, then government bonds should be used to determine the relevant risk-free rate.

Stage 2: If government bonds are available, but they do not meet the IRTS criteria, then they should be adjusted for their deficiencies relating to these criteria. The adjusted rates should approximate government bond rates that, according to CEIOPS, meet the risk-free criteria. The adjusted rates should be used to determine the IRTS.

Stage 3: If government bonds are not available or if government bond rates cannot be adjusted to meet the IRTS criteria for different reasons, other financial instruments should be used to derive the risk-free interest rates. They should be as similar to government bonds as possible and their rates should be adjusted for credit risk and any similar risks.

Illiquidity Premium

The IRTS could be increased to reflect cases where the liability cannot be cancelled at short notice without any penalty. The addition to the IRTS is usually referred to as the *(il)liquidity premium;* see Section 6.2.4 for a general discussion. The inclusion of the illiquidity premium would, according to CEIOPS (2009f14), lead to a significant decrease of TPs and would lower the level of protection of policyholders. Among other reasons against the inclusion of illiquidity premium is that the decreased value of TPs would not be sufficient to meet the insurance obligations because undertakings may not be able to earn the illiquidity premium in a risk-free manner in practice.

CEIOPS (2009f13) discussed the following proposals regarding changes to the SCR standard formula that need to be made to address the risk inherent in the illiquidity premium:

- According to an economic approach and in order to achieve the confidence level set out in the FD text, the downward stress scenario of the interest rate risk submodule should be modified as follows: Liabilities that are discounted with the illiquidity premium should incur an additional stress of the size of the illiquidity premium.
- The underlying idea of the illiquidity premium is that illiquid liabilities can be covered with illiquid bonds and that these assets are not needed to pay other obligations. If the SCR calculation takes diversification between risks relation to illiquid and liquid obligations into account, then this contradicts the underlying idea of the illiquidity premium. Because diversification implies that a loss relating to the liquid obligations can be paid with the assets covering the illiquid obligations. Owing to the illiquidity of these assets, this is not possible. Therefore, no diversification between the risks relation to illiquid and liquid obligations should be taken into account in the calculation of the SCR standard formula.
- The allowance for using an illiquidity premium would increase the amount of basic OFs, being necessary to assess the quality of such increase and its relevant tier.

The IRTS for Euro and Other Currencies

The European Central Bank, ECB, determines an IRTS for the euro on a regularly basis. The term "structure" is derived from AAA-rated bonds issued in euro by a euro area central government. CEIOPS believes that this government bond term "structure" satisfies all of the risk-free rate criteria and therefore is the most appropriate risk-free term structure for the euro. The three-stage process described above should be seen as the benchmark for other currencies.

Liquidity and Long Maturities

Any appropriate IRTS is necessarily constructed from a finite number of data points of sufficient liquidity. Hence, we need to *interpolate* between these data points and *extrapolate* beyond the last available data point of sufficient liquidity. The discount factor increases with the time to maturity. This means that any extrapolation of the risk-free curve beyond the last available data point significantly impacts the present value of long-term insurance liabilities. Therefore, the technique of extrapolation needs to adhere to the desired risk-free criteria set out in this advice, with the exception of liquidity.

CEIOPS (2009f14) discusses four techniques that could be used for extrapolation:

- *Simple extrapolation techniques:* They require no deeper analysis of the fundamentals or shape of the curve. In its purest form, the simple extrapolation technique assumes that the final liquid data point is also the long-term interest rate level. From the final liquid point onward, the curve is, therefore, a horizontal line. It is easy, objective, robust, and reliable, but it has disadvantages. The extrapolation curve is dependent on the final data point. This could be handled by using an extremely liquid data point.

- *Macroeconomic extrapolation techniques:* These techniques involve identifying, for example, by the use of economic theory, a long-term equilibrium interest rate and to interpolate in-between the last available data point and the long-term interest rate. Two such approaches are discussed in Section 6.2.3. The techniques are stable in the long end of the curve. The long-term level is based on market consistency if a thorough and well-founded macroeconomic analysis is applied and the equilibrium level will be updated as new information becomes available.

- *Parameterization techniques:* They emphasize smoothing and provide an objective construction of the term "structure" if the parameterization technique is fixed. They could be based on economic assumptions. This category of extrapolation techniques is the one most used currently in market practice. There are many types of parameterization techniques: for example, constant forward rates, the Svensson method, the one-factor, or Vasicek class of models.

- *Constant or variable spread methods:* They are alternative methods for non-Euro currencies: first an appropriate extrapolation technique for the Euro is defined; then the rates for the other currencies are extrapolated by using the Euro curve plus:
    - In the case of a constant spread method, the constant spread between the Euro and the relevant currency for the last available liquid data point of the relevant currency.
    - In the case of a variable spread, the spreads might be derived by fitting a curve to the spreads observed in the nonextrapolated part of the curve.

CEIOPS (2009f14) recognizes the importance of the choice of the extrapolation technique and thus does not prescribe any method for extrapolating the interest rate curve now. Instead, during Level 3 process, CEIOPS will develop a set of principles for the choice of an appropriate extrapolation method and will, based on these principles, choose for each currency the method deemed to be most appropriate.

In the late 2009 CEIOPS sets up a task force, with representatives from stakeholders (e.g., CEA, Groupe Consultatif), to discuss the illiquidity premium, macroeconomic extrapolation, and the choice of risk-free interest rates. The proposals were given to the European Commission for Level 2 implementing measures.

24.4 SEGMENTATION

Undertakings are not necessarily required to use the same segmentation for the purposes of determining best estimate, RM, SCR, MCR, and statutory reporting. The segmentation used for different purposes should depend upon what is best for that purposes; CEIOPS (2009b02, 2009f02).

Due to the diversity of products sold and because undertakings have the best understanding of their business and how to segment it, the advice given by CEIOPS are only with regard to the LoB that are *prescribed*. Undertakings are required to further segment their business in the manner needed to derive appropriate assumptions for the calculation of the BE. They should segment their obligations into LoB as defined below. The principle of substance over form should be followed in determining how contracts with obligations from different LoBs should be treated.

Where a contract covers risks across nonlife and life insurance, these contracts should be unbundled into their life and nonlife parts.

In general, business is managed in more granular homogeneous risk groups than the minimum segmentation presented here; CEIOPS (2009f02). A homogeneous risk group is defined by CEIOPS as a set of insurance obligations that are managed together and that have similar risk characteristics in terms of, for example, underwriting policy, claims settlement patterns, risk profile of policyholders, likely policyholder behavior, product features (including guarantees), future management actions, and expense structure. The risks in each group should be sufficiently similar to allow for a reliable valuation of TPs (including a meaningful statistical analysis). The classification is undertaking-specific.

Unbundling of insurance obligations is of major importance to achieve an appropriate and suitable segmentation for the assessment of TPs.

In the case of cross-border activities, risks belonging to the same LoB may have substantially different characteristics, for example, due to different local regulations, market practices, or social and economic conditions. Therefore, in the case of cross-border activities, the undertaking shall first segment its insurance obligations by country and then according to the requirements of this advice.

24.4.1 Segmentation for Life TP

As stated above, the starting point in calculation of life TP should be a policy-by-policy approach. It is also possible to group the policies into homogeneous risk groups, but a contract covering life insurance business should always be unbundled according to the four top-level segmentations defined below. Hence the homogeneous risk groups used should be unbundled.

With regard to the second level of segmentation, unbundling should be applied to life insurance contracts where those contracts

- Cover a combination of risks relating to different LoBs
- Could be constructed as stand-alone contracts covering each of the different risks

Unbundling may not be required where only one of the risks covered by a contract is material. In this case, the contract may be allocated according to the major risk driver. Note that the principle of substance over form should be applied in order to determine how each of the unbundled components of a given contract should be allocated to different LoBs.

Life insurance and reinsurance business shall be segmented into 16 LoBs defined as follows.

Level 1:

- Contracts with profit participation clauses
- Contract where policyholder bears the investment risk
- Other contracts without profit participations clauses
- Accepted reinsurance

At Level 2, they should be further segmented into

- Contracts where the main risk driver is *death*
- Contracts where the main risk driver is *survival*
- Contracts where the main risk driver is *disability/morbidity risk*
- *Savings contracts*, that is, contracts that resemble financial products providing no or negligible insurance protection relative to the aggregated risk profile

24.4.2 Segmentation for Nonlife TPs

Where a contract covers risks across different nonlife insurance LoBs, these contracts should be unbundled into the appropriate LoB. Subject to the principle of proportionality, contracts covering risks across different nonlife insurance with one major risk driver might not require unbundling, but might be allocated according to the major risk driver.

CEIOPS proposes the following LoBs for accepted nonlife insurance:

- *Accident* which includes obligations caused by accident or misadventure but excludes obligations considered as workers' compensation insurance
- *Sickness* which includes obligations caused by illness, but excludes obligations considered as workers' compensation insurance
- *Workers' compensation* which includes obligations covered with workers' compensation insurance that insures accident at work, industrial injury and occupational diseases
- *Motor vehicle liability* which covers all liabilities arising out of the use of motor vehicles operating on the land including carrier's liability
- *Motor, other classes* which covers all damage to or loss of land motor vehicles, land vehicles other than motor vehicles and railway rolling stock
- *Marine, aviation, and transport* which covers all damages or loss to river, canal, lake and sea vessels, aircraft, and damage to or loss of goods in transit or baggage, irrespective of the form of transport
- *Fire and other damages* which cover all damage to or loss of property other than motor, marine aviation, and transport due to fire, explosion, natural forces, including storm, hail or frost, nuclear energy, land subsidence, and any event such as theft

- *Third-party liability* which covers all liabilities arising out of use of aircraft, ships, vessels, or boats, including carrier's liability and all other liabilities
- *Credit and suretyship* which covers insolvency, export credit, instalment credit, mortgages, agricultural credit, and direct and indirect suretyship
- *Legal expenses* which covers legal expenses and cost of litigation
- *Assistance* which covers assistance for persons who get into difficulties while traveling, while away from home, or while away from their habitual residence
- *Miscellaneous nonlife insurance* which covers employment risk, insufficiency of income, bad weather, loss of benefits, continuing general expenses, unforeseen trading expenses, loss of market value, loss of rent or revenue, indirect trading losses other than those mentioned before, and other financial losses (not trading), as well as any other risk of nonlife insurance business not covered by the LoBs mentioned before

Proportional nonlife reinsurance should be segmented separately according to the nonlife insurance segmentation described above.

Nonproportional nonlife treaty and nonproportional nonlife facultative reinsurance shall be split into

- Property business
- Casualty business
- Marine aviation and transport business

24.4.3 Segmentation for Health TPs

Health insurance obligations shall be segmented into

- Health insurance obligations pursued on a similar technical basis to that of life
- Insurance (*SLT Health*)
- Health insurance obligations pursued on a similar technical basis to that of nonlife insurance (*Non-SLT Health*)

SLT Health obligations should be further segmented according to the segmentation for life insurance obligations described above and Non-SLT Health obligations according to the segmentation for nonlife insurance obligations described above (accident, sickness, and workers' compensation).

24.5 SIMPLIFICATIONS AND PROXIES

Earlier thoughts, discussions, and calibrations are discussed in Appendix G, Section G.5. The (draft) final advice is given by CEIOPS (2009d07, 2009e14, 2010a11).

Solvency II is based on a principles-based approach to the valuation of Tps, meaning that the regulatory requirements relating to the valuation process would generally not prescribe any specific approaches to carrying out the valuation. Instead, there will typically be a range of different approaches that are available to the undertaking, which then has to select a valuation methodology that is appropriate with regard to valuation principles established. The principle of proportionality, see Section 22.2, requires that the undertaking should be allowed to choose and apply a valuation method which is

- Suitable to achieve the objective of deriving a market-consistent valuation according to the Solvency II principles; but
- Not more sophisticated than is needed in order to reach this objective

Thus, the valuation method chosen should be compatible with the FD valuation principles and should be proportionate to the nature, scale, and complexity of the risks. If simplified approaches are used in the valuation of the TPs this could introduce additional uncertainty or model error. The degree of such model error in the measurement of TPs is linked to the reliability and suitability of the valuation.

Simplifications

There will be a range of different valuation methods available for the undertaking, differing in their degree of complexity and sophistication. Following the proportionality principle will enable the undertaking to simplify a given valuation method in case where the simplified method is still proportionate to the underlying risks. In this case, the term "simplified method" would refer to a situation where a specific valuation technique has been simplified in line with the proportionality principle. The term could also be used to refer to a valuation method that is considered to be simpler than more complex "commonly used" methods. There will be no categorization of simplifications on Level 2.

Proxies

Under certain circumstances it will be unavoidable for the undertaking to only have insufficient company-specific data of appropriate quality to apply a reliable statistical actuarial method for the determination of TPs. It is therefore important to develop valuation techniques that would substitute a lack of company-specific data by, for example, using external information. In the FD text, such techniques are referred to as approximations.

Solvency II envisages a principle-based approach to the valuation of TPs. Hence, the regulatory requirements relating to the valuation process would generally not prescribe any specific approaches to carrying out the valuation. This means that the undertaking should be flexible to reflect the specificities of its business and risk profile, taking into account the principle of proportionality as laid out in Section 22.2. Supervisory guidelines on Level 3 would be developed. CEIOPS also expects that national and international associations of the actuarial profession will issue guidelines on technical issues, including the valuation of TPs.

CEIOPS (2009d07) discussed when it would be appropriate to introduce external thresholds guiding the use of simplified methods for the valuation of the TPs. Any such threshold would be regarded as a cutoff point below which it would be regarded as justifiable to use simplifications. Thresholds would either apply to the scale of the undertaking risks or the degree of model error inherent in the valuation methods. It would also be useful to distinguish between thresholds that broadly apply to all methods, or a specific class of methods, or those that applies to individual methods. This view gives rise to the following threshold table:

	Relating to	
Thresholds	**Scale of Risk**	**Model Error**
Broadly applied	*Type 1*	*Type 2*
Individually applied	*Type 3*	*Type 4*

The usual interpretation of thresholds as providing a cutoff point below which it would be regarded as justifiable to use simplified valuation techniques could be problematic. Hence, CEIOPS considers that rather than as a criterion for the allowance for specific simplified techniques it may therefore be more appropriate to use thresholds as a criterion for their rejection, which means that where the threshold is exceeded it would be considered that the degree of model error in the calculation is material, so that the use of the simplified method would not seem appropriate. However, where the threshold is not exceeded, the undertaking would need to conduct further analysis and assessment before it can be decided whether an application of the simplified method would be proportionate with regard to its risk profile.

Simplifications for the RRs are discussed in Section 24.3.1.6.1 and for RMs in Section 24.3.2.4.

The following two chapters discuss simplified methods and techniques that could be used with respect to the calculation of BEs of nonlife and life insurance business, and also sets out the circumstances in which use of such simplifications could be considered to be appropriate; see CEIOPS (2009e14, 2010a11).

24.5.1 Nonlife Insurance Specific

Simplification 1: Outstanding Reported Claims Provisions

This simplification applies to the calculation of the BE of reported claims by means of considering the number of claims reported and the average cost thereof. Therefore a simplification is applicable when it does not deliver material model error in the estimate of frequency, severity, and its combination.

$$\textit{Calculation:}\ \sum_{i=1}^{n}[N_i \cdot A_i] - P_i,$$

where N_i: number of claims reported, incurred in year i, known; A_i: average cost of claims closed in year i, unknown but using the average claims cost, multiplied by a factor taking care of future inflation and discounting; and P_i: payments for claims incurred in year i, known.

Simplification 2: Outstanding Reported Claims Provisions

This approach consists in the simple sum of estimates of each claim reported at the date of reference of the valuation. The allowance of a simplified method based on a "case-by-case approach" should be assessed carefully according to the features of the claims portfolio and the undertaking internal structure and capabilities.

Calculation: This method should start estimating each individual provision for a single claim upon up-to-date and credible information and realistic assumptions. Furthermore

- This estimate should take account of future inflation according a reliable forecast of the time-pattern of the payments
- The future inflation rates should be market consistent and suitable for each LoB and company
- Individual valuations should be revised as information is improved
- Furthermore, where back testing evidences a systematic bias in the valuation, this should be offset with an appropriate adjustment according the experience gained with claims settlement in previous years and the expected future deviations
- Undertakings should complete the valuation resulting from this method with an *incurred but not reported* (IBNR) provision and an *unallocated loss adjustment expenses* (ULAE) provision

Simplification 3: IBNR

This simplification applies to the calculation of the BE of IBNR claims by means of an estimation of the number of claims that would expected to be declared in the following years and the cost thereof.

Calculation: The final estimate of this TPis derived from the following expression, where just for illustrative purposes a three-year period of observation has been considered (the adaptation of the formula for longer series is immediate):

$$\text{IBNR}_t = C_t \cdot N_t = C_t \cdot R_t \cdot \text{Av}_t,$$

where C_t is the average cost of IBNR claims, after taking into account inflation and discounting. This cost should be based on the historical average cost of claims reported in the relevant accident year. Since a part of the overall cost of claims comes from provisions, a correction for the possible bias should be applied.

$$N_t = R_t \cdot \text{Av},$$

$$\text{where Av} = \frac{N_{t-1}/p_1 + N_{t-2}/p_2 + N_{t-3}}{R_{t-1} + R_{t-2} + R_{t-3}} \text{(3 years is only for illustration)}$$

N_{t-i}: number of claims IBNR at the end of the year $t - i$, independently of the accident year (to assess the number of IBNR claims all the information known by the undertaking till the end of the year t should be included)

p_1: percentage of IBNR claims at the end of year $t-3$ that have been reported during the year $t-2$

p_2: percentage of IBNR claims at the end of year $t-3$ that have been reported during the years $t-2$ and $t-1$

R_{t-i}: claims reported in year t, independently of accident year

Simplification 4: IBNR

This simplification should apply only when it is not possible to apply reliably Simplification 3. In this simplification, the BE of nonreported claims (IBNR) is estimated as a percentage of the provision for reported outstanding claims.

$$\textit{Calculation: } \mathrm{IBNR}_k = \mathrm{factor}_k * \mathrm{PCO}_{\mathrm{rep},k},$$

where $\mathrm{PCO}_{\mathrm{rep},k}$: provision for reported claims outstanding and factor_k: factor specific for each LoB k.

Simplification 5: ULAE

This simplification estimates the provision for claims settlement expenses (ULAE) as a percentage of the claims provision. It should be applied to each LoB k.

$$\textit{Calculation: } \mathrm{ULAE}_k = R_k \cdot \left[\mathrm{IBNR}_k + a \cdot \mathrm{PCO}_{\mathrm{rep},k}\right],$$

where R_k: simple average of R_i (e.g., over the last two exercises) and R_i= Expenses/(gross claims+subrogations); IBNR_k: provision for IBNR; $\mathrm{PCO}_{\mathrm{rep},k}$: provision for reported claims outstanding; and a: percentage of claim provisions (i.e., set as 50% in QIS4).

Simplification 6: Premium Provision

This simplification estimates the BE of the premium provision when the undertaking is not able to derive a reliable estimate of the expected future claims and expenses derived from the business in force. It should be applied to each LoB k.

$$\begin{aligned}\textit{Calculation: } & (\mathrm{UP}_{\mathrm{Pro-rate}} + \mathrm{Adj})/(1 + \mathrm{RF}_{1y}/3) \cdot \mathrm{BE} \\ & = \{\text{Provision for unearned premiums} + \text{Provisions for unexpired risks}\}/(1 + RF_{1y}/3),\end{aligned}$$

where RF_{1y} is the risk-free interest rate 1 year term.

24.5.2 Life Insurance Specific

Simplifications—Biometric Risk Factors

Biometric risk factors (RFs) are underwriting risks covering any of the risks related to human life such as mortality/longevity rate, morbidity rate, and disability rate. The possible simplifications are to

- Neglect the expected future changes in biometric RFs, using static mortality, and morbidity/disability tables instead of dynamics ones

- Assume independency between biometric RFs (i.e., between mortality rates and morbidity/disability rates)
- Use cohort or period data to analyze biometric RFs;
- Apply current tables in use adjusted by suitable multiplier function. The construction of reliable mortality, morbidity/disability tables and the modeling of trends could be based on current (industry standard or other) tables in use adjusted by suitable multiplier function. Industry-wide and other public data and forecasts should provide useful benchmarks for suitable multiplier functions.

An undertaking's selection of technique will depend on materiality and different complexities will be appropriate for different types of product.

Simplifications—Surrender Options

The surrender option gives the policyholders the right to terminate the contracts before maturity and to receive the surrender value. The surrender value is commonly predetermined according to some principles. The surrender option is a very important element for undertakings and should be taken into account when valuing the TPs. It could have significant financial effect for instance on uncharged expenses and uncharged costs for options and guarantees.

Different possible simplification approaches are possible, for example,

- Assume that surrenders occur independently of financial/economic factors
- Assume that surrenders occur independently of biometric factors
- Assume independency in relation to management actions
- Assume that surrenders occur independently of the undertaking-specific information
- Use a table of surrender rates that are differentiated by factors such as age, time since policy inception, product type, and so on
- Model the surrender as a hazard process either with a nonconstant or constant intensity

For with-profit contracts, the surrender option and the minimum guarantees are clearly dependent. Furthermore, management actions will also have a significant impact on the surrender options that might not easily be captured in a closed formula.

Simplifications—Financial Options and Guarantees

Life insurance contracts have usually implicitly or explicitly built in different kinds of financial options or guarantees. If they are nonhedgeable, a mark-to-model approach could be used (deterministic approach, simulation techniques, or closed forms such as Black–Scholes formula.

Simplifications—Investment Guarantees

Some unit-linked products guarantee a minimum benefit at maturity in absolute term or as an annual constant guaranteed rate of return at the issue of the contract. At maturity, the policyholder will receive an amount corresponding to the index but not less than the guaranteed amount.

The random payout at maturity $\tilde{V}(T)$is given by the formula

$$\tilde{V}(T) = \max\left\{\tilde{S}(T); G(T)\right\} = \begin{cases} G(T) + \max\left\{\tilde{S}(T) - G(T); 0\right\} & \text{Call option} \\ \tilde{S}(T) + \max\left\{G(T) - \tilde{S}(T); 0\right\} & \text{Put option} \end{cases},$$

where $\tilde{S}(T)$ is the random account value at time T and $G(T)$ the guaranteed payment at maturity. The time value of investment guarantees IG assuming put–call parity for European options (assuming thus implicitly the Black–Scholes framework), using risk-free discount rate and taking the expectation of the random payout at maturity with respect to the risk-neutral measure, can be written as

$$\text{IG} = \begin{cases} \text{Call option} = \text{Guarantee} + \text{Intrinsic value of extra benefits} + \text{Option Time Value} \\ \text{Put option} = \text{Underlying Assets} + \text{Intrinsic value of guarantees} \\ \qquad\qquad + \text{Option Time value} \end{cases}$$

The time value of the investment guarantee shows the expected amount that should be held in addition to the underlying assets to be able to deliver the benefits due to the investment guarantee.

In the **call option** approach, the *Intrinsic value of extra benefits* corresponds to the amount the call option is in-the-money if it would be exercised immediately and the *Option Time Value* captures the potential to receive further extra benefits in the future due to the random fluctuations of the underlying assets.

In the **put option** approach, the *Intrinsic value of guarantee* corresponds to the amount the guarantee is in-the-money if it would be exercised immediately and the *Option time value* captures the potential for the cost to change in value (guarantee to bite further) in the future, as the guarantee move (related to the variability of the underlying assets) into or out-of-the-money.

Possible simplifications for investment guarantees are, for example, to

- Assume nonpath dependency in relation to management actions, regular premiums, and cost deductions (e.g., management charges, etc.)
- Use representative deterministic assumptions of the possible outcomes for determining the intrinsic values of extra benefits
- Assume deterministic scenarios for future premiums (when applicable), mortality rates, expenses, surrender rates, and so on
- Apply formulaic simplified approach for the time values if they are not considered to be material
- Apply stochastic simulation techniques to group of contracts instead of individual policies

Simplifications—Other Options and Guarantees

Commonly life insurance contracts include many types of options and guarantees. Hence it is impossible to give detailed valuation approaches that would be suitable for all possible options and guarantees. As an interim approach, one could ignore those options and guarantees that are not material. *Ad hoc* valuation should consist of the following steps:

1. Analyze the characteristics of the option or the guarantee and how it would probably effect the CFs
2. Analyze the amount the option or the guarantee is expected to be currently in-the-money
3. Determine how much the cost of the option or the guarantees is expected to vary as the time passes
4. Estimate the probability that the cost of the option or the guarantee would become more costly/less costly in the future

Simplifications—Distribution of Future Discretionary Benefits

The management discretion and the wording of insurance contracts have a large influence on the valuation of TPs for with-profit business. An accurate assessment and a detailed documentation of the mechanism for distribution of extra benefits form the cornerstones of the valuations of extra benefits. Possible simplifications for distribution of extra benefits are

- Assume that economic conditions will follow a certain pattern, not necessarily stochastic, appropriately assessed
- Assume that the business mix of undertaking's portfolio will follow a certain pattern, not necessarily stochastic, appropriately assessed

Simplifications—Expenses and Other Charges

Undertakings shall take into account, among others, all expenses that will be incurred in servicing insurance and reinsurance obligations when calculating TPs. The estimation of the best estimate assumptions for expenses should be based on the analysis of the undertaking's own experience. The possible simplification for expenses is to use an assumption built on simple models using information from current and past expense loadings to project future expense loadings, including inflation.

Simplifications—Other Charges and Issues

Simplifications for charges for embedded options could be based on assumptions such as the other charges are a constant share of extra benefits or a constant charge from the policy fund.

Future CFs should take into account the time value of money. Possible simplifications for the payments to the policyholders and beneficiaries is to assume that

- The projection period is one year and that
- CFs to the policyholders occur either at the end of the year or in the middle of the year

Possible simplification for the payments of the premiums are to assume that future premiums are paid independently of the financial markets and undertakings specific information or alternatively.

Possible simplifications in relation to fund/account value projections (which is important for valuing financial options and guarantees) are to

- Group assets with similar features/use representative assets or indexes
- Assume independency between assets, for instance, between equity rate of return and interest rate

24.6 OTHER LIABILITIES

The (draft) final advice is discussed in CEIOPS (2009b10, 2009f11).

Contingent liabilities are defined in IAS 37 as

a. A possible obligation that arises from past events and whose existence will be confirmed only by the occurrence or nonoccurrence of one or more uncertain future events not wholly within the control of the entity; or

b. A present obligation that arises from past events but is not recognized because (1) it is not probable that an outflow of resources embodying economic benefits will be required to settle the obligation; or (2) the amount of the obligation cannot be measured with sufficient reliability.

According to IAS 37, contingent liabilities shall not be recognized in a company's accounts. For Solvency II purposes, CEIOPS believes that the principles established in IAS 37 are reasonably compatible with Article 74 of the FD. If there is any probability of an outflow of future economic benefits, the company shall reassess that responsibility and recognize a liability in accordance.

Useful definitions of *deferred tax liabilities* are given in IAS 12, where it is stated that they are the amounts of income taxes payable in future periods with respect to taxable temporary differences, that is, between the carrying amount of a liability in the statement of financial position and its tax base.

For Solvency II purposes, the valuation of deferred tax liabilities shall only be taken into account when they are linked to specific and identifiable assets on the solvency balance sheet.

Unused tax losses and tax credits should be valued at nil if there is no link to any specific identifiable liability on the Solvency II balance sheet.

Deferred tax liabilities should not be discounted and should be measured at the tax rates expected to apply when the liability is settled.

Other financial liabilities and amounts payables are, according to IAS 39, only recognized when an undertaking becomes a party to the contractual provisions of the instrument. On initial recognition, they are measured at fair value through profit or loss. After initial recognition, they are measured at amortized cost using an effective interest method, except for

- Financial liabilities at fair value through profit or loss
- Financial liabilities arising from a transfer of a financial asset not qualifying for recognition or when continuing involvement approach applies
- Financial guarantee contracts (measured at the highest of the amount determined in accordance with IAS 37 and initially recognized less cumulative amortization recognized in accordance with IAS 18)
- Commitments to provide a loan at a below-market interest rate (measured at the highest of the amount determined in accordance with IAS 37 and initially recognized less cumulative amortization recognized in accordance with IAS 18)

According to IFRS, any changes in a company's own credit standing is reflected in the fair value of its financial liabilities.

For Solvency II noninsurance liabilities valuation, CEIOPS recommends that undertakings apply an approach that combines the use of the risk-free rate for some liabilities and consideration of own credit standing at inception for other liabilities according to the features of the item being valued. Based on this approach, liabilities would be valued using the relevant risk-free interest rate.

Post employment benefit plans are either classified as defined contribution plans or defined benefit plans. The IFRS treat the defined contribution plan as either a liability, after deducting any contribution already paid, or as an expense. Accounting for defined benefit plans involves making a reliable actuarial estimate of the benefit employees have earned in current and prior periods, discounting the benefit, determining the fair value of any plans assets, and the total amount of actuarial gains and losses to be recognised.

CEIOPS recommends IAS 19 to be used as a proxy for an economic valuation until any revision has been made by IASB.

CHAPTER 25

European Solvency II Project

Eligible Own Funds and Investments

FIGURE 6.1 DEPICTS the concept of own funds. The capital, and its quality, which is backing-up the capital requirements are important issues. The investments done by the undertaking to make this backing-up accepted is also a key issue. These issues are discussed in this chapter. Investments are based on the "prudent person principle"; see Appendix E, Section E.5. In Section 25.7 we discuss the requirements for investing in ABS, that is, repackaged loans.

25.1 EXTRACTS ("RECITALS") FROM THE FRAMEWORK DIRECTIVE PREAMBLE

The recitals to the FD give us the ideas behind the FD. In each of Chapters 22 through 26 we are quoting the recitals that give the background to the topics discussed. Recitals not quoted in these chapters will be discussed in Appendix D, Section D.8. The number to the left of the recital is the number in the preamble. The headings are given for "easy reading."

25.1.1 Eligible Own Funds

(47) In accordance with that approach, capital requirements should be covered by own funds, irrespective of whether they are on or off the balance-sheet items. Since not all financial resources provide full absorption of losses in the case of winding-up and on a going-concern basis, own-fund items should be classified in accordance with quality criteria into three tiers, and the eligible amount of own funds to cover capital requirements should be limited accordingly. The limits applicable to own-fund items should only apply to determine the solvency standing of insurance and reinsurance undertakings, and should not further restrict the freedom of those undertakings with respect to their internal capital management.

(48) Generally, assets which are free from any foreseeable liabilities are available to absorb losses due to adverse business fluctuations on a going-concern basis and in the case of winding-up. Therefore the vast majority of the excess of assets over liabilities, as valued in accordance with the principles set out in this Directive, should be treated as high-quality capital (Tier 1).

(49) Not all assets within an undertaking are unrestricted. In some Member States, specific products result in ring-fenced fund structures which give one class of policyholders' greater rights to assets within their own fund. Although those assets are included in computing the excess of assets over liabilities for own-fund purposes they cannot in fact be made available to meet the risks outside the ring-fenced fund. To be consistent with the economic approach, the assessment of own funds needs to be adjusted to reflect the different nature of assets, which form part of a ring-fenced arrangement. Similarly, the Solvency Capital Requirement calculation should reflect the reduction in pooling or diversification related to those ring-fenced funds.

(50) It is current practice in certain Member States for insurance companies to sell life insurance products in relation to which the policyholders and beneficiaries contribute to the risk capital of the company in exchange for all or part of the return on the contributions. Those accumulated profits are surplus funds, which are the property of the legal entity in which they are generated.

(51) Surplus funds should be valued in line with the economic approach laid down in this Directive. In this respect, a mere reference to the evaluation of surplus funds in the statutory annual accounts should not be sufficient. In line with the requirements on own funds, surplus funds should be subject to the criteria laid down in this Directive on the classification in tiers. This means, *inter alia*, that only surplus funds which fulfil the requirements for classification in Tier 1 should be considered as Tier 1 capital.

(52) Mutual and mutual-type associations with variable contributions may call for supplementary contributions from their members (supplementary members' calls) in order to increase the amount of financial resources that they hold to absorb losses. Supplementary members' calls may represent a significant source of funding for mutual and mutual-type associations, including when those associations are confronted with adverse business fluctuations. Supplementary members' calls should therefore be recognized as ancillary own-fund items and treated accordingly for solvency purposes. In particular, in the case of mutual or mutual-type associations of shipowners with variable contributions solely insuring maritime risks, the recourse to supplementary members' calls has been a long-established practice, subject to specific recovery arrangements, and the approved amount of those members' calls should be treated as good-quality capital (Tier 2). Similarly, in the case of other mutual and mutual-type associations where supplementary members' calls are of similar quality, the approved amount of those members' calls should also be treated as good-quality capital (Tier 2).

25.1.2 Finite Reinsurance and SPVs

(90) Due to the special nature of finite reinsurance activities, Member States should ensure that insurance and reinsurance undertakings concluding finite reinsurance contracts or pursuing finite reinsurance activities can properly identify, measure, and control the risks arising from those contracts or activities.

(91) Appropriate rules should be provided for special purpose vehicles which assume risks from insurance and reinsurance undertakings without being an insurance or reinsurance

undertaking. Recoverable amounts from a special purpose vehicle should be considered as amounts deductible under reinsurance or retrocession contracts.

(92) Special purpose vehicles authorised before 31 October 2012 should be subject to the law of the Member State having authorised the special purpose vehicle. However, in order to avoid regulatory arbitrage, any new activity commenced by such a special purpose vehicle after 31 October 2012 should be subject to the provisions of this Directive.

(93) Given the increasing cross-border nature of insurance business, divergences between Member States' regimes on special purpose vehicles, which are subject to the provisions of this Directive, should be reduced to the greatest extent possible, taking account of their supervisory structures.

(94) Further work on special purpose vehicles should be conducted taking into account the work undertaken in other financial sectors.

25.1.3 Investments

(67) As a matter of principle, the new risk-based approach does not comprise the concept of quantitative investment limits and asset eligibility criteria. It should, however, be possible to introduce investment limits and asset eligibility criteria to address risks which are not adequately covered by a sub-module of the standard formula.

(71) Insurance and reinsurance undertakings should have assets of sufficient quality to cover their overall financial requirements. All investments held by insurance and reinsurance undertakings should be managed in accordance with the "prudent person" principle.

(72) Member States should not require insurance or reinsurance undertakings to invest their assets in particular categories of assets, as such a requirement could be incompatible with the liberalisation of capital movements provided for in Article 56 of the Treaty.

(73) It is necessary to prohibit any provisions enabling Member States to require pledging of assets covering the technical provisions of an insurance or reinsurance undertaking, whatever form that requirement might take, when the insurer is reinsured by an insurance or reinsurance undertaking authorised pursuant to this Directive, or by a third-country undertaking where the supervisory regime of that third country has been deemed equivalent.

(74) The legal framework has so far provided neither detailed criteria for a prudential assessment of a proposed acquisition nor a procedure for their application. A clarification of the criteria and the process of prudential assessment is therefore needed to provide the necessary legal certainty, clarity, and predictability with regard to the assessment process, as well as to the result thereof. Those criteria and procedures were introduced by provisions in Directive 2007/44/EC. As regards insurance and reinsurance, those provisions should therefore be codified and integrated into this Directive.

(75) Maximum harmonisation throughout the Community of those procedures and prudential assessments is therefore critical. However, the provisions on qualifying holdings should not prevent the Member States from requiring that the supervisory authorities are to be informed of acquisitions of holdings below the thresholds laid down in those provisions, so long as a Member State imposes no more than one additional threshold below 10% for that purpose. Nor should those provisions prevent the supervisory authorities from

providing general guidance as to when such holdings would be deemed to result in significant influence.

25.2 REFERENCE TO THE FD

The main legal basis for the (draft) final advice given by CEIOPS on the classification and eligibility of own funds, CEIOPS (2009d08, 2009f19), is primarily found in Articles 97 and 99 of the FD. Other Articles that are relevant for own funds are Articles 88, 89, 93–96, and 98. The provisions given in the FD should be read in connection with the Recitals given in Section 25.1.

The supervisory approval of ancillary own funds, CEIOPS (2009b04), has its legal basis in Article 92 of the FD. Also, Articles 87–90 are relevant for background to the advice. Ancillary own funds are discussed in Section 25.3.2.

Special purpose vehicles, SPVs, and finite reinsurance are dealt with in Article 211 of the FD. Authorization of SPVs is discussed in CEIOPS (2009b11, 2009f12). The requirements for using SPVs are also discussed in op. cit., see also Section 25.4.

Ring-fenced funds are defined and discussed in Section 25.5; CEIOPS (2009e06, 2010a06). The treatment of participations for own funds purposes are discussed in Section 25.6; CEIOPS (2009e05, 2010a05).

The investments rules, following the prudent person rule, of the FD are given in Articles 132–135. Earlier thoughts, discussions, and calibrations are discussed in Appendix E, Section E.5, which is the reference to this issue.

25.3 OWN FUNDS

Earlier thoughts, discussions, and calibrations are discussed in Appendix E, Section E.4.

An undertaking's eligible own funds, or "available capital" or the "capital base," corresponds to the financial resources that can serve as a buffer against risks and absorb losses where necessary. The amount of own funds are divided between

- Basic own funds (items on the balance sheet)
- Ancillary own funds (off-balance-sheet items)

According to their loss-absorbing capacity, these items are classified into three tiers, depending on their nature and to the extent to which they meet certain quality criteria.

The development and calibration of own funds are discussed in Appendix E, Section E.4.

One key lesson that CEIOPS learned from the crises, CEIOPS (2009a, 2009d08, 2009f19), was that own funds must be available in times of stress to fully absorb losses. According to CEIOPS, they have to be built up when the undertaking is not in stress. In terms of FD Article 93, there are own funds that *fully possess* the characteristic of permanent availability. Article 94 requires Tire 1 own-fund items to *substantially possess* this characteristic, meaning that they have lower quality than those that fully possess this characteristic. CEIOPS has construed this in terms of increasing the amount and quality of Tier 1, increasing the quality of Tier 2, and decreasing the amount and increasing the quality of Tier 3 own funds.

CEIOPS final advice on classification and eligibility of own funds are found in CEIOPS (2009f19). The general advice concludes that

- Tier 1 should possess the highest quality of own funds and should substantially absorb losses to enable an undertaking to continue as a going concern.
- The proportion of Tier 1 items in the eligible own funds must be *significantly higher* than one-third of the total amount of eligible own funds (Article 98 in FD: higher than one third).
- The average quality of own funds should be increased as described above.
- For Tier 1, this could be achieved by restricting the Tier 1 capital to ordinary share capital, or the equivalent mutual/mutual-type capital, and nonrestricted reserves. Further restrictions were also discussed (restrictions on hybrid capital instruments, subordinated liabilities to those absorbing losses first in a going concern).
- Tier 1 classified capital must be fully paid in.
- There must be certain minimum qualitative requirements for capital to be included in own funds.

We follow CEIOPS' headings in their final advice; CEIOPS (2009f19).

25.3.1 CEIOPS Proposed Limit Structure

- In relation to compliance with the SCR:
  - Proportion of Tier 1 > Proportion of eligible Tier 2 > Proportion of eligible Tier 3.
  - Proportion of Tier 1 items in eligible own funds is at least 50% of the total amount of eligible own funds.
  - Proportion of Tier 3 items in eligible own funds is set to a maximum of 15%.
  - Any inclusion of high-quality hybrids should be restricted to no more than 20% of Tier 1.
- In relation to compliance with MCR:
  - Proportion of Tier 1 > Proportion of eligible Tier 2 Basic own funds.
  - Proportion of Tier 1 items in eligible own funds is at least 80% of the total amount of eligible own funds.

25.3.2 Minimum Characteristics for Own Funds

- Eligible own funds:
  - All CFs on OF items should be subject to supervisory approval once the SCR is breached (both coupon and principal payments).

- Tier 3 basic own funds:
  - Should demonstrate features to ensure that subordination is effective and not just a nominal requirement.
  - Should not be freely redeemable, when an undertaking's solvency position is deteriorated.
  - Coupons should not be freely payable, when an undertaking's solvency position is deteriorated.
- Other issues:
  - Own-fund items should not be allowed to cause, or accelerate, an undertaking to go into insolvency.
  - Any redemption, conversion, or exchange of capital instruments should be subject to prior supervisory approval.
- Sufficient Duration:
  - The average duration of own-fund items should not be significantly lower than the average duration of an undertaking's liabilities. The assessment of sufficient duration of own-fund items should be done on a reporting date basis, and be part of the risk management and the ORSA.
  - Capital instruments:
    - Should be included in own funds on the basis of their issue date.
    - The first contractual possibility of repayment. The time horizon must be the expected duration, or anticipated duration over the next twelve months. To ensure that capital is of sufficient duration, they should have benchmark minimum maturities: CEIOPS recommends:
      - 10 years for Tier 1
      - 5 years for Tier 2
      - 3 years for Tier 3

25.3.3 Supervisory Approval of Assessment and Classification

The approach to supervisory approval should be principle-based. CEIOPS recommends allowing room for the criteria below to be elaborated on as part of Level 3 guidance.

Supervisory approval should be granted before undertakings are allowed to include an own-fund item not covered by the lists given below in Sections 25.3.1 and 25.3.2. The request for approval should include *at least* the following details:

- Amount to be used
- Legal form of the element to be included

- Counterparty (belonging to the same group or not)
- The capacity of the own-fund items to absorb losses either on a going-concern basis (Tier 1) or in a winding-up (Tier 2 and 3)
- If the item is fully paid or called up
- Duration of the item
- Existence of requirements or incentives to redeem the instrument
- Existence of mandatory fixed charges
- Subordination in winding up
- Duration of insurance and reinsurance obligations

A three-step process has been proposed by CEIOPS for the approval of own-fund items not included in the lists. The undertakings are recommended to use the same process internally.

Step 1: The supervisory authority, taking into account the legal enforceability and the characteristics of the item, assesses to what extent it possesses the characteristics of permanent availability and subordination. In addition, the supervisory authority assesses whether the duration of the item is compatible with the maturity of undertaking's insurance and reinsurance obligations. For undated items without a call, this assessment is unnecessary.

Step 2: The supervisory authority assesses to what extent the item possesses the features of absence of incentive to redeem, mandatory servicing costs, and encumbrances.

Step 3: The supervisory authority assesses whether the inclusion of the item is compatible with the quantitative limits envisaged by implementing measures to cover SCR and the MCR.

25.3.4 Basic Own Funds

Basic Own Funds, Tier 1-Capital Instruments

CEIOPS suggests the following list:

- Paid-in ordinary share capital.
- Paid-in equivalent of ordinary share capital of mutual and mutual-type undertakings.
- Other paid-in capital instruments, including preference shares that absorb losses first or rank *pari passu* with capital instruments that absorb losses first (in a going concern).
    - Instruments that automatically convert to ordinary share capital, or to the equivalent of ordinary share capital of mutual and mutual-type undertakings, as and when the undertaking needs to absorb losses, and in any case when the undertaking breaches its SCR.

- Instruments subject to write down as long as losses persist, as and when the undertaking needs to absorb losses, and in any case when the undertaking breaches its SCR.

The following six characteristics have been developed by CEIOPS; see, for example, Appendix E, Section E.4. Tier 1 own funds should posses all of them.

i. *Subordination:* The item must be the most deeply subordinated in a winding-up.

ii. *Loss absorbency:* The item must be fully paid in, must be the first instrument to absorb losses or rank *pari passu* with an instrument that absorbs first losses, and must not hinder recapitalization.

iii. *Sufficient duration:* The item should not have a legal maturity of less than 10 years at the issue date. The item must be contractually locked-in on a breach of the SCR where redemption is only permitted in exceptional circumstances, if the item is replaced by an own-fund item of equivalent or higher quality and subject to the consent of the supervisory authority.

iv. *Free from requirements or incentives to redeem:* There must be no incentives to redeem the item. The item must only be redeemable at the option of the undertaking (i.e., not at the option of the holder) and any redemption should be subject to the approval of the supervisory authority.

v. *Free from mandatory fixed charges:* At all times, coupons/dividends must be able to be cancelled and must at a minimum be cancelled on a breach of the SCR after which they can only be paid in exceptional circumstances and subject to the consent of the supervisory authority. Undertakings should have full discretion over the amount of payment; coupons/dividends must not be at a fixed rate, and there should be no preference as to income or return of capital.

vi. *Absence of encumbrances:* The instrument must be free from encumbrances and therefore should not be connected with any other transaction which, when considered with the own-fund item, could undermine the characteristics of that item. Examples of potential encumbrances include, but are not limited to, rights of set off, restrictions, charges or guarantees. Where an investor subscribes for capital in an undertaking and at the same time that undertaking has provided financing to the investor, only the net financing provided by the investor is considered as eligible own funds.

The principle behind the requirement that own-fund item should not hinder recapitalization is that Tier 1 capital instruments must absorb losses first. Own-fund items should not hinder recapitalization, that is, investors are the first to be called upon to recapitalize the undertaking. If Tier 1 capital instruments include hybrid capital instruments or subordinated liabilities, the requirement mean that the instrument absorb losses in a going concern through appropriate mechanisms reducing the potential future outflows to the holders of the instruments.

Basic Own Funds, Tier 1-Other Own-Fund Items

These items are referred to in the FD text as "excess of assets over liabilities"; see recitals 48, 49, and Article 88. In this Article it is stated that these items should be considered as basic own funds.

CEIOPS suggests the following list:

- *Reserves:* to the extent that they are available to *substantially* absorb losses at any time arising from any segment of liabilities or from any risks. They include
  - Retained earnings
  - Share premium account
  - Surplus funds falling under Article 91(2)
  - Revaluation reserves
  - Other reserves
- Paid in subordinated mutual member accounts

Less

- Own shares, or units of equivalent capital of mutual and mutual-type undertakings, held by the undertaking. Taking the economic approach and applying the principle of substance over form, where there is evidence of a group of connected transactions whose economic effect is the same as the holding of "own shares," the assets that those transactions generate for the undertaking shall be deducted from its own funds, to the extent necessary to guarantee that own funds reliably represent the net financial position of its shareholders, further to other allowed items.

CEIOPS exclude the following reserves from Tier 1 "excess of assets over liabilities":

- *Restricted reserves:* such as certain equalization reserves, legal reserves, and statutory reserves, which should only be eligible for inclusion in own funds in relation to the risks they cover.
- The difference *between* the value of technical provisions calculated as an ongoing concern basis, *and* the amounts that the original undertaking shall have to pay to its policyholders to honor their rights according to the contracts in force in the case of winding up with no transfer of portfolios. Where such a difference exists, it should be classified as Tier 3. This calculation shall allow for the amounts that policyholders are legally or contractually obliged to pay to the undertaking in a situation of winding up. ("winding-up gap")
- Deferred tax assets, which should be excluded from own funds or be classified in Tier 3.

- Intangible assets not valued at zero and not subject to a capital charge should be excluded from own funds or included as Tier 3 and subject to an intangible assets risk module.
- Expected future profits, defined as the actual value of any type of profit included, either explicit or implicitly, in the future inflows considered in the calculation of the best estimate. Implicit profits shall be assessed using appropriate actuarial methods.

Basic Own Funds, Tier 2-Capital Instruments
CEIOPS suggests the following list:

- Called-up ordinary share capital.
- Other called-up capital instruments that absorb losses first or rank *pari passu*, in going concern, with capital instruments that absorb losses first.
- Other capital instruments, including preference shares that do not have the conversion features required for Tier 1 but that display the characteristics below.
- Other capital instruments, including preference shares, not subject to write down as long as losses persist, but that display the characteristics below.

Characteristics set up that Tier 2 basic own funds should display:

i. *Subordination:* The item must be effectively subordinated in a winding-up.

ii. *Loss absorbency:* The item does not need to be fully paid in, but can simply be called up, and must absorb losses to a degree. The undertaking must be able to defer coupon payments once the SCR has been breached.

iii. *Sufficient duration:* The item should not have a legal maturity of less than 5 years at the issue date. The item must be contractually locked-in on a breach of the SCR where redemption is only permitted if the item is replaced by an own-fund item of equivalent or higher quality and subject to the consent of the supervisory authority.

iv. *Free from requirements or incentives to redeem:* There may be moderate incentives to redeem the item. The item must only be redeemable at the option of the undertaking (i.e., not at the option of the holder) and any redemption should be subject to the approval of the supervisory authority.

v. *Free from mandatory fixed charges:* Coupons/dividends must at a minimum be deferred for an indefinite term on a breach of the SCR after which they can only be paid subject to the consent of the supervisory authority.

vi. *Absence of encumbrances:* The instrument must be free from encumbrances and therefore should not be connected with any other transaction which, when considered with the own-fund item, could undermine the characteristics of that item. Examples of potential encumbrances include, but are not limited to, rights of setoff, restrictions, charges, or guarantees. Where an investor subscribes for capital in an undertaking and at the same time that undertaking has provided financing to the investor, only the net financing provided by the investor is considered as eligible own funds.

Basic Own Funds, Tier 2-Other Own-Fund Items
No items are included in this category.

Basic Own Funds, Tier 3-Capital Instruments
The role of Tier 3 basic own funds is to provide loss absorbency in a winding-up situation in order to adequately protect policyholders and beneficiaries. This is particularly important as Tier 3 can represent a very substantial proportion of the undertaking's own funds unless its eligibility is limited.

CEIOPS suggests that other capital instruments, including. preference shares, which do not display the characteristics required for Tier 1 and Tier 2, are included.

Tier 3 basic own funds must posses at least some of the characteristics required for Tier 1 and Tier 2 eligibility, but to a lesser degree. They should be prevented from having characteristics that undermine effective subordination.

Tier 3 basic own funds must be free from encumbrances. There should be no redemption of Tier 3 basic own funds, or coupon payments, on a breach of the SCR (i.e., during the ladder of supervisory intervention), unless the supervisory authority determines that redemption is necessary to facilitate a recapitalization. Tier 3 basic own funds should have a minimum maturity (e.g., 3 years).

If intangible assets is regarded as Tier 3, then CEIOPS wants to introduce an intangible assets risk module as a parallel to the five top risk modules under Basic SCR; CEIOPS (2009f19). This is discussed in Chapter 26.

Basic Own Funds, Tier 3-Other Own-Fund Items
CEIOPS suggests the following list:

- The difference between technical provisions (Articles 75–86) on a going-concern basis and the amounts the undertaking should have to pay to its policyholders in case of winding up
- Deferred tax assets, if not excluded from own funds
- Expected future profit

25.3.5 Ancillary Own Funds

The assessment of the amount of an ancillary own-fund item requires supervisory judgement to be approved; CEIOPS (2009b04, 2009f04). The approval of ancillary own funds will have a principle-based approach at Level 2 implementing measures to be elaborated at Level 3 Guidance.

The approval process has three steps both for the supervisory authority and for the undertaking that seeks approval.

Step 1. The supervisory authority, taking into account the legal enforceability and the characteristics of the item, assesses to what extent it possesses the characteristics of permanent availability and subordination. In addition, the supervisory authority assesses whether the duration of the item is compatible with the maturity of the undertaking's insurance and reinsurance obligations. For undated items without a call, this assessment is unnecessary.

Step 2. The supervisory authority assesses to what extent the item possesses the features of absence of incentive to redeem: mandatory servicing costs and encumbrances.

Step 3. The supervisory authority assesses whether the inclusion of the item is compatible with the quantitative limits envisaged by implementing measures to cover the SCR and the MCR.

In the three-step process, the supervisory authority needs to consider some criteria setup. The information should be provided by the undertaking to the supervisor, who will use it as part of its documentation.

- The status of the counterparties concerned, in relation to their ability and willingness to pay
    - Ability to pay: consider both the default risk and the risk of delay in the transfer of funds
    - Willingness to pay: for example, the amount recoverable
- The recoverability of the funds, taking account of the legal form of the item, as well as any conditions that would prevent the item from being successfully paid in or called up.
- Any information on the outcome of past calls that undertakings have made for such ancillary own funds, to the extent that information can be reliably used to assess the expected outcome of future calls.

It is the undertaking's responsibility to inform the supervisory authority of any significant changes in the recoverability of ancillary own funds, and to provide the supervisory authority, as soon as possible, with the relevant documentation to support this. The supervisor may approve the amount of ancillary own funds item for a specified period of time. When Solvency II is fully implemented, CEIOPS will revisit the issue of an appropriate time frame for approval.

The undertaking is required to publicly disclose, as part of Pillar III, detailed information on ancillary own funds.

Ancillary Own Funds, Tier 2

Tier 2 ancillary own-fund items should be callable own funds of the highest quality and demonstrably absorb unexpected losses to enable an undertaking to continue as a going concern. They would represent own-fund items that, if called up and paid in, would be classified as Tier 1.

CEIOPS suggests, CEIOPS (2009f19), the following list:

- Ordinary share capital on demand
- Equivalent of ordinary share capital, callable on demand, of mutual and mutual-type undertakings
- Supplementary member calls of mutual or mutual-type undertakings, within the next 12 months, that can be made on demand, where the call generates Tier 1 own funds and is clear of encumbrances
- Letter of credit and guarantees that are held in trust for the benefit of insurance creditors by an independent trustee and provided by credit institutions authorized in accordance with Level 1 text 2006/48/EC
- Other capital instruments, callable on demand, that absorb losses first or rank *pari passu*, in a going concern, with capital instruments that absorb losses first, for example
  - Instruments that automatically convert to ordinary share capital (or equivalent for mutual companies), as and when the undertaking needs to absorb losses, and in any case when the undertaking breaches its SCR or
  - Instruments subject to write down as long as losses persist, as and when the undertaking needs to absorb losses, and in any case when the undertaking breaches its SCR

Ancillary Own Funds, Tier 3

Tier 3 ancillary own funds do not need to be subject to any specific requirements, due to the proposed limit structure, and taking into account the supervisory approval of ancillary own funds.

CEIOPS suggests the following list:

- Callable preference shares classified in Tier 2 or Tier 3
- Other callable capital instruments classified in Tier 2 or Tier 3

25.4 SPECIAL PURPOSE VEHICLES, SPVs

According to Article 13(26), an SPV is defined as

> any undertaking, whether incorporated or not, other than an existing insurance or reinsurance undertaking, which assumes risks from insurance or reinsurance undertakings and which fully funds its exposure to such risks through the proceeds of a debt issuance or any other financing mechanism where the repayment rights of the providers of such debt or financing mechanism are subordinated to the reinsurance obligations of such an undertaking.

An undertaking can use an SPV to transfer insurance risks through a contract, in a similar way as it would cede insurance risk to a reinsurance undertaking. The undertaking reinsures risks to the SPV and may transfer an amount of supplementary assets, or pays an adequate premium, necessary to offer the investors a rate of return appropriate to the risk. This rate of return is calculated as a percentage of the amount to be raised from the market. The undertaking would then take credit for the risks ceded to the SPV as reinsurance recoverables calculated.

The scope of authorization and the conditions to be included in all contract issued are discussed in (2009f12). SPVs should only be considered for authorization in a Member State if (a) and either (b) or (c) of the following three scope issues are fulfilled:

a. The transaction has the structure of an SPV as defined in Article 13(26) and meets the requirements established below; *and either*

b. The SPV assumes risk from an undertaking through a reinsurance contract; *or*

c. The SPV assumes insurance risks from an undertaking transferred through a contract that is "reinsurance like."

If, in the authorization process, the undertaking has clearly stated the aim of reusing the SPV, along with the details of the reuse, then detailed follow-up discussions with the supervisor shall not be necessary when the envisaged reuse occurs if the same circumstances apply as at authorization of the SPV and the SPV is acting within its Articles of incorporation. Where an SPV is to be located in an EEA jurisdiction other than where the undertaking is located, the decision for authorization of the SPV shall be taken by the supervisory authority in the jurisdiction where the SPV is to be established. Where separate undertakings within a group use an SPV, that SPV shall be established in such a way that the SPV is protected from the impact of a related undertaking within a group being wound up; CEIOPS (2009f12).

There are five principles that should be included in the mandatory conditions of the contracts issued in the relation to the establishment of the authorization. Also, governance and reporting SPVs are discussed. For more details, on these subjects we refer to CEIOPS (2009f12).

Principle 1: Fully funded

Principle 2: Investors have a subordinated claim on SPV assets

Principle 3: Prudent person

- Assets should reflect the duration of underlying liabilities
- Assets should be of high quality and counterparty exposures should be sufficiently diversified
- Derivatives should be used only for risk reduction/efficient portfolio management

Principle 4: Effective risk transfer

Principle 5: Nonrecourse

The documentation requirements include, for example,

- A copy of the proposed contract between the SPV and the undertaking and a statement containing a description of that contract, accompanied by or including satisfactory information about the identities and qualifications therein
- A copy of the SPV's memorandum and Articles, or proposed memorandum and Articles of association
- A description of, for example, any terms and conditions for payments under the contract, the aggregate limit of the relevant contract, compliance with the fully funded principle and any stress tests results
- Actuarial review of underlying business, independent of the undertaking
- Prospectus/Offering Circular or Private Placement Memorandum—if any
- Overall risk management plan including details as to how the SPV will continue to be fully funded during the term of the contract
- Risk implications of the SPV's investment strategy
- Details of any intended hedging instruments, such as interest rate swaps or currency contracts
- Capital including size, growth, potential investor concentration, and management share of the capital base
- A contingency plan explaining what will occur if, for example, the fully funded principle is breached, a disagreement arises
- Details of Directors/Management fitness and probity
- Details on how the SPV meets its system of governance requirements (especially risk management and internal control) as set out in this paper
- Investment authority and guidelines for assets held in Trust, along with details of any leverage permitted within these guidelines

The authorization may also require other documents. In an appendix to CEIOPS (2009f12), the requirements for undertakings using SPVs are discussed. The requirements were not within the scope of Article 211 on SPVs but were seen as important issues for supervisors and thus included in op. cit.

It is the responsibility of the administrative or management body of the undertaking to ensure that all mandatory conditions are present within the contractual arrangements at the time of the SPV's authorization. The requirements are given below.

- *Effects of the fully funded concept on the undertaking:* The undertaking, through its system of governance, must continuously monitor the fully funded requirement. The maximum reinsurance credit taken for an SPV should be capped at the amount equal to *min*(the aggregate maximum liability transferred; the aggregate value of the assets of the SPV). Any fall in the value of assets within the SPV should be mirrored by a fall in the reinsurance asset within the undertaking.

- *Risks remaining within the undertaking:* Any remaining risk from the SPV should be fully taken into account through the undertakings risk management system.
- *Alignment of interests between the undertaking and the SPV:* To ensure, for example,
  - That claims management processes operate effectively;
  - To provide a discipline on the underwriting of risks, that is, the undertaking cannot just transfer risks it may not have fully understood or properly managed to an SPV; and
  - The SPV is established and subsequently run in an appropriate manner for all the interested parties.
- *Transparency:*
  - Full disclosure within the Solvency and Financial Condition Report, SFCR, and the annual accounts should be made regarding its reinsurance within the SPV.
- *Fit and proper requirements for the undertaking:* Before the undertaking enters into an SPV transaction, the supervisory authority should assess whether the administrative or management body of the undertaking has the appropriate modeling and risk management understanding to fully comprehend the risks being transferred to the SPV and the consequences of such actions.

25.5 RING-FENCED FUNDS

The treatment of ring-fenced funds was discussed in CEIOPS (2009e06) and the final advice CEIOPS (2010a06); see also recital 49 above.

Where capital is not totally transferable and therefore not able to meet losses as they occur across the whole undertaking, there are two relevant aspects that need to be carefully analyzed, in order to reflect the economic effect of ring-fencing and the potential implications for the measurement of own funds and capital requirements.

The first aspect relates to the *availability* of own funds within an undertaking in the presence of ring-fenced funds and the measurement of the extent to which own funds held within the ring-fenced fund (restricted own funds) contribute toward the coverage of the total SCR of the undertaking. The second aspect relates to the *calculation* of the undertakings SCR when ring-fenced funds are in place. The problem lies on the fact that the existence of ring fencing may reduce the overall level of diversification between risks, that is, reduces the extent to which losses and profits observed in and out of the ring-fenced fund compensate each other.

A ring-fenced fund arises as a result of an arrangement where; CEIOPS (2010a06):

a. There is a barrier to the sharing of profits/losses arising from different parts of the undertaking's business leading to a reduction in pooling/diversification related to that ring fenced fund or

b. Own funds (restricted own funds) can only be used to cover losses on a defined portion of the undertaking's (re)insurance portfolio or with respect to particular policyholders or in relation to particular risks such that those restricted own funds are only capable of fulfilling the criteria in Article 93(1) (a) and/or (b) in respect of that defined portion of the portfolio, or with respect to those policyholders or those risks or

c. Both (a) and (b) apply

Certain types of business do not fall within the scope of ring-fenced funds, for example, unit-linked and reinsurance business.

Depending on the specific provisions of product regulation and national, insolvency, and contract law, as well as the products that exist in different Member States' ring-fenced funds may arise, which give rise to the following treatments: (a) There is a need to make an adjustment to the calculation of the undertaking's SCR or (b) there is a need to make an adjustment to restrict own funds arising in respect of the ring-fenced fund; or (c) both SCR and own funds need to be adjusted.

In the presence of ring-fenced funds that have restricted own funds, or the potential to have restricted own funds, the restricted availability should be reflected through an adjustment to own funds.

The adjustment should be based on these principles:

- If the ring-fenced fund has sufficient own funds to cover the notional SCR for that ring-fenced fund, then any surplus over the notional SCR cannot be used to cover risks in the rest of the firm and should be excluded.

- Notwithstanding legal and contractual requirements, if the ring-fenced fund does not have sufficient capital to meet the notional SCR for that ring-fenced fund, then the deficit should be covered by own funds outside the ring-fenced fund that could be transferred to meet the deficit.

The calculation also needs to address the treatment of future transfers attributable to shareholders in respect of profit-sharing arrangements where benefits to policyholders are reflected in technical provisions. These future transfers should not form part of the own funds of the ring-fenced fund when calculating the ring-fencing restriction.

CEIOPS also outlines the steps for appropriate adjustment to the eligible own funds in practice; see CEIOPS (2010a06).

25.6 PARTICIPATIONS

The treatment of participations for own funds purposes is discussed in CEIOPS (2009e05, 2010a05). First we have to define what is meant by participations, and also what is meant with related undertakings.

The FD defines participation (from the view of the participating undertaking, that is, a parent or similar) as the ownership, direct or by way of control, of 20% or more of the voting rights or capital of an undertaking. However, Article 92 makes clear that the definition of participations for the purposes of these implementing measures should be the definition that is set out in the third subparagraph of Article 212(2). This broader definition states that supervisory authorities shall also consider as participation the holding, directly or indirectly, of voting rights or capital in an undertaking over which, in the opinion of the supervisory authorities, a significant influence is effectively exercised.

Related undertakings comprise subsidiaries, participations, or an undertaking linked with another undertaking by a relationship that could be summed up as "unified management."

The treatment of participations must ensure that the supervisors have a meaningful picture of the solvency position of each solo undertaking. Therefore CEIOPS considers that the issue of double gearing needs to be addressed at both the solo and the group levels, and this is fundamental as to maintain the integrity of the solo solvency calculation.

The following objectives are also relevant when considering the treatment of participations; CEIOPS (2009e05, 2010a05):

- Ensuring that the capital held in each solo entity is commensurate with the risks run in that entity—this requires supervisors to have the ability to identify where capital and risks reside
- Limiting systemic risk
- Avoiding the contagion of risks within a group through subsidiaries/participations
- Avoiding incentives to regulatory arbitrage through group structuring

Contagion risk will impact on all undertakings; so, even though the assessment may be done at group level, this is also relevant at the solo level.

The treatment of participations, *included in the scope of group supervision*, should be, for example, defined below.

Financial and credit institutions: The own funds arising from participations in financial and credit institutions should not be recognized as eligible own funds for the purpose of the SCR and MCR of the participating undertaking. Hence, own funds of a participation undertaking will not be available to absorb losses of the parent in any possible situation, and particularly in times of crisis. Any holdings in subordinated claims and other instruments in the participation will also be excluded. Where the participation is an intermediate holding company, this should be treated as a financial institution.

Insurance and reinsurance companies: The amount of own funds held by the participation to meet its SCR should be treated as a restricted item, and excluded from the participating entity's eligible own funds. Any inherent goodwill in the valuation should be excluded from own funds of the participating undertaking. An alternative approach is discussed in Section 26.7.2.

Financial nonregulated undertakings: The same approach used for regulated financial and credit institutions should be adopted for financial nonregulated undertakings.

Nonfinancial nonregulated undertakings: These undertakings have a standard equity risk charge approach applied to them.

Financial and credit institutions, (re) insurers and Financial nonregulated undertakings: According to CEIOPS (2010a05), there are no major views. Three different views are proposed, of which two are described below:

- Three members view: Participations in (re)insurers and financial nonregulated undertakings included in the scope of group supervision should be treated as equity investments at the solo level and therefore be subject to an equity risk charge approach. Two of these Members also believe that the same approach should apply to participations in financial and credit institutions.
- Two members view: According to Article 111(m), the equity risk shock should be reduced, in order to take into account the likely reduction in volatility due to the strategic nature of the related undertakings and the influence exercised by the participating undertaking. An appropriate reduction is 50% of the standard shock.

In the case of participations *excluded from the scope of group supervision*, the solo supervisor shall consider whether the circumstances leading to that exclusion also apply to the assessment of the solo solvency position of the relevant undertaking. If the solo supervisor concludes that these circumstances do apply, and that the loss absorbency of the own funds derived from the participation is affected, then the amount should not be recognized as eligible own funds. This assessment should be carried out in respect of participations in (re)insurers, financial and credit institutions, and financial nonregulated undertakings.

In the case of nonregulated nonfinancial undertakings, CEIOPS proposed that a standard equity risk charge approach should be used, subject to the criteria in Article 111(m). The equity risk charge will need to take into account whether a nonregulated nonfinancial undertaking should ever be considered a strategic investment.

25.7 INVESTMENTS

The general principle for investing assets is the prudent person principle, discussed in more detail in Appendix E, Section E.5. One "clarification" of the prudent person principle is the requirements on so-called repackaged loans. They are repackaged and sold on by an originator to a third party. These loans are generically referred to ABS. CEIOPS (2009e01, 2010a01) wants to ensure that insurance undertakings take sound and sensible investment decisions—in line with the prudent person principle—and to achieve a cross-sectoral consistency to avoid regulatory inconsistency across financial sectors. Therefore, the proposed requirements and principles closely follow the amended Capital Requirements Directive (CRD); CRD (2009).

From the CRD adopted in 2006, OJ (2006), we have the following definitions that are useful for the following discussion:

"*Securitization*" means a transaction or scheme, whereby the credit risk associated with an exposure or pool of exposures is tranched, having the following characteristics:

a. Payments in the transaction or scheme are dependent upon the performance of the exposure or pool of exposures and
b. The subordination of tranches determines the distribution of losses during the ongoing life of the transaction or scheme;

"*Tranche*" means a contractually established segment of the credit risk associated with an exposure or number of exposures, where a position in the segment entails a risk of credit loss greater than or less than a position of the same amount in each other such segments, without taking account of credit protection provided by third parties directly to the holders of positions in the segment or in other segments;

"*Securitization position*" shall mean an exposure to a securitization;

"*Originator*" means either of the following:

a. An entity which, either itself or through related entities, directly or indirectly, was involved in the original agreement which created the obligations or potential obligations of the debtor or potential debtor giving rise to the exposure being securitized or

b. An entity that purchases a third party's exposures onto its balance sheet and then securitizes them;

"*Sponsor*" means a credit institution other than an originator credit institution that establishes and manages an asset-backed commercial paper program or other securitization scheme that purchases exposures from third-party entities.

ABS loans are loans that are repackaged into tradeable securities and other financial instruments, and backed (collateralized) by a specified pool of underlying assets. They can take different forms and are assigned different risk classes (tranches): super senior, senior, mezzanine, and equity. Super senior and senior tranches are considered as safest. Interest and principal payments are made in order of seniority. Hence, junior tranches, that is, mezzanine and equity, give highest coupon payments and interest rates.

The originator of an ABS earns a commission at the time of issue and earns management fees during the life of the ABS. The originator, which is often called an SPV, is usually an investment bank. One incentive for creating ABS is to remove risky assets from the balance sheet by having another institution taking over the credit risk. The originator receives cash in return; CEIOPS (2009e01).

CEIOPS observe that the ability to earn substantial fees from originating and securitizing loans, coupled with the absence of any residual liability, skews the incentives of originators in favor of loan volume rather than loan quality. This is a structural flaw in the debt-securitization market and it is widely accepted that this contributed greatly to the credit bubble of 2000 as well as the credit crisis, and the prevailing banking crisis, of 2008; CEIOPS (2009e01).

Insurance undertakings have not traditionally purchased large volumes of these products. However, the proposed requirements could be seen as preventive measures. For investment to take place, CEIOPS propose seven principles that need to adhere to. The principles to be applied to tradeable securities should take into consideration the principle of substance over form. This is because it is not possible to anticipate the specific nature that these products may take in future years. There is also a grandfathering rule for these arrangements stating that these rules will apply from December 31, 2014 for existing arrangements of October 31, 2012.

25.7.1 Principle 1: Originators' Retained Interest

Undertakings should only invest in repackaged loans or similar financial arrangements if the originator, sponsor, or original lender has explicitly disclosed to the undertaking in the contract that it will retain, on an ongoing basis, a net economic interest, which in any event should not be less than 5%. For this purpose, retention of a net economic interest should mean

a. Retention of not less than 5% of the nominal value of each of the tranches sold or transferred to the investors; or

b. In the case of securitizations of revolving exposures, retention of originators' interest of not less than 5% of the nominal value of the securitised exposures; or

c. Retention of randomly selected exposures, equivalent to not less than 5% of the nominal amount of the securitized exposures, where these would otherwise have been securitized in the securitization, provided that the number of potentially securitised exposures is not less than 100 at origination; or

d. The same or a less severe risk profile and not maturing any earlier than those transferred or sold to investors, so that the retention equals in total not less than 5% of the nominal value of the securitized exposures.

Exemptions on the requirements for the originator in relation to *Principle 1* may be authorized by supervisory authorities on a case-by-case basis on the request of the undertaking and will only include some specific transactions.

25.7.2 Principle 2: Criteria for Sponsor and Credit Institutions

Prior to an undertaking investing in repackaged loans or similar financial arrangements, it is required that the undertaking shall ensure that the sponsor and originator credit institutions

- Base credit granting (such as the issuance of loans or mortgages) on sound and well-defined criteria and clearly establish the process for approving, amending, renewing, and refinancing loans to the exposures to be securitized as they apply to exposures they hold;
- Ensure that an originator operates effective systems to manage the ongoing administration and monitoring of its various credit risk-bearing portfolios and exposures, including for identifying and managing problem loans and for making adequate value adjustments and provisions;
- Diversify each credit portfolio given its target market and overall credit strategy; and
- Maintain documentation to include its policy for credit risk, including its risk appetite and provisioning policy and shall describe how it measures, monitors, and controls that risk.

25.7.3 Principle 3: Transparency and Disclosure of the Underlying

Prior to an undertaking investing in tradable securities and other financial instruments based on repackaged loans, the undertaking should ensure that the sponsor and the originator credit institutions disclose to the undertaking the level of their commitment, as under *Principle 1*, to maintain a net economic interest in the securitization. The originator and the sponsor should also disclose any features of the holding that could undermine the concept of its retained interest such as commission payments associated with the transaction.

25.7.4 Principle 4: Skill, Care and Diligence

Undertakings investing in repackaged loans or similar financial arrangements need to be able to properly identify, measure, monitor, manage, control, and report the risks of these products and shall pay particular attention to assessing the ALM risk, concentration risk and investment risk arising from these products. Undertakings who invest in these products shall also have due consideration within their internal investment policy.

25.7.5 Principle 5: Monitoring Procedures

Undertakings shall establish formal monitoring procedures commensurate with the risk profile of their investments in repackaged loans or similar financial arrangements to monitor on an ongoing basis and, in a timely manner, performance information on the exposures underlying their securitization positions. Undertakings need to have access to relevant information to be able to perform this analysis.

25.7.6 Principle 6: Stress Tests (Including Using Financial Models)

Undertakings, where appropriate, should perform stress tests as part of their ORSA and wherever else appropriate to ensure the adequate management, monitoring, and control of the investments in accordance with Article 132 of the Level 1 text. Where an undertaking holds a material value of tradable securities and other financial instruments based on repackaged loans, CEIOPS will expect these to be subject to their own stress tests simultaneously taking into account that the dynamic effect of the stress test scenario on the rest of their business. The results of any stress tests should be documented and mitigating actions taken as appropriate.

25.7.7 Principle 7: Formal Policies, Procedures, and Reporting

Before investing in repackaged loans or similar financial arrangements, and as appropriate thereafter, an undertaking shall be able to demonstrate to its competent supervisory authorities that for each of its individual securitization positions, it has a comprehensive and thorough understanding of and has implemented formal policies and procedures appropriate to its investment portfolio. These formal policies and procedure shall commensurate with the risk profile of their investments in securitized positions.

Undertakings shall also ensure that there is an adequate level of internal reporting to administrative or management body so that they are aware of material investment made in repackaged loans and that the risks from these products are adequately managed.

Undertakings shall also include appropriate information on their investments in these products, and their risk management procedures in this area, in the supervisory reporting and public disclosure requirements.

PART E

European Solvency II Standard Formula: Final Advice

I think and think for months and years. Ninety-nine times, the conclusion is false. The hundredth time I am right

Albert Einstein, (1879–1955)
Swiss-German-US physicist

THE NEXT PART of the European Solvency II system is the standard formula for the SCR. The earlier developments are described in Part F.

The standard formula SCR, as described in Chapters 26 through 34, are based on the Level 1 texts, that is, the Framework Directive, and the Level 2 implementing measures, that is, detailed measures complementing the Framework Directive, proposed by CEIOPS in its draft proposals and its final advice as published up to January 2010. The last final advices given by CEIOPS were published in March 2010. Based on all these advices the European Commission developed the Level 2 implementing measures, mainly as EU regulation. These will be published in late 2011.

When we talk about an *undertaking* in both Part D and Part E chapters, we mean both an insurance undertaking and a reinsurance undertaking, unless otherwise is explicitly mentioned.

> In this context, it is crucially important that stakeholders comment constructively on all key aspects of the draft advice and that they also suggest concrete alternative solutions if they do not agree with what CEIOPS are proposing.
>
> The industry has a particular role to play in this respect, and I would urge financial services firms to seize all opportunities to engage in an active discussion with CEIOPS at technical level whenever you feel that the tentative direction taken would be deviating from the political agreement embedded in the Framework Directive.
>
> Once the final advice from CEIOPS has been received on the numerous issues on which the Commission has requested advice over the coming months, we will start drafting the implementing measures. The Commission will do so in close

co-operation with experts from Member States, who will see to it that the substance of what they have agreed in the level 1 text is well preserved.

I can assure you that, in developing the level 2 measures, my services will be careful to actively consult stakeholders and prepare a thorough impact assessment. Of particular importance here will be the Fifth Quantitative Impact Study, which will provide evidence on the likely impact of the draft implementing measures and will help ensure that their final version is properly calibrated.

Mr Charlie McCreevy

European Commissioner for Internal Market and Services.
Speech at the Financial Risk Solutions 10th Anniversary Dinner
Dublin, 28 September, 2009.

CHAPTER 26

Solvency II

Standard Formula Framework

WE BEGIN THIS MAIN CHAPTER—the standard formula is the main issue of this book's "practical part"—with the recitals that are important for the standard formula framework. Then we briefly address the Framework Directive (FD) and those articles that are the most important ones.

In Section 26.3, the general issues of the standard formula are discussed. The modular approach and dependence structures are presented in Section 26.4 and the adjustments to the Basic SCR that are possible to do are described in Section 26.5. Risk mitigation techniques are illustrated in Section 26.6. Limiting issues like ring fenced funds and participations are discussed in Section 26.7. The use of undertaking specific parameters is the subject of Section 26.8 and simplifications are discussed in Section 26.9.

26.1 EXTRACTS FROM THE FRAMEWORK DIRECTIVE PREAMBLE ("RECITALS")

The recitals to the FD give us the ideas behind the FD. Chapters 22 through 26 and Chapter 34 will quote the recitals, given the background to the topics discussed. Recitals not quoted in these chapters are presented in Appendix D, Section D.8. The number to the left of the recital represents the number in the preamble.

(60) The supervisory regime should provide for a risk-sensitive requirement, which is based on a prospective calculation to ensure accurate and timely intervention by supervisory authorities (the SCR), and a minimum level of security below which the amount of financial resources should not fall (the MCR). Both capital requirements should be harmonized throughout the Community in order to achieve a uniform level of protection for policyholders. For the good functioning of this Directive, there should be an adequate ladder of intervention between the SCR and the MCR.

(61) In order to mitigate undue potential procyclical effects of the financial system and avoid a situation in which insurance and reinsurance undertakings are unduly forced to raise additional capital or sell their investments as a result of unsustained adverse movements in financial markets, the market risk module of the standard formula for the SCR should

include a symmetric adjustment mechanism with respect to changes in the level of equity prices. In addition, in the event of exceptional falls in financial markets, and where that symmetric adjustment mechanism is not sufficient to enable insurance and reinsurance undertakings to fulfill their SCR, provision should be made to allow supervisory authorities to extend the period within which insurance and reinsurance undertakings are required to reestablish the level of eligible own funds covering the SCR.

(62) The SCR should reflect a level of eligible own funds that enables insurance and reinsurance undertakings to absorb significant losses and that gives reasonable assurance to policyholders and beneficiaries that payments will be made as they fall due.

(63) In order to ensure that insurance and reinsurance undertakings hold eligible own funds that cover the SCR on an ongoing basis, taking into account any changes in their risk profile, those undertakings should calculate the SCR at least annually, monitor it continuously, and recalculate it whenever the risk profile alters significantly.

(64) In order to promote good risk management and align regulatory capital requirements with industry practices, the SCR should be determined as the economic capital to be held by insurance and reinsurance undertakings in order to ensure that ruin occurs no more often than once in every 200 cases or, alternatively, that those undertakings will still be in a position, with a probability of at least 99.5%, to meet their obligations to policyholders and beneficiaries over the following 12 months. That economic capital should be calculated on the basis of the true risk profile of those undertakings, taking account of the impact of possible risk-mitigation techniques, as well as diversification effects.

(65) Provision should be made to lay down a standard formula for the calculation of the SCR to enable all insurance and reinsurance undertakings to assess their economic capital. For the structure of the standard formula, a modular approach should be adopted, which means that the individual exposure to each risk category should be assessed in a first step and then aggregated in a second step. Where the use of USPs allows for the true underwriting risk profile of the undertaking to be better reflected, this should be allowed, provided such parameters are derived using a standardized methodology.

(66) In order to reflect the specific situation of small and medium-sized undertakings, simplified approaches to the calculation of the SCR in accordance with the standard formula should be provided for.

26.2 REFERENCE TO THE FD

In the FD, Section 4 of Chapter 6 in the first title is devoted to the capital requirement. It includes articles 100–135, including SCR (articles 100–111), MCR (articles 128–131), and (partial) internal models (articles 100–102 and 112–127). Articles 132–135 treat investments.

26.3 GENERAL ISSUES

In this chapter we look at the proposed capital requirement based on a standard formula. The developments from early thoughts are given in Appendix H. Thanks to the Lamfalussy procedure; there will probably never be a final solution as there will always be a possibility to change a parameter or a dependence structure (correlation) in a timely manner.

In Chapters 27 through 33 we look closer at the main different risk modules. In this chapter, we look at the general thoughts behind the target capital requirement, SCR, aggregation of main risks, and so on. In Chapter, 34 we consider the MCR.

The subsections will be based on the (draft) final advice given by CEIOPS.

26.3.1 Purpose of the SCR

The SCR should deliver a level of capital that enables an insurance undertaking to absorb significant unforeseen losses over a specified time horizon and gives reasonable assurance to policyholders that payments will be made as they fall due.

The concept of SCR shares many features with economic capital in value-based management. Commercially, an undertaking will define its risk appetite and the public rating it wishes to achieve. It then determines the economic capital that will be necessary to limit its probability of insolvency to a defined level.

26.3.2 Risk Measure

A quantitative solvency assessment could be based on a simplified balance sheet, consisting of assets, liabilities and available capital, that is, the excess of assets over liabilities. Changes in the level of available capital will depend on the risks to which an undertaking is exposed over the time horizon of the solvency assessment. Because the future development of assets and liabilities is unknown, the future level of available capital will behave stochastically. It may be described by a probability distribution, which measures the likelihood of all possible outcomes.

A "risk measure" is, in general terms, a function that assigns an amount of capital to a risk distribution; see Chapter 14. Commonly used risk measures are VaR and TVaR. VaR assesses the probability of ruin at a specified quantile, for example, 99.5%. According to the FD and Article 101 (3), the SCR should be calibrated to correspond to the VaR.

26.3.3 Confidence Level

The level of prudence, or confidence, for the SCR will be used to calibrate the standard formula. It will also be a required design feature of internal models. The choice of the level of confidence is, according to Article 101 (3) of the FD, 99.5%, that is, a ruin probability of 0.5%. It does not imply that a ruin event will occur once in every 200 years, or that, on an annual basis, 1 in every 200 undertakings will fail. The causes of ruin in one undertaking may have a wider impact, leading to clusters of insurer failures.

26.3.4 Time Horizon

The time horizon for the SCR should reflect

- The frequency with which results are produced
- The ability of undertakings to take timely and effective management action
- The ability of supervisors to respond to a breach of the requirement

Given periodic reporting cycles, it seems sensible that, generally, a time horizon of one year should be applied to the SCR calculation. In Article 101 (3) of the FD, the time horizon is set to one year. It shall cover existing business, as well as new business expected to be written over the following 12 months.

26.3.5 Going Concern versus Runoff/Winding Up Assumptions

The purpose of regulatory capital requirements for solvency purposes is twofold. On the one hand, it aims at ensuring that the insurer is sufficiently capitalized during the defined time horizon as a going concern. On the other, regulatory capital should also provide for a successful run-off of an insurance undertaking in a ruin situation. Therefore, regulatory capital has aspects of both the going concern and run-off situations.

Over the one-year time horizon, new business may change the risk profile of an insurance undertaking. As the undertaking should be regarded as a going concern until an insolvency event, capital requirements should generally reflect new business. Additional legal considerations should be studied, especially for those Member States where law requires precise and clear definitions for imposing higher individual solvency requirements to address projections of future business.

In a ruin situation, the issue of costs specifically linked with the run-off of the insurers' business arises. However, to some extent, it could be argued that some costs associated with run-off are already reflected in TPs.

Article 101 (2) of the FD states that the presumption is that the undertaking will pursue its business as a going concern.

26.3.6 Risk Classification

The pillar I quantitative requirements is designed to address the main financial risks to which an insurance undertaking is exposed. As a general principle, a Pillar I treatment may be applied to any risk that is susceptible to quantification or limitation.

By contrast, Pillar II requirements should consider all risks, even if they cannot be quantified. It should be noted that there is no unique way of breaking down risks into categories. A categorization that provides a good fit to the risk profile of one undertaking may be less appropriate in other circumstances. This will depend largely on *the nature, scale, and complexity of the business*, that is, the proportionality principle, undertaken by an individual undertaking. In addition, the practicability criterion means that some subcategories of risk could be treated in a Pillar I internal model, but not in a Pillar I standard formula.

In Article 101 (4) of the FD, it is stated that SCR should *at least* cover the following risks:

a. Nonlife underwriting risk; see Chapter 31 for details

b. Life underwriting risk; see Chapter 32 for details

c. Health underwriting risk; see Chapter 33 for details

d. Market risk; see Chapter 27 for details

e. Credit risk; see Chapter 28 for details

f. Operational risk; see Chapter 29 for details

Operational risk shall include legal risks, but exclude risks arising from strategic decisions as well as reputation risks.

CEIOPS has also proposed, CEIOPS (2009f19), that where basic own funds allow for the value of intangible assets than the calculation of the capital requirement should include a risk module for intangible assets risk. This module is discussed in Section 26.10.

26.4 THE MODULAR APPROACH AND DEPENDENCE STRUCTURE

According to Article 103 of the FD, the SCR should be determined as

$$\mathrm{SCR} = \mathrm{BSCR} + \mathrm{CR}_{\mathrm{OR}} - \mathrm{Adj}.$$

where BSCR is the Basic SCR, $\mathrm{CR}_{\mathrm{OR}}$ is the capital charge for operational risk; see Chapter 29, and Adj is an adjustment for the risk-absorbing effect of future profit sharing and deferred taxes; see Section 26.5.

BSCR is the SCR before any adjustments, combining capital charges for five major risk categories. The following input information was required, including all subrisk modules; see Article 103 and Annex IV of the FD:

$\mathrm{CR}_{\mathrm{NL}}$: the nonlife underwriting risk module; see Chapter 31 for details about the calculation of this capital charge.

$\mathrm{CR}_{\mathrm{LR}}$: the life underwriting risk module; see Chapter 32 for details about the calculation of this capital charge.

$\mathrm{CR}_{\mathrm{HR}}$: the health underwriting risk module; see Chapter 33 for details about the calculation of this capital charge.

$\mathrm{CR}_{\mathrm{MR}}$: the market risk module; see Chapter 27 for details about the calculation of this capital charge.

$\mathrm{CR}_{\mathrm{CR}}$: the credit risk: default risk module; see Chapter 28 for details about the calculation of this capital charge.

$\mathrm{CR}_{\mathrm{IAR}}$: the intangible assets risk module; see Section 26.10 for details and about the calculation of this capital charge.

The *BSCR* is determined as

$$\mathrm{BSCR} = \mathrm{CR}_{\mathrm{IAR}} + \sqrt{\sum_{\mathrm{rxc}} \rho_{\mathrm{rc}} \cdot \mathrm{CR}_{\mathrm{r}} \cdot \mathrm{CR}_{\mathrm{c}}}.$$

This could be illustrated as in Figure 26.1.

Table 26.1 has the dependence structure used in the formula above. The structure was first given by the Annex IV in the FD, but changed by CEIOPS (2009e12) in its discussion on correlations. The main reason for this change was the new health risk module. The dependence structure was not discussed in the final advice; CEIOPS (2010a09).

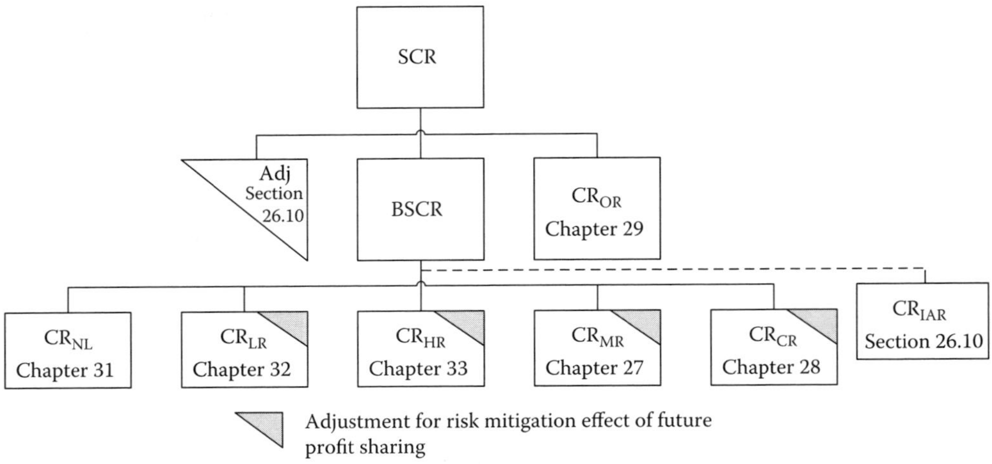

FIGURE 26.1 The modular structure of the standard formula as given in the FD and in the advice from CEIOPS. The different main risk modules are discussed in later chapters; see the figure. Adjustments are discussed in Section 26.5.

26.4.1 Standard Formula Dependence

Only for a restricted class of distributions the aggregation with linear correlation coefficients produces the correct result. In the literature, numerous examples can be found where linear correlations do not well reflect the dependence between distributions and the use of linear correlations lead to wrong or even absurd aggregation results; see, for example, Embrecht et al. (2002) and Pfeifer and Strassburger (2008), but also Chapter 15. Two main reasons for this aggregation problem are

- The dependence between the distributions is not linear; for example, there are tail dependencies

TABLE 26.1 Dependence Structures was First Defined in the FD, But Changed by CEIOPS

ρ_{rc}	CR_{NL}	CR_{LR}	CR_{HR}	CR_{MR}	CR_{CR}
CR_{NL}	1	0	0.25	0.25	0.50
CR_{LR}		1	0.75	0.25	0.25
CR_{HR}			1	0.25	0.25
CR_{MR}				1	0.50
CR_{CR}					1

Source: Adapted from CEIOPS. 2009e12. Draft L2 Advice on SCR Standard Formula—Correlation Parameters. CEIOPS-CP-74-09. November 2. Available at www.ceiops.org.

- The shape of the marginal distributions is significantly different from the normal distribution; for example, the distributions are skewed

Both characteristics are shared by many of the main risks and subrisks in the Solvency II model. A discussion on the aggregation technique is done in CEIOPS (2009e12, 2010a09). They have observed that under the financial crisis, mainly 2007–2009, market parameters such as credit spreads, property prices, equity prices, and currency exchange rates, simultaneously showed strong adverse changes. These changes are not observed under normal conditions. It also became apparent that a change in one parameter had a reinforcing effect on the deterioration of other parameters.

The underlying distributions are not normal, but usually skewed and often truncated by reinsurance or hedging.

Therefore CEIOPS (2009e12) has expressed the choice of dependence parameters as the choice of the correlation factors should attempt to avoid misestimating the aggregate risk. In particular, linear correlations are in many cases not an appropriate choice for the correlation parameter. Instead, the correlation parameters should be chosen in such a way as to achieve the best approximation of the 99.5% VaR for the aggregated capital requirement. In mathematical terms, this approach can be described as follows: for two risks X and Y with $E(X) = E(Y) = 0$, the correlation parameter ρ should minimize the aggregation error

$$\left|\mathrm{VaR}\left[(X+Y)^2\right] - \mathrm{VaR}[X^2] - \mathrm{VaR}[Y^2] - 2 \cdot \rho_{XY} \cdot \mathrm{VaR}[X] \cdot \mathrm{VaR}[Y]\right|$$

Several risks are assumed to be independent and hence a correlation of zero is chosen. CEIOPS (2009e12) shows that this is not always the best choice and if the underlying distributions are not normal positive correlation parameters may be the appropriate choice. In particular, the correlation parameters should deviate from linear correlation coefficients if the best approximation of the 99.5% VaR for the aggregated capital requirement do not achieve this, and should allow for any tail dependence between risks.

The dependence structure within the different main risk modules is discussed in Chapters 27, 31 through 33.

In the future CEIOPS will collect appropriate data from undertakings in the Member States to support any revision of the dependence structures ("the correlation factors").

26.5 ADJUSTMENTS

Adjustments to the BSCR are treated in articles 103, 108, and 111. The adjustments that could be done are deduction from the loss-absorbing capacity of technical provisions and deferred taxes. This has been discussed in CEIOPS (2009d16, 2009f26).

The adjustments for the loss-absorbing capacity of technical provisions and deferred taxes should be based on a balance sheet excluding the risk margin of the technical provisions.

26.5.1 Loss-Absorbing Capacity of Technical Provisions

Two alternative approaches could be used for adjustments: the modular approach and the single equivalent scenario approach. The latter is also called a killer scenario. In both approaches a gross calculation is made as a first step. Any reference here to gross SCR means an SCR calculated gross of the risk mitigation effect of profit sharing (and not gross of reinsurance).

A lower boundary SCR, calculated under the assumption that the undertaking passes on the impact of shocks to policyholders rather than absorb the loss themselves using own funds, has also to be determined under both approaches.

Under the *modular approach*, CEIOPS (2009f26), the net and gross calculation of SCR should be defined as follows.

- *Net calculation:* The insurer is able to vary its assumptions on future bonus rates in response to the shock being tested, based on reasonable expectations and having regard to plausible management actions.
- *Gross calculation:* In the calculation of the net SCR for each (sub)module, undertakings calculate a stressed balance sheet and compare it to the unstressed balance sheet that was used to calculate own funds. Therefore, for each (sub) module undertakings can derive the best estimate value of the technical provisions relating only to future discretionary benefits from both balance sheets. The change in these provisions measures the impact of the risk mitigation. For each submodule, this difference should be added to the net SCR used to derive the gross SCR.

Under the *single equivalent scenario approach*, the basic SCR is calculated using a single scenario under which all of the risks covered by the standard formula occurred simultaneously (the killer scenario approach). The process involved the following steps:

- The capital charge for each risk is calculated under the assumption that the insurer is not able to vary its assumptions on future bonus rates in response to the shock being tested (*gross calculation*).
- The gross capital charges are used as inputs to determine the single equivalent scenario based on the relative importance of each of the subrisks to the undertaking. Undertakings have the option to determine the single equivalent scenario using net capital charges as inputs if this is felt to more accurately reflect the relative importance of each risk.
- The undertaking then consider the management actions that should be applied in such a scenario and, in particular, whether their assumptions about future bonus rates would change if such a scenario is to occur. It is to be noted that therefore the management actions that would be applied if all stresses occur simultaneously may not be the same as those that would be applied if the stresses occur individually as in the modular approach.
- The change in the undertaking's net asset value is then calculated on the assumption that all the shocks underlying the single equivalent scenario occurred simultaneously and that the undertaking makes an operational loss equal to the capital charge in

respect of operational risk. The management actions identified above as well as the loss-absorbing capacity of deferred taxes are taken into account.

- The adjustment to the basic SCR for the loss-absorbing capacity of future discretionary benefits is determined by deducting the SCR for operational risk and the SCR calculated under the single equivalent scenario from the gross SCR.

The single equivalent scenario approach could be clarified by the use of the following notations, discussed in Annex 7 of the technical specifications to QIS 4; QIS4 (2008). See also the appendices in CEIOPS (2009d16, 2009f26).

Let SCR_D be the diversified SCR calculated as $\text{SCR}_D = \sqrt{\sum_{i=1}^{k}\sum_{j=1}^{k}\rho_{ij}C_iC_j}$, where ρ_{ij} is the dependence between risks i and j, and $i, j = 1, \dots, k$ and C_i denotes the capital charge from risk $i = 1, \dots k$. The new stress weight for risk $j = 1, \dots, k$ is defined as

$$w_j = \frac{1}{\sqrt{D_0}} \cdot \sum_{i=1}^{k} \rho_{ij}C_i, \quad \text{where } D_0 = \sum_{j=1}^{k} w_jC_j.$$

The new stress tests, the killer stress test, for risk module $j = 1, \dots, k$, are defined as $\text{original\_stress\_test}_j \times w_j$.

CEIOPS defines future discretionary benefits by distinguishing between guaranteed and discretionary benefits:

- *Guaranteed benefits:* This represents the value of future cash flows, which does not take into account any future declaration of future discretionary bonuses. The cash flows take into account only those liabilities to policyholders or beneficiaries to which they are entitled at the valuation date.
- *Conditional discretionary benefits:* This is a liability based on declaration of future benefits influenced by legal or contractual declarations and performance of the undertaking/fund. It could be linked with IFRS definition of "discretionary participation features" as additional benefits that are contractually based on
    - The performance of a specified pool of contracts or a specified type of contract or a single contract
    - Realized and/or unrealised investment return on a specified pool of assets held by the issuer
    - The profit or loss of the company, fund or other entity that issues the contract
- *Pure discretionary benefit:* This represents the liability based on the declaration of future benefits that are in discretion of the management. It could be linked with IFRS definition of "discretionary participation features" as additional benefits whose amount or timing is contractually at the discretion of the issuer.

Both types of discretionary benefits could potentially be considered to have loss-absorbing capacity. CEIOPS' paper on assumptions about future management actions, CEIOPS (2009b07, 2009f07), states that in order to be taken into account for the calculation of the technical provisions, the management actions must be *objective, realistic, and verifiable.* The advice also sets out the detail as to how an undertaking might meet these criteria; see also Section 24.3.1.4.

In applying stress scenarios to calculate gross SCR and net SCR the following rules should be considered; CEIOPS (2009f26, para 3.46):

Calculation of ΔTP (with-profits)	Gross SCR Calculation	Net SCR Calculation
ΔGuaranteed benefits	Yes	Yes
ΔConditional discretionary benefits	Only allow changes directly due to the impact of the risk under stress Disregard legal or contractual rules of the profit-sharing mechanism	Yes Apply any legal or contractual rules of the profit-sharing mechanism
ΔPure discretionary benefits	Only allow changes directly due to the impact of the risk under stress	Yes Allow effect of management actions under stress

The adjustment for loss-absorbing capacity of technical provisions should account for risk-mitigating effects in relation to the following main risk modules:

- Life underwriting risk
- Health—similar to lift technique—underwriting risk
- Market risk
- Counterparty default risk

26.5.2 Loss-Absorbing Capacity of Deferred Taxes

The calculation of this adjustment should be consistent with the calculation for loss-absorbing capacity of technical provisions. The loss-absorbing capacity of deferred taxes can be taken account for in the single equivalent scenario approach.

The adjustment for loss-absorbing capacity of deferred taxes is based on the difference between the value of deferred taxes as included in other liabilities on the balance sheet and the value of deferred taxes under the single equivalent scenario approach.

If the adjustment for loss absorbency of technical provisions is calculated using the modular approach, a further adjustment should be made to reflect the loss-absorbing capacities of deferred taxes. The adjustment should be calculated as follows:

- The BSCR should be calculated on the basis that the current, prestress, liability in respect of deferred taxes is excluded from the current balance sheet.

- The capital requirement for operational risk should be added to the BSCR. The outcome is reduced by the adjustment for the loss-absorbing capacity of technical provisions. The result of this calculation is called SCR shock.
- The liability in respect of deferred taxes should then be calculated under the assumption that the undertaking made an immediate loss equal to the SCR shock.

The adjustment to the BSCR equals the change in the deferred tax liability. Only decreases in deferred taxes should be taken account for.

26.6 RISK MITIGATION TECHNIQUES

Risk mitigation techniques are discussed in CEIOPS (2009b06, 2009f06, 2009d14, 2009f24, 2009b11, 2009f12). They are techniques that enable an undertaking to transfer part or all of their risk to another party. Allowance for mitigation techniques are treated in Article 111. Special purpose vehicles, SPV, are treated in Article 211.

26.6.1 Allowance for Financial Mitigation Techniques

A financial risk mitigation technique is a financial contract whose future value or future cash flows vary in opposite direction and equivalent, or sufficiently similar, amount to the variations of the future value or future cash flows of the assets or liabilities considered by the undertaking in its solvency assessment.

The use of financial mitigation techniques is discussed in CEIOPS (2009b06, 2009f06). The effect of financial risk mitigation techniques on the SCR is only recognized if the following two conditions are satisfied, according to the FD text:

- Credit risk and other risks arising from the use of such techniques are properly reflected in the SCR
- The instrument provides for an effective transfer of risk from the undertaking to a third party

Securitization is a mitigation technique to transfer out financial risks. However, CEIOPS' advice does not apply when the financial risks are transferred with underwriting risks and such financial risks have been assumed as part of the liabilities derived from an insurance contract, and furthermore they are not significant.

The use of a financial risk mitigation technique should be based on an intended decision to mitigate the undertaking's risk profile according the targeted overall risk management policy, where both qualitative and quantitative features shall be considered. According the principles set out below, the allowance for financial risk mitigation techniques in the calculation of the SCR with the standard formula is restricted to instruments and excludes processes and controls the undertaking has in place to manage the investment risk.

CEIOPS has set up five principles for allowing the use of financial mitigation techniques.

Principle 1: Economic effect over legal form Techniques that have material impact on an undertaking's risk profile should be recognized and treated equally, regardless of their legal form and accounting treatment.

Principle 2: Legal certainty, effectiveness, and enforceability The financial mitigation instruments together with the action and steps taken and procedures and policies implemented by the undertaking should be such as to result in risk mitigation arrangements that are legally effective and enforceable in all relevant jurisdictions.

Principle 3: Liquidity and ascertainability of value To recognize an instrument, the technique should have a value over time sufficiently reliable to provide appropriate certainty as to the risk mitigation achieved. The liquidity of the technique should follow these statements:

- The undertaking should have written internal policy regarding liquidity requirements that financial mitigation techniques should meet according to the risk management policy.
- Financial mitigation techniques considered to reduce the SCR have to meet the liquidity requirements established by the undertaking.
- The liquidity requirements shall guarantee an appropriate coordination of the liquidity features of the hedged assets or liabilities, the liquidity of the financial risk mitigation technique, and the overall policy of the undertaking regarding liquidity risk management.

Principle 4: Credit quality of the provider of the financial mitigation instrument The provider of a financial mitigation instrument should have an adequate credit quality to guarantee with appropriate certainty that the undertaking will receive the protection in the cases specified by the contract.

For the standard formula SCR, only providers of financial mitigation instruments with a rating of BBB or better should be allowed.

Principle 5: Direct, explicit, irrevocable, and unconditional features The financial mitigation techniques can reduce the SCR only if

- *Direct feature:* they provide the undertaking with a direct claim on the protection provider;
- *Explicit feature:* they contain an explicit reference to specific exposures or a pool of exposures, so that the extent of the cover is clearly defined and incontrovertible;
- *Irrevocable feature:* they are not subject to any clause, the fulfillment of which is outside the direct control of the undertaking, that would allow the protection provider to unilaterally cancel the cover or that would increase the effective cost of protection as a result of certain developments in the hedged exposure; and
- *Unconditional feature*: they are not subject to any clause outside the direct control of the undertaking that could prevent the protection provider from its obligation to pay out in a timely manner in the event that a loss occurs on the underlying exposure.

The reduction of the standard formula SCR based on mitigation of credit exposures by using *credit derivatives* is only allowed when the undertaking has in force generally applied procedures for this purposes and considers generally admitted criteria. To recognize a credit derivative contract, its cover must at least cover

- Failure to pay the amounts due under the terms of the underlying obligation that are in effect at the time of such failure
- Bankruptcy, insolvency, or inability of the obligor to pay its debts, or its failure or admission in writing of its inability generally to pay its debts as they fall due, and analogous events
- Restructuring of the underlying obligation, involving forgiveness or postponement of principal, interest, or fees, which results in a credit loss event

Admissible *collateral* for the standard formula SCR calculation must protect the undertaking against the same events listed above for credit derivatives. A collaterized transaction is a transaction in which the undertaking has a (potential) credit exposure that is hedged in whole or in part by collateral posted by a counterparty or by a third party on behalf of the counterparty. If the liabilities of the counterparty are covered by strictly *segregated assets* under arrangements that ensure the same degree of protection as a collateral that meets the above-mentioned requirements, then the segregated assets shall be treated as if they were collateral with an independent custodian.

26.6.2 Allowance for Reinsurance Mitigation Techniques

The use of reinsurance mitigation techniques is discussed in CEIOPS (2009d14, 2009f24).

In order to allow for reinsurance risk mitigation, the arrangement must satisfy the principles for risk transfer to a third party as set out below. They are similar to the principles of financial mitigation techniques. As it is difficult to anticipate the specific nature that these risk transfers may take in future years, and in order not to constrain innovation and risk management, CEIOPS has proposed high-level principles hat would facilitate the ongoing development and evolution of reinsurance risk mitigation techniques within a predefined supervisory framework.

Principle 1: Effective risk transfer The risk mitigation technique shall effectively transfer risk from the undertaking. The undertaking needs to be able to show the extent to which there is an effective transfer of risk in order to ensure that any reduction in SCR or increase in available capital resulting from its reinsurance arrangements is commensurate with the change in risk that the insurer is exposed to. The transfer of risk from the undertaking to the third party shall be effective in all circumstances in which the undertaking may wish to rely upon the transfer.

Principle 2: Economic effect over legal form Reinsurance risk mitigation techniques shall be recognized and treated equally, regardless of their legal form or accounting treatment, provided that their economic or legal features meet the requirements for such recognition. The economic effect of the transaction shall be considered over the legal form.

The design of the standard formula SCR shall allow for the changed risk profile by reflecting the economic substance of the arrangements that implement the technique. Thus, in principle, the SCR shall reflect

- A reduction in requirements commensurate with the extent of risk transfer and
- An appropriate treatment of any corresponding risks that are acquired in the process.

Principle 3: Legal certainty, effectiveness, and enforceability The reinsurance contracts used to provide the risk mitigation together with the action and steps taken, and procedures and policies implemented by the insurance undertaking, should be such as to result in risk mitigation arrangements that are legally effective and enforceable in all relevant jurisdictions. To the extent that the effectiveness or ongoing enforceability cannot be verified or the mitigation technique is not documented, the benefits of the mitigation technique should not be recognized in the SCR calculation, but the calculation shall recognize any additional risks in accordance with the formula. The standard formula SCR should, to the extent practicable, allow for the possibility that reinsurance protection will not be renewed on expiry or will be renewed on adverse terms.

Principle 4: Liquidity and valuation The design of the standard SCR calculation should recognize reinsurance risk mitigation techniques in such a way that there is no double counting of risk mitigation effects.

Where the reinsurance risk mitigation techniques reduce risk, the capital requirement should be no higher than if there were no recognition in the SCR of such reinsurance risk mitigation techniques. Where the reinsurance risk mitigation techniques actually increase risk, the SCR should be increased

Principle 5: Credit quality of the provider of the reinsurance risk mitigation instrument Undertakings shall consider the credit quality of the providers of reinsurance risk mitigation contractual arrangements and shall only take into account effective risk transfer having regard to the credit quality.

26.6.3 Treatment of Special Purpose Vehicles

The credit of an SPV within the undertaking should be equal or less than the value of the assets recoverable from the SPV. The recoverable amounts from an SPV should be considered by the undertaking as amounts deductible under reinsurance or retrocession contracts.

An SPV should be fully funded at all times and is therefore not required to calculate an individual MCR or an SCR; see CEIOPS (2009b11, 2009f12).

26.7 LIMITING ISSUES

26.7.1 Ring-Fenced Funds

The identification of ring-fenced funds is discussed in Section 25.5. CEIOPS discusses the treatment of ring-fenced funds in CEIOPS (2009e06) and in its final advice.

The procedure would be equivalent using the equivalent scenario approach; CEIOPS (2010a06).

The SCR should be calculated according to the following procedure, assuming that the modular approach is used to calculate the adjustment for loss absorbency of technical provisions; CEIOPS (2010a06):

a. When performing the calculation of each individual capital charge, the corresponding impact at the level of subportfolios of assets and liabilities (those relevant to capture the effect of each ring-fenced fund) shall be computed;

b. Where positive effects are observed at the level of a ring-fenced fund, the gross capital charge at such level should take into account any potential increase of liabilities (e.g., additional distribution of profits to policyholders) even though the overall impact of the shock on the undertaking is negative. In practice, this can only happen in those cases of bidirectional scenarios (interest rate risk, currency risk, and lapse risk) where positive effects calculated at the level of a ring-fenced fund can be observed.

c. In parallel, the capital charges at the level of each ring-fenced fund should be calculated as the net of the mitigating effect of future discretionary benefits. Where the ring-fenced fund relates to the existence of profit-sharing mechanisms, the assumptions on the variation of future bonus rates should be realistic, with due regard to the impact of the shock at the level of the ring-fenced fund and to any contractual, legal, or statutory clauses of the profit-sharing mechanism. The relevant (downward) adjustment for the loss absorbency capacity of technical provisions should not exceed, in relation to a particular ring-fenced fund, the amount of future discretionary benefits within the ring-fenced fund;

d. For each of gross/net, the total capital charge for the individual risk is given by the sum of the capital charges calculated at the level of each ring-fenced fund and that calculated at the level of the remaining subportfolio of business;

e. For each of gross/net, the total capital charges for each individual risk are then aggregated using the usual procedure of the standard formula to derive the total SCR.

26.7.2 Participations

The treatment of participations is discussed in CEIOPS (2009e05, 2010a05). CEIOPS discusses an alternative approach in treating participations in an insurance entity within the scope of group supervision as compared to the one presented in Section 25.6. Definitions of participating undertakings and related undertakings are given there.

The treatment of participations as part of the SCR calculation should deliver a treatment that mitigates the risks identified in Section 25.6 and in particular addresses double gearing. If participations were included in the equity risk submodule with no specific adjustment, this would not be achieved due to the effect of correlations in the successive levels of the

SCR formula. Therefore, there are two possible alternatives to achieve the desired outcome; CEIOPS (2010a05):

a. Position a "participations risk module" on the top of the calculations of the BSCR, in such a manner that

$$\text{Basic\_SCR} = \text{SCR\_market\_default\_life\_health\_non-life} + \text{SCR\_intangible\_assets} + \text{SCR\_participations}$$

This might be the most effective approach and the easiest formula to facilitate workable and accountable risk management practices, and supervisory review.

b. Position the SCR treatment of participations risks within the equity risk submodule but allowing for appropriate correction factors, firstly addressing double gearing and, secondly, to provide an appropriate treatment of diversification with other risks considered in the calculation of the Basic SCR.

For this risk assessment in the participating undertaking, that is, the "parent," use can be made of the risk assessment in the related undertaking, for example, a subsidiary. If the related undertaking is an insurance undertaking subject to EEA supervision, then it has to calculate its solvency capital requirement SCR_{rel}. A straightforward approach for the measurement of the risk in the participation ("parent") is to base the *equity capital charge* of the participating undertaking on SCR_{rel}. In case of an x% participation in the related undertaking, the equity capital charge of the participating undertaking for this participation could be calculated as

$$\text{CR}^{\text{Part}}_{\text{MR,ER}} = x\% \cdot \text{SCR}_{\text{rel}}.$$

One drawback, according to CEIOPS (2009e05, 2010a05), is that the diversification between the equity capital charge and other subrisks may be overstated. To correct this, they propose to include an add-on factor $f > 1$:

$$\text{CR}^{\text{Part}}_{\text{MR,ER}} = f \cdot x\% \cdot \text{SCR}_{\text{rel}}.$$

The add-on factor is proposed to be set equal to 1.2; CEIOPS (2009e05, 2010a05).

26.8 UNDERTAKING-SPECIFIC PARAMETERS

A subset of the parameters of the standard formula may, after approval of the supervisory authority, be replaced by USPs. If the supervisor finds it inappropriate to calculate the capital requirement in accordance with the standard formula they may require a company to replace a subset of the parameters by USPs. One reason could be that the risk profile of the undertaking deviates significantly from the assumptions underlying the standard formula. If so, the supervisory authority shall require the undertaking to use either USPs or (partial) internal models. If this is not applicable, a capital add-on may be imposed. This is discussed in CEIOPS (2009e13, 2010a10).

The following subset of parameters may be replaced by USP. Standardized methods for USPs are discussed in different chapters mentioned below.

- Nonlife underwriting risk—reserve and premium risk parameters; see Section 31.2.1
- Life underwriting risk—revision risk shock parameter; see Section 32.6.1
- Health underwriting risk—SLT risk—health revision risk shock parameter; see Section 33.2.6.1
- Health underwriting risk—Non-SLT risk—reserve and premium risk parameters; see Section 33.3.1.1

USPs may be used to replace different subsets of parameters within the particular risk modules stated above. For all other parameters undertakings shall use the values of standard formula parameters; CEIOPS (2010a10). Calibration of the USPs shall be carried out *at least* annually.

Parameters based on simplifications are not considered as standard parameters and may therefore not be changed for USPs.

The use of USPs is based on an approval procedure and if an undertaking wishes to go back from the use of USPs to the standard formula parameters, they have to get an approval from the supervisory authority.

The approval process consists of the following; CEIOPS (2009e13, 2010a10).

- The undertakings shall demonstrate as best as possible that the calibration of the standard formula parameters does not appropriately reflect their risk profile and that the use of USPs leads to a more appropriate result.
- Supervisors shall be satisfied that USPs are not being used to "cherry-pick" the areas that give the lowest SCR. Where USPs have only been used for some LOBs, undertakings shall explain why.
- Supervisors shall be satisfied that the USPs have been calibrated following the standardized methods laid down in this advice and meets the following criteria:
  - The risks covered by the USPs are conceptually the same as those covered by the standard formula parameters,
  - The underlying assumptions behind the standard formula parameters and behind USPs are the same,
  - The standard methodology provided should enable a robust and reliable estimation of the USPs,
  - The data used to estimate such USPs comply with the criteria set out in the advice.

- Where the supervisory authority requires supplementary information to make the assessment and verify the suitability of the USPs, approval shall also be subject to the availability of such additional information.
- The supervisory authority should assess, on the basis of the information supplied by the undertaking, whether the data and revised calibration are relevant to the undertaking and whether the data are sufficient to justify the revised calibration.
- Should supervisors require undertakings to replace the subset of parameters specified above by USPs; their decision shall state the reasons. The supervisor shall be required to
    - Explain as best as possible why they consider the standard formula parameters inadequate
    - Where the undertaking concerned is not able to comply with such decision, supervisory authorities shall provide alternative actions, in line with the Level 1 Text and relevant Level 2 implementing measures

26.9 SIMPLIFICATIONS

The undertaking is responsible to determine the SCR by using appropriate methods selecting from the following list, taking into account nature, scale, and complexity of the risks:

- Full internal model
- Standard formula using
    - Partial internal model
    - USPs
    - Explicit as described in Chapters 26 through 33
    - Simplification

The proportionality is discussed in more detail in Section 22.3. In assessing whether the standard calculation or a simplified calculation could be considered proportionate to the underlying risks, see CEIOPS (2009e15, 2010a12), the insurer should have regard to the following steps.

Step 1: Assessment of nature, scale, and complexity The insurer should assess the nature, scale, and complexity of the risks. This is intended to provide a basis for checking the appropriateness of specific simplifications carried out in the subsequent step.

Step 2: Assessment of the model error In this step, the insurer shall assess whether a specific simplification can be regarded as proportionate to the nature, scale, and complexity of the risks analysed in the first step.

Specific simplifications for the risk submodules of the SCR calculation are discussed in the following sections.

26.10 INTANGIBLE ASSETS RISK MODULE

Where basic own funds allow for, that is, increase with, the value of intangible assets, CEIOPS (2009f19) recommends that the risks inherent in intangible assets should be considered in the standard calculation of the SCR. For this purpose, it is considered that all risks intangible assets are exposed to should be taken into account, including not only market risks, but also internal risks, inherent to the specific nature of these elements. Any inclusion of intangible assets within the market risk module would allow for diversification benefits with the market risk submodules and with the other four top modules of the BSCR (counterparty default, life, nonlife, and health).

The capital requirement for intangible assets shall be calculated as follows:

$$\mathrm{CR}_{\mathrm{IAR}} = f_{\mathrm{IA}} \cdot IA_{\mathrm{FV}} - T3_{\mathrm{DCP}},$$

where $f_{\mathrm{IA}} = 1$, $\mathrm{IA}_{\mathrm{FV}}$ is the fair value of intangible assets, and $T3_{\mathrm{DCP}}$ is the amount of intangible assets not considered as own funds due the trigger of the limit of Tier 3 ("double counting prevention"). According to a reasonable method, the effect of triggering the Tier 3 limit should be allocated among the different eligible items of own funds included in this tier.

CHAPTER 27

Solvency II Standard Formula

Market Risk

MARKET RISK arises from the level of volatility of market prices of financial instruments. Exposure to market risk is measured by the impact of movements in the level of financial variables, such as stock prices, interest rates, real estate prices, and exchange rates.

27.1 GENERAL FEATURES

The market risk and its capital charge are treated in different articles of the FD, namely 104, 105, 106, 111, and also in Articles 13 and 132. The development of the market risk module is discussed in Appendices I and J. CEIOPS has discussed the general structure of the market risk module in CEIOPS (2009d09, 2009f20). Dampeners are discussed in CEIOPS (2009e02, 2009e07, 2010a07), and in Section 27.8. Assets that are allocated to policies where the policyholders bear the investment risk are excluded from the module only to the extent that the risk is passed on to policyholders. Where assets held include an investment in a subsidiary undertaking, which manages all or part of the insurance undertaking's investments on its behalf, the undertaking shall take into account the underlying assets held by the subsidiary undertaking.

27.1.1 Standard Formula

The structure of the market risk module is given in Annex IV (4) of the FD. Figure 27.1 shows the structure of the risk module and its subrisks.

CR_{MR} denotes the capital charge from the market risk module. The capital charges of its subrisks modules are the following:

- $CR_{MR,IR}$: the interest rate risk module
- $CR_{MR,ER}$: the equity risk module
- $CR_{MR,PR}$: the property risk module
- $CR_{MR,CR}$: the currency risk module

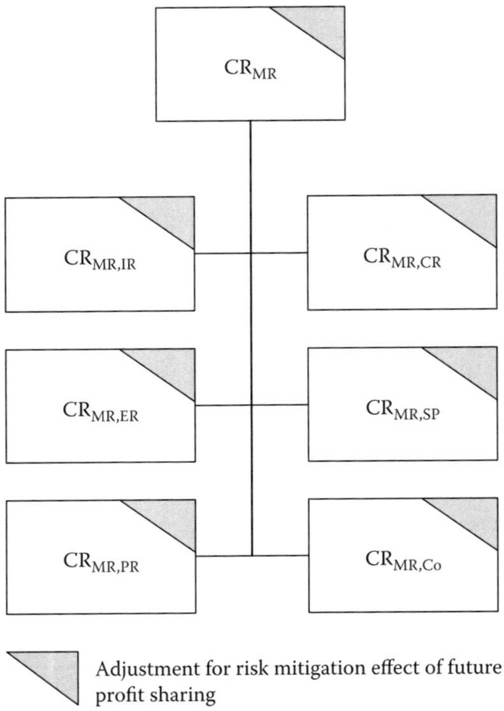

FIGURE 27.1 The modular structure of the capital charge for the market risk.

- $CR_{MR,SP}$: the spread risk module
- $CR_{MR,CO}$: the concentration risk module

The capital charge will be calculated as

$$CR_{MR} = \sqrt{\sum_{rxc} \rho_{rc} \cdot CR_{MR,r} \cdot CR_{MR,c}},$$

where r and c are the rows and columns of the dependence matrix for the market risk. This matrix is shown in Table 27.1.

The dependence structure shown in Table 27.1 is discussed in CEIOPS (2010a09). The main reason for the new structure as compared to the dependence structures used in the earlier Quantitative Impact Studies, QIS, is the financial crisis. For CEIOPS it became apparent that the dependence structure in market risk changed in stressed situation. It was also observed that the risks had a reinforcing effect on each other; CEIOPS (2009e12).

In CEIOPS (2010a09), it was proposed to introduce a two-sided correlation between interest rate risk and equity/property/spread risks in the standard formula:

- In case the insurer is exposed to a *fall in interest rate risk (fIR)*, a correlation parameter of 50% between interest rate risk and equity/property/spread risks should be applied to aggregate the respective capital charges;

TABLE 27.1 Dependence Matrix for the Market Risk's SubRisks as Proposed by CEIOPS

ρ_{rc}	$CR_{MR,IR}$	$CR_{MR,ER}$	$CR_{MR,PR}$	$CR_{MR,CR}$	$CR_{MR,SP}$	$CR_{MR,Co}$
$CR_{MR,IR}$	1	fIR:0.50 rIR: 0	fIR:0.50 rIR: 0	0.50	fIR:0.50 rIR: 0	0.50
$CR_{MR,ER}$		1	0.75	0.50	0.75	0.50
$CR_{MR,PR}$			1	0.50	0.50	0.50
$CR_{MR,CR}$				1	0.50	0.50
$CR_{MR,SP}$					1	0.50
$CR_{MR,Co}$						1

Source: Adapted from CEIOPS. 2010a09. CEIOPS' Advice for L2 Implementing Measures on SII: SCR Standard Formula—Correlations (former Consultation Paper no. 74). CEIOPS-DOC-70-10. January 29. Available at www.ceiops.org.

Note: fIR: fall in Interest Rate risk, rIR: rise in Interest rate risk.

- In case the insurer is exposed to a *rise in interest rate risk (rIR)*, a correlation parameter of 0% between interest rate risk and equity/property/spread risks should apply.
- The correlation parameter then results from the decisive risk for the undertaking. Therefore the application of the two-sided correlation depends on whether a fall or rise in interest rates is the crucial factor.

Net Capital Charge

For all subrisks, a net value of the capital charge is calculated:

$nCR_{MR,i}$: Capital charge for market subrisk i, including the risk-absorbing effect of future profit sharing.

The total net market risk capital charge was then calculated using the dependence matrix above:

$$nCR_{MR} = \sqrt{\sum_{rxc} \rho_{rc} \cdot nCR_{MR,r} \cdot nCR_{MR,c}}.$$

These net charges will be used to calculate a risk adjustment on the top level; see Chapter 26.

27.1.2 Delta-NAV Approach

Most of the market risk capital charges are based on stress tests using the so-called delta-NAV (net asset value) approach. This approach considers the change in the net value of assets minus liabilities due to the stress test. The change in NAV should be based on a balance sheet that excludes the risk margin of the technical provisions. This approach is based on the assumption that the risk margin does not change materially under the stress test.

Where a delta-NAV (or ΔNAV) approach is used, the impact of hedging instruments should be allowed for as part of the submodule: use of the delta-NAV calculation ensures the impact of the stress scenario on the hedging instrument is captured alongside the impact on all other assets and liabilities. See Section 26.6 and the discussion there regarding the use of financial risk mitigation instruments.

27.1.3 Investment Funds

It has been proposed to CEIOPS that it would be important to have the treatment of *collective investment vehicles*, and other investments packaged as funds, included in the market risk module.

Therefore CEIOPS proposes to properly assess the market risk inherent in these instruments and it will be necessary to examine their economic substance.

If possible this should be achieved by applying *a look-through approach* in order to assess the risks applying to the assets underlying the investment vehicle. Each of the underlying assets would then be subjected to the relevant submodule stresses and capital charges calculated accordingly. The same look-through approach will also be applied for other indirect exposures.

Where an investment fund is invested in other investment funds, that is, where a number of iterations of the look-through approach is required, the number of iterations shall be sufficient to ensure that all material market risk is captured.

27.1.4 SPV Notes

Notes issued by an SPV and held by an undertaking are assets. Due to their intrinsic legal and financial architecture, SPV notes are one of the investments where the requirements of the FD text are complex to meet. However, any inappropriate application may endanger the solvency position of an undertaking holding significant exposures in SPV notes; CEIOPS (2009f20).

It is important to have in mind that the same issuance of notes uses to have different tranches with substantially different risk profiles.

Non-intra-group SPV notes: Those materializing a securitization of risks of entities external to the group the undertaking belongs to should be treated as follows:

- Non-intra-group SPV notes having mostly the features of fixed income bonds, authorized and subject to the requirements set out in Chapter 25, and rated BBB (stable) or better: Their risks shall be considered in the "spread risk," "interest rate risk," and concentration submodules according its rating.
- Non-intra-group SPV notes having mostly the features of fixed income bonds, authorized and subject to the requirements set out in Chapter 25, rated below BBB (stable) or unrated: Their risks shall be considered in the "equity risk" submodule as nontraded equities, unless they are traded actively in a financial market.

- *Other non-intra-group SPV notes, including those having significant features of equities (i.e., equity tranche notes):* Their risks shall be considered in the "equity risk" submodule. For this purpose the SPV notes shall be considered as nontraded equities, unless they are traded actively in a financial market.

Intra-group SPV notes: Those materializing in a major part a securitization of risks of entities belonging to the group of the undertaking; CEIOPS (2009f12):

- SPVs could be used by more than one undertaking within the same group, to transfer risk to outside this group. However an SPV should only be used by one group and not by a number of undertakings from different groups. . .
- Where separate undertakings within a group use an SPV, that SPV should be established in such a way that the SPV is protected from the impact of a related undertaking within a group being wound up ("such as a bankruptcy remote vehicle"). . .
- An important mandatory condition for authorizing an intragroup SPV is that the undertaking cannot use an internal SPV (i.e., one where no element of finance is raised externally) to achieve a regulatory capital reduction at the group level in the absence of any financing external to the group.
- In the absence of external financing, only the solo undertaking who has the contract with the SPV may take regulatory capital relief for the SPV.
- CEIOPS considers that, as separate undertakings within a group using an SPV should ensure that the SPV is structured in such a way that the SPV is protected from the impact of a related undertaking within a group being wound up, the same principle should apply to intragroup SPVs.
- The requirements should be assessed in relation to the solo undertaking.

An undertaking holding intragroup SPV bears various and complex risks—not only underwriting risks, but might also bear operational risks, market risks, counterparty default risks, and additionally group-specific risks. Therefore CEIOPS considers that such risks cannot be captured appropriately in the standard formula. Hence, in order to ascertain that the SCR will reflect the risks assumed by the undertaking, it is expected that an undertaking aiming to assume a material exposure in intragroup SPV notes will develop appropriate *internal model* to capture each of the risks associated to the holding of the SPV notes. Otherwise, the undertaking's SPV notes should receive similar treatment as in the case of external SPV notes, being considered in any case as nontraded assets.

27.2 INTEREST RATE RISK

Let ΔNAV be the net value of assets minus liabilities, and $\{\Delta\text{NAV}|\text{upwardshock}\}$ and $\{\Delta\text{NAV}|\text{downwardshock}\}$ are the changes in the net value of asset and liabilities due to revaluing all interest–rate-sensitive instruments using altered term structures. The intention is that the calibration of the interest rate shock will capture changes in level, slope, and curvature of the term structure.

The capital charge will be a combination of two predefined factors in *the term structure* (the upward shock and the downward shock) and specific alterations in the interest rate *implied volatility*. The volatility shocks are usually only relevant where insurers' asset portfolios and/or their insurance obligations are sensitive to changes in interest rate volatility, for example, where liabilities contain embedded options and guarantees. Hence, for many nonlife obligations the interest rate volatility stress will be immaterial and on that basis could be ignored.

However, in general we will have four predefined scenarios; see CEIOPS (2009e08, 2010a08).

- UU: Interest rate shock: *upward* + implied volatility experience: *upward parallel shift*
- UD: Interest rate shock: *upward* + implied volatility experience: *downward parallel shift*
- DU: Interest rate shock: *downward* + implied volatility experience: *upward parallel shift*
- DD: Interest rate shock: *downward* + implied volatility experience: *downward parallel shift*

Let

$$\begin{aligned}
\mathrm{CR}_{\mathrm{MR,IR}}^{\mathrm{UU}} &= \left\{\Delta\mathrm{NAV} \mid \text{upward shock} + \text{upward parallel shift}\right\},\\
\mathrm{CR}_{\mathrm{MR,IR}}^{\mathrm{UD}} &= \left\{\Delta\mathrm{NAV} \mid \text{upward shock} + \text{downward parallel shift}\right\},\\
\mathrm{CR}_{\mathrm{MR,IR}}^{\mathrm{DU}} &= \left\{\Delta\mathrm{NAV} \mid \text{downwardshock} + \text{upward parallel shift}\right\},
\end{aligned}$$

and

$$\mathrm{CR}_{\mathrm{MR,IR}}^{\mathrm{DD}} = \left\{\Delta\mathrm{NAV} \mid \text{downward shock} + \text{downward parallel shift}\right\}$$

then the interest rate capital charge will be

$$\mathrm{CR}_{\mathrm{MR,IR}} = \max\left\{0; \mathrm{CR}_{\mathrm{MR,IR}}^{\mathrm{UU}}; \mathrm{CR}_{\mathrm{MR,IR}}^{\mathrm{UD}}; \mathrm{CR}_{\mathrm{MR,IR}}^{\mathrm{DU}}; \mathrm{CR}_{\mathrm{MR,IR}}^{\mathrm{DD}}\right\}.$$

When calculating the capital charges for the four scenarios above, a correlation of 0% should be assumed between the volatility stress event and the term structure stress event in each term structure/volatility stress pair.

The $\{\Delta\mathrm{NAV}|\cdot\}$ are the changes in ΔNAV due to revaluation of all interest rate sensitive assets and liabilities based on

- Specified alterations to the interest rate term structures in combination with
- Specified alterations to interest rate volatility

TABLE 27.2 The Term Structure Stresses in % for Interest Rates

Maturity in Years	Up (%)	Down (%)
0.25	70	−75
0.5	70	−75
1	70	−75
2	70	−65
3	64	−56
4	59	−50
5	55	−46
6	52	−42
7	49	−39
8	47	−36
9	44	−33
10	42	−31
11	39	−30
12	37	−29
13	35	−28
14	34	−28
15	33	−27
16	31	−28
17	30	−28
18	29	−28
19	27	−29
20	26	−29
21	26	−29
22	26	−30
23	26	−30
24	26	−30
25	26	−30
30	25	−30

Source: Adapted from CEIOPS. 2010a08. CEIOPS' Advice for L2 Implementing Measures on SII: SCR Standard Formula—Calibration of Market Risk Module (former Consultation Paper no. 70). CEIOPS-DOC-66-10. January 29. Available at www.ceiops.org.

The upward and downward shocks were determined by CEIOPS (2010a08) as in Table 27.2.

The absolute change of interest rates in the downward scenario should at least be one percentage point. Where the unstressed rate is lower than 1%, the shocked rate in the downward scenario should be assumed to be 0%. The downward stress can be defined by the following formula:

$$r' = \max\left\{\min\left[(1 + \text{stress factor}) \cdot r; r - 1\%\right]; 0\right\}$$

where r is the unstressed and r' the stressed rate.

The volatility stresses are given in Table 27.3; CEIOPS (2010a08).

TABLE 27.3 The *Additive* Volatility Stresses in % for Interest Rates

Volatility (additive) stress: Up	12%
Volatility (additive) stress: Down	−3%

Example To calculate the capital charge $CR^{UU}_{MR,IR}$ applying for an interest rate of term 10 years and given a current 10-years rate of r% and volatility of v% the company has to calculate the change in NAV (ΔNAV) on moving to stressed interest rate of $(1 + 0.42)^* r$ and stressed volatility of $(v + 12)$% and then combine them with a correlation of 0.

27.2.1 Simplifications

Captives could use simplifications for the interest rate risk; see CEIOPS (2009e16, 2010a13). Assets are grouped into maturity buckets for a simplified duration:

Maturity of Assets	Simplified Duration
<1 year	0.5 year
1–3 years	2 years
3–5 years	4 years
5–10 years	7 years
>10 years	12 years

The effect of the interest rate shocks on the market value of interest rate sensitive assets MV_i, grouped in maturity intervals i, is calculated as follows:

$$\text{Interest rate risk asset up} = \sum_i MV_i \cdot Dur_i \cdot rate_i \cdot Shock_{i,up}$$

$$\text{Interest rate risk asset down} = \sum_i MV_i \cdot Dur_i \cdot rate_i \cdot Shock_{i,down}$$

where MV_i is the market value of interest-rate-sensitive assets I, Dur_i is the simplified duration of maturity interval i, $rate_i$ is the risk-free rate for simplified duration of maturity interval I, and $Shock_{i,up}$ is the relative upward shock of interest rate for simplified duration of maturity interval i, and $Shock_{i,down}$ relative downward shock of interest rate for simplified duration of maturity interval i.

The simplified calculation should be done separately for assets of different currency.

For the shocks on liabilities, captives should in a first step asses the duration of the liabilities per LOB. In a second step, the relevant term structure is used to calculate the change in the best estimate BE_k as follows:

$$\text{Interest rate risk best estimate up} = -\sum_{LOB:k} BE_k \cdot Dur_k \cdot rate_k \cdot Shock_{k,up}$$

$$\text{Interest rate risk best estimate down} = -\sum_{LOB:k} BE_k \cdot Dur_k \cdot rate_k \cdot Shock_{k,down}$$

where BE_k is the best estimate of LOB k, Dur_k is the modified duration of the best estimate in LOB k, $rate_k$ is the risk-free rate for modified duration Dur_k, $Shock_{k,up}$ is the relative upward shock of interest rate for modified duration Dur_k, and $Shock_{k,down}$ is the relative downward shock of interest rate for modified duration Dur_k.

The simplified calculation should be done separately for liabilities of different currency.

27.3 EQUITY RISK

Equity risk arises from the level or volatility of market prices for equities. Exposure to equity risk refers to all assets and liabilities whose value is sensitive to changes in equity prices.

The standard capital charge is discussed in CEIOPS (2009e07, 2010a07) and covers two categories: equities listed in EEA and OECD countries (global equities) and other equities. This is the same approach used in QIS4.

There are two possible ways to calculate the equity risk capital charge: the standard approach, discussed here, and a "duration dampener" approach discussed in Section 27.8.2. For the standard approach there is also a symmetric dampener adjustment mechanism to be applied to the standard capital stress SS below. This is discussed in Section 27.8.1.

The calculation is made in three steps:

1. Equity Level capital requirement
2. Equity Volatility capital requirement
3. Aggregation

27.3.1 Level Equity Capital Requirement

For each category of equity

$i = 1$: Global, G (equities listed in EEA and OECD countries)
$i = 2$: Other, O (compromising equities not listed in EEA or OECD countries, nonlisted and private equities, hedge funds, commodities, and other alternative investments)

the capital charge is calculated as

$$CR^{L}_{MR,ER,i} = \{\Delta NAV | \text{equity shock down}_i; 0\}$$

where ΔNAV is the net value of equity assets minus equity liabilities after an *equity shock* of index i, $i = 1, 2$, and L stands for Level. The equity shock, or standard capital stress, proposed by CEIOPS is given in Table 27.4a. As they have published both a majority view and a minority view, it will be up to the European Commission to decide. (One of CEIOPS' members supported a 32% stress).

The equity shock used in the calculation is an adjusted standard capital stress as defined in Section 27.8.1; see Equation 27.1.

For simplicity we denote the two capital charges (Global equity and Other equity) by $CR(G; L)$ and $CR(O; L)$, respectively.

TABLE 27.4a The Standard Capital Stress, SS, or Level Equity Shocks Proposed by CEIOPS

	Global	**Other**
Level equity shock_i	Majority view: 45% Minority view: 39%	55%

Source: Adapted from CEIOPS. 2010a07. CEIOPS' Advice for L2 Implementing Measures on SII: Equity Risk Sub-Module (former Consultation Paper no. 69). CEIOPS-DOC-65-10. January 29. Available at www.ceiops.org.

TABLE 27.4b The Equity Volatility Shocks Proposed by CEIOPS

	Upward	**Downward**
Volatility equity shock_i	50%	−15%

Source: Adapted from CEIOPS. 2010a07. CEIOPS' Advice for L2 Implementing Measures on SII: Equity Risk Sub-Module (former Consultation Paper no. 69). CEIOPS-DOC-65-10. January 29. Available at www.ceiops.org.

27.3.2 Volatility Equity Capital Requirement

Firms should assess whether they are affected by equity volatility or not, and if so if they are affected by an increase or a decrease or both; CEIOPS (2010a07, para 3.111).

CEIOPS (2010a07) proposed an equity volatility stress consisting of a relative volatility stress of 50% in the upward direction. If a downward stress is seen relevant, then by assuming that the relative strengths of the up and down stresses are similar for 5-year options as for 1-month options CEIOPS proposed a downward relative stress of around 15%.

The Volatility equity shock proposed by CEIOPS is summarized in Table 27.4b.

For each category of equity

$i = 1$: Global, G

$i = 2$: Other, O

the capital charges are calculated as separate volatility stress (using simplifying notations) for Global and Other, respectively. We use the upward and downward shocks from Table 27.4b and get

$$\begin{aligned}
\mathrm{CR}(G;\mathrm{Vol}^{\mathrm{Up}}) &= \Delta\mathrm{NAV} \mid \text{Volatility shock Up}_i \\
\mathrm{CR}(G;\mathrm{Vol}^{\mathrm{Down}}) &= \Delta\mathrm{NAV} \mid \text{Volatility shock Down}_i \\
\mathrm{CR}(O;\mathrm{Vol}^{\mathrm{Up}}) &= \Delta\mathrm{NAV} \mid \text{Volatility shock Up}_i \\
\mathrm{CR}(O;\mathrm{Vol}^{\mathrm{Down}}) &= \Delta\mathrm{NAV} \mid \text{Volatility shock Down}_i,
\end{aligned}$$

where ΔNAV is the net value of equity assets minus equity liabilities after a *volatility shock* of index i, $i = 1,2$, and Vol stands for Volatility.

The overall volatility capital charges for Global equity and Other equity are calculated as Global equity:

$$\begin{aligned}\mathrm{CR}(G;\mathrm{Vol}^{\mathrm{Up}},L) &= \Delta\mathrm{NAV} \mid \text{Volatility shock } \mathrm{Up}_i, \text{Level}\\ &= \sqrt{\mathrm{CR}(G;\mathrm{Vol}^{\mathrm{Up}})^2 + \mathrm{CR}(G;L)^2 + 2\cdot\rho_{L,\mathrm{Up}}\cdot\mathrm{CR}(G;\mathrm{Vol}^{\mathrm{Up}})\cdot\mathrm{CR}(G;L)}\end{aligned}$$

$$\begin{aligned}\mathrm{CR}(G;\mathrm{Vol}^{\mathrm{Down}},L) &= \Delta\mathrm{NAV} \mid \text{Volatility shock } \mathrm{Down}_i, \text{Level}\\ &= \sqrt{\mathrm{CR}(G;\mathrm{Vol}^{\mathrm{Down}})^2 + \mathrm{CR}(G;L)^2 + 2\cdot\rho_{L,\mathrm{Down}}\cdot\mathrm{CR}(G;\mathrm{Vol}^{\mathrm{Down}})\cdot\mathrm{CR}(G;L)}\end{aligned}$$

Other equity:

$$\begin{aligned}\mathrm{CR}(O;\mathrm{Vol}^{\mathrm{Up}},L) &= \Delta\mathrm{NAV} \mid \text{Volatility shock } \mathrm{Up}_i, \text{Level}\\ &= \sqrt{\mathrm{CR}(O;\mathrm{Vol}^{\mathrm{Up}})^2 + \mathrm{CR}(O;L)^2 + 2\cdot\rho_{L,\mathrm{Up}}\cdot\mathrm{CR}(O;\mathrm{Vol}^{\mathrm{Up}})\cdot\mathrm{CR}(O;L)}\\ \mathrm{CR}(O;\mathrm{Vol}^{\mathrm{Down}},L) &= \Delta\mathrm{NAV} \mid \text{Volatility shock } \mathrm{Down}_i, \text{Level}\\ &= \sqrt{\mathrm{CR}(O;\mathrm{Vol}^{\mathrm{Down}})^2 + \mathrm{CR}(O;L)^2 + 2\cdot\rho_{L,\mathrm{Down}}\cdot\mathrm{CR}(O;\mathrm{Vol}^{\mathrm{Down}})\cdot\mathrm{CR}(O;L)},\end{aligned}$$

where L stands for Level equity shock and the dependence are set as $\rho_{L,\mathrm{Up}} = 0.75$ and $\rho_{L,\mathrm{Down}} = 0$, that is, the firms may assume a correlation coefficient of 0.75 between equity level and upward equity volatility and a correlation coefficient of 0 between equity level and downward equity volatility.

27.3.3 Aggregation

The total capital charge including volatility equity shock and level equity shock is calculated as

$$\mathrm{CR}(G) = \max\left\{\mathrm{CR}(G;\mathrm{Vol}^{\mathrm{Up}},L), \mathrm{CR}(G;\mathrm{Vol}^{\mathrm{Down}},L)\right\} \text{ for Global equity and}$$

$$\mathrm{CR}(O) = \max\left\{\mathrm{CR}(O;\mathrm{Vol}^{\mathrm{Up}},L), \mathrm{CR}(O;\mathrm{Vol}^{\mathrm{Down}},L)\right\} \text{ for Other equity.}$$

The overall capital charge is calculated using a dependence between the Global index and Other index of $\rho_{G,O} = 0.75$, that is,

$$\mathrm{CR}_{\mathrm{MR,ER}} = \sqrt{\mathrm{CR}(G)^2 + \mathrm{CR}(O)^2 + 2\cdot\rho_{G,O}\cdot\mathrm{CR}(G)\cdot\mathrm{CR}(O)}$$

27.4 PROPERTY RISK

The capital charge is calculated as, CEIOPS (2009d09, 2009f20),

$$\mathrm{CR}_{\mathrm{MR,PR}} = \{\Delta\mathrm{NAV} \mid \text{property shock}\},$$

where the *property shock* is the immediate effect expected in the event of a fall in real estate benchmarks, taking account of all the participant's individual direct and indirect exposures to property prices.

CEIOPS assumes that it would not be proportionate to explicitly test changes in the volatility of property prices as part of the standard formula approach. However, these factors are implicitly taken into account in the calibration of the shock scenarios.

The property shock was determined by CEIOPS (2010a08) as given in Table 27.5.

The following investments shall be treated as property, and hence their risks considered in the property risk submodule:

- Land, buildings, and immovable-property rights
- Direct or indirect participations in real estate companies that generate periodic income or which are otherwise intended for investment purposes
- Property investment for the own use of the insurance undertaking

Otherwise, the following investments shall be treated as equity and their risks considered in the equity risk submodule:

- An investment in a company engaged in real estate management, or
- An investment in a company engaged in real estate project development or similar activities, or
- An investment in a company that took out loans from institutions outside the scope of the insurance group in order to leverage its investments in properties.

Collective real estate investment vehicles should be treated like other collective investment vehicles with a look-through approach.

TABLE 27.5 Stress Test for Properties

Property Sector	Stress
All properties	−0.25

Source: Adapted from CEIOPS. 2010a08. CEIOPS' Advice for L2 Implementing Measures on SII: SCR Standard Formula—Calibration of Market Risk Module (former Consultation Paper no. 70). CEIOPS-DOC-66-10. January 29. Available at www.ceiops.org.

27.5 CURRENCY RISK

For each currency, the capital charge for currency risk is determined from the result of two predefined scenarios:

$$\mathrm{CR}^{\mathrm{up}}_{\mathrm{MR,CR}} = \{\Delta\mathrm{NAV} \mid \text{upward currency shock}\} \quad \text{and}$$
$$\mathrm{CR}^{\mathrm{down}}_{\mathrm{MR,CR}} = \{\Delta\mathrm{NAV} \mid \text{downward currency shock}\},$$

where the upward and downward shocks are respectively the immediate effect expected on the net value of asset and liabilities in the event of a rise and fall, respectively in value of all other currencies against the local currency in which the undertaking prepares its local regulatory accounts, taking account of all the participant's individual currency positions and its investment policy (e.g., hedging arrangements, gearing etc.).

The total capital charge for currency risk is determined as

$$\mathrm{CR}_{\mathrm{MR,CR}} = \sum_{\text{each\_currency}} \max\left\{\mathrm{CR}^{\mathrm{up}}_{\mathrm{MR,CR}};\mathrm{CR}^{\mathrm{down}}_{\mathrm{MR,CR}}\right\}.$$

For currencies pegged to Euro there will be different stress tests, see Table 27.6. The currency upward and downward shocks were determined by CEIOPS (2010a08) as in Table 27.6.

27.6 SPREAD RISK

The capital charge we are seeking is denoted $C_{\mathrm{MR,SP}}$ and is a part of the Market risk module in the standard formula framework.

The spread risk submodule should cover the credit risk of

- Bonds (including deposits with credit institutions)
- Loans guaranteed by mortgages
- Structured credit products, such as asset-backed securities and collateralized debt obligations
- Credit derivatives, such as credit default swaps, total return swaps, and credit-linked notes

TABLE 27.6 Currency Stress Tests

Currency	Up	Down
Danish Krone against Euro, Lithuanian litas, or Estonian kroon	0.0225	−0.0225
Estonian Kroon against Euro or Lithuanian litas	0	0
Latvian lats against Euro, Lithuanian litas, or Estonian kroon	0.01	−0.01
Lithuanian litas against Euro or Estonian kroon	0	0
Latvian lats against Danish Krone	0.035	−0.035
All other currency pairs	0.25	−0.25

Source: Adapted from CEIOPS. 2010a08. CEIOPS' Advice for L2 Implementing Measures on SII: SCR Standard Formula—Calibration of Market Risk Module (former Consultation Paper no. 70). CEIOPS-DOC-66-10. January 29. Available at www.ceiops.org.

In relation to credit derivatives, only the credit risk that is transferred by the derivative is covered in the spread risk submodule. Investments for the benefit of life-insurance policyholders who bear the investment risk are included only to the extent that the undertaking assumes spread risk. No capital charge shall apply to borrowings by or guaranteed by national government of an OECD or EEA state, issued in the currency of the government, or issued by a multilateral development bank as listed in the Capital Requirements Directive, CRD, or issued by an international organization listed in the CRD.

CEIOPS proposes that this submodule applies to at least the following classes of bonds:

- Investment grade corporate bonds
- High yields corporate bonds
- Subordinated debt
- Hybrid debt

The spread risk module is applicable to all types of asset-backed securities as well as to all the tranches of structured credit products such as collateralized debt obligations. This class of securities includes transactions of schemes whereby the credit risk associated with an exposure or pool of exposures is tranched, having the following characteristics:

- Payments in the transaction or scheme are dependent upon the performance of the exposure or pool of exposures;
- The subordination of tranches determines the distribution of losses during the ongoing life of the transaction or scheme.

The spread risk submodule should further cover in particular credit derivatives, for example, credit default swaps, total return swaps, and credit linked notes that are not held as part of a recognized risk mitigation policy. As indicated above, the spread risk submodule will also be applicable to all tranches of structured credit products such as collateralized debt obligations. In addition, traditional forms of asset backed securities, that is, commercial and residential mortgage backed securities, home equity loans, credit card receivables, auto loans, student loans as well as whole-business securitizations, infrastructure finance notes, and other covered bonds are also addressed by this submodule.

Instruments sensitive to changes in credit spreads may also give rise to other risks, which should be treated accordingly in the appropriate modules. For example, the counterparty default risk associated with the counterparty should be addressed in the counterparty default risk module, rather than in the spread risk submodule; CEIOPS (2009d09, 2009f20).

Government bonds are exempted from the application of this submodule. The exemption concerns borrowings by the national government, or guaranteed by the national government, of an OECD or EEA state, issued in the currency of the government.

The capital charge, $CR_{MR,SP}$, is based on two factor-based approaches, one assuming a rise in credit spread and the second a fall in credit spreads. The charge will be the maximum of these two calculations. Migration and default risk are addressed implicitly in both the calibration of the factors and in the movements in credit spreads. For credit derivatives, the capital charge is based on a stress-test approach.

The capital charge for the spread risk is split up into four components, one for bonds, one for structured credit products, one for credit derivatives, and one for mortgage loans (real estate). Hence we have the following capital charge:

$$C_{\text{MR,SP}} = C_{\text{SP,Bo}} + C_{\text{SP,Struc}} + C_{\text{SP,Cd}} + C_{\text{SP,Re}}.$$

For bonds:

$$C_{\text{SP,Bo}} = \sum_i \text{EAD}_i \cdot \text{dur}_i \cdot f(g_i) + \Delta\text{Liab}_{\text{UL}}$$

where EAD_i is the exposure at default, that is, the credit risk exposure i as determined by reference to market values, dur_i is the duration of credit risk exposure i, $f(g_i)$ is a function of the rating class (g) of the credit risk exposure i, calibrated to deliver a shock consistent with VaR 99.5%; see Table 27.7, and $\Delta\text{Liab}_{\text{UL}}$ is the overall impact on the liability side for policies where the policyholders bear the investment risk with embedded options and guarantees of the stressed scenario, with a minimum value of 0 (sign convention: positive sign means losses). The stressed scenario is defined as a drop in value on the assets (except government bonds issued by an EEA or OECD government in its local currency) used as the reference to the valuation of the liabilities by $f(g_i)$, for example, for a BBB-rated asset with a duration of 4 years, this means a drop by 10.0%.

For Structured Credit Products:

As ratings of structured credit instruments were considered to be one of the reasons for the financial turmoil during 2007–2009, they are seen as not fit for the determination of this capital charge. Instead ratings of the underlying assets are used which represents a look-through approach to the ultimate risks of a securitized asset. The specific characteristics of

TABLE 27.7 The Function *f* Determined by Rating

$f(g_i)$	g_i	**Duration Floor**	**Duration Cap**
AAA	0.013	1	–
AA	0.015	1	–
A	0.018	1	–
BBB	0.025	1	–
BB	0.045	1	8
B or lower	0.075	1	6
Unrated	0.030	1	–

Source: Adapted from CEIOPS. 2010a08. CEIOPS' Advice for L2 Implementing Measures on SII: SCR Standard Formula—Calibration of Market Risk Module (former Consultation Paper no. 70). CEIOPS-DOC-66-10. January 29. Available at www.ceiops.org.

the structured credit instrument, and especially the subordination of the tranche, fit into this approach.

$$C_{\text{SP,Struc}} = \sum_i \frac{[\text{EAD}_i \cdot G(\text{RD}_i, T_i) \cdot (1 - R(\text{RD}_i) - \text{AP}_i]}{\text{DP}_i - \text{AP}_i}$$

Where EAD_i is the exposure at default, that is, the credit risk exposure i as determined by reference to market values. $G(\text{RD}_i,T_i)$ is a function of the rating class (RD) and tenure (T) of the credit risk exposure within a securitized asset pool which is calibrated to deliver a shock consistent with VaR 99.5%; see Table 27.8; $R(\text{RD}_i)$ is a function of the rating class of the credit risk exposure within a securitized asset pool which is calibrated to deliver a shock consistent with VaR 99.5%; see Table 27.9; AP_i stands for structured credit products, the attachment point of the tranche held; DP_i stands for structured credit products, the detachment point of the tranche held; T_i stands for structured credit products, the average tenure of the assets securitized; and RD_i stands for structured credit products, a vector of the rating distribution in the asset pool securitised.

When calculating $C_{\text{Sp,Struc}}$, a cap of 100% of MV_i and a floor of 10% of MV_i should be used.

If the originator of a structure credit product does not comply with the 5% net retention rate foreseen in the CRD (2006/48/EC), the capital charge for the product should be 100%, regardless of the seniority of the position. If a look-through on the level of securitized assets is not possible, the same stress as for the "equity, other" category should be applied to the structured product for which the look-through is not possible; CEIOPS (2010a08).

TABLE 27.8 The Function *G* of the Rating Class and Tenure

$G(\text{RD}_i, T_i)$	AAA	AA	A	BBB	BB	B	CCC or Lower	Unrated
0–1.9 y	0.008	0.016	0.047	0.081	0.209	0.415	0.659	0.097
2–3.9 y	0.016	0.031	0.081	0.147	0.341	0.597	0.833	0.176
4–5.9 y	0.023	0.050	0.109	0.202	0.430	0.682	0.884	0.242
6–7.9 y	0.035	0.074	0.140	0.252	0.504	0.733	0.907	0.302
8 + y	0.047	0.097	0.171	0.302	0.562	0.771	0.919	0.362

Source: Adapted from CEIOPS. 2010a08. CEIOPS' Advice for L2 Implementing Measures on SII: SCR Standard Formula—Calibration of Market Risk Module (former Consultation Paper no. 70). CEIOPS-DOC-66-10. January 29. Available at www.ceiops.org.

Note: y: years.

TABLE 27.9 The Function *R* (Recovery Rate) of the Rating Class

$R(\text{RD}_i)$	AAA	AA	A	BBB	BB	B	CCC or Lower	Unrated
Recovery rate	0.50	0.45	0.40	0.35	0.30	0.25	0.20	0.35

Source: Adapted from CEIOPS. 2010a08. CEIOPS' Advice for L2 Implementing Measures on SII: SCR Standard Formula—Calibration of Market Risk Module (former Consultation Paper no. 70). CEIOPS-DOC-66-10. January 29. Available at www.ceiops.org.

For Credit Derivatives:
For credit derivatives, the capital charge $C_{SP,CD}$ is determined as the change in the value of the derivative (i.e., as the decrease in the asset or the increase in the liability) that would occur following

a. A widening of credit spreads by 600% if overall this is more onerous or

b. A narrowing of credit spreads by 75% if this is more onerous

A notional capital charge should then be calculated for each event. The capital charge should then be the higher of these two notional changes.

For Mortgage Loans:
The capital charge for the spread risk of exposures secured by real estate is determined as follows:

$$CR_{SP,Re} = 0.08 \cdot \sum_i \left[RW_{i,\sec} \cdot \mathrm{Sec}_i + RW_{i,un\sec} \cdot \max\left(\mathrm{Exp}_i - \mathrm{Sec}_i; 0\right)\right]$$

where Exp_i is the total mortgage exposure to borrower i, Sec_i denotes the fully and completely secured part of the exposure to borrower i, calculated as the part of the exposure covered by real estate collateral after application of the haircut, $RW_{i,\sec}$ is the risk weight associated with the fully and completely secured part of the exposure to borrower i, and $RW_{i,un\sec}$ is the risk weight associated with the unsecured part of to exposure to borrower i.

The fully and completely secured part of the exposure is that part of the mortgage exposure that is covered by a real estate collateral, after application of a haircut to that collateral value. It should also meet the conditions given in Directive 2006/48/EC (appendix VI Section 9). The haircut to be applied to the value of the real estate collateral is 25% for residential real estate and 50% for commercial real estate. Therefore, the fully and completely secured part of the exposure is equal to 75% of the value of residential real estate collateral, and 50% of the value of commercial real estate collateral; CEIOPS (2010a08).

For *residential property,* a risk weight of 35% applies to the fully and completely secured part of exposure i in the following circumstances:

- Exposures or any part of an exposure fully and completely secured, to the satisfaction of the competent authorities, by mortgages on residential property, which is, or shall be, occupied or let by the owner, or the beneficial owner in the case of personal investment companies.
- Exposures fully and completely secured, to the satisfaction of the competent authorities, by shares in Finnish residential housing companies, operating in accordance with the Finnish Housing Company Act of 1991 or subsequent equivalent legislation, in respect of residential property which is, or shall be, occupied or let by the owner.
- Exposures to a tenant under a property leasing transaction concerning residential property under which the insurer is the lessor and the tenant has an option to purchase,

provided that the competent authorities are satisfied that the exposure of the insurer is fully and completely secured by its ownership of the property.

If the secured part of exposure i does not fall within the circumstances stated in the previous paragraph, or if the conditions given in Directive 2006/48/EC, Appendix VI Section 9, are not met, it cannot be treated as fully and completely secured. In that case, a risk weight of 100% will be applied. The unsecured part of exposure i also receives a risk weight of 100%.

For *commercial property*, a risk weight of 100% is applied to both the fully and completely secured part and the unsecured part. A risk weight of 50% is applied to the fully and completely secured part only if the conditions given in Directive 2006/48/EC, Appendix VI Section 9, are met. Fully and completely secured exposures receive a risk weight of 0% if these exposures are guaranteed by an OECD or EEA government, and if these exposures are in the currency of the government. This applies to both residential and commercial real estate.

27.6.1 Simplifications

If the following is provided, simplifications may be used; see CEIOPS (2009e15, 2010a12):

- The simplification is proportionate to the nature, scale, and complexity of the risks that the undertaking faces.
- The average credit rating for long duration bonds (10 year and above) is not less than one rating below the credit rating for short duration bonds (5 years or below).
- The standard calculation of the spread risk submodule is not an undue burden for the undertaking.

Simplification for bonds:

$$C_{\mathrm{SP,Bo}}^{\mathrm{Simp}} = \mathrm{MV}_{\mathrm{Bo}} \cdot \mathrm{Dur}_{\mathrm{Bo}} \cdot \sum_i \%\mathrm{MV}_{\mathrm{Bo},i} \cdot f(g_i) + \Delta\mathrm{Liab}_{\mathrm{UL}},$$

Simplifications for structured credit products:

$$C_{\mathrm{SP,St}}^{\mathrm{Simp}} = \mathrm{MV}_{\mathrm{St}} \cdot \mathrm{Dur}_{\mathrm{St}} \cdot \sum_i \%\mathrm{MV}_{\mathrm{St},i} \cdot G(\mathrm{RD}_i, T_i)$$

Simplifications for credit derivatives:

$$C_{\mathrm{SP,Cd}}^{\mathrm{Simp}} = \mathrm{Dur}_{\mathrm{Cd}} \cdot \sum_i \%\mathrm{MV}_{\mathrm{Cd},i}$$

and

$$C_{\mathrm{SP}}^{\mathrm{Simp}} = C_{\mathrm{SP,Bo}}^{\mathrm{Simp}} + C_{\mathrm{SP,St}}^{\mathrm{Simp}} + C_{\mathrm{SP,Cd}}^{\mathrm{Simp}},$$

where $\mathrm{MV}_{\mathrm{Bo}}$ is the total market value of nongovernment bond portfolio, $\mathrm{MV}_{\mathrm{St}}$ is the total market value of structured credit portfolio, $\mathrm{Dur}_{\mathrm{Bo}}$ is the modified duration of nongovernment bond portfolio, $\mathrm{Dur}_{\mathrm{St}}$ is the modified duration of structured credit portfolio, $\mathrm{Dur}_{\mathrm{Cd}}$

is the modified duration of credit derivatives portfolio, $\%MV_{Bo,i}$ is the proportion of non-government bond portfolio held at rating i, $\%MV_{St,i}$ is the proportion of structured credit portfolio held at rating i, $\%MV_{Cd,i}$ is the proportion of credit derivatives portfolio held at rating i, f is the function defined as in the standard calculation; see above, G is the function defined as in the standard calculation; see above, ΔLiab_{UL} is the the overall impact on the liability side for policies where the policyholders bear the investment risk with embedded options and guarantees of the stressed scenario, with a minimum value of 0 (sign convention: positive sign means losses). The stressed scenario is defined as a drop in value on the assets by

$$MV_{Bo} \sum_i \%MV_{Bo,i} \cdot f(\text{dur}_i) + \Delta\text{Liab}_{UL}.$$

CEIOPS (2010a12) does not include any simplification for mortgage loans.

Captives may assume all assets to be submitted to the spread risk module are rated BBB. For structured bonds, credit derivatives, and bonds with a lower rating than BBB the standard calculation of the spread risk module needs to be applied; CEIOPS (2009e16, 2010a13).

27.7 CONCENTRATION RISK

The capital charge we are seeking is denoted $C_{MR,Co}$ and is a part of the Mmarket risk module in the standard formula framework. The calibration to this submodule capital charge is found in Annex A to CEIOPS (2009d09, 2009f20).

The scope of this submodule is to cover assets considered in equity, interest rate, spread risk, and property risk submodules within the market risk module, but excluding assets covered by the counterparty default risk module in order to avoid any overlap between both elements of the standard calculation of the SCR.

The capital charge is calculated as

$$C_{MR,Co} = \sqrt{C_{Co,Fi}^2 + C_{Co,Pr}^2 + 2\rho_{Fi,Pr} \cdot C_{Co,Fi} \cdot C_{Co,Pr}},$$

where $C_{Co,Fi}$ is the capital charge for financial concentration risk, $C_{Co,Pr}$ is the capital charge for property concentration risk, and $\rho_{Fi,Pr}$ is the dependence applied to the submodules of properties and equity risk given in Table 27.1.

27.7.1 Financial Concentration Risk

The financial concentration risk per name is calculated as (name i)

$$C_{Co,Fi,i} = A_{xl} \cdot XS_i \cdot g_i + \Delta\text{Liab}_{ul}.$$

A_{xl} is the amount of total assets considered in this submodule (see above) and excluding those where the policyholder bears the investment risk.

The excess exposure is calculated as

$$\text{XS}_i = \max\left\{0; \frac{\text{EAD}_i}{A_{xl}} - \text{CT}\right\},$$

TABLE 27.10 The Concentration Threshold, CT, and the Parameter g_i

Rating	Credit Quality Step	Threshold CT	g_i
AAA	1	3%	0.12
AA	1	3%	0.12
A	2	3%	0.21
BBB	3	1.5%	0.27
BB or lower, unrated	4–6, -	1.5%	0.73

Source: Adapted from CEIOPS. 2009f20. CEIOPS' Advice for L2 Implementing Measures on SII: SCR—Standard formula—Market risk (former Consultation Paper no. 47). CEIOPS-DOC-40-09. November 10. Available at www.ceiops.org.

where EAD_i is the net exposure at default to counterparty i and CT is a concentration threshold. Where an undertaking has more than one exposure to a counterparty then EAD_i is the aggregate of those exposures at default and rating g_i should be a weighted rating determined as the rating corresponding to a weighted average credit quality step calculated as average of the credit quality steps of the individual exposures to that counterparty, weighted by the net exposure at default in respect of that exposure to that counterparty.

As $A_{xl} \cdot \text{XS}_i = \max\{0; \text{EAD}_i - A_{xl} \cdot \text{CT}\}$, we see that we compare the net exposure at default with a percentage of the asset hold of the counterparty. The lower the rating, the higher part of the excess exposure is charged.

$\Delta\text{Liab}_{\text{UL}}$ is the overall impact on the liability side for policies where the policyholders bear the investment risk with embedded options and guarantees of the stressed scenario, with a minimum value of 0 (sign convention: positive sign means losses). The stressed scenario is defined as a drop in value on the assets for counterparty i used as the reference to the valuation of the liabilities by $\text{XS}_i^* g_i$.

For names that can only be found on the assets used as the reference to the valuation of the liabilities, the risk concentration charge per name i is calculated as $C_{\text{Co,Fi},i} = \Delta\text{Liab}_{\text{ul}}$.

The concentration threshold and the parameter g_i are given in Table 27.10.

The financial concentration risk capital charge should be adjusted by the overall impact on technical provisions with future profit features of this submodule, provided the undertaking is able to assess such impact with the same requirements applied to the calculation of best estimate values, and preventing that double counting of this effect is allowed with other submodules or modules.

The capital charge from financial concentration risk is calculated as

$$C_{\text{Co,Fi}} = \sqrt{\sum_i C^2_{\text{Co,Fi},i} + 0.5 \cdot \sum\sum_{i \neq j} C_{\text{Co,Fi},i} \cdot C_{\text{Co,Fi},j}},$$

where a dependence between name i and j is assumed to be 0.25.

Investments in a single undertaking for collective investment in transferable securities (UCITS) are exempted from the concentration risk submodule if the maximum share of the

UCITS assets that are invested in a single body does not exceed (UCITS i)

$$CT_{UCITS,i} = CT \cdot \frac{A_{xl}}{MV_{UCITS,i}},$$

where $CT_{UCITS,i}$ is the concentration threshold for UCITS i,$MW_{UCITS,i}$ is the market value of the undertaking's investment in UCITS i,CT is the concentration threshold of the submodule, and A_{xl} is the comparative measure of the submodule.

Investments in *mortgage covered bonds and public sector covered bonds* should be covered when the following requirements are met:

- The asset has an AAA credit quality
- The portfolio of mortgages backing the asset is diversified into a sufficiently high number of borrowers
- There is no evidence of high correlation or connection among the default of one or few borrowers
- The covered bond meets the requirements defined in Article 22(4) of the UCITS directive 85/611/EEC

The threshold is set to 15%.

27.7.2 Property Concentration Risk

Undertakings shall identify the exposures in a single property higher than 10% of "total assets" considered in this submodule according to the text above. For this submodule, the undertaking shall take into account both properties directly owned and those indirectly owned, that is, funds of properties, and both ownership and any other real exposures, mortgages or any other legal right regarding properties. Properties located in the same building or sufficiently nearby shall be considered a single property.

Exposures exceeding the threshold shall deliver a capital requirement calculated *applying the formula reflected in this submodule for financial investments rated as AA*; see Section 27.2.1.

Capital requirements for different properties shall be aggregated assuming a correlation factor 0 between the requirements for each property. The capital charge is thus calculated as

$$C_{Co,Pr} = \sqrt{\sum_i C^2_{Co,Pr,i}}.$$

27.7.3 Simplifications

Captives may use simplifications as discussed in CEIOPS (2009e16, 2010a13) based on intra-group asset pooling arrangements. The exemption is based on the condition that there exist legally effective formal provisions where the captive's liabilities can be offset by intragroup exposures it may hold on the entities of the group. The threshold shall be 15% where the following requirements are met:

- The credit institution or cash-pooling entity of the group has a rating of AA
- The credit institutions do not belong to the same group and have no other dependency between each other

A look-through approach to intragroup asset pooling arrangements may be applied for the calculation of the market risk module, if the account of the captive undertaking meets the requirements stated for segregated assets in CEIOPS (2009f06).

27.8 DAMPENER

The volatility of asset prices has been debated in relation to the new solvency regime, as a sharp decline in those prices could lead to breaches of the capital requirement level. This could be a start of a vicious cycle. Within Solvency II this is addressed by several measures of which a dampener approach is one that was tested in QIS4.

The use of a dampener approach, the Pillar 2 dampener, in terms of a longer recovery period is discussed in Section 22.6.2; see also CEIOPS (2009e02).

The use of an equity dampener is discussed in Section 27.8.1. A discussion on the general effects of a dampener approach is given by Lechkar and van Welie (2008).

27.8.1 Equities: Symmetric Dampener Adjustment

In calibrating the symmetric adjustment mechanism based on Article 106 in the FD, CEIOPS (2009e07, 2010a07) has considered the following objectives:

- Allow sufficient time for undertakings to rebalance their profile in a stressed scenario
- Avoid unintended procyclical effects (in particular, a rise in the equity charge in the middle of a crisis)
- Ensure that the equity charge remains sufficiently risk sensitive
- Prevent fire sales of assets
- Avoid undertakings having to adjust their risk profile frequently solely as a result of movements in the equity capital charge
- Avoid any incentive to invest in one or the other asset class
- Allow the adjustment to be set independently of the standard equity stress

CEIOPS' calibration of the symmetric adjustment mechanism is based on the following variables and formulation:

Standard capital stress, SS: Equity shocks as defined in Table 27.4a.

Adjusted capital stress, AS: New equity shocks after applying the symmetric adjustment.

LA: Limited adjustment as defined below.

The Unlimited adjustment term is

$$\mathrm{UA} = \frac{I_t - 1/n(\sum_{s=t-1}^{t+n})I_s}{1/n(\sum_{s=t-1}^{t+n})I_s},$$

where the adjustment term I_t is the value of the MSCI Developed index at time t. The UA is subject to a band of $\pm 10\%$ either side of the standard capital stress, that is,

$$\mathrm{LA} = \begin{cases} -10 & \text{if} \quad \mathrm{UA} < -10 \\ \mathrm{UA} & \text{if} \quad -10 < \mathrm{UA} < 10 \\ +10 & \text{if} \quad \mathrm{UA} > 10 \end{cases}.$$

Hence,

$$\mathrm{AS} = \mathrm{SS} + \mathrm{LA}\, x\, \beta, \tag{27.1}$$

where the β is calculated from a regression of the index level on the weighted average index level and is set equal to 1.

The averaging period is set to one year.

27.8.2 Equities: Duration Dampener

The duration approach was introduced in Article 304 of the FD. If the average duration of the liabilities, corresponding to “retirement liabilities” and where all assets and liabilities are ring-fenced, managed, and organized separately from other activities, exceeds an average of 12 years, then the duration approach can be applied.

Subject to Member State option and supervisory approval, undertakings may apply a 22% capital charge to the equities held corresponding to the liabilities falling under the scope of Article 304. This capital charge should be integrated in the SCR standard formula:

$$\mathrm{CR}_{\mathrm{MR,ER,Tot}} = \mathrm{CR}_{\mathrm{MR,ER}} + \mathrm{CR}_{\mathrm{MR,ER,Dur}},$$

where $\mathrm{CR}_{\mathrm{MR,ER}}$ is the equity capital charge as calculated in Section 27.3 and $\mathrm{CR}_{\mathrm{MR,ER,Dur}}$ is the equity capital charge using the duration approach.

CHAPTER 28

Solvency II Standard Formula

Credit Risk

THE COUNTERPARTY default risk module should reflect possible losses due to unexpected default, or deterioration in the credit standing, of the counterparties and debtors of insurance and reinsurance undertakings over the forthcoming 12 months. The counterparty default risk module should cover risk-mitigating contracts, such as reinsurance arrangements, securitizations, and derivatives, and receivables from intermediaries, as well as any other credit exposures that are not covered in the spread risk submodule. For the purpose of the counterparty default risk module, reinsurance should include financial reinsurance (CEIOPS, 2009b03).

28.1 GENERAL ISSUES

In the standard formula for the total capital requirement, SCR, the credit risk is split up into the counterparty default risk (Section 28.1), the (credit) spread risk (Section 27.6), and the concentration risk (Section 27.7). The two latter risks are in the SCR standard formula considered as parts of the market risk.

The credit risk, counterparty default risk, and its capital charge are treated in different articles of the Framework Directive, namely 111, but also in 13, 104, 105, and 109. The development of the counterparty default risk is discussed in Appendix J. CEIOPS has discussed the general structure of the counterparty default risk module in CEIOPS (2009b03) and CEIOPS (2009d13) and in its final advice, CEIOPS (2009f03).

28.2 COUNTERPARTY DEFAULT RISK

As a consequence of the definition of the scope of the spread risk submodule (see Section 27.6), the counterparty default risk module should cover the credit risk of

- risk-mitigating contracts, such as reinsurance arrangements, securitizations, and derivatives,
- receivables from intermediaries, and

- any other credit exposures that are not covered in the spread risk submodule, in particular (list not exhaustive): policyholder debtors, cash at bank, deposits with ceding institutions, capital, initial funds, letters of credit as well as any other commitments received by the undertaking that have been called up but are unpaid, and guarantees, letters of credit, letters of comfort that are provided by the undertaking as well as any other commitments that the undertaking has provided and that depend on the credit standing of a counterparty. In relation to credit derivatives, the credit risk that is transferred by the derivative should not be covered in the counterparty default risk module as it is already covered in the spread risk submodule (CEIOPS, 2009f03).

28.2.1 Standard Formula

Figure 28.1 shows the structure of the counterparty default risk module and its two subrisks.

$\mathrm{CR_{CR}}$ denotes the capital charge for the counterparty default risk module. The capital charges of its subrisks modules are

- $\mathrm{CR_{CR,T1}}$: Type 1 counterparty default exposures, defined below
- $\mathrm{CR_{CR,T2}}$: Type 2 counterparty default exposures, defined below

The capital charge (see CEIOPS, 2009f03) will be calculated with a low diversification effect ($\rho_{\mathrm{T1,T2}} = 0.75$) as

$$\mathrm{CR_{CR}} = \sqrt{\mathrm{CR^2_{CR,T1}} + \mathrm{CR^2_{CR,T2}} + 2 \cdot 0.75 \cdot \mathrm{CR_{CR,T1}} \cdot \mathrm{CR_{CR,T2}}}.$$

Type 1 exposures are exposures in relation to

- Reinsurance arrangements
- Securitizations and derivatives
- Any other risk-mitigating contracts
- Cash at bank

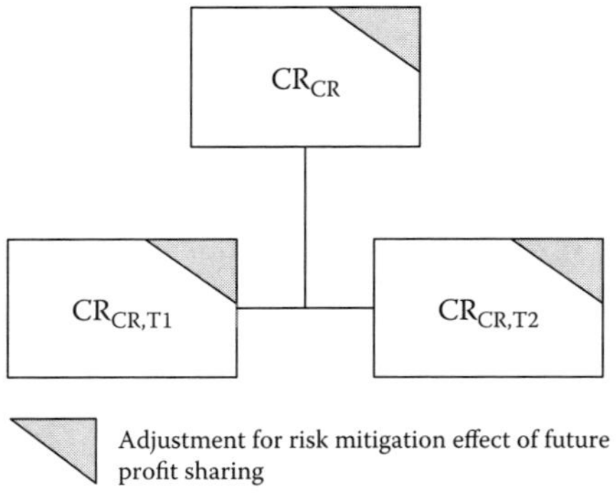

FIGURE 28.1 The modular structure of the capital charge for the counterparty default risk.

- deposits with ceding institutions if the number of independent counterparties does *not exceed* a certain threshold
- capital, initial funds, letters of credit as well as any other commitments received by the undertaking that have been called up but are unpaid if the number of independent counterparties does *not exceed* a certain threshold
- guarantees, letters of credit, letters of comfort provided by the undertaking and other commitments that are provided by the undertaking and that depend on the credit standing of a counterparty

Type 2 exposures are all other exposures that are within the scope of the module, in particular

- Receivables from intermediaries
- Policyholder debtors
- Deposits with ceding institutions if the number of independent counterparties *exceeds* a certain threshold
- Capital, initial funds, letters of credit as well as any other commitments received by the undertaking that have been called up but are unpaid if the number of independent counterparties *exceeds* a certain threshold

Threshold to distinguish between Type 1 and Type 2 exposures: If the number of independent counterparties in relation to deposits with ceding institutions *does not exceed* 15, these exposures should be treated as type 1 exposures. The same should apply to called up but unpaid commitments.

28.2.2 Type 1 Capital Charge

The model that was used for QIS 4 was criticized because of its inconsistency in the determination of the risk factors. CEIOPS therefore decided to use the ter Berg model described in Section 18.3.

Given the probabilities of default, PD, and losses-given-default, LGD, of the counterparties in the portfolio of Type 1 exposures, the ter Berg model provides an estimate of the variance of the portfolio's loss distribution. This estimate is used to calculate the capital requirement for Type 1 exposures as follows:

$$\mathrm{CR}_{\mathrm{CR,T1}} = \min\left\{\sum_i \mathrm{LGD}_i; k_{1-\alpha}\cdot\sigma_Z\right\},$$

where LGD_i is the loss-given-default for Type 1 exposure of counterparty and $k_{1-\alpha}\cdot\sigma_Z$, given by Equation 18.15a, is the product of the quantile and the standard deviation of the loss distribution of the Type 1 exposures. The standard deviation is calculated as follows (*cf.* Equation 18.14c and Section 18.3 for the definition of the variables Z and Y_*):

$$\sigma_Z = \sqrt{\sum_{c=1}^{C}\sum_{d=1}^{C} u_{cd}\cdot Y_c\cdot Y_d + \sum_{c=1}^{C} v_c\cdot Y_{2c} - \left(\sum_{c=1}^{C} w_c\cdot Y_c\right)^2},$$

where c and d are the rating classes and u_{cd}, v_c, w_c are parameters that depend on the counterparties PD, that is, are a function of its rating. The variables Y_c and Y_{2c} are calculated as

$$Y_c = \sum_{i\in c}\mathrm{LGD}_i \quad \text{and} \quad Y_{2c} = \sum_{i\in c}\mathrm{LGD}_i^2.$$

Following Section 18.3 and letting $b_c = \mathrm{PD}_c/[1 + r\cdot(1-\mathrm{PD}_c)]$, where $r = \alpha/\tau$ is the ratio between the two parameters in the ter Berg model (this ratio is calibrated by CEIOPS, see below) and PD_c is the probability of default for rating class c, the parameters can be written as

$$u_{cd} = \frac{r\cdot b_c\cdot b_d\cdot(1-b_c)\cdot(1-b_d)}{r\cdot b_c\cdot b_d + b_c + b},$$

$$v_c = \mathrm{PD}_c\cdot(1-\mathrm{PD}_c) + (\mathrm{PD}_c - b_c)^2 - \frac{r\cdot b_c^2\cdot(1-b_c)^2}{2 + r\cdot b_c^2},$$

and

$$w_c = \mathrm{PD}_c - b_c = \frac{r\cdot\mathrm{PD}_c\cdot(1-\mathrm{PD}_c)}{1 + r\cdot(1-\mathrm{PD}_c)}.$$

The following parameters should be used (CEIOPS, 2009d13):

- The ratio between the two parameters used in the ter Berg model, α and τ (see Section 18.3), is set to $r = \alpha/\tau = 4$, see CEIOPS (2009d13, 2009f03). This calibration proposal is based on expert opinion.
- The quantile factor $k_{0.995}$ should be set to

$$k_{0.995} = \begin{cases} 3 & \text{if } \sigma_Z \leqslant 0.03\cdot\sum_i \mathrm{LGD}_i \\ 5 & \text{else.} \end{cases}$$

The calculation of LGD is discussed in Section 28.3 and a simplification for the PD determination based on own funds and SCR is given in Section 28.4.

28.2.3 Type 2 Capital Charge

The Type 2 exposures usually relate to unrated counterparties and an undertaking's portfolio usually consists of a larger number of such exposures. In most cases the default risk originating from these exposures is very small compared to the overall risks. Therefore, CEIOPS has decided to use a simple factor-based approach.

The value of a Type 2 exposure towards a counterparty is equal to the corresponding asset value. The value might be netted with liabilities towards the same legal entity to the extent that they could be set off in the case of the default of the legal entity. Also, the value of the exposure might be reduced by the value of collateral for the exposure.

The capital charge for counterparty default risk of Type 2 exposures is calculated as

$$\mathrm{CR}_{\mathrm{CR,T2}} = x \cdot E + y \cdot E_{p-d},$$

where E is the sum of the values of Type 2 exposures, except for receivables from intermediaries that are due for more than T months, E_{p-d} is the sum of the values of receivables from intermediaries that are due for more than T months (past-due), x is the risk factor for Type 2 exposures, and y is the risk factor for past-due receivables from intermediaries.

The following parameters should be used (CEIOPS, 2009f03):

x	0.15
y	0.90
T	3 months

28.3 CALCULATION OF LGD

The LGD of an exposure could conceptually be defined to be the loss of basic own funds that the insurer would incur if a counterparty defaults. As the size of a potential loss usually changes over time and a failure of the counterparty is more likely when the potential loss is high, the LGD should be determined for the case of a stressed situation. A straightforward approach to measure the additional loss owing to the stressed situation is the approximation of the risk-mitigating effect of reinsurance arrangement or derivative in the SCR calculation. In order to determine the LGD, the value of the current exposure should be increased by the risk-mitigating effect of the contract.

In case of default, typically a part of the exposure can still be collected. In order to allow for the potential recovery of the counterparty, the LGD is amended by a factor $(1-\mathrm{RR})$ where RR denotes the recovery rate of the counterparty. The recovery rate may be different for reinsurance arrangements and securitizations, on the one hand, and for derivatives, on the other hand.

The loss-given-default for Type 1 exposure of the counterparty depends on the kind of exposure.

For a reinsurance arrangement or a securitization, the LGD_i should be calculated as

$$\mathrm{LGD}_i = \max\left\{0; (1-\mathrm{RR}_{\mathrm{re}}) \cdot \left[(\mathrm{Rec}_i - \mathrm{Col}_i) - \mathrm{RM}_{\mathrm{re},i}\right]\right\},$$

where the variables are defined as follows:

$\mathrm{RR}_{\mathrm{re}}$ is the recovery rate for reinsurance arrangements.

Rec_i is the best estimate of recoverables from the reinsurance contract (or SPV) i plus any other debtors arising out of the reinsurance arrangement or SPV securitization. Might be netted with liabilities towards the same legal entity to the extent that they could be set off in case of default of the legal entity.

Col_i is the market value of collateral in relation to the reinsurance arrangement or SPV securitization i. Collateral for a reinsurance arrangement could be a reinsurance depot at the cedant undertaking.

$\text{RM}_{\text{re},i}$ is the risk-mitigating effect on underwriting risk of the reinsurance arrangement or SPV securitization i. This is an approximation of the difference *between* the hypothetical capital requirement for underwriting risk under the condition that the reinsurance arrangement or the SPV securitization is not taken into account in its calculation (CR^{gross}) *and* the capital requirement for underwriting risk, without any amendments (CR^{net}). Denote this amount by $\text{RM}_{\text{re},i,\text{uw}} = \text{CR}_{\text{uw}}^{\text{gross}} - \text{CR}_{\text{uw}}^{\text{net}}$. This is discussed in detail in CEIOPS (2009d13, 2009f03).

If an SPV also transfers market risk, the risk-mitigating effect $\text{RM}_{\text{re},i}$ should be given by the aggregation between $\text{RM}_{\text{re},i,\text{uw}}$ and the difference *between* the (hypothetical) capital requirement for market risk under the condition that the risk-mitigating effect of the SPV is not taken into account in its calculation *and* the capital requirement for market risk (without any amendments). The latter is denoted by $\text{RM}_{\text{re},i,\text{MR}}$. The dependence is assumed to be the same as that between underwriting and market risk modules, that is, $\rho = 0.25$. Hence, we get (see CEIOPS, 2009f03)

$$\text{RM}_{\text{re},i} = \sqrt{\text{RM}^2_{\text{re},i,\text{uw}} + \text{RM}^2_{\text{re},i,\text{MR}} + 2 \cdot 0.25 \cdot \text{RM}_{\text{re},i,\text{uw}} \cdot \text{RM}_{\text{re},i,\text{MR}}}.$$

For a *derivative*, the LGD_i should be calculated as

$$\text{LGD}_i = \max\left\{0; (1 - \text{RR}_{\text{fi}}) \cdot \left[(\text{MV}_i - \text{Col}_i) - \text{RM}_{\text{fi},i}\right]\right\},$$

where RR_{fi} is the recovery rate for derivatives, MV_i is the market value of the derivative i and Col_i is the market value of collateral in relation to the derivative i.

$\text{RM}_{\text{fi},i}$ is the risk-mitigating effect on market risk of the derivative i. This is an approximation of the difference *between* the hypothetical capital requirement for market risk under the condition that the risk-mitigating effect of the derivative is not taken into account in its calculation (CR^{gross}) *and* the capital requirement for market risk, without any amendments (CR^{net}). This is discussed in detail in CEIOPS (2009d13).

The following recovery rates should be used (CEIOPS, 2009f03):

	Recovery Rate (%)
RR_{re}	50
RR_{fi}	10

If the counterparty has tied up an amount for collateralization commitments for reinsurance arrangements, both on and off balance sheet, including commitments to other parties, greater than 60% of the assets on its balance sheet, then the recovery rate is assumed to be 10% rather than 50%.

The calculation of the risk-mitigating effect, RM, is discussed in CEIOPS (2009d13, 2009f03). The determination of the risk-mitigating effects $RM_{re,i}$ and $RM_{fi,i}$ is based on the calculation of two capital requirements:

- The (hypothetical) capital requirement for underwriting and market risk under the condition that the risk-mitigating effect of the reinsurance arrangement, SPV or derivative of a particular counterparty is not taken into account in its calculation. These values are only determined for the purpose of the counterparty default risk module.
- The capital requirements for underwriting risk and market risk without any amendments for these modules. They are available as soon as the calculations of the particular modules have been made.

Life reinsurance and derivatives: The gross capital requirements in relation to counterparty i are determined by a recalculation of the modules that are affected by the risk-mitigating contracts with that counterparty.

- *Scenario-based modules:* the scenario outcome should be reassessed assuming that the risk-mitigating contract with counterparty i will not provide any compensation for the losses incurred under the scenario.
- *Factor-based modules:* the volume measures that allow for the risk-mitigating effect of the contract need to be reassessed. In particular, in the concentration risk submodule of the market risk, the exposure measures should be calculated without allowance for risk-mitigating effects of contracts with counterparty i.

If a module of the SCR did not allow for the risk-mitigating effect of the risk-mitigating contract with counterparty i in the calculation of the net capital requirement, the net and gross capital requirements coincide and $RM_{re,i}$ and $RM_{fi,i}$ are zero.

Nonlife reinsurance: If the reinsurance treaties with a counterparty affect only one nonlife line of business, then the difference $CR_{NL}^{gross} - CR_{NL}^{net}$ should be approximated by the following term:

$$\sqrt{\begin{aligned}&\left(\mathrm{NL}_{\mathrm{CAT}}^{\mathrm{gross}} - \mathrm{NL}_{\mathrm{CAT}}^{\mathrm{net}}\right)^2 + \left(3 \cdot \sigma_{\mathrm{PR},k} \cdot \left[P_k^{\mathrm{gross}} - P_k^{\mathrm{net}}\right]\right)^2 \\ &\quad + \left(3 \cdot \sigma_{\mathrm{RR},k} \cdot \mathrm{Rec}\right)^2 + 9 \cdot \sigma_{\mathrm{PR},k} \cdot \left[P_k^{\mathrm{gross}} - P_k^{\mathrm{net}}\right] \cdot \sigma_{\mathrm{RR},k} \cdot \mathrm{Rec}\end{aligned}}$$

where $\mathrm{NL}_{\mathrm{CAT}}^{\mathrm{gross}} - \mathrm{NL}_{\mathrm{CAT}}^{\mathrm{net}}$ is the counterparty's share of CAT losses, $P_k^{\mathrm{gross}} - P_k^{\mathrm{net}}$ is the reinsurance premium of the counterparty in the affected line of business, Rec is the reinsurance recoverables in relation to the counterparty in the affected line of business, $\sigma_{\mathrm{PR},k}$ is the standard deviation for premium risk in the affected line of business as used in the premium and reserve risk submodule, and $\sigma_{\mathrm{RR},k}$ is the standard deviation for reserve risk in the affected line of business as used in the premium and reserve risk submodule.

Simplifications are also proposed in CEIOPS (2009d13, 2009f03). For derivatives, the difference $CR^{gross} - CR^{net}$ may be replaced by the sum of the differences $CR^{gross}_{MR,sr} - CR^{net}_{MR,sr}$, where sr stands for submodule risk.

For life reinsurance a simplification could be made using the sum of the differences $CR^{gross}_{LR,sr} - CR^{net}_{LR,sr}$ of the submodules risks (sr) affected. For proportional life reinsurance, a further simplification could be used: $CR^{gross}_{LR,sr} - CR^{net}_{LR,sr} \approx CR^{net}_{LR} \cdot (BE^{gross}/BE^{net} - 1)$.

For nonlife reinsurance, the simplifications could be done in two steps. First calculate $CR^{gross}_{NL} - CR^{net}_{NL}$ for all reinsurance counterparties together. In the second step, we approximate the share of a single counterparty i by

$$\left(CR^{gross}_{NL} - CR^{net}_{NL}\right)_i \approx \left(CR^{gross}_{NL} - CR^{net}_{NL}\right) \cdot \frac{Rec_i}{Rec_{total}},$$

where Rec stands for recoverable (from counterparty i and the overall reinsurance recoverables).

There are many further simplifications if we look at $CR^{gross}_{NL} - CR^{net}_{NL}$ and then approximate this difference, see CEIOPS (2009d13). See also the discussion in Appendix J, Section J.2.1.

28.4 PROBABILITY OF DEFAULT

PD for Type 1 exposures: it is necessary to assign a PD or at least a rating class to each counterparty. In CEIOPS (2009d13, 2009f03), this is discussed. It is supposed that there are a finite number of rating classes such that an average PD can be assigned to the rating classes and one of the rating classes can be assigned to each counterparty with a type 1 exposure.

The assignment of a PD should follow these steps:

- *Rated:* If the counterparty is rated by a credit rating agency (CRA) that meets certain quality requirements, the credit rating should be used to derive a PD. In order to make use of credit ratings for the determination of the PD, two elements should to be specified:

    - Recognition of the CRAs whose credit ratings can be used in the standard formula.
    - For each recognized CRA, an assignment of probabilities of default to the rating classes used by the CRA. This assignment should distinguish between different kinds of rated instruments and counterparties.

  The credit ratings used in the standard formula should meet the highest standards and should be registered according to the new regulation on how credit rating agencies should be recognized. In the case where there are more than one ratings available for a counterparty, the second highest should be used.

- *Unrated:* If the counterparty is an insurance undertaking that is subject to Solvency II supervision and up-to-date information about the solvency position of the undertaking are there, then PD, depending on the SCR and the eligible own funds to meet the SCR (OF), could be defined as follows. If a counterparty does not meet the MCR, a default probability of 30% should be assigned.

OF/SCR (%)	PD (%)
>200	0.025
>175	0.050
>150	0.10
>125	0.20
>100	0.50
>90	1.00
>80	2.00
⩽80	10.00

A PD of 10% should be assigned to counterparties that are not rated by a recognized CRA and no solvency ratio rating can be assigned to them, either because they are not under Solvency II supervision or they do not meet the requirement of the solvency ratio rating. Unrated banks are treated as if their rating was BBB. The same holds for unrated counterparties under supervision equivalent to Solvency II.

If an undertaking has more than several counterparties that are not independent, for example because they belong to one group, then it is necessary to assign a PD to the whole set of dependent counterparties. This overall PD should be average probability of the counterparties weighted with the corresponding LGD.

28.5 OTHER ISSUES

Independence of counterparties: An economic approach should be taken in order to decide whether counterparties are independent or not, meaning, for example, that counterparties that belong to the same corporate group should usually not be considered to be independent for the purpose of the counterparty default risk module.

Collaterals: If a collateral is posted in relation to the exposure, the custodian holding the collateral is independent of the counterparty and the requirements defined for collaterals in the advice on financial risk mitigation techniques, see Section 26.6, are met, then the loss-given-default, for a type 1 exposure, or the value of the exposure, for a type 2 exposure, may be reduced by the risk-adjusted value of the collateral. The risk-adjusted value of the collateral is calculated as, see CEIOPS (2009f03),

$$\text{Collateral} = 0.80 \cdot \{\text{Market Value}_{\text{Col}} - \text{MR}_{\text{Col}}\},$$

where Market $\text{Value}_{\text{Col}}$ is the market value of the collateral assets.

MR_{Col} is the adjustment for market risk. The reduction of the market value of the collateral according to the equity, property, credit spread and currency risk submodules should be determined and aggregated according to the correlation matrix of the market risk module. For the calculation of the currency risk submodule, the currency of the collateral is compared to the currency of the secured credit exposure. For reasons of practicability, the interest rate risk and concentration risk are neglected. If the collateral assets are bank deposits, which are not subject to the credit spread risk, the adjustment should be increased by the capital requirement for counterparty default risk of the deposits.

Simplified collaterals: A simplification may be applied, if it is proportionate to the nature scale and complexity of the risks inherent in the collateral arrangement: Collateral $= 0.70 \cdot$ Market $\text{Value}_{\text{Col}}$.

A simplification may be applied where the collateral assets are bankruptcy remote and there is no credit risk present: Collateral $= 0.85 \cdot$ Market $\text{Value}_{\text{Col}}$.

Segregated assets: If the liabilities of the counterparty are covered by strictly segregated assets that meet the risk mitigation requirements, see Section 26.6, then the segregated assets should be treated like collaterals in the calculation of the CR_{CR}.

Letter of credit: If a letter of credit, not approved as ancillary own funds, is provided to secure a credit exposure and meets the risk mitigation requirement set up in Section 26.6, then the counterparty of the credit exposure may be replaced by the provider of the letter of credit in the calculation of the CR_{CR}.

Netting: The LGD, in the case of a type 1 exposure, or the value of the exposure, in the case of a type 2 exposure, may be netted with liabilities towards the same legal entity to the extent that they could be set off in the case of default of the legal entity.

Pools: Pool arrangements are important under the counterparty risk module. Pools can be organized and managed in many different ways. These are discussed in CEIOPS (2009f03).

CHAPTER 29

Solvency II Standard Formula

Operational Risk

OPERATIONAL RISK is the risk of loss arising from inadequate or failed internal processes, people, and systems or from external events. Operational risk also includes legal risks.

Operational risk shall include legal risks and exclude risks arising from strategic decisions, as well as reputation risks. The operational risk module is designed to address operational risks to the extent that these have not been explicitly covered in other risk modules.

29.1 GENERAL FEATURES

The operational risk and its capital charge are treated in different articles of the Framework Directive, namely 111, 101, 103, and 107 and also in 13 and 49. The development of the operational risk is discussed in Appendix K. CEIOPS has discussed the general structure of the operational risk module in CEIOPS (2009d15) and CEIOPS (2009f25).

29.2 STANDARD FORMULA

The operational risk module of the SCR standard formula does not differ significantly from the QIS4 proposal since it was based on volume measures generally available and, at the same time, was more closely aligned with the drivers of the main operational risks; see Appendix K.

It is assumed that there are no diversification effects, that is, the capital charge C_{OR} is just added to the Basic SCR; see Figure 29.1.

The capital charge for operational risk is

$$\mathrm{CR}_{\mathrm{OR}} = \min\left[0.30 \cdot \mathrm{BSCR}; \mathrm{CR}_{\mathrm{OR,TPE}}\right] + 0.25 \cdot \mathrm{Exp}_{\mathrm{UL}},$$

where $\mathrm{CR}_{\mathrm{OR,TPE}}$ is the basic operational risk charge for all businesses other than UL business (gross of reinsurance), defined as

$$\mathrm{CR}_{\mathrm{OR,TPE}} = \max\left\{\mathrm{CR}_{\mathrm{OR,TP}}; \mathrm{CR}_{\mathrm{OR,Pr}}\right\},$$

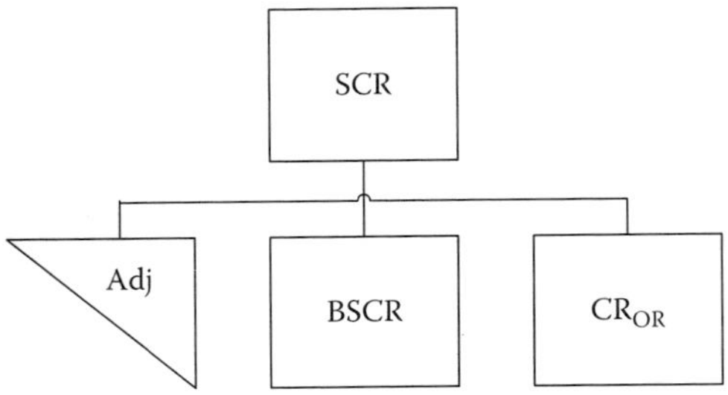

FIGURE 29.1 The structure of the modular approach for the standard formula of SCR and the place of the operational risk capital charge.

where

$$\begin{aligned} CR_{OR,TP} &= 0.006 \cdot \left(TP_L + TP_{H,SLT} - TP_{UL}\right) + 0.036 \cdot \left(TP_{NL} + TP_{H,NonSLT}\right) \\ &\quad + \max\{0; 0.006 \cdot (\Delta TP_L - \Delta TP_{UL})\} + \max\{0; 0.036 \cdot \Delta TP_{NL}\}, \end{aligned}$$

and

$$\begin{aligned} CR_{OR,Pr} &= 0.055 \cdot \left(Earn_L + Earn_{H,SLT} - Earn_{UL}\right) + 0.038 \cdot \left(Earn_{NL} + Earn_{H,nonSLT}\right) \\ &\quad + \max\{0; 0.055 \cdot (\Delta Earn_L - \Delta Earn_{UL})\} + \max\{0; 0.038 \cdot \Delta Earn_{NL}\} \end{aligned}$$

TP: Technical provisions. All TPs should have a floor equal to zero.

TP_L: Total life insurance TPs–gross of reinsurance—with a floor at zero. This also includes UL business and life-like obligations on nonlife contracts such as annuities.

$TP_{H,SLT}$: TPs corresponding to health insurance corresponding to Health SLT insurance-gross of reinsurance—with a floor at zero.

TP_{UL}: Total life insurance TPs for UL business–gross of reinsurance—with a floor at zero.

TP_{NL}: Total nonlife insurance TPs–gross of reinsurance—with a floor at zero (excluding life-like obligations of nonlife contracts such as annuities).

$TP_{H,nonSLT}$: TPs corresponding to health insurance corresponding to Health non-SLT–gross of reinsurance—with a floor at zero.

Δ: Change in TPs from year $t-1$ to t, for TP increases that have exceeded an increase of 10%. Furthermore, no offset shall be allowed between life and nonlife Δ.

Pr: Premiums.

$Earn_L$: Total earned life premium–gross of reinsurance—including UL business.

$Earn_{SLT}$: Total earned premiums corresponding to health insurance corresponding to Health SLT–gross of reinsurance.

$Earn_{UL}$: Total earned life premium for UL business–gross of reinsurance.

$Earn_{NL}$: Total earned nonlife premium–gross of reinsurance.

$Earn_{H,nonSLT}$: Total earned health insurance premium corresponding to health insurance corresponding to Health non-SLT–gross of reinsurance.

Δ: Change in earned premiums from year $t-1$ to t, for TP increases that have exceeded an increase of 10%. Furthermore, no offset shall be allowed between life and nonlife Δ.

BSCR: the Basic SCR, calculated over all quantified risks; see Chapter 26.

Exp_{UL}: The amount of annual expenses–gross of reinsurance—incurred in respect of UL business. Administrative expenses should be used, excluding acquisition expenses; the calculation should be based on the latest past years expenses and not on future projected expenses.

CHAPTER 30

Solvency II Standard Formula

Liquidity Risk

The liquidity risk is not included in the calculation of the capital charges for the quantifiable risks. This is mainly due to lack of simple and explicit formulas; *cf.* Section 20.1.

The liquidity risk should be included in the company's own risk and solvency assessment procedure, ORSA.

Due to the financial turmoil (2008–2010) and suggestions arising from QIS 4, there was a discussion on whether the liquidity risk should be treated as a quantitative risk and whether it should be included in the market risk module.

CEIOPS has concluded that, for the moment, this risk is better captured in Pillars 2 and 3; see CEIOPS (2009d09) and CEIOPS (2009f20, para 4.6). In their paper *Lessons learned from the crisis,* CEIOPS (2009a) have discussed the treatment of liquidity risk. Their conclusion was that the liquidity risks need more attention and therefore there need to be a higher reporting frequency in a stressed situation to the board of directors (weekly or daily reporting). Specific information should be requested by supervisors for their understanding of the risks.

Liquidity risk for banks: Quantitative rules on the liquidity risk set up by the Bank for International Settlements (BIS) will probably be a complement to the qualitative rules set out in the Capital Requirements Directive (CRD) for banks. It is also very likely that any such rules set up by CEBS for future amendments of the CRD will influence the work by CEIOPS on capital requirements for insurance undertakings. The reader is referred to Chapter 20 for a discussion of proposals from BIS.

CHAPTER 31

Solvency II Standard Formula

Nonlife Underwriting Risk

31.1 GENERAL FEATURES

The nonlife underwriting risk and its capital charge are treated in different articles of the Framework Directive, namely 111, but also in 101, 104, and 105. The development of the nonlife underwriting risk is discussed in Appendix M. CEIOPS has discussed the general structure of the nonlife underwriting risk module in CEIOPS (2009d10) and (2009f21), its correlations in CEIOPS (2009e12) and (2010a09), and its calibration in CEIOPS (2009e09). The final calibration was done in connection with the development of the Technical Specifications of QIS5, during the spring 2010.

Reserve risk stems from two sources: on the one hand, the absolute level of the claims provisions may be misestimated. On the other hand, because of the stochastic nature of future claims payouts, the actual claims will fluctuate around their statistical mean value. The underlying risk is assumed to follow a lognormal distribution.

Some of the stochastic effects relate to individual claims so that they are generally less significant for large portfolios. Others relate to economic conditions and other factors that affect the whole portfolio so the law of large numbers does not apply to them.

Premium risk is understood to relate to *future claims arising during and after the period until the time horizon for the solvency assessment.* The risk is that expenses plus the volume of losses (incurred and to be incurred) for these claims (comprising both amounts paid during the period and provisions made at its end) are higher than the premiums received (or if allowance is made elsewhere for the expected profits or losses on the business, that the profitability will be less than expected). The underlying risk is assumed to follow a lognormal distribution.

Premium risk is present at the time the policy is issued, before any insured events occur. Premium risk also arises because of uncertainties prior to issue of policies during the time horizon. These uncertainties include the premium rates that will be charged, the precise terms and conditions of the policies, and the precise mix and volume of business to be written.

Premium risk relates to policies to be written (including renewals) during the period, and to unexpired risks on existing contracts.

Catastrophic (CAT) risks stem from extreme or irregular events that are not sufficiently captured by the charges for premium and reserve risk.

31.1.1 Changes as Compared to QIS4

In CEIOPS (2009d10) and (2009f21), the following main changes are presented:

Proportional risk mitigation reinsurance is fully reflected through the use of net volume measures, via the design of the nonlife premium and reserve risk formula, and the nonlife catastrophe submodule. An average level of risk mitigating effect of *nonproportional reinsurance* is implicitly allowed for in the calibration of the nonlife premium and reserve risk module.

Geographical diversification: After QIS 4 and consultation, CEIOPS has decided not to apply geographical diversification for nonlife business across the globe. CEIOPS establishes that while this change is crucial for reinsurers and cross-border groups, it was seen as introducing unnecessary complexity at the solo level, in view of the materiality of the reduction in capital requirement they could obtain from the calculation.

Multiyear contracts: CEIOPS considers that a risk that was missed under the QIS4 approach was the risk relating to the change in the premium provisions, which is set up for multiyear contracts. If the outstanding period of an insurance contract is longer than 1 year at the valuation date t, then a part of the premium provision covers the CFs that relate to cover given from $t + 1$ onward. After 1 year (at $t + 1$), a new premium provision needs to be set up. This premium provision may be higher, apart from discounting effects, because the assumptions about the CFs may have changed from t to $t + 1$.

Undertaking-specific estimate: During QIS4 the standard deviation for premium risk for each LoB was derived as a credibility mix of an undertaking-specific estimate and a market-wide estimate. The industry welcomed the inclusion of such an approach. However, after careful consideration, CEIOPS believes that the mechanic estimation of the standard deviation from loss ratios is not sufficiently robust and reliable unless the credibility factors are very low. Moreover, the loss ratios for the estimation may not be appropriate for reasons such as changes in the composition of the portfolio, product changes, portfolio transfers, change of reinsurance, etc. CEIOPS considers that the drawbacks of using such an approach outweigh the benefits, and providing solutions would make the standard formula overly complicated; therefore the approach has not been retained.

Complex risk mitigation arrangements: CEIOPS is aware that the standard formula may not be able to consider the full risk mitigation effect of particular risk mitigation arrangements such as nonproportional reinsurance. CEIOPS has consulted extensively on this issue and has reached the conclusion that this is an area that may be further analyzed as part of the implementing measure of Article 111(e). However CEIOPS also considers that the standard formula is already very complex and introducing further complexity may not be welcomed. CEIOPS would encourage undertakings with complex risk mitigation arrangements to use partial internal models.

Distinction between CAT risks and other risks: QIS4 feedback, as well as other industry papers, has highlighted the importance of making a clear distinction between catastrophe risks and other risks in order to ensure that there is no double counting of catastrophe risk in the capital requirements. CEIOPS agrees with this. However, CEIOPS believes that the assumptions underlying the estimation of the premium risk capital charge, implicitly allow for double counting. The selection of a lognormal distribution for the underlying exposure was made under the assumption that such distribution would not capture extreme events and therefore would avoid double counting.

Expected profit or loss in the standard formula is not allowed for.

31.1.2 Standard Formula

The structure of the nonlife underwriting risk module is given in Annex IV (2) of the Framework Directive. Figure 31.1 shows the structure of the risk module and its subrisks.

CR_{NL} denotes the capital charge from the nonlife underwriting risk module. The capital charges of is subrisk modules are

- $CR_{NL,RP}$: the reserve and premium risk module
- $CR_{NL,CAT}$: the catastrophe risk module

CEIOPS (2009e12, 2010a09) discussed the dependence between the two subrisks, but also the dependence structure between conceptually independent risks. They maintained that even if the premium and reserve risks and the catastrophe risk are conceptually independent, a positive correlation factor may be appropriate as, for example, the underlying distributions are skewed and also truncated.

It is therefore assumed that dependence between the CAT-risk and the combined reserve and premium risks should be set to $\rho_{RP,CAT} = 0.25$; CEIOPS (2009e12, 2010a09). Hence the total capital charge is defined as

$$C_{NL} = \sqrt{C_{NL,RP}^2 + C_{NL,CAT}^2 + 0.5 \cdot C_{NL,RP} \cdot C_{NL,CAT}}.$$

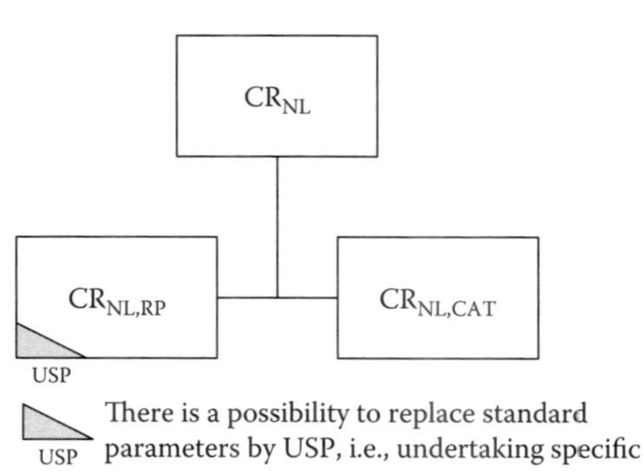

FIGURE 31.1 Modular structure of the capital charge for the nonlife underwriting risk.

TABLE 31.1 Segmentation in LoB, Which Should be Used

LoB	LoB
1	Motor, third-party liability
2	Motor, others
3	MAT
4	Fire/property
5	Third-party liability
6	Credit
7	Legal expenses
8	Assistance
9	Miscellaneous
10	NP Reinsurance (property)
11	NP Reinsurance (pasualty)
12	NP Reinsurance (MAT)

Source: Adapted from CEIOPS. 2009f21. CEIOPS' Advice for L2 Implementing Measures on SII: SCR-Standard formula-Non Life Underwriting Risk (former Consultation Paper no. 48). CEIOPS-DOC-40-09. November 10. Available at www.ceiops.org.

Note: Faculty and proportional reinsurance are classified according to Classes 1–10. For nonproportional reinsurance, a split into three different LoBs has been proposed.

31.1.2.1 Segmentation

The calculation of capital charge for both the premium risk and the reserve risk should be done for eachLoB, as defined in Table 31.1.

The calibration of the parameters used in the standard formula is given by CEIOPS (2009e09).

31.2 RESERVE RISK AND PREMIUM RISK MODULES

The capital requirement for the combined premium and reserve risks is determined as

$$C_{\text{NL,RP}} = \rho_{0.995}(\sigma)_{\text{VaR}} \cdot V_{\text{RP}},$$

where V_{RP}: a volume measure, which is a combination of volume measures from reserve and premium volumes. See Steps 1 and 2; σ: combined standard deviation, resulting from the combination of the reserve and premium risk standard deviations. See Steps 1 and 2; and $\rho_{0.995}(\sigma)$ is a charge function, defined by Equation 21.9b.

The charge function $\rho_{0.995}(\sigma)$ is set such that, assuming a lognormal distribution of the underlying risk, a capital charge consistent with the VaR 99.5% confidence is produced. This could be approximated by $\rho_{0.995}(\sigma) \approx 3 \cdot \sigma$.

The calculation is made in two steps:

1. For each LoB, the standard deviation and volume measures is calculated for both the reserve risk and the premium risk.

2. The standard deviation and the volume measure for the reserve risk and the premium risk in the individual LoBs are aggregated to derive an overall volume measure V and an overall standard deviation σ.

Let

$V_{R,k}$: the volume measure of the reserve risk

$V_{P,k}$: the volume measure of the premium risk

$\sigma_{R,k}$: the standard deviation for the reserve risk

$\sigma_{P,k}$: the standard deviation for the premium risk.

Step 1: Volume and Volatility Measures per LoB
For each LoB, k, we define the volume measures as

Step 1.1: Volume Measure—Reserve Risk
$V_{R,k} = R_k^0$, where we have used the notation used in Section 21.1.4.
R_k^0 =BE for claims outstanding in LoB k (the opening reserve)

Step 1.2: Volume Measure—Premium Risk

$$V_{P,k} = \max\left\{P_k^{01}; \tilde{P}_k^{01}; P_k^{-1}\right\} + M_k^{PP},$$

where we have used the notation used in Section 21.1.4.

P_k^{01} = estimate of the net *written* premium in LoB k in the forthcoming year
$\tilde{P}_k^{01}$ = estimate of the net *earned* premium in LoB k in the forthcoming year
P_k^{-1} = the net *written* premium in LoB k during the previous year
M_k^{PP} = expected present value of net claims and expense payments that relate to claims incurred after the following year and covered by existing contracts for each LoB k; Multiyear contracts. The term relates purely to part of the premium provision brought forward, whereas the other term is a proxy for premiums to be written or premiums to be earned, noting that the risks relating to these are rather different and only partly overlap. It is not intended to cover random events after the year but changes in provisions on claims after the year as a result of new information.

If the insurer has committed to its regulator that it.he/she will restrict premiums written over the period so that the actual premiums written (or earned) over the period will not exceed its estimated volumes, the volume measure is determined only with respect to estimated premium volumes, so that in this case

$$V_{P,k} = \max\left\{P_k^{01}; \tilde{P}_k^{01}\right\} + M_k^{PP}.$$

Step 1.3: Volatility Measure—Reserve Risk
For each LoB, k, we define the volatility measures as

$$\sigma^2_{R,k} = s^2_{R,k},$$

where $s_{R,k}$ = the standard deviation for reserve risk for LoB k; see Table 31.2.

Step 1.4: Volatility Measure—Premium Risk
For each LoB, k, we define the volatility measures as

$$\sigma^2_{P,k} = s^2_{P,k},$$

where $s_{P,k}$ = the standard deviation of the premium risk for LoB k; see Table 31.2.

The volatility for the reserve and premium risks for each LoB is calculated by

$$\sigma^2_k = \frac{\left(V_{R,k} \cdot \sigma_{R,k}\right)^2 + \left(V_{P,k} \cdot \sigma_{P,k}\right)^2 + 2\alpha \cdot V_{R,k} \cdot \sigma_{R,k} \cdot V_{P,k} \cdot \sigma_{P,k}}{V^2_k},$$

where $V_k = V_{R,k} + V_{P,k}$ is the volume measure within each LoB k and α is a factor representing on overall assumption between premium risk and the reserve risk. The dependence-factor α is set to 0.5 as in QIS4.

Step 2: Overall Volume and Volatility Measures
At this step, we aggregate the volume measures and the volatility measures using the dependency structure in Table 31.3.

The overall volume measure is defined as the sum of the volume measures for the reserve and premium risks summed up over all LoBs and taken care of the diversification effect:

$$V_{RP} = \sum_k V_k,$$

where $V_k = V_{R,k} + V_{P,k}$ is the volume measure within each LoB k.

The overall volatility measure is defined by

$$\sigma^2 = \frac{1}{V^2} \sum_{j,k} \rho_{jk} \cdot V_j \cdot V_k \cdot \sigma_j \cdot \sigma_k,$$

TABLE 31.2 Draft Standard Deviations for the Market-Wide Reserve Risk ($s_{R,k}$) and for the Market-Wide Premium Risk ($s_{P,k}$)

LoB	1	2	3	4	5	6	7	8	9	10	11	12
$s_{R,k}$	0.125	0.125	0.175	0.15	0.20	0.20	0.125	0.125	0.20	0.30	0.30	0.30
$s_{P,k}$	0.10	0.10	0.20	0.125	0.175	0.20	0.075	0.10	0.20	0.30	0.30	0.30

Source: Adapted from CEIOPS. 2009e09. Draft L2 Advice on SCR Standard Formula-Calibration of Non-Life Underwriting Risk. CEIOPS-CP-71-09. November 2. Available at www.ceiops.org.
Note: The calibrations of these parameters are given by CEIOPS. The LoBs are given in Table 31.1.

TABLE 31.3 Dependence Matrix Used for QIS 4

LOB	1	2	3	4	5	6	7	8	9	10	11	12
1: M 3rd	1	0.5	0.5	0.25	0.5	0.25	0.5	0.25	0.5	0.25	0.25	0.25
2: M other		1	0.25	0.25	0.25	0.25	0.5	0.5	0.5	0.25	0.25	0.25
3: MAT			1	0.25	0.25	0.25	0.25	0.5	0.5	0.25	0.25	0.5
4: Fire				1	0.25	0.25	0.25	0.5	0.5	0.5	0.25	0.5
5: Gen liab					1	0.5	0.25	0.25	0.25	0.25	0.25	0.25
6: Credit						1	0.5	0.25	0.5	0.25	0.5	0.25
7: Legal exp							1	0.25	0.5	0.25	0.5	0.25
8: Assist.								1	0.5	0.5	0.25	0.25
9: Misc.									1	0.25	0.25	0.5
10: Reins-p										1	0.25	0.25
11: Reins-c											1	0.25
12: Reins-MAT												1

Source: Adapted from CEIOPS. 2009e12. Draft L2 Advice on SCR Standard Formula-Correlation Parameters. CEIOPS-CP-74-09. November 2. Available at www.ceiops.org.

Note: The LoBs are defined in Table 31.1.

where the dependencies between LoB, ρ_{jk}, are given in the table below and the sum is taken over all LoBs.

31.2.1 Undertaking-Specific Parameters

In QIS4, the premium risk for each LoB was derived as a credibility mix of market-wide parameters and undertaking-specific parameters, USP. This option was deleted from the standard formula described above. However, the approach will be allowed for both the reserve risk and the premium risk; CEIOPS (2009e13, 2010a10). This is discussed in general terms in Section 26.8.

31.2.1.1 USP for Reserve Risk

Let $\sigma^2_{\mathrm{M,R},k} = s^2_{\mathrm{R},k}$ be the market-wide parameters as defined in Section 31.2 and $\sigma^2_{\mathrm{U,R},k}$ be the corresponding USP (U stands for USP and R for the reserve risk). Then the reserve risk parameters should be calculated as

$$\sigma_{\mathrm{R},k} = c \cdot \sigma_{\mathrm{U,R},k} + (1 - c) \cdot \sigma_{\mathrm{M,R},k},$$

where c is a credibility weight given in Table 31.4.

TABLE 31.4 Credibility Weights (c) According to the Length of Time Series with Own Data (N_k)

Data	N_k	5	6	7	8	9	10	11	12	13	14	15+
ID1	c	0.34	0.43	0.51	0.59	0.67	0.74	0.81	0.87	0.92	0.96	1
ID2	c	0.34	0.51	0.67	0.81	0.92	1	1	1	1	1	1
ED1	c	0.30	0.34	0.38	0.42	0.46	0.50	0.53	0.56	0.58	0.61	0.63
ED2	c	0.30	0.38	0.46	0.53	0.58	1	1	1	1	1	1

Source: Adapted from CEIOPS. 2010a10. CEIOPS'–Advice for L2 Implementing Measures on SII: SCR standard formula-Undertaking-Specific Parameters (former Consultation Paper no. 75). CEIOPS-DOC-71-10. January 29. Available at www.ceiops.org.

Note: Data ID1 is for internal data (third-party liability, motor vehicle liability, and credit and suretyship) and ID2 is for internal data for all other LoB. Data ED1 are for external data (third-party liability, motor vehicle liability, and credit and suretyship) and data ED2 are for external data for all other LoBs. When using of a mix of internal and external data, ED1 and ED2 should be used.

The estimated USP should be complemented with a reserve risk component for unexpected extreme events and model risk:

$$\sigma_{U,R,k} = \sqrt{\sigma_{U,R,k}^2 + \tau^2},$$

where τ is a fixed parameter, reflecting risks missing in the USP. CEIOPS (2010a10) has decided not to fix the amount at this stage. $\sigma'_{U,R,k}$ is calculated by one of the methods described below.

We define the following terms:

$R_{k,i,j}$: the BE for claims outstanding by LoOB k, for accident year i and development year j

$C_{k,i,j}$: the incremental paid claims by LoB k, for accident year i and development year j

$V_{R,Y,k} = \sum_{i+j=Y+1} R_{k,i,j}$ is the volume measure by LoB k and calendar year (Y) and

$$R_{Y,k} = \sum_{\substack{i+j=Y+2 \\ i \neq' Y+1}} R_{k,i,j} + \sum_{\substack{i+j=Y+2 \\ i \neq' Y+1}} C_{k,i,j}$$

is the BE for outstanding claims and incremental paid claims for the exposures covered by the volume measure, but in one year's time by LoB k and calendar year (Y).

Method 1:

$$\sigma'_{U,R,k} = \frac{\hat{\beta}_k}{\sqrt{V_{R,k}}},$$

where $V_{R,k}$ = BE for claims outstanding in LoB k, and

$$\hat{\beta}_k^2 = \frac{1}{N_k - 1} \sum_Y \frac{\left(R_{Y,k} - V_{R,Y,k}\right)^2}{V_{R,Y,k}}$$

is a constant of proportionality for the variance of the BE for claims outstanding in one year plus the incremental claims paid over the one year by LoB k. N_k is the number of observations when data pairs are available.

Method 2: Merz–Wüthrich Method

$$\sigma'_{\mathrm{U,R,k}} = \frac{\sqrt{\mathrm{MESP}}}{V_{\mathrm{R,k}}}.$$

This approach is based on the mean squared error of prediction (MSEP) of the claims development result over the one year and fitting a model to these results. The mean squared errors are calculated using the approach detailed in Merz and Wüthrich (2008); see also Bühlmann et al. (2009).

Method 3: Merz–Wüthrich Method

$$\sigma'_{\mathrm{U,R,k}} = \frac{\sqrt{\mathrm{MESP}}}{\mathrm{CLV}_{\mathrm{R,k}}}.$$

The approach is similar to Method 2, besides that the denominator is calculated using chain ladder (CL); see Merz and Wüthrich (2008); see also Bühlmann et al. (2009).

31.2.1.2 USP for Premium Risk

Let $\sigma^2_{\mathrm{M,P,k}} = s^2_{\mathrm{P,k}}$ be the market-wide parameters as defined in Section 31.2 and $\sigma^2_{\mathrm{U,P,k}}$ be the corresponding USP (U stands for USP and P for the premium risk). Then the premium risk parameters should be calculated as

$$\sigma_{\mathrm{P,k}} = c \cdot \sigma_{\mathrm{U,P,k}} + (1 - c) \cdot \sigma_{\mathrm{M,P,k}},$$

where c is a credibility weight given in Table 31.4.

To allow for the expense risk, an add-on to the USP estimate should be done:

$$\sigma_{\mathrm{U,P,k}} = \sigma'_{\mathrm{U,P,k}} \cdot \left(1 - \frac{\mathrm{Exp}^{t-1}_{\mathrm{k}}}{C^{t-1}_{\mathrm{k}}}\right),$$

where σ' is calculated by one of the methods described below. *Exp* is the expense payments of the last year excluding allocated claims handling expenses and C is the net claims payments of the last year.

Let

$U_{Y,k}$ is the ultimate after one year by accident year Y and LoB k,

$V_{P,Y,k}$ is the earned premium by accident year Y and LoB k,

$V_{P,k}$ is the calculation measure for premium risk as defined above.

N_k is the number of data points available for LoB k.

An estimate of the expected loss ratio by LoB k is defined as

$$\hat{\mu} = \frac{\sum_Y U_{Y,k}}{\sum_Y V_{P,Y,k}}.$$

Method 1:

$$\sigma'_{U,P,k} = \frac{\hat{\beta}_k}{\sqrt{V_{P,k}}},$$

where

$$\hat{\beta}_k^2 = \frac{1}{N_k - 1} \sum_Y \frac{\left[U_{Y,k} - V_{P,Y,k} \cdot \hat{\mu}_k\right]^2}{V_{P,Y,k}}.$$

Method 2:

$$\sigma'_{U,P,k} = \frac{\hat{\beta}_k}{\sqrt{V_{P,k}}}$$

This estimator is similar to Method 1, but the parameters $\hat{\beta}_k$ are the ones that maximize the log likelihood

$$\log L = \sum_Y \left[-\log(S_{Y,k}) - \frac{\left(\log(U_{Y,k}) - M_{Y,k}\right)^2}{2 \cdot S_{Y,k}^2} \right],$$

where we have defined

$$S_{Y,k}^2 = \log\left(1 + \frac{\beta_k^2}{V_{P,Y,k} \cdot \mu_k^2}\right) \quad \text{and} \quad M_{Y,k} = \log(V_{P,Y,k} \cdot \mu_k) - \frac{1}{2} \cdot S_{Y,k}^2.$$

Method 3: based on SST

$$\sigma'_{U,P,k} = \frac{\sqrt{Var_k(S_N)}}{V_{P,k}},$$

where $Var_k(S_N)$ is the variance of the sum of claim size (random), and the number of claims N is also random. V is the volume measure known at the beginning of the year. It is also assumed that the number of claims, given a random parameter, follows a Poisson distribution. It is also assumed that the claims size, given a random parameter, follows a distribution F and that the claims size and the number of claims are conditionally independent.

This Bayesian approach is described in SST (2006b) and Gisler (2009).

31.2.2 Simplifications

For nonlife premium and reserve risks, a factor-based approach is used, no simplifications are considered, except for captives, as this approach should be acceptable also for less sophisticated undertakings; CEIOPS (2009e15) and (2010a12).

Simplifications for captives are discussed in CEIOPS (2009e16) and (2010a13). For each LoB k, you can calculate the capital charge as

$$C_{\mathrm{NL,RP,k}} = 0.9 \cdot \sqrt{V_{\mathrm{RR,k}}^2 + V_{\mathrm{PR,k}}^2 + 2 \cdot 0.5 \cdot V_{\mathrm{RR,k}} \cdot V_{\mathrm{PR,k}}},$$

and the total capital requirement as

$$CR_{\mathrm{NL,RP}} = \sqrt{\sum_k CR_{\mathrm{NL,RP,k}}^2 + 0.35 \cdot \sum_{\substack{r,c \\ r \neq c}} CR_{\mathrm{NL,RP,r}} \cdot CR_{\mathrm{NL,RP,c}}},$$

where r, c denote a pair of LoBs and

$V_{\mathrm{RR,k}}$ volume measure for the reserve risk for LoB k, as defined above and $V_{\mathrm{PR,k}}$ volume measure for the premium risk for LoB k, as defined above.

The risk mitigating effect of an aggregate limit can be taken into account by modifying the volume measure for premium risk of an LoB in the calculation above as follows:

$$V'_{\mathrm{PR,k}} = \min\left\{\frac{Agg_{\mathrm{k}}}{0.9}; V_{\mathrm{PR,k}}\right\},$$

where Agg_k is the aggregate limit for LoB k. This aggregate limit shall represent the net retention per LoB, after reinsurance, taken into account the limits stated in acceptance as well as in reinsurance treaties, increased by a possible reinstatement premium.

The formulas above will be updated once the calibration for the nonlife underwriting risk module is done in connection with the development of the Technical Specifications of QIS5 during the spring 2010.

31.3 NONLIFE CAT RISK

The CAT-risk submodule should be calculated according to a standardized scenario approach, see Section 31.3.1, and an alternative method (the factor approach), see Section 31.3.2.

The capital charge per event is calculated as $CR_{\mathrm{NL,CAT}}^{\mathrm{event}} = \max\{CR_{\mathrm{NL,CAT}}^{\mathrm{SS}}; CR_{\mathrm{NL,CAT}}^{\mathrm{AM}}\}$.

The capital requirement for catastrophe risk for an LoB should not exceed the aggregate limit for a specific LoB. The aggregate limit shall represent the net retention per LoB, after reinsurance, taking into account the limits stated in acceptance as well as in reinsurance treaties, increased by a possible reinstatement premium. The aggregation of the capital charges for different CAT events and relevant countries are illustrated in CEIOPS (2009d10) and (2009f21).

31.3.1 Standard Scenarios Approach

The capital charge of the standard formula CAT risk module should result from the application of standardized scenarios: $CR_{\mathrm{NL,CAT}}$ = the undertakings loss relative to the standardized scenarios specified by CEIOPS at a later stage.

This approach will be more detailed and comprehensive compared to the second method used under QIS 4; see Appendix M.

Scenarios shall be constructed for at least the following minimum set of events:

Events	**Lines of business affected**
Natural perils	
Storm	Fire and property; Other motor
Flood	Fire and property; Other motor
Earthquake	Fire and property; Other motor
Hail	Fire and property; Other motor
Man Made disasters	
Major fires, explosions	Fire and property
Major MAT disaster	MAT
Major motor third party liability disasters	Motor third party liability
Major third party liability disaster	Third party liability

For credit and health CEIOPS shall define relevant pan European events.

Pan European Events	Lines of business affected
Credit (Recession, large exposures)	Credit
Health SLT/Non-SLT (Pandemic, mass accident, recession, etc.).	Health SLT/Non-SLT

For each standardized scenario the undertaking should calculate the net amount of the gross share of the total loss according to the following formula.

Where the XL cover follows a proportional cover:

$$\mathrm{Max}\{[L \cdot MS \cdot QS] - XLC; 0\} + \mathrm{Min}\{[L \cdot MS \cdot QS]\,; XLF\} + \mathrm{Reinst},$$

where a proportional cover follows an XL cover:

$$\mathrm{Max}\,\{[L \cdot MS] - XLC; 0\} \cdot QS + \mathrm{Min}\,\{[L \cdot MS]\,; XLF\} \cdot QS + \mathrm{Reinst},$$

where L = the total gross loss amount. The total gross loss amount of the catastrophe will be provided as part of the information of the scenario; MS = the market share. This proportion might be determined with reference to exposure estimates, historical loss experience, or the share of total market premium income received. The total market loss amount of the catastrophe will be provided as part of the information of the scenario; QS = quota share retention. Allowance must be made for any limitations, for example, event limits that are frequently applied to QS treaties; XLC = the upper limit of the XL program that is applicable in case of the scenario event; XLF = the XL retention of the XL program that is applicable in case of the scenario event; $REINST$ = the reinstatement premium or premiums (in case of scenarios with a succession of 2 or more identical events).

The capital charge of the standardized scenario is denoted $CR^{SS}_{NL,CAT}$. It should be net of reinsurance recoverables. Each standardized scenario shall be designed, quantified, and reported in line with a common set of definitions. This will increase harmonization. Scenarios should, where relevant be calibrated based on sums insured by CRESTA zones (see www.cresta.org). The calibration of the standard scenarios will be provided by CEIOPS in mid-2010.

31.3.2 Alternative Approach

Undertakings will use a simpler factor-based alternative method in two cases: as an interim solution while the scenarios are being developed, and on a permanent basis, in cases that will be clearly specified. One such case is that the factor method shall be the default method for the miscellaneous LoB. This method was called Method 1 in QIS 4; see Appendix M.

When undertakings have selected the standard formula, they shall apply the factor-based alternative method in two cases:

- When a standardized scenario is not relevant and a partial internal model is not proportionate
- For the miscellaneous LoB

Circumstances in which the factor method shall be used instead of a standardized scenario are

- Undertakings with nonmaterial exposures outside the EU, in relation to these exposures
- Undertakings writing nonmaterial nonproportional reinsurance, in relation to these exposures
- The scenario is not applicable, but it has exposure

CEIOPS has improved the calibration of the factor method, as compared with the QIS4 calculation, by introducing the following changes; CEIOPS (2009e09):

- The factor has been calibrated gross of reinsurance. This allows undertakings to apply their respective reinsurance program in order to estimate the net amount
- The factor has been calibrated by peril for the property LoB, in order to introduce further segmentation at an LoB level

TABLE 31.5 CAT-Risks Factors for the Alternative Approach

Events	LoB Affected	Factor, c_k
Storm	4: Fire and property; 2: Other motor	1.75
Flood	4: Fire and property; 2: Other motor	1.13
Earthquake	4: Fire and property; 2: Other motor	1.20
Hail	2: Other motor	0.30
Major fires, explosions	4: Fire and property	1.75
Major MAT disaster	3: MAT	1.00
Major motor, third-party liability disasters	1: Motor third party liability	0.40
Major third-party liability disaster	5: Third-party liability	0.85
Miscellaneous	9: Miscellaneous	0.40
NPL property	10: NPL property	2.50
NPL casualty	11: NPL casualty	2.50
NPL MAT	12: NPL MAT	2.50

Source: Adapted from CEIOPS. 2009e09. Draft L2 Advice on SCR Standard Formula-Calibration of Non-Life Underwriting Risk. CEIOPS-CP-71-09. November 2. Available at www.ceiops.org.

Note: The final calibration was done in connection with the development of the Technical Specifications of QIS5, during the spring 2010.

The standard capital charge, according to the alternative approach, is given by

$$C_{\text{NL,CAT}}^{\text{AM}} = \sqrt{\sum_{k \neq 3,4,10,12} (c_k \cdot P_k)^2 + (c_3 \cdot P_3 + c_{12} \cdot P_{12})^2 + (c_4 \cdot P_4 + c_{10} \cdot P_{10})^2},$$

where P_k = estimate of net written premium for LoB k in the forthcoming year and the factors c_k are defined in Table 31.5.

In line with its final advice, CEIOPS (2009f21) has decided to provide a set of CAT-factors for captives, CEIOPS (2009e09). They are given in Table 31.6.

TABLE 31.6 CAT-Risk Factors for the Alternative Approach for Captives

LoB Affected	Factor, c_k
1: Motor, third-party liability	2.25
2: Other motor	5.40
3: MAT	9.20
4: Fire and property (other damage)	4.50
9: Miscellaneous	9.20

Source: Adapted from CEIOPS. 2009e09. Draft L2 Advice on SCR Standard Formula-Calibration of Non-Life Underwriting Risk. CEIOPS-CP-71-09. November 2. Available at www.ceiops.org.

Note: The final calibration was done in connection with the development of the Technical Specifications of QIS5, during the spring 2010.

31.3.3 Simplifications

The standard formula catastrophe risk module shall result from the application of standardized scenarios. The scenarios should be constructed in such a way that they are proportionate to the risks that they attempt to capture. Where this is not possible, simplifications should be introduced. CEIOPS will publish advice on the standardized scenarios at a later stage; CEIOPS (2009e15). They will also look at simplifications for captives.

CHAPTER 32

European Solvency II Standard Formula

Life Underwriting Risk

ANY STANDARDIZED APPROACH for life insurance will need to take account of the characteristics given by IAA (2004) and listed in Section 21.2. It will require the classification of all life insurance business into defined product types.

The calibration of the life underwriting parameters shall capture changes in the level and trend of the parameters only. It is assumed that the volatility risk component is implicitly covered by the level, trend, and catastrophe risk components.

32.1 GENERAL FEATURES

The life underwriting risk and its capital charge are treated in different articles of the Framework Directive, namely 111, but also in 105 and 109. The development of the life underwriting risk is discussed in Appendix N. CEIOPS has discussed the general structure of the life underwriting risk module in CEIOPS (2009d11) and (2009f22), and the dependence structures in CEIOPS (2009e12).

32.1.1 Standard Formula

The structure of the life underwriting risk module is given in Annex IV (3) of the Framework Directive. Figure 32.1 shows the structure of the risk module and its subrisks.

CR_{LR} denotes the capital charge from the life underwriting risk module. The capital charges of its subrisks modules are

- $CR_{LR,MR}$: mortality risk
- $CR_{LR,LO}$: longevity risk
- $CR_{LR,DR}$: disability risk
- $CR_{LR,LR}$: lapse risk

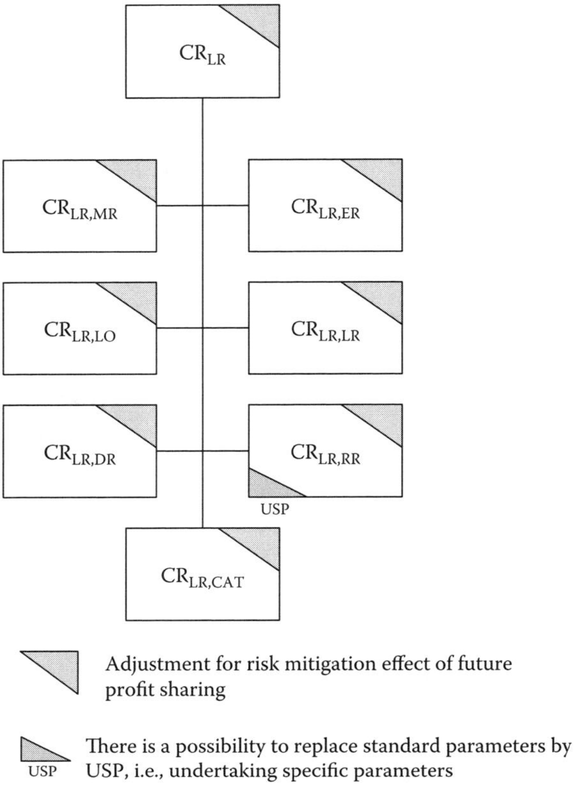

FIGURE 32.1 Modular structure of the capital charge for the life underwriting risk.

- $\mathrm{CR}_{\mathrm{LR,ER}}$: expense risk
- $\mathrm{CR}_{\mathrm{LR,CAT}}$: CAT risk
- $\mathrm{CR}_{\mathrm{LR,RR}}$: revision risk

The capital charge will be calculated as

$$CR_{\mathrm{LR}} = \sqrt{\sum_{rxc} \rho_{\mathrm{rc}} \cdot CR_{\mathrm{LR,r}} \cdot CR_{\mathrm{LR,c}}},$$

where r and c are the rows and the columns of the dependence matrix for the life underwriting risk. This matrix is given by Table 32.1.

The choice of the dependence structure given in Table 32.1 was based on expert opinion; CEIOPS (2009e12). In op. cit. they discuss mainly the dependence between mortality and longevity risks and expense risk versus disability, lapse, and revision risks. The dependence between the other risks is probably low or close to zero. Hence, by the discussion in CEIOPS (2009e12) and (2010a09), these correlations are set to either 0 or 0.25.

32.1.1.1 Risk Measures

The standard approach of the life underwriting risk stresses are based on the delta-NAV (or ΔNAV) approach as defined in Section 27.1.1, that is, the change in

TABLE 32.1 Dependence Matrix for the Life Underwriting Risk's Subrisks

	$CR_{LR,MR}$	$CR_{LR,LO}$	$CR_{LR,DR}$	$CR_{LR,LR}$	$CR_{LR,ER}$	$CR_{LR,CAT}$	$CR_{LR,RR}$
$CR_{LR,MR}$	1	–0.25	0.25	0	0.25	0.25	0
$CR_{LR,LO}$		1	0	0.25	0.25	0	0.25
$CR_{LR,DR}$			1	0	0.50	0.25	0
$CR_{LR,LR}$				1	0.50	0.25	0
$CR_{LR,ER}$					1	0.25	0.50
$CR_{LR,CAT}$						1	0
$CR_{LR,RR}$							1

Source: Adapted from CEIOPS. 2010a09. CEIOPS' Advice for L2 Implementing Measures on SII: SCR Standard Formula—Correlations (former Consultation Paper no. 74). CEIOPS-DOC-70-10. January 29. Available at www.ceiops.org.

value of assets minus liabilities. Where not applicable, scenario calculations will be used.

Furthermore, where a ΔNAV approach is used, the revaluation of technical provisions should allow for any relevant adverse changes in option take-up behavior of policyholders in this scenario.

32.1.1.2 Segmentation

Segmentations for technical provisions are discussed by CEIOPS (2009b02) and (2009f02). As for the calculation of life technical provision, a starting point should be a policy-by-policy approach. For the calculation of the capital requirement it should also be possible to group the policies into homogeneous risk groups, but a contract covering life insurance business should always be unbundled according to the implicit segmentation according to the different subrisk modules. The granularity must be decided by the company depending on its portfolio.

32.1.1.3 Net Capital Charge

For all subrisks, a net value of the capital charge is calculated:

$nCR_{LR,i}$ is the capital charge for life subrisk i, including the risk absorbing effect of future profit sharing.

The total net life underwriting risk capital charge was then calculated using the dependence matrix above:

$$nCR_{LR} = \sqrt{\sum_{rxc} \rho_{rc} \cdot nCR_{LR,r} \cdot nCR_{LR,c}}.$$

These net charges will be used to calculate a risk adjustment on the top level; see Section 26.1.

32.2 MORTALITY RISK

Mortality risk is associated with insurance obligations, such as term assurance or endowment policies, where an undertaking guarantees to make a single or recurring series of payments in the event of the death of the policyholder during the policy term.

It is applicable for obligations contingent on mortality risk, that is, where the amount currently payable on death exceeds the technical provisions held and, as a result, an increase in mortality rates is likely to lead to an increase in the technical provisions.

The capital charge for mortality risk is intended to reflect the uncertainty in mortality parameters as a result of changes in the level, trend, and volatility of mortality rates and capture the risk that more policyholders than anticipated die during the policy term.

This risk is normally captured by increasing the mortality rates either by a fixed amount or by a proportion of the base mortality rates. The calibration (of the increase) should capture the impact of each of the above factors (level, trend, and volatility); CEIOPS (2009d11) and (2009f22).

The capital charge for life mortality risk is defined as the result of a scenario as

$$CR_{\text{LR,MR}} = \sum_i \{\Delta NAV_i \mid \text{Mortshock}\},$$

where subscript i denotes each policy where the payment of benefits, either lump sum or multiple payments, was contingent on mortality risk. The other terms are

ΔNAV: the change in the net value of assets minus liabilities

Mortshock: a permanent 15% *increase* in mortality rates for each age

The life mortality scenario would be calculated under the condition that the assumptions on future bonus rates, reflected in the valuation of future discretionary benefits in technical provisions, remained unchanged before and after the shocks being tested.

Additionally, the result of the scenario was determined under the condition that the participant was able to vary its assumptions in future bonus rates in response to the shock being tested. The resulting capital charge is $nCR_{\text{LR,MR}}$.

32.2.1 Simplifications

There are also *simplifications* that could be used, CEIOPS (2009e15) and (2010a12), provided that

- The simplification is proportionate to the nature, scale, and complexity of the risks that the undertaking faces.
- The assumed 10% increase in mortality rates underlying the simplification for each annual increase in age is consistent with the mortality assumption used in the calculation of the best estimate liability.

- The capital requirement for mortality risk under the simplified calculation is less than 5% of the overall SCR before adjustment for the loss-absorbing capacity of technical provisions and deferred taxes. For this comparison, the overall SCR can be calculated by means of the simplified calculation for the mortality risk capital requirement.
- The standard calculation of the mortality risk submodule is not an undue burden for the undertaking.

$$\text{Mortality capital requirement} = (\text{Total CaR}) * q\,(\text{firm-specific}) * n \\ * 0.15 * 1.1^{((n-1)/2)},$$

where CAR is the Capital-at-Risk, n is the modified duration of liability cash flows, q is the expected average death rate over the next year weighted by sum assured, and PMI, Projected Mortality Increase $= 1.1^{((n-1)/2)}$.

32.3 LONGEVITY RISK

Longevity risk is associated with insurance obligations, such as annuities, where an undertaking guarantees to make recurring series of payments until the death of the policyholder and where a decrease in mortality rates leads to an increase in the technical provisions, or with obligations, such as pure endowments, where an undertaking guarantees to make a single payment in the event of the survival of the policyholder for the duration of the policy term.

It is applicable for insurance obligations contingent on longevity risk, that is, where there is no death benefit or the amount currently payable on death is less than the technical provisions held and, as a result, a decrease in mortality rates is likely to lead to an increase in the technical provisions.

The risk that a policyholder lives longer than anticipated is longevity risk. Longevity risk is particularly significant as a result of an increasing life expectancy among policyholders in most developed countries.

The capital charge for longevity risk is intended to reflect the uncertainty in mortality parameters as a result of changes in the level, trend, and volatility of mortality rates and capture the risk of policyholders living longer than anticipated.

The capital charge for life longevity risk is defined as a result of a longevity scenario as follows:

$$CR_{\text{LR,LO}} = \sum_i \left\{\Delta NAV_{\text{i}} \mid \text{Longevityshock}\right\},$$

where subscript i denotes each policy where the payment of benefits, either lump sum or multiple payments, was contingent on longevity risk. The other terms are

ΔNAV: the change in the net value of assets minus liabilities

Longevity shock: a permanent 25% *decrease* in mortality rates for each age

The life longevity scenario should be calculated under the condition that the assumptions on future bonus rates, reflected in the valuation of future discretionary benefits in technical provisions, remained unchanged before and after the shocks being tested.

Additionally, the result of the scenario was determined under the condition that the participant was able to vary its assumptions in future bonus rates in response to the shock being tested. The resulting capital charge is $nCR_{\text{LR,LO}}$.

32.3.1 Simplifications

There are also *simplifications* that could be used, CEIOPS (2009e15) and (2010a12), provided that the conditions given in Section 32.2.1 are met.

$$\text{Longevity capital requirement} = 25\% * q * (1.1)^{((n-1)/2)} * n * (\text{Best estimate provisions for contracts subject to longevity risk}),$$

where n is the modified duration of liability cash flows and q is the expected average death rate over the next year weighted by sum assured.

32.4 DISABILITY RISK

Morbidity or disability risk is associated with all types of insurance compensating or reimbursing losses, for example, loss of income, caused by illness, accident, or disability (income insurance), or medical expenses due to illness, accident, or disability (medical insurance), or where morbidity acts as an acceleration of payments or obligations that fall due on death. It is applicable for obligations contingent on a definition of disability.

The capital charge for disability risk was defined as a result of a disability scenario as follows:

$$CR_{\text{LR,DR}} = \sum_{i} \left\{ \Delta NAV_{\text{i}} \mid \text{Disabilityshock} \right\},$$

where subscript i denotes each policy where the payment of benefits, either lump sum or multiple payments, was contingent on disability risk. The other terms are

ΔNAV: the change in the net value of assets minus liabilities

Disability shock: an increase of 35% in disability rates for the next year, together with a permanent 25% increase over best estimate in disability rates at each age in following years. Where applicable, add also a permanent decrease of 20% in morbidity/disability recovery rates.

The life disability scenarios should be calculated under the condition that the assumptions on future bonus rates, reflected in the valuation of future discretionary benefits in technical provisions, remained unchanged before and after the shocks being tested.

Additionally, the result of the scenario was determined under the condition that the participant was able to vary its assumptions in future bonus rates in response to the shock being tested. The resulting capital charge is $nCR_{\text{LR,DR}}$.

32.4.1 Simplifications

There are also *simplifications* that could be used, CEIOPS (2009e15) and (2010a12), provided that the conditions given in Section 32.2.1 are met.

$$\begin{aligned}\text{Disability capital requirement} = {} & (\text{total disability capital at risk})_1 * i(\text{firm-specific})_1 * 0.50 \\ & + (\text{total disability capital at risk})_2 * i(\text{firm-specific})_2 \\ & * 0.25 * 1.1^{((n-1)/2)} * n\end{aligned}$$

where n is the modified duration of liability cash flows; i_1, i_2 are the expected movements from healthy to sick over the first (next) year and the second year, respectively, weighted by sum assured or annual payment as appropriate for the product in question; and PDI, Projected Disability Increase $= 1.1^{((n-1)/2)}$.

An undertaking using the above simplification should ensure that the assumed 10% increase in disability rates for each annual increase in age is consistent with the disability assumption used in the calculation of the best estimate liability. For products where benefits consist of a series of payments payable until death or recovery of the policyholder, there is also a risk that the duration of the claim is higher than anticipated. The above simplification should therefore be extended to a combined stress that considers a 20% decrease in termination rates in addition to the increase in inception rates described above.

Disability capital requirement in respect of the risk that duration of claims is greater than expected $= 20\% * t * (1.1)((n-1)/2) * n *$ (Best estimate provisions for contracts subject to disability claims),

where n is the modified duration of liability cash flows, t is the expected termination rate, that is, movement from sick to healthy/dead over the next year, and PDI, Projected Disability Increase $= 1.1^{((n-1)/2)}$.

32.5 EXPENSE RISK

Expense risk arises from the variation in the expenses incurred in servicing insurance or reinsurance contracts. It is likely to be applicable for all insurance obligations.

The capital charge for expense risk was determined as follows:

$$CR_{\mathrm{LR,ER}} = \left\{\Delta NAV \mid \text{Expenseshock}\right\},$$

where ΔNAV is the change in the net value of assets minus liabilities,

Expense shock: All future expenses are higher than best estimate anticipations by 10%, and the rate of expense inflation is 1% per annum higher than anticipated; but for policies with adjustable loadings, 75% of these additional expenses can be recovered from second year onward through increasing the charges payable by policyholders. Policies with adjustable loadings were those for which expense loadings or charges may be adjusted within the next 12 months.

The life expense risk scenario should be calculated under the condition that the assumptions on future bonus rates, reflected in the valuation of future discretionary benefits in technical provisions, remained unchanged before and after the shock being tested.

An expense payment should not be included in the scenario if its amount is already fixed at the valuation date (for instance, agreed payments of acquisition provisions).

There may be some firms that have outsourced their expense activities, such as IT systems, policyholder complaints handling and so on. This can produce a different set of risks that, if significant, may mean a capital charge in respect of expense risk outside of the standard formula calibration.

Additionally, the result of the scenario was determined under the condition that the participant was able to vary its assumptions in future bonus rates in response to the shock being tested. The resulting capital charge is $nCR_{\text{LR,ER}}$.

32.5.1 Simplifications

There are also *simplifications* that could be used, CEIOPS (2009e15) and (2010a12), provided that

- The simplification is proportionate to the nature, scale, and complexity of the risks that the undertaking faces.
- The capital requirement for expense risk under the simplified calculation is less than 5% of the overall SCR before adjustment for the loss-absorbing capacity of technical provisions and deferred taxes. For this comparison, the overall SCR can be calculated by means of the simplified calculation for the expense risk capital requirement.
- The standard calculation of the expense risk submodule is not an undue burden for the undertaking.

$$\text{Expense risk capital requirement} = (\text{Renewal expenses in the 12 months prior to valuation date}) * \left[n(\exp)^{*}0.10 + \left\{\frac{(1+k)^{n(\exp)}-1}{k} - \frac{(1+i)^{n(\exp)}-1}{i}\right\}\right],$$

where $n(\exp)$ is the average (in years) period over which risk runs off, weighted by renewal expenses, i is the expected inflation rate (the same inflation as used in calculation of the best estimate), and k is the stressed inflation rates, that is, $i + 1\%$.

32.6 REVISION RISK

In the context of the life underwriting risk module, revision risk was intended to capture the risk of adverse variation of an annuity's amount, as a result of an unanticipated revision of the claims process. This was meant to impact only on annuities that are genuinely reviewable. Annuities whose amount is linked to earnings or another index such as prices or that vary in deterministic value on change of status should not be classified as genuinely reviewable for these attributes. This risk should be applied only to annuities and to those benefits that can be approximated by a life annuity arising from nonlife claims, including accident insurance, but excluding workers compensation that are allocated to the life underwriting risk module.

The capital charge for revision risk was determined as follows:

$$CR_{\text{LR,RE}} = \{\Delta NAV \mid \text{Revision shock}\},$$

where ΔNAV is the change in the net value of assets minus liabilities and *Revision shock* is a 3% increase in the annual amount payable for annuities exposed to revision risk. The impact should be assessed considering the remaining run-off period.

On the computation of this risk charge, participants should only consider the impact on those nonlife annuities for which a revision process is possible to occur during the next year, for example, annuities where there are legal or other eligibility restrictions should not be included.

32.6.1 Undertaking-Specific Parameters

A subset of the parameters of the standard formula may, after approval of the supervisory authority, be replaced by USPs. This is discussed in general terms in Section 26.8. The capital charge for the revision risk can be calculated by using USP standardized methods. This is discussed in CEIOPS (2009e13) and (2010a10).

Revision risk is intended to capture the risk of adverse variation of an annuity's amount, as a result of an unanticipated revision of the claims process. This risk should be applied only to annuities and to those benefits that can be approximated by a life annuity arising from nonlife claims (in particular, life assistance benefits from workers' compensation LOB). The *undertaking-specific shock* for revision risk is restricted only to workers' compensation or to annuities that are not significantly subject to inflation risk. This restriction stems from the assumption in calculation procedure that the number and severity of revisions are independent. In case of inflation the number and severity are usually dependent because the value of inflation determines which annuities will be revised and the severity of this revision.

On the computation of this risk charge, it shall be considered that the impact on those annuities for which a revision process is possible to occur during the next year; for example, annuities where there are legal or other eligibility restrictions should not be included. Unless the future amounts payable are fixed and known with certainty, all those benefits that can be approximated by a life annuity (life assistance) are also subject to revision risk. In order to derive undertaking-specific parameters for revision risk, undertaking concerned shall use time series of annual amounts of individual annuities (life assistance benefits) in payment in consecutive years, during the time horizon in which they are subject to revision risk.

For each calendar year t,

- Identify the set of annuities that were exposed to revision risk during the whole year
- Include those individual annuities that were exposed only during a part of the year, but where an upward revision has effectively occurred in that period

- Annuities that entered or exited the books during the period, for example, new claims, death of the beneficiary, should be excluded

Input data are the historical average relative change of individual annuities, the historical standard deviation of relative change of individual annuities estimated by means of the standard estimator, the estimate of percentage of individual annuities for which a revision process is possible to occur during the forthcoming year, and the historical standard deviation of percentage of individual annuities for which a revision process occurred, estimated by means of the standard estimator.

Statistical fitting techniques should be applied, to estimate a probability distribution, and validated by goodness-of-fit tests. After this, the mean, $\bar{X}_{\text{Rev}}$, and standard deviation should be calculated. The revision shock is then calculated as

$$\hat{C}_{\text{LR,RE}} = \frac{VaR_{0.995}(X_{\text{Rev}})}{\bar{X}_{\text{Rev}}} - 1.$$

The Value-at-Risk, VaR(X_{Rev}), should be derived by simulation; see CEIOPS (2010a10) for details.

32.6.2 Simplifications

There are also *simplifications* that could be used; CEIOPS (2009e15) and (2010a12).

Revision capital requirement = 3% $*$ Total net technical provisions for annuities exposed to revision risk.

32.7 LAPSE RISK

This submodule should capture the adverse change in value of insurance liabilities, resulting from changes in level or volatility of the rates of policy lapses, terminations, renewals, and surrenders; see Article 105 (3f) in the Framework Directive.

For the assessment of lapse risk a prespecified stress test can easily be applied. The capital requirement is of the form of the difference between a special valuation of policy liabilities and the normal valuation. For the special valuation, the lapse assumption is multiplied by a specified factor greater or less than one. For some policies, an increase in lapse rates will result in an increase in policy liabilities, and for other policies, liabilities will increase when assumed lapses decrease.

A lapse case, which cannot be addressed in a factor-based approach are those products for which the lapse risk does not act uniformly over the products life, such as lapses at early durations that may reduce the company's exposure to later risks for some policies and not for others.

Lapse risk relates to the loss, or adverse changes in the value of insurance liabilities, resulting from changes in the level or volatility of the rates of policy lapses, terminations, changes to paid-up status (cessation of premium payment), and surrenders. The standard formula allows for the risk of a permanent change of the rates as well as for the risk of a mass lapse event.

The capital charge for lapse risk should be based on three scenarios: a permanent increase of lapse rates, a permanent decrease of lapse rates, and a mass lapse event. It is defined as follows:

$$CR_{\mathrm{LR,LA}} = \max\left\{\mathrm{Lapse}_{\mathrm{down}}; \mathrm{Lapse}_{\mathrm{up}}; \mathrm{Lapse}_{\mathrm{mass}}\right\},$$

where $\mathrm{Lapse}_{\mathrm{down}}$ is the capital charge for the risk of a permanent decrease of rates of lapsation, $\mathrm{Lapse}_{\mathrm{up}}$ is the capital charge for the risk of a permanent increase of rates of lapsation, and $\mathrm{Lapse}_{\mathrm{mass}}$ is the capital charge for the risk of a mass lapse event.

The capital charges were calculated based on a policy-by-policy comparison of surrender value and best estimate provision. The surrender strain of a policy was defined as the difference between the amount currently payable on surrender and the best estimate provision held. The amount payable on surrender should be calculated from net of any amounts recoverable from policyholders or agents, for example, net of any surrender charge that may be applied under the terms of the contract.

The three capital charges are calculated as below.

$$\mathrm{Lapse}_{\mathrm{down}} = \sum_i \left[\Delta NAV_{\mathrm{i}} \mid \mathrm{Lapseshock}_{\mathrm{down}}\right],$$

where i denotes each policy, ΔNAV is the change in the net value of assets minus liabilities (not including the loss-absorbing effect of future discretionary benefits and taxation), and Lapse $\mathrm{shock}_{\mathrm{down}}$ is the reduction of 50% in the assumed option take-up rates in all future years for all policies without a positive surrender strain. Affected by the reduction are options to fully or partly terminate, decrease, restrict, or suspend the insurance cover. Where an option allows the full or partial establishment, renewal, increase, extension, or resumption of insurance cover, the 50% reduction should be applied to the rate that the option is not taken up. The shock should not change the rate to which the reduction is applied to by more than 20% in absolute terms.

$$\mathrm{Lapse}_{\mathrm{up}} = \sum_i \left[\Delta NAV_i \mid \mathrm{Lapseshock}_{\mathrm{up}}\right],$$

where i denotes each policy, ΔNAV is the change in the net value of assets minus liabilities (not including the loss-absorbing effect of future discretionary benefits and taxation), and Lapse $\mathrm{shock}_{\mathrm{up}}$ is the increase of 50% in the assumed option take-up rates 33/51 in all future years for all policies with a positive surrender strain. Affected by the increase are options to fully or partly terminate, decrease, restrict, or suspend the insurance cover. Where an option allows the full or partial establishment, renewal, increase, extension, or resumption of insurance cover, the 50% increase should be applied to the rate that the option is not taken up. The shocked rate should not exceed 100%.

The shocked take-up rate should have the following restrictions:

$$R_{\mathrm{up}}(R) = \min\{1.5 \cdot R; 1\},$$

and

$$R_{\text{down}}(R) = \min\{\max[0.5 \cdot R; R - 0.20]; 0\},$$

where R_{up} is the shocked take-up rate in $\text{Lapseshock}_{\text{up}}$, R_{down} is the shocked take-up rate in $\text{Lapseshock}_{\text{down}}$, and R is the take-up rate before shock.

$\text{Lapse}_{\text{mass}}$ is defined as 30% of the sum of surrender strains over the policies where the surrender strain is positive. The result reflects the loss, which is incurred in a mass lapse event. For nonretail business, the capital requirement for the risk of a mass lapse event $\text{Lapse}_{\text{mass}}$ should be defined as 70% of the sum of surrender strains over the policies where the surrender strain is positive.

In order to determine $nCR_{\text{LR,LA}}$, the results of the scenarios was also calculated under the condition that the undertaking was able to vary its assumptions on future bonus rates in response to the shock being tested. If the scenario that gave the maximum net calculation did not coincide with the scenario that gave the maximum gross calculation, the definition of $CR_{\text{LR,LA}}$ should be changed in order to ensure consistency with the net calculation.

For instance, if $\text{Lapse}_{\text{down}} = 10$, $\text{Lapse}_{\text{up}} = 20$, $\text{Lapse}_{\text{mass}} = 30$, $n\text{Lapse}_{\text{down}} = 9$, $n\text{Lapse}_{\text{up}} = 5$, and $n\text{Lapse}_{\text{mass}} = 8$, then $\text{Life}_{\text{lapse}}$ should be chosen to be 10 but not 30.

32.7.1 Simplifications

There are also *simplifications* that could be used; see CEIOPS (2009d11) and (2009f22).

If it is proportionate to the nature, scale, and complexity of the risk, the comparison of surrender value and best estimate provision in the above calculations may be made on the level of homogeneous risk groups, or at finer granularity, instead of a policy-by-policy basis.

In particular, if the conditions are met, this simplification may be applied, if technical provisions are not calculated on a policy-by-policy basis.

A simplified calculation of $\text{Lapse}_{\text{down}}$ and Lapse_{up} may be made, if the following conditions are met:

a. The simplified calculation is proportionate to nature, scale, and complexity of the risk.

b. The undertaking is small or the capital charge for lapse risk under the simplified calculation is less than 5% of the overall SCR before adjustment for the loss-absorbing capacity of technical provisions.

The simplified calculations are defined as follows:

$$\text{Lapse}_{\text{down}} = 0.5 \cdot l_{\text{down}} \cdot n_{\text{down}} \cdot S_{\text{down}},$$

and

$$\text{Lapse}_{\text{up}} = 0.5 \cdot l_{\text{up}} \cdot n_{\text{up}} \cdot S_{\text{up}},$$

where $l_{\text{down}}; l_{\text{up}}$ is the estimate of the average rate of lapsation of the policies with a negative/positive surrender strain, $n_{\text{down}}; n_{\text{up}}$ is the average period (in years), weighted by surrender strains, over which the policy with a negative/positive surrender strain runs off, and $S_{\text{down}}; S_{\text{up}}$ is the sum of negative/positive surrender strains.

32.8 LIFE CAT RISK

The CAT risks stem from extreme or irregular events that are not sufficiently captured by the charges for the biometric risks, lapse risk, and expense risk. These are one-time shocks from the extreme, adverse tail of the probability distribution that are not adequately represented by extrapolation from more common events and for which it is usually difficult to specify a loss value, and thus an amount of capital to hold. For example, a contagious disease process or a pandemic may affect many persons simultaneously, nullifying the usual assumption of independence among persons.

The mortality catastrophe module is restricted to insurance obligations that are contingent on mortality, that is, where an increase in mortality leads to an increase in technical provisions.

Scenarios may be used to model extreme events where the assumptions of the analytic model break down, or to take into account risks that are not covered by analytic models, particularly systemic risk. Mixing two different techniques may actually reduce modeling risk associated with a standard formula.

Possible scenarios include

- Severe epidemic: for example, Spanish Flu in 1918
- Natural catastrophe: for example, earthquake
- Terrorist attack: for example, events of 9/11

A more restricted range might be applied to take account of relative data availability. For example, one might include periodic natural catastrophes and epidemic, but exclude extreme, episodic events, such as terrorist activity.

The capital charge for life CAT risk was determined as follows:

$$CR_{\mathrm{LR,CAT}} = \{\Delta NAV \mid \mathrm{Lifeshock}_{\mathrm{CAT}}\},$$

where the $\mathrm{Lifeshock}_{\mathrm{CAT}}$ is an absolute 2.5 per mille increase in the rate of policyholders dying over the following year (e.g., from 1.0 per mille to 3.5 per mille).

Participants are requested to calculate the capital charge for life CAT risk should be calculated under the condition that the assumptions on future bonus rates, reflected in the valuation of future discretionary benefits in technical provisions, remained unchanged before and after a life CAT event.

Additionally, participants were also requested to determine the result of the scenario under the condition that the participant was able to vary its assumptions in future bonus rates in response to the shock being tested. The resulting capital charge is $nCR_{\mathrm{LR,CAT}}$.

32.8.1 Simplifications

There are also *simplifications* that could be used; CEIOPS (2009e15) and (2010a12).

The following formula could be used as a simplification for the Life catastrophe risk submodule: the input data are required for each policy where the payment of benefits, either lump sum or multiple payments, is contingent on either mortality or disability:

$$CR_{\mathrm{LR,CAT}} = \sum_i 0.0015 \cdot CaR_i,$$

where subscript i denotes each policy where the payment of benefits, either lump sum or multiple payments, is contingent on either mortality or disability, and where CaR_i is determined as

$$CaR_i = SA_i + AB_i \cdot Af - TP_i,$$

and TP_i is the technical provision (net of reinsurance) for each policy i; SA_i: for each policy i: where benefits are payable as a single lump sum, the Sum Assured (net of reinsurance) on death or disability. Otherwise, zero; AB_i: For each policy i: where benefits are not payable as a single lump sum, the Annualized amount of Benefit (net of reinsurance) payable on death or disability. Otherwise, zero. Af: Average annuity factor for the expected duration over which benefits may be payable in the event of a claim.

CHAPTER 33

Solvency II Standard Formula

Health Underwriting Risk

HEALTH INSURANCE obligations are all types of insurance compensating or reimbursing losses, for example, loss of income, caused by illness, accident, or disability (income insurance), or medical expenses due to illness, accident, or disability (medical insurance); CEIOPS (2009d12) and (2009f23).

33.1 GENERAL FEATURES

The health underwriting risk and its capital charge are treated in different articles of the Framework Directive, namely 111, but also in 104 and 105. The development of the health underwriting risk is discussed in Appendix O. CEIOPS has discussed the general structure of the health underwriting risk module in CEIOPS (2009d12) and (2009f23), and its calibration in CEIOPS (2009e10). The final calibration was done in connection with the development of the Technical Specifications of QIS5, during the spring 2010.

CEIOPS has concluded that the standard formula developed at the European level cannot capture all features of national private health insurance systems. Given the social impact of health insurance, the risks underlying health insurance contracts should be adequately evaluated in the standard formula. In order to recognize the impact of the underlying Social Security systems in the calibration of the standard formula, undertakings having health insurance obligations will be allowed to use internal/external data. It is important to take into account how the Social Security system impacts the risk profile of undertakings with health activities. Such undertakings use simplifications in assessing their premium and reserve risk volatilities, and simplifications are not considered under the scope of USPs.

33.1.1 Standard Formula

CEIOPS has decided to split Health underwriting risks into three categories; CEIOPS (2010a09):

- Health insurance obligations pursued on a *similar technical basis* to that of life insurance (SLT Health)

- Health insurance obligations *not* pursued on a *similar technical basis* to that of life insurance (N-SLT Health)
- Health Catastrophe risks (Health CAT)

Figure 33.1 shows the structure of the health underwriting risk module and its subrisks.

The capital charge will be calculated as (with simplified notation)

$$CR_{\text{HR}} = \sqrt{\sum_{rxc} \rho_{\text{rc}} \cdot CR_{\text{HR,r}} \cdot CR_{\text{HR,c}}},$$

where the dependence is given in Table 33.1.

The capital charges including the risk-absorbing effects of future profit sharing, that is, the loss-absorbing capacity of TPs, are given by nCR_{SLT} and $CR_{\text{-}SLT}$. The capital charge for

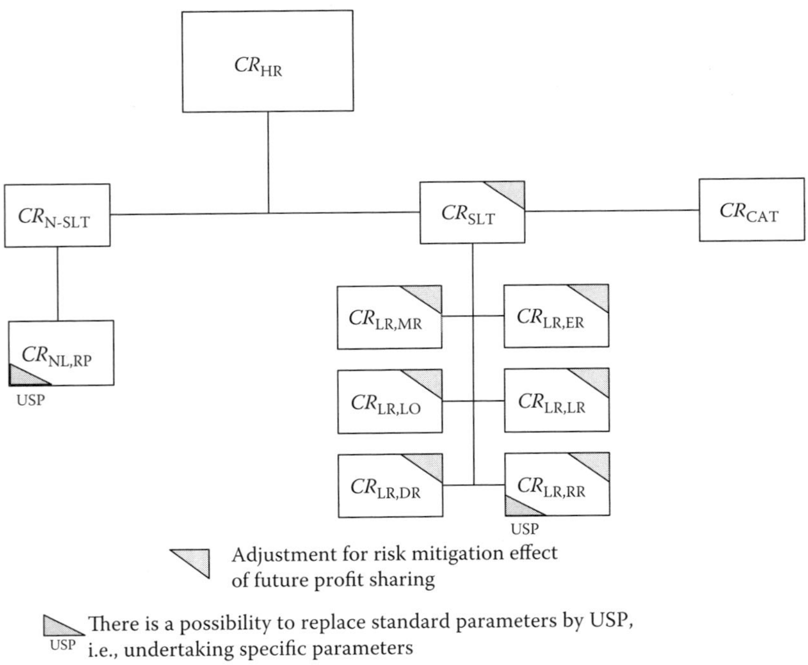

FIGURE 33.1 An overall description of the modular structure of the capital charge for the health underwriting risk and its subrisks. (Adapted from CEIOPS. 2009f23. CEIOPS' Advice for L2 Implementing Measures on SII: SCR-Standard formula—Health Underwriting Risk (former Consultation Paper no. 50). CEIOPS-DOC-43-09. November 10. Available at www.ceiops.org and discussion in CEIOPS. 2010a09. CEIOPS' Advice for L2 Implementing Measures on SII: SCR Standard Formula—Correlations (former Consultation Paper no. 74). CEIOPS-DOC-70-10. January 29. Available at www.ceiops.org.)

TABLE 33.1 Dependence Matrix for the Three Main Subrisks of the Health Underwriting Risk Module

ρ_{rc}	N-SLT $CR_{\text{N-SLT}}$	SLT CR_{SLT}	CAT CR_{CAT}
$CR_{\text{N-SLT}}$	1	0.75	0.25
CR_{SLT}		1	0.25
CR_{CAT}			1

Source: Adapted from CEIOPS. 2010a09. CEIOPS' Advice for L2 Implementing Measures on SII: SCR Standard Formula—Correlations (former Consultation Paper no. 74). CEIOPS-DOC-70-10. January 29. Available at www.ceiops.org.

CR_{SLT}: Capital charge for health insurance obligations pursued on a similar technical basis to that of life insurance;

$CR_{\text{N-SLT}}$: Capital charge for health insurance obligations *not* pursued on a similar technical basis to that of life insurance; and

CR_{CAT}: Capital charge for CAT risks.

nCR_{HR} is determined as, see CEIOPS (2009f23),

$$\begin{aligned} nCR^2_{\text{HR}} &= CR^2_{\text{N-SLT}} + nCR^2_{\text{SLT}} + CR^2_{\text{CAT}} + 2 \cdot \rho_{\text{N-SLT,SLT}} \cdot CR_{\text{N-SLT}} \cdot nCR_{\text{SLT}} \\ &\quad + 2 \cdot \rho_{\text{CAT,SLT}} \cdot CR_{\text{CAT}} \cdot nCR_{\text{SLT}} + 2 \cdot \rho_{\text{CAT,N-SLT}} \cdot CR_{\text{CAT}} \cdot CR_{\text{N-SLT}} \end{aligned}$$

and the dependence is given as above.

33.1.2 Simplifications

The simplifications proposed for the life subrisks discussed in Section 32.1 can be applied; CEIOPS (2009e15) and (2010a12).

33.2 SLT HEALTH UNDERWRITING RISK

The capital charge for the SLT Health underwriting risk is determined by combining the capital charges from the subrisk modules as

$$CR_{\text{SLT}} = \sqrt{\sum_{rxc} \rho^{\text{SLT}}_{\text{rc}} \cdot CR_{\text{LR,r}} \cdot CR_{\text{LR,c}}}$$

where r and c are the rows and columns of the dependence matrix, Table 33.2, for the SLT Health subrisks pursued on a similar technical basis to life insurance (LR); see CEIOPS (2010a09).

TABLE 33.2 Dependence Matrix for the Subrisks of the SLT Health Underwriting Risk Module

ρ_{rc}^{SLT}	Mortality $CR_{LR,MR}$	Longevity $CR_{LR,LO}$	Disability $CR_{LR,DR}$	Expense $CR_{LR,ER}$	Lapse $CR_{LR,LR}$	Revision $CR_{LR,RR}$
$CR_{LR,MR}$	1	−0.25	0.25	0.25	0	0
$CR_{LR,LO}$		1	0	0.25	0.25	0.25
$CR_{LR,DR}$			1	0.5	0	0
$CR_{LR,ER}$				1	0.5	0.5
$CR_{LR,LR}$					1	0
$CR_{LR,RR}$						1

Source: Adapted from CEIOPS. 2010a09. CEIOPS' Advice for L2 Implementing Measures on SII: SCR Standard Formula—Correlations (former Consultation Paper no. 74). CEIOPS-DOC-70-10. January 29. Available at www.ceiops.org.

The capital charges including the risk-absorbing effects of future profit sharing, that is, the loss-absorbing capacity of TPs, is given by

$$nCR_{HR,SLT} = \sqrt{\sum_{rxc} \rho_{rc}^{SLT} \cdot nCR_{LR,r} \cdot nCR_{LR,c}},$$

where r and c are the rows and columns of the dependence matrix, Table 33.1 and $nCR_{LR,r}$ and $nCR_{LR,c}$ are the corresponding capital charges including the risk-absorbing effects.

Except for CAT risk, CEIOPS has decided that the risk drivers of the SLT Health underwriting risk module should be developed consistently with those of the Life underwriting risk module:

- SLT Health Mortality risk
- SLT Health Longevity risk
- SLT Health Disability/morbidity risk (only income insurance)
- SLT Health Expense risk
- SLT Health Lapse risk
- SLT Health Revision risk (with a larger scope)

The computation of the capital charges including the loss-absorbing capacity of TPs should be computed as set out in CEIOPS (2009d16); see also Section 26.5.1.

33.2.1 Mortality Risk

The SLT Health mortality risk covers the risk of loss, or of adverse change in the value of insurance liabilities, resulting from changes in the level, trend, or volatility of mortality rates, where an increase in the mortality rate leads to an increase in the value of insurance liabilities; CEIOPS (2009d12), (2009f23), and CEIOPS (2009e10).

The calculation of $CR_{\text{LR,MR}}$ and $nCR_{\text{LR,MR}}$ are computed as in Section 32.2.

33.2.2 Longevity Risk

The SLT Health longevity risk covers the risk of loss, or of adverse change in the value of insurance liabilities, resulting from the changes in the level, trend, or volatility of mortality rates, where a decrease in the mortality rate leads to an increase in the value of insurance liabilities; CEIOPS (2009d12), (2009f23), and CEIOPS (2009e10).

The calculation of the $CR_{\text{LR,LO}}$ and $nCR_{\text{LR,LO}}$ are computed as in Section 32.3.

33.2.3 Disability Risk

The SLT Health Disability/morbidity risk covers the risk of loss, or of adverse changes in the value of insurance liabilities, resulting from changes in the level, trend, or volatility of the frequency or the initial severity of the claims, due to changes

- In the disability, sickness and morbidity rates
- In medical inflation

The disability/morbidity risk includes the recovery, which is the risk of loss, or of adverse changes in the value of insurance liabilities, resulting from the changes in the level, trend, or volatility of the recovery rates where a decrease in the recovery rate (moving from sick or disabled to full revalidation) leads to an increase in the value of insurance liabilities; CEIOPS (2009f23), and CEIOPS (2009e10).

The capital charge for SLT Health disability/morbidity risk is determined as

$$CR_{\text{LR,DR}} = CR_{\text{Med}} + CR_{\text{Inc}},$$

where CR_{Med}: the capital charge for medical insurance and CR_{Inc}: the capital charge for income insurance.

The corresponding capital charges, including the risk-absorbing effects, are given by

$$nCR_{\text{LR,DR}} = nCR_{\text{Med}} + nCR_{\text{Inc}}.$$

33.2.3.1 SLT Health Disability Risk for Medical Insurance

A large part of the risk in medical expense insurance is independent from the actual health status. Hence, for medical insurance, the determination of the disability/morbidity capital charge cannot be based on disability or morbidity probabilities; *cf.* CEIOPS (2009d12) and (2009f23).

The calculations of the capital charges are scenario-based. Input information is the effect of two specific scenarios on the net value of assets minus liabilities; see Section 27.1.1 on the delta-NAV approach.

As a first step, we calculate an upward and a downward shock:

$$CR_{\text{Med}}^{\text{Up}} = \left\{\Delta\text{NAV} \mid \text{claim\_shock\_up}\right\},$$

and

$$CR_{\text{Med}}^{\text{Down}} = \{\Delta\text{NAV} \mid \text{claim\_shock\_down}\},$$

where the claim shocks are defined as given in Table 33.3.

The claim shock down the scenario needs only to be analyzed for policies that include a premium adjustment mechanism that foresees an increase of premiums if claims are higher than expected and a decrease of premiums if claims are lower than expected. Otherwise, undertakings should assume that the result of the scenario claim shock down is zero.

The same calculations are also made under the condition that the assumptions on future bonus rates may be changed in response to the shock. These capital charges are

$$nCR_{\text{Med}}^{\text{Up}} = \left\{\Delta\text{NAV} \mid \text{claim\_shock\_up}\right\},$$

and

$$nCR_{\text{Med}}^{\text{Down}} = \{\Delta\text{NAV} \mid \text{claim\_shock\_down}\},$$

where the claim shocks are defined as in Table 33.2. The relevant scenario is the most adverse one taking into account the loss-absorbing capacity of TPs:

$$nCR_{\text{Med}} = \max\left\{nCR_{\text{Med}}^{\text{Up}}; nCR_{\text{Med}}^{\text{Down}}\right\}.$$

Then the capital charge for medical insurance is set to

$$CR_{\text{Med}} = \begin{cases} CR_{\text{Med}}^{\text{Up}} & \text{if } nCR_{\text{Med}}^{\text{Up}} > nCR_{\text{Med}}^{\text{Down}} \\ CR_{\text{Med}}^{\text{Down}} & \text{if } nCR_{\text{Med}}^{\text{Up}} < nCR_{\text{Med}}^{\text{Down}} \\ \max\left\{CR_{\text{Med}^{\text{Up}}}; CR_{\text{Med}}^{\text{Down}}\right\} & \text{if } nCR_{\text{Med}}^{\text{Up}} = nCR_{\text{Med}}^{\text{Down}} \end{cases}.$$

TABLE 33.3 Capital Charge is Computed by Analyzing the Scenarios Claim Shock Up and Claim Shock Down, Respectively

Scenario	Permanent Absolute Change of Claim Inflation	Permanent Relative Change of Claims
Claim shock up	+1%	+10%
Claim shock down	−1%	−10%

Source: Adapted from CEIOPS. 2009f23. CEIOPS' Advice for L2 Implementing Measures on SII: SCR-Standard formula—Health Underwriting Risk (former Consultation Paper no. 50). CEIOPS-DOC-43-09. November 10. Available at www.ceiops.org.

The calibration of the shocks is discussed in CEIOPS (2009d12), (2009f23), and on CEIOPS (2009e10).

33.2.3.2 SLT Health Disability Risk for Income Insurance

For income insurance, the determination of the capital requirement for disability/morbidity risk is based on disability or morbidity probabilities. Considering that the risk in income insurance depends on the health status, the SLT Health disability/morbidity risk for income insurance should be treated in the same way as disability/morbidity risk in the Life underwriting risk module; CEIOPS (2009d12), (2009f23), and CEIOPS (2009e10).

The calculation of CR_{Inc} and nCR_{Inc} are computed as in Section 32.4.

33.2.4 Expense Risk

The SLT Health expense risk covers the risk of loss, or of adverse changes in the value of insurance liabilities, resulting from changes in the level, trend, or volatility of the expenses incurred in servicing insurance or reinsurance contracts. Expense risk arises if the expenses anticipated when pricing a guarantee are insufficient to cover the actual costs accruing in the following year. All expenses incurred have to be taken into account; CEIOPS (2009d12), (2009f23), and CEIOPS (2009e10).

The calculation of $CR_{\text{LR,ER}}$ and $nCR_{\text{LR,ER}}$ are computed as in Section 32.5.

33.2.5 Lapse Risk

The SLT Health Lapse risk covers the risk of loss, or of adverse change in the value of insurance liabilities, resulting from changes in the level or volatility of the rates of policy lapses, terminations, renewals, and surrenders; CEIOPS (2009d12), (2009f23), and CEIOPS (2009e10).

The calculation of $CR_{\text{LR,LR}}$ and $nCR_{\text{LR,LR}}$ are computed as in Section 32.7.

33.2.6 Revision Risk

The SLT Health Revision risk covers the risk of loss, or of adverse changes in the value of insurance liabilities resulting from fluctuations in the level, trend, or volatility of the revision rates applied to benefits, due to changes in either:

- The legal environment (or court decision); only future changes approved or strongly foreseeable at the calculation date under the principle of constant legal environment,
- The state of health of the person insured (sick to sicker, partially disabled to fully disabled, and temporarily disabled to permanently disabled); CEIOPS (2009f23) and CEIOPS (2009e10).

The calculation of $CR_{\text{LR,RR}}$ and $nCR_{\text{LR,RR}}$ are computed as in Section 32.6.

33.2.6.1 Undertaking-Specific Parameters

A subset of the parameters of the standard formula may, after approval of the supervisory authority, be replaced by USPs. This is discussed in general terms in Section 26.8. The capital charge for the revision risk can be calculated by using USP standardized methods as described in Section 32.6.1. This is also discussed in CEIOPS (2009e13) and (2010a10).

33.2.7 Life CAT Risk

The SLT Health CAT risk covers at least the risk of loss, or of adverse changes in the value of insurance liabilities, resulting from the significant uncertainty of pricing and provisioning assumptions related to outbreaks of major epidemics, as well as the unusual accumulation of risks under such extreme circumstances; CEIOPS (2009d12). These risks will be included in the Health CAT-risk module.

CEIOPS has decided that the CAT risk exposure for both SLT Health and Non-SLT Health should be treated in the same way as the Non-life CAT risk module (according to the same methodologies). Health CAT risk is required in both the SLT and the Non-SLT subrisk modules because the life CAT risk does not apply to health business.

SLT Health CAT scenarios, for example, include

- Pandemic, for example, bird flu
- Mass accident
- Polio type debilitating disease effects
- Biohazard from an insecure laboratory
- Terrorist action (e.g., pathogen released, terrorist action with delayed effects)
- Concentrated office block accident
- Sudden downturn in the economy (e.g., with an impact on the disability/morbidity inception rate)

The calculation of $CR_{\text{LR,CAT}}$ and $nCR_{\text{LR,CAT}}$ are computed as in Section 31.3 and according to CEIOPS (2009e10) will the calibration of the health CAT be provided by CEIOPS in mid-2010.

33.3 NON-SLT HEALTH UNDERWRITING RISK

The Non-SLT Health underwriting risk arises from the underwriting of health insurance obligations, not pursued on a similar technical basis to that of life insurance, following from both the perils covered and processes used in the conduct of business; CEIOPS (2009d12) and (2009f23).

The capital charge for the Non-SLT Health underwriting risk is determined by combining the capital charges from two subrisk modules, premium and reserve risks and CAT risks, as

$$CR_{\text{N-SLT}} = CR_{\text{N-SLT,RP}},$$

where $CR_{\text{N-SLT,RP}}$ is the capital charge for N-SLT Health premium and reserve risk.

33.3.1 Reserve Risk and Premium Risk

This module combines a treatment for the three main sources of underwriting risk: premium risk, reserve risk, and expense risk. Premium risk is understood to relate to future claims arising during and after the period until the time horizon for the solvency assessment. The risk is that expenses plus the volume of losses (incurred and to be incurred) for these claims (comprising both amounts paid during the period and provisions made at its end) are higher than the premiums received (or if allowance is made elsewhere for the expected profits or losses on the business that the profitability will be less than expected); CEIOPS (2009d12).

The capital charge is calculated, for N-SLT Health insurance obligations, as for the nonlife reserve risk and the premium risk discussed in Section 31.2:

$$C_{\text{N-SLT,RP}} = \rho(\sigma_{\text{N-SLT}}) \cdot V_{\text{N-SLT}}.$$

The calculation follows the same steps as in Section 31.2. The only difference is the number of LoBs defined for the N-SLT Health module. For each LoB, the dependence between the reserve risk and the premium risk, the same factor as in QIS 4 is kept, that is, $\rho_{\text{RR,PR}} = 0.5$; CEIOPS (2009e10).

Three LoBs are used. These are accident, sickness, and workers compensation. The following standard deviations and dependence structure as given in Table 33.4 should be used, CEIOPS (2009e10).

33.3.1.1 *Undertaking-Specific Parameters*

A subset of the parameters of the standard formula may, after approval of the supervisory authority, be replaced by USPs. This is discussed in general terms in Section 26.8. The capital charge for the reserve and premium risks can be calculated by use of USP standardized methods as described in Section 31.2.1. This is also discussed in CEIOPS (2009e13) and (2010a10).

TABLE 33.4 Standard Deviations for the Reserve Risk and Premium Risk, and the Dependence Structure Used

LOB	s_R	s_P	Accident	Sickness	Workers Comp
Accident	0.175	0.10	1	0.50	0.50
Sickness	0.125	0.075		1	0.50
Workers Comp	0.125	0.10			1

Source: Adapted from CEIOPS. 2009e10. L2 Advice on SCR Standard Formula—Calibration of Health Underwriting Risk. CEIOPS-CP-72-09. November 2. Available at www.ceiops.org; CEIOPS. 2010a09. CEIOPS' Advice for L2 Implementing Measures on SII: SCR Standard Formula—Correlations (former Consultation Paper no. 74). CEIOPS-DOC-70-10. January 29. Available at www.ceiops.org.

33.3.2 Nonlife CAT Risk

The CAT risk capital charge covers the risk of loss, or of adverse changes in the value of insurance liabilities, resulting from the significant uncertainty of pricing and provisioning assumptions related to outbreaks of major epidemics, as well as the unusual accumulation of risks under such extreme circumstances [Article 105 (4) c) in Level 1 text]. The CAT risk exposure for both SLT Health and Non-SLT Health should be treated in the same way as nonlife CAT risk module, that is, according to the same methodologies; CEIOPS (2009d12). These risks will be included in the Healthy CAT risk module.

The N-SLT Health CAT scenarios should at least include

- Terrorism, mostly for group contracts
- Pandemic
- Stagflation (as this touches upon death spiral territory, it may not be well captured in the premium and risk component)

The capital charge $CR_{\text{N-SLT,CAT}}$ is calculated as the CAT-risk module for nonlife underwriting risk in Section 31.3 and according to CEIOPS (2009e10) will the calibration of the health CAT be provided by CEIOPS in mid-2010.

CHAPTER 34

Solvency II Standard Formula

Minimum Capital Requirement

THE EUROPEAN SOLVENCY II capital requirement is, as we know, based on a two-level system, namely an SCR and an MCR.

The MCR could be articulated as a measure based on a time horizon and a probability of ruin at which the risks to new policyholders would be unacceptable, even on the short term *or* the point at which it ceases to be economically rational for the undertaking to be recapitalized to the level of the SCR.

As the MCR will trigger the most serious supervisory intervention, its calculation needs to be simple, robust, and objective.

34.1 GENERAL FEATURES

The recitals to the FD give us the ideas behind the Articles in the FD. Each of the Chapters 22 through 26 and Chapter 34 quote the recitals that give the background to the topics discussed. The number to the left before the recital is the numbering in the FD.

See also recital (60) in Section 26.1.

(69) When the amount of eligible basic own funds falls below the Minimum Capital Requirement, the authorisation of insurance and reinsurance undertakings should be withdrawn where those undertakings are unable to reestablish the amount of eligible basic own funds at the level of the Minimum Capital Requirement within a short period of time.

(70) The Minimum Capital Requirement should ensure a minimum level below which the amount of financial resources should not fall. It is necessary that that level be calculated in accordance with a simple formula, which is subject to a defined floor and cap based on the risk-based Solvency Capital Requirement in order to allow for an escalating ladder of supervisory intervention, and that it is based on the data which can be audited.

The MCR is treated in different Articles of the FD, namely 128 and 129, and also in Article 74. The development of the MCR is discussed in Appendix P. CEIOPS has discussed the general structure of the MCR module in CEIOPS (2009d17, 2009f27), and its calibration in CEIOPS (2009e11).

34.2 STANDARD FORMULA

The MCR must always be higher than an absolute floor, absolute minimum capital requirement (AMCR), as defined in Article 129.1 (d):
(d) It shall have an absolute floor of

i. 2.200.000 EUR for nonlife insurance undertakings, including captive insurance undertakings, except in the case where all or some of the risks included in one of the classes 10–15 listed in point A of Annex 1 are covered, in which case it shall not be less than 3.200.000 EUR,

ii. 3.200.000 EUR for life insurance undertakings, including captive insurance undertakings,

iii. 3.200.000 EUR for reinsurance undertakings, except in the case of captive reinsurance undertakings, in which case the MCR shall not be less than a minimum of 1.000.000 EUR,

iv. The sum of the amounts set out in points (i) and (ii) for insurance undertakings as referred to in Article 73(5).

The above floor values expressed in euro will be adjusted according to inflation.

The MCR will be calculated using what has been labeled a *combined approach*, which is done in two steps.

Step 1: Calculate an MCR_{L} by using a *linear formula* explained in Section 34.3. It is a simple factor-based formula based on basic volume measures such as written premiums, TPs, capital-at-risk, deferred taxes and administrative expenses.

Step 2: The MCR_{L} is then squeezed into a corridor *if needed*: a floor equal to $MCR_{\mathrm{Floor}} = 0.25 * SCR$ and a cap equal to $MCR_{\mathrm{Cap}} = 0.45 * SCR$.

These procedures are described in Figure 34.1.

Although both the SCR and the MCR linear formula are calibrated to a VaR measure subject to a given confidence level, some important structural differences between the two should be noted; CEIOPS (2009f27):

- The linear formula is a simple factor-based measure, while the level of complexity of the SCR calculation is typically higher, involving nonlinear calculations and scenario analysis
- In particular, as opposed to the SCR, the linear formula includes no allowance for diversification effects
- The linear formula is retrospective (e.g., previous year actual volume measures) whereas the SCR is prospective (e.g., next year projected volume measures)

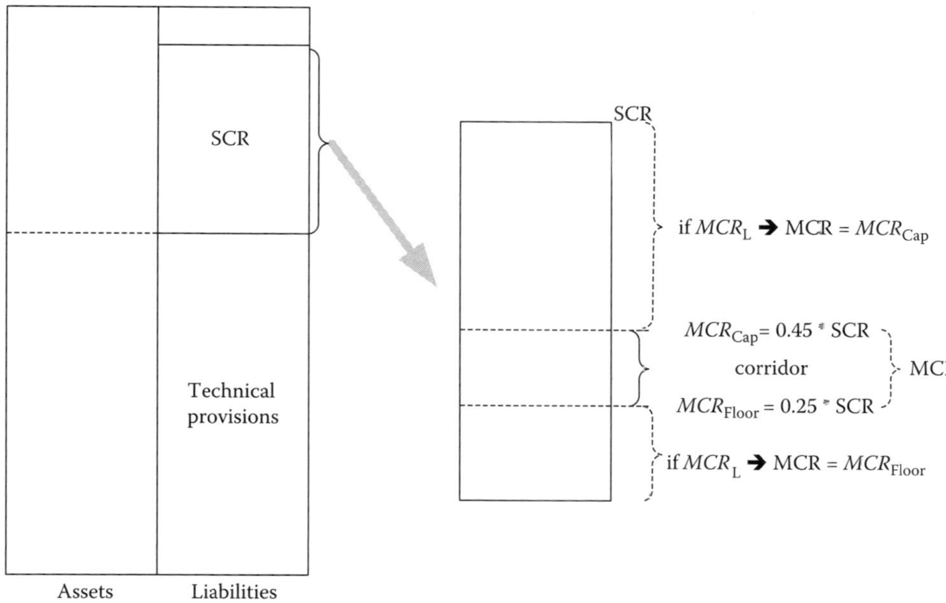

FIGURE 34.1 A description of the procedure to determine the MCR (always >AMCR).

The reason for having a corridor cap is that there must be a reasonable difference between the SCR level and the highest possible value of MCR as there will be a ladder of intervention between the SCR and the MCR. As MCR is an ultimate level, it must also be reasonable higher than the TP.

If an undertaking has an approved internal model, then the corridor used for calculating its MCR is determined by the internal model SCR result.

NOTE: For no longer than 2 years after the entry into force of Solvency II, the supervisory authority has the power to require that the corridor is calculated from the SCR standard formula.

Let MCR_{Comb} be the combined MCR resulting from "squeezing" the MCR_{L} to the corridor. Then the MCR is defined as

$$MCR = \max\{MCR_{\text{Comb}}; AMCR\}.$$

The absolute floors should apply to the notional nonlife and life MCR as follows; CEIOPS (2009f27):

Old composites: The notional nonlife MCR should not be lower than the nonlife absolute floor and the notional life MCR should not be lower than the life absolute floor.

New composites: The amount of the absolute floor is not defined in the Level 1 text. Since the overwhelming majority of new composites are life undertakings that have taken up accessory nonlife activities, CEIOPS has considered that the economic reality of these undertakings is best reflected if they are treated like life insurance undertakings. The same absolute floor should apply to the notional life MCR of a new composite undertaking, whereas a zero absolute floor should apply to its notional nonlife MCR.

The MCR_{Comb} can be written as

$$MCR_{\text{Comb}} = \min\left[\max\left\{MCR_{\text{L}}; MCR_{\text{Floor}}\right\}; MCR_{\text{Cap}}\right], \tag{34.1}$$

where

$$MCR_{\text{Floor}} = 0.25 \cdot SCR \quad \text{and} \quad MCR_{\text{Cap}} = 0.45 \cdot SCR$$

34.3 MCR LINEAR FORMULA

The MCR linear formula is calculated as the sum of four components:

$$MCR_{\text{L}} = MCR_A + MCR_B + MCR_C + MCR_D,$$

where MCR_A: MCR for nonlife activities practiced on a nonlife technical basis; MCR_B: MCR for nonlife activities technically similar to life; MCR_C: MCR for life activities practiced on a life technical basis; and MCR_D: MCR for life activities—supplementary obligations practiced on a nonlife technical basis.

The volume measures as referred to above in the linear formula, in particular TPs, written premiums, and capital-at-risk,

- Should be allocated between the above four components without double counting
- Should be subject to a floor of zero

The TP net of reinsurance is the difference between the gross TP and the reinsurance recoverables, where the recoverables should not include recoverables from finite reinsurance.

The TP volume measures should be without the risk margin (RM).

CEIOPS has decided to exclude finite reinsurance from the volume measures for the MCR in order to ensure that the linear MCR is robust and produces a result in line with the calibration objective of 85% VaR.

The premiums net of reinsurance are the premiums received from the policyholders less the reinsurance premiums paid for reinsurance contracts that correspond to these policyholder premiums. The reinsurance premiums should not include payments of reinsurance premiums for finite reinsurance.

34.3.1 MCR_A for Nonlife Activities Practiced on a Nonlife Technical Basis

The linear formula component MCR_A for nonlife business with activities on a nonlife technical basis is calculated by the following formula:

$$MCR_A = \sum_{j \in \text{LOB}} \max\left\{\alpha_j \cdot TP_j; \beta_j \cdot P_j\right\},$$

where TP_j: technical provisions (not including the RM) for each LoB, net of reinsurance, subject to a minimum of zero; and P_j: written premiums in each LoB over the last 12-month period, net of reinsurance, subject to a minimum of zero.

In Table 34.1, the parameters α and β together with the segmentation used for MCR are given; CEIOPS (2009e11). The final calibration was done in connection with the development of the Technical Specifications of QIS5, during the spring 2010.

34.3.2 MCR_B for Nonlife Activities Technically Similar to Life

The calculation of the MCR_B for nonlife activities technically similar to life insurance should be the same as the calculation for life activities, with the same segmentation and the same factors as described below in component MCR_C; see below.

Examples of nonlife activities that are similar in nature to life insurance include long-term health insurance and nonlife annuities.

34.3.3 MCR_C for Life Activities Practiced on a Life Technical Basis

The linear formula component MCR_C for life business with activities on life technical basis is calculated by the following formula, *cf.* CEIOPS (200d17, 2009e11)

$$MCR_C = \max\{\alpha_{C.1.1} \cdot TP_{C.1.1} + \alpha_{C.1.2} \cdot TP_{C.1.2};\ \gamma \cdot TP_{C.1.1}\} + \sum_{i \in \{C.2.1, C.2.2, C.3\}} \alpha_i \cdot TP_i$$
$$+ \beta_{C.4} \cdot CaR_{C.4}$$

where TP stands for the technical provisions excluding the RM, net of reinsurance, subject to a minimum of zero and $CaR_{C.4}$ stands for capital-at-risk, that is, the sum of financial strains for each policy on immediate death or disability where it is positive. The financial strain on immediate death or disability is the amount currently payable on death or disability of the insured and the present value of annuities payable on death or disability of the insured less

TABLE 34.1 The Segmentation (LoB) and Parameters for MCR_A Calculation

LoB No. j	LoB	α_j	β_j
1	Motor vehicle liability	0.18	0.14
2	Motor, other classes	0.18	0.14
3	Marine, aviation, transport (MAT)	0.25	0.29
4	Fire and other property damage	0.22	0.18
5	Third-party liability	0.29	0.25
6	Credit and suretyship	0.29	0.29
7	Legal expenses	0.18	0.11
8	Assistance	0.18	0.14
9	Miscellaneous	0.29	0.29
10	NP reinsurance (property)	0.42	0.42
11	NP reinsurance (casualty)	0.42	0.42
12	NP reinsurance (MAT)	0.42	0.42
13	Accident	0.25	0.14
14	Sickness	0.18	0.12
15	Workers compensation	0.18	0.16

Source: Adapted from CEIOPS. 2009e11. Draft L2 Advice on MCR—Calibration. CEIOPS-CP-73-09. November 2. Available at www.ceiops.org.
Note: Segmentations, except 10–12, include proportional reinsurance.

the net TPs, not including the RM, and less the increase in reinsurance recoverables, which is directly caused by death or disability of the insured.

Especially, we have $TP_{C.1}$: technical provisions, net best estimate, for guaranteed benefits relating to with-profits contracts; $TP_{C.2}$: technical provisions, net best estimate, for discretionary bonuses relating to with-profits contracts; and TP_i: technical provisions, excluding the RM, net of reinsurance and subject to a minimum of zero for each segment i other than with-profits business, according to the granularity defined in Table 34.2.

34.3.4 MCR_D for Life Activities: Supplementary Obligations Practiced on a Nonlife Technical Basis

The calculation of the linear formula component MCR_D should follow the same segmentation, the same factors, and the same volume measures with respect to supplementary nonlife and health obligations.

34.4 COMPOSITE UNDERTAKINGS

For composite undertakings, a notional nonlife and life *MCR*, $NMCR_{NL}$ and $NMCR_{Life}$ has to be calculated. These capital requirements have to be covered by eligible basic OFs with respect to the nonlife and life activity.

TABLE 34.2 The Segmentation and Parameters for MCR_C

LOB	Segment	α_j	β_j	γ
Contracts With-Profit (WP) Participation Clauses				
C.1.1	TP for guaranteed benefits	0.077		0.025
C.1.2	TP for future discretionary benefits	–0.130		
Contracts where the Policyholder Bears the Investment Risk (e.g., unit-linked)				
C.2.1	TP for contracts without guarantees	0.008		
C.2.2	TP for contracts with guarantees	0.028		
Contracts Without Profit Participation Clauses				
C.3	TP for contracts without profit participation clauses	0.032		
Capital-at-Risk Segmentation (CaR)				
C.4	CaR for contracts with an outstanding term of 5 years or more		0.0014	

Source: Adapted from CEIOPS. 2009e11. Draft L2 Advice on MCR—Calibration. CEIOPS-CP-73-09. November 2. Available at www.ceiops.org.

Note: The final calibration was done in connection with the development of the Technical Specifications of QIS5, during the spring 2010.

The *linear formula* result of a composite undertaking is split between notional nonlife and life components as follows:

$$NMCR_{L:\text{NL}} = MCR_A + MCR_B \quad \text{and} \quad NMCR_{L:\text{Life}} = MCR_C + MCR_D.$$

For the calculation of the corridor for notional nonlife and life MCR we need a notional split of the SCR into its nonlife and life parts. These is done by taking the SCR without any add-on and multiply it by the notional and the actual SCR:

$$NSCR_{\text{NL}} = SCR_{\text{No:add\_on}} \times \frac{NMCR_{L:\text{NL}}}{MCR_L} \quad \text{and} \quad NSCR_{\text{Life}} = SCR_{\text{No:add\_on}} \times \frac{NMCR_{L:\text{Life}}}{MCR_L}.$$

In calculating the combined notional MCR for nonlife and life business, we use Equation 34.1 and change the notation to "notional notation" where SCR is changed for the notional SCR, that is, $NSCR_{\text{NL}} + \text{Add\_on}_{\text{NL}}$ and $NSCR_{\text{Life}} + \text{Add\_on}_{\text{Life}}$ for nonlife and life business respectively.

The $\text{Add\_on}_{\text{NL}}$ and $\text{Add\_on}_{\text{Life}}$ are any capital add-on imposed with respect to the undertakings activity.

The notional MCR is defined as

$$NMCR_{\text{NL}} = \max\{NMCR_{\text{Comb:NL}}; AMCR_{\text{NL}}\}, \quad \text{and}$$
$$NMCR_{\text{Life}} = \max\{NMCR_{\text{Comb:Life}}; AMCR_{\text{Life}}\}$$

where $AMCR_{\text{NL}}$: for *old composite undertakings* the nonlife absolute floor should apply, and for new composite undertakings a zero absolute floor should apply; and $AMCR_{\text{Life}}$: the life absolute floor.

34.5 OTHER ISSUES

34.5.1 Deferred Taxes

According to Article 129 of the FD, deferred tax liabilities can be used in the calculation of the linear MCR. The objective of the inclusion of the deferred tax liability in the calculation would be to capture the loss-absorbing capacity of this balance sheet item. The SCR calculation allows for this effect by an adjustment to the BSCR.

CEIOPS has decided that the inclusion of deferred tax liabilities in the MCR linear formula would not lead to any significant regulatory benefit. Therefore it is excluded from the MCR calculation.

34.5.2 Quarterly Calculation

As the MCR has to be calculated quarterly, and the SCR according to the FD at least once a year, CEIOPS recognize that the SCR has to be calculated on a quarterly basis. Also the OFs eligible to cover the MCR should be calculated on a quarterly basis, as this is to ascertain whether or not the MCR has been breached.

For the quarterly calculation of the SCR, CEIOPS has proposed to use a simplification if the SCR calculation is done by the standard formula.

This simplification of the SCR calculation consists of a *partial recalculation* of the last reported SCR. A partial recalculation means that only those submodules of the SCR whose main risk drivers have changed significantly since the last calculation has to be recalculated.

However, CEIOPS has decided that no simplifications will be allowed in the following cases:

- *Significant change in risk profile*: If there has been a significant change in the risk profile of an undertaking since the last reported SCR
- *Proximity of intervention point*: If the undertaking falls below the following capital thresholds, indicating that there is an increased probability of MCR level intervention in the forthcoming period:
    - The undertaking has breached the SCR,
    - The undertaking has breached the MCR, or
    - The undertaking does not hold eligible Tier 1 and Tier 2 basic OFs covering at least 200% of the MCR, without taking into account the absolute floor.

PART F

Backgrounds and Calibrations

No sensible decision can be made any longer without taking into account not only the world as it is, but the world as it will be . . .

Isaac Asimov (1920–1992)
U.S. "science writer, science-fiction writer"

WE DESCRIBE how the European Solvency II has been developed by the EC and CEIOPS during the first decade in the twenty-first century. Much of the development has been based on the different Quantitative Impact Studies conducted during 2005–2008 (QIS1–QIS4). The descriptions made are mainly based on official documents from the organizations involved. I have used, for example, the technical specifications and the calibration documents to the different QIS for the explanation. Therefore, the texts are close to the original texts published.

In Appendix D, the Solvency II (SII) process is described. The general ideas behind SII are discussed in Appendix E. In Appendices F and G, the valuation processes are described. The standard formula framework is described in Appendix H. In the subsequent appendices, we look at the development of the different risk modules: market risk is discussed in Appendix I, credit risk in Appendix J, operational risk in Appendix K, and liquidity risk in Appendix L. The underwriting risks are discussed in the following appendices: nonlife in Appendix M, life in Appendix N, and health in Appendix O. The MCR is discussed in Appendix P. When we talk about an undertaking in these chapters, we mean both an insurance undertaking and a reinsurance undertaking, unless otherwise explicitly mentioned.

APPENDIX A

Some Statistical Clarifications

A.1 CONDITIONAL VARIANCES AND COVARIANCES

Let $\mathbb{F} = \sigma\{\mathbb{M}, \Theta, \mathbb{T}\}$ be a σ-algebra, where $\mathbb{M}$ is a finite set of models, Θ is a finite set of parameters in the models, and $\mathbb{T}$ is a finite set of trends. To simplify the discussion, we may assume that the set of trends is a part of the set of models, that is, $\mathbb{T} \subset \mathbb{M}$. In this case, we have that $\mathbb{F} = \sigma^*\{\mathbb{M}, \Theta\}$. This will be discussed in more detail in Section A.2 later.

Conditional Variances:

$$V(Y) = E_{\mathbb{F}}[V(Y|\mathbb{F})] + V_{\mathbb{F}}[E(Y|\mathbb{F})]. \tag{A.1.1}$$

Proof In the proof, we use the fact that $V(Y|\mathbb{F}) = E\left[Y^2|\mathbb{F}\right] - (E[Y|\mathbb{F}])^2$.

$$\begin{aligned}
E_{\mathbb{F}}[V(Y|\mathbb{F})] &= E_{\mathbb{F}}\left\{E\left[Y^2|\mathbb{F}\right] - (E[Y|\mathbb{F}])^2\right\} \\
&= E_{\mathbb{F}}\left\{E\left(Y^2|\mathbb{F}\right)\right\} - E_{\mathbb{F}}\left\{E(Y|\mathbb{F})^2\right\} = E\left(Y^2\right) - E_{\mathbb{F}}\left\{E(Y|\mathbb{F})^2\right\} \\
&= E\left(Y^2\right) - E(Y)^2 - \left\{E_{\mathbb{F}}\left[E(Y|\mathbb{F})^2\right] - E(Y)^2\right\} \\
&= V(Y) - \left\{E_{\mathbb{F}}\left[E(Y|\mathbb{F})^2\right] - E_{\mathbb{F}}[E(Y|\mathbb{F})]^2\right\} = V(Y) - V_{\mathbb{F}}[E(Y|\mathbb{F})].
\end{aligned}$$

This gives us Equation A.1.1.

From the proof, we also see that the variance of the conditional expectation can be seen as a weighted mean of $[E_{\mathbb{F}}(Y|\mathbb{F}) - E(Y)]^2$:

$$V_{\mathbb{F}}[E(Y|\mathbb{F})] = E_{\mathbb{F}}\left\{[E_{\mathbb{F}}(Y|\mathbb{F}) - E(Y)]^2\right\}. \tag{A.1.2}$$

□

Conditional Covariances:

$$\operatorname{cov}(X, Y) = E_{\mathbb{F}}[\operatorname{cov}(X, Y|\mathbb{F})] + \operatorname{cov}_{\mathbb{F}}[E(X|\mathbb{F}), E(Y|\mathbb{F})]. \tag{A.1.3}$$

The proof is very much similar to the proof of Equation A.1.1.

Proof As before, we use the fact that $\text{cov}(X, Y|\mathbb{F}) = E(XY|\mathbb{F}) - E(X|\mathbb{F})\,E(X|\mathbb{F})$.

$$\begin{aligned}
E_{\mathbb{F}}\{\text{cov}(X, Y|\mathbb{F})\} &= E_{\mathbb{F}}\{E(XY|\mathbb{F}) - E(X|\mathbb{F})\,E(X|\mathbb{F})\} \\
&= E_{\mathbb{F}}[E(XY|\mathbb{F})] - E_{\mathbb{F}}[E(X|\mathbb{F})\,E(Y|\mathbb{F})] = E(XY) \\
&\quad - E_{\mathbb{F}}[E(X|\mathbb{F})\,E(Y|\mathbb{F})] \\
&= \text{cov}(X, Y) - \{E_{\mathbb{F}}[E(X|\mathbb{F})\,E(Y|\mathbb{F})] - E(X)E(Y)\} \\
&= \text{cov}(X, Y) - \{E_{\mathbb{F}}[E(X|\mathbb{F})\,E(Y|\mathbb{F})] - E_{\mathbb{F}}[E(X|\mathbb{F})]\,E_{\mathbb{F}}[E(Y|\mathbb{F})]\} \\
&= \text{cov}(X, Y) - \text{cov}_{\mathbb{F}}[E(X|\mathbb{F}), E(Y|\mathbb{F})],
\end{aligned}$$

which gives us Equation A.1.3. Of course, putting $Y = X$ in Equation A.1.3 gives us Equation A.1.1. □

A.2 MEAN SQUARE ERROR

We now look at the *mean square error* (MSE). Let $\hat{\theta} = \hat{\theta}(X_1, X_2, \ldots, X_n)$ be an estimator of a parameter θ and $X_1, X_2, \ldots, X_n$ be a random sample of size n from a pdf with parameter $\theta, f_\theta(\cdot)$, and $\theta \in \Theta$, the parameter space.

MSE is the expectation of the squared-error loss in estimating θ by $\hat{\theta}$:

$$E\left[\left(\hat{\theta} - \theta\right)^2\right] = E\left[\left(\hat{\theta} - E\left(\hat{\theta}\right)\right) + \left(E\left(\hat{\theta}\right) - \theta\right)\right]^2 = (\text{bias})^2 + V\left(\hat{\theta}\right), \tag{A.2.1}$$

where $(\text{bias})^2 = E\left[\left(\theta - E\left(\hat{\theta}\right)\right)^2\right]$

and

$$V\left(\hat{\theta}\right) = E\left[\left(\hat{\theta} - E\left(\hat{\theta}\right)\right)^2\right].$$

The last term in Equation A.2.1 can be written as, *cf.* Section A.1,

$$V\left(\hat{\theta}\right) = E_{\mathbb{F}}\left(V\left(\hat{\theta}|\mathbb{F}\right)\right) + V_{\mathbb{F}}\left(E\left(\hat{\theta}|\mathbb{F}\right)\right)$$

and hence the MSE equals

$$E\left[\left(\hat{\theta} - \theta\right)^2\right] = (\text{bias})^2 + E_{\mathbb{F}}\left(V\left(\hat{\theta}|\mathbb{F}\right)\right) + V_{\mathbb{F}}\left(E\left(\hat{\theta}|\mathbb{F}\right)\right) \tag{A.2.2}$$

A.3 EULER'S THEOREM AND COROLLARY

We will sketch the proof of Euler's theorem and its corollary.

A.3.1 Euler's Theorem

If a function $f: R^v \to R$ is linearly homogeneous of degree 1, then:

$$f(x_1, x_2, \ldots, x_v) = \sum_{i=1}^{v} \frac{\partial f}{\partial x_i} x_i. \tag{A.3.1}$$

Proof Let λ be a scalar. Then by the definition of linear homogeneity of degree 1, we have $\lambda f(x) = f(\lambda x)$, where $x = (x_1, x_2, \ldots, x_v)$. Differentiate both sides of $\lambda f(x) = f(\lambda x)$ with respect to x_i and by the chain rule, we obtain

$$\frac{\partial \lambda f(x)}{\partial f(x)} \cdot \frac{\partial f(x)}{\partial x_i} = \frac{\partial f(\lambda x)}{\partial(\lambda x_i)} \cdot \frac{\partial(\lambda x_i)}{\partial x_i}.$$

As $\frac{\partial \lambda f(x)}{\partial f(x)} = \lambda$ and $\frac{\partial(\lambda x_i)}{\partial x_i} = \lambda$, we obtain $\lambda \cdot \frac{\partial f(x)}{\partial x_i} = \frac{\partial f(\lambda x)}{\partial(\lambda x_i)} \cdot \lambda$ and hence

$$\frac{\partial f(x)}{\partial x_i} = \frac{\partial f(\lambda x)}{\partial(\lambda x_i)}. \tag{A.3.2}$$

In the next step, we will differentiate $\lambda f(x) = f(\lambda x)$ with respect to λ. This gives us

$$\frac{\partial \lambda f(x)}{\partial \lambda} = \frac{\partial f(\lambda x)}{\partial \lambda} = \sum_{i=1}^{v} \frac{\partial f(\lambda x)}{\partial(\lambda x_i)} \cdot \frac{\partial(\lambda x_i)}{\delta \lambda}. \tag{A.3.3}$$

The left-hand side of Equation A.3.3 is simply equal to $f(x)$. The sum at the right-hand side of Equation A.3.3 is, using Equation A.3.2, equal to

$$\sum_{i=1}^{v} \frac{\partial f(x)}{\partial x_i} \cdot x_i \quad \text{as} \quad \frac{\partial(\lambda x_i)}{\partial \lambda} = x_i.$$

This completes the proof. □

A.3.2 Corollary

If a function $f: R^v \to R$ is linearly homogeneous of degree 1, then:

$$\sum_{i=1}^{v} \frac{\partial^2 f(x)}{\partial x_i \partial x_j} \cdot x_i = 0 \quad \text{for any } j.$$

Proof By Euler's theorem, we have $f(x) = \sum_{i=1}^{v} \frac{\partial f}{\partial x_i} x_i$. Differentiating with respect to x_j yields

$$\frac{\partial f(x)}{\partial x_j} = \frac{\partial^2 f(x)}{\partial x_1 \partial x_j} x_1 + \cdots + \frac{\partial^2 f(x)}{\partial x_j \partial x_j} x_j + \frac{\partial f(x)}{\partial x_j} + \cdots + \frac{\partial^2 f(x)}{\partial x_v \partial x_j} x_v.$$

This could be rewritten as

$$\frac{\partial f(x)}{\partial x_j} = \sum_{i=1}^{v} \frac{\partial^2 f(x)}{\partial x_i \partial x_j} x_i + \frac{\partial f(x)}{\partial x_j}.$$

This completes the proof. □

A.3.3 Quadratic Approximation

For Equation 12.14,

$$f^2(x_1, x_2, \ldots, x_v) = f^2(\xi_1, \xi_2, \ldots, \xi_v) + \sum_{i=1}^{v} \frac{\partial f^2}{\partial x_i} (x_i - \xi_i)$$
$$+ \frac{1}{2} \sum_{i=1}^{v} \sum_{j=1}^{v} \frac{\partial^2 f^2}{\partial x_i \partial x_j} (x_i - \xi_i)(x_j - \xi_j) \tag{A.3.4}$$

we will use Euler's theorem and the previous corollary to show Equation 12.15.

Note that

$$\sum_{i=1}^{v} \frac{\partial f}{\partial x_i} \xi_i = \sum_{i=1}^{v} \frac{\partial f}{\partial \xi_i} \xi_i \cdot \frac{\partial \xi_i}{\partial x_i} = f(\xi) \tag{A.3.5}$$

in a neighborhood $x_i = \xi_i$ as the right-hand term in the middle equation is equal to 1 and the equality follows from Euler's theorem A.3.1.

We also use

$$\left.\frac{\partial f^2}{\partial x_i}\right|_{x=\xi} = 2f(\xi) \cdot \left.\frac{\partial f}{\partial x_i}\right|_{x=\xi} \tag{A.3.6}$$

and

$$\left.\frac{\partial^2 f^2}{\partial x_i \partial x_j}\right|_{x=\xi} = \left.\frac{\partial}{\partial x_j}\left(\frac{\partial f^2}{\partial x_i}\right)\right|_{x=\xi} = \left.\frac{\partial}{\partial x_j}\left(2f(x)\frac{\partial f}{\partial x_i}\right)\right|_{x=\xi}$$
$$= 2\left.\frac{\partial f}{\partial x_j}\frac{\partial f}{\partial x_i}\right|_{x=\xi} + 2f(\xi) \cdot \left.\frac{\partial^2 f}{\partial x_i \partial x_j}\right|_{x=\xi}. \tag{A.3.7}$$

Expanding Equation A.3.4 and from Equations A.3.5, A.3.6, and A.3.7, we obtain

$$f^2(\xi_1, \xi_2, \ldots, \xi_v) = \frac{1}{2} \sum_{i=1}^{v} \sum_{j=1}^{v} \frac{\partial^2 f^2}{\partial x_i \partial x_j} \xi_i \xi_j$$

and by changing x for ξ, we obtain the desired result.

Approximations for Skewness

THE GENERALIZED CORNISH–FISHER expansion, see for example, Johnson and Kotz (1970), has as its aim to approximate an α-quantile of a distribution function F in terms of the α-quantile of Θ, which is a base distribution with density function ϕ. Usually, as in this case, Θ is the standard normal distribution. The key to this series expansion in terms of derivatives of F and Θ is the Lagrange's inversion theorem. The expansion is made in several steps.

The Lagrange's inversion theorem states that if a function $s|->t$ is implicitly defined by $t = c + s \cdot h(t)$ and h is analytic* in c, then an analytic function $f(t)$ can be developed into a power series in a neighborhood of $s = 0$ $(t = c)$: $f(t) = f(c) + \sum_{r=1}^{\infty} \frac{s^r}{r!} D^{r-1}\left[f' \cdot h^r\right](c)$, where D denotes the differentiation operator† $Df(x) = f'(x) = \mathrm{d}f(x)/\mathrm{d}x$, see, for example, Johnson and Kotz (1969). For a given probability $c = \alpha, f = \Theta^{-1}$, and $h = (\Theta - F) \circ \Theta^{-1}$ we have

$$x = z + \sum_{r=1}^{\infty} (-1)^r \frac{1}{r!} D_{(r-1)}\left[\alpha^r\right](z), \tag{B.1}$$

where $x = F^{-1}(\alpha)$, $z = \Theta^{-1}(\alpha)$, $s = 1$,
$\alpha = \dfrac{(F - \Theta)}{\phi}$, $\quad D_{(r)} = \left(D + \frac{\phi'}{\phi}\right)\left(D + 2\dfrac{\phi'}{\phi}\right)\cdots\left(D + r\dfrac{\phi'}{\phi}\right)$, and $D_{(0)} = 1$, the identity operator. Equation B.1 is the generalized Cornish–Fisher expansion.

The next step is to choose a specific base distribution, Θ. Here we assume the standard normal distribution. The third step is to expand α into a series expansion. Here we use a Gram–Charlier series expansion and reorder the terms. The idea behind this is to develop the ratio of the moment generating function of the "general distribution" F, $M(t) = Ee^{t\Delta Y}$, and the moment generating function of the standard normal distribution, $M_{\Theta}(t) = e^{t^2/2}$, into a power series at 0: $M(t)e^{-t^2/2} = \sum_{h=0}^{\infty} c_k t^k$, where c_k is the Gram–Charlier coefficient expressing the cumulants of the distributions. Assume that the general distribution function

*A function is analytic if it is differentiable at every point of its domain.

†If the function $f(x)$ can be expressed in terms of a Taylor series, then $f(x+h) = \sum_{j=0}^{\infty} (h^j/j!) D^j f(x)$ and the operator acting on $f(x)$ can be written as $\sum_{j=0}^{\infty} (hD)^j/j! \equiv e^{hD}$.

is normalized such that its mean value is zero and variance is one. As the cumulants of the standard normal distributions are $\kappa_1 = 0$, $\kappa_2 = 1$, $\kappa_j = 0$, $j > 2$ the cumulants in the Gram–Charlier coefficients will express those of the general distribution function. Fourier inversion gives us the corresponding series for the probability density:

$$f(x) = \sum_{k=0}^{\infty} c_k(-1)^k \phi^{(k)}(x). \tag{B.2}$$

Let $\phi(x) = \frac{1}{\sqrt{2\pi}}e^{-x^2/2}$, then

$D\phi(x) = \phi'(x) = -x\phi(x), D^2(x) = \phi''(x) = \left(x^2 - 1\right)\phi(x), \phi'''(x) = \left(3x - x^3\right)\phi(x)$,

$\phi^{(4)}(x) = (x^4 - 6x^2 + 3)\phi(x), \phi^{(5)}(x) = -(x^5 - 10x^3 + 15x)\phi(x)$, and so on. The result is obviously a polynomial in x multiplied by the standard normal density function. The Hermite polynomials are defined as before. Hence, the derivatives of the standard normal density can be expressed in terms of Hermite polynomials H_k: $(-1)^k\phi^{(k)}(x) = \phi(x)H_k(x)$. This gives us the formal Cornish–Fisher expansion. Expressing Equation B.2 in terms of the Hermite polynomials gives us

$$f(x) = \sum_{k=0}^{\infty} c_k H_k(x)\phi(x) = \phi(x) + \sum_{k=3}^{\infty} c_k H_k(x)\phi(x), \tag{B.3}$$

where we have used the fact that $c_1 = c_2 = 0$.

The Gram–Charlier coefficients and Hermite polynomials are given in Table B.1.

Using the coefficients and polynomials from Table B.1, Equation B.3 could be rewritten and reordered. In most applications, only the third and fourth cumulants are used. This gives us the Gram–Charlier expansion

$$f(x) \approx \phi(x) - \kappa_3\frac{(x^3 - 3x)}{6}\phi(x) + \kappa_4\frac{\left(x^4 - 6x^2 + 3\right)}{24}\phi(x). \tag{B.4}$$

TABLE B.1 Some Gram–Charlier Coefficients and Hermite Polynomials up to Order 8

Gram–Charlier Coefficients	Hermite Polynomials
$c_0 = 1$	$H_0 = 1$
$c_1 = 0$	$H_1 = x$
$c_2 = 0$	$H_2 = x^2 - 1$
$c_3 = \kappa_3/6$	$H_3 = x^3 - 3x$
$c_4 = \kappa_4/24$	$H_4 = x^4 - 6x^2 + 3$
$c_5 = \kappa_5/120$	$H_5 = x^5 - 10x^3 + 15x$
$c_6 = (\kappa_6 + 10\kappa_3)/720$	$H_6 = x^6 - 15x^4 + 45x^2 - 15$
$c_7 = (\kappa_7 + 35\kappa_4\kappa_3)/5040$	$H_7 = x^7 - 21x^5 + 105x^3 - 105x$
$c_8 = \left(\kappa_8 + 56\kappa_5\kappa_3 + 35\kappa_4^2\right)/40320$	$H_8 = x^8 - 28x^6 + 210x^4 - 420x^2 + 105$

Source: Data from Kendall, Sir Maurice and Alan Stuart. 1977. The Advanced Theory of Statistics. Distribution Theory. Volume 1, Fourth Edition. Charles Griffin & Company Ltd, London & High Wycombe.

Adding two more terms, including the squared skewness gives us the Edgeworth expansion. There is no general superiority of the Edgeworth expansion over the Gram–Charlier.

The first two terms in Equation B.4 are used to form the first-order normal power approximation. Including also the last term in Equation B.4 gives us the second-order normal power approximation.

From expansions given in Equation B.3 we obtain for the cumulative distribution function

$$F(x) \approx \Theta(x) + \sum_{k=3}^{\infty} c_k H_{k-1}(x)\phi(x) = \Theta(x) + \kappa_3 \frac{(x^2 - 1)}{6}\phi(x) + \kappa_4 \frac{(x^3 - 3x)}{24}\phi(x). \tag{B.5}$$

To get the VaR at the $1 - \alpha$ percentile we have used Equation B.5. As we are going to consider the ratios of these expansions we use only the first-order expansion. Multiplying Equation B.4 with x, integrating from $\text{VaR}_{1-\alpha} = z_{1-\alpha}$ to ∞, and dividing by $1 - (1 - \alpha) = \alpha$, we obtain the desired results for the tail expectation, TailVaR:

$$\frac{1}{\alpha}\int_{z_{1-\alpha}}^{\infty} xf(x)\,\mathrm{d}x = \frac{1}{\alpha}\int_{z_{1-\alpha}}^{\infty} x\phi(x)\,\mathrm{d}x + \frac{1}{\alpha}\frac{\kappa_3}{6}\int_{z_{1-\alpha}}^{\infty} (x^4 - 3x^2)\phi(x)\,\mathrm{d}x, \tag{B.6a}$$

$$+\frac{1}{\alpha}\kappa_4 \int_{z_{1-\alpha}}^{\infty} \frac{(x^5 - 6x^3 + 3x)}{24}\phi(x)\,\mathrm{d}x. \tag{B.6b}$$

Equation B.6a can be written as the first-order TailVaR expansion (an outline of the proof is given next):

$$k_{2,1-\alpha}(\gamma)_F = k_{2,1-\alpha} + \gamma \frac{k_{1-\alpha}^3}{6} k_{2,1-\alpha}, \tag{B.7a}$$

where $k_{2,1-\alpha} = 1/R(z_{1-\alpha})$ is the TailVaR for a standard normal distribution and $R(z_{1-\alpha}) = \alpha/\phi(z_{1-\alpha})$ is the Mill's ratio for the standard normal distribution. $k_{1-\alpha}(\gamma)$ is the corresponding percentile or VaR measure for the standard normal distribution. This result was first established by Giamouridis (2006), as a correction of the results in Christoffersen and Goncalves (2005). Giamouridis (2006) also showed that a second-order TailVaR expansion, including the kurtosis γ_2, is given by

$$k_{2,1-\alpha}(\gamma)_F = k_{2,1-\alpha} + \gamma \frac{k_{1-\alpha}^3}{6} k_{2,1-\alpha} + \frac{\gamma_2}{24}\left(k_{1-\alpha}^4 - 2k_{1-\alpha}^2 - 1\right) k_{2,1-\alpha}. \tag{B.7b}$$

Proof of Equation B.7a The integrand in the right most term in Equation B.6a can be written as

$$\left(x^4 - 3x^2\right)\phi(x) = \left(x^4 - 6x^2 + 3 + 3x^2 - 3\right)\phi(x) = \left[\phi^{(4)}(x) + 3\phi''(x)\right]\phi(x) \tag{B.8}$$

by using the above derivatives. In addition, $-x\phi(x) = \phi'(x)$ shows that Equation B.6a can be written as:

$$\frac{1}{\alpha}\int_{z_{1-\alpha}}^{\infty} xf(x)\,\mathrm{d}x \approx -\frac{1}{\alpha}\int_{z_{1-\alpha}}^{\infty} \phi'(x)\,\mathrm{d}x + \frac{1}{\alpha}\frac{\kappa_3}{6}\int_{z_{1-\alpha}}^{\infty}\left[\phi^{(4)}(x) - 3\phi''(x)\right]\mathrm{d}x$$

$$= -\frac{1}{\alpha}\left[\phi(x)\right]_{z_{1-\alpha}}^{\infty} + \frac{1}{\alpha}\frac{\kappa_3}{6}\left[\phi'''(x) + 3\phi'(x)\right]_{z_{1-\alpha}}^{\infty}$$

$$= -\frac{1}{\alpha}\left[\phi(x)\right]_{z_{1-\alpha}}^{\infty} + \frac{1}{\alpha}\frac{\kappa_3}{6}\left[\left(3x - x^3\right)\phi(x) - 3x\phi(x)\right]_{z_{1-\alpha}}^{\infty}$$

$$= -\frac{1}{\alpha}\left[\phi(x)\right]_{z_{1-\alpha}}^{\infty} - \frac{1}{\alpha}\frac{\kappa_3}{6}\left[x^3\phi(x)\right]_{z_{1-\alpha}}^{\infty}$$

$$= -\frac{1}{\alpha}\lim_{t\to\infty}\phi(t) + \frac{\phi(z_{1-\alpha})}{\alpha} + \frac{1}{\alpha}\frac{\kappa_3}{6}\lim_{t\to\infty}\left(-t^3\right)\phi(t) + \frac{\kappa_3}{6}z_{1-\alpha}^3\frac{\phi(z_{1-\alpha})}{\alpha}.$$

The first and third terms to the right go to zero as $t \to \infty$ and $\phi(z_{1-\alpha})/\alpha = 1/R(z_{1-\alpha})$. With the notations used before, we have Equation B.7. □

List of Different Papers Published by CEIOPS

Web site: http://www.ceiops.org/

CONSULTATION PAPERS

2004

Consultation Paper No. 1—Consultation practices (July 21)

Consultation Paper No. 2—Guidelines for coordination committees (August 2)

Consultation Paper No. 3—Implications of IAS/IFRS introduction for prudential supervision (November 29)

2005

Consultation Paper No. 4—Answers to the first wave of Calls for Advice—Solvency II (February 25)

Consultation Paper No. 5—Occupational Pensions Protocol (February 28 plus October 27)

Consultation Paper No. 6—Recommendation on possible need for amendments—Insurance Groups Directive (March 3)

Consultation Paper No. 7—Answers to the second wave of Calls for Advice—Solvency II (July 4)

Consultation Paper No. 8—Insurance Mediation Protocol (June 30)

Consultation Paper No. 9—Answers to the third wave of Calls for Advice—Solvency II (December 9)

Consultation Paper No. 10—Developing CEIOPS' Medium-Term Work Programme (November 11)

Consultation Paper No. 11—Recommendation on independence and accountability (December 9)

Consultation Paper No. 12—Treatment of "Deeply Subordinated Debt" (December 9)

2006

Consultation Paper No. 13—Advice on insurance undertakings' Internal risk and capital requirements, supervisors' evaluation procedures and harmonized supervisors' powers and tools (July 4)

Consultation Paper No. 14—Draft Advice on subgroup supervision, diversification effects, cooperation with third countries and issues related to the MCR and the SCR in a group context (July 4)

Consultation Paper No. 15—Draft Advice to the European Commission on Supervisory Reporting and Public Disclosure in the Framework of the Solvency II project (November 6)

Consultation Paper No. 16—Draft Advice to the European Commission in the Framework of the Solvency II project on Pillar II issues relevant for reinsurance (November 6)

Consultation Paper No. 17—Draft Advice to the European Commission in the Framework of the Solvency II project on Pillar II capital add-ons for solo and group undertakings (November 3)

Consultation Paper No. 18—Draft Advice to the European Commission in the Framework of the Solvency II project on Supervisory powers—further advice (November 3)

Consultation Paper No. 19—Draft Advice to the European Commission in the Framework of the Solvency II Project on Safety Measures (Limits on Assets) (November 10)

Consultation Paper No. 20—Draft Advice to the European Commission in the Framework of the Solvency II Project on Pillar I Issues—Further Advice (November 10 plus Supplement December 13)

2007

3L3 Consultation Paper—Impact Assessment Guidelines for EU Level 3 Committees (May 24)

Consultation Paper No. 21—Establishment of a mediation mechanism between insurance and pensions supervisors (July 5)

Consultation Paper No. 22—General protocol relating to the collaboration of the insurance supervisory authorities of the Member States of the European Union (November 6)

3L3 Consultation Paper—3L3 Medium Term Work Program (November 22)

Consultation Paper No 23—Interim Report on Proxies (December 21)

2008

Consultation Paper No. 24—Draft Advice on the Principle of Proportionality in the Solvency II Framework Directive Proposal (February 25)

Consultation Paper No. 25—Draft Advice on aspects on the Framework Directive Proposal related to Insurance Groups (February 25)

3L3 Consultation Paper—CESR, CEBS, CEIOPS Joint Public Consultation on a Common Understanding on the Information on the Payer Accompanying a Funds Transfer (March 26)

IWCFC Consultation Paper—IWCFC Recommendations on Capital for Financial Conglomerates (July 7)

3L3 Consultation Paper—CESR, CEBS, CEIOPS Joint Guidelines for the prudential assessment of acquisitions and increase of holdings in the financial sector required by Directive 2007/44/EC (July 11)

2009

1st Set of Draft Advice on Level 2 Implementing Measures

March 2009 cover letter for Consultation Papers 26–37 (March 26)

Consultation Paper No. 26—Draft Level 2 Advice on Technical Provisions—Methods and statistical techniques for calculating the best estimate

Consultation Paper No. 27—Draft Level 2 Advice on Technical Provisions—Segmentation

Consultation Paper No. 28—Draft Level 2 Advice on SCR Standard Formula—Counterparty default risk

Consultation Paper No. 29—Draft Level 2 Advice on Own Funds—Criteria for supervisory approval of ancillary own funds

Consultation Paper No. 30—Draft Level 2 Advice on Technical Provisions—Treatment of future premiums

Consultation Paper No. 31—Draft Level 2 Advice on SCR Standard Formula—Allowance of financial mitigation techniques

Consultation Paper No. 32—Draft Level 2 Advice on Technical Provisions—Assumptions about future management actions

Consultation Paper No. 33—Draft Level 2 Advice on System of Governance

Consultation Paper No. 34—Draft Level 2 Advice on Transparency and Accountability

Consultation Paper No. 35—Draft Level 2 Advice on Special Purpose Vehicles

Consultation Paper No. 36—Draft Level 2 Advice on Valuation of Assets and "other Liabilities"

Consultation Paper No. 37—Draft Level 2 Advice on the Procedure to be followed for the approval of an Internal Model

Consultation Paper No. 38—Budapest Protocol (April 7)

Consultation Paper—Review of the Financial Conglomerates Directive. Joint with CEBS (May 28)

2nd Set of Draft Advice on Level 2 Implementing Measures

July 2009 cover letter for Consultation Papers 37 Addendum plus 39–62 (July 2)

Addendum to CEIOPS-CP-37–09 Draft L2 Advice on the procedure to be followed for the approval of a group internal model

Consultation Paper No. 39—Draft L2 Advice TP—Best estimate

Consultation Paper No. 40—Draft L2 Advice on TP—Risk free interest rate

Consultation Paper No. 41 Draft L2 Advice TP—Calculation as a whole

Consultation Paper No. 42—Draft L2 Advice—Risk margin

Consultation Paper No. 43—Draft L2 Advice on TP—Standards for data quality

Consultation Paper No. 44—Draft L2 Advice on TP—Counterparty default adjustment

Consultation Paper No. 45—Draft L2 Advice on TP—Simplifications

Consultation Paper No. 46—Draft L2 Advice on own funds—Classification and eligibility

Consultation Paper No. 47—Draft L2 Advice on SCR Standard Formula—Market risk

Consultation Paper No. 48—Draft L2 Advice on SCR Standard Formula—Nonlife underwriting risk

Consultation Paper No. 49—Draft L2 Advice on SCR Standard Formula—Life underwriting risk

Consultation Paper No. 50—Draft L2 Advice on SCR Standard Formula—Health underwriting risk

Consultation Paper No. 51—Draft L2 Advice on SCR Standard Formula—Counterparty default risk

Consultation Paper No. 52—Draft L2 Advice on SCR Standard Formula—Reinsurance mitigation

Consultation Paper No. 53—Draft L2 Advice on SCR Standard Formula—Operational risk

Consultation Paper No. 54—Draft L2 Advice on SCR Standard Formula—Loss absorbing capacity of TP

Consultation Paper No. 55—Draft L2 Advice on SCR Standard Formula—MCR calculation

Consultation Paper No. 56—Draft L2 Advice on Tests and Standards for internal model approval

Consultation Paper No. 57—Draft L2 Advice on capital add-on

Consultation Paper No. 58—Draft L2 Advice on supervisory reporting and disclosure

Consultation Paper No. 59—Draft L2 Advice on remuneration (released July 21)

Consultation Paper No. 60—Draft L2 Advice on group solvency assessment

Consultation Paper No. 61—Draft L2 Advice on intragroup transactions and risk concentration

Consultation Paper No. 62—Draft L2 Advice on cooperation and colleges of supervisors

3rd Set of Draft Advice on Level 2 Implementing Measures

November 2009 cover letter for Consultation Papers 63–77 and 79 (November 2)

Consultation Paper No. 63—Draft L2 Advice on repackaged loans investment

Consultation Paper No. 64—Draft L2 Advice on the extension of the recovery period—Pillar II dampener

Consultation Paper No. 65—Draft L2 Advice on partial internal models

Consultation Paper No. 66—Draft L2 Advice on the group solvency for groups with centralized risk management

Consultation Paper No. 67—Draft L2 Advice on SCR Standard Formula and Own Funds—Treatment of participations

Consultation Paper No. 68—Draft L2 Advice on SCR Standard Formula and Own Funds—Treatment of ring-fenced funds

Consultation Paper No. 69—Draft L2 Advice on SCR Standard Formula—Design of the equity risk submodule

Consultation Paper No. 70—Draft L2 Advice on SCR Standard Formula—Calibration of market risk

Consultation Paper No. 71—Draft L2 Advice on SCR Standard Formula—Calibration of nonlife underwriting risk

Consultation Paper No. 72—Draft L2 Advice on SCR Standard Formula—Calibration of health underwriting risk

Consultation Paper No. 73—Draft L2 Advice on MCR—Calibration

Consultation Paper No. 74—Draft L2 Advice on SCR Standard Formula—Correlation parameters

Consultation Paper No. 75—Draft L2 Advice on SCR Standard Formula—Undertaking specific parameters

Consultation Paper No. 76—Draft L2 Advice on Technical Provisions—Simplifications

Consultation Paper No. 77—Draft L2 Advice on SCR Standard Formula—Simplifications

Consultation Paper No. 78—Draft L2 Advice on assessing 3rd country equivalence (released November 30)

Consultation Paper No. 79—Draft L2 Advice on simplifications for captives

2010

Consultation Paper No. 80—Draft Level 3 Guidance on Solvency II: Preapplication process for internal models (March 8)

ISSUES PAPER

2007

Issues Paper on risk management and other corporate issues (July 17)

Policy on Harmonization of contents and formats for public disclosure and supervisory reporting (November 1)

2008

Issues Paper Own Risk and Solvency Assessment under the Solvency II Framework Directive Proposal (May 26)

Issues Paper on Supervisory Review Process and Undertakings' Reporting Requirements (August 18)

Issues Paper on Implementing Measures on System of Governance (November 4)

APPENDIX D

European Solvency II Project

THE SOLVENCY II PROJECT can be divided into three phases:

1. 1999/2000–2003: the learning phase
2. 2003–2008/2009: the framework directive phase
3. 2008/2009–2012/2013: the implementing phase

The first two phases are discussed here in detail and the third is briefly described.

D.1 PHASE I: LEARNING PHASE: 1999/2000–2003

The EU working document (EC 1999) sketched out some further reflections from the Müller report (1997) that included a review of the existing solvency system and a study on failures among European insurance companies, which could be considered. These reflections later became the Solvency II project. At the 23rd meeting of the Insurance Committee (now EIOPC), it was agreed that a more fundamental and wide-ranging review of the overall financial position of an insurance undertaking should be done. This review would also include risk classes that have not been considered at that time, that is, investment risk (market and credit), operational risk, liquidity risk, and ALM risk.

In a paper entitled *The review of the overall financial position of an insurance undertaking (Solvency II review)*, MARKT (1999), the Commission services outlined the new project for the first time. They found six key aspects to discuss further.

1. *Technical provisions:* There was inadequate harmonization of nonlife TPs. The "Manghetti Group" of the Insurance Committee had a crucial role to play in harmonizing the provisions; Manghetti-report (2001). Subjects that were discussed were, for example, discounting the provisions, setting up equalization reserves and IBNR/IBNER reserves. The situation seemed more satisfactory in life insurance. A description of the reserving approaches made in nine EU countries is given in Wolthuis et al. (1997).

2. *Assets/Investment Risk:* Many Member States were using stress tests (or resilience tests). Also, the Müller report identified investment risk as important. One reinsurer had indicated that 60–85% of the variation in the financial results could be attributed to volatility in investments.

   Another issue to be discussed was asset valuation rules. Increased transparency was wanted.

3. *Asset liability management:* Certain ALM features were already established in the existing rules. Therefore it would be interesting to make an examination of how ART,* derivative products, catastrophe bonds, and so on could be used to control and transfer risks. Liquidity requirements were also of interest.

4. *Reinsurance:* No explicit account had been taken to the difference in the quality of reinsurance arrangements. A report was commissioned and the result and a more general study on a possible EU framework for reinsurance supervision were of vital importance.

5. *Solvency margin requirement—Methodology:* A methodology that more accurately reflects the true risks encountered was of interest. It had to remain simple, feasible, robust, and transparent, but also reflect the reduction in risk owing to portfolio diversification.

   It would also be of interest to review risk models in other jurisdictions and investigate risk capital requirements in other financial sectors. This was essential for financial conglomerates.

6. *Accounting system:* The accounting system links the previous five key factors. The accounting liabilities can have fiscal repercussions. IASC (now IASB) had started its insurance accounting project. Relevant issues were the focus on insurance contracts (and not undertakings) and the approach in valuing liabilities and assets.

At the end of the 1990s the Insurance Conference (now CEIOPS) set up a WG to continue the work done by the earlier Angerer Group on TPs in nonlife insurance. The group, which was chaired by Mr Giovanni Manghetti, focused on the concept of ultimate cost† and discounting. They presented a report in 2001, the Manghetti-report (2001).

The Commission services also discussed several external sources from different stakeholders such as IASB, IAIS, CEA, and so on. In its work plan the services proposed to submit a paper with recommendations to the Insurance Committee in 2002/2003. At the end of December 2000 the Commission contracted KPMG to conduct a study on background knowledge and updates on market developments in relation to its Solvency II project. A report was published in May 2002; KPMG (2002); see also Section D.1.3.

* ART: Alternative Risk Transfer. The Commission Services presented a study on ART made by Tillinghast Towers-Perrin 2000; see EC (2000).

† See Article 28 of the Insurance Accounting Directive, IAD, EEC (1991).

During 2001 the Commission services produced three notes to the Solvency Subcommittee of the Insurance Committee on (1) a presentation of the proposed work; MARKT (2001a), (2) banking rules: relevant for the insurance sector; MARKT (2001b), and (3) an RBC system; MARKT (2001c). In the notes, different questions were asked, for example, if internal models (IM) and RBC[‡] systems are to be used, are the banking rules, that is, the Basel system, and the corresponding RBC system of interest? The services also summarized the Basel project and the RBC systems of the United States, Australia, and Canada.

The outline of the proposed work in MARKT (2001a) are summarized by headlines below. The Commission papers on different subjects are given within parenthesis.

1. Qualitative criteria for a solvency system
2. Risk modeling adapted to each firm (IMs) (=> MARKT, 2002a)
3. RBC solvency systems (=> MARKT, 2001c)
4. European solvency system and supplementary rules
5. Solo solvency requirements and insurance groups
6. Study on the banking prudential system (=> MARKT, 2001b)
7. Accounting developments
8. Study on the risks to be taken into account
9. Nonlife technical provisions (=> MARKT, 2002f)
10. Life technical provisions (=> MARKT, 2002e)
11. Rules on the admissibility of assets covering technical provisions
12. Investment risk
13. Asset–liability management (life, => MARKT, 2002e)
14. Reinsurance
15. Qualitative rules

In its review of the work, MARKT (2002b), the Commission services stated that the project have two parts: the first part consists of examining the issues relating to the general design of the system (and to gather knowledge) while the second part focuses on the detailed arrangements for taking account of risks in the new system. In notes to the IC Solvency Subcommittee the services discussed current work done by IAIS and IAA/Groupe Consultatif, MARKT (2002c), and the Lamfalussy procedure; MARKT (2002d).

[‡] RBC: Risk-Based Capital.

D.1.1 Lamfalussy Procedure

In 2001, the European Council endorsed the so-called Lamfalussy procedure for regulation and supervision of the European securities markets. The approach was described in a report of the Committee of Wise Men, chaired by Baron Alexandre Lamfalussy, on the regulation of European Security Markets.

The procedure was to get a more flexible and efficient regulatory approach and permitting more rapid decision making and improved supervisory convergence at EU level.

In the Solvency II project, there was an agreement to seek a solvency margin regime that better reflects the true risks and was easier to change when the financial environment changes. This would call for a more detailed regulation that would not be adopted by a directive or a regulation (the primary level), but would be implemented under a comitology regime (the secondary level).

The Lamfalussy procedure is a four-level approach:

Level 1: The Commission adopts a proposal for a directive (or a regulation) containing framework principles. Once the Parliament and the Council agrees on the framework, the detailed implementing measures are developed in Level 2.

Level 2: After consulting the Level 2 committee, European Insurers and Occupational Pensions Committee (EIOPC), the Commission will *request* a Level 3 committee on advice: CEIOPS. The CEIOPS prepares this advice in consultation with market participants, for example, Groupe Consultatif and CEA, and submits it to the Commission. A formal proposal is then made by the Commission and submitted to EIOPC that must vote on the proposal within three months. After that, the Commission adopts the measure.

Level 3: The CEIOPS works on joint interpretations and recommendations, consistent guidelines, and common standards (e.g., like the IAIS or IAA Standards). It should also undertake peer review and compare regulatory practice to ensure consistent implementation and application.

Level 4: The Commission checks the Member States compliance with the EU legislation and may take legal action.

At this time, neither EIOPC nor CEIOPS existed in the forms they have today.

The two working groups within the Solvency Subcommittee (life and nonlife insurance) published their reports to the Insurance Committee; see MARKT (2002e) and MARKT (2002f).

The Conference of European Insurance Supervisors contributed to the first phase of the project by setting up a WG to study recent insolvencies and near insolvencies. This "London Group," or "Sharma group" after its chairman, published a report ("the Sharma report") in 2002; see Sharma (2002).

A summary and a winding-up of the first phase of the project is given in MARKT (2002h); see also MARKT (2002g).

D.1.2 Summary of Phase 1

The winding-up paper, MARKT (2002h), consists of three parts. The first is a recapitulation of the first phase, the second draws lessons, and the third outlines a future prudential supervisory system.

The Solvency Subcommittee of the EU Insurance Committee started four different projects during this first phase. All these projects can be seen as parts in a learning process:

- The "learning project" was done by KPMG, see Section 2.3.1.3
- The life group looked at rules for calculating mathematical provisions and asset–liability management methods
- The nonlife group looked at rules for calculating TPs
- The Insurance Conference, now CEIOPS, was initiated to study insolvency and near insolvency within EU ("the London Group" or "the Sharma Group")

The Insurance Conference (now CEIOPS) set up a WG to draw up principles for insurance undertakings and supervisory authorities in assessing an undertaking's internal control systems ("the Madrid Group"). A report was published at the end of 2003; the Madrid report (2003).

The working documents for the project were circulated widely and comments from Member States and organizations such as CEA and Groupe Consultatif were taken on board in Commissions work.

It was learned that there are three meanings for the term *solvency*. One is the direct one relating to the *solvency margin*, that is, a set of rules for calculating an MCR and the available solvency capital. The next concept relates to the set of rules intended to ensure that a company is *financially sound*. The last concept, known as *overall solvency*, corresponds to a company's financial soundness, taking account of the external environment and the conditions under which it operates.

The risks were also discussed. It was stated that the capital requirement (CR) in itself could not be the sole measure of the risk exposure, but that a solvency system must include other rules for measuring and limiting the risks. The CR could serve as a binding minimum threshold for a company to remain in the market and the minimum margin could serve as an early warning system. Another approach could be to determine what capital is required to provide against business fluctuations.

The harmonization within the European market was seen important and the adoption of the Lamfalussy procedure was another step in this direction. International standards were also seen as an important part in the work, for example, new accounting standards set up by IASB and new standards from IAIS and IAA.

A new idea for the future solvency system was the introduction of the concept of the *target capital* (or desirable capital), which would replace the solvency margin concept. At this stage it was seen as a soft threshold. Later on the concept was changed to the SCR.

One of the main issues that the Commission borrowed from the bank sector was the three-pillar system as proposed in Basel II.

The *first pillar* contains the system's quantitative requirements, at least rules on provisions and investments and the latter's capital. The two WGs on life and nonlife provisions reviewed these topics.

For the investment rules, the third directives had laid down the principle of prudent financial management (*prudent man principle*). This principle would be further clarified by quantitative rules on the diversification and spread, ALM, and possible extension of the coverage rules on items other than TPs on the liability side of the BS. The "prudent person rule," as it is called in EU directives, is clarified in the directive on supervision of institutions for occupational retirement provisions (IORP), COM (2003).

The CR in a prudential system could be viewed as

- The target capital (for maintaining an acceptable probability of default)
- An early warning system
- The absolute minimum capital

These three concepts give rise to different margin requirements and different kinds of intervention rules (by the supervisory authorities).

The *second pillar* contains the SRP as set out within the Basel II project. The review process includes internal control and rules for sound risk management (RM). The harmonization and the introduction of the Lamfalussy procedure give a common framework for prudential supervision, which include

- A common framework for assessing corporate governance
- Compilation of common statistics
- Harmonization of the main early warning indicators
- Devising reference scenarios (stress tests)
- On-site inspection
- A common validation framework for IMs
- A device for sharing information and coordination in crisis

The *third pillar* contains market discipline, that is, greater transparency and harmonization of accounting rules. The main factors contributing to market discipline are the financial markets and rating agents.

There should be a close link between the third-pillar measures and those of the first and second pillars.

D.1.3 KPMG Report (KPMG, 2002)

The KPMG report was published in May 2002. Its purpose was to sum up background knowledge and update developments in relation to the Solvency II project. The following summary is based on its executive summary.

The current system is based on three interconnected pillars:

- Assets
- TPs
- The solvency margin (based on fixed ratios)

The main limitations of the approach are the narrow scope of risks considered and the insensitivity of CRs with respect to company-specific risk profiles.

There was a need for a level playing field across the global financial sector and a trend toward convergence of prudential rules for different sectors. Changes in international accounting system give further pressure on the European solvency system.

Key risks to the financial position of the insurance undertakings are

- Insurance risk (underwriting risk and TPs)
- Asset risks (market values, interest rates and inflation, exchange rate, and commodity risk)
- Liability risk factors
- Credit risk (mainly in relation to reinsurer security and bond portfolios)
- Liquidity risk
- ALM risk
- operational risks

It is also important to recognize the interaction between these risks.

The development in the banking sector was helpful in getting insights. IMs must be a basis for decision making and it must be possible to quantify the risks involved and provide a value as a result. The models have to be validated.

TPs were also discussed. There was difference between the existing prudent valuation of the provisions and the "fair valuation" discussed in the IFRS. The disclosure of the TPs consists of, for example, the disclosure of used methodologies and assumptions made, the sensitivity of the calculations to changes in assumptions, and the details of run-off development. Additional safety margins as equalization reserves were also studied.

The study also looked at stress tests (resilience tests) on the assets side of the BS. Various risk reduction techniques were discussed, for example, reinsurance, ALM, and portfolio diversification.

Other topics that were studied were new accounting rules from IASB, the role of rating agencies, and different solvency margin methodologies. In the latter case, a comparison was made between the current EU system and RBC systems used in other parts pf the world.

Risk categories in an risk-based system should, as a minimum, include

- Underwriting risk (exposures less reinsurance)
- Market risk
- Credit risk, and also
- Operational risk and
- Asset–liability mismatch risk

KPMG also proposed the use of the three-pillar banking system.

D.1.4 Life Report (MARKT, 2002e)

In 2001 the Insurance Committee Solvency Subcommittee decided to set up two WGs, of which the Life WG was supposed to study rules for calculating mathematical provisions and ALM methods.

The WG was composed of Member-State experts and one representative from the Groupe Consultatif. To acquire common knowledge, the Members described the characteristics of their different markets. By this knowledge they tried to find common concerns and also common European solutions. The questions they discussed were

- Guaranteed interest rates
- Annuities and mortality risk
- Profit-sharing clauses
- UL products
- Options embedded in the contracts

For each question the members discussed principles and methods for mathematical provisions, and also the principle of premium sufficiency. The discussion on ALM methods followed the same line and the group discussed ideas for improvement.

The group believed that the directives contain the most necessary prudence principles, and that other types of principles could be created or strengthened in the directives. Therefore, they suggested that two supplemented principles should be developed:

- A principle of prudence in the choice of mortality table (corresponding to the existing prudence principle in the choice of interest rate)
- A principle of asset diversification applied to UL products

Other prudence principles suggested by the group were

- A principle aimed at protection of policyholders and fair conduct of business:
  1. With-profit products: perhaps a general principle of fair-sharing of the profits should be established.
  2. Unit or index-linked products: necessary with disclosure principles to ensure that policyholders are aware of the risks in these products.
     - Principles regarding RM and supervisory review: one way would be to introduce a requirement for companies to use appropriate prospective tools for their ALM, which could be used as the supervisory basis.

One method that could be used in improving the calculation of the mathematical provisions would be to focus on the interest rate used in the calculations. One way to proceed is to make reference to current market interest rates and another way would be to introduce a "resilience provision."

If the last principle above would be laid down at the EU level, then it would be necessary to harmonize or at least coordinate the supervisory methods. First, the supervisors need to have *benchmarks*, for example, maximum interest rate, reference mortality tables, and for ALM reference, adverse scenarios. Second, the supervisors have a role in *monitoring*, for example, by exchanging indicators and statistical data, and third, the supervisors need to have supervisory *powers* defined.

D.1.5 Nonlife Report (MARKT, 2002f)

The nonlife WG set up by the Insurance Committee Solvency Subcommittee in 2001 studied the major issue:

- Rules for calculating TPs: outstanding claims and equalization

As the corresponding life group, it composed of Member-State experts and one representative of the Groupe Consultatif. The report was to be seen as complements to the "Manghetti-report" on TPs in nonlife insurance, Manghetti-report (2001), and the KPMG report; see Section 2.3.1.3.

D.1.5.1 Provisions for Outstanding Claims

The group explored the possibility of comparing levels of prudence of provisions for outstanding claims using statistical indicators (quantitative approach) and also to have a discussion with transnational insurance groups, who had experience of different provisioning (qualitative approach), to get a better understanding of these issues. They found that the level of prudence in provisions was a more complex issue than it was usually presented.

The group found that the supervisors lack a common set of data for the analysis of claims runoff, and therefore it would be useful to require companies to provide statistical data according to a common layout. Such a database would be the basis for a common SRP (*monitoring* tool).

The group believed that it was better to let the provisioning converge toward a common level of prudence and in this respect set up principles and guidance for sound claims management and provisioning practices.

D.1.5.2 Provisions for Equalization

Equalization provisions are used as a buffer for bad years or catastrophes, and they are an additional buffer to the solvency. The group believed that it would make sense to take these provisions into account when assessing the solvency position. It should nevertheless be noted that under IFRS equalization reserves do not exist. If Solvency II and IFRS should be coordinated, it must be decided whether or not such reserves would be permitted to smooth the results over time or for tax reasons.

There was a large diversity in the size of equalization provisions. One way to promote further harmonization would be to explicitly link the provisions to the volatility of the business written.

D.1.6 The Sharma Report (Sharma, 2002)

When the EC started its project on insurance regulation, the Solvency II project, in May 2001, the Insurance Conference (now CEIOPS) was asked to give input and make recommendations. They set up a WG of insurance supervisors, called the London Working Group (LWG) as Paul Sharma from the U.K. Financial Services Authority (FSA) chaired it. Its report that was published in December 2002 is usually named *the Sharma report;* see Sharma (2002).

The goals of the LWG were to understand the risks to solvency of insurance undertakings and how better to monitor the undertaking's RM. They followed four main lines in their work:

- Risk classification and causal chain mapping
- Surveys on actual failures and near misses during 1996–2001; an update of the Müller report
- A total of 21 detailed case studies were discussed
- Diagnostic and preventive tools questionnaire

All the case studies showed *a chain of multiple causes*, but the most obvious causes were *inappropriate risk decisions, external trigger event,* and *resulting adverse financial outcomes*. The study also showed that these causal chains started with *underlying internal causes*: *problems with management, shareholders, or other external controllers*. The problems included

- Incompetence
- Operating outside their area of expertise

- Lack of integrity or conflicting objectives
- Weakness in the face of inappropriate group decisions

The underlying internal problems led to inadequate internal controls and decision-making processes which resulted in inappropriate risk decision. The LWG concluded that supervision would be most effective when there are tools to tackle the full causal chain.

For each of the 21 case studies the LWG made a risk map based on the template. The risk-models can be presented in three different levels:

- A summary level showing the broad categories
- A detailed level applied to the specific cases
- A detailed level showing the relevant supervisory tools and lessons learnt; see Annex E in Sharma (2002)

Based on the questionnaires on actual failures, near misses, and case studies, the LWG identified 12 risk types according to similarities in the causal chain rather than the effects.

1. The parent sets inappropriate policy in pursuit of group objectives (strategic investments); see the risk map in Annex E1 in Sharma (2002).

   The insurer's parent undertaking set an aspect of policy which had a detrimental effect on the insurance undertaking because they had objectives other than prudent management of the insurance undertaking. Group management overrode the local management.

2. The parent sets inappropriate policy through poor understanding of insurance; see the risk map in Annex E2 in Sharma (2002).

   The insurer's parent undertaking had a noninsurance focus and because they lacked proper understanding of the insurance business they set an aspect of policy that had a detrimental effect on the insurance undertaking.

3. *Mutual insurer faces conflicting objectives*; see the risk map at Annex E3 in Sharma (2002).

   A mutual insurer's management may have social or other objectives besides prudent management of insurance business, for example, in one case, this led the management to invest in other activities for the benefit of their members.

4. Business risk: Large insurer faces merger integration issues; see the risk map at Annex E4 in Sharma (2002).

   It would be hard for the large composite firms to manage efficiently, particularly those that have grown up through a series of acquisitions and mergers.

5. Cross-border management of insurance group; see the risk map at Annex E5 in Sharma (2002).

Poor management centrally can affect the conduct of operations in more than one Member State.

6. Life insurer—high expectations/long-term interest rate guarantees; see the risk map at Annex E6 in Sharma (2002).

   The management sets policies that gamble on future economic conditions. Interest rate guarantees can contain long-term options that could be expensive to service or high expectations of discretionary bonuses with low reserves.

7. *Stagnating insurer diversifies*; see the risk map at Annex E7 in Sharma (2002).

   Stagnating undertakings can move into noncore business with little experience.

8. Underwriting risk: a niche player with an evolving market; see the risk map at Annex E8 in Sharma (2002).

   The management takes naïve approach ignoring developments in their market changing the nature of their risks taken on.

9. Insurer matches liabilities with correlated investments; see the risk map at Annex E9 in Sharma (2002).

   The correlation between the risk profiles of assets and liabilities is ignored.

10. Firms have inappropriate distribution strategies; see the risk map at Annex E10 in Sharma (2002).

    Inappropriate strategy concerning agents and brokers can have adverse effects on the insurer, for example, poor customer service or high distribution costs not linked to portfolio outcomes.

11. *Catastrophe/inadequate reinsurance planning*; see the risk map at Annex E11 in Sharma (2002).

    An undertaking may find that it has insufficient reinsurance when catastrophic losses occur.

12. *Outsourcing of key functions*; see the risk map at Annex E12 in Sharma (2002).

    An undertaking may outsource a key activity and fail to maintain proper control over it.

The LWG found that almost all of the 21 case studies shared the same root causes: *poor or inexperienced management leading to inadequate decision making or internal controls.*

Based on a questionnaire on diagnostic and preventive tools the LWG set up toolkits for the supervision of the insurance undertakings. These toolkits are overlaid on the risk-map structure; see Figure 5.1 in Sharma (2002).

The toolkits are

Toolkit 1: Underlying causes—internal (management, governance, and ownership)

Toolkit 2: Underlying or Trigger causes—external (general and insurance specific)

Toolkit 3: Inadequacy of failed internal processes, people or systems

Toolkit 4: Risk decision and outcomes: investments, credit, and ALM risks

Toolkit 5: Risk decision and outcomes: underwriting and TPs

Toolkit 6: Risk decision and outcomes: reinsurance risk

Toolkit 7: Risk decision and outcomes: other risks and general tools

Toolkit 8: Curative Tools

For each of the first seven toolkits the LWG have listed what they expect of firms, their supervisory aims, tools that have been used, examples of other tools in use, and ideas for new tools or uses of tools.

The LWG also gave general considerations on a new supervisory approach. There should be a right balance between supervisory control and prevention and maintaining an insurer's freedom of operations. In order to assure the right mixture, a prudential regime should address risk in three main ways.

- Capital adequacy and solvency regime
- A broad range of tools needed to cover the full causal chain
- Internal factors

The LWG also list future work for the Insurance Conference (now CEIOPS), for example, harmonize common early warning signals, improve focus on RM and internal controls and develop a framework and guidance on asset–liability matching.

D.2 PHASE II: FRAMEWORK DIRECTIVE PHASE: 2003–2009

The first phase of the Solvency II project was a learning phase, in the sense that a series of studies was made to build the foundations for the project. During the first two years of the second phase the Commission Services produced some additional documents with questionnaires. At the Insurance Committee's June meeting, 2004 a road map for the development of the future work was presented, including a first wave of specific calls for advice from CEIOPS. In December the same year, the second wave of specific calls for advice was released and the third, and the last, wave was released together with a new road map during spring 2005.

Before summarizing the road map we will briefly discuss the various documents and questionnaires that have preceded it.

Regarding the winding-up paper of phase I, MARKT (2002h), the Commission Services made a questionnaire to facilitate discussion, MARKT (2003c). The general questions made in the paper were on the design of a prudential system, the function of the CR, the harmonization of European insurance supervision, the consistency of rules between sectors (insurance and bank), and international developments (IAA, IAIS, etc.).

In a paper to the Insurance Committee, MARKT (2003d), and including MARKT (2003a) as an Annex, the Commission Services made recommendations on the design of a prudential supervisory system. The new system is supposed to provide the supervisory authorities with appropriate qualitative and quantitative tools to assess the overall solvency of an insurance undertaking. The Basel II three-pillar approach should be the starting point. The system should be a risk-based approach, encouraging undertakings to measure and manage their risks. There should be consistency between the financial sectors and a more efficient supervision of insurance groups and financial conglomerates. This would lead to harmonization of the supervisory methods. The Lamfalussy comitology techniques would be used to make the new solvency system effective and flexible.

The following is a list of names of the persons at the European Commission, Internal Market and Services DG, Financial Institutions, Insurance and Pensions Unit, who have been working on Solvency II project between 2002 and 2009:

David Deacon (Head of Unit, 1999–2004)
Karel Van Hulle (Head of Unit, 2004–2009)

Joao de Abreu
Manuel Altemir Mergelina
Ines Alpert
Ben Carr
Ramón Carrasco
Pauline de Chatillon
Paulina Dejmek
Ivo van Es
Olivier Fliche
Manuel de Frutos
Alexandros Iatrou
Adam Jacobs
Bertrand Labilloy
Jung-Duk Lichtenberger
Ulf Linder
Janne Lipponen
Evelyne Massé
Axel Oster
Vesa Ronkainen
Susanne Rosenbaum
Teresa Rubino
Kristina Summanen
Dominique Thienpont
Mike Thom

Thanks to Yvette Chrissantonis and Valérie Kupferman, who were the assistants working in the project, and Kristina Summanen for this most exhaustive list of persons working with the Solvency II project.

D.2.1 Recommendations for the First Pillar

The new system was proposed to evolve two regulatory capital requirements:

- An *SCR*, reflecting the economic capital an undertaking would need to operate with a low probability of failure. It should reflect most risks to which a company is exposed. This would be the principal instrument for the supervisory process.

The new system should allow the companies to use *internal risk models* for the calculation of the SCR.

- An *MCR*, or *safety net*, should be established to constitute a basic trigger level for ultimate supervisory action. It should be calculated in a simple and objective way, as supervisory actions at this level may need court decisions in certain jurisdictions.

The asset risk should be captured in a more explicit way in the calculation of SCR. If all risks an insurance company faces are considered in the risk capital calculation it is not necessary that there are rules for assets. When insurers take all potential risks into account, they should be free to choose the way that they invest these assets.

Other main components of the first pillar are the treatment of the TPs, that is, in terms of a BE plus a risk margin, as well as a definition of intervention levels.

D.2.2 Recommendations for the Second Pillar

It is essential to have quantitative tools and a strengthened SRP in place. Several supervisory areas could be harmonized, such as principles for internal control, sound risk and financial management, asset/liability matching, and criteria for the structure of the reinsurance program of the undertaking.

In the second pillar risk categories not captured by the quantitative approach under pillar one will be assessed, for example liquidity risk (and perhaps the operational risk).

There should also be minimum criteria for on-site inspections and defined intervention powers, responsibilities of the supervisory authorities and transparency of supervisory action.

D.2.3 Recommendations for the Third Pillar

Transparency and disclosure requirements will also be an important part of a new system. Reporting requirements would be coordinated in order to reduce administrative burden on undertakings.

The recommendations were welcomed by the Insurance Committee (IC); MARKT (2003e). The decision taken was that the Commission should proceed and prepare a more detailed paper on how the project would be progressed in the future. This Update document was presented to the Insurance Committee in November 2003, MARKT (2003f). But before that, at the IC Solvency Subcommittee in October 2003, a Commission Services document, MARKT (2003b), was discussed. This document sets up a proposed structure for an FD, relating to the codified life directive, COM (2002c).

D.3 FURTHER STEPS

D.3.1 Creation of CEIOPS and Stakeholders' Action

In the Update document, mandates for the new committee of European supervisory authorities CEIOPS are discussed. The Commission on 5 November 2003 officially created CEIOPS. Consultations with the insurance industry and the actuarial profession (through the Groupe Consultatif) are also important issues for the coming work. CEIOPS decided to create a number of technical subgroups to perform the work related to the Solvency II project (e.g., Pillar I: life, Pillar I: nonlife, Pillar II, Pillar III, and group/conglomerates/cross-sector issues). These groups were fully operational from summer 2004.

The Groupe Consultatif set up shadow WGs to be able to cooperate with the CEIOPS WGs. These groups were operational during spring 2004. In the beginning of autumn 2004 the first concrete questions from CEIOPS WGs were delivered to GC's WGs.

CEA changed its working structure in that they set up a Steering Group to enhance input to and dialogue with the Solvency II project key drivers. In the beginning of 2005 CEA also set up working groups to deal with similar issues. In a pilot project, carried out by CEA, a comparative study between different solvency systems was made; see CEA (2005).

EIOPC, the successor body to the Insurance Committee, got a key role in the development of the project, for example, discussion on the overall solvency confidence level, the level of prudence in the TPs, and the target capital level.

D.3.2 Organization and Basic Architecture of Solvency II

In February 2004 the Commission Services published a paper on the organization of the future work, Pillar I work areas, and suggestions of further work on Pillar II for CEIOPS, MARKT (2004a).

It was proposed to split the work between different levels:

- *The FD* The work would not differ from the usual drafting of directives, that is, the Commission Services will draft texts and consult the EIOPC. A Commission solvency working group (CSWG), now a subcommittee to the Insurance Committee, prepared the work of EIOPC.
- *Implementing regulation: comitology* Once the new committee architecture (Lamfalussy procedure) was in place, detailed work in these areas was formally delegated to CEIOPS, via a mandate established by the Commission (until then the Commission gave *specific calls for advice* from CEIOPS).
- *Role of the Commission and the regulatory committee* Implementing measures have the same legal value as a directive. The Commission has the monopoly of regulatory initiatives at both levels. When the new EIOPC (as a regulatory committee) and CEIOPS (as a supervisory committee) were formally in place, the new structure became
  - The Commission, after consulting EIOPC, requests advice from CEIOPS.
  - CEIOPS will determine the most appropriate way to fulfil its mandates from the Commission.

    - In the light of the advice given by CEIOPS, the Commission will draw up its proposal of implementing measures. This proposal may be subject to an additional consultation. The EP can render an opinion on this draft. The Commission then presents it to the EIOPC, who votes on it.
    - Depending on the results of this vote, different procedures can be followed until final adoption.

- *Role of CEIOPS* CEIOPS will provide technical advice on implementing measures.
- *Role of other parties—Importance of transparency* The Commission Services considered the dialogue with the insurance industry, actuaries, and others to be essential (i.e., stakeholders). Transparency was an important element in the new committee architecture.
- *Role of the EP* In the extension of the new committee architecture there will be a need to involve the Parliament much more closely in the technical analysis than in the past.

The document discussed general issues and gave questions to be answered by interested parties. Discussion was made on accounting environment issues, IAIS standards, type of harmonization, and so on. Other hot topics discussed were the TPs in life and nonlife insurance, BE and MVM level of prudence, discount rates, and so on.

In calculating the SCR the Commission Services suggested a whole spectrum of available approaches:

- An ESA
- A national SA that would result from the European one with calibration of national parameters
- An IM that would wholly or partly substitute for the SA

One way of motivating companies to develop IMs is that the resulting solvency capital level may be lower in the IM than in the SA.

Other issues that were discussed were suitable risk measures (e.g., VaR and TailVaR[*]) and a suitable time horizon for the SCR. Should the calculations be made on a going-concern basis or on a run-off or a winding-up basis? Classification of risk factors and the structure of the SA, as well as dependencies and correlations, were the other issues discussed. One appendix dealt with suggested requests on Pillar II issues for CEIOPS.

There would be several types of measures regarding the prudent asset management in insurance companies, such as the investment risk in the SCR, and the "prudent-person" approach including ALM.

[*] In the Second wave of specific Calls for Advice, MARKT (2004d) and (2004e), the TailVaR was proposed for IMs!

The paper also included proposals for amended articles in the FD, such as Article 22 in the codified life directive, COM (2002c) and Article 20 in the third nonlife directive, EEC (1992)—underlined text was new:

"The assets covering the TPs *and the capital requirements* shall take account of the type of business carried on by an assurance undertaking in such a way as to secure the safety, yield and marketability of its investments, which the undertaking shall ensure are diversified and adequately spread. *To this end an assurance undertaking shall have an appropriate investment plan.*"

Other issues discussed were the calculation of the minimum solvency margin, the guarantee fund, and solvency control levels.

D.3.3 Road Map

At the Insurance Committee's meeting on June 30, 2004 the Commissions Services made a proposal for a road map for the future work and also gave a first wave of specific calls for advice from CEIOPS; MARKT (2004a2). The Insurance Committee adopted the road map and the first wave, except for occupational pensions, that became a separate subject addressed at a later date. In October, the draft proposal for a second wave of specific calls for advice from CEIOPS was published, MARKT (2004b). Other stakeholders, such as Group Consultatif and CEA, were also asked to give comments on the requests before the Insurance Committee took it, MARKT (2004c).

- The Commission would prepare a draft proposal for an FD, before mid-2006. In the beginning of 2005 it was stated that this proposal would not be issued until October 2006. Later it was postponed until July 2007.
- The detailed Solvency II legislation would be subsequently adopted through implementing measures under the comitology procedure with a target completion date of 2008 or 2009.
- A framework on the policy principles and guidelines for consulting CEIOPS and other stakeholders was drawn up. CEIOPS and other stakeholders were also consulted by means of three waves of specific calls for advice; see Section D.3.3.2 for a list.

D.3.3.1 Basic Architecture

In 2004 the EC published a first proposed framework for the solvency system introducing the basic architecture of Solvency II, MARKT (2004a1). The framework was published for consultation among different stakeholders.

The Framework for Consultation, FfC, covered the totality of the Solvency II project. It set out the policy principles and guidelines within which CEIOPS was asked to develop its advice for the Solvency II project. In addition to the FfC, CEIOPS was requested to provide advice on detailed aspects of the new solvency system through "specific calls for advice," which was annexed to the FfC; see Sectoin D.3.3.2 below. CEIOPS was invited to develop work on Solvency II topics on its own initiative, in accordance with the policy principles and guidelines, and seeking inspiration from earlier Commission Services' papers.

This basic architecture paper was amended both in 2005, MARKT (2005b) and in 2006, MARKT (2006b).

The latest version of the framework from 2006 is discussed below. For each point we give comments on the developments made during the years 2004–2006.

General Features

1. The new solvency system should provide supervisors with the appropriate tools and powers to assess the "overall solvency" of all institutions* based on a prospective and risk-oriented approach. It should not only consist of quantitative elements, but also cover qualitative aspects that influence the risk-standing of the undertaking (managerial capacity, internal risk control, risk monitoring processes, etc.). The new solvency system is likely to result in changes to most of the present insurance Directives.

2. This solvency system defined in a broader sense should take its starting point in a three-pillar structure *inspired by Basel II/Capital Requirements Directive (CRD): quantitative requirements (Pillar I), supervisory activities (Pillar II), and supervisory reporting and public disclosure (Pillar III)*. This implies that special considerations are made concerning the interaction between quantitative and qualitative supervision, as well as concerning the role of disclosure. The importance of the SRP in Pillar II is to be highlighted. *It should be noted that the scope and magnitude of the individual pillars do not need to be identical to Basel II.*

   Comments: The text in *italic* was new from 2005. In the first text it said "... quantitative capital requirements, an SRP and disclosure requirements."

3. In Pillar I the new solvency system contains two capital requirements *with different purposes and calculated accordingly*: the SCR and the MCR. *The SCR may not be lower than the MCR.*

   Comments: The text in *italic* was new from 2005. It was stressed that the SCR could not be lower than the MCR.

4. The solvency system should be designed in such a way that it ~~encourages and~~ gives an incentive to the supervised institutions to measure and properly manage their risks. In this regard, common EU principles on RM and supervisory review should be developed. Furthermore, the SCR should cover the quantifiable risks to which a supervised institution is exposed. This risk-oriented approach implies the recognition of IMs (either partial or full), provided these improve the institution's RM, better reflect its true risk profile than under the standard formula, and can be appropriately validated.

   Comments: The cancelled text was omitted from 2005.

5. An IM can result in a higher or lower amount for the SCR than the amount based on the standard formula, subject to a floor (the MCR). Supervisors may require undertakings

* Life assurance, nonlife insurance, and reinsurance undertakings as defined in the relevant Directives.

for which the activities deviate substantially from the assumptions underlying the standard formula to develop an IM.

Comments: This text, introducing IMs, was new from 2005.

6. The main focus of the Solvency II project is on capital requirements and supervisory review at the level of the individual legal entity. However, issues related to insurance groups and financial conglomerates also have to be addressed, including the implications for existing legislation [e.g., the Insurance Groups Directive (98/78/EC) and the Financial Conglomerates Directive (2002/87/EC)].

    6.1. The application of IMs in a group or conglomerate context is a key issue in this regard, as well as possible diversification benefits/costs and how to allocate these.

    6.2. It should be recognized that management decisions are increasingly being taken by the ultimate parent company. The relevant rules therefore need to be set at the relevant level. This includes an appropriate split of responsibilities between the supervisors involved in the supplementary supervision of financial groups, in particular concerning the validation of IMs.

    6.3. Solo supervision remains the responsibility and task of the national supervisor. Any rules on the adequate distribution of capital should be reinforced, to ensure that at the solo level sufficient capital is available. However, this does not exclude any streamlining of the supervisory activities for groups subject to supervision in several Member States.

    Comments: 6.1 through 6.3 were new from 2005. In the first text from 2004, the head text ended with "and the application of IMs in a group or conglomerate context."

7. The Solvency II regime contains prudential valuation standards for assets and liabilities of insurance undertakings, as well as rules on supervisory reporting and public disclosure. In order to ensure convergence of valuation rules, supervisory reporting, and public disclosure, as well as to limit the administrative burden for supervised institutions, the Solvency II rules should be compatible with accounting rules elaborated by the IASB. The following clarifications of this approach can be given:

    7.1. IASB is currently working on Phase II of its "Insurance Contracts" project, but it will take significant time until the final standard is presented. It is therefore likely that the Solvency II rules on valuation, reporting, and disclosure will be elaborated without having an adopted IFRS in place. The likely outcome (*cf.* paragraph 7.2) of the IASB work should, however, be taken into account. Additions and adjustments to the IASB accounting rules may be proposed, provided specific reasons are given. The Solvency II rules may be adjusted when the IASB has finalized Phase II of its "Insurance Contracts" project.

    7.2. The following elements are likely to be part of a future IFRS on insurance: use of a prospective asset–liability valuation approach; the valuation methodologies should make optimal use of information provided by the financial markets;

CFs related to assets and liabilities should be discounted; and for TPs, the "best estimate" 2 as well as the RM should be disclosed.

7.3. Presently only listed EU insurance companies are required to present consolidated financial statements according to endorsed IAS/IFRS, although Member States may broaden the field of application. The Solvency II regime will not require full use of IAS/IFRS by all insurance undertakings. Certain prudential valuation rules, and reporting and disclosure rules, may however be similar to IAS/IFRS rules, and in those cases a wider use of IAS/IFRS-like methods by insurance undertakings is not unlikely. Certain simplifications may apply for certain types of insurance undertakings, but the general principles should apply to all insurance undertakings.

Comments: The original text from 2004 was not as comprehensive as the text adopted from 2005 depending on the development within IASB.

8. The aim to attain *an appropriate level of harmonisation that is at the same time higher than the present one* should be reflected by solvency rules that do not need additional requirements. The new solvency system should provide for uniform application *and sufficient consumer protection whilst supporting fair competition.*

   Comments: Instead of "an appropriate level of harmonization" the first text introduced the concept of "maximum harmonization." The text in *italic* was new from 2005.

9. The solvency system sets a uniform level of prudence, both for TPs and for the SCR.

10. In order to ensure consistency across financial sectors, the general layout of the solvency system should, to the extent necessary and possible, be compatible with the approach and rules used in the banking field. Products containing similar risks should, in principle, be supervised in the same way and should be subject to the same capital adequacy or solvency requirements. The new solvency system should be constructed in a way that facilitates efficient supervision of insurance groups and financial conglomerates and avoids regulatory arbitrage between and within financial sectors. *However, the use of a more accurate or appropriate approach for measuring risks should be encouraged.*

    Comments: The text in *italic* was new from 2005.

11. Further international convergence is promoted through compatibility of the new solvency system with work of the IAIS and the Groupe Consultatif Actuariel Européen/IAA. *The IAIS Framework for insurance supervision and Cornerstones for the formulation of regulatory financial requirements provide a valuable basis for the development of a new system.*

    Comments: The text in *italic* was new from 2005.

12. In order to evaluate the impact of the new solvency system *on insurers*, one or more QIS will have to be made. Coherent data are a fundamental requirement to perform such

studies. The ~~corresponding~~ data collection requirements, methodology, and timing for these QIS all need analysis. *CEIOPS has been asked to perform these quantitative impact studies in cooperation with the industry. The Commission Services will provide an Impact Assessment to accompany the Proposal for a Directive, which will have a broader scope and less detail than the QIS.*

Comments: The text in *italic* was new from 2005. The cancelled text was omitted from 2005.

13. Guarantee schemes are a last resort for policyholders and beneficiaries to be indemnified for their loss. The calibration of the new solvency system should not take into consideration the existence of a guarantee scheme. The new solvency system should provide enough safety and confidence in the insurance industry without relying on guarantee schemes.

    Comments: The text was new from 2005.

14. Solvency II should not be overly costly for small undertakings but does not intend to have an entirely special treatment of them. Small undertakings should respect the same basic principles as all other institutions. However, in certain areas, it may be necessary to provide for specific rules for them.

    Comments: The text was new from 2005.

15. Solvency II should take due account of the particularities of reinsurance undertakings.

    Comments: The text was new from 2005.

Pillar I, Quantitative Requirements

16. An increased level of harmonization for EPs is a cornerstone of the new solvency system. TPs need to be established in order for the undertaking to fulfill its (re)insurance obligations toward policyholders and beneficiaries, taking account of expenses. In line with the IAIS Cornerstones and expected IASB developments, TPs have to be prudent, reliable, and objective, and allow comparison between (re)insurers. They should make optimal use of and be consistent with information provided by the financial markets and generally available data on insurance technical risks. They are the sum of a BE[*] and an RM.

    16.1. The BE equals the expected PV of future CFs, using the relevant risk-free yield curve, based on current and credible information and realistic assumptions. The use of realistic assumptions implies that surrender value floors should not be applied to the calculation of TPs.

    16.2. The RM covers the risks linked to the future liability CFs over their whole time horizon. It should be determined in a way that enables the (re)insurance obligations to be transferred or put into runoff. Such an approach protects

[*] In international fora also "current" and "central" estimate are used. However, this terminology may not always have the same meaning.

policyholder rights and takes account of the uncertainty of valuation of the BE. Two possible ways to calculate this RM should be considered as working hypotheses. It can be calculated as the difference between the 75th percentile of the underlying probability distribution until runoff and the BE. However, the RM should as a minimum be equal to half a standard deviation in order to take account of strongly skewed distributions. Alternatively, the RM can be calculated based on the cost of providing SCR capital to support the business-in-force until runoff. Further quantitative impact information should be collected to assess the merits of the two methods. It would be preferable, if the same method would apply to all undertakings, irrespective of size, sophistication, or type of business. However, the specificities of life and nonlife insurance may require different treatments. Small undertakings could be allowed to use a simplified version of that method. If in certain circumstances it would prove to be appropriate to use different methods for determining the RM, this will be based on objective criteria and not subject to Member States' discretion.

Comments: This issue has been extensively developed between the years. In the first text from 2004 it was stated that "An increased level of harmonisation for TPs is a cornerstone of the new solvency system. To that effect it is recommended to set a quantitative benchmark for the prudence level in TPs. The relationship between TPs in the new solvency system and the future accounting regime and an appropriate level of prudence for TPs require analysis." 16.1 and 16.2 were new from 2006.

17. The SCR reflects a level of capital that enables an institution to absorb significant unforeseen losses and that gives reasonable assurance to policyholders *and beneficiaries. When an undertaking does not fulfil the SCR, it shall re-establish the amount of capital covering the SCR in due time, based on a concrete and realisable plan submitted to the supervisor for approval. The parameters in the SCR should be calibrated in such a way that the quantifiable risks to which an institution with a diversified portfolio of risks is exposed are taken into account and based on the amount of economic capital corresponding to a ruin probability of 0.5% (Value at Risk of 99.5%) and a one year time horizon. This percentage reflects a working hypothesis. Ruin occurs when the amount of admissible assets is lower than the amount of technical provisions as defined in paragraph 16. The methods used to check that this level is effective must be defined.* The SCR should be based on a going-concern basis. *These principles shall apply regardless of whether a standard formula or an internal model is used.*

    Comments: The text in *italic* was new from 2005 and extended the original text extensively.

18. The standard formula to calculate the SCR can be based on a variety of methods, for example, a factor-based formula, probability distribution-based formula, scenarios, or combinations thereof. The most suitable standard formula taking into account the specificities of life, nonlife, and reinsurance business, which consequently vary according to the specific sector concerned, requires analysis.

19. The MCR reflects a level of capital below which ultimate supervisory action would be triggered. *The level of the MCR will be set once quantitative impact studies have been performed.* It shall be calculated in a more simple and robust manner than the SCR as this kind of action may need authorization by national courts. To facilitate and stabilize the transition to the new overall solvency system, the MCR should be constructed in a straightforward manner such as under the present "Solvency I Directives," while maintaining a sufficient level of prudence. *It will have an absolute floor.*

    Comments: The text in *italic* was new from 2005. It introduced an absolute floor put on the MCR.

20. The risks addressed in the capital requirements should be based on the IAA risk classification and include underwriting risk, credit risk, market risk, operational risk, and liquidity risk. Additions and adjustments to the IAA risk classification could be made, provided specific reasons are given. To the extent these risks are not quantifiable they will be taken into account in Pillar II.

21. IMs may be used to replace the standard formula of the SCR if the IM has been validated for this purpose. The validation criteria and the validation process should be developed and harmonized. *Partial use of models may also be authorised, if these models fulfil validation conditions, including compatibility with the standard formula.* ~~The possibility to extend this option to group-wide internal models requires analysis.~~

    Comments: The text in *italic* was new from 2005. The cancelled text was omitted from 2005.

Pillar II, Supervisory Activities

22. *The supervisory activities should aim to identify institutions with financial, organizational, or other features susceptible to producing a higher risk profile. Such institutions can be required to hold a higher solvency capital than under the SCR and/or to take measures to reduce the risks incurred. In addition,* it should increase the level of harmonization of supervisory methods, tools, and powers *by developing common standards and methods, for example, for the validation process of IMs. The scope therefore goes beyond that of the SRP defined in Basel II. The supervisory activities also include increased cooperation between supervisors combined with peer reviews.*

    Comments: The text in *italic* was new from 2005 and developed the original text.

Pillar III, Supervisory Reporting and Public Disclosure

23. Harmonized reporting from insurance undertakings to their supervisors ("supervisory reporting") will be an important part of the future regulatory architecture in the EU. Supervisory reporting goes beyond the notion of financial reporting rules, and includes different types of information that a supervisor needs to perform his functions. This information is normally not in the public domain. In addition, transparency and disclosure of information by undertakings to the public will serve to reinforce market mechanisms and discipline ("public disclosure").

    Comments: The text was new from 2005.

24. In the Basel II framework, Pillar III requirements deal only with public disclosure, and supervisory reporting is only harmonized to a limited extent. While remaining compatible, it is deemed necessary to explicitly include both areas in the Solvency II exercise.

 Comments: The text was new from 2005.

25. *Supervisory reporting and* disclosure requirements should be in line with those elaborated by the IAIS and IASB in order to reduce the administrative burden for supervised institutions. They should also be compatible with disclosure requirements in the banking sector. Additions and adjustments could be proposed, provided specific reasons are given. Confidentiality aspects linked to disclosure requirements need careful consideration.

 Comments: The text in *italic* was new from 2005. In the original text from 2004 it was stated that "Disclosure requirements enhance market discipline and complement requirements under Pillars I and II."

It was also stated in the framework text that the solvency requirements of IORPs were linked to those of life assurance undertakings by way of Article 17 of Directive 2003/41/EC. However, solvency requirements for IORPs were an important and separate subject that was to be looked at again at the time of the reviews that was foreseen in the IORP Directive.

D.3.3.2 Three Waves of Specific Calls for Advice

As stated earlier, at the Insurance Committee's meeting on June 30, 2004 the Commissions Services made a proposal for a road map for the future work and also gave a first wave of specific calls for advice from CEIOPS; MARKT (2004a2). The Insurance Committee adopted the first wave of specific calls. In October, the draft proposal for a second wave of specific calls for advice from CEIOPS was published, MARKT (2004b). Other stakeholders, such as Group Consultatif and CEA, were also asked to give comments on the requests before the Insurance Committee took it, MARKT (2004c).

- The Commission would prepare a draft proposal for an FD, before mid-2006. In the beginning of 2005 it was stated that this proposal would not be issued until October 2006. Later it was postponed until July 2007.

- The detailed Solvency II legislation would be subsequently adopted through implementing measures under the comitology procedure with a target completion date of 2008 or 2009.

- A framework on the policy principles and guidelines for consulting of CEIOPS and other stakeholders was drawn up. CEIOPS and other stakeholders were also consulted by means of three waves of specific calls for advice.
 1. *The first wave of specific calls:* Pillar II issues (published in July 2004). Technical reports should be transmitted no later than 30 June 2005 (or 31 March depending on

the request; see below). Progress reports should be given at four-monthly intervals, the first by 31 October 2004; MARKT (2004a2).

The first wave of requests includes the following issues (Pillar II)

Request 1: Internal control and RM

Request 2: Supervisory review process (general)

Request 3: Supervisory review process (quantitative tools)

Request 4: Transparency of supervisory action

Request 5: Investment Management rules

Request 6: Asset–Liability Management

2. *The second wave of specific calls:* Pillar I issues (published in December 2004). Technical reports should be transmitted no later than 31 October 2005. Progress reports should be given at four-monthly intervals, the first by 28 February 2005; MARKT (2004c).

   The second wave of requests include the following issues (Pillar I)

   Request 7: Technical provisions in life assurance

   Request 8: Technical provisions in nonlife assurance

   Request 9: Safety measures

   Request 10: Solvency capital requirement: standard formula (life and nonlife)

   Request 11: Solvency capital requirement: internal models (life and nonlife) and their validation

   Request 12: Reinsurance (and other risk mitigation techniques)

   Request 13: Quantitative impact study and data-related issues

   Request 14: Powers of the supervisory authorities

   Request 15: Solvency control levels

   Request 16: Fit and proper criteria

   Request 17: Peer reviews

   Request 18: Group and cross-sectoral issues

3. *The third wave of specific calls:* Pillar III issues (draft published in February 2005); MARKT (2005).

   The third wave of requests include the following issue (Pillar III)

   Request 19: Eligible elements to cover the capital requirements

   (Request 20: Independence and accountability of supervisory activities[*])

   Request 20: Cooperation between supervisory authorities

   Request 21: Supervisory reporting and public disclosure

[*] Request 20 was removed before the third wave of requests was sent to CEIOPS. Requests 21–24 were renumbered as 20–23.

Request 22: Procyclicality

Request 23: Small- and medium-sized enterprises

D.3.3.3 Brief Summary

In the framework for discussion a new concept was introduced instead of the earlier TCR[*] the *SCR*. The "target" had now become "a requirement"! The question whether the SCR should be a soft or hard level was also discussed. The decision would influence the composition of trigger levels and perhaps also the confidence level!

The new solvency system would be based on a risk-oriented approach and provide the supervisors with appropriate tools and powers to assess the "overall solvency" of the life insurance, nonlife insurance, and reinsurance undertakings.

The new solvency system would be defined in a three-pillar approach: quantitative capital requirements (Pillar I), qualitative requirements, for example, an SRP (Pillar II) and transparency and disclosure requirements (Pillar III). RM is important. The supervised institutions would be encouraged to manage their risks properly. Issues related to insurance groups and financial conglomerates would also have to be addressed, including the application of IMs in a group or conglomerate.

In order to ensure convergence in financial and regulatory reporting, as well as to limit the administrative burden, supervisory reporting would be compatible with accounting rules elaborated by the IASB, especially techniques and methods used to calculate TPs.

A general goal was to get *maximum harmonisation* in the new solvency system. It would also be compatible with the approach and rules of the banking sector.

There would be a uniform level of prudence, both for TPs and for the SCR.

International convergence is promoted. Work done by IAIS and Groupe Consultatif (and IAA) would be considered.

QIS have to be made.

Pillar I Features

One cornerstone of the new system would be an increased level of harmonization for TPs. The SCR should reflect the risks the company faces. The ESA to calculate the SCR could be based on a variety of methods, for example, on factor-based approaches, probability distribution-based approaches, scenarios, or a combination of these.

The MCR reflects the level of capital below which ultimate supervisory action would be triggered. It should be calculated in a simple way.

The risk factors would be based on the IAA risk classification, that is, the Basel II risks (market risk, credit risk, and operational risk) and the underwriting risk (insurance risk). IMs may replace the ESA to the SCR if IMs have been validated for this purpose.

[*] The question whether the TCR should be seen as a soft or hard level have been discussed. In answer to a direct question to the Commissions Services at a meeting at the CEA in the spring of 2004 it was said to be hard. This answer reflects the change from TCR to SCR!

Pillar II Features

The SRP would increase the level of harmonization of supervisory methods, tools, and processes.

Pillar III Features

Disclosure requirements enhance market discipline and complement requirements under Pillar II and Pillar I. They should be in line with those of IAIS and IASB.

D.4 STEPS TOWARD A SOLVENCY DIRECTIVE

Two main lines of work have been carried out:

- *Level 1 workstream:* The work with the FD (Level 1 in the Lamfalussy procedure; see Section D.3.1)
- *Level 2–3 workstream:* Consultations for implementing measures (Level 2–3 in the Lamfalussy procedure; see Section D.3.1)

To describe the work development within the Commission and by EIOPC, we used extracts from the minutes of the EIOPC meetings to give a "live picture" of the development.

D.4.1 EIOPC Meeting December 2005

In 2005 the Commission introduced a document (not public) about recasting and codifying the Life and Non-life directives. At EIOPC' meeting in December 2005, MARKT (2006a), this was discussed. From the fourth paragraph of these minutes we have the following extract:

> *A. Recasting of the Insurance Directives*
>
> The Commission introduced the document MARKT/2524/05* and its Annex, a working document recasting and codifying the Life and Non-life Insurance Directives. The consolidated version of the Insurance Groups Directive had been circulated to EIOPC members as a separate document. Detailed discussions about those documents were to take place in the working group on 14 December 2005. The Commission briefly recalled the background for the codification work. It was pointed out that the current structure of the Insurance Directives had been maintained but was likely to be adapted when the Solvency II changes were to be incorporated into the document. Provisions that had been identified as obsolete would need to be discussed with Member States. Moreover, EIOPC members were informed about the Commission's intention to incorporate into the codification document the recently adopted Reinsurance Directive. The Insurance Groups Directive (98/78/EC) and the Winding-up Directive (2001/17/EC) would be included in separate appendixs.
>
> Several Member States congratulated the Commission for the good work it had carried out. One delegation asked whether the timetable for Solvency II could be

* Not published by EIOPC.

adapted, given the significant work already undertaken on the codification. One delegation noted that the wording of the merged articles could be improved. Another delegation observed that the wish to keep terminology consistent had occasionally been pushed too far. One Member State was opposed to the idea of including the Insurance Groups Directive and the Winding-up Directive in the codification document. Seven Member States expressed an explicit wish to include the Insurance Groups and Winding-up Directives. One delegation asked about the implications for the Insurance Accounts Directive. Two delegations inquired about a possible inclusion of the Insurance Mediation Directive (2002/92/EC).

The Commission thanked delegations for their comments and noted that a majority of the Member States that had spoken seemed to be in favour of including the Insurance Groups Directive as well as the Winding-up Directive in the codification exercise. This corresponded to the Commission's preferred option. Policy issues such as the implications for the Insurance Accounts Directive would need to be considered further. There was no intention to include the Insurance Mediation Directive, since this Directive was of a different nature. A new working document including provisions related to reinsurance, winding-up and insurance groups would be circulated shortly and discussed in the Solvency Working Group on 14 December.

B. Impact assessment

The Commission staff reminded EIOPC members that a document setting out the structure of the impact assessment report had been sent out in the summer, and briefly presented a new document containing the further work strands identified for the impact assessment report and further consultation (namely an Open Hearing was foreseen for June 2006). A detailed timetable was also communicated. EIOPC members were asked for input as to who should be consulted. The work programme and timetable were approved. Several Member States stressed the need to involve small insurers in the quantitative impact studies. The Commission staff pointed out that the impact assessment report, which had to accompany the Level 1 directive proposal (which would not contain all the technical details) was separate from the impact studies, for which several rounds would be needed to calibrate the detailed formulas. The impact assessment report would use the quantitative impact studies that were available up to the time of its finalization.

From this discussion we can summarize the fact that the FD now was supposed to be a codified text from existing directives. At the end it showed up to include 13 different old insurance directives.

As a complement to the QISs conducted by CEIOPS an IA study had to be done.

At the same meeting, CEIOPS presented its final results of the second wave of call for advice (CfA).

C. CEIOPS' presentation on the final results of the second wave of calls for advice

The CEIOPS representative reported on the considerable amount of work accomplished in a short period of time under the first and second waves. Major highlights were the work on the calculation of technical provisions, safety measures, internal models, quantitative impact studies (QIS), the fit and proper criteria, the supervisory review process as well as group and cross-sectoral issues. There had also been a public hearing organised by CEIOPS in September and the industry appeared to be satisfied with the general direction of the work.

There were of course some open issues from the second wave on which work was still continuing. Though technical in nature these were also political issues.

One such issue arising under calls for advice Nos 7 and 8 related to the calculation of technical provisions. Various percentiles would be tested. However there was also discussion on the cost of capital approach. Another open issue related to the calculation of the MCR, where there were three possible approaches, one similar to the current regime, another linked to the SCR and the third involving a margin over liabilities. Discussions were ongoing on asset requirements, where some saw no need for these, while others felt that some quantitative limits were still needed. There were also two views on the validation of internal models (call for advice No 18 with a link to No 20), one taking the CRD as the starting point, while the other view favoured a more bottom-up approach and was reluctant to see the group level dictate to the solo level. On this and the other disputed issues, work to seek a compromise was being pursued.

CEIOPS also reported on its work on QIS 1 relating to technical provisions. Work had begun in October 2005 and was due to end in February 2006. QIS 2 would then begin in May. The CEIOPS representative felt that four QIS rounds would probably be needed. He was encouraged by the significant number of participating entities, including small and medium sized insurers.

The Commission representative noted that the second wave was at the heart of the future regime. It was understandable that some issues were still open. The Commission would be sending further questions to CEIOPS.

One delegation thought that where further questions were put to CEIOPS it should be given a working hypothesis on which to base its work.

A number of delegations were of the opinion that where there were different political approaches on various substantive issues it would be appropriate for the EIOPC to confront such issues, discuss them and give a steer. It was suggested that the Commission produce a paper on those issues for the next meeting of the Committee.

The Commission noted that the distinction between technical and political issues was not an easy one to make. Working group meetings would look at first drafts and would identify issues for further discussion at the EIOPC at its next meeting. Such issues should indeed be discussed in the EIOPC but only after a preparatory stage.

D.4.2 EIOPC Meeting in April 2006

In EIOPC's 3rd meeting in April 2006 the IA and the work with CEIOPS QIS2 were reported. In paragraph 7 of the minutes of the meeting, MARKT (2006c), the first part gave a report on the progress of the work done by the Commission and CEIOPS:

> The Commission services reported that work on Solvency II had accelerated in recent months and paid tribute to the enormous efforts put in by all those involved in this work both from the CEIOPS Secretariat and from all the Member States involved. As a result of good cooperation, Solvency II was moving forward, though the timetable remained extremely challenging.
>
> Since the last EIOPC meeting on 2 December 2005 the Solvency Working Group had met three times. On 14 December 2005 there had been a discussion on the draft new codified Directive (MARKT/2524/05). Following this discussion, a new version had been presented at the following meeting on 15–16 February 2006. Nine Member States had sent in written comments. Work was currently ongoing with the Legal Service on the codified Directive.
>
> On 15–16 February 2006 new draft articles reflecting the first wave of calls for advice were discussed (document MARKT/2501/06). Written comments from six Member States had been received. There had also been a presentation by the CRO Forum and the CEA on the cost of capital approach.
>
> The meeting on 20 March 2006 discussed technical provisions and the cost of capital approach on the basis of document MARKT/2502/06. One Member State had sent in written comments on this document. It was on the basis of the discussion on 20 March and the answers provided to the questions raised at that meeting that the Commission had drafted its proposed changes to the Framework for Consultation, which constituted the next agenda subitem.
>
> The Commission Services noted that documents 2501/06 and 2502/06 were not public and accordingly the Member States' comments on those papers would not be made public. The Commission asked, however, whether Member States who had commented on these working papers could agree to their answers being circulated to the other Member States.
>
> CEIOPS' answer to the third wave of calls for advice had been due by the end of February but owing to the complexity of the issues, CEIOPS would only approve the final answers to the calls for advice at its 25–26 April Members meeting. The EIOPC would discuss these answers at its 6 July meeting.
>
> The next Solvency Working Group meeting would be on 5–6 May and would discuss new Solvency II draft text on issues covered by to the second wave as well as elements of the Impact assessment report. The Commission informed the Committee that the codified Directive would not be discussed at that meeting since there were on-going discussions with the revisors of the Legal Service. The codification would undergo considerable drafting changes, but the substance would remain unchanged. It was planned that a new version of the codified directive would be ready in the autumn.
>
> The following meeting would be held on 22–23 June, back-to-back with the Solvency II Public Hearing taking place on 21 June.
>
> The CEIOPS representative reported on the work on preparing the final answer to the 3rd wave of calls for advice. The Copenhagen meeting on 6 December had

approved the consultation paper. Following a shortened consultation period, final approval was envisaged at the Extraordinary Members Meeting on 25 and 26 April, after which the answers would be forwarded to the Commission.

In 2006, CEIOPS would publish further consultation papers and refine its existing work. This had always been planned and would also be in line with the Commission request set out in its January letter. The CEIOPS representative outlined the further papers to be produced on Pillar 1, Pillar 2 and group issues. The plan was to issue papers for consultation in June 2006, with final advice being scheduled for October. The time-frame would be different for the advice that depended on the results of QIS2. In this case, consultation papers would be issued in October after the results of QIS2 were available.

CEIOPS believed it was making good progress, but noted that there were high-level questions (e.g. on the prudential margin) where political guidance was still needed. The revised Framework, for discussion at this meeting of the Committee, was a useful step in this direction.

The next discussion was regarding an FfC and proposed changes.

The Commission gave a brief introduction to document Markt/2511/06 and its annex (Framework for Consultation) and explained that the proposed changes to the Framework for Consultation related mainly to paragraph 16 on technical provisions. In particular the principle of market valuation of technical provisions and the specification of the risk margin and its calculation methods should provide the additional guidance requested by CEIOPS.

A majority of members that intervened could accept the amendments, although some additional clarification was sought and text suggestions provided. Two members took the view that the amendments were not really needed, but could broadly accept the text. Only one member opposed the amendments, clarifying that prudence in technical provisions was of the utmost importance and that further guidance on the risk margin could only be provided after QIS 2.

The principle of market valuation was thought to be somewhat artificial by one member, as a real market for technical provisions does not exist. It pointed out that there was no consensus within the industry and that mutuals would also have difficulty with this concept. It requested an alignment on IAIS Cornerstone V in this respect. Two other members, however, requested the text to be reinforced on this point and to clarify that market valuation implied that hedgeable risks were valued at the market value of the respective hedging instrument.

Three members pointed out that the Best Estimate is the key issue. It should be calculated in a reliable manner. Further guidance by the actuarial profession should be sought. One of these members, supported by another member, agreed that realistic assumptions implied that surrender value floors should not be applied to the calculation of technical provisions. However, it was pointed out that overall, sufficient financial resources should be available to cover the surrender value obligation.

Concerning the risk margin, members that expressed themselves on this point could agree that the Cost of Capital method would be tested, although three members believed this method not to be adequate or prudent. Two members suggested including a third method based on scenarios. A large number of members that intervened took the view that eventually one method should be applied, as a preferred option. This would be particularly important for insurance groups, doing life and non-life business. However, it was noted that a simplified method could be useful for smaller undertakings. One member took the view that more than one method could be acceptable if they were economically equivalent. Another member stressed that no Member State option should be given in case several methods would be accepted.

One member, although agreeing with the amendments, pointed out that, to the extent technical provisions would be higher than a Best Estimate plus Risk Margin based on market consistent valuation, the surplus should be allowed to cover the SCR. This member would send a note to the Commission explaining this point in more detail.

The Chairman concluded that there was a lot of agreement on the proposed amendments to the Framework for Consultation. Although agreeing with the comments made on hedgeable risks, the Commission believed that this terminology was not yet fully defined. The Commission concurred with those members that stressed the importance of the calculation of the Best Estimate. The Groupe Consultatif had offered its help on this issue and the Commission would further communicate with the Groupe on this matter. A further alignment of the Framework with IAIS papers was welcomed. The Commission would slightly redraft the Framework for Consultation and send the redraft to the members. However, it was noted that this would not lead to any delay in CEIOPS' work, as they could now go ahead with the current text. It was also agreed that the comments provided by members would be circulated to other member.

The IA and the second QIS were also discussed.

At the last meeting of EIOPC, the Commission Services had presented a paper providing an update on the impact assessment of Solvency II (MARKT 2525/05). It had identified a number key pieces of additional work that needed to be carried out as part of the impact assessment including: quantitative impact studies; impact on the macro-economy and financial stability; impact on insurance products and markets; impact on supervisory authorities, consumers and SMEs; and market failure analysis. The paper had also recognised the need for further consultation work. In particular, the Commission Services proposed organising a public hearing and conducting interviews with a range of stakeholders as well as setting up an inter-service steering group. The Commission had also promised to keep the Committee up to date with developments on the impact assessment work.

Regarding quantitative impact studies, CEIOPS had now completed QIS1 and had published summary results. QIS2 was currently being finalised and would test both the percentile approach and the cost-of-capital approach. The Commission had given a presentation to CEIOPS FSC setting out its requirements regarding QIS2 at the joint Pillar 1/FSC meeting on 14 February in Paris and had also given the same presentation to the Solvency Working Group on 15 February. CEIOPS planned to submit the QIS2 report to the Commission in the autumn.

Work on the assessment of the macro-economic effects of the introduction of Solvency II including the impact on financial stability was involving the ECB, DG ECFIN, the Joint Research Centre. Three tri-lateral meetings had been held so far.

As regards the impact on insurance products and markets, CEA/AISAM/ACME would produce a typography of European Insurance Markets and would also be conducting an exception based analysis of the impact on insurance products and markets.

To gauge the impact on supervisory authorities, consumers and SMEs, CEIOPS had created a task force COMPASS to look at the impact on supervisory authorities. The Commission Services had made a presentation to FIN-USE on 22 March and was currently working on a document looking at key issues from a consumer perspective.

Concerning market failure analysis, the first elements of the impact assessment report would be presented to the Solvency Working Group Experts meeting on 4–5 May. This would cover procedural issues, the problem definition for the market failure analysis, the objectives and the policy options.

Further Consultations would involve a public hearing to be held on 21 June 2006. This would focus on the impact on insurance products and markets; consumers and SMEs and international developments. The GNAIE was performing a comparative study of Solvency II and the US system and both the Commissioner and the Director General were due to speak at the meeting. Discussions with various stakeholders had already been conducted including: insurance undertakings, think tanks, strategic consultants, rating agencies and non-EU supervisory authorities. A questionnaire was being prepared.

An Inter-Service Steering Group (ISG) had been set up with participation from a wide number of other DGs including ECFIN, EMPL, COMP, JLS, ENTR, SANCO and SEC GEN. The Inter-Service Steering Group met for the first time on 8 December 2005. It was also intended to set up a focus group within DG MARKT.

The Next milestones were the Solvency Working Group on 4–5 May and the 21 June Public Hearing in Brussels. A further update would be provided at the EIOPC meeting on 6 July. Two Member States raised the issue of calibration of QIS2. In particular, concern was raised regarding the 40% capital charge for equity exposures. The Commission said it would raise the issue at the next meeting of CEIOPS Financial Stability Committee due to take place on 11th April 2006.

D.4.3 EIOPC Meeting July 2006

In the fourth EIOPC meeting in July 2006, MARKT (2006d), the Commission gave an oral presentation on the progress of the project. Paragraph 3.1 in the minutes gave a summary of the presentation.

The Commission first reported on the continuing good progress of the project in general, in line with the framework for consultation and the roadmap. The Solvency Working Group had held a number of meetings. The following points were noteworthy. The Commission observed that discussions were sometimes made more difficult by the fact that it was not allowed to formally present draft legislative texts as such for discussion in the Working Group. It was clear that there was a need to discuss the use of VAR or Tail-VAR and CEIOPS had been asked to produce a note. The Commission was also continuing its thinking on the practical implementation of the project and questions of timing of entry into force.

The Commission also referred to the inclusion of Solvency II under the other business agenda item at the informal Financial Service Committee to be held on the following day in Barcelona. The Commission was firmly of the view that the technical discussions should be kept within the EIOPC. Although political discussions elsewhere were not excluded it was important to avoid any confusion.

The Commission reported on the successful public hearing which had brought together some 250 people and had heard unanimous support for the need for a new solvency regime. The four panel discussions and general debate had raised a number of issues. More attention might have to be paid to specialist or monoline insurers in the context of the standard formula since they could not invoke any diversification effects. The linkage with IAS was also noted and the need not to produce a disincentive for investment in equity capital. While the importance of Pillar 1 was clear it had to be remembered that failure was often the result of management failings, which pointed to the equal importance of Pillar 2. The importance of the consumer angle was noted, together with improved communication with policyholders. As the Commissioner had pointed out, Solvency II should set an example for regulatory convergence. Lastly, the issue was raised of whether the risk of herding behaviour would increase due to the higher level of harmonisation of Pillar 2 powers under Solvency II.

The Commission would publish the proceedings of the public hearing on its website.

The Commission updated the Committee on impact assessment work. The public hearing was part of this, pointing to areas where more work might be needed. A draft elements paper had been prepared for the Solvency Working Group and had prompted many comments from Member States. This would probably lead to a more extensive paper later in the year.

There were five key strands to the impact assessment work. CEIOPS was mainly responsible for the work in relation to insurance undertakings and the supervisory authorities. On insurance products and markets CEA and AISAM were helping and were working on an extensive draft typology of the EU insurance industry. Many insurers had shown interest in responding to their questionnaire. On the macroeconomic and financial stability aspects, probably the most difficult parts of the impact assessment (with little academic literature), help was being provided by the Commission's Joint Research Centre and the European Central Bank. With regard to consumer issues the help of FIN-USE was being used.

The Commission also referred to its intention to carry out interviews of insurance undertakings and called on Member States to provide by end July the names of up to three companies as potential candidates for interview in Brussels. The Commission would try to interview at least a representative sample.

The Commission further reported on work on the Framework for Consultation. There had been a discussion on the Framework at the April meeting of the Committee. A number of Member States had submitted comments and most of these had been taken on board. The new amended Framework was to be found in document Markt/2515/06, which had been published on DG Markt's website and sent to CEIOPS and members of the EIOPC. The changes related predominantly to Paragraph 16 on technical provisions. The main changes could be summarised as follows: there should be a market consistent valuation of technical provisions, which implied no surrender value floors; technical provisions were the sum of a Best Estimate (BE) and a Risk Margin (RM); the BE was the expected present value of future cash flows using a risk free yield curve to discount; the RM could be based either on a 75% percentile approach or using the Cost of Capital approach; however the preferable outcome would be a single calculation method for the RM.

Progress on the recast of the insurance directives was also reported. This work was being conducted with the revisers of the Commission's Legal Service. Many changes had been made to the structure of the text (shorter articles, more modern language) but there were no changes of substance. The document now had to be put into the Legiswrite format (marking up changes) and it remained the Commission's intention to present to the Committee a new consolidated document in the autumn.

The Chairman then gave the floor to CEIOPS for an update of its work. The CEIOPS representative referred to the annual report that had been distributed and to its sixth progress report to the Commission.

At its 29 June meeting CEIOPS had issued its first set of post 3rd wave consultation papers, thus coming back as planned and as requested by the Commission to some issues already covered in the three waves. These papers were mainly concerned with Pillar 2 aspects, namely internal risk assessment requirements, supervisory evaluation procedures and supervisory tools and powers. Another paper focussed on group issues. These papers concentrated on principles for inclusion and were much shorter. CEIOPS had had to shorten the consultation period (12 September deadline) but was organising a public hearing in Frankfurt on 7 September. The final advice was expected in October. In October it was also planned to issue a second set of post 3rd wave consultation papers, mainly relating to key Pillar 1 issues in the light of the results of QIS 2. There would thus be further advice on technical provisions, capital requirements, including risk mitigation tools, on internal models and on eligible capital elements.

The October advice would include advice on the SCR standard formula but, consistent with the scope of QIS 2, would most likely deal with the structure of the standard formula rather than calibration.

The June CEIOPS meeting had also discussed the organisation of future work. It was planned to hold a 2-day members meeting on 12 and 13 March 2007 to approve the advice on the consultation papers issued in October 2006 but also to finalise the specifications for a further round of QIS studies. There would at least be a 3rd QIS running from April to June 2007, which would focus more on calibration. CEIOPS should then issue further advice on Pillar 1, including the standard formula. After a further consultation paper the final advice should emerge in spring 2008.

The CEIOPS work assisting the Commission with the impact assessment should be finalised after the summer break on the basis of the questionnaire circulated to members.

The Chairman of the CEIOPS COMPASS Task Force on convergence and impact assessment reported on this important work to gauge the impact of Solvency II on supervisors, from the angles of cost, staffing and competence requirements. The aims were twofold, namely in the short term to provide the Commission with input on the major changes affecting the supervisory authorities. A questionnaire had been sent to 30 countries and 27 had answered. There were a number of common concerns relating, for example, to the organisation of supervision of internal models and to the competences to be required of supervisors. The medium term work of the Task Force would identify educational needs to ensure a smooth and convergent implementation. Work on training needs would begin in September and should lead to a report end 2006 or early 2007.

Several delegations praised the Commission for the quality of its work and the very successful public hearing.

Two delegations expressed some reticence concerning the FSC discussing Solvency II, while the Member State that had drafted the FSC document reminded the Committee that the FSC had itself asked for a discussion and stressed that there was no intention to have a parallel debate.

Other points made by delegations on Solvency II concerned the need to cover the important question of equity investment in the impact assessment, the need for the EIOPC to tackle matters on which agreement proved impossible in CEIOPS and the possible need for additional meetings. One delegation requested the Commission to state its reasons when it departed from CEIOPS advice. Another underlined the importance for CEIOPS to study how to supervise internal models in a co-ordinated way. One delegation called for a review of the organisation of CEIOPS groups now that the inventory of problems was becoming clear.

In response to points raised the CEIOPS representative recalled that CEIOPS could, if necessary, give advice on a qualified majority basis.

The Commission representative agreed that additional meetings could be held if necessary and if they could be properly prepared, but noted that excessive staff turnover was a complicating factor.

Other issues that were discussed at the meeting were the relation between the Insurance Accounts Directive, IAD, and Solvency II (paragraph 3.2), and deeply subordinated debt (DSD) (paragraph 4).

Relationship between IAD and Solvency II

The Commission introduced the important but complex question of the relationship between the Insurance Accounts Directive (IAD) and the Solvency II project. A number of the Directives that were to be incorporated in the new Solvency II text made reference to the Accounts Directive. The question the Commission was raising related to whether the Accounts Directive should undergo an in-depth revision or whether the necessary prudential rules on the calculation of technical provisions and valuation of admissible assets should be introduced in the Solvency II text. In the latter case the IAD would remain untouched at least until the completion of the IASB's ongoing work. This debate also raised the question of the calculation of different sets of figures for solvency and for accounting purposes.

The Commission was advocating that the IAD should not be amended now and that the necessary rules should be included in Solvency II. The consequence was that two different sets of rules would continue to apply for solvency and accounting purposes. The eventual aim was the convergence of solvency and financial reporting but this depended on the second phase of the IASB work and was a long-term objective. If Member States were reluctant about the maintenance of two reporting structures, the Commission requested their views on a possible option, to be introduced into the IAD, for insurance undertakings to make use in the preparation of their financial statements of the rules on the calculation of technical provisions and valuation of admissible assets to be contained in the Solvency II prudential directive.

All the Member States that took the floor supported the Commission recommendation not to amend the IAD at this stage but to include the necessary rules as part of Solvency II.

The Commission representative noted that the message from the Committee was very clear. At the same time he saw a slight contradiction between the view expressed by several delegations of the need not to overburden the industry and the apparent acceptance of the burden involved in having two different reporting frameworks. The Commission's suggested possible option would involve a short-term fix. The problem was on the liabilities side, given that the assets side already allowed for considerable flexibility. The Commission would continue to think about this option, though not as a matter of urgency, and also asked CEIOPS to think further on this issue.

Deeply subordinated debt

The Commission reminded the Committee that one Member State had presented a paper on deeply subordinated debt (DSD) at the first meeting of the EIOPC on 29 June 2005. It had been argued that this type of instrument could be as safe as own funds and it was suggested that the Life and Non-Life Directives could be amended via comitology to formally admit this type of instrument as an element, up to a certain percentage, of the solvency margin.

Most of the Member States that spoke at that meeting had taken an open position and it had been decided to seek advice from CEIOPS. That advice had been provided in May. CEIOPS considered that the proposal was technically feasible but did not offer any opinion as to whether the use of comitology would be the right way to make such a change.

The Commission noted that major insurance companies and the mutuals seemed to be in favour of an extension of the list of eligible elements to include deeply subordinated debt more fully than is the case in the current Directives. The question centered on whether we should do it now through the comitology procedure or later through the forthcoming Solvency II Directive.

The Commission requested a clear view from Member States orally at the meeting and to be confirmed in writing on the technical aspects of the proposal and on whether they favoured the use of the comitology procedure to make such an adjustment. The Commission services would then be in a position to make a recommendation to the Commissioner on the best way forward.

The CEIOPS representative noted that CEIOPS preferred to give technical advice rather than to express a real preference in this matter. He underlined, however, the need to avoid inconsistencies with the banking sector rules on own funds.

A tour de table revealed a number of Member States in favour of using the comitology route now, while a larger number preferred to address this issue as part of Solvency II.

However, the overall picture remained unclear as many Member States continued to suggest that their position remained open. Several delegations stressed the need for a holistic approach covering all kinds of hybrid capital instruments and understood that CEBS and CEIOPS were working on this. There was also considerable concern about retaining consistency with the banking sector in order to avoid regulatory arbitrage.

Summing up, the Commission stated that it believed, subject to the view of the Legal Service, that comitology could be used to make this adaptation and was prepared to undertake the work if there was a clear qualified majority among the Member States. This was still not clear even after the tour de table and the Commission called on Member States to give a clear yes or no answer to the use of comitology for the inclusion of DSD by the end of July.

D.4.4 EIOPC Meeting November 2006

In its 5th meeting in November 2007 EIOPC got a general update of the work done by the Commission and by CEIOPS; MARKT (2007a).

The Commission reported on its efforts since the previous EIOPC meeting. A number of meetings had been held with Member States, inter alia on group issues and the recast of the insurance directives. Work had also progressed on the recitals of the recast directive and a major exercise would have to be carried out to cross check all references. A document on minimum capital requirements had been discussed with Member States. Work also continued on the impact assessment and the Commission thanked the CEA for its input on the impact on products and markets. The ECB was providing input on the macroeconomic aspects and DG ECFIN and CEIOPS were

> also contributing. The Commission had started a series of interviews with individual insurance undertakings, which were proving very interesting, and had participated in a number of conferences in various Member States. Further meetings with Member States were planned for 20 and 21 December, 29 and 30 January 2007, 22 and 23 February and 26 and 27 March. The Commission would have to finalise the drafting work on the framework directive at the end of March, which was already very late in terms of the anticipated adoption of the proposal by the Commission in mid-July and its immediate transmission to the Council and Parliament.
>
> Two Member States praised the Commission for its work and good contacts with the industry. The question was raised about stepping up the frequency of meetings and mention was made of a possible extra EIOPC meeting devoted to Solvency II in the first quarter of 2007. The Commission promised it would do its best but noted that it was difficult to increase the pace of meetings given the necessary preparatory work.
>
> CEIOPS reported on its ongoing work following its responses to the three waves of advice. It would provide a first set of additional advice in early December. At its last members meeting (end October) it had discussed the results of QIS2 (to be dealt with in a separate agenda item) and would issue a public report in early December. In the light of the QIS2 results it had issued further draft advice covering all three pillars. These included a major paper on key Pillar 1 issues, including technical provisions and capital requirements. This advice was open for comment until 12 January and a public hearing was scheduled for 10 January. Some issues were resolved by consensus; others had been put to a vote. On other issues it had not been possible to find a common CEIOPS position. In such cases the options were explained, setting out the pros and cons and the support they enjoyed. CEIOPS referred in particular to the common position on the calculation of technical provisions, which represented a milestone and a great achievement. This was a qualified majority position and care had been taken to explain the position of the minority. In March 2007 CEIOPS expected to deliver final advice on the above issues. There would probably be additional consultation papers plus the preparation of QIS3 (to run from April to June 2007). CEIOPS expected to come back on MCR/SCR calibration and planned a consultation paper in October 2007 with final advice in spring 2008. A first report on the impact on supervisory authorities was in written consultation and the final report would be sent to the Commission in December.
>
> The Commission thanked CEIOPS again for its efforts.

From the beginning onward the Commission had been striving to have TailVaR as a risk measure. The industry had been lobbying for VaR. The choice of risk measure was discussed at the meeting.

> The Commission presented the cover note it had prepared, as well as the two background documents annexed to this note (CEIOPS letter and CEA working document). Three amendments to the current Framework for Consultation were under consideration by the Commission with respect to: firstly, the definition of ruin, to

make this definition more comprehensive; secondly, the risk measure (VaR vs. TailVaR); finally, the possibility to allow for the use of alternative risk measures where an insurance or reinsurance company developed an internal model (flexibility principle).

After this introduction, the Chair invited the Members of the Committee to express their views on these three issues.

A vast majority of Member States expressed their preference for the new definition of ruin suggested by CEIOPS, because it better reflected the financial standing of an insurance or reinsurance company (in particular by taking account of "other liabilities," which sometimes took precedence over insurance claims in the case of winding-up). Three Member States suggested that the definition of ruin should however refer to "eligible assets," rather than "assets."

Two Member States opposed the proposed new definition of ruin, as the new solvency regime should strictly focus on policyholders' protection and should not aim at protecting other creditors.

Regarding the choice of a risk measure, the opinions of Member States appeared to be split: a number of Member States underlined the theoretical merits of TailVaR, while an equal number of them explained they preferred to use VaR, since it seemed to be more suitable in practice (in particular, for SMEs). Some Members said they were indifferent or had not made any decision yet. Almost all Members of the Committee pointed out that such a risk measure should appropriately capture large exposures (e.g., natural catastrophes) while being computable by all companies, which obviously made the choice between VaR and TailVaR very difficult.

Most Member States supported the introduction of a flexibility principle for companies using an internal model. One Member State opposed the proposed amendment to the extent it hampered the harmonisation of standards and practices throughout the Community.

The Chair concluded that the Commission would soon circulate a draft Amended Framework for Consultation, in order to change the definition of ruin (and adopt the CEIOPS proposal) and include the flexibility principle. Regarding the risk measure, it needed to further consider this issue, taking full account of practical concerns.

Also the results of QIS2 were presented by CEIOPS.

A CEIOPS representative provided an advance presentation of the QIS2 report. He noted that QIS1 had been focussed on technical provisions, while QIS2 had a much broader scope but did not involve a final calibration. The emphasis on methodology and the multiple options had led to complexity and many complaints. For provisions, for example, they had tested both the percentile and the cost of capital approaches. 514 companies from 23 countries took part (at least in part). The main results were that QIS2 did not suggest big problems with capitalisation. Most undertakings met the virtual requirements though the impact on small companies was bigger than on the larger ones. For most companies, MCR was less than 75% of SCR, but because of profit-sharing (in SCR and not in MCR) for some companies the MCR exceeded the SCR. In extreme cases MCR actually exceeded SCR. Industry showed a slight preference for the cost of capital approach but for those that did both calculations the difference was very small as compared with the 75th percentile. For most companies the solvency ration remained above 100%. QIS2 also provided information on which risks were most important: for life firms it was market risk, for non-life underwriting risk and for composites a mix of the two.

Other comments received called for no charge for concentration risk in the standard formula, suggested that group diversification was not sufficiently addressed and were against the Basel approach to credit risk. The equity risk charge was felt to be prohibitive in some cases. The module for underwriting risk in life assurance did not take account of volatility suggesting in this case that the calibration was too low. There was criticism of the operational risk part of the standard formula.

The conclusion was that it was necessary to rethink certain risk modules and that the calibration for all modules would have to be looked at again. For QIS3 it would be possible to put in parameters that could be defended empirically. A lesson for QIS3 was that there was a danger of losing support if the size of QIS3 was similar to number 2. Therefore decisions had to be taken. A single QIS3 was expected without options. An even greater participation of small firms was needed.

The Commission thanked CEIOPS for this presentation and all their hard work and invited Member States' comments.

One Member State regretted that this was the first time that the Level 2 Committee was looking at the formula that was at the heart of Solvency II. Several Member States referred to the equity holdings where the charge was the same whether insurers' investment horizon was one month or 30 years. There was concern too at the effect on smaller companies and it was necessary to keep markets open and competitive. For some the k factor was a source of problems.

One Member State referred to the MCR/SCR relationship, which was wrong for many firms. It was necessary to bear in mind what the MCR was for, i.e., as the point at which firms should go into run-off.

On the equity risk charge CEIOPS accepted that 40% was on the high side but felt that any figure lower than 35% could not be justified empirically. 99.5% over one year gave 35–40%. It was recognised that equity was unduly punished because of the correlation structure but it was harder to say that the 40% scenario was way out of line.

Concern was raised concerning the paper presented by two Member States on group issues.

The Chairman of CEIOPS raised the question of amending the Framework for Consultation (FfC). Two previous amendments were felt in CEIOPS to be good Lamfalussy practice in that they reflected common progress and an evolving learning curve. However, the new paper on group supervision was very different and was an attempt to shift the learning curve and was not a natural outcome of the internal process. The CEIOPS Chairman did not comment on the merits of the proposal, which had not been discussed in CEIOPS. Rather he raised procedural concerns and feared that such a radical change could prevent CEIOPS from finalising its advice in what was an already tight timetable.

Certain Member States voiced their opposition to the group supervision paper and feared its practical effects. Others felt that there was insufficient time now to consider all of its ramifications. It was noted that the directive could be amended in the future.

Other delegations supported the ideas in the paper and felt that the proposal could not simply be ignored or that procedural considerations should determine the process.

The CEIOPS Chairman noted that if the FfC was not amended, CEIOPS would continue to follow it as it currently stood.

A new comitology arrangement was presented and discussed; *cf.* the Lamfalussy procedure in Section D.1.1.

The Commission reported on the new comitology procedure, namely the regulatory procedure with scrutiny introduced by Council Decision 2006/512/EC amending Council Decision 1999/468/EC. A consolidated version of the Decision had been published in Official Journal C 255/4 of 21.10.2006.

The role of the EIOPC was not affected and its competences were not touched. However, the EIOPC would in due course be called on to play a major role in the comitology area under Solvency II.

The new Decision maintained the old regulatory procedure under which the European Parliament had limited control rights over comitology measures but added the new scrutiny procedure under which the Parliament would now have a substantial role to check whether the Commission had not acted ultra vires and had respected the permitted scope for comitology measures and the principles of subsidiarity and proportionality.

DG Markt already had substantial experience of Lamfalussy comitology, notably in the securities sector, and had tried to facilitate matters by trying to take account of Parliament and Member State concerns already in the comitology procedure before any vote.

In return for the greater powers under the new procedure Parliament had agreed to remove the sunset clauses whereby implementing powers in certain directives lapsed after a specified time.

A number of priority legislative acts were being adapted to introduce the new procedure (four in the insurance field).

D.4.5 EIOPC Meetings in February and July 2007

The 6th CEIOPS meeting was held on February 21, 2007. No minutes of the meeting have been published. It was a special meeting on Solvency II and at the meeting an update on progress and timetable going forward was discussed, as well as a presentation of the main policy choices in the proposal. Also risk mitigation, with a focus on securitization, and the treatment of the third-country reinsurer and composites were discussed.

Its 7th meeting was held in Brussels on July 4; MARKT (2007b). The meeting was largely devoted to the Solvency II IA work.

D.4.6 EIOPC 8th–12th Meetings November 2007–April 2009

The 8th meeting of EIOPC was held in November 2007; MARKT (2008a). The Commission made a presentation on the proposed outline for a Level 2 IA works for Solvency II implementing measures. The Member States expressed a measure of support for the general approach that was adopted. This Level 2 IA was conducted during 2009.

The Commission also gave an oral report on the preparations for QIS4 that was conducted in spring 2008; see Section D.4.2.

> The Commission representative first referred to the recent publication by CEIOPS of the QIS3 summary report on its website and thanked CEIOPS warmly for its good work in the QIS3 exercise. The level of participation in QIS3 had been remarkable, since more than 20% of EU companies had taken part in the exercise.
>
> From the Commission's point of view, there were already three important lessons to be drawn from QIS3, so as to make QIS4 even more successful:
>
> Firstly, the data submitted on groups were disappointing. It was necessary to see significant improvement in both the level of submission to CEIOPS' centralized database and the quality and completeness of data submitted in QIS4, so that the impact of the Level 2 measures could be properly assessed and tested.
>
> Secondly, further work was needed on simplifications for SMEs for technical provisions and the calculation of the SCR in order to give SMEs a clear indication of what the proportionality principle meant in practice.
>
> Thirdly, QIS3 also illustrated that further analysis and data needed to be collected under Solvency II regarding the use of partial and full internal models to calculate the SCR.
>
> From a practical point of view, the QIS3 summary report contained a number of very helpful recommendations (e.g., more guidance, more user-friendly spreadsheets, and increased contact with the industry). The Commission would work closely with CEIOPS and stakeholders to try to put them into practice for QIS4.
>
> Regarding the way forward on QIS4, the Commission, in its 19 July letter to CEIOPS on further work on Solvency II, had set out a revised Solvency II Roadmap; one of its sections was dedicated to QIS4. Basically, there were three phases: development of the draft QIS4 specifications, consultation on the draft QIS4 specifications, and running of the QIS4 exercise.
>
> The Commission had asked CEIOPS to provide it with the QIS4 draft technical specifications by 20 December 2007; these should include simplifications for technical provisions and the calculation of the SCR as well as the use of entity specific parameters.
>
> In contrast to the previous QIS exercises, the Commission was politically responsible for the consultation. It planned to publish the draft QIS4 specifications produced by CEIOPS on its website on 21 December 2007. The consultation would run from

> 21 December 2007 to 15 February 2008. People would have to connect to the website to comment on the draft QIS4 specifications. Work was in hand with DG MARKT's IT team to develop the website and structure the on-line consultation (by main issues, e.g., SCR, MCR, etc.). It was also intended to meet with the industry (i.e., CEA, AISAM – ACME) to ensure that the consultation and the QIS4 exercise ran as smoothly as possible.
>
> Meanwhile, the Commission intended to organise three events to consult on the draft QIS4 specifications:
>
> - There would be a Public Hearing on 28 January 2008 to discuss QIS4 technical specifications, where the Commission proposed to focus on 3 key topics: the proportionality principle and SMEs, groups and internal models.
> - It was also proposed to have an extraordinary EIOPC meeting on 1 February 2008 to discuss QIS4 technical specifications with Member States.
> - It was suggested to hold a meeting with Stakeholders on 11 February 2008 in order to examine detailed technical aspects.
>
> Once the consultation was closed, the Commission would collate comments and classify them into political and technical issues. It would pass on the technical issues to CEIOPS, but would process the political comments from all stakeholders, including Ministries of Finance and Members of Parliament. With the help of CEIOPS, the draft QIS4 specifications would be updated accordingly, reflecting the consultation process.
>
> It was planned to finalise the technical specifications towards the end of March, and issue a Call for Advice asking CEIOPS to run QIS4 between April and July 2008, and to publish the results in November 2008. In practice, there would thus not be any significant change as compared to QIS3 with respect to that phase: CEIOPS would develop the QIS4 spread-sheets over the first half of April (so-called "pre-test"), and afterwards supervisory authorities would assist the participants if they had questions on the QIS4 specifications (Q&A process); companies would have to return their results to supervisory authorities, as usual.
>
> As could be seen, CEIOPS was very much involved in QIS4 and its invaluable help was greatly appreciated.

An extra 9th meeting was held on February 1, 2008. At the meeting the TSs of QIS4 were discussed. No minutes from the meeting have been published.

At the 10th meeting held on June 27, 2008, updates on QIS4 and the IA were given; MARKT (2008b). There was also a discussion on IORP and solvency.

A draft document on the harmonization of solvency rules applicable to IORP had been circulated and was discussed. It was important to distinguish between different types of IORPs.

> Accordingly, the consultation first of all concentrated on those IORPs that were covered by Article 17 of the IORP Directive: these IORPs underwrote their own liabilities/provided guarantees and were therefore required to have regulatory own funds,

i.e., additional assets above the technical provisions to serve as a buffer. For these regulatory own funds, Article 17(2) of the IORP Directive referred to the Solvency I regime under Directive 2002/83/EC concerning life assurance ('Life Assurance Directive'). As the latter Directive would cease to exist with the adoption of the Solvency II Directive, the question arose whether and to what extent this regime should be replaced by solvency rules similar or equivalent to the Solvency II rules. In addition, the consultative document also dealt in its Chapter 4 with those IORPs that engaged in cross-border business. The IORPs envisaged could be IORPs covered by Article 17 of IORP Directive as well as other IORPs. The extension of the scope of the draft consultation document was justified by the fact that the European Commission wanted to see to what extent the differences in the solvency regimes for IORPs that operated on a cross-border basis were creating internal market problems.

A number of Member States fully supported the draft Consultative document presented by the Commission Services and most of the other Member States did not raise any concerns. Some Member States emphasised particularly that the consultation was largely driven by the Solvency II question and that it should not lead to a wider debate.

Two Member States suggested that the consultation should be extended to cover all IORPs, including the IORPs that operated only on a domestic basis. It was underlined that, while financing arrangements may be different, insurance companies could offer comparable pension products with the same guarantees to the consumer. If IORPs that were not subject to Article 17 of the IORP Directive were excluded from the consultation this might potentially give rise to an unlevel playing field issue. Moreover, it was also suggested that the rules for cross-border and domestic IORPs should be similar within the single market.

It was clarified, however, that the consultation would address cases where such IORPs competed with insurance undertakings on a cross-border basis. Further, it was also noted that many IORPs are not always in competition with insurance companies (although it is not true for many IORPs) where they exist solely to provide a mechanism for an employer to defer remuneration into the post retirement period for its own employees only, not to provide pensions services generally in the market. Some Member States then expressed the view that if the consultation were to cover all IORPs, including those that operated only domestically, the consultation should also cover domestic arrangements falling outside the IORP Directive, in particular pay-as-you-go and book reserve systems.

While a few Member States expressed concern that the focus of the European Commission proposal would limit the consultation to IORPs in only a very few Member States, some delegates suggested that the cross-border perspective potentially covered all IORPs. In the meeting it was also mentioned that potentially every IORP operated on a cross-border basis.

Some delegates also pointed out that the objective of the consultation should be made clear from the very start. If the consultation gave rise to an expectation that a Solvency II type regime would be introduced for IORPs, there was a risk that the industry and employers would expect a significant rise in the cost of occupational pension provision.

Other representatives suggested that if Solvency I applied to IORPs then it would be logical that Solvency II should also apply to IORPs. The Chairman, however, reminded the meeting that Solvency II was much more sophisticated and that the innovations (e.g., 99.5% VaR over one year) would not necessarily be appropriate or proportionate for IORPs.

A number of Member States also asked for a clarification on how the link between the IORP Directive and the current Solvency II Directive Proposal would be broken. The Commission Services confirmed that the aim was clearly to maintain the status quo, i.e., keep Solvency I applicable for IORPs covered by Article 17 of the Directive. However, the amendment proposed by Mr Skinner, MEP in his report, i.e., keep the reference in the IORP Directive to the Solvency I provisions in the Consolidated Life Directive was not possible from a legal perspective and an alternative approach, i.e., an amendment of the IORP Directive to include the wording of the Solvency I provisions in the Consolidated Life Directive was currently being discussed with the Secretariat of the European Parliament.

The Chairman concluded the discussion by asking Member States for written suggestions on the draft consultative document by 4 July 2008, as it was intended that the consultation should be launched as soon as possible.

An update on the QIS4 was provided by the Commission Services.

QIS4 preliminary results would be presented by CEIOPS in Parliament's ECON Committee in mid-September 2008, and final results would be published on 19 November 2008.

CEIOPS provided an update on work done in QIS4, the level of participation in the exercise, and future work.

The Commission Services also gave an update on the IA work for Solvency II implementing measures.

A seminar would be held at the Commission's premises at the end of July 2008 between CEIOPS and Commission representatives involved in the Solvency II level 2 Impact Assessment. A call for tender would be launched in autumn 2008 for a study on specific areas for the C annexes of the Impact Assessment report. CEIOPS made some general comments about steps already taken and its future contribution to the level 2 Impact Assessment work.

One Member State made some comments and suggestions on the CEIOPS paper on policy issues and options. The Commission Services invited written comments on this paper by July 14th.

A follow up on the Impact Assessment work for level 2 measures would be provided at the November 2008 EIOPC meeting.

The minutes from the 10th meeting was changed on the request by one Member State regarding the IORP item at the 11th meeting held on November 26, 2008.

At this meeting, the financial crisis and turmoil was discussed. It was maintained that the market turbulences have had limited impacts on the European insurance and pensions markets so far. The exposures to AIG and Lehman Brothers had been relatively minimal. However, as important investors, insurers, and pensions funds participated in general market developments and were therefore impacted by indirect effects. It was pointed out that overreactions by the industry should be avoided. CEIOPS believed that robust RM and internal control mechanisms are necessary tools.

In March 2009, CEIOPS published a document regarding the lessons learned from the crises; CEIOPS (2009a).

The Commission informed about the state-of-play regarding the occupational pension issue. At the beginning of September the document for consultation had been launched. A public hearing was held in Brussels on May 27, 2009.

The results of QIS4 were presented, and the ground for Level 2 implementing measures was discussed. The Commission Services acknowledged that it was difficult at this stage to outline with certainty a timetable for the Level 2 implementing measures, but they considered it helpful to outline what the timing might be. It is based on the Commission Proposal and therefore assumes an implementation deadline of October 31, 2012. The Commission Services aim to adopt all implementing measures by April 2011 in order to give all stakeholders involved enough time to prepare for the new regime.

The Commission Services also presented a draft CfA to CEIOPS on its contribution to the IA work on Level 2 implementing measures for Solvency II. It included a list of policy issues and options that should be analyzed as part of the Level 2 IA and is annexed to the draft CfA.

At the 12th meeting of EIOPC, April 3, 2009, an amendment to the minutes of the 11th meeting was requested by the French delegation regarding hybrid capital.

Also, the financial crisis was discussed. There was an intensive discussion on issues raised in a Note sent out to EIOPC members a month before the meeting. A Commission representative presented a summary of the most recent developments as far as the response to the financial crisis was concerned. Questions were raised regarding the rational for the separation between micro- and macrolevel supervisory bodies. Some member states also identified the differences between insurance and banking!

On these questions the chairman emphasized the current international focus on macroprudential supervisory issues in, for example, the United States, the G-20, and so on and the usefulness of having specialized bodies to carry out these tasks. The Chairman also pointed out that insurers had taken an intellectual leadership from bankers in finding solutions like Solvency II, which is more sophisticated than the CRD (based on Basel II). Therefore the insurance industry will have to be listened to, whatever precise structures are decided on for the supervisory architecture. The Chairman also indicated that early redemptions of life assurance policies might be caused by their miss-selling, as they create unreasonable expectations.

The outcome of the de Larosière Group was described as a realistic one. However, there was a lack in reference to pensions. With the creation of the new European Systemic Risk Council, ESRC, and the European System of Financial Supervisors (ESFS) the link between micro- and macrolevel is very important. The CEIOPS General Secretary also mentioned that a new insurance authority, the successor of CEIOPS, would need far more than 15 staff which CEIOPS has.

The CEIOPS (2009a) document regarding the lessons learned from the crises was presented and discussed. The CEIOPS Secretary General also mentioned the potential need to modify correlations at Level 2 taking into account the lessons learned from the crisis. Procyclicality will be an important issue.

The Chairperson also mentioned that the issue as to what extent the same valuation for Solvency II purposes can be used for financial reporting purposes would be considered.

The Chairman reported on the Level 1 discussions, and negotiations in the EP and in the Council. Solvency II also discussed regarding, for example, the future timetable. The Czech Presidency reported also on the compromise reached in the Council and in the trialogue. Also, a question was raised whether the Level 2 implementing measures would be adopted in the form of a Directive or regulation. The legal form of the upcoming measures would be discussed in Commission's reconvened "Expert Group on Insurance: Solvency" on May 6. The choice of the legal form of the implementing measures will determine the time needed for their implementation. It was important to note that only amended provisions of the recast part of the Solvency II Directive would need to be newly introduced into national law. The Commission Services intended to prepare a list indicating the changes under the recast exercise.

Two meetings in EIOPC were tentatively scheduled for the rest of 2009 (June 13–17 and November 14–26). The interested reader is referred to EIOPC Web site: http://ec.europa.eu/internal_market/insurance/committee_en.htm.

D.5 DRAFT FD

In its work on the draft FD, the Commission issued three waves of Calls for Advice to CEIOPS, regarding different aspects of the new solvency system; see Section D.3.3.2. The Commission also set out policy guidelines and principles to guide CEIOPS in its task in the "FfC," published on July 14, 2004; MARKT (2004a1). The FfC was amended during the process and a new version was published on July 15, 2005; MARKT (2005b). A third version of the text was published on June 2, 2006; MARKT (2006b).

Following extensive internal discussions, interaction with many of the key stakeholders, and subsequent public consultations, CEIOPS sent its final answers to the three waves for Calls for Advice to the Commission on 30 June 2005, CEIOPS (2005b), 1 November 2005, CEIOPS (2005c) and 3 May 2006, CEIOPS (2006a), respectively.

The Commission hold a public hearing on Solvency II on June 21, 2006. A total of 250 people participated in the hearing. During the whole process the Commission was working on both formulating new solvency articles in line with the FfC and in line with the answers that the Commission got from its Calls for Advice, but also at a later stage consolidating old insurance directive texts.

The EIOPC set up a subcommittee on solvency, called the Solvency Working Group (labeled IWG, insurance working group, by EIOPC). It has been working as a discussion forum for the EC in its development of the proposal for an FD since 2005 and it consisted of experts from the regulatory authorities in the Member States. The IWG has also prepared the meetings that EIOPC has had regarding Solvency II issues and other solvency issues. Different stakeholders such as CEA have also presented their views on different issues. In 2005, the IWG had two meetings and discussed a first draft directive text from the EC (COM) at their meeting in May (the layout of the directive).

In 2006, the IWG conducted eight meetings, both one-day and two-day meetings, with the COM to discuss different issues of the FD and also the consolidation of older insurance directives. In these meetings the following main issues regarding the directive texts were discussed.

February 15–16, 2006: draft codified directive, IA, and solvency issues

March 20, 2006: specific solvency issues

May 4–5, 2006: impact assessment, specific solvency issues

June 22–23, 2006: text on group issues and relating to the first and second CfA

September 18, 2006: strategic issues, group issues, and other important issues

October 16, 2006: draft directive texts

November 8, 2006: texts on MCR and eligible elements of capital

In November 2006, the HM Treasury and the FSA of the United Kingdom published a discussion paper on insurance groups; HM Treasury (2009). This started an intensive debate on group issues.

December 20–21, 2006: texts on TPs, SCR (standard formula and IMs), MCR, ALM and investment rules, eligible elements of capital, and group issues.

In spring 2007, the IWG conducted four meetings. They discussed the following directive issues and texts.

January 29–30, 2007: Pillar I, II, and III draft texts

February 22–23, 2007: revised proposal texts on Pillar I, II, and III issues and group issues

March 26–27, 2007: new revised proposal texts on Pillar I, II, and III issues, group issues, and exclusions from the scope and transitions

April 16, 2007: the full draft proposal for an FD

After that meeting the COM decided to finalize the proposal for the FD on its own and not distribute the final version until July 10 (2007).

The EC adopted the proposal concerning the "taking-up and pursuit of the business of Insurance and Reinsurance," to introduce a new solvency capital and supervision framework for insurers and reinsurers at its meeting on July 10, 2007. It was published on July 19, COM (2007), and an amended version was published on February 26, 2008; COM (2008).

The press release that was published on July 10, 2007 after the adoption of the proposed FD is reproduced below.

"Solvency II": EU to take global lead in insurance regulation

The European Commission has proposed a ground-breaking revision of EU insurance law designed to improve consumer protection, modernise supervision, deepen market integration and increase the international competitiveness of European insurers. Under the new system, known as 'Solvency II,' insurers would be required to take account of all types of risk to which they are exposed and to manage those risks more effectively. In addition, insurance groups would have a dedicated 'group supervisor' that would enable better monitoring of the group as a whole. The Commission aims to have the new system in operation in 2012. This proposal is part of the Commission's Better Regulation strategy and its firm commitment to simplify the regulatory environment and cut red tape. It will mean replacing 14 existing directives with a single directive. The proposal now passes to the European Parliament and Council for consideration.

Internal Market and Services Commissioner Charlie McCreevy said: "This is an ambitious proposal that will completely overhaul the way we ensure the financial soundness of our insurers. We are setting a world-leading standard that requires insurers to focus on managing all the risks they face and enables them to operate much more efficiently. It's good news for consumers, for the insurance industry and for the EU economy as a whole."

What "Solvency II" would introduce

The new system would introduce more sophisticated solvency requirements for insurers, in order to guarantee that they have sufficient capital to withstand adverse events, such as floods, storms or big car accidents. This will help to increase their financial soundness. Currently, EU solvency requirements only cover insurance risks, whereas in future insurers would be required to hold capital also against market risk (e.g., a fall in the value of an insurer's investments), credit risk (e.g., when debt obligations are not met) and operational risk (e.g., malpractice or system failure). All these risk types pose material threats to insurers' solvency but are not covered by the current EU system.

Insurers would also be required to focus on the active identification, measurement and management of risks, and to consider any future developments, such as new business plans or the possibility of catastrophic events, that might affect their financial standing. Under the new system, insurers would need to assess their capital needs in light of all risks by means of the 'Own Risk and Solvency Assessment', while the 'Supervisory Review Process' (SRP) would shift supervisors' focus from compliance monitoring and capital to evaluating insurers' risk profiles and the quality of their risk management and governance systems.

In addition, the new system would enable insurance groups to be supervised more efficiently, through a 'group supervisor' in the home country that would have specific responsibilities to be exercised in close cooperation with the relevant national supervisors. This would entail a more streamlined approach to supervision that would recognise the economic realities of such groups. The introduction of group supervisors would ensure that group-wide risks are not overlooked and would enable groups to operate more efficiently, while providing policyholders with a high level of protection. Groups that are sufficiently diversified may also be allowed to lower their capital requirements under certain conditions.

Background

The aim of EU solvency rules is to ensure that insurance undertakings are financially sound and can withstand adverse events, in order to protect policyholders and the stability of the financial system as a whole. However, the current EU solvency system is over 30 years old and financial markets have developed dramatically in recent years, leading to a large discrepancy between the reality of the insurance business today and its regulation. Also, many Member States have introduced their own additional rules at national level, leading to a range of different regulatory requirements across the EU, which ultimately undermine the Single Market and especially hinder insurance groups. 'Solvency II' would replace this patchwork of different rules, ensuring a level playing field and a uniform level of consumer protection. It is in line with international discussions within the International Association of Insurance Supervisors (IAIS).

More information is available at: http://ec.europa.eu/internal_market/insurance/solvency_en.htm.

In order to help prepare the ground for the development of implementing measures, that is, Level 2 and Level 3 measures, once the Solvency II proposal had been adopted by the Council and the Parliament, the Commission Services sent CEIOPS a CfA along with detailed TSs on March 31, 2008 asking them to run a fourth QIS) between April and July 2008. After the CfA was sent to CEIOPS at the end of March an extensive Public Consultation based on the draft technical specifications developed by CEIOPS was conducted. During the consultation, a public hearing on Solvency II and QIS4 was held on January 28, 2008 in Brussels. The Commission got over 1600 comments on the draft technical specifications.

From the publication of the proposal for an FD, the "Solvency II directive," there have been two main workstreams during the 1.5 years up to the end of 2008, one within the EP and one within the Council. When both workstreams came to a conclusion a third workstream was started, the trialogue consisting of a negotiation board with people from the EP and the Council. The trialogue was headed by Karel van Hulle at the EC.

During the negotiations, two main political issues were raised. First, we had the proposal of a duration approach for equity risks that aroused as an issue in the discussion within the Council, and second, the proposal remove the group support regime which was part of the proposal for an FD; Articles 234–247. The outcome of the discussions within the EP and the Council was divergent.

D.5.1 Workstream of the EP

An EP *rapporteur* is a primary agenda setter for the parliament's position in the legislative process where the EP has the power of approving and amending a legislative proposal submitted by the Commission (the sole legislative initiator of the EU).

Once the EP has received a Commission's legislative proposal, the EP president assigns a committee named as *the committee responsible* in the relevant policy area to present a report on the proposal to the plenary. The committee responsible appoints a rapporteur to write the report on its behalf and votes on it with some amendments if necessary before presenting to the plenary. Other relevant committees are also allowed to present to the EP plenary their "opinions" on the Commission's proposal.

The first rapporteur for the Solvency II project was Peter Skinner, appointed by the ECON Committee ("economic and monetary affairs") within the EP. Also, the JURI committee ("legal affairs") appointed a rapporteur, Sharon Bowles, for the project.

Both rapporteurs published their draft amendments to the proposed FD during spring 2008; Skinner (2008) and Bowles (2008a). A final version from rapporteur Bowles was published in the beginning of July; Bowles (2008b). The amendments were the outcome of many discussions between members within the ECON and JURI committees, but also on discussions with different stakeholders, for example, CEA, AMICE. Also, different lobby organizations and national insurance associations were lobbying directly with the rapporteurs and the MEPs. In total, there were more than 800 proposed amendments. Some article had more than 10–15 amendments proposed.

Based on the different suggestions, a report from Peter Skinner was adopted by the EP's ECON Committee on October 7, 2008. The final report, with the adopted amendments was published on October 16; EP (2008). This proposal did not include the equity duration issue, the equity dampener, or the removal of the group support regime.

The EP's Socialist Party, PES, organized a meeting on January 28, 2009 to discuss and to facilitate an agreement within the group on the Solvency II main political issues. The most important socialist MEPs were present, for example, the first rapporteur Peter Skinner, the ECON chair Pervenche Berés, the PES coordinator Elisa Ferreira, and Udo Bullman. Also the liberal MEP Ieke van den Burg was present together with representatives from the insurance industry, the CEA.

D.5.2 Workstream of the Council

The work by the Council was set up directly after the release of the proposed FD in July 2007 by the Portuguese Presidency. A new WG, Working Party on Financial Services, WPFS, was set up by the Council. The members of the WPFS were more or less the same as in the old IWG set up by the EIOPC for the work with the Commission. Some of the meetings were only for national experts, some only for the attachés in Brussels, and some with both experts and attachés.

During the Portuguese Presidency, July 1–December 31, 2007: the WPFS met 7–8 times and discussed different articles and proposed amendments.

During the Slovenian Presidency, January 1–June 30, 2008: the WPFS met 10 times and continued the work started under the Portuguese Presidency. They succeeded to make compromises on most of the articles, except on equity and group support issues. Twelve Member States opposed on the group support regime. In its meeting on June 12, the Commission produced a paper on equity risk.

During the French Presidency, July 1–December 31, 2008: the WPFS was now reformed as a Working Party on Insurance, WPI. The WPI had 10 meetings during the French Presidency. The main focus during the initial talks was on group issues and especially on the group support regime. In its meeting on September 16, the Presidency included the equity duration as an issue for the first time. A proposal for a recital to a new Article 105 was proposed: *The Solvency Capital Requirement should reflect a level of eligible own funds that enables insurance and reinsurance undertakings to absorb significant losses and that gives reasonable assurance to policyholders and beneficiaries that payments will be made as they fall due. The calibration of the capital charge shall properly take into account the long holding period of assets that is typical in insurance business, in particular for certain types of assets, such as equity, and shall not discourage undertakings from holding participations in financial and non-financial firms.*

Different solutions to the group support regime were discussed. At its meeting on September 26, a new proposal from the French Presidency was presented. It included a new article, Article 105, regarding the introduction of a dampener for avoiding procyclical effects. A proposed recital to this new article said: *To avoid the exacerbation of procyclical effects, a macro-economic dimension should be introduced in the supervisory rules and practices.*

First, the equity risk submodule should include a symmetric adjustment mechanism (equity dampener) to avoid that insurance and reinsurance undertakings would be unduly forced to raise additional capital or sell equities as a result of an adverse movement in equity markets and that they unduly buy equities as a result of a favorable movement in equity markets.

Second, in the event of exceptional falls in financial markets and where this adjustment mechanism is not sufficient to enable insurance and reinsurance undertakings to comply with their SCR, provisions are also made to allow supervisory authorities to extend the time period which undertakings have to r-establish the level of EOFs covering the SCR.

The Presidency did not succeed in finding a compromise. There was a feeling that the Presidency was nearly giving up. According to the plan the ECOFIN, that is, the committee of the Ministers of Finance should discuss the proposal from the Council at the same day that the EP's ECON vote, that is on October 7.

Before the ECOFIN discusses a new directive or proposal, it is always discussed within the COREPER II, representing the "EU ambassadors." COREPER, from French "Comité des représentants permanents," is the Committee of Permanent Representatives in EU. The COREPER consists of the head (COREPER II) or deputy head (COREPER I) of the mission from the EU Member States in Brussels. COREPER II deals largely with political, financial, and foreign policy issues. With the Permanent Representatives in EU, there are also attachés, that is, people officially assigned to the staff of a diplomatic mission to serve in a particular capacity.

Due to the fact that the COREPER II meeting at the very beginning of October did not come up to any agreement on the Solvency II proposal, the ECOFIN on October 7 decided to send back the act to the COREPER for them to find a solution on the group support issue and also on the equity issue. The French Presidency decided to present a compromise including the equity proposals but excluding the group support regime.

At the COREPER II meeting on November 19, the compromise was adopted as a "general approach." A General Approach,* in the Council of Ministers is an informal agreement within the Council, sometimes by qualified majority, before the Parliament has given its opinion on first reading. Such an agreement speeds up work and facilitates an agreement on first reading. On the other hand, the Commission gives no definitive undertaking to the Council owing to the absence of an opinion from the Parliament. Once the Council has received the Parliament's opinion, the Council prepares a political agreement.

At the ECOFIN meeting on December 2, the amended proposal for an FD was adopted as an A item, that is, without discussion, a General Approach. However, the EC made it clear that they could not support the compromise as the group support regime was carved out and there was no final solution on the equity issue. Three Member States declared that they were making a reservation against the proposal.

The last WPI meeting during 2008 was held on December 9 with only the attachés and no experts. At the meeting, the Member States had the first opportunity to comment on a draft table containing a comparison between the draft directive text, proposed amendments by the EP, and the compromise text adopted by ECOFIN. The comparison was made by two types of information

a. Classification of relevance (by letters A to D)

b. Possible compromise text prepared by the French Presidency. This was related to those parts of the Solvency II proposal that were touched by the EP amendments.

D.5.3 Trialogue

At the beginning of December 2008, we had the following positions regarding the proposals for an amended FD:

European Commission: wants the group support regime, not the equity proposal (open for discussion on the equity proposal)

European Parliament: wants the group support regime, not the equity proposal

Council: carve out of group support regime; wants to have the equity proposal.

The "trialogue" is a debate and negotiation meeting, in a constellation between the Council represented by the Presidency, the EC, and the EP, to try and sort out any substantial

* http://ec.europa.eu/codecision/index_en.htm

differences on a new code of access to EU documents. The trialogue was first launched as a new committee body in January 2001. This is the place where the real negotiations take place. The objective of the trialogue meetings is to put forward a compromise text.

The first trialogue was scheduled for December 16, 2008, but it was postponed to January 22, 2009.

During the Czech Presidency, January 1–June 30, 2009: The Council Working Group, CWG, with only attachés, had two meetings in January 2009, three meetings in February, and six meetings in March.

There were two trialogue meetings in January 2009, on 22nd and 28th, three in February 2009, on 4th, 13th, and 19th. In March 2009, there were three meetings: on 3rd, 11th, and 17–18th.

At the two January meetings of the CWG, the tabled comparison that was presented by the French Presidency at the last attaché meeting in December was discussed.

At the *first trialogue* the EP representatives (the ECON rapporteur and the shadow rapporteurs) presented a united front on the basis of their mandate, that is, the approved ECON report from/October 2008. The discussion was more of procedural nature and also set the timeline and process going forward.

At the *second trialogue* there was an agreement to discuss five political issues further: group support, anticyclical mechanisms (including the equity issue), cooperation among supervisors, third-country issues, and the quality of OFs. There was also an agreement to establish a technical committee for nonpolitical issues.

At the *third trialogue* it was reported that the technical committee had met several times and agreed on some 96 articles of the draft directive. The political issues were also discussed.

The *fourth trialogue* was a short meeting and the Czech Presidency explained the results of its questionnaire to the member state and the fact that it leaves the Presidency without a mandate to put up a compromise. At the meeting, the EP was clear that it would not discuss the equity issue without a compromise put forward on the group support regime. The Czech Presidency produced a "nonpaper" with draft compromise texts on the group support regime and on the equity issue.

At the CWG's meeting on February 18, it was decided to give the Czech Presidency mandate to negotiate on the group support regime. This was seen as a major step forward and was a thaw in the negotiations.

At the *fifth trialogue* meeting on February 19, the Czech Presidency presented its proposal for a compromise on the two major issues. It was decided that the EP was going to analyze the proposal and come back with its own proposal.

During the very serious economic and financial crises in 2008, the President of the EC, José Manuel Barrosso, in October mandated Jacques de Larosière to chair a "*High-Level Group on supervision*" to give advice on the future of European financial regulation and supervision. The de Larosière report was presented on February 25 2009; see de Larosière (2009). The conclusion of the group put further pressure on the negotiations within the trialogue, as in their recommendation five the Group considered that

- The Solvency II directive must be adopted

- The final agreement should include a group support regime, to be coupled with sufficient safeguards for host Member States, a binding mediation process between supervisors and the setting-up of harmonized insurance guarantee schemes

The EP's answer to the proposal from the Czech Presidency was first discussed at the attaché meeting (CWG) on March 2.

At the *sixth trialogue* meeting on March 3, it was shown that it was difficult to reach an agreement on possible compromises on the group support regime among the Member States. There was a group of 15 countries opposing the proposal. They also presented its reaction to the EP's answer to the Czech Presidency compromise, showing its rejection of moving beyond the first compromise. The meeting did not achieve much progress. One problem was the Presidency's limited margin to act given by the Council members. The Presidency decided to come up with a new compromise proposal to break the deadlock.

At the *seventh trialogue* meeting on March 11, the Presidency informed the EP and the EC about their discussion with Member States on a possible and feasible solution of a review clause for the group issues. Most members showed willingness for such a solution. The EP's representatives were also open for this proposal. This would mean that the group support regime would be lifted out from the FD as proposed by the Presidency's proposal.

There was also a proposal for the duration approach, limiting it for pension's activities.

These compromises were discussed at the attaché meeting (CWG) on March 12.

The technical subcommittee of the CWG finalized its work by finding a compromise for the technical issues at the mid of March; the "global compromise."

The last trialogue meeting on March 17 did not come up with any agreements after 11 h negotiation. The meeting continued the day after (March 18). The outstanding issue at this time was the duration approach. A compromise was reached on a Pillar I dampener. The Presidency debriefed the attachés at their meeting the same day. The EP sent its final position on group support and procyclicality. They suggested that it should be brought forward to the Coreper for a final decision and agreement. The final position included the review clause with reference to the original FD proposal (which includes the group support regime), a Pillar I dampener, and an optional duration approach limited to pension's activities (ring-fenced and no cross-border business). The EP also wanted an average duration of the pension liabilities of 15 years. The Presidency had suggested 10 years.

The final position of the EP was given by the first rapporteur late on Friday, March 20. The Coreper II-meeting was planned for Wednesday, March 25 (the EU ambassadors meeting).

A draft consolidated FD was sent to the Coreper on Monday, March 23. It was based on the negotiations in the last trialogue (March 17–18), the attaché meeting (March 18), and the final position of the EP.

The Coreper meeting on March 25 was positive to the proposal and to support the text except a couple of issues: changing the average pension liability duration from 15–12 years and changing the MCR corridor for captives. The EP agreed to the first issue, and at an extra Coreper meeting on March 26 an informal agreement was reached. At the Coreper 1 April, the consolidated FD was formally endorsed as an A item, that is, approved without discussion.

Press release from the Czech Presidency 26 March 2009:

Czech Presidency finalises negotiations on Solvency II Directive
The Czech Presidency has succeeded to finalise the negotiations concerning the Solvency II Directive which had been discussed for the past two years.
This turning point has only been reached after heavy and difficult negotiations in the Council and with the European Parliament.

The final compromise solution was approved, on a preliminary basis, by the European Parliament and supported by the COREPER today. The two main sensitive issues which represented the hardest obstacles were solved. Concerning the first one, the group support regime, both institutions have agreed not to include it in the directive and come back to it in a few years.

The second issue, the treatment of equity risk, includes the "duration approach." This is to be used only after being transposed into Member State legislation and for clearly defined life insurance products. A formal approval by the COREPER is expected next Wednesday, the vote in the European Parliament plenary is expected later this month.

The European insurance industry welcomed the compromise. The following press release was sent from CEA 26 March 2009.

CEA welcomes informal agreement on Solvency II Framework Directive
Pleased that compromise has been reached by European Parliament and Council

Brussels, 26 March 2009: The CEA, the European insurance and reinsurance federation, is delighted that informal but unanimous agreement has been reached at political level in Europe on the text of the proposed Solvency II Framework Directive after prolonged negotiations.

"This is a decisive step towards the new, enhanced regulatory regime that we have been seeking for Europe's insurers," said Michaela Koller, CEA director general. "We are happy that the timetable for implementing the Directive is on track. Solvency II is an important and timely piece of legislation and any delay would have been most unfortunate in the current economic climate."

The CEA, however, feels that carving out group support from the text agreed means that Europe has missed the opportunity to introduce a tool that would have met the need for the efficient and effective supervision of multinational groups which was highlighted last month in the De Larosière Group's report on financial supervision. "The industry looks forward to Europe taking this step as soon as possible," said Alberto Corinti, deputy director general of the CEA.

The text of the Framework Directive agreed informally today by the Committee of Permanent Representatives is expected to be formally endorsed next week. The European Parliament will put the Directive to a plenary vote on 22 April. Formal adoption of the Framework Directive could then take place during the 5 May Economic and Financial Affairs (ECOFIN) Council.

"The CEA stands ready to continue contributing to the work on the Level Two implementing measures of the Directive, to ensure that the best possible framework for the supervision of Europe's insurers is achieved," said Corinti.

The FD was adopted by the EP at its Plenary Vote April 22, 2009 and by the European Finance Ministries at the ECOFIN Council May 5.

D.5.4 Adoption in the EP

The FD was adopted by 593 votes in favor, 80 against and 3 abstentions on Wednesday, April 22, 2009. It would have sufficed with 393 votes for an absolute majority.

Amendment 145, which was adopted, comprised the draft Directive as amended and agreed by the trialogue negotiations with the Council. In a news flash from CEA the final compromise was summarized as

- The review clause (Art. 246) calls on the EC to present, by October 31, 2014 (2 years after the transposition date of October 2012), a report on the application of the provisions regarding group supervision, accompanied with legislative proposals if appropriate. The report should focus in particular on the cooperation of supervisory authorities, the colleges of supervisors, the legal status of CEIOPS, and supervisory practices on setting the capital add-ons.
- The review clause also states that by October 2015 (3 years after the transposition date of October 2012), the Commission shall make an assessment of the benefit of enhancing group supervision, and maintains the reference to the original EC proposal and the ECON report of October 16, 2008.
- The Pillar I dampener issue (Article 105.ter) and an average duration of the liabilities corresponding to the retirement business of 12 years (Article 305b). The latter is subject to a revision clause (305.b.2) whereby the EC is requested to submit to the EP and EIOPC, by October 2015 (3 years after the transposition date of October 2012) a report, and a legislative proposal if appropriate, on the application of the duration approach. This report shall address in particular cross-border effects of the use of this approach in a view to preventing regulatory arbitrage from insurance and reinsurance undertakings.
- Recitals calls on the EC to present a proposed revision of the Insurance Mediation Directive before the end of 2010; a revision of the IORP Directive "as quickly as possible," and requesting that the EC and CEIOPS develop a system of solvency rules for pension provision; recommending the EC to put forward legislation to tackle the shortcomings identified regarding the provisions related to supervisory cooperation, in line with the de Larosière report.

From the discussion we cite the first rapporteur Peter Skinner, the Commissioner Charlie McCreevy, and the shadow rapporteur Sharon Bowles, and also the negative voice of the MEP Godfrey Bloom.

> *Peter Skinner*, rapporteur: Madam President, you caught me slightly unawares as I had not seen the complete change of the timetable today, but I am very grateful for the chance to address the Chamber about a very important issue of the financial services industry, that is, the insurance and reinsurance industry, what we have done with the Solvency II report and how we have finally brought this now to Parliament

in order to be able to establish what I think will be a very concrete basis for regulation across the European Union.

It is, of course, something, which we are coming back to. There was Solvency I, and I am grateful to Mr Ettl when previously in Parliament we discussed this at some length and we managed to come up with some basis. But now we have to modernise, and the insurance industry is amongst many financial services industries, which have to be at the forefront of change. It is clear, with the financial crisis and everything that has gone along with it, that the insurance industry is something, which cannot be left alone.

There are several measures which come about from Solvency II which I think have helped to make this one of the leading reports, which will be a global leader at that. Amongst them is the issue of management of risk. I think it is not enough now that regulators simply tick boxes to establish whether or not the industry that they are meant to be guarding and protecting on behalf of the consumer is doing the right thing. It is essential that the day-to-day business of insurance and reinsurance companies is actually watched, managed and monitored by regulators over a period of time.

It is by this process and this process alone that we will be able to establish a proper and appropriate form of regulation. It is the reporting of companies: yes, they will be doing things to tell regulators what they are doing, but regulators will have to be involved. And across 27 Member States: not each individual Member State with their own separate rules now, considering what they can apply in terms of that regulation, but indeed they will be applying a standard formula of regulation across the European Union, which will lead, frankly, to the better consumer platform of protection that we expect.

Similarly, companies will manage to get economies of scale from this regulation, because now they will be reporting only in one way to each of the regulators. What they produce, what they have to say, what they do and how they report, will not just be to one regulator but it could be to a college of regulators, especially for groups, because, as insurance companies cross borders, it is now important that regulators team up and work together to ensure that the appropriate levels of reporting, the appropriate levels of figures and what information is supplied, are brought to bear to make sure that the markets are best protected.

It was during the discussion with the Council that Parliament saw some interesting and perhaps sometimes even deliberate ploys to move national industries one way or another, so I cannot pretend that this has not been a very difficult dossier to try to negotiate with the Council: it has. Parliament has pushed the Council a long way. It pushed it further than I think the Council established and really wanted to go under the last two presidencies, so I am very proud and pleased to have worked with the team that I have in order to be able to get the Council to move.

Unfortunately we will not have the kind of group support that we initially envisaged that we should, but because we are able to insert a review clause in this directive, we will be able to come back to group support and, three years after the introduction of this particular directive, I am hoping – and I expect the Commissioner to tell me that he will anticipate doing this as well – to be able to bring back group support in one way or another, specifically to match the economic side of this particular approach.

We want a regulation that is risk-based and principle-based, but one that will also support the capacity of the industry and one that will promote the very best instincts of regulators across the European Union and abroad. I will just finish on this one note. We must also challenge regulators elsewhere in the world and recognise country-to-country regimes only. I hope the Commissioner will agree with me about that.

Charlie McCreevy, Member of the Commission: Madam President, today's debate takes place at a time when we are facing the greatest challenge to the European economy in modern times. Action is needed urgently: vigorous, targeted and comprehensive action in order to restore confidence, growth and jobs and to repair the financial system, to rebuild stability for the future, to promote trade and investment and to better protect our citizens – in short, to deliver an effective and stable financial system.

Based on the Commission's communication of the beginning of March, the Spring European Council set out a strong EU action plan for the future – a strategy to address the regulatory gaps in the financial sector, to restore incentives and to reform supervision to match the single EU financial market. In a few weeks' time the Commission will present its views on the road towards building a state-of-the-art supervisory framework in Europe. These will be discussed by the heads of state or government in June. The Commission is ready to put concrete measures on the table in the autumn.

Clearly, global problems also require global solutions. The EU initiative to agree a coordinated global response to the financial crisis has been very successful. At the London meeting, G20 leaders made extensive commitments to address the weak points of the financial system in a coordinated manner, to jointly build a new financial architecture while defending an open, global economy.

The situation in the EU financial sector is serious. But a lot has already been done, and I am glad to note that the Commission, the European Parliament and the Council have reacted quickly and cooperated closely to respond to the crisis. We are about to successfully conclude the adoption of three key measures: firstly, the regulation on credit-rating agencies; secondly, the recast of Solvency II, as well as, thirdly, the revision of the Third and Sixth Company Law Directives on domestic mergers and divisions.

Firstly, the agreement reached on a regulation on credit-rating agencies will help address one of the problems that contributed to this crisis and thus will offer some prospect of restoring market confidence. The proposal adopted by the Commission last November sets some clear objectives for improving integrity, transparency, responsibility and good governance of the credit-rating agencies. The thrust of the initial proposal is preserved in this regulation, which will in particular secure the analytical independence of credit-rating agencies, the integrity of the rating process and an adequate management of conflicts of interest that existed before in the rating process. Moreover, a comprehensive supervisory regime will be put in place. European regulators will supervise the conduct of credit-rating agencies and take enforcement action where necessary.

On the issue of supervision, I have been vocal about the need to strengthen supervisory cooperation. I have therefore no difficulty in agreeing on the need to push forward in this crucial domain. Therefore, in order to ensure consistency and coherence in all relevant financial sector regulation, the Commission agrees, on the basis of the recommendations of the de Larosière report, to examine the need to strengthen the provisions of this regulation with regard to supervisory architecture.

On the issue of the treatment of credit ratings issued in third countries, the outcome of the G20 summit has changed the global situation. All G20 members have agreed on regulating credit-rating agencies through the introduction of mandatory registration and oversight regime. That is why I agree with the solution agreed in the negotiations between the Council and Parliament on the treatment of ratings issued in third countries.

I am pleased to note that the ambitious goals set by the Commission proposal have been kept. The Commission is very pleased with the outcome of the codecision process.

Let me now turn to Solvency II. I would like to thank the rapporteur, Mr Skinner, and Parliament for their work and their willingness to compromise in order to reach agreement in a single reading on this important subject. Such an outcome will be widely welcomed by the EU insurance industry, by supervisors and by stakeholders in general.

However, I also have to admit that I am disappointed with certain aspects of the compromise. The deletion of the group support regime, which I consider one of the most innovative aspects of the Commission's proposal, means that we will not be able to modernise – as much as we wanted – the supervisory arrangements for insurers and reinsurers operating on a cross-border basis.

I also remain concerned that some of the amendments regarding the treatment of equity risk could result in the introduction of an imprudent regime for investment in risk-based capital. This is particularly the case for the amendments, which introduce the so-called duration approach as a Member State option. The Commission will pay close attention to ensure that the implementing measures brought forward in this regard are prudentially sound.

Nonetheless, the Commission will support the agreement between Parliament and the Council, if it is endorsed by your vote. The current Solvency regime is over 30 years old. Solvency II will introduce an economic risk-based regime that will deepen integration of the EU insurance market, enhance policyholder protection and increase the competitiveness of EU insurers.

As confirmed recently by CEIOPS in their report on lessons learned from the financial crisis, we need Solvency II more than ever as a first response to the present financial crisis. We need regulation that requires companies to properly manage their risks, that increases transparency and that ensures that supervisory authorities cooperate and coordinate their activities more effectively. Solvency II will bring about a regime for the insurance industry that can serve as a model for similar reforms internationally.

The introduction of a review clause specifically mentioning the group support regime will enable the Commission to come back to this issue. I expect that progress in a number of different areas, connected to the recommendations of the de Larosière report, will have created a more favourable environment for reforms related to cross-border cooperation between home and host supervisors.

I now turn to the Weber report. Thanks to the efficient work of the rapporteur, Ms Weber, it has been possible to identify a compromise on simplified reporting and documentation requirements, in the case of mergers and divisions of public limited liability companies which will maintain a very significant part of the savings potential of the original Commission proposal, which amounts to EUR 172 million per year.

Measurements and studies carried out in the context of reduction of administrative burdens show that company law is one of the most burdensome areas of the EU acquis. For several reasons, administrative burdens hit SMEs harder than bigger companies. An expert report from 2007 estimates that small enterprises spend 10 times the amount that large enterprises have to spend in order to comply with information obligations imposed by legislation. Ten times, I repeat. At the same time, small businesses are the backbone of our European economy, and they are currently facing very difficult economic times.

In the current difficult and challenging economic situation we cannot afford such impediments. Instead we must strengthen our effort to ease the burden on our companies. In its resolution of 12 December 2007, the European Parliament welcomed the Commission's determination to reach the goal of a 25% reduction in administrative burdens on undertakings at EU and national level by 2012 and underlined that it would examine legislative proposals in this light. Today, only seven months after the proposal was put forward by the Commission, I am very pleased with this compromise, even though the Commission had gone even further in its original proposal. I look forward to Parliament endorsing this compromise, which will rapidly bring significant benefits to companies, especially to SMEs. And we should not stop there. Simplification and reducing red tape will remain at the heart of the Commission's agenda.

Sharon Bowles, the rapporteur for the opinion of the Committee on Legal Affairs: Madam President, I welcome the agreement for Solvency II and, like others, I regret the relegation of group support to a future review and the eventual inability on the part of the Council to explore with us ways to make it workable, taking into account some well founded concerns. In both the Committee on Legal Affairs and the Committee on Economic and Monetary Affairs I looked at what happens to the movement of capital at times of group stress, such as near-insolvency, and it is certainly not as straightforward as the Commission draft or insurance industry representatives portrayed.

However, there are instruments available that could achieve the objective and we recommended Level 2 measures, but now we are left to look for the future for ways to maximise safe, economic use of capital in a group. I hope that Member States will rise to the challenge when it comes to looking for better solutions on winding-up.

Turning now to some of the things in the package, changes to Article 27 specified that supervisory authorities must have relevant expertise and capacity. I made the original amendment in part with the report on Equitable Life in mind, but in the context of the financial crisis it has a wider resonance and I have achieved similar inclusions in the capital requirements and credit-rating proposals.

It must be absolutely clear that taking a risk-based approach is not a soft option. Proper understanding of models and underlying assumptions should be a more intensive way of supervising than tick-boxes. Stress tests must challenge beyond the comfort zone of assumptions, and correlation factors should remain under active review.

Group supervision is now made an inclusive process, not winner-takes-all for the group supervisor, although there has to be responsibility at a single end point ultimately. The role of CEIOPS is augmented, and it is worth stressing that it was the discussion around Solvency II that led the way towards mainstream thinking on enhanced roles for the Level 3 committees. Importantly, it is also clarified that there should not be conflicts between the mandate of a national supervisor and its role within CEIOPS.

These amendments were somewhat prescient when made quite some time ago, but have shown their worth as the financial crisis has developed. As the rapporteur has said, the Parliament team has done well and in the context of Solvency II; so has the Czech Presidency.

Not all MEP's were positive to the solvency II directive:

Godfrey Bloom, on behalf of the IND/DEM Group.—Madam President, I have spent 40 years in financial services, so I think perhaps I know a little bit about what I am talking about here.

Let me just say a little about the UK Financial Services Authority (FSA), which will guide us onto the target of how mistakes are made. The FSA in the United Kingdom has a rule book of half a million words. Nobody understands it - least of all the FSA. The FSA interprets its own rule book in secret; they keep the fines that they impose to beef up their own salaries and pensions; there is no court of appeal. I have written to Commissioner McCreevy on this subject and it drives a coach and horses through Articles 6 and 7 of his own Human Rights Act. There is no court of appeal. There is no legal recourse at all if they get it wrong. The general public has been given the impression that if a regulation has an FSA stamp on it it cannot go wrong. There is no concept of caveat emptor.

Now it is going to be, it would appear, subsumed by some sort of EU overseer, consisting no doubt of ignorant bureaucrats, Scandinavian housewives, Bulgarian mafia and Romanian peg-makers. Frankly, I think you are going to get on really well with each other.

D.5.5 Adoption in the ECOFIN Council

At the 2940th ECOFIN Council meeting on May 5, 2009, in Brussels, according to a press release, the ECOFIN Council took note of progress on the following four financial services dossiers:

- Solvency requirements for insurance companies (draft "Solvency II" directive)
- Credit-rating agencies (draft regulation)
- Electronic money (draft directive)
- Cross-border payments in the EU (draft regulation)

On all four dossiers the Presidency has reached agreement with the EP in first reading, enabling adoption by the Council at a forthcoming meeting, once the texts have been finalized.

This meant that the texts would be formally adopted at a forthcoming meeting once they have undergone jurist/linguist scrutiny and translated to all languages. This was only a technical and legal translation formality. Coreper adopted the final version at its meeting on November 4 and at the ECOFIN Council on November 10, 2009 the FD was adopted.

The Solvency II FD was published in *Official Journal* December 17, 2009; see OJ (2009). The Directive entered into force on January 5, 2010.

D.6 CONSULTATIONS FOR IMPLEMENTING MEASURES: CEIOPS' WORKSTREAM

Based on the general approach laid out in the road map for the Solvency II project, see Section D.3.2 above, the European supervisory committee CEIOPS produced a lot of CPs, issue papers, and quantitative impact studies.

CEIOPS' consultation and transparency are two constituent elements of CEIOPS' functioning and of the Lamfalussy process; see Section D.1.1. On its Web site, www.ceiops.org, they establish that *the creation of a robust regulatory framework for supervision and the adoption of effective supervisory practices rely both on a clear and complete knowledge of market situations and needs, and on a wide sharing of the regulatory and supervisory policy.*

In preparing its advice to the EC, and also in drafting its own recommendations, guidelines, and standards, CEIOPS consulted interested stakeholders. Disclosure has been the prestige word for CEIOPS in its work and its fulfilling of its duties, within any limits of secrecy and confidentiality constraint in the supervisory activity.

Based on its first consultation, CP 1, regarding the consultation practices CEIOPS drafted its Public Statement of Consultation Practices (CEIOPS-DOC-01/05), adopted in February 2005.

CEIOPS has been committed to consult, both before and after the drafting of each CP, market stakeholders such as market participants, consumers, and end users, in different ways:

1. *Predraft papers:* On a continuous basis at the level of Working Groups through informal discussions at an early stage with those most likely to be directly affected. In this phase,

CEIOPS informs interested parties about the work under way and find practical ways to facilitate external inputs while a draft document or statement is under preparation.

2. *Draft papers for consultation and hearings:* On the basis of wide, formal consultations, also via the Web site and by holding public hearings, once a draft document or statement (CPs) has been finalized.

3. *Inputs from the Consultative Panel:* By using inputs from the Consultative Panel, whose tasks are mainly to oversee CEIOPS' policy, work program, and consultations practice.

4. *Input on specific projects:* By inviting stakeholders' inputs and reactions to specific projects (issue papers, surveys, and questionnaires).

Between 2004 and 2008 CEIOPS produced 25 numbered *CPs* and some non-numbered. The list of CEIOPS papers is given in Appendix C. These papers were sent out and commented by different interested stakeholders, such as CEA, the body of the European insurance industry, and also by different insurance companies. During the same period five *Issues papers* were published; see also Appendix C.

The aim of the *public hearings* has been to give all stakeholders the opportunity for a direct dialogue with CEIOPS' representatives in commenting on draft texts.

In order to fulfill the transparency requirements of the Lamfalussy Procedure, CEIOPS has been required to consult extensively and in an open and transparent manner with providers of insurance and occupational pension's funds products, market participants, consumers and end-users. In 2003 a *consultative panel* of experts from the industry, end-users and consumers was established. The panel has at least three meeting a year with representatives from CEIOPS and gives suggestions for CEIOPS work programme, gives comments on draft and interim results. The panel has been composed of a limited number of independent high level persons, committed to the objectives of the European Union. They have been appointed on a personal basis and do not represent national positions or sectoral interests.

In order not to lose tempo in work and after consultations, CEIOPS has conducted four QISs. In the different QISs, CEIOPS has tested different approaches to valuation of assets and liabilities and also formulas for the SCR and MCR calculations. The outcome of the results has been used for calibration of new studies, and also as an input for the EC in its work on the FD. The different QISs that have been conducted are

- QIS0: preparatory field study (2005—spring)

- QIS1: valuation of assets and liabilities (2005: October 17–December 31) QIS1 had the focus on testing the level of prudence in current TPs by benchmarking them against best estimate liabilities (BEL) plus an RM. The firms' ability to perform the requested calculations was also looked at by CEIOPS. The result from QIS1 was published on March 17, 2006; see CEIOPS Web site.

- QIS2: valuation and standard formula for SCR, MCR (2006: May 3–July 31) QIS2 had the broader focus on assessing the SCR, the MCR, the valuation of assets and liabilities, and the definition of available capital ("OFs"). The main goals were to provide information on the practicality of the calculation methods involved, to provide information about the impact on the BS, and to provide both quantitative and qualitative information on the different approaches used. The calibration of the capital requirements (SCR and MCR) was quite tentative. There was no guidance provided on which items of capital could be included as "eligible capital." The result from QIS2 was published on December 7, 2006; see CEIOPS Web site.

- QIS3: A full SCR and MCR calculation (2007: April 2–June 29) The main goal of QIS3 was to study the potential effect of new proposals by CEIOPS for both insurance undertakings and groups of undertakings. CEIOPS advice to the EC on Pillar I issues, together with some development work done, formed the basis for the revised proposals. They had also taken into account useful comments that were received from insurance industry.

  The goals of QIS3 were fourfold. Firstly, CEIOPS hoped that the exercise would give further information about the practicability and suitability of the calculations involved. Secondly, they looked for quantitative information about the possible impact on the BSs, and the amount of capital that might be needed, if the approach and the calibration that was set out in the QIS3 specification were to be adopted as the Solvency II standard. Thirdly, they looked for information about the suitability of the suggested calibrations for the calculation of the SCR and MCR. Fourthly, they looked for information about the effect of applying the QIS3 specification to insurance groups

  The result from QIS3 was published in November 2007; see CEIOPS Web site.

- QIS4: A full solvency study, including group calculation (2008: April 1–July 31). This study was conducted by the Commission with the help of CEIOPS. The TS was sent out for consultation at the end of December 2007 for a period of less than 2 months. The March 26, 2008 the EC sent a CfA to CEIOPS asking them to run the QIS.

  Key changes made in the TS were in the following areas: calculation of MCR, equities, deferred taxes, future premiums in TPs, determination of risk-free interest rate, hybrid capital instrument, participations, calibration of nonlife underwriting risk, and use of entity-specific data, and groups.

  The TS was divided into six main parts: Valuation of assets and liabilities, OFs, SCR: standard formula, SCR: IMs, MCR, and group issues. The result from QIS4 was published in November 2008; see CEIOPS Web site.

Details of these studies and the calibrations of the standard formula are discussed in Appendices E to P of this book.

Other documents, reports, standards, and letters are published on CEIOPS Web site.

D.7 PHASE III: IMPLEMENTING PHASE: 2009–2012

In a letter from Director General at Internal Market and Services DG, EC, Jörgen Holmquist to CEIOPS' chair Thomas Steffen the further work on Solvency II was planned. The letter was sent the same day as the proposal for an FD was published. Apart from the planning for the next 1.5 years, there was also a planning for the years behind that period.

There was a planning for CEIOPS to give the FA on potential future Level 2 and Level 3 measures. Two groups of issues were presented. The first relates to quantitative requirements such as the valuation of assets and liabilities, including TPs, the approach to OFs, the detailed technical design, and calibration of the SCR and the design of the MCR. The work should also include detailed consideration of the simplifications to be applied both with respect to the valuation of assets and liabilities, including TPs, and the SCR as well as the use of entity-specific parameters in the underwriting risk modules. The second group of issues relates to full and partial IMs, including the application and approval processes; governance requirements, regulatory supervisory reporting and public disclosure requirements, and the use of capital add-ons.

Any FA from CEIOPS should be in line with the adopted FD and should take into account the results of QIS3 and QIS4 as well as international developments in the supervisory and accounting areas.

Dead-line for this was set to October 2009.

An IA was also planned for the Level 2 implementing measures on the above issues.

The letter included an Annex I with a list of implementing measures in the proposal for a directive. In the "implementing measures" articles of the proposal for an FD, that is, the comitology articles, the choice of *may* or *shall* reflects whether the Commission retains some discretion with respect to adopting such measures, or actually undertakes to do so. Annex II gave a Solvency II Road map for CEIOPS, the Commission, and so on.

Solvency II Road map sent to CEIOPS in July 2007

2009		**2010**			
1st Semester	**2nd Semester**	**1st Semester**	**2nd Semester**	**2011**	**2012**
Adoption of the Directive	CEIOPS: FA	COM: proposal for implementing measures + IA report	Implementing measures adopted CEIOPS: level 3 guidance finalized	Industry and Member States preparation for entry into force of the new regime	Industry and Member States preparation for entry into force of the new regime. Regime into force 18 months after implementing measures adopted

Source: Adapted from COM. 2007b. Letter from Jörgen Holmquist, European Commission, to the chairman of CEIOPS, Thomas Steffen, regarding further work on Solvency II, Markt/H2/BC/sv D(2007) 10830, Brussels, July 19. Available at http://ec.europa.eu/internal_market/insurance/solvency/index_en.htm.

Mr Holmquist ended the letter by noting that in addition to the issues listed in the letter, CEIOPS was of course free to start work on other areas where it believes that further guidance would help to foster supervisory convergence and facilitate harmonized implementation of the new regime.

In November 2008, the EIOPC set up a tentative Workplan for the years 2009–2012. The Workplan has not been published. The main issues on the agenda are summarized below.

Level 1—Framework Directive

2009 April: Adoption of the Framework Directive

2009 November–2010January: Publication in the EU *Official Journal*

Level 2—Implementing measures

2009 May–2010 January: Informal discussions in EIOPC, "Solvency Expert Group," within EP, with stakeholders and intensive discussions with the Legal Service and other DGs

2010 February–April: Finalizing the drafting of the Level 2 measures

2010 May–June: The adoption process of the Commission (COM) and formal adoption by COM

2010 July–November: Formal discussions in EIOPC followed by the official opinion from EIOPC

2010 December–2011 April: Discussions in the Council and the EP following the opinion of EIOPC → adoption of the Level 2 implementing measures by COM

2011 May–2012 September: Getting prepared for the entry into force of Solvency II in October 2012

Translations to the languages of the member States

2009 May–2010 March: Set up cooperation between DG Translation and Technical Experts. Organize seminar(s) with DG Translation to start thinking about technical terms.

2010 April–September: Translation work on the Level 2 implementing measures.

2010 October–November: Finalizing the translation of the implementing measures and the summary of the impact assessment report.

Quantitative Impact Study—QIS5

2009 December–2010 January: CEIOPS finalizes the QIS5 draft technical specifications and Commission publicly consults on this draft with CEIOPS' technical support.

2010 February–March: The COM and CEIOPS finalize the QIS5 technical specifications.

2010 April–July: The QIS5 is running. Later changed to August–October/November.

2010 September: Preliminary results.

2010 November: Final QIS5 results.

CEIOPS work, excluding QIS5

2009 March/April: Launch 1st wave of consultations on its draft final advice on implementing measures.

2009 April–May: Following political agreement on Level 1 CEIOPS should update a list of policy issues and options.

2009 June/July: Launch 2nd wave of consultations on its draft FA on implementing measures

2009 October/November: The FA on implementing measures is delivered (including its IA contribution)

2009 November–2010 October: Works with Level 3 Guidelines.

2010 November–2012 September: Getting prepared for the entry into force of Solvency II.

In a new letter from the Director General Jörgen Holmquist to the chairman of CEIOPS Thomas Steffen in June 2009, an update of how the EC saw on the rest of the project (after the adoption of the FD) and especially the work on Level 2 implementing measures and Level 3 supervisory guidance.

The new Solvency II timetable set out by the EC is summarized below.

Level 1—Framework Directive

2009 October–December: Publication in the EU *Official Journal*

Level 2—Implementing measures and Level 3—Supervisory guidelines

2009 May–2010 June: Informal discussions in EIOPC, "Solvency Expert Group," within the EP, with stakeholders and intensive discussions with the Legal Service and other DGs

2009 October: CEIOPS publishes the FA on first and second rounds of CP and launches third round

2010 January: FA on the third round of CPs

2010 July–September: finalizing the drafting of the Level 2 measures

2010 October–November: Commission's adoption process. Formal adoption of Level 2 implementing measures will be made

2011 January–May: Formal discussion in EIOPC and official opinion from EIOPC

2011 June–October: Discussions in Council and in Parliament. Adoption of Level 2 implementing measures by COM will be made

2011 December: Recommended delivery date for CEIOPS final Level 3 supervisory guidelines

Translations to the languages of the Member States

2009 May–2010 March: Set up cooperation between DG Translation and Technical Experts. Organize seminar(s) with DG Translation to start thinking about technical terms

2010 April–October: Preparatory translation work on the Level 2 implementing measures

2010 November–2011 March: Translation work starts

2011 April–May: Finalizing the translation of the implementing measures and the summary of the impact assessment report

Quantitative Impact Study—QIS5

2009 November–2010 February: CEIOPS finalizes the QIS5 draft technical specifications

2010 March: Draft QIS5 sent to COM

2010 April–May: Commission publicly consults on draft QIS5 technical specifications with CEIOPS' technical support

2010 mid-June: COM publishes the final QIS5 technical specifications

2010 end-July: CEIOPS publishes final spreadsheets for QIS5

2010 August–October/November: The QIS5 is running. Solo companies submit results by the end of October, Groups by mid-November

2011 January–March: CEIOPS prepare report

2011 April: Final QIS5 report is published

Third-country equivalence

2009 December: CEIOPS launches consultation on third-country assessments

2010 March: CEIOPS delivers the FA on third-country assessments

2010 April–June: Informal discussion within EIOPC, EP, and with third countries

2011 March: CEIOPS launches consultation on the initial wave of individual assessments of third-country equivalence

2011 June–July: CEIOPS delivers the FA on the initial wave of individual assessments of third-country equivalence

2011 August–October: Informal discussion within EIOPC, EP, and with third countries

2011 November–December: COM finalizes drafting proposal

2012 January-February: Commission's adoption process

2012 March–June: Formal discussions in EIOPC

Impact assessment (IA)

2009 July–December: COM coordinates the various IA work-streams

2009 September: Progress report by Contractor

2009 December: Interim report delivered by Contractor

2010 January–March: COM starts drafting the IA Main Report and Executive Summary

2010 March: Final report draft delivered by Contractor

2010 April–June: COM finalizes the draft IA package

2010 June: Final report's final version delivered by Contractor

2010 July: Positive opinion from COM IA Board on the draft IA package

2010 August–October: Processing IA Board's comments. Final IA package will be published

The COM also planned for a public hearing in Brussels during the 1st half of 2010 to get comments on future Level 2 implementing measures, QIS5 draft technical specifications, and the IA.

The draft final advice and the FA on Level 2 implementing measures are discussed in Chapters 23 (asset valuation), 24 (liability valuation), 25 (investments and eligible own funds) and 26 through 34 (the capital requirements).

D.7.1 CEIOPS' 1st Launch of Draft Final Advices

On March 26, 2009, after the Members meeting, CEIOPS' launched 12 draft final advises for consultation; see Appendix C and CEIOPS (2009b01)–(2009b12). These drafts were developed on the basis of the General approach on the Solvency II Directive proposal adopted by the ECOFIN Council on December 2, 2008 ("Level 1 text"). The consultation period was set to 2 months except for two papers; see below.

CP-26: Draft L2 Advice on TP—Methods and statistical techniques for calculating the BE. This CP aims at providing advice with regard to actuarial and statistical methodologies for the calculation of the BE as requested in Article 85(a) of the Level 1 text. *Consultation period:* April 1–June 1; see Chapter 31.

CP-27: Draft L2 Advice on TP—Segmentation. This CP aims at providing advice with regard to the LoBs on the basis of which insurance and reinsurance obligations are to be segmented in order to calculate TPs as requested in Article 85(e) of the Level 1 text. The advice covers the minimum level of segmentation – that is, LoBs—that undertakings need to consider when calculating their TPs. *Consultation period:* April 1–May 3; see Chapter 31.

CP-28: Draft L2 Advice on SCR Standard Formula—Counterparty default risk. This CP aims at providing advice with regard to the treatment of counterparty default risk in the standard formula for the SCR as requested in Article 109 of the Level 1 text. The objective of this paper is to give draft advice on the scope of the module and the calculation of the capital requirement for counterparty default risk. *Consultation period:* April 1–June 1; see Chapter 32.

CP-29: Draft L2 Advice on Own Funds—Criteria for supervisory approval of ancillary OFs. This CP aims at providing advice with regard to supervisory approval of ancillary OFs as requested in Article 92 of Level 1 text. The objective of the paper is to provide a framework that specifies the criteria for granting supervisory approval. *Consultation period:* April 1–June 1; see Chapter 31.

CP-30: Draft L2 Advice on TP—Treatment of Future Premiums. This CP aims at providing advice with regard to the treatment of future premiums in the assessment of TPs as requested in Article 85 of the Level 1 text. The advice covers the recognition of an insurance or reinsurance obligation as well as on the boundaries of these obligations. *Consultation period:* April 1–June 1; see Chapter 31.

CP-31: Draft L2 Advice on SCR Standard Formula—Allowance of financial mitigation techniques. This CP aims at providing advice with regard to financial mitigation techniques as requested in Article 109(1), letter (f), of the Level 1 text. This advice develops the qualitative treatment of financial mitigation techniques. *Consultation period:* April 1–June 1; see Chapter 32.

CP-32: Draft L2 Advice on TP—Assumptions about Future management actions. This CP aims at providing advice with regard to the use of management actions in the assessment of the TPs as requested in Article 85(a) of the Level 1 text. The objective of the paper is to propose a framework that identifies the circumstances in which it is appropriate for undertakings to take account of future management actions in the calculation of their TPs. *Consultation period:* April 1–June 1.

CP-33: Draft L2 Advice on System of Governance. This CP aims at providing advice for Level 2 measures with regard to the System of Governance as requested in Article 49 of the Level 1 text. It also includes material that could be considered for the future when developing Level 3 guidance. The text covers the most important issues to be regulated to ensure appropriate governance standards within insurance and reinsurance undertakings. *Consultation period:* April 1–June 8.

CP-34: Draft L2 Advice on Transparency and Accountability. This CP aims at providing advice for Level 2 measures with regard to Transparency and Accountability, as requested in Article 30 of the Level 1 text, which provides that "the supervisory authorities shall conduct their tasks in a transparent and accountable manner with due respect for the protection of confidential information." It also presents some ideas as to how the accessibility of information could be improved through Level 3 guidance. *Consultation period:* April 1–June 8.

CP-35: Draft L2 Advice on Valuation of Assets and "Other Liabilities." This CP aims at providing advice for Level 2 implementing measures with regard to the valuation of assets and liabilities other than TPs in accordance with Article 74 of the Level 1 text. *Consultation period:* April 1–June 8; see Chapter 31.

CP-36: Draft L2 Advice on Special Purpose Vehicles. This CP aims at providing advice for Level 2 measures with regard to Special Purpose Vehicles (SPVs), as required in Article 209 of the Level 1 text. The advice addresses the authorization, regulatory requirements, and scope of supervisory review that relate to the establishment of SPVs under Solvency II. It also includes material that could be considered for Level 3 guidance. *Consultation period:* April 1–June 8.

CP-37: Draft L2 Advice on the procedure to be followed for the approval of an IM. This CP aims at providing advice for Level 2 implementing measures with regard to the procedure to be followed for the approval of an IM in accordance with Article 112(1) of the Level 1 text. Specificities related to the approval of group IMs as set out in Article 229 of the Level 1 text will be provided as an addendum to this CP later in 2009. *Consultation period:* April 1–June 1.

CEIOPS also published a paper on the lessons learnt from the financial crisis, CEIOPS (2009a). CEIOPS' analysis revealed that the financial crisis highlights the need for a further refinement of the existing Solvency II calibrations, both at module and submodule levels in Levels 2 and 3. According to CEIOPS, the current experience demonstrated that in real crisis situations, only high-quality capital elements could truly be a first line of defence, in the sense of absorbing losses without the insurer being taken into full bankruptcy.

As to the financial sector at large, CEIOPS had found that governance, RM, and internal controls in the insurance sector needed to be further strengthened. These elements are key to Solvency II's underlying philosophy as a risk-sensitive system. It is not just about risk measurement and quantification, but also very much about effective governance and RM.

Focusing on the insurance sector, CEIOPS' report had also supplementary findings and reflections. These related to capital and solvency issues, the need for the scope of regulation and supervision to focus more on consolidated entities rather than only solo entities, insurers' exposures to special purpose vehicles, Credit Default Swaps (CDS), and the issue of procyclicality of regulatory regimes.

These findings would be reflected in the work on Level 2 and Level 3 implementing measures.

D.7.2 CEIOPS' 2nd Launch of Draft Final Advices

Late on July 2, 2009, CEIOPS published 24 new CPs. Consultation period: until September 11, 4 p.m. CET.

Addendum to CP-37: Draft L2 Advice on the procedure to be followed for the approval of a group IM. This addendum aims at providing advice for Level 2 implementing measures with regard to the procedure to be followed for the approval of a group IM in accordance with Articles 112(1) and Article 229 of the Level 1 text.

CP-39: Draft L2 Advice TP—Best Estimate. This CP aims at providing advice with regard to actuarial and statistical methodologies for the calculation of the BE as requested in Article 85(a) of the Level 1 text. A first part of the advice on this Article has been released as CP-26. The objective of this paper is to further elaborate on the appropriate methodologies for the calculation of the BE.

CP-40: Draft L2 Advice on TP—Risk-free interest rate. This CP aims at providing advice with regard to the relevant risk-free interest rate term structure to be used in the assessment of TPs as requested in Article 85(b) of the Level 1 text.

CP-41: Draft L2 Advice TP—Calculation as a whole. This CP aims at providing advice with regard to the circumstances in which TPs shall be calculated as a whole (and not as a sum of a BE and an RM), as required in Article 85(c) of the Level 1 text.

CP-42: Draft L2 Advice—Risk Margin (RM). This CP aims at providing advice with regard to the calculation of the RM as requested in Article 85(d) of the Level 1 text. The objective of this paper is to specify the overall structure of the calculation of the RM, including the following aspects: the definition of the reference undertaking, including the assumption this undertaking has to fulfill, the stipulation (calibration) of the CoC rate, and the projection of the future SCRs related to the reference undertaking.

CP-43: Draft L2 Advice on TP—Standards for data quality. This CP aims at providing advice with regard to the standards that should be met with respect to ensuring the appropriateness, completeness, and accuracy of the data used in the calculation of TPs, and with the specific circumstances in which it would be appropriate to use approximations, as requested in Article 85(f) of the Level 1 text.

CP-44: Draft L2 Advice on TP—Counterparty default adjustment. This CP aims at providing advice with regard to the methods to be used when calculating the counterparty default adjustment to recoverables from reinsurance contracts and SPVs as requested in Article 85(g) of the Level 1 text.

CP-45: Draft L2 Advice on TP—Simplifications. This CP aims at providing advice with regard to simplified methods and techniques to calculate TPs in order to ensure that actuarial and statistical methodologies are proportionate to the nature, scale, and complexity of the risks, as requested in Article 85(h) of the Level 1 text. Concrete simplifications will be consulted upon in a third set of advice.

CP-46: Draft L2 Advice on Own funds—Classification and eligibility. This CP aims at providing advice with regard to the classification of OFs (Article 97) and the eligibility of OFs (Article 99) in the Level 1 text.

CP-47: Draft L2 Advice on SCR Standard Formula—Market Risk. This CP aims at providing advice with regard to the design and structure of the market risk module of the SCR standard formula, as required by Article 109 of the Level 1 text. The objective of this paper is to give draft advice on the structure and design of interest rate risk, spread risk, currency risk, property risk, and concentration risk submodules. The equity risk submodule, the calibration of the modules (except concentration risk), and correlations will be covered in the third set of advice, together with advice on potential simplified approaches in the light of the proportionality principle.

CP-48: Draft L2 Advice on SCR Standard Formula—Nonlife underwriting risk. This CP covers advice with regard to the design of the nonlife underwriting risk module in the SCR standard formula, in particular the methods, assumptions, and standard parameters to be used when calculating this risk module, as required in Article 109(c) of the Level 1 text. Calibration of the module, the use of USPs, correlations, and simplifications will be included in the third set of advice.

CP-49: Draft L2 Advice on SCR Standard Formula—Life underwriting risk. This CP aims at providing advice with regard to the design, structure, and calibration of the life underwriting module in the SCR standard formula as requested in Article 109 of the Level 1 text.

Calibration of the module, correlations, and simplifications will be included in the third set of advice.

CP-50: Draft L2 Advice on SCR Standard Formula—Health underwriting risk. This CP aims at providing advice with regard to the health underwriting risk module in the SCR standard formula as requested in Article 109 of the Level 1 text. Calibration of the module, correlations, and simplifications will be included in the third set of advice.

CP-51: Draft L2 Advice on SCR Standard Formula—Counterparty Default risk. This CP aims at providing further advice with regard to the treatment of counterparty default risk in the SCR standard formula as requested in Article 109 of the Level 1 text. This CP complements CEIOPS' CP no. 28.

CP-52: Draft L2 Advice on SCR Standard Formula—Reinsurance mitigation. This CP covers the advice with regard to the qualitative criteria that reinsurance and securitization arrangements must meet in order to ensure that there has been effective underwriting risk transfer to a third party, for allowance in the SCR standard formula. The paper is complementary to the draft advice CP 31.

CP-53: Draft L2 Advice on SCR Standard Formula—Operational Risk. This CP aims at providing advice with regard operational risk in the SCR standard formula, as required in Article 109(1) of the Level 1 text.

CP-54: Draft L2 Advice on SCR Standard Formula—loss-absorbing capacity of TPs. This CP aims at providing advice with regard to the calculation of the adjustment for loss-absorbing capacity of TPs in the SCR standard formula as requested in Article 109(g) of the Level 1 text.

CP-55: Draft L2 Advice on SCR Standard Formula—MCR calculation. This CP aims at providing advice with regard to the calculation of the MCR as requested in Article 128 of the Level 1 text. Calibration of the module will be included in the third set of advice.

CP-56: Draft L2 Advice on Tests and Standards for Internal Model Approval. This CP aims at providing advice with regard to the use test, statistical quality, calibration, P&L attribution, validation, documentation and external models and data, as requested in Article 125 of the Level 1 text. It also includes draft advice with regard to specificities for Internal Models on the System of Governance and specificities related to the requirements for approval of group IMs are included. Specific requirements for the approval of a partial IM will be included in the third set of advice.

CP-57: Draft L2 Advice on Capital add-on. This CP aims at providing advice for Level 2 measures with regard to the power to set a capital add-on as provided for in Article 37 of the general approach on the Level 1 text. According to Article 37(6) of the Level 1 text the EC shall adopt implementing measures laying down further specifications for the circumstances under which a capital add-on may be imposed and the methodologies for the calculation thereof.

The advice also develops the requirements for setting a capital add-on in a group context, further to Article 232 of the Level 1 text.

CP-58: Draft L2 Advice on Supervisory reporting and disclosure. This CP aims at providing advice with regard to supervisory reporting and public disclosure, as requested in Article 35(6), Article 55, and Article 260 of the Level 1 text, along with supporting material that could be used to develop Level 3 guidance.

CP-59: Draft L2 Advice on remuneration. A CP will be released in due course providing additional, standalone, advice with regard to remuneration issues. It will serve as a basis for implementing measures on this specific issue in the framework of Article 49 of the Level 1 text.

CP-60: Draft L2 Advice on Group Solvency Assessment. This CP aims at providing advice with regard to the technical principles and methods set out in Article 218–227 and the application of Articles 228–231 as requested in Article 232 of the Level 1 text. The advice covers the scope of the group solvency assessment and the calculation methods. It includes advice on assessment of available elements of OFs, including the fungibility and transferability of OFs. The advice also addresses issues related to third-country entities and/or groups.

CP-61: Draft L2 Advice on IGTs and risk concentration. As a lesson to be learned from the crisis, this CP aims at providing advice with regard to the important role risk concentration and intragroup transactions can play in the financial well-being of a group. This own initiative paper develops CEIOPS' initial views on the supervision of these aspects under

Solvency II, including suggestions for Level 3 guidance and potentially, for the development of Level 2 implementing measures if requested by the EC.

CP-62: Draft L2 Advice on Cooperation and Colleges of supervisors. This CP aims at providing advice with regard to the coordination of group supervision along with supporting material that could be used to develop Level 3 guidance as requested in Articles 252(3) and 253(2) of the Level 1 text.

D.7.3 CEIOPS' Third Launch of Draft Final Advices

Late on November 2, 2009, CEIOPS published 16 new CPs. Consultation period: until December 11, 12.00 CET.

CP-63: Draft L2 Advice on repackaged loans investment. This Paper aims at providing advice for Level 2 measures with regard to investments in repackaged loans or similar financial arrangements as required by Article 135 of the Level 1 text. The aim of the implementing measure is to ensure cross-sectoral consistency and remove any potential misalignment of interests between the originators (the companies issuing the financial instrument) and investors in such financial instruments.

CP-64: Draft L2 Advice on the extension of the recovery period—Pillar II dampener. This Paper aims at providing advice for Level 2 measures with regard to the extension of the recovery period by the supervisory authority, in the event of an exceptional fall in financial markets, in cases where the SCR of an undertaking is no longer complied with, as required in Articles 143 in conjunction with Article 138 of the Level 1 text.

CP-65: Draft L2 Advice on partial IMs. This Paper aims at providing advice to take account of the limited scope of the application of the partial IM as requested in Article 112.2 of the Level 1 text. It addresses the scope of partial IMs, specific provisions for the approval of partial IMs, and in particular how results of partial IMs may be integrated into the results of the standard formula. Furthermore, the Paper contains advice on the concept of major business unit, the integration of risks not covered in the standard formula, and adaptations to be made to standards set out in Articles 118–123.

CP-66: Draft L2 Advice on the group solvency for groups with centralized RM. This Paper aims at providing advice for Level 2 measures with regard to the group solvency for groups with centralized RM as requested by Article 245 of the Solvency II Level 1 text. The Paper differentiates between centralized RM on the one hand and group-wide RM, including consistent implementation of the RM in all undertakings forming part of a group, on the other hand.

CP-67: Draft L2 Advice on SCR Standard Formula and Own Funds—Treatment of Participations. This Paper aims at providing advice with regard to the solo treatment of participations in credit and financial institutions for the determination of OFs and the approach to be used with respect to related undertakings in the calculation of the SCR, in particular the equity risk submodule of the SCR, as requested by Articles 92 and 109(ja).

CP-68: Draft L2 Advice on SCR Standard Formula and Own Funds—Treatment of Ring Fenced Funds. This Paper aims at providing advice with regard to the adjustments that should be made to reflect the lack of transferability of OFs and the reduced scope for risk diversification at the level of the SCR related to ring-fenced funds, as requested in Articles 99(b) and 109(1) (fa) of the Solvency II Level 1 text.

CP-69: Draft L2 Advice on SCR Standard Formula—Design of the Equity Risk Submodule. This Paper aims at providing advice with regard to the design and calibration of the equity risk submodule as requested in Article 109(b) of the Solvency II Level 1 text. It includes advice on the symmetric adjustment mechanism according to Article 105(a) and the duration dampener foreseen by Article 305(b) of the Level 1 text.

CP-70: Draft L2 Advice on SCR Standard Formula—Calibration of Market Risk. This Paper aims at providing advice with regard to the calibration of the market risk module for the SCR standard formula, as requested by Article 109(b) of the Solvency II Level 1 text. This Paper follows CP47, which was published in June 2009, and for which the consultation period closed on September 11, 2009. This Paper gives draft advice on the calibration of the interest rate risk, spread risk, currency risk, and property risk. The calibration of the concentration risk sub module has already been covered in former CP47. The calibration of the equity risk submodule and the correlations between the market risk submodules and between the market risk module and other modules are being covered in separate.

CP-71: Draft L2 Advice on SCR Standard Formula—Calibration of Nonlife Underwriting Risk. This Paper aims at providing advice with regard to the calibration of the nonlife underwriting risk module for the SCR standard formula, as requested by Article 109(b) of the Solvency II Level 1 text. CEIOPS will provide further advice on the calibration of catastrophe Standardised Scenarios under the nonlife catastrophe risk submodule in the course of 2010. The Advice should be read in conjunction with previous advice consulted upon in former CP48.

CP-72: Draft L2 Advice on SCR Standard Formula—Calibration of Health Underwriting Risk. This Paper aims at providing advice with regard to the calibration of the health underwriting risk module as requested by Article 109(b) of the Solvency II Level 1 text. CEIOPS will provide further advice on the Standardized Scenarios required under the SLT Health catastrophe risk submodule and the non-SLT Health catastrophe risk submodule in the course of 2010. The Advice should be read in conjunction with previous advice consulted upon in CP50.

CP-73: Draft L2 Advice on MCR—Calibration. This Paper aims at providing advice with regard to the calculation of the MCR as requested in Article 128 of the Solvency II Level 1 text. This Paper follows CP55, which was released for consultation in July 2009. It gives draft advice on the calibration of the MCR, in particular on the calibration of the linear function referred to in Article 127(1b) of the Level 1 text.

CP-74: Draft L2 Advice on SCR Standard Formula—Correlation Parameters. This Paper aims at providing advice with regard to the choice of the correlation parameters applied in the SCR standard formula to aggregate capital requirements on module and submodule level as requested in Article 109(1c) of the Solvency II Level 1 text.

CP-75: Draft L2 Advice on SCR Standard Formula—Undertaking Specific Parameters. This Paper provides advice as requested by Articles 109(1) (h) and (i) of the Solvency II Level 1 text with regard to the subset of standard parameters in the life, nonlife, and health underwriting risk modules that may be replaced by USPs. The Paper covers the standardized methods to be used by a (re)insurance undertaking to calculate those USPs, and the criteria for supervisory approval relating to the completeness, accuracy, and appropriateness of the data. The Paper should be read in conjunction with advice on which CEIOPS consulted in CP43, CP48, CP49, and CP50.

CP-76: Draft L2 Advice on Technical Provisions—Simplifications. This Paper aims at providing advice with regard to simplified methods and techniques to calculate TPs in order to ensure that actuarial and statistical methodologies are proportionate to the nature, scale, and complexity of the risks, as requested in Article 85(h) of the Solvency II Level 1 text. The advice integrates previous advice on which CEIOPS consulted in CP45; hence part of the advice has been finalized following previous consultation. For this reason, stakeholders are asked not to comment on the paragraphs identified in the introduction of the Paper. This Paper should be read in conjunction with the advice on which CEIOPS consulted in CP39 and CP43.

CP-77: Draft L2 Advice on SCR Standard Formula—Simplifications. This Paper aims at providing advice with regard to simplified calculations for specific submodules and risk modules as well as the criteria the (re)insurance undertakings shall be required to meet in order to be entitled to use the simplifications, as requested in Article 109(j) of the Solvency II Level 1 text.

CP-79: Draft L2 Advice on simplifications for captives. This Paper aims at providing advice with regard to simplified calculations for the calculation of the SCR for captives, as requested in Article 109(j) of the Solvency II Level 1 text. It elaborates on possible simplifications for the calculation of the SCR for captives, due to their specific business model.

CP-78: Draft L2 Advice on Solvency II: Technical criteria for assessing third-country equivalence in relation to Articles 172, 227, and 260.

Published on November 30, 2009. Consultation period: until February 5, 2010, 12.00 CET.

D.7.4 CEIOPS' Final Advices

On November 10, 2009, CEIOPS published 34 FAs on the issues that had been out for consultation during the 1st and 2nd wave earlier in 2009. CEIOPS' feedback on the 1st and 2nd waves of CPs was given in CEIOPS (2009f35). Web site: www.ceiops.org

The following are the first FA:

CEIOPS-DOC-21-09: CEIOPS' Advice for L2 Implementing Measures on SII: Elements of actuarial and statistical methodologies for the calculation of the BE (*former CP no. 26*).

CEIOPS-DOC-22-09: CEIOPS' Advice for L2 Implementing Measures on SII: Segmentation for the calculations of TPs (*former CP no. 27*).

CEIOPS-DOC-23-09: CEIOPS' Advice for L2 Implementing Measures on SII: SCR—Standard formula—Counterparty default risk (*former CP no. 28 and 51*).

CEIOPS-DOC-24-09: CEIOPS' Advice for L2 Implementing Measures on SII: Supervisory approval of ancillary OFs (*former CP no. 29*).

CEIOPS-DOC-25-09: CEIOPS' Advice for L2 Implementing Measures on SII: Technical Provisions—Treatment of future premiums (*former CP no. 30*).

CEIOPS-DOC-26-09: CEIOPS' Advice for L2 Implementing Measures on SII: SCR—Standard formula—Allowance of risk mitigation techniques (*former CP no. 31*).

CEIOPS-DOC-27-09: CEIOPS' Advice for L2 Implementing Measures on SII: Technical Provisions—Assumptions about future management actions (*former CP no. 32*).

CEIOPS-DOC-28-09: CEIOPS' Advice for L2 Implementing Measures on SII: Procedure to be followed for the approval of an IM, including the addendum on group specificities (*former CP no. 37*).

CEIOPS-DOC-29-09: CEIOPS' Advice for L2 Implementing Measures on SII: System of Governance (*former CP no. 33*).

CEIOPS-DOC-30-09: CEIOPS; Advice for L2 Implementing Measures on SII: Transparency and accountability (*former CP no. 34*).

CEIOPS-DOC-31-09: CEIOPS' Advice for L2 Implementing Measures on SII: Valuation of assets and "other liabilities" (*former CP no. 35*).

CEIOPS-DOC-32-09: CEIOPS' Advice for L2 Implementing Measures on SII: Special Purpose Vehicles (*former CP no. 36*).

CEIOPS-DOC-33-09: CEIOPS' Advice for L2 Implementing Measures on SII: Technical Provisions—Actuarial and statistical methodologies to calculate the BE (*former CP no. 39*).

CEIOPS-DOC-34-09: CEIOPS' Advice for L2 Implementing Measures on SII: Technical Provisions—Risk free interest rate (*former CP no. 40*).

CEIOPS-DOC-35-09: CEIOPS' Advice for L2 Implementing Measures on SII: Calculation of Technical Provisions as a whole (*former CP no. 41*).

CEIOPS-DOC-36-09: CEIOPS' Advice for L2 Implementing Measures on SII: Risk Margin (*former CP no. 42*).

CEIOPS-DOC-37-09: CEIOPS' Advice for L2 Implementing Measures on SII: Technical Provisions—Standard for data quality (*former CP no. 43*).

CEIOPS-DOC-38-09: CEIOPS' Advice for L2 Implementing Measures on SII: Technical Provisions—Counterparty default adjustment (*former CP no. 44*).

CEIOPS-DOC-39-09: CEIOPS' Advice for L2 Implementing Measures on SII: Classification and eligibility of OFs (*former CP no. 46*).

CEIOPS-DOC-40-09: CEIOPS' Advice for L2 Implementing Measures on SII: SCR—Standard formula—Market risk (*former CP no. 47*).

CEIOPS-DOC-41-09: CEIOPS' Advice for L2 Implementing Measures on SII: SCR—Standard formula—Nonlife underwriting risk (*former CP no. 48*).

CEIOPS-DOC-42-09: CEIOPS' Advice for L2 Implementing Measures on SII: SCR—Standard formula—Life underwriting risk (*former CP no. 49*).

CEIOPS-DOC-43-09: CEIOPS' Advice for L2 Implementing Measures on SII: SCR—Standard formula—Health underwriting risk (*former CP no. 50*).

CEIOPS-DOC-44-09: CEIOPS Advice for L2 Implementing Measures on SII: SCR—Standard formula—Reinsurance mitigation (*former CP no. 52*).

CEIOPS-DOC-45-09: CEIOPS' Advice for L2 Implementing Measures on SII: SCR—Standard formula—Operational Risk (*former CP no. 53*).

CEIOPS-DOC-46-09: CEIOPS' Advice for L2 Implementing Measures on SII: SCR—Standard formula—loss-absorbing capacity of TPs (*former CP no. 54*).

CEIOPS-DOC-47-09: CEIOPS' Advice for L2 Implementing Measures on SII: MCR Calculation (*former CP no. 55*).

CEIOPS-DOC-48-09: CEIOPS' Advice for L2 Implementing Measures on SII: Tests and standards for IM approval (*former CP no. 56*).

CEIOPS-DOC-49-09: CEIOPS' Advice for L2 Implementing Measures on SII: Capital add-ons (*former CP no. 57*).

CEIOPS-DOC-50-09: CEIOPS' Advice for L2 Implementing Measures on SII: Supervisory reporting and disclosure (*former CP no. 58*).

CEIOPS-DOC-51-09: CEIOPS' Advice for L2 Implementing Measures on SII: Remuneration (*former CP no. 59*).

CEIOPS-DOC-52-09: CEIOPS' Advice for L2 Implementing Measures on SII: Group solvency assessment (*former CP no. 60*).

CEIOPS-DOC-53-09: CEIOPS' Advice for L2 Implementing Measures on SII: Intragroup transactions and risk concentration (*former CP no. 61*).

CEIOPS-DOC-54-09: CEIOPS' Advice for L2 Implementing Measures on SII: cooperation and colleges of supervisors (*former CP no. 62*).

The 29 January 2010 CEIOPS' published the 2nd set of Finale Advice, FA. There were 13 FA published. Three FAs were postponed for the TSs for QIS5 that was sent to the EC at the end of March 2010.

CEIOPS-DOC-59-10: Repackagd Loans Investment (*former CP no. 63*).

CEIOPS-DOC-60-10: CEIOPS' Advice for L2 Implementing Measures on SII: Extension of the Recovery Period (*former CP no. 64*).

CEIOPS-DOC-61-10: CEIOPS' Advice for L2 Implementing Measures on SII: Partial IMs (*former CP no. 65*).

CEIOPS-DOC-62-10: CEIOPS' Advice for L2 Implementing Measures on SII: Supervision of Group Solvency for Groups with Centralized Risk Management (*former CP no. 66*).

CEIOPS-DOC-63-10: CEIOPS' Advice for L2 Implementing Measures on SII: Treatment of participations (*former CP no. 67*).

CEIOPS-DOC-64-10: CEIOPS' Advice for L2 Implementing Measures on SII: Ring-fenced Funds (*former CP no. 68*).

CEIOPS-DOC-65-10: CEIOPS' Advice for L2 Implementing Measures on SII: Equity risk submodule (*former CP no. 69*).

CEIOPS-DOC-66-10: CEIOPS' Advice for L2 Implementing Measures on SII: SCR Standard Formula—Calibration of Market Risk Module (*former CP no. 70*).

CEIOPS-DOC-70-10: CEIOPS' Advice for L2 Implementing Measures on SII: SCR Standard Formula—Correlations (*former CP no. 74*).

CEIOPS-DOC-71-10: CEIOPS' Advice for L2 Implementing Measures on SII: SCR standard formula—Undertaking-specific parameters (*former CP no. 75*).

CEIOPS-DOC-72-10: CEIOPS' Advice for L2 Implementing Measures on SII: Technical Provisions—Simplified methods and techniques to calculate TPs (*former CP no. 76*).

CEIOPS-DOC-73-10: CEIOPS' Advice for L2 Implementing Measures on SII: SCR standard formula—Simplified calculations in the standard formula (*former CP no. 77*).

CEIOPS-DOC-74-10: CEIOPS' Advice for L2 Implementing Measures on SII: SCR standard formula—Simplifications/Specifications for captives (*former CP no. 79*).

D.8 FD PREAMBLE (RECITALS)

Some of the recitals (2) to (142) from the FD as adopted by the EP and the ECOFIN Council 2009 are quoted here. The numbering is adjusted to the final version of the FD text. Recitals not quoted here are given in Chapters 22 through 26 and Chapter 34 as explanatory texts. The number to the left before the recital is the numbering in the preamble. The headings are given for "easy reading." The recitals given here are exactly as they were on October 15 and have been updated by juridical and linguistic people. This text was approved on October 16, just before its publication in *Official Journal*; OJ (2009).

Exclusion of certain undertakings, authorizations, and so on.

(4) It is appropriate that certain undertakings which provide insurance services are not covered by the system established by this Directive due to their size, their legal status, their nature—as being closely linked to public insurance systems—or the specific services they offer. It is further desirable to exclude certain institutions in several Member States, the business of which covers only a very limited sector and is restricted by law to a specific territory or to specified persons.

(5) Very small insurance undertakings fulfilling certain conditions, including gross premium income below EUR 5 million, are excluded from the scope of this Directive. However, all insurance and reinsurance undertakings which are already licensed under the current Directives should continue to be licensed when this Directive is implemented. Undertakings that are excluded from the scope of this Directive should be able to make use of the basic freedoms granted by the Treaty. Those undertakings have the option to seek authorisation under this Directive in order to benefit from the single licence provided for in this Directive.

(6) It should be possible for Member States to require undertakings that pursue the business of insurance or reinsurance and which are excluded from the scope of this Directive to register. Member States may also subject those undertakings to prudential and legal supervision.

Motor Insurance

(7) Council Directive 72/166/EEC of 24 April 1972 on the approximation of the laws of Member States relating to insurance against civil liability in respect of the use of motor vehicles, and to the enforcement of the obligation to insure against such liability1; Seventh

Council Directive 83/349/EEC of 13 June 1983 based on the Article 54(3)(g) of the Treaty on consolidated accounts2; Second Council Directive 84/5/EEC of 30 December 1983 on the approximation of the laws of the Member States relating to insurance against civil liability in respect of the use of motor vehicles3; Directive 2004/39/EC of the European Parliament and of the Council of 21 April 2004 on markets in financial instruments4; and Directive 2006/48/EC of the European Parliament and of the Council of 14 June 2006 relating to the taking-up and pursuit of the business of credit institutions5 lay down general rules in the fields of accounting, motor insurance liability, financial instruments and credit institutions and provide for definitions in those areas. It is appropriate that certain of the definitions laid down in those directives apply for the purposes of this Directive.

(12) Directive 2000/26/EC of the European Parliament and of the Council of 16 May 2000 on the approximation of the laws of the Member States relating to insurance against civil liability in respect of the use of motor vehicles (Fourth motor insurance Directive) lays down rules on the appointment of claims representatives. Those rules should apply for the purposes of this Directive.

Reinsurance and Captives

(8) The taking-up of insurance or of reinsurance activities should be subject to prior authorisation. It is therefore necessary to lay down the conditions and the procedure for the granting of that authorisation as well as for any refusal.

(9) The directives repealed by this Directive do not lay down any rules in respect of the scope of reinsurance activities that an insurance undertaking may be authorised to pursue. It is for the Member States to decide to lay down any rules in that regard.

(10) References in this Directive to insurance or reinsurance undertakings should include captive insurance and captive reinsurance undertakings, except where specific provision is made for those undertakings.

(13) Reinsurance undertakings should limit their objects to the business of reinsurance and related operations. Such a requirement should not prevent a reinsurance undertaking from pursuing activities such as the provision of statistical or actuarial advice, risk analysis or research for its clients. It may also include a holding company function and activities with respect to financial sector activities within the meaning of Article 2(8) of Directive 2002/87/EC of the European Parliament and of the Council of 16 December 2002 on the supplementary supervision of credit institutions, insurance undertakings and investment firms in a financial conglomerate1. In any event, that requirement does not allow the pursuit of unrelated banking and financial activities.

(89) In order to take account of the international aspects of reinsurance, provision should be made to enable the conclusion of international agreements with a third country aimed at defining the means of supervision over reinsurance entities which conduct business in the territory of each contracting party. Moreover, a flexible procedure should be provided for to make it possible to assess prudential equivalence with third countries on a Community basis, so as to improve liberalisation of reinsurance services in third countries, be it through establishment or cross-border provision of services.

Products

(76) In view of the increasing mobility of citizens of the Union, motor liability insurance is increasingly being offered on a cross-border basis. To ensure the continued proper functioning of the green card system and the agreements between the national bureaux of motor insurers, it is appropriate that Member States are able to require insurance undertakings providing motor liability insurance in their territory by way of provision of services to join and participate in the financing of the national bureau as well as of the guarantee fund set up in that Member State. The Member State of provision of services should require undertakings which provide motor liability insurance to appoint a representative in its territory to collect all necessary information in relation to claims and to represent the undertaking concerned.

(77) Within the framework of an internal market it is in the interest of policy holders that they should have access to the widest possible range of insurance products available in the Community. The Member State in which the risk is situated or the Member State of the commitment should therefore ensure that there is nothing to prevent the marketing within its territory of all the insurance products offered for sale in the Community as long as they do not conflict with the legal provisions protecting the general good in force in that Member State and in so far as the general good is not safeguarded by the rules of the home Member State.

(78) Provision should be made for a system of sanctions to be imposed when, in the Member State in which the risk is situated or the Member State of the commitment, an insurance undertaking does not comply with any applicable provisions protecting the general good.

(79) In an internal market for insurance, consumers have a wider and more varied choice of contracts. If they are to benefit fully from that diversity and from increased competition, consumers should be provided with whatever information is necessary before the conclusion of the contract and throughout the term of the contract to enable them to choose the contract best suited to their needs.

(80) An insurance undertaking offering assistance contracts should possess the means necessary to provide the benefits in kind which it offers within an appropriate period of time. Special provisions should be laid down for calculating the Solvency Capital Requirement and the absolute floor of the Minimum Capital Requirement which such undertaking should possess.

(81) The effective pursuit of Community co-insurance business for activities which are by reason of their nature or their size likely to be covered by international co-insurance should be facilitated by a minimum of harmonisation in order to prevent distortion of competition and differences in treatment. In that context, the leading insurance undertaking should assess claims and fix the amount of technical provisions. Moreover, special cooperation should be provided for in the Community co-insurance field both between the supervisory authorities of the Member States and between those authorities and the Commission.

(82) In the interest of the protection of insured persons, national law concerning legal expenses insurance should be harmonised. Any conflicts of interest arising, in particular, from the fact that the insurance undertaking is covering another person or is covering a person in respect of both legal expenses and any other class of insurance should be precluded

as far as possible or resolved. To that end, a suitable level of protection of policy holders can be achieved by different means. Whichever solution is adopted, the interest of persons having legal expenses cover should be protected by equivalent safeguards.

(83) Conflicts between insured persons and insurance undertakings covering legal expenses should be settled in the fairest and speediest manner possible. It is therefore appropriate that Member States provide for an arbitration procedure or a procedure offering comparable guarantees.

(84) In some Member States, private or voluntary health insurance serves as a partial or complete alternative to health cover provided for by the social security systems. The particular nature of such health insurance distinguishes it from other classes of indemnity insurance and life insurance insofar as it is necessary to ensure that policy holders have effective access to private health cover or health cover taken out on a voluntary basis regardless of their age or risk profile. Given the nature and the social consequences of health insurance contracts, the supervisory authorities of the Member State in which a risk is situated should be able to require systematic notification of the general and special policy conditions in the case of private or voluntary health insurance in order to verify that such contracts are a partial or complete alternative to the health cover provided by the social security system. Such verification should not be a prior condition for the marketing of the products.

(85) To that end, some Member States have adopted specific legal provisions. To protect the general good, it should be possible to adopt or maintain such legal provisions insofar as they do not unduly restrict the right of establishment or the freedom to provide services, it being understood that such provisions should apply in an identical manner. Those legal provisions may differ in nature according to the conditions in each Member State. The objective of protecting the general good may also be achieved by requiring undertakings offering private health cover or health cover taken out on a voluntary basis to offer standard policies in line with the cover provided by statutory social security schemes at a premium rate at or below a prescribed maximum and to participate in loss compensation schemes. As a further possibility, it may be required that the technical basis of private health cover or health cover taken out on a voluntary basis be similar to that of life insurance.

Taxation

(86) Host Member States should be able to require any insurance undertaking which offers, within their territories, compulsory insurance against accidents at work at its own risk to comply with the specific provisions laid down in their national law on such insurance. However, such a requirement should not apply to the provisions concerning financial supervision, which should remain the exclusive responsibility of the home Member State.

(87) Some Member States do not subject insurance transactions to any form of indirect taxation, while the majority apply special taxes and other forms of contribution, including surcharges intended for compensation bodies. The structures and rates of such taxes and contributions vary considerably between the Member States in which they are applied. It is desirable to prevent existing differences leading to distortions of competition in insurance services between Member States. Pending subsequent harmonisation, the application of the tax systems and other forms of contribution provided for by the Member States in which

the risk is situated or in the Member State of the commitment is likely to remedy that problem and it is for the Member States to make arrangements to ensure that such taxes and contributions are collected.

(88) Those Member States not subject to the application of Regulation (EC) No 593/2008 of the European Parliament and of the Council of 17 June 2008 on the law applicable to contractual obligations (Rome I) should, in accordance with this Directive, apply the provisions of that Regulation in order to determine the law applicable to contracts of insurance falling within the scope of Article 7 of that Regulation.

Winding-Up

(117) Since national legislation concerning reorganisation measures and winding-up proceedings is not harmonised, it is appropriate, in the framework of the internal market, to ensure the mutual recognition of reorganisation measures and winding-up legislation of the Member States concerning insurance undertakings, as well as the necessary cooperation, taking into account the need for unity, universality, coordination and publicity for such measures and the equivalent treatment and protection of insurance creditors.

(118) It should be ensured that reorganisation measures which were adopted by the competent authority of a Member State in order to preserve or restore the financial soundness of an insurance undertaking and to prevent as far as possible a winding-up situation, produce full effects throughout the Community. However, the effects of any such reorganisation measures as well as winding-up proceedings vis-a-vis third countries should not be affected.

(119) A distinction should be made between the competent authorities for the purposes of reorganisation measures and winding-up proceedings and the supervisory authorities of the insurance undertakings.

(120) The definition of a branch for insolvency purposes, should, in accordance with existing insolvency principles, take account of the single legal personality of the insurance undertaking. However, the legislation of the home Member State should determine the manner in which the assets and liabilities held by independent persons who have a permanent authority to act as agent for an insurance undertaking are to be treated in the winding-up of that insurance undertaking.

(121) Conditions should be laid down under which winding-up proceedings which, without being founded on insolvency, involve a priority order for the payment of insurance claims, fall within the scope of this Directive. Claims by the employees of an insurance undertaking arising from employment contracts and employment relationships should be capable of being subrogated to a national wage guarantee scheme. Such subrogated claims should benefit from the treatment determined by the law of the home Member State (lex concursus).

(122) Reorganisation measures do not preclude the opening of winding-up proceedings. Winding-up proceedings should therefore be able to be opened in the absence of, or following, the adoption of reorganisation measures and they may terminate with composition or other analogous measures, including reorganisation measures.

(123) Only the competent authorities of the home Member State should be empowered to take decisions on winding-up proceedings concerning insurance undertakings. The decisions should produce their effects throughout the Community and should be recognised by all Member States. The decisions should be published in accordance with the procedures of the home Member State and in the *Official Journal* of the EU. Information should also be made available to known creditors who are resident in the Community, who should have the right to lodge claims and submit observations.

(124) All the assets and liabilities of the insurance undertaking should be taken into consideration in the winding-up proceedings.

(125) All the conditions for the opening, conduct and closure of winding-up proceedings should be governed by the law of the home Member State.

(126) In order to ensure coordinated action amongst the Member States the supervisory authorities of the home Member State and those of all the other Member States should be informed as a matter of urgency of the opening of winding-up proceedings.

(127) It is of utmost importance that insured persons, policy holders, beneficiaries and any injured party having a direct right of action against the insurance undertaking on a claim arising from insurance operations be protected in winding-up proceedings, it being understood that such protection does not include claims which arise not from obligations under insurance contracts or insurance operations but from civil liability caused by an agent in negotiations for which, according to the law applicable to the insurance contract or operation, the agent is not responsible under such insurance contract or operation. In order to achieve that objective, Member States should be provided with a choice between equivalent methods to ensure special treatment for insurance creditors, none of those methods impeding a Member State from establishing a ranking between different categories of insurance claim. Furthermore, an appropriate balance should be ensured between the protection of insurance creditors and other privileged creditors protected under the legislation of the Member State concerned.

(128) The opening of winding-up proceedings should involve the withdrawal of the authorisation to conduct business granted to the insurance undertaking unless this has already occurred.

(129) Creditors should have the right to lodge claims or to submit written observations in winding-up proceedings. Claims by creditors resident in a Member State other than the home Member State should be treated in the same way as equivalent claims in the home Member State without discrimination on grounds of nationality or residence.

(130) In order to protect legitimate expectations and the certainty of certain transactions in Member States other than the home Member State, it is necessary to determine the law applicable to the effects of reorganisation measures and winding-up proceedings on pending lawsuits and on individual enforcement actions arising from lawsuits.

Implementation

(131) The measures necessary for the implementation of this Directive should be adopted in accordance with Council Decision 1999/468/EC of 28 June 1999 laying down the procedures for the exercise of implementing powers conferred on the Commission1.

(132) In particular, the Commission should be empowered to adopt measures concerning the adaptation of Annexes and measures specifying in particular the supervisory powers and actions to be taken and laying down more detailed requirements in areas such as the system of governance, public disclosure, assessment criteria in relation to qualifying holdings, calculation of technical provisions and capital requirements, investment rules and group supervision. The Commission should also be empowered to adopt implementing measures granting to third countries the status of equivalence with the provisions of this Directive. Since those measures are of general scope and are designed to amend non-essential elements of this Directive, inter alia, by supplementing it with non-essential elements, they must be adopted in accordance with the regulatory procedure with scrutiny laid down in Article 5(a) of Decision 1999/468/EC.

(133) Since the objectives of this Directive cannot be sufficiently achieved by the Member States and can therefore, by reason of their scale and effects, be better achieved at Community level, the Community may adopt measures, in accordance with the principle of subsidiarity as set out in Article 5 of the Treaty. In accordance with the principle of proportionality, as set out in that Article, this Directive does not go beyond what is necessary in order to achieve those objectives.

(134) Council Directive 64/225/EEC of 25 February 1964 on the abolition of restrictions on freedom of establishment and freedom to provide services in respect of reinsurance and retrocession1; Council Directive 73/240/EEC of 24 July 1973 abolishing restrictions on freedom of establishment in the business of direct insurance other than life insurance; Council Directive 76/580/EEC of 29 June 1976 amending Directive 73/239/EEC on the coordination of laws, regulations and administrative provisions relating to the taking-up and pursuit of the business of direct insurance other than life assurance3; and Council Directive 84/641/EEC of 10 December 1984 amending, particularly as regards tourist assistance, First Directive (73/239/EEC) on the coordination of laws, regulations and administrative provisions relating to the taking-up and pursuit of the business of direct insurance other than life4 have become obsolete and should therefore be repealed.

(135) The obligation to transpose this Directive into national law should be confined to those provisions which represent a substantive change as compared with the earlier Directives. The obligation to transpose the provisions which are unchanged is provided for in the earlier Directives.

(136) This Directive should be without prejudice to the obligations of the Member States relating to the time-limits for transposition into national law of the Directives set out in Annex VI, Part B.

Insurance Guarantee Schemes

(137) The Commission will review the adequacy of existing guarantee schemes in the insurance sector and make an appropriate legislative proposal.

IORP

(138) Article 17(2) of Directive 2003/41/EC of the European Parliament and of the Council of 3 June 2003 on the activities and supervision of institutions for occupational retirement

provision1 refers to the existing legislative provisions on solvency margins. Those references should be retained in order to maintain the status quo. The Commission should conduct its review of Directive 2003/41/EC under Article 21(4) thereof as quickly as possible. The Commission, assisted by CEIOPS, should develop a proper system of solvency rules concerninginstitutions for occupational retirement provision, whilst fully reflecting the essential distinctiveness of insurance and, therefore, should not prejudge the application of this Directive to be imposed upon those institutions.

Further Issues

(139) Adoption of this Directive changes the risk profile of the insurance company vis-a-vis the policy holder. The Commission should as soon as possible and in any event by the end of 2010 put forward a proposal for the revision of Directive 2002/92/EC of the European Parliament and of the Council of 9 December 2002 on insurance mediation1, taking into account the consequences of this Directive for policy holders.

(140) Further wide-ranging reforms of the regulatory and supervisory model of the EU financial sector are greatly needed and should be put forward swiftly by the Commission with due consideration of the conclusions presented by the group of experts chaired by Jacques de Larosiere of 25 February 2009. The Commission should propose legislation needed to tackle the shortcomings identified regarding the provisions related to supervisory coordination and cooperation arrangements.

(141) It is necessary to seek advice from CEIOPS on how best to address the issues of an enhanced group supervision and capital management within a group of insurance or reinsurance undertakings. CEIOPS should be invited to provide advice that will help the Commission to develop its proposals under conditions that are consistent with a high level of policy holder (and beneficiary) protection and the safeguarding of financial stability. In that regard CEIOPS should be invited to advise the Commission on the structure and principles which could guide potential future amendments to this Directive which may be needed to give effect to the changes that may be proposed. The Commission should submit a report followed by appropriate proposals to the European Parliament and the Council for alternative regimes for the prudential supervision of insurance and reinsurance undertakings within groups which enhance the efficient capital management within groups if it is satisfied that an adequate supportive regulatory framework for the introduction of such a regime is in place. In particular, it is desirable that a group support regime operate on sound foundations based on the existence of harmonised and adequately funded insurance guarantee schemes; a harmonised and legally binding framework for competent authorities, central banks and ministries of finance concerning crisis management, resolution and fiscal burden-sharing which aligns supervisory powers and fiscal responsibilities; a legally binding framework for the mediation of supervisory disputes; a harmonised framework on early intervention; and a harmonised framework on asset transferability, insolvency and winding-up procedures which eliminates the relevant national company or corporate law barriers to asset transferability. In its report, the Commission should also take into account the behaviour of diversification effects over time and risk associated with being part of a group,

practices in centralised group RM, functioning of group IMs as well as supervision of intragroup transactions and risk concentrations.

(142) In accordance with point 34 of the Interinstitutional agreement on better law-making1, Member States are encouraged to draw up, for themselves and in the interest of the Community, their own tables illustrating, as far as possible, the correlation between this Directive and the transposition measures, and to make them public.

D.9 FD STRUCTURE

The structure of the FD in terms of Titles, Chapters, and Sections are given here. We have not included the subsections.

Title I: General Rules on the Taking-Up and Pursuit of Direct Insurance and Reinsurance

Article 1–177

Chapter I—Subject Matter, Scope and Definitions: Art 1–13

Section 1: Subject matter and scope: 1–2 Section 2: Exclusions from scope: 3–12

Section 3: Definitions: 13

Chapter II—Taking-Up of Business: Art 14–26

Chapter III—Supervisory Authorities and General Rules: Art 27–39

Chapter IV—Conditions Governing Business: Art 40–72

Section 1: Responsibility of the Administrative or Management Body: 40

Section 2: System of Governance: 41–50

Section 3: Public Disclosure: 51–56

Section 4: Qualifying holdings: 57–63

Section 5: Professional secrecy, exchange of information, and promotion of supervisory convergence: 64–71

Section 6: Duties of auditors: 72

Chapter V—Pursuit of Life and Nonlife Insurance Activity: Art 73–74

Chapter VI—Rules Relating to the Valuation of Assets and Liabilities, Technical Provisions, Own Funds, Solvency Capital Requirement, Minimum Capital Requirement and Investment Rules: Art 75–135

Section 1: Valuation of assets and liabilities: 75

Section 2: Rules relating to technical provisions: 76–86

Section 3: Own funds: 87–99

Section 4: Solvency capital requirement: 100–135

Chapter VII—Insurance and Reinsurance Undertakings in Difficulty or in An Irregular Situation: Art 136–144

Chapter VIII—Right of Establishment and Freedom to Provide Services: Art 145–161

Section 1: Establishment by insurance undertakings: 145–146

Chapter II—Final Position: Art 218–246

Section 1: Group solvency: 218–243

Section 2: Risk concentration and intra-group transactions: 244–245

Section 3: Risk management and internal control: 246

Chapter III—Measures to Facilitate Group Supervision: Art 247–259

Chapter IV—Third Countries: Art 260–264

Chapter V—Mixed-Activity Insurance Holding Companies: Art 265–266

Title IV: Reorganization and Winding-Up of Insurance Undertakings

Article 267–296

Chapter I—Scope and Definitions: Art 267–268

Chapter II—Reorganization Measures: Art 269–272

Chapter III—Winding-up proceedings: Art 273–284

Chapter IV—Common Provisions: Art 285–296

Title V: Other Provisions

Article 297–304

Title VI: Transitional and Final Provisions

Article 305–312

Chapter I—Transitional Provisions: Art 305–307

Chapter II—Final Provisions: Art 309–312

Annex I–VII

I. Classes of nonlife insurance

II. Classes of life insurance

III. Legal forms of undertakings

IV. SCR standard formula

V. Groups of non-life insurance classes for the purposes of Article 157

VI. Part A: Repealed Directives, Part B: List of time limits

VII. Correlation table (between articles)

APPENDIX E

European Solvency II

General Ideas

THE GENERAL DEVELOPMENT of the European Solvency II system and the basic architecture has been discussed in Section 2.3 and in Appendix D. The general ideas in the final FD are discussed in Chapter 22.

The undertakings participating in QIS4, conducted during 2008, considered the architecture of the Solvency II system as sound, and also indicated that they needed more guidance on the calculation of TPs, the valuation of some assets and of other liabilities, the calculation of the SCR, and the assessment of eligible capital; CEIOPS (2008e).

Lessons made from the financial crisis, mainly during 2007–2009, are discussed by CEIOPS (2009a). They concluded that the overall architecture of the system was sound, but nevertheless some improvements could be done to get Solvency II operable in both normal and stressed situations. These considerations were included in the final advice that CEIOPS published during late 2009. Some of the main issues discussed by CEIOPS (2009a) are the following:

- Further refinement of the existing calibrations, both at main-module and submodule levels
    - Strengthened dependency structure in the standard formula
    - Only high-quality capital elements for loss absorbing
- Strengthened governance
- Strengthened risk management
    - Require a strong own risk assessment
- Liquidity risk needs more attention
- Strengthened internal controls
- The scope of regulation and supervision needs focusing on consolidated entities rather than on solo entities

- A holistic approach to insurers' exposures to SPVs
- Consistent treatment of CDS
- Rigorous stress testing under Pillar 2 (SRP)
- "Through-the-cycle" or dynamic reserving (equalization mechanism)

Here we discuss the following general issues:

- The proportionality principle: Section E.1
- Internal models: Section E.2
- Groups: Section E.3
- EOFs: Section E.4
- Investments: Section E.5

E.1 PROPORTIONALITY PRINCIPLE

Owing to the level of sophistication of the new Solvency II regime, one of the key questions has been how to gear it to the *nature, scale, and complexity of the risks* to which an insurance undertaking is exposed, particularly, but not only, with regard to small- and medium-sized undertakings (SMEs). This gives a brief sum up of the proportionality principle, which is discussed in, for example, CEIOPS (2008b).

CEIOPS addressed this issue during the third wave of Calls for Advice, in answer to Call for Advice 23. In this Call for Advice, the Commission Services asked CEIOPS to advise on whether a specific treatment of small undertakings was necessary. In its answer, CEIOPS (2006a) recognized the importance of the proportionality principle in the application of the Solvency II regime and also found that "policyholders should not expect a lower degree of protection simply because their cover is provided by a smaller undertaking. In addition, CEIOPS recognised that size in itself may not be an adequate proxy for the risk to which an undertaking is exposed. Undertakings within the scope of the Directive should not be classified differently on the basis of size."

In the draft framework directive (FD) published in July 2007 and amended in February 2008, COM (2007) and COM (2008), particular care was taken to ensure that the new solvency regime is not too burdensome for low risk-profile undertakings, which are often SMEs. Appropriate treatment of these undertakings has to be achieved through the application of the *principle of proportionality*.

The FD established the proportionality principle as a general principle that applies throughout the Directive, highlighting it in several provisions and leaving its concrete implementation to Level 2 measures and Level 3 guidance.

The proportionality principle is a generally acknowledged principle of the due course of law and is not therefore comprehensively defined in the FD. It applies throughout the directive. It has two aspects:

- Proportionality has to be taken into account when implementing the requirements laid down in the FD
- Supervision has to be carried out in a proportionate manner

Proportionality does not mean the introduction of automatic and systematic simplifications for certain undertakings. The principle will be applied where it would be disproportionate to the

- nature
- scale
- complexity

of undertakings' business to apply general quantitative and qualitative rules without relief. The individual risk profile should be the primary guide in assessing the need to apply the proportionality principle.

The principle of proportionality applies to all the provisions in the directive and, as a consequence, to all future implementing measures. The proportionality principle should be applied in a coherent way across the three pillars as well as the group provisions.

Where simplifications with respect to quantitative Pillar I requirements are applied, its implications on the supervisory review process, on the insurer's ORSA, and on disclosure requirements under Pillar III have to be considered.

Moreover, the proportionality principle is being applied regardless of whether the principle of proportionality is explicitly mentioned in a provision or not. The mention of the principle of proportionality in certain articles should not lead to the conclusion *per se* that it does not apply or applies less where it is not explicitly mentioned.

All Solvency II provisions need to be suitable and necessary to achieve their objective as well as appropriate in light of the nature, scale, and complexity of an undertaking's risk profile. The two aspects have to be taken into consideration and put into relation to each other: the purpose that is to be achieved and the means employed to serve this purpose. In order to be considered proportionate a measure has to be, at least, suitable and necessary to achieve its objective appropriately. A measure is necessary if there is no less onerous method available that is equally or even better suited to serve the objective. Appropriateness requires that the drawbacks of a measure are not totally disproportionate to the benefits it reaps.

Proportionality works in two ways: It justifies simpler and less burdensome ways of meeting requirements for low risk-profile portfolios, and also increases the likelihood that undertakings in fulfilling requirements will need to apply more sophisticated methods and techniques for more complex risk portfolios.

In the explanatory memorandum (preamble, recitals) of the FD, COM (2007) and (2008), the importance of the principle of proportionality is explicitly linked to the need to avoid excessive strain on SMEs. This does, however, not mean that size is the only relevant factor when the principle is considered. There is a link between the proportionality requirement to the nature, scale, and complexity of the risks inherent in the business, which is consistent with the general risk-based approach of Solvency II where the size of an undertaking on its own is not a relevant risk-based criterion.

In considering the *nature* of the risks, supervisors will take into account the underlying risk profiles of the classes of business, an undertaking is written, for example, whether it is a long- or short-tailed business, or whether it is a low-frequency and high-severity business, or consists of high-frequency and low-severity risks. The specific nature of risks inherent to the reinsurance business and to the captives business should also be taken into account.

Complexity is linked to the nature of the business as certain kinds of business may dictate the use of more demanding methods or an advanced system of governance. In particular, a more sophisticated risk management system may be regarded in order to deal properly with all risks the undertaking faces. However, it may also be introduced via the investment strategy of the undertaking or because the insurer chooses to employ challenging methods or processes in some areas that require a commensurate degree of complexity in other areas of the undertaking. It is also linked to the complexity in the evaluation of the commitments, for example, unlimited motor liability, or investment in a complex option, or annuities (as opposed to a lump sum), or nonproportional reinsurance (as opposed to a straightforward direct insurance business).

Through *scale* a size criterion is introduced. Relating to the valuation of assets, liabilities, or risks, this criterion resembles a materiality requirement and the approach applied should ensure an appropriate relative or absolute approximation of the theoretically correct value. This shall be calculated with regard to the provisions of the FD. Relating to Pillar II, cost–benefit analysis can also be seen as a scale issue, applied, for example, to governance processes.

In assessing what is proportionate, the focus must be on *the combination of all three criteria* to arrive at a solution that is adequate to the risk an undertaking is exposed to. For instance, a business may well be of small scale but could still include complex risk profiles or, on the contrary, it may be of large scale with a simple risk profile. In the first case, it should not be allowed to use simplified methods, while the possibility may be considered in the second case under very specific circumstances.

So, while it will probably tend to be the SMEs that will find relief in the application of the proportionality principle through simpler ways of meeting supervisory requirements, it is actually imprecise to talk of proportionality as size based. Consequently, CEIOPS will not find a definition for SMEs or to develop a set of simplified requirements to be applied only to these SMEs.

The proportionality principle was mentioned explicitly eight times in the draft FD, COM (2007) and (2008). In relation to

- *Pillar I:* TPs and the SCR standard formula

- *Pillar II:* supervision in establishment and freedom to provide services in a recital, supervisory powers, the system of governance and the supervisory review process—frequency, scope
- *Pillar III:* supervisory reporting

The TPs are split between a BE and an RM. With respect to the BE, both simplifications and proxies have been suggested in the QIS4 technical specifications, QIS4 (2008).

Proxies have been developed in case there is insufficient company-specific data of appropriate quality to apply a reliable statistical actuarial method. Therefore proxies can be regarded as special types of simplified methods that are positioned at the "lower end" of continuum of methods that could be applied. The simplifications and proxies proposed for QIS4 were the joint work of CEIOPS' Financial Requirements Expert Group, FinReg, and the Coordination Group set up between CEIOPS and the Groupe Consultatif; see CEIOPS-GC (2008).

The development of proxy methods for the valuation of TPs acknowledges the fact that, even after the introduction of the Solvency II regime, sufficient statistical data will not always be available, for example, in the case where an insurer sells products within a new LoB or where the portfolio is too small to allow the build-up of credible historic claims data. In the draft FD, COM (2007) and (2008), this issue is recognized and it is possible to use case-by-case methods for the valuation of TPs, which can be regarded as a special case of proxy techniques, where insurers have insufficient data of appropriate quality to apply an appropriate actuarial technique.

Simplifications and proxies for the BE are discussed in Chapter 24. Simplifications for the capital charges in the SCR calculation are discussed in Chapters 26 through 33.

FinReg asked CEA for advice on applying the proportionality principle. In its answer, CEA (2008b) outlined its holistic view on the principle, suggesting a two-stage approach to apply proportionality:

Stage 1: Use *nature and complexity* as a filter to identify where simplifications are likely to be appropriate

Stage 2: Use *scale thresholds* to limit the use of certain simplifications where this is needed to mitigate risk from any inaccuracy inherent in the approaches

Nature: Risks that are more predictable are easier to model and there is less chance that they will be understated. In such instances it is easier to develop proxies and benchmark runoff patterns that contain relatively small calibration margins to compensate for the extra uncertainty associated with not having entity-specific experience. As the level of uncertainty gets bigger it becomes increasingly more difficult to be sure that a particular calibration margin will be appropriate, making it more important to get specialist actuarial help on a case-by-case basis. Typically, there is more uncertainty over underwriting risk for nonlife business than is the case for life because there are less data on which to base best estimate assumptions, and future trend risk is usually more significant.

Complexity: Complexity is associated with the level of calculation sophistication and/or level of expertise needed. Where a particular simplification materially does not capture the complexity of the business, it may not be possible to alleviate this by incorporating a degree of conservatism in its calibration because the simplified and sophisticated approaches are just too different.

Complexity and nature can be interrelated. For example, on nonlife business, uncertainty over future claims experience may require more sophisticated and complex approaches, for example, requiring specialist actuarial knowledge in order to determine appropriate BEL.

However, relating to how to interpret nature and complexity, the use of simplifications is more appropriate for some risks and LoBs than others. The CEA therefore saw the nature and complexity criteria as being used as *a filter* to identify when simplifications are likely to be appropriate and to avoid excessive risk from model error.

Size: CEA believes that similar levels of protection should be provided to policyholders regardless of the absolute size of their company and whether or not they are using proportionality. This requires a *relative* rather than absolute *scale measure.*

In order to illustrate this, CEA considered two companies that are identical, except that one is five times the size of the other in all aspects. Suppose for the larger company it is decided that a particular simplification can be used for a particular LoB, but that its use should not exceed 10% of some volume measure in order to avoid unacceptable risks to its policyholders. If scales were to have an absolute as well as relative component, this would imply that the smaller company should get preferential treatment, that is, be allowed to use it for a threshold greater than 10%. However, given that the companies are identical in all aspects except absolute size, this would mean that the policyholders in the smaller company were at greater risk than those in the larger company, which the CEA is against. Hence, the conclusion is that scale thresholds have to be relative and not absolute.

Scale thresholds provide a means of comparing different simplifications and a common measure of model error. The illustrative Table E.1 below demonstrates this with the common measure being derived as the model error per 1% of the overall SCR/BEL. Note that the figures are only illustrative and also that there is no need to calculate the absolute model error. *A relative measure of 1 denotes the maximum amount of model error that would be acceptable,* for example, that obtained when using a particular simplification at the threshold limit.

Model error is akin to random noise and as such the model error associated with different simplifications is likely to be uncorrelated. Other things being equal, there is less overall model error risk associated with using a number of different simplifications to a lesser

TABLE E.1 Illustration of the Scale Threshold

Simplification	Individual Scale Threshold (% Overall SCR/ BEL)	Measure of Model Error per 1% of the Overall SCR/BEL
A	50%	1/50
B	20%	1/20
C	5%	1/5

Source: From CEA. 2008b. Initial thoughts on the use of simplifications. CEA Working Paper. ECO 8370. December 22. Available at http://www.cea.eu. With permission.

extent than using one simplifications to a greater extent because of diversification effects. Such diversifications effects therefore need to be allowed for. In theory a dependence matrix approach could be used to allow for diversification effects, but this is unlikely to feasible because the very large number of possible simplification combinations would make the dependence matrix to be too large.

A pragmatic approach would be to simply add the measure of model error where a high positive dependence is expected, for example, where the same or very similar simplifications are used, and then to use a sum of squares approach to combine this with the measure of model error from other uncorrelated simplifications.

To illustrate how such an aggregation approach might work consider an example using the illustrative figures in Table E.1 where 25% of the overall SCR was calculated using simplification A ($25 * 1/50 = 1/2$) and 10% of the overall SCR calculated using simplification B ($10 * 1/20 = 1/2$). Since the two different simplifications are used, the overall measure of model error would be $\sqrt{0.5^2 + 0.5^2} = 0.71$ showing that there is further scope to use simplifications as the overall measure of model risk is less than the sum.

E.2 INTERNAL MODELS

With the approval from the local supervisor the (partial) internal models may be used. For a given subrisk or risk module, a company can exchange the standard formula for its own methodology to create a model for the capital charge. If only one or more subrisks are changed, they are called partial internal model. They combine elements of internal models and enable companies to incrementally improve their risk management.

The use of partial or fully internal models for the risk measurement has the intention of giving the company incentives to evaluate and control their risks more accurate.

The goals of QIS4 for internal models are, as outlined in QIS4 (2008, TS.XIV),

- To collect reliable and comparable *quantitative data* from partial and full internal models that were currently used by firms for assessing their capital needs. These data would assist CEIOPS in conducting a range of statistical analyses and then providing advice to the EC and other stakeholders when discussing the calibration of the standard model and its likely impact on the Solvency II regime;
- To collect high-level *qualitative information* from insurance undertakings that use internal models for assessing their capital needs to influence the qualitative aspects of the implementation measures; and
- To collect *general information* from all insurance undertakings to assess the current and potential future status of internal modeling in Europe.

To achieve the first goal, firms had to have assessed the quality and comparability of the data against high-level principles. Therefore, firms had to concentrate on comparing the results and the modeling aspects of the standard formula with those derived from their internal models. Key areas to address in this context were the modeling requirements of the draft FD and the data that had been used when firms calibrated their models. It was also important to understand the differences in assumptions and definitions between those underlying existing models in firms and those anticipated under Solvency II. To this end and to the extent that this was practicable, the estimates derived from internal models had to be compatible with the overall calibration objectives for the standard formula.

The importance of qualitative issues was highlighted by the second goal. CEIOPS did not seek detailed qualitative information at the time of QIS4, but it was likely that further analysis would be undertaken at a later stage of the process.

Finally, the QIS4 exercise also served as a tentative mapping exercise of the current development stage of internal models used by market participants, and to indicate to what extent insurance undertakings planned to use an internal model for calculating their SCR or to use partial models to calculate the eligible capital for some modules of the SCR.

Approximately 50% of the solo undertakings participating in QIS4 also provided information on internal models. However, only 10% of the undertakings provided quantitative results for the IM calculations. Some undertakings (13%) reported that the standard formula works well and that they did not intend to develop any IM. Most of the companies (63%) reported that they would probably develop partial and/or full internal models due to better risk management and governance. The SCR calculated from internal models resulted in a lower capital requirement than from the standard formula. Some of the risk modules generated, on the average, lower capital requirement than under the standard formula (e.g., interest rate risk, longevity risk, lapse risk, and premium and reserve risk), while other risk modules generated higher capital requirement (e.g., operational risk, equity risk, and property risk); CEIOPS (2008e).

In the light of the financial crisis, CEIOPS (2009a), it was shown that internal models are valid management tools and are supported by CEIOPS. However, the lessons learnt from the crisis showed that some areas had to be addressed in the final advice for Level 2 implementing measures. For example, the quality of data used in the model must be an issue for both internal purposes and also for the approval of a model. Model errors need to be accounted for.

E.3 GROUP ISSUES

Earlier, the nonlife and life directives ("Solvency 0" and Solvency I) provided an insurance undertaking, a "solo" company, to have own capital as a protection of their solvency (solvency margin). The *Insurance Group Directive (IGD)*, COM (1998), and the *Financial Conglomerate Directive (FCD)*, COM (2002b), prevented these requirements from being circumvented by groups of insurance undertakings or financial conglomerates which could otherwise allow the same regulatory capital to be used more than once to cover the solvency requirement of an insurance undertaking belonging to a group or conglomerate (*double gearing*).

These two directives ensured that a group's or conglomerate's insurance or reinsurance undertakings have a financial situation that was taken care of by an *adjusted solvency*.

With an *insurance group* we mean two or more insurance undertakings (direct insurer, reinsurer, etc.) that act as a *group*. A member of a group could be a participating undertaking (at least 20% or more of the voting rights or capital are held); see COM (1998, Article 1).

The IGD was one of the insurance directives that were consolidated in the new "Solvency II directive."

Groups were specifically addressed for the first time in QIS3 in 2007. The QIS4 developed those initial specifications in order to test the methods that were set out in the Directive.

Group supervision and group support were the most "controversial" elements in the draft FD. The local supervisor will play a vital role in the day-to-day supervision of a nondomestic company or subsidiary. The group supervisor will be the "home regulator" for the group, which safeguards that the group holds capital and provides *group support*. This would have allowed for a much more efficient and economic approach to the use of capital. Unfortunately, the group support text in the draft FD was taken out and a review clause was included; see Appendix D.

The group support would have been particularly important, as it would have allowed the freedom to exploit the diversification of assets. The parent undertaking would have to make any required capital available as agreed between the local supervisor and the group supervisor, and certainty of this was essential. Moreover, the proposal foresaw the possibility of providing to a subsidiary group support from funds present in another subsidiary. Company law aspects could have implied that transfers was not that easy to affect and also not as timely as requested. Insurance supervisors are not qualified to make a value judgement on the legal possibility to transfer funds between companies, possibly established in different jurisdictions.

The *group solvency*, defined as the difference between EOFs and the SCR could be calculated in three ways, based on the existing legislation (Solvency I); also, see the following:

- Accounting consolidation-based method (Method 1, the default method), where EOFs and consolidated SCR were based on consolidated accounts.
- Deduction and aggregation method (Method 2, the alternative method), where the SCR is the sum of the SCRs of all group companies.
- A combination of Methods 1 and 2, where the exclusive application of Method 1 would not be appropriate.

If supervisors do not decide differently, Method 1 had to be applied (default). The alternative method, that is, Method 2 did not provide diversification benefits to groups.

The consolidated SCR (Method 1) and the SCRs of group companies (Method 2) could be calculated either by the standard formula or by an approved (partial) internal model. Capital add-ons are possible where the risk profile of the group was not adequately reflected in the standard formula or (partial) internal model.

In case the group wants to use an (partial) internal model to calculate the consolidated group SCR, the supervisory authorities concerned shall cooperate to decide whether or not to grant permission. A joint decision should be reached within 6 months. In the absence of a joint decision within this period, the group supervisor makes its own decision. CEIOPS could be consulted at the request of the group or any of the concerned supervisors, which extends the period by 2 months.

The competent authority responsible for group supervision, that is, the current Lead Supervisor appointed by each Coordination Committee, managed the QIS4 process for each of their groups.

The specifications in QIS4 were designed for the purposes of QIS4 only and did not necessarily reflect the final solutions for Solvency II.

Data had to be valued in accordance with the QIS4 valuation specifications; see Appendices F and G. The annual accounts could be taken as a starting point, adjusted for material differences with QIS4 valuation standards. Where this was not possible, material differences had to be noted.

QIS4 had four main objectives with respect to groups: (1) to establish a functional standard SCR formula and OFs calculation for groups, (2) to establish an appropriate recognition of group level diversification effects in the standard SCR formula for groups, (3) to collect information on the use of internal group models, and (4) to collect information on the group support regime.

The diversification effects varied considerably between different groups. On average it was 21% when measured as the difference between the sum of the solo SCRs and the group SCR. This was mainly an effect of the diversity of their business. Only a small number of groups provided any quantitative information on the amount of group support that was allowed to be used as part of OFs. The group support was classified as Tier 2 capital. See CEIOPS (2008e) for more details.

In the light of the financial crisis, CEIOPS (2009a), it was shown that Solvency II needed to clarify the treatment of holdings, in particular regarding their supervision, as well as the treatment of non-EU parts of an insurance group. CEIOPS requires that the insurers integrate their internal control procedures within a group, including the internal audit and risk management functions. The focus needs to be on consolidated entities rather than on solo entities.

Two methods for calculating the *group capital requirement* were decided in the FD:

- *Default method—Accounting consolidation:* This method recognizes diversification benefits between different group entities, including between EEA and non-EEA (re)insurance entities and with-profit businesses. The calculation also takes into account any participation in (re)insurance entities. No diversification benefits are recognized for noninsurance participations or participations with no control relationship. All worldwide (re)insurance undertakings of the group, including any non-EEA (re)insurance undertakings, should be taken into account in the calculations. Groups may take into account geographical diversification benefits where these are permitted in the Standard Formula for the solo entity.

- *Alternative method—Deduction and aggregation method:* This calculates the required capital as the sum of solo SCRs of each group entity. Under QIS4, this was calculated from the output of the solo spreadsheets. In order to distinguish intragroup effects from diversification effects when comparing this alternative method with the default method adjustments are needed to eliminate market and counterparty risk charges on IGTs.

Groups that were participating in QIS4 were requested to calculate the following:

- The group capital requirement and the group OFs according to the following two methods:
    - The standard SCR formula and OFs calculation applied to the consolidated group position, that is, the default accounting consolidation method;
        - In addition, groups that had non-EEA entities and/or with-profits life business were also invited to show the results of two variations on this method to assess the extent of the diversification benefits arising from non-EEA entities (Variation 1) and from with-profit businesses (Variation 2).
    - The sum of the solo SCRs and OFs of each group entity, that is, the alternative method, the deduction, and aggregation method;
        - In order to produce an accurate group position in the Solvency II context, this method needs to be adjusted to eliminate market and counterparty risk charges on IGTs.
        - The unadjusted sum of solo SCRs of each group entity will also be calculated from the output of the solo spreadsheets in order to distinguish intragroup effects from diversification effects when comparing this method with the accounting consolidation method.
- The group capital requirements and capital resources under the regime currently in force for (re)insurance entities, as calculated under the IGD or FCD, as appropriate, see above.

In this sequel, we only look at the two main methods, the default and the alternative, used in QIS4.

E.3.1 Default Method: Accounting Consolidation

The component of Group SCR with respect to the insurance* entities in the group is termed $SCR_{wwconso}$. This component is calculated by applying the standard formula approach, as outlined in Chapters 26 through 33, of the SCR to the group insurance business as if it were a single entity.

The BS for the group insurance business, including both EEA and non-EEA entities, should therefore be calculated based on QIS4 specifications.

The total capital requirement for the group is then calculated as the sum of the

- consolidated SCR ($SCR_{wwconso}$)
- SCR for other financial sectors (SCR_{ofs})
- SCR for noncontrolled participations (SCR_{ncp})

*With insurance entities we mean both direct insurance companies and/or reinsurance companies,

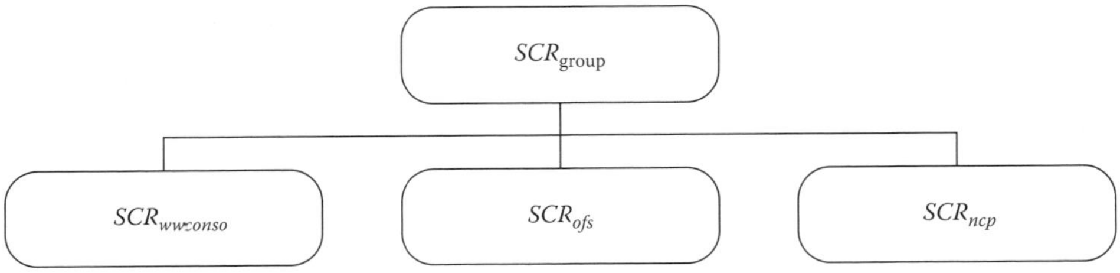

FIGURE E.1 The group capital requirement using the default method: Accounting consolidation. $SCR_{wwconso}$: the consolidated SCR from the group. SCR_{ofs}: the capital requirement from other financial sectors, and SCR_{ncp}: the capital requirement for noncontrolled participations.

This can then be shown as a bottom-up aggregation of the SCR components as shown in Figure E.1.

The contribution of participations held in *other financial sectors* to the capital requirement of the group should be the other financial sector's requirements. When participations in another financial sector form a group for which a specific capital requirement exists, the latter, instead of the sum of the requirements of each solo entity, should be used. This will form SCR_{ofs} which was added to $SCR_{wwconso}$ without recognition of any diversification effects.

When the group's participation in an EEA insurer can be regarded as *a relationship of control,* the contribution to the group SCR for this participation would follow a *look-through approach*. Control means the relationship between a parent undertaking and a subsidiary, as set out in Article 1 of EEC (1983), or a similar relationship between any natural or legal person and an undertaking. This means that the "parent" has a majority of the shareholders' or members' voting rights, or has the right to appoint or remove a majority of the members of the administrative, management, or supervisory body and is at the same time a shareholder in or member of that undertaking, or has the right to exercise a dominant influence over an undertaking of which it is a shareholder or member, or is a shareholder in or member of an undertaking.

This is consistent with the full consolidation of the participation in the accounts or the proportional consolidation (if there is jointly shared control of the participation). In case of a full-consolidated participation, minority interests would in turn contribute to cover part of the group SCR, with some limitations.

When the group's participation in an EEA insurer is *greater or equal than 20%* but without a relationship of control, the contribution to the group SCR in respect of the participation would be calculated as the group's share in the participation multiplied by the solo SCR of this participation. This would be consistent with the equity method consolidation where such participation would be accounted for at equity value in the group's consolidated accounts. The contribution of these participations to the group SCR would be the sum of the above-mentioned calculations. If the solo SCR of the current year was not available, then the previous SCR could be used adjusted for the annual movement in premiums.

When the group's interest in an EEA insurer is *lower than 20%,* the contribution to the group SCR with respect to the participation was calculated by applying the equity risk charge to the value of the participation.

The contribution of the EEA insurance undertakings in which the group has no relationship of control will form SCR_{ncp}, that is, the SCR of noncontrolled participations.

Some changes had to be made to the different subrisks discussed in Chapters 26 through 33. They were

- *Nonlife underwriting risk:* The method is the same as at solo level, except that a Herfindahl–Hirchmann index had to be calculated to take geographical diversification into account.
- *Life underwriting risk:* The method is the same as at solo level.
- *Credit risk—counterparty default risk:* The group's LGD had to be calculated as the sum of all of the solo LGDs for a particular counterparty. As IGTs had to be eliminated, they were set to zero in the group calculation; QIS4 (2008, TS.XVI.B.11).
- *Market risk:* The method is the same as at solo level, except for
  - Interest rate risk—$CR_{MR,IR}$: The effect of interest rate shocks could be calculated on the consolidated approach by working with the components of the capital charge calculated for each solo entity or with-profit fund. The calculation needs to take into account that upward and downward shocks on interest rate cannot happen at the same time. The capital charge for the group can then be expressed as follows: the maximum of either the sum of capital charges based on an upward shock or the sum of capital charges based on a downward shock.
  - Currency risk—$CR_{MR,CR}$: The currency risk for non-EEA countries had to apply on the net asset value minus the capital requirement of the subsidiary or the subgroup, that is, if the net asset value of the subsidiary is 100 and its capital requirement is 80, the currency risk applies only to $100 - 80 = 20$. As for interest rate risk above, the effect of foreign exchange rate shocks can be calculated on the consolidated approach by working with the components of the capital charges calculated for each solo entity or with-profit fund. The calculation needs to take into account that upward and downward shocks on exchange rate cannot happen at the same time. The capital charge for the group can then be expressed in the same way as for the interest rate risk above.
- *Operational risk:* This was calculated as the sum of the solo operational risk charges. This included the 30% overall cap specified in the standard formula. In addition, groups should apply the standard formula for the operational risk module to the consolidated business of the group. Groups had to calculate SCR_{op} in this way at the consolidated level but without the 30% cap. The results of the two latter calculations had to be reported as additional information.

Adjustment for the loss-absorbing effect of TPs: Participants' attention was drawn to the fact that the loss-absorbing effect of TPs could be limited to certain parts of the group because of contractual or legal constraints. When calculating the adjustment for the loss-absorbing effect of TPs at group level, participants should ensure that the assumptions they make were consistent with any such contractual or legal constraints in this regard.

Adjustment for the loss-absorbing effect of deferred taxes: Where the taxation regime applicable to insurance groups did not allow them to benefit from tax integration for all the entities part of the group, for example, cross-border groups, groups could use the following simplification to assess the adjustment for the loss-absorbing effect of deferred taxes at group level:

$$Adj_{\mathrm{DT}}^{\mathrm{Group}} = \sum_i Adj_{\mathrm{DT},i}^{\mathrm{Group}} \cdot \frac{SCR^{\mathrm{Group}}}{\sum_i SCR_i^{\mathrm{solo}}}$$

where index i covers all the entities of the group included in the calculation of $SCR_{wwconso}$ and

$Adj_{\mathrm{DT},i}^{\mathrm{solo}}$ is the solo Adjustment for the loss-absorbing effect of deferred taxes of entity I, SCR^{Group} is the consolidated group SCR, and SCR_i^{solo} is the solo SCR of entity i.

Group own funds: Had to be calculated by applying the solo capital specifications to the group as a whole. The BS of the group, including both EEA and non-EEA entities, should therefore be calculated based on QIS4. Adjustments must be made to eliminate double use of EOFs and to limit the use of eligible elements of OFs to the group undertaking in which they were held when those elements of OFs cannot effectively be made available to cover the SCR of other group undertakings.

Total share capital (shareholders' capital + minority interests): Group OFs should be calculated on the basis of groups' on- and off-balance sheet positions valued in accordance with the solo valuation specifications. The QIS4 participants could use their statutory accounts as a starting point that had to be adjusted for valuation differences. In particular, participants had to specify the amount of the adjustment due, for example, to the difference between the value of TP and investments calculated according to the QIS4 valuation standards and any different standards used for individual group undertakings.

A minority interest's share in any surplus assets of a group entity in which it holds an interest is not necessarily available for use elsewhere in a group. Therefore, a minority interest's share in any surplus OFs should only be included in group OFs up to the minority interest's proportional share in the group entity's SCR. In the draft FD, eligible minority interests were calculated on the basis of the percentage of participation of these minority interests multiplied by the required solvency margin of the entities in which these minority interests hold—directly or indirectly—participations. In the Solvency II framework, under the consolidated approach, the solvency capital requirement for the group will not be the sum of the solo requirements due to the recognition of some diversification benefits. Thus, it would not be possible to calculate directly the contribution of a solo entity to the group SCR. However a proxy contribution with respect to minority interest j could be calculated, resulting from the following formula:

$$Contr_j = SCR \cdot \frac{SCR_j}{\sum_i SCR_i}$$

where index i covers all the group entities included in the calculation of $SCR_{wwconso}$.

Hybrid capital: These capital items, mainly noncumulative preference shares and subordinated debt, could not be considered as transferable if not issued or guaranteed by the ultimate parent of the group. In essence, this depends on the rights of the subscribers on the revenues of these instruments.

Participations in non-EEA insurance entities: Groups had to calculate OFs in non-EEA undertakings separately from OFs in EEA undertakings on a Solvency II basis. Surpluses should be identified separately. EOFs in non-EEA undertakings were clearly available to meet the SCR of the undertaking in which they were held but a final decision had not yet been made on the extent to which surplus OFs in non-EEA undertakings could be considered transferable and hence contribute to overall available group capital. Participants were therefore requested to provide information on any legal or other barriers to the free transfer of surpluses from the non-EEA jurisdictions in which they hold capital surpluses.

Participations in other financial sector entities: For information purposes, groups had to report the amount of any surplus OFs in other financial sector entities that were "consolidated" in the calculation. Surpluses should be identified separately. With regard to participations that were "not consolidated" in the group's solvency assessment, OFs with respect to such participations had to be deducted from the OFs of the group. Relevant sectoral valuation rules should be applied.

With-profit business: A firm could contain items of EOFs and/or profit-sharing mechanisms within the TPs, which could only be used to cover the liabilities for a limited set of policyholders, for example, where a firm writes with-profit business, or protected cell or statutory lines/social insurance with participation. A set of assets, liabilities, and OFs that is so restricted is termed a "fund." EOFs that were only available to cover losses in one entity could be included in the calculation of the group OFs subject to a limit. This limit should be the SCR of the related entity. Participants were requested to provide information on any legal or other barriers to the free transfer of surpluses from the with-profits funds in which they hold capital surpluses.

Nontransferable assets: The sum of nontransferable assets valued as OFs, for example, minority interests, could not exceed the SCR in which these assets were located, with a specific reduction due to the diversification effects recognized in the consolidated group SCR. Here, the contribution to group OFs from entity j, Contr_j, was limited according to the following formula

$$\text{Contr}_j = SCR \cdot \frac{SCR_j}{\sum_i SCR_i}$$

where index i covers all the entities of the group included in the calculation of $SCR_{wwconso}$.

E.3.2 Alternative Method: Deduction and Aggregation Method

The required capital was calculated as the sum of each individual SCR for each entity in the group, including non-EEA entities, minority interests, and cross-sectoral or other participations.

In order to produce an accurate group position in the Solvency II context, the solo SCRs had to be adjusted to eliminate double counting of market and counterparty risk requirements on IGTs. The aim was to calculate the contribution of each EEA solo entity to the SCR of the group by summing the "solo-adjusted SCRs" and adding the capital requirements of other entities where an adjusted SCR calculation cannot be readily calculated, for example, this may include non-EEA entities, minority interests, or cross-sectoral or other participations. This can be expressed in the following formula:

$$SCR_{\text{group}} = \sum SCR_{\text{solo-adjusted}} + CR_{\text{ot}},$$

where the "solo-adjusted SCR" is defined as the SCR calculated at each solo entity level with the elimination of IGTs. This elimination had to be carried out at each submodule level and the solo-adjusted SCR equalled the total SCR of the entity multiplied by the percentage used for the consolidated accounts.

CR_{ot} is defined as the sum of the capital requirements for all other group businesses where a "solo-adjusted SCR" cannot be readily calculated.

In practice, the "solo-adjusted" SCR would be calculated for CR_{MR} and CR_{CR} in the following manner:

- CR_{MR}: The idea was to say that the shocks prescribed in a scenario-based approach did not affect the IGTs. With a factor-based approach, there was a zero charge for intragroup assets.
- CR_{CR}: The capital charge stemming from default risk of intragroup cedants, that is, risks transferred into another entity of the group, should be taken to be equal to zero.

Groups could take into account materiality considerations in calculating the adjustment for IGTs. In that case, participants had to explain what materiality rule was used, as well as its rationale. Participants may wish to focus on the most material IGTs, for example, financial reinsurance arrangements, loans, and so on. Where participants could not calculate the solo-adjusted SCR for each single entity in the group, they could calculate an overall adjustment to the sum of the solo SCRs instead, for those entities for which an adjusted solo SCR could not be calculated separately.

Group OFs: These were calculated as the OFs of the ultimate participating insurance undertaking or insurance holding company, see, for example, Chapter 25, plus its proportional share of the OFs in each group entity. That share was equal to the one used for the consolidated accounts. In order to eliminate the potential for double gearing, the OFs in each group entity had to be based on an assessment of the solo OFs after the deduction of participants and subsidiaries and removal of other intragroup arrangements. As under this option no diversification benefits were being considered in assessing the group SCR, there would be no adjustments in the capital resources reflecting diversification benefits.

E.4 ELIGIBLE OWN FUNDS

Here we discuss the development and calibration of OFs that has been one of the building blocks in Solvency II. The final advice from CEIOPS is discussed in Chapter 25.

The EOFs are available and are discussed in Chapter 10 from a general point of view, and also from the view of the draft FD. Ideas that had been discussed within the EC were also discussed. The objective of QIS4 was to collect further information. Further information was needed because the QIS3 specifications were limited to the high-level principles set out in the FD proposal, which was interpreted rather broadly when classifying OFs into tiers. So as to remedy this problem, QIS4 TSs included much more detailed guidance on how those high-level principles could be implemented in practice.

QIS4 specifications for OFs essentially focus on the implementation of the tiering structure set out in the draft FD, COM (2007) and (2008), based on a further specification of those principles. To enhance the participants' awareness of the potential changes in OFs as compared to the solvency regime that was currently in force, the QIS4 spreadsheets will automatically compute and indicate the SCR and MCR coverage ratios, based on participants' own classification of OF items into tiers and on the quantitative limits set out in the draft.

E.4.1 Principles

The main concern about a particular eligible element was to what extent it meets the characteristics set forth in the draft FD. In QIS4, elements were classified in relation to how well and when they absorb losses compared to paid-up ordinary share capital or paid-up initial fund. There was a broad spectrum of capital instruments that were potentially eligible in OFs. These included equity instruments with debt-like features, and debt instruments with equity-like features. Member States referred to these instruments using different terms: some considered subordinated liabilities to be hybrid capital instruments, while others consider subordinated liabilities to be distinct from hybrid capital instruments. This specification referred to both hybrid capital instruments and subordinated liabilities; but participants were reminded that what is ultimately relevant is the extent to which a particular instrument holds the qualitative characteristics required for classification in a particular tier.

For QIS4 purposes, the following applied:

- The excess of assets over liabilities was classified as a *Tier 1 item*, with specification of any elements of the excess of assets over liabilities that may be subject to restricted loss absorption
- A hybrid capital instrument, regardless of its legal form, could be *a Tier 1, Tier 2, or Tier 3 item*
- A subordinated liability could be a *Tier 1, Tier 2, or Tier 3 item*
- A promise to provide OFs could be a *Tier 2 or Tier 3 item*

The definition of *subordination* for Solvency II purposes was similar to the definition used for accounting purposes, that is, capital items should not only be subordinated to policyholders' interests but also to all liabilities that are not explicitly "subordinated." Another relevant issue relates to the transferability of OFs within a company, in particular when ring-fenced structures had been introduced.

As the Solvency II project has prudential supervision as its exclusive purpose. it is neutral and agnostic with regard to any issue concerning general financial statements or tax issues. Therefore the QIS4 should not be understood as impacting existing accounting or taxation rules.

The treatment of *ring-fenced fund structures* described here was especially developed for QIS4. It was acknowledged that the treatment of such funds under the Solvency II framework had to be further analyzed following QIS4, once additional information had been collected during the QIS4 exercise.

Where part of the business of participants is segregated from the rest of their operations in a ring-fenced fund, they should follow the guidance.

Ring-fenced fund should be understood as a contractual or legal arrangement whereby part of the assets or eligible surplus of the company are strictly segregated from the rest of the company's investments or resources and can only be used to meet the insurance obligations with respect to which the ring-fenced fund has been established.

As a consequence, the OFs held within the ring-fenced fund, that is, the excess of the segregated assets over the insurance obligations concerned, could only absorb the losses stemming from the risks associated with the ring-fenced insurance portfolio. The OFs held within the ring-fenced fund were not available to meet the company's other obligations and could not be transferred from the ring-fenced fund to support the rest of the activity, on a going-concern basis. Consequently, when assessing the solvency of the company as a whole, it might seem appropriate to adjust the amount of OFs eligible to cover the SCR in order to take account of the nontransferability of the OFs held within ring-fenced funds.

The QIS4 results, CEIOPS (2008e), showed an increase of total OFs across countries of 27% in comparison with the OFs reported earlier. The increase is explained by

1. Valuation adjustments following the move to market-consistent valuation
2. The reclassification of equalization reserves from TPs to OFs
3. The inclusion of hybrid capital instruments, subordinated liabilities, and ancillary OFs

The general criteria for the classification of OFs were supported. The impacts of deferred taxes on the amount of OFs were unclear as the participating companies used different approaches. Surplus funds represented a significant amount in three countries (Sweden, Germany, and Denmark). Companies were also reporting other reserves with full or restricted loss-absorbing capacity. Life insurers writing with-profit business reported most ring-fenced funds.

One main lesson from the financial crisis discussed by CEIOPS (2009a) was the Solvency II needed to ensure a sufficiently high quality of capital that could guarantee its loss-absorbing capacity (especially under stressed situations). The final advice of CEIOPS published in late 2009 will reflect this.

E.4.2 Classification of Own Funds into Tiers

For QIS4 purposes, CEIOPS used the characteristics of tiering set out in the draft FD, and also to further specifying the meaning of the term *sufficient* in the perpetuality characteristic and the meaning of the term *to a substantial degree* in the draft FD.

The work resulted in a detailed list of OF items given in QIS4 (2008, TS.V.K). The list set forth, per tier, separately for *basic own funds* and *ancillary own funds*, the relevant characteristics and the interpretation of the characteristics or key features.

For QIS4 purposes, participants were requested to provide the amount of each eligible element of capital included in the list mentioned above. When working on the list, CEIOPS concluded that the characteristics could be made more operational by proceeding as follows:

- Distinguish more clearly between loss absorbency on winding-up and loss absorbency in going concern
- Merge subordination with loss absorption on winding-up
- Distinguish the different elements of mandatory servicing costs

As a result, CEIOPS developed in the list of tiers, six characteristics that are broadly in line with the draft FD:

1. Subordination of total amount on winding-up
2. Full loss-absorbency in going concern
3. Undated or of sufficient duration (perpetuality)
4. Free from requirements/incentives to redeem the nominal amount
5. Absence of mandatory fixed charges
6. Absence of encumbrances

The term *to a substantial degree* applied to characteristics 3–6 in QIS4. CEIOPS view is that characteristics 3–6 should be viewed as features to be taken into account when assessing the loss-absorbency features in characteristics 1 and 2. To be more precise,

- Inclusion in Tier 1 capital:
    - A hybrid capital, instrument, or subordinated liability had to be able to be written down or converted into equity in times of stress, notwithstanding a possible later write-up in case of subsequent profits.
- Inclusion in Tier 2 capital:
    - Any payment (principal or coupon) on a hybrid capital instrument or subordinated liability had to be able to be deferred in times of stress until the financial position is restored.
    - The receipts of a promise to provide OFs had to be certain.

The precise level of losses that would trigger conversion or write-down of hybrid capital instruments and subordinated liabilities was still under discussion. For QIS4 purposes, participants were requested to classify items according to whether conversion or write-down was a contractual provision.

Given the multiplicity of the actual form that hybrid capital instruments, subordinated liabilities, and promises to provide OFs can take, participants were requested to provide a specification of each line item qualified as "other," providing brief details of which characteristics those items possess. The verification of the perpetuality characteristic (Key Features 3) for each capital item, using minimum durations (e.g., 5 years or 10 years) as a reference, was still under consideration. The use of minimum durations from the issue date could simplify the assessment of this characteristic and could also enhance cross-sector consistency, given the current banking framework. However, fixed minimum durations from the issues date would not be sufficiently risk sensitive. For QIS4 purposes, participants were requested to classify in Tier 1 those instruments with a maturity from the issue date of at least 10 years, and in Tier 2 those instruments with a maturity from the issue date of at least 5 years. But participants could also provide additional information on the remaining duration of those instruments from the reporting date, as well as information on the duration of their insurance liabilities, in order to allow for detailed analysis.

For all ancillary OF items, that is, off-balance sheet items, the characteristics and key features had to apply to the basic own fund item that arises once the ancillary own fund item had been called up.

Examples that were given in QIS4 (2008, TS.V.F) are listed below.

Basic Own Funds, Tier 1

- The excess of assets over liabilities, determined in accordance with QIS4 valuation principles.
- The BS items that contribute to this difference are mentioned in the list of tiers. Each item, and the amount, must be stated separately.
- A net surplus on an insurer's scheme for employee benefits, such as postretirement benefits, is not included in the excess of assets over liabilities unless the net surplus can absorb losses for the benefit of policyholders, because the insurer has a legal claim on the net surplus and can cash the net surplus to settle policyholder claims.
- Budgeted supplementary calls that mutual undertakings can make on their members are eligible for inclusion in the excess of assets over liabilities.
- Subordinated mutual member accounts.
- Noncumulative perpetual preference shares.
- Noncumulative fixed-term preference shares with a minimum duration of at least 10 years from the issue date.

- Other hybrid capital instruments that fulfill the criterion of loss-absorbency in going concern. The instrument must be undated or have a minimum maturity of at least 10 years from the issue date. Any interest step-ups must not apply before 10 years from the issue date and must not exceed the highest of 100 basis points or 50% of the initial credit spread.
- Subordinated liabilities that fulfill the criterion of loss-absorbency in going concern. The instrument must be undated or have a minimum maturity of at least 10 years from the issue date. Any interest step-ups must not apply before 10 years from the issue date and must not exceed the highest of 100 basis points or 50% of the initial credit spread.

Basic Own Funds, Tier 2

- Cumulative perpetual preference shares.
- Cumulative fixed-term preference shares with a minimum maturity of at least 5 years from the issue date.
- Other hybrid capital instruments that are either undated or have a minimum maturity of at least 5 years from the issue date. Any interest step-ups must not apply before 5 years from the issue date and must not exceed the highest of 100 basis points or 50% of the initial credit spread.
- Subordinated liabilities that are either undated or that have a minimum maturity of at least 5 years from the issue date. Any interest step-ups must not apply before 5 years from the issue date and must not exceed the highest of 100 basis points or 50% of the initial credit spread.

Basic Own Funds, Tier 3

- Cumulative fixed-term preference shares with a minimum maturity of less than 5 years from the issue date
- Other hybrid capital instruments that are either undated or have a minimum maturity of less than 5 years from the issue date
- Subordinated liabilities that are either undated or have a minimum maturity of less than 5 years from the issue date

Ancillary Own Funds, Tier 2

- Unpaid common shares; unpaid initial fund.
- Unpaid noncumulative preference shares.
- Unpaid and callable hybrid capital instruments eligible for inclusion in Tier 1.
- Letters of credit and guarantees, in accordance with the draft FD.
- Supplementary member calls of Protection and Indemnity Associations in accordance with the draft FD.

- Part of the amount of unbudgeted supplementary member calls by mutual undertakings. These calls are subject to recovery risk, as the callable amount might not be fully received following a call. It is also possible that receipt is delayed so that the claim is not available immediately to cover losses. As a consequence, only part of unbudgeted supplementary member calls can be classified in Tier 2 ancillary own funds, being calls, the recoverability of which is considered certain. For QIS4, 40% of the maximum callable amount specified in the statutes of the mutual company can be classified in Tier 2 ancillary own funds, and the rest in Tier 3 ancillary own funds.

- Other commitments with equivalent loss absorption to ancillary own fund items mentioned specifically in the draft FD.

Ancillary Own Funds, Tier 3

- Unpaid cumulative preference shares

- Unpaid and callable hybrid capital instruments eligible for inclusion in Tier 2 or Tier 3

- Letters of credit and guarantees not eligible for inclusion in Tier 2

- Supplementary member calls of mutual undertakings not eligible for inclusion in Tier 2

- Other commitments not eligible for inclusion in Tier 2

E.5 INVESTMENTS

In the first wave of Calls for Advice from the EC, MARKT (2004a2), they asked CEIOPS on advice about investment management rules covering the capital requirements, that is, the MCR and the SCR. Analysis was needed if different rules were to be applied for different purposes such as TPs, MCR, and SCR. The Commission also required CEIOPS' advice on the requirements concerning the (minimum) content of an appropriate investment plan as well as its relationship to general business planning, internal control, and risk management processes, including quantitative aspects, for example, ALM and capital adequacy, as well as qualitative aspects, for example, prudent person. Practical issues concerning reviews, proportionality, and so on. had to be taken into account. CEIOPS gave an outline of the investment issues in its answer; CEIOPS (2005b).

Also in its answer to the Call for Advice 9 on safety measures, CEIOPS (2005c) discussed investments, quantitative limits on assets, and the prudent persons' role.

Under existing Directives that precede the Solvency II directive (see Section 2.3 for more details), TPs have to be covered by assets that are secure, liquid, and well diversified. This is largely a principle-driven approach, although specific limits are set to achieve investment diversification.

As an example, the equivalent of no more than 5% of total gross Ps can be invested in unsecured loans to achieve an appropriate mixture of investments, and no more than 1% in any single unsecured loan to achieve an appropriate spread of investments.

In preliminary proposals for the draft FD, it was envisaged that the current Articles on the prudent management of assets* would be strengthened as follows:

The assets covering the TPs and the SCR shall take account of the type of business carried on by an assurance undertaking in such a way as to secure the safety, yield, and liquidity of its investments, which the undertaking shall ensure are diversified and adequately spread. To this end, an assurance undertaking shall have an appropriate investment plan.

CEIOPS argued that an insurer that fails to meet its MCR could not afford to take any risks with its investments. Any insurer that fails to meet its SCR will be expected to take action to restore its position and one of the actions it might decide to take is to rearrange its investments to reduce risks and thus the SCR. If this "derisking" response can be generally anticipated, and the extent of "derisking" increases as available capital nears the MCR, the risks associated with investments should reduce and there is no need for the MCR itself to reflect investment risks.

The present rules for investment diversification may be said to offer a partial treatment for specific risk. Specific risk is that which can be addressed by holding a diversified portfolio. This is because it originates from factors associated with the individual issuer or counterparty that are not correlated with general economic trends.

TPs, MCR, and SCR have to a very large extent a common purpose—that is, enabling the undertaking to fulfill its obligations as they fall due.

CEIOPS considered that, in principle, the same limits should be applied to investments covering TPs, MCR, and SCR, but there may be a need for some exceptions for some types of business, for example, in a winding-up situation. The same limits should be applied regardless of whether the SCR is calculated using the standard formula, partial internal models, or full internal models.

An argument supporting a uniform treatment of assets was the fact that the MCR/SCR is a risk buffer that may have to be used up during the solvency assessment time horizon; in such a situation, the insurer should always be able to cover its liabilities with assets of sufficient quality. This could constitute a problem if assets of minor quality were allowed to cover the SCR/MCR. Applying the same rules at all three levels, that is, TPs, MCR, and SCR, has the added advantage of simplicity; CEIOPS (2005c).

Some types of insurance business may require a special treatment because some of the risks are borne by the policyholder—for example, in UL products. The SCR should cover any risks retained by the insurance undertaking.

Insurance undertakings should be required to have an overall investment strategy for all their assets, including derivatives, set by the Board of Directors, which should be updated as often as necessary (at least annually). Senior management should describe how this strategy will be implemented through an investment policy, including investment planning or investment procedures.

* Article 22 in the Codified Life Directive and Article 20 in the third nonlife Directive; *cf.* Section 2.3.

CEIOPS had the view that investment and ALM should be considered in a coherent manner. The overall investment strategy and policy should therefore adequately reflect ALM considerations.

The characteristics of liabilities are the driving force in developing investment policies for an insurer. The nature of the insurance business and the nature, terms, and conditions of the policies written require the establishment of TPs, and the investment in assets which are appropriate to the liabilities. The design and underwriting of products, and thus the resulting liabilities of an insurer, could not be considered in isolation from its investment activities. In order to ensure that it can meet its contractual liabilities to policyholders, an insurance undertaking would manage its assets in a sound and prudent manner, taking account of the profile of its liabilities, its solvency position, and its complete risk-return profile.

This forms the essence of the undertaking's ALM.

The undertaking's investment policy should describe the types and limits of investments that reflect its risk profile and financial strength. Details should be given if the policy is different for assets covering TPs, assets covering solvency requirements, and free assets. Special attention should be paid to financial derivatives and structured products that have the effect of financial derivatives. Within the framework of applicable laws and regulations, the investment policy, plans, and procedures, which would be communicated to all staff involved in investment activities, would, in principle, address the following main elements; CEIOPS (2005b):

- The determination of the asset allocation, including ALM considerations—that is, the short-term and long-term asset mix over the main investment categories.
- The establishment of limits for the allocation of assets by geographical area, markets, sectors, counterparties, and currency.
- The return on investments that should be achieved, given decisions on the distribution of profits to policyholders and shareholders (including where discretion is exercised in relation to with-profits life business).
- The degree of sensitivity to investment risks (including matching, RMs, and capital requirements) and the results of using quantitative tools in previous years (e.g., stress tests and/or scenarios).
- The extent to which the holding of some types of assets is ruled out or restricted where, for example, the sale of the asset could be difficult due to the illiquidity of the market or where independent (i.e., external) verification of pricing is not available.
- The use of financial derivatives as part of the general portfolio management process or of structured products that have the economic effect of derivatives.
- Key staff involved with investment activities have the appropriate levels of skills, experience, and integrity.

- The framework of accountability for all asset transactions.
- The nature of outsourcing and requirements for the safe keeping of assets (custodial arrangements).
- The methodology and frequency of the performance measurement and analysis.

Insurance undertakings would be required to have ALM tools that show the managers of an insurance undertaking how their decisions interrelate and impact on the financial results. Better understanding of these interrelationships enables better strategic choices, such as

- Can the undertaking reallocate its investments to boost returns, reduce risk, or both?
- How much reinsurance should the undertaking purchase? What type?
- How fast should the undertaking aim to grow?
- Should the undertaking exit certain business lines and enter others?
- Will potential acquisitions add value to the insurance undertaking?
- Does the insurance undertaking have enough capital to ensure its continued solvency?

CEIOPS supported the inclusion of high-level general requirements on insurance undertakings to manage their assets and liabilities appropriately, and to invest appropriately. Senior management would be required to exercise judgement in a sound and prudent manner.

The balance between this, specific limits and capital requirements, that is, quantitative and qualitative requirements, could be termed a prudent person plus approach. CEIOPS suggests that the future regime uses a "prudent person plus" approach. The idea of the "prudent person plus" approach is the combination of three different types of requirements:

- The risk-based SCR
- Qualitative requirements on the management of assets and liabilities
- Eligibility criteria for assets, asset–liability mismatches, and limits on asset concentrations; This issue is discussed in Section 25.2.

In its response to the first wave of Call for Advice, CEIOPS supported the inclusion of high-level general requirements for insurance undertakings to manage their assets and liabilities appropriately; CEIOPS (2005b).

Prudent management would be supported by risk-sensitive capital requirements, which would take into account asset and liability risks and the degree of an insurance undertaking's asset–liability mismatch. The SCR should therefore serve as an important tool in addressing the risk to which an undertaking is exposed, and encourage good risk management and internal control. As a result, the SCR would need to be supported by a safety net, consisting of rules on the eligibility of assets and limits on risk concentrations. This safety net would be necessary, irrespective of whether the SCR is calculated using the standard formula, partial internal models, or full internal models; CEIOPS (2005c).

The concept of prudent person approach was also included in the IORP Directive (Directive 2003/41/EC on the activities and supervision of institutions for occupational retirement provision).

APPENDIX F

European Solvency II

Asset Valuation

HERE WE DISCUSS the development and calibration of asset valuation that has been one of the building blocks in Solvency II. The final advice from CEIOPS is discussed in Chapter 23.

According to QIS2 conducted in the first summer of 2006, assets should be valued at their MV, taking account of any bid–ask spread. In cases where there is no readily available MV, alternative approaches may be adopted, but these would still be consistent with any relevant market information. For tradable assets, this should be an estimate of the realizable value, CEIOPS (2006d).

In QIS3 (CEIOPS, 2007a), CEIOPS developed the valuation of assets. Assets should be valued at their MV. Where reliable, observable market prices in deep and liquid markets existed, asset values would be set equal to these market prices. For long positions on assets, appropriate quoted market price would be the bid price taken at the valuation date, while for short positions, it would be the offer price.

If a market price was observable but is not reliable, for instance, due to illiquidity, reasonable proxies for valuation could be used, taking into account the degree of unreliability and illiquidity of the asset in an adequate manner. Participants were asked to provide a description of the proxies used.

In cases where there was no readily available market value, an alternative approach should be adopted, but this would still have to be consistent with any relevant market information. For tradable assets, this would be an estimate of the realizable value.

Illiquid or nontradable assets should be valued on a prudent basis, fully taking into account the reduction on value due to the credit and liquidity risks attached.

- In absence of any sufficient evidence, the value of these assets should not be higher than their acquisition cost reduced by the estimated profit margin charged by the seller at that moment, and the depreciation due to the use or obsolescence of the asset.

- In absence of any sufficient evidence, intangible assets, furniture, fittings, data process equipment, and similar assets with a significant risk of depreciation in case of realization should be valued at nil.

If independent and reliable expert opinions were available, these should be considered in the valuation.

A valuation of assets at their MV is taken as a reference standard for the calculation of investments and eligible capital. For illiquid (and hard-to-value) instruments, it should be noted that they normally have a value lower than, otherwise similar, more marketable instruments.

For QIS4, the following hierarchy of high-level principles was proposed for the valuation of both assets and liabilities; QIS4 (2008):

1. Wherever possible, a firm must use *mark-to-market* methods in order to measure the economic value of assets and liabilities.

2. Where this is not possible, *mark-to-model* procedures should be used. Marking to model is any valuation, which has to be benchmarked, extrapolated, or otherwise, calculated from market input. When marking to model, undertakings will use as much observable and market-consistent inputs as possible.

3. Firms may opt to follow the guidance in the annexed tables to QIS4 Technical Specifications, QIS4 (2008), to determine where the treatment under IFRS is considered an allowable proxy for economic value for the purposes of QIS4. Where possible, this guidance may also be applied to local GAAP.

4. Under the following circumstances, the national accounting figures could be used, even though these might not reasonably be regarded as a proxy for economic value:
   - Where a firm can demonstrate that an asset or liability is not significant in terms of the financial position and the performance of the entity as determined under the applicable financial reporting framework and the solvency assessment. Participants should refer to the materiality principle set out in their applicable financial reporting framework to determine what was deemed significant or not, and apply the same principle for solvency purposes.

   - When the calculation of an economic value is unjustifiable and impractical in terms of the costs involved and the benefits derived.

When participants have ring-fenced funds in place, which separate part of the resources from the rest of the business, the calculation of the liabilities and assets for each ring-fenced fund should include all CFs in and out of that fund. For example, interfund CFs should be considered as assets of the fund, which receives them, and as a liability of the fund of origin. When preparing accounts for the whole undertaking, the transactions between funds should be netted off.

The attention of participants was drawn to the two following points:

F.1 INTANGIBLE ASSETS (INCLUDING GOODWILL)

For solvency purposes, the economic value of most intangible assets was considered to be nil or negligible, since they very rarely have a cashable value. Therefore, for the purpose of QIS4, all intangible assets should be valued at nil.

Participants should, however, provide the following additional quantitative information.

1. The accounting value ascribed to the following four intangible asset categories:
   a. Goodwill on acquisition of participations
   b. Goodwill on acquisition of business
   c. Brand names
   d. Other intangible assets (please specify their nature)
2. For intangible assets in a–d that have an economic value that is cashable, participants should provide the economic value of that intangible asset. In these cases, participants also had to provide a detailed description of the valuation method and valuation assumptions used, the valuation process and the valuation governance followed, and the difference (if any) with the accounting value.

F.2 DEFERRED TAXES

Solvency II has prudential supervision as its exclusive purpose and is therefore neutral and agnostic with regard to any issue concerning general accounting or taxation As Solvency II is not introducing any amendments in insurance accounting nor the valuation basis used for tax purposes, the difference stemming from the prudential revaluation of TPs for Solvency II purposes does not correspond to a one-off profit in the accounts and therefore does not create a one-off tax liability. Thus, participants should not include in their solvency BS a deferred tax liability specifically related to the change in value of TPs arising from the move from Solvency I to Solvency II. However, the economic approach underpinning Solvency II implies that all expected future cash-outflows and cash-inflows should be recognized in the solvency BS, including those related to taxes applicable under the fiscal regime currently in force in each country.

Where the figures used for QIS4 differ from the figures used for general purpose accounting, participants were invited to explain how those QIS4 figures were derived, for example,

- Evaluated through the use of a purposefully designed system expand on reliability and experience thereof or
- Roughly evaluated on the basis of more reliable, less economic figures, for example, slight amortization of a relatively recent economic valuation or
- Rough estimate

If applicable, participants were asked to also indicate whether these figures were already used for another purpose in the conduct of business, that is, other than for QIS4.

Guidance for (1) *and* (2): *marking to market and marking to model* Where an MV was already available because it has been calculated or assessed for purposes other than accounting, it should be reported within QIS4. It was recognized that a number of BS items, including most marketed investments, would have an economic value readily available through market appraisals, which may or may not be conducted for accounting purposes.

It was understood that when marking to market or marking to model, participants would verify market prices or model inputs for accuracy and relevance and have in place appropriate processes for collecting and treating information and for considering valuation adjustments.

Participants were also invited to provide additional information on the following:

- The identification of those assets and liabilities, which are marked to market and those which are marked to model
- Where relevant, the characteristics of the models and the nature of input used when marking to model
- Any differences between the economic values obtained and the accounting figures (in aggregate, by category of assets and liabilities)

Guidance for (3): *adjustments for relevant BS items under IFRS* Considering that some undertakings in the EU already use IFRS as a basis for their financial reporting, and because IFRS is the only common European accounting standard, some tentative views on the extent to which IFRS BS figures could be used as a reasonable proxy for economic valuations under Solvency II had been provided in the QIS4 specifications.

These views were presented in the tables included in the Technical Specifications; QIS4 (2008). In these tables, the Commission had identified the items for which IFRS valuation rules might be considered consistent with economic valuation, and for other items, adjustments to IFRS standards were proposed in order to bring the value of the item closer to an economic valuation approach. Firms using local GAAP had to apply the principles and adjustments indicated in those tables to their local GAAP standards, where feasible and appropriate.

If firms considered that other adjustments to the accounting figures should be provided for, then they had to identify and explain those adjustments. This analysis should not be considered as setting any interpretations of IFRS standards. Furthermore, this analysis did not preempt future conclusions on the possible need for solvency adjustments under IFRS.

Guidance for (4): *use of accounting figures not regarded as economic values* When accounting figures were used, which could not be regarded as economic values, participants had to be able to demonstrate that

a. The difference between the economic value and the accounting value was unlikely to be significant and/or
b. That the explicit calculation of an economic value entails excessive costs

The use of stochastic simulation or a deterministic approach for both asset and liability valuation was discussed in QIS4 (QIS4, 2008, pp. 39–41).

From the results of QIS4, it could be seen that there were a broad support for the general design of the valuation approach. It did not create any major difficulties for most participants in those countries applying IFRS accounting principles; (CEIOPS, 2008e). There was also a strong desire to the Solvency II economic valuation approach and the international accounting standards (IAS) (i.e., IFRS phase II) to develop consistently. Further guidance was requested regarding the use of mark-to-market approach. It was also mentioned that the IFRS valuation increased the volatility in both assets and liabilities leading to volatility in OFs held by the undertakings. There were diverging views and responses on both the treatment of deferred taxes and the intangible assets.

One of the main drivers of the financial crisis, mainly during 2007–2009, has been directly linked to a combination of lack of disclosure and inappropriate valuation of complex products. As pointed out by CEIOPS (2009a), the principle of not buying or selling an asset you cannot value or understand was constantly left aside. Drivers of this trend were the reliance on external ratings, inappropriate valuation techniques, and wrong remuneration incentives. The fair valuation was seen as an issue for the financial turmoil, especially in illiquid or nonactive markets. CEIOPS considered that the economic valuation of both assets and liabilities should remain valid (and necessary) for the Solvency II project.

APPENDIX G

European Solvency II

Liability Valuation

HERE WE DISCUSS the development and calibration of liability valuation that has been one of the building blocks in Solvency II. The final advice from CEIOPS is discussed in Chapter 24.

From the results of QIS4 it could be seen that there were a broad support for the general design of the valuation approach. It did not create any major difficulties for most participants in those countries applying IFRS accounting principles; CEIOPS (2008e). There was also a strong desire to the Solvency II economic valuation approach and the International accounting standards (i.e., IFRS phase II) to develop consistently. Further guidance was requested regarding the use of mark-to-market and mark-to-model approaches. It was also mentioned that the IFRS valuation increased the volatility in both assets and liabilities leading to volatility in OFs held by the undertakings.

The fair valuation was seen as an issue for the financial turmoil, especially in illiquid or nonactive markets. However, CEIOPS (2009a) considered that the economic valuation of both assets and liabilities should remain valid (and necessary) for the Solvency II project. In the work on the final advice given by CEIOPS in the late 2009 the lack of market for insurance liabilities and the complexities involved in their valuation was considered.

Early in 2004 the European Commission Services asked CEIOPS for advice on, among other things, how to valuate liabilities and especially the TSs. We start by looking at these initial thoughts and the advice given by CEIOPS; see Section G.1. These advices were later developed in the different QISs. We discuss the valuation of TPs as they have evolved during the QISs in Section G.2. Life technical provision in Section G.3, nonlife in Section G.4, simplifications and proxies in Section G.5, and the RM for nonhedgeable risks/obligations in Section G.6. Other liabilities are briefly discussed in Section G.7.

G.1 INITIAL THOUGHTS

The insurance liabilities consist of TPs and "other liabilities." CEIOPS gave its view on TPs in the answers to the Call for Advice (CfA) 7, TPs in life assurance, and to the CfA 8, TPs in nonlife insurance; see CEIOPS (2005c). The CfAs were published in MARKT (2004c). Life insurance TP is discussed in Section G.1.1 and nonlife technical provision in Section G.1.2.

G.1.1 Life Insurance Technical Provision

In MARKT (2004c), the Commission Services asked CEIOPS to advise on TPs issues. The provisioning standards to be included in the implementing measures could be supplemented by actuarial guidance and thus close c-operation with actuarial profession would be necessary.

G.1.1.1 CfA 7: TPs for Life Insurance

At this initial stage of the process, there was no distinction between hedgeable and nonhedgeable risks or obligations. These concepts were introduced in QIS2 in 2006.

G.1.1.2 Best Estimate

Expected PV or "BE" provisions should be unambiguously defined in a way that is IASB compatible and takes international developments into account (in particular, the work underway in the IAIS and the Groupe Consultatif/IAA).

G.1.1.3 Risk Margin

To take into account the uncertainty of valuation, and to protect policyholders, for example, when the solvency margin has disappeared, the portfolio must be transferred or put into runoff, RMs must be set in addition to expected values. Relevant factors, such as insurance risks, lapses, expenses, and so on, and methods for establishing RMs as well as the appropriate level of aggregation should be addressed. The methods should be, to the greatest extent possible, compatible with IASB developments and the SCR calculation methodology in order to minimize additional workload and costs for insurance companies. However, non-IFRS companies and SMEs required specific consideration, and actuarial rules should allow different approaches and approximations ranging from traditional deterministic methods to sophisticated stochastic modeling. After an appropriate technical structure for RMs has been found, an analysis of the goals of provisioning from the points of view of solvency, supervision, accounting, and feasibility, is needed. Finally, a proposal regarding the possible benchmark level of prudence of the TPs with field-testing results had to be communicated to the Services.

G.1.1.4 Risk-Free Interest Rate

CEIOPS was requested to provide advice on how to define the risk-free market interest rate for discounting the estimated future CFs applicable both inside and outside of the Euro zone. In addition, methods to establish TPs for the investment-related parts of life assurance contracts, that is, interest rate guarantees, bonuses, and embedded options, should be developed. Ultimately, valuation methods of profit-sharing policies should be explicit and promote fairness and transparency to clients and other stakeholders. However, proper recognition has to be given to national differences in insurance products. CEIOPS had also to address the new prudential aspects that arise when the guaranteed interest rate differs from the discount rate. Moreover, valuation techniques of mathematical finance should be investigated and their use encouraged particularly when assessing risks and prices of certain options embedded in life assurance contracts (both Pillar I and II measures should be considered). Insurance contracts, including UL contracts, as well as investment contracts should be studied.

G.1.1.5 Other Issues

Further analysis and development of open issues, for example, the use of retrospective methods, the definition of book of contracts, unit of calculation, the surrender value floor, and the provisioning for expenses, should be carried out in the light of international accounting and actuarial developments.

Finally, it was imperative for supervision to ensure that any changes in provisioning methods (both in national and company levels) did not lead to unfair sharing of profits or distortion of the distribution of bonuses. This applied to existing portfolios as well as between different generations of policyholders. CEIOPS's advice should include analysis if certain transitory measures (e.g., regarding provisioning rules for certain "old" contracts) are considered necessary.

G.1.1.6 CEIOPS' Advice

The objective of CEIOPS answer, CEIOPS (2005c), was to suggest adequate amendments to the existing rules and to introduce implementing measures to the extent that detailed provisions were not included in the Directive. The aim of the amendments would be to introduce principles assuring sufficient TPs and to achieve a greater degree of harmonization throughout the EU with regard to provisioning rules and practices.

In answering to the CfA7, the following issues had been identified and would be addressed through subsequent work by CEIOPS.

- Quantitative standard for TPs
- Segmentation
- Discounting
- Investment-related parts
- Profit sharing and potential sharing
- Surrender value floor
- Expenses
- Retrospective methods
- Small undertakings
- Recognition of national differences

The liabilities to be valued needed to be defined. The term "insurance liabilities" has been used for TPs so that the terminology was consistent with that of IASB. However, the concept will apply for Solvency II purposes to insurance-policy-related obligations, whether or not these are obligations under an "insurance contract" as defined for IASB purposes.

The insurance liabilities would encompass provisions in respect of all cash-outflows that an insurer will incur in fulfilling its obligations toward policyholders and other beneficiaries. For example, this will include tax, expenses, or fees under service agreements. Allowance needed to be made for contractual bonus payments reflecting the manner in which they will be determined. Where discretionary, noncontractual, or constructive liabilities to policyholders exist, these should be provisioned realistically.

Adequate allowance would also need to be included in an insurance undertaking's BS for all liabilities other than those falling to be provided for as part of the insurance liabilities.

The application of a common set of capital requirements would likely produce different views of insurer strength for each accounting system used because of the different ways accounting systems can define liability and asset values. In the view of the IAA, these definitions may create a hidden surplus or deficit that must be appropriately recognized for the purpose of solvency assessment. It is vital for prudential supervision that the valuation methods used by undertakings were adequate to make an undertaking's financial risk profile clear.

TPs needed to be determined on a basis compatible with IASB methodology, but not necessarily identical. IASB methodology may determine liabilities from a shareholder perspective and may thus explicitly allow for reductions in liabilities as an insurer's creditworthiness declines. This was not seen as appropriate in a regulatory framework, which must consider the policyholder perspective. Regulatory valuation of insurance liabilities would not include any reduction to reflect the shareholder option to default on obligations. Whatever the solution adopted under IAS, it may be required to adjust the IASB provisions, perhaps through an explicit additional liability shown in the regulatory returns, or through other means to achieve the same effect. Such an adjustment need not make regulatory TP incompatible with IASB methodology, as such.

Solvency II should aim at a methodology that would be transparent in the valuation of insurance liabilities. The resulting provision could cover both

- The expected PV of the liability CF, given insights at the time the insurance liabilities are being determined.
- An RM. This RM should be set prudently, but not so prudently as to act as a disincentive for the private industry to underwrite insurance risk.

Further analysis was necessary to determine the extent to which methodologies for the regulatory valuation of technical liabilities should be prescribed by the supervisor.

The expected PV could relate to individual contracts. But any RM may be applied at a higher level of aggregation, for example, homogeneous risk groups. A single provisioning philosophy should underlie the methodologies adopted in practice for determining insurance liabilities for all policies, regardless of the

- *Nature:* for example, profit sharing, nonprofit sharing, UL
- *Premium features:* for example, regular premium, single premium, reviewable premiums

- *Contract conditions:* for example, guarantees, options, and so on
- Contract duration

It was proposed by CEIOPS that policy options and guarantees should be explicitly provisioned. This included both financial options, guarantees embedded within the product, and other forms. Financial options and guarantees would be provisioned in a manner consistent with market-based values. Others, such as lapse options, or the ability to convert the policy from a short term to a whole of life policy on preagreed terms (without evidence of continued good health) needed to be provisioned on the basis of expected present values of CFs and also considering further comments mentioned below in the paragraph on the required margin for risk. The PV of expected CFs could be determined including allowance for BEs of all relevant factors, for example, levels of antiselection in exercising options, or realistically assessed profits on policy lapsation.

Some allowance could be made for more adverse levels of option take-up, higher or lower rates of discontinuance, and so on. But it would be appropriate within the context of the SCR to disallow assumptions that a given policy may lapse if this would lead to a reduced insurance liability. The precise treatment of option take-up rates within best estimate CFs, insurance liabilities, and the SCR was to be considered later.

Valuing insurance liabilities by using the BE approach could lead to the recognition of profits or losses at inception of an insurance contract. On this issue, it was stated in the Basis for conclusions IFRS 4 that: "Assets and liabilities arising from insurance contracts should be measured at fair value. In the absence of market evidence to the contrary, the estimated value of an insurance liability shall not be less, but may be more, than the entity would charge to accept new contracts with identical contractual terms and remaining maturity from new policy holders. It follows that an insurer would not recognise a net gain at inception of an insurance contract, unless such market evidence is available."

For with-profit contracts and in a number of other cases, any surplus at inception might not be recognized as profit because of requirements in national law to distribute surplus to policyholders. In these cases, calculating the BE of guaranteed benefits could lead to a number of practical difficulties. A pragmatic approach would be to value the contract on a tariff basis at inception, although valuation postinception would require further consideration.

Alternatively, any difference between the "fair value" of a policy and the valuation of a policy based on guaranteed benefits discounted using a risk-free interest rate could be interpreted as a TP corresponding to future or potential bonus, rather than as "profit."

G.1.1.7 Segmentation: Homogenous Risk Groups

Assessing the probability distributions of future CFs requires a classification of URs into groups with similar characteristics, known as homogenous risk groups. This classification should be based in part on information from historical data on the liabilities portfolio, the undertaking's specific circumstances, and relevant data from the insurance industry.

Conceptually, any RM related to the quantitative level of prudence should be calculated at the level of the insurance undertaking as a whole. However, in practice, a valuation of liabilities would require a classification of URs into homogenous risk groups.

G.1.1.8 Discounting

The PV of the expected CFs is equal to the realistic value of an investment with identical CFs to these expected CFs that will be paid with certainty. Such an investment replicates the expected CFs of the liabilities. A more efficient way to arrive at this PV is via discounting. An undertaking's insurance liabilities CFs might be discounted using the term structure of interest rates that has to be based on the effective yields on default-free capital market instruments.

A risk-free interest rate curve might be determined using the yield on the highest-quality government bond issues available in the currency of denomination of the liability. Where the risk-free yield cannot be determined with sufficient confidence, for example, because long duration bond issues are not available in a particular currency, then reference could be made to proxy measures, such as the swap curve. However, in such circumstances, there should be appropriate adjustments for the differences observed between government bond yields and the observed proxy yields.

According to CEIOPS it may even be more efficient if the nominal term structure of interest rates for the various currencies in the EU is prescribed such as the Euro spot rate term structure. Various central banks publish such term structures. This would allow insurers to value their liabilities using a prescribed term structure. Transparency is fostered by this method. If the capital market does not have the financial instruments available to replicate the longer-term liabilities of insurers, prudent long-term spot rates to arrive a prudent extension to the market-induced term structure of interest rates may be prescribed. Another option was not to impose the discount rate to be used but to set principles for this rate, notwithstanding the possibility for the supervisor to advice on a market-related rate for a given book of business.

However, valuing TPs with a "risk-free" interest rate curve may not allow for "sufficient prudence" TPs according to some members of CEIOPS. It might weaken the position of policyholders, if the provisions plus margins for risk did not reach the present level of provisions. An alternative may be that the supervisor prescribes or approves the term structure including a margin of prudence to calculate the insurance liabilities. But using margins in the discount rate for valuing liabilities may not be a satisfactory means of achieving an explicit level of prudence. For example, using a discount rate of 60% of the yield of a specified bond in fact implies using varying implicit RMs because the actual margin changes as interest rates change and is therefore not properly risk related.

G.1.1.9 Profit Sharing and Potential Sharing

With-profits benefits, profit sharing, could be conditional or unconditional. In this context, two types of liabilities are identified:

- Liabilities that are enforceable contractual liabilities and
- Discretionary or constructive liabilities

An example of an enforceable contractual obligation is an unconditional with-profits benefit where the amount of the benefit is linked only to an objective financial event so that the amount can be ascertained immediately. In modeling the CFs, an undertaking must take account of the fact that the amount of the benefit depends directly, for example, on corporate profits, investment yields, or objective external returns. An example of such an option is the guarantee of minimum annual returns for with-profits insurance. Another example is the right to extend the contract on preagreed terms and/or rates. Such options could affect the CF from an obligation and thus have a value.

Constructive liabilities include liabilities that result from an established pattern of past practice and that have created a valid expectation that they will be fulfilled in the future. They also include liabilities where policyholders may have been led to anticipate some form of benefit, for example, by reference to communications made by the insurance undertaking, whether or not those communications were directly to policyholders.

An example of a discretionary or constructive liability is a conditional with-profits benefit where the amount is determined wholly or partly by a decision of the Board. A conditional CF for insurers is the profit sharing that depends on a Board decision on allocating operating profit to policyholders. It is generally specified that there is profit sharing but the amount is not certain in advance. Usually there is a link with actual investment results. But the relationship between these results, the profit sharing, and the timing of the allocation is not set out unambiguously.

G.1.1.10 Surrender Value Floor

Under the present supervision legislation in the EU, there has been a minimum value for the provision to be maintained for insurance liabilities. The minimum insurance liabilities provision is any guaranteed surrender value of commitments in each contract. The insurer will need to select contract discontinuance assumptions when the entity is exposed to risk from the potential use of the option that the policyholder has to withdraw or persist, or to select the timing or the amount of such contract termination. Discontinuance can take the form of ceasing premium payments or terminating the contract. Discontinuance may give rise to the payment of the surrender value, to the granting of a paid-up policy, or to lapse without value.

To determine the surrender value payable on withdrawal, the insurer usually would take the following into account:

- Market assumptions assumed in the projection
- Any guaranteed surrender or transfer value scale
- Constructive obligations incorporated within the contract

G.1.1.11 Reinsurance

It is desirable to have valuation standards for gross provisions and net provisions. When determining net liabilities, an insurance undertaking should have regard to its gross liabilities, before reinsurance and other types of risk reduction. Appropriate allowance will need to be made for risk arising from insufficient reinsurance.

G.1.2 Nonlife Insurance Technical Provisions

G.1.2.1 CfA 8: TPs for Nonlife Insurance

In MARKT (2004c), the Commission Services asked CEIOPS to advise on rules to value nonlife TPs, with the aim of establishing TPs that are sufficient to cover the liabilities with a quantified level of confidence. Given the diversity of nonlife business lines and the many factors impacting on ultimate claims settlement, the Commission was aware of the methodological difficulties involved with this approach. To take into account the uncertainty of the valuation and to protect policyholders, for example, when the solvency margin has disappeared and the portfolio must be transferred or put into runoff, RMs must be set in addition to the expected PV or "BE" provisions. The Commission Services proposed that the following aspects should be taken into account in the analysis.

- Required quantitative prudence level: how should it be fixed, to what amounts exactly should it be applied.
- Actual quantitative prudence level: how should it be evaluated, at what level of aggregation, and so on.
- Obligation to use several appropriate actuarial methods among the generally accepted methods—some or all of these methods may have to be defined.
- If considered appropriate, criteria to decide on final amount of TPs when methods give differing results, for example, the criteria could be to choose the amount that is given by several methods. These criteria must not be so strict that it prevents the exercise of actuarial judgement and discretion.
- Introduction of a detailed annual report on the valuation of technical liabilities making explicit the actuarial and economic assumptions made, their evolution over time, and the reasons for the benefits or losses of liquidation. This report should also disclose the RMs above the expected value included in the TPs and the justification of the discount rate used.
- Appropriateness of introducing guidelines on reserving (levels of aggregation, allowance for risk interdependency, etc.).
- Harmonization of the data collected (implying a prior definition of the terms used), including runoff triangles of undiscounted TPs used to assess provisioning adequacy and claims paid triangles.
- Adaptation of the estimates of the TPs (whenever there is significant information and at least at periodic intervals) and mechanisms to ensure estimates of TPs are responsive to major external events, for example, rapid inflation or catastrophic events.
- Treatment of non-IFRS companies and SMEs, avoiding undue complexity.

- To promote solvency on a long-term basis, should a compulsory equalization mechanism be maintained regardless of the taxation and accounting regimes (this could, for example, be incorporated in the SCR calculation)?

CEIOPS was invited to give advice on how best to develop common EU actuarial standards to achieve greater consistency in provisioning practices. As stated in the Framework for consultation, CEIOPS was asked to work with actuarial bodies, in particular the Groupe Consultatif. The advice must be compatible with the likely accounting developments, unless duly justified. The advice should take into account insurance-specific standards and recommendations adopted in other international fora (namely, IAIS and OECD).

G.1.2.2 CEIOPS' Advice

There was a wide diversity in approaches to valuing TPs across the EU. One of the objectives would be to harmonize these approaches. The simplest way of introducing such a quantitative standard would be to define the desired outcome in quantitative terms and leave the precise method of calculation to the discretion of individual undertakings, subject to best practice and professional standards for treating homogenous risk groups. However, this may do little to correct the disparities in practice evident in the current system; CEIOPS (2005c).

When statistical methods could be used, the evaluation of the BE and the RM depends on the type of method used. It is probably not desirable to impose a single calculation method: different methods may be more or less adapted to different situations. CEIOPS said that regardless of the calculation mechanics, undertakings themselves would retain the ultimate responsibility for ensuring the adequacy of their provisions.

CEIOPS did not see it possible to list the cases where statistical methods can be applied or not in a regulation. The decision to apply or not to apply statistical methods for achieving the regulatory level of prudence in the provisions must be part of the actuarial analysis required in the undertaking's provisioning procedures. In any event, guidance and requirements on the type of methods that are acceptable to supervisors would be needed. In particular, these should address small- and medium-sized companies, which will not always have resources to develop sophisticated methods for evaluating their RM.

Any introduction of a quantitative standard on provisions, even expressed at a very general level, must provide for

- Cases for which statistical data will be inadequate or of doubtful applicability (although these cases cannot be identified ex ante in regulation).
- The need to supplement the general quantitative principle with minimum requirements for statistical techniques used by undertakings agreed at European level.

Given that no single calculation method will be appropriate to all circumstances, the SRP related to the quantitative standard on provisions will have to take into account, as far as possible, the specificity of each undertaking.

At this stage of the Solvency II project it was not yet decided how the RM should be calculated.

G.1.2.3 Segmentation

Three different options were discussed by CEIOPS.

- *Individual claim estimation:* Determining the level of confidence for an individual claim will require considerable judgement. In most cases, aggregating the results of individual claims estimates will result in TPs with very prudent margins, as diversification benefits will not be taken into account. Under this option, the articulation of the quantitative standard and the current case-by-case principle is, of course, straightforward.
- *Line of business:* This might correspond to the most current actuarial practice. Even if the percentile were to be calculated on another level, separate reporting by LoB would be needed to facilitate transfers, and so on. Under this option, the provisioning process would be more likely to be integrated into the underwriting risk management of the undertaking. But not all diversification benefits will be taken into account as it would not consider, for example, diversification between different LoBs.
- *Whole entity:* Whole portfolio diversification benefits will be taken into account. However, the result of statistical methods applied to nonhomogeneous categories of claims may be questioned. The risk of encouraging poor underwriting and provisioning procedures should also be highlighted.

G.1.2.4 Reinsurance

The quantitative standard should be envisaged for provisions gross or net of reinsurance.

Net provisions: Providing that the level of reinsurance is adequate, statistical methods should in general be easier to apply to provisions net of reinsurance and more relevant. However, this may not always be the case: for example, a significant change in the reinsurance program might be an obstacle to the application of common actuarial techniques. Net provisions need to be considered because they represent economic exposure.

Gross provisions: Reinsurance does not exempt the direct insurer from its commitments toward policyholders. The insurer still faces a risk of counterparty on the amount of ceded provisions. To have a global view of the risks linked to reserving, it is necessary to consider gross provisions. For those reasons, harmonization should also be sought at gross level. However, in practice, a more significant part of gross provisions might fall out of the scope of application of statistical methods.

G.1.2.5 Treatment of Future CFs and Discounting

Specificity in nonlife insurance is that the main uncertainty lies in the amount of claims. Nonlife insurance largely covers short-tail business. In addition, even in the case of long-tail claims, the final amount remains the major source of uncertainty, because the randomness of the final amount increases with the time horizon—due to inflation, juridical risks, and so on. Thus the additional complexity arising from discounting, and the difficulty in setting a realistic rate, could simply increase modeling error, rather than improving the realism of the solvency framework.

Discounting of nonlife TPs has not been widely practiced within the EU. Although the Directives permit discounting under certain circumstances, few Member States have been tolerating this option, and even in these States, few insurance companies make use of this possibility. However, in order to better reflect economic reality, the timing of future CFs would need to be taken into account, resulting in the use of discounted provisions.

Although the existing European regulation obliges insurers to take into account inflation in their estimation of ultimate cost, one of the main benefits, according to CEIOPS, of discounting would be to force insurers to consider claims inflation more explicitly. This would suggest the need for a requirement on how to consider inflation explicitly. Discounting would also be more consistent with the market-consistent valuation of assets.

According to CEIOPS, to ensure a level playing field, the calculation of TPs, including the relevant discount rate, should reflect the location and the currency of the liability.

G.1.2.6 Provision for Claims Outstanding

In the Accounting Directive (91/674/EEC) the provision for claims outstanding was defined as the total estimated ultimate cost to an insurance undertaking of settling all claims arising from events that have occurred up to the end of the financial year, whether reported or not, less amounts already paid in respect of such claims. In line with this definition, Article 60 of the Accounting Directive states the principle of case-by-case estimation of notified claims completed by an evaluation of IBNR.

Case-by-case estimation is closely linked to the nature of contracts and local law. In some cases, local law may impose the minimum amount to be reserved. It should also be added that, in some Member States, this way of envisaging provisions was also the basis for winding-up regulations, granting exclusive access to a subset of assets meeting TPs. The case-by-case principle, which is a necessary reference for a prudential framework, should actually not be envisaged as in contradiction with statistical methods.

In any event, the insurer should understand, and be able to explain to the supervisor, differences between case-by-case estimates and the provisions it establishes. If the two methods are in conflict, either the statistical method or the case-by-case estimation may not be adequate. However, the statistical method may result in a lower estimate since it may take diversification effects between individual contracts into account, which is out of scope of the case-by-case estimation. The undertaking is responsible for using appropriate and reliable techniques to value its provisions. This should be reviewed as part of the SRP. Therefore, the most natural way of introducing the quantitative benchmark would be to define it as a principle supplementing the case-by-case principle.

G.1.2.7 Premium Provisions

The Accounting Directive provides for two components of the premium provisions.

- *The provision for unearned premiums:* which "shall comprise the amount representing that part of gross premiums written which is to be allocated to the following financial year or to subsequent financial years."
- *The provision for unexpired risks:* "that is the amount set aside in addition to unearned premiums in respect of risks to be borne by the insurance undertaking after the end of the financial year, in order to provide for all claims and expenses in connection with insurance contracts in force in excess of the related unearned premiums and any premiums receivable on those contracts."

It could be argued that the split between provision for unearned premiums and provision for unexpired risk is not suitable and may even be confusing. In a rather straightforward manner all claim payments arising from future events insured under existing policies up until their next renewal are related to risks being unexpired at the BS date (the reporting date).

G.2 QIS1–QIS4 VALUATION OF TP

In late 2005, the first Quantitative Impact Study, QIS1, was conducted. It focused only on the valuation of TPs. We will indicate these first thoughts in the following discussions where we will start the discussion on the valuation of TPs with general ideas from CEIOPS. Then we look closer at the valuation of TPs for life insurance, Section G.2.2, and for nonlife insurance in Section G.2.3. In these appendices we focus on BEs. RMs are discussed in Section G.6. In QIS4, a general valuation approach for both assets and liabilities were developed; see QIS4 (2008) and Appendix F for details.

G.2.1 General Principles

In QIS1, CEIOPS (2005d), for each segment, TPs would have to be shown on the following bases:

- Current basis
- BE
- 75th percentile
- 90th percentile

TPs both net of reinsurance and gross of reinsurance had to be shown on each of these bases.

Participants were asked to cover their entire portfolio in the valuation. The coverage was allowed to be partial if a business line or subportfolio was considered immaterial, or if insufficient data existed. The value of excluded TPs (on the current basis) had to be given separately.

Specific for nonlife insurance was that the BE, the 75th percentile, and the 90th percentile had to be shown on both a discounted and an undiscounted basis. The standard deviation of the probability distribution of the liability CF had to be given separately.

For life insurance, the total difference between the BE and the PV of the guaranteed surrender or transfer values of each contract should be given separately.

The approach to valuate the TPs for QIS2 was set out in CEIOPS (2006d) and could be summarized as

- Market-consistent values for risks where hedges are readily available, for example, financial risks (hedgeable risks)
- BE+RM approach to the 75th percentile for other risks, for example, some insurance risks (nonhedgeable risks). RMs are discussed in Section G.6

Where participants were unsure of the distinction between hedgeable and nonhedgeable risks, or where market-consistent values cannot be derived, the BE+RM approach would be followed. TPs should be shown both gross and net of reinsurance.

The segments/LoBs were not necessarily mutually exclusive. Business should therefore be allocated according to its predominant characteristic. Additionally, overall estimates for life, health, and nonlife TPs should be provided. The summation approach could include some allowance for diversification benefits, provided that sound actuarial techniques were used and potential regulatory restrictions were taken into account.

The liability valuation according to QIS3 is outlined in the TSs, CEIOPS (2007a). We discuss only the main considerations and leave the more details for life and nonlife insurance valuation to the presentation of the QIS4 approach in Sections G.2.2 and G.2.3.

As for QIS2, in QIS3 there was a split of the TPs or the obligations into hedgeable and nonhedgeable parts. For nonhedgeable risks the valuation should correspond to the explicit sum of

- A BE, Section G.2.3 in general terms, plus
- An RM, the latter being determined according to a CoC approach; *cf.* Section G.6

The main change in QIS3 as compared with QIS2 was that the CoC approach became the main approach adopted.

However, for long-tailed nonlife business, alternative methods were envisaged. This may also include risks that are of a financial nature, whenever there was no hedgeable price from deep, liquid and transparent markets including an implicit allowance for additional uncertainty.

If from a nonhedgeable risk a hedgeable subrisk could be separated for which there is a reliable hedgeable price on a sufficiently deep, liquid and transparent market, then the market value of this hedgeable subrisk could be used in the valuation. If for a nonhedgeable risk there is a hedge available that was traded on a financial market but was incomplete and would only to some extent eliminate the risks associated with a liability, then the valuation of the BE would be done by a reference to the market value of the incomplete hedge increased with an appropriate valuation of the expected basis risk.

Nonhedgeable financial risks include, for instance, different kinds of embedded financial options and guarantees in life insurance contracts that are not traded on a financial market, risks where the duration exceeds a reasonable extrapolation from durations traded on the financial market, traded financial instruments that, however, are not available in sufficient quantities, and so on.

If within a CF an option or guarantee can be completely separated and as such be perfectly hedged on a deep, liquid and transparent market the separated risk would be classified as hedgeable. On the other hand, if the CF contains nonhedgeable financial—due to incomplete markets—or nonfinancial risks—due to options and guarantees on mortality and expenses for instance—that cannot be hedged on a deep, liquid and transparent market, they would be valued by inter/extrapolating directly observable market prices or as a BE plus RM valuation. It should be noted that on a deep, liquid and transparent market a perfect hedge has no basis risk. Note also that an RM has to be added to the BE.

For both QIS3 and QIS4, CEIOPS proposed a two-step approach for hedgeability and nonhedgeability. We use the latter two-step definition here:

1. The first step focuses on the split of the obligations into hedgeable and nonhedgeable and
2. The second step focuses on how an explicit RM for nonhedgeable CFs is to be calculated

The valuation of the TPs should cover both hedgeable and nonhedgeable (re)insurance obligations.

Even if it would be desirable, the values of hedgeable and nonhedgeable risks might not be separable under all circumstances, for example, a risk-neutral probability method has been used.

The general principles for QIS4, QIS4 (2008), were

- Participants had to value TPs at the amount for which they could be transferred, or settled, between knowledgeable willing parties in an arm's length transaction.
- The calculation of TPs was based on their current exit value.
- The calculation of TPs had to make use of and be consistent with the information provided by the financial markets and generally available data on insurance technical risk.
- The TPs were established with respect to all obligations toward policyholders and beneficiaries of insurance contracts.
- TPs had to be calculated in a prudent, reliable, and objective manner. No reduction in TPs should be made to take account of the creditworthiness of the undertaking itself. Prudence should not be understood as a requirement that TPs should include any implicit or explicit margin above the RM required to bring the value of the TP to the current exit value.
- The value of the TPs was equal to the sum of a BE and an RM.
  - The BE and the RM had to be valued separately; with the exception of hedgeable (re)insurance obligations.

- Separate calculations of the BE and the RM were not required, where future CFs associated with insurance obligations can be replicated using financial instruments for which a market value is directly observable. In this case, the value of TPs should be determined on the basis of the market value of those financial instruments.
- In certain specific circumstances, the BE element of TPs may be negative, for example, for some individual contracts. This was acceptable and participants should not set to zero the value of the BE with respect to those individual contracts.

In line with the principle set out above, where the future CFs associated with (re)insurance obligations could be replicated using financial instruments, those obligations were considered as "hedgeable" and separate calculations of the BE and RM were not required; see Section G.2.2.

Conversely, where (re)insurance obligations were considered as "nonhedgeable" because the future CFs associated with those obligations could not be replicated using financial instruments, separate calculations of the BE and RM were required. Nonhedgeable insurance obligations were still to be valued on a market-consistent basis as set out in Section G.2.3. In particular, where financial markets provide for relevant, credible, and up-to-date information for valuation purposes, this could be duly taken into account.

If within a contract an option, guarantee, or other part of the contract can be completely separated and as such be perfectly hedged on a deep, liquid and transparent market, the separate benefit was classified as a hedgeable component and was valued as set out in Section G.2.2.

Where there was an unsure distinction between hedgeable and nonhedgeable CFs, or where market-consistent values cannot be derived, the nonhedgeable approach would be followed, that is, separate calculations of BE and RM. The respective values of hedgeable and nonhedgeable (re)insurance obligations had to be separately disclosed. For nonhedgeable (re)insurance obligations, the RM had also to be separately disclosed.

The use of stochastic simulation or a deterministic approach for both asset and liability valuation was discussed in QIS4, QIS4 (2008, pp. 39–41). Small- and medium-sized companies felt that the lack of appropriate data was an obstacle for the use of stochastic valuation methods; CEIOPS (2008e).

The design of the proposed methods for the calculation of the TPs was in general supported by the companies participating in QIS4; CEIOPS (2008e). That included the simplifications and proxies also.

G.2.2 Hedgeable Risks or Obligations

For QIS2, CEIOPS (2006d), it was proposed that financial guarantees and options would be considered on a market-consistent basis. TPs for financial guarantees and options would be derived using risk-neutral discount rates applying at the BS date. An allowance for the time value of hedgeable guarantees and options was also considered, which brought in a range of potential future levels of interest rate.

In QIS3, CEIOPS (2007a), and QIS4, QIS4 (2008), it was assumed that if a risk can be perfectly hedged or replicated on a sufficient deep, liquid and transparent market, the hedge or the replicating portfolio provided a directly observable price, that is, mark-to-market. Reasonable inter/extrapolations from directly observable prices are also permitted. Deep, liquid, and transparent markets are defined as markets where participants can rapidly execute large-volume transactions with little impact on prices.

A perfect hedge or replication is one that completely eliminates all risks associated with the liability. In practice, perfect hedges are expected to be relative rare. Circumstances where perfect hedges could possibly be derived include, for instance, some options and guarantees embedded in life insurance contracts, some UL, for example, equity-indexed, life insurance contracts, CFs where there is no uncertainty in the amount and timing, and so on.

For a perfect hedge or replication, the nonarbitrage principle implies that the market-consistent value of the hedgeable risk would be equal to the market value of the relevant hedge or replicating portfolio.

A market was defined to be deep, liquid, and transparent if it meets the following requirements:

a. Market participants could rapidly execute large-volume transactions with little impact on prices.

b. Current trade and quote information was readily available to the public.

c. The properties specified in a. and b. were expected to be permanent.

Basis risk originates from differences between the exposure to undertakings liabilities and the contract terms of what may be purchased from the market.

G.2.3 Nonhedgeable Risks or Obligations: BE

In QIS1, the expected PV of future CFs was used. The estimate should be based on policy-by-policy data and the expected cash flows should be based on actuarial assumptions that were deemed to be realistic for the book of business in question, that is, each element sampled from a distribution believed to be reasonable and realistic with regard to all the available information. Assumptions had to be made based on a participant's experience for the probability distributions for each RF, but taking into consideration market or industry data where own experience was limited or not sufficiently credible.

To the extent practical, expected CFs had to reflect expected demographic, legal, medical, technological, social, or economic developments. For example, any foreseeable trends in life expectancy had to be taken into account.

The BE should be separately disclosed and the expected PV of future CFs should be used. In QIS2, in principle, the estimate should be based on policy-by-policy data, but reasonable actuarial methods and approximations could also be used.

The realistic valuation of assets and liabilities means that all potential future CFs that would be incurred in meeting liabilities to policyholders needed to be identified and valued. The PV of contract loadings and the PV of expected expenses would be recognized explicitly in the CF projection. Any shortfall would need to be recognized as an additional liability. Expenses that would be incurred in the future to service an insurance contract are CFs for which a provision had to be calculated. Participants would select assumptions with respect to future expenses arising from commitments made on, or prior to, the valuation date.

All future administrative costs, including investment management, commissions, claims expenses, and overheads had to be considered. Expense assumptions would include an allowance for future cost escalation. This would concern the types of cost involved. The allowance for inflation would be consistent with the economic assumptions made. For disability income and other similar types of business, claims expenses could be a significant factor. Expenses related to future deposits or premiums would usually be taken into consideration. Participants would also consider their own analysis of expenses, future plans, and relevant market data.

More specific details on BE of life and nonlife nonhedgeable risks for QIS2 were discussed in CEIOPS (2006d).

In QIS3, CEIOPS (2007a), the BE should be assessed using at least two different methods that could be considered reliable and relevant. The most appropriate method, or combination of methods, would then be used to value the BE.

A most appropriate method is a technique which is part of best practice and which captures the nature of the liability most adequately in a prudent, reliable, and objective manner. In deriving the BE, all potential future CFs that would be incurred in meeting liabilities to policyholders need to be identified and valued.

The BE equals the expected PV, probability weighted averages, of all future potential CFs (distributional outcomes), based on current and reliable information and entity-specific assumptions.

A projection horizon, which was long enough to capture all material CFs arising from the contract or groups of contracts being valued, should be used. If the projection horizon does not extend to the term of the last policy or claim payment, the firm should ensure that the use of a shorter projection horizon does not significantly affect the results.

The main valuation issue within an incomplete market is which of the possible prices should be picked for the valuation. The selection procedure is dependent on the user of the information, the user's preferences, and the user's attitude toward risk. Therefore, the most appropriate approach should be chosen for the valuation of the BE.

The reference to different methods within the valuation of the BE implicitly also concerns fitting distributions to statistical samples, such as, for instance, mortality and morbidity, which are used within the valuation of BE. However, since changes to mortality occurs slowly on a rather long-term basis, alternative methods and approaches to this kind of samples would be expected to be carried out less frequently than annually. Hence life firms often only have one available method.

G.2.3.1 General Assumptions

Appropriate assumptions for future inflation had to be built into the CF projections. Care should be taken to identify the type of inflation to which particular CFs were exposed. For some CFs, the link could be to consumer prices, but others are linked to salary inflation, which tends to exceed consumer price inflation.

For QIS3, CEIOPS (2007a), the expected CFs should be based on assumptions that were realistic for the book of business in question. Assumptions should be based on a participant's experience for the probability distributions for each RF, but taking into consideration market or industry data where own experience is limited or not sufficiently credible. Such realistic assumptions should neither be deliberately overstated nor be deliberately understated when performing professional judgements on factors where no credible information is available.

According to QIS4, QIS4 (2008), the BE was defined to be equal to the probability-weighted average of future CFs, taking account of the time value of money, using the relevant risk-free interest rate term structure.

The calculation of BE had to be based on current and credible information and realistic assumptions and to be performed using adequate actuarial methods and statistical techniques. The CF projection used in the calculation of the BE had to take into account of all the cash inflows and cash outflows required to settle the obligations over their lifetime. It should be calculated gross, without deduction of the amounts recoverable from reinsurance contracts and SPVs.

Where the CFs associated with the insurance obligations contained nonhedgeable financial, due to incomplete markets, or nonfinancial risks, due to options and guarantees on mortality and expenses for instance, that, when combined in a single insurance contract, could not be hedged or replicated using instruments on a deep, liquid and transparent market, the obligations could be valued by inter/extrapolating from directly observable market prices.

Market-consistent valuation techniques could be used to set the assumptions for, say, financial risks within a nonhedgeable contract and, for the remaining risks, the nonfinancial risks in this example, valued using BE assumptions. The RM would then be determined according to a CoC approach. The CoC calculation in QIS4 excluded market risk as this would otherwise double-count margins that were implicitly included in market prices.

Not all financial risks can be hedged or replicated using instruments traded on a deep, liquid, and transparent market. For instance, different kinds of embedded financial options and guarantees in life insurance contracts may include risks where there is a nontraded underlying, or risks where the duration exceeds a reasonable extrapolation from durations traded on the financial market, or risks relating to traded financial instruments that are not available in sufficient quantities, and so on. Where this was the case and if the remaining risk was considered material, alternative methods to find a "hedgeable cost" could be used to adjust market information and capture an additional market-consistent RM. Even if it would be desirable, the values of hedgeable and nonhedgeable risks might not be separable under all circumstances.

The BE was set equal to the expected PV of all future potential CFs, probability-weighted average of distributional outcomes, based on current and credible information, having due regard to all available information and reflecting the characteristics of the underlying insurance portfolio. Entity-specific information could be used in the calculation to the extent it enabled participants to better reflect the characteristics of their insurance portfolio, for example, entity-specific information regarding claims management and expenses.

The BE had to be assessed using a relevant and reliable actuarial method. Ideally, the method retained by participants could be part of actuarial best practice and could capture the technical nature of the insurance liabilities most adequately.

The local GAAP numbers should not be used as an input for the BE for QIS4 purposes, unless local GAAP standards actually delivered a valuation of the TPs, which was in line with the valuation principles, that is, current exit value, market consistency, BE plus explicit RM.

In many cases, the valuation of TPs in accordance with Solvency II was likely to be different from local GAAP figures. In line with the BE definition, the projection horizon used in the calculation had to cover the full lifetime of the insurance portfolio. In practice, the projection horizon used by participants had to be long enough to capture all significant CFs arising from the contract or groups of contracts being valued. And if the projection horizon did not extend to the term of the last policy or claim payment, participants had to ensure that the use of a shorter projection horizon did not significantly affect the results. Insurers had to describe which actuarial method they used to determine the BE and whether they used various actuarial methods.

Simplifications for BE calculations are discussed in Section G.5.

G.2.3.2 Discounting

For QIS1, participants were supplied with data on the term structure of interest rates that applied at the BS date for different EEA-currencies. These duration-dependent, risk-neutral discount rates were derived from the spot-rate term structure for each different currency, adjusted for credit risk. CFs were to be discounted at the risk-neutral discount rate applicable for that duration. Where the given rate structure provided no data for duration, the interest rate had to be interpolated or extrapolated in a suitable fashion. The resulting discount rates would be disclosed. For nonlife business, the TPs also had to be calculated at a discount rate of 0%.

In QIS2, the CFs would be discounted at the risk-neutral discount rate applicable for the relevant duration. Participants were supplied with data from CEIOPS on the term structure of interest rate for different EEA currencies, together with the US Dollar, Japanese Yen, and Swiss Franc. Where the given rate structure did not provide any data for a duration, the interest rate could be interpolated or extrapolated in a suitable fashion.

For QIS3, CEIOPS (2007a), the CFs would be discounted at the risk-free discount rate applicable for the relevant maturity at the valuation date. These would be derived from the risk-free interest rate term structure at the valuation date.

The use of risk-adjusted discount rates, so-called deflators, was allowed for CFs linked to financial variables, provided that the underlying estimation process led to results equivalent to those that would be obtained if the CFs were projected using risk neutral probabilities and discounted with the risk-free interest rate term structure.

The participants would use the term structure of interest rate supplied by CEIOPS for different EEA currencies, together with the US Dollar, Japanese Yen, and Swiss Franc.

The creditworthiness of the undertaking was intended to have no influence on the value of the TP. Thus, if participants needed to use term structures for other currencies that were not supplied by CEIOPS, they would derive them based on the following rationale:

- The risk-free interest rates relating to bullet maturities should be credit risk-free.
- Taking into account yields on government bonds could set the risk-free interest rates, where available and appropriate.
- In some markets it could, however, be more appropriate due to illiquidity or/and insufficient selection of maturities to use swap rates as proxies for risk-free interest rates. If so, appropriate considerations related to possible illiquidity or insufficient credit quality in the swap rates was to be given.

For QIS4 purposes, the prescribed risk-free interest rate term structure for the Euro was derived from swap rates. The methodology of its derivation could be found in an annex to QIS4 (2008). Yield curves for other EEA currencies and certain other currencies that were consistent with the methodology of the Euro curve were provided as well. Participants were expected to use a similar approach for nonspecified currencies. A number of companies participating in the QIS4 expressed concerns about the use of the swap rates and used adapted curves instead. No conclusion could be drawn from QIS4 on the choice between swap rates or government bonds as the basis for deriving the risk-free interest rate; CEIOPS (2008e).

If for certain currencies, a swap market did not exist, the government bonds could be used to determine the risk-free interest rate term structure. To determine that alternative risk-free interest rate term structure, a model that was close to the model used by the European Central Bank had to be applied.

In addition, a participant could deviate from the prescribed term structure and apply an interest rate term structure that was derived by the participant itself. Creditworthiness of the undertaking would not have any influence on the interest rate term structure derived by the participant. The participant was requested to disclose the term structure, as well as the reason for the deviation, and was invited to indicate the impact on the best-estimate TPs of the internal interest rate curve as compared to the prescribed interest rate term structure.

G.2.3.3 Expenses

In QIS1, the realistic valuation of assets and liabilities meant that all potential future CFs which would be incurred in meeting liabilities to policyholders needed to be identified and valued. The PV of contract loadings and the PV of expected expenses had to be recognized explicitly in the CF projection. Any shortfall is needed to be recognized as an additional liability. Expenses that would incur in future to service an insurance contract are CFs for which a provision should be calculated. Participants would select assumptions with respect to future expenses arising from commitments made on or prior to the valuation date. All future administrative costs including investment management, commissions, claims expenses, and overheads had to be considered.

Expense assumptions had to include an allowance for future cost escalation. This was with regard to the types of cost involved. The allowance for inflation had to be consistent with the economic assumptions made. For disability income, and other similar types of business claims, expenses may be a significant RF.

Expenses related to future deposits or premiums should usually be taken into consideration.

Participants had to consider allowance for expenses in relation to their own analysis of expenses, future plans, and relevant market data.

For QIS3, CEIOPS (2007a), and QIS4, QIS4 (2008), the expenses that would incur in the future to service an insurance contract are CFs for which a provision should be calculated. For the valuation, firms would make assumptions with respect to future expenses arising from commitments made on or prior to the valuation date.

All future administrative costs, including investment management, commissions, claims expenses, and an appropriate amount of overheads would be considered. Expense assumptions would include an allowance for future cost increases. These would take into account the types of cost involved. The allowance for inflation would be consistent with the economic assumptions made. For disability income and other similar types of business, claims expenses can be a significant factor. Expenses related to future deposits or premiums should be taken into consideration.

Firms would consider their own analysis of expenses, future plans, and relevant market data. But this would not include economies of scale where these have not yet been realized. Whenever the PV of expected future contract loadings was taken as a starting point any shortfall relative to future expenses that will have to be incurred in the future to service an insurance contract would be recognized as an additional liability.

To the extent that future deposits or renewal premiums were considered in the evaluation of BE, expenses relating to those future deposits and renewal premiums would usually be taken into consideration as well. Expenses related to the CFs due to future premiums were excluded if the latter were excluded from the evaluation of the BE. Whenever the PV of expected future contract loadings was taken as a starting point, any shortfall relative to future expenses that would have to be incurred in the future to service an insurance contract had to be recognized as an additional liability (and the opposite).

G.2.3.4 Taxation

To the extent that taxation payments needed to be made in order to meet policyholder liabilities, tax would be allowed for on the basis that currently applied, except where changes had been agreed to be introduced in which case the adjustments to the tax regime should be reflected in the calculations.

Taxation payments required meeting policyholder liabilities would be allowed for on the basis that currently applies. In cases where changes to taxation requirements had been agreed, but not yet implemented, the pending adjustments would be reflected in the calculations. In certain reassurances, the timing of recoveries and the time of direct payments might markedly diverge, and this would be taken into account when valuing the TPs, for example, when discounting CFs.

In a minority of the Member States of EU, taxation payments were charged to the policyholder. Where this was the case, participants were required to apply the following guidance. First of all, the assessment of the expected CFs underlying the TPs had to include the tax liabilities assumed to be charged to the policyholder. If this were the case, the undertaking's tax liabilities would be included as "other liabilities" within the BS. This would allow for the notional recharge of tax liabilities to policyholders. When valuing the BE, the recognition of taxation and compulsory contributions to the policyholders had to be consistent with the amount and timing of the taxable profits and losses that were expected to be incurred in the future. In cases where changes to taxation requirements had been agreed, but not yet implemented, the pending adjustments would be reflected. In all other cases, participants would assume that the taxation system remains unaffected by the introduction of Solvency II.

G.2.3.5 Reinsurance and SPVs

From QIS1, it was stated that in certain reassurances, the timing of recoveries and the time of direct payments might markedly diverge. In calculating TPs net of reinsurance, participants should assume that its reinsurer will not default.

In calculating TPs net of reinsurance, participants had to assume that the reinsurer would not default.

In certain classes of business, nonreinsurance recoveries may be material, their timing markedly diverging from that of direct payments. Nevertheless, participants could assume for QIS2 that their counterparts will not default. No reduction in liabilities would be made on account of the creditworthiness of the undertaking itself.

For QIS3, CEIOPS (2007a), the BE should be valued on both gross and net of reinsurance.

For QIS4, QIS4 (2008), the BE of the insurance liabilities of the participants had to be calculated on gross of reinsurance contracts and SPVs arrangements. Therefore, the amounts recoverable from reinsurance contracts and SPVs had to be shown separately, on the asset side of participants' BS, as *reinsurance and SPV recoverables*. The value of reinsurance recoverables had to be adjusted in order to take account of expected losses due to counterparty default, whether this aroused from insolvency, dispute, or another reason. A similar principle applied to CFs from an SPV.

In certain types of reinsurance, the timing of recoveries and that of direct payments might markedly diverge, and this had to be taken into account when valuing reinsurance and SPV recoverables. Recoverables should also fully take into account cedents' deposits. In particular, if the deposit exceeded the BE claim on the reinsurer, the recoverable would be negative.

The adjustment for counterparty default should be based on an assessment of the probability of default of the counterparty and average loss resulting from such a default, LGD. The assessment had to take into account the duration of the reinsured liabilities also.

The assessment of the PD and the LGD of the counterparty had to be based on current, reliable, and credible information. Among the possible sources of information were credit spreads, rating judgements, information relating to the supervisory solvency assessment, and the financial reporting of the counterparty. The assessment of the PD would implicitly take into account that the PD may increase under adverse scenarios. If the PD of the counterparty significantly depends on the amount payable to the insurance or reinsurance undertaking under the reinsurance contract or SPVs, the average PD should be used. The average probability should be weighted with the product of the amount payable and the probability that the amount will be payable.

The assessment of the PD should take into account the fact that the probability increases with the time horizon of the assessment. If no reliable estimate of the LGD were available, 50% of the value of the amounts recoverable would be used. Note that information such as credit spreads may already include an implicit allowance for the LGD.

If no reliable estimate of the PD was available, the PD of the counterparty according to the default risk submodule of the SCR standard formula could be used for a time horizon of 1 year. For a time horizon of t years, the probability $1 - (1 - PD)^t$ could be used, where PD is the probability for a time horizon of 1 year.

As far as recoverables were covered by collateral or a letter of credit, the PD of the collateral or the letter of credit occurring at the same time as the default of the counterparty, along with its LGD could replace the PD and the LGD of the counterparty in the calculation of the expected loss. The adjustment for expected loss had to be calculated separately for each counterparty. However, if the estimates of the PD and the LGD of several counterparties coincide, no separate calculation was necessary under the simplified approach.

A simplified approach was possible. It is described in Section G.5.3.

Companies participating in QIS4, and also some supervisors, raised concern about the valuation of nonproportional reinsurance accepted and reinsurance recoverables; CEIOPS (2008e).

In its discussion on the financial crisis, CEIOPS (2009a) concluded that the main finding that needed to be translated to its final advice is the need for a holistic approach to undertakings, including SPVs and risks off the balance. This is also linked to transparency and disclosure.

G.2.3.6 Future Premiums from Existing Contracts

For QIS3, CEIOPS (2007a), an appropriate allowance for future premiums was given. However, future premiums in exceedance of the necessary level to support the obligations under an existing contract would not be taken into account. Contractual recurring premiums under the contracts would be taken into account, but no allowance would be given for expected renewal premiums that were not included within the current insurance contract and that both parties are free to refuse.

Where a contract includes options or guarantees that provide rights under which the policyholder can obtain a further contract on favorable terms, for example, renewal with restrictions on repricing or further underwriting, then the value of these guarantees and options would be included in the valuation of TPs.

Any uncertainty surrounding future premiums would be reflected through an appropriate probability assumption, consistent with the probability assumptions applied to other CFs. Thus future premiums would be included in the determination of future CFs with an appropriate assessment of the future expected persistency.

According to QIS4, QIS4 (2008), the CFs included in the BE of the insurance liability would only include CFs associated with the current insurance contracts and any existing ongoing obligation to service policyholders. This would not include expected future renewals that were not included within the current insurance contracts.

Recurring premiums had to be included in the determination of future CFs, with an assessment of the future persistency based on actual experience and anticipated future experience. Where a contract included options and guarantees that provided rights under which the policyholder could obtain a further contract on favorable terms, then these options or guarantees had to be included in the valuation of the insurance liability arising under the existing contract. Where no such restrictions on repricing or underwriting exist, there was no ongoing obligation to service policyholders. In particular, future premiums should be included in the determination of future CFs when

- The payment of future premiums by the policyholder was legally enforceable
- Guaranteed amounts at settlement were fixed at subscription date

Clearer definitions and more guidance on the treatment of future premiums were requested by undertakings that participated in QIS4; CEIOPS (2008e).

G.3 LIFE TECHNICAL PROVISIONS

The QIS' participants could use credible and relevant discontinuance experience to the extent practical. Where a discretionary surrender value was paid on discontinuance, the estimates should allow for the payment the insurer would reasonably make in the scenario under consideration.

It was important to consider policyholder options to change the terms of the contract. CF projections should take account of the proportion of policyholders that were expected to take up options. This could depend on financial conditions at the time the option crystallizes, which will affect the value of the option. Nonfinancial conditions would also be considered—for example, deterioration in health could be expected to impact take-up rates of guaranteed insurability options.

TPs were also to be estimated as if the policyholder were sure to surrender the contract when this is unfavorable to the insurer. "Surrender" also refers to the transfer of the contract and the transfer values that may be specified in nonsurrenderable policies, when these policies give policyholders the option to transfer the contract from one insurer to another. For each contract, the participant was also requested to compare the BE with the highest PV of the surrender values of the contract, or the PV of the nearest surrender value if more practicable, and total up the differences when the second term of the comparison is higher.

To the extent possible, the participant in QIS2 was requested to give an approximation of the contribution of the surrender risk in the 75th/90th percentile RMs, and to briefly describe the approach followed.

Future management actions had to be reflected in the projected CFs. The assumptions used should reflect the actions that management would reasonably expect to carry out in the circumstances of each scenario, such as changes in asset allocation, changes in bonus rates or product charges, or the way in which an MV adjustment is applied. Allowance would be made for the time taken to implement actions. In considering the reasonableness of the projected management actions, participants would consider their obligations to policyholders, whether through policy wordings, marketing literature, or other statements that give rise to policyholder expectations of how management will run the business.

TPs had to include amounts with respect to guaranteed, statutory, and discretionary benefits. Assumptions for these would follow the general principles for management actions set out above. In considering future bonuses it was likely that insurers would take into consideration recent bonus rates, especially where the undertaking's policy was to smooth changes in bonus rates. Where undertakings differentiate their bonuses between policy type or risk group, this would be reflected in the assumptions on future bonus rates. Where material to the results, participants would take into consideration the expected apportionment between annual and final bonuses.

An aspect influencing the participant's realistic assessment of future bonus rates would be the extent of reserves already held by the undertaking for future bonuses, and the way in which the undertaking plans to distribute such reserves. Another aspect is the participant's expectations about future rates of investment return, as informed by market indicators. The margins available from future premiums payable under in-force contracts would also come into consideration, as well as RMs included within the TPs and the competitive position that an insurer plans to take.

The approach to bonus and profit-sharing assumptions would also take into consideration any constraints arising from legal restrictions or profit-sharing clauses in policy conditions. Undertakings would assume that, in applying such clauses, the approach to calculating profits for profit-sharing purposes would not change from that which currently applies.

For unit and index-linked business, the same CF projection approach would be used as for other products. All CFs arising from the product would be considered, including expenses, death benefits, and charges receivable by the undertaking. Where insurers have the right to increase charges the assumed increase in charging would be consistent with the management action that would be carried out. Participants would assume UL funds perform on a market-consistent basis.

Simplifications for the estimation of life TPs are discussed in Section G.5.4.

G.3.1 Segmentation

The valuation of liabilities could require a classification of URs into homogenous risk groups according to the broad nature of the risks assumed under the contract. Examples in QIS1 included

- Homogenous mortality risk (e.g., life cover risk, annuitant mortality)
- Morbidity risks (e.g., disability income, critical illness)

- Lapse risks (e.g., regular premium contracts)
- Profit-sharing/non-profit-sharing business
- Guaranteed/nonguaranteed business

Segmentation may also be driven by local factors, such as the tax treatment of business.

For QIS4, the participants was asked to segment its portfolio in the following way for reporting purposes—this was the segmentation suggested in QIS2 and QIS3:

First level of segmentation:

- Contracts with-profit participation clauses
- Contracts where the policyholder bears the investment risk
- Other contracts without-profit participation clauses
- Accepted reinsurance

For the valuation of the RM, each of the first-level segments had to be further disaggregated into risk drivers in the following way:

Second level of segmentation:

- Death protection contracts
- Survivorship protection contracts
- Contracts where the main risk driver is disability/morbidity risk
- Saving contracts, that is, contracts that resemble financial products providing no or negligible insurance protection relative to the aggregated risk profile

The segments/LoB described in the first and second levels of segmentation were not necessarily mutually exclusive. Business would therefore be allocated according to its predominant characteristics, for example, the allocation of endowment policies should depend on the relative significance of the death and survivorship benefits and where endowment policies with the same sum assured on death as on survival, are managed separately, these should be classified in the 4th subsegment as a "savings product."

Amounts for health contracts with features similar to life business should be disclosed separately.

In the QIS it was explicitly said that for life TPs, relevant *risk factors* had to include at least the following:

- Mortality rates
- Morbidity rates
- Longevity
- Lapse rates

- Option take-up rates
- Expense assumptions

Mortality, longevity, and morbidity assumptions would be assessed separately for different risk groups. Where a participant assumes correlation of risks between different risk groups, the assumptions made and the rationale had to be disclosed. Assumptions on the volatility of mortality, longevity, and morbidity experience would also be disclosed.

The segmentation for life insurance valuation in QIS4 and QIS3 were assumed as in QIS2. In QIS4, some companies had problems calculating the BE based on the proposed segmentation.

G.3.2 Grouping of Contracts

To start with, the valuation should to be based on policy-by-policy data, but reasonable actuarial methods and approximations were allowed in QIS4. In particular the projection of future cash flows based on suitable specimen policies can be permitted, that is, the use of a model office; see Section G.3.2.1.

Due to the principle of proportionality the reasonable actuarial methods and approximation could be used if

- The grouping of policies for valuing the costs of guarantees, options or smoothing, and their representation by representative policies (*model points*) is acceptable, provided that it can be demonstrated that the grouping of policies does not materially misrepresent the underlying risk and does not significantly misstate the costs.
- The grouping of policies should not inappropriately distort the valuation of TP, by, for example, forming groups containing life policies with guarantees that are "in the money" and life policies with guarantees that are "out of the money."
- Sufficient validation should be performed to be reasonably sure that the grouping of life policies has not resulted in the loss of any significant attributes of the portfolio being valued. Special attention should be given to the amount of guaranteed benefits and any possible restrictions (legislative or otherwise) for a firm to treat different groups of policyholders fairly (e.g., no or restricted subvention between homogeneous groups).

The grouping of contracts was also discussed for QIS3 in CEIOPS (2007h).

G.3.3 Behavior of Policyholders and Management

It is important to consider whether the presence of *policyholder options* could materially change the economic nature of the risks covered under the terms of the contract if exercised, that is, where they have an option enabling this.

In such circumstances, and where the effect of doing was expected to be material, CF projections should take account of the proportion of policyholders that were expected to take up the option. Expectations should be found on appropriate statistical analysis. This may depend on financial conditions at the time the option crystallizes, which would affect the value of the option. Nonfinancial conditions should also be considered, for example, deterioration in health could be expected to have an impact on take-up rates of guaranteed insurability options.

When credible and relevant discontinuance experience will be available, firms should make use of it. Where a discretionary surrender value is paid on discontinuance, the estimates should allow for the payment the insurer would reasonably make in the scenario under consideration. When assessing the experience of policyholders' behavior, appropriate attention have be given to the fact that the behavior when an option is out of or barely in the money should not be considered a reliable indication of likely policyholders' behavior when an option is significantly in the money.

Appropriate considerations should also be given for an increasing future awareness of policy options as well as policyholders' possible reactions to a reduced solvency of a firm. In general, policyholders' behavior should not be assumed to be independent of financial markets, a firm's treatment of customers, or publicly available information, unless proper evidence to support the assumption can be observed.

Future *management actions* could be reflected in the projected CFs and any items taken into account should be consistent with the firm's current principles and practices to run the business. Any assumptions used should reflect the actions that management would reasonably expect to carry out in the circumstances of each scenario, such as changes in asset allocation, changes in rates of extra benefits or product charges, or the way in which a market value adjustment is applied. Allowance should be made for the time taken to implement actions. Participants should use reasonable assumptions in incorporating management actions into projections of CFs such that the mitigating effects of the management actions are not overstated.

In considering the sensibility of projected management actions, firms should also consider their obligations to policyholders, whether through policy wordings, marketing literature, or other statements that give rise to policyholder expectations of how management will run the business. The reflection of management actions in the valuation would normally require that the assumptions used, the calculations carried out, the numerical results obtained, and the performed sensitivity analyses are based on objective, reasonable, and verifiable bases. The applied principles and practices should normally also be maintained in time unless there is sufficient evidence about the necessity of their updating.

Management actions have to be calculated using the same methods and assumptions in a risk neutral valuation as in a real-world valuation. For a given scenario, each valuation should have identical management actions. The risk neutral valuation and real-world valuation may either use a different set of scenarios or place different weights on the same scenarios.

Companies participating in QIS4 said that they would welcome clearer definitions and more guidance on the valuation of options and guarantees; CEIOPS (2008e).

G.3.4 With-Profit Business

When calculating TPs, participants have to take account of all payments to policyholders and beneficiaries, including *future discretionary bonuses*, which they expect to make, whether or not these payments are contractually guaranteed, unless those payments are defined as Surplus Funds (Article 91 of the Framework Directive).

For QIS4, the term "guaranteed benefits" included any benefits to which policyholders were already individually and unconditionally entitled to at the valuation date, including extra benefits from realized profits, irrespective of how the benefits are described, for example, vested, declared, or allotted.

Discretionary benefits included all payments to policyholders and beneficiaries in addition to those guaranteed benefits. The amount of future discretionary benefits could be influenced by legal or contractual restrictions, market practice, and/or management actions. In any case, all future discretionary bonuses, except surplus funds, should be accounted for in the calculation of TPs.

For with-profit contracts, all participants were requested to split the amount of their BE into the three following items:

1. Guaranteed and allocated benefits, that is, the sum of
   a. Allocated extra benefits that the policyholders are individually and unconditionally entitled
   b. Allocated extra benefits that the policyholders are collectively and unconditionally entitled
   c. Guaranteed future benefits (e.g., linked with contractual clauses that guarantee an absolute minimum for bonus rates)
2. Other future benefits that relate to a legal or contractual obligation, that is, the sum of
   a. Future benefits in excess of previous items that are linked with a legal obligation (e.g., firms must give to their policyholders a minimum share of their profits)
   b. Future benefits in excess of previous items that are linked with a contractual obligation (e.g., firms may guarantee in their contracts a minimum share of their profits)
3. Future discretionary benefits in excess of previous items (e.g., firms must apply a certain bonus rate above legal and contractual obligations in order to stay competitive)

Any constraints arising from legal restrictions or profit-sharing clauses in policy conditions had to be taken into consideration. It had to be assumed that, in applying such clauses, the approach to calculating profits for profit-sharing purposes would not change from that which applies currently.

Assumptions for distributing extra benefits should follow the general principles for management actions and a firm's principles and practices to run the business. Firms could take into consideration recent levels of extra benefits, especially where their policy is to smooth changes in rates of extra benefits.

Where firms differentiate their extra benefits between policy types or risk groups, this had to be reflected in the assumptions on the level of future extra benefits. Where material to the results, firms had to take into consideration the expected apportionment between annual and final extra benefits.

The valuation of extra benefits had to be consistent with the future return on assets assumed to back the liabilities.

If a firm's principles and practices for distributing extra benefits were expected to lead to payments that were in excess of what can be generated from the reserve held for the policy or group of policies, any such amounts would be taken into account unless otherwise stated. These amounts could be related to realized or unrealized profits and they might be subject to a different and a primary loss-absorbing nature in adverse circumstances compared to those extra benefits generated from the policy fund.

However, CFs arising from realized profit reserves appearing in the BS where they may be used to cover any losses that may arise and where they have not been made available for distribution to policyholders, that is, *surplus funds* should be excluded from the valuation of TPs. In particular, this relates to certain profit-sharing systems where insurance companies have established surplus funds. Surplus fund systems can be found, for example, in Austria, Denmark, Germany, and Sweden. As a consequence, existing surplus funds, which currently appear on the accounting BS of those insurance companies, should not be regarded as TPs, to the extent they meet the above-mentioned conditions.

In some products the smoothing of extra benefits in time imposes a so-called *soft guarantee* that could have more or less restrictions attached to it. These should be given appropriate attention.

In some cases, such as *extra benefits*, options, and guarantees, the valuation of TP is intrinsic on the assets held by the firm. The assets assumed in such circumstances could be chosen accordingly to one or several combinations of the following principles:

- The actual assets held to back a specific liability (assuming a segmented investment portfolio).
- The assets considered most reasonable to back the specific liability and that attribute future investment returns to that fund.
- A proportion of the assets allocated in accordance with the cover of TPs.
- A proportion of the assets allocated in accordance with the general investment portfolio.

The valuation of extra benefits, including any projections or assumptions on future returns of the firm's asset portfolio, should be consistent with information provided by the financial markets and generally available data on insurance and reinsurance technical risks, that is, market consistency. The assumptions on future asset returns underlying the valuation of extra benefits should not exceed the level given by the forward rates derived from the risk-free interest rates.

Where the extra benefits include options and guarantees dependent on the return on assets, see Section G.3.6 on the market cost of hedging the option or guarantee. In the absence of financial options or guarantees, the assumptions on future asset returns underlying the valuation of extra benefits should be consistent with the forward rates derived from the risk-free interest rates.

The level of *extra benefits* should be consistent with the future return on investments assumed (these should be consistent with forward rates derived from the risk-free interest rates) and possible management actions. Even if the valuation of extra benefits would induce path dependencies, these might be disregarded or only partly addressed. Possible path dependencies should, however, be qualitatively assessed.

Regarding the amount of the extra benefits that are "in the money," a historical average distribution ratio, reflecting past management actions, applied to the appropriate risk-free forward rate could be used. If extra benefits were also distributed from a guarantee related to mortality or expenses, these could be taken into account as an increment of the distribution ratio related to investment returns and hence these did not have to be stochastically modeled. If the firm aims at extra benefits in excess of those that were generated from the policy fund, these could be taken into account by an appropriate increment of the distribution ratio to reflect the amount distributed from excess assets.

Companies participating in QIS4 said that they would welcome clearer definitions and more guidance on the valuation of guarantees and the calculation of future discretionary benefits; CEIOPS (2008e).

G.3.5 Linked Business

The same CF projection approach would be used for unit and index-linked business. Firms had to assume also that UL funds perform on a market-consistent basis. All CFs arising from the product should be considered, including expenses, death benefits, and charges receivable by the insurer. Where firms have the right to increase charges, assumptions on increased charging should be consistent with the general principles for management actions.

As a simplification, the income from the policy charges may often be expressed on a basis that could be valued as a percentage of the current unit fund, valued as the current face value of the units, or series of fixed payments, which can be discounted using the forward rates derived from the risk-free interest rates. A full stochastic model is often not needed to value UL business market consistently.

Applying the outlined valuation principles also for unit- and index-linked business, the TP could in some cases be less than the current value of the fund value reflecting the excess of future charges over expected expenses.

G.3.6 Other Issues

The CF projections for health insurance business should take account of claims inflation and premium adjustment clauses. It could be assumed that the effects of claims inflation and premium adjustment clauses cancel out each other in the CF projection, provided this approach undervalues neither the BE, nor the risk involved with the higher CFs after claims inflation and premium adjustment.

Nonlife insurance methodologies had to be applied to pure risk insurance belonging to the insurance classes "accident" and "health." However, where the characteristics of the contracts clearly require a different treatment, in line with life insurance valuation methodologies, participants had to treat these contracts as life insurance.

The costs of options and guarantees had to be valued on a market-consistent basis including both the intrinsic and the time value. Considerations regarding the effects of policyholder behavior and other nonfinancial factors should also be taken into account in the valuation of options and guarantees.

The costs of any option and guarantee may be valued by using one or more of the following four methods:

- If the risk from the option or guarantee is hedgeable, the market costs of the hedge or replicating portfolio of the option or guarantee should be used.
- A stochastic approach using, for instance, a market-consistent asset model (includes both closed form and stochastic simulation approaches).
- A series of deterministic projections with attributed probabilities.
- A deterministic valuation based on expected CFs in cases where this delivers a market-consistent valuation of the TP, including the cost of options and guarantees.

Generally, dynamic hedging strategies should not be assumed in the valuation of options and guarantees unless it forms an integrated part of a firm's principles and practices to run the business.

A guarantee is defined as a benefit that is the maximum of either a quantity related in some way to the value of the underlying assets or a guaranteed amount (which may be time dependent and increasing on future valuation dates when extra benefits are added). A guarantee thus defines the possibility to receive extra benefits in excess of the guaranteed benefits. In financial terms, a guarantee is linked to option valuation.

For a with-profit life insurance contract with an investment guarantee, the intrinsic value represents the amount at which the extra benefits are "in the money" at the valuation date. The intrinsic value can be estimated by using representative deterministic assumptions of possible future financial outcome. The time value of the guarantee captures the potential for the cost to change in value in the future, as the guarantee moves "into" or "out of the money." Thus, under certain economic scenarios where amounts above the intrinsic value are required to meet policyholder's payments, the average additional cost of these events forms the time value of the guarantee. Where the option or guarantee is capable of being hedged, then the cost of the guarantee or option would be the market cost of hedging the option or guarantee.

For the time value of an *investment guarantee*, it could be assumed that a Black–Scholes or any other market-consistent framework holds. A stochastic simulation approach may be required to accurately capture policyholder behavior and management actions but with further assumptions, a closed-form solution may be used as a simplification.

If a firm charges for the cost of guarantees, options, or smoothing in the determination of extra benefits, then when calculating the credit for those charges, the projected future levels of such charges should be separately assessed and be consistent with the firm's principles and practices to run the business.

Applying one or several of the following approaches could assess other charges than expenses:

- If the charges are fixed in some way (e.g., they are a fixed percentage of future regular premiums or fund value), then it may be sufficient to discount the expected future charges at the appropriate risk-free interest rate.
- If the future charges are to be reassessed periodically in the light of the future cost of guarantees, options, or smoothing, possibly net of residual accrued past charges and costs, then the valuation of them should allow for future changes to the charges if appropriate and material.
- Especially if a firm can exercise discretion, the reasonability of the projected charges should be considered. A firm should consider the actual costs of guarantees, options, or smoothing and the firm's possible obligations to policyholders, whether through policy wordings, marketing literature, or other statements that give rise to policyholder expectations of how the management will run the business.

The outlined general valuation approach was expected by CEIOPS to be used for *small insurers or portfolios.* However, for some factors, elements, or procedures, more pragmatic approaches could be accepted. The general valuation objective for small insurers or portfolios is that the valuation approach should not materially alter the overall valuation result and systematically underestimate the true liability. The valuation approach for small insurers or portfolios should therefore reflect the main characters of the underlying liability to be valued and produce reasonable proxies for BE values.

It should be noted that the simplifications for small firms or portfolios were in principle equally well applicable for larger insurers and larger portfolios especially where risks are not considered to be significant following the principle of proportionality.

Assumptions should generally reflect both past experiences and any foreseeable trend. A more pragmatic approach would be allowed, where this distinction was not explicitly made, but was nevertheless qualitative explored. Thus more approximate methods sets a reasonable BE where the historical experience and the trends were not separated and therefore some prudence was expected to be included in the estimate in order to cover model and parameter uncertainties. The prudence level set should, however, not be such that it includes prudence related to adverse deviations.

Concerning *mortality assumptions,* a birth-year cohort approach did not need to be followed, even if it normally would be appropriate to do so. Moreover, any biometric risk could be considered to be independent from any other variable.

Generally, where there was considerable variation in the cost of options and guarantees relative to time and the conditions prevailing at that time, single deterministic scenarios cannot capture the BE costs in a reliable way. Since policyholder's option to surrender, and commonly also investment guarantees, can be seen to constitute a material part of the valuation approach and of the overall liability, they need to be explicitly modeled. However, a pragmatic approach leading to approximate closed-form formulas could be adopted.

Concerning *policyholders' option to surrender*, surrendering is often dependent on financial markets and firm-specific information. However, for the purpose of QIS4, it was assumed that the process of surrendering was independent of financial markets and firm-specific information. This assumption simplified the modeling and enabled the process to be modeled, for instance, with the use of hazard rates. Care has to be taken to define the surrender intensity in an actuarially sound manner.

Other options and guarantees should also be qualitatively assessed. This includes identifying them and an assessment of key drivers, including any possible changes in value as time passes, triggering events and possible impacts on the firm. If considered material, other options and guarantees could be given a subjective *ad hoc* cost approximation given by an expected intrinsic amount increased with an amount that equals the expected probability that the option will move more "into the money" as time passes times the expected costs given that the event will occur.

In general, *future premiums* are not paid independently of the financial market or a firm's solvency position. This creates complicated path-dependent structures. It may be assumed that future premiums are paid independently of the financial market and the firm's solvency position. Possible path dependencies should, however, be qualitatively assessed.

In general, expected *future expenses* should be explicitly recognized in the CF projection. A pragmatic approach could be to recognize as a liability the future expense loadings expected to incur increased with possible historical deficiencies in the expense loadings.

G.4 NONLIFE TECHNICAL PROVISIONS

G.4.1 Segmentation

In QIS1, the insurance values should be indicated in each of the LoBs defined in the present accounting Directive, that is, Article 63 of the Council Directive on the annual accounts and consolidated accounts of insurance undertakings, 91/674/EEC. If values were not provided on this basis, participants had to describe the method they used to determine an appropriate segmentation. They would also explain why this segmentation is more appropriate to their business.

In QIS2 values for nonlife insurance should be indicated in each of the LoBs defined in Article 63 of the Council Directive on the annual accounts and consolidated accounts of insurance undertakings (91/674/EEC):

- Accident and health
- Motor, third-party liability
- Motor, other classes
- MAT: Marine, aviation, and transport
- Fire and other property damage

- Third-party liability
- Credit and suretyship
- Legal expenses
- Assistance
- Miscellaneous nonlife insurance
- Reinsurance. In the case of facultative reinsurance cover, business may be allocated to the other segments, if this is more reflective of how an insurer's accounting systems operate in practice.

In both QIS1 and QIS2 the TPs to be tested comprise:

- The provision for claims outstanding
- Premium provisions, unearned premium provision, provision for unexpired risks

The valuation of the provision for claims outstanding and the premium provisions would generally be carried out separately. However, if such a separate treatment was not practical, for example, where business is written on an underwriting year basis, and a split between covered but not incurred (CBNI) and incurred but not settled (IBNS) claims would be artificial, participants could value these provisions together.

For QIS3 and QIS4, the nonlife the segment "accident and health" was split into workers' compensation, health insurance, and "others/default." Proportional nonlife reinsurance was treated as direct insurance, that is, it was allocated to one of the LOBs listed.

Nonproportionality reinsurance was split into property, casualty, and MAT business.

Simplifications for the estimation of nonlife Ps are discussed in Section G.5.5.

The principle of *substance over form* should be followed in determining how contracts are to be treated, whether with respect to an allocation within nonlife insurance, or with respect to an allocation between life and nonlife insurance. In practice, certain types of liabilities, although stemming from claims covered by nonlife insurance contracts, may be similar in nature to liabilities commonly observed in life insurance business. These claims should be valued based on their technical nature, that is, life insurance principles.

For those nonlife LoB affected by this issue, participants were asked to disclose separately the BE of liabilities similar in nature to "standard" applicable nonlife principles and the BEL where life principles need to be used. Analogously, certain types of liabilities stemming from claims classified under life insurance business could be better approximated (in terms of technical nature) by nonlife valuation principles. These claims should be valued using relevant and applicable nonlife insurance principles. Participants should disclose separately the BEL that were so valued.

The segmentation used in QIS4 posed problems to some of the participating companies; CEIOPS (2008e). Many companies were using a more granular segmentation in their internal systems.

G.4.2 Best Estimate

As stated above, the valuation of BE for claims outstanding provisions and for premium provisions would generally be carried out separately. However, if a separate treatment was not practical, participants could value these provisions together. Participants were encouraged to perform the valuation of BE on the basis of homogeneous groups of risks, which could be more granular than the above segmentation, following actuarial best practice. Results had to be disclosed on the basis of the above segmentation. To the extent possible, insurers should describe on what basis the groupings were made.

Participants had to use statistical methods compatible with current actuarial "best practice" and had to take into account all factors that might have a material impact on the expected future claims experience. Typically, this would require the use of claims data on an occurrence/accident year basis or an underwriting year basis for the run-off triangles.

"Goodness-of-fit" tests should be applied to all statistical methods considered. The results from this analysis would be taken into account together with the estimate of future trends, the relevance of past data (particularly the inclusion of exceptional events), and other elements of actuarial judgment in determining the BE provisions. .

In a letter to Groupe Consultatif in September 2008, CEIOPS asked the Groupe Consultatif to closely cooperate with the joint CEIOPS-Groupe Consultatif Coordination Group for establishing a taxonomy of actuarial methods, which could be used to calculate TPs for nonlife insurance liabilities and which might be used under Solvency II in accordance with the probability-weighted average of future CFs, taking account the time value of money.

In its answer, Groupe Consultatif, GC (2008), maintained that achieving consistent TPs is one of the core challenges for the quantitative measures within Solvency II, even if there is wide support in principle for market-consistent valuation with the TPs being split into two components: the BE and the RM.

The focus of the report was the consistency of BE valuation. It was evident from QIS4 that it was difficult to assess the consistency of methodology applied. This difficulty in the view of Groupe Consultatif is due to a number of factors:

- Accounting practices (local GAAPs) vary across Europe and do not always produce TPs that are fit for comparison. The local GAAP figures are currently not aligned to the Solvency II valuation principles, and not everyone has had the time fully to assess the implication of reserving on the new basis.
- There is no standardized definition of a reserving "best estimate" across Europe.
- Historical differences in coverage and in definitions of business classes.
- Lack of appropriate data and systems for some organizations and countries.
- The fact that BE valuation is not an exact science but should incorporate professional judgments.

With the introduction of Solvency II, companies will have to prepare for a move toward more advanced reserving techniques, greater transparency, and communication of uncertainties to the Administrative or Management Body and external stakeholders including supervisors.

To get a view of the current practice for valuation of BE for nonlife insurance companies, Groupe Consultatif decided to make a country comparison. On the basis of the comparison, GC (2009), Groupe Consultatif concluded that

- There is no standardized definition of a reserving "best estimate" across Europe.
- Accounting practices (local GAAPs) vary across Europe and do not always produce TPs that are fit for comparison. The local GAAP figures are currently not aligned to the Solvency II valuation principles.
  - The underlying claims reserving principles vary by country but tend to be based on a prudent approach.
  - The premium reserves are currently based on accounting practice and are set without consideration of the mean of the discounted CFs.
  - No consistent approach to the calculation of loss adjustment expenses exist.
  - There are historical differences in coverages and in definitions of business classes.
- The common industry practice across Europe is to hold undiscounted reserves, with only one country so far having introduced mandatory discounting for all LoBs.
- There is no country within Europe which, to our knowledge, has introduced a requirement to hold explicit RMs. However, RMs/prudence are implicitly included within the reserves. These implicit margins are typically build up through a combination of
  - Conservative (case) reserving.
  - No recognition for the time value of money.
  - In some cases restriction on reserves imposed through regulation.
- For a number of countries, actuarial guidance is available on reserving (techniques, reporting, etc.), although this does not exist in the majority of cases.
- The requirement for an Appointed Actuary for nonlife business only exists in 3 out of the 14 countries within this comparison. A few countries do, however, require an Appointed Actuary for certain LoBs.
- It is not unusual for the majority of countries that signed actuarial reports or statements may be required internally by management, board or for audit purposes, in their effort to fulfill existing (regulatory) demands of corporate governance.

Given the diversity of current practices across Europe, it is evident that every country has a different starting point and their own changes as well as challenges to deal with when moving to a Solvency II reserving basis. It also highlights the difficulty in developing sensible market benchmarks and potential "default methods" across Europe at least in the short to medium term.

G.4.3 Premiums Provisions

Premium provisions substitute current unearned premium provisions and unexpired risk provisions (stand-ready obligations). Premium provisions relate to the coverage period when the insurer provides the service of accepting and managing the risks to its policyholders. During the coverage period, the insurer is at risk of insured events occurring with varying severity.

The calculation of the BE of the premium provision relates to all future claim payments arising from future events post the valuation date that would be insured under the insurer's existing policies that have not yet expired, administrative expenses, and to all expected future premiums. Premium provision is determined on a prospective basis taking into account the expected cash inflows and cash outflows and time value of money.

Applying appropriate methodologies and underlying models and using assumptions that were deemed to be realistic for the LoB or homogenous groups of risk should determine the expected CFs.

G.4.4 Postclaims Technical Provisions: Outstanding Claims Provisions

Postclaims TPs relate to the settlement period between claims being incurred and claims being settled. During the settlement period, the insurer is at risk due to uncertainties regarding, for example, the number of claims not yet reported (IBNR claims), the stochastic nature of claim sizes, and the timing of claim payments (reflecting the claims handling processes and the potential reopening of claims) as well as uncertainties related to, for example, changes in the legal environment.

For claims with low uncertainty, both in timing and amount, generally claims that are settled in a short term, either the result of their individual valuation (case by case) or the result of sound statistical methods may be assumed as reasonable proxies of their BE, provided the entity has checked that the alternative used has produced consistent estimates with the actual results obtained in back testing.

For claims with significant uncertainty, in either timing or amount (generally claims that are settled in a medium or long term), the BE should in principle be valued using relevant actuarial methods based on runoff triangles. To guarantee that the insurer controls model and parameter errors, some general principles were suggested:

- The BE should in general be assessed using at least two different methods that could be considered reliable and relevant. Two methods are considered different when they are based both on different actuarial techniques and different sets of assumptions, therefore cross-checking each other if there is some model or parameter error. Judgment should then be used to choose the most appropriate method. A most appropriate method is a technique which is part of best practice and which captures the nature of the liability most adequately.

- If the available data do not offer a robust behavior to be integrated directly into run-off triangles and treated through generally accepted actuarial methods, the participant will try to adjust the historical data using objective and verifiable criteria, maintaining in any case homogeneity of different series used.

- If this adjustment were not possible or reliable, a case-by-case assessment is preferable to the application of too heterogeneous methods or to inconsistent sets of data.

However, if it is considered that the claims handlers consistently under- or overestimate claims, this should be reflected in the overall BE provision.

G.5 SIMPLIFICATIONS AND PROXIES

Simplifications and proxies were introduced in QIS4 according to the proportionality principle set out in the draft Framework Directive, COM (2007). The proportionality principle is also discussed in Appendix E, Section E.1. "Less advanced approaches" for life TPs was also discussed for QIS3 in CEIOPS (2007h).

For QIS4, CEIOPS provided the industry a valuation tool based on the chain ladder approach. The tool, an Excel spreadsheet, was developed by the German industry organization GDV. A description of the tool was given in CEIOPS (2008c).

G.5.1 Proxies

Proxies for the valuation of TPs could be used where there was insufficient company-specific data of appropriate quality to apply a reliable statistical actuarial method for the determination of the BE. Proxies could be regarded as special types of simplified methods that are positioned at the "lower end" of continuum of methods that could be applied.

A Bayesian method to estimate the predictive distribution for outstanding loss liabilities was discussed by Meyers (2009).

Under a future Solvency II regime, proxy methods would be needed whenever a lack of sufficiently credible own data cannot be avoided. This could be the case, for example,

- For entirely new types of insurance in the market that will not have any historical data to act as a guide (e.g., cyber risks).

- For classes of business that are being written for the first time by an insurer.

- Where due to legislative or significant underwriting changes the characteristics of the terms of the insurance contracts are changed in such a manner that historical data are rendered useless.

- When the insurer, or the class of business in question, is too small to allow the build-up of credible historical claims data.

Under Solvency II, proxies could be used to determine TPs, if

- The proxy is compatible with the general principles underlying the valuation of TPs under Solvency II and
- The use of the proxy is proportionate to the underlying risks

An appropriate valuation of TPs under the Solvency II principles, including the use of proxies, will require sufficient actuarial expertise. Consistent with this, the Framework Directive Proposal required insurers to provide an actuarial function to ensure the appropriateness of the methodologies and underlying models used as well as the assumptions made in the calculation of TPs.

Section TS.IV of the Technical Specification, QIS4 (2008), contained a description of a range of proxy valuation techniques for TPs, including criteria under which the proxies could be applied. When applied with sufficient actuarial expertise and professional judgment, these techniques, or parts of these techniques, could under certain circumstances be regarded as sound actuarial techniques. It is inappropriate to overrely on any proxy method. A method could at a specific point in time produce a sensible estimate, but changing the circumstances may render its accuracy and validity of limited use. Therefore, to the extent this was practicable, participants would not rely on a single proxy method thought to be appropriate, but rather consider a range of approaches before making a final decision on which method they would take.

G.5.2 Simplification: Reinsurance Recoverables

A simplified calculation of the expected loss was allowed if the following conditions were met:

- The expected loss according to the simplified calculation is less than 5% of the recoverables before adjustment for counterparty default.
- The approximation is proportionate to the nature, scale, and complexity of the risks supported by the undertaking, in particular there are no indications that the simplified formula significantly underestimate the expected loss.

The simplified calculation should be made as follows:

$$EL = -LGD_{\%} \cdot BE_{\text{Rec}} \cdot \max\{Dur_{\text{mod}}; 0\} \cdot \frac{PD}{1 - PD},$$

where EL is the adjustment for expected loss; $LGD_{\%}$ is the relative LGD of the counterparty, for instance, 50%, if no reliable estimate of the lLGD was available; BE_{Rec} is the BE of recoverables not taking account of expected loss due to default of the counterparty; Dur_{mod} is the modified duration of the recoverables; and PD is the probability of default of the counterparty.

The adjustment for expected loss should be calculated separately for each counterparty. If the estimates of the PD and the LGD of several counterparties coincide, no separate calculation was necessary under the simplified approach.

G.5.3 Simplification: Life Insurance BE

"Less advanced approaches" for life TPs were discussed for QIS3 in CEIOPS (2007h). The following is a summary of some of the possible less advanced approaches illustrated in op. cit. that could be applied in the estimation of the BEL:

- Grouping of contracts (this is discussed above for QIS4)
  - Choose suitable model points within a specimen policy instead of modeling each contract in the portfolio separately.
- Biometric assumptions
  - Neglect the reflection of the so-called trend forecast
  - Apply a period instead of a cohort approach for mortality
  - Use assumptions that are currently available in the valuation of TPs appropriately scaled to obtain an approximation for the BE
  - Use a deterministic approach and neglect any possible need for stochastic simulation
  - Assume independence from any other variable
- Surrender option
  - Assume that surrenders occur independent of financial and biometric risks
  - Assume that surrenders occur independent of firm-specific information
  - Model the surrender as a hazard process either with a nonconstant or constant intensity
- Financial options and guarantees
  - Approximate guarantees and options by assuming a Black–Scholes type of environment
- Investment guarantees
  - Assume nonpath dependency
  - Focus on intrinsic values
  - Apply formulaic simplified approaches for the time values if they are considered to be material
- Other options and guarantees
  - Focus on material other options and guarantees
  - Approximate, for instance, by grouping investment, mortality, and expense guarantees into one single investment guarantee
  - In the absence of a well-defined valuation approach, use subjective *ad hoc* approaches if the options and guarantees are considered to be of material importance

- Distribution of extra benefits
  - Assume nonpath dependency
  - Assume a constant distribution ratio of extra benefits reflecting past practices. Apply the distribution ratio to the overall valuation to determine the amount of extra benefits and the time values of guarantees
  - Alternatively, approximate the amount of available extra benefits for distribution to policyholders as the difference of liabilities currently held and the PV of expected future guaranteed benefits adjusted with appropriate considerations to future expenses needed to service the insurance contracts.
- Expenses and other charges
  - Use information from current expense loadings, future projected expense loadings, and past expense analysis
  - Assume other charges to be a constant reduction of extra benefits or a constant charge from the policy fund
- Other issues
  - Chose a projecting period equal to 1 year
  - Assume CFs to occur at the end or in the middle of the time intervals
  - Assume that future premiums are paid independently of the financial market and firm-specific information or alternatively neglect the premiums

The simplification used in QIS4 was based on a profit-sharing life insurance system from Italy. Given that the profit-sharing mechanism followed a similar approach, it could be extended to other profit-sharing systems. In particular, the simplification could be used for countries where the revaluation clauses of the sum insured were defined in the insurance contracts or in the national law. Moreover, an additional simplification was proposed, for policies where annual bonuses were determined by an insurer's decision.

Following the proportionality principle, only participants with a low risk profile could use the simplification. In the application, the assets portfolio should have a small component of equity investment, that is, the simplified formula should be limited to funds where the percentage invested in equity was lower than 20%, and should not contain financial derivatives.

The following input information was required separately for each fund and at least for different minimum guaranteed rates and for different maturities:

S_0: the total sum insured at the valuation date

T: the average maturity of the policies

R: the technical interest rate

Δ: the minimum guaranteed spread over r

B: the participation coefficient; for policies where annual bonuses are determined by an insurer's decision, the same approach could be used for deriving an assessment of Future Discretionary Benefits. In this case, β could be set equal to the average participation coefficient over the last three years.

w_E: the fraction of the fund invested in equity.

The simplification delivers the following output:

BE: best estimate of with-profit contracts

FDB: value of future discretionary benefits

In order to calculate the BE of the TPs of a profit-sharing policy, let us consider a benefit Y_T to be paid at date T. The benefit will be determined as follows:

$$Y_T = S_0 \cdot \prod_{t=1}^{T} (1 + R_t),$$

where R_t is the revaluation rate in year $t = 1, 2, \ldots, T$; R_t is a function $R_t = m(I_t)$of the return I_t on the investments in year t;

As a simple example: $m(I_t) = \max\left\{\frac{\beta \cdot I_t - r}{1+r}; \delta\right\}$.

By this rule, the value of the minimum guaranteed benefit is

$$\text{BE}_{\text{guaranteed}} = S_0 \cdot (1 + \delta)^T \cdot v_T,$$

where v_T is the risk-free discount factor for maturity T.

The intrinsic value (IV) of Y_T is defined as $\text{IV} = S_0 \cdot v_t \cdot \prod \left[1 + m\left(f_t\right)\right]$, where $f_t = (v_t/v_{t-1}) - 1$ is the forward rate for the period $[t-1, t]$ derived from the risk-free interest rate term structure. As is known, IV provides an underestimation of the BE of Y_T (the difference being the time value of Y_T).

Therefore, the simplification for the BE is equal to

$$\text{BE} \approx S_0 \cdot v_t \cdot \prod [1 + m(f_t^*)],$$

where f_t^* is a projection rate obtained by incrementing the forward rate: $f_t^* = f_t + \Delta f_t$,

Considering that the calibration of the increment Δf_t shall take into account the nature, scale, and complexity of the risks borne by insurance undertakings, f_t^* is calculated as follows:

$$f_t^* = f_t + \frac{\sigma_B \cdot (1 - w_E) + \sigma_E \cdot w_E}{\sqrt{t}},$$

where $\sigma_B = 2,5\%$ and $\sigma_E = 15\%$.

The value of Future Discretionary Benefits is, $FDB = \text{BE} - \text{BE}_{\text{guaranteed}}$.

G.5.4 Simplification: Nonlife Insurance BE

As a simplified approach for the premium provisions, an "Expected Loss Based Proxy" with a combined ratio estimated from the firm's own data and other information could be used to derive a BE for the premium provision. A description of such a method was given in QIS4 (2008, TS.IV.F).

A simplified approach for the claims TPs would be to use a "case-by-case" estimation to stipulate the BE for claim amounts related to the reported but not settled claims (the RBNS provisions). However, the "case-by-case" estimation of RBNS provisions must be supplemented by a simplified method for stipulating the claim amounts related to incurred but not reported claims (IBNR claims).

In cases like this, a simplified method for calculating the IBNR provisions could be given by a prespecified percentage applied to the sum of cumulated claims payments and the RBNS provisions or as the difference between the estimated overall claims costs and the sum of cumulated claims payments and the RBNS provisions. With this approach, the stipulation of the IBNR provisions must be carried out per occurrence/ accident year (or underwriting year).

A simplified method for calculating the IBNR claims could be based on the total of paid claims and the RBNS amount, for example, as a given percentage of this total) or on an estimate of the total claims costs (e.g., as a residual given by the difference between the estimated overall claims cost and the total of paid claims and the RBNS amount).

G.5.5 Proxies: Nonlife BE

Before QIS4, CEIOPS and Groupe Consultatif set up a coordination group to discuss nonlife proxies. Behind this group different local supervisory authorities, together with actuary societies, set up working groups to consider different national methods used. The group published its report in connection with QIS4; CEIOPS-GC (2008). Section TS.IV of the Technical Specification, QIS4 (2008), contained a description of a range of these proxy valuation techniques for the TPs.

Table G.1 gives an overview of the proxies that could be used in QIS4. The proxies were applied to claims provision and premium provision, but also to discounting and "gross-to-net" reinsurance.

Following the "substance over form" principle, annuities arising from nonlife insurance contracts are to be treated as life insurance obligations for solvency valuation purposes.

For each of the proxy techniques that require market parameters as an input data, a short description of the type of data required was discussed in CEIOPS (2008a).

It should be noted that the testing of explicit proxy techniques for the valuation of TPs was a new feature in QIS4. Whereas a number of markets have already developed tentative calibrations for selected proxy market parameters, the QIS4 exercise was important to initiate further technical discussions with stakeholders involved in the Solvency II process on the availability of market data and the appropriateness and feasibility of potential calibration techniques. An adequate calibration of market parameters for proxies would be central to improve the consistency and comparability of valuation techniques for TPs across Europe.

TABLE G.1 When are Proxies Allowed for in Nonlife Best Estimation?

Applied to Proxy	Claims Provision	Premium Provision	Discounting	Gross-to-Net
Market development patterns	√			√
Average severity/frequency	√		√	√
Bornhuetter–Ferguson	√			√
Case by case	√		√	√
Expected loss		√		√
Simplified application of standard statistical techniques	√			√
Premium based		√		
Claims handling costs	√			

Source: Adapted from QIS4. 2008. QIS 4 Technical Specifications. European Commission. MARKT/2505/08. March 31. Available at http://www.ceiops.eu/content/view/118/124.

Note: The table gives a hint where proxies could be used.

The following gives an overview on where market parameters were needed for an application of the proxy techniques that were tested under QIS4; proxy and required market parameter:

- *Market development pattern proxy:* development factors that reflect the average evolution of gross paid claims
- *Frequency–severity proxy:* expected severity of claims, as a market average
- *Bornhutter–Ferguson-based proxy*: initial market-based ultimate loss ratio
- *Discounting proxy:* market average of modified duration
- *Risk margin proxy:* market factors to determine RM as percentage of BE; see Section G.6.3
- *Factor-based claims-handling-costs proxy:* market factors for claims handling costs

Each of the market parameters given above applied at the level of an individual LoB in an individual market. This did not preclude the possibility that the same parameter was used across several LoBs/markets where this would seem adequate considering the diversity of the risk profiles of the insurers operating in these segments.

It should be noted that a number of proxy techniques to be tested under QIS4 were not relying on market data, and were therefore not contained in the list above, among them, for example, case-by-case proxies that rely on case-by-case reserves and other insurer-specific input.

Proxies based on market development patterns had been suggested by almost all of the national proxy expert groups that discussed the issue before QIS4. By combining the use of statistical loss-reserving techniques with market data, they would allow insurers a gradual transition to more sophisticated modeling techniques when more insurer-specific claims data becomes available over time.

G.6 RISK MARGIN FOR NONHEDGEABLE RISKS/OBLIGATIONS

G.6.1 General Considerations

In QIS2 technical specifications, CEIOPS (2006d), the required RM for nonhedgeable risks was defined as the difference between the expected value and the value needed to achieve a given, overall, entity-wide level of confidence, including uncertainty over the assumed distributions, the 75th percentile. Participants could calculate the RM using sound actuarial techniques, for example, by stochastically simulating the variation in CFs, based on random variation in the RFs, to determine an appropriate distribution. The approach to calculating the 75th percentile would generally reflect the same considerations that apply when calculating the BE.

Following discussions held within CEIOPS and EIOPC on a standard for TPs, it was anticipated that the Commission was going to issue the following clarification to the Amended Framework for Consultation:

The RM covers the risks linked to the future liability CFs over their whole time horizon. Two possible ways to calculate the RM should be considered as working hypotheses. It can be calculated as the difference between the 75th percentile of the underlying probability distribution until runoff and the BE... Alternatively, the RM can be calculated based on the cost of providing SCR capital to support the business-in-force until runoff. Further quantitative impact information should be collected to assess the merits of the two methods.

To enable an assessment of the two methods, participants were strongly encouraged to provide estimates of TPs according to the *cost of capital approach*. Estimates provided under this approach were only supplementary information to facilitate a comparison.

From QIS3, it was decided to use only a CoC methodology in the determination of the RM for nonhedgeable risks; CEIOPS (2007a).

For the purposes of the calculation of the CoC margin, it was assumed that as a result of an economic loss incurred during the solvency time horizon the undertaking became insolvent at the end of the current year and had no available capital left. It was further assumed that, at time $t = 1$, the portfolio of assets and liabilities was taken over by another undertaking and that the acquiring or purchasing undertaking—the reference undertaking—needed to be compensated for the additional SCR which it had to put up during the whole runoff of the portfolio.

The CoC risk margin (RM was then defined as the cost of the PV, at $t = 0$, of future SCR which the reference undertaking would have to put up during the runoff of the portfolio of assets and liabilities for the in-force book of business at time $t = 1$. As the reference undertaking, that is, the undertaking that received the transferred obligations, the ceding undertaking would be taken, that is, it shall be assumed that the insurer, at time $t = 1$, transfers its obligations to itself.

For the purpose of QIS4, participants were requested to perform their SCR calculations on the basis of the standard formula, when calculating the RM, even if it should be possible to use the output of an approved internal model to perform the SCR calculation under the future Solvency II framework. On an optional basis, participants who had developed a full or partial internal model were also invited to communicate the result of their RM calculations based on these models, provided that the results using the standard formula were also communicated.

Where the RM calculation was based on the standard formula, it should be calculated net of reinsurance. In other words, a single net calculation of the RM should be performed, rather than two separate calculations, that is, one for the RM of the TPs and one for the RM of reinsurance and SPV recoverables. Where participants calculate the RM using an IM, they could either perform one single net calculation or two separate calculations.

The risk modules that need to be taken into account in the CoC calculations were operational risk, underwriting risk with respect to existing business, and counterparty default risk with respect to ceded reinsurance. It was assumed that related to the insurance and reinsurance obligations there did not arise any market risk or risk of default of the counterparties to financial derivative contracts.

Renewals and future business should be considered only to the extent that they have been included in the current BEL. Participants were requested to differentiate calculations on different segments:

- For life insurance, the value of the RM should be reported separately for each segment as defined in Section G.3.1.
- For nonlife insurance, the value of the RM should be reported separately for each LoB as defined in Section G.4.1.

To obtain the overall value of TPs, participants had to assume that no diversification benefits aroused from the grouping of TPs calculated per segment.

All participants had to assume that the CoC rate was 6%.

Most of all companies participating in QIS4 supported the CoC approach, but many also expressed concern regarding the complexity in the calculation methods. The 6% CoC rate was questioned by many companies.

G.6.1.1 General Description of the CoC Methodology

The steps to calculate the RM under a CoC methodology can be summarized as follows. It was assumed that the valuation date is the beginning of year 0, that is, $t = 0$:

- For each insurance/reinsurance segment find an SCR for year $t = 0$ and for each future year throughout the lifetime of the obligations in that segment. SCR for year 0 corresponds to the capital requirement that the firm should hold today with the exception that only part of the risks are considered. The risks to be taken into account are operational risk, underwriting risk with respect to existing business, and counterparty default risk with respect to reinsurance ceded as described in Chapters 26 through 33.
- Multiply each of the future SCRs by the CoC rate to get the cost of holding the future SCRs.

- Discount each of the amounts calculated on the previous step using the risk-free yield curve at $t = 0$. The sum of the discounted values corresponds to the RM to be attached to the BE of the relevant liabilities at $t = 0$.
- The total amount of RM is the sum of the RMs in all the segments.

The main practical difficulty of the method is deriving the SCR for future years. A sophisticated approach requires the projection of the RFs underlying the liabilities for the whole runoff period. However, simpler approximations may be envisaged. This was discussed in CEIOPS (2007a). Confer also the discussion made in Section 8.5.

While the SCR at $t = 0$ is fully calculated, that is, the capital requirements covering the one-year solvency period immediately starting at the valuation date which would not be included on the calculation of the RM of TPs, the design of the future SCR depends on the assumptions chosen for the implementation of the CoC method.

The operational risk capital charge could always be calculated using the SCR standard formula. The formula uses as input parameters earned premiums gross of reinsurance and BEs of TPs, comprising both premium provision and outstanding claims provision, gross of reinsurance. There was also an upper limit with respect to the basic SCR. These input data had to be estimated for each respective year in each segment. Participants were reminded that the BEs are valued at the time value of money of the development year in question, consistent with the use of the interest rate term structure at the valuation date.

G.6.2 Simplifications

There are two ways to calculate simplifications. The first main approach is given in Sections G.6.2.1 through G.6.2.5. If the companies could not use these simplifications, they could use the approaches given in Section G.6.2.6.

G.6.2.1 Credit Risk: Counterparty Default Risk

Counterparty default risk charge (RC) with respect to reinsurance ceded could be calculated directly from the definition for each segment and each year. If the exposure to the default of the reinsurers did not vary considerably throughout the development years, the risk charge could be approximated by applying reinsurers' share of BEs to the level of risk charge that was observed in year 0.

According to the standard formula, the counterparty default risk for reinsurance ceded was assessed for the whole portfolio instead of separate segments. If the risk of default in a segment was deemed to be similar to the total default risk or if the default risk in a segment was negligible, then the risk charge could be arrived at by applying reinsurers' share of BEs to the level of the total capital charge for reinsurers' default risk in year 0.

G.6.2.2 Nonlife Underwriting Risk

Underwriting risk charge for nonlife business, other than catastrophe risk, could be calculated directly from the formula using BE for outstanding claims provision net of reinsurance, other than annuities, and earned premiums net of reinsurance as input parameters. Renewals and future business were not taken into account. For simplicity, it could be assumed that the undertaking-specific estimate of the standard deviation for premium risk remained unchanged throughout the years.

Underwriting risk charge for catastrophe risk was taken into account only with respect to the insurance contracts that exist at $t = 0$. If no better estimate of the catastrophe risk charge for a segment in year y was accessible, then the size of the risk charge could be assumed to be in direct proportion to the earned premiums net of reinsurance in that segment.

If it was not possible to differentiate the catastrophe risk charges in between segments, then it can be assumed that the exposure was proportionate to the net earned premiums.

Usually the periods of insurance were not very long in nonlife insurance so that the earned premiums differ from zero only for the first few years. This provides for a further simplification. Since there does not exist any premium or catastrophe risk for the years when earned premiums are zero, the underwriting risk module for nonlife consists only of the reserve risk. The risk charge for the reserve risk in a segment is simply of the form constant times the BE of the outstanding claims provision net of reinsurance.

G.6.2.3 Health Underwriting Risk

In short-term health insurance, the lifetime of the obligations is short by definition. Typically the capital charge for the first 12 months will suffice ($t = 0$). If there were obligations that are not negligible beyond the first year, simplifications similar to those in nonlife underwriting risk could be used. For simplicity, it may be assumed that the overall standard deviation σ remains the same over time.

Similarly, the underwriting risk charge for the workers' compensation module should be calculated using the guidelines proposed for nonlife underwriting risk. However, the workers' compensation annuities risk charge would be calculated using the methods proposed for the life underwriting risk charge.

G.6.2.4 Life Underwriting Risk

As an approximation, the future SCRs for submodules could be calculated using the simplified SCR approaches in Chapter 32. Future SCRs would then be calculated using inputs projected into the future required to calculate the simplified SCRs.

G.6.2.5 Risk-Absorbing Effect of Future Profit Sharing

Undertakings had to project the SCR net of the risk-absorbing effect of profit sharing for the purpose of calculating the RM. Profit sharing could be ignored where this was largely a result of risks that have been excluded from the projection (e.g., market risk).

Alternatively, the effect of profit sharing could be approximated by calculating the SCR at future periods calculated gross of the profit-sharing effect multiplied by the ratio of the SCR net of profit-sharing effect at $t = 0$, excluding market risk, divided by the SCR gross of profit-sharing effect at $t = 0$, excluding market risk.

G.6.2.6 Alternative Simplifications

The simplified calculations should be made per segment. They could only be applied if the standard formula was applied to calculate the SCR. For those segments that include risks calculated by the nonlife, life and/or health methods mentioned below, the overall RM was calculated by combining the results from the simplifications by means of the aggregation method of the SCR standard formula.

G.6.2.6.1 Nonlife Underwriting Risk The CoC RM for an LoB could be determined using the formula

$$CoCM \approx CoC \cdot \left[SCR_k^{tf}(0) + \left(Dur_{\text{mod},k} - 1\right) \cdot \left(3 \cdot \sigma_{R,k} \cdot R_k^0 + 0.02 \cdot R_k^{gr} + CR_{Re,k}\right)\right]$$

where $CoCM$ is the CoC margin; CoC is the CoC rate; $SCR_k^{tf}(0)$ is the current SCR for the LoB k, excluding market risk and default risk for financial derivatives; $Dur_{\text{mod},k}$ is the modified duration of R_k^0; $\sigma_{R,k}$ is the standard deviation for reserve risk of the LoB k, as defined in the SCR standard formula premium and reserve risk module; R_k^0 is the net BE provision for claims outstanding in the LoB k; R_k^{gr} is the gross BE provision for claims outstanding in the LoB k; and $CR_{Re,k}$ is the current capital charge for reinsurance default risk assigned to the LoB k.

If the portfolio of the LoB k contained treaties with a contract period that exceeds the following year, an amendment of the above result for the premium and catastrophe risk for the time after the following year had to be made.

In order to simplify the determination of $SCR_k^{tf}(0)$, the current SCR for premium and reserve risk in the LoB could be estimated as follows:

$$CR_{P,k} \approx 3 \cdot \sqrt{\left(\sigma_{P,k} \cdot P_{\text{k}}^{\text{existing}}\right)^2 + \left(\sigma_{R,k} \cdot R_k^0\right)^2 + 2 \cdot \alpha \cdot \sigma_{P,k} \cdot P_k^{\text{existing}} \cdot \sigma_{R,k} \cdot R_k^0},$$

where $\sigma_{P,k}$ is the standard deviation for premium risk of the LoB k, as defined in the SCR standard formula premium and reserve risk module; P_k^{existing} is the net earned premium in the individual LoB k during the forthcoming year relating to contracts closed before the valuation date; and $\alpha = 0.5$ is the correlation factor between premium risk and reserve risk as specified I the premium and reserve risk submodule.

G.6.2.6.2 Life Underwriting Risk The CoC RM for a segment is determined using the formula

$$CoCM \approx CoC \cdot Dur_{\text{mod},k} \cdot SCR_k^{tf}(0),$$

where $CoCM$ is the CoC margin; CoC is the CoC rate; $Dur_{\text{mod,k}}$ is the modified duration of the BE provision in the segment k, net of reinsurance; and $SCR_k^{tf}(0)$ is the current SCR for the segment k, excluding market risk and default risk for financial derivatives.

In order to determine $SCR_k^{tf}(0)$, a recalculation of the life underwriting SCR restricted to the segment may be necessary. This may be simplified by redistributing the subrisk charges (mortality, longevity, etc.) for the whole portfolio to the segments proportionally to appropriate exposure measures.

The following exposure measures may be taken into consideration:

Subrisk	Exposure Measure
Mortality	(Capital at risk) (duration of treaties under mortality risk)
Longevity	BE of treaties under longevity risk
Disability	(Capital at risk) (duration of treaties under disability risk)
Lapse	(BE of treaties under lapse risk)—(surrender values of treaties under lapse risk)
Expenses	(renewal expenses) duration
Revision	BE of annuities exposed to revision risk
CAT	Capital at risk of treaties under mortality and disability risk

The formula is based on the assumption that the relative loss-absorbing capacity is constant over the runoff of the portfolio. Amendments to the estimation should be made if this assumption did not hold. For example, when the simplified calculation was applied, attention had to be given to the appropriate allowance for the loss-absorbing capacity of future discretionary benefits.

G.6.2.6.3 Health Underwriting Risk The CoC RM for health insurance that is practiced on a similar technical basis to that of life assurance is determined using the formula

$$CoCM \approx CoC \cdot \sum_{t \geq 0} SCR_k^{tf}(0) \frac{L(t)}{(1+r_t)^t \cdot L(0)},$$

where $CoCM$ is the CoC margin; CoC is the CoC rate; $SCR_k^{tf}(0)$ is the current SCR for the LoB k, excluding market risk and default risk for financial derivatives; $L(t)$ is the expected benefits, allowing for claim inflation, paid in year t; and r_t is the risk-free interest rate for the maturity t.

The formula is based on the assumption that the relative loss-absorbing capacity is constant over the runoff of the portfolio. Amendments to the estimation shall be made if this assumption did not hold. For example, when the simplified calculation is applied, attention had to be given to the appropriate allowance for the loss-absorbing capacity of future discretionary benefits.

The RM for health short term and workers compensation general modules has to be calculated using the guidelines proposed for nonlife underwriting RM.

Workers' compensation annuities RM should be calculated using the methods proposed for life underwriting.

G.6.2.7 Overall SCR Simplifications

Alternatively to the simplifications provided in Sections G.6.2.1 through G.6.2.6, companies could derive future SCR values for each segment assuming that the ratio of SCR for that segment at $t = 0$, incorporating only the appropriate risks, over the BE at $t = 0$, or other exposure measure deemed appropriate as a reflection of the underlying risks, *is constant throughout the whole run-off period* of liabilities. This is the approach discussed in Section 8.5.

For example, the calculation of future SCRs for the profit-sharing business may be based on a projection of guaranteed benefits if this is appropriate.

G.6.3 RM proxies

In QIS4, CEIOPS considered a proxy for the RM to be calculated by applying a percentage figure to the BE amount, calculated by using a proxy method; CEIOPS (2008a). These percentages would be indicated per LoB.

The entry values were fixed by CEIOPS and reflected the average payout patterns. This proxy was intended to be used when all other methods to determine the CoC RM, including the simplifications described above, were not available due to a lack of data.

On the basis of the QIS3 results, a first quantitative analysis of the ratios between the CoC margin and the BE for individual insurers (per LoB/market) had been conducted. On the basis of this analysis, a tentative calibration had been derived and included for testing in the QIS4 technical specifications; QIS4 (2008, TS.IV.N).

However, given the difficulties met by undertakings in calculating the BE and the absence of detailed guidance on this calculation given by CEIOPS in the previous QIS exercises, further technical work has to be done to decide whether adequate average ratios for the calibration of the proxy could be determined. This should also include an analysis of the statistical spread of individual ratios around the average in order to assess the quality of the estimation of the RM achieved by this proxy.

G.7 OTHER LIABILITIES

In accordance with QIS2 technical specifications, CEIOPS (2006d), the total of liabilities, other than TPs, would be disclosed according to local valuation practices.

In QIS3, no adjustment in the valuation of other liabilities would be made on account of the creditworthiness of the undertaking itself. Other liabilities that were tradable in a deep, liquid market were valued at the amount for which they could be transferred, or settled, between knowledgeable willing parties in an arm's length transaction. Obligations that were not tradable in a deep, liquid market would be valued on a prudent basis at the PV of the future CFs allowing for all aspects that affect those CFs, such as the right to early repayment, the right of conversion, and by being consistent with information provided by the financial markets. Reasonable simplifications were allowed.

In QIS4, QIS4 (2008, TS.III.B), other liabilities are defined as

Provisions: A provision is a liability of uncertain timing or amount. A provision should be recognized when and only when

a. An entity has a present obligation (legal or constructive) as a result of a past event

b. It is probable (i.e., more likely than not) that an outflow of resources will be required to settle the obligation

c. A reliable estimate can be made of the amount of the obligation

Financial Liabilities

Financial liabilities at fair value through profit or loss: Only recognized when an entity becomes a party to the contractual provisions of the instrument.

Other financial liabilities and amounts payable: Only recognized when an entity becomes a party to the contractual provisions of the instrument.

"Other Liabilities"
Deferred tax liabilities: Income taxes include all domestic and foreign taxes based on taxable profits and withholding taxes payable by a group entity.

Current tax liabilities: Income taxes include all domestic and foreign taxes based on taxable profits and withholding taxes payable by a group entity.

Employee Benefits
Short-term employee benefits: Employee benefits falling due within 12 months after the period in which employee services were rendered.

Postemployment benefits (including pension commitments): Employee benefits other than termination benefits payable after completion of employment. Postemployment benefit plans are classified as defined contribution plans and *defined benefit plans*.

Other long-term employee benefits: Other employee benefits not falling due within 12 months after the end of the period in which employee services were rendered.

Termination benefits: Benefits payable as a result of either

a. An entity's decision to terminate an employee's employment or

b. An employee's decision to accept voluntary redundancy

The treatment of these liabilities is discussed in op. cit.

APPENDIX H

European Solvency II

Standard Formula Framework

HERE WE LOOK at the proposed capital requirement (CR) within the EU Solvency II project, based on a standard approach, and its calibration. The final proposal from CEIOPS is discussed in Chapter 26.

We follow the development from early thoughts that CEIOPS made, based on suggestions by the EC, to a more or less final solution. A final solution will probably never be found. Thanks to the Lamfalussy procedure, there will always be a possibility to change a parameter or a dependence structure (correlation) the Level 2 implementing measures.

In Appendices I through O we look closer at the main different risk modules, and their calibration, proposed by the Commission and CEIOPS. In this appendix, we look at the general thoughts behind the "target capital requirement" (TCR): the solvency capital requirement (SCR), aggregation of main risks, and so on. In Appendix P we discuss the different approaches to assess an MCR.

We start with the advices from CEIOPS, CEIOPS (2005), based on the comments to the CP on the CfA 10 from the EC.

We then look at the different solutions that have been proposed and tested by CEIOPS in the QISs that the EC requested CEIOPS to conduct in order to assess the practicability, implications, and possible impact on different alternative considered. The first QIS, QIS1, was conducted in 2005. It mainly focused on the valuation of assets and liabilities. The first QIS to include a proposal for a standard formula for the CR was QIS2 conducted during 2006.

The subappendices are therefore based on the answers to CfA 10 given in Section H.1, the QIS2 testing in Section H.2, the QIS3 testing in Section H.3, and the QIS4 testing in Section H.4.

CEIOPS has also considered the effects of the financial turmoil during 2007–2009 and how it could effect the FA on the Solvency II implementing measures that were delivered at the end of 2009; CEIOPS (2009a). CEIOPS considered that the overall architecture of the Solvency II system was sound. However, further refinement on the existing calibrations that had been made up to QIS4 had to be done both for the main modules and their submodules. The dependency structure underlying the standard formula had to be strengthened. According to CEIOPS, the crisis had shown that the diversification benefits may have been overstated in QIS4. Different risks were reviewed in light of the financial turmoil, for example, the custodian risk that was not modeled in QIS4.

Stéphane Loisel pointed out in JRMS (2008, p. 75) that the way correlations are defined in Solvency II (up to QIS4) did not take into account "correlation crisis" that could occur after a catastrophe.

H.1 EARLY THOUGHTS AND IDEAS FROM 2005

The early thoughts about the structure of the SCR were given in CEIOPS (2005) and the presentation below is taken from that discussion.

In a preliminary proposal for the draft Directive, the European Commission Services (COM) suggested to introduce a new article for the definition of the SCR.

For illustration and discussion on these ideas, COM gave the following tentative example, suggesting that the issues in square brackets would receive further analysis and consideration:

"To be able to absorb significant unforeseen losses and to give [*reasonable*] assurance to policyholders, an insurance undertaking shall hold at all times solvency capital which is adequate having regard to its overall risk profile. Solvency capital requirement shall cover the relevant [*underwriting, credit, market, liquidity, operational and other*] risks. It shall be calibrated so that the probability of failure of an undertaking within [*one*] year is sufficiently low [*"1/200" or "as defined in implementing measures."*].

The Commission invited CEIOPS to offer advice on the appropriate EU standards for calculating the SCR. The specific CfA would also be examined in conjunction with other relevant specific CfAs, and should address Pillar II issues as well as Pillar I issues.

H.1.1 Purpose of the SCR

The SCR should deliver a level of capital that enables an insurance undertaking to absorb significant unforeseen losses over a specified time horizon and give reasonable assurance to policyholders that payments will be made as they fall due.

The concept of SCR shares many features with economic capital in value-based management. Commercially, an undertaking will define its risk appetite and, where applicable, the public rating it wishes to achieve. It then determines the economic capital that will be necessary to limit its probability of insolvency to a defined level.

H.1.2 Risk Measure

A quantitative solvency assessment could be based on a simplified BS, consisting of assets, liabilities, and available capital, that is, the excess of assets over liabilities. Changes in the level of available capital will depend on the risks to which an undertaking is exposed over the time horizon of the solvency assessment. Because the future development of assets and liabilities is unknown, the future level of available capital will behave stochastically. It may be described by a probability distribution, which measures the likelihood of all possible outcomes.

A "risk measure" is, in general terms, a function that assigns an amount of capital to a risk distribution. Commonly used risk measures are VaR and TVaR.

VaR assesses the probability of ruin at a specified quantile, for example, 99.5%. By contrast, TVaR considers both the probability and the severity of losses in the event that specified quantile is breached.

On a fundamental level, TVaR would encourage supervisors, undertakings, and policyholders to consider the consequences of a potential default, rather than focusing simply on the probability of insolvency. To some extent, a comparison could be drawn with the advanced approach to credit risk in the banking sector, where undertakings consider both the PD and the LGD.

In many business lines, insurance undertakings may be subject to infrequent, high-impact losses, that is, catastrophic losses. The risk distribution will then feature a "fatter tail" than the normal distribution. Because TVaR reflects these losses, it creates an incentive for the insurance undertakings to improve their identification, management, monitoring, and control of low-frequency, high-severity risks. For different reasons explained, TVaR is the risk measure that the IAA Insurer Solvency Assessment Working Party suggested to use for the purpose of setting solvency requirements; IAA (2004).

While the standard formula would be calibrated to simulate the effects of a particular risk measure, undertakings operating under this approach would not be expected to perform a VaR or TVaR calculation themselves. The effects could be simulated using a prespecified, formulaic calculation. IMs could deliver requirements that are closer to an undertaking's "true" VaR/TVaR result.

The most significant disadvantage associated with TVaR is the scarcity of data, which could lead to increased modeling error. A formula based on TVaR might be difficult to generalize in such a way as to provide a good fit for the majority of insurance undertakings, that is, the standard formula would over- or underestimate CRs in many cases because it would be calibrated using tail data that may not be representative.

CEIOPS noted that the COM was proposing the use of VaR as a general principle for calculating the SCR. For IMs, the Commission Services acknowledged in one of the CfAs that more advanced modeling techniques could be used, including the use of TVaR as a risk measure.

CEIOPS stressed the importance of a common underlying philosophy for the SCR, applicable to both the standard formula and IMs. Using different risk measures would impact the incentives to move from the standard formula to IMs and would lead to unpredictable results in the context of partial models. Since IMs strive for a more accurate mapping of the business and therefore would be more likely to address also the consequences of tail events, the use of TVaR would smooth the transition from the standard formula to IMs and facilitate partial use.

H.1.3 Confidence Level

The level of prudence, or confidence, for the SCR will be used to calibrate the standard formula. It may also be a required design feature of IMs. The choice of such a level of confidence will have to reflect the overall prudential objectives of Pillar I requirements on insurers.

The impact of the chosen confidence level will need to be assessed using quantitative analysis. CRs are unlikely to increase in a linear manner as the confidence level is raised. Raising the absolute level of the survival ratio from 99.5% to 99.9% could potentially lead to a much greater marginal increase in CRs than a move from 95% to 99.5%. Choosing to calibrate the SCR to a defined ruin probability will not necessarily lead to smooth or predictable results. For example, a confidence level of 99.5% (i.e., a ruin probability of 0.5%) does not imply that a ruin event will occur once in every 200 years, or that, on an annual basis, 1 in every 200 undertakings will fail. The causes of ruin in one undertaking may have a wider impact, leading to clusters of insurer failures.

H.1.4 Time Horizon

The time horizon for the SCR should reflect

- The frequency with which results are produced
- The ability of undertakings to take timely and effective management action
- The ability of supervisors to respond to a breach of the requirement

Given periodic reporting cycles, it seems sensible that, generally, a time horizon of 1 year should be applied to the SCR calculation.

However, CEIOPS noted at this time that a longer time horizon or a degree of variation within the time horizon might be necessary to enable any formula or IM to describe more realistically how quickly the insurer or its supervisor would be able to react.

H.1.5 Unacceptable Level of Capital

The unacceptable level of capital is the "ruin" situation that the SCR is designed to "avoid," to, for example, a 99.5% confidence level. Traditionally, "ruin" might be characterized as the point at which assets no longer exceed liabilities. However, an undertaking's liabilities may extend beyond the 1-year time horizon for the SCR. At the end of the time horizon (and the beginning of the new one), assets may still exceed liabilities, but capital could represent a much lower confidence level.

The SCR should also reflect the capital required at the end of the time horizon to properly address the runoff of an undertaking's liabilities. This does not mean that available capital should be sufficient to cover the full runoff period of the liabilities. But the liabilities should be transferable to a third party at the end of the time horizon, or policyholders should have reasonable assurance that their claims would be covered. The risk margin (RM) in TPs provides a reasonable proxy for these aims. The SCR should therefore deliver the amount of capital necessary to ensure, with a 99.5% confidence level, that assets will exceed TPs, and other liabilities, as estimated over the period to the end of the specified time horizon.

H.1.6 Going Concern Versus Runoff/Winding up Assumptions

The purpose of regulatory CRs for solvency purposes is twofold. On the one hand, it aims at ensuring that the insurer is sufficiently capitalized during the defined time horizon as a going concern. On the other, regulatory capital should also provide for a successful runoff of an insurance undertaking in a ruin situation. Therefore, regulatory capital has aspects of both the going-concern and runoff situations.

Over the 1-year time horizon, new business may change the risk profile of an insurance undertaking. As the undertaking should be regarded as a going concern until an insolvency event, CRs should generally reflect new business. Additional legal considerations should be studied, especially for those Member States where law requires precise and clear definitions for imposing higher individual solvency requirements to address projections of future business.

In a ruin situation, the issue of costs specifically linked with the runoff of the insurers' business arises. However, to some extent, it could be argued that some costs associated with runoff are already reflected in TPs. Other runoff costs might be addressed when determining the level of confidence in the SCR calculation.

CEIOPS also noted that the valuation principles for assets and liabilities underlying the calculation of the SCR should be compatible with IFRS to the greatest possible extent.

H.1.7 Risk Classification

To the extent possible, the Pillar I quantitative requirements should be designed to address the main financial risks to which an insurance undertaking is exposed. As a general principle, a Pillar I treatment may be applied to any risk which is susceptible to quantification or limitation. However, risks may be excluded from an explicit requirement in Pillar I if, for example,

- On average, the risk is considered marginal
- Simplifying assumptions can be made
- A standardized risk treatment would not be practicable

By contrast, Pillar II should consider all risks, even if they cannot be quantified. The relative emphasis on Pillar II requirements in the solvency framework will depend on the adequacy of Pillar I treatments. It should be noted that there is no unique way of breaking down risks into categories. A categorization that provides a good fit to the risk profile of one undertaking may be less appropriate in other circumstances. This will depend largely on *the nature, scale, and complexity of the business*, that is, the proportionality principle, undertaken by an individual undertaking. In addition, the practicability criterion means that some subcategories of risk could be treated in a Pillar I IM, but not a Pillar I standardized formula.

Based on the work of the IAA (2004), the risks faced by a typical insurance undertaking could be categorized under five major headings:

- *Underwriting risk:* specific insurance risk arising from the underwriting of insurance contracts, associated with both the perils covered and the processes followed in the conduct of the business; these risk modules, depending on whether they are nonlife,

life, or health risks, are discussed in Appendices M, N, and O, respectively. The FA is presented in Chapters 31 through 33.

- *Market risk:* risk arising from the level or volatility of the market prices of financial instruments; This risk module is discussed in Appendix I. The FA is presented in Chapter 27.
- *Credit risk:* the risk of default and change in the credit quality of the issuers of securities, counterparties (notably reinsurers), and intermediaries to whom an undertaking has an exposure; this risk module is discussed in Appendix J. The FA is presented in Chapter 28.
- *Operational risk:* risk of loss resulting from inadequate or failed internal processes, people, and systems or from external events; This risk module is discussed in Appendix K. The FA is presented in Chapter 29.
- *Liquidity risk:* exposure to loss in the event that insufficient liquid assets will be available to meet the CF requirements of policyholder obligations as they fall due. This risk module is discussed in Appendix L. The FA is presented in Chapter 30.

In view of the ongoing work at that time by the Joint Forum, liquidity risk was added as a separate risk category. The risk of asset liability mismatch is also significant, particularly in life insurance business. ALM risk can manifest itself through all of these risk categories and therefore its quantifiable aspects should be addressed as part of the SCR. But it is important that CRs are supported by an appropriate framework for assessing ALM under Pillar II.

H.1.8 Risk Dependencies

Generally, CEIOPS may assume that not all risks will occur at the same time. Due to diversification effects, the overall CR might be smaller than the sum of the capital components for the individual risks. Simple addition of the components could therefore overstate the appropriate amount of capital.

A commonly used mathematical tool to analyze risk dependencies is linear correlation. Linear correlation is based on the assumption of linear dependency between risks and has the advantage of familiarity, as well as being relatively easy to compute. However, its effectiveness is limited, particularly as insurance risks are not generally subject to a normal distribution. For risks that follow a heavily skewed distribution, a linear correlation assumption may underestimate CRs. In addition, normally uncorrelated risks may become highly correlated in extreme circumstances.

Notwithstanding its theoretical deficiencies, linear correlation, *together with a simplified form of tail correlation*, may provide a starting point (and practical expedient) for the standard formula. However, it would be important to keep note of any dependencies that would not be addressed properly by this treatment.

This idea was in line with the proposal made by IAA (2005) and which was discussed in Section 15.3.

An important associated question is the degree of granularity in the analysis. Breaking down risks into many categories and then assuming independence may underestimate the capital needs if, in reality, some of the categories turn out to be positively correlated. Conversely, using too few categories will result in requirements that are insensitive to the different risks posed by LoBs with very different characteristics.

In principle, an adequate overall CR should reflect the underlying distributions of the individual risks, rather than just the resulting capital components. In practice, this is likely to require approximation, using reasonable assumptions about the underlying distributions and dependencies.

H.2 QIS2

The standard formula calculation was divided into modules, following the risk classification set out in Section H.1. This is illustrated in Figure H.1.

The main modules and its subrisk modules as described in Figure 29.1 are as follows.

CR_{NL}: the nonlife underwriting risk module. Its subrisk modules are

- $CR_{NL,PR}$: the premium risk module
- $CR_{NL,RR}$: the reserve risk module
- $CR_{NL,CAT}$: the catastrophe risk module

These subrisks are discussed in Appendix M.

CR_{LR}: the life underwriting risk module. Its subrisk modules are

- $CR_{LR,MR}$: the mortality risk module
- $CR_{LR,MO}$: the morbidity risk module
- $CR_{LR,LO}$: the longevity risk module
- $CR_{LR,DR}$: the disability risk module
- $CR_{LR,LA}$: the lapse risk module
- $CR_{LR,ER}$: the expense risk module

These subrisks are discussed in Appendix N.

CR_{HR}: the health underwriting risk module. Its subrisk modules are

- $CR_{HR,ER}$: the expense risk module
- $CR_{HR,MR}$: the excessive loss/mortality/cancellation risk
- $CR_{HR,AR}$: the epidemic/accumulation risk

These subrisks are discussed in Appendix O.

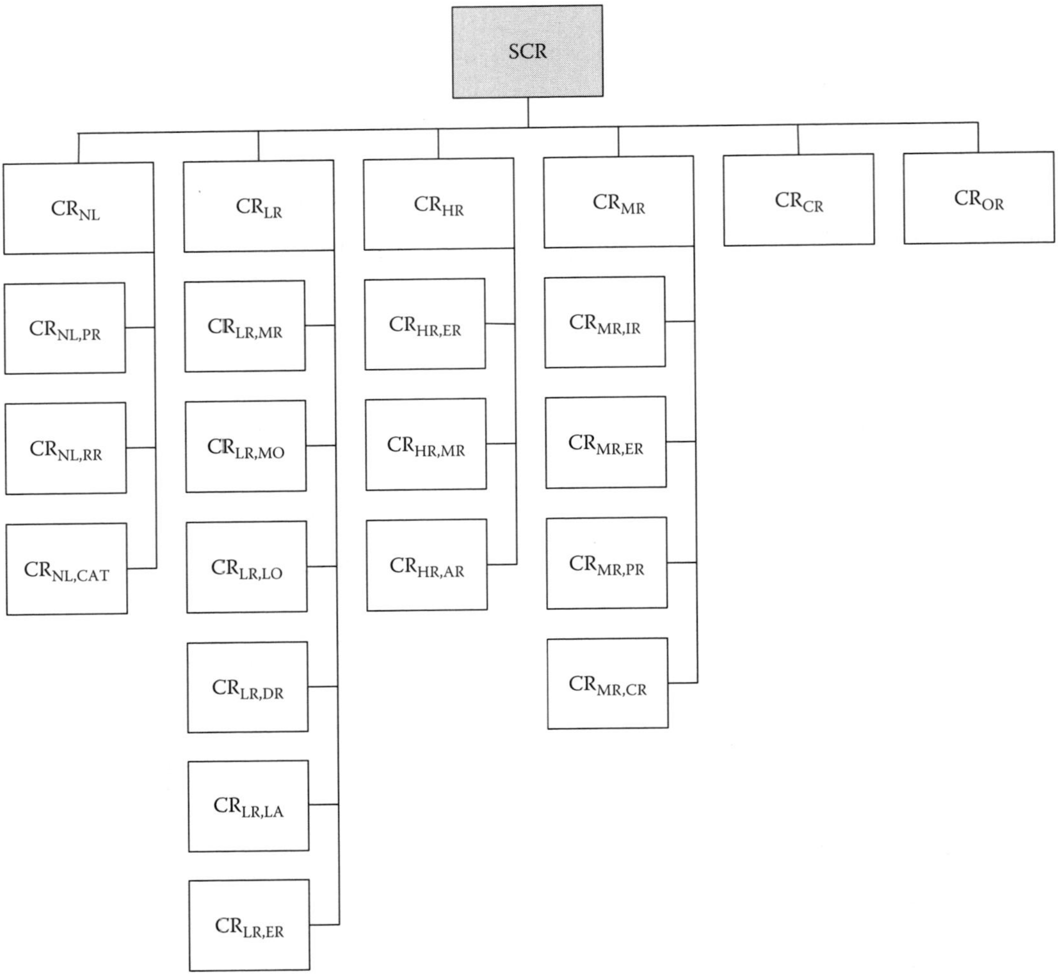

FIGURE H.1 The modular structure of the standard formula proposed for QIS2. The different modules are described below. (Adapted from CEIOPS. 2006d. Quantitative Impact Study No. 2: Technical Specification, CEIOPS-PI-08/06. Available at www.ceiops.org.)

CR_{MR}: the market risk module. Its subrisks modules are

- $CR_{MR,IR}$: the interest rate risk module
- $CR_{MR,ER}$: the equity risk module
- $CR_{MR,PR}$: the property risk module
- $CR_{MR,CR}$: the currency risk module

These subrisks are discussed in Appendix I.

CR_{CR}: the credit risk modules

This risk is discussed in Appendix J.

CO_{OR}: the operational risk module.

This risk is discussed in Appendix K.

In QIS2 the participants were requested to test a number of different modeling approaches for the same risk. Generally, that meant to test both a relatively simple, "robust" approach and a more sophisticated, "risk-sensitive" approach. Applying these different approaches would enable CEIOPS to form a view on the most appropriate treatments and, in particular, the right balance between risk-sensitivity and complexity.

In each risk module, the results of a single approach are singled out to serve as a "placeholder" risk capital charge. This enables the construction of an overall "placeholder SCR" charge for comparative purposes.

The parameters and assumptions used reflect only an initial and very tentative calibration made in QIS2. Parameters and shocks have been selected with the aim of approximating for 1/200 year events. However, the focus for this exercise was on methodological/design issues. CEIOPS also recognized that further calibration work would be required at a later stage in the Solvency II project to fully reflect the prudential objectives set out in the answer to CfA No. 10; see CEIOPS (2005).

General Approach to Risk Mitigation

A broad assumption was made that the effect of risk mitigation techniques should be given full recognition in reducing the relevant risk capital charges. However, the risk of risk mitigation failure should be addressed through an explicit charge for counterparty risk, as part of credit risk. Implicitly, the operational risk charge also addresses the risk of risk mitigation failure. No additional criteria apply for hedging instruments.

In life insurance, provisions relating to future discretionary profit sharing may have significant risk absorption abilities. This was reflected in the calculations by following a three-step approach as follows:

- *In the first step*, capital charges for the individual modules were calculated before allowing for the risk mitigating effects of future profit sharing. This implies that, for these calculations, the valuation of TPs was restricted to guaranteed and statutory benefits, whenever this valuation is used under the factor-based and scenario-based treatments set out below.

- *In the second step*, the CRs for each of the major risk modules were aggregated by applying a dependence matrix, thus allowing for diversification effects across those risk modules.

- *In the third step*, an offset to the overall CR obtained in Step 2 equals to a certain proportion of the amount in TPs relating to future discretionary benefits was included to derive the final SCR value. The determination of this proportion would need to reflect the degree to which future discretionary benefits may be used to absorb risk.

H.2.1 Overall QIS2 SCR Calculation

The placeholder capital charge for the SCR included an allowance for the *risk absorption ability of future profit sharing*, and—for nonlife insurance—the *expected profit or loss from next year's business*, such that

$$\text{SCR} = BSCR - RPS - NL\_PL,$$

where *BSCR*: the BSCR; *RPS*: the reduction for profit sharing; and *NL_PL*: for nonlife insurance, the expected profit or loss arising from next year's business.

Where nonlife business was profitable, the assumption of continued new business might reduce the SCR. Therefore CEIOPS also wanted to consider the effect of omitting CR_{NL} and *NL_PL* from the placeholder formula.

H.2.1.1 BSCR: Basic Solvency CR

The BSCR is calculated before any adjustments for profit sharing or the expected profit or loss arising from next year's business; *cf.* Figure 29.1. It is calculated as

$$BSCR = \sqrt{\sum_{rxc} \rho_{rc} \cdot CR_r \cdot CR_c},$$

where *r* and *c* are the rows and columns of the dependence matrix given in Table H.1, CR_{NL}: the capital requirement/charge for the nonlife underwriting risk; See Appendix M; CR_{LR}: the capital requirement/charge for the life underwriting risk; see Appendix N; CR_{HR}: the capital requirement/charge for the health underwriting risk; see Appendix O; CR_{MR} the capital requirement/charge for the market risk; see Appendix I; CR_{CR}: the capital requirement/charge for the credit risk; and see Appendix J; CR_{OR}: the capital requirement/charge for the operational risk; see Appendix K.

The participants in QIS2 were also asked to calculate two additional BSCRs:

- Assuming full independence, that is, $\rho = 0$ for all combinations
- Assuming no diversification effect, that is, $\rho = 1$ for all combinations

TABLE H.1 A Proposed Structure of the Dependence Matrix for the Main Risks in QIS2

ρ_{rc}	CR_{NL}	CR_{LR}	CR_{HR}	CR_{MR}	CR_{CR}	CR_{OR}
CR_{NL}	1	L	ML	ML	M	M
CR_{LR}		1	L	ML	ML	ML
CR_{HR}			1	ML	ML	ML
CR_{MR}				1	MH	M
CR_{CR}					1	ML
CR_{OR}						1

Source: Adapted from CEIOPS. 2006d. Quantitative Impact Study No. 2: Technical Specification, CEIOPS-PI-08/06, Available at www.ceiops.org.

Note: L: low dependence, ML: medium–low dependence, M: medium dependence, MH: medium–high dependence, and H: high dependence.

H.2.1.2 RPS: Reduction for Profit-Sharing Using the k-Factor

The following input information was required:

TPbenefits: total amount in the placeholder valuation of TPs relating to future discretionary benefits

k: risk-absorbing proportion of *TPbenefits*

The factor k ranges between 0 and 1 and was intended to reflect the extent to which future discretionary profit sharing may be used to absorb future losses under adverse circumstances. Generally, this would depend on a range of aspects, including

- The extent to which legal or statutory restrictions impede the use of future discretionary benefits to absorb losses
- The nature of agreed management actions in adverse circumstances
- The degree of policyholder expectations on future profit sharing
- The extent to which cross-subsidy would be allowed across policies or across different funds

The k-factor would be set by the participating undertakings using their own assumptions, taking into account any aspect that has a material impact on the degree to which amounts in TPs relating to discretionary benefits may be used to cover losses under adverse circumstances. National supervisors could also provide additional guidance, taking into account the legal environment and general practices in their markets.

In some cases, life insurance undertakings have a complex fund structure consisting of a number of nonprofit and with-profit funds. Typically, to each fund a separately managed pool of assets and liabilities is associated. Profit-sharing rules may be different in different funds, restricting or disallowing a sharing of profits and risks across funds. In such cases, undertakings should set the factor k such that it is consistent with the structure of funds in their portfolio.

In cases where amounts relating to future discretionary benefits were excluded from the "placeholder" valuation of TPs and treated instead as part of available capital, the k-factor needed to be set to zero to avoid a double counting of such amounts both as available capital and as a risk mitigant under the SCR calculation.

The reduction of the overall capital charge with respect to future profit sharing was defined as

$$\text{RPS} = k \cdot \text{TPbenefits}.$$

H.2.1.3 NL_PL: Nonlife Expected Profit or Loss

For nonlife insurance business, the determination of the overall capital charge also took into account the expected profit or loss NL_PL arising from next year's business.

The following input information was required:

P_k: estimate of the net earned premium in the forthcoming year in each of the LoBs k

$P_{k,y}$: earned net premiums in each of the LoBs k and for historic years y (to the extent available, not more than 5 years)

$CR_{k,y}$: net combined ratios in each of the LoBs k and for historic years y (to the extent available, not more than 5 years)

PCO: the net provision for claims outstanding for the overall business

PCO_k: the net provision for claims outstanding in each of the LoBs k

In each of the LoBs, the estimate P_k of the net earned premium in the forthcoming year should be determined as

- The undertaking's estimate of the net earned premium volume for the forthcoming year, in cases where the undertaking estimates that this will exceed previous year's net earned premiums by more than 5%
- In other cases, 105% of the previous year's net earned premiums

The combined ratio $CR_{k,y}$ is the ratio for year y of expenses and incurred claims in a given LoB k over earned premiums, determined at the end of year y. The earned premiums would exclude prior year adjustments, the expenses should be those attributable to the premiums earned other than claims expenses, and incurred claims should exclude the runoff result, that is, they should be the total losses occurring in year y of the claims paid, including claims expenses, during the year and the provisions established at the end of the year. Alternatively, if it is more practicable, participants may calculate the combined ratio as the sum of the expense ratio and the claims ratio, where the expense ratio is the ratio of expenses, other than claims expenses, to written premiums and the expenses are those attributable to the written premiums.

Under this approach, the expected profit or loss arising from next year's premiums NL_PL_{prem} was defined as

$$NL\_PL_{\text{prem}} = (1 - \mu) \cdot P$$

where μ: the estimate of the expected value of the combined ratio for the overall nonlife business and where P was defined as $P = \sum_k P_k$.

The estimate μ was set to $\mu = \sum_k (P_k/P) \cdot \mu_k$ where μ_k: company-specific estimate of the expected value of the combined ratio in the individual LoBs k, and μ_k was defined as the premium-weighted average of historic combined ratios:

$$\mu_k = \sum_k \left(\frac{P_{k,y}}{\sum_y P_{k,y}} \right) \cdot CR_{k,y}.$$

Here, the summation was run over at least 3, but not more than 5 years. In the case where less than 3 years of historic data are available, μ_k is set as 100%.

The expected surplus or deficit NL_PL_{res} arising from next year's runoff result was defined as $NL\_PL_{\text{res}} = \mu \cdot PCO$, where μ is the estimate of the expected value of the (relative) runoff result for the overall business in the forthcoming year and μ is defined as

$$\mu = \sum_k \frac{PCO_k}{PCO} \cdot \mu_k,$$

where μ_k is the estimate of the expected value of the (relative) runoff result in the forthcoming year in each of the LOBs k. The estimate μ_k is defined as follows:

$$\mu_k = \alpha \cdot \frac{RM_k}{PCO_k}$$

where α is the proportion of the claims provision PCO_k that was expected to be paid out in the forthcoming year; RM_k is the risk margin in the claims provision PCO_k; the parameter α could be approximated by $\alpha = 1/D$, where D is the mean duration of the claims provision PCO_k, but where the firm can make a more accurate estimate it should attempt to do so.

The expected profit or loss NL_PL arising from next year's business may be determined as

$$NL\_PL = NL\_PL_{\text{prem}} + NL\_PL_{\text{res}}.$$

Alternatively, the expected profit or loss NL_PL arising from next year's business could also be determined by following the approach set out for the determination of NL_PL_{prem}, but where instead of combined ratios excluding the runoff result "full" combined ratios including the runoff result are used.

H.3 QIS3

The standard formula calculation was divided into modules. Following the comments given to QIS2, the risk classification was changed in accordance to the modules illustrated in Figure H.2.

Special considerations apply for participants writing composite business or which had one or more funds in life insurance business where the assets of such funds are not transferable to other parts of the undertaking's business. These are discussed later in this appendix.

The principle of substance over form would be followed in determining how risks are to be treated. For instance, where claims were payable in the form of an annuity, agreed claims would normally be part of CR_{LR}, unless the impact of the associated risk on the risk capital charges for the individual risk modules could be expected to be negligible.

For the purposes of the SCR standard formula calculation specified in Appendices I through O, TPs would be valued in accordance with the specifications laid out in CEIOPS (2007a). To avoid any circularity in the calculations, any reference to TPs within the calculations for the individual SCR modules was to be understood to exclude the CoC RM.

The parameters and assumptions used for the calculation of the SCR were intended to reflect a VaR risk measure, calibrated to a confidence level of 99.5%, and a time horizon of 1 year.

To ensure that the different modules of the standard formula were calibrated in a consistent manner, these calibration objectives have been applied to each individual risk module,

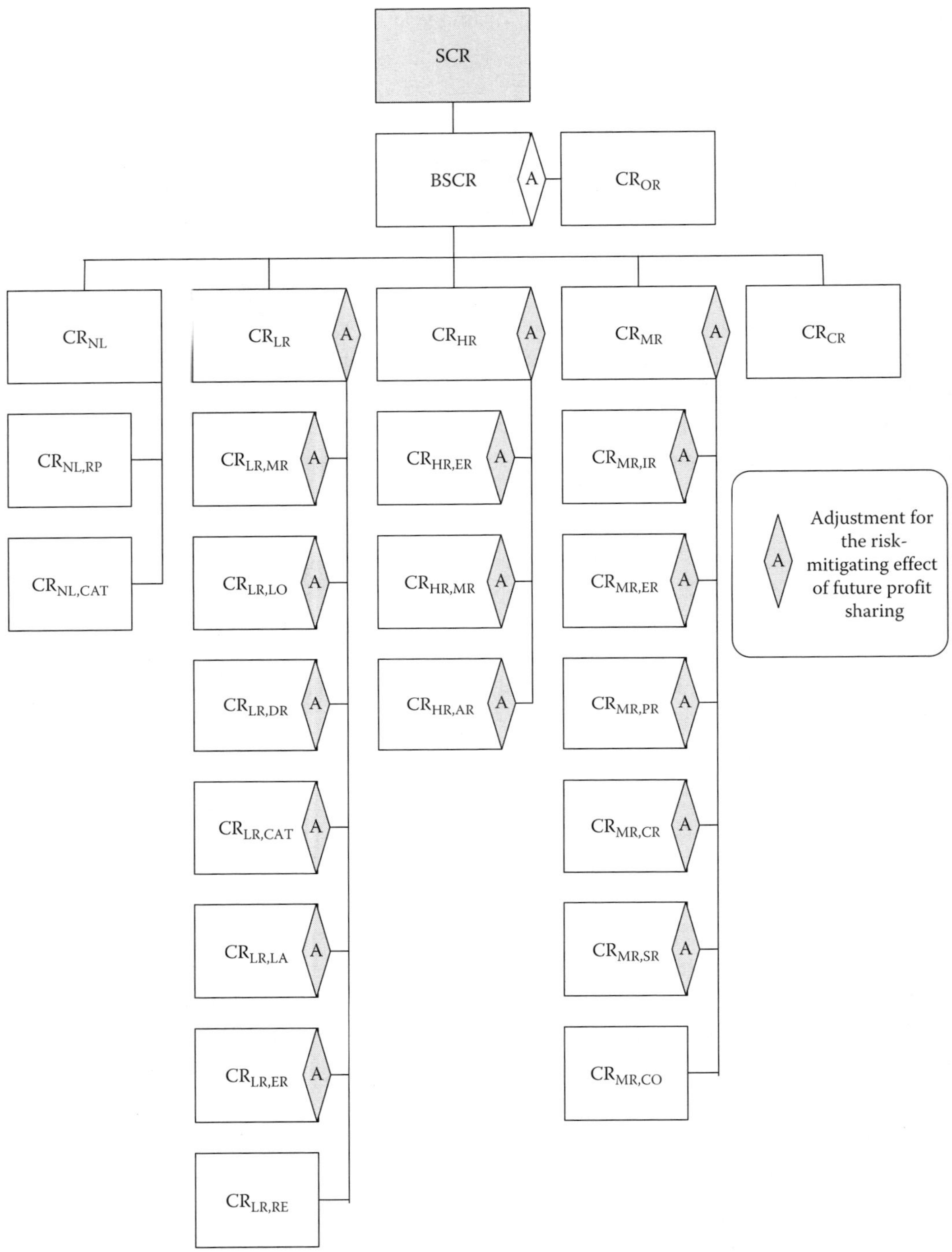

FIGURE H.2 The modular structure of the standard formula proposed for QIS3. (Adapted from CEIOPS. 2007a. QIS 3 Technical Specifications, Part I: Instructions, CEIOPS-FS-11/07 & Part II: Background Information, CEIOPS-FS-12/07. April. Available at www.ceiops.org.)

while also taking account of any model error arising from the particular technique chosen to assess that risk.

For the aggregation of the individual risk modules to an overall SCR, linear correlation techniques were applied. The setting of the dependence structures was intended to reflect potential dependencies in the tail of the distributions, as well as the stability of any dependence assumptions under stress conditions.

Segmentation for Nonlife Insurance Business

The analysis of nonlife underwriting risk would require a segmentation of the participant's nonlife insurance business into individual LoBs. For direct insurance and facultative reinsurance, the LoBs are defined as in Article 63 of the Council Directive on the annual accounts and consolidated accounts of insurance undertakings 91/674/EEC, namely

- Accident and health—workers' compensation
- Accident and health—health insurance
- Accident and health—others/default
- Motor, third-party liability
- Motor, other classes
- Marine, aviation, and transport
- Fire and other property damage
- Third-party liability
- Credit and suretyship
- Legal expenses
- Assistance
- Miscellaneous nonlife insurance

For reinsurance business, written either by reinsurers or direct insurers, a split between proportional and nonproportional reinsurance and between property and casualty reinsurance should be made as follows:

- For proportional reinsurance, the same LoBs should be used as for direct insurance
- Nonproportional reinsurance business should be split into three LoBs, namely
    - Property business
    - Casualty business
    - Marine, aviation, and transport business

Market Risk on Assets in Excess of the SCR ("Free Assets")

Under the *simplified balance sheet* concept underlying the SCR, consideration was required as to whether one should apply market stresses to assets in excess of the SCR. This issue was highlighted in the feedback received from QIS2. As in QIS2, the specifications for the SCR standard formula followed the approach to include CRs on all assets, including assets in excess of the SCR. CEIOPS believed that this idea was consistent with the *simplified balance sheet* concept, under which available capital was defined as the excess of (all) assets over liabilities.

However, in the context of the feedback from the QIS2 exercise, some stakeholders had argued that market risk should only apply to the assets that are backing TPs and the SCR, and that excess capital should not lead to increased CRs for market risk. In its future technical work, CEIOPS would further consider this issue. Therefore, participants were invited to supply, as additional information, an overall SCR estimate where the assets to be taken into consideration are limited to those required backing the total of the TPs and the SCR, that is, there is no capital charge with respect to free assets in excess of the SCR. Such an estimate may, for example, be derived by an iterative calculation as follows.

First, undertakings could calculate an initial SCR, SCR1, based on including the full balance sheet, that is, all assets, as is specified as the "placeholder" SCR in the specifications; CEIOPS (2007a). In a second iteration, free assets not needed to cover either the SCR or TPs may be excluded, leading to a smaller result for the SCR, SCR2. This calculation should be repeated until the coverage ratio is not significantly different from 100%.

H.3.1 Overall SCR Calculation

The SCR is determined as

$$\mathrm{SCR} = \mathrm{BSCR} + \mathrm{CR}_{\mathrm{OR}}$$

where *BSCR* is the Basic SCR; CR_{OR} is the capital charge for operational risk; see appendix K.

The Basic SCR Calculation

BSCR is the SCR before any adjustments, combining capital charges for five major risk categories. The following input information was required, including all subrisk modules:

$\mathrm{CR}_{\mathrm{NL}}$: the nonlife underwriting risk module. Its subrisk modules are

- $\mathrm{CR}_{\mathrm{NL,RP}}$: the reserve and premium risk module
- $\mathrm{CR}_{\mathrm{NL,CAT}}$: the catastrophe risk module

These subrisks are discussed in Appendix M.

CR_{LR}: the life underwriting risk module. Its subrisk modules are

- $\mathrm{CR}_{\mathrm{LR,MR}}$: the mortality risk module
- $\mathrm{CR}_{\mathrm{LR,LO}}$: the longevity risk module
- $\mathrm{CR}_{\mathrm{LR,DR}}$: the disability risk module

- $CR_{LR,CAT}$: the catastrophe risk module
- $CR_{LR,LA}$: the lapse risk module
- $CR_{LR,ER}$: the expense risk module
- $CR_{LR,RR}$: the revision risk module

These subrisks are discussed in Appendix N.

CR_{HR}: the health underwriting risk module. Its subrisk modules are

- $CR_{HR,ER}$: the expense risk module
- $CR_{HR,MR}$: the excessive loss/mortality/cancellation risk
- $CR_{HR,AR}$: the epidemic/accumulation risk

These subrisks are discussed in Appendix O.

CR_{MR}: the market risk module. Its subrisks modules are

- $CR_{MR,IR}$: the interest rate risk module
- $CR_{MR,ER}$: the equity risk module
- $CR_{MR,PR}$: the property risk module
- $CR_{MR,CR}$: the currency risk module
- $CR_{MR,SR}$: the spread risk module
- $CR_{MR,CO}$: the concentration risk module

These subrisks are discussed in Appendices I and J.

CR_{CR}: the credit risk: default risk module

This risk is discussed in Appendix K.

Additional input was required:

- *FDB*: total amount in TPs corresponding to future discretionary benefits
- KC_{LR}: the risk mitigating effect of future profit sharing for LUR
- KC_{HR}: the risk mitigating effect of future profit sharing for health underwriting risk
- KC_{MR}: the risk mitigating effect of future profit sharing for market underwriting risk

The BSCR was determined as

$$BSCR = \sqrt{\sum_{rxc} \rho_{rc} \cdot CR_r \cdot CR_c} - \min\left\{\sqrt{\sum_{rxc} \rho_{rc} \cdot KC_r \cdot KC_c}; FDB\right\}.$$

TABLE H.2 The Dependence Matrix for the Main Risks in QIS3

ρ_{rc}	CR_{NL}	CR_{LR}	CR_{HR}	CR_{MR}	CR_{CR}
CR_{NL}	1	0	0	0.25	0.5
CR_{LR}		1	0.25	0.25	0.25
CR_{HR}			1	0.25	0.25
CR_{MR}				1	0.25
CR_{CR}					1

Source: Adapted from CEIOPS. 2007a. QIS 3 Technical Specifications, Part I: Instructions, CEIOPS-FS-11/07 & Part II: Background Information, CEIOPS-FS-12/07. April. Available at www.ceiops.org.

Table H.2 presents the dependence used in the formula above.

H.3.2 General Approach to Risk Mitigation

A broad assumption was made that the effect of risk mitigation techniques should be given adequate recognition in reducing the relevant risk capital charges. Risk mitigation was taken to include both traditional and nontraditional risk transfer instruments on the asset side (e.g., financial hedging) and on the liability side (e.g., hedging instruments, reinsurance).

The SCR would allow for the effects of risk mitigation through

- A reduction in requirements commensurate with the extent of risk transfer
- Appropriate treatment of any corresponding risks that are acquired in the process

To simplify the overall treatment of risk mitigation in the context of the standard formula calculation of the SCR, the following two effects were separated:

- The extent of the risk transfer was recognized in the assessment of the individual risk modules.
- The acquired counterparty risks, for example, in the case of reinsurance, in the event of the reinsurer's default, were captured in the counterparty default risk module.

Implicitly, the operational risk charge also addresses the risk of risk mitigation failure.

Requirements on the Recognition of Risk Mitigation Instruments

The underlying impact on risk associated with risk mitigation would be treated consistently, regardless of the legal form of the protection. Risk mitigation arrangements would be legally effective and enforceable in all relevant jurisdictions. Risk mitigation arrangements would provide appropriate assurance as to the risk mitigation achieved, with regard to the approach used to calculate the extent of risk transfer and the degree of recognition in the SCR.

In an annex to CEIOPS (2007a), a tentative set of principles on financial risk mitigating tools was laid out which could be used to define minimum requirements on the allowance of such tools with respect to a standard formula calculation of the SCR. These principles were inspired by requirements in the banking sector on the credit quality of the provider of the risk mitigation instrument, and some CEIOPS members believed that they might usefully complement the advice on risk mitigation instruments that CEIOPS has given.

CEIOPS at the time of QIS3 had not yet reached a final position on the issue, and participants were invited to comment on the appropriateness of these principles in the context of a standard formula calculation of the SCR.

In cases where participants apply risk mitigation instruments for the calculation of the QIS3 standard formula SCR which do not fulfill these principles, and where such mitigating instruments have a significant impact on the SCR, it was recommended that the participant indicates which of the principles were violated, and gave an estimation of the impact of the instruments out of the scope on the SCR estimate.

H.3.3 Composites (Insurers Carrying Out Both Life and Nonlife Business)

The EC had pointed out that the application of the standard solo SCR formula to composites produced a lower capital requirement than the separated application of the same formula to two insurers, one specialized in life insurance and the other one in nonlife insurance, both of them jointly having the same activity as the composite. To solve this issue some proposals were analyzed, although for the time being at the QIS3 there was not a sufficiently definitive decision on this issue.

Under QIS2, composites were allowed to make one global calculation. A major issue with respect to QIS3 was that introducing substantial changes in QIS2 specifications regarding composites would overburden this type of insurers. This was particularly true for composite insurers belonging to groups. For some countries, where the composites had an important market share, this argument weighs heavily on the decision.

For QIS3 purposes composite insurers would calculate a single standard solo SCR and a single MCR, applying the formulas reflected in the relevant Appendices (I–O). In the qualitative part of the QIS3 exercise, composites were allowed to offer information about the following:

- Alternatives to solve the problem described regarding composites when compared with separated entities.
- On voluntary basis, comparison of standard solo SCR with the capital requirement obtained by
    a. In the first step, calculating solo SCR for each type of business separately.
    b. In the second step, adding both capital requirements obtained in (a), following, for example, QIS3 specifications regarding Groups SCR.
- Participant's views on how to deal with composites when developing IMs.
- Additionally, CEIOPS should be able, on the basis of the information collected in the nonlife and life risk modules, to approximate the proposal of the EC by modifying the dependence matrix between both risk modules.

H.3.4 Adjustments for Risk Mitigating Properties of Future Profit Sharing

For with-profits business in life insurance, the specification of the standard formula calculation as set out in the specifications, CEIOPS (2007a), took into account the risk absorption ability of future profit sharing. This was achieved by a three step "bottom-up" approach as follows:

1. The first step was to calculate the capital requirements for individual subrisks—for example, interest rate risk—under two different assumptions:
   - That the insurer was able to vary its assumptions on future bonus rates in response to the shock being tested, based on reasonable expectations and with regard to plausible management decisions ($nCR_{\text{MR,IR}}$)
   - That the insurer was not able to vary its assumptions on future bonus rates in response to the shock being tested ($gCR_{\text{MR,IR}}$)

   The difference between the two capital requirements ($gCR_{\text{MR,IR}} - nCR_{\text{MR,IR}}$) is termed KC ($KC_{\text{MR,IR}}$).

   Performing these two calculations for different risks reflects the fact that the ability to vary policyholder benefits will depend on the nature of the shock to which the insurer is exposed. For example, the potential for risk mitigation might be more significant in the case of yield curve movements than a shock to property values.
2. The second step was to aggregate capital requirements for risks within the same category (equity, interest rate, property, etc.) using the relevant dependence matrices. To preserve the coherence of the modular approach, the aggregation uses the capital requirements produced assuming no change in the assumptions used to estimate policyholder benefits. For instance, the capital requirement gCR_{MR} for market risk is derived by combining $gCR_{\text{MR,IR}}$, $gCR_{\text{MR,ER}}$, and so on. The KCs should also be combined using the same dependence matrices.
3. The third and final step was to repeat the aggregation process for the major risk categories. gCR_{MR} was combined with gCR_{LR} and all the other risk modules using the relevant dependence matrix. KC_{LR}, KC_{HR}, and KC_{MR} were combined by also using the relevant dependence matrix. The resulting "BSCR" was reduced by the minimum of the aggregated KC and the total amount of TPs corresponding to future discretionary benefits.

For reasons of simplification, the adjustment at the level of individual SCR risks was restricted to the submodules of the market risk, with the exception of the submodule for risk concentrations, LUR, with the exception of the submodule for revision risk, and health underwriting risk modules.

If a participant wishes to simplify the process—particularly in cases where the risk mitigating effect was not expected to be material—it could simply declare the calculation "net" of the risk mitigating effects of future profit sharing to be equal to the "gross" calculation, that is, it may put $gCR_{\text{MR,IR}} = nCR_{\text{MR,IR}}$ and $KC_{\text{MR,IR}} = 0$.

H.3.5 Ring-Fenced Funds

Where an undertaking had one or more funds where the assets of such funds were not transferable to other parts of the undertaking's business, the SCR standard formula for the fund should be calculated as if that fund were a separate insurer.

Similarly, a calculation should be carried out for the undertaking's remaining business as if it were a separate insurer. The overall SCR should be calculated as the sum of the SCR for each fund and for the remaining business.

H.4 QIS4

For QIS4, the QIS3 calibration had not been substantially changed. The differences in calibration between QIS3 and QIS4 are made in the following risk modules and submodules:

- LUR: Lapse risk, Life catastrophe risk. See Appendix N.
- Nonlife underwriting risk: number of historic years, LoB standard deviations, Geographical diversification, NL Cat Method 1. See Appendix M.
- Health underwriting risk: Accident and Health Short Term, Health workers compensation. See Appendix O.
- Market risk: Concentration risk. See Appendix I.
- Counterparty default: Counterparty default. See Appendix J.

QIS4 was proposed by the EC and the TSs was published in QIS4 (2008). The work to conduct QIS4 was made by CEIOPS. The proposed standard formula was published as a part of the TSs published in QIS4 (2008), and the calculation was as in the earlier QIS' divided into modules. Following the comments given to QIS3, the risk classification was changed in accordance with the modules illustrated in Figure H.3.

The main issues from QIS3 were treated in a similar way in QIS4, that is, the principle of substance over form would be followed in determining how risks are to be treated. For instance, where claims were payable in the form of an annuity, agreed claims would normally be part of CR_{LR}, unless the impact of the associated risk on the risk capital charges for the individual risk modules could be expected to be negligible.

For the purposes of the SCR standard formula calculation specified in Appendices I through O, TPs would be valued in accordance with the specifications laid out in QIS4 (2008). To avoid any circularity in the calculations, any reference to TPs within the calculations for the individual SCR modules was to be understood to exclude the CoC RM.

The participants could, within the design of the standard formula, replace one or a subset of its parameters by parameters specific to the undertaking concerned when calculating the underwriting risk modules. This option could be used, provided that the participant used for the calculation of its own specific parameters the same standardized methods adopted for the calculation of the standard parameters, including distributional assumptions. If another distribution was used, then a partial IM would be required.

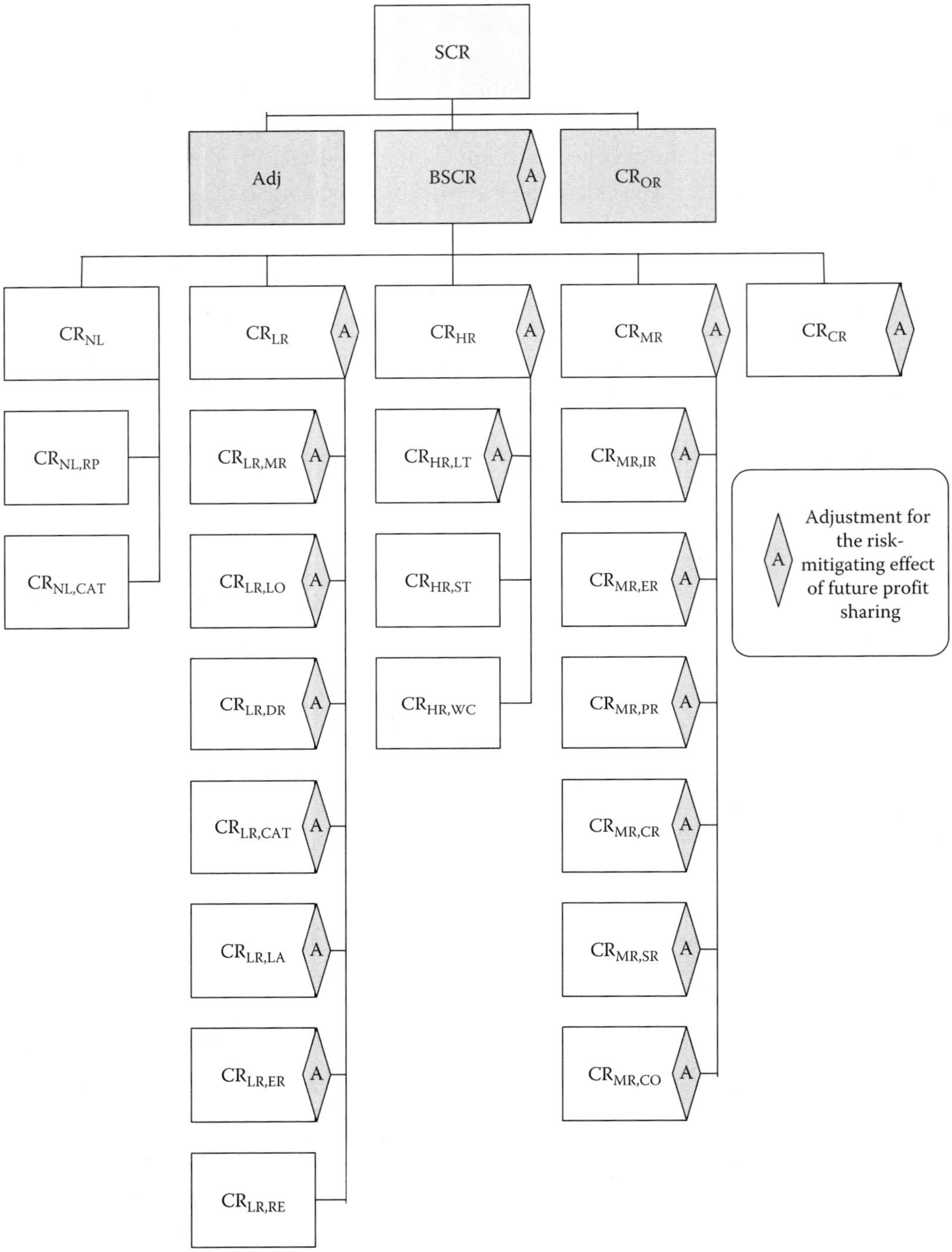

FIGURE H.3 The modular structure of the standard formula proposed for QIS4. (Adapted from QIS4. 2008. QIS 4 Technical Specifications. European Commission, MARKT/2505/08, March 31. Available at http://www.ceiops.eu/content/view/118/124/.)

The scope of application of undertaking-specific parameters are: given the availability of explicit standardized methods, the use of undertaking-specific parameters was limited in QIS4 to certain parameters of the nonlife and health ("Accident and health short term" and "Workers compensation") underwriting risk modules. For the calculation of the standard parameters for these modules, CEIOPS had used a lognormal assumption. However, this did not preclude a more extensive application to other parameters, including in the "life" and "long-term health" module in the future. Insurance and reinsurance undertakings could replace one or several of the following parameters of the nonlife and health underwriting risk modules by parameters specific to their business: the standard deviation for reserve risk in individual LoB $\sigma_{RR,k}$ and the standard deviation for premium risk in individual LoB $\sigma_{PR,k}$.

The parameters and assumptions used for the calculation of the SCR were intended to reflect a VaR risk measure, calibrated to a confidence level of 99.5%, and a time horizon of 1 year.

To ensure that the different modules of the standard formula were calibrated in a consistent manner, these calibration objectives had been applied to each individual risk module, while also taking account of any model error arising from the particular technique chosen to assess that risk.

For the aggregation of the individual risk modules to an overall SCR, linear correlation techniques were applied. The setting of the dependence structures was intended to reflect potential dependencies in the tail of the distributions, as well as the stability of any dependence assumptions under stress conditions.

Market Risk on Assets in Excess of the SCR ("Free Assets")

In QIS3, participants were invited to supply as additional information, an overall SCR estimate, where the assets to be taken into consideration were limited to those required to back the total of the TPs and the SCR. This means that no capital charge would be applied with respect to free assets in excess of the SCR. As this testing proposal is not in line with the Framework Directive Proposal, which was based on a TBSA. Therefore, QIS4 did not include this testing proposal.

Valuation of Intangible Assets for Solvency Purposes

For the calculation of the SCR, when participants were requested to input the value of assets, in accordance with the specifications on the valuation of assets and liabilities other than TPs, intangible assets, including goodwill, were to be valued at nil.

H.4.1 Segmentation

Segmentation for Nonlife Insurance Business

For direct insurance and facultative reinsurance, the LoBs used were the same as in QIS3:

- Accident and health—workers' compensation
- Accident and health—health insurance
- Accident and health—others/default

- Motor, third-party liability
- Motor, other classes
- Marine, aviation, and transport
- Fire and other property damage
- Third-party liability
- Credit and suretyship
- Legal expenses
- Assistance
- Miscellaneous nonlife insurance

A new treatment of health and accident risks was introduced in QIS4. The first three LoBs were for SCR calculation purposes classified in the SCR health underwriting risk module.

For reinsurance business, written either by reinsurers or direct insurers, a split between proportional and nonproportional reinsurance and between property and casualty reinsurance was made as in QIS3.

The health underwriting risks for different types of health business in the EEA would be allocated in QIS4 in the following way:

SCR Health Module

- *Long-term* submodules were used for long-term health business practiced on a similar technical basis to that of life insurance with additional restrictions according to National Law (as sold in Germany and Austria).
- *Short-term* submodules were used for the health and accident LoBs that was part of the nonlife business.
- *Workers' compensation* submodule was used for the workers compensation and similar LoBs. This was used for both long-term and short-term types of workers' compensation. Workers' compensation includes many of the risks of the life underwriting risk (LUR) module (such as longevity, revision, and expenses). However, the appropriate instructions on how these were combined for the workers' compensation LoB are included within the workers' compensation submodule.

SCR Life Underwriting Module

- Disability submodule was used for long-term health business other than that included within the long-term submodule within Health.
- Catastrophe submodule was used for all LoBs (long-term and short-term, including workers' compensation) which were exposed to the policyholder experiencing mortality, morbidity, or disability underwriting risk.

Dutch Health Insurance

Compulsory health insurance for Dutch citizens was discussed in an Annex to QIS4 (2005).

UK Alternative Disability Risk Submodule within Life Underwriting

An alternative to the disability risk submodule within the life underwriting risk module was discussed in an Annex to QIS4 (2005).

H.5 OVERALL SCR CALCULATION

The SCR is determined as

$$SCR = BSCR + CR_{\mathrm{OR}} - Adj$$

where *BSCR*: the Basic SCR, CR_{OR}: the capital charge for operational risk; see Appendix K, and *Adj*: an adjustment for the risk-absorbing effect of future profit sharing and deferred taxes, see Section H.4.4.

The Basic SCR Calculation

BSCR is the SCR before any adjustments, combining capital charges for five major risk categories. The following input information was required, including all subrisk modules:

$\mathrm{CR}_{\mathrm{NL}}$: the nonlife underwriting risk module. Its subrisk modules are

- $\mathrm{CR}_{\mathrm{NL,RP}}$: the reserve and premium risk module
- $\mathrm{CR}_{\mathrm{NL,CAT}}$: the catastrophe risk module

These subrisks are discussed in Appendix M.

$\mathrm{CR}_{\mathrm{LR}}$: the life underwriting risk module. Its subrisk modules are

- $\mathrm{CR}_{\mathrm{LR,MR}}$: the mortality risk module
- $\mathrm{CR}_{\mathrm{LR,LO}}$: the longevity risk module
- $\mathrm{CR}_{\mathrm{LR,DR}}$: the disability risk module
- $\mathrm{CR}_{\mathrm{LR,CAT}}$: the catastrophe risk module
- $\mathrm{CR}_{\mathrm{LR,LA}}$: the lapse risk module
- $\mathrm{CR}_{\mathrm{LR,ER}}$: the expense risk module
- $\mathrm{CR}_{\mathrm{LR,RR}}$: the revision risk module

These subrisks are discussed in Appendix N.

CR_{HR}: the health underwriting risk module. Its subrisk modules are

- $\mathrm{CR}_{\mathrm{HR,LT}}$: the long-term health risk underwriting module
- $\mathrm{CR}_{\mathrm{HR,ST}}$: the accident and health short term underwriting risk module

- $\text{CR}_{\text{HR,WC}}$: the workers compensation underwriting risk

These subrisks are discussed in Appendix O, and also in Appendices M and N.

CR_{MR}: the market risk module. Its subrisks modules are

- $\text{CR}_{\text{MR,IR}}$: the interest rate risk module
- $\text{CR}_{\text{MR,ER}}$: the equity risk module
- $\text{CR}_{\text{MR,PR}}$: the property risk module
- $\text{CR}_{\text{MR,CR}}$: the currency risk module
- $\text{CR}_{\text{MR,SR}}$: the spread risk module
- $\text{CR}_{\text{MR,CO}}$: the concentration risk module

These subrisks are discussed in Appendices I and J.

CR_{CR}: the credit risk: default risk module

This risk is discussed in Appendix J.

Additional input was required:

FDB: total amount in TPs corresponding to *future discretionary bonuses (FDBs)*

nCR_{LR}: the capital charge for LUR including the risk-absorbing effect of future profit sharing

nCR_{HR}: the capital charge for health underwriting risk including the risk-absorbing effect of future profit sharing

nCR_{MR}: the capital charge for market risk including the risk-absorbing effect of future profit sharing

nCR_{CR}: the capital charge for credit risk, that is, the counterparty default risk, including the risk-absorbing effect of future profit sharing

Adj_{FDB}: the adjustment for the risk-absorbing effect of future profit sharing

Adj_{DT}: the adjustment for the risk-absorbing effect of deferred taxes

nBSCR: net BS CR

The basic solvency capital requirement, *BSCR*, the *Adj*, and the *nBSCR* were determined as

$$BSCR = \sqrt{\sum_{rxc} \rho_{rc} \cdot CR_r \cdot CR_c},$$

$$Adj = Adj_{FDB} - -Adj_{DT}, \quad \text{and}$$

$$nBSCR = BSCR - -Adj_{FDB}.$$

TABLE H.3 The Dependence Matrix for the Main Risks in QIS4

ρ_{rc}	CR_{NL}	CR_{LR}	CR_{HR}	CR_{MR}	CR_{CR}
CR_{NL}	1	0	0.25	0.25	0.5
CR_{LR}		1	0.25	0.25	0.25
CR_{HR}			1	0.25	0.25
CR_{MR}				1	0.25
CR_{CR}					1

Source: Adapted from QIS4. 2008. QIS 4 Technical Specifications. European Commission, MARKT/2505/08, March 31. Available at http://www.ceiops.eu/content/view/118/124/.

Table H.3 shows the dependence structure used in the above formula.

The adjustment for the risk-absorbing effect of future profit sharing is defined as follows:

$$Adj_{FDB} = \min\left\{\sqrt{\sum_{rxc}\rho_{rc}\cdot CR_r\cdot CR_c} - \sqrt{\sum_{rxc}\rho_{rc}\cdot nCR_r\cdot nCR_c}; FDB\right\},$$

The adjustment for the risk-absorbing effect of deferred taxes is defined as follows:

$$Adj_{DT} = \left\{\left|\Delta DeferredTaxes\right| |CRshock\right\},$$

that is, the absolute value of the reduction in the value of deferred taxes under the scenario CR shock, where $\left|\Delta DeferredTaxes\right|$: the absolute value of the reduction in deferred taxes and *CRshock*: the immediate loss of basic OFs of the amount $BSCR - Adj_{FDB} + CR_{Op}$.

H.5.1 Simplifications in SCR Calculation

According to the proportionality principle introduced in the proposed FD, the undertakings could use simplified methods and techniques to calculate the SCR, using actuarial methods and statistical techniques that were *proportionate to the nature, scale, and complexity of the risks* they support.

Simplified methods could be employed in the valuation of SCR where the produced result was not material or not materially different from that which would result from a more accurate valuation process. If the criteria outlined in the following paragraph were satisfied, that is, the criteria were expected to be met, simplified actuarial methods and statistical techniques could be used.

Simplified actuarial methods and statistical techniques could be used if

- The types of contracts written for each LoB was not complex, for example, path dependency did not have a significant effect; for example: life contract that did not include any options or guarantees, nonlife insurance that did not include options for renewals; and
- The LoB was simple by nature of the risk, for example, insured risks were stable and predictable; for example, term assurance, insurance of damage to land motor vehicles and

- Any additional nature and complexity standards set out for each SCR calculation were met and
- The SCR that was valued in the simplified approach was not material in absolute terms or relative to the overall size of the total SCR. For the purpose of QIS4, the following guidance on materiality for when simplifications could be used for the SCR was outlined:
  - The SCR resulting from the simplified approach was no more than 50 million Euro for life business, and 10 million Euro for nonlife business; for composites, the same threshold applied on each part of the business, life, and nonlife; or
  - The value of each capital charge (prediversification) determined with simplified methods for each risk was no more than 10% of the total SCR; and
  - The sum of the capital charges (prediversification) determined with simplified methods was no more than 30% of the total SCR.

If a participant, for example, a captive (re)insurer, did not meet the threshold indicated, but nevertheless thought it would be allowed to apply a simplified approach because of the specificities of its situation, it could do so provided that it

1. Explained the reasons for this
2. Indicated the criteria it considered relevant in its situation

The participant was also invited to carry out the more accurate calculation to allow CEIOPS to benchmark the simplified calculation.

Additional criteria can be set for a specific simplification, in relation to the nature of the simplification itself. For example, in order to determine the interest rate submodule in the market risk, a simplified calculation according to the duration approach could be used on the value of assets and liabilities other than TPs if they had no embedded options and the convexity of the curve did not lead to a material error.

H.5.2 Adjustments for Risk-Absorbing Properties

H.5.2.1 Adjustments for Risk-Absorbing Properties of Future Profit Sharing

For with-profits business in life insurance, the specification of the standard formula calculation took into account the risk absorption ability of future profit sharing. This was achieved by a three-step "bottom-up" approach in the following way:

1. The first step was to calculate the capital requirements for individual subrisks under two different assumptions:
   - That the insurer was able to vary its assumptions on future bonus rates in response to the shock being tested, based on reasonable expectations and with regard to plausible management decisions (e.g., interest rate risk: $nCR_{MR,IR}$).

- That the insurer was not able to vary its assumptions on future bonus rates in response to the shock being tested (e.g., interest rate risk: $CR_{MR,IR}$).

  Performing these two calculations for different risks reflects the fact that the ability to vary policyholder benefits would depend on the nature of the shock to which the insurer was exposed. For example, the potential for risk mitigation might be more significant in the case of yield curve movements than a shock to property values.

2. The second step was to aggregate both kinds of capital requirement separately, using the relevant dependence matrices. The results were two overall capital charges, excluding operational risk, one derived from capital charges including the risk-absorbing effect of future profit sharing on the submodule level (aggregate of the nCRs), and one derived from capital charges disregarding this effect, that is, the BSCR.
3. The final step was to determine an adjustment Adj_{FDB} to the BSCR by comparing both overall capital charges. Generally, the adjustment was given by the difference between the Basic SCR and the aggregate of the nCRs.

The adjustment for the loss-absorbing capacity of FDBs itself can never exceed the total value of FDBs. Hence

$$Adj_{FDB} = \min\{BSCR - \text{aggregate of } nCRs; FDB\}.$$

This upper bound to the adjustment was necessary to prevent double counting of risk-absorbing effects on the submodule level in the determination of the capital charge. More details can be found in Appendices I through O for the different risk modules.

A participant could simplify the process—particularly in cases where risk-absorbing effect was not expected to be material—by declaring the calculation including the risk-absorbing effects of future profit sharing to be equal to the calculation excluding the risk-absorbing effects of future profit sharing, that is, it could put $nCR_{MR,IR} = CR_{MR,IR}$).

H.5.2.2 Adjustments for Risk-Absorbing Properties of Deferred Taxation

The specification of the standard formula calculation should take account of the risk-absorbing capacities offered by deferred taxation. Participants should take the following approach:

- First, the liability for deferred taxation within the current (prestress) BS should be excluded from the prestress BS.
- The BSCR should be calculated as shown below. All references to change in net asset value should be interpreted to exclude the potential change in deferred taxes.
- The CR for operational risk should be calculated as in Appendix K.
- The liability for deferred taxes should be recalculated based on the assumption that the undertaking makes an immediate loss equal to the value of the calculated SCR.

The change in the liability for deferred taxes, that is, the current liability for deferred taxes minus the recalculated liability for deferred taxes after the SCR, should be added to the adjustment for loss absorbency capacity of TPs such as it can be used to reduce the SCR.

Where the TPs included tax liabilities charged to policyholders, this would be included within the calculation of the BSCR in order to accurately value options and guarantees. Where this was the case, participants would ensure that the undertaking's liability for taxation in the 99.5% event (following the SCR) and the amount passed to policyholders underlying the SCR are consistent.

Participants may use a range of simplifications to calculate the allowance for deferred and future taxation within the TPs and the adjustment for loss absorbency as a result of deferred taxes within the SCR.

H.5.2.3 Simplification and an Alternative Method

Simplification

When an undertaking used the simplified method based on the *Italian profit sharing life insurance system* described in QIS4 (2008) to calculate theBE, they could apply the following formula to evaluate the adjustment for the risk-absorbing effect of future profit sharing: $Adj = +0,1FDB$.

An Alternative Method

Firms could also use the following scenario-based method to calculate the adjustment for the loss-absorbing capacity of TPs and deferred taxes, where they suspect that those effects were not linearly correlated between risk modules as assumed in the calculation.

The approach involved replacing the application of the SCR Standard Formula by a single scenario test covering all the risks included in the SCR Standard Formula. The particular combination of simultaneous shocks to be used by the participant was determined using the spreadsheet provided for this purpose as part of the QIS4 package. This single scenario was referred to as *the single equivalent scenario*. The single equivalent scenario derived a linear approximation of the BSCR standard formula, taking into account the specific risk profile of the company, and used this approximation to identify a scenario underlying the SCR.

In order to calculate the adjustment for the loss-absorbing capacity of TPs and deferred taxes using the single equivalent scenario, participants would carry out the following five steps:

1. The participant should first calculate the capital charge for each subrisk module in the SCR standard formula using the relevant Appendices I through O. The calculation should be calculated assuming that assumptions about future bonus rates, reflected in the valuation of future discretionary benefits in TPs, remain unchanged before and after the shocks being tested (i.e., the submodule SCRs).
2. The participant should then determine the single equivalent scenario it should apply using the spreadsheet provided for this purpose as part of the QIS4 package, by introducing in the "input" sheet the capital charges calculated in Step 1. First, the spreadsheet calculated the relative importance, that is, the weightings, of each of the subrisks in the

participant's overall SCR. Then the spreadsheet used those weights to determine what simultaneous shocks, for example, interest rates, equity, and so on, should be used by the participant in the single equivalent scenario. The single equivalent scenario to be used was automatically displayed in the output tab of the spreadsheet. Note that since the relative importance of each of the subrisks will vary from company to company, the single equivalent scenario applied will also vary from company to company.

3. The participant should consider what management actions they would take in the single equivalent scenario and in particular how their assumptions regarding future bonus rates would change in the event that such a scenario would occur.
4. The participant should then calculate the change in the undertaking's net asset value in the face of the equivalent scenario, taking into account management actions identified in Step 3 as well as the loss-absorbing capacity of deferred taxes. The calculation of the change in net asset value should be performed on the assumption that all the shocks making up the single equivalent scenario occur simultaneously and that the undertaking makes an operational risk loss equal to CR_{op} within the equivalent scenario, in order to ensure that the loss-absorbing capacity of deferred taxes is properly captured.
5. Finally, the participant should calculate the "Adjustment for the loss-absorbing capacity of TPs and deferred taxes" as follows:

$$Adj = BSCR + CR_{op} - SCR_{net}.$$

   SCR_{net} is the change in the undertaking's net asset value in the face of the equivalent scenario calculated in Step 4.

For the calculation of some risk modules (interest rate risk, currency risk, and lapse risk), undertakings were required to consider both an increase and a decrease in parameters. Undertakings should satisfy themselves that the direction of the change in parameters continues to be appropriate within the scenario test. This may be done by further sensitivity testing or by another method, for example, considering more than one scenario.

H.5.3 Risk Mitigation in SCR

The General Approach to Risk Mitigation

The effect of risk mitigation techniques would be given adequate recognition in reducing the relevant risk capital charges. Risk mitigation includes both traditional and nontraditional risk transfer instruments on the asset side, for example, financial hedging, and on the liability side, for example, hedging instruments, reinsurance.

The SCR should allow for the effects of risk mitigation through a reduction in requirements commensurate with the extent of risk transfer and appropriate treatment of any corresponding risks that are acquired in the process.

To simplify the overall treatment of risk mitigation in the context of the standard formula calculation of the SCR, these two effects were separated as follows:

- The extent of the risk transfer was recognized in the assessment of the individual risk modules

- The acquired counterparty risks, for example, in the case of reinsurance, in the event of the reinsurer's default, were captured in the counterparty default risk module

Implicitly, the operational risk charge also addresses the risk of risk mitigation failure.

Requirements on the Recognition of Risk Mitigation Tools

The underlying impact on risk associated with risk mitigation would be treated consistently, regardless of the legal form of the protection. Risk mitigation arrangements would be legally effective and enforceable in all relevant jurisdictions. Risk mitigation arrangements would provide appropriate assurance as to the risk mitigation achieved, with regard to the approach used to calculate the extent of risk transfer and the degree of recognition in the SCR.

A set of principles on financial risk mitigating tools, therefore, excluding reinsurance, was laid out, which could be used to define minimum requirements on the allowance of such tools with respect to a standard formula calculation of the SCR. These principles were inspired by requirements in the banking sector on the credit quality of the provider of the risk mitigation instrument. The allowance for risk mitigating effects in the standard formula SCR was restricted to instruments and excludes processes and controls the firm had in place to manage the investment risk. For example, where a firm had a dynamic investment strategy, for example, delta-hedging or CF matching, a firm would calculate the capital charge assuming that they continued to hold their current assets during the change in financial conditions.

Principle 1: Economic Effect over Legal Form

For standard SCR purposes, financial risk mitigating instruments that have a material impact on an insurance undertaking's risk profile, should be recognized and treated equally, regardless of their legal form or accounting treatment, provided that their economic or legal features do not oppose to the principles and rules required for such recognition.

Where financial risk mitigation instruments were recognized in the SCR calculation, any material new risks should be identified and the capital required at the 99.5th confidence level quantified. This includes any basis risk between the firm's liability and the risk mitigation instrument. The additional capital required should be included within the SCR.

Principle 2: Legal Certainty, Effectiveness, and Enforceability

The instruments used to provide the financial risk mitigation together with the action and steps taken and procedures and policies implemented by the insurance undertaking should be as resulting in risk mitigation arrangements, which are legally effective and enforceable in all relevant jurisdictions.

The insurance undertaking should take all appropriate steps, for example, a sufficient legal review, to ensure and confirm the effectiveness and continuing enforceability of the financial risk mitigation arrangement and to address related risks. In case the full effectiveness or continuing enforceability could not be verified, the risk mitigation instrument should not be recognized in the SCR calculation.

Undocumented or deficiently documented financial risk mitigation instruments should not be considered, not even on a partially sufficient basis, for standard SCR purposes.

Principle 3: Liquidity and Ascertainability of Value
To be eligible for recognition, the financial risk mitigation instruments relied upon should have a value over time sufficiently reliable to provide appropriate certainty as to the risk mitigation achieved.

Regarding liquidity, QIS4 specifications did not contain any concrete requirement, but only the following two general statements were made

- The insurer should have written guidance regarding liquidity requirements that financial risk mitigation instruments should meet, according to the objectives of the own insurer's risk management policy.
- Financial risk mitigation instruments considered to reduce the SCR should meet the liquidity requirements established by the own entity.

The standard SCR calculation should recognize financial risk mitigation techniques in such a way that there was no double counting of mitigating effects. Where the risk mitigation instrument reduced risk, the capital requirement should be no higher than if there were no recognition in the standard SCR of such mitigation instruments. Where the risk mitigation instrument actually increased risk, then the SCR should be increased.

Principle 4: Credit Quality of the Provider of the Risk Mitigation Instrument
Providers of financial risk mitigation should have an adequate credit quality to guarantee with appropriate certainty that the insurer would receive the protection in the cases specified by the contracting parties. Credit quality should be assessed using objective techniques according to generally accepted practices.

As a general rule, when the insurer applied the standard calculation for a certain risk module, only financial protection provided by entities rated BBB or better would be considered in the assessment of SCR. In the event of the default, insolvency or bankruptcy of the provider of the financial risk mitigation instrument, the financial risk mitigation instrument should be capable of liquidation in a timely manner or retention. The degree of dependence between the value of the instruments relied upon for risk mitigation and the credit quality of their provider should not be undue, that is, material positive.

Principle 5: Direct, Explicit, Irrevocable, and Unconditional Features
Financial mitigating instruments only can reduce the capital requirements if

- They provide the insurer a direct claim on the protection provider (*direct feature*).
- They contain explicitly reference to specific exposures or a pool of exposures, so that the extent of the cover is clearly defined and incontrovertible (*explicit feature*).
- They do not contain any clause, the fulfillment of which is outside the direct control of the insurer, that would allow the protection provider unilaterally to cancel the cover or that would increase the effective cost of protection as a result of certain developments in the hedged exposure (*irrevocable feature*).

- They do not contain any clause outside the direct control of the insurer that could prevent the protection provider from being obliged to pay out in a timely manner in the event that a loss occurs on the underlying exposure (*unconditional feature*).

Special Features Regarding Credit Derivatives

Reduction of standard SCR based on the mitigation of credit exposures by using credit derivatives was allowed when the insurer had in force generally applied procedures for this purposes and considered generally admitted criteria. Requirements set out in other financial sectors for the same mitigating techniques were considered as generally applied procedures and admitted criteria.

In order for a credit derivative contract to be recognized, the credit events specified by the contracting parties must at a minimum cover

- *Failure* to pay the amounts due under terms of the underlying obligation that are in effect at the time of such failures (with a grace period that was closely in line with the grace period in the underlying obligation).
- *Bankruptcy*, insolvency, or inability of the obligor to pay its debts, or its failure or admission in writing of its inability generally to pay its debts as they become due, and analogous events.
- *Restructuring* of the underlying obligation, involving forgiveness or postponement of principal, interest, or fees that results in a credit loss event.

Collateral

A collaterized transaction is one in which insurers have a credit exposure or potential credit exposure and it is hedged in whole or in part by a collateral posted by a counterparty or by a third party on behalf of the counterparty. In addition to the general requirements for legal certainty, the legal mechanism by which a collateral is pledged or transferred must ensure that the insurer has the right to liquidate or take legal possession of it, in a timely manner. Insurers must have clear and robust procedures for the timely liquidation of collateral to ensure that any legal conditions required for declaring the default of the counterparty and liquidating the collateral are observed, and that collateral can be liquidated promptly.

Unless it becomes impossible according to market conditions, admissible collateral for standard SCR purposes must protect the insurer against the same events listed in this paper for credit derivatives.

European Solvency II Standard Formula

Market Risk

HERE WE DISCUSS the development and calibration of the market risk module and its capital charge that has been one of the building blocks in Solvency II. The final advice from CEIOPS is discussed in Chapter 27.

In the light of the financial crisis, mainly during 2007–2009, CEIOPS (2009a) decided to review the calibration and dependency structure between the different submodules of the market risk. For Pillar II, concrete requirements regarding the submissions of information from insurer to supervisors would be necessary.

I.1 GENERAL FEATURES

I.1.1 Background

As an answer to the EC's CfA 10, MARKT (2004c), CEIOPS gave a background to the modeling of the standard formula and its risk modules; see CEIOPS (2005).

I.1.2 QIS2–QIS4

I.1.2.1 QIS2, CEIOPS (2006d)

Under QIS2, CEIOPS (2006d), the capital charge was comprised of the capital requirements from the following four subrisk modules:

$$CR_{MR,IR} = \text{Interest rate risk}$$

$$CR_{MR,ER} = \text{Equity risk}$$

$$CR_{MR,PR} = \text{Property risk}$$

$$CR_{MR,CR} = \text{Currency risk}$$

Under QIS2, the capital requirements for interest rate risk, equity risk, property risk, and currency risk were combined using Table I.1:

TABLE I.1 Dependence Matrix for the Market Risk's Subrisks Proposed for QIS2

	$CR_{MR,IR}$	$CR_{MR,ER}$	$CR_{MR,PR}$	$CR_{MR,CR}$
$CR_{MR,IR}$	1	0.75	0.75	0.25
$CR_{MR,ER}$		1	1	0.25
$CR_{MR,PR}$			1	0.25
$CR_{MR,CR}$				1

Source: Adapted from CEIOPS. 2006d. Quantitative Impact Study No. 2: Technical Specification, CEIOPS-PI-08/06. Available at www.ceiops.org.

The total capital charge for the market risk was calculated as

$$CR_{MR} = \sqrt{\sum_{rxc} \rho_{rc} \cdot CR_{MR,r} \cdot CR_{MR,c}},$$

where the indices r and c are the rows and columns in the above dependence matrix.

In QIS2, an allowance for the risk absorption ability of future profit sharing was built into the formula by an RPS, which led to a reduction of the "basic" SCR capital requirement. This reduction became known as the "k-factor" approach since it was computed as the product of the total amount in the placeholder valuation of TPs relating to future discretionary benefits and the factor k, the risk-absorbing proportion of the TPs. Note that the "k-factor" approach was consistent with the suggestions for the standard formula made by the industry, CEA (2006b). No reduction in the different subrisk modules of the market risk was allowed.

Participants in QIS2 raised a number of concerns with the specific values used in this matrix. They suggested in particular that the dependencies between

- Interest rate risk and equity risk
- Interest rate risk and property risk
- Property risk and equity risk

were significantly higher than might be expected and did not recognize the diversification of portfolios. Some participants also noted that the matrix approach did not provide sufficient incentive for insurers to pursue investment strategies that were well diversified across different asset classes.

In spite of the concerns raised by QIS2 participants, CEIOPS observed that the dependence assumptions used to combine market risks needed to reflect risk dependencies under stressed conditions (consistent with the overall definition of the SCR). Although the degree of dependence between interest rate risk and equity risk would be comparatively low in "normal" market conditions, the relationship may change significantly under the manner of combined stress that the SCR is designed to test.

CEIOPS recognized that on market risk the QIS2 approach did not give due recognition for diversification effects and that some of the dependence assumptions would need to be revised downward.

A minority of CEIOPS Members considered that the analysis on market risk dependencies performed for the Dutch Financial Assessment Framework (FTK) was broadly consistent with the need for a simple, robust approach to aggregation and calibration identified earlier in this section. They would prefer the use of the QIS2 market risk dependence assumptions as a starting point for QIS3.

The two credit risk modules spread and concentration risks were included in the market risk as a consequence of the comments given by QIS2 participants. These two risks are discussed in Appendix J, Sections J.3 and J.4, respectively:

$$CR_{MR,SP} = \text{Credit spread risk}$$

$$CR_{MR,Co} = \text{Concentration risk}$$

In CP 20, CEIOPS (2006b), it was proposed that the capital requirements for six market subrisks were combined using Table I.2.

The comparatively low dependence assumptions for spread risk and concentration risk reflected features of the design of these new modules:

- As concentration risk quantifies the diversifiable, residual risk of individual exposures, it could be considered independent from the other market risks.
- In times of *flight to quality*, credit spreads tend to widen when other asset classes such as equity are depressed as well, suggesting a positive dependence between spread risk and other market risks.

However, the assumption of a zero dependence between the concentration risk module and the other market risk modules was a first suggestion in CEIOPS (2006b).

I.1.2.2 QIS3, CEIOPS (2007a)

As proposals for QIS3 developed, CEIOPS decided to replace the top-level RPS module of QIS2 with adjustments at the level of individual SCR subrisks.

TABLE I.2 Dependence Matrix for the Market Risk's Subrisks Proposed for Coming QIS3

	$CR_{MR,IR}$	$CR_{MR,ER}$	$CR_{MR,PR}$	$CR_{MR,CR}$	$CR_{MR,SP}$	$CR_{MR,Co}$
$CR_{MR,IR}$	1	0.75	0.75	0.25	0.25	0
$CR_{MR,ER}$		1	1	0.25	0.25	0
$CR_{MR,PR}$			1	0.25	0.25	0
$CR_{MR,CR}$				1	0.25	0
$CR_{MR,SP}$					1	0
$CR_{MR,Co}$						1

Source: Adapted from CEIOPS. 2006b. Draft Advice to the European Commission in the Framework of the Solvency II project on Pillar I issues—further advice. Consultation Paper 20, CEIOPS-CP-09/06, November 10. Available at www.ceiops.org.

The first step would be to calculate the capital requirements for individual risks—for example, interest rate risk—under two different assumptions:

- That the insurer is able to vary its assumptions on future bonus rates in response to the shock being tested; $nCR_{MR,IR}$
- That the insurer is not able to vary its assumptions on future bonus rates in response to the shock being tested $gCR_{MR,IR}$

Performing these two calculations for different risks reflects the fact that the ability to vary policyholder benefits would depend on the nature of the shock to which the insurer is exposed. For example, the potential for risk mitigation might be more significant in the case of yield curve movements than, say, a shock to property values.

The difference between the two capital requirements $gCR_{MR,IR} - nCR_{MR,IR}$ would be termed $KC_{MR,IR}$. If insurers wished to simplify the process, particularly in cases where the risk mitigating effect was not expected to be material, they could simply declare the first calculation to be equal to the second, that is, such that $KC_{MR,IR} = 0$.

In CP 20, CEIOPS (2006b), it was suggested that a second step would be to aggregate capital requirements for risks within the same category (equity, interest rate, property, etc.) using the relevant dependence matrix. To preserve the coherence of the modular approach, the aggregation would use the capital requirements produced assuming no change in the assumptions used to estimate policyholder benefits. For instance, the capital requirement gCR_{MR} for market risk would be derived by combining $gCR_{MR,IR}$, $gCR_{MR,ER}$, and so on.

Note that CEIOPS in the draft version of the TSs to QIS3 proposed that KC_{MR} should be calculated as $KC_{MR} = \max\{KC_{IR}; KC_{ER}; KC_{PR}; KC_{CR}\}$ in accordance with CP 20. The aggregated capital requirement gCR_{MR} would be reduced by the largest value of KC derived for the underlying risk modules, such that $nCR_{MR} = gCR_{MR} - KC_{MR}$.

In the final QIS3 specification, CEIOPS (2007a), that both the CR_{MR} and KC_{MR} should be calculated using the dependence matrix given in Table I.3:

$$CR_{MR} = \sqrt{\sum_{rxc} \rho_{rc} \cdot CR_{MR,r} \cdot CR_{MR,c}}$$

TABLE I.3 Dependence Matrix for the Market Risk's Subrisks Used for QIS3

ρ_{rc}	$CR_{MR,IR}$	$CR_{MR,ER}$	$CR_{MR,PR}$	$CR_{MR,CR}$	$CR_{MR,PR}$	$CR_{MR,Co}$
$CR_{MR,IR}$	1	0	0.50	0.25	0.25	0
$CR_{MR,ER}$		1	0.75	0.25	0.25	0
$CR_{MR,PR}$			1	0.25	0.25	0
$CR_{MR,CR}$				1	0.25	0
$CR_{MR,SP}$					1	0
$CR_{MR,Co}$						1

Source: Adapted from CEIOPS. 2007a. QIS 3 Technical Specifications, Part I: Instructions, CEIOPS-FS-11/07 & Part II: Background Information, CEIOPS-FS-12/07. April. Available at www.ceiops.org.

and

$$KC_{MR} = \sqrt{\sum_{rxc} \rho_{rc} \cdot KC_{MR,r} \cdot KC_{MR,c}},$$

where r and c are the rows and columns of the matrix.

Note that CEIOPS decided that the KC for the concentration risk was set to zero, that is, $KC_{MR,Co} = 0$.

In QIS3, the dependence matrix that was used, CEIOPS (2007a), is given by Table I.3.

I.1.2.3 QIS4, QIS4 (2008)

For policies where the policyholders bear the investment risk, such as UL policies, the undertaking would remain exposed to market risks where the value of the charges taken from these policies is dependent on fund performance. Exposure to interest rates would occur where fixed charges are received in the future. The value of any options and guarantees embedded within these contracts may also be exposed to market risk.

Where an undertaking has purchased derivatives, provided they accord with the risk mitigating principles of QIS4 (2008, TS.VII), the risk mitigating/increasing effect should be considered within each sub-module (e.g., currency forwards should be considered alongside the insurers other exposures within the currency risk submodule). Where the financial instrument does not accord to the risk mitigating principles of op. cit., their risk mitigating effect should be excluded from the calculation of the SCR.

Risk exposures of collective investment schemes should be allocated to submodules on a look-through basis if possible and on a best effort basis otherwise. Where a collective investment scheme is not sufficiently transparent to allow a reasonable best effort allocation, reference should be made to the investment mandate of the scheme. It should be assumed that the scheme invests in accordance with its mandate in such as manner as to produce the maximum overall charge. For example, it should be assumed that the scheme invests in currencies other than the undertaking's reporting currency to the maximum possible extent permitted by the investment mandate. It should be assumed that the scheme invests assets in each rating category, starting at the lowest category permitted by the mandate, to the maximum extent. If a scheme may invest in a range of assets exposed to the risks assessed under this module, then it should be assumed that the proportion of assets in each exposure category is such that the overall charge is maximized.

As a third choice to the look-through and mandate-based methods, participants should consider the collective investment scheme as an equity investment and apply the global equity risk charge (if the assets within the collective investment scheme are predominately listed) or other risk charge (if the assets within the collective investment scheme are predominately unlisted).

The same dependence matrix as for QIS3 was used. In order to test the impact of a different dependence situation, as has been observed during recent and earlier periods of market turmoil, CEIOPS decided to carry out a sensitivity analysis by testing a new dependence factor between equity and interest rate risk, leading to an alternative SCR value. To this end, the dependence factor of 0 between $CR_{MR,IR}$ and $CR_{MR,ER}$ was replaced by a positive dependence factor of 0.25 in the scenario of a downward movement of the interest rate and by a negative dependence factor of −0.25 in the case of an upward movement of the interest rate.

Risk-absorbing effects was taken into account in QIS4 in another way as compared to QIS3. For all subrisks, including the concentration risk, a net value of the capital charge is calculated.

$nCR_{MR,i}$: Capital charge for market subrisk i, including the risk-absorbing effect of future profit sharing.

The total net market risk capital charge is then calculated using the dependence matrix above:

$$nCR_{MR} = \sqrt{\sum_{rxc} \rho_{rc} \cdot nCR_{MR,r} \cdot nCR_{MR,c}}.$$

These net charges will be used to calculate a risk adjustment on the top level; see Appendix H.

The results from QIS4 showed that the market risk module accounts for the majority of capital requirements in life insurance business and to a lesser extent in nonlife business; CEIOPS (2008e).

I.1.3 Calibration

The calibration of the market risk module within the SCR standard formula is based on an economic approach, that is, the maximum use of historical observed changes in market rates and market prices.

For the different QISs, the following SCR features are taken into consideration:

- Risk measure: VaR
- Confidence level: 99.5%
- Time horizon: 1 year

I.1.3.1 QIS2

The dependence assumptions in QIS2 were based on the work that was done for the domestic FTK in the Netherlands; see CEIOPS (2006b). The following observations were made in that work: "The correlation between interest rates and shares (and variable yield securities) is unstable over time; consequently, the standardized method uses a robust estimate, allowing for the parameter uncertainty in that correlation. A degree of diversification is assumed between variable-yield securities and interest rates, being a correlation of $\rho = 0.8$ between the effects of the interest rate scenario and the scenarios for variable-yield securities."

Using a rolling-window technique, the authorities in the Netherlands estimated the distribution of the 12-month dependence between interest rate risk and "variable-yield" risk. They determined that the dependence corresponding to the 99.5% confidence level was 0.8. For QIS2 purposes, this was rounded down to 0.75 for both the interest rate–equity and the interest rate–property relationship. Also following the Dutch analysis, perfect dependence was assumed between equity and property risk.

I.1.3.2 QIS3

The analysis of the magnitude of market shocks was built upon the advice that CEIOPS had submitted to the EC.

The analysis used a "top-down" approach for the calibration of the dependence coefficient between interest rate and equity risk. This approach is chosen in order to ensure that the overall CR_{MR} risk charge, given the calibration of the individual shocks to a VaR 99.5% standard, is again consistent with a VaR 99.5% standard.

The dependence between interest rate and equity risk was calibrated using data from the following sources:

- MSCI Developed Markets index, total returns, available from 1972, yearly data (*Source*: Datastream)
- German zero rates, different maturities available from 1972, yearly data (*Source*: Bundesbank)

In order to carry out the "top-down" approach, some simplified model assumptions were made. For example, the analysis assumed a constant asset mix (30% equity and 70% bonds) and works with a constant positive 10-year duration gap between the duration of liabilities and the duration of bond investments and liabilities.

The historical returns of this model portfolio were assumed to follow a normal distribution. Figure I.1 compares the empirical data with this normality assumption. Figure I.1 shows that the normality assumption seems appropriate for the balanced portfolio.

Based on the normality assumption, the 99.5% confidence level corresponds to an 18% overall shock for market risk.

Total Return Model Portfolio	
Arithmetical mean	7.6%
Geometrical mean	7.1%
Standard deviation	9.8%
99.5 percentile; normal distribution	18.0%

Given this 18% overall shock for the model portfolio, the corresponding dependence between interest rate and equity risk can be derived. The calibration work for the individual shocks for interest rate and equity risk is used to determine the corresponding dependence of the model portfolio.

The analysis assumed for equity risk a 35% shock (compare the results given in Section I.3.3), and applied for interest rate risk the stress factors from Table I.8 stated in Section I.2.3.

$$\text{Stress factor for } CR_{MR}\ (\rho = 0.25) = 0.168$$

$$\text{Stress factor for } CR_{MR}\ (\rho = 0.50) = 0.184$$

$$\text{Stress factor for } CR_{MR}\ (\rho = 0.75) = 0.198$$

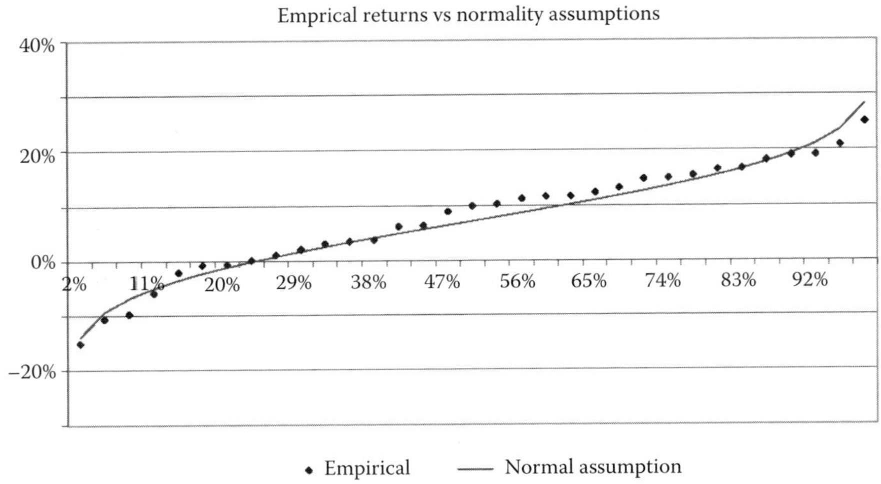

FIGURE I.1 Historical total returns of a balanced portfolio (30% equity, 70% bonds, and 10 years duration gap) compared to the normality assumption. Data from MSCI and Bundesbank 1972–2005. (Adapted from CEIOPS. 2007f. Calibration of the market risk module. CEIOPS-FS-05/07. February 19. Available at www.ceiops.org and CEIOPS. 2007g. Calibration of the underwriting risk, market risk and MCR. CEIOPS-FS-14/07. April. Available at www.ceiops.org, © Committee of European Insurance and Occupational Pensions Supervisors.)

From these results we can easily find that "the exact" dependence corresponding to the overall 99.5% shock. It equals 0.44 $[(0.18 - 0.168) * (0.5 - 0.25)/(0.184 - 0.168) = 0.4375]$.

Table I.4 shows the split of the resulting CR_{MR} of the model portfolio into the interest rate and equity risk parts.

A sensitivity analysis showed that the result is extremely sensitive to the assumed percentage invested in equity and used duration gap. Table I.5 reveals that the dependence between interest rate and equity risk ranges between 0.75 and 0.10.

Given this great uncertainty, CEIOPS proposed for QIS3 purposes the use of a rounded number of 0 for the dependence between interest rate and equity risk. This also allows symmetry between the assumptions about the direction of movements in the value of

TABLE I.4 The Split of the Resulting Market Risk of the Model Portfolio into the Subrisks Interest Rate and Equity

	Weight	**Implied Shock**	**Stress Factor for CR_{MR}**
Bonds	30%	15%	10.7%
Equity	70%	35%	10.5%
Portfolio	100%		18.0%

Source: Adapted from CEIOPS. 2007f. Calibration of the market risk module. CEIOPS-FS-05/07. February 19. Available at www.ceiops.org. and CEIOPS. 2007g. Calibration of the underwriting risk, market risk and MCR. CEIOPS-FS-14/07. April. Available at www.ceiops.org.

TABLE I.5 The Derived Dependence Between Interest Rate and Equity Risk for Different % in Equity and Duration Gap Assumptions

	Percentage Invested in Equity		
Duration Gap	**25%**	**30%**	**35%**
5 years	0.39	0.23	0.10
10 years	0.56	0.44	0.36
15 years	0.75	0.60	0.50

Source: Adapted from Table 6.1 in CEIOPS. 2007f. Calibration of the market risk module. CEIOPS-FS-05/07. February 19. Available at www.ceiops.org. and CEIOPS. 2007g. Calibration of the underwriting risk, market risk and MCR. CEIOPS-FS-14/07. April. Available at www.ceiops.org.

equities during an interest rate increase scenario and an interest rate decrease scenario. In addition, it may be noted that the main interest rate risk for insurers is likely to be a fall in interest rates. In a falling interest rate scenario, it could be arguable that the direction of movement in equities is just as likely to be upward as downward.

I.1.3.3 QIS4

No changes were made at the top level of the market risk module in QIS4, except for the tested dependencies between interest rate risk and market risk.

A review of the dependency between equity risk and interest rate risk was asked for after the QIS4; CEIOPS (2008e).

I.2 INTEREST RATE RISK

Interest rate risk exists for all assets and liabilities of which the value is sensitive to changes in the term structure of interest rates or interest rate volatility and which are not allocated to policies where the policyholders bear the investment risk. In any event, these are fixed-income investments, insurance liabilities, and financing instruments (loan capital) and interest rate derivatives. The value of assets and liabilities sensitive to interest rate changes can be determined using the (prescribed) term structure of interest rates ("zero rates"). This term structure can, of course, change over the period of a year.

I.2.1 Background

As an answer to the EC's CfA 10, MARKT (2004c), CEIOPS gave a background to the modeling of the standard formula and its risk modules; see CEIOPS (2005).

Interest rate risk exists for all investments and liabilities whose value is sensitive to changes in the term structure of interest rates or interest rate volatility. In any event, these are fixed-income investments, insurance liabilities, and financing instruments, loan capital, and derivatives with a value dependent on interest rates. The value of investments and liabilities sensitive to interest rate changes may be established from the (prescribed) term structure of interest rates ("zero rates"). This term structure can, of course, change over the period of a year.

The value of the changes in the risk-free interest rate could be modeled with some interest rate model that should be chosen according to the criterion of predictive power. The parameters of such a model would be fixed by supervisors using historic time series and allowing for current market assessments. One possibility may be the Cox–Ingersoll–Ross model, see Section 17.1.3, whose parameters are the drift (mean reversion factor), the volatility and the mean reversion level (long-term average).

Then the development of the long-term risk-free interest rate is given by

$$dr = \kappa(\mu - r)dt + \sigma \cdot \sqrt{r}dW,$$

where W denotes a Brownian motion. For determining the required risk, capital movements of the yield curve may be analyzed including parallel shifts, twists at the short end, and fluctuations in the middle range. For simplicity, parallel shifts are considered and the change in interest rate is chosen to be the difference between the current level and the quantile (0.5% for a drop, 99.5% for a rise) of the distribution with respect to the time horizon of one year.

I.2.1.1 Scenario-Based Approach

In a scenario-based approach, the aim of the interest rate stress test is to establish the sensitivity of the surplus to movements in the term structure and the desired solvency for interest rate risk derived from it. The stress test to be computed relate to a general "rise" and "fall" in the term structure of interest rates as used to discount the insurance liabilities. The value of the liabilities and investments is again determined comprehensively, assuming the prescribed higher or lower term structure of interest rates. The scenario with the largest loss has to be computed. This applies to both scenarios if it cannot be said with certainty in advance whether an interest rate increase or fall is the more unfavorable for the financial position of the institution. The change in the surplus, that is, the difference between the value of the investments and the value of the liabilities, is established for each scenario. The greatest loss is included when determining the capital charge for interest rate risk.

Yield curves movements (rising and falling) for the standard formula may be prescribed. Two aspects may have to be taken into account in determining the stress tests. Firstly, the volatility in "zero rates" for long periods is relatively less than for short periods. The higher the initial "zero rate," the larger the expected change. Both characteristics are commonly observed empirically.

I.2.1.2 Factor-Based Approach

Alternatively, a factor-based approach could be used to model interest rate risk. The concept of modified duration may be applied for the actual assessment. After a drop in interest rates (negative), the market values of fixed-income securities and of liabilities increase. The capital requirement to cope with the interest rate shock amounts to

$$CR_{IR} = \max\left\{0; -\Delta \cdot \left(MV_{TP} \cdot D_{\mathrm{TP}}^{\mathrm{mod}} - MV_{\mathrm{FI}} \cdot D_{\mathrm{inv}}^{\mathrm{mod}}\right)\right\},$$

where MV denotes market values, CR the capital requirement and D^{mod} the modified duration, and Δ the drop in interest rates. The indices refer to TPs and fixed income (FI).

For a rise in interest rates, the concept of modified duration yields the capital requirement

$$CR_{IR} = \max\left\{0; \Delta \cdot \left(MV_{FI} \cdot D_{inv}^{\text{mod}} - MV_{TP} \cdot D_{TP}^{\text{mod}}\right)\right\},$$

where the index *FI* refers to fixed income.

Since a drop and a rise in interest rates cannot occur simultaneously, risk capital should be taken as the maximum of both.

I.2.2 QIS2–QIS4

I.2.2.1 QIS2, CEIOPS (2006d)

Under QIS2, CEIOPS (2006d), the capital charge was based on a factor-based approach as follows:

$$CR_{MR,IR} = \max\left\{0; \left(MV_{FI} \cdot GD_{FI}\left(r, s_{\text{up}}\right) - TP \cdot GD_{TP}\left(r, s_{\text{up}}\right)\right); \left(MV_{FI} \cdot GD_{FI}\left(r, s_{\text{down}}\right) - TP \cdot GD_{TP}\left(r, s_{\text{down}}\right)\right)\right\},$$

where MV_{FI}: the net market value of interest-rate-dependent assets and financial instruments not allocated to policies where the policyholders bear the investment risk; *TP*: total technical provisions not allocated to policies where the policyholders bear the investment risk; GD_x: the generalized duration of "$x = FI$": the interest-rate-dependent assets and financial instruments; and "$x = TP$": the technical provisions.

The term structure $r(t)$ and the stresses $s_{\text{up}}(t)$ and $s_{\text{down}}(t)$ were prescribed as in Table I.6. The formula looks quite general, but an important approximation enters the equation in the computation of the generalized duration, as is shown in Section I.2.1.

Both the up stress $s_{\text{up}}(t)$ and the down stress $s_{down}(t)$ are constant over five maturity buckets, see Table I.6.

The altered term structures can be derived by multiplying the current term structure by the relevant stress factor. For example, the n-year spot rate in 12 months time in the up stress situation corresponds with

$$R_{12}(t) = R_0(t) \cdot \left[1 + s_{\text{up}}(t)\right],$$

where, $R_0(t)$ is the current t-year spot rate based on the term structure data supplied, and $s_{\text{up}}(t)$ follows from Table I.6. Example, from CEIOPS (2007f), (2007g): For a 4% 10-year

TABLE I.6 Two Separate Shocks were Performed under QIS2

Maturity in years:	1–3	4–6	7–12	13–18	>18
Relative change $s_{\text{up}}(t)$	0.75	0.50	0.40	0.35	0.30
Relative change $s_{\text{down}}(t)$	–0.40	–0.35	–0.30	–0.25	–0.20

Source: Adapted from CEIOPS. 2006d. Quantitative Impact Study No. 2: Technical Specification, CEIOPS-PI-08/06. Available at www.ceiops.org.

Note: One up stress and one down stress were done. The altered term structure was arrived at by multiplying the current interest rate curve, prescribed by CEIOPS, by $(1+s_{\text{up}}(t))$ and $(1+s_{\text{down}}(t))$.

interest rate, the upward stressed interest rate corresponds to 5.6% [4% × (1+,0.4)], while the downward stressed interest rate equals to 2.8% [4% × (1 − 0.3)].

Alternatively, those undertakings that cannot easily perform the separation into five pools of CFs for the TPs, the subdivision could be based on the BEs instead.

Participants were also asked to test the change in net asset value that would occur, given a predefined scenario, that is, to get a second alternative of the capital charge. Let *NAV* be the net value of assets minus liabilities and $\Delta NAV|$ upward shock and $\Delta NAV|$ downward shock be the changes in the net value of asset and liabilities due to revaluing all interest-rate-sensitive instruments using altered term structures.

$$CR_{MR,IR} = \max\left\{0;\, \Delta NAV|\text{upward shock};\, \Delta NAV|\text{downward shock}\right\}.$$

The altered term structures can be derived by multiplying the current interest rate curve by $(1 + s_{\text{up}})$ and $(1 + s_{\text{down}})$ as above.

I.2.2.2 QIS3, CEIOPS (2007a)

The value of assets and liabilities sensitive to interest rate changes can be determined using the prescribed term structure of interest rates ("zero rates") that was enclosed with the QIS3 specifications. This term structure can, of course, change over the period of a year.

In QIS3, the number of buckets was increased from 5 in Table I.6 to 20 given in Table I.8. The last bucket was >20 years with the stress factors from bucket 20.

Let ΔNAV be the net value of assets minus liabilities and $\Delta NAV|$ upward shock and $\Delta NAV|$ downward shock be the changes in the net value of asset and liabilities due to revaluing all interest-rate-sensitive instruments using altered term structures. Where an undertaking is exposed to interest rate movements in more than one currency, the capital charge for interest rate risk should be calculated based on the same relative change on all relevant yield curves.

The capital charge for the interest rate risk is defined as

$$CR_{MR,IR} = \max\left\{0;\, \Delta NAV \mid \text{upward shock};\, \Delta NAV \mid \text{downward shock}\right\}.$$

I.2.2.3 QIS4, QIS4 (2008)

The value of assets and liabilities sensitive to interest rate changes can be determined using the prescribed term structure of interest rates ("zero rates") that was enclosed with the QIS4 specifications. This term structure can, of course, change over the period of a year.

Let ΔNAV be the net value of assets minus liabilities and $\Delta NAV|$ upward shock and $\Delta NAV|$ downward shock be the changes in the net value of asset and liabilities due to revaluing all interest-rate-sensitive instruments using altered term structures. Where an undertaking is exposed to interest rate movements in more than one currency, the capital charge for interest rate risk should be calculated based on the same relative change on all relevant yield curves.

Let $CR^{\text{up}}_{MR,IR} = \left\{\Delta NAV|\text{upward shock}\right\}$ and
$CR^{\text{down}}_{MR,IR} = \{\Delta NAV|\text{downward shock}\}.$

The scenarios for interest rate risk should be calculated under the condition that the assumptions on future bonus rates (reflected in the valuation of future discretionary benefits in TPs) remain unchanged before and after the shocks being tested.

Additionally, the result of the scenarios should be determined under the condition that the participant is able to vary its assumptions in future bonus rates in response to the shock being tested. The resulting capital charges are $nCR^{up}_{MR,IR}$ and $nCR^{down}_{MR,IR}$, that is, the capital charges for interest rate risk after upward shock and downward shock including the risk-absorbing effect of future profit sharing.

The capital charge for interest rate risk is derived from the type of shock that gives rise to the highest capital charge including the risk-absorbing effect of future profit sharing:

If

$$nCR^{up}_{MR,IR} > nCR^{down}_{MR,IR},$$

then

$$nCR_{MR,IR} = \max\left\{0; nCR^{up}_{MR,IR}\right\}$$

and

$$CR_{MR,IR} = CR^{up}_{MR,IR} \quad \text{if } nCR_{MR,IR} > 0 \quad \text{and} = 0 \text{ otherwise.}$$

If

$$nCR^{up}_{MR,IR} \leq nCR^{down}_{MR,IR},$$

then

$$nCR_{MR,IR} = \max\left\{0; nCR^{down}_{MR,IR}\right\}$$

and

$$CR_{MR,IR} = CR^{down}_{MR,IR} \quad \text{if } nCR_{MR,IR} > 0 \quad \text{and} = 0 \text{ otherwise.}$$

There was also a *simplification* that could be tested under QIS4:

In order to determine the interest rate scenario effect on the value of assets and liabilities, a simplified calculation could be used whereby changes in value are estimated as the yield curve change multiplied by the relevant modified duration separately for the assets and for the liabilities. The condition to be met for using this simplification is that the CFs of the item are not interest rate sensitive; in particular, the item has no embedded options. This simplification could be used for assets, nonlife TPs, and other liabilities. This simplification should not be used for life TPs.

The shocks are parallel yield stress, at all durations of

- Downward shock: −40%
- Upward shock: +55%

One suggestion that was raised after the QIS4 calculation was done was the possibility to introduce sensitivity to the changes in the shape of the yield curve for the interest rate module; CEIOPS (2008e).

I.2.3 Calibration

I.2.3.1 QIS2, CEIOPS (2006b) QIS3, CEIOPS (2007f, 2007g)

Both the factor-based approach and the scenario-based approach from QIS2 used a series of stress factors for interest rate risk that were constant over five maturity buckets; see Table I.6.

In the scenario-based approach, which was the placeholder, the participants were asked to perform two separate shocks, recalculating the net value of assets and liabilities by revaluing all interest-rate-sensitive instruments using altered term structures. The altered term structures could be derived by multiplying the current interest rate curve (prescribed by CEIOPS for QIS2) by $(1 + s_{\text{up}}(t))$ and $(1 + s_{\text{down}}(t))$. The capital requirement was the larger of the two changes in the net asset value in response to these shocks (or zero if the shocks actually generated an improvement in net asset value).

CEIOPS also tested a duration approach in QIS2 which broadly approximated the scenario using formulaic requirements.

Given a CF $C = (C(1), C(2), \ldots)$, the relative change in its market value subject to a small change of the term structure $r(t)$ can be approximated using the first-order Taylor approximation:

$$\frac{MV[C, r(1+s)] - MV[C, r]}{MV[C, r]} \approx -\frac{1}{MV[C, r]} \cdot \sum_t \frac{s(t) \cdot r(t)}{1 + r(t)} \cdot t \cdot d(t) \cdot C(t) := GD_C\,(r, s).$$

Using the generalized duration effectively restricts the calculation of the capital charge to a first-order Taylor approximation. From its definition, the generalized duration can take positive and negative values (unlike the modified duration, see below, that can only take positive values).

If the term structure $r(t)$ and the stress factor $s(t)$ are constant in t, then the right-hand side of the equation becomes $s \cdot r \cdot D_C^{\text{mod}}$, given that the modified duration is defined as

$$D_C^{\text{mod}} = -\frac{1}{MV(C)} \cdot \frac{1}{1+r} \cdot \sum_t t \cdot d(t) \cdot C(t) \cdot s.$$

Both the up stress $s_{\text{up}}(t)$ and the down stress $s_{\text{down}}(t)$ are constant over five maturity buckets, see Table I.6, which allows approximating the generalized duration by pooling CFs (or assets) in the five maturity buckets:

$$GD_s(r, s) \approx \cdot \sum_{bucket:b} r_b \cdot s_b \cdot D_b^{\text{mod}} \cdot \frac{MV_b}{MV(C)},$$

where r_{b} is the average interest rate for bucket b, s_{b} the stress for bucket b, the modified duration of the CF pooled in bucket b, and MV_{b} the market value of the CF pooled in bucket b.

For those undertakings that cannot easily perform the separation into the five pools of CFs, the following is to be considered an optional approximation:

$$GD_C(r,s) \approx r_b \cdot s_b \cdot D_C^{\text{moc}},$$

where the chosen interest rate and stress depends on the bucket b, into which the duration of the CF C falls. C stands for either assets or liabilities.

The stress factors for changes in interest rates were calibrated using two data sources:

- Monthly data from 1972 onward on German government bond zero rates, for maturities between 1 and 10 years (*Source*: Bundesbank)
- Daily data from 1997 onward on European zero swap rates, for maturities 1 year–5 years–10 years–15 years–20 years–25 years–30 years (*Source*: Datastream)

The observed data showed that in general higher interest rates were associated with higher absolute changes in interest rates. The lognormal model exhibits this property and the calibration of the lognormal model appeared more robust than the normal model. The Black–Karasinski and Cox–Ingersoll–Ross models were also considered; *cf.* Section 17.1.3. However, these mean reversion models were not used because, based on the observed data, the mean reversion assumption did not hold. The resulting shocks were also highly dependent on the exact model chosen.

The lognormal model treats proportionate changes in interest rates as a lognormal process; so it has been assumed that the distribution of the t-year spot rate in 12 months is given by

$$R_{12}(t) = R_0(t) \cdot e^X, \tag{I.1}$$

where X is assumed to be distributed as $N(\mu_t, \sigma_t^2)$.

For y sufficiently close to zero, $\ln(1+y)$ is approximately y; hence, Equation I.1 could be rearranged to

$$X = \log\left[\frac{R_{12}(t)}{R_0(t)}\right] = \log\left\{1 + \left[\frac{R_{12}(t) - R_0(t)}{R_0(t)}\right]\right\} \approx \left[\frac{R_{12}(t) - R_0(t)}{R_0(t)}\right], \tag{I.2}$$

showing that the lognormal model assumes that the absolute change in interest rates, $[R_{12}(t) - R_0(t)]$, linearly depends on the level of interest rate, $R_0(t)$.

The annualized standard deviations are given for the maturities in Table I.7.

For interest rates having maturities longer than 10 years (often described as the long end of a term structure), no long-dated data series were available. To determine stress factors for these long end interest rates that are consistent with the short end stress factors, the factors were fitted on information from both data sources. For the long end interest rates, the annualized standard deviations are determined by using a constant volatility ratio (yearly data/daily data).

The stress factors for interest rate risk were estimated in Table I.8 and Figure I.2.

TABLE I.7 The Annualized Standard Deviations s_G of German Zero Rates Yearly Data from 1972 to 2006, and s_E of EMU Swap Rates Daily Data from 1997 to 2006

	Time to Maturity													
	1 year	2 years	3 years	4 years	5 years	6 years	7 years	8 years	9 years	10 years	15 years	20 years	25 years	30 years
s_G	0.27	0.23	0.21	0.20	0.19	0.17	0.17	0.16	0.15	0.15				
s_E	0.20	0.21			0.17					0.14	0.14	0.12	0.12	0.12

Source: Adapted from CEIOPS. 2007f. Calibration of the market risk module. CEIOPS-FS-05/07. February 19. Available at www.ceiops.org/

TABLE I.8 The Resulting Stress Factors for Interest Rate Risk

t years	$s_{up}(t)$ Maturity	$s_{down}(t)$
1	0.94	−0.51
2	0.77	−0.47
3	0.69	−0.44
4	0.62	−0.42
5	0.56	−0.40
6	0.52	−0.38
7	0.49	−0.37
8	0.46	−0.35
9	0.44	−0.34
10	0.42	−0.34
11	0.42	−0.34
12	0.42	−0.34
13	0.42	−0.34
14	0.42	−0.34
15	0.42	−0.34
16	0.41	−0.33
17	0.40	−0.33
18	0.39	−0.32
19	0.38	−0.31
20+	0.37	−0.31
21	0.37	−0.31
22	0.37	−0.31
23	0.36	−0.31
24	0.36	−0.31
25	0.36	−0.31
26	0.36	−0.31
27	0.36	−0.31
28	0.36	−0.31
29	0.36	−0.31
30	0.36	−0.31

Source: Adapted from CEIOPS. 2007f. Calibration of the market risk module. CEIOPS-FS-05/07. February 19. Available at www.ceiops.org and CEIOPS. 2007g. Calibration of the underwriting risk, market risk and MCR. CEIOPS-FS-14/07. April. Available at www.ceiops.org.

Note: The 20 first buckets are used in QIS3.

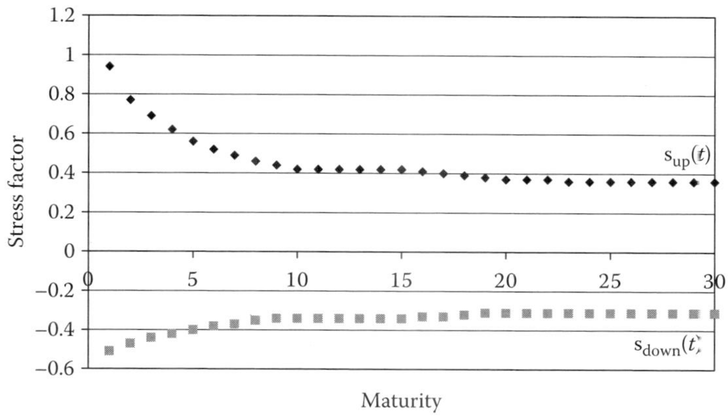

FIGURE I.2 Estimated stress factors for the interest rate risk using a lognormal model. (Adapted from Table I.8.)

The annualized standard deviations for different maturities were calculated, showing (for both data sources) higher standard deviations for shorter maturities than longer maturities. The same result was observed when using data on three non-euro EU countries: Denmark, Sweden, and the United Kingdom. The comparison was based on the following data:

- Relevant zero swap rates (*Source:* Datastream)
- Maturities 2 years–5 years–10 years–20 years–30 years
- Overlapping time periods starting from 1997
- Data on a daily basis

Table I.9 shows for each country and maturity, the corresponding standard deviation.

For all three added countries, the standard deviations were more or less consistent with the EMU results, with the exception of the short end of the UK term structure that showed

TABLE I.9 The Annualized Standard Deviation for the Euro-zone (EMU), Denmark (DEK), Sweden (SEK), and United Kingdom (GBP)

	Time to Maturity				
Annualized Standard Deviation	**2 year**	**5 year**	**10 year**	**20 year**	**30 year**
EMU	0.21	0.17	0.14	0.12	0.12
DEK	0.19	0.16	0.14	0.13	0.13
SEK	0.18	0.17	0.14	NA	NA
GBP	0.15	0.14	0.13	0.13	0.13

Source: Adapted from CEIOPS. 2007f. Calibration of the market risk module. CEIOPS-FS-05/07. February 19. Available at www.ceiops.org and CEIOPS. 2007g. Calibration of the underwriting risk, market risk and MCR. CEIOPS-FS-14/07. April. Available at www.ceiops.org.

Note: Swap rates, daily data from 1997 to 2006; Datastream.

lower results. However, for all countries, the short end of the term structure exhibits higher standard deviations than the long end.

In QIS3, the number of buckets was increased from 5 in Table I.6 to 20 given in Table I.9. The last bucket was >20 years with the stress factors from bucket 20.

I.2.3.2 QIS4

No changes were made.

I.3 EQUITY RISK

Equity risk arises from the level or volatility of market prices for equities. Exposure to equity risk refers to all assets and liabilities whose value is sensitive to changes in equity prices.

A distinction can be made between systematic risk and idiosyncratic risk, which comes from inadequate diversification. The systematic risk refers to the sensitivity of the equity's returns to the returns of market portfolios and cannot be reduced by diversification (undiversified risk).

The equity subrisk module is intended to capture the systematic risk, whereas the undiversified risk is captured in the concentration subrisk module.

I.3.1 Background

As an answer to the EC's CfA 10, MARKT (2004c), CEIOPS gave a background to the modeling of the standard formula and its risk modules; see CEIOPS (2005).

I.3.1.1 Factor-Based Approach

In a factor-based approach to modeling equity risk, share values may be assumed to follow a lognormal distribution. Its mean value (yield) and standard deviation (volatility) can be derived from historical data and, if appropriate, be modified so as to allow for the current market situation and trends. The RF then would be a suitable quantile, for example, 0.5%, of the chosen lognormal distribution.

I.3.1.2 Scenario-Based Approach

In a scenario-based approach, given the overall equity position, the institution must ascertain the effect on the surplus of the value change described below in the benchmark used. To ascertain the capital charge for equity risk in a scenario approach, the relevant risk position consists of the value of all long and short positions in shares and all financial instruments whose value is influenced wholly or partly by share prices, such as options, futures, convertibles, equity notes, and total return swaps. Liabilities from UL insurance and the assets covering them are to be considered simultaneously when determining the relevant risk position.

The scenario for equity risk may distinguish between shares listed on mature markets, emerging markets shares, and private equity (unlisted shares). From empirical observations, the two latter categories are riskier than the former. For example, a fall of 40% may be assumed for mature markets shares and 45% for emerging markets shares and private equity.

As a result of empirically observed (and theoretically expected) higher positive dependence between the three types of share, the outcomes for the three subcategories may be added together in the standard formula.

I.3.2 QIS2–QIS4

I.3.2.1 QIS2, CEIOPS (2006d)

Two different approaches were tested in QIS2. The first was a factor-based approach:

$$CR_{MR,ER} = \{\Delta MV_{EQ} \mid \text{equity shock}\} - \{\Delta MV_{EQ,Link} \mid \text{equity shock}\},$$

where MV_{EQ}: the market value of the overall equity exposure; and $MV_{EQ,Link}$: the market value of the equity exposures where the policyholders bear the investment risk, for example, linked business.

The equity shock is the immediate effect expected in the event of a 40% fall in all individual equities, also considering the effect on derivatives and short positions. If participants liked to take account of the effect of short positions and derivatives, they could do so in the following way: the change in value should be calculated on the basis of the change in value of the underlying instrument. However, no consideration should be given to management actions or active trading strategies.

The second approach was a scenario approach:

$$CR_{MR,ER} = \{\Delta NAV \mid \text{equity shock}\},$$

where ΔNAV is the net value of equity assets minus equity liabilities. The *equity shock* is the immediate effect expected in the event of a 40% fall in equity benchmarks, for example, Eurostoxx, taking account of all the participant's individual direct and indirect exposures to equity prices. The equity shock takes account of the specific investment policy including, for example, hedging arrangements, gearing, and so on.

I.3.2.2 QIS3, CEIOPS (2007a)

The QIS2 technical specification document provided a 40% stress factor for equity risk. Equity risk arises from the level or volatility of market prices for equities. The stress factor for equity risk can be calibrated at different levels of granularity, that is, at the global or region level. For reasons of simplicity, the level of granularity used for QIS2 was set at the global level. For QIS3, this hypothesis was refined and equity risk was divided into two parts. The first part contains equity invested in indices of development markets and the second component comprises the remaining elements, for example, emerging markets, nonlisted equities, and alternative investments.

The equity subrisk module used indices as risk proxies, which means that the volatility and dependence information is derived from these indices. It was assumed that all equities could be allocated to an index of the prescribed set.

The capital charged was based on a scenario test. It was assumed that the equity portfolio of the insurer has the same exposure to systematic risk as the index, that is, the proxy, itself and is therefore assumed that the beta is 1.

It is assumed that hedging and risk transfer mechanisms were taken into account when calculating the capital charge. As a general rule it was assumed that hedging instruments would only be allowed with the average protection level over the next year. Hedging instruments not in force at the BS date were not allowed.

For each index

1. Global
2. Other (compromising emerging markets, nonlisted equities, and alternative investments)

the capital charge is calculated as

$$CR_{MR,ER,i} = \max\{\Delta NAV \mid \text{equity shock}_i; 0\},$$

where ΔNAV is the net value of equity assets minus equity liabilities after an *equity shock* of index $i, i = 1, 2$. The equity shock used in QIS3 is given in Table I.10.

The overall capital charge was calculated using a dependence between the Global index and Other index of $\rho = 0.75$, that is,

$$CR_{MR,ER} = \sqrt{CR^2_{MR,ER,1} + CR^2_{MR,ER,2} + 1.5 \cdot CR_{MR,ER,1} CR_{MR,ER,2}}.$$

A duration approach was also tested. For more details, see Appendix I.3.2.

I.3.2.3 QIS4, QIS4 (2008)

In QIS4, the only main change was the equity shock for index 2, Other, which were increased to 45%, that is, we had the following shocks as in Table I.11A:

TABLE I.10 Equity Shocks used in QIS3

	Global	**Other**
Equity shock$_i$	32%	40%

Source: Adapted from CEIOPS. 2007a. QIS 3 Technical Specifications, Part I: Instructions, CEIOPS-FS-11/07 & Part II: Background Information, CEIOPS-FS-12/07. April. Available at www.ceiops.org.

TABLE I.11A Equity Shocks and Volatility Factors used in QIS4

	Global	**Other**
Equity shock$_i$ or volatility factor f_i	32%	45%

Source: Adapted from QIS4. 2008. QIS 4 Technical Specifications. European Commission. MARKT/2505/08. March 31. Available at http://www.ceiops.eu/content/view/118/124/.

I.3.2.3.1 Treatment of Participations and Subsidiaries at Solo Level The treatment of participations and subsidiaries at solo level is explained in Annex SCR 1 in the Technical Specifications; QIS4 (2008). Three options were given and the QIS4 participants were requested to test option 1, the default option, and options 2. The third option was on an optional basis. For options 1 and 2 where the *parent* owns a participation in another undertaking, or has a subsidiary, this participation or subsidiary was valued on an economic basis. If a fair value treatment under IAS 39 was applied, this was considered as an acceptable proxy.

I.3.2.3.2 Option 1: Differentiated Equity Stress Approach For all participations and subsidiaries, participants had to treat these holdings in the SCR calculation as if they were an equity investment as described below (i.e., by calculating a differentiated capital charge for equity risk).

The shocks are calculated as above. For participations and subsidiaries (e.g., ownership of more than 20%) in insurance and financial undertakings included in the scope of consolidated or supplementary supervision, the equity shock would be reduced to 16% for "Global" firms and to 22.5% for the "Other" participations.

The same reduction would apply for other participations and subsidiaries in

- Noninsurance and nonfinancial undertakings that are taken into consideration within the consolidated or supplementary supervision.
- Insurance and financial undertakings that are not included in the scope of consolidated or supplementary supervision and do not exceed the 10% of the participating undertaking's OFs.

In cases where the mother owns more than 20% of another insurance or financial undertaking which (1) is not included in the scope of consolidation or supplementary supervision and (2) where the value of that participation or subsidiary exceeds 10% of the participating undertaking's OFs the calculation of the regulatory capital requirement of the parent shall be carried out using deduction and aggregation method.

I.3.2.3.3 Option 2: Across the Board Approach Under this option, all participations and subsidiaries were treated as if they were a "standard" equity investment when calculating the SCR capital charge for equity risk. They are not granted any specific treatment with respect to equity risk.

I.3.2.3.4 Option 3: Look-Through Approach On an optional basis, participants could replace their solo SCR calculation, with the group SCR calculation for the subgroup formed by the participant itself (the "parent") and its subsidiaries and participations. Where this method is followed, undertakings should follow the default method for groups. Under this option, both the parent's OFs and SCR had to be replaced with the OFs and group SCR of the subgroup.

The calculations for equity risk should be carried out under the condition that the assumptions on future bonus rates, reflected in the valuation of future discretionary benefits in TPs, remain unchanged before and after the shocks being tested.

Additionally, the overall result of the calculation was determined under the condition that the participant was able to vary its assumptions in future bonus rates in response to the shock being tested. The resulting capital charge is $nCR_{MR,IR}$.

There are also *simplifications* that could be tested under QIS4:

The determination of the capital charge $CR_{MR,ER,i}$ with respect to an individual index i could be carried out by taking into account hedging and risk transfer mechanisms using a two-step process.

1. The first step relates to the level of the individual equity. If there are hedging instruments for single equities, they have to be taken into account at the level of the single equity. The hedge reduces the stress with the change in MV of the instrument itself. The impact has to be determined by the company itself.

   The calculations within this first step would be carried out as follows:

   For each index i, the MV of individual equities allocated to i in the event of the stress scenario *equity shock*$_i$ would be calculated, taking into account hedging instruments. The "stressed" MVs would be calculated as follows:

$$EQ\_\text{stress}_{i,j} = MV_{EQ,i,j} \cdot (1 - f_i) + MV_{\text{Hedge},i,j},$$

   where $EQ\_\text{stress}_{i,j}$: market value of equity j, allocated to index i, after stress; $MV_{EQ,i,j}$ is the market value of equity j allocated to index i; f_i: prescribed volatility factor of the index i, see Table I.11A; and $MV_{\text{Hedge},i,j}$: the change in market value of hedges per individual equity i, j under stress

2. In a second step, hedging instruments for subportfolios for example, indices or special funds would be taken into account. The risk mitigation would be reflected by the change in MV of the hedging instrument per index, which stands for the subportfolio. If there would be a global hedge for all equity positions in force, it would be allocated on an MV weighted basis to the relevant equity indices (excluding alternative investments).

Within this second step, the changes in MV for all equities under index i would be aggregated to a capital charge taking into account hedging instruments for equity risk for the individual index i as follows:

$$\Delta CR_{EQ,i} = \sum_j \left[MV_{EQ,i,j} - EQ\_\text{stress}_{i,j}\right] - \Delta MV_{\text{Hedge},i},$$

where $\Delta CR_{EQ,i}$: the risk capital charge for equity risk for index i; and $\Delta MV_{\text{Hedge},i}$: the change in market value of hedges per individual index i under stress.

The overall value of equities under stress would be derived by combining the "ChangeInEquityValue" $\Delta CR_{EQ,i}$ for the individual indices using the dependence described above to provide the "AggregateChangeInEquityValue" ΔCR_{EQ}. This should be converted into a revised stress test and this stress test should be applied to the liabilities:

$$\text{RevEQ\_stress} = \frac{\Delta CR_{EQ}}{MV_{EQ+\text{Hedges}}},$$

where *RevEQ_stress*: the revised equity stress test; and $MV_{EQ+\text{Hedges}}$: the current market value of all investments in equities and hedges.

Finally, the capital charge, $CR_{EQ,i}$, is calculated as the change in the net asset value of the undertaking as follows:

$$CR_{EQ,i} = \max\left\{\Delta L - \Delta CR_{EQ}; 0\right\},$$

where ΔL: the change in the value of the liabilities following a change in the value of equities/hedges of the revised equity stress test *RevEQ_stress.*

A dampener alternative for equity risk was introduced in QIS4. For details on the dampener see Section I.3.5.

In QIS4, many of the participating companies (and their supervisors) stated that the 32% calibration of the equity stress was too low for a 99.5% calibration and instead suggested a figure around 40%; CEIOPS (2008e). The views on the different approaches to the treatment of participations were mixed.

I.3.3 Calibration

I.3.3.1 QIS2, CEIOPS (2006b) QIS3, CEIOPS (2007f, 2007g)

The design of both the factor approach and the scenario approach assumed for simplicity that equity risk could be tested using a single, global shock. However, the magnitude of the initial shock chosen for QIS2 was identified as problematic by many of the stakeholders.

Additionally, concern was expressed that the treatment did not reflect the use of equities to match longer-term liabilities, resulting in overdominance of the SCR by the equity risk component.

The QIS2 shock was calibrated using quarterly data from the MSCI Developed Markets index on total returns over the period 1970–2005 (*Source*: Datastream). The index covered 23 indices from developed markets, excluding private equity investments and, by definition, emerging markets. Individual country weights were calculated on the basis of their market capitalization. Total returns were estimated on the assumption that dividends are reinvested in the index on the day the security is quoted exdividend.

As a first step, global returns were assumed to follow a normal distribution, although the observed data exhibited negative skew and a negative fat tail.

There are different options to deal with thickness of negative tails.

- *Extreme value theory*, where the assumption is made that the distribution of the tail converges to a limit distribution, cf. extreme value copulas in Section 13.4.3.1
- *Log-linear estimation methods*, where the tail in the historical probability distribution is extrapolated using linear regression for the historically worst outcomes
- *The Gumbel distribution* is a special case of the Generalized Extreme Value distribution and is especially suitable for light tailed, *cf.* Section 13.4.1
- Fitting a Generalized Pareto Distribution (GPD) to the tail of the equity return distribution, which can be estimated using maximum likelihood (ML) estimators or with regression models

The geometric mean for the source data equals 10.1% (arithmetic mean = 11.5%) and the standard deviation corresponds to 16.9%. Assuming normally distributed equity returns, the shock corresponding to the 99.5% confidence level is 33.4%. However, after tail correction, the 99.5% confidence level corresponds to a shock of approximately 35%, depending on the exact correction method chosen:

- VaR 99.5%, Log-linear: 34.3%
- VaR 99.5%, Gumbel: 34.6%
- VaR 99.5%, GPD via ML: 33.3%
- VaR 99.5%, GPD via regression (R): 37.9%
- VaR 99.5%, Normal, arithmetic mean: 32.0%
- VaR 99.5%, Normal, geometric mean: 33.4%

The equivalent result for TailVaR 99% was a shock of approximately 37.5%.

In CP 20, CEIOPS (2006b), further developments were discussed. CEIOPS believed that the equity risk treatments could continue to be developed in advance of QIS3 on the basis of the QIS2 proposal, with a focus on refining the calibration.

One particular aspect that required further attention was the interaction between equity and currency risk. The analysis presented above was based on hedged returns. The QIS2 currency risk module considered only direct exposures, since, the indirect impact may be technically difficult to quantify, for example, a euro-denominated stock will be impacted indirectly by a move in the US dollar via the issuer's unhedged dollar activities. Alternatively, unhedged returns, include movements in exchange rates, could be used. However, this would lead to double counting if equity positions are included in the currency risk module.

Institutional investors considered the MSCI Developed Markets index as one of the main benchmarks for worldwide equity investments. Unfortunately, the MSCI data are only available from 1970. Consequently, important pre-WWII market declines were not included

TABLE I.11B The Dependence between Five Major Equity Markets for Three Different Time Periods: (a) 1900–2005, (b) 1970–2005, and (c) 1985–2005

	France	Germany	Japan	United Kingdom	United States
			1900–2005		
France	1	0.31	0.16	0.38	0.24
Germany		1	0.12	0.16	0.16
Japan			1	0.16	0.15
United Kingdom				1	0.51
United States					1
			1970–2005		
France	1	0.75	0.51	0.46	0.58
Germany		1	0.41	0.49	0.62
Japan			1	0.32	0.37
United Kingdom				1	0.64
United States					1
			1985–2005		
France	1	0.82	0.68	0.77	0.67
Germany		1	0.47	0.70	0.73
Japan			1	0.56	0.39
United Kingdom				1	0.83
United States					1

Source: Adapted from CEIOPS. 2007f. Calibration of the market risk module. CEIOPS-FS-05/07. February 19. Available at www.ceiops.org and CEIOPS. 2007g. Calibration of the underwriting risk, market risk and MCR. CEIOPS-FS-14/07. April. Available at www.ceiops.org.

Note: Data from Ibbotson Associated.

in the MSCI world index. On the other hand, due to the globalization of financial markets, the dependency between the markets of the developed countries has increased. This effect is clearly observed in Table I.11B. The table shows the dependencies between five major equity markets for three different time periods.

These dependencies are calculated on the basis of Dimson total return indices (Data source: Ibbotson Associates, now Morningstar), dated from 1900 and on a yearly basis. The Dimson indices lead to approximately equal results:

1. Geometric mean equals to 11.2%
2. The standard deviation corresponds to 16.5%
3. Based on normality the 99.5 confidence level corresponds to a shock of 31.2%
4. After fat tail correction, the 99.5 confidence level corresponds to a shock of approximately 35%, depending on the exact correction method chosen

CEIOPS therefore performed analysis on a data series covering the period 1900–2000.

Table I.12 shows the outcome of applying the analyses at different levels of granularity.

In line with expectations, the more diversified the index, the lower was the shock. In this context, investing purely in European indices will lead on average to a 3% increase, while global ex-Europe investments show slightly lower increases compared to the global index.

TABLE I.12 The Scenario Outcome using Different Granularity

Region/Country	Global	Europe	Global, ex-Europe
Time period	1970–2006	1970–2006	1970–2006
Frequency	Quarterly	Quarterly	Quarterly
VaR 99.5%, Normal	33%	38%	34%
VaR 99.5, Log-linear	34%	38%	34%
VaR 99.5%, Gumbel	35%	36%	35%

Source: Adapted from CEIOPS. 2007f Calibration of the market risk module. CEIOPS-FS-05/07. February 19. Available at www.ceiops.org and CEIOPS. 2007g Calibration of the underwriting risk, market risk and MCR. CEIOPS-FS-14/07. April. Available at www.ceiops.org.

Also the appropriateness of a global shock was considered further. Alternative proposals for the standard formula were suggested that different factors should be applied to take account of the volatility of returns experienced in different markets. Due to a lack of historical data, it seemed unlikely that this could reach the level of granularity of different shocks for exposures in different Member States. But it could distinguish between developed and emerging markets, or between different global regions. However, CEIOPS' first analysis suggests that increased granularity could actually increase the magnitude of the equity risk shock (Table I.13).

The result could be seen as intuitive, given that increasing the diversification of the index reduces the level of the shock.

CEIOPS was also about to consider the practical implementation of the scenario approach—for example, whether the equity benchmarks need to be specified more clearly, for example, MSCI Developed Markets Index, EuroStoxx, and so on. This also presented the question of how "nonmarket" portfolios would map across to the benchmarks under the stress conditions.

There was a concern that a large risk weight on equities may have the unwanted effect of European insurers reducing their equity holding significantly. This could be addressed in two ways:

- If the dependence parameter between interest rate risk and equity risk was reduced to a relatively low level, for example, 25%, then moderate equity holdings would lead to a larger than proportional increase of the market risk SCR.

TABLE I.13 Data Series used for the Equity Risk Module

Data Series	Region	Global	Europe	Germany
	Time Period	1970–2006	1970–2006	1970–2006
	Frequency	Quarterly	Quarterly	Monthly
Corrected VaR 99.5	GPD ML	33%	36%	N/A
	GPD R	38%	41%	49%

Source: Adapted from CEIOPS. 2007f. Calibration of theMarket RiskModule. CEIOPS-FS-05/07. February 19. Available at www.ceiops.org. and CEIOPS. 2007g. Calibration of the Underwriting Risk, Market Risk and MCR. CEIOPS-FS-14/07. April. Available at www.ceiops.org.

Note: Data from Datastream.

- Lower risk weights could be used if the proportion of all assets in equities was relatively small, while the dependence parameter between interest rate risk and equity risk could continue to be set cautiously, for example, 75%.

Both solutions would penalize concentrations in the equity asset class. In the second solution, the equity SCR would represent an estimate of the risk contribution to the portfolio, rather than a standalone estimate of the risk.

I.3.4 Equity Duration

Some of the Member States advocated a different solution to the treatment of equity risk. They noted that, in the long run, equities typically provide better returns than bonds and provide good cover against various types of inflation. It would therefore be appropriate to consider equity risk in conjunction with the liabilities that the assets are being used to match.

Volatility of equities is important in the short term but is not significant in the long term. Some theoretical and empirical studies suggest that a duration of 25 years could be attributed to equities. Accordingly, the treatment of equity risk might differentiate between cases where the liability side is instable or of short duration and cases where the liability side is stable and of long duration, such as life annuities. Even on the nonlife side, there are examples where the duration of liabilities is, effectively, long and stable, for example, through tacit renewal of insurance contracts.

In QIS2, there were examples, particularly for nonlife insurers, where the capital resulting from the equity risk treatment resulted in a disproportionately high contribution to the final SCR. But the solution did not seem to be a universal reduction in the 40% shock applied under QIS2. This would not be prudent for insurers with a high proportion of the BS invested in equities. Under some implementations of Solvency I, Member States apply a 0% loading on equities that represent less than two-thirds of policyholder liabilities, and 100% for any equities above this amount. This could, of course, be replaced with a more finely graduated approach.

An alternative approach, where the magnitude of the equity shock depends on the expected holding period of the equity position and the overall concentration of its investments in equities, was tested. The shocks might be set as follows:

$$CR_{MR,ER,D} = EQ_{\text{load}} \cdot \rho VaR_{EQ},$$

where ρVaR_{EQ} is a proxy of VaR 99.5% estimated as 70% of the weighted average one-year volatility of the insurer's equity portfolio and EQ_{load} is defined in Table I.14.

Objective criteria would be needed to determine the expected holding period. However, many CEIOPS Members are strongly opposed to such a concept that they consider inconsistent with the design of the SCR and in particular, the one-year time horizon for assessing risk.

The 70% factor applied to the weighted average one-year volatility of the insurer's equity portfolio would produce results consistent with the 99.5% VaR observed for the EuroStoxx

TABLE I.14 The Equity Loading Factors for the Model Proposed by Some Member States

EQ_{load}		Equities as a Proportion of Policyholder Liabilities		
		15–25%	**25–40%**	**40%+**
Expected holding period	Less than 2 years	1.00	1.00	1.00
	2–5 years	0.70	0.80	0.90
	More than 5 years	0.60	0.70	0.80

Source: Adapted from CEIOPS. 2006b. Draft Advice to the European Commission in the Framework of the Solvency II project on Pillar I issues—further advice. Consultation Paper 20, CEIOPS-CP-09/06, November 10. Available at www.ceiops.org.

Note: It is assumed to exclude equities and liabilities where the policyholder bears the investment risk.

50 equity index. However, the assumption, and the equity loadings given in the table above, would needed further testing and refinement as part of the following QIS3.

Basing the size of the equity shock on the volatility of the equity portfolio provides incentives for insurers to manage their equity risk effectively, rather than applying a "one-size-fits-all" shock. Although the practicalities of such an approach would require careful consideration, it seemed that information on the volatility of traded equities could be readily available—and where unknown, a default (prescribed) value could be used.

I.3.4.1 QIS3, CEIOPS (2007a)

The French duration approach was tested on an optional basis in QIS3. For each index in QIS3, the capital charge is calculated as

$$CR_{MR,ER,i} = \max\left\{\Delta NAV \mid \text{equity shock}_i; 0\right\},$$

where the equity shock is the immediate effect on the net value of assets and liabilities (ΔNAV) in the event of a fall in equities, taking account of the participant's individual direct or indirect exposure to equity prices.

The duration of the equity portfolio of backing insurance liabilities would be set equal to the average duration of the total portfolio of insurance liabilities. The duration of the equity portfolio backing insurance liabilities other than insurance liabilities would be considered to be less than 2 years. The fall of equities used in this optional approach was set according to Table I.15.

TABLE I.15 Equity Fall Used in the Optional Duration Approach in QIS3

Duration of the Portfolio	Equity Fall
0–2 years	36%
2–5 years	33%
5–10 years	23%
>10 years	13%

Source: Adapted from CEIOPS. 2007a. QIS 3 Technical Specifications, Part I: Instructions, CEIOPS-FS-11/07 & Part II: Background Information, CEIOPS-FS-12/07. April. Available at www.ceiops.org.

I.3.5 Dampener as Alternative to the Equity Risk

A "dampener" formula was tested as an alternative for QIS4 for the "global" market equity risk component. In QIS4 (2008), a background document, provided by French authorities (Annex SCR 8—TS.XVII.J) was published as an alternative approach to assess the capital charge for equity risk, incorporating an equity dampener.

The theoretical basis of a "dampener" is that the probability that the value of the equity indices increase is small when this value is high, and high when this value is low.

The value of an equity index is split into a trend component, and a cyclical component. The cyclical component is the difference between the mean of the value of the equity index in the last 10 trading days before the day when the SCR is calculated and the mean of the value of the equity index in the last year (around 250 trading days) before the day when the SCR is calculated.

The dampener effect only applies to those liabilities with a duration of more than 3 years.

In this option, the calculation for $CR_{MR,ER,1} = \max\left\{\Delta NAV|\text{equity shock}_1; 0\right\}$, the capital charge for equity risk with respect to index 1(Global)) is replaced by the following calculation:

$$CR_{MR,ER,1} = MV_{EQ,1} \cdot [\alpha \cdot [F(k) + G(k) \cdot c_t] + (1 - \alpha) \cdot 0.32],$$

where $MV_{EQ,1}$: the market value of the equity portfolio 1, that is, the Global; α: the share of the TPs accounting for more than 3-year commitments. $F(k)$ and $G(k)$ are coefficients defined in Table I.16.

The factor c_t is a cyclical component defined as

$$c_t = \bar{Y}_t^{10} - \bar{Y}_t^{261},$$

where $\bar{Y}_t^{10}$ is the mean of the 10 trading days before the day when the SCR is calculated; $\bar{Y}_t^{261}$ is the mean of the last year, approximately 261 trading days, before the day when the SCR is calculated, and

$$\bar{Y}_t^N = \frac{\sum_{i=0}^{N-1} \log Y_{t-i}}{N}$$

is the mean of the last N trading days of the equity index before the day where the SCR is calculated, and $logY_t$ is the value of the natural logarithm of the equity index Y_t at time t.

TABLE I.16 Coefficients $F(k)$ and $G(k)$ for the Dampener Formula

Duration of the Liabilities (k)	**$F(k)$**	**$G(k)$**
3–5 years	29%	0.20
5–10 years	26%	0.11
10–15 years	23%	0.08
>15 years	22%	0.07

Source: Adapted from QIS4. 2008. QIS 4 Technical Specifications. European Commission, MARKT/2505/08, March 31. Available at http://www.ceiops.eu/content/view/118/124/.

The equity index to be considered for this QIS4 option is the MSCI Developed Markets index.

The value of c_t as of December 31, 2007 was: $c_t = -0,013$

A *simplification* to the Dampener approach was also introduced:

Undertakings could use as an approximation for the duration of their liabilities an average, weighted by the share of the TPs held by LoB that they write, of market average durations per LoB. This was specified in national guidance.

The duration dampener resulted in a reduction of approximately 10% in equity risk capital charge in QIS4; CEIOPS (2008e). This approach was opposed by many supervisors and undertakings owing to the lack of theoretical and empirical justifications. The effects of using a dampener has been discussed by Lechkar and van Welie (2008) and they concluded that the duration dampener approach will not provide necessary confidence that assets will cover liabilities within the one-year time horizon.

I.4 PROPERTY RISK

Property risk arises from the level or volatility of market prices of property.

I.4.1 Background

As an answer to the EC's CfA 10, MARKT (2004c), CEIOPS gave a background to the modeling of the standard formula and its risk modules; see CEIOPS (2005).

I.4.1.1 Factor-Based Approach

Changes in the value of real estate may be modeled using a factor-based approach, where the RFs are calibrated according to a lognormal distribution. Its parameters (yield and volatility) can be derived from suitable market indices. Risk capital is deduced in the same way as for equity.

I.4.1.2 Scenario-Based Approach

Alternatively, a scenario-based approach could be used to model property risk. For the total real estate position, and taking account of the investment policy, the institution has to determine the effect on the surplus of a fall of, for example, 20% in the real estate benchmark used. The position in real estate is the value of all long and short positions in real estate and all financial instruments, such as real estate derivatives, whose value is influenced wholly or partly by the value of real estate.

Under any approach, the standard formula may not distinguish between direct and indirect real estate in the real estate portfolio or the real estate investment subcategories for reasons of simplicity.

I.4.2 QIS2–QIS4

I.4.2.1 QIS2, CEIOPS (2006d)

For reasons of simplicity, QIS2 did not make any distinction between direct and indirect real estate or between different types of real estate investment (offices, retail, residential, etc.).

For the purpose of QIS2, no differentiation was made between property investments which may have equity-type characteristics, for example, freehold ownership of a property, and those with more bond-like characteristics, for example, property rented for a fixed period at agreed rents.

Two different approaches were tested in QIS2. The first was a factor-based approach:

$$CR_{MR,PR} = 0.2 \cdot MV_{PR},$$

where MV_{PR} is the market value of the overall property position not allocated to policies where the policyholders bear the investment risk.

The second approach was a scenario approach:

$$CR_{MR,PR} = \{\Delta NAV \mid \text{property shock}\},$$

where the *property shock* is the immediate effect expected in the event of a 20% fall in real estate benchmarks, taking account of all the participant's individual direct and indirect exposures to property prices. The property shock takes account of the specific investment policy including, for example, hedging arrangements, gearing and so on.

I.4.2.2 QIS3, CEIOPS (2007a)

The scenario approach from QIS2 was used in QIS3, that is, the capital charge is calculated as

$$CR_{MR,PR} = \{\Delta NAV \mid \text{property shock}\},$$

where the *property shock* is the immediate effect expected in the event of a 20% fall in real estate benchmarks, taking account of all the participant's individual direct and indirect exposures to property prices. The property shock takes account of the specific investment policy including, for example, hedging arrangements, gearing, etc.

A duration approach was also tested. For more details, see Section I.4.2.

I.4.2.3 QIS4, QIS4 (2008)

No changes from QIS3 were made.

The scenario for property risk should be calculated under the condition that the assumptions on future bonus rates, reflected in the valuation of future discretionary benefits in TPs, remain unchanged before and after the shock being tested.

Additionally, the result of the scenario was determined under the condition that the participant was able to vary its assumptions in future bonus rates in response to the shock being tested. The resulting capital charge is $nCR_{MR,PR}$.

I.4.3 Calibration

I.4.3.1 QIS2, CEIOPS (2006b) QIS3, CEIOPS (2007f, 2007g)

For reasons of simplicity, it was assumed in QIS2 that property returns are normally distributed. Choosing a more sophisticated model might have given a better fit, but insufficient data exist to model the negative tail of the distribution very precisely.

The aim was to use a simple and transparent model to produce reasonable estimators for the lower percentiles. The stress factor was calibrated using the following data indices.

Source for data was the Investment Property Databank, IPD, except for the Dutch data that were taken from ROZ-IPD. The French, German, and Swedish IPD data lack long-dated information on these property markets. Consequently, the corresponding analyses do not include a full property cycle.

- The Netherlands (NL), 1977–2005
- France (FR), 1998–2005
- Germany /GER), 1996–2005
- Sweden (SWE), 1997–2005
- The United Kingdom (UK), 1971–2005

The indices were based on annualized total returns (capital growth + income) of direct investments in real estate. The total returns were based on valuation data, such as surveyors' estimates, rather than actual market prices, and therefore reflected a degree of smoothing over time. Since transaction prices are important, an unsmoothed return also needs to be considered. This can be derived mechanistically from the unsmoothed data.

In the IPD Pan-European property index, the individual country returns are grossed up according to the estimated value of the investment market in each country. Table I.17 shows market size earlier at the end 2005.

Since transaction prices are important, an "unsmoothed" property returns can be derived from the observable smoothed data; see Fisher and Geltner (2000).

The following simple desmoothing mechanism is applied:

$$\hat{R}_t = \omega \cdot R_t + (1 - \omega) \cdot \hat{R}_{t-1},$$

where $\hat{R}_t$: the smoothed property return at time t; R_t: the unsmoothed property return at time t; ω: the weight, given in Table I.16.

TABLE I.17 Invested Market Size in Million Euros, at the End of 2005

	Market Size	**Weight**	**ω**
France	138,714	14%	0.4388
Germany	259,375	26%	0.1046
Netherlands	68,032	7%	0.4696
Sweden	58,212	6%	0.5523
United Kingdom	462,469	47%	0.6522
Sum:	986, 802	100%	

Source: Adapted from CEIOPS. 2007f. Calibration of the Market Risk Module. CEIOPS-FS-05/07. February 19. Available at www.ceiops.org and CEIOPS. 2007g. Calibration of the Underwriting Risk, Market Risk and MCR. CEIOPS-FS-14/07. April. Available at www.ceiops.org.

Note: Data from IPD. The weights ω are discussed below.

TABLE I.18 The Mean, Standard Deviation and 99.5% VaR for Property Risk

Country	Mean	Standard Deviation		99.5% shock
		Smoothed	Unsmoothed	
France	10.5%	3.4%	7.6%	8.92%
Germany	3.6%	1.7%	9.3%	20.36%
Netherlands	9.4%	5.1%	8.4%	12.20%
Sweden	9.9%	7.2%	11.4%	19.40%
United Kingdom	12.4%	10.3%	16.0%	28.87%

Source: Adapted from CEIOPS. 2006b. Draft Advice to the European Commission in the Framework of the Solvency II Project on Pillar I Issues—Further Advice. Consultation Paper 20, CEIOPS-CP-09/06. November 10. Available at www.ceiops.org.

In other words, it was assumed that the IPD data is smoothed as a weighted average of last year's smoothed and this year's actual return. The weight, ω, can be found using the existing autocovariance in the observed data. ω is estimated from the slope coefficient of the regression of the smoothed returns on their values lagged one year, with the estimation undertaken on a rolling basis. Compared to the other countries, the German returns showed a relatively high level of autocovariance, leading to a relatively low value for ω.

Based on the specific unsmoothing mechanism used, the standard deviations of the unsmoothed property returns were determined, see Table I.18.

Since the historical series varied considerably in length, covariances were estimated using the shortest common subset of returns, thereby discarding some of the information in the longer time series. Instead of using the market-weighted basket of the five countries, the 99.5% was conservatively rounded to 20%.

I.4.3.2 QIS4, QIS4 (2008)

No changes from QIS3 were made.

I.4.4 Property Duration

As for the equity risk, some Member States suggested the use of a duration approach also for the property risk.

A minority of CEIOPS Members advocate this approach where the magnitude of the property risk shock depends on the average duration of the insurer's liabilities and the overall concentration of its investments in property. The shocks might be set as given in Table I.19.

The shocks reflect a proposal to CEIOPS by the Fédération Française des Sociétés des Assurances, FFSA. However, the values of *Property fall* would require further refinement and testing as part of the coming QIS3. Many CEIOPS members were strongly opposed to this type of model which they considered inconsistent with the design of the SCR (in particular, the one-year time horizon for assessing risk).

I.4.4.1 QIS3, CEIOPS (2007a)

The same approach as for the equity risk is used for the property risk on an optional basis, but with different values of the property fall tested. They are given in Table I.20.

TABLE I.19 Property Fall in the Context of a "Duration Approach"

Average Duration of Liabilities (Years)	Property as a Proportion of Policyholder Liabilities			
	<15%	15–25%	25–40%	40%+
1	20%	20%	20%	50%
2	14.5%	14.5%	20%	50%
3	11%	11%	20%	50%
4	8%	10%	20%	50%
5	7%	10%	20%	50%
6	6%	10%	20%	50%
7	5.5%	10%	20%	50%
8	5%	10%	20%	50%
9	4.5%	10%	20%	50%
10+	4%	10%	20%	50%

Source: Adapted from CEIOPS. 2006b. Draft Advice to the European Commission in the Framework of the Solvency II project on Pillar I issues—further advice. Consultation Paper 20, CEIOPS-CP-09/06, November 10. Available at www.ceiops.org.

Note: The proposal excluded equities and liabilities where the policyholder bears the investment risk.

TABLE I.20 Property Fall used in the Optional Duration Approach in QIS3

Duration of the Portfolio	Property Fall
0–2 years	18%
2–5 years	17%
5–10 years	12%
>10 years	7%

Source: CEIOPS. 2007a. QIS 3 Technical Specifications, Part I: Instructions, CEIOPS-FS-11/07 & Part II: Background Information, CEIOPS-FS-12/07. April. Available at www.ceiops.org.

I.5 CURRENCY RISK

Currency risk arises from the level or volatility of currency exchange rates.

I.5.1 Background

As an answer to the EC's CfA 10, MARKT (2004c), CEIOPS gave a background to the modeling of the standard formula and its risk modules; see CEIOPS (2005).

Currency risk relates to bonds, real estate, and liabilities and will be considered, provided a given threshold on the basis of current values is exceeded. Due to difficulties in denomination, further analysis is needed to determine whether currency risk of shares should be addressed. Currency risk could be addressed through a *scenario-based approach*. For the total foreign currency position, and taking account of the applicable investment policy, the institution has to determine the effect on the surplus of a fall in value of all other currencies against the euro of, for example, 25%.

Alternatively, in a *factor-based approach*, RFs may be derived for different currencies assuming normal distributions.

I.5.2 QIS2–QIS4

I.5.2.1 QIS2, CEIOPS (2006d)

For each currency other than the local currency, the currency position is the difference in the TPs for liabilities in that currency and the assets in that currency.

Two different approaches were tested in QIS2. The first was a factor-based approach:

$$CR_{MR,CR} = 0.25 \cdot MV_{FX},$$

where MV_{FX} is the market value of the overall property net foreign currency position.

The second approach was a scenario approach:

$$CR_{MR,CR} = \{\Delta NAV \mid \text{currency shock}\},$$

where the *currency shock* is the immediate effect expected in the event of a 25% change, more onerous of a rise or fall, in value of all other currencies against the local currency in which the undertaking prepares its local regulatory accounts, taking account of all the participant's individual currency positions and its investment policy, for example, hedging arrangements, gearing, and so on.

I.5.2.2 QIS3, CEIOPS (2007a)

In QIS3 the scenario approach was adopted:

$$CR_{MR,CR} = \{\Delta NAV \mid \textit{currency shock}\},$$

where the *currency shock* is the immediate effect expected in the event of a 20% change, more onerous of a rise or fall, in value of all other currencies against the local currency in which the undertaking prepares its local regulatory accounts, taking account of all the participant's individual currency positions and its investment policy, for example, hedging arrangements, gearing, and so on.

I.5.2.3 QIS4, QIS4 (2008)

In QIS4, the capital charge for currency risk was determined from the result of two predefined scenarios:

$$CR^{Up}_{MR,CR} = \{\Delta NAV \mid \text{upward currency shock}\},$$
$$CR^{Down}_{MR,CR} = \{\Delta NAV \mid \text{downward currency shock}\},$$

where the upward and downward shocks are respectively the immediate effect expected on the net value of asset and liabilities in the event of a 20% change, rise and fall respectively in value of all other currencies against the local currency in which the undertaking prepares its local regulatory accounts, taking account of all the participant's individual currency positions and its investment policy (e.g., hedging arrangements, gearing, etc.).

The scenario for currency risk should be calculated under the condition that the assumptions on future bonus rates (reflected in the valuation of future discretionary benefits in technical provisions) remain unchanged before and after the shock being tested.

The size of the shock applied in the calculation for an Exchange Rate Mechanism II, ERM II*, Member State currency versus the Euro should reflect the maximum fluctuations set under ERM II. For QIS4 purpose, the following shock had to be considered:

- 2,25% for the Danish krona (DKK)
- 15% for the Estonian kroon (EEK), the Latvian lats (LVL), the Lithuanian litas (LTL), and the Slovak koruna (SKK)

Additionally, the result of the scenarios should be determined under the condition that the participant was able to vary its assumptions in future bonus rates in response to the shock being tested. The resulting capital charges are nMktfxUp and nMktfxDown.

The capital charge for currency risk is derived from the type of shock that gives rise to the highest capital charge including the risk-absorbing effect of future profit sharing:

If

$$nCR_{MR,CR}^{\text{up}} > nCR_{MR,CR}^{\text{down}},$$

then

$$nCR_{MR,CR} = nCR_{MR,CR}^{\text{up}}$$

and

$$CR_{MR,CR} = CR_{MR,CR}^{\text{up}}$$

If

$$nCR_{MR,CR}^{\text{up}} \leq > nCR_{MR,CR}^{\text{down}},$$

then

$$nCR_{MR,CR} = nCR_{MR,CR}^{\text{down}}$$

and

$$CR_{MR,CR} = CR_{MR,CR}^{\text{down}}.$$

I.5.3 Calibration

I.5.3.1 QIS2, CEIOPS (2006b) QIS3, CEIOPS (2007f, 2007g)

The calibration of the stress factor in QIS2 was based on work performed for the domestic FTK in the Netherlands. This considered the exchange rates of seven currencies against the

* Exchange Rate Mechanism II—see http://www.ecb.int/ecb/legal/107663/1350/html/index.en.html

Euro. A currency basket was constructed using an estimation of currency positions held by Dutch financial institutions, with weighting as follows:

- 35% United States dollar (USD)
- 24% British sterling (GBP)
- 13% Argentine peso (ARP)
- 8% Japanese yen (JPY)
- 7% Swedish krona (SEK)
- 7% Swiss Franc (CHF)
- 6% Australian dollar (AUD)

The comparatively high weighting for the Argentine peso reflected its use as a proxy for all currency exposure to emerging markets. Note that the analysis includes the period 1992–2001 in which the peso was fixed to the US dollar. Some undertakings where the base currency is linked to the Euro excluded the currency risk on exposures in euro in order to reflect the rather modest risk.

The source data were monthly exchange rates for the period 1958–2006, Data source: Datastream, using a synthetic euro for the period before the currency's adoption. Given the monthly frequency of the data, the holding period in the calculation of the risk measure needed to be scaled up to a one-year risk evaluation. This adjustment assumes that the monthly distributions are statistically independent.

The annualized standard deviations of the seven exchange rates versus the Euro were calculated. For reasons of simplicity, it was assumed that the relative changes in exchange rates are normally distributed. Figure I.3 shows the lower percentiles of the observed distribution of the monthly relative changes in the dollar–euro exchange rate. Looking at the negative tail of the distribution, the normality assumption seems to be acceptable. Of course, choosing a more sophisticated model might give a better fit. However, the purpose here is to use a simple and transparent model to produce reasonable estimators for the lower percentiles.

The 99.5% shocks were estimated as in Table I.21a.

For this specific currency basket, the 99.5% confidence level corresponds to a shock of approximately 17%. However, if the period of the Bretton Woods agreement was excluded, the 99.5% shock corresponds to 20%. The Bretton Woods agreement from 1944 consisted of an obligation for each country, which had assigned it, to adopt a monetary policy that maintained the exchange rate of its currency within a fixed value, plus or minus 1%, in terms of gold and the ability of the IMF to bridge temporary imbalances of payments. The system collapsed in 1971, following the United States' suspension of convertibility from dollars to gold. This created the unique situation, whereby the United States dollar became the reserve currency for the states which had signed the agreement.

The same analysis was performed using British Sterling as the base currency. For simplicity, the Euro was assumed to represent 24% of the currency basket (i.e., directly replacing

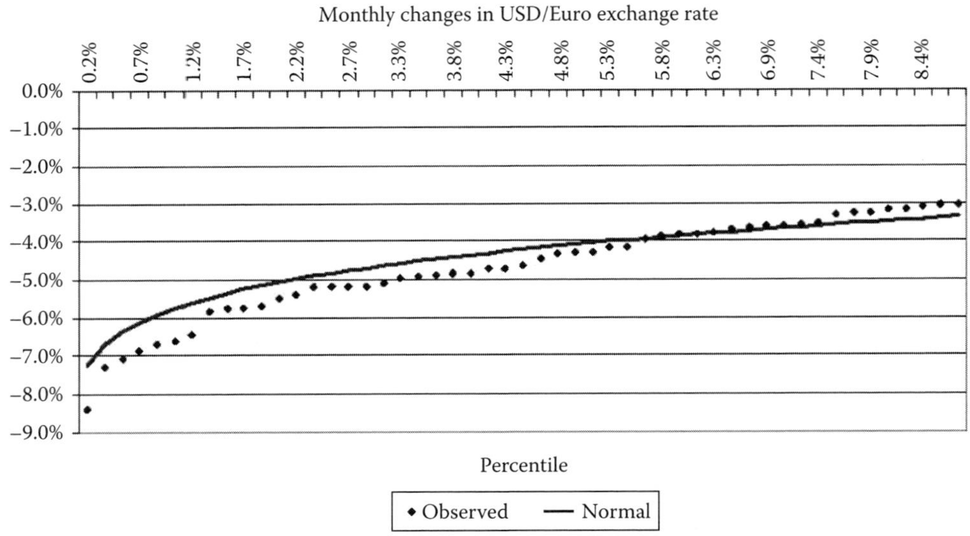

FIGURE I.3 The negative tail of the distribution of monthly changes in USD/Euro exchange rate. (Adapted from CEIOPS. 2007f. Calibration of the market risk module. CEIOPS-FS-05/07. February 19. Available at www.ceiops.org and CEIOPS. 2007g. Calibration of the underwriting risk, market risk and MCR. CEIOPS-FS-14/07. April. Available at www.ceiops.org; © Committee of European Insurance and Occupational Pensions Supervisors.)

TABLE I.21a The Standard Deviation and the 99.5% VaR for the Euro versus the Seven Different Currencies

Euro versus	USD	GBP	ARP	JPY	SEK	CHF	AUD	Basket
σ	9%	7%	37%	9%	6%	6%	11%	7%
99.5% shock	22%	18%	95%	23%	15%	14%	28%	17%

Source: Adapted from CEIOPS. 2006b. Draft Advice to the European Commission in the Framework of the Solvency II Project on Pillar I Issues—Further Advice. Consultation Paper 20, CEIOPS-CP-09/06. November 10. Available at www.ceiops.org.

Note: Data from Datastream 1958–2006.

sterling), with all other weights held the same. The 99.5% shocks were estimated as in Table I.21b.

Assuming the same currency basket, the standard deviation for British sterling is higher and the corresponding 99.5% shock is approximately 21%.

TABLE I.21b The Standard Deviation and the 99.5% VaR for the GBP versus the Seven Different Currencies

GBP versus	USD	EUR	ARP	JPY	SEK	CHF	AUD	Basket
Σ	9%	7%	37%	10%	8%	9%	14%	8%
99.5% shock	23%	18%	96%	26%	21%	23%	37%	21%

Source: Adapted from CEIOPS. 2006b. Draft Advice to the European Commission in the Framework of the Solvency II Project on Pillar I Issues—Further Advice. Consultation Paper 20, CEIOPS-CP-09/06. November 10. Available at www.ceiops.org.

Note: Data from Datastream 1958–2006.

TABLE I.22a The Standard Deviation and the 99.5% VaR for the Basket versus Euro Depending on the Basket Chosen

		USD		GBP		ARP	
Basket Versus Euro	**No change**	**More**	**Less**	**More**	**Less**	**More**	**Less**
Weight	—	50%	0%	50%	0%	20%	0%
σ	7%	7%	8%	6%	8%	9%	5%
99.5% shock	17%	17%	20%	15%	21%	23%	12%

Source: Adapted from CEIOPS. 2006b. Draft Advice to the European Commission in the Framework of the Solvency II project on Pillar I issues—further advice. Consultation Paper 20, CEIOPS-CP-09/06, November 10. Available at www.ceiops.org.

Note: Data from Datastream 1958–2006.

TABLE I.22b The Standard Deviation and the 99.5% VaR for the Basket versus GBP depending on the basket chosen

		USD		EUR		ARP	
Basket Versus GBP	**No change**	**More**	**Less**	**More**	**Less**	**More**	**Less**
Weight	—	50%	0%	50%	0%	20%	0%
σ	8%	8%	10%	7%	9%	10%	7%
99.5% shock	21%	20%	25%	19%	24%	25%	17%

Source: Adapted from CEIOPS. 2006b. Draft Advice to the European Commission in the Framework of the Solvency II project on Pillar I issues—further advice. Consultation Paper 20, CEIOPS-CP-09/06, November 10. Available at www.ceiops.org.

Note: Data from Datastream 1958–2006.

CEIOPS recognized that the implied level of shock depended on the precise currency basket chosen; therefore, sensitivity analysis was performed by varying the weights of the three largest currencies (all other weights moved in the same proportion as the original basket) (Table I.22a).

The key observation was the impact of inclusion or exclusion of the Argentine peso from the currency basket on the level of the overall shock. This is an intuitive result, given ARP has the highest standard deviation and because of its specific dependency structure with the other currencies.

The same analysis was performed using British sterling as the base currency, producing shocks as in Table I.22b.

Since all the modeled currency baskets produced shocks in the range 12–25%, the QIS2 shock was cautiously set at 25%.

I.5.3.2 QIS4, QIS4 (2008)

Regarding the calibration, no changes from QIS3 were made.

APPENDIX J

European Solvency II Standard Formula

Credit Risk

HERE WE DISCUSS the development and calibration of the counterparty default risk module and its capital charge that has been one of the building blocks in Solvency II. The final advice from CEIOPS is discussed in Chapter 28.

In the standard formula for the capital requirement, SCR, the credit risk is split up into the counterparty default risk (Section J.1), the (credit) spread risk (Section J.2), and the concentration risk (Section J.3). The latter two risks are in the SCR standard model and are considered as parts of the market risk

When we refer to QIS3 and QIS4 in the following, we mean the technical specifications QIS3 (2007) and QIS4 (2008), respectively.

As an answer to the EC's CfA 10, MARKT (2004c), CEIOPS gave a background to the modeling of the standard formula and its risk modules; see CEIOPS (2005).

In the light of the financial crisis, mainly during 2007–2009, CEIOPS (2009a) decided to review the calibration of the credit risk. As a step toward a better regulation, the EC has forwarded a directive regarding the regulation of credit rating agencies (CRAs). As losses above 50% is due to financial derivatives, the formula used to calculate LGD has to be reassessed. Concentration risk, discussed in Section J.3, was also reviewed (including contagion lines within the financial sector).

Custodian risks, defined as the risk of loss of securities held in custody occasioned by the insolvency, negligence, or fraudulent action of the custodian or subcustodian, was also reviewed in the work for the final advice.

Background Thoughts

A combination of rating agency analysis, to establish the rating buckets, and market information reflected in the credit spread could be used to model credit risk. A separate credit spread multiplier could be applied for each rating bucket. An example of such an approach has been chosen by the British FSA:

$$C_i = \sqrt{CS_i} \cdot Dur_i \cdot f_j,$$

where CS_i: the credit spread for corporate bond i; Dur_i: the duration of corporate bond i; and f_j: multiplier for rating bucket j, to which corporate bond i has been assigned.

As an alternative, one could take the following approach:

The RF is derived from a variable that represents rating quality, like a PD estimate. The PD estimate could be backed out of credit spread data or rating information; whichever is deemed more reliable in the specific segment. Next, a credit portfolio risk model or the beta-distribution model of the GDV could be used to convert the PD into a capital requirement. This capital requirement could be corrected for tenor by applying a maturity correction. In the end, the results of these steps can be combined in a single table or formula.

If no credit rating exists, alternatively, an approach could be developed using only credit spreads to reflect the market's perception of credit quality. Higher credit spreads will be more volatile and therefore should result in a higher capital requirement. The maturity of an exposure will also determine the effect of credit-spread changes on the value of surplus.

The simplest form uses a fixed factor on the credit spread of an instrument. Using the first-order Taylor approximation of the relative change in bond value could approximate capital requirements.

$$C_i \approx CS_i \cdot Dur_i \cdot f,$$

where C_i: the capital charge for corporate bond i; CS_i: the credit spread for corporate bond i; Dur_i: the duration of corporate bond i; and f: the fixed factor prescribed by the supervisor.

CEIOPS' example: With a fixed factor of 0.6 a corporate bond with duration of 5 years and a credit spread of 80 basis points would lead to a capital requirement of 2.4% of the bond's market value.

J.1 QIS2 PROPOSAL

In QIS2, CEIOPS (2006d) the credit risk was defined as the risk of default and change in the credit quality of the issuers of securities, counterparties (including reinsurers and other recoveries), and intermediaries to whom an undertaking has an exposure. Exposures to counterparties should take account of the availability of risk mitigants, such as collateral.

A rating-based approach was proposed. It was calculated as

$$CR_{CR} = \sum_i g_i \cdot RD_i \cdot EAD_i,$$

where g_i: a risk weight depending on the rating, see Table J.1; RD_i: the effective duration of credit risk exposure i, with a minimum value of 1 year and a maximum value of 5 years; and EAD_i: the nominal size of the credit risk exposure I as determined by reference to market values, that is, the exposure at default.

In cases where there was no readily available MV of a credit risk exposure i, alternative approaches could be adopted to determine EAD_i, for example, in the case of insurance-related recoveries, the BE of the credit risk exposure, but these must still be consistent with any relevant market information.

In the case of a reinsurance exposure, the duration of the exposure would be an estimate of the modified duration of the projected payments to the cedant under the terms of the reinsurance contract. In principle, this would be the modified duration of relevant CFs in stressed conditions assumed to underlie the SCR. For the purposes of QIS2, it was acceptable to have in regard only the projected CFs included in the determination of the change in TPs due to reinsurance. For yearly renewed reinsurance this meant that the duration could be assumed to be 1 year.

Exposures to reinsurance counterparties should take account of the availability of risk mitigants, such as collateral. Participants should consider the net exposure to the reinsurance. These should be treated as follows:

- Where the reinsurer is rated, the risk weight function should be applied
- Where the reinsurer is unrated but would be subject to the requirements of the Reinsurance Directive, including supervision by an EEA competent authority, the exposure should be assigned to bucket IV (BBB)
- In other cases, the exposure should be assigned to bucket VI

Where collateral and other risk mitigants are recognized as credit risk mitigants, other risks arising from those mitigants should be taken into account in determining the SCR. For example, asset risks associated with collateral provided should be assessed in the same way as other assets.

A second formula was also tested in QIS2. It was based on ratings and credit spread information:

$$CR_{CR} = \sum_{rating:\, i} \sqrt{CSE_i} \cdot h_i \cdot D_i \cdot EAD_i$$

where for each credit risk exposure i; CSE_i: the credit spread equivalent; h_i: a risk weight depending on the rating, see Table J.1; and D_i: the effective duration.

TABLE J.1 Risk Weights g and h for Different Ratings

Rating$_i$	**CEIOPS Rating Buckets**	g	h
AAA	I—Extremely strong	0.008%	3%
AA	II—Very strong	0.056%	5.25%
A	III—Strong	0.66%	6.75%
BBB	IV—Adequate	1.312%	9.25%
BB	V—Speculative	2.032%	15%
B	VI—Very speculative	4.446%	24%
CCC or lower	VII—Extremely speculative	6.95%	24%
Unrated (except reinsurance)	VIII—unrated	1.6%	12.5%

Source: Adapted from CEIOPS. 2006d. Quantitative Impact Study No. 2: Technical Specification, CEIOPS-PI-08/06. Available at www.ceiops.org.

For traded exposures, the term CSE_i could be provided by estimating the credit spread directly. For nontraded exposures, an equivalent to the credit spread may be

- Inferred from the ratings or
- Inferred using the product of conservative estimates of the PD and the LGD

J.2 COUNTERPARTY DEFAULT RISK

The capital charge that we seek is denoted C_{CR}. The proposed method is based on the Basel II IRB model adjusted in a proposal by Sachs (2007). We first follows Sachs' proposal and then look closer at the calibrations made by CEIOPS. His starting point is not the traditional bank default risk, but the default risk of a reinsurer.

Following Sachs (2007), the starting point is the determination of the possible exposure, that is, the maximum loss to the insurer if the reinsurer defaults. The exposure depends on a second event. Is there an insurance event and will the insurer claim its cover from the reinsurer? This is called the "wrong way risk," which means the unfavorable dependency of exposure and default event. The insurance event has usually not a direct influence of the default of the insurer, but it may have some affect on the reinsurer default risk. Sachs looks at two distinct cases.

- *There is no insurance event for the insurer and the reinsurer defaults* (no wrong way risk) → the insurer would need to repurchase coverage (possibly at a higher cost than before)
- *There is an insurance event for the insurer and the reinsurer defaults* (wrong way risk) → the exposure would then correspond to the amount of TPs for the ceded business corresponding to the reinsurance receivables (from that particular reinsurer)

A reasonably conservative exposure measure could be "Exposure $=$ Volume$_{TP}-$ Collateral."

The PD is modeled as the one-factor model in Equation 18.3b in Section 18.1 and if the asset value falls below some threshold, the default point, assumed to be economically in default (does not mean that it is in legal default). We can now use the portfolio theory briefly discussed in Section 18.1, that is, for a portfolio of infinitely many loans, we get the Vasicek distribution (18.4). The distribution that depends only on two parameters: the PD and the correlation coefficient ρ.

In reality, we do not have infinitely many reinsurers available on the market. If we had, the model would have been a perfect description of the reality. What would be an appropriate correlation in that case? In the Basel II model, it is assumed as a well-diversified loan portfolio. The correlation is modeled as an inverse function of the PD having an upper correlation of 24% and a lower limit of 12%. In industry models as discussed in Chapter 18 such as Moody's KMV, CreditMetrics and CreditRisk+ allow for much higher correlations. In the KMV database, only 5% of the companies' exceed a correlation coefficient of 45%. KMV uses an upper limit of 65%, see Zeng and Zhang (2001, panel C) and CEIOPS (2007c).

Sachs (2007), laying down the fact that reinsurance companies tend to be quite large but also quite correlated, proposes to use $\rho_\infty = 0.5$ as a conservative lower limit in a theoretical portfolio of infinitely many reinsurers. In practice, we need to determine the correlation for a small number of n different reinsurers as well. Sachs (2007) suggests using

$\rho(n) = \rho_\infty + \frac{1}{n}(1 - \rho_\infty)$, with $\rho_\infty = 0.5$. For a single reinsurer the correlation would be one, if $n = 2$, then $\rho(2) = 0.75$. This approximation assumes that all n reinsurers in the portfolio take the same share of risk. As an enhancement, Sachs (2007) proposes to change the term $1/n$ with the Herfindahl–Hirschman index (HHI), defined in Equation 18.26a. Hence, we have the following approximation of the correlation in the Vasicek model:

$$\rho(n) = 0.5 + 0.5 \cdot HHI, \tag{J.1}$$

where the HHI is defined as $HHI = \sum_{i=1}^{n} p_i^2$, where p_i is the relative size of the firm. If all firms are equal in size, then $p_i = 1/n$, and we obtain the first suggestion mentioned above.

The LGD is an important part in the default risk assessment. It is the fraction of exposure that is lost in the event of a default. The LGD could be much less than 100% of the exposure due to third-party guarantees, and so on. Sachs (2007) proposes that the definition of LGD should be based on the TVaR and VaR given by

$$LGD = \frac{TVaR_{1-\alpha}(X) - VaR_{1-\alpha}(X)}{1 - VaR_{1-\alpha}(X)},$$

where TVaR and VaR are as defined in Chapter 14. But, as there is no economic framework available, the proposal has to be based on empirical default and recovery rates ($=1-$LGD). Sachs uses data from Standard & Poor's and proposes that LGD should be set to 50%. Data from both Standard & Poor's and Fitch gives the similar results.

Sachs (2007) proposes that the cost for taking credit risk being the sum of expected and unexpected loss. It includes a factor for taking the risk, the return on economic capital (RoEC). For solvency purposes, one is interested in the total amount of capital, that is, the sum of expected and unexpected events. This is given by setting $RoEC = 1$ in the cost formula $Cost = PD + RoEC \cdot [C(PD, \rho, 1 - \alpha) - E(X)]$ taken in some base units. C is a measure like VaR or TailVaR. Sachs shows that if the PD is larger than the confidence level, then the entire portfolio is at risk. For some rating classes, the diversification effect is material.

J.2.1 QIS3 and QIS4 Models

Reference to QIS3 is made in CEIOPS (2007a) and to QIS4 in QIS4 (2008).

The capital charge for the counterparty default risk is defined as

$$C_{CR} = \sum_j C_{CR,j}, \tag{J.2}$$

where the sum adds up the individual capital charges for reinsurance exposures, financial derivatives, receivables from intermediaries, as well as any other credit exposures to get the capital requirement for counterparty credit risk. Each capital charge is factor based, that is, a factor times a volume measure. The volume measures used are the LGD.

For each counterpart i, we need the following:

LGD_i: the loss-given-default of reinsurance, financial derivatives, intermediary, or any other credit exposures if counterparty i default. (In QIS3 the LGD was called replacement cost, RC). This is the volume measure.

PD_i: probability of default of counterpart i, which is used in the factor that is multiplied to the volume LGD.

The capital charge, for each counterparty, is defined as

$$C_{CR,i} = LGD_i \cdot \Phi\left[\frac{\Phi^{-1}(PD_i) - \sqrt{\rho} \cdot \Phi^{-1}(0.995)}{\sqrt{1-\rho}}\right], \tag{J.3}$$

that is, the product of the LGD (volume) and the Vasicek distribution for a given PD and a correlation less than 1 ("factor"). For the Vasicek distribution (model) see Chapter 18.

If the correlation is i, then the capital charge is equal to $LGD_i \cdot \min\{100 \cdot PD_i; 1\}$.

The correlation is defined by Equation J.1 above and the HHI is calculated with the relative sizes defined by $p_i = LGD_i / \sum_{j=1}^{n} LGD_j$.

The LGDss are defined separately for reinsurance (and SPVs and for financial derivatives.

J.2.1.1 LGD: Reinsurance

$$LGD_i = 50\% \cdot \max\left\{(\text{Recoverables} - \text{Collateral})_i + \left(C_{UW,i}^{\text{gross}} - C_{UW,i}^{\text{net}}\right); 0\right\}.$$

The current exposure is written as the (positive) sum of the "current exposure" plus the risk mitigating effect of reinsurance in the capital requirement. CEIOPS has also assumed that even if there is a default, the reinsurer will usually be able to meet a larger part of its obligations. CEIOPS used studies published by, for example, Fitch Ratings and Standard & Poor's (2006) to make the conservative choice of 50%; *cf.* also Sachs (2007).

Current Exposure

Recoverables: BE of recoverables from the reinsurance contract or SPV as defined in QIS4 Technical Specifications. This is the amount that the company is expected to get from the reinsurer or SPV.

Collateral: collateral covering the loss in relation to the counterparty. If the collateral is held by the counterparty itself, t should be set to nil. If the collateral bears any default risk, it should be included in the module calculation like receivables from intermediaries and other credit exposures. This could, for example, be premiums retained at the direct insurer from the counterparty as a security.

Risk mitigating effect of reinsurance in the capital requirement:

$C_{UW,i}^{\text{gross}} - C_{UW,i}^{\text{net}} = \Delta C_{UW,i}$: the difference between the gross and net, of reinsurance, capital charges for the underwriting risks calculated according to the standard formula for counterparty i. In both cases, we disregard any loss-absorbing capacity of future bonuses and deferred taxes.

J.2.1.2 LGD: Financial Derivatives

$$LGD_i = 50\% \cdot \max\left\{(\text{Market value} - \text{Collateral})_i + \left(C_{MR,i}^{\text{gross}} - C_{MR,i}^{\text{net}}\right); 0\right\},$$

where the 50% is taken from Sachs (2007), see above.

Current Exposure

Market value: value of the financial derivative (as defined in the FD, Article 75).

Collateral: collateral covering the loss in relation to the counterparty. If the collateral is held by the counterparty itself, t should be set to nil. If the collateral bears any default risk, it should be included in the module calculation like receivables from intermediaries and other credit exposures.

Risk Mitigating Effect of FDs in the Capital Requirement

$C_{MR,i}^{\text{gross}} - C_{MR,i}^{\text{net}}$: the difference between the gross and net, of reinsurance, capital charges for the market risks calculated according to the standard formula for counterparty i. In both cases, we disregard any loss-absorbing capacity of future bonuses and deferred taxes.

J.2.1.3 LGD: Intermediary Risks/Credit Exposures

In relation to the intermediary risk and any other credit exposures, the LGD is the BE of the credit to intermediaries and any other credit exposures, respectively.

J.2.1.4 Probability of Default

A PD estimate is derived from external ratings according to Table J.2. The rating scale used by Standard & Poor's is given for illustrative purposes. In cases where several ratings are available for a given credit exposure, the second-best rating should be applied. Unrated insurers and reinsurers not subject to Solvency II regulation would be treated as rating class 6 (CCC). Unrated insurers and reinsurers subject to Solvency II regulation would be treated as rating class 3 (BBB).

TABLE J.2 PD Estimates from External Ratings

Rating$_i$	**AAA**	**AA**	**A**	**BBB Unrated Subject to SII Regulation**	**BB**	**B**	**CCC-Unrated Not subject to SII regulation**
Credit quality step	1	1	2	3	4	5	6, -
Counterparty solvency ratio, SR:	>2	>1.6	>1.3	>1.0	>0.7	>0.5	⩽0.5
PD$_i$ in %	0.002	0.01	0.05	0.24	1.20	6.04	30.41

Source: Adapted from CEIOPS. 2007c. QIS3 Calibration of the credit risk. CEIOPS-FS-23/07, April. Available at www.ceiops.org.

The rating scale of Standard & Poor's is only for illustrative purposes.

The counterparty SR is discussed in the context of internal reinsurance.

In case of reinsurance ceded to an unrated reinsurer is part of the same group (internal reinsurance), the PD of counterparty i is replaced, for the share of the reinsurance that is retroceded outside the group to a counterparty k by the PD of counterparty k. In this case, the PD of counterparty i will still be used for the share of the reinsurance kept in retention by reinsurer i.

For intragroup reinsurance that does not meet the requirements specified in the previous paragraph, a regulatory rating should be used to determine the PD of the intragroup counterparty. The PD depends on the solvency ratio (ratio of OFs and SCR) according to Table J.2.

The calculation of the capital charge for counterparty default risk should be derived under the condition that the assumptions on future bonus rates (reflected in the valuation of future discretionary benefits in PD) remain unchanged before and after a presumed change in default counterparty to risk mitigating contracts such as reinsurance and financial derivatives.

Additionally, the result of the calculation should be determined under the condition that the participant is able to vary its assumptions in future bonus rates in response to the shock being tested. The resulting capital charge is nC_{CR}.

J.2.1.5 Simplifications

If it was seen as proportionate to the underlying risk, participants could determine the LGDi and the requirements Defi on the level of rating classes rather than on the level of counterparties.

In Question & Answer's to QIS4, CEIOPS proposed an approximation for the nonlife reinsurance counterparty risk; CEIOPS (2008d). The risk mitigating effect could be written as $\Delta C_{UW,i} = C^{\text{gross}}_{UW,i} - C^{\text{net}}_{UW,i} = V_i^G \cdot \rho\left(\sigma_i^G\right) - V_i^N \cdot \rho\left(\sigma_i^N\right)$, where we have used the fact that the capital charge could be written as a volume times a factor. From Equation M.1, $\rho(\sigma) \approx 3 \cdot \sigma$, we see that this could be rewritten approximately as $\Delta C_{UW,i} \approx V_i^G \cdot 3 \cdot \sigma_i^G - V_i^N \cdot 3 \cdot \sigma_i^N$. Extend this with $C^N_{UW} \approx V^N \cdot 3 \cdot \sigma^N$ and we get the following approximation:

$$\Delta C_{UW,i} \approx C^N_{UW} \frac{V_i^G \cdot 3 \cdot \sigma_i^G - V_i^N \cdot 3 \cdot \sigma_i^N}{V^N \cdot 3 \cdot \sigma^N} = C^N_{UW} \frac{V_i^G \cdot \sigma_i^G - V_i^N \cdot \sigma_i^N}{V^N \cdot \sigma^N}.$$

Assume that $\sigma_i^G = \sigma_i^N = \sigma^N$. From this we get the following approximation

$$\Delta C_{UW,i} \approx \frac{C^N_{UW}}{V^N} \left[V_i^G - V_i^N\right].$$

The volume measure could be written as the sum of the reserve risk volume (R) and the premium risk volume (P): $V_i = V_{Ri} + V_{Pi}$, and hence

$$\begin{aligned} \Delta C_{UW,i} &\approx \frac{C^N_{UW}}{V^N} \cdot \left[\left(V^G_{Ri} - V^N_{Ri}\right) + \left(V^G_{Pi} - V^N_{Pi}\right)\right] \\ &= \frac{C^N_{UW}}{V^N} \cdot \left[(\text{Recoverables}_i) + (\text{CededRe}_i)\right]. \end{aligned}$$

The recoverables is the difference between the gross and net (BEs) reserves and the Ceded reinsurance ("CededRe") is the difference between gross and net premiums. CEIOPS proposed the use of earned premiums in this case. If $\sigma_i^G > \sigma_i^N$, then we will have an underestimate of the risk mitigating effect. With a reversed sign we will have an underestimate.

J.2.2 Calibration

External data are used to calibrate the estimates of the PD given in Table J.1, and the base correlation of 0.5; *cf.* Sachs (2007) and above.

We start with the calibration of the PD, which is discussed in CEIOPS (2007c). The default rates are extracted from the following public reports:

- Annual Global Corporate Default Study: Corporate Defaults Poised to Rise in 2005 (Standard & Poor's, 2005). Data used for the study contain 11,605 companies. The rating histories of these companies were analyzed for the time frame 1980 until 2005. These companies include industrials, utilities, financial institutions, and insurance companies around the world with long-term local currency ratings.
- Default and Recovery Rates of Corporate Bond Issuers, 1920–2005 (Moody's, 2005). The study is based on Moody's database of ratings and defaults for corporate bond issuers. In the beginning of 2005 over 500 corporate issuers held a long-term bond rating. The whole database includes over 16,000 corporate issuers for the time between 1920 and 2005.

CEIOPS has used both data from Standard & Poor's and Moody's for smoothing data. We will only discuss the smoothing using the data from Standard & Poor's as the approach is similar for Moody's.

First, the empirical default rates for all 17 rating categories and the resulting PDs using exponential smoothing are shown. Column 3 includes linear regression after taking the logarithm, where we have assumed that $PD = e^{a_0+a_1RC}$, where RC is the rating category. Columns 4 and 5 show the results from nonlinear regression, where we have assumed that: $PD = 1/1 + e^{-a_0-a_1RC}$. In the last column, the PD estimates are obtained by a weighting $1/PD^2$. In the case of a default rate less than 0.03%, the weights used for smoothing are set to a minimum PD of 0.03%. This is in common with the treatment of default risk in banking sector for corporates. In order that the fitted PD estimates and the empirical default rates are in line for the linear interpolation, it is also assumed that there is an additional rating category (CCC) between B− and C.

The estimates of the PDs are shown in Figure J.1.

CEIOPS concluded that it did not seem appropriate to use all 17 rating categories. The same analysis was made for the default rates of consolidated rating categories. Data are given in Table J.3 and the estimated PDs are shown in Figure J.2.

The default rates published by Standard & Poor's are calculated on a database that includes also the issuers in the rating category "NR." These are issuers that had ratings withdrawn. A conservative approach for estimating PDs by external default rates is to eliminate these issuers from the data and recalculate the default rates. These recalculated

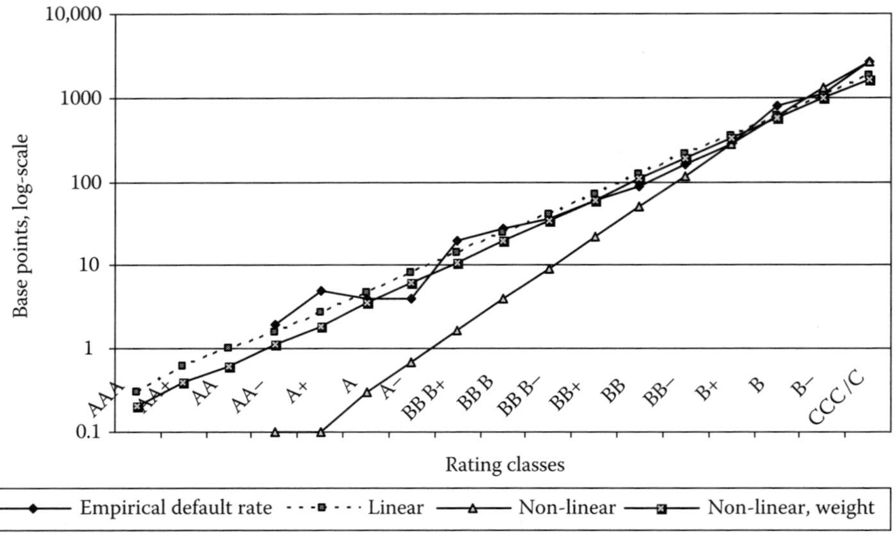

FIGURE J.1 Plot of external default rates and smoothed PDs for the 17 Standard & Poor's rating categories. 1 base point = 0.0001. Base points in log scale. (Adapted from CEIOPS. 2007c. QIS3 Calibration of the credit risk. CEIOPS-FS-23/07, April. Available at www.ceiops.org; © Committee of European Insurance and Occupational Pensions Supervisors.)

TABLE J.3 Standard & Poor's Rating 17 Categories, Corresponding Default Rates and Smoothed PDs in Base Points (1 bp = 0.0001)

Standard & Poor's Rating Category	Empirical Default Rate	Linear Regression	Nonlinear Regression	
			Unweighted	**Weighted**
AAA	0	0,3	0,0	0,2
AA+	0	0,6	0,0	0,4
AA	0	1,0	0,0	0,6
AA–	2	1,6	0,1	1,1
A+	5	2,8	0,1	1,9
A	4	4,8	0,3	3,5
A–	4	8,2	0,7	6,1
BBB+	20	14,1	1,7	10,9
BBB	28	24,1	4,0	19,4
BBB–	36	41,3	9,4	34,4
BB+	59	70,7	21,9	60,9
BB	87	121,2	51,0	107,8
BB–	162	207,6	118,5	189,9
B+	286	355,6	272,5	332,8
B	778	609,2	614,6	576,6
B–	1122	1043,6	1327,4	980,9
CCC/C	2702	1787,8	2634,8	1619,9

Source: Adapted from CEIOPS. 2007c. QIS3 Calibration of the credit risk. CEIOPS-FS-23/07, April. Available at www.ceiops.org.

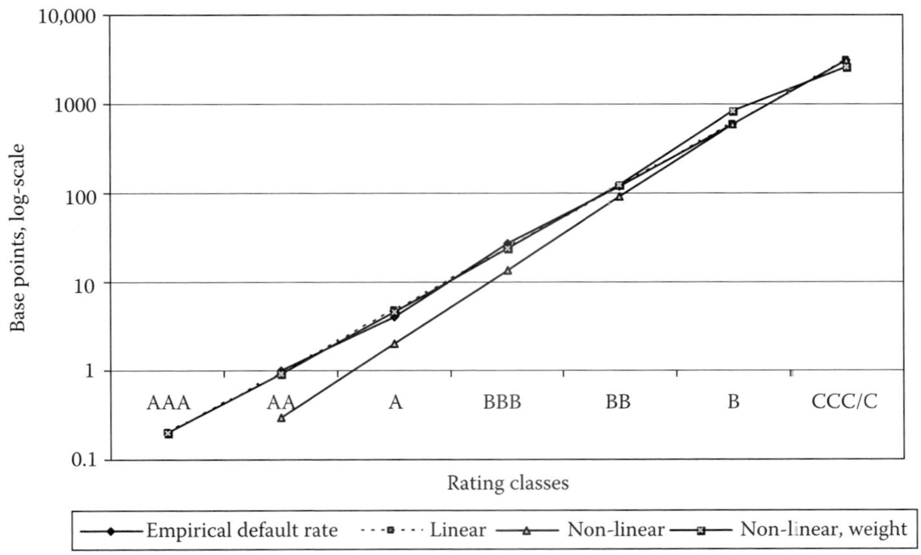

FIGURE J.2 Plot of external default rates and smoothed PDs for the 7 consolidated Standard & Poor's rating categories. 1 base point = 0.0001. Base points in log scale. (Adapted from CEIOPS. 2007c. QIS3 Calibration of the credit risk. CEIOPS-FS-23/07, April. Available at www.ceiops.org; © Committee of European Insurance and Occupational Pensions Supervisors.)

rates are, in general, greater than those of the conventional default rate calculation. It shows that the behavior of the default rates is nearly similar excluding them as if they had been included. Data in Table J.4 and Figure J.2 exclude these issuers with ratings withdrawn.

A similar study was carried out for default rates published by Moody's, and the results were similar. Therefore CEIOPS made the assumption that Standard & Poor's 7 rating categories are comparable to Moody's 7 rating categories (Table J.5). In Table J.4, the empirical default

TABLE J.4 Standard & Poor's 7 Consolidated Rating Categories, Corresponding Default Rates and Smoothed PDs in Base Points (1 bp = 0.0001)

Standard & Poor's Rating Category	Empirical Default Rate	Linear Regression	Nonlinear Regression	
			Unweighted	Weighted
AAA	0	0,2	0,0	0,2
AA	1	0,9	0,3	0,9
A	4	4,7	2,0	4,6
BBB	27	23,8	13,7	24,2
BB	120	119,9	92,7	126,3
B	591	603,9	600,9	833,5
CCC/C	3041	3040,7	3039,6	2633,5

Source: Adapted from CEIOPS. 2007c. QIS3 Calibration of the credit risk. CEIOPS-FS-23/07, April. Available at www.ceiops.org.

TABLE J.5 Moody's and Standard & Poor's 7 Rating Categories, Corresponding Default Rates, Aggregated Rates, and Smoothed PDs in Base Points (1 bp = 0.0001)

Moody's		Standard & Poor's			Estimates		
				M + S%P		Nonlinear Regression	
Rating Category	Empirical Default Rate	Rating Category	Empirical Default Rate	Aggregated Data	Linear Regression	Unweighted	Weighted
Aaa	0	AAA	0	0	0	0	0
Aa	1	AA	1	1	1	1	1
A	2	A	4	3	4	4	4
Baa	21	BBB	27	24	22	20	21
Ba	131	BB	120	125	110	111	115
B	569	B	591	580	555	586	595
Caa-C	2098	CCC/C	3041	2570	2797	2569	2555

Source: Adapted from CEIOPS. 2007c. QIS3 Calibration of the credit risk. CEIOPS-FS-23/07, April. Available at www.ceiops.org.

ratings from the two rating institutes are shown together with their aggregated default ratings and their estimates. They are also illustrated in Figure J.3.

R^2, the coefficient of determination, is a measure of the global fit of the linear model. $R^2 = 1$ indicates that the model fits ideal and $R^2 = 0$ indicates a nonlinear relationship. The values of the different interpolations were between 0.9561 and 0.9989. Because of this and

FIGURE J.3 Plot of aggregated default rates and smoothed PDs for the 7 Standard & Poor's rating categories. 1 base point = 0.0001. Base points in log scale. For illustration we have used Moody's rating categories; *cf.* Table J.5. (Adapted from CEIOPS. 2007c. QIS3 Calibration of the credit risk. CEIOPS-FS-23/07, April. Available at www.ceiops.org; © Committee of European Insurance and Occupational Pensions Supervisors.)

the aim of the standard formula to provide a "prudent approach," CEIOPS suggested the estimated PDs given in Table J.1.

In one of CEIOPS CPs (CP 20), it was suggested to use a base correlation of 0.5 for the theoretical case of infinitely many issuers and to adjust the correlation with the use of the HHI following Equation J.1; *cf.* the discussion above.

The model used in QIS4 is discussed and criticized in Hürlimann (2008b) where the TailVaR measure is proposed instead of the VaR.

The calculation of the counterparty default risk charge made in QIS4 was considered to be too complex by many participating companies; CEIOPS (2008e). This was particularly for the nonlife reinsurance counterparties. The capital charge for the exposure to unrated counterparties was criticized as being prohibitively high and inconsistent with current practice.

J.3 CREDIT SPREAD RISK

The capital charge we are seeking is denoted $C_{MR,SP}$ and is a part of the market risk module in the standard formula framework.

J.3.1 QIS3 and QIS4 Models

Reference to QIS3 is made in CEIOPS (2007a) and to QIS4 in QIS4 (2008).

From QIS4 we note the following: Spread risk is the part of risk originating from financial instruments that is explained by the volatility of credit spreads over the risk-free interest rate term structure. It reflects the change in value due to a move of the yield curve relative to the risk-free term structure. Assets that are allocated to policies where the policyholders bear the investment risk should be excluded from this risk module. However, as these policies may have embedded options and guarantees, an adjustment (calculated using a scenario-based approach) is added to the formula to take into account the part of the risk that is effectively borne by the insurer.

For the purposes of determining the SCR for spread risk, companies should assume the more onerous (in aggregate) of a rise or fall in credit spreads. It is not required to assume different directional movements in credit spreads when determining the different components of the spread risk submodule. Currently, default and migration risks are not explicitly built in the spread risk module. However, the spread risk module will include parts of these risks implicitly via the movements in credit spreads. The credit indices used for the calibration rebalance on a monthly basis and, consequently, the change of their constituents, due to downgrades or upgrades, has a monthly frequency as well. Hence, the impact of intramonth downgrades/upgrades will partly be reflected in the movements of credit spreads.

Government bonds are exempted from an application of this module. The exemption relates to borrowings by the national government, or guaranteed by the national government, of an OECD or EEA state, issued in the currency of the government.

The spread risk module is applicable to all tranches of structured credit products like ABSs and collateralized debt obligations (CDOs). In general, these products include transactions or schemes, whereby the credit risk associated with an exposure or pool of exposures is tranched, having the following characteristics: (a) payments in the transaction or scheme are dependent on the performance of the exposure or pool of exposures; and (b) the subordination of tranches determines the distribution of losses during the ongoing life of the transaction or scheme.

The spread risk module further covers credit derivatives, for example, CDSs, total return swaps (TRS), credit-linked notes (CLN) that are not held as part of a recognized risk mitigation policy.

The spread risk module was developed for testing under QIS3 in 2007. In CP 20, CEIOPS (2006b), indicated that as a starting point, the capital charge would be calculated as

$$C_{MR,SP} = \sum_i EAD_i \cdot m(dur_i) \cdot f\left(g_i\right), \tag{J.4a}$$

where the sum is taken over the credit risk exposures i; EAD_i: the exposure at default; the nominal size pf credit risk exposure i as determined by reference to market values; dur_i: the effective duration of credit risk exposure i. If the bond has no embedded options, or behaves like an option-free bond, effective duration could be estimated using modified duration which equals the Macaulay duration divided by 1 plus the yield-to-maturity of the bond; CEIOPS (2006a); and g_i: the external rating of credit exposure i.

In QIS4, the capital charge for the spread risk was split up into three components, one for bonds, one for structured credit products, and one for credit derivatives. Hence, we have in QIS4 the following capital charge:

$$C_{MR,SP} = C_{SP,Bo} + C_{SP,Struc} + C_{SP,Cd}, \tag{J.4b}$$

where

$$C_{SP,Bo} = \sum_i EAD_i \cdot m(dur_i) \cdot f\left(g_i\right) + \Delta Liab_{UL} \tag{J.5a}$$

$$C_{SP,Struc} = \sum_i EAD_i \cdot m(dur_i) \cdot g\left(g_i\right). \tag{J.5b}$$

$C_{SP,Cd}$: for the credit derivatives, the capital charge is determined as the change in the value of the derivative (i.e., as the decrease in the asset or the increase in the liability) that would occur following (a) a widening of credit spreads by 300% if overall this is more onerous or (b) a narrowing of credit spreads by 75% if this is more onerous. A notional capital charge should then be calculated for each event. The capital charge for derivatives should then be the higher of these two notional charges.

$\Delta Liab_{UL}$: is the overall impact on the liability side for policies where the policyholders bear the investment risk with embedded options and guarantees of the stressed scenario, with a minimum value of 0 (sign convention: positive sign means losses). The stressed scenario is defined as a drop in value on the assets (except government bonds used as the reference to the valuation of the liabilities by $m(dur_i) * f(g_i)$ (e.g., for a BBB-rated asset with a duration of 4 years, this means a drop by 5%).

TABLE J.6 Risk Weights f and g as Functions of External Rating as Proposed in CP 20 and used in QIS3 and QIS4

		Risk Weights, %			
		CP 20	QIS3	QIS4	
Rating, g_i	**Credit Quality Step**	$f(g_i)$	$f(g_i)$	$f(g_i)$, **Bonds**	$g(g_i)$, **Structured Products**
AAA	1	0.04	0.25	0.25	2.13
AA	1	0.28	0.25	0.25	2.55
A	2	3.30	1.03	1.03	2.91
BBB	3	6.56	1.25	1.25	4.11
BB	4	10.16	3.39	3.39	8.42
B	5	22.23	5.60	5.60	13.35
CCC	6	34.96	11.20	11.20	29.71
Unrated	–	8.00	2.00	2.00	100

Source: Adapted from CEIOPS. 2006a. Answers to the European Commission on the third wave of Calls for Advice in the framework of the Solvency II project. CEIOPS-DOC-03/06, May. Available at www.ceiops.org.

The calculation of the capital charge for spread risk should be derived under the condition that the assumptions on future bonus rates (reflected in the valuation of future discretionary benefits in TPs) remain unchanged before and after a presumed change in spread levels.

For CDOs companies should ensure that the rating reflects the nature of the underlying risks associated with collateral assets. For example, in the case of a CDO-squared, the rating should take into account the risks associated with the CDO tranches held as collateral, that is, the extent of their leveraging and the risks associated with the collateral assets of these CDO tranches.

The function f produces a risk weight as given in Table J.6 and is calibrated to deliver a shock consistent with VaR 99.5%. The function m is a function of the duration.

For QIS3, CEIOPS analyzed a cap of duration in the capital charge (J.4) in order to allow for the nonlinearity of the risk. These caps are given in Table J.7.

TABLE J.7 Duration Caps used in QIS3 and QIS4

	$m(dur_i) =$		
Rating, g_i	**QIS3**	**QIS4, Bonds**	**QIS4, Structured**
BB	$\min(dur_i;8)$	$\min(dur_i;8)$	$\max\{\min(dur_i;5),1\}$
B	$\min(dur_i;6)$	$\min(dur_i;6)$	$\max\{\min(dur_i;4),1\}$
CCC or lower	$\min(dur_i;4)$	$\min(dur_i;4)$	$\max\{\min(dur_i;2.5),1\}$
Unrated	$\min(dur_i;4)$	$\min(dur_i;4)$	1
"Otherwise"	dur_i	dur_i	$\max\{dur_i;1\}$

Source: Adapted from QIS3. 2007. QIS 3 Technical Specification. CEIOPS-DOC-01/07. Available at http://www.ceiops.eu/content/view/118/124/; see also CEIOPS (2007a) and QIS4. 2008. QIS 4 Technical Specifications. European Commission. MARKT/2505/08. March 31. Available at http://www.ceiops.eu/content/view/118/124/.

The calculation of the capital charge for spread risk should be derived under the condition that the assumptions on future bonus rates, reflected in the valuation of future discretionary benefits in TPs, remain unchanged before and after a presumed change in spread levels.

Additionally, the result of the calculation should be determined under the condition that the participant is able to vary its assumptions in future bonus rates in response to the shock being tested. The resulting capital charge is $nC_{MK,SP}$. On assessing the scope of variation of the assumptions on future bonus rates, firms should have regard to the size of shock (as measured by the amount of the spread risk charge) and of the resulting financial position, and then reflect on what would be a realistic level of reduction of future bonuses (subject to any legal or contractual constraints) as a trade-off between improvement of the solvency position, managing policyholders expectations, and deteriorating commercial objectives.

In QIS4, it was possible to use a *simplification* provided that

- The average credit rating for long-duration bonds (10 year and above) was not less than one rating below the credit rating for short-duration bonds (5 years or below)
- The general criteria set out in QIS4 (2008) for simplifications were followed

$$CR_{MR,SP} = CR^{\text{bonds}}_{MR,SP} + CR^{\text{struct}}_{MR,SP} + CR^{cd}_{MR,SP},$$

where for bonds: $CR^{\text{bonds}}_{MR,SP} = MV^{*}Dur^{\text{bonds}*}\Sigma(\%Mv_i^{\text{bonds}}f(g_i)) + \Delta Liab_{UL}$; for structured credit products: $CR^{\text{struct}}_{MR,SP} = MV^{*}Dur^{\text{struct}*}\Sigma(\%Mv_i^{\text{struct}}\, g(g_i))$; and for credit derivatives: $CR^{cd}_{MR,SP} = \Sigma(Mv_i^{cd})^{*}Dur^{cd}$, where $MV =$ the total market value of nongovernment bond portfolio; $MV^{\text{struct}} =$ the total market value of structured credit products portfolio; $Dur^{\text{bonds}} =$ the modified duration of nongovernment bond portfolio; $Dur^{\text{struct}} =$ the modified duration of structured credit portfolio; $Dur^{cd} =$ the modified duration of credit derivatives portfolio; $\%Mv_i^{\text{bonds}} =$ the proportion of nongovernment bond portfolio held at rating i; $\%Mv_i^{\text{struct}} =$ the proportion of structured credit portfolio held at rating i; and $\%Mv_i^{cd} =$ the proportion of credit derivatives portfolio held at rating i and where $\Delta Liab_{UL}$ equaled the overall impact on the liability side for policies where the policyholders bear the investment risk with embedded options and guarantees of the stressed scenario, with a minimum value of 0 (sign convention: positive sign means losses). The stressed scenario was defined as a drop in value on the assets, except government bonds, used as the reference to the valuation of the liabilities by $MV^{*}Dur^{\text{bonds}}*\Sigma(\%Mv_i^{\text{bonds}}f(g_i))$.

The risk weights $f(g_i)$ and $g(g_i)$ are taken as for the nonsimplified approach.

J.3.2 Calibration

External data from Moody's are used to calibrate the 99.5% shock on bond spreads. CEIOPS used both Moody's median bond spread data series (1991–2006, monthly data) and Moody's long-term credit spreads (1950–2006, monthly data).

All data series are based on information from global credit portfolios. Separate series were available for several rating classes. The credit spread was measured versus US Treasury, since at time being CEIOPS did not had excess to long-dated data series versus EU risk-free term structures. However, except for the last few years, US credits will highly dominate the global credit portfolios.

The observed data showed that in general higher credit spreads were associated with higher absolute changes in credit spreads. The lognormal model exhibits this property. The lognormal model treats proportionate changes in credit spreads as a lognormal process, so it has been assumed that the credit spread in 12 months is given by

$$CS_{12} = CS_0 \cdot e^X,$$

where X is distributed $N(\mu, \sigma^2)$.

For y sufficiently close to zero, $\ln(1 + y)$ is approximately y; hence, the above formula could be rearranged into

$$X = \ln\left(\frac{CS_{12}}{CS_0}\right) = \ln\left[1 + \left(\frac{CS_{12} - CS_0}{CS_0}\right)\right] \approx \frac{CS_{12} - CS_0}{CS_0}$$

showing that the lognormal model assumes that the absolute change in credit spreads, $[CS_{12} - CS_0]$, linearly depends on the current credit spread level, CS_0.

Relative changes in credit spreads for data from 1991 to 2006 are shown in Table J.8.

It has been decided to use a factor-based approach and CEIOPS used the following linear approximation

$$\Delta EAD_i \approx -EAD_i \cdot dur_i \cdot \Delta y,$$

where Δy is the resulting change in interest rates after the shock in the credit spread. Using the assumption made above, we can rewrite this approximation as

$$\Delta EAD_i \approx -EAD_i \cdot dur_i \cdot CS_i \cdot F(g_i),$$

where $F(g_i)$ can be found from Table J.8. CEIOPS therefore assumed the credit spread shock for bonds to be 70%.

As credit spreads might not always be available, one solution would be to exclude them from the formula and make use of the average credit spread of each rating class. This would

TABLE J.8 Relative Changes in Credit Spreads Based on Monthly Data 1991–2006

Rating Bucket	Mean, %	Standard Deviation, %	99.5 Shock, %
AAA	5	30	72
AA	4	28	70
A	4	28	68
BBB	4	27	67
BB or lower	−2	29	77

Source: Adapted from Table 2.2 in CEIOPS. 2007c. QIS3 Calibration of the credit risk. CEIOPS-FS-23/07, April. Available at www.ceiops.org.

Note: Calculations made by CEIOPS.

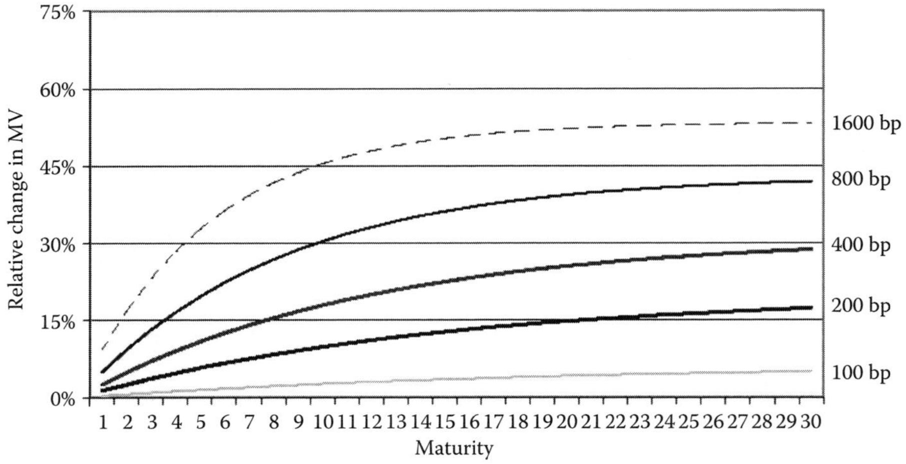

FIGURE J.4 The linear approximation in duration of the relative change in market values versus the "true" change for different maturities and credit spread levels. *Source*: CEIOPS (2007c): QIS3 Calibration of the credit risk. CEIOPS-FS-23/07, April. Website: www.ceiops.org; copyright Committee of European Insurance and Occupational Pensions Supervisors.

give us the capital charge for credit spread risk as

$$C_{MR,SP} = \sum_i EAD_i \cdot dur_i \cdot f(g_i).$$

For small credit spreads or durations, the described linear approximation works well, but where both durations and spreads are large, it might not. For coupon paying bonds, the duration is not equal, and will be lower than, the maturity of the bond. Therefore, the linear relation in duration will not be linear in maturity. More important is that the combination of long-maturity and high credit spreads, the approximation seems to too conservative. The differences between the linear approximation and the "true" relative change in MVs are shown in Figure J.4. For long maturities (and consequently, long durations as well) and high credit spread levels, the linear approximation is significantly higher.

CEIOPS' solution to this nonlinearity problem was to cap the product of the duration and the credit spread in the model above. This produced the function $m(dur_i)$ given above.

In an unpublished working document for CEIOPS QIS Task Force (TF), a member of that TF showed that the assumed bond shock of 70% as shown above was too low for structured credit products. For such products and the calibration, the shock was assumed to be 300%. To arrive in the g-function in Equation J.5b given in Table J.6, the same procedure as above was used.

J.4 CONCENTRATION RISK

The capital charge we are seeking is denoted $C_{MR,Co}$ and is a part of the market risk module in the standard formula framework.

As an answer to the EC's CfA 10, MARKT (2004c), CEIOPS gave a background to the modeling of the standard formula and its risk modules; see CEIOPS (2005).

Concentration effects could be taken into account by adjusting RFs or increasing volume measures. Since concentration results in more "dangerous" distributions, adjusted RFs might be based on higher moments, for example, skewness. For the sake of simplicity of the standard formula, however, the choice of increasing volume measures appears more practicable. For example, a certain percentage of the amount of bonds of a single issuer which exceeds a given limit might be added to the corresponding volume measure.

J.4.1 QIS3 and QIS4 Models

Reference to QIS3 is made in CEIOPS (2007a) and to QIS4 in QIS4 (2008).

The capital charge is calculated as

$$C_{MR,Co} = \sqrt{\sum_i C^2_{Co,i}}, \tag{J.6}$$

where under

$$\text{QIS3} : C_{Co,i} = A_{xUL} \cdot XS_i \cdot \left(g_0 + g_1 \cdot XS_i\right) \tag{J.7a}$$

and

$$\text{QIS4} : C_{Co,i} = A_{xUL} \cdot XS_i \cdot g_i + \Delta Liab_{ul}. \tag{J.7b}$$

A_{xUL}: the amount of total assets excluding those where the policyholder bears the investment risk.

$$XS_i = \max\left\{0; \frac{EAD_i}{A_{xUL}} - CT\right\},$$

where EAD_i is the net exposure at default to counterparty i and CT is a concentration threshold.

Where an undertaking has more than one exposure to a counterparty, then EAD_i is the aggregate of those exposures at default and rating g_i should be a weighted rating determined as the rating corresponding to a weighted average credit quality step calculated as average of the credit quality steps of the individual exposures to that counterparty, weighted by the net exposure at default in respect of that exposure to that counterparty.

$\Delta Liab_{UL}$: the overall impact on the liability side for policies where the policyholders bear the investment risk with embedded options and guarantees of the stressed scenario, with a minimum value of 0 (sign convention: positive sign means losses). The stressed scenario is defined as a drop-in value on the assets for counterparty i used as the reference to the valuation of the liabilities by $XS_i^* g_i$.

The concentration threshold and the QIS3 parameters g_0 and g_1 and QIS4 parameter g_i are given in Table J.9. As $A_{xUL} \cdot XS_i = \max\{0; EAD_i - A_{xUL} \cdot CT\}$, we see that we compare the net exposure at default with a percentage of the asset hold of the counterparty.

TABLE J.9 The Concentration Threshold, CT, and the QIS3 Parameters g_0 and g_1 and QIS4 Parameter g

Rating	Credit Quality Step	Threshold CT(%)	QIS3 g_0	QIS3 g_1	QIS4 g_i
AAA	1	5	0.184	0.040	0.15
AA	1	5	0.184	0.040	0.15
A	2	5	0.268	−0.016	0.18
BBB	3	3	0.386	−0.042	0.30
BB or lower, unrated	4–6, –	3	0.923	−0.431	0.73

Source: Adapted from QIS3. 2007. QIS 3 Technical Specification. CEIOPS-DOC-01/07. Available at http://www.ceiops.eu/content/view/118/124/; see also CEIOPS (2007a) and QIS4. 2008. QIS 4 Technical Specifications. European Commission. MARKT/2505/08. March 31. Available at http://www.ceiops.eu/content/view/118/124/.

From QIS4 we note the following: Market risk concentrations present an additional risk to an insurer because of

- Additional volatility that exists in concentrated asset portfolios and
- The additional risk of partial or total permanent losses of value due to the default of an issuer

Assets that are allocated to policies where the policyholders bear the investment risk should be excluded from this risk module. However, as these policies may have embedded options and guarantees, an adjustment (calculated using a scenario-based approach) is added to the formula to take into account the part of the risk that is effectively borne by the insurer.

For the sake of simplicity and consistency, the definition of market risk concentrations is restricted to the risk regarding the accumulation of exposures with the same counterparty. It does not include other types of concentrations (e.g., geographical area, industry sector, etc.).

In case an undertaking owns shares representing more than 20% of the capital of another insurance or financial undertaking which (1) is not included in the scope of consolidation or supplementary supervision and (2) where the value of that participation or subsidiary exceeds 10% of the participating undertaking's OFs, these shares are exempted from the application of the concentration risk module when using option 1; see Section J.3.3, for the treatment of participations (deduction–aggregation method). In line with this approach and when using option 3 for the treatment of participations (look-through approach), the concentration risk module should not be applied.

Government bonds are exempted from the application of this module. The exemption concerns borrowings by the national government, or guaranteed by the national government, of an OECD or EEA state, issued in the currency of the government.

Bank deposits with a term of less than 3-months terms, of up to 3 million Euros, in a bank that has a minimum credit rating of AA are also exempted from an application of this module.

All entities that belong to the same group should be considered as a single counterparty for the purposes of this submodule.

The net exposure at default to an individual counterparty *i* shall comprise the asset classes of equity and fixed income (including hybrid instruments, e.g., junior debt, mezzanine CDO tranches, etc.).

Financial derivatives on equity and defaultable bonds should be properly attributed (via their "delta") to the net exposure, that is, an equity put option reduces the equity exposure to the underlying "name" and a single-name CDS ("protection bought") reduces the fixed-income exposure to the underlying "name." The exposure to the default of the counterparty of the option or the CDS is not treated in this module, but in the counterparty default risk module. Also, collaterals securitizing bonds should be taken into account. Similarly, a look-through approach should be applied to assets representing reinsurers' funds withheld by a counterpart.

Exposures via investment funds or such entities whose activity is mainly the holding and management of an insurer's own investment need to be considered on a look-through basis. The same holds for CDO tranches and similar investments embedded in "structured products."

This capital charge is calculated for concentration risk under the condition that the assumptions on future bonus rates (reflected in the valuation of future discretionary benefits in TPs) remain unchanged before and after a presumed change in volatility and/or default level of concentrated assets.

Additionally, the result of the calculation should be determined under the condition that the participant is able to vary its assumptions in future bonus rates in response to the shock being tested. The resulting capital charge is $nC_{MR,Co}$.

J.4.2 Calibration

The calibration has been categorized into three main steps:

Step 1: Designing a well-diversified portfolio

Step 2: Concentration exposures in the initial portfolio

Step 3: Concentration exposures for names rated AA, A, BBB, and BB and worse and different sectors

Step 1: *The starting point is the design of a well-diversified portfolio of investments in individual names with the following characteristics*

a. The portfolio has a mix, representative of EU average insurers' portfolios of investments in bonds and equities. The mix proposed is 70–30% corresponding bonds–equities, respectively (see Figure 11A, p. 12, Financial Stability Report Conglomerates 2005–2006).

b. Within each of these two groups, a sector-distribution of investments is built, also according to an EU expected average, as follows:

 1. Investment in bonds: We have assumed that 25% of bonds portfolio is invested in risk-free bonds, and the rest (75%) is invested in different sectors and ratings as described below.

 2. Investment in equities: To the extent that this exercise is assumed as the starting point of a well-diversified portfolio, consequently, it should replicate some equity index sufficiently representative and well-known. The selected index is Eurostoxx 50, and the period used to record data on prices of each of its element, ranges from January 11, 1993 until November 30, 2006. The length of this period guarantees sufficient historical data to derive VaR 99.5% with a high degree of reliability. Some elements of the selected index have been removed, since their records of data prices are only available for a significantly shorter period than that mentioned above.

J.4.2.1 Description of Bonds Portfolio

In order to avoid the effect of the change in Macaulay Duration (as the life of the bond expires) and the renewal of investment 12, and what is more important, to reflect the whole risk belonging to each sector/rating it was decided that

1. Bonds used in the computation are notional bonds, all of them issued at 5% rate and pending 5 years to maturity. At any moment of the simulation, each bond maintain these features (which could be accepted as representative average features of the bonds existing in insurance portfolios).

2. To capture and summarize market information about each sector/rating, notional bonds described in point (1) are valued with Bloomberg corporate yield curves, according the corresponding sector/rating. The following table lists these yield curves:

Interest Rates Data	
1	F888 EUR BANK AAA
2	F462 INDS AA+
3	F890 BANK AA
4	F580 UTIL AA
5	F892 BANK A
6	F583 UTIL A
7	F465 INDUS A
8	F898 BANK BBB
9	F625 TELEF A
10	F468 INDUS BBB
11	F469 INDUS BBB
12	F682 TELEF BBB+
13	F470 INDUS BB

J.4.2.2 Description of Equities Portfolio

To obtain a well-diversified portfolio, after selecting the components of Eurostoxx 50 mentioned above, other additional names have been added to complete all the buckets of the cross-table resulting from, on the one dimension, rating categories considered, and, on the other dimension, economic sectors included in this exercise.

Weights for each name in this initial portfolio depend on the following features:

1. When calculating BB concentration risk polynomial: we use names ranged from B to AAA
2. When calculating BBB, A, and AA-AAA concentration risk polynomials: we use the names ranging from BBB to AAA with the relevant adjustment in their initial weight
3. Besides, the level or quality of diversification of these two starting portfolios has being checked by calculating their HHI, see Section 14.4.1, and comparing with their minimum possible value:

	Eurostoxx 50	**Eurostoxx 37**	**Eurostoxx 34**
HHI	0.0256	0.0335	0.0349
$1/n$	0.0200	0.0270	0.0294

This table shows in all cases the HHI for each portfolio is only 0.005 higher than the minimum possible value, which confirms that the selected portfolios are actually well diversified.

Finally, the calibration exercise has calculated the historic 1-year VaR 99.5% of a mixed portfolio (30% invested in the equities portfolio, and 70% invested in the bonds portfolio). The asset allocation in 30–70 between equities and bonds are motivated from the observations made in, for example, CEIOPS (2006b). This measure is calculated twice:

First, taking into account all the names and its corresponding yield curves as listed above, $VaR_{99.5\%}=12.85\%$.

Second, excluding BB names and its corresponding yield curve, as listed above. $VaR_{99.5\%}=10.88\%$.

In both cases, risk-free bonds are priced with the German sovereign curve.

As one can appreciate, there is sufficient rationale to calibrate firstly BB polynomial using the whole portfolio and afterward, in a second step, to calibrate BBB, A and AA-AAA polynomial with a less volatile portfolio.

Step 2: *Concentrating exposures in the initial portfolio*

First of all, we have established a bijective correspondence between each equity name and one of the interest rates curves listed above, taking into account its sector/rating. This means that when we concentrate the whole portfolio, we concentrate at the same time the investment in the selected equity and its correspondent notional bond.

1. The exercise begins selecting a concrete name with a certain rating, (i.e., a bank rated AAA) and its correspondent notional bond (Banks AAA). Then, we increase in steps of 1% its total weight in respect of the whole portfolio, obviously reducing simultaneously the participation of the rest of counterparties (to isolate purely the effect of concentration on the selected name).

2. Increases of concentration levels range from the starting weight up to the starting weight plus 71%, (as mentioned above, using 1% steps). For each level of concentration, we calculate the difference between the historic 1-year VaR 99.5% of the starting portfolio and the historic 1-year VaR 99.5% of the resulting concentrated portfolio, and this difference is considered a raw proxy of an eventual concentration charge (it is called *Variation VaR.*)

3. For QIS3**:** Points of raw-concentrations charges obtained in the successive increases of concentration for each name are drawn, interpolating a second-degree polynomial, and then deriving the parameters g_0 and g_1. Thus, for each level of rating i, we will have

$$C_{Co,i} = A_{xUL} \cdot XS_i \cdot \left(g_0 + g_1 \cdot XS_i\right).$$

4. For QIS4: Points of raw-concentrations charges obtained in the successive increases of concentration for each name are drawn, interpolating a straight line, and then deriving the parameter g_i. Thus, for each level of rating, we will have

$$C_{Co,i} = A_{xUL} \cdot XS_i \cdot g_i.$$

Step 3: *The same procedure is repeated for names rated AA, A, BBB, and BB or worse and different sectors*

Note that the initial investment in risk-free bonds remains unchanged. Therefore, concentration exercise refers to the whole equity portfolio and 75% of the bonds portfolio.

Tables in CEIOPS (2007c) and CEIOPS (2008a) compare 1-year historical VaR 99'5% for the starting portfolio versus the extreme 1-year historical VaR 99'5% (portfolios with a concentration increase of 71% above the initial weight).

Once this point is reached and the graphs obtained are analyzed, the interpolation of a second-degree polynomial is carried out taking into account the worst-behaved names. This criterion is necessary to guarantee the consistency of the calibration exercise with the rationale grounding the standard SCR formula, which focuses on stressed scenarios.

Under QIS3, finally, g_0 and g_1 parameters of each polynomial for each rating are estimated using a conventional minimum squares method and included a judgmental adjustment to consider:

- The method has focused on nonsystemic volatility risk (the first type of risk captured in this submodule), assuming that due to the long period analyzed, outputs reflect partially the second source of risk which should be captured in concentration sub-module (risks of losses due to a default in the concentrated exposure). Nevertheless, the method may not achieve a full reflection of this second risk.

- After QIS3, some adjustments in the coefficients resulting from this calibration may be necessary to coordinate, on the one hand, the correlation implicit in this method among concentration risk and other market risks (equity, interest rate risk, and spread risk), and, on the other hand, the correlation coefficients used in QIS3.

Under QIS4, finally, *g* parameters for each rating are estimated using a conventional minimum squares method.

J.4.3 Options in QIS 4

Participants were requested to test options 1 and 2 as described below. The default option to be used in the calculation of BSCR is option 1. In addition, on an optional basis, participants may test option 3.

For options 1 and 2 where an undertaking (called "the parent" below) owns a participation in another undertaking, or has a subsidiary, this participation or subsidiary should be valued on an economic basis. If a fair value treatment under IAS 39 is applied, this is considered as an acceptable proxy.

J.4.3.1 Option 1: "Differentiated Equity Stress" Approach

All participations and subsidiaries, except those falling under the Participating rule mentioned below, should treat these holdings in the SCR calculation as if they were an equity investment as described in the following, that is, by calculating a differentiated capital charge for equity risk.

Under this option, when calculating the equity risk module in a first step, for each index i, a capital charge is determined as the result of a predefined stress scenario for index i as follows:

$$CR_{MR,ER,i} = \max\left(\Delta NAV \mid \text{Shock}_{Eq,i}; 0\right),$$

where $\text{Shock}_{Eq,i}$ = Prescribed fall in the value of index I depending on the confidence level and standard deviation of the index i; and $CR_{MR,ER}$ = Capital charge for equity risk with respect to index i, and where the equity shock scenarios for the individual indices are specified as follows:

	Global	Other
$\text{Shock}_{Eq,i}$ =	32%	45%

For participations and subsidiaries, for example, ownership of more than 20% in insurance and financial undertakings included in the scope of consolidated or supplementary supervision the equity shock will be reduced to 16% for "Global" firms and to 22,5% for the "Other" participations.

The same reduction should be applied for other participations and subsidiaries in

- Noninsurance and nonfinancial undertakings that are taken into consideration within the consolidated or supplementary supervision.
- Insurance and financial undertakings that are not included in the scope of consolidated or supplementary supervision and do not exceed the 10% of the participating undertaking's OFs.

Participating: In cases where the mother owns more than 20% of another insurance or financial undertaking which (1) is not included in the scope of consolidation or supplementary supervision and (2) where the value of that participation or subsidiary exceeds 10% of the participating undertaking's OFs, the calculation of the regulatory capital requirement of the parent shall be carried out using deduction and aggregation method.

Concerning the concentration risk module, participations and subsidiaries are exempted only if they fall under the definition set out in the Participating rule mentioned above.

J.4.3.2 Option 2: "Across the Board" Approach

Under this option, all participations and subsidiaries are treated as if they were a "standard" equity investment when calculating the SCR capital charge for equity risk. They are not granted any specific treatment with respect to equity risk.

Therefore, when calculating the equity risk module, all equity investments, including participations and subsidiaries, should be subject to the "standard" equity shock scenarios (with no adjustment) as follows:

	Global	**Other**
$\text{Shock}_{Eq,i} =$	32%	45%

No exemption is applied from the application of the concentration risk module.

J.4.3.3 Option 3: "Look-Through" Approach

On an optional basis, participants may replace their solo SCR calculation, with the group SCR calculation for the subgroup formed by the participant itself (the "parent") and its subsidiaries and participations. Where this method is followed, undertakings should follow the default method set out in the Default method—Accounting consolidation for the calculation of OFs and SCR.

Under this option, both the parent's OFs and the SCR are to be replaced with the OFs and the group SCR of the subgroup.

Where a consolidated approach is taken, participants should follow the guidance on nonlife underwriting risk, counterparty default risk, LUR, market risk, and operational risk.

APPENDIX K

European Solvency II Standard Formula

Operational Risk

HERE, WE WILL DISCUSS the development and calibration of the operational risk module and its capital charge that has been one of the building blocks in Solvency II. The final advice from CEIOPS is discussed in Chapter 29.

Operational risk is the risk of loss arising from inadequate or failed internal processes, people, and systems or from external events. Operational risk also includes legal risks. Reputation risks and risks arising from strategic decisions do not count as operational risks.

As an answer to the European Commission's Call for Advice 10, MARKT (2004c), CEIOPS gave a background to the modeling of the standard formula and its risk modules (see CEIOPS, 2005). CEIOPS made a presentation of the methods proposed in Basel II; see Chapter 19.

In the light of the financial crisis, mainly during 2007–2009, CEIOPS (2009a) decided to review the operational risk to better measure its potential impact during stressed situations.

K.1 STANDARD FORMULA

The challenges of setting capital requirements for operational risk are well documented. The greatest obstacle is the lack of data, arising from the absence of established approaches to classify and quantify losses (see also Section 19.1). In those Member States that have encouraged the holding of capital specifically to address operational risk, the methods used by insurers vary widely in their sophistication. But operational risk could be a potentially material threat to policyholder protection; therefore some attempt should be made to quantify it under the SCR.

The introduction of the operational risk was based on the similar introduction in Basel II.

K.1.1 QIS2 (CEIOPS, 2006d)

The approach to test was given by a simple, but robust capital charge formula as follows:

$$
\begin{aligned}
\mathrm{CR_{OR}} = \max\{&0.006 \times \mathrm{TP_L} + 0.03 \times \mathrm{TP_{NL}} + 0.03 \times \mathrm{TP_H};\\
&0.06 \times \mathrm{Earn_L} + 0.03 \times \mathrm{Earn_{NL}} + 0.03 \times \mathrm{Earn_H}\},
\end{aligned}
$$

where $\mathrm{TP_L}$: total life insurance technical provisions 14 (gross of reinsurance); $\mathrm{TP_{NL}}$: total nonlife insurance technical provisions (gross of reinsurance); $\mathrm{TP_H}$: total health insurance technical provisions (gross of reinsurance); $\mathrm{Earn_L}$: total earned life premium (gross of reinsurance); $\mathrm{Earn_{NL}}$: total earned nonlife premium (gross of reinsurance); and $\mathrm{Earn_H}$: total earned health insurance premium (gross of reinsurance).

In the case of linked business with no policyholder guarantees, this may be reduced to one-tenths of the technical provisions and the earned life premium.

K.1.2 QIS3 (CEIOPS, 2007a)

In QIS3, the capital charge was changed in accordance with the proposal in CEIOPS (2006b). The capital charge was set to

$$
\begin{aligned}
\mathrm{CR_{OR,TPE}} &= \max\{0.006 \times \mathrm{TP_L} + 0.03 \times \mathrm{TP_{NL}} + 0.03 \times \mathrm{TP_H};\\
&\qquad 0.06 \times \mathrm{Earn_L} + 0.03 \times \mathrm{Earn_{NL}} + 0.03 \times \mathrm{Earn_H}\},\\
\mathrm{CR_{OR}} &= \min\left[f_{\mathrm{OR}} \times \mathrm{BSCR}; \mathrm{CR_{OR,TPE}}\right] = \min\left[0.30 \times \mathrm{BSCR}; \mathrm{CR_{OR,TPE}}\right],
\end{aligned}
$$

where f_{OR} is a predefined operational loading factor equal to 0.30; and BSCR: the basic SCR, calculated over all quantified risks (see Appendix H), and $\mathrm{CR_{OR,TPE}}$ is the capital charge in QIS2, see above.

K.1.3 QIS4 (QIS4, 2008)

The capital charge for operational risk is given by

$$
\mathrm{CR_{OR}} = \min\left[0.30 \times \mathrm{BSCR}; \mathrm{CR_{OR,TPE}}\right] + 0.25 \times \mathrm{Exp_{UL}},
$$

where $\mathrm{CR_{OR,TPE}}$: basic operational risk charge for all business other than UL business (gross of reinsurance), defined as

$$
\begin{aligned}
\mathrm{CR_{OR,TPE}} = \max\{&0.006 \times (\mathrm{TP_L} - \mathrm{TP_{UL}}) + 0.03 \times \mathrm{TP_{NL}} + 0.03 \times \mathrm{TP_H};\\
&0.06 \times (\mathrm{Earn_L} - \mathrm{Earn_{UL}}) + 0.03 \times \mathrm{Earn_{NL}} + 0.03 \times \mathrm{Earn_H}\},
\end{aligned}
$$

where $\mathrm{TP_L}$: total life insurance technical provisions 14 (gross of reinsurance); $\mathrm{TP_{UL}}$: total life insurance technical provisions for unit-linked business (gross of reinsurance); and

TP_{NL}: total nonlife insurance technical provisions (gross of reinsurance). It concerns all the LoBs in nonlife, excluding the risks related to annuities in LoBs:

- Accident and health—workers' compensation
- Accident and health—health insurance
- Accident and health—others not included under first two items

TP_H : total health insurance technical provisions (gross of reinsurance). It concerns the risks related to both long-term health insurance and annuities in LoBs:

- Accident and health—workers' compensation
- Accident and health—health insurance
- Accident and health—others not included under first two items

$Earn_L$: total earned life premium (gross of reinsurance); $Earn_{L-UL}$: total earned life premium for UL business (gross of reinsurance); and $Earn_{NL}$: total earned nonlife premium (gross of reinsurance). It concerns all the LoBs in nonlife excluding the risks related to annuities in LoBs:

- Accident and health—workers' compensation
- Accident and health—health insurance
- Accident and health—others not included under first two items

$Earn_H$ is the total earned health insurance premium (gross of reinsurance). It concerns the risks related to both long-term health insurance and to annuities in LoBs:

- Accident and health—workers' compensation
- Accident and health—health insurance
- Accident and health—others not included under first two items

BSCR the basic SCR, calculated over all quantified risks (see Appendix H); Exp_{UL} the amount of annual expenses, gross of reinsurance, incurred in respect of UL business. Administrative expenses should be used (excluding acquisition expenses), calculation should be based on the latest years expenses and not on future projected expenses.

Companies participating in QIS4 had diverged views whether this capital charge was calibrated adequately (CEIOPS, 2008e). Further improvements were needed, for example, the full dependency between operational risks and the other risks (no diversification effect). The standard formula was not reflecting the wide spectrum of operational risks that can be materialized. The capital 30% was too high.

K.2 CALIBRATION

K.2.1 QIS2 (CEIOPS, 2006b)

The initial calibration of the factors or the standard formula was taken from a proposal by the German Insurance Association adjusted to take account of business where a material portion of the overall risk is borne by policyholders (see Section 19.3).

K.2.2 QIS3 (CEIOPS, 2006b)

QIS2 results showed a very wide degree of dispersion in the capital requirements for operational risk, suggesting that the proposal was too simplistic to capture differences in the management of this risk between insurers. In some cases, the operational risk charge completely dominated the SCR, while other insurers reported that their internal modeling approaches implied the need for higher capital requirements.

Despite the problems encountered during QIS2, CEIOPS remained of the view that the SCR should take account of operational risk. Deferring the problem entirely to the insurer's ORSA would not provide much incentive to improve the identification and management of operational risks.

CEIOPS was encouraged after QIS2 by industry attempts to improve the understanding and consistency of data on operational risk losses, which would eventually support more sophisticated approaches to operational risk (including partial models).

As with QIS2, the standard formula treatment for operational risk should be based on a simple, robust formula. But in the Consultation Paper 20 (CEIOPS 2006b), it was proposed to consider the following revision to the operational risk charge as it develops proposals for QIS3:

$$\begin{aligned} \mathrm{CR}_{\mathrm{OR,TPE}} &= \max\{0.006 \times \mathrm{TP}_{\mathrm{L}} + 0.03 \times \mathrm{TP}_{\mathrm{NL}} + 0.03 \times \mathrm{TP}_{\mathrm{H}}; \\ &\qquad 0.06 \times \mathrm{Earn}_{\mathrm{L}} + 0.03 \times \mathrm{Earn}_{\mathrm{NL}} + 0.03 \times \mathrm{Earn}_{\mathrm{H}}\}, \\ \mathrm{CR}_{\mathrm{OR}} &= \min\left[f_{\mathrm{OR}} \times \mathrm{BSCR}; \mathrm{CR}_{\mathrm{OR,TPE}}\right], \end{aligned}$$

where f_{OR}: a predefined operational loading factor <1; and BSCR: the basic SCR, calculated over all quantified risks (see Appendix H).

The aim of the loading factor is to avoid excessive dominance of the overall SCR by the component for operational risk.

CEIOPS recognized that a loading approach had significant weaknesses. One particular difficulty is the potential exaggeration of risk mitigation effects in other modules. For example, an insurer might use credit derivatives to reduce its exposure to credit risk. The SCR charge for credit risk is therefore lower, which feeds through to a lower operational risk charge—when, arguably, operational risk has actually increased. Equally, there may be instances where an insurer had a very limited exposure to other risk categories, for example, some forms of linked business, and that it would be quite appropriate for operational risk to be the largest contributor to the SCR.

Tentatively, CEIOPS proposed that the operational loading factor is set in the range of 25–50%. Further attempt was made before QIS3 to establish the consistency of all

CR_{OR} factors with the soundness standard for the SCR, although this will be extremely difficult given the absence of historical operational risk data. CEIOPS also analyzed whether a differential treatment is needed in cases where operational risk is the dominant risk.

As CR_{OR} would no longer be part of the BSCR, and hence the capital requirements for operational risk would no longer reflect any diversification effects.

K.2.3 QIS4 (QIS4, 2008)

The capital (0.30 × BSCR) restricting CR_{OR} to a percentage of the other capital requirements (BSCR) is provided in Article 106(3) of the FD Proposal (COM, 2007).

APPENDIX L

European Solvency II Standard Formula

Liquidity Risk

The liquidity risk is not included in the calculation of the capital charges for the quantifiable risks. This is mainly due to the lack of simple and explicit formulas (*cf.* Chapter 20).

The liquidity risk should be included in the company's ORSA procedure.

In Consultation Paper 7 (CEIOPS, 2005), the liquidity risk is defined as the exposure to losses in the event that insufficient liquid assets will be available to meet the CF requirements of policyholder obligations as they fall due. In view of the work performed by the Joint Forum, CEIOPS proposed that liquidity risk should be added as a separate risk category.

CEIOPS noticed that ALM coordinates the CFs on the asset and liability side of the BS, and thus proves to be an effective tool for reducing the liquidity risk in both life and nonlife insurance. For a large portfolio of life business, CFs would be reasonably predictable on a one-year time horizon because of the law of large numbers.

Effective liquidity planning should be addressing most sources of liquidity risk and can be tested under Pillar II. Other sources of liquidity risk may be considered implicitly under Pillar I through the assessment of other risk categories. For example, in life business, an increase in lapse rates could be assessed through its impact on an undertaking's market and UR exposures.

In the light of the financial crisis, mainly during 2007–2009, CEIOPS (2009a) decided to look closer on the liquidity risk and to see if they needed to deal with it as a Pillar I issue, that is, include a capital charge in the standard formula.

L.1 QIS4 (QIS4, 2008)

The QIS4 specifications did not contain any concrete requirement, but only the following two general statements:

1. The insurer should have written the guidance regarding liquidity requirements that financial risk mitigation instruments should meet, according to the objectives of the own insurer's risk management policy.
2. Financial risk mitigation instruments considered to reduce the SCR have to meet the liquidity requirements established by the own entity.

APPENDIX M

European Solvency II Standard Formula

Nonlife Underwriting Risk

HERE, WE WILL DISCUSS the development and calibration of the nonlife UR module and its capital charge that has been one of the building blocks in Solvency II. The final advice from CEIOPS is discussed in Chapter 31.

M.1 GENERAL FEATURES

M.1.1 Background

As an answer to the EC's Call for Advice 10 (MARKT 2004c), CEIOPS gave a background to the modeling of the standard formula and its risk modules (see CEIOPS, 2005).

CEIOPS assumed the time horizon to be one business year. The forthcoming business year, at the point of time of the solvency assessment, is referred to as the current year. The risk charge (RC) for UR is therefore derived from the properties of $\text{UR}^{\text{technical}}$, the underwriting result of the undertaking in the current year, which is regarded as a random variable

If you are excluding investment yields on claims' provisions and premiums, one can say that:

$$\text{UR}^{\text{technical}} = \text{EP}^{\text{CY}} - \text{E}^{\text{CY}} - \text{IL}^{\text{CY}} + \text{RunOff},$$

where: EP^{CY}: the earned premiums in the current year; E^{CY}: the expenses related to the current year; IL^{CY}: the incurred losses for claims arising in the current year; and RunOff: the claim's provision runoff result in the current year.

Here, earned premiums are defined as written premiums adjusted by the change in premium provisions, comprising both the provision for unearned premiums and the provision for unexpired risks.

Clearly, the split between premium risk and reserve risk corresponds to

$$\text{UR}^{\text{technical}} = \text{UR}^{\text{premium}} + \text{UR}^{\text{reserve}},$$

where $UR^{premium} = EP^{CY} - E^{CY} - IL^{CY}$, that is, the part of the UR relating to future claims arising from coverage provided on existing contracts and $UR^{reserve} = RunOff$, is the part that relates to reserve risk.

One can assume that the risk capital charge for UR is determined according to ruin probability α and risk measure $\rho_{1-\alpha}$.

For example, one may choose $\rho_{1-\alpha} = VaR_{1-\alpha}$ for Value-at-Risk, or $\rho_{1-\alpha'} = TVaR_{1-\alpha}$ for Tail Value-at-Risk. In this context, $1-\alpha$ or $1-\alpha'$ corresponds to the confidence level ensuring the degree of prudence that CEIOPS wishes to achieve.

The risk capital charge for UR is given by

$$CR^{technical} = \rho_{1-\alpha}(UR^{technical}).$$

At this stage, there was no decision about which risk measure to use (VaR or TVaR).

Considering premium and reserve risk separately, risk capital charges can be defined as

$$CR_{NL,PR} = \rho_{1-\alpha}\left[EP^{CY} - \left(E^{CY} + IL^{CY}\right)\right] \quad \text{and} \quad CR_{NL,RR} = \rho_{1-\alpha}\left[RunOff\right],$$

where $CR_{NL,PR}$: the risk capital charge for premium risk; and $CR_{NL,RR}$ the risk capital charge for reserve risk.

Expressing these charges in terms of the volume measures EP^{CY} and PCO_0, the provision for claims outstanding at the beginning of the current year, one has that:

$$CR_{NL,PR} = \rho_{1-\alpha}\left(1 - CR^{CY}\right) \times EP^{CY} \text{ and } CR_{NL,RR} = \rho_{1-\alpha}\left(RunOff^{\%}\right) \times PCO_0,$$

where $CR^{CY} = (E^{CY} + IL^{CY})/EP^{CY}$ is the combined loss ratio of the insurer in the current year and $RunOff^{\%} = RunOff/PCO_0$ is the (relative) runoff result, expressed in percentage of the outstanding provision at the beginning of the current year.

Translating these theoretical equations into a factor-based, standardized formula requires the following:

- Analysis at the level of individual undertakings
- Generalized analysis that can be applied across the industry

M.1.1.1 Reserve Risk

In general terms, to be able to compute $\rho_{1-\alpha}$, you need to know the probability distribution of the random variable $RunOff^{\%}$. On a practical level, it may be assumed that this distribution is of a type that is completely specified by its first two moments. For example, one may assume that $RunOff^{\%}$ follows a shifted lognormal, or gamma, distribution. Then, $\rho_{1-\alpha}$ may be determined once the following has been specified:

- The type of the distribution
- Its expected value μ
- Its variance σ^2

Assuming that the supervisor sets the type of distribution of RunOff%, the determination of its expected value and variance allows for a wide range of approaches, which vary in their degree of personalization, as follows:

- All parameters are set by the supervisor; the result would be a table of industry-wide factors for reserve risk that can be applied to the insurers' provisions in each segment or
- The expected value and/or variance of the distribution are computed using company-specific data or
- The expected value and/or variance of the distribution are computed using a mixture of company-specific data and data that are set by the supervisor

The decision about the degree of personalization requires a trade-off between accuracy and practicability of the determination of the risk within the limits of the standard formula.

The third alternative is an intermediate approach that uses limited portfolio-specific data to measure the portfolio-specific risk in a reliable and practicable way. For example, the variance of the distribution may be regarded as a function of the size of the portfolio, $f(n)$:

$$\sigma^2\left(\text{RunOff}^{\%}\right) = f(n).$$

The supervisor would provide the function f, and the size n of the portfolio would be determined individually by the insurer. The size of the portfolio could be measured, for example, by the number of risks in the portfolio at the beginning of the time horizon. This approach would combine an assumption on the volatility of the distribution, which is specific to the business line and independent of the single company with the diversification effects caused by the size of the portfolio, which is specific to the company.

M.1.1.2 Premium Risk

The premium risk could be treated in a similar way as the reserve risk. Hence, if the distribution was completely specified by its first two moments, the risk parameter could be determined once the following has been specified:

- The type of the distribution
- Its expected value μ
- Its variance σ^2

This specification can differ in the degree of company-specific information that is evaluated. The level of premium risk strongly depends on the expected value of the combined ratio CR^{CY}.

An estimated combined ratio above 1 increases the risk by the estimated loss, whereas an estimated combined ratio below 1 decreases the risk by the estimated profit. Since insurers tend to differ in the expected value of the combined ratio of their business, it seems advisable to personalize this parameter of CR^{CY}. This could be achieved by estimating the expected value of the combined ratio of the insurer using the historical ratios.

As to the personalization of the variance of the combined ratio, a number of different approaches seem possible. The determination of the variance may be left completely, partly or not at all to the insurer. The decision requires a trade-off between accuracy and practicability of the determination of the variance within the limits of the standard formula. A completely company-specific determination of the variance would make a high demand on the insurer. It would require the insurer to observe the volatility of the relevant business for a long period of time and to make an actuarial analysis on those data. This approach may not seem feasible for the standard formula.

The opposite approach would be to uniformly fix the assumed variance of the combined ratio of a business line for all companies. It would not take into account the differences between the insurers in the volatility of their combined ratios. In particular, the standard formula would not differentiate between the volatility of large and small portfolios.

An intermediate approach would be to determine the variance of the combined ratio by referring to data that are partly specific to the company and partly independent from the company. Given a coefficient of variation for the claim size, the insurer may calculate the coefficient of variation of the loss ratio by estimating the expected number of claims of its portfolio.

While this approach is quite ambitious, a more pragmatic attempt to personalize the determination of the variance is to regard the variance as a function of the size of the portfolio, $f(n)$:

$$\sigma^2 \left(\mathrm{CR}^{\mathrm{CY}}\right) = f(n).$$

The supervisor would provide the function f, and the size n of the portfolio would be determined individually by the insurer. The size of the portfolio could be measured by the number of risks in the portfolio at the beginning of the time horizon. This approach would combine an assumption on the volatility of the combined ratio, which is specific to the business line and independent of the single company with the diversification effects caused by the size of the portfolio, which is specific to the company.

Also scenario-based approaches were discussed (see CEIOPS, 2005).

M.1.1.3 Segmentation

In general terms, an assessment of UR involves an estimation of the variability of the underwriting result of the undertaking. This requires underlying data that are sufficiently homogeneous with respect to emergence, development, and statistical pattern of claims.

For a heterogeneous product, such as commercial multiperil or miscellaneous liability insurance, experience may be segregated into more homogeneous groupings. A suitable segmentation of the book of business might be explicitly defined within the formula, or some flexibility could be allowed so that national particularities can be taken into account. A standard classification that is more closely aligned with actual undertaking behavior should have positive consequences for risk management.

Both premium and reserve risk may be analyzed on the basis of homogenous segments of the portfolio to take the particularities of single segments into account. Such a segmented approach to UR would present the problem of how to aggregate individual RCs. Simply adding up the individual charges would neglect diversification effects between different LoBs. This may lead to an overestimation of the required risk capital.

There are two approaches to deal with this problem:

- One may determine premium, or reserve, risk capital charges for each segment and calculate the overall premium, or reserve, risk capital charge using capital aggregation methods.
- One may determine only the first two moments of the distribution of the premium, or reserve, risk for each segment and calculate the first two moments of the overall premium, or reserve, risk using a dependence matrix for the second moments. Assuming the overall premium, or reserve, risk to have a specific two-parametric probability distribution, one may then calculate the overall premium, or reserve, risk capital charge.

M.1.2 High-Level QIS2–QIS4

In CEIOPS, 20th consultation paper (CEIOPS, 2006b), it is explained that the nonlife UR is a specific insurance risks arising from insurance contracts. These risks should be based on the technicalities of the insurance business: The insurance undertaking has to ensure future payment commitments and the volume of such payments must be calculated in advance. The following explanation has been used in QIS3 (CEIOPS, 2007a; QIS3, 2007), and QIS4 (QIS4, 2008).

The UR relates to the uncertainty about the results of the insurer's underwriting. This includes uncertainty about the following:

- The amount and timing of the eventual claim settlements in relation to existing liabilities
- The volume of business to be written and the premium rates at which it will be written
- The premium rates that would be necessary to cover the liabilities created by the business written

The nonlife UR component of the solvency capital requirement is intended to cover the excess losses that might occur over the 12 months following the date at which it is evaluated on existing provisions and new business. Losses mean the underwriting losses in excess of those expected or the expected profit-less actual outcome at the end of the period.

In the standard formula, there are two risk modules under the heading of the nonlife underwriting risk:

- Premium and reserve risk
- Catastrophe risk (CAT risk)

CEIOPS identifies the two main sources of UR as the premium risk and reserve risk; see CEIOPS (2006b), CEIOPS (2007a), and QIS4 (2008). They are explained in the following way:

Premium risk is understood to relate to *future claims arising during and after the period until the time horizon for the solvency assessment.* In this risk, expenses plus the volume of losses (incurred and to be incurred) for these claims (comprising both amounts paid during the period and provisions made at its end) are higher than the premiums received (or if allowance is made elsewhere for the expected profits or losses on the business, then the profitability will be less than expected).

Premium risk is present at the time the policy is issued, before any insured events occur. Premium risk also arises because of uncertainties prior to issue of policies during the time horizon. These uncertainties include the premium rates that will be charged, the precise terms and conditions of the policies, and the precise mix and volume of business to be written.

Premium risk relates to policies to be written (including renewals) during the period, and to unexpired risks on existing contracts.

Reserve risk stems from two sources: On the one hand, the absolute level of the claims' provisions may be misestimated. On the other hand, because of the stochastic nature of future claims payouts, the actual claims will fluctuate around their statistical mean value.

Some of the stochastic effects relate to individual claims so that they are generally less significant for large portfolios. Others relate to economic conditions and other factors that affect the whole portfolio, so the law of large numbers does not apply to them.

The catastrophe risk is explained in CEIOPS (2006b) as follows.

CAT risks stem from extreme or irregular events that are not sufficiently captured by the charges for premium and reserve risk.

In QIS3 (CEIOPS, 2007a) and QIS4 (QIS4, 2008), it was added that to avoid double counting, the calibration of the scenarios and market losses should allow for the parts of catastrophe risks that are covered by premium risk.

Consultation Paper 20 (CEIOPS, 2006b): When considering possible catastrophe losses over the following 12 months, the intention is that the CAT charge should represent the average effect on the net asset value of the insurer of the 1% of scenarios, including multiple catastrophes that cause the greatest fall in net assets (i.e., 1% TVaR).

QIS3 (CEIOPS, 2007a): For the modeling of nonlife CAT risk in QIS3, regional CAT scenarios were considered that were specified by the local regulator. Additionally, a list of European (transregional) scenarios was prescribed.

QIS4 (QIS4, 2008): The CAT-risk submodule was proposed to be calculated according to two alternative main methods or to an optional method, see Section M.5. This was welcomed by companies participating in QIS4 (CEIOPS, 2008e).

M.1.3 QIS2

It is assumed that there is a dependence of 0.5 between the reserve and premium risk, and zero dependence between the CAT risk and the reserve and premium risks. Hence, the total capital charge is defined as:

$$C_{NL} = \sqrt{C_{NL,RR}^2 + C_{NL,PR}^2 + C_{NL,CAT}^2 + C_{NL,RR}C_{NL,PR}}.$$

The main change that CEIOPS made as a proposal in their Consultation Paper 20 (CEIOPS, 2006b) was to change the charge calculation. As the dependence between the reserve and premium risks is higher between LoB levels than on the total level, the aggregation between LoBs was made by using dependencies on the LoB's level.

The insurance industry made a lot of lobbying on having a VaR approach instead of the TVaR. This was also the approach used in the third quantitative impact study that was conducted during the summer 2007.

M.1.4 QIS3 + QIS4

It is assumed that there is a zero dependence between the CAT risk and the combined reserve and premium risks. Hence, the total capital charge is defined as

$$C_{NL} = \sqrt{C_{NL,RP}^2 + C_{NL,CAT}^2}.$$

M.2 RESERVE AND PREMIUM RISK MODULES

Two main measures are introduced in Solvency II. For each risk, a volume measure and a volatility measure are defined.

In a study on nonlife long-tail business AISAM–ACME* (2007) declared that the calibration of the reserve risk needs to reflect strictly the one-year time horizon rather than the classical full runoff approach. They also found that the magnitude of the UR parameters within QIS3 seemed to be consistent with a full runoff approach rather than the one-year time horizon volatility. For the companies in this study, the calibrated reserve RC was two to three times higher than it would have been under the one-year time horizon.

In the sequel, we will follow how CEIOPS has developed the standard formula for the capital charge of the reserve and premium risks.

M.2.1 QIS2 Model (2006)

In the second quantitative impact study conducted during 2006 (CEIOPS, 2006b, 2006d), the LoB was used (Table M.1).

The TVaR was used as a risk measure for both the two risk types. The modeling used a lognormal assumption and used the capital charge, $\rho_{0.995}(\sigma)_{\text{TVaR}} \times V$, as defined in Equations 21.10a and 21.10b.

* AISAM and ACME have merged into AMICE, Association of Mutual Insurers and Insurance Cooperatives in Europe. Web site: www.insurance-mutuals.org

TABLE M.1 LoB Used in QIS2

LoB	LoB
1	Accident and health
2	Motor, third-party liability
3	Motor, other
4	MAT
5	Fire
6	General, third-party liability
7	Credit
8	Legal expenses
9	Assistance
10	Miscellaneous
11	Reinsurance

Source: Adapted from CEIOPS. 2006d. Quantitative Impact Study No. 2: Technical Specification, CEIOPS-PI-08/06. Available at www.ceiops.org.

Reserve risk | **Premium risk**

General capital charge; see Equations 21.10a and 21.10b

Reserve risk:

$$\rho_{0.995}\,(\sigma_{\mathrm{R}})_{\mathrm{TVaR}} \times V_{\mathrm{R}}$$

Premium risk:

$$\rho_{0.995}\,(\sigma_{\mathrm{P}})_{\mathrm{TVaR}} \times V_{\mathrm{P}}$$

For the reserve risk the final capital charge is calculated as

$$C_{\mathrm{NL,RR}} = \max\{0; \rho_{0.995}\,(\sigma_{\mathrm{R}})_{\mathrm{TVaR}} \times V_{\mathrm{R}} - \mathrm{ES}_{\mathrm{R}}\},$$

where ES_{R} = the expected surplus arising from the next year's runoff result

For the premium risk the final capital charge is calculated as

$$C_{\mathrm{NL,PR}} = \max\{0; \rho_{0.995}\,(\sigma_{\mathrm{P}})_{\mathrm{TVaR}} \times V_{\mathrm{P}} - \mathrm{ES}_{\mathrm{P}}\}$$

where ES_{R} is the expected surplus or deficit arising from next year's premiums

Volume measures

Reserve risk: V_{R} = net technical provision

Premium risk: V_{P} = net earned premium

The volume measures are first calculated for each LoB, k, and then summed up:

$$V_{\mathrm{R}} = \sum_k V_{\mathrm{R},k} \qquad\qquad V_{\mathrm{P}} = \sum_k V_{\mathrm{P},k}$$

The expected surplus are defined as

$$\begin{aligned}\mathrm{ES}_{\mathrm{R}} &= \mu V_{\mathrm{R}} = \frac{\sum_k \mu_k V_{\mathrm{R},k}}{V_{\mathrm{R}}} V_{\mathrm{R}} \\ &= \sum_k \mu_k V_{\mathrm{R},k} = \sum_k \alpha_k \frac{RM_k}{V_{\mathrm{R},k}} V_{\mathrm{R},k} \\ &= \sum_k \alpha_k RM_k \approx \sum_k \frac{1}{D_k} RM_k\end{aligned}$$

$$\begin{aligned}ES_{\mathrm{P}} &= (1-\mu)V_{\mathrm{P}} = \left(1 - \frac{\sum_k \tilde{\mu}_k V_{\mathrm{P},k}}{V_{\mathrm{P}}}\right) \\ &\times V_{\mathrm{P}} = \left(V_{\mathrm{P}} - \sum_k \tilde{\mu}_k V_{\mathrm{P},k}\right)\end{aligned}$$

$$\text{Definition: } \mu_k = \sum_{y:\text{years}} \mathrm{CR}_{k,y} \frac{V_{\mathrm{P},k,y}}{\sum_y V_{\mathrm{P},k,y}}$$

where μ is an estimate of the expected value of the relative runoff result for the overall business in the forthcoming year, which could be estimated by the weighted sum of the relative runoff result for the forth-coming year per LoB.
This μ_k is defined as a proportion (α_k) of the claims provision that is expected to be paid out in the forthcoming year *times* the share of the risk margin to the claims provision.
The proportion (α_k) could be approximated by the reciprocal to the mean duration of the claims provision.

where μ is an estimate of the expected value of the combined ratio of the overall business and is set equal to a weighted sum over the different LoB k.

The $\tilde{\mu}_k$is defined as μ_k, with the distinction that the summation is run over at least 3, but not more than five years. If the number of available historic combined ratios is <3 the $\tilde{\mu}_k$is set equal to 1.

Volatility functions and measures

Reserve risk: $\rho_{0.995}\,(\sigma_{\mathrm{R}})_{\mathrm{TVaR}}$

Premium risk: $\rho_{0.995}\,(\sigma_{\mathrm{P}})_{\mathrm{TVaR}}$ Combined ratio, CR, is used.

The volume measures, σ_*, are first calculated for each LoB, k, and then summed up

The estimate of σ_{R} is given by

$$\sigma_{\mathrm{R}}^2 = \frac{1}{V_{\mathrm{R}}^2}\sum_{j,k} r_{jk} V_{R,j} V_{\mathrm{R},k} \sigma_{Rj} \sigma_{Rk},$$

where the summation is made over all combinations of the LoBs j and k. The dependencies r_{jk} used are given in Table M.3.

The standard deviations are market-wide estimates of the standard deviation of the runoff result in the individual LOBs. Like in the GDV model, see Section 21.1.2, we can write $\sigma_{Rk} = sf \times f_{Rk}$, where sf is a size factor and f_{Rk} a variability factor depending on the LoB k, see Table M.2 below.

Two different approaches was tested. One was based on a market-wide estimate of σ_{P}, σ_{PM}, and one based on an undertaking-specific estimate of σ_{P}, σ_{PU}. Both are estimated from

$$\sigma_{\mathrm{P}}^2 = \frac{1}{V_{\mathrm{P}}^2}\sum_{j,k} r_{jk} V_{\mathrm{P},j} V_{\mathrm{P},k} \sigma_{\mathrm{P}j} \sigma_{\mathrm{P}k},$$

where the volatilities are either based on market-wide data or on undertaking-specific data. The dependencies are given in Table M.3.
Market-specific data
Like in the GDV model, see Section 21.1.2, we can write $\sigma_{\mathrm{P}k} = sf \times f_{\mathrm{P}k}$, where sf is a size factor and f_{Pk} a variability factor depending on the LOB k, see Table M.2 below.

TABLE M.2 Volatility Factors for the Reserve Risk and the Premium Risk

LoB	1	2	3	4	5	6	7	8	9	10	11
f_{Rk}	0.15	0.15	0.075	0.15	0.10	0.20	0.10	0.10	0.20	0.20	0.20
f_{Pk}	0.05	0.125	0.075	0.15	0.10	0.25	0.10	0.15	0.10	0.15	0.15

Source: Adapted from CEIOPS. 2006d. Quantitative Impact Study No. 2: Technical Specification, CEIOPS-PI-08/06. Available at www.ceiops.org.

Note: For the LoB k, see Table M.1.

The size factor is calculated as

$$sf = \begin{cases} 1 & \text{if } 100\,\text{mEuro} \le V_{R,k,\text{Gross}} \\ \dfrac{10}{\sqrt{10^{-6} \times V_{R,k,\text{Gross}}}} & \text{if } 20\,\text{mEuro} < V_{R,k,\text{Gross}} < 100\,\text{mEuro} \\ \dfrac{10}{\sqrt{20}} & \text{if } V_{R,k,\text{Gross}} < 20\,\text{mEuro} \end{cases}$$

The volatility factors are defined as in Table M.2.

Undertaking-specific data

The undertaking-specific standard deviation is a credibility weighted function:

$$\sigma^2_{PU,k} = c_k \times \sigma^2_{CR,k} + (1 - c_k) \times sf^2 \times f^2_{Pk},$$

where c_k is a credibility function defined as

$$c_k = 0.2 \times \max\{0, Y_k - 10\},$$

Y_k is the number of available historic combined ratios for each LoB k. If $Y_k > 10$ the estimate of $\sigma^2_{PU,k}$ is fully based on the data from the undertaking.

$$\sigma^2_{CR,k} = \frac{1}{Y_k - 1} \sum_{y:\text{year}} V_{P,k,y} \left(CR_{k,y} - \mu_{k,y}\right)^2,$$

and CR is the undertaking-specific combined ratios for the historic years.

TABLE M.3 Dependence Matrix for the 11 Different LoBs Used in QIS2

LoB	1	2	3	4	5	6	7	8	9	10	11
1. A&H	1	0.25	0	0	0	0.25	0	0.5	0	0	0
2. Motor, third-party liability		1	0.5	0	0	0	0	0.25	0	0	0
3. Motor, other			1	0.5	0.5	0	0	0	0.5	0	0.5
4. MAT				1	0.25	0	0	0	0.5	0	0.5
5. Fire					1	0	0	0	0.5	0	0.5
6. General liability						1	0.75	0.5	0	0	0.5
7. Credit							1	0.75	0	0	0
8. Legal expenses								1	0	0	0
9. Assistance									1	0	0
10. Miscellaneous										1	0
11. Reinsurance											1

Source: Adapted from CEIOPS. 2006d. Quantitative Impact Study No. 2: Technical Specification, CEIOPS-PI-08/06. Available at www.ceiops.org.

Note: See also Table M.1 for the LoB.

TABLE M.4 LoB Used in QIS3

LoB	LoB
1	A&H: worker's comp
2	A&H: health
3	A&H: others/default
4	Motor, third-party liability
5	Motor, other
6	MAT
7	Fire/property
8	Third-party liability
9	Credit
10	Legal expenses
11	Assistance
12	Miscellaneous
13	NP reinsurance (property)
14	NP reinsurance (casualty)
15	NP reinsurance (MAT)

Source: Adapted from CEIOPS. 2007a. QIS 3 Technical Specifications, Part I: Instructions, CEIOPS-FS-11/07 & Part II: Background Information, CEIOPS-FS-12/07. April. Available at www.ceiops.org.

Note: A&H, accident and health. Faculty and proportional reinsurance is classified according to Classes 1–10. For nonproportional reinsurance a split into three different LoB was used.

The different volatility measures for the whole business are summed up using the dependency structure in Table M.3.

M.2.2 QIS3 Model (2007)

In the third quantitative impact study conducted during the summer 2007 (CEIOPS, 2007a; QIS3, 2007) used the LoB in Table M.4.

The VaR was used as a risk measure for both the two risk types. The modeling used a lognormal assumption and the capital charge for the combined reserve and premium risks was set to$C_{NL,RP} = \rho_{0.995}(\sigma)_{VaR} \times V$; see Equations 21.9a and 21.9b. It was also shown that the factor function ρ could be approximated by:

$$\rho_{0.995}(\sigma)_{VaR} \approx 3\sigma. \tag{M.1}$$

The calculation was made in two steps:

1. For each LoB, the standard deviation and volume measures were calculated for both the reserve risk and the premium risk
2. The standard deviation and the volume measure for the reserve risk and the premium risk in the individual LoB are aggregated to derive an overall volume measure V and an overall standard deviation σ

Step 1: Volume and volatility measures per LoB

Reserve risk	**Premium risk**
	Loss ratio, LR, is used.

For each LoB, k, we define the volume measures as

Reserve risk:

$V_{R,k} = R_k^0$,

where we have used the notation used in notation used in Section 21.1.4. R_k^0 = the net provisions for claims outstanding in LoB k (the opening reserve)

Premium risk:

$V_{P,k} = \max\left\{P_k^{01}; \tilde{P}_k^{01}; 1.05 \times P_k^{-1}\right\}$,

where we have used the notation used in Section 21.1.4. P_k^{01} = estimate of the net *written* premium in LoB k in the forthcoming year; $\tilde{P}_k^{01}$ = estimate of the net *earned* premium in LoB k in the forthcoming year; and P_k^{-1} = of the net *written* premium in LoB during the previous year.

For each LoB, k, we define the volatility measures as

Reserve risk:

$\sigma_{R,k}^2 = s_{R,k}^2$,

where $s_{R,k}$ = standard deviation for reserve risk for LoB k; see Table M.5.

Premium risk:

$\sigma_{P,k}^2 = c_k \times \sigma_{U,k}^2 + (1 - c_k) \times \sigma_{M,k}^2$

where c_k is a credibility function defined as

$$c_k = \begin{cases} \dfrac{Y_k}{Y_k + 4} & \text{if } Y_k \geq 7 \text{ years}, \\ 0 & \text{otherwise.} \end{cases}$$

The figure 4 is a credibility constant. $\sigma_{M,k}$ = the market-wide estimate of the standard deviation of the premium risk is given in Table M.5. $\sigma_{U,k}$ = the undertaking estimate of the standard deviation of the premium risk is defined as

$$\sigma_{U,k}^2 = \frac{1}{(Y_k - 1)\, V_{P,k}} \sum_{y:\text{year}} V_{k,y}^{\tilde{P}} (\text{LR}_{k,y} - \mu_{k,y})^2$$

where $V_{k,y}^{\tilde{P}}$ = the net earned premiums for historical years $-1, -2, \ldots, -n$; $LR_{k,y}$ = the loss-ratio in LoB k, for years $-1, -2, \ldots, -n$; and $\mu_k = \sum_{y:\text{years}} \text{LR}_{k,y} \times \dfrac{V_{k,y}^{\tilde{P}}}{\sum_y V_{k,y}^{\tilde{P}}}$ is a company-specific estimate of the expected value of the loss ratio in LoB k weighted by historical net earned premiums.

TABLE M.5 Standard Deviation for the Market-Wide Reserve Risk ($S_{R,k}$) and for the Market-Wide Premium Risk ($\sigma_{M,k}$)

LoB	1	2	3	4	5	6	7	8	9	10	11	12	13	14	15
$s_{R,k}$	0.15	0.075	0.15	0.125	0.075	0.15	0.10	0.15	0.10	0.10	0.10	0.15	0.15	0.20	0.20
$\sigma_{M,k}$	0.075	0.03	0.05	0.10	0.10	0.125	0.10	0.10	0.125	0.05	0.075	0.125	0.15	0.15	0.15

Source: Adapted from CEIOPS. 2007a. QIS 3 Technical Specifications, Part I: Instructions, CEIOPS-FS-11/07 & Part II: Background Information, CEIOPS-FS-12/07. April. Available at www.ceiops.org.

Step 2: Overall volume and volatility measures

At the last step, we aggregate the volume measures and the volatility measures using the dependency structure in Table M.6.

The overall volume measure is defined as the sum of the volume measures for the reserve and premium risks summed up over all LoB:

$$V_{RP} = \sum_k (V_{RR} + V_{PR})$$

The overall volatility measure is defined by

$$\sigma^2 = \frac{1}{V^2} \sum_{j,k} r_{jk} V_j V_k a_j a_k,$$

where the dependencies between LoB, ρ_{RP}, are given in Table M.6 and the factors a_k is defined as

$$a_k = \begin{cases} \sigma_{R,k} & \text{if Reserve risk} \\ \sigma_{P,k} & \text{if Premium risk} \end{cases}.$$

TABLE M.6 Dependence Matrix Used for QIS3

LoB	1	2	3	4	5	6	7	8	9	10	11	12	13	14	15
1. A&H-worker's compensation	1	0.5	0.5	0.25	0.25	0.25	0.25	0.5	0.25	0.5	0.25	0.5	0.25	0.25	0.25
2. A&H-health		1	0.5	0.25	0.25	0.25	0.25	0.25	0.25	0.25	0.25	0.5	0.25	0.25	0.25
3. A&H-default			1	0.25	0.25	0.25	0.25	0.25	0.25	0.5	0.25	0.5	0.25	0.25	0.25
4. Motor, third-party liability				1	0.5	0.5	0.25	0.5	0.25	0.5	0.25	0.5	0.25	0.25	0.25
5. Motor, other					1	0.25	0.25	0.25	0.25	0.5	0.5	0.5	0.25	0.25	0.25
6. MAT						1	0.25	0.25	0.25	0.25	0.5	0.5	0.25	0.25	0.5
7. Fire							1	0.25	0.25	0.25	0.5	0.5	0.5	0.25	0.5
8. General liability								1	0.5	0.25	0.25	0.25	0.25	0.25	0.25
9. Credit									1	0.5	0.25	0.5	0.25	0.5	0.25
10. Legal expenses										1	0.25	0.5	0.25	0.5	0.25
11. Assistance											1	0.5	0.5	0.25	0.25
12. Miscellaneous												1	0.25	0.25	0.5
13. Reinsurance (property)													1	0.25	0.25
14. Reinsurance (casualty)														1	0.25
15. Reinsurance (MAT)															1

Source: Adapted from CEIOPS. 2007a. QIS 3 Technical Specifications, Part I: Instructions, CEIOPS-FS-11/07 & Part II: Background Information, CEIOPS-FS-12/07. April. Available at www.ceiops.org.

Note: The LoBs are defined in Table M.4.

$$r_{jk} = \begin{pmatrix} \rho_{RP} & \alpha\rho_{RP} \\ \alpha\rho_{RP} & \rho_{RP} \end{pmatrix},$$

and α is a factor representing on overall assumption between premium and reserve risk. In QIS3, it was set equal to 0.5.

M.2.3 QIS4 Model (2008)

In the forth quantitative impact study that was conducted during the summer 2008 (QIS4, 2008), the same LoB as in QIS3 was used; see Table M.4. But, as the accident and health LoB were taken out of the nonlife risk module and presented under the Health risk module, we have to renumber the LoBs. Old number 4 is now 1, and so on.

The main change from QIS3 was the introduction of a new second step in the calculation of the volume and volatility measures: geographical diversification.

The calculation was made in two steps:

1. For each LoB, the standard deviation and volume measures was calculated for both the reserve risk and the premium risk
2. For each LoB, geographical diversification is determined
3. The standard deviation and the volume measure for the reserve and premium risks in the individual LoB are aggregated to derive an overall volume measure V and an overall standard deviation σ

Step 1: Volume and volatility measures per LoB

Reserve risk	**Premium risk**
	Loss ratio, LR, is used.

For each geographical area we do the calculation set out below.
We have dropped the index for geographical area.
For each LoB, k, we define the volume measures as

$V_{R,k} = R_k^0$,	$V_{P,k} = \max\left\{P_k^{01}; \tilde{P}_k^{01}; 1.05 \times P_k^{-1}\right\}$,
where we have used the notation used in Section 21.1.4. R_k^0 = the net provisions for claims outstanding in LoB k (the opening reserve)	where we have used the notation used in Section 21.1.4. P_k^{01} = estimate of the net *written* premium in LoB k in the forthcoming year; $\tilde{P}_k^{01}$ = estimate of the net *earned* premium in LoB k in the forthcoming year; and P_k^{-1} = of the net *written* premium in LoB k during the previous year. If the insurer has committed to its regulator that it will restrict premiums written over the period so that the actual premiums

written (or earned) over the period will not exceed its estimated volumes, the volume measure is determined only with respect to estimated premium volumes, so that in this case:

$$V_{P,k} = \max\left\{P_k^{01}; \tilde{P}_k^{01}\right\}.$$

For each LoB, k, we define the volatility measures as

$$\sigma_{R,k}^2 = s_{R,k}^2,$$

where $S_{R,k}$= standard deviation for reserve risk, for LoB k; see Table M.7.

$$\sigma_{P,k}^2 = c_k\sigma_{U,k}^2 + (1 - c_k)\,\sigma_{M,k}^2$$

where c_k is a credibility function defined in Table M.8.

$\sigma_{M,k}$ = the market-wide estimate of the standard deviation of the premium risk is given in Table M.7. $\sigma_{U,k}$ = the undertaking estimate of the standard deviation of the premium risk is defined as

$$\sigma_{U,k}^2 = \frac{1}{(Y_k - 1)\,V_{P,k}} \times \sum_{y:\text{year}} \tilde{P}_k^y \left(\mathrm{LR}_{k,y} - \mu_{k,y}\right)^2$$

where $\mathrm{LR}_{k,y}$ = the loss-ratio in LoB k, for years $-1, -2, \ldots, -n$

$\mu_k = \sum_{y:\text{years}} \mathrm{LR}_{k,y} \dfrac{\tilde{P}_k^y}{\sum_y \tilde{P}_k^y}$ is a company-specific estimate of the expected value of the loss ratio in LoB k weighted by historical net earned premiums.

The volatility for the reserve and premium risks for each LoB is calculated by

$$\sigma_k^2 = \frac{\left(V_{R,k}\sigma_{R,k}\right)^2 + \left(V_{P,k}\sigma_{P,k}\right)^2 + 2\alpha V_{R,k}\sigma_{R,k}V_{P,k}\sigma_{P,k}}{V_k^2},$$

where $V_k = V_{R,k} + V_{P,k}$ is the volume measure within each LoB k and α is a factor representing on overall assumption between premium and reserve risk. In QIS4, as well in QIS3, it was set equal to 0.5.

TABLE M.7 Standard Deviation for the Market-Wide Reserve Risk ($s_{R,k}$) and for the Market-Wide Premium Risk ($\sigma_{M,k}$)

LoB	1	2	3	4	5	6	7	8	9	10	11	12
$s_{R,k}$	0.12	0.07	0.10	0.10	0.15	0.15	0.10	0.10	0.10	0.15	0.15	0.15
$\sigma_{M,k}$	0.09	0.09	0.125	0.10	0.125	0.15	0.05	0.075	0.11	0.15	0.15	0.15

Source: Adapted from QIS4. 2008. QIS 4 Technical Specifications. European Commission. MARKT/2505/08. March 31. Available at http://www.ceiops.eu/content/view/118/124/.

Note: LoB number 1 is number 4 in Table M.4, number 2 is number 5 in that table.

TABLE M.8 Credibility Factors c_k Depending on LoB k and the Maximum Value of the Number of Historical Years Used

c_k	**Number of historical years of data available (excl. the first three years after the LoB was first written)**														
Max. no. years															
LoB	1	2	3	4/*1*	5/*2*	6/*3*	7/*4*	8/*5*	9/*6*	10/*7*	11/*8*	12/*9*	13/*10*	14/*11*	15/*12*
15	0	0	0	0	0	0	0.64	0.67	0.69	0.71	0.73	0.75	0.76	0.78	0.79
10	0	0	0	0	0.64	0.69	0.72	0.74	0.76	0.79	—	—	—	—	—
5	0	0	0.64	0.72	0.79	—	—	—	—	—	—	—	—	—	—

Source: Adapted from QIS4. 2008. QIS 4 Technical Specifications. European Commission. MARKT/2505/08. March 31. Available at http://www.ceiops.eu/content/view/118/124/.

Note: The LoB here includes the three accident and health risks; see Table M.4: The numbers in *italics* show the new numbering of the nonlife LoB. The maximum of years to use is 15 for LoBs: *1,5,6,11*. Ten years for LoBs: *3,9,12*, and five years for LoBs: *2,4,7,8,10*. (This is valid for the 12 LoBs in the nonlife risk module.)

Step 2: Geographical Diversification

Diversification was not allowed for LoB's miscellaneous (no. 9) and credit and suretyship (no. 6).

The Herfindahl–Hirshmann index is calculated for premium and reserve risks for each LoB (*cf*. Section 14.4.1):

$$H_{\mathrm{RP},k} = \frac{\sum_{g:\mathrm{geo}} \left(V_{\mathrm{R},k,g} + V_{\mathrm{P},k,g}\right)^2}{\left[\sum_{g:\mathrm{geo}} V_{\mathrm{R},k,g} + V_{\mathrm{P},k,g}\right]^2},$$

where the sum is taken *over* prespecified geographical areas.

Premiums and provisions have to be allocated between, in total, 54 geographical areas:

- Each country of the EEA
- Switzerland
- The rest of Europe
- Asia (excluding Japan and China)
- Japan
- China
- Oceania (excluding Australia)
- Australia
- Canada
- United States
- Mexico
- Rest of North and Central America
- Each country of South America
- Africa

The volume measure is defined as

$$V_k = 0.75\left(V_{R,k} + V_{P,k}\right) + 0.25H_{RP,k}\left(V_{R,k} + V_{P,k}\right).$$

Step 3: Overall Volume and Volatility Measures

At the last step, we aggregate the volume measures and the volatility measures using the dependency structure in Table M.9.

The overall volume measure is defined as the sum of the volume measures for the reserve and premium risks summed up over all LoBs and taken care of the diversification effect:

$$V_{RP} = \sum_k V_k$$

The overall volatility measure is defined by

$$\sigma^2 = \frac{1}{V^2}\sum_{j,k} \rho_{jk} V_j V_k \sigma_j \sigma_k,$$

where the dependencies between LoB, ρ_{jk}, are given in the table below and the sum is taken over all LoBs.

The overall capital charge is then defined as

$$C_{NL,RP} = \rho_{0.995}(\sigma)_{VaR}\, V_{RP},$$

where the charge function is defined by Equation 21.9b.

TABLE M.9 Dependence Matrix Used for QIS4

LoB	1	2	3	4	5	6	7	8	9	10	11	12
1. Motor, third-party liability	1	0.5	0.5	0.25	0.5	0.25	0.5	0.25	0.5	0.25	0.25	0.25
2. Motor, other		1	0.25	0.25	0.25	0.25	0.5	0.5	0.5	0.25	0.25	0.25
3. MAT			1	0.25	0.25	0.25	0.25	0.5	0.5	0.25	0.25	0.5
4. Fire				1	0.25	0.25	0.25	0.5	0.5	0.5	0.25	0.5
5. General liability					1	0.5	0.5	0.25	0.5	0.25	0.5	0.25
6. Credit						1	0.5	0.25	0.5	0.25	0.5	0.25
7. Legal expenses							1	0.25	0.5	0.25	0.5	0.25
8. Assistance								1	0.5	0.5	0.25	0.25
9. Miscellaneous									1	0.25	0.25	0.5
10. Reinsurance (property)										1	0.25	0.25
11. Reinsurance (casualty)											1	0.25
12. Reinsurance (MAT)												1

Source: Adapted from QIS4. 2008. QIS 4 Technical Specifications. European Commission. MARKT/2505/08. March 31. Available at http://www.ceiops.eu/content/view/118/124/.

Note: The LoBs are defined in Table M.4 where LoB number 1 equals number 4 in that table and so on.

In QIS4, the geographical diversification that was introduced was seen as introducing unnecessary complexity at solo level (CEIOPS, 2008e).

M.2.4 Calibration

Under QIS2, the "market-wide" standard deviations σ, for the reserve and premium risks on the level of an individual LOB, were determined as

$$\sigma_{*k} = f_{*k} \times sf,$$

where $_*$ indicates either the reserve risk or the premium risk; sf was a size factor equal for both the reserve and premium risks, dependant on the size of the volume measure; and f was a constant factor specific for the LoB and risk.

The assumption behind this is based on the fact that the variance could be estimated by linear regression, $\sigma^2 \approx a + b \times V^2$ with the constant a equal to zero, the slope $\sqrt{b} = f_{*k}$, and the volume measure is taken as a size factor defined in given buckets of size. When the firms "volume" is zero, there is no business, and hence, in a going concern, the volatility is zero.

M.2.4.1 Pre-QIS3 Calibration

In consultation caper 20 (CEIOPS, 2006b), it is explained that the design of the size factors in QIS2 assumed that the variance of the *outturn* of an LoB was proportionate to the square root of the premium income or the claims provision (with very small and very large accounts being treated differently). This as assumed as appropriate, if the portfolios of small and large insurers are similar and the risks are all independent.

Both these assumptions were not true. In general, small insurers do not write the larger risks, and they generally need and therefore tend to buy more reinsurance protection. The risks are therefore not independent. They may be affected by, for example, economic conditions such as inflation, the weather, changes in society, and changes in legislation or judicial decisions. Some of these risks apply after the claims have been incurred as well as before. Such risks are largely not diversifiable within an LoB (CEIOPS, 2006b, 5.331–5.333).

The risk that is not diversifiable would be expected to be directly proportional to the size of the portfolio, while the diversifiable risk should be proportional to the square root (at least if the risks are similar). The variance of the *outturn* could be defined as

$$\sigma^2_{O*k} = f^2_{*k} V^2_{*k} + nd_{*k} V_{*k} = \left[f^2_{*k} + \frac{nd_{*k}}{V_{*k}} \right] V^2_{*k}, \tag{M.2}$$

where V_{*k} is a volume measure for either the reserve risk or the premium risk, f and nd are constant factors specific for the LoB and risk (f for the diversifiable and nd for the nondiversifiable parts).

While the assumption that all risks are similar does not hold, this might still be a reasonable approximation. We redefine the slope, that is, the variability factor f above as an estimate of the systematic standard deviation, that is, $f_{*k} = S_{*k}$. Then the market-wide standard deviations of the reserve and premium risks are

$$\sigma_{*k} = \sqrt{S_{*k}^2 + \frac{nd_{*k}}{V_{*k}}}. \tag{M.3}$$

This was a proposal set out in the Consultation Paper 20 (CEIOPS, 2006b), and if this formula was to be adopted, CEIOPS assumed that it would be logical to aggregate the components $S^2 \times V^2$ across LoBs and premium and reserve risk with a covariance matrix and the components $nd \times V$ assuming independence. This has the advantage that it allows the "law of large numbers" to apply across LoBs as well as within them. The resulting expression for the variance of the outturn would be as follows:

$$\sigma_{\mathrm{O}}^2 = \frac{1}{V^2}\left\{\sum_{i,j} r_{ij}S_iS_jV_iV_j + \sum_i nd_iV_i\right\},$$

where the summation runs over all indices *i,j* of the form (prem,*k*) or (res,*k*), $k =$ LoB. V denotes the overall volume measure, that is, the sum of the reserve and premium risk volumes.

CEIOPS also indicates that when selecting dependence coefficients, allowance should be made for tail dependence. To allow for this, the dependence used should be higher than simple analysis of relevant data would indicate. For the above formula, the dependencies should be increased to reflect the diversification implicit in the second summation (CEIOPS, 2006b, 5.336).

M.2.4.2 Estimation of Market-Wide Parameters S_{*k}^2 *and* nd_{*k}

Calibration of the market-wide parameters S_{*k}^2 and nd_{*k} is discussed in CEIOPS (2007d).

Premium risk: The calibration was carried out by comparing undertaking's specific estimations of the volatility for premium risk with the market-wide estimate of the standard deviations defined by Equation M.3. Ideally, the parameters s_k and nd_k can be chosen such that the market-wide estimate is near to the undertaking's specific estimate, so that

$$\sigma_{\mathrm{P}k}^2 = S_{\mathrm{P}k}^2 + \frac{nd_{\mathrm{P}k}}{V_{\mathrm{P}}} \approx Sd_{\mathrm{P}k,U}^2,$$

that is,

$$nd_{\mathrm{P}k} + V_{\mathrm{P}}S_{\mathrm{P}k}^2 \approx V_{\mathrm{P}}Sd_{\mathrm{P}k,U}^2,$$

for each individual insurer (U) and for each relevant volume measure V. Note that both the market-wide estimate σ_k and the undertaking-specific standard deviation depend on the size of the volume measure V.

Therefore, the parameters S_k and nd_k were determined using least-squares optimization, that is, they were chosen such that the sum of the squares of the residuals is minimized:

$$VS = \sum_U \left[\left(nd_k + S_k^2V\right) - Sd_{k,U}^2V\right]^2.$$

To carry out this approach, the volume measure V has to be specified. Since the undertaking's specific estimations of the premium risk volatility were derived using the volatility of historic loss ratios

$$\mathrm{LR}_{k,U}^{-1}, \mathrm{LR}_{k,U}^{-2}, \ldots, \mathrm{LR}_{k,U}^{-n},$$

V was determined as the average gross premium income for the individual insurer in the LoB k, over the period that was used to derive the estimate $\mathrm{sd}_{k,U}$. For each insurer, an observation period between 11 and 15 years was used.

For the determination of the undertaking's specific estimate $\mathrm{sd}_{k,U}$ of the standard deviation, the following set of assumption was used:

- The historic net loss ratios $\mathrm{LR}_{k,U}^{-1}, \mathrm{LR}_{k,U}^{-2}, \ldots, \mathrm{LR}_{k,U}^{-n}$ are uncorrelated
- The standard deviation of the loss ratio LR_k of the insurer during the observation period is proportional to

  $$\sqrt{S_{\mathrm{P}k}^2 + \frac{nd_{\mathrm{P}k}}{P_{ky}}}$$

  that is, there exists a proportionality constant a_{kU} such that

  $$\sigma\left(\mathrm{LR}_k^y\right) = a_{k,U}\sqrt{S_{\mathrm{P}k}^2 + \frac{nd_{\mathrm{P}k}}{P_{ky}}}$$

  for each $y = -1, -2, \ldots, -n$, where LR_k^y = net loss ratios in the LoB for historic years $y = -1, -2, \ldots, -n$. In the calibration exercise for the German market, combined ratios including costs were used and $P_{k,y}$ = earned gross premium of the insurer in the LoB and for historic year y.

Under these assumptions, an unbiased estimator of the proportionality constant a_{kU} is given by:*

$$\hat{a}_{kU} = \sqrt{\frac{1}{(n_k - 1)} \sum_{y:\text{year}} w_y \left(\mathrm{LR}_k^y - \mu_k\right)^2},$$

where w_y are weights depending on historic years y, n_k is the number of historic years (at most 15), and μ_k is a company-specific estimate of the expected value of the loss ratio in the LoB k, defined as the weighted average of historic loss ratios

$$\mu_k = \frac{\sum_y w_y \times \mathrm{LR}_k^y}{\sum_y w_y},$$

* This follows from a simple algebraic calculation using the assumptions in the previous paragraph.

where the weights w_y are defined as

$$w_y = \frac{1}{s_{\mathrm{P}k}^2 + \dfrac{nd_{\mathrm{P}k}}{P_{ky}}}.$$

From the assumptions on the standard deviation of the loss ratios for individual insurers U, it follows that the standard deviation corresponding to the volume measure V is given by:

$$\mathrm{sd}_{kU} = a_{kU}\sqrt{\mathrm{s}_k^2 + \frac{nd_k}{V}}.$$

Therefore, given the estimator of a_{kU} as above, the individual standard deviation can be estimated as:

$$\mathrm{sd}_{kU} = \sqrt{\frac{1}{(n_k - 1)\,w} \sum_{y:\mathrm{year}} w_y \left(\mathrm{LR}_k^y - \mu_k\right)^2},$$

where $w = \dfrac{1}{s_k^2 + \dfrac{nd_k}{V}}$.

Following this rationale, the estimator of sd_{kU} was used in the least squares optimization to derive the parameters s_k and nd_k.

Calibration results for the seven LoBs were discussed in CEIOPS (2007d). Note that nd_k is written as d in the following Figures M.1a–g.

Reserve risk: This risk is not discussed in CEIOPS (2007d).

Due to comments on the Consultation Paper 20 and a draft proposal for QIS3, as described above, CEIOPS made some changes to the final QIS3 version (CEIOPS, 2007a, 2007e).

M.2.4.3 QIS3: Market-Wide Factors for Premium and Reserve Risks

For premium risk, the calibration of the market-wide factors was carried out by analyzing different undertaking's specific estimations of the volatility for premium risk, derived from historic loss ratios, using data from the German insurance market. The following approach was chosen:

Ideally, the market-wide estimate σ_{M} can be chosen such that it is near to the undertaking's specific estimate σ_{U}, that is,

$$\sigma_{\mathrm{M}} \approx \sigma_{\mathrm{U}} \text{ so that } V\sigma_{\mathrm{M}} \approx V\sigma_{\mathrm{U}}$$

for each individual insurer, U, and for each relevant volume measure V.

Therefore, the market-wide estimate σ_{M} was determined using least-squares optimization, that is, it was chosen such that the sum of the squares of the residuals is minimized:

$$VS = \sum_{U} (V\sigma_{\mathrm{M}} - V\sigma_{\mathrm{U}})^2,$$

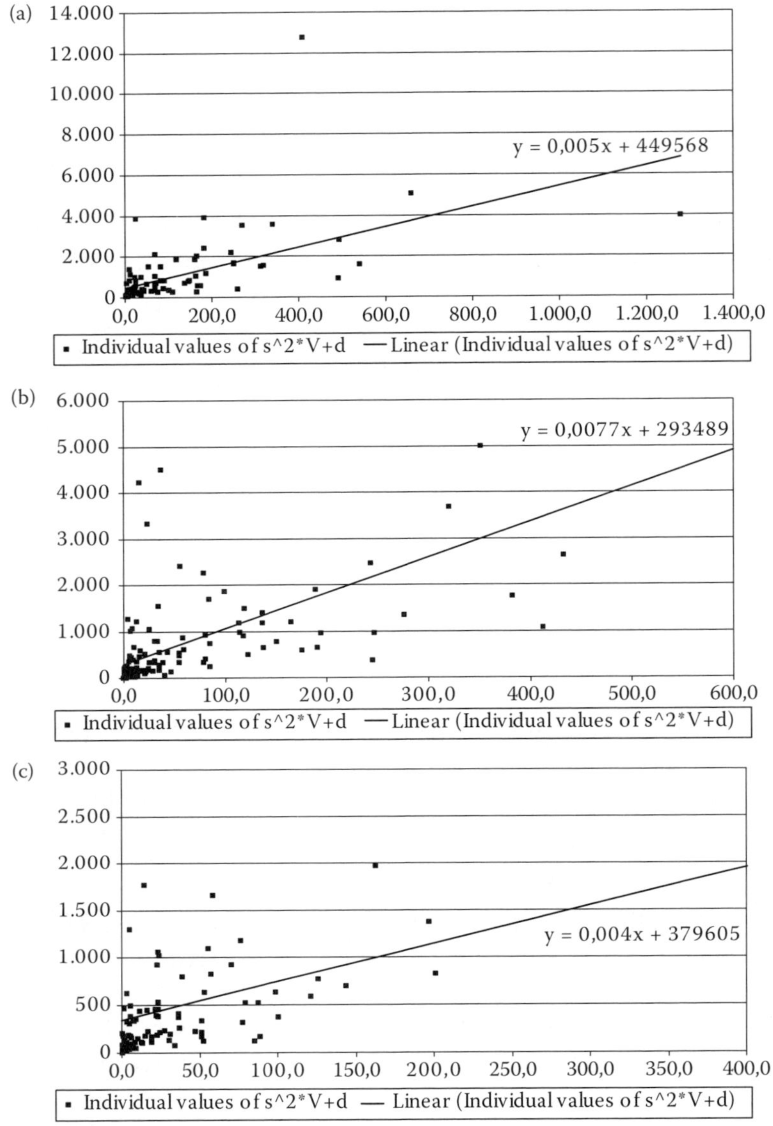

FIGURE M.1 (a) Calibration results for motor third-party liability based on data from 95 companies (1988–2002). Parameter estimates: $d = 449,568, \mathrm{s}^2 = 0.50\%$. (b) Calibration results for fire and property based on data from 125 companies (1988–2002). Parameter estimates: $d = 293,489, \mathrm{s}^2 = 0.77\%$. (c) Calibration results for third-party liability based on data from 101 companies (1988–2002). Parameter estimates: $d = 379,605, \mathrm{s}^2 = 0.40\%$.

where the volume measure V was determined as the average gross premium income for the individual insurer in the LoB, over the period that was used to derive the insurer's individual estimate of the standard deviation. The constant in the linear regression is set to equal zero. The results of performing a linear regression analysis are described in Table M.10.

In CEIOPS (2007e), the calculation is only illustrated by the linear regression analysis for LoB 4, motor third-party liability (see Figure M.2).

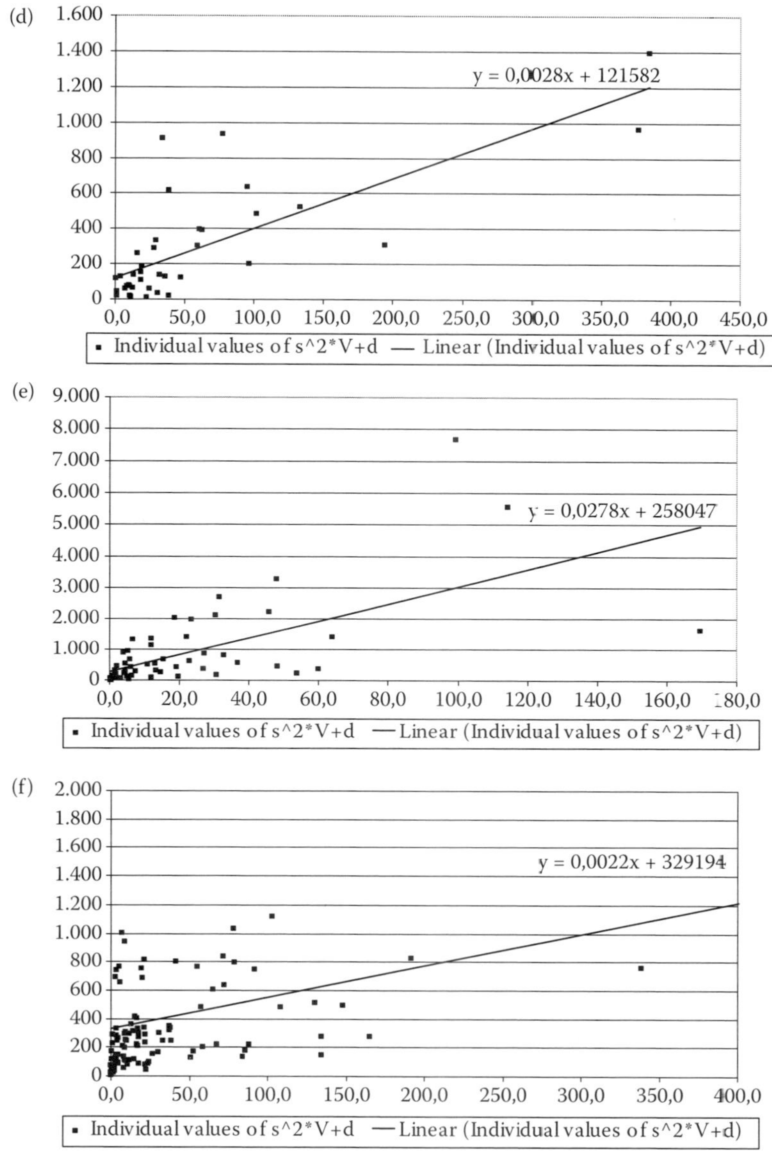

FIGURE M.1 (Continued) (d) Calibration results for legal expenses based on data from 38 companies (1988–2002). Parameter estimates: $d = 121,582$, $s^2 = 0.28\%$. (e) Calibration results for MAT based on data from 57 companies (1988–2002). Parameter estimates: $d = 258,047$, $s^2 = 2.78\%$. (f) Calibration results for accident and health based on data from 112 companies (1988–2002). Parameter estimates: $d = 329,194$, $s^2 = 0.22\%$.

Motor Third-Party Liability: To allow for a more refined calibration, the "accident and health" LoB was split into three sub-LoBs:

- Accident and health—worker's compensation business (LoB no. 1)
- Accident and health—health (LoB no. 2)

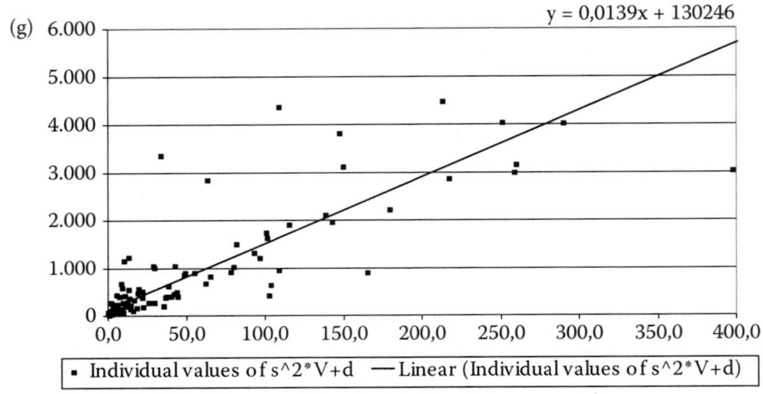

FIGURE M.1 (Continued) (g) Calibration results for motor and other classes based on data from 97 companies (1988–2002). Parameter estimates: $d = 130,246$, $s^2 = 1.39\%$. (Adapted from CEIOPS. 2007d. Calibration of QIS3 SCR standard formula for non-life underwriting risk. CEIOPS, SCR Subgroup, 2007-02-19; © Committee of European Insurance and Occupational Pensions Supervisors.)

TABLE M.10 Results from a Linear Regression Analysis Based on German Data

LoB,*k*	**No. Firms**	**No. LR**	**Statistical σ (%)**	**Chosen σ (%)**
1–3. A& health	112	1609	6	5[a]
4. Motor third party	96	1364	10	10
5. Motor, other	98	1394	10	10
6. MAT	58	815	13	12.5
7. Fire/property	127	1821	9	10
8. Third-party liability	99	1421	6	10
10. Legal expenses	30	430	5	5

Source: Adapted from CEIOPS. 2007e. Calibration of the UR modules for QIS3. CEIOPS. April 1. Available at www.ceiops.org.

[a] This factor was used for LoB 3, accident and health—other.

- Accident and health—other (LoB no. 3)

The calibration of the premium risk factor (RF) for *worker's compensation* business was carried out using statistical data from the Portuguese insurance market, following a similar approach as described above.

The calibration of the premium RF for *health business* was derived from a statistical analysis on basis of data from the French insurance market, also taking into account the results of a study conducted by Swiss Re (Sigma 2006).

The calibration of the premium RFs for the *remaining LoBs* (credit and suretyship, assistance, miscellaneous nonlife insurance, and the three nonproportional reinsurance classes) was chosen judgmentally, taking into account the feedback received from QIS2.

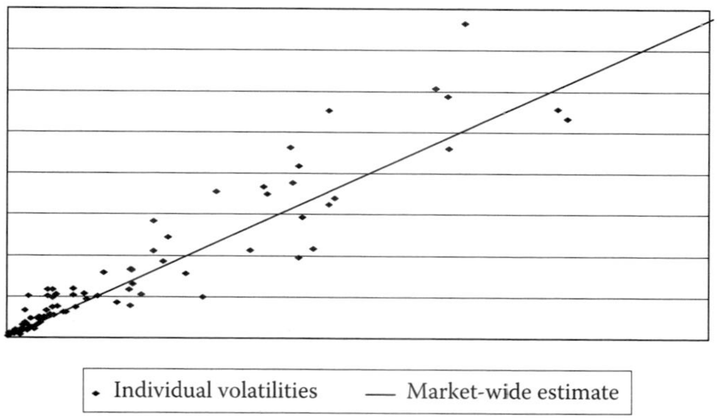

FIGURE M.2 Calibration results for motor third-party liability classes based on data from 96 companies (1988–2002). Parameter estimate: $s^2 = 0.01\%$. Some of the largest insurance companies were left out from the diagram. (Adapted from CEIOPS. 2007e. Calibration of the underwriting risk modules for QIS3. CEIOPS. 1 April. Available at www.ceiops.org; © Committee of European Insurance and Occupational Pensions Supervisors.)

For reserve risk, the calibration RF for the following LoBs:

- Motor, third-party liability
- Third-party liability

was carried out using statistical data from the U.K. insurance market, following a similar approach as was used for the determination of the premium RFs.

For *health business*, as in the case of premium risk, the calibration of the reserve RF was derived from a statistical analysis on basis of data from the French insurance market.

For the *remaining LoBs*, the reserve RFs were assessed judgmentally, using as a starting point the base reserve RFs applied in QIS2. Taking into consideration results from the British ICAS submissions into U.K. FSA and feedback from the QIS2 exercise, the QIS2 base reserve factors were adjusted downwards in some cases (assistance and miscellaneous nonlife insurance).

M.2.4.4 QIS3 Calibration of Credibility Constant for Premium Risk

The calibration of the credibility constant is discussed in CEIOPS (2007e) and we follow the discussion made there. The variance for premium risk in the individual LoB is derived as a credibility mix of an undertaking-specific estimate (U) and the market-wide estimate (M) as follows:

$$\sigma_{\mathrm{P},k}^2 = c_k \sigma_{U,k}^2 + (1 - c_k)\,\sigma_{M,k}^2,$$

where $\sigma_{\mathrm{P},k}$: the resulting estimate of the standard deviation for premium risk; LoB kc_k: credibility factor for LoB k; $\sigma_{U,k}$: undertaking-specific estimate of the standard deviation for premium risk, LoB k; and $\sigma_{M,k}$: market-wide estimate of the standard deviation for premium risk, LoB k, calibrated as described above.

The credibility factor c_k is defined as

$$c_k = \begin{cases} \dfrac{Y_k}{Y_k + b_k} & \text{if } Y_k \geq 7 \text{ years,} \\ 0 & \text{otherwise,} \end{cases}$$

where b_k: a credibility constant depending on the LoB k; and Y_k: the number of historic loss ratios available for undertaking U in LoB k; at most 15.

The calibration of the credibility factor was derived by an application of the classical Bühlmann–Straub credibility model. For this, the following set of assumptions on the loss ratios $LR_{k,U}^{-1}, LR_{k,U}^{-2}, \ldots, LR_{k,U}^{-n}$ of individual insurers U were used:

a. Conditionally, for fixed structure parameter Θ_U, $LR_{k,U} = \left(LR_{k,U}^{-1}, LR_{k,U}^{-2}, \ldots, LR_{k,U}^{-n}\right)$ are independent and there exist functions $\mu(\Theta_U)$ and $\sigma(\Theta_U)$ such that

   i. $E\left[LR_{k,U}^{y}|\Theta_U\right] = \mu(\Theta_U)$

   ii. $V\left[LR_{k,U}^{y}|\Theta_U\right] = \dfrac{\sigma^2(\Theta_U)}{P_{k,Uy}}$

   iii. $\mu_4\left[LR_{k,U}^{y}|\Theta_U\right] = \dfrac{3\sigma^4(\Theta_U)}{P_{k,Uy}^2}$

   where $E\left[LR_{k,U}^{y}|\Theta_U\right]$ = the expected value of $LR_{k,U}^{y}$, given Θ_U; $V\left[LR_{k,U}^{y}|\Theta_U\right]$ = the variance of $LR_{k,U}^{y}$, given Θ_U; $\mu_4\left[LR_{k,U}^{y}|\Theta_U\right]$ = the fourth central moment of $LR_{k,U}^{y}$, given Θ_U; and P_{k,U_y} = the earned gross premium of the insurer U in the LoB k, and in historic year y; and n = the number of historic loss ratios for insurer U.

b. The pairs $\left(\Theta_U, LR_{k,U}^{-1}\right)$, $U = 1, \ldots, N$, are independent and the risk parameters Θ_U, $U = 1, \ldots, N$ are independent and identically distributed (iid).

Under these assumptions, it can be shown that the credibility factor c_k is given by (see Centeno, 1989):

$$c_k = \frac{Y_{k,U} - 1}{Y_{k,U} - 1 + 2\dfrac{E\left[\sigma^4(\Theta_U)\right]}{V\left[\sigma^2(\Theta_U)\right]}} = \frac{V\left[\sigma^2(\Theta_U)\right]}{V\left(\Sigma_U^2\right)},$$

where

$$\Sigma_U^2 = \frac{1}{Y_{k,U} - 1} \sum_{y:\text{year}} P_{k,Uy}\left(LR_{k,U}^{y} - \mu_{k,U}\right)^2.$$

Using unbiased estimators for the structural parameters $\xi = V\left(\Sigma_U^2\right)$ and $\phi = V\left[\sigma^2(\Theta_U)\right]$ constructed from the statistical data (*cf.* Section 4 in Centeno 1989), estimates of the credibility factor c_k in the individual LoBs were determined. From this, values for the credibility constants b_k were derived. Depending on the individual LoB, these ranged from 3 to 5.

In the light of this analysis and for reasons of simplification, the credibility constant b_k was set equal to 4.0 for each LoB k.

M.2.4.5 QIS3 Rational for Aggregation Formula

The aggregation is discussed in CEIOPS (2007e) and we follow the discussion made there. The rationale for the aggregation formula is as follows:

For an individual insurer, the overall risk variable for premium and reserve risk is described by the sum of the risk variables for the individual LoB k, where for each LoB a distinction is made between premium and reserve risks. Hence,

$$C = \sum_k \left[C_{\mathrm{R},k} + C_{\mathrm{P},k}\right],$$

where C: the overall risk variable for reserve and premium risks of the entire portfolio; k: LoB k; $C_{\mathrm{R},k}$: the risk variable for the systematic part of the reserve risk in LoB k; and $C_{\mathrm{P},k}$: the risk variable for the premium risk in LoB k.

Hence,

$$\sigma^2\left[X\right] = \sigma^2\left[\sum_k \left[C_{\mathrm{R},k} + C_{\mathrm{P},k}\right]\right] = \sum_{i,j} \rho^*_{ij} \times \sigma\left(X_i\right) \times \sigma\left(X_j\right),$$

where the indices i and j goes over all combinations of (P,i) and (R,j), i and j are LoBs, and ρ^*_{ij} ρ_{ij} is the dependency between LoBs i and j. $\sigma\left(X_i\right)$ is the standard deviation of the risk variable X_i.

The overall standard deviation that needs to be derived is defined relative to the overall volume measure V, so that

$$\sigma\left(X\right) = \sigma \times V.$$

Likewise, the standard deviations for premium and reserve risks that are determined on the level of individual LoBs k are defined relative to the volume measures for premium and reserve risks, so that:

$$\sigma\left(X_{\mathrm{P},k}\right) = \sigma_{\mathrm{P},k} \times V_{\mathrm{P},k},$$

and

$$\sigma\left(X_{\mathrm{R},k}\right) = \sigma_{\mathrm{R},k} \times V_{\mathrm{R},k}$$

for each LoB k, where $\sigma_{\mathrm{P},k}$: the standard deviation of the premium risk in the individual LoB k, relative to the volume measure; and $\sigma_{\mathrm{R},k}$: the standard deviation of the reserve risk in the individual LoB k, relative to the volume measure.

Hence σ is given by

$$\sigma = \frac{\sigma\left[X\right]}{V} = \frac{1}{V}\sqrt{\sum_{i,j} \rho^*_{ij} V_i \sigma_i V_j \sigma_j}.$$

M.2.4.6 QIS3 Dependency Structures

The settings of dependency structures are discussed in CEIOPS (2007e) and we follow the discussion made there. Within the *grand dependence matrix*, coefficients have to be set between the following pairs of indices:

- Between (P,k1) and (P,k2)
- Between (P,k1) and (R,k2)
- Between (R,k1) and (R,k2)

for each pair (k1,k2) of individual LoBs.

In view of the insufficiency of currently available data, the setting of these dependency structures would necessarily include a certain degree of judgment. This is also so because, when selecting the coefficients, allowance should be made for nonlinear tail dependence, which is not captured under a "pure" linear correlation approach. As an example, two risk variables X and Y may have zero linear correlation, but may nonetheless be dependent "in the tail," that is, in the occurrence of adverse events. In fact, such a situation is not uncommon for variables related to insurance risk. In such cases, the dependence matrix used in the standard formula to aggregate the risk capital charges for the two risks should be set to capture such tail dependence, that is, the related coefficient should be set higher than zero. To allow for this, the coefficients used should be higher than the simple analysis of relevant data would indicate.

For reasons of simplification the *grand dependence matrix* was determined such that:

- For any two LoBs k1 and k2, the dependence within premium and reserve risk coincide, that is, the coefficients between (P,k1) and (P,k2) and between (R,k1) and (R,k2) are the same
- For each individual LoB k, the dependence between premium and reserve risk is set as 50%
- The dependencies between (P,k1) and (R,k2) for different LoBs k1 and k2 are determined as 50% of the dependence between (P,k1) and (P,k2)

Therefore, the *grand dependence matrix*, GCM, was specified as:

$$GCM = \begin{pmatrix} \rho & \alpha \times \rho \\ \alpha \times \rho & \rho \end{pmatrix},$$

where GCM: the dependence matrix for premium (P) and reserve (R) risk, arranged in such a way that the first (respectively, the last) 15 rows and columns refer to the indices of the form (P,k) [respectively, to the indices of the form (R,k)]; ρ: the dependence matrix for premium risk; and α: factor representing on overall assumption between premium and reserve risks (set as 50%).

As a starting point for the determination of the dependence matrix for premium risk, an analysis of the statistical dependence structure between individual LoBs (on the level of individual insurers) in the German insurance market was carried out.

For a given pair (k1,k2) of LoBs k1 and k2, the analysis used historic loss ratios of individual insurers in k1 and k2 during the period 1988–2002.

For each individual insurer (where at least 10 historic loss ratios in each of k1 and k2 were available), an insurer-specific estimate of the dependence between k1 and k2 was derived. An overall estimate of the dependence between k1 and k2 was then determined as the average of these insurer-specific dependencies.

Moreover, to visualize the dependency between the loss ratios in k1 and k2, diagrams were produced showing the standardized residuals of the loss ratios of the individual insurers (with respect to loss ratios in k1 in the x-coordinate, and with respect to loss ratios in k2 in the y-coordinate).

For example, in the case of the dependency between:

- k1: Motor, third-party liability and
- k2: Third-party liability

an average overall dependence of 28% (using the data of 89 firms and 1.269 loss ratios) was derived, together with the following plot of standardized residuals (Figure M.3):

It was interesting to see that the plot clearly shows a nonlinear dependency between k1 and k2, with an accumulation of residuals in the lower-left corner, that is, the dependency increases under the occurrence of adverse events.

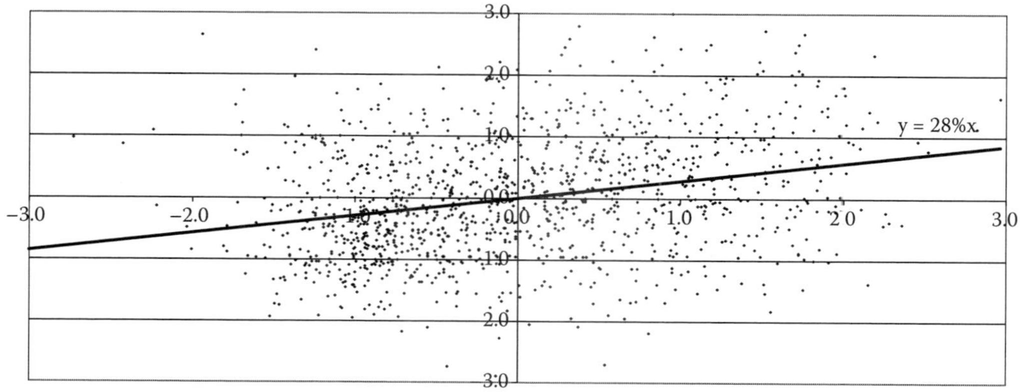

FIGURE M.3 Calibration of residuals for motor third-party liability versus third-party liability classes based on data from 89 companies (1988–2002) and 1,269 loss ratios. Parameter estimate: $\rho = 0.28$. (Adapted from CEIOPS. 2007e. Calibration of the underwriting risk modules for QIS3. CEIOPS. 1 April. Available at www.ceiops.org; © Committee of European Insurance and Occupational Pensions Supervisors.)

Taking into account results from internal models of selected insurers, the general considerations regarding the selection of dependencies derived a final choice of the coefficients of the matrix. These results were not published by CEIOPS.

M.2.4.7 QIS4 Calibration

The nonlife UR module was more or less similar to the final QIS3 model. However, some new considerations were made. We follow CEIOPS (2008a).

Number of historic years: A lot of work was carried out by working groups on Best Estimate at national level within different supervisory authorities and by the Groupe Consultatif. Due to this work, recognition of a differentiation regarding the number of historic years between the various nonlife LoBs could be introduced in the QIS4 specification. The table in TS.XIII.B.13 (QIS4 2008), was designed using the recommendations from the French supervisory authority (see ACAM, 2007).

LoB's standard deviations: Following the feedback from QIS3, the factors used within the SCR nonlife UR module were adjusted to better reflect the relative and overall riskiness of different LoBs. Based on the QIS3 calibration of premium risk in the German market, the recalibration reflects information collected through QIS3 on internal models, the results from current regulatory regimes, and other market information from several Members States (United Kingdom, Portugal, and the Netherlands). Results from over 46 firms were used to recalibrate the factors.

Geographical diversification: During the QIS3 exercise, CEIOPS received comments that suggested that the proposed geographical diversification for groups, which was based on the location of the entities' headquarters, was not enough risk sensitive. It has also been highlighted that an entity operating in different countries with branches, or under the Freedom to Provide Services regime, should also benefit from geographical diversification.

In the model tested in QIS4, see Section M.2.3, Step 2, the minimum diversification factor equals 0.75. Hürlimann (2009) shows that this factor is not optimal and he also explains the background for the proposed diversification factor.

We can assume that the geographical decomposition of the volume measure in volume measures of an LoB into n geographical regions is $V = \sum_{j=1}^{n} V_j$. Also assume that the diversification can be measured by the intra-portfolio correlation coefficient $Q = \sum_{i=1}^{n}\sum_{j=1}^{n} \rho_{ijw} w_i w_j \in [-1, 1]$, where $w_i = V_i / V$. Here ρ_{ij} is the correlation coefficient and the weight w_i is the portfolio weight in the geographical region $i, i = 1, \ldots, n$.

Assume that CR is the original capital charge. Then we can adjust for diversification by letting CR not taking diversification into account:

$$\mathrm{CR}^{\mathrm{Adj}} = \frac{1}{2}(1 + Q)\,\mathrm{CR}.$$

If we have perfect positive dependence between the regions, then $Q = 1$ and hence the adjusted capital charge equals the original capital charge. In a similar way, if we have perfect negative dependence between the regions, the capital charge vanishes.

If we assume a linear dependence structure between perfect dependence and independence such that the correlation coefficients are given by

$$\rho_{ij} = \frac{1}{2} + \frac{1}{2}\delta_{ij}, \quad \delta_{ij} = \begin{cases} 1 & i = j \\ 0 & i \neq j \end{cases},$$

then we obtain $Q = \frac{1}{2}(1 + H)$ and $H = \sum_{j=1}^{n} w_j^2$, where H denotes the Herfindahl–Hirschman index. A motivation for this is given by Hürlimann (2008b). This implies that

$$\mathrm{CR}^{Adj} = (0.75 + 0.25H)\,\mathrm{CR}.$$

For example, Hürlimann (2009) shows that if the risk measure is proportional to the standard deviation of the risk, then the absolute minimum value would be 0.707 instead of 0.75, that is, an additional reduction of the diversification effect of 4.3%.

Consequently, CEIOPS proposed a revised structure of geographical diversification and extended it to solo entities. In QIS4, geographical diversification is calculated with a concentration index (Herfindahl–Hirschman index; *cf.* Section 14.4.1) based on the location of risks (for premiums and reserves) for each LoB, except for credit and suretyship and miscellaneous.

The diversification benefit was capped to 25% for the concerned LoBs. That cap was seemed reasonable by CEIOPS for the standard formula considering the limited number of data gathered by CEIOPS on well-diversified groups during the QIS3 process.

CAT risk: Following the QIS3 feedback, many firms expressed the view that the methodology for calculating the nonlife CAT-risk module produced results that were inconsistent with their own assessment of risk. The standard approach, Method 1, were configured on the basis of QIS3 returns and benchmarked against market practice for a range of more than 20 insurers under the U.K. FSA supervision to ensure a reasonable calibration.

Gisler (2009) has compared the Solvency II nonlife insurance risk model (from QIS4) with the corresponding model used in the Swiss Solvency Test (SST). He discussed the common parts and what was the difference.

M.3 NONLIFE CAT RISK

M.3.1 QIS2

From a "market loss" approach, QIS2 had a capital charge defined as:

$$C_{\mathrm{NL,CAT}} = \max\left\{f_U \times \mathrm{MS}_U \times \mathrm{ML} - X_2; 0\right\} + \min\left\{f_U \times \mathrm{MS}_U \times \mathrm{ML}; X_1\right\},$$

where f_U: the retention factor of the reinsurance program of the undertaking; MS_U: the market share of the undertaking measured by the sum of gross written premiums in the LoB affected by the CAT risk considered to the total gross premiums written in the LoB for the entire market; ML: the market loss; X_1 and X_2: the lower and upper bounds of a CAT-XL (excess of loss) layer in the reinsurance program of the undertaking (if applicable) (*cf.* Section 22.1.2.1).

TABLE M.11 CAT-Risks Factors for the Standard Approach

LoB, *k*	c_k
1. Motor, third-party liability	0.15
2. Motor, other	0.075
3. MAT	0.50
4. Fire/property	0.75
5. Third-party liability	0.15
6. Credit and suretyship	0.60
7. Legal expenses	0.02
8. Assistance	0.02
9. Miscellaneous	0.25
10. NP-reinsurance (property)	1.50
11. NP-reinsurance (casualty)	0.50
12. NP-reinsurance (MAT)	1.50

Source: Adapted from QIS4. 2008. QIS 4 Technical Specifications. European Commission. MARKT/2505/08. March 31. Available at http://www.ceiops.eu/content/view/118/124/.

The national regulator had the freedom to define CAT events and also to modify the design of the market loss approach.

M.3.2 QIS3

For the modeling of the CAT risk in QIS3, regional CAT scenarios were considered that were specified by the local regulator. Additional to this was a list of European transregional scenarios prescribed.

$\mathrm{CR}_{\mathrm{NL,CAT}}$: the undertakings loss relative to the local market and the specified CAT events.

M.3.3 QIS4

The CAT risk submodule was proposed to be calculated according two alternative main methods or to an optional method.

Method 1: Standard Approach If no regional scenarios are provided, a standard formula is applied. The standard capital charge, according to the standard formula, is given by:

$$C^2_{\mathrm{NL,CAT}} = \sum_{k \neq 3,4,10,12} (c_k P_k)^2 + (c_3 P_3 + c_{12} P_{12})^2 + (c_4 P_4 + c_{10} P_{10})^2 ,$$

where P_k: estimate of net written premium for LoB k in the forthcoming year and the factors c_k are defined in Table M.11.

Method 2: Scenarios If regional scenarios are available, provided by the local supervisor (the supervisor of the relevant territory, not necessarily the insurer's own supervisor), they replace the standard formula of Method 1. Regional scenarios include natural catastrophes and man-made catastrophes. They were done in a similar way as in QIS3.

$\mathrm{CR}_{\mathrm{NL,CAT}}$: the undertakings loss relative to the local market and the specified CAT events.

The materiality threshold was set to 25% of the cost of the most severe scenario. Therefore, participants would take into account:

1. The most severe scenario and
2. Additionally, any other scenario whose cost exceeds 25% of the cost of the most severe scenario

For each of the scenarios specified by the local regulator, participants had to estimate the cost CAT_i of the scenario, that is, the effect on the net value of assets and liabilities, if the cost exceeds the materiality threshold.

For the regional scenarios, the calculation of CAT_i should follow the specifications set out by the local regulator. This was either based on a scenario-based approach or a market loss approach.

Where more than one regional scenario, in the same national market, is relevant for a participant, the aggregation of the results of the calculation for each of the regional scenarios should follow the specifications determined by the local supervisor.

Nonproportional nonlife reinsurance business and nonlife insurance and reinsurance business that were located in areas outside of the European Economic Area—for which areas no regional scenarios are provided—were not allowed for in this approach. Instead, to the extent to which the business may give rise to catastrophe risk, participants should quantify the risk by means of a partial internal model. The capital charge for this risk should be added to the capital charge derived under the approach described.

Method 3: Optional—Personalized Scenarios In addition, undertakings could, on an optional basis, use personalized catastrophe scenarios according to the classes of business written and geographic concentration, and explaining the appropriate definition for calculation purposes (Method 3). The *catastrophe personalization* can include partial or full internal model output, including where available commercial catastrophe model output, but is equally applicable where modeling was not carried out.

Occurrence basis: Cat scenarios were defined on the basis of the occurrence of a single event, for example, single windstorm, flood, earthquake, fire, and explosion. The scenarios to be selected were those that the firm considers will exceed the materiality threshold, which is 25% of the most severe scenario.

Annual basis: Annual basis was to be used when assessing the effect reinsurance treaties on the nonlife CAT risk exposure. For most firms, the SCR calibration in line with a 99.5% confidence level over a one-year time horizon was likely to involve the occurrence of not one catastrophic event, but a series of catastrophic events over the forthcoming 12 months. In most reinsurance treaties, distinct catastrophic events are subject to separate retentions, as well as different reinstatements and associated costs. Therefore, when participants have to simulate a series of events to derive CAT_{NL}, they would take into account the impact of those separate retentions, reinstatements and associated costs on their nonlife CAT risk exposure.

Personalized CAT scenarios should always reflect the full 12-month exposure. For example, for flood risk, a single scenario might include the financial impact of multiple events during a single 12-month period.

Some examples of man-made scenarios were outlined in QIS4 (2008):

Motor Third Party

- Car falls onto railway line, causing a train to derail—multiple deaths and injuries
- Mont Blanc type event, for example, in Mont Blanc or Channel Tunnel
- Level crossing accident including a train (possibly including a suicide, even though it would be an intentional act)
- Petrol tanker crash or collision, causing noxious fumes and poisoning
- Lorry accident involving another form of public transport with long-term injuries, not death
- An accident involving nuclear material on a train

Motor (Other)

- Driver's own debilitating injuries following an accident
- Epidemic of cars being taken without owner's consent and damaged

MAT

- Oil rig event like Piper Alpha
- A number of high-worth vehicles in transit or on another transport are destroyed
- Aviation collision (see QIS3)
- Aviation crash or collision during take-off or landing, including collision with airport buildings
- Piracy—damage to property, theft, injury, and death
- LPG ship collision with a cruise ship carrying high net worth individuals

Fire

- Terrorism event (from QIS3)
- Total loss to the largest single property risk (QIS3)
- Buncefield type event (http://www.buncefieldinvestigation.gov.uk/index.htm)

- Flixborough (http://www.hse.gov.uk/comah/sragtech/caseflixboroug74.htm)
- Oil wells exploding
- National Galleries, historic sites destroyed
- Conflagration across a city center (covering property [some historic], petrol stations, etc.)

Third-Party Liability

- New latent claims
- New individual diseases—from chemicals, and so on.
- Enron, Parmalat-type events leading to D&O and E&O claims
- Construction: ship building (e.g., insolvency half way through building), builders, and architects
- Pharmaceuticals: for example, Thalidomide http://news.bbc.co.uk/1/hi/uk/2031459.stm)
- Large medical claim, for example, baby born with brain damage through medical malpractice and, in particular, a series of such claims arising from consistent errors
- Pollution

Credit

- Total impact of single policy, for example, credit or bond guarantees of single largest policyholder
- Wider recession, effect of interest rates, for example, 1930 global recession

Assistance

- Terrorism at Olympic Games (QIS3)

Miscellaneous

- Very successful product sold with extended waivers that are all claimed upon because of a latent defect in the product

APPENDIX N

European Solvency II Standard Formula

Life Underwriting Risk

Here, we will discuss the development and calibration of the life underwriting risk (LUR) module and its capital charge that has been one of the building blocks in Solvency II. The final advice from CEIOPS is discussed in Chapter 32.

This concerns specific risks arising from the underwriting of life insurance contracts, associated with both the perils covered and the processes followed in the conduct of the business.

N.1 GENERAL FEATURES

N.1.1 Background

As an answer to the European Commission's Call for Advice (CfA) 10 (MARKT, 2004c), CEIOPS gave a background to the modeling of the standard formula and its risk modules (see CEIOPS, 2005).

The modeling approaches to be used in the SCR standard formula required considerable further analysis at the stage the development process was in. A first indication of possible approaches was given by CEIOPS (2005). A factor-based approach to mortality, lapse, and expense risks was first suggested. Such an approach would require

- To specify the volume measures that are specific to UR
- To determine the coefficients applicable to these volume measures
- To specify the degree of "personalization" that should be reflected in the coefficients
- To clarify how the approach chosen may be combined with a segmentation of the book of business of the insurer

As a supplement to this factor-based approach, scenario techniques were discussed for life CAT risks and for the lapse risk. This is discussed in Sections N.8.1 and N.6.1, respectively.

N.1.1.1 Choice of Volume Measure

Considering the split between mortality, lapse, and expense risks, and the different nature of these three subrisks, it seems advisable to choose three volume measures for UR, that is, one measure specific to mortality risk, one to lapse risk, and one specific to expense risk. These are discussed in Sections N.2.1, N.6.1, and N.7.1, respectively.

N.1.1.2 Choice of Coefficients

A choice of coefficients within a factor-based model for UR needs to reflect the use of a limited—one year—time horizon, but with full allowance for changes in the expectation over that period of future CFs to be reserved for at the end of that period. It is to be based on an analysis of the insurers' underwriting result during the time horizon; a loss of capital occurs if the underwriting result is negative.

The underwriting result of the insurer will strongly depend on the valuation of the technical provisions. The principles for this valuation in the context of the calculation of the capital RC for UR should be compatible with the rules on the calculation of the technical provisions to be developed as part of the future solvency framework. To simplify, it is assumed that the time horizon is one business year. The forthcoming business year (at the point of time of the solvency assessment) is referred to as *the current year*. The RC for UR is therefore derived from the properties of $\mathrm{UR}^{\text{technical}}$, the underwriting result of the undertaking in the current year, which is regarded as a random variable.

One can assume that the risk capital charge for UR is determined according to ruin probability α and risk measure $\rho_{1-\alpha}$.

For example, one may choose $\rho_{1-\alpha} = \mathrm{VaR}_{1-\alpha}$ for Value-at-Risk, or $\rho_{1-\alpha'} = \mathrm{TVaR}_{1-\alpha}$ for Tail Value-at-risk. In this context, $1-\alpha$ or $1-\alpha'$ corresponds to the confidence level ensuring the degree of prudence, that CEIOPS wishes to achieve.

The risk capital charge for UR is given by:

$$\mathrm{CR}^{\text{technical}} = \rho_{1-\alpha}(\mathrm{UR}^{\text{technical}}).$$

At this stage, there was no decision about risk measure to use.

N.1.1.3 Degree of Personalization

Translating these theoretical equations into a factor-based, standardized formula requires:

- Analysis at the level of individual undertakings and
- Generalized analysis that can be applied across the industry

This is carried out for the mortality, lapse, and expense risks, which are discussed in Sections N.2.1, N.6.1, and N.7.1, respectively.

N.1.1.4 Aggregation

Mortality, lapse, and expense risk may be analyzed on the basis of homogenous segments of the portfolio to take the particularities of the single segments into account. Such a segmented approach to UR would present the problem of how to aggregate individual RCs. Simply adding up the individual charges would neglect diversification effects between different homogenous risk groups. This may lead to an overestimation of the required risk capital.

There are two approaches to deal with this problem:

- One may determine mortality (or lapse/expense) risk capital charges for each segment and calculate the overall mortality (or lapse/expense) risk capital charge using capital aggregation methods or
- One may determine only the first two moments of the distribution of the mortality (or lapse/expense) risk for each segment and calculate the first two moments of the overall mortality (or lapse/expense) risk using a dependence matrix for the second moments. Assuming the overall mortality (or lapse/expense) risk to have a specific two-parametric probability distribution, one may then calculate the overall mortality (or lapse/expense) risk capital charge.

For example, the IAA follows the first approach (see Section 15.3). The advantage of this approach is that for the calculation of the capital charges of the single segments, the underlying probability distribution of the risk can be chosen according to the particularities of the segment. The disadvantage of this approach is that a standardized aggregation of the risk capital charges of the segments is problematic. To be in a position to aggregate them in a mathematically precise manner, the complete dependence structure of the risks has to be known. This is rarely the case.

According to the second alternative, it is not necessary for the supervisor to set a probability distribution for the risk on the level of the individual segment. This may help to reduce the model error of the determination of the risk capital, since the moments of the risks can be aggregated precisely once the dependence structure between those risks are known. Moreover, it may be easier to make an adequate assumption on the type of the distribution on the level of the diversified overall risk than on the level of the segment risk. However, this approach takes only the first two moments of the probability distribution on the segment level into account.

N.1.1.5 Segmentation

In general terms, an assessment of UR involves an identification of factors that influence the variability of the underwriting result of the undertaking. This requires a classification of URs into groups with similar characteristics, known as homogenous risk groups. This classification must be based in part on information from historical data on the liabilities portfolio, the institution's specific circumstances, and relevant data from the insurance industry.

The LUR groups to be used need further examination. It is advisable to identify these groups on a European level, but national specificities resulting in country-specific groups may be taken into account. A suitable segmentation of the book of business might be explicitly defined within the formula, or some flexibility could be allowed so that national particularities can be taken into account. A standard classification that is more closely aligned with actual undertakings' behavior should have positive consequences for risk management.

N.1.2 QIS2–QIS4

N.1.2.1 QIS2 (CEIOPS 2006d)

LUR is split into

- Biometric risks, comprising
  - $CR_{LR,MR}$: mortality risk
  - $CR_{LR,LO}$: longevity risk
  - $CR_{LR,MO}$: morbidity risk
  - $CR_{LR,DR}$: disability risk
- $CR_{LR,LR}$: lapse risk
- $CR_{LR,ER}$: expense risk

Under QIS2, the capital requirements for the six subrisks were combined using the dependence matrix in Table N.1.

The total capital charge for the LUR was calculated as

$$CR_{LR} = \sqrt{\sum_{r \times c} \rho_{rc} \times CR_{LR,r} \times CR_{LR,c}},$$

where the indices r and c are the rows and columns in the above dependence matrix, respectively.

TABLE N.1 Dependence Matrix for the LUR's Subrisks Proposed for QIS2

	$CR_{LR,MR}$	$CR_{LR,LO}$	$CR_{LR,MO}$	$CR_{LR,DR}$	$CR_{LR,LR}$	$CR_{LR,ER}$
$CR_{LR,MR}$	1	0	0.5	0.25	0	0.5
$CR_{LR,LO}$		1	0	0	0.5	0.5
$CR_{LR,MO}$			1	1	0	0.5
$CR_{LR,DR}$				1	0	0.5
$CR_{LR,LR}$					1	0.5
$CR_{LR,ER}$						1

Source: Adapted from CEIOPS. 2006d. Quantitative Impact Study No. 2: Technical Specification, CEIOPS-PI-08/06. Available at www.ceiops.org

N.1.2.2 QIS3 (CEIOPS 2006b, 2007a)

In QIS2, for mortality, longevity, and disability risk, two risk components had been tested: one for trend risk and one for volatility risk. However, compared to the trend risk, the volatility risk proved to be considerably lower. Thus for QIS3, it was decided to reduce the complexity of the design of the UR module by maintaining the trend risk components only, assuming the volatility risk components to be implicitly covered by the trend and catastrophe risk components.

For mortality, longevity, disability, lapse, and expense risks, CEIOPS proposed in its consultation paper 20 (CEIOPS, 2006b) to apply a scenario-based approach. Conversely, for the catastrophe risk component, CEIOPS proposed a factor-based approach.

The components for mortality, longevity, disability, lapse, expense, and catastrophe risks are then aggregated together through the application of a dependence matrix. In consultation paper 20 (CEIOPS, 2006b), participants raised the concern that this matrix would not present a consistent set of assumptions on dependence structures between pairs of risks*: whereas morbidity risk and disability risk were assumed to be 100% correlated, the dependence assumptions for these risks with respect to the remaining risks were not identical. Also, some participants pointed out that for some contracts, the distinction between morbidity and disability risk would be problematic.

Concerning the CAT risks surcharges within the mortality, disability, and morbidity modules, concerns were raised whether the simple factor-based treatment of such risks would be adequate. Also, it was pointed out that adding these CAT surcharges to the other surcharges for mortality (respectively, disability/morbidity) risk would not adequately reflect that CAT risks could be assumed to be independent from the other sources of risk.

For QIS3, CEIOPS suggested to combine disability and morbidity risks into one module with invalidity/morbidity probability as the underlying risk driver. Also, CEIOPS proposed to comprise the treatment of CAT risks in LUR into a new "CAT-risk" module (see Section N.8).

Based on the *principle of substance over form*, agreed claims arising from nonlife business payable in the form of annuity should normally be part of the life underwriting risk charge, CR_{LR}. In particular, the risk of revision is applicable only to this type of annuities.

LUR was proposed to consist of the following seven subrisks:

- $CR_{LR,MR}$: mortality risk
- $CR_{LR,LO}$: longevity risk
- $CR_{LR,DR}$: disability risk
- $CR_{LR,LR}$: lapse risk
- $CR_{LR,ER}$: expense risk

*That is, the matrix is not positive-definite (*cf.* Section 15.3.4).

TABLE N.2 Dependence Matrix for the LUR's Subrisks Proposed for QIS3

	$CR_{LR,MR}$	$CR_{LR,LO}$	$CR_{LR,DR}$	$CR_{LR,LR}$	$CR_{LR,ER}$	$CR_{LR,CAT}$	$CR_{LR,RR}$
$CR_{LR,MR}$	1	0	0.5	0	0.25	0	0
$CR_{LR,LO}$		1	0	0.25	0.25	0	0.25
$CR_{LR,DR}$			1	0	0.5	0	0
$CR_{LR,LR}$				1	0.5	0	0
$CR_{LR,ER}$					1	0	0.25
$CR_{LR,CAT}$						1	0
$CR_{LR,RR}$							1

Source: Adapted from CEIOPS. 2007a. QIS 3 Technical Specifications, Part I: Instructions, CEIOPS-FS-11/07 & Part II: Background Information, CEIOPS-FS-12/07. April. Available at www.ceiops.org

- $CR_{LR,CAT}$: CAT risk
- $CR_{LR,RR}$: revision risk

The dependence matrix is shown in Table N.2.

N.1.2.3 QIS4 (QIS4 2008)

The same subrisks as in QIS3 were used in this model. The dependence matrix was proposed to be as in Table N.3.

Risk-absorbing effects were taken into account in QIS4 in another way when compared with QIS3. For all subrisks, excluding the revision risk, a net value of the capital charge is calculated.

$nCR_{LR,i}$: Capital charge for market subrisk i, including the risk-absorbing effect of future profit sharing.

The total net LUR capital charge was then calculated using the above dependence matrix:

$$nCR_{LR} = \sqrt{\sum_{r \times c} \rho_{rc} \times nCR_{LR,r} \times nCR_{LR,c}}.$$

These net charges will be used to calculate a risk adjustment on the top level, see Appendix H.

TABLE N.3 Dependence Matrix for the LUR's Subrisks Proposed for QIS4

	$CR_{LR,MR}$	$CR_{LR,LO}$	$CR_{LR,DR}$	$CR_{LR,LR}$	$CR_{LR,ER}$	$CR_{LR,CAT}$	$CR_{LR,RR}$
$CR_{LR,MR}$	1	−0.25	0.5	0	0.25	0	0
$CR_{LR,LO}$		1	0	0.25	0.25	0	0.25
$CR_{LR,DR}$			1	0	0.5	0	0
$CR_{LR,LR}$				1	0.5	0	0
$CR_{LR,ER}$					1	0	0.25
$CR_{LR,CAT}$						1	0
$CR_{LR,RR}$							1

Source: Adapted from QIS4. 2008. QIS 4 Technical Specifications. European Commission. MARKT/2505/08. March 31. Available at http://www.ceiops.eu/content/view/118/124/.

Some of the participating companies in QIS4 reported that lapse risk gave too high capital charge (CEIOPS, 2008e). The allocation of contracts between life, nonlife, and health risk modules was not clear.

N.2 MORTALITY RISK

N.2.1 Background

As an answer to the European Commission's CfA 10 (MARKT, 2004c), CEIOPS gave a background to the modeling of the standard formula and its risk modules (see CEIOPS, 2005).

N.2.1.1 Choice of Volume Measure

With regards to mortality risk, depending on the product design, two natural candidates for a volume measure are the technical provision, if the risk of longevity is relevant, and the capital at risk for term insurance at the beginning of the solvency assessment time horizon.

The valuation of the technical provision for the purposes of calculating capital requirements for UR should be compatible with the rules on the calculation of the technical provisions to be developed as part of the future solvency framework.

If the technical provision was defined as a sum of the best estimate and an RM that reflects the volatility of the claims, it may also be possible to choose the best estimate as the volume measure. However, such a choice would need to be reflected in the degree of volatility that is taken into account in the definition of the coefficient applicable to the volume measure.

N.2.1.2 Choice of Coefficients

Expressing the capital charges in terms of the volume measures TP_0 and CaR_0, the technical provision, and the capital at risk at the beginning of the current year, one has that

$$\mathrm{CR_{MR}} = \max\left\{\beta \times \mathrm{TP_0}; \gamma \times \mathrm{CaR_0}\right\},$$

where $\beta = \rho_{1-\alpha}\left(\mathrm{UR}^{\mathrm{mortality}}/\mathrm{TP_0}\right)$ and $\gamma = \rho_{1-\alpha}\left(\mathrm{UR}^{\mathrm{mortality}}/\mathrm{CaR_0}\right)$ denotes the quantile of the (relative) mortality result in the current year, expressed in percentage of the corresponding volume measure at the beginning of the current year.

N.2.1.3 Degree of Personalization

Translating these theoretical equations into a factor-based, standardized formula requires:

- Analysis at the level of individual undertakings and
- Generalized analysis that can be applied across the industry

Defining $\mathrm{UR}\%^{\mathrm{mortality}}$ as the quotient of $\mathrm{UR}^{\mathrm{mortality}}$ and the corresponding volume measure TP_0 or CaR_0, it can be seen that on an abstract level, one needs to choose the coefficient β or γ applicable to our volume measure as

$$\beta = \rho_{1-\alpha}\left(\mathrm{UR}_{\%}^{\mathrm{mortality}}/\mathrm{TP_0}\right) \text{ and } \gamma = \rho_{1-\alpha}\left(\mathrm{UR}_{\%}^{\mathrm{mortality}}/\mathrm{CaR_0}\right),$$

respectively, where ρ is a given risk measure and α is the ruin probability.

In general terms, to be able to compute the coefficient β according to the above formula, one needs to know the probability distribution of the random variable $\mathrm{UR\%}^{\mathrm{mortality}}$. On a practical level, it may be assumed that this distribution is of a type that is completely specified by its first two moments. Then, β may be determined once the following has been specified;

- The type of the distribution
- Its expected value μ
- Its variance σ^2

Assuming that the supervisor sets the type of distribution of $\mathrm{UR\%}^{\mathrm{mortality}}$, the determination of its expected value and variance allows for a wide range of approaches, which vary in their degree of personalization:

- All parameters are set by the supervisor; the result would be a table of industry-wide factors β and γ for mortality risk that can be applied to the insurers' provisions in each segment or
- The expected value and/or variance of the distribution are computed using company-specific data or
- The expected value and/or variance of the distribution are computed using a mixture of company-specific data and data that are set by the supervisor

The decision about the degree of personalization requires a trade-off between accuracy and practicability of the determination of the risk within the limits of the standard formula.

The third alternative is an intermediate approach that uses limited portfolio-specific data to measure the portfolio-specific risk in a reliable and practicable way. For example, the variance of the distribution may be regarded as a function of the size of the portfolio:

$$\sigma^2(\mathrm{UR\%}^{\mathrm{mortality}}) = f(n).$$

The supervisor would provide the function f, and the size n of the portfolio would be determined individually by the insurer. The size of the portfolio could be measured, for example, by the number of risks in the portfolio at the beginning of the time horizon. This approach would combine an assumption on the volatility of the distribution, which is specific to the homogenous risk group and independent of the single company with the diversification effects caused by the size of the portfolio, which is specific to the company. The Dutch supervisory authority has chosen this approach for the Dutch Financial Assessment Framework. The reliability and practicability of the determination of the coefficient β will depend on the rules for the valuation of technical liabilities within the solvency assessment framework. Therefore, it seems premature to make a definite advice on the degree of personalization at this stage.

According to CEIOPS (2006b) the volatility risk should be defined as the risk of random fluctuations of actual mortality rates during the solvency time horizon around the expected mortality rates. The uncertainty risk should be defined as the risk that the models used to estimate mortality rates are misspecified or that the parameters within the models are misestimated. It should also comprise the risk that the risk structure, that is, parameters, can change over time or be uncertain for other reasons. For example, a new medical breakthrough (e.g., cure for cancer) could change the assumptions on future mortality rates.

The life mortality risk charge (RC) should capture volatility and uncertainty risk only to the extent these risks have not already been addressed in the valuation of technical provisions.

N.2.2 QIS2–QIS4

N.2.2.1 QIS2 (CEIOPS, 2006d)

The treatment of mortality risk was split into the following risk components:

- Volatility risk
- Uncertainty risk
- CAT risk

Uncertainty risk comprises accumulation, trend, and parameter risks, to the extent these were not already reflected in the valuation of technical provisions.

Define the following variables:

q_x: the average probability of death. The insurer, using sound actuarial methods and approximations, should determine the average probability of death q_x. For example, q_x may be determined as the ratio of total actual claims paid and claims related expenses (in the most recent business year) over the sum of the capital at risk in the portfolio.

CaR: the sum of the net capital at risk in the portfolio.

N: the number of insurance contracts.

TP_M: the sum of net technical provisions.

TP_i: the technical provisions for policy i.

D_i: the amount payable on immediate death for policy i.

Both factor- and scenario-based approaches were tested in QIS2:

$$\mathrm{CR}_{\mathrm{LR,MR}} = \mathrm{CR}_{\mathrm{Mort,Vol}} + \mathrm{CR}_{\mathrm{Mort,Trend}} + \mathrm{CR}_{\mathrm{Mort,CAT}}$$

where $\mathrm{CR}_{\mathrm{Mort,Vol}}$: either the factor-based risk capital for volatility risk or the results of the mortality scenario for volatility risk; $\mathrm{CR}_{\mathrm{Mort,Trend}}$: either the factor-based risk capital for trend/uncertainty risk or the results of the mortality scenario for trend/uncertainty; and $\mathrm{CR}_{\mathrm{Mort,CAT}}$: the risk capital for mortality CAT risks.

N.2.2.1.1 Volatility Risk The risk capital charge for volatility risk under the factor-based approach was defined as follows:

$$\mathrm{CR}_{\mathrm{Mort,Vol}} = 2.58 \times \sigma_{\mathrm{Mort}} \times \mathrm{CaR},$$

where σ_{Mort}: the estimate of the standard deviation in the loss distribution for mortality risk and this is estimated as

$$\sigma_{\mathrm{Mort}} = \sqrt{\frac{q_x\left(1 - q_x\right)}{N}}.$$

The life mortality scenario for volatility risk was defined as

$$\mathrm{CR}_{\mathrm{Mort,Vol}} = \sum_i \{\Delta\mathrm{NAV}_i \mid \mathrm{Mortshock}_{\mathrm{Vol}}\},$$

where subscript i denotes each policy where the payment of benefits, either lump sum or multiple payments, was contingent on death. The other terms represent: ΔNAV: the change in the net value of assets minus liabilities; and $\mathrm{Mortshock}_{\mathrm{Vol}}$: a 10% increase in mortality rates for each age during the next business year.

N.2.2.1.2 Trend/Uncertainty Risk In the factor-based approach, the risk capital charge for uncertainty risk is defined as follows:

$$\mathrm{CR}_{\mathrm{Mort,Trend}} = 0.002 \times \mathrm{TP}_M.$$

The life mortality scenario for trend/uncertainty risk was defined as

$$\mathrm{CR}_{\mathrm{Mort,Trend}} = \sum_i \{\Delta\mathrm{NAV}_i \mid \mathrm{Mortshock}_{\mathrm{Trend}}\},$$

where subscript i denotes each policy where the payment of benefits, either lump sum or multiple payments, was contingent on mortality risk. The other terms represent: ΔNAV: the change in the net value of assets minus liabilities; and $\mathrm{Mortshock}_{\mathrm{Trend}}$: a (permanent) 20% increase in mortality rates for each age.

N.2.2.1.3 CAT Risk The CAT RC for mortality risk was defined as follows:

$$\mathrm{CR}_{\mathrm{Mort,CAT}} = \sum_i [0.003 \times \max\{\mathrm{TP}_i; D_i\}],$$

where the subscript i denotes each policy where the payment of benefits, either lump sum or multiple payments, was contingent on mortality risk.

N.2.2.2 QIS3 (CEIOPS, 2007a)

The treatment of mortality risk was intended to reflect uncertainty risk. Uncertainty risk comprises trend and parameter risks, to the extent these are not already reflected in the valuation of technical provisions.

It is applicable to the class of insurance contracts contingent on mortality risk, that is, where the amount currently payable on death exceeds the technical provisions held, and therefore an increase in mortality rates is likely to lead to an increase in technical provisions.

The capital charge for life mortality risk was defined as the result of a scenario as

$$\text{CR}_{\text{LR,MR}} = \sum_{i} \{\Delta\text{NAV}_i \mid \text{Mortshock}\},$$

where the subscript i denotes each policy where the payment of benefits, either lump sum or multiple payments, was contingent on mortality risk. The other terms represent: ΔNAV: the change in the net value of assets minus liabilities; and Mortshock: a permanent 10% increase in mortality rates for each age.

The life mortality scenario would be calculated under the condition that the assumptions on future bonus rates, reflected in the valuation of future discretionary benefits in technical provisions, remained unchanged before and after the shocks being tested.

The results of the mortality scenario could also be approximated by a factor-based approach as follows:

$$\text{CR}_{\text{LR,MR}} = 0.0015 \times \text{CaR},$$

where CaR is the sum of the capital at risk, net of reinsurance, in the portfolio, that is, the sum of the amounts currently payable on death less the (net of reinsurance) technical provisions held for each policy.

The factor 0.0015 represents 10% of the assumed average probability of death times an average duration. To simplify the calculation, no differentiation has been assumed for different time periods into the future.

N.2.2.3 QIS4 (QIS4, 2008)

Mortality risk is intended to reflect the uncertainty in trends and parameters, to the extent these are not already reflected in the valuation of technical provisions. It is applicable to the insurance contracts contingent on mortality risk, that is, where the amount currently payable on death exceeds the technical provisions held, and therefore an increase in mortality rates is likely to lead to an increase in technical provisions.

For those contracts that provide benefits both in the cases of death and survival, one of the following two options should be chosen and applied consistently to all contracts in the various lines of business concerned:

- Option 1: Contracts where the death and survival benefits are contingent on the life of the same insured person(s) should not be unbundled. This can be relaxed to include persons who are considered to belong to the same cohort, that is, same relative age and health conditions. For all the remaining contracts, unbundling is required, that is, Option 2 is applicable. For these contracts, the mortality scenario should be applied fully allowing for the netting effect provided by the "natural" hedge between the death benefits component and the survival benefits component. Note that a floor of zero applies at the level of contract if the net result of the scenario is favorable to the (re)insurer.

- Option 2: All contracts are unbundled into two separate components: one contingent on the death and other contingent on the survival of the insured person(s). Only the former component is taken into account for the application of the mortality scenario.

The capital charge for life mortality risk was defined as the result of a scenario as

$$\mathrm{CR}_{\mathrm{LR,MR}} = \sum_i \{\Delta\mathrm{NAV}_i \mid \mathrm{Mortshock}\},$$

where subscript i denotes each policy where the payment of benefits, either lump sum or multiple payments, was contingent on mortality risk. The other terms represent: ΔNAV: the change in the net value of assets minus liabilities; and Mortshock: a permanent 10% increase in mortality rates for each age.

The life mortality scenario would be calculated under the condition that the assumptions on future bonus rates, reflected in the valuation of future discretionary benefits in technical provisions, remained unchanged before and after the shocks being tested.

Additionally, the result of the scenario was determined under the condition that the participant was able to vary his/her assumptions in future bonus rates in response to the shock being tested. The resulting capital charge is $n\mathrm{CR}_{\mathrm{LR,MR}}$.

There are also *simplifications* that could be tested under QIS4, provided that:

1. There is no significant change in the capital at risk over the policy term of the contract
2. The general criteria for simplifications are followed

$$\text{Mortality capital requirement} = (\text{Total CaR}) \times q(\text{firm-specific}) \times n \times 0.10 \times 1.1^{((n-1)/2)},$$

where n is the modified duration of liability CFs; q the expected average death rate over the next year weighted by sum assured; and PMI (projected mortality increase) $= 1.1^{((n-1)/2)}$.

N.2.3 Calibration

N.2.3.1 QIS2 Results (CEIOPS, 2006b)

N.2.3.1.1 The Factor-Based Approach In QIS2, the main capital charge for mortality risk was calculated by a factor-based approach. For volatility risk, this approach derived an estimate of the standard deviation in the loss distribution for mortality risk. This estimate used the average probability of death and the number of contracts in the portfolio as input parameters. For uncertainty risk, multiplying the volume of technical provisions with a market-wide risk factor derived the main capital charge.

QIS2 participants raised concerns regarding the following treatments:

- The risk capital charge for mortality uncertainty should be calculated by reference to sum at risk rather than provisions, since a higher savings component of the contract should not automatically lead to an increase in the capital charge.
- Outstanding duration is an important element in assessing the uncertainty and trends in future mortality rates; however, it does not impact the capital charge.
- The RC for volatility risk does not reflect the part of volatility in the result that is driven by variations in policy size. However, this effect may be material in relation to the volatility arising from the number of lives component.
- For the volatility risk capital charge, the number of insured heads (rather than the number of contracts) should be used.

It was also assumed that the formula should reflect that volatility risk, and uncertainty risk could be assumed to be independent.

N.2.3.1.2 The Scenario-Based Approach QIS2 also tested a scenario-based approach for mortality risk. For volatility risk, this assumed a nonpermanent 10% increase in mortality rates for each age during the solvency time horizon. For uncertainty risk, a permanent 20% increase in mortality rates for each age was considered.

Participants raised the following concerns regarding the scenario-based treatment:

- Some insurers did not have the capacity to perform the necessary calculations.
- It was questioned whether a separate scenario calculation for volatility would be necessary, if this would always be a small proportion of the scenario calculation for uncertainty risk (as the former assumes a change in experience only during the solvency time horizon, whereas the latter assumes a permanent change).
- Volatility risk will depend on the degree of diversification within the portfolio, which can be expected to be significantly higher in large portfolios than in small portfolios. However, the assumed shock for volatility risk makes no allowance for portfolio size.
- For contracts with reviewable rate options, an assumption of a permanent shock to assumed future mortality rates may not be appropriate.

N.2.3.1.3 Quantitative Results The following observations could be made:

- On average, the capital charges for mortality risk under the scenario-based approach were significantly higher than under the factor-based approach.
- In the factor-based treatment for volatility risk, changes in the average probability of death had far less impact on the estimation of the standard deviation, and hence on

the capital charge, than changes in portfolio size; at the same time, the calculations to derive estimates for average death probabilities were relatively complex.

N.2.3.2 QIS3 (CEIOPS, 2007e, 2007g)

It was decided to only use a scenario-based approach for the mortality risk and its trend part as the volatility part was considerably lower.

For mortality risk, CEIOPS used information derived from a study published in 2004 by Wyatt about the 99.5% assumptions over a 12-month time horizon that firms were proposing to make for the British Individual Capital Adequacy Standards (ICAS) submissions in the United Kingdom (see CEIOPS, 2007g). This indicated a range of 10–35%, with an average of around 23%. However, it is thought that this assumption may cover both the trend and the volatility risk, as well as possibly CAT risk. ICAS submissions were believed to have included a fairly low level of mortality trend shock. This may reflect the likelihood that the probability distribution for mortality is skewed, with a current trend toward improving mortality.

It was also relevant to note that many firms may not allow explicitly for future improvements when assessing the best estimate mortality rates for insured lives. Accordingly, in the light of this information, it was suggested that the 20% mortality shock in QIS2 should now be reduced to 10% for the purpose of QIS3.

N.2.3.3 QIS4 (QIS4, 2008)

For QIS4, the QIS3 calibration has not been substantially changed. As a consequence, the QIS3 calibration paper above should still be used as a reference for the QIS4 calibration.

N.3 LONGEVITY RISK

N.3.1 Background

As an answer to the European Commission's CfA 10 (MARKT, 2004c), CEIOPS gave a background to the modeling of the standard formula and its risk modules (see CEIOPS, 2005). Longevity risk was not a separate risk module in this discussion.

N.3.2 QIS2–QIS4

N.3.2.1 QIS2 (CEIOPS, 2006d)

The treatment of longevity risk is split into the following risk components:

- Volatility risk
- Uncertainty risk

Uncertainty risk comprises trend and parameter risks, to the extent these are not already reflected in the valuation of technical provisions.

The following input information was required for the class of insurance contracts contingent on longevity risk, that is, where a decrease in mortality rates leads to an increase in technical provisions:

q_x: the average probability of death. The average probability of death q_x should be determined by the participating undertakings using sound actuarial methods and approximations. For example, q_x would be determined as the ratio of the actual release of reserve for insurance contracts contingent on longevity risk (in the most recent business year) over total technical provisions (net of any benefits payable on immediate death) for those contracts.

PR: total of net technical provisions, net of any benefits payable on immediate death.

N: the number of insurance contracts.

TP_L: the sum of net technical provisions.

Both factor- and scenario-based approaches were tested in QIS2:

$$\mathrm{CR}_{\mathrm{LR,LO}} = \mathrm{CR}_{\mathrm{Long,Vol}} + \mathrm{CR}_{\mathrm{Long,Trend}}$$

where $\mathrm{CR}_{\mathrm{LongVol}}$ is either the factor-based risk capital for volatility risk or the results of the longevity scenario for volatility risk; and $\mathrm{CR}_{\mathrm{Long,Trend}}$ is either the factor-based risk capital for trend/uncertainty risk or the results of the longevity scenario for trend/uncertainty.

N.3.2.1.1 Volatility Risk The risk capital charge for volatility risk under the factor-based approach was defined as follows:

$$\mathrm{CR}_{\mathrm{Long,Vol}} = 2.58 \times \sigma_{\mathrm{Long}} \times \mathrm{PR},$$

where σ_{Long}: the estimate of the standard deviation in the loss distribution for mortality risk and this is estimated as

$$\sigma_{\mathrm{Mort}} = \sqrt{\frac{q_x\left(1 - q_x\right)}{N}}.$$

The life longevity scenario for volatility risk was defined as

$$\mathrm{CR}_{\mathrm{Long,Vol}} = \sum_i \left\{\Delta\mathrm{NAV}_i \mid \mathrm{Longevityshock}_{\mathrm{Vol}}\right\},$$

where subscript i denotes each policy where the payment of benefits, either lump sum or multiple payments, was contingent on survival. The other terms represent: ΔNAV: the change in the net value of assets minus liabilities; and $\mathrm{Longevityshock}_{\mathrm{Vol}}$: a 10% decrease in mortality rates for each age during the next business year.

N.3.2.1.2 Trend/Uncertainty Risk In the factor-based approach, the risk capital charge for uncertainty risk is defined as follows:

$$\mathrm{CR}_{\mathrm{Long,Trend}} = 0.005 \times \mathrm{TP}_L.$$

The life longevity scenario for trend/uncertainty risk was defined as

$$\mathrm{CR}_{\mathrm{Long,Trend}} = \sum_i \{\Delta\mathrm{NAV}_i \mid \mathrm{Longevityshock}_{\mathrm{Trend}}\},$$

where subscript i denotes each policy where the payment of benefits, either lump sum or multiple payments, was contingent on survival. The other terms represent: ΔNAV: the change in the net value of assets minus liabilities; and $\mathrm{Longevityshock}_{\mathrm{Trend}}$: a (permanent) 20% decrease in mortality rates for each age.

N.3.2.2 QIS3 (CEIOPS, 2007a)

The treatment of longevity risk was intended to reflect uncertainty risk. Uncertainty risk comprises trend and parameter risks, to the extent these are not already reflected in the valuation of technical provisions.

It is applicable to the class of insurance contracts contingent on longevity risk, that is, where there is no death benefit or where the amount currently payable on death is less than the technical provisions held, and therefore a decrease in mortality rates is likely to lead to an increase in technical provisions.

The capital charge for longevity risk was defined as a result of a longevity scenario as follows:

$$\mathrm{CR}_{\mathrm{LR,LO}} = \sum_i \{\Delta\mathrm{NAV}_i \mid \mathrm{Longevityshock}\},$$

where the subscript i denotes each policy where the payment of benefits, either lump sum or multiple payments, was contingent on longevity risk. The other terms represent: ΔNAV: the change in the net value of assets minus liabilities; and Longevityshock: a permanent 25% decrease in mortality rates for each age.

The life longevity scenario should be calculated under the condition that the assumptions on future bonus rates, reflected in the valuation of future discretionary benefits in technical provisions, remained unchanged before and after the shocks being tested.

The results of the longevity scenario for trend/uncertainty risk may be approximated by a factor-based approach as follows:

$$\mathrm{CR}_{\mathrm{LR,LO}} = 0.06 \times \mathrm{P\_Release},$$

where P_Release is the total of net technical provisions, net of any benefits payable on immediate death.

The factor 0.06 represents an estimate of the effect of a 25% permanent decrease (or of a 2.5% p.a. rate of improvement) in mortality rates.

N.3.2.3 QIS4 (QIS4, 2008)

Longevity risk was intended to reflect the uncertainty in trends and parameters, to the extent these are not already reflected in the valuation of technical provisions. It is applicable to the class of insurance contracts contingent on longevity risk, that is, where there is no death benefit, or where the amount currently payable on death is less than the technical provisions held, and therefore a decrease in mortality rates is likely to lead to an increase in technical provisions. The provision for disability claims in payment should be included within the longevity risk module.

For those contracts that provide benefits both in the cases of death and survival, the procedure set for the mortality risk in Section N.2.2 should be applied in an analogous and consistent manner.

The capital charge for longevity risk was defined as a result of a longevity scenario as follows:

$$\mathrm{CR}_{\mathrm{LR,LO}} = \sum_i \{\Delta\mathrm{NAV}_i \mid \mathrm{Longevityshock}\},$$

where subscript i denotes each policy where the payment of benefits, either lump sum or multiple payments, was contingent on longevity risk. The other terms represent: ΔNAV: the change in the net value of assets minus liabilities; and Longevityshock: a permanent 25% decrease in mortality rates for each age.

The life longevity scenario should be calculated under the condition that the assumptions on future bonus rates, reflected in the valuation of future discretionary benefits in technical provisions, remained unchanged before and after the shocks being tested.

Additionally, the result of the scenario was determined under the condition that the participant was able to vary its assumptions in future bonus rates in response to the shock being tested. The resulting capital charge is $n\mathrm{CR}_{\mathrm{LR,LO}}$.

There are also *simplifications* that could be tested under QIS4, provided that:

1. The average age of policyholders within the portfolio that are 60 years or over
2. The general criteria for simplifications are followed

$$\text{Longevity capital requirement} = 25\% \times q \times (1.1)^{((n-1)/2)} \times n\,(\text{technical provisions for contracts subject to longevity risk}),$$

where n is the modified duration of liability CFs, and q the expected average death rate over the next year weighted by sum assured.

N.3.3 Calibration

N.3.3.1 QIS2 Results (CEIOPS, 2006b)

The feedback arising from QIS2 on longevity risk was similar to that for mortality risk; see Section N.2.3. QIS2 participants pointed out that specific to longevity risk in the scenario approach an assumption of an X% per annum improvement in longevity (i.e., reduction in mortality rates) might be more suitable than a one-off permanent decrease in mortality rates.

N.3.3.2 QIS3 (CEIOPS, 2007e, 2007g)

It was decided to only use a scenario-based approach for the mortality risk and its trend part as the volatility part was considerably lower.

For longevity risk, CEIOPS had regard to information derived from a study published in 2004 by Watson Wyatt about the 99.5% assumption over a 12-month time horizon that firms were proposing to make for the British ICAS submissions in the United Kingdom, for the reduction in mortality rates, when expressed as a single uniform permanent decrease in mortality (see CEIOPS, 2007g; Watson Wyatt, 2004b). This indicated a range of 5–35%, with an average of around 18% as the decrease in mortality that was assumed. ICAS submissions in the United Kingdom are believed to have shown however an assumed decrease of around 25% in mortality rates, to cover longevity risk, and this was understood to be consistent with some external expert advice that has been received.

This was also consistent with the observed experience in recent years, with an accelerating rate of improvement in longevity for retired persons (see, e.g., Willets et al., 2004). For example in Table 6.11b of op. cit., it is shown that the quanta jump in mortality between the various standard tables that were utilized in the United Kingdom for annuitant mortality over the last 25 years. Each of these standard mortality tables included an allowance for future improvements in longevity based on the best estimates made at that time. However, as can be seen, each of these standard tables has underestimated the actual rates of improvement that have occurred.

These improvements were believed to be attributable to a combination of factors, such as significant medical advances, particularly in the treatment of heart disease and cancer, a reduction in the number of smokers, and better living conditions. There is a wide range of views among academics about the potential for further significant improvements in longevity.

However, the above-suggested 25% reduction in mortality rates would be equivalent to around another three years expected life for a person aged 65, and would be consistent with the quantum shifts in longevity trends seen in recent years.

How should a scenario test to calculate, for example, the volatility component be performed? One way to illustrate how CEIOPS did this is to start with the following formula, where q_x is the average mortality rate. This is an approach that was discussed before the final QIS3 solution and it illustrates the thinking.

$$
\begin{aligned}
2.58 \times \sigma \times \text{CaR}\sqrt{q_x \times N} &= 2.58 \times \sigma \times \text{CaR} \times q_x/\sqrt{q_x \times N} \\
&= 258\% \times \sigma \times \text{CaR} \times q_x/\sqrt{q_x \times N} \\
&= 10 \times 25.8\% \left[\sigma \times \text{CaR} \times q_x\right]/\sqrt{q_x \times N} \\
&= 10 \times \{25\%\,\text{shock}\}/\sqrt{q_x \times N},
\end{aligned}
$$

that is, a factor 10 times a 25% shock. CEIOPS did not think that a 258% shock was possible to communicate, so they introduced a factor 10 times a 25% shock instead.

N.3.3.3 QIS4 (QIS4, 2008)

For QIS4, the QIS3 calibration has not been substantially changed. As a consequence, the above QIS3 calibration paper should still be used as a reference for the QIS4 calibration.

N.4 MORBIDITY RISK

N.4.1 Background

As an answer to the European Commission's CfA 10 (MARKT, 2004c), CEIOPS gave a background to the modeling of the standard formula and its risk modules (see CEIOPS, 2005). Morbidity risk was not a separate risk module in this discussion.

N.4.2 QIS2

N.4.2.1 QIS2 (CEIOPS, 2006d)

The treatment of morbidity risk is split into the following risk components:

- Volatility risk
- Uncertainty risk
- CAT risk

Uncertainty risk comprises accumulation, trend, and parameter risks, to the extent these are not already reflected in the valuation of technical provisions.

The following input information was required for the class of insurance contracts contingent on health status:

CaR: the sum of the (net) capital at risk

i_x: the average morbidity probability

N: the number of insurance contracts

TP_{Morb}: the sum of net technical provisions

SA_i: the sum assured for each policy i, where benefits are payable as a single lump sum, otherwise zero

AB_i: the annualized amount of benefit payable for each policy i, where benefits are not payable as a single lump sum, otherwise zero

Both factor- and scenario-based approaches were tested in QIS2:

$$CR_{LR,MO} = CR_{Morb,Vol} + CR_{Morb,Trend} + CR_{Morb,CAT}$$

where $CR_{Morb,Vol}$: either the factor-based risk capital for volatility risk or the results of the morbidity scenario for volatility risk; $CR_{Morb,Trend}$: either the factor-based risk capital for trend/uncertainty risk or the results of the morbidity scenario for trend/uncertainty; and $CR_{Morb,CAT}$: the risk capital for morbidity CAT risks.

N.4.2.1.1 Volatility Risk The risk capital charge for volatility risk under the factor-based approach was defined as follows:

$$CR_{Morb,Vol} = 2.58 \times \sigma_{Morb} \times CaR,$$

where σ_{Morb}: the estimate of the standard deviation in the loss distribution for morbidity risk and this is estimated as

$$\sigma_{\text{Morb}} = \sqrt{\frac{i_x(1 - i_x)}{N}}.$$

The life morbidity scenario for volatility risk was defined as

$$\text{CR}_{\text{Morb,Vol}} = \sum_i \left\{\Delta\text{NAV}_i \mid \text{Morbidityshock}_{\text{Vol}}\right\},$$

where the subscript i denotes each policy where the payment of benefits, either lump sum or multiple payments, was contingent on health status. The other terms represent: ΔNAV: the change in the net value of assets minus liabilities; and $\text{Morbidityshock}_{\text{Vol}}$: a 10% increase in morbidity rates for each age during the next business year.

N.4.2.1.2 Trend/Uncertainty Risk *n* the factor-based approach, the risk capital charge for uncertainty risk is defined as follows:

$$\text{CR}_{\text{Morb,Trend}} = 0.002 \times \text{TP}_{\text{Morb}}.$$

The life morbidity scenario for trend/uncertainty risk was defined as

$$\text{CR}_{\text{Morb,Trend}} = \sum_i \left\{\Delta\text{NAV}_i \mid \text{Morbidityshock}_{\text{Trend}}\right\},$$

where the subscript i denotes each policy where the payment of benefits, either lump sum or multiple payments, was contingent on health status. The other terms represent: ΔNAV: the change in the net value of assets minus liabilities; and $\text{Longevityshock}_{\text{Trend}}$: a (permanent) 25% increase in assumed rates for morbidity for each age, including probability of remaining sick or disabled.

N.4.2.1.3 CAT Risk The CAT RC for morbidity risk was defined as follows:

$$\text{CR}_{\text{Morb,CAT}} = \sum_i [0.001 \times \text{SA}_i + 0.005 \times \text{AB}_i],$$

where the subscript i denotes each policy where the payment of benefits, either lump sum or multiple payments, was contingent on health status.

N.4.2.2 QIS3 (CEIOPS, 2006b)

The morbidity risk was aggregated into subrisk "disability risk," Section N.5, in QIS3.

N.5 DISABILITY RISK

N.5.1 Background

As an answer to the European Commission's CfA 10 (MARKT, 2004c), CEIOPS gave a background to the modeling of the standard formula and its risk modules (see CEIOPS, 2005). Disability risk was not a separate risk module in this discussion.

N.5.2 QIS2–QIS4

N.5.2.1 QIS2 (CEIOPS, 2006d)

The treatment of disability risk is split into the following risk components:

- Volatility risk
- Uncertainty risk
- CAT risk

Uncertainty risk comprises accumulation, trend, and parameter risks, to the extent these are not already reflected in the valuation of technical provisions.

The following input information was required for the class of insurance contracts contingent on a definition of disability:

CaR: the sum of the (net) capital at risk

i_x: the average disability probability

N: the number of insurance contracts

TP_{Dis}: the sum of net technical provisions

SA_i: the sum assured for each policy i, where benefits are payable as a single lump sum, otherwise zero

AB_i : the annualized amount of benefit payable for each policy i, where benefits are not payable as a single lump sum, otherwise zero

Both factor- and scenario-based approaches were tested in QIS2:

$$\text{CR}_{\text{LR,DR}} = \text{CR}_{\text{DR,Vol}} + \text{CR}_{\text{DR,Trend}} + \text{CR}_{\text{DR,CAT}}$$

where $\text{CR}_{\text{DR,Vol}}$: either the factor-based risk capital for volatility risk or the results of the disability scenario for volatility risk; $\text{CR}_{\text{DR,Trend}}$: either the factor-based risk capital for trend/uncertainty risk or the results of the disability scenario for trend/uncertainty; and $\text{CR}_{\text{Morb,CAT}}$: the risk capital for disability CAT risks.

N.5.2.1.1 Volatility Risk The risk capital charge for volatility risk under the factor-based approach was defined as follows:

$$\text{CR}_{\text{DR,Vol}} = 2.58 \times \sigma_{\text{DR}} \times \text{CaR},$$

where σ_{DR}: the estimate of the standard deviation in the loss distribution for disability risk and this is estimated as

$$\sigma_{\text{DR}} = \sqrt{\frac{i_x\,(1 - i_x)}{N}}.$$

The disability scenario for volatility risk was defined as

$$\mathrm{CR}_{\mathrm{DR,Vol}} = \sum_i \{\Delta \mathrm{NAV}_i \mid \mathrm{Disabilityshock}_{\mathrm{Vol}}\},$$

where the subscript i denotes each policy where the payment of benefits, either lump sum or multiple payments, was contingent on a definition of disability. The other terms represent: ΔNAV: the change in the net value of assets minus liabilities.; and $\mathrm{Disabilityshock}_{\mathrm{Vol}}$: a 10% increase in disability rates for each age during the next business year.

N.5.2.1.2 Trend/Uncertainty Risk In the factor-based approach, the risk capital charge for uncertainty risk is defined as follows:

$$\mathrm{CR}_{\mathrm{DR,Trend}} = 0.002 \times \mathrm{TP}_{\mathrm{Dis}}.$$

The life disability scenario for trend/uncertainty risk was defined as

$$\mathrm{CR}_{\mathrm{DR,Trend}} = \sum_i \{\Delta \mathrm{NAV}_i \mid \mathrm{Disabilityshock}_{\mathrm{Trend}}\},$$

where the subscript i denotes each policy where the payment of benefits, either lump sum or multiple payments, was contingent on a definition of disability. The other terms represent: ΔNAV: the change in the net value of assets minus liabilities; and $\mathrm{Disabilityshock}_{\mathrm{Trend}}$: a (permanent) 25% increase in assumed rates for disability for each age, including probability of remaining sick or disabled.

N.5.2.1.3 CAT Risk The CAT RC for disability risk was defined as follows:

$$\mathrm{CR}_{\mathrm{DR,CAT}} = \sum_i [0.001 \times \mathrm{SA}_i + 0.005 \times \mathrm{AB}_i],$$

where the subscript i denotes each policy where the payment of benefits, either lump sum or multiple payments, was contingent on a definition of disability.

N.5.2.2 QIS3 (CEIOPS, 2006b)

The treatment of disability risk was intended to reflect uncertainty risk. Uncertainty risk comprises trend and parameter risks, to the extent these are not already reflected in the valuation of technical provisions.

It is applicable to the class of insurance contracts where benefits are payable contingent on a definition of disability.

The capital charge for disability risk was defined as a result of a disability scenario as follows:

$$\mathrm{CR}_{\mathrm{LR,DR}} = \sum_i \{\Delta \mathrm{NAV}_i \mid \mathrm{Disabilityshock}\},$$

where the subscript i denotes each policy where the payment of benefits, either lump sum or multiple payments, was contingent on disability risk. The other terms represent: ΔNAV: the change in the net value of assets minus liabilities; and Disabilityshock: an increase of 35% in disability rates for the next year, together with a permanent 25% increase over the best estimate in disability rates at each age in following years.

The life disability scenarios should be calculated under the condition that the assumptions on future bonus rates, reflected in the valuation of future discretionary benefits in technical provisions, remained unchanged before and after the shocks being tested.

The results of the disability scenario may be approximated by a factor-based approach as follows:

$$\mathrm{CR}_{\mathrm{LR,DR}} = 0.01 \times \mathrm{CaR},$$

where the capital at risk term, CaR, could be approximated by

$$\mathrm{CaR} = \sum_i [\mathrm{SA}_i + \mathrm{AB}_i \times \mathrm{Af} - \mathrm{TP}_i],$$

where for each policy i, SA_i: the sum assured (net of reinsurance) on disability, where benefits are payable as a single lump sum; otherwise, zero; AB_i: the annualized amount of benefit (net of reinsurance) payable on disability, where benefits are not payable as a single lump sum; otherwise, zero; TP_i: the net of reinsurance held technical provisions; and Af: average annuity factor for the expected duration over which benefits may be payable in the event of a claim.

The factor 0.01 would represent 40% of the assumed average probability of disability times an average duration. To simplify the calculation, no differentiation was assumed for different future time periods.

N.5.2.3 QIS4 (QIS4, 2008)

The treatment of disability risk was intended to reflect uncertainty risk in trends and parameters, to the extent these are not already reflected in the valuation of technical provisions. It is applicable to the class of insurance contracts where benefits are payable contingent on a definition of disability. Disability also includes morbidity or sickness, that is, policies with lump sum or annual benefits that are payable contingent on some definition of sickness.

The capital charge for disability risk was defined as a result of a disability scenario as follows:

$$\mathrm{CR}_{\mathrm{LR,DR}} = \sum_i \{\Delta\mathrm{NAV}_i \mid \mathrm{Disabilityshock}\},$$

where the subscript i denotes each policy where the payment of benefits, either lump sum or multiple payments, was contingent on disability risk. The other terms represent: ΔNAV: the change in the net value of assets minus liabilities; and Disabilityshock: an increase of 35% in disability rates for the next year, together with a permanent 25% increase over the best estimate in disability rates at each age in following years.

The life disability scenarios should be calculated under the condition that the assumptions on future bonus rates, reflected in the valuation of future discretionary benefits in technical provisions, remained unchanged before and after the shocks being tested.

Additionally, the result of the scenario was determined under the condition that the participant was able to vary its assumptions in future bonus rates in response to the shock being tested. The resulting capital charge is $n\mathrm{CR}_{\mathrm{LR,DR}}$.

There are also *simplifications* that could be tested under QIS4, provided that:

1. There is no significant change in the capital at risk over the policy term of the contract
2. The general criteria for simplifications are followed

$$\begin{aligned}\text{Disability capital requirement} = {} & (\text{total disability sum at risk}) \times i(\text{firm-specific}) \\ & \times 0.35 \times 1.1^{((n-1)/2)} \times n\end{aligned}$$

where n is the modified duration of liability CFs, i the expected movements from healthy to sick over the next year weighted by sum assured/annual, and PDI (projected disability increase) $= 1.1^{((n-1)/2)}$.

N.5.3 Calibration

N.5.3.1 QIS2 results (CEIOPS, 2006b)

The feedback arising from QIS2 on disability (and morbidity) risk was similar to that for mortality risk; see Section N.2.3.

N.5.3.2 QIS3 (CEIOPS, 2007e, 2007g)

For disability risk, CEIOPS had regard to information derived from a study published in 2004 by Watson Wyatt about the 99.5% assumption over a 12-month time horizon that firms were proposing to make for the British ICAS submissions in the United Kingdom, for the increase in sickness and disability rates (see CEIOPS, 2007g; Watson Wyatt, 2004b). This indicated a wide range of between 10% and 60%, with an average of around 40%, as the level of increase that was assumed in the number of new sickness and disability claims.

It is likely that much of that potential variation would be attributable to short-term factors such as epidemics, and also to the effect of the economic cycle that can increase the number of longer-term claims. For example, significant variations in the numbers of new longer-term sickness claims between different 4-year periods can be observed from different reports published by the U.K. actuarial profession. However, there could also be more permanent changes as a result of a resurgence of diseases such as tuberculosis, or that are attributable to a lack of relevance or credibility of the data on which the best estimates are based.

Since comments from firms claimed the 40% average increase in disability rates as being slightly high, it is proposed to adopt a factor of 35% for the increase in sickness and disability rates over the next 12 months, reducing to a 25% increase over the best estimate thereafter. The latter figure represented an allowance for both a permanent increase in sickness rates, over the assumed best estimate, and the effect on claims of the economic cycle.

N.5.3.3 QIS4 (QIS4, 2008)

For QIS4, the QIS3 calibration has not been substantially changed. As a consequence, the above QIS3 calibration paper should still be used as a reference for the QIS4 calibration.

N.6 LAPSE RISK

N.6.1 Background

As an answer to the European Commission's CfA 10 (MARKT, 2004c), CEIOPS gave a background to the modeling of the standard formula and its risk modules (see CEIOPS, 2005).

N.6.1.1 Choice of Volume Measure

With regards to lapse risk there are two primary effects of unanticipated lapse rates. The first involves the payment of surrender or termination values. The relationship of the amount of a surrender payment to the value of the liability being held in respect of a particular policy is of great importance. When a policy lapses, the insurer pays the surrender value and "receives" the actuarial reserve that is released by the policy's termination. If surrender values are lower than policy reserves, the insurer is at risk from lapse rates that are lower than expected, particularly if high lapse rates were anticipated in the pricing of a product. The case that surrender values exceed policy reserves results in higher lapse rates being unfavorable to the insurer. However, if according to IAIS, technical provisions must not be lower than surrender values, there is no risk in an increase in lapse rates.

The second primary effect of unanticipated lapse rates is that the insurer may not realize the expected recovery from future premiums of initial policy acquisition expenses. These acquisition expenses may be recognized implicitly in financial statements through the use of modified net-level premium valuation methods. These implicit methods currently do not include any provision for unfavorable variations in lapse rates. Under a best estimate plus an RM valuation approach, these unfavorable variations should be partly included in the RM.

A capital requirement with respect to the first type of lapse risk requires the division of an insurance company's policies into two classes: first those policies for which the technical provisions TP are greater than surrender values S, and second those policies for which $S > \text{TP}$. This suggests choosing $S - \text{TP}$ and $\text{TP} - S$, as volume measures for the first type of lapse risk.

For the second type of lapse risk, the technical provision seems to be the appropriate volume measure. Within a best estimate plus an RM valuation approach, the technical provision will need to include a provision for the impact of unfavorable variations in lapse rates on the expected recovery of the acquisition expenses.

N.6.1.2 Choice of Coefficients

Regarding a capital requirement with respect to the first type of lapse risk, the insurance company's policies should be divided into two classes: those policies for which technical provisions TP are greater than surrender values S, and those policies for which $S > \text{TP}$.

Capital at risk denotes the difference between the payment falling due when the contract is triggered and the technical provision for that contract. The capital requirements would then be of the form:

$$\mathrm{CR}_{\mathrm{LR,LA}} = j(\mathrm{TP}_0 - S_0)\,, \quad \mathrm{TP}_0 > S_0,$$

$$\mathrm{CR}_{\mathrm{LR,LA}} = k(\mathrm{TP}_0 - S_0)\,, \quad \mathrm{TP}_0 < S_0,$$

respectively, for appropriately chosen factors j and k.

N.6.1.3 Degree of Personalization

Translating these theoretical equations into a factor-based, standardized formula requires:

- Analysis at the level of individual undertakings
- Generalized analysis that can be applied across the industry

Given the practicability of the determination of the coefficients for lapse risk j and k could be set by the supervisor; the result would be a table of industry-wide factors j and k for lapse risk that can be applied to the insurers' provisions in each segment.

N.6.1.4 Scenario Techniques

For the assessment of lapse risk, a prespecified stress test can easily be applied. The capital requirement is of the form of the difference between a special valuation of policy liabilities and the normal valuation. For the special valuation, the lapse assumption is multiplied by a specified factor greater or less than one. Since for some policies, an increase in lapse rates will result in an increase in policy liabilities, and for other policies liabilities will increase when assumed lapses decrease.

A lapse case, which cannot be addressed in a factor-based approach, are those products for which lapse risk does not act uniformly over the products life, such as lapses at early durations that may reduce the company's exposure to later risks for some policies and not for others.

N.6.2 QIS2–QIS4

N.6.2.1 QIS2 (CEIOPS, 2006d)

Lapse risk relates to an unanticipated (higher or lower) rate of policy lapses, terminations, changes to paid-up status (cessation of premium payment), and surrenders.

The following input information was required:

TP: technical provision

RB: total amount of claims against policyholders and insurance agents and Zillmer/agents' and other intermediaries' commission claw-back claims

Both factor- and scenario-based approaches were tested in QIS2:

The factor-based approach is

$$\mathrm{CR}_{\mathrm{LR,LA}} = 0.005 \times \mathrm{TP} + 0.1 \times \mathrm{RB}.$$

and the scenario-based approach is

$$\mathrm{CR}_{\mathrm{LR,LA}} = \sum_i \{\Delta\mathrm{NAV}_i \mid \mathrm{Lapseshock}\},$$

where i denotes each policy. The other terms represent: ΔNAV: the change in the net value of assets minus liabilities; and Lapseshock: the more adverse of a 50% increase or 50% decrease was assumed rates of lapsation at each duration, subject to a minimum change of 3% per annum. This means that if an assumed lapsation rate is required to be reduced from a rate already below 3% per annum, it should be reduced to zero.

N.6.2.2 QIS3 (CEIOPS, 2007a)

The capital charge for lapse risk was defined as follows:

$$\mathrm{CR}_{\mathrm{LR,LA}} = \sum_i \{\Delta\mathrm{NAV}_i \mid \mathrm{Lapseshock}\},$$

where i denotes each policy. The other terms represent: ΔNAV: the change in the net value of assets minus liabilities; and Lapseshock: the greater each year of

1. A 50% increase in the assumed rates of lapsation or
2. An increase in absolute terms of 3% per annum in the assumed rate of lapsation, for policies where the surrender value currently exceeds the technical provisions held; together with a 50% reduction in the assumed rates of lapsation for policies where the surrender value is currently less than the technical provisions held.

The life lapse risk scenario should be calculated under the condition that the assumptions on future bonus rates, reflected in the valuation of future discretionary benefits in technical provisions, remained unchanged before and after the shock being tested.

The results of the lapse scenario could be approximated by use of a factor-based approach as follows:

$$\mathrm{CR}_{\mathrm{LR,LA}} = 0.2 \times (\mathrm{Surrender\_release} + \mathrm{Surrender\_strain}),$$

where Surrender_release: The sum of the differences (where positive) between the technical provisions held for policies that can be lapsed or surrendered, and the amount currently payable on surrender; and Surrender_strain: The sum of the differences (where positive) between the amounts currently payable on surrender, and the technical provisions held for policies that can be lapsed or surrendered.

N.6.2.3 QIS4 (QIS4, 2008)

Lapse risk relates to the loss, or adverse change, in the value of insurance liabilities, resulting from changes in the level or volatility of the rates of policy lapses, terminations, changes to paid-up status (cessation of premium payment), and surrenders. The standard formula allows for the risk of a permanent change of the rates as well as for the risk of a mass lapse event.

The capital charge for lapse risk was defined as follows:

$$\mathrm{CR}_{\mathrm{LR,LA}} = \max\left\{\mathrm{Lapse}_{\mathrm{down}}; \mathrm{Lapse}_{\mathrm{up}}; \mathrm{Lapse}_{\mathrm{mass}}\right\},$$

where $\mathrm{Lapse}_{\mathrm{down}}$: capital charge for the risk of a permanent decrease of rates of lapsation; $\mathrm{Lapse}_{\mathrm{up}}$: capital charge for the risk of a permanent increase of rates of lapsation; and $\mathrm{Lapse}_{\mathrm{mass}}$: capital charge for the risk of a mass lapse event

The capital charges were calculated based on a policy-by-policy comparison of surrender value and the best estimate provision. The surrender strain of a policy was defined as the difference between the amount currently payable on surrender and the best estimate provision held. The amount payable on surrender should be calculated as the net of any amounts recoverable from policyholders or agents, for example, net of any surrender charge that may be applied under the terms of the contract.

$$\mathrm{Lapse}_{\mathrm{down}} = \sum_i \left[\Delta\mathrm{NAV}_i \mid \mathrm{Lapseshock}_{\mathrm{down}}\right],$$

where i denotes each policy; ΔNAV: change in the net value of assets minus liabilities (not including the loss-absorbing effect of future discretionary benefits and taxation); and $\mathrm{Lapseshock}_{\mathrm{down}}$: reduction of 50% in the assumed rates of lapsation in all future years for policies where the surrender strain is expected to be negative

$$\mathrm{Lapse}_{\mathrm{up}} = \sum_i \left[\Delta\mathrm{NAV}_i \mid \mathrm{Lapseshock}_{\mathrm{up}}\right],$$

where i denotes each policy; ΔNAV: change in the net value of assets minus liabilities (not including the loss-absorbing effect of future discretionary benefits and taxation); and $\mathrm{Lapseshock}_{\mathrm{up}}$: increase of 50% in the assumed rates of lapsation in all future years for policies where the surrender value is expected to be positive.

$\mathrm{Lapse}_{\mathrm{mass}}$ is defined as 30% of the sum of surrender strains over the policies where the surrender strain is positive. The result reflects the loss, which is incurred in a mass lapse event.

To determine $n\mathrm{CR}_{\mathrm{LR,LA}}$, the results of the scenarios was also calculated under the condition that the undertaking was able to vary its assumptions on future bonus rates in response to the shock being tested. If the scenario that gave the maximum net calculation did not coincide with the scenario that gave the maximum gross calculation, the definition of $CR_{LR,LA}$ should be changed to ensure consistency with the net calculation.

For instance, if $\text{Lapse}_{\text{down}} = 10$, $\text{Lapse}_{\text{up}} = 20$, $\text{Lapse}_{\text{mass}} = 30$, $n\text{Lapse}_{\text{down}} = 9$, $n\text{Lapse}_{\text{up}} = 5$, and $n\text{Lapse}_{\text{mass}} = 8$, then $\text{Life}_{\text{lapse}}$ should be chosen to be 10 but not 30.

There are also *simplifications* that could be tested under QIS4, provided that: If it is proportionate to the nature, scale, and complexity of the risk, the comparison of surrender value and the best estimate provision in the above calculations may be made on the level of homogeneous risk groups, or at finer granularity, instead of a policy-by-policy basis.

In particular, if the conditions are met this simplification may be applied if technical provisions are not calculated on a policy-by-policy basis. A simplified calculation of $\text{Lapse}_{\text{down}}$ and Lapse_{up} may be made if the following conditions are met:

1. The simplified calculation is proportionate to nature, scale, and complexity of the risk.
2. The undertaking is small or the capital charge for lapse risk under the simplified calculation is <5% of the overall SCR before adjustment for the loss-absorbing capacity of technical provisions.

The simplified calculations are defined as follows:

$$\text{Lapse}_{\text{down}} = 0.5 \times l_{\text{down}} \times n_{\text{down}} \times S_{\text{down}}$$

and

$$\text{Lapse}_{\text{up}} = 1.5 \times l_{\text{up}} \times n_{\text{up}} \times S_{\text{up}},$$

where l_{down}; l_{up}: estimate of the average rate of lapsation of the policies with a negative/positive surrender strain; n_{down}; n_{up}: average period (in years), weighted by surrender strains, over which the policy with a negative/positive surrender strain runs off; and S_{down}; S_{up}: sum of negative/positive surrender strains.

N.6.3 Calibration

N.6.3.1 QIS2 Results (CEIOPS, 2006b)

N.6.3.1.1 The Factor-Based Approach In QIS2, the main capital charge for lapse risk was calculated by a factor-based approach. This approach used technical provisions and the total amount of claims against policyholders and insurance agents as volume measures.

QIS2 participants raised the following concerns regarding this approach:

- It can be expected that under the future solvency valuation of technical provisions, no "Zillmerising" will be allowed, so that the materiality of claims against insurance agents should diminish.
- The treatment is unsuitable for certain types of product, for example, annuity business where there is no lapse option.

- Technical provisions would not seem to be an appropriate exposure measure for lapse risk.

N.6.3.1.2 The Scenario-Based Approach QIS2 also tested a scenario-based treatment of lapse risk. This required undertakings to assess the impact of the most adverse of a 50% increase or 50% decrease in assumed rates of lapsation at each duration, subject to a minimum change of 3% per annum.

Participants commented that the specification of this scenario was not sufficiently clear, and questioned the appropriateness of requiring minimum absolute changes.

N.6.3.2 QIS3 (CEIOPS, 2007e, 2007g)

For lapse risk, CEIOPS had regard to information derived from a published study in 2004 by Watson Wyatt about the 99.5% assumption over a 12-month time horizon that firms were proposing to make for the British ICAS submissions in the United Kingdom, for the changes in lapse rates, either upwards or downwards. This indicated a range of 35–60% in the rate of assumed lapses, with an average of around 50% as the change in the rate of lapses that was assumed (see CEIOPS, 2007g; Watson Wyatt, 2004b).

However, CEIOPS were also conscious that a 50% increase in the rate of lapses might underestimate the potential increase, if lapse rates are currently quite low. Consequently, CEIOPS proposed to apply a minimum increase of 3% per annum in the assumed lapse rate.

N.6.3.3 QIS4 (QIS4, 2008)

Some aspects of the approach used in QIS3 were criticized by participants: The 75% shock in the CAT module was considered to be too high and participants identified a potential for double-counting as two submodules cover the risk of increase in lapse rates. Moreover, a simplification of the approach was asked for.

In response to this feedback, the following changes were made:

The 75% shock of the mass lapse scenario was reduced to 30%. The new calibration is an expert estimate based on past mass lapse events in the German life insurance market. The scope of application of the shock was extended from linked policies to all policies.

To avoid double-counting, only the more adverse of the mass lapse shock and the scenario of permanent 50% increase in lapse rates was used to determine the capital charge.

The scenario component of an increase in absolute terms of 3% was removed for reasons of simplification.

N.7 EXPENSE RISK

N.7.1 Background

As an answer to the European Commission's CfA 10 (MARKT, 2004c), CEIOPS gave a background to the modeling of the standard formula and its risk modules (see CEIOPS, 2005).

N.7.1.1 Choice of Volume Measure

A detailed understanding of the insurer's expense structure and expense drivers is a key element when determining the expense risk. Using a prospective valuation approach of assets and liabilities means that all possible future CFs will have to be identified and valued. Expenses that will have to be made in future to service an insurance contract are one of those CFs for which a provision will have to be calculated.

The IAA observes that the insurer normally selects assumptions with respect to the future expenses associated with obligations arising from commitments the entity has made on, or prior to, the valuation date, including overheads. When setting expense assumptions, it may be useful to differentiate between:

The entity's strategy for determining the level of service provided to policyholders, and its approach to claims' management, if applicable

The entity's efficiency in providing that level of service, implementing its approach to claims' management, if applicable

Usually all future administrative costs and consequent commissions would need to be considered. Where future deposits or premiums are factors in the determination of the liabilities, expenses related to the deposits or premiums would usually be taken into consideration. In addition, where appropriate, the expenses of administering investments normally would be taken into consideration too.

N.7.1.2 Choice of Coefficients

A methodology for determining the expense risk capital requirement could involve looking at the expenses of a company in aggregate and simply estimating the capital charge as

$$\mathrm{CR}_{\mathrm{LR,ER}} = t \times E_{\mathrm{total}},$$

where t is an appropriately chosen factor and E_{total} is the provision for all expenses that will have to be made in future to service an insurance contract.

N.7.1.3 Degree of Personalization

Translating these theoretical equations into a factor-based, standardized formula requires:

- Analysis at the level of individual undertakings
- Generalized analysis that can be applied across the industry

The coefficient t could be set by the supervisor; the result would be a table of industry-wide factors t for expense risk that can be applied to the insurers' provisions in each segment.

N.7.2 QIS2–QIS4

N.7.2.1 QIS2 (CEIOPS, 2006d)

Expense risk arises from the variation in the expenses associated with the insurance contracts.

The following input information was required:
$E_{\text{fixed:the}}$ total annual amount of the fixed expenses of the undertaking

Both factor- and scenario-based approaches were tested in QIS2:
The factor-based approach is

$$\text{CR}_{\text{LR,ER}} = 0.1 \times E_{\text{fixed}}$$

and the scenario-based approach is

$$\text{CR}_{\text{LR,ER}} = \left\{\Delta\text{NAV} \mid \text{Expenseshock}\right\},$$

where ΔNAV: the change in the net value of assets minus liabilities; and Expenseshock: all future expenses are higher than the best estimate anticipations by 10% and the rate of expense inflation was 1.5% per annum higher than anticipated.

N.7.2.2 QIS3 (CEIOPS, 2007a)
The capital charge for expense risk was determined as follows:

$$\text{CR}_{\text{LR,ER}} = \left\{\Delta\text{NAV} \mid \text{Expenseshock}\right\},$$

where ΔNAV: the change in the net value of assets minus liabilities; and Expenseshock: All future expenses are higher than the best estimate anticipations by 10%, and the rate of expense inflation is 1% per annum higher than anticipated; but for policies with adjustable loadings, 75% of these additional expenses can be recovered from year 2 onwards through increasing the charges payable by policyholders. Policies with adjustable loadings were those for which expense loadings or charges may be adjusted within the next 12 months.

The life expense risk scenario should be calculated under the condition that the assumptions on future bonus rates, reflected in the valuation of future discretionary benefits in technical provisions, remained unchanged before and after the shock being tested.

The results of the revised scenario could be approximated by a factor-based approach as follows:

$$\text{CR}_{\text{LR,ER}} = 0.12 \times f_{\text{fixed}} \times E_{\text{fixed}} + 0.03 \times f_{\text{adj}} \times E_{\text{adj}},$$

where f_{fixed}: average outstanding duration of business with fixed; E_{fixed}: total annual amount of the expenses for business with fixed loadings; f_{adj}: average outstanding duration of business with adjustable loadings; and E_{adj}: total annual amount of the expenses for business with adjustable loadings.

N.7.2.3 QIS4 (QIS4, 2008)
Expense risk arises from the variation in the expenses incurred in servicing insurance or reinsurance contracts.

The capital charge for expense risk was determined as follows:

$$\text{CR}_{\text{LR,ER}} = \left\{\Delta\text{NAV} \mid \text{Expenseshock}\right\},$$

where ΔNAV: the change in the net value of assets minus liabilities; and Expenseshock: All future expenses are higher than the best estimate anticipations by 10%, and the rate of expense inflation is 1% per annum higher than anticipated; but for policies with adjustable loadings, 75% of these additional expenses can be recovered from year 2 onwards through increasing the charges payable by policyholders. Policies with adjustable loadings were those for which expense loadings or charges may be adjusted within the next 12 months.

The life expense risk scenario should be calculated under the condition that the assumptions on future bonus rates, reflected in the valuation of future discretionary benefits in technical provisions, remained unchanged before and after the shock being tested.

An expense payment should not be included in the scenario, if its amount is already fixed at the valuation date (for instance agreed payments of acquisition provisions).

Additionally, the result of the scenario was determined under the condition that the participant was able to vary its assumptions in future bonus rates in response to the shock being tested. The resulting capital charge is $n\mathrm{CR}_{\mathrm{LR,ER}}$.

There are also *simplifications* that could be tested under QIS4, provided that:

$$\text{Expense risk capital requirement} = (\text{Renewal expenses in the 12 months prior to valuation date}) \times n(\text{expenses}) \times (0.1 + 0.005 \times n(\text{expenses})),$$

where n(expenses) = average (in years) period over which risk runs off, weighted by renewal expenses.

N.7.3 Calibration

N.7.3.1 QIS2 Results (CEIOPS, 2006b)

N.7.3.1.1 The Factor-Based Approach In QIS2, the main capital charge for lapse risk was calculated by a factor-based approach. This approach used the total annual amount of the fixed expenses of the undertaking as the volume measure.

QIS2 participants raised the following concerns regarding this approach:

- It would seem spurious to consider only fixed expenses, as variable expenses are just as likely to suffer from cost inflation. A restriction of considerations to fixed expenses would only be appropriate if a closed book scenario is predicted and hence diseconomies of scale need to be considered.
- Considering the long-term nature of life insurance business, a RC of 10% of one year's expenses may underestimate the true expense risk.

N.7.3.1.2 The Scenario-Based Approach QIS2 also tested a scenario-based treatment of lapse risk. This required undertakings to consider the scenario that all future expenses are higher than the best estimate anticipations by 10% and the rate of expense inflation is 1.5% per annum higher than anticipated.

N.7.3.2 QIS3 (CEIOPS, 2007e, 2007g)

For expense risk, CEIOPS had regard to information derived from a published study in 2004 by Watson Wyatt about the 99.5% assumption over a 12-month time horizon that firms were proposing to make for the British ICAS submissions in the United Kingdom, for the potential increase in the level of expenses (see CEIOPS, 2007g; Watson Wyatt, 2004b). This indicated a range of 5–50% in the rate of assumed lapses, with an average of around 26% as the increase in the level of expenses that was assumed. ICAS submissions in the United Kingdom are believed to have shown though an assumed increase of around 10% in the level of expenses in the following year, together with an increase of between 1% and 2% per annum in the rate of future expense inflation.

Increases of 10% or more in expense levels have certainly been observed in the accounts of undertakings from one year to the next. An increase of between 1% and 1.5% in the rate of inflation would also be consistent with the postulated movement in nominal rates of interest that are proposed for the interest rate risk component of the market risk module. Given that real rates of interest are likely to be fairly stable, this suggests that a 1% per annum increase in the rate of expected future inflation would be a consistent financial assumption. Accordingly, it was proposed that undertakings should assess the impact of a scenario of a 10% increase in the level of expenses in the following year, together with a 1% per annum increase in the assumed rate of inflation.

However, in the case of some policies, the undertakings are entitled to adjust the expense loadings or charges. Making allowance for this opportunity for those policies, it is assumed that 75% of the additional expenses can be recovered from year 2 onward through increasing the charges payable by policyholders.

N.7.3.3 QIS4 (QIS4, 2008)

For QIS4, the QIS3 calibration has not been substantially changed. As a consequence, the above QIS3 calibration paper should still be used as a reference for the QIS4 calibration.

N.8 LIFE CAT RISK

The CAT risks stem from extreme or irregular events that are not sufficiently captured by the charges for the biometric, lapse, and expense risks. These are one-time shocks from the extreme, adverse tail of the probability distribution that are not adequately represented by extrapolation from more common events and for which it is usually difficult to specify a loss value, and thus an amount of capital to hold. For example, a contagious disease process or a pandemic may affect many persons simultaneously, nullifying the usual assumption of independence among persons.

N.8.1 Background

As an answer to the European Commission's CfA 10 (MARKT, 2004c), CEIOPS gave a background to the modeling of the standard formula and its risk modules (see CEIOPS, 2005). Life CAT risk was not a separate risk module in this discussion.

Scenario techniques for life CAT risks were discussed in CEIOPS (2005).

N.8.1.1 Scenario Techniques

A factor-based approach to modeling UR is based on certain probabilistic assumptions on the frequency and severity of claims. Typically, a parametric family of distributions is chosen to model the future occurrence of loss. Parameters are fitted to statistical data that are collected from historical experience. The major part of such claims' experience relates to "normal" circumstances, where a certain regularity and smoothness in claims' patterns may be observed.

Extreme or irregular events may either be absent from the data, or may have to be "smoothed out" in the calibration process. By nature of their construction, factor-based models may be less able to predict extreme, catastrophic events. One response to this issue might be the provision of a separate treatment for catastrophic UR.

Scenarios may be used to model extreme events where the assumptions of the analytic model break down, or to take into account risks that are not covered by analytic models, particularly systemic risk. Mixing two different techniques may actually reduce modeling risk associated with a standard formula.

Possible scenarios include:

- Severe epidemic: for example, Spanish Flu in 1918
- Natural catastrophe: for example, earthquake
- Terrorist attack: for example, events of 9/11

A more restricted range might be applied to take account of relative data availability. For example, one might include periodic natural catastrophes and epidemic, but exclude extreme, episodic events, such as terrorist activity.

Taking account of extreme events implies a degree of domestic variability to reflect climatic and geographical differences. Given such specificities, it may be questioned whether UR scenarios are a plausible candidate for Pillar I. An alternative might be to require undertakings to define and test their own scenarios that could then be reviewed under Pillar II. This would offer alternatives to the imposition of capital charges—for example, the development of risk mitigation programs. However, the credibility of the standard formula might be undermined if quantifiable, highly visible (albeit extreme) risks are not addressed in Pillar I.

N.8.2 QIS2–QIS4

In QIS2, the CAT risk was part of mortality, morbidity, and disability risk submodules. From QIS3, it was seen as a separate subrisk.

N.8.2.1 QIS3 (CEIOPS, 2007a)

Life CAT risks stem from extreme or irregular events, for example, a pandemic, which are not sufficiently captured by the charges for the other LUR submodules. The treatment considers catastrophe risk in relation to both biometric and lapse risks.

The following input data were required for each policy where the payment of benefits, either lump sum or multiple payments, was contingent on either mortality or disability:

CaR: the sum of the (net) capital at risk

TP_i: the net technical provisions for policy i

SA_i: the sum assured for each policy i, where benefits are payable on death or disability as a single lump sum, otherwise zero

AB_i: the annualized amount of net benefit payable on death or disability for each policy i, where benefits are not payable as a single lump sum, otherwise zero

Af: average annuity factor for the expected duration over which benefits may be payable in the event of claim

Additionally, the following input data were required for the class of linked policies, which can be lapsed or surrendered:

SS_L is the sum of the differences, where positive, between (a) the amount currently payable on surrender and (b) the technical provisions held. The amount payable on surrender should be calculated as the net of any amounts recoverable from policyholders or agents, for example, net of any surrender charge that may be applied under the terms of the contract.

The CAT RC for LUR was determined as follows:

$$CR_{LR,CAT} = \sqrt{CR^2_{Mort+Dis,CAT} + CR^2_{Lapse,CAT}},$$

where $CR_{Mort+Dis,CAT} = 0.0015 \times CaR$, that is, the results of the calculation for mortality and disability catastrophe risk. The CaR is determined as

$$CaR = \sum_i [SA_i + AB_i \times Af - TP_i].$$

Subscript i denotes each policy where the payment of benefits, either lump sum or multiple payments, is contingent on either mortality or disability; and $CR_{Lapse,CAT} = 0.75 \times SS_L$, that is, the results of the calculation for lapse catastrophe risk.

The capital charge for life CAT risk should be calculated under the condition that the assumptions on future bonus rates, reflected in the valuation of future discretionary benefits in technical provisions, remain unchanged before and after a life CAT event.

N.8.2.2 QIS4 (QIS4, 2008)

The capital charge for life CAT risk was determined as follows:

$$CR_{LR,CAT} = \{\Delta NAV \mid Lifeshock_{CAT}\},$$

where the $Lifeshock_{CAT}$ is a combination of the following events all occurring at the same time:

- An absolute 1.5 per mille increase in the rate of policyholders dying over the following year (e.g., from 1.0 to 2.5 per mille).

- An absolute 1.5 per mille increase in the rate of policyholders experiencing morbidity over the following year. Where appropriate, undertakings should assume that one-third of these policyholders experience morbidity for 6 months, one-third for 12 months, and one-third for 24 months from the time at which the policyholder first becomes sick.

Participants are requested to calculate the capital charge for life CAT risk, which should be calculated under the condition that the assumptions on future bonus rates, reflected in the valuation of future discretionary benefits in technical provisions, remained unchanged before and after a life CAT event.

Additionally, participants were also requested to determine the result of the scenario under the condition that the participant was able to vary its assumptions in future bonus rates in response to the shock being tested. The resulting capital charge is $n\text{CR}_{\text{LR,CAT}}$.

There are also *simplifications* that could be tested under QIS4. The following formula could be used as a simplification for the life catastrophe risk submodule: the input data are required for each policy where the payment of benefits, either lump sum or multiple payments, is contingent on either mortality or disability:

$$\text{CR}_{\text{LR,CAT}} = \sum_i 0.0015 \times \text{CaR}_i,$$

where the subscript i denotes each policy where the payment of benefits, either lump sum or multiple payments, is contingent on either mortality or disability, and where CaR_i is determined as

$$\text{CaR}_i = \text{SA}_i + \text{AB}_i \times \text{Af} - \text{TP}_i$$

and TP_i: technical provision (net of reinsurance) for each policy i; SA_i: for each policy i, where benefits are payable as a single lump sum, the sum assured (net of reinsurance) on death or disability; otherwise, zero; AB_i: for each policy i, where benefits are not payable as a single lump sum, the annualized amount of benefit (net of reinsurance) payable on death or disability; otherwise, zero; and Af: average annuity factor for the expected duration over which benefits may be payable in the event of a claim.

N.8.3 Calibration

N.8.3.1 QIS2 Results (CEIOPS, 2006b)

When considering possible catastrophe losses over the following 12 months, the intention was that the CAT charge should represent the average effect on the net asset value of the undertaking of the 1% of scenarios, including multiple catastrophes, which cause the greatest fall in net assets.

A number of catastrophe scenarios were agreed by supervisors. They intended to be representative scenarios for 1/100 events, that is the average cost of the worst event in 100 years, that is, TailVaR. The catastrophes may include catastrophes concerning biometric risks (e.g., a pandemic) and possibly also events that can have retrospective effect on existing liabilities: for example, a sudden increase in prices or an increase in inflationary expectations. A clear outline of the range of the scenarios to be considered would need to be defined to ensure a consistent approach.

The capital charge for each scenario was estimated by the insurer by evaluating the effect of it, taking into account the peculiarities of its business. Insurers also had the opportunity to estimate the impact of a specified scenario by using a market-loss approach, that is, by estimating the market-wide loss for the scenario, which may be specified by the supervisor, and deriving the entity-specific capital charge by using its market share.

N.8.3.2 QIS3 (CEIOPS, 2007e, 2007g)

Mortality and disability catastrophe risk reflects the potential effect of epidemics and other hazards such as large fires, earthquakes, war, and terrorism. The epidemic risk is thought to outweigh all these other potential risks at present and, consequently, is the only catastrophe considered further here. The particular risk at the forefront of thinking by the WHO at that time was of an avian flu pandemic, thought here are of course other potential diseases such as SARS or Ebola that needs to be borne in mind as well.

N.8.3.2.1 Mortality Catastrophe Risk It is known that there are recurring serious flu epidemics every 20–30 years. The most intense of these epidemics in living memory was the 1918 epidemic. This was believed to have been a significant mutation of earlier flu viruses that caused serious illness and resulted in a mortality rate of as much as 1–2% of the population in some countries, focused mainly on those aged between 20 and 40, but with lower rates of mortality in other age groups. Over Europe as a whole, the additional death rate was around 5 per mille. It is not entirely clear why the death rate varied between countries or why the 20–40 age group was particularly badly affected. One theory, however, is that many of the deaths resulted from an overreaction of the immune system, which is at its strongest level for individuals in this age group, and which caused the lungs of infected people to be overwhelmed by a form of pneumonia.

The H5N1 flu strain, at this time, was of considerable concern to the WHO. It was fairly difficult to transmit to humans and there has been around a 50% or higher death rate from known cases of people infected people by this virus. It was also believed that it may only be a matter of time before the virus mutated or combined with a human flu virus to become contagious between humans. It would then spread very rapidly within the global population, though the prognosis for people infected with the mutated virus was at that time unknown. There are a number of possible mitigants such as vaccines and antiviral drugs. However, it would take some time to develop an effective and widely available specific vaccine for this virus, and this might well not be available to prevent a widespread outbreak. It was also known that the H5N1 virus was developing some resistance to the most commonly available antiviral drug.

CEIOPS assumed, as a starting point, that the 1918 epidemic represents a 1 in 200 year event, and then that this would suggest an additional number of deaths of around 5 per mille. However, it may be reasonable to allow some reduction for the medical advances that have taken place since then, albeit that there was still considerable uncertainty about the potential effectiveness of these mitigating measures in combating a mutated H5N1 (or any other) virus.

Accordingly, it was proposed to assume a capital component equal to 1.5 per mille of capital at risk for mortality catastrophe risk.

N.8.3.2.2 Disability Catastrophe Risk For sickness and disability catastrophe risk, a major epidemic is also likely to be the main risk factor. However, we were already assuming an increase of 35% in the level of new claims to calculate the capital component for sickness and disability risk, and a serious flu epidemic would be more likely to cause death within a week or two than a lengthy sickness claim.

Accordingly, it was proposed to assume an additional level of claims of 1.5 per mille.

N.8.3.2.3 Lapse Catastrophe Risk A further "catastrophe" risk to consider was that of a sudden adverse policyholder reaction in the event of either a loss of reputation of a firm, or some other operational difficulty, resulting in a sizeable number of surrenders or lapses of (UL) policies. This risk may not be fully reflected in the operational risk component. The most effective means of covering this risk would be to hold capital to cover the difference between the surrender value of the policies and the technical provision held. This would also achieve some equivalence with the banking and investment sectors, where firms are not allowed to anticipate on the BS the expected profits from future management charges. This could, however, be offset by the lapse risk component in respect of UL policies and possibly by part of the operational risk component.

Accordingly, it was proposed to include a capital component for 75% of the difference between the surrender value of the policies and the technical provision held. This parameter will be reviewed further in the light of the QIS3 results.

N.8.3.3 QIS4 (QIS4, 2008)

The QIS4 calibration of the mortality and disability catastrophe risk was unchanged compared to QIS3. The capital charge was calculated as 1.5‰ of the capital at risk. The calibration was supported by a study of Swiss Reinsurance Company. Based on an epidemiological model, for a pandemic with a level of severity expected once every 200 years, the excess mortality within an insurance portfolio was estimated to be between 1 and 1.5 deaths per 1000 lives in most developed countries.

N.9 REVISION RISK

N.9.1 Background

As an answer to the European Commission's CfA 10 (MARKT, 2004c), CEIOPS gave a background to the modeling of the standard formula and its risk modules (see CEIOPS, 2005). Revision risk was not a separate risk module in this discussion. It was introduced in QIS3.

N.9.2 QIS2–QIS4

Revision risk was intended to capture the risk of adverse variation of an annuity's amount, as a result of an unanticipated revision of the claims' process. This is meant to impact only on annuities that are genuinely reviewable. Annuities whose amount is linked to earnings or prices or to some other index, or that vary in deterministic value on the change of status was not classified as genuinely reviewable for these attributes.

This risk should be applied only to annuities arising from nonlife claims that are allocated to the LUR module according to a principle set out in QIS3.

It should be noted that the revision risk was not included in QIS2, but a number of CEIOPS members highlighted the importance of its inclusion for a proper assessment of the specificities of the risks stemming from nonlife annuities. The design and calibration of the risk were thus at an early stage in QIS3. The objective was to gather market information on the appropriateness of the inclusion of the revision risk for the various markets and lines of business, including the relative size of the initial tentative calibration.

N.9.2.1 QIS3 (CEIOPS, 2007a)

The capital charge for revision risk was determined as follows:

$$\mathrm{CR}_{\mathrm{LR,RE}} = \{\Delta\mathrm{NAV} \mid \mathrm{Revisionshock}\},$$

where ΔNAV is the change in the net value of assets minus liabilities; and Revisionshock is a 3% increase in the annual amount payable for annuities exposed to revision risk. The impact should be assessed considering the remaining runoff period.

On the computation of this RC, participants should only consider the impact on those nonlife annuities for which a revision process is possible to occur during the next year, for example, annuities where legal or other eligibility restrictions should not be included.

N.9.2.2 QIS4 (QIS4, 2008)

In the context of the LUR module, revision risk was intended to capture the risk of adverse variation of an annuity's amount, as a result of an unanticipated revision of the claims' process. This was meant to impact only on annuities that are genuinely reviewable. Annuities whose amount is linked to earnings or another index such as prices, or that vary in deterministic value on change of status should not be classified as genuinely reviewable for these attributes. This risk should be applied only to annuities and to those benefits that can be approximated by a life annuity arising from nonlife claims, including accident insurance, but excluding workers' compensation that are allocated to the LUR module.

The capital charge for revision risk was determined as follows:

$$\mathrm{CR}_{\mathrm{LR,RE}} = \{\Delta\mathrm{NAV} \mid \mathrm{Revisionshock}\}$$

where ΔNAV: the change in the net value of assets minus liabilities; and Revisionshock: a 3% increase in the annual amount payable for annuities exposed to revision risk. The impact should be assessed considering the remaining runoff period.

On the computation of this RC, participants should only consider the impact on those nonlife annuities for which a revision process is possible to occur during the next year, for example, annuities where legal or other eligibility restrictions should not be included.

There are also *simplifications* that could be tested under QIS4.

$$\text{Revision capital requirement} = 3\% \times \text{Total net technical provisions for annuities exposed to revision risk.}$$

N.9.3 Calibration

N.9.3.1 QIS3 (CEIOPS, 2007e, 2007g)

For revision risk, the 3% increase scenario was calibrated using historical data for pensions in payment for the workers' compensation line of business in Portugal.

CEIOPS fitted a binomial compound distribution to the historical data, assuming a binomial distribution for the frequency process and a lognormal distribution to model the severity of revision. The aggregate loss distribution was derived using Monte-Carlo simulation for different portfolio sizes. All pensions were assumed to be independent and their annual amount was assumed to be constant. Different assumptions were considered for pensions homologated and pensions not yet defined, the latter with higher frequency and severity volatilities.

The 3% scenario corresponds to the 99.5% quantile of the aggregate loss distribution for an average-sized portfolio comprising pensions at different legal stages in "average" proportions.

It was recognized that further study is needed to assess the adequacy of the scenario, namely by increasing the number of historical years, the number of firms providing data, and, probably, differentiating the scenarios per type of pension, for example, pensions not yet homologated are more prone to revision risk. Also, the analysis will need to be extended to cover other lines of business and markets.

N.9.3.2 QIS4 (QIS4, 2008)

No changes were made in comparison to QIS3.

APPENDIX O

European Solvency II Standard Formula

Health Underwriting Risk

HERE, WE WILL DISCUSS the development and calibration of the health underwriting risk (UR) module and its capital charge that has been one of the building blocks in Solvency II. The final advice from CEIOPS is discussed in Chapter 33.

O.1 GENERAL FEATURES

This module is intended to cover UR for all health and workers' compensation guarantees. Different structures have been tested in QIS2–QIS4.

O.2 QIS2 PROPOSAL

In QIS2 (CEIOPS, 2006d), this risk module set out specifications for UR in health insurance that is practiced on a similar technical basis to that of life insurance.

Health UR was split into the three components:

- $CR_{HR,ER}$: expense risk
- $CR_{HR,MR}$: excessive loss/mortality/cancellation risk
- $CR_{HR,AR}$: epidemic/accumulation risk

The following input information was required:

H_{exp}: placeholder capital charge for health expense risk
H_{xs}: placeholder capital charge for health excessive loss/mortality/cancellation risk
H_{ac}: placeholder capital charge for health epidemic/accumulation risk
EH_{exp}: expected result in health expense risk
EH_{xs}: expected result in health excessive loss/mortality/cancellation risk

The overall capital charge for the health UR was calculated by starting with the following formula:

$$\mathrm{CR}_{\mathrm{HR},1} = \sqrt{\left(H_{\mathrm{exp}} + \mathrm{EH}_{\mathrm{exp}}\right)^2 + (H_{\mathrm{xs}} + \mathrm{EH}_{\mathrm{xs}})^2 + \left(H_{\mathrm{exp}} + \mathrm{EH}_{\mathrm{exp}}\right) \cdot (H_{\mathrm{xs}} + \mathrm{EH}_{\mathrm{xs}})}$$
$$+ H_{\mathrm{ac}} - \left(\mathrm{EH}_{\mathrm{exp}} + \mathrm{EH}_{\mathrm{xs}}\right),$$

where the dependence between the two first subrisks was assumed to be 0.5 (see Table O.1), and then setting the final capital charge to

$$\mathrm{CR}_{\mathrm{HR}} = \max\left\{\mathrm{CR}_{\mathrm{HR},1}; 0\right\}.$$

O.2.1 Health Expense Risk

Expense risk arises if the expenses anticipated in the pricing of a product are insufficient to cover the actual costs accruing in the accounting year. There are numerous possible causes of such a shortfall; therefore, all cost items of private health insurers have to be taken into account. To ensure comparability among the financial years, all annual results will be related to the gross premiums earned in the specific financial year.

It was calculated as

$$H_{\mathrm{exp}} = 2.58 \times \sigma_{\mathrm{exp}} \times \mathrm{GP}_{\mathrm{ay}} - \mathrm{EH}_{\mathrm{xp}} = \left[2.58 \times \sigma_{\mathrm{exp}} - \mu_{\mathrm{exp}}\right] \times \mathrm{GP}_{\mathrm{ay}},$$

where σ_{exp}: the standard deviation of the expense result over the previous 10-year period; $\mathrm{GP}_{\mathrm{ay}}$: gross premium earned for the accounting year; $\mathrm{EH}_{\mathrm{xp}} = \mu_{\mathrm{exp}} \times \mathrm{GP}_{\mathrm{ay}}$; and $\mu_{\mathrm{exp}} =$ the mean value of the expense result in the last three financial years.

O.2.2 Health Excessive Loss/Mortality/Cancellation Risk

This "multirisk" should cover the following:

- The excessive loss risk or per capita loss risk results if the actual per capita loss is greater than the loss assumed in the pricing of the product.
- The mortality risk exists if the actual funds from provisions for increasing age becoming available due to death are lower than those assumed in the pricing of the product.
- The cancellation risk exists if the actual funds from provisions for increasing age becoming available due to cancellations are lower than those assumed in the pricing of the product.

It was calculated as

$$H_{\mathrm{xs}} = 2.58 \times \sigma_{\mathrm{xs}} \times \mathrm{GP}_{\mathrm{ay}} - \mathrm{EH}_{\mathrm{xs}} = [2.58 \times \sigma_{\mathrm{xs}} - \mu_{\mathrm{xs}}] \times \mathrm{GP}_{\mathrm{ay}},$$

where σ_{xs}: the standard deviation of the H_{xs} result over the previous 10-year period; $\mathrm{GP}_{\mathrm{ay}}$: gross premium earned for the accounting year; $\mathrm{EH}_{\mathrm{xs}} = \mu_{\mathrm{xs}} \times \mathrm{GP}_{\mathrm{ay}}$; and $\mu_{\mathrm{xs}} =$ the mean value of the H_{xs} result in the last three financial years.

O.2.3 Health Epidemic/Accumulation Risk

The capital charge was calculated as

$$H_{ac} = 0.01 \times \mathrm{CE}_{ay} \times \frac{\mathrm{GP}_{ay}}{\mathrm{HGP}_{ay}},$$

where CE_{ay}: the claims' expenditure for the accounting year; GP_{ay}: gross premium earned for the accounting year; and HGP_{ay}: total gross premium earned for the accounting year in the health insurance market.

O.2.4 QIS2 Experience

The experience from QIS2 was presented in CEIOPS (2006b). The participants noted that the treatment of the expected result in health expense risk and excessive loss/mortality/cancellation risk within health UR module was not consistent with the non-life UR module. The nonlife UR was confined to measuring the "volatility-related" parts of premium and reserve risks, whereas the expected profit/loss arising from the nonlife business was considered in a module at a "top-level" adjustment to the SCR (see Appendix H). It was thus suggested for QIS3 that the treatment of expected losses/profits in health UR was aligning with the corresponding treatment in nonlife business. This means the following:

- The treatment of health expense risk and excessive loss/mortality/cancellation risk in health UR is confined to the assessment of excess losses (i.e., losses that occur in excess of the expected result)
- The expected result in health expense risk and excessive loss/mortality/cancellation risk is considered (in an analogous way to the expected profit/loss in nonlife business) as a top-level adjustment to the SCR

O.2.5 Health Expense Risk and Health Excessive Loss/Mortality/Cancellation Risk

The QIS2 participants generally supported the modeling approach. However, concerns were raised with respect to the following two points:

- CEIOPS did not specify how the capital charge should be derived in cases where the results from the preceding 10 years would not be available
- The treatment of the risks should be confined to the "volatility-related" part of this risk, whereas the expected result should be used as a "top-level" adjustment to the SCR

O.2.6 Health Epidemic/Accumulation Risk

The QIS2 participants generally supported the modeling approach. However, concerns were raised whether the 1% RF would not be too low, considering, for example, the prospect of a bird flu epidemic.

O.3 QIS3 PROPOSAL

In the Consultation Paper 20 (CEIOPS, 2006b), the health risk module was renamed to a "special" risk module, referring to types of business where the treatment of UR follows a different structure, such as the form of actuarial health insurance common in Austria and Germany.

Health UR was the risk arising from the underwriting of health insurance contracts, associated with both the perils covered and the processes followed in the conduct of the business. It concerns health insurance that is practised on a similar technical basis to that of life assurance.

Participants in QIS2 noted that the treatment of the expected result in health expense risk and excessive loss/mortality/cancellation risk within health UR module was not consistent with the nonlife UR module. $\mathrm{CR_{NL}}$ was confined to measuring the "volatility-related" parts of premium and reserve risks, whereas the expected profit/loss arising from the nonlife business was considered in the $\mathrm{EP_{NL}}$ module as a "top-level" adjustment to the SCR (see Appendix H).

In QIS3 (CEIOPS 2007a), the same description of the UR was made as in QIS2. Health UR is split into the three components:

$\mathrm{CR_{HR,ER}}$: expense risk

$\mathrm{CR_{HR,MR}}$: claim/mortality/cancellation risk

$\mathrm{CR_{HR,AR}}$: epidemic/accumulation risk

Combining the capital charges for the health subrisks using a dependence matrix as follows derived the capital charge for the health UR:

$$\mathrm{CR_{HR}} = \sqrt{\rho_{r\times c} \times \mathrm{CR}_r \times \mathrm{CR}_c}.$$

Table O.1 gives the dependence matrix.

TABLE O.1 Dependence Matrix for the Health UR Used for QIS2 and QIS3

	$\mathrm{CR_{HR,ER}}$	$\mathrm{CR_{HR,MR}}$	$\mathrm{CR_{HR,AR}}$
$\mathrm{CR_{HR,ER}}$	1	0.5	0
$\mathrm{CR_{HR,MR}}$		1	0
$\mathrm{CR_{HR,AR}}$			1

Source: Adapted from CEIOPS. 2006d. Quantitative Impact Study No. 2: Technical Specification, CEIOPS-PI-08/06. Available at www.ceiops.org; CEIOPS. 2007a. QIS 3 Technical Specifications, Part I: Instructions, CEIOPS-FS-11/07 & Part II: Background Information, CEIOPS-FS-12/07. April. Available at www.ceiops.org.

O.3.1 Health Expense Risk

The capital charge for health expense risk was determined as follows:

$$CR_{HR,ER} = \lambda_{exp} \times \sigma_{exp} \times GP_{ay}, \qquad (O.1)$$

where λ_{exp}: expense RF that was set to deliver a health expense risk charge consistent with a VaR 99.5% standard; the expense RF was set to 2.58; σ_{exp}: the standard deviation of the expense result over the previous 10-year period; and GP_{ay}: gross premium earned for the accounting year.

O.3.2 Special Treatment for Small and New Health Insurance Undertakings

In some cases, especially for rather young undertakings, expense results are only available for a short time period, that is, the standard deviation of the expense result cannot be determined directly on the basis of the previous 10-year period. Furthermore, expense results relating to the first years after starting-up of an undertaking might not be representative for future expense results. In those cases, the standard deviation for the expense result should be estimated as follows:

$$\sigma_{exp} = \frac{1}{4} \times \left(10 - n'\right) \times f_{exp} + \frac{1}{4} \times \left(n' - 6\right) \times \sigma_{exp}(n),$$

where $n' = \min\{\max[n; 6]; 10\}$; n: number of recent accounting years, where the gross premium earned continuously exceeded 3 Mio Euro (at most 10); $\sigma_{exp}(n)$: the standard deviation of the expense result over the previous n-year period; and f_{exp}: parameter that will be used to estimate σ_{exp} for small companies.

This means that for $n \geq 7$, the company's individual standard deviations $\sigma_{exp}(n)$ are taken into account; if $n < 7$, the estimate will be determined solely by the parameter f_{exp}, which is independent of the undertaking's individual standard deviations. The parameter f_{exp} is set to 2%.

The capital charge for health expense risk should be calculated under the condition that the assumptions on future bonus rates, reflected in the valuation of future discretionary benefits (FDBs) in technical provisions, remained unchanged before and after the assumed increase in expense costs.

O.3.3 Health Claim/Mortality/Cancellation Risk

This "multirisk" covers:

- The claim risk or per capita loss risk arising in cases where the actual per capita loss is greater than the loss assumed in the pricing of the product.
- The mortality risk exists if the actual funds from provisions for increasing age becoming available due to death are lower than those assumed in the pricing of the product.

- The cancellation risk exists if the actual funds from provisions for increasing age becoming available due to cancellations are lower than those assumed in the pricing of the product.

It was calculated as

$$\mathrm{CR}_{\mathrm{HR,MR}} = \lambda_{\mathrm{cl}} \times \sigma_{\mathrm{cl}} \times \mathrm{GP}_{\mathrm{ay}}, \tag{O.2}$$

where λ_{cl}: health RF that was set to deliver a health claim/mortality/cancellation risk charge consistent with a VaR 99.5% standard; the expense RF was set to 2.58; σ_{cl}: the standard deviation of the health risk over the previous 10-year period; and $\mathrm{GP}_{\mathrm{ay}}$: gross premium earned for the accounting year.

O.3.4 Special Treatment for Small and New Health Insurance Undertakings

In some cases, especially for rather young undertakings, claim/mortality/cancellation results are only available for a short time period, that is, the standard deviation of these results cannot be determined directly on the basis of the previous 10-year period. Furthermore, health results relating to the first years after starting-up of an undertaking might not be representative for future health results. In those cases, the standard deviation for the health result should be estimated as follows:

$$\sigma_{\mathrm{cl}} = \frac{1}{4} \times \left(10 - n'\right) \times f_{\mathrm{cl}} + \frac{1}{4} \times \left(n' - 6\right) \times \sigma_{\mathrm{cl}}\left(n\right),$$

where $n' = \min\{\max[n; 6]; 10\}$; n: number of recent accounting years, where the gross premium earned continuously exceeded 3 Mio Euro (at most 10); $\sigma_{\mathrm{cl}}(n)$: the standard deviation of the health result over the previous n-year period; and f_{cl}: parameter that will be used to estimate σ_{cl} for small companies.

This means that for $n \geq 7$, the company's individual standard deviations $\sigma_{\mathrm{exp}}(n)$ are taken into account; if $n < 7$, the estimate will be determined solely by the parameter f_{exp}, which is independent of the undertaking's individual standard deviations. The parameter f_{exp} is set to 2%.

The capital charge for health claim/mortality/cancellation risk should be calculated under the condition that the assumptions on future bonus rates, reflected in the valuation of FDBs in technical provisions, remained unchanged before and after the assumed occurrence of a claim/mortality/cancellation event.

O.3.5 Health Epidemic/Accumulation Risk

Epidemic/accumulation risk concerns the risks arising from the outbreaks of major epidemics, for example, a severe outbreak of influenza. Such events also typically lead to accumulation risks, since the usual assumption of independence among persons would be nullified.

The capital charge was calculated as

$$\mathrm{CR}_{\mathrm{HR,AR}} = \lambda_{\mathrm{ar}} \times \mathrm{CE}_{\mathrm{ay}} \times \frac{\mathrm{GP}_{\mathrm{ay}}}{\mathrm{HGP}_{\mathrm{ay}}}, \tag{O.3}$$

where λ_{ac}: a health RF set as 6.5%; CE_{ay}: the claims' expenditure for the accounting year in the health insurance market; GP_{ay}: gross premium earned for the accounting year; and HGP_{ay}: total gross premium earned for the accounting year in the health insurance market.

The specified factor was higher than the factor used under QIS2 (1%) to adequately reflect health epidemic/accumulation risk, as well as the adjusted dependence assumptions for health epidemic/accumulation risk with respect to the other health risk submodules.

The capital charge for health epidemic risk should be calculated under the condition that the assumptions on future bonus rates, reflected in the valuation of FDBs in technical provisions, remained unchanged before and after the assumed occurrence of an epidemic event.

O.3.6 QIS3 Calibration

The calibration of the health risk modules is discussed in CEIOPS (2007e, 2007f).

O.3.7 Health Expense Risk

The capital charge for health expense risk was determined as in Equation O.4, that is,

$$CR_{HR,ER} = \lambda_{exp} \times \sigma_{exp} \times GP_{ay}, \tag{O.4}$$

where λ_{exp}: expense RF that was set to deliver a health expense risk charge consistent with a VaR 99.5% standard; the expense RF was set to 2.58; σ_{exp}: the standard deviation of the expense result over the previous 10-year period; and GP_{ay}: gross premium earned for the accounting year.

To determine the appropriate value for the factor λ_{exp}, an analysis of the empirical distribution of the health expense results for the German market was carried out. An application of standard statistical testing tools, such as Kolmogorow–Smirnow and Shapiro–Wilk tests, yielded the result that it would be appropriate to assume that the expense result follows a normal distribution.

Since the calibration follows a VaR 99.5% standard, this led to setting the factor λ_{exp}, as 2.58. The standard deviation of the expense risk result is generally calculated on the basis of undertaking's specific data and therefore does not require an explicit calibration. However, in cases where the undertaking's specific data are insufficient to determine σ_{exp}, the standard deviation is estimated as a convex combination of the undertaking's specific estimate and a market-wide parameter f_{exp}. The calibration of the parameter f_{exp} was carried out on the basis of the data of 43 health insurance undertakings in the German market. Ideally, these parameters can be chosen such that the market-wide estimate is near to the undertaking's specific estimate, so that

$f_{exp} \approx \sigma_{exp}$, that is, $P \times f_{exp} \approx P \times \sigma_{exp}$ for each individual insurer and for each relevant volume measure P.

Therefore, the parameter f_{exp} was determined using least-squares optimization, that is, it was chosen such that the sum of the squares of the residuals is minimized:

$$VS = \sum_i \left[P \times f_{exp} - P \times \sigma_{exp}\right]^2.$$

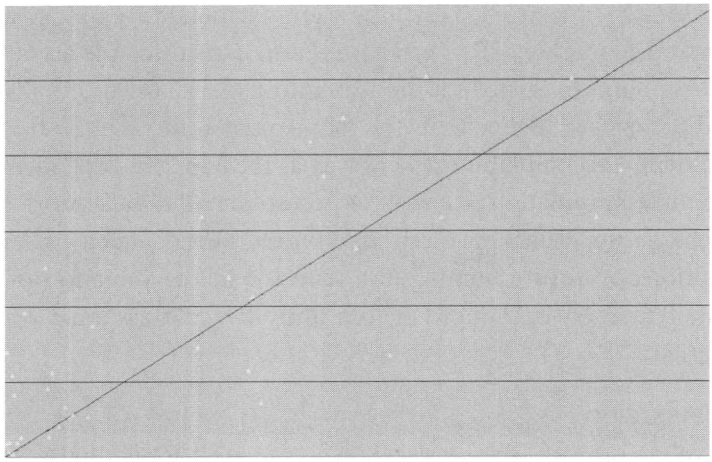

FIGURE O.1 The linear regression analysis for health expense risk module. The white dots are $P \times \sigma_{exp}$ and the regression line is the result for $P \times f_{exp}$. (Adapted from CEIOPS. 2007g. Calibration of the Underwriting Risk, Market Risk and MCR. CEIOPS-FS-14/07. April. Available at www.ceiops.org; copyright Committee of European Insurance and Occupational Pensions Supervisors.)

The result of performing this linear regression analysis is shown in Figure O.1. On the basis of this analysis, the factor f_{exp} was set to 2%.

O.3.8 Health Claim/Mortality/Cancellation Risk

The capital charge for this health subrisk was determined as in Equation O.5, that is,

$$\mathrm{CR}_{\mathrm{HR,MR}} = \lambda_{\mathrm{cl}} \times \sigma_{\mathrm{cl}} \times \mathrm{GP}_{\mathrm{ay}}, \tag{O.5}$$

where λ_{cl}: health RF that was set to deliver a health claim/mortality/cancellation risk charge consistent with a VaR 99.5% standard; the expense RF was set to 2.58; σ_{cl}: the standard deviation of the health risk over the previous 10-year period; and GP_{ay}: gross premium earned for the accounting year.

To determine an appropriate value for the factor λ_{cl}, an analysis of the empirical distribution of the health results for the German market was carried out. As in the case of the expense risk, an application of standard statistical testing tools yielded the result that it would be appropriate to assume that the health result follows a normal distribution. Hence, the factor λ_{cl} was set as 2.58.

As for expense risk, in cases where the undertaking's specific data are insufficient to determine σ_{cl}, the standard deviation was estimated as a convex combination of the undertaking's specific estimate and a market-wide factor f_{cl}.

The calibration of the factor f_{cl} was carried out using the same linear regression approach as for expense risk. The result of performing a linear regression analysis is shown in Figure O.2.

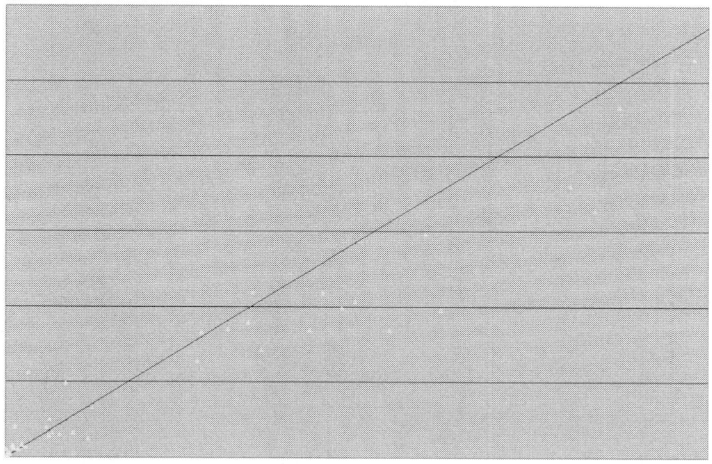

FIGURE O.2 The linear regression analysis for health claim/mortality/cancellation risk module. The white dots are $P \times \sigma_{cl}$ and the regression line is the result for $P \times f_{cl}$. (Adapted from CEIOPS. 2007g. Calibration of the Underwriting Risk, Market Risk and MCR. CEIOPS-FS-14/07. April. Available at www.ceiops.org; copyright Committee of European Insurance and Occupational Pensions Supervisors.)

O.3.9 Epidemic/Accumulation Risk

The capital charge for this health subrisk was determined as in Equation O.3.

In accordance with international guidelines issued by the WHO to respond to threats and occurrences of pandemic influenza, the German state and German Länder had worked out a national influenza pandemic plan under the leadership of the Robert Koch Institute, RKI*. As part of this plan, it was estimated for Germany that, under an assumed influenza infection rate of 50% and within a time period of eight weeks, 25% of the population would seek medical consultation and 0.75% of the population would require clinical treatment, not regarding additional therapeutically and prophylactic measures.

The likelihood for the occurrence of such a scenario was considered to lie within the 99.5% quantile used for the SCR, and therefore this scenario was taken into account to determine λ_{ay}. On the basis of this analysis, the factor λ_{ay} was set to 6.5%.

O.4 QIS4 PROPOSAL

This module is intended to cover UR for all health and workers' compensation guarantees and is split into three submodules:

- $CR_{HR,LT}$: long-term health that is practised on a similar technical basis to that of life assurance, which exists only in Germany and Austria
- $CR_{HR,ST}$: short-term health
- $CR_{HR,WC}$: workers' compensation

*The Robert Koch Institute: www.rki.de

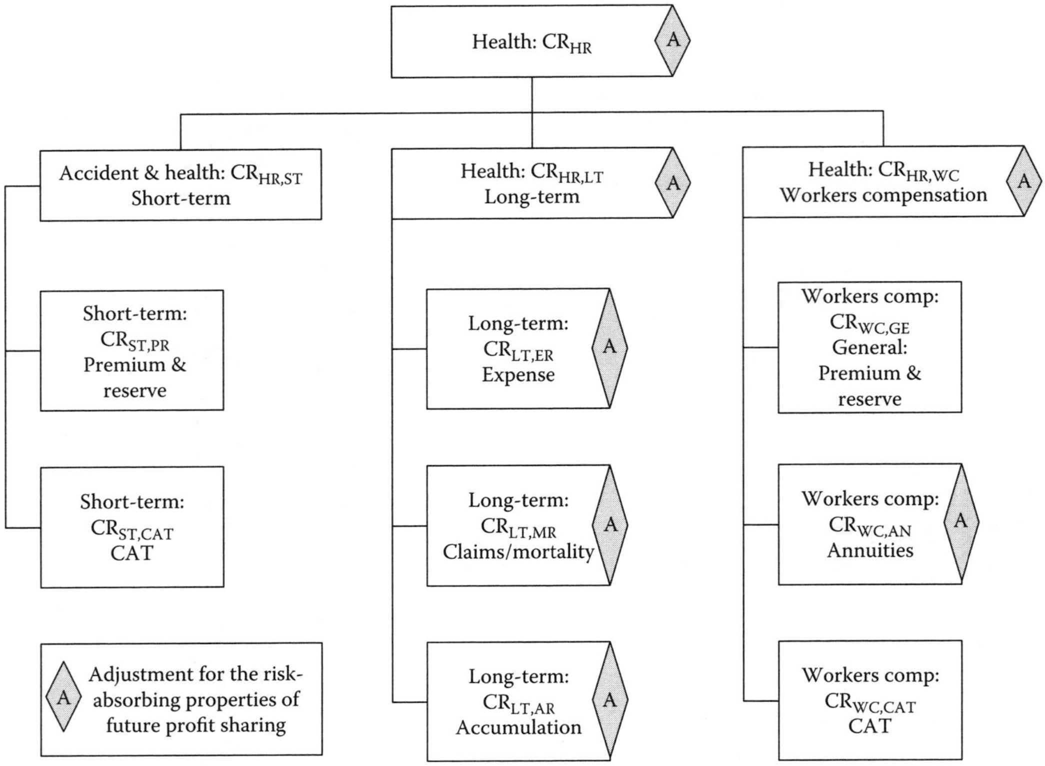

FIGURE O.3 An overall description of the health UR module and its subrisk modules. (Adapted from QIS4. 2008. QIS 4 Technical Specifications. European Commission. MARKT/2505/08. March 31. Available at http://www.ceiops.eu/content/view/118/124/.)

Note that it differs from the health module in QIS3 in that its scope was wider since it includes health lines of business previously included in the nonlife underwriting module. Hence, we will have two levels of subrisks. This is illustrated in Figure O.3.

We have also changed the denotations of the "old health risk" submodules as they are now subrisks to the long-term health risk: The old $\mathrm{CR}_{\mathrm{HR,ER}}$ is now denoted $\mathrm{CR}_{\mathrm{LT,ER}}$, old $\mathrm{CR}_{\mathrm{HR,MR}}$ is now $\mathrm{CR}_{\mathrm{LT,MR}}$, and $\mathrm{CR}_{\mathrm{HR,AR}}$ is now $\mathrm{CR}_{\mathrm{LT,AR}}$.

Combining the capital charges for the health submodules using a dependence matrix, the capital charge for the health UR is derived:

$$\mathrm{CR}_{\mathrm{HR}} = \sqrt{\sum_{r\times c} \rho_{rc} \times \mathrm{CR}_{\mathrm{HR},r} \times \mathrm{CR}_{\mathrm{HR},c}},$$

where the dependencies are given in Table O.2, and r and c are the rows and columns of the matrix, respectively.

The risk-absorbing effect of future profit sharing is taken care of for the long-term health and the workers' compensation risk modules. The capital charge for health UR including

TABLE O.2 Dependence Matrix for the Health UR Used for QIS4

	$\mathrm{CR_{HR,LT}}$	$\mathrm{CR_{HR,ST}}$	$\mathrm{CR_{HR,WC}}$
$\mathrm{CR_{HR,LT}}$	1	0	0
$\mathrm{CR_{HR,ST}}$		1	0.5
$\mathrm{CR_{HR,WC}}$			1

Source: Adapted from QIS4. 2008. QIS 4 Technical Specifications. European Commission. MARKT/2505/08. March 31. Available at http://www.ceiops.eu/content/view/118/124/.

the risk-absorbing effect of future profit sharing is determined by

$$\mathrm{CR_{HR}} = \sqrt{\sum_{r\times c} \rho_{rc} \times n\mathrm{CR}_{\mathrm{HR},r} \times n\mathrm{CR}_{\mathrm{HR},c}},$$

where $n\mathrm{CR_{HR,ST}} = \mathrm{CR_{HR,ST}}$ and $n\mathrm{CR_{HR,WC}} = \mathrm{CR_{HR,WC}}$.

Companies participating in QIS4 mainly welcomed the new structure, but some undertakings were still unsure about the classifying particular insurance contracts according to the new structure (CEIOPS, 2008e).

O.4.1 Health Long-Term UR Module

This module was concerned with UR in health insurance that is practised on a similar technical basis to that of life insurance with additional restrictions according to National Law as sold in Germany and Austria. The URs, including morbidity and disability, within other forms of health insurance practised on a similar technical basis to that of life insurance should be measured using the life underwriting module or the workers' compensation.

Health long-term UR is split into the three components (*cf.* QIS3 above):

- $\mathrm{CR_{LT,ER}}$: the capital charge for the expense risk
- $\mathrm{CR_{LT,MR}}$: the capital charge for the claim/mortality/cancellation risk and
- $\mathrm{CR_{LT,AR}}$: the capital charge for the epidemic/accumulation risk

Combining the capital charges for the long-term health subrisks using the dependence matrix given in Table O.3 gave us the capital charge for the health UR:

$$\mathrm{CR_{HR,LT}} = \sqrt{\rho_{r\times c} \times \mathrm{CR}_{\mathrm{LT},r} \times \mathrm{CR}_{\mathrm{LT},c}}.$$

The capital charges including the risk-absorbing effects of future profit sharing were given by $n\mathrm{CR_{LT,ER}}$, $n\mathrm{CR_{LT,MR}}$, and $n\mathrm{CR_{LT,AR}}$, respectively. The capital charge for the long-term health UR including the risk-absorbing effects of future profit sharing was given by:

$$n\mathrm{CR_{HR,LT}} = \sqrt{\rho_{r\times c} \times n\mathrm{CR}_{\mathrm{LT},r} \times n\mathrm{CR}_{\mathrm{LT},c}},$$

where the dependencies are provided in Table O.3.

TABLE O.3 Dependence Matrix for the Long-Term Health UR Used for QIS4

	$CR_{HR,ER}$	$CR_{HR,MR}$	$CR_{HR,AR}$
$CR_{HR,ER}$	1	0.5	0
$CR_{HR,MR}$		1	0
$CR_{HR,AR}$			1

Source: Adapted from QIS4. 2008. QIS 4 Technical Specifications. European Commission. MARKT/2505/08. March 31. Available at http://www.ceiops.eu/content/view/118/124/.

O.4.1.1 Health Expense Risk

Expense risk arises if the expenses anticipated in the pricing of a product are insufficient to cover the actual costs accruing in the accounting year. There are numerous possible causes of such a shortfall; therefore, all cost items of private health insurers have to be taken into account. To ensure comparability among the financial years, all annual results will be related to the gross premiums earned in the specific financial year.

The capital charge for the long-term health expense risk was determined as follows:

$$CR_{LT,ER} = \lambda_{exp} \times \sigma_{hexp} \times P_{ay},$$

where λ_{exp}: the expense RF that was set to deliver a health expense risk charge consistent with a VaR 99.5% standard; in QIS4, it was set to 2.58; σ_{hexp}: the gross earned premium weighted standard deviation of the expense result in relation to the gross premium over the previous 10-year period; and P_{ay}: the gross premium earned for the accounting year.

The capital charge was also calculated taking the risk-absorbing effect of future profit sharing. It was denoted as $nCR_{LT,ER}$.

O.4.1.2 Special Treatment for Small and Recently Established Health Insurance Companies

In some cases, especially for rather recent established undertakings, expense results are only available with a short history, that is, the standard deviation of the expense result cannot be determined directly on the basis of the previous 10 years. Furthermore, expense results relating to the first years after start-up might not be representative of future expense results. In those cases, the gross earned premium weighted standard deviation for the expense result would be estimated as follows:

$$\sigma_{hexp} = \frac{1}{4} \times \left(10 - n'\right) \times f_{exp} + \frac{1}{4} \times \left(n' - 6\right) \times \sigma_{hexp}\left(n\right),$$

where $n' = \min\{\max\{n; 6\}; 10\}$; n: the number of recent accounting years, where the gross premium earned continuously exceeded 3 Mio Euro (at most 10). The number would not allow for the first three years after start up of business; $\sigma_{hexp}(n)$: the gross earned premium weighted standard deviation of the expense result over the previous n-year period; and f_{exp}: the parameter that will be used to estimate σ_{hexp} for small companies; in QIS4, it was set to 0.02.

This means that for $n \geq 7$, the company's individual standard deviations $\sigma_{\text{hexp}}(n)$ are taken into account; if $n < 7$, the estimate will be determined solely by the parameter f_{exp} which is independent of the undertaking's individual standard deviations.

The capital charge for health expense risk should be calculated under the condition that the assumptions on future bonus rates (reflected in the valuation of FDBs in technical provisions) remain unchanged before and after the assumed increase in expense costs.

Additionally, the result of the scenario was determined under the condition that the participant was able to vary its assumptions in future bonus rates in response to the shock being tested. The resulting capital charge is $n\text{CR}_{\text{LT,ER}}$.

O.4.1.3 Health Claim/Mortality/Cancellation Risk

This risk covers:

- Claim risk or per capita loss risk arising in cases where the actual per capita loss is greater than the loss assumed in the pricing of the product.
- Mortality risk arising when the actual funds from TPs becoming available due to death are lower than those assumed in the pricing of the product.
- Cancellation risk arising when the actual funds from TPs becoming available due to cancellations are lower than those assumed in the pricing of the product.

The capital charge for long-term health claim/mortality/cancellation risk was determined as follows:

$$\text{CR}_{\text{LT,MR}} = \lambda_{\text{mr}} \times \sigma_{\text{hmr}} \times P_{\text{ay}},$$

where λ_{mr}: the health RF that was set to deliver a health claim/mortality/cancellation risk charge consistent with a VaR 99.5% standard; in QIS4, it was set to 2.58; σ_{hmr}: the gross earned premium weighted standard deviation of the expense result in relation to the gross premium over the previous 10-year period; and P_{ay}: the gross premium earned for the accounting year.

The capital charge was also calculated taking the risk-absorbing effect of future profit sharing. It was denoted as $n\text{CR}_{\text{LT,MR}}$.

O.4.1.4 Special Treatment for Small and Recently Established Health Insurance Companies

In some cases, especially for rather recent established undertakings, expense results are only available for a short time period, that is, the standard deviation of the expense result cannot be determined directly on the basis of the previous 10 years. Furthermore, health results relating to the first years after start-up might not be representative for future health results. In those cases, the gross earned premium weighted standard deviation for the health result should be estimated as follows:

$$\sigma_{\text{hmr}} = \frac{1}{4} \times \left(10 - n'\right) \times f_{\text{mr}} + \frac{1}{4} \times \left(n' - 6\right) \times \sigma_{\text{hmr}}\,(n)\,,$$

where $n' = \min\{\max\{n; 6\}; 10\}$; n: the number of recent accounting years, where the gross premium earned continuously exceeded 3 Mio Euro (at most 10). The number would not allow for the first three years after start up of business; $\sigma_{\text{hmr}}(n)$: the gross earned premium weighted standard deviation of the expense result over the previous n-year period; and f_{mr}: the parameter that will be used to estimate σ_{hmrp} for small companies; in QIS4, it was set to 0.03.

This means that for $n \geq 7$, the company's individual standard deviations $\sigma_{\text{hmrp}}(n)$ are taken into account; if $n < 7$, the estimate will be determined solely by the parameter f_{mr}, which is independent of the undertaking's individual standard deviations.

The capital charge for health claim/mortality/cancellation risk was calculated under the condition that the assumptions on future bonus rates, reflected in the valuation of FDBs in technical provisions, remain unchanged before and after the assumed occurrence of a claim/mortality/cancellation event.

Additionally, the result of the scenario was determined under the condition that the participant was able to vary its assumptions in future bonus rates in response to the shock being tested. The resulting capital charge is $n\text{CR}_{\text{LT,MR}}$.

O.4.1.5 Health Epidemic/Accumulation Risk

Epidemic/accumulation risk concerns the risks arising from the outbreaks of major epidemics, for example, a severe outbreak of influenza. Such events typically also lead to accumulation risks, since the usual assumption of independence among persons would be nullified.

The capital charge was calculated as

$$\text{CR}_{\text{LT,AR}} = \lambda_{\text{ar}} \times \text{CE}_{\text{ay}} \times \frac{\text{GP}_{\text{ay}}}{\text{HGP}_{\text{ay}}},$$

where λ_{ac}: a health RF set in QIS4 as 6.5%; CE_{ay}: the claims' expenditure for the accounting year in the health insurance market; GP_{ay}: gross premium earned for the accounting year; and HGP_{ay}: total gross premium earned for the accounting year in the health insurance market.

The capital charge for health epidemic risk was calculated under the condition that the assumptions on future bonus rates, reflected in the valuation of FDBs in technical provisions, remain unchanged before and after the assumed occurrence of an epidemic event.

Additionally, the result of the scenario should be determined under the condition that the participant is able to vary its assumptions in future bonus rates in response to the shock being tested. The resulting capital charge is $n\text{CR}_{\text{LT,AR}}$.

O.4.2 Accident and Health Short-Term UR Module

This module covers the premium and reserve risks and catastrophe risk of short-term health and accident lines of business:

$\text{CR}_{\text{ST,RP}}$: the capital charge for reserve and premium risk

$\text{CR}_{\text{ST,CAT}}$: the capital charge for catastrophe risk

TABLE O.4 Dependence Matrix for the Short-Term Health UR Used for QIS4

	$CR_{ST,RP}$	$CR_{ST,CAT}$
$CR_{ST,RP}$	1	0
$CR_{ST,CAT}$		1

Source: Adapted from QIS4. 2008. QIS 4 Technical Specifications. European Commission. MARKT/2505/08. March 31. Available at http://www.ceiops.eu/content/view/118/124/.

The capital charge for short-term health and accident UR was derived by combining the capital charges for the short-term health and accident subrisks using the dependence matrix given by Table O.4 as follows:

$$CR_{HR,ST} = \sqrt{\rho_{r\times c} \times CR_{ST,r} \times CR_{ST,c}},$$

O.4.2.1 Accident and Health Short-Term Premium and Reserve Risks

This module covers the premium and reserve risks of short-term health and accident lines of business. Premium and reserve risks are defined as set out in Appendix M for the nonlife UR.

$CR_{ST,RP,SH}$: the reserve and premium risks for short-term health insurance

$CR_{ST,RP,AO}$: the reserve and premium risks for short-term accident and others insurance

The calculation follows that set out in Appendix M for the premium and reserve risk calculation of $CR_{NL,PR}$ using the following parameter: Maximum $n_k = 5$.

The standard deviation for reserve risk and the market-wide estimate of the standard deviation for premium risk in the individual lines of business were determined as in Table O.5.

The dependence matrix was specified as in Table O.6.

O.4.2.2 Accident and Health Short-Term CAT Risk

The accident and health short-term CAT risk is defined as set out in Appendix M for the nonlife UR, using the following methods.

Method 1

$$CR_{ST,CAT} = \sqrt{(C_1 \times P_1)^2 + (C_2 \times P_2)^2},$$

TABLE O.5 Standard Deviation for Reserve and Premium Risks in the Individual Lines of Business

LOB, *k*	$CR_{ST,RP,SH}$	$CR_{ST,RP,AO}$
$\sigma_{R,k}$	7.5%	15%
$\sigma_{M,P,k}$	3%	5%

Source: Adapted from QIS4. 2008. QIS 4 Technical Specifications. European Commission. MARKT/2505/08. March 31. Available at http://www.ceiops.eu/content/view/118/124/.

TABLE O.6 Dependence Matrix between the Short-Term Reserve and Premium Risks for Health and Accident, and Others Insurance

	$CR_{ST,RP,SH}$	$CR_{ST,RP,AO}$
$CR_{ST,RP,SH}$	1	0.5
$CR_{ST,RP,AO}$		1

Source: Adapted from QIS4. 2008. QIS 4 Technical Specifications. European Commission. MARKT/2505/08. March 31. Available at http://www.ceiops.eu/content/view/118/124/.

where P_1 and P_2 are the estimates of the net written premium in the individual line of business (LoB) *short-term health and accident* and *others*, respectively, during the forthcoming year, and $C_1 = C_2 = 0.1$.

Method 2: Scenarios

Some examples of scenarios are outlined below for health:

- Pandemic, for example, the bird flu
- Polio-type debilitating disease effects
- Biohazard from an insecure laboratory
- Terrorist pathogen released
- Terrorist action with delayed effects (e.g., poisoning a water supply with a difficult to detect and slow working poison)
- Concentrated office block accident (similar to the workers' compensation scenario)

O.4.3 Workers' Compensation UR Module

This module was concerned with UR in workers' compensation LoB.

In general, workers' compensation insurance covers a diversity of liability profiles, related to short-term or long-term sick leave whatever the cause of the sickness should be for instance:

a. Standard nonlife type of liabilities, including medical treatments and lump-sum indemnity payments. Due to their characteristics, these claims have commonly a short- to medium-term perspective; in several markets, the scope of workers' compensation insurance is limited to this first type of liabilities.

b. Annuities payable to injured workers and beneficiaries.

c. Regular and recurrent benefits on a (generally) long-term basis, specifically aimed to provide life assistance to an injured worker with a significant level of incapacity, for example, medical treatments on a continuous basis, replacement of artificial limbs, salary to a person providing assistance, and so on. In the subsequent paragraphs, these liabilities will be referred to as "*life assistance liabilities.*" The main difference with

TABLE O.7 Description of the Scope of the Charges in Terms of Capture of the Risks

	Past Events, Reported or Not Reported	Future Events
Standard nonlife type of liabilities	$CR_{WC,GE}$	$CR_{WC,GE}$
Annuities and life assistance	$CR_{WC,AN}$	$CR_{WC,GE}$

Source: Adapted from QIS4. 2008. QIS 4 Technical Specifications. European Commission. MARKT/2505/08. March 31. Available at http://www.ceiops.eu/content/view/118/124/.

benefits in (a) is the expectation that the flow of benefits will continue on a regular basis until the death of the beneficiary.

In Table O.7 the scope of each of the charges in terms of capture of the risks per type of liability and per timing of the event that triggers the benefits are identified.

The following subrisks are included:

$CR_{WC,GE}$: the capital charge capturing the premium risk and the reserve risk (the latter relating only to the "standard nonlife type of liabilities")

$CR_{WC,AN}$: the capital charge capturing the risks stemming from liabilities paid in the form of annuities and "life assistance" liabilities

$CR_{WC,CAT}$: the capital charge for catastrophe risk

The capital charges for the workers' compensation UR were derived by combining the capital charges for the workers' compensation subrisks using a dependence matrix as follows:

$$CR_{HR,WC} = \sqrt{\sum_{r\times c} \rho_{rc} \times CR_{WC,r} \times CR_{WC,c}},$$

where the dependence structure is given in Table O.8.

O.4.3.1 Workers' Compensation "General": Premium and Reserve Risks

This submodule covered premium and reserve risks (the latter related only to the "standard nonlife type of liabilities") resulting from the underwriting of workers' compensation insurance contracts. Some forms of workers' compensation insurance covers expose the undertaking to life (mortality/longevity) type catastrophes and the capital associated with

TABLE O.8 Dependencies between the Workers' Compensation Subrisks

	$CR_{WC,GE}$	$CR_{WC,AN}$	$CR_{WC,CAT}$
$CR_{WC,GE}$	1	0.5	0
$CR_{WC,AN}$		1	0
$CR_{WC,CAT}$			1

Source: Adapted from QIS4. 2008. QIS 4 Technical Specifications. European Commission. MARKT/2505/08. March 31. Available at http://www.ceiops.eu/content/view/118/124/.

this risk, for these covers, needs to be assessed (by inclusion in life-CAT) as part of the life UR charge.

Premium and reserve risks were to be understood as set in the premium and reserve risks in Appendix M. The assessment of the premium risk for workers' compensation LoB will not differentiate between the type of liabilities—annuities, life assistance, and standard nonlife type of liabilities—that may stem from future claims. In the context of the workers' compensation LOB, reserve risk was intended to only cover the incurred claims that correspond to the "standard nonlife type of liabilities" classification.

The module delivered the following information:

$\mathrm{CR}_{\mathrm{WC,GE}}$: the capital charge capturing the premium and the reserve risks (the latter relating only to the "standard nonlife type of liabilities") as defined in Appendix M with the use of the parameters below.

R^0_{WC}: the net provision for claims outstanding in the workers' compensation LOB ($k = 1$, i.e., "WC"), relating to the "standard nonlife type of liabilities": $V_R = R^0_{WC}$.

Premium input included the total amount of workers' compensation premiums. We set the risk-absorbing effect as $nCR_{\mathrm{WC,GE}} = CR_{\mathrm{WC,GE}}$.

The calculation was computed as in the nonlife premium and reserve risk calculation, see Appendix M, using the following parameter:

$$\text{Maximum } n_k = 5,$$

where the relative weight of longer-term annuity annuities and life assistance was significant, for example, the corresponding best estimate is higher than 50% of the total best estimate of Workers' compensation LOB k, participants should use the maximum parameter $n_k = 15$.

The standard deviation for reserve risk was $\sigma_{\mathrm{R}} = 10\%$ and the market-wide estimate of the standard deviation for premium risk was $\sigma_{\mathrm{M,P}} = 7\%$.

The overall volume measure V is determined as follows:

$$V = V_{\mathrm{P}} + V_{\mathrm{R}},$$

where V_{P} and V_{R} are the volume measures for premium and reserve risks.

The overall standard deviation σ is determined as follows:

$$\sigma = \sqrt{\frac{1}{V^2}\left(\sigma_{\mathrm{P}}^2 \times V_{\mathrm{P}}^2 + \sigma_{\mathrm{R}}^2 \times V_{\mathrm{R}}^2 + \sigma_{\mathrm{P}} \times \sigma_{\mathrm{R}} \times V_{\mathrm{P}} \times V_{\mathrm{R}}\right)}.$$

O.4.3.2 Workers' Compensation: Annuities

This submodule covered the risks underlying Workers' compensation benefits paid in the form of annuities and life assistance liabilities. It intended to cover liabilities originated from events already incurred at the valuation date. Regarding life assistance, it was assumed that the best estimate of these liabilities could be approximated using an annuity factor applied to an "average" expected annual amount of benefits (note that this annual amount is subject to a certain degree of uncertainty).

This submodule was split into

- $\text{CR}_{\text{AN,LO}}$: the capital charge for longevity risk
- $\text{CR}_{\text{AN,RE}}$: the capital charge for revision risk
- $\text{CR}_{\text{AN,DR}}$: the capital charge for disability risk
- $\text{CR}_{\text{AN,ER}}$: the capital charge for expenses risk

Combining the capital charges for the relevant subrisks using the dependence matrix given in Table O.9 derived the capital charge for the UR underlying annuities and life assistance:

$$\text{CR}_{\text{WC,AN}} = \sqrt{\sum_{r \times c} \rho_{rc} \times \text{CR}_{\text{AN},r} \times \text{CR}_{\text{AN},c}}.$$

The four submodules were also calculated including the risk mitigating effects of future profit sharing: $n\text{CR}_{\text{AN,LO}}$, $n\text{CR}_{\text{AN,RE}}$, $n\text{CR}_{\text{AN,DR}}$, and $n\text{CR}_{\text{AN,ER}}$.

O.4.3.2.1 Annuities Longevity Risk The capital charge for longevity risk should be calculated along the methodology set in life- longevity risk. The longevity shock to be applied was a permanent 25% decrease in mortality rates for each age.

The capital charge $\text{CR}_{\text{AN,LO}}$ should be calculated under the condition that the assumptions on future bonus rates, reflected in the valuation of FDBs in technical provisions, remain unchanged before and after the shocks being tested.

$n\text{CR}_{\text{AN,LO}}$ was calculated under the condition that the assumption that the participant was able to vary its assumptions on future bonus rates in response to the shock being tested. The capital charge for longevity risk was calculated along the methodology set in life-longevity risk, excluding the risk-absorbing effect of future profit sharing.

TABLE O.9 Dependence Structure between the Subrisks in the Workers' Compensation Annuity Risk Module

	$\text{CR}_{\text{AN,LO}}$	$\text{CR}_{\text{AN,RE}}$	$\text{CR}_{\text{AN,DR}}$	$\text{CR}_{\text{AN,ER}}$
$\text{CR}_{\text{AN,LO}}$	1	0	0	0.25
$\text{CR}_{\text{AN,RE}}$		1	0	0.25
$\text{CR}_{\text{AN,DR}}$			1	0.5
$\text{CR}_{\text{AN,ER}}$				1

Source: Adapted from QIS4. 2008. QIS 4 Technical Specifications. European Commission. MARKT/2505/08. March 31. Available at http://www.ceiops.eu/content/view/118/124/.

O.4.3.2.2 Annuities Disability Risk The capital charge $CR_{AN,DR}$ should be calculated along the methodology set in life-disability risk, gross of the risk mitigating effect of future profit sharing (no mitigating effect).

The disability shock to be applied was an increase of 35% in disability rates for the next year, together with a permanent 25% increase, over the best estimate, in disability rates at each age in following years.

$nCR_{AN,DR}$ was calculated under the condition that the participant was able to vary its assumptions on future bonus rates in response to the shock being tested. The capital charge for disability risk should be calculated along the methodology set in life-disability risk, excluding the risk-absorbing effect of future profit sharing.

O.4.3.2.3 Annuities Revision Risk In the context of workers' compensation LOB, revision risk captures the risk of adverse variation of an annuity's amount, as a result of an unanticipated revision of the claims' process, and for those benefits that can be approximated by a life annuity (life assistance), the uncertainty underlying the "average" annual amount assumed in the computation of the best estimate.

On the computation of this risk charge, $CR_{AN,RE}$, participants should consider the impact on those annuities for which a revision process was possible to occur during the next year, for example, annuities where there are legal or other eligibility restrictions should not be included. Unless the "average" annual amount was fixed and known with certainty, all those benefits that can be approximated by a life annuity (life assistance) are also subject to revision risk.

The revision shock to be applied was:

- *Revisionshock* for annuities: a 2% increase in the annual amount payable
- *Revisionshock* for life assistance benefits: a 5% increase in the annual amount payable

The impact should be assessed considering the remaining runoff period.

$nCR_{AN,RE}$ should be calculated under the condition that the assumptions on future bonus rates, reflected in the valuation of FDBs in technical provisions, remain unchanged before and after the shocks being tested.

In addition, $nCR_{AN,RE}$ should be calculated under the condition that the assumption that the participant was able to vary its assumptions on future bonus rates in response to the shock being tested.

O.4.3.2.4 Annuities Expense Risk The capital charge for expense risk should be calculated along the methodology set in life-expense risk, excluding the risk-absorbing effect of future profit sharing

The expense shock to be applied was:

- *Expenseshock*: all future expenses are higher than best estimate anticipations by 10%, and the rate of expense inflation is 1% per annum higher than anticipated.

In addition, $n\mathrm{CR}_{\mathrm{AN,ER}}$ should be calculated under the condition that the assumption that the participant was able to vary its assumptions on future bonus rates in response to the shock being tested.

O.4.3.3 Workers' Compensation: CAT Risk

Workers' compensation catastrophe risk was defined as set in the nonlife catastrophe risk description in Appendix M.

The capital charge was denoted by $\mathrm{CR}_{\mathrm{WC,CAT}}$. The calculation is computed as set in nonlife catastrophe risk calculation, using the following methods.

Method 1

$$\mathrm{CR}_{\mathrm{WC,CAT}} = C \times P,$$

where P was the estimate of the net written premium in the individual LCB (1: WC) during the forthcoming year, and $C = 0.07$.

Method 2: Scenarios

Some examples of man made scenarios are outlined below:

- An industrial disease could be very costly and affect a large number of people
- A large concentrated accident or terrorist incident involving a large workforce for one firm or in one area

O.4.4 QIS4 Calibration

Insurers in various countries writing *accident and health* and *workers' compensation* types of business found their activities difficult to fit with the QIS3 split of the nonlife UR module. For this reason, the health module was restructured for QIS4.

Owing to the restructuring of the health module, a 0.25 dependence factor between the capital charge for nonlife UR, $\mathrm{CR}_{\mathrm{NL}}$, and the capital charge for health UR, $\mathrm{CR}_{\mathrm{HR}}$, has been introduced.

O.4.4.1 Accident and Health Short Term

Calibration for the new accident and health short-term submodule was identical to the calibration used for LoB 2 and LoB 3 of the QIS3 Nonlife UR specification, see Appendix M.

O.4.4.2 Health Workers' Compensation

Market-wide factor for premium risk: The calibration was based on the analysis of historical data from Portugal. It is based on a similar approach to that used for the premium risk in other nonlife LOBs, that is, it was based on the observation of the volatility of historic loss ratios. These historic loss ratios reflect the volatility of claims falling into the standard nonlife type of liabilities category, as well as that of annuities and life assistance liabilities.

Market-wide factor for reserve risk: The calibration was based on the analysis of historical data from Portugal. The factor was derived from the observation of the impact of applying a stress test to the development pattern of the runoff triangle (claims paid), corresponding to a VaR 99.5% one-year scenario. The data used comprises only the standard nonlife type of liabilities.

Longevity risk: No specific analysis was made. The same shock was assumed as for the life underwriting risk (LUR) module.

Revision risk: The calibration procedure is detailed in Appendix N. The only addition was the use of a more granular approach, to derive separate factors for annuities and life assistance liabilities.

Expense risk: No specific analysis was made. The same shock was assumed as for the LUR module.

European Solvency II Standard Formula

Minimum Capital Requirement

HERE, WE WILL DISCUSS the development and calibration of the MCR and its capital charge that has been one of the building blocks in Solvency II. The final advice from CEIOPS is discussed in Chapter 34.

The European Solvency II capital requirement (CR) is based on a two-level system, that is, we have a targetSCR and an ultimateMCR. The latter level is called the PCR by IAIS (see IAIS, 2008c, 2008d and Section 6.1). A description of the two levels is given in Figure 6.2. The two-level approaches are in line with the proposal by IAIS.

P.1 BACKGROUND

The Commission's Framework for Consultation (FfC) (COM, 2006) envisaged two main CRs under Pillar I:

- The SCR
- The MCR

The MCR was intended to provide a "safety net." This means that, on an ongoing basis, the MCR does not necessarily represent an adequate level of capital. But a level of capital below the MCR is clearly unacceptable, even over the short term. While temporary breaches of the SCR might be tolerated—subject to the prospect of restorative action—a breach of the MCR could not be tolerated. A discussion on different MCR issues is given in CEIOPS (2005).

The MCR could be articulated and calibrated as

- A measure based on a time horizon and a probability of ruin at which the risks to new policyholders would be unacceptable, even on the short term

- The point at which it ceases to be economically rational for the undertaking to be recapitalized to the level of the SCR, either by a parent or by a third party

While the first interpretation could be relatively straightforward, it places less emphasis on the ability of undertakings to be recapitalized. Under the alternative interpretation, the MCR would seek to ensure a solvent runoff in a situation when there is no incentive to recapitalize an undertaking. This might prove more realistic, but would be more difficult to implement.

Alternatively, the MCR could be regarded as a rule-of-thumb measure designed primarily to ensure continuity with the existing Solvency I CR, without the need for reference to any specific risk measure, assuming that the existing requirements provide a reasonable reflection of the point below which the risk for policyholders becomes unacceptable.

In the Commission's FfC and its Calls for Advice 9, the MCR gets the following design priorities:

- Simple and straightforward calculation
- Robustness
- Objectivity
- Smooth transition

As the MCR will trigger the most serious supervisory intervention, its calculation needs to be simple, robust, and objective.

According to CEIOPS, while a combination of factor- and scenario-based approaches could be feasible for the calculation of the SCR, a factor-based approach would be more suitable for the purposes of the MCR, since the advantages of factor-based approaches are consistent with the above priorities.

The MCR should not be a volatile measure. Some undertakings may seek to maintain large capital buffers over and above the requirement to mitigate the risk of serious supervisory intervention arising from the breach.

To provide an effective safety net, internal models would not be allowed to replace, or affect, the calculation of the MCR.

For smooth transition, the presumption might be toward the retention of existing requirements, unless an alternative could be shown to be demonstrably preferable. In the case of the adoption of an alternative approach, the existing requirements may be retained for a transitional period.

In addition to the priorities set by the Commission, CEIOPS suggested, when judging the merits of an MCR approach, to consider also the following preferences:

- Risk sensitivity
- Suitability for interim calculations

- Reference to audited/auditable data only
- Consistency with the valuation standards for assets and liabilities and the calculation of the SCR

There is a familiar trade-off between risk sensitivity and the need for simplicity. It should be recognized that any genuinely simple formula would have limitations. Therefore, pressure would always develop to refine the formula to address those limitations. Adding new items in an iterative process may eventually result in the loss of simplicity, transparency, and objectivity. The MCR could be optimized for simplicity while the SCR could be optimized for risk sensitivity.

The insurer must meet its CRs not only at the balance sheet date, but also at all times. When the SCR is breached, both the insurer and the supervisor must be able to constantly monitor the undertaking's solvency situation, to enable timely reaction. Therefore, the structure of the MCR formula should enable interim calculations at any point of time in the year.

The need in certain Member States to support MCR-level intervention with court decisions raises the expectation that only audited data are used. There is a tension between this, the need for timely intervention, and the possible need for interim calculations. The scope of data that are audited, auditable, or are in the annual accounts or supervisory reporting may vary according to the Member State. However, the data requirements for the MCR would need to be reasonably simple, both to reduce the administrative burden on insurers and to allow a straightforward verification.

Regarding the calculation of the MCR, CEIOPS considered the following alternatives:

- Adopting a calculation based on the existing Solvency I requirements
- Using the SCR standard formula as a reference
- Establishing a simple RM over and above liabilities

P.1.1 A Formula Based on the Existing Solvency I

A formula based on the existing Solvency I requirement has, according to CEIOPS, the advantage of continuity with the existing regime, which also reduces the risk of modeling error associated with the innovative and yet unproven SCR. Transitional costs for undertakings, as well as the expenses of developing and testing a formula would also be minimized.

Against this, an MCR based on Solvency I would import the disadvantages of the existing requirements into Solvency II. It is difficult to identify an underlying theoretical basis for the present requirements, so it would also be difficult to demonstrate how the MCR achieved the purpose set out in our working definition. Even if additional risk categories were added, there would be a lack of consistency with the treatment of underwriting risk under the SCR. However, it may not be essential for the MCR to measure risks precisely in order to provide an effective safety net. Although the theoretical basis of the current formula may have been forgotten, several working groups of supervisors have recognized that the solvency margin worked well, at least as a safety net.

Since the definition of the TPs is going to change and, for some Member States, changes in asset valuation will impact the level of available capital, even when the existing Solvency I calculations are retained, the combined requirement of TPs plus the MCR is going to change. This also means that, even if the existing calculations are adopted, a review of the factors via Quantitative Impact Studies (QIS) had to be performed.

Assuming that the existing formula was retained without major changes, the following adjustments need to be/might be considered:

- Adjustments to ensure IASB compatibility
- Adjustments aimed to correct minor anomalies without added complexity
- Shortcuts aiming for more simplicity, where this can be performed without significant loss of risk sensitivity; considering also that the SCR will be able to capture risks more elaborately

The use of some approximations could be allowed to facilitate interim calculations. However, if the existing formulae are retained for a transitional period only, it was suggested to keep adjustments to the absolute minimum necessary, since marginal improvements would not justify the additional costs of a double reform. In its present form, the nonlife formula captures mainly underwriting risk. While the core formula uses BS and profit-and-loss items, the differentiation of the factors according to business lines leads to data requirements that are normally not part of the BS or profit-and-loss. The differentiation does not always follow the present EU classification.

There was a concern whether the nonlife formula was conducive for interim calculations. While some members argued that the formula is easy to calculate throughout the year and does not represent an undue burden, other members hold the view that the data requirement of the nonlife formula was not well aligned to interim calculations. In particular, because it takes into account the history of the last three years, it cannot reflect recent changes in the nature of the portfolio. A related complication was the treatment of portfolio transfer situations, where, to reflect the change of the portfolio, reference had to be made to revenue data of another insurer (or several other insurers).

A number of changes might be considered to improve its suitability for interim calculations, and to better deal with portfolio transfer situations. Illustrative examples include:

- Changing the reflection of reinsurance
- Replacing the 3-year claims index by a provision index (this would be a move toward the margin over liabilities approach): this would also make unnecessary the additional requirement that the MCR must not fall at a faster rate than claims' provisions
- Using just one definition of premiums as a basis for the premium index, in lieu of the maximum of gross premiums earned and gross premiums written, and using 12-month rolling accounting figures

The existing life formula attempts to capture underwriting risk and investment risk. Note that the margin only depends on whether the contracts involve an investment risk to the undertaking, not on the level of investment risk. The result of the calculation depends on the definition of life TPs. If the formula is retained, recalibration of the factors may be suggested by the results of field testing. The potential volatility of results also needs to be field-tested, for example, the sensitivity of the MCR to interest rate changes. The differentiation of the factors according to contract types and the reliance on capital at risk data leads to data requirements that are normally not part of the BS or profit-and-loss.

P.1.2 A Simple Calculation Based on the SCR Standard Formula

The advantage of a simple calculation based on the standard formula of the SCR is that it would be fully integrated into the new risk-based framework and would be consistent with the overall prudential objectives of the new regime. By making sure that the insurance undertakings' CR is more closely aligned with the risks they face, the new framework would reinforce financial stability and promote the competitiveness of European industry. However, the feasibility of this approach is fully dependent on the progress, testing, and eventual success of the SCR standard formula. Therefore, if this approach is adopted, a transitional period between the introduction of the SCR and the adaptation of the MCR is suggested. For the duration of this period, the MCR defined in Solvency I could apply.

The simplest approach would be to fix the MCR as a fraction of the SCR, for example, $MCR = 0.5 \times SCR$. Note that SCR here means the standard SCR, as internal models are not to be allowed for the calculation of the MCR. This shortcut, however, would not be aligned with the theoretical basis of the SCR (VaR or TVaR based on a probability of ruin), and would not deliver a common level of prudence. Moreover, in this case, the calculation of the MCR would not be more simple and robust than the calculation of the SCR.

Another option would be a simplified version of the standard SCR formula that would concentrate on the most important risk categories, possibly using a more straightforward technique for aggregation, and calibrated to a lower level of prudence than the SCR. Scenario-based elements of the SCR formula might be replaced with factor-based items for the purposes of the MCR.

P.1.3 MCR as an RM Over and Above Liabilities

Making the MCR a simple proportion of liabilities was another relevant approach. The formula could be recalculated with ease, at least approximately, throughout the year. All liabilities should be covered in principle, including liabilities under contracts that under international accounting standards are required to be treated as financial contracts. Any liability to which minimal uncertainty attaches, for example, current liabilities for agreed amounts, might be excluded, provided that this does not increase the complexity of the calculation disproportionately to the increase in accuracy.

In life insurance, an MCR expressed as a margin over liabilities would be similar, but not identical to the Solvency I CR. A precondition for an MCR calculated as a margin over liabilities is the harmonization of TPs. In this approach, the main concern was the ability of the insurer to meet its liabilities in a runoff situation. Risks associated with new business are not reflected, assuming that an insurer who fails to meet the MCR will not be permitted to write new business.

However, the MCR is relevant not only for those undertakings that already breached it, but also for the undertakings that meet the MCR. There was a concern that the liabilities basis alone will not adequately reflect the true risk, for example, in the case of short-term business with a low amount of provisions, or in the case of new/rapidly growing undertakings. If field-testing shows that this is a serious shortcoming, a combination of the liabilities result with another volume measure, for example, like the premiums, or with two volume measures (premiums and claims, as in the current solvency margin) might address this problem, although at the cost of some added complexity. Consideration needs to be given whether a maximum approach (as in the current solvency margin) would appear preferable to any additive approach.

For short-term claims business, the level of the overall liabilities would be driven mainly by the provision for unearned premiums. This may lead to undesirable effects in cases where the level of unearned premiums fluctuates significantly throughout the year, for example, where a large proportion of the insurer's business consists of contracts with similar periods of cover.

Operational risk is hard to quantify and for the purpose of the MCR only a rough and ready allowance can be made. The amount of the liabilities might be a suitable proxy for the potential exposure. This allows operational risk to be covered by a loading to the factors that are applied to the liabilities for the purpose of calculating the liability risk. Alternatively, a measure based on premiums could be developed.

P.1.4 Investment Risk in the MCR

Under all approaches, there was an issue whether the MCR should include an allowance for investment risk. To provide advice on this point, the following should be considered:

- Whether the investment risk was sufficiently material for those undertakings that have breached the SCR, taking account of the potentially shorter time horizon of supervisory intervention, to justify the increased complexity
- Whether it was possible to properly reflect investment risk by way of a simple factor-based calculation (e.g., indirectly by applying a ratio on liabilities)
- The extent to which other safety nets (e.g., investment rules) reduce the need to reflect investment risk in the MCR

It may be argued that an insurer who fails to meet its MCR cannot afford to take any risks with its investments. Any insurer who fails to meet its SCR will be expected to take action to restore its position and one of the actions it might decide to take is to rearrange its investments to reduce risks and so the SCR. If this "de-risking" response can be generally anticipated, and the extent of "de-risking" increases as available capital nears the MCR, the risks associated with investments should reduce and there is no need for the MCR itself to reflect investment risks.

Conversely, such a "de-risking" response might not be generally anticipated for a life insurer carrying on with-profits insurance because of commitments to policyholders. Such an insurer may choose other remedial actions to restore its financial resources that do not reduce the risks associated with its investments. Thus, an MCR for life insurance may need to reflect some investment risks.

P.1.5 Interplay with the SCR

The MCR will be a floor for the SCR. However, in some cases, the result suggested by the SCR calculation will be lower than the MCR, in other cases the SCR might be above, but very close to, the MCR.

Only field testing will reveal how frequently this happens. While adjusting the factors in the formulae may reduce the number of such situations, it would be unlikely that they can be avoided entirely. Even if the MCR formula was based on the standard SCR, the SCR estimate suggested by an internal model may fall below the MCR. The future regime, however, would need to avoid abrupt shifts from the "no intervention" control level to the level of "ultimate intervention": the relative levels of the SCR and the MCR should be calibrated so that the SCR represents a meaningful margin over the MCR for most insurers.

Possible approaches to deal with this problem are:

- Setting a floor to the SCR equal to MCR $\times j$ (where $j \geq 1$)
- Basing the MCR on the SCR (either as a fixed percentage of the SCR, or by omitting some of its components) or
- Changing the definition of "ruin" under the SCR so that it is based on having sufficient capital at the end of the time horizon to cover both the TPs and the prospective MCR calculated at that time

This discussion made by CEIOPS reflects the uncertainty in measuring the capital floor MCR before different approaches were tested.

P.2 QIS2

In QIS2 (see CEIOPS, 2006b, 2006d), the MCR largely followed the same modular approach as the standard formula, but with the following key differences:

- To retain a degree of simplicity, there were no adjustments for the risk-absorbing. capacity of profit-sharing liabilities or the expected profitability of nonlife business
- To support the use of audited/auditable data only, factor-based approaches were tested in each of the modules
- There was no explicit charge for operational risk
- All dependence assumptions were prescribed

The initial calibration of the MCR in QIS2 was set using two methods:

- In most cases, the output from the equivalent modules of the standard formula was reduced by 50%.
- In some cases, factors in the standard formula approaches were recalibrated to the equivalent of 90% TVaR.

Transitional MCR Based on Solvency I

CEIOPS' advice on Call for Advice (CfA) No. 9 suggested that a formula based on the Solvency I requirements should be used to calculate the MCR for a set transitional period. However, this will need to reflect the Solvency II methodology for valuing TPs.

For nonlife business, an assumption is made that changes to the valuation basis have a limited effect on the results of the formulaic Solvency I requirements. For life business, the exercise adopts a shortcut whereby the calculations are performed on a gross of reinsurance basis.

The Following Approach for Life Insurance was to be Tested

$$\begin{aligned} \text{TMCR}_{\text{life}} = 0.5 \times [&0.04 \times \text{TP}_1 + 0.003 \times \text{CR}_1 + 0.001 \times \text{CR}_2 + 0.0015 \times \text{CR}_3 \\ &+ \text{CRC}_{\text{sup}} + 0.04 \times \text{TP}_{\text{health}} + \text{CRC}_{\text{health}} + 0.04 \times \text{TP}_{\text{red}} + 0.01 \times \text{A}_{\text{ton}} \\ &+ 0.04 \times \text{TP}_{\text{L1}} + 0.01 \times \text{TP}_{\text{L2}} + 0.25 \times \text{CE}_{\text{L3}} + 0.003 \times \text{CR}_{\text{L}}] \end{aligned}$$

where

$$\text{CR}_1 = \text{CCR}_1 \left(1 + \frac{\text{CTP}_1 - \text{TP}_1}{\text{CCR}_1 + \text{CCR}_2 + \text{CCR}_3}\right),$$

$$\text{CR}_2 = \text{CCR}_2 \left(1 + \frac{\text{CTP}_1 - \text{TP}_1}{\text{CCR}_1 + \text{CCR}_2 + \text{CCR}_3}\right),$$

$$\text{CR}_3 = \text{CCR}_3 \left(1 + \frac{\text{CTP}_1 - \text{TP}_1}{\text{CCR}_1 + \text{CCR}_2 + \text{CCR}_3}\right),$$

$$\text{TP}_{\text{L1}} = \text{CTP}_{\text{L1}} \left(1 + \frac{\text{TP}_{\text{L}} - \text{CTP}_{\text{L}}}{\text{CTP}_{\text{L}}}\right),$$

$$\text{TP}_{\text{L2}} = \text{CTP}_{\text{L2}} \left(1 + \frac{\text{TP}_{\text{L}} - \text{CTP}_{\text{L}}}{\text{CTP}_{\text{L}}}\right),$$

$$\text{CR}_{\text{L}} = \text{CCR}_{\text{L}} + \text{CTP}_{\text{L}} - \text{TP}_{\text{L}},$$

and CTP_1: current mathematical provisions for nonlinked life assurance; CTP_{L}: current TPs for all linked assurance; CTP_{L1}: current TPs for linked assurance, insofar as the insurer bears an investment risk; CTP_{L2}: current TPs for linked assurance, insofar as the insurer bears no investment risk but the allocation to cover management expenses is fixed for a period exceeding 5 years; CCR_1: current capital at risk for nonlinked life assurance, other than temporary assurance on death of a maximum term of 5 years; CCR_2: current capital at risk for temporary assurance on death of a maximum term of 3 years; CCR_3: current capital at risk for temporary assurance on death of a term more than 3 years but not more than 5 years; CCR_L: current capital at risk for linked assurance insofar as the insurer covers a death risk; CRC_{sup}: Solvency I required capital for supplementary insurance; CRC_{health}: Solvency I required capital for permanent health insurance not subject to cancellation; CE_{L3}: last year's net administrative expenses pertaining to linked assurance, insofar as the insurer bears no investment risk and the allocation to cover management expenses is not fixed for a period exceeding 5 years.

The following inputs were also required using the placeholder valuation requirements of this specification:

TP_l: TPs for nonlinked life assurance

TP_L: TPs for linked life assurance

TP_{health}: TPs for permanent health insurance

TP_{red}: TPs for capital redemption operations

A_{ton}: market-consistent value of tontine assets

The Following Approach for Nonlife Insurance was to be Tested

The CRC_{NL}, that is, the current capital required under Solvency I for the nonlife business was to be calculated. A simple factor-based approach was to be tested:

$$TMCR_{NL} = 0.5 \times CRC_{NL}.$$

The companies were asked to also calculate the double TMCR.

Post-Transition MCR

In its answer to CfA No. 9, CEIOPS gave the following working hypothesis for the MCR: *CEIOPS will develop a simple factor-based formula for the MCR by simplifying the SCR, possibly by retaining its most significant items, by using a more straightforward technique for aggregation and by calibrating the factors to a lower level of confidence.*

For a placeholder calculation of the MCR, CEIOPS asked the participants to test the following:

- The results of the relatively simple, "robust" modeling approaches for the SCR, reduced by applying fixed factors
- The placeholder dependence assumptions used in the SCR

This meant that many of the outputs from the calculation of the SCR risk modules in QIS2 were reused for the MCR. The MCR calculation is divided into components as shown in Figure P.1.

For QIS2, the operational risk is not included in the calculation of the MCR.

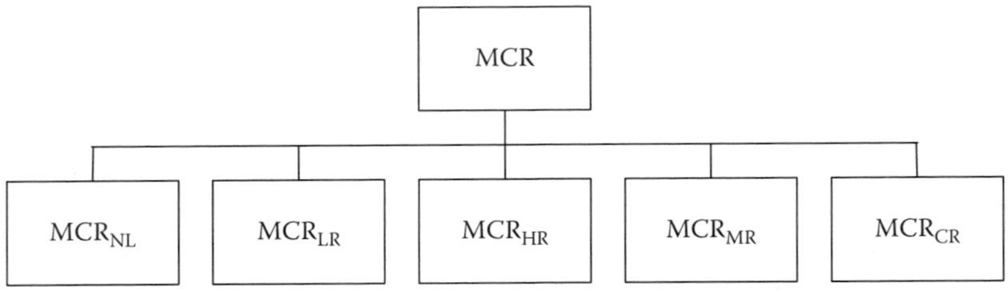

FIGURE P.1 The five MCR modules used for calculating the MCR. (Adapted from CEIOPS. 2006d. Quantitative Impact Study No. 2: Technical Specification, CEIOPS-PI-08/06, Available at www.ceiops.org.)

The capital charges for the main risk components should be combined using a dependence matrix as follows:

$$\text{MCR} = \sqrt{\sum_{r\times c} \rho_{\text{rc}} \times \text{MCR}_r \times \text{MCR}_c},$$

where the dependence structures are given in Table P.1 and the MCR_r and MCR_c are the capital charges for the five individual MCR risks modeled in Figure P.1.

This calculation should also be combined with two calculations assuming the following:

- Full diversification effects between the main subrisks are assumed, that is, $\rho = 0$ for all combinations of the dependencies.
- No diversification effects between the main subrisks are assumed, that is, $\rho = 1$ for all combinations of the dependencies.

Market MCR$_{MR}$

The capital charges for the market subrisks should be combined as above using the dependence matrix given in Table P.2.

For interest rate risk ($\text{MCR}_{\text{MR,IR}}$), the MCR uses the factor-based approach under the SCR calculation, see Appendix I. However, a different shock was applied in accordance with Table P.3.

The following approximation would also be used in all cases:

$$D_C^{\text{gen}}(r,s) \approx r_b \times s_b \times D_C^{\text{mod}}.$$

TABLE P.1 A Proposed Structure of the Dependence Matrix for the Main Risks in QIS2

	MCR$_{NL}$	**MCR$_{LR}$**	**MCR$_{HR}$**	**MCR$_{MR}$**	**MCR$_{CR}$**
MCR$_{NL}$	1	0	0	0.25	0.5
MCR$_{LR}$		1	0	0.25	0.25
MCR$_{HR}$			1	0.25	0.25
MCR$_{MR}$				1	0.75
MCR$_{CR}$					1

Source: Adapted from CEIOPS. 2006d. Quantitative Impact Study No. 2: Technical Specification, CEIOPS-PI-08/06. Available at www.ceiops.org

TABLE P.2 A Proposed Structure of the Dependence Matrix for the Main Risks in QIS2

	$MCR_{MR,IR}$	$MCR_{MR,ER}$	$MCR_{MR,PR}$	$MCR_{MR,CR}$
$MCR_{MR,IR}$	1	0.75	0.75	0.25
$MCR_{MR,ER}$		1	1	0.25
$MCR_{MR,PR}$			1	0.25
$MCR_{MR,CR}$				1

Source: Adapted from CEIOPS. 2006d. Quantitative Impact Study No. 2: Technical Specification, CEIOPS-PI-08/06. Available at www.ceiops.org

The MCR charges for equity, property, and currency risks are each calculated by applying a fixed factor to the "simple" treatments used under QIS2 calculation of the CR, see Appendix I.

$$MCR_{MR,X} = 0.5 \times CR_{MR,X},$$

where CR stands for the CR calculated in Appendix I, and X is either ER, PR, or CR.

Credit MCR_{CR}

The ratings-based approach Appendix I would be used with application of a fixed factor, so that:

$$MCR_{CR} = 0.5 \times CR_{CR}.$$

Life MCR_{LR}

The MCR charge for LUR was given by applying a fixed factor to the "simple" treatment used for CR in Appendix N. For the biometric, lapse, and expense risks, we have

$$MCR_{LR,X} = 0.5 \times CR_{LR,X}.$$

The overall MCR charge MCR_{LR} for LUR was calculated from the subrisk charges in the same way as in the CR_{LR}, see Appendix N.

Health MCR_{HR}

The MCR charge for health underwriting risk was given by applying a fixed factor to the "simple" treatment used for CR in Appendix M. For the biometric, lapse, and expense risks, we have

$$MCR_{HR} = 0.5 \times CR_{HR}.$$

TABLE P.3 Market Shock used for the MCR Calculation in QIS2

Maturity t (years)	**1–3**	**3–6**	**6–12**	**12–18**	**+18**
Relative change $s^{up}(t)$	0.3	0.25	0.2	0.15	0.15
Relative change $s^{down}(t)$	−0.25	−0.20	−0.15	−0.1	−0.1

Source: Adapted from CEIOPS. 2006d. Quantitative Impact Study No. 2: Technical Specification, CEIOPS-PI-08/06. Available at www.ceiops.org

Nonlife MCR_{NL}

The calculation should follow the same structure as the CR_{NL} calculation, except for the following changes:

- In each of the LoBs k, the estimate P_k of the net earned premium in the forthcoming year should be determined as the previous year's net earned premiums.
- In the premium and reserve subrisk modules, the function $\rho(x)$ should be re-specified and calculated at the 90% quantile.
- In the CAT subrisk module, the market-loss parameter referred to in Appendix M would generally be consistent with a TailVaR risk measure calibrated to a confidence level of 90%.

P.2.1 QIS2 Experience

CEIOPS were aware of that the approach used in QIS2 was driven by pragmatic considerations and that a more coherent approach to calibration would be required for the final MCR.

As a result, the comments made after QIS2 led CEIOPS to conclude that different MCR proposals should be considered for future QIS exercises.

It is clear that there needs to be a sufficient gap between the MCR and the SCR, to allow the ladder of supervisory actions elaborated in Pillar II. The MCR would be a floor for the SCR. However, the results of QIS2 show that in some Member States, the SCR result may be lower in too many cases than the MCR. In other cases, the SCR might be above, but very close to, the MCR. This was largely because, unlike the QIS2 version of the MCR, the SCR included adjustments for reduction for profit sharing (RPS) in life insurance and expected profitability in nonlife insurance.

Under QIS2, the standard SCR included an RPS, expressed as $k \times \mathrm{TP}_{\mathrm{benefits}}$, where $\mathrm{TP}_{\mathrm{benefits}}$ was the TP relating to future discretionary benefits, and k is a factor between 0 and 1 that is intended to reflect the extent to which future discretionary profit sharing may be used to absorb future losses under adverse circumstances. Generally, this depends on a range of aspects specific to the country and the insurer. No adjustment for the loss-reduction potential of discretionary profit sharing was included in the QIS2 version of the MCR.

QIS2 experience suggested that, while the RPS may be zero or negligible in some Member States, at the same time it is quite significant in other Member States, effectively closing the gap between the MCR and the SCR. Therefore, most CEIOPS members believed that ignoring the k-factor in the MCR could lead to unacceptable differences in the shape of the "supervisory ladder" in different countries. However, some CEIOPS members considered that including the k-factor in the MCR would make it complex and difficult to present before a court. To ensure that the SCR remains above the MCR, these members consider that the k-factor should be retained as an eligible element of capital rather than being deducted from the SCR.

As with the SCR, most CEIOPS' members believed that the recognition of the loss-reduction potential of discretionary profit sharing could be resolved in an overall framework including the CRs and available capital. However, the approach used for the MCR would need to be sufficiently simple, robust, and objective suggesting a degree of simplification is necessary.

If the adjustment for the expected profitability of nonlife business was included in the MCR, the corresponding treatment would need to be sufficiently simple, robust, and objective. However, some CEIOPS members considered that the recognition of profit sharing in the MCR does not ensure that MCR is lower than SCR. In effect, formulaic MCR and SCR calculations will differ in their requirements and deductions, which may lead to the SCR lower than the MCR even when the latter recognizes profit sharing.

The structure for the MCR as shown in Figure P.2 might be consistent with the considerations made above. In the above proposal, the health module has got a new name, Special, and we have also added two new modules for runoff expenses (Exp_{RO}) and CAT risks (MCR_{CAT}). The AMCR is an absolute MCR expressed in Euros and decided by the framework directive (FD). Based on the QIS2 results, the credit risk is omitted as it was not seen as of material importance for the MCR.

The MCR is proposed to be calculated as

$$\text{MCR} = \max\left\{\text{AMCR}; \sqrt{\text{MCR}^2_{\text{assets}} + \text{MCR}^2_{\text{liabilities}}} + \text{Exp}_{\text{RO}}\right\},$$

where

$$\text{MCR}_{\text{assets}} = \text{MCR}_{\text{MR}}$$

and

$$\text{MCR}_{\text{liabilities}} = \text{MCR}_{\text{NL}} + \text{MCR}_{\text{LR}} + \text{MCR}_{\text{CAT}}.$$

It is assumed that the asset risks and liability risks are independent and the liability risks are fully correlated between themselves.

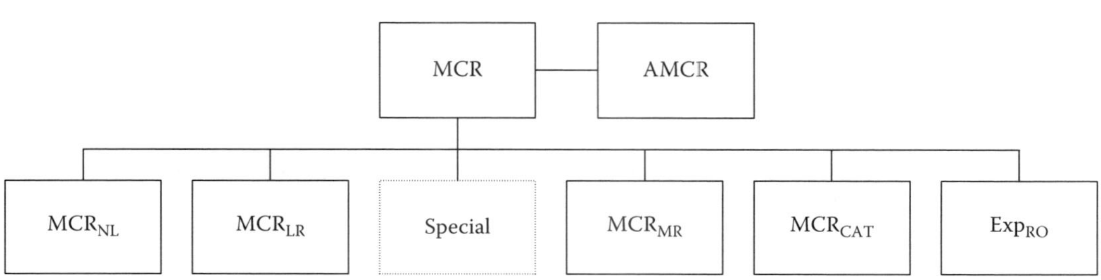

FIGURE P.2 A proposal for the MCR structure. (Adapted from CEIOPS. 2006b. Draft Advice to the European Commission in the Framework of the Solvency II Project on Pillar I Issues—Further Advice. Consultation Paper 20, CEIOPS-CP-09/06. November 10. Available at www.ceiops.org.)

The MCR market risk charge (RC) could be calculated by the following simple formula:

$$MCR_{MR} = (\alpha \times EQU) + (\beta \times RE) + (\chi \times FI),$$

where EQU: the market value of the overall equity and UCITS exposure; RE: the market value of the property exposure; FI: the market value of fixed income assets; and α, β, and χ are fixed coefficients.

The calculation would be performed on the basis of the total BS, but assets covering UL liabilities would be excluded.

For the calculation of the MCR nonlife underwriting RC, a factor-based formula similar to the Solvency I capital charge could be suggested. However, a component based on the TPs might also be included to address situations where the existing requirements do not always provide a good risk proxy.

The calculation could proceed as follows:

$$MCR_{NL} = \max\{\beta \times TP_{NL}; \chi \times P_{NL}; \delta \times X_{NL}\},$$

where TP_{NL}: a volume measure based on nonlife TPs; P_{NL}: a volume measure based on nonlife premiums; X_{NL}: a volume measure based on past nonlife claims; and β, χ, and δ are fixed coefficients.

For the calculation of the MCR life underwriting RC, a factor-based formula similar to the Solvency I capital charge could be applied. However, for better risk sensitivity, a distinction could be made between those (nonunit linked) liabilities that are subject to longevity risk, and those that are not.

The calculation would proceed as follows:

$$MCR_{LR} = \alpha \times TP_{UL} + \beta_1 \times TP_{L1} + \beta_2 \times TP_{L2} + \delta \times CaR,$$

where TP_{UL}: life TPs where the investment risk is borne by the policyholder; TP_{L1}: TPs (net of reinsurance) relating to the part of the life portfolio that is subject to longevity risk; TP_{L2}: TPs (net of reinsurance) relating to the part of the life portfolio that is not subject to longevity risk; CaR: the capital at risk of the life portfolio (net of reinsurance); and α, β_1, β_2, and δ are fixed coefficients.

It seems that the MCR should reflect catastrophe risk and the mitigating effect of reinsurance on the exposure to such risk. However, there are also concerns that CAT risk could not be reflected in the MCR in a robust, reliable, and practicable way, or that the inclusion of the CAT risk module might distort the interplay with the SCR. It was therefore suggested that the issue was revisited after QIS3.

The MCR CAT RC could be directly derived as:

$$MCR_{CAT} = X_{PML},$$

where X_{PML} is the total cost of the probable maximum loss (PML) to which the insurer can be exposed, net of the current reinsurance cover.

This definition of the CAT risk would be forward-looking, as it was the current reinsurance cover that was used to determine the net amount of the PML. This approach could also encourage better risk management as the undertakings need to estimate their PML. However, it is potentially difficult to audit and may be complex to estimate for some risks (e.g., motor insurance).

The capital charges in the SCR adopted a going-concern perspective while the MCR is geared toward winding up. As a consequence, Exp_{RO} provides an allowance for runoff expenses.

The TPs already include a part of the runoff expenses, as they should be sufficient to ensure the settlement of liabilities to the policyholders and include future administrative costs. The runoff expenses CR was intended to ensure the payment of additional unexpected expenses.

The allowance for runoff expenses in the MCR could be calculated by the following simple formula:

$$\text{Exp}_{\text{RO}} = h \times \text{Exp}_{t-1} \times \text{Dur}_{\text{TP}},$$

where Exp_{t-1}: the total annual claims' handling expenses including an allowance for overheads, all costs required to continue to service the existing business and associated discontinuance costs (e.g., redundancy payments, closure costs) at the time of the most recently reported TPs (by default, TPs at the end of the previous financial year) to the extent that these are not already included in the calculation of TPs; and Dur_{TP}: the estimated duration of the TPs at the time of the MCR's calculation, and h is a fixed constant.

The function applies a simple loading to expenses to reflect additional expenses that could occur during run off. This is projected forward to reflect the potential length of the runoff period, although no discounting is applied to retain simplicity. The rationale for using Exp_{t-1} rather than current expenses is to secure the "auditability" of the MCR, but the supervisor could request interim MCR calculations based on more up-to-date estimates. The use of the current estimate for Dur_{TP} is necessary to avoid penalizing the insurer for provisions that have unwound between $t - 1$ and the point of the MCR's calculation. This is still prudent because it assumes that the insurer's capital position would immediately deteriorate to the point of ruin, whereas the SCR should ensure the adequacy of an insurer's assets to meet TPs throughout its one-year time horizon. Alternatively, YPs at the end of the previous financial year could replace Exp_{t-1} as a risk proxy.

P.2.1.1 The Compact Approach

Some of CEIOPS members were concerned that the "modular" approach described above will not deliver a clear hierarchy of regulatory requirements (with an associated ladder of supervisory actions), in which the SCR should be above the MCR. They were also concerned that the specific "modular" proposal outlined above oversimplifies the relationship between asset and liability risks. These members advocate a "compact" MCR for development and testing under QIS3. A compact approach means taking a percentage of the SCR as the MCR.

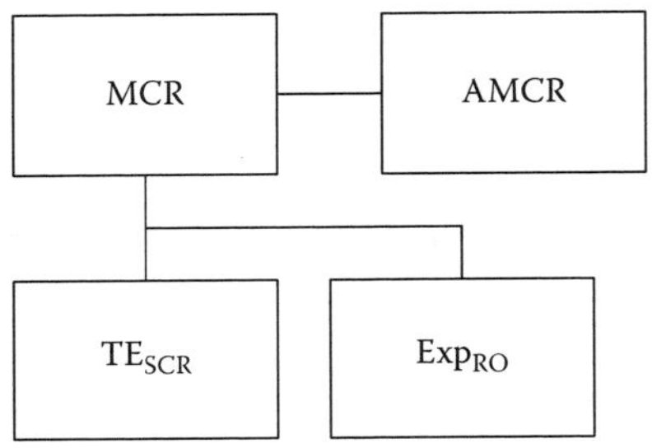

FIGURE P.3 The structure of the compact approach. (Adapted from CEIOPS. 2006b. Draft Advice to the European Commission in the Framework of the Solvency II Project on Pillar I Issues—Further Advice. Consultation Paper 20, CEIOPS-CP-09/06. November 10. Available at www.ceiops.org.)

It was important to note that other CEIOPS members were strongly opposed to the specific "compact" proposal suggested, which they considered being inconsistent with the MCR's role as a safety net and the design criteria set by the Commission's Amended FfC.

The following simple structure given in Figure P.3 would be consistent with the compact approach.

The MCR would be calculated as:

$$\text{MCR} = \max\left\{\text{AMCR}; \text{TE}_{\text{SCR}}; \text{Exp}_{\text{RO}}\right\},$$

where AMCR: absolute MCR; TE_{SCR}: the timing error; the risk that the insurer and the supervisor are unable to identify the breach (or potential breach) of a requirement sufficiently quickly in order to successfully execute restorative measures; and Exp_{RO}: the runoff expenses; the SCR adopts a going-concern perspective and therefore does not explicitly address additional costs that would arise in a runoff situation (e.g., staff redundancy costs or costs of breaking a lease for office premises). In principle, TPs should already include an allowance for such expenses, but, in practice, they may be difficult to project with any accuracy.

The timing error, TE_{SCR}, could be calculated as $TE_{\text{SCR}} = g \times SCR_{t-1}$, where g is a prespecified coefficient <1. The rationale for using SCR_{t-1}, rather than the current SCR, is to secure the "auditability" of the MCR. Clearly, this is at the expense of accuracy—in particular, this approach will be less well suited to cases where, for example, an insurer's business is growing rapidly, or the insurer has taken steps to de-risk its business, thereby reducing its current SCR.

P.3 QIS3

The MCR in QIS3 (CEIOPS, 2007a) used the following module structure as described in Figure P.4.

The different risk modules of QIS3 are as follows:

MCR_{NL}: the nonlife underwriting risk

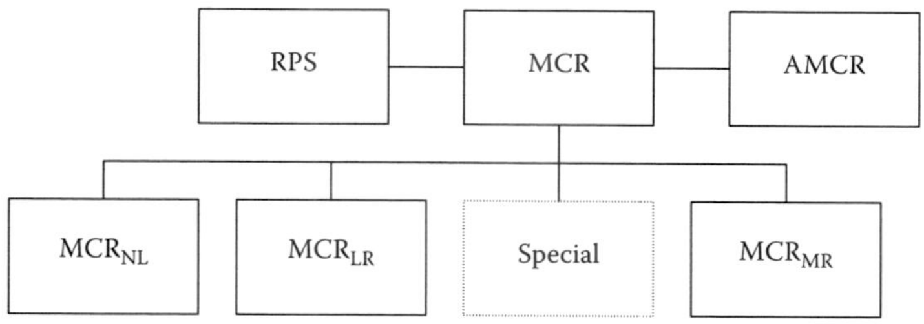

FIGURE P.4 The MCR module structure in QIS3. (Adapted from CEIOPS 2007a. QIS 3 Technical Specifications, Part I: Instructions, CEIOPS-FS-11/07 & Part II: Background Information, CEIOPS-FS-12/07. April. Available at www.ceiops.org.)

MCR_{LR}: the LUR

MCR_{HR}: the health underwriting risk as a special risk component

MCR_{MR}: the market risk

RPS: Reduction for profit sharing

AMCR: the absolute MCR

The MCR is calculated as

$$\text{MCR} = \max\{\text{BMCR};\text{AMCR}\},$$

where BMCR is the base MCR calculated as $\text{BMCR} = \sqrt{\sum_{r \times c} \rho_{rc} \times \text{MCR}_r \times \text{MCR}_c} - \text{RPS}$, and r and c are the rows and columns in the dependence matrix given in Table P.4, respectively.

The Reduction for Profit Sharing, RPS

This component reflects the loss-reduction potential of future nonguaranteed bonuses. The scope of the reduction includes both life and health insurance business. The approach specified below does not represent a final position on part of CEIOPS. The calculation assumes that, in the context of the MCR, a risk reduction factor (k-factor) of 100% can be assumed; however, on the contrary, the reduction is capped by a surrender value limit. The scope of the reduction includes both life and health insurance business.

TABLE P.4 Dependence Matrix for the Calculation of Market MCR under QIS3

	MCR_{NL}	MCR_{LR}	MCR_{MR}	MCR_{HR}
MCR_{NL}	1	0	0.25	0
MCR_{LR}		1	0.25	0.25
MCR_{MR}			1	0.25
MCR_{HR}				1

Source: Adapted from CEIOPS. 2007a. QIS 3 Technical Specifications, Part I: Instructions, CEIOPS-FS-11/07 & Part II: Background Information, CEIOPS-FS-12/07. April. Available at www.ceiops.org.

It is calculated as:

$$\mathrm{RPS} = \sum_i \min\left[\max\left\{0; \mathrm{TP}_{\mathrm{wp},i} - \mathrm{TP}_{\mathrm{surrender},i}\right\}; \mathrm{TP}_{\mathrm{benefits},i}\right],$$

where $\mathrm{TP}_{\mathrm{wp},i}$: the sum of TPs for with-profits fund i; including the element relating to guaranteed benefits and the element relating to future nonguaranteed bonuses; $\mathrm{TP}_{\mathrm{surrender},i}$: the surrender value of benefits guaranteed under contracts (i.e., excluding any discretionary benefits) for with-profits fund i; and $\mathrm{TP}_{\mathrm{benefits},i}$: the element of TPs relating to future nonguaranteed bonuses for with-profits fund i; as calculated within QIS3

For the purposes of this calculation, a with-profit fund means a group of with-profit contracts, which are treated as a unit together with the underlying liabilities and a segregated portfolio of backing assets that are not normally available to cover other liabilities.

The Market Risk Component, MCR_{MR}

Two alternatives were tested on an equal footing, without specifying a placeholder. For both alternatives, the calculation would be performed on the basis of the total BS, but assets covering UL liabilities shall be excluded.

The two alternative calculations are:

$$\mathrm{MCR}_{\mathrm{MR1}} = \sqrt{[0.12 \times \mathrm{EQU} + 0.08 \times \mathrm{RE}]^2 + [0.054 \times \mathrm{FI}_{\mathrm{L}} + 0.027 \times \mathrm{FI}_{\mathrm{NL}}]^2}$$

and

$$\mathrm{MCR}_{\mathrm{MR2}} = \sqrt{[\mathrm{MCR}_{\mathrm{ER}} + \mathrm{MCR}_{\mathrm{PR}}]^2 + \mathrm{MCR}_{\mathrm{SR}}^2 + \mathrm{MCR}_{\mathrm{IR}}^2},$$

where EQU: the market value of the overall equity and UCITS exposure; RE: the market value of the property exposure; and FI_{X}: the market value of fixed income assets related to life or nonlife business (including fixed income UCITS) including government bonds (X: L = life and NL = nonlife)

$$\mathrm{MCR}_{\mathrm{ER}} = 0.12 \times \mathrm{EQU}$$

$$\mathrm{MCR}_{\mathrm{PR}} = 0.08 \times \mathrm{RE}$$

$$\mathrm{MCR}_{\mathrm{SR}} = 0.025 \times \mathrm{FI}^*$$

and

$$\mathrm{MCR}_{\mathrm{IR}} = \max\left\{0; \mathrm{FI} \times D_{\mathrm{FI}}^{\mathrm{mod}} \times r\left(D_{\mathrm{FI}}^{\mathrm{mod}}\right) \times s^{\mathrm{up}} - \mathrm{TP} \times D_{\mathrm{TP}}^{\mathrm{mod}} \times r\left(D_{\mathrm{TP}}^{\mathrm{mod}}\right) \times s^{\mathrm{up}}; \mathrm{FI} \times D_{\mathrm{FI}}^{\mathrm{mod}} \times r\left(D_{\mathrm{FI}}^{\mathrm{mod}}\right) \times s^{\mathrm{down}} - \mathrm{TP} \times D_{\mathrm{TP}}^{\mathrm{mod}} \times r\left(D_{\mathrm{TP}}^{\mathrm{mod}}\right) \times s^{\mathrm{down}}\right\}$$

The modified duration of a CF $C(t)$ is calculated from the duration as:

$$D_{\mathrm{C}}^{\mathrm{mod}} = \frac{1}{1 + r\left(D_{\mathrm{C}}\right)} \times D_{\mathrm{C}}$$

and the interest rate shock parameters are determined as:

$$s^{\text{up}} = 0.18$$
$$s^{\text{down}} = -0.20,$$

UCITS exposures should be split between the EQU, RE, and FI (L, NL, or *) exposure on a look-through basis where this is possible. Otherwise, they should be split consistently with their classification; if no simple classification is available, they should be counted as equity.

The CFs used to determine the duration of TPs should be consistent with the CFs used to determine the best estimate (BE).

FI: the market value of fixed income assets ($\text{FI}_L + \text{FI}_{NL}$).

FI*: the market value of fixed income assets (including fixed income UCITS), excluding government bonds. The exemption relates to borrowings by the national government, or guaranteed by the national government, of an OECD or EEA state, issued in the currency of the government.

TP: the market value of TPs.

D_{FI}: the mean duration of the discounted CFs relating to fixed income assets.

D_{TP}: the mean duration of TPs.

$r(t)$: the term structure of interest rates (prescribed).

The Nonlife Underwriting Risk Component, MCR_{NL}

The MCR nonlife underwriting risk component was calculated by the following function:

$$\text{MCR}_{\text{NL}} = \max\left\{\sqrt{H_{\text{P}}}; 0.65\right\} \times \left[\sum_i \alpha_i \times P_i\right] + \max\left\{\sqrt{H_{\text{PCO}}}; 0.65\right\} \times \left[\sum_i \beta_i \times \text{PCO}_i\right],$$

where H_{P}: the Herfindahl indices for premiums; and H_{PCO}: the Herfindahl indices for claims provisions; which serves as a proxy measures for diversification between LoBs:

$$H_{\text{P}} = \frac{\sum_i P_i^2}{\left(\sum_i P_i\right)^2} \quad \text{and} \quad H_{\text{PCO}} = \frac{\sum_i \text{PCO}_i^2}{\left(\sum_i \text{PCO}_i\right)^2}.$$

PCO_i: the MCR_{NL} TP volume measure for QIS3 purposes: total provisions for claims outstanding for line of business i, net of reinsurance; and P_i: the earned premiums in line of business i during the previous year, net of reinsurance

The coefficients α and β are given in Table P.5.

The LUR Component, MCR_{LR}

The MCR life underwriting risk component was calculated by the following function:

$$\text{MCR}_{\text{LR}} = \sqrt{\text{MCR}^2_{\text{LR,MR}} + \text{MCR}^2_{\text{LR,LR}}} + \text{MCR}_{\text{UL}},$$

TABLE P.5 Coefficients α and β used for the Nonlife MCR Module

LoB	1	2	3	4	5	6	7	8	9	10	11	12	13	14	15
α_i	10	6.5	6.5	13	13	16.5	13	13	16.5	6.5	10	16.5	19.5	19.5	19.5
β_i	19.5	19.5	19.5	19.5	10	19.5	13	26.5	13	13	26.5	26.5	26.5	26.5	26.5

Source: Adapted from CEIOPS. 2007a. QIS 3 Technical Specifications, Part I: Instructions, CEIOPS-FS-11/07 & Part II: Background Information, CEIOPS-FS-12/07. April. Available at www.ceiops.org.

Note: Values are given in percentages.

where a distinction is made between mortality risk (LR,MR), longevity risk (LR,LR), and UL contracts. The calculation of the subcomponents is as follows:

$$\text{MCR}_{\text{LR,MR}} = 0.00025 \times \text{CaR},$$
$$\text{MCR}_{\text{LR,LR}} = 0.0015 \times \text{TP}_{\text{long}},$$
$$\text{MCR}_{\text{UL}} = 0.12 \times \text{Exp}_{\text{UL}},$$

where CaR: the sum of the net of reinsurance capital at risk in the portfolio, that is, the sum of the amounts currently payable on death less the (net of reinsurance) TP held for each policy that gives rise to a financial strain on immediate death of the insured; TP_{long}: the sum of net TPs net of any benefits payable on immediate death in respect of contracts which give rise to a financial surplus on immediate death of the insured; and Exp_{UL}: the last year's net administrative expenses relating to UL business.

The Special Risk Component: The Health Underwriting Risk, MCR_{HR}

This module is concerned with underwriting risk in health insurance that is practised on a similar technical basis to that of life assurance. The capital charge was calculated as:

$$\text{MCR}_{\text{HR}} = 1.28 \times \frac{\rho}{\sqrt{N_{\text{health}}}} \times \text{BE},$$

with $\rho = 0.5$; N_{health}: the number of health-insured persons; and BE: the sum of the annual gross benefits (settled or not) for the policyholders and the annual expenses of the insurance company related to health insurance business that occurred in the accounting year. BE subsumes all the claims and expenses associated with claims risk, mortality risk, cancellation risk and expense risk.

For the AMCR three different floors expressed in Euros was tested: 1 million EUR, 2 million EUR, and 3 million EUR.

Additional Quantitative Information that was Requested in QIS3

To assist the design of transitional arrangements and the setting of the MCR floor, and to enable the testing of the CEA's alternative MCR proposal, that is, MCR = % × SCR, the following information should be disclosed:

1/3 × SCR: one-third of the SCR of the participant, calculated according to the standard formula

$1/3 \times SCR_{IM}$: one-third of the SCR of the participant, calculated according to an internal model—if available

RSM: the Solvency I required solvency margin of the participant at the reference date (calculated according to the current valuation base)

MGF: the Solvency I minimum guarantee fund of the participant at the reference date

P.3.1 Calibration of QIS3

The calibration of the MCR is discussed in CEIOPS (2007g).

P.3.1.1 Market Risk

The calibration of the MCR market risk module generally adopted the methodology and the results of the calibration of the SCR market risk module, discussed in Appendix I. However the reduced granularity of the MCR also meant that, in certain cases, broad average assumptions had to be used.

P.3.1.1.1 Alternative 1: MCR_{MR1} The calibration was benchmarked on a member state market, taking into account about 400 undertakings, using the results of QIS1 and QIS2; this aimed at adjusting BSs to reflect changes in the valuation of liabilities and assets.

Equity Component The MCR equity RC was calibrated using normally distributed returns with a 10% return and 16.9% volatility. This led to an 11.7% capital charge, rounded to 12%.

Property Component Regarding property risk, it seemed that the QIS2 methodology overestimated return rates, due to estimations based on the 1998–2005 period for the reference market; such estimations excluded the crisis that took place in the beginning of the 1990s on the property reference market. Those estimates would have led to a negative capital charge on a 90% VaR basis. For that reason, a 7% return and 12% volatility were chosen.

Fixed-Income Component The fixed-income charge covered the interest rate risk and its initial calibration follows the calibration of the QIS2 interest rate risk module. The liability side was taken into account, not on a company-by-company basis but on a market-wide basis: a mean rate of 4% was chosen, combined with a 2-year duration for nonlife business, and a 7-year duration for life business. A 20% shock was applied and led to the coefficients in the specifications: 2.7% for nonlife business and 5.4% for life business.

P.3.1.1.2 Alternative 2: MCR_{MR2}

Equity and Property Components For the equity component and the property component, the calibrations were the same as in Alternative 1 above.

Credit Spread Component The risk weight applied to FI* was derived from the SCR spread risk module, assuming a 5-year duration and A-rated bonds, and the factor was adjusted for the 90% VaR level shock and rounded.

Interest Rate Risk Component The modeling of the upward and downward interest rate shocks as a function of maturity followed the approach used to calibrate the SCR standard formula interest rate risk submodule; the calibration of the MCR took into account the shocks corresponding to the 90% VaR level. The average values for medium maturities (the former middle maturity bucket) of 7–12 years were chosen, leading to $s^{\mathrm{up}} = 0.18$ and $s^{\mathrm{down}} = -0.20$. Reference to the shorter durations and their higher shock factors was deliberately avoided, because of the concern that this would close the gap between the MCR and the SCR.

Table P.6 provides the 90%-level stress factors for interest rate risk, from which the above parameters have been derived.

P.3.1.2 Nonlife Underwriting Risk

No independent model was built to calibrate $\mathrm{MCR_{NL}}$; the present calibration aimed to approximate the 90%-level SCR equivalent (SCR_90) in a simpler algebraic structure. The model used to determine the SCR_90 charge was the same as in the SCR standard formula nonlife premium and reserve risk submodule, yet with a treatment of premium risk relying solely on market-wide volatility factors.

The factors for each line of business, k, and for both premium and reserve risk were derived as

$$\alpha_k = \rho_{90}\left(\sigma_{\mathrm{PR},k}\right) \text{ and } \beta_k = \rho_{90}\left(\sigma_{\mathrm{RR},k}\right),$$

where ρ_{90} is the 90% VaR equivalent of the risk measure function ρ used in the standard formula NLprem submodule; and where $\sigma_{\mathrm{PR},k}$ and $\sigma_{\mathrm{RR},k}$ are the volatility factors used in the SCR for each line of business.

The 0.65 floor applied to the concentration/diversification factors was selected to minimize the sum of the squares of the residuals:

$$\sum\left[\frac{\mathrm{MCR_{NL}}}{\mathrm{SCR\_90}} - 1\right]^2,$$

for a sample of insurer data. A real-life sample of 80 insurers from two countries was used. However, since the data matching the QIS3 segmentation were not always available, the calculation involved some rough estimates. Because of this, a simulated sample was also set up to supplement the calibration.

The two approaches led to very similar results. In terms of MCR-to-SCR ratios, the calibration exercise indicated that, under the present choice of parameters, the ratio of the SCR_90 charge to the (nonpersonalized) SCR nonlife premium and reserve RC generally falls close to 47%.

TABLE P.6 Stress Factors, at 90% Level, for the Interest Rate Risk

Maturity n (years)	$s^{up}(n)$	$s^{down}(n)$
1	0.37	–0.31
2	0.31	–0.28
3	0.28	–0.26
4	0.25	–0.25
5	0.23	–0.23
6	0.21	–0.22
7	0.20	–0.21
8	0.19	–0.21
9	0.18	–0.20
10	0.17	–0.20
11	0.17	–0.20
12	0.17	–0.20
13	0.17	–0.20
14	0.17	–0.20
15	0.17	–0.20
16	0.17	–0.19
17	0.17	–0.19
18	0.16	–0.19
19	0.16	–0.18
20	0.16	–0.18
21	0.15	–0.18
22	0.15	–0.18
23	0.15	–0.18
24	0.15	–0.18
25	0.15	–0.18
26	0.15	–0.18
27	0.15	–0.18
28	0.15	–0.18
29	0.15	–0.18
30	0.15	–0.18

Source: Adapted from QIS4. 2008. QIS 4 Technical Specifications. European Commission, MARKT/2505/08, March 31. Available at http://www.ceiops.eu/content/view/118/124/.

For the samples used in the calibration exercise, the ratio of the MCR nonlife underwriting RC to the corresponding standard formula charge (with no personalization applied) generally fell between 35% and 55%.

Figures P.5 and P.6 illustrate the relationship between MCR_{NL} and SCR_90.

P.3.1.3 Life Underwriting Risk

The mortality and longevity components were calculated on the same technical basis as in the factor-based CR_{LR} proxies, with a calibration of 90% VaR instead of 99.5% VaR. The

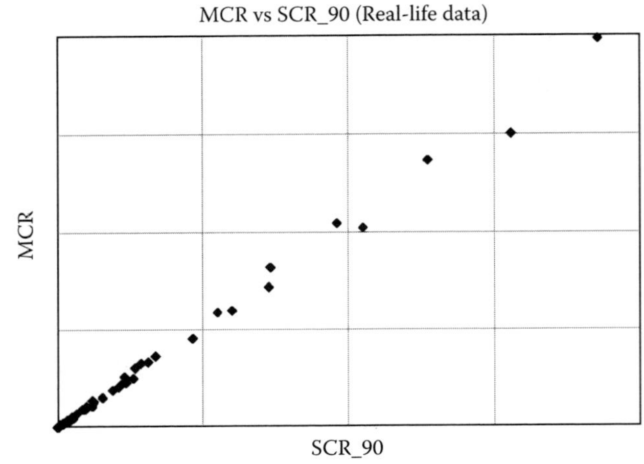

FIGURE P.5 The relation between the MCR and SCR_90 for "real-data." (Adapted from CEIOPS. 2007g. Calibration of the underwriting risk, market risk and MCR. CEIOPS-FS-14/07. April. Available at www.ceiops.org; © Committee of European Insurance and Occupational Pensions Supervisors.)

definition of the UL charge was an initial one meant for QIS3 purposes only. Parallel to the review of the standard formula operational RC, CEIOPS decided to revise this component after QIS3.

The present MCR life underwriting risk formula did not take into account disability and morbidity risks. The inclusion of a RC reflecting these risks would be considered after QIS3.

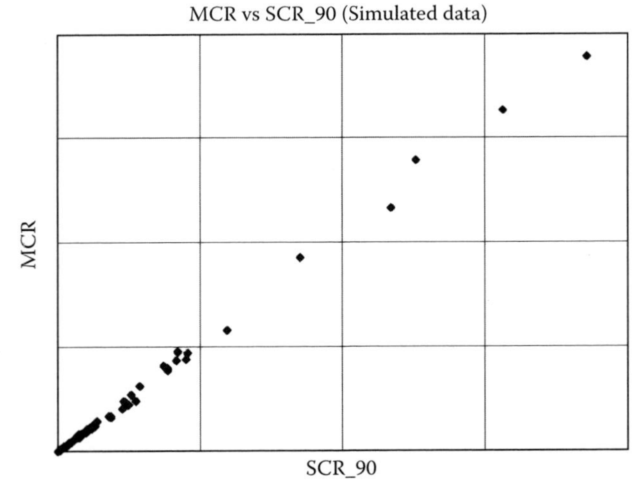

FIGURE P.6 The relation between the MCR and SCR_90 for "simulated-data." (Adapted from CEIOPS. 2007g. Calibration of the underwriting risk, market risk and MCR. CEIOPS-FS-14/07. April. Available at www.ceiops.org; © Committee of European Insurance and Occupational Pensions Supervisors.)

P.3.1.4 Health Underwriting Risk

The calibration was based on market data regarding the following variables:

I_k: the number of sampled risks of the kth insurer

A_k: the overall number of risks of the kth insurer

SumClaim_k: the sum of sampled claims of the kth insurer, that is, $\sum_{i \in I_k} x_{i,k}$

QSumClaim_k: the sum of sampled squares of claims of the kth insurer, that is, $\sum_{i \in I_k} x_{i,k}^2$

s_k: the estimated standard deviation of the random variable supposed to describe the claim per person and accounting year of the kth insurer

The purpose of the calibration was to determine the coefficient ρ in the MCR health formula:

$$\text{MCR}_{\text{HR}} = c \times \frac{\rho}{\sqrt{A_k}} \times \text{SumClaim}_k,$$

in such a way that the MCR capital charge, together with TPs, provides a 90% confidence level that the available capital for a risk (unbiased randomly chosen from all risks on the market) will stay above the TP reserved for that risk.

The factor ρ is formally determined from the equation

$$\frac{\sum_k A_k \times \Pr\left[X < c \times \rho \times \dfrac{\text{SumClaim}_k}{I_k \times s_k}\right]}{\sum_k A_k} = \frac{\sum_k A_k \times \Phi\left[c \times \rho \times \dfrac{\text{SumClaim}_k}{I_k \times s_k}\right]}{\sum_k A_k} = 0.9$$

where X denotes an N(0,1) distributed random variable and Φ denotes the cumulative distribution function of N(0,1). The equality was solved for ρ by a Newton or fixed-point procedure.

Applying the above methodology to data obtained from a member state market, estimates of ρ fell between 2.1 and 7.6. For QIS3, a calibration of $\rho = 5$ was chosen.

P.4 QIS4

The technical specifications for QIS4 are presented in QIS4 (2008). Two approaches were tested in QIS4:

1. A *linear approach* simplifying the modular approach tested in QIS3. It was building on the margin over liabilities, that is, a percentage of TPs, approach, but made it more risk-sensitive by adding other volume measures. However, asset-side volume measures were excluded from this variant of the linear approach.

2. A *combined approach*, given by the linear MCR approach in combination with a cap of 50% and a floor of 20% of the SCR, whether calculated using the standard formula or an internal models.

Let

MCR_{linear}: the linear MCR, that is, the sum of the linear MCRs for each type of business undertaken by the participant, before applying any cap or floor.

$MCR_{combined}$: the combined MCR of the participant, as calculated by the combined approach, after applying the cap and the floor (50% and 20% of the SCR, respectively) to the linear MCR.

MCR: the final MCR of the participant, as calculated by applying the absolute minimum floor to the combined approach.

The Linear Approach

Participants were *first* requested to calculate the components of their linear MCR before applying any cap or floor, depending on the type of business they write, namely:

MCR_{NL}: the linear MCR for nonlife business

MCR^*_{NL}: the linear MCR for nonlife business similar to life business

MCR_{Life}: the linear MCR for life business

MCR^*_{Life}: the linear MCR for supplementary nonlife business underwritten in addition to life insurance

Then in the *second* step, the overall linear MCR of the participant was set equal to the sum of the components of the linear MCR:

1. For nonlife participants: $MCR_{linear} = MCR_{NL} + MCR^*_{NL}$.
2. For life participants: $MCR_{linear} = MCR_{Life} + MCR^*_{Life}$.
3. For composite participants, which conduct both life and nonlife business:

$$MCR_{linear} = MCR_{NL} + MCR^*_{NL} + MCR_{Life} + MCR^*_{Life}.$$

The Combined Approach

Then in the *third* step, the combined MCR was calculated, by applying the cap and the floor (50% and 20% of the SCR, respectively) to the linear MCR.

$$MCR_{combined} = \{\min\left[\max\left(MCR_{linear};\ 0.2 \times SCR\right);\ 0.5 \times SCR\ \right]\},$$

where SCR is the SCR of the participant as calculated in accordance with Appendices H through O.

In the last step, an absolute floor is applied to the combined MCR:

$$MCR = \max\{MCR_{combined}; AMCR\},$$

where AMCR is the absolute floor of the MCR: 1 million EUR for nonlife insurance undertakings and for reinsurance undertakings, 2 million EUR for life insurance undertakings, or 1 million EUR + 2 million EUR = 3 million EUR for composite undertakings.

Notional Nonlife and Life Linear MCR (for Composite Undertakings)

Composite participants were also requested to report the following outputs:

$NMCR_{NL}$: the notional nonlife linear MCR of the participant

$NMCR_{Life}$: the notional life linear MCR of the participant.

They are calculated as

$$NMCR_{NL} = MCR_{NL} + MCR^*_{NL}$$

and

$$NMCR_{Life} = MCR_{Life} + MCR^*_{Life}.$$

Linear MCR for Nonlife Business

The MCR for nonlife business was calculated by the following function, where the sum is taken over lines-of-business:

$$MCR_{NL} = \sum_{LOB:k} \max\,[\alpha_k \times TP_k;\ \beta_k \times P_k],$$

where the factors α_{lob} and β_{lob} are determined in Table P.7, and

TP_k: TPs, excluding the RM, for each line of business k, net of reinsurance, subject to a minimum of zero.

P_k: written premiums in each line of business k at the reporting date, net of reinsurance, subject to a minimum of zero.

Certain nonlife LoBs may include claims that are similar in nature to life insurance business. Health insurance practised on a similar technical basis to that of life insurance and nonlife annuities have been identified as such activities whose MCR is specified in this subsection.

The MCR for nonlife business similar to life is calculated by the following function:

$$MCR^*_{NL} = \alpha_h \times TP_h + \alpha_a \times TP_a,$$

where the factors were determined as $\alpha_h = 0.013$ and $\alpha_a = 0.025$, and TP_h: TPs, excluding the RM, net of reinsurance, subject to a minimum of zero for health insurance that is practised on a similar technical basis to that of life insurance; and TP_a: TPs, excluding the RM, other than TP_h that are disclosed separately as nonlife liabilities valued according to life insurance principles, for example, TPs for nonlife annuities, net of reinsurance, subject to a minimum of zero.

TABLE P.7 Factors used for the Linear Nonlife MCR Calculation for the 15 Nonlife LoB

	Lines-of-Business (LoB)	α_{lob}	β_{lob}
1	A&H—workers' compensation	0.13	0.09
2	A&H—health insurance	0.10	0.04
3	A&H—others/default	0.20	0.06
4	Motor, third-party liability	0.16	0.12
5	Motor, other classes	0.09	0.12
6	Marine, aviation, transport	0.13	0.16
7	Fire and other property damage	0.13	0.13
8	Third-party liability	0.20	0.16
9	Credit and suretyship	0.20	0.20
10	Legal expenses	0.13	0.06
11	Assistance	0.13	0.10
12	Miscellaneous	0.13	0.14
13	NP reinsurance—property	0.20	0.20
14	NP reinsurance—casualty	0.20	0.20
15	NP reinsurance—MAT	0.20	0.20

Source: Adapted from QIS4. 2008. QIS4 Technical Specifications. European Commission, MARKT/2505/08, March 31. Available at http://www.ceiops.eu/content/view/118/124/.

Linear MCR for Life Business

The MCR for life business is calculated by the following function:

$$
\begin{aligned}
MCR_{\text{Life}} = \max\{&\alpha_{\text{WP\_guaranteed}} \times \text{TP}_{\text{WP\_guaranteed}} \\
&+ \alpha_{\text{WP\_bonus}} \times \text{TP}_{\text{WP\_bonus}};\ \gamma \times \text{TP}_{\text{WP\_guaranteed}}\} + \\
&+ \sum_{i\{\text{non}-\text{WP}\}} \alpha_i \times \text{TP}_i + 0.25 \times \text{Exp}^*_{\text{ul}} + \sum_j \beta_j \times \text{CaR}_j
\end{aligned}
$$

The factors relating to TPs for with-profits business were determined as follows: $\alpha_{\text{WP\_guaranteed}} = 0.035$, $\alpha_{\text{WP\_bonus}} = -0.09$, and $\gamma = 0.015$.

The factors α_i applied to TPs other than with-profits business, following the segmentation of life TPs, are given in Table P.8.

The following variables are used in the formula above:

$\text{TP}_{\text{WP\_guaranteed}}$: TPs, net BE, for guaranteed benefits relating to with-profits contracts.

TABLE P.8 Factors α_i Applied to TPs Other Than With-Profits Business

	Risk Driver	
First-Level Segment	Death or Savings	Survivorship or Morbidity
UL	0.005	0.0175
Nonprofit	0.01	0.035
Reinsurance accepted	See below	See below

Source: Adapted from QIS4. 2008. QIS4 Technical specifications. European Commission, MARKT/2505/08, March 31. Available at http://www.ceiops.eu/content/view/118/124/.

TP_{WP_bonus}: TPs, net BE, for discretionary bonuses relating to with-profits contracts.

TP_i: TPs, excluding the RM, net of reinsurance, subject to a minimum of zero for each segment i other than with-profits business according to the granularity defined in Table P.8.

CaR_j: capital at risk, that is, the sum of the amounts currently payable on death or disability and the present value of annuities payable on death or disability less the TP held for each policy that gives rise to a financial strain on immediate death or disability of the insured, calculated net of reinsurance for each segment j according to the granularity defined in Table P.9.

Exp^{*}_{ul}: only with respect to nonretail UL business and management of group pension funds where the policyholder takes the investment risk: the amount of last year's net administrative expenses.

Reinsurance accepted should be apportioned according to the segmentation of direct classes, using the same factors as for direct business. UL products with a guarantee on survival would use a 0.0175 factor.

The participants would also calculate the MCR of supplementary nonlife insurance underwritten in addition to life insurance. The MCR of such classes was calculated in a manner technically similar to the nonlife MCR.

The following input information was required:

TP_{lob}: TPs, excluding the RM, for each line of business, net of reinsurance, subject to a minimum of zero

P_{lob}: written premiums in each line of business at the reporting date, net of reinsurance, subject to a minimum of zero

The MCR for supplementary nonlife business is calculated by the following function:

$$\mathrm{MCR}^{*}_{\mathrm{Life}} = \sum_{\mathrm{LOB:k}} \max\left(\alpha_k \times \mathrm{TP}_k;\ \beta_k \times \mathrm{P}_k\right),$$

where the factors α_k and β_k are identical to the nonlife MCR factors defined above.

The QIS4 combined approach was better received by both companies participating and the majority of supervisors the modular approach tested in QIS3 (CEIOPS, 2008e).

TABLE P.9 Factors β_j Applied to Capital-at-Risk are Determined According to the Table

j	Outstanding Term of Contract (years)	β_j
1	5 or more	0.00125
2	3–5	0.0009
3	3 or less	0.0005

Source: Adapted from QIS4. 2008. QIS 4 Technical Specifications. European Commission. MARKT/2505/08. March 31. Available at http://www.ceiops.eu/content/view/118/124/.

P.4.1 Calibration of QIS4

Late in 2007, CEIOPS published a discussion paper on the pros and cons of different approaches to calculate the minimum CR (CEIOPS, 2007i). They first discussed and compared the modular approach, the percentage of the SCR standard formula approach, and the classical percentage of the TP approach (similar to the original solvency calculation for life insurance within the EU. From the feedback, they instead proposed two other approaches to be tested for QIS4. These were the combined and the linear approaches.

In the calibration of the MCR for the QIS4 technical specifications, several countries' QIS3 data were used (see CEIOPS, 2008a). We will briefly look at the calibration results that CEIOPS obtained.

P.4.1.1 General Assumptions

Following the FD proposal, the MCR should be calibrated to a confidence level in the range of 80–90% VaR over a one-year period. In developing and testing the calibration for QIS4, a percentage of the SCR as a proxy calibration target had been used. It was recognized that there was no linear relationship between 80% or 90% VaR, and 99.5% VaR through all distributions. In CEIOPS' calibration exercise, following the lognormal assumptions underlying the SCR standard formula, the 25%·SCR to 45%·SCR interval was taken as a rough equivalent of the 80–90% VaR range, and the midpoint of this interval—that is, 35% of the SCR—was used as a proxy calibration target.

P.4.1.2 Nonlife Business

The nonlife MCR premium and TPs factors as described in Table P.7 was derived from the same underlying lognormal assumptions as in the SCR standard formula for the nonlife premium and reserve risk, respectively.

Starting from the market-wide standard deviation parameters $\sigma_{R,k}$ and $\sigma_{M,k}$, a $\rho(\sigma)$-function corresponding to 90% VaR, roughly $\rho(\sigma) \approx 1.3\sigma$, was used to derive the α_k and β_k factors. Factors corresponding to 90% VaR over a one-year time horizon, that is, the high end of the 80–90% target interval of the FD proposal, were chosen to implicitly compensate for the fact that this calibration approach did not take into account risks other than premium and reserve risks.

The α_h factor given above for long-term health insurance provisions was calibrated to 35% of the observed QIS3 SCR to TPs ratio on one local market. The α_a factor was calibrated using nonlife annuity data on a local market, reflecting the middle point between the 80% VaR and the 90% VaR calibration (yielding a 0.0197 and a 0.0306 factor, respectively, on that local market).

These calibrations were back-tested on QIS3 data. The testing took into account the proposed QIS4 changes in SCR premium and reserving RFs. For composite firms, a proxy nonlife SCR was calculated to allow a separate nonlife MCR to SCR comparison.

The results of the back testing are given in Table P.10.

One significant outlier group with high MCR-to-SCR ratios that was identified in the testing were health insurers on a local market where a market-wide mandatory equalization system was in place. All but one other nonlife MCR-to-SCR ratios observed in the testing

TABLE P.10 Back-Testing for 460 Insurers in 19 Countries Resulted in these "Nonlife MCR-to-SCR Ratios"

MCR-to-SCR Ratio (Nonlife), %	Number of Firms
Lower than 10	20
10–20	60
20–30	126
30–40	125
40–50	88
50–60	20
60–70	9
70–80	2
80–90	8
90–100	0
>100	2

Source: Adapted from CEIOPS. 2008a. QIS4 Background Document: Calibration of SCR, MCR, and Proxies. CEIOPS-DOC-02/08 (draft January 31). April 1. Available at www.ceiops.org.

were lower than 70%, with three-quarter of the results falling between 20% and 50%. Given these results, the factors that were proposed for the nonlife business generally provided a satisfactory interplay between the SCR and the MCR.

P.4.1.3 Life Business

The calibration was developed in two steps.

1. Initially, the calibration was derived via least-squares linear fitting for 35% of the SCR on the QIS3 data of one local market, taking into account the following adjustments:

- The counterparty default risk was removed from the SCR, as this risk component was concentrated in a small number of firms, and was difficult to reproduce by a linear formula.
- The lapse catastrophe component was removed from the SCR, given the change of methodology in QIS4.
- The SCR was adjusted to exclude "free assets," so that the calibration of the MCR reflected the financial position of a company with little to no "free assets" above the TPs and the SCR. The rationale for this adjustment was that the MCR being tested was unaffected by assets.

Where QIS3 data were insufficient to yield a reasonable factor, expert adjustments were applied to the fitting results to obtain a calibration. These included the TPs charge for the with-profit death, disability, and survivorship; UL death, disability, and survivorship; nonprofit death, disability, and savings classes; and the capital-at-risk charge for remaining contract term of less than 5 years.

TABLE P.11 Back-Testing for 286 Insurers in 18 Countries Resulted in these "Life MCR-to-SCR Ratios"

MCR-to-SCR Ratio (life)	Number of Firms
Lower than 10%	33
10–20	63
20–30	62
30–40	51
40–50	26
50–60	21
60–70	12
70–80	5
80–90	6
90–100	2
>100	5

Source: Adapted from CEIOPS. 2008a. QIS4 Background Document: Calibration of SCR, MCR, and Proxies. CEIOPS-DOC-02/08 (draft January 31). April 1. Available at www.ceiops.org.

2. As a second step, this calibration was back-tested on QIS3 data. The testing took into account the proposed QIS4 changes regarding lapse catastrophe risk. For composite firms, a proxy life SCR was calculated to allow a separate life MCR to SCR comparison.

A major issue that emerged from the testing related to the different risk absorption characteristics of future profit sharing on different markets. On those markets where future discretionary benefits have high-risk absorbency, there was a strong negative dependence between discretionary bonus provisions and risks, justifying a negative factor.

On some of these markets, the initial calibration, with a zero factor for discretionary bonus provisions and a 2.5–3.5% factor on provisions for guaranteed benefits, resulted in high MCR-to-SCR ratios in the testing. For one specific market, a –26% factor for discretionary bonus provisions was suggested instead, while a 6.8% factor would apply to provisions for guaranteed benefits. The latter factor reflected the risks of a firm that has no discretionary bonus provisions to absorb losses.

However, on other markets the relationship between future discretionary bonuses and risk mitigation was less straightforward. It was raised that future discretionary bonuses may actually have a higher risk profile, for example, through riskier investments on some markets. On such markets, the factors suggested above could lead to negative MCR results—stopped only by the absolute floor.

Therefore taking the linear MCR, given above, refined the initial approach. This refined approach for the with-profits segment was suggested as a middle ground between the two types of market identified, that is, guaranteed benefits and discretionary bonuses. It recognized future profit sharing as a risk-mitigating factor; however, it also included a floor equal to 1.5% of TPs for guaranteed benefits to avoid extremely low results. Thus, the capital charge would remain in a band between 1.5% and 3.5% of the guaranteed part of provisions.

Then a second round of back-testing, including QIS3 data of 286 firms in 18 countries, focusing on the refined approach for with-profits contracts, led to the life MCR-to-SCR ratios as given in Table P.11.

References

Aas, Kjersti. 2005. *The Basel II IRB Approach for Credit Portfolios: A Survey*. SAMBA/33/05. Norsk Regnesentral, Oslo, October.

Aas, Kjersti, Xeni K. Dimakos, and Anders Øksendal. 2005. *Risk Capital Aggregation*. SAMBA/40/05. Norsk Regnesentral, Oslo, December.

Aase, Knut K. and Svein-Arne Persson. 2003. New econ for life actuaries. *ASTIN Bulletin*, Vol. 33, No. 2, pp. 117–124 (incl. Comments by Hans Bühlmann).

Abbink, Michel and Matt Saker. 2002. Getting to grips with fair value. Presented to *The Staple Inn Actuarial Society* on March 5, 2002. Available at http://www.sias.org.uk

ACAM. 2007. *Work Group Report on the Best Estimate in Non-Life insurance*. Autcrité de contrôle des assurances et des mutuelles, ACAM, November 21. Available at http://www.acam-france.fr

AISAM–ACME. 2007. *AISAM-ACME Study on Non-Life Long Tail Liabilities: Reserve Risk and Risk Margin Assessment Under Solvency II*. October 17. Joint report from AISAM and ACME. Available at http://www.amice-eu.org

Albrecht, Peter. 2004a. Risk measures. In Teugels, Jozef L. and Bjørn Sundt, Editors. 2004. *Encyclopaedia of Actuarial Science*. John Wiley & Sons, Chichester, ISBN: 0-470-84676-3.

Albrecht, Peter. 2004b. Risk-based capital allocation. In Teugels, Jozef L. and Bjørn Sundt, Editors. 2004. *Encyclopaedia of Actuarial Science*. John Wiley & Sons, Chichester, ISBN: 0-470-84676-3.

Alexander, Carol and Jacques Pézier. 2003. *On the Aggregation of Market and Credit Risks*. ISMA Centre Discussion Papers in Finance 2003-13. ISMA Centre, University of Reading, UK.

Algo Risk. 2007. *Replicating Portfolios in Algo Risk*. Using the Algo Risk Optimizer to replicate a set of scenario dependent liability cash flows. Algorithmics, September. Available at http://www.algorithmics.com

Alworth, Julian S. 1993. *The Valuation of US Dollar Interest Rate Swaps*. BIS Economic Papers, No. 35, January. Bank for International Settlements, Basel.

Amato, Jeffery D. and Eli M. Remolona. 2003. The credit spread puzzle. *BIS Quarterly Review*, December. Available at http://www.bis.org/forum/research.htm (free of charge).

Antonio, David, Stephen Carlin, John Hibbert, Colin Holmes, Zhuoshi Liu, David Roseburgh, and Steffen Sorensen. 2009. A framework for extrapolation of long-term interest rates. *Life & Pensions*, July, pp. 39–44. Available at http://www.life-pensions.com

AON. 2005. *Liability Driven Investment, Intouch Opinion*. Part I, February, Part II May. Available at http://www.aon.com

APRA. 2000. *A Proposed Reform to the Prudential Supervision of General Insurance Companies in Australia*. Policy Discussion Paper, APRA, April. Available at http://www.apra.gov.au

Artzner, Phillipe. 1999. Application of coherent risk measures to capital requirements in insurance. *North American Actaurial Journal*, Vol. 3, No. 2, pp. 11–25.

Artzner, Phillipe, Freddy Delbaen, Jean-Marc Eber, and David Heath. 1999. Coherent measures of risk. *Mathematical Finance*, Vol. 9, No. 3, pp. 203–228.

ASB. 1988. *Concerning Cash Flow Testing For Life and Health Insurance Companies.* Actuarial Standard of Practice, No. 7, Actuarial Standards Board, Washington DC, October (No longer in effect).

ASB. 2002. *Analysis of Life, Health, or Property/Casualty Insurers Cash Flow.* Actuarial Standard of Practice, No. 7, Actuarial Standards Board, Washington DC, June.

Aven, Terje. 2003. *Foundations of Risk Analysis.* John Wiley & Sons, Chichester, ISBN: 10-0-471-49548-4.

Bangia, Anil, Francis X. Diebold, Til Schuermann, and John D. Stroughair. 1999. *Modeling Liquidity Risk: With Implications for Traditional Market Risk Measurement and Management.* Report, Wyman & Company, Oliver, April 26.

Barnett, Glen and Ben Zehnwirth. 2000. Best estimates for reserves. *Proceedings of the Casualty Actuarial Society*, Vol. LXXXVII, Part 2, No. 167, pp. 245–321.

Barrie and Hibbert. 2008. *A Framework for Estimating and Extrapolating the Term Structure of Interest Rates.* Exposure Draft, Version 1.0, September. Available at www.barrhibb.com

Baumgartner, Gabi, Hans Bühlmann, and Michael Koller. 2004. Multidimensional valuation of life insurance policies and fair value. *SAV Bulletin*, No. 1, pp. 27–64.

Baxter, Martin and Andrew Rennie. 1996. *Financial Calculus: An Introduction to Derivative Pricing.* Cambridge University Press, Cambridge, ISBN: 0 521 55289 3.

Behan, Donald F. 1993. Cash-flow testing as a managing tool. *Record of Society of Actuaries*, Vol. 19, No. 3, pp. 1711–1724.

Bellini, Fabio and Camilla Caperdoni. 2007. Coherent distortion risk measures and higher-order stochastic dominances. *North American Actuarial Journal*, Vol. 11, No. 2, pp. 35–42.

Benjamin, Bernard. 1977. *General Insurance.* Heinemann, London (published for the Institute of Actuaries and the Faculty of Actuaries).

Bennet, C. 2004. *Dictionary of Insurance*, 2nd edition. FT Prentice Hall, Pearson Education Ltd, Harlow.

ter Berg, Peter. 2008. Portfolio modeling of counterparty reinsurance default risk. *Life & Pensions*, April, pp. 29–33. Available at http://www.life-pensions.com

Bhatia, Mohan. 2009. *An Introduction to Economic Capital.* Risk Books, Incisive Media, London, ISBN: 978-1-906348-09-0.

Bikker, Jacob A. and Katharina Haaf. 2002. Measures of competition and concentration in the banking industry: A review of the literature. *Economic and Financial Modelling*, Vol. 9, pp. 53–98.

BIS. 1999. *Market Liquidity. Research Findings and Selected Policy Implications.* Report on a Study Group established by the CGFS. Basle, May 3. Available at http://www.bis.org

BIS. 2004. *International Convergence of Capital Measurement and Capital Standards: A Revised Framework.* Basel Committee on Banking Supervision, Bank for International Settlements, June. Available at http://www.bis.org

BIS. 2005. *An Explanatory Note on the Basel II IRB Risk Weight Functions.* July. Basel Committee on Banking Supervision, Bank for International Settlements. Available at http://www.bis.org

BIS. 2006a. *International Convergence of Capital Measurement and Capital Standards: A Revised Framework Comprehensive Version.* June. Basel Committee on Banking Supervision, Bank for International Settlements. Available at http://www.bis.org

BIS. 2006b. *Studies on Credit Risk Concentration.* Working Paper No. 15. November. Basel Committee on Banking Supervision, Bank for International Settlements, Basel.

BIS. 2008. *Principles for Sound Liquidity Risk Management and Supervision.* June. Bank for International Settlements. Available at http://www.bis.org/publ/bcbs138.htm

BIS. 2009. International framework for liquidity risk measurement, standards and monitoring. Consultative Document. December. Bank for International Settlements. Available at http://www.bis.org

Björk, Tomas. 1998. *Arbitrage Theory in Continuous Time.* Oxford University Press, New York, ISBN: 0-19-877518-0.

Blum, Peter, and Michel Dacorogna. 2004. DFA—Dynamic financial analysis. In Teugels, Jozef L. and Bjørn Sundt, Editors. 2004. *Encyclopaedia of Actuarial Science.* John Wiley & Sons, Chichester, ISBN: 0-470-84676-3.

Bohman, Harald. 1974. Rule of the thumb for the determination of a sufficient risk reserve. *Scandinavian Actuarial Journal*, pp. 237–240.

Bohman, Harald. 1977. A rule of the thumb and its verification. *Scandinavian Actuarial Journal*, Supplement, pp. 25–36.

Bouro, G., G. Pavesi and G. Zucchiatti. 1980. Observations on the method of calculation of the solvency margin; 21st International Congress of Actuaries, in Italian; reprinted in English in *l'Argus International*, 1981, No. 22, pp. 65–71.

Bowles, Sharon. 2008a. Draft Opinion on the Committee on Legal Affairs for the Committee on Economic and Monetary Affairs on the amended proposal for a directive of the European Parliament and of the Council on the taking-up and pursuit of the business of Insurance and Reinsurance (recast). [COM(2008)0119-C6-0231/2007-2007/0143(COD)], Committee on Legal Affairs. 2007/0143(COD). 2.7.2008, European Parliament.

Bowles, Sharon. 2008b. Opinion on the Committee on Legal Affairs for the Committee on Economic and Monetary Affairs on the amended proposal for a directive of the European Parliament and of the Council on the taking-up and pursuit of the business of Insurance and Reinsurance (recast). [COM(2008)0119-C6-0231/2007-2007/0143(COD)], Committee on Legal Affairs. 2007/0143(COD). 1.4.2008, European Parliament.

Brehm, Paul J., Geoffrey R. Perry, Gary G. Venter, and Susan E. Witcraft, Editors. 2007. *Enterprise Risk Analysis for Property and Liability Insurance Companies.* Guy Carpenter and Company LLC, New York, ISBN: 0-615-13356-8. Available at http://ERMbook.guycarp.com

Brender, Allan. 2004. Risk-based capital requirements. In Teugels, Jozef L. and Bjørn Sundt, Editors. 2004. *Encyclopaedia of Actuarial Science.* John Wiley & Sons, Chichester, ISBN: 0-470-84676-3.

van Broekhoven, Henk. 2002. Market value of liability mortality risk: A practical model. *North American Actuarial Journal*, Vol. 6, pp. 95–106.

Buchwalder, Markus, Hans Bühlmann, Michael Merz, and Mario V. Wüthrich. 2006. The mean square error of prediction in the chain ladder reserving method (Mack and Murphy revisited). *ASTIN Bulletin*, Vol. 36, No. 2, pp. 521–542.

Buchwalder, Markus, Hans Bühlmann, Michael Merz, and Mario V. Wüthrich. 2007. Valuation portfolio in non-life insurance. *Scandinavian Actuarial Journal*, No. 2, pp. 108–125.

Bühlmann, Hans. 1987. Actuaries of the third kind? *ASTIN Bulletin*, Vol. 17, No. 2, pp. 137–138.

Bühlmann, Hans. 2002. New math for life actuaries. *ASTIN Bulletin*, Vol. 32, No. 2, pp. 209–211.

Bühlmann, Hans. 2003. On teaching actuarial science. Guest Editorial. *British Actuarial Journal*, Vol. 9, III, pp. 491–492.

Bühlmann, Hans. 2004. Multidimensional valuation. *Finance*, Vol. 25, pp. 15–29.

Bühlmann, Hans and Michael Merz. 2007. The valuation portfolio. *Journal of the Swiss Association of Actuaries*, Hefte 1, pp. 69–84.

Bühlmann, Hans, Massimo DeFelice, Alois Gisler, Franco Moriconi, and Mario V. Wüthrich. 2009. Recursive credibility formula for chain ladder factors and the claims development result. *ASTIN Bulletin*, Vol. 39, No. 1, pp. 275–306.

Cairns, Andrew J.G. 2004a. Interest-rate modeling. In Teugels, Jozef L. and Bjørn Sundt, Editors. 2004. *Encyclopaedia of Actuarial Science.* John Wiley and Sons, Chichester, ISBN: 0-470-84676-3.

Cairns, Andrew J.G. 2004b. Interest-rate risk and immunization. In Teugels, Jozef L. and Bjørn Sundt, Editors. 2004. *Encyclopaedia of Actuarial Science.* John Wiley & Sons, Chichester, ISBN: 0-470-84676-3.

Cairns, Andrew J.G. 2004c. *Interest-Rate Models: An Introduction.* Princeton University Press, Princeton and Oxford, ISBN: 0-691-11893-0.

Cairns, Andrew J.G. 2004d. Hedging and risk management. In Teugels, Jozef L. and Bjørn Sundt, Editors. 2004. *Encyclopaedia of Actuarial Science*. John Wiley & Sons, Chichester, ISBN: 0-470-84676-3.

Cairns, Andrew J.G., David Blake, and Kevin Dowd. 2008. Modelling and measurement of mortality risk: A review. *Scandinavian Actuarial Journal*. No. 2–3, pp. 79–113.

Campagne, C. 1957. *Minimum Standards of Solvency for Insurance Firms*. OEEC, Insurance Sub-Committee, TFD/PC/565, Paris, November 6. Provisional Translation.

Campagne, C. 1961a. *Minimum Standards of Solvency for Insurance Firms*. OEEC, Insurance Sub-Committee, TP/AS(61)1, Paris, March 11. English version of Campagne (1961b).

Campagne, C. 1961b. *Standard minimum de solvabilité applicable aux enterprises d'assurances*. Report of the OECE, March 11. Reprinted in Het Verzekerings-Archief deel XLVIII, 1971–1974.

Campagne, C., B.H. de Jongh, and J.N. Smit. 1947. *Bijdrage tot de Wiskundige Theorie van de Bedrijfsreserve en het Eigenbehoud in de Brandverzekring (Contribution to the Mathematical Theory of the Stabilization Reserve and the Net Retention in Fire Insurance)*. S-GravenhageRRijksuitgeverij, pp. 131– (in both Dutch and English).

Campagne, C., van der Loo, and A.J. Yntema. 1948. *Contribution to the Method of Calculating the Stabilization Rreserve in Life Assurance Business*. Gedenboek Verzekeringskamer 1923–1948, Staatsdrukkerij en uitgeverijbedrijf, Den Haag, pp. 338–378 (in both Dutch and English).

CAS. 1995. *Dynamic Financial Analysis, Property/Casualty Insurance Companies Handbook*. Release 1.0 (Final), September. Casualty Actuarial Society, Arlington, VA.

CEA. 2005. *Solvency Assessment Models Compared: Essential Groundwork for the Solvency II Project*. Produced by the CEA and Mercer Oliver Wyman in cooperation with all European Insurance Markets, Brussels. February.

CEA. 2006a. *CEA Document on Cost of Capital*. CEA, Brussels, April 21. Available at http://www.cea.eu

CEA. 2006b. *CEA Working Document on the Standard Approach for Calculating the Solvency Capital Requirement*. CEA, Brussels, March 22. Available at http://www.cea.eu

CEA. 2007a. *CEA Working Paper on the Total Balance Sheet Approach*. CEA, February 23. Available at http://www.cea.eu

CEA. 2007b. *Solvency II: Frequently Asked Questions*. CEA Briefing Note. February. Available at http://www.cea.eu

CEA. 2007c. *Solvency II: Understanding the Process*. CEA Briefing Note. February. Available at http://www.cea.eu

CEA. 2008a. Simplifications for calculating risk margins using cost-of-capital approach. CEA. February 15. Available at http://www.cea.eu

CEA. 2008b. *Initial Thoughts on the Use of Simplifications*. CEA Working Paper, ECO 8370. December 22. Available at http://www.cea.eu

CEA-CRO Forum. 2006. *Solutions to Major Issues for Solvency II*. Joint submission by the CRO Forum and CEA, Brussels. February 17.

CEA-GC. 2007. *Solvency II Glossary*. Comité Européen des Assurances and Groupe Consultatif Actuariel Européen, Brussels. March.

CEA-Tillinghast. 2006. *Solvency II: Introductory Guide*. CEA and Towers Perrin & Tillinghast. June. Available at http://www.cea.eu

Cech, Christian. 2006. *Copula-Based Top-Down Approaches in Financial Risk Aggregation*. No. 32, Working Paper Series by the University of Applied Sciences of bfi Vienna, Vienna. December.

CEIOPS. 2005a. Draft Answers to the European Commission on the "Second Wave" of Calls for Advice in the Framework of the Solvency II Project. CEIOPS-CP-04/05. May. Available at www.ceiops.org

CEIOPS. 2005b. Answers to the European Commission on the "first wave" of Calls for Advice in the Framework in the Solvency II Project. CEIOPS-DOC-03/05. June. Available at www.ceiops.org

CEIOPS. 2005c. Answers to the European Commission on the Second Wave of Calls for Advice in the Framework in the Solvency II Project. CEIOPS-DOC-07/05. October. Available at www.ceiops.org

CEIOPS. 2005d. QIS1 Specification, Technical provisions. CEIOPS-FS-11/05. October. Available at www.ceiops.org

CEIOPS. 2006a. Answers to the European Commission on the Third Wave of Calls for Advice in the Framework of the Solvency II Project. CEIOPS-DOC-03/06. May. Available at www.ceiops.org

CEIOPS. 2006b. Draft Advice to the European Commission in the Framework of the Solvency II Project on Pillar I Issues—Further Advice. Consultation Paper 20, CEIOPS-CP-09/06. November 10. Available at www.ceiops.org

CEIOPS. 2006c. Financial Conditions and Financial Stability in the European Insurance and Occupational Pension Fund Sector 2005–2006. CEIOPS-FS-14/06S. December. Available at www.ceiops.org

CEIOPS. 2006d. Quantitative Impact Study No. 2: Technical Specification, CEIOPS-PI-08/06. Available at www.ceiops.org

CEIOPS. 2007a. QIS 3 Technical Specifications, Part I: Instructions, CEIOPS-FS-11/07 & Part II: Background Information, CEIOPS-FS-12/07. April. Available at www.ceiops.org

CEIOPS. 2007b. Risk Management and Other Corporate Issues. Issues Paper. CEIOPS-PII-11/07, July 17. Available at www.ceiops.org

CEIOPS. 2007c. QIS3 Calibration of the Credit Risk. CEIOPS-FS-23/07. April. Available at www.ceiops.org

CEIOPS. 2007d. Calibration of QIS3 SCR Standard Formula for Non-Life Underwriting Risk. CEIOPS, SCR Subgroup, 2007-02-19.

CEIOPS. 2007e. Calibration of the Underwriting Risk Modules for QIS3. CEIOPS. April 1. Available at www.ceiops.org

CEIOPS. 2007f. Calibration of the Market Risk Module. CEIOPS-FS-05/07. February 19. Available at www.ceiops.org

CEIOPS. 2007g. Calibration of the Underwriting Risk, Market Risk and MCR. CEIOPS-FS-14/07. April. Available at www.ceiops.org

CEIOPS. 2007h. QIS 3 Technical Specifications. Annexes. April. CEIOPS-FS-13/07. Available at www.ceiops.org

CEIOPS. 2007i. Architecture of the MCR. Pros and Cons of Different Approaches. CEIOPS-DOC-22/07. December 17. Available at www.ceiops.org

CEIOPS. 2008a. QIS4 Background Document: Calibration of SCR, MCR and Proxies. CEIOPS-DOC-02/08 (draft January 31). April 1. Available at www.ceiops.org

CEIOPS. 2008b. Advice to the European Commission on the Principle of Proportionality in the Solvency II Framework Directive Proposal. CEIOPS-DOC-24/08. May. Available at www.ceiops.org

CEIOPS. 2008c. QIS4 Best Estimate Valuation Tool: User's Guide and Functionality Overview. CEIOPS-DOC-20g/08. May 5. Available at www.ceiops.org

CEIOPS. 2008d. Quantitative Impact Study 4, Questions & Answers, QIS4—List of Methodological Issues Raised by participants and supervisors. CEIOPS-DOC-12/08. Version July 8. Available at www.ceiops.org

CEIOPS. 2008e. CEIOPS' Report on its fourth Quantitative Impact Study (QIS4) for Solvency II. CEIOPS-SEC-82/08. November. Available at www.ceiops.org

CEIOPS. 2008f. CEIOPS' Issues Paper on the Own Risk and Solvency Assessment. CEIOPS-IGSRR-09/08. May. Available at www.ceiops.org

CEIOPS. 2009a. Lessons Learned from the Crisis (Solvency II and Beyond). CEIOPS-SEC-17/08. March 19. Available at www.ceiops.org

CEIOPS. 2009b01. Draft L2 Advice on TP—Methods and Statistical Techniques for Calculating the Best Estimate. CEIOPS-CP-26-09. March 26. Available at www.ceiops.org

CEIOPS. 2009b02. Draft L2 Advice on TP—Segmentation. CEIOPS-CP-27-09. March 26. Available at www.ceiops.org

CEIOPS. 2009b03. Draft L2 Advice on SCR Standard Formula—Counterparty Default Risk. CEIOPS-CP-28-09. March 26. Available at www.ceiops.org

CEIOPS. 2009b04. Draft L2 Advice on Own Funds—Criteria for Supervisory Approval of Ancillary Own Funds. CEIOPS-CP-29-09. March 26. Available at www.ceiops.org

CEIOPS. 2009b05. Draft L2 Advice on TP—Treatment of Future Premiums. CEIOPS-CP-30-09. March 26. Available at www.ceiops.org

CEIOPS. 2009b06. Draft L2 Advice on SCR Standard Formula—Allowance of financial mitigation techniques. CEIOPS-CP-31-09. March 26. Available at www.ceiops.org

CEIOPS. 2009b07. Draft L2 Advice on TP—Assumptions About Future Management Actions. CEIOPS-CP-32-09. March 26. Available at www.ceiops.org

CEIOPS. 2009b08. Draft L2 Advice on System of Governance. CEIOPS-CP-33-09. March 26. Available at www.ceiops.org

CEIOPS. 2009b09. Draft L2 Advice on Transparency and Accountability. CEIOPS-CP-34-09. March 26. Available at www.ceiops.org

CEIOPS. 2009b10. Draft L2 Advice on Valuation of Assets and "Other Liabilities." CEIOPS-CP-35-09. March 26. Available at www.ceiops.org

CEIOPS. 2009b11. Draft L2 Advice on Special Purpose Vehicles. CEIOPS-CP-36-09. March 26. Available at www.ceiops.org

CEIOPS. 2009b12. Draft L2 Advice on the Procedure to be Followed for the Approval of an Internal Model. CEIOPS-CP-37-09. March 26. Available at www.ceiops.org

CEIOPS. 2009b13. Addendum to CEIOPS-CP-37-09 Draft L2 Advice on the Procedure to be Followed for the Approval of a Group Internal Model. Addendum to CEIOPS-CP-37-09. July 2. Available at www.ceiops.org

CEIOPS. 2009c. Insurance Linked Securities Report. CEIOPS-DOC-17/09. June. Available at www.ceiops.org

CEIOPS. 2009d01. Draft L2 Advice Technical Provisions. Article 85a. Actuarial and statistical methodologies to calculate the best estimate. CEIOPS-CP-39-09. July 2. Available at www.ceiops.org

CEIOPS. 2009d02. Draft L2 Advice Technical Provisions. Article 85b. Risk-free interest rate term structure. CEIOPS-CP-40-09. July 2. Available at www.ceiops.org

CEIOPS. 2009d03. Draft L2 Advice Technical Provisions. Article 85c. Circumstances in which technical provisions shall be calculated as a whole. CEIOPS-CP-41-09. July 2. Available at www.ceiops.org

CEIOPS. 2009d04. Draft L2 Advice Technical Provisions. Article 85d. Calculation of the Risk Margin. CEIOPS-CP-42-09. July 2. Available at www.ceiops.org

CEIOPS. 2009d05. Draft L2 Advice Technical Provisions. Article 85f. Standards for Data Quality. CEIOPS-CP-43-09. July 2. Available at www.ceiops.org

CEIOPS. 2009d06. Draft L2 Advice Technical Provisions. Article 85g. Counterparty default adjustment to recoverables from reinsurance contracts and SPVs. CEIOPS-CP-44-09. July 2. Available at www.ceiops.org

CEIOPS. 2009d07. Draft L2 Advice Technical Provisions. Article 85h. Simplified methods and techniques to calculate technical provisions. CEIOPS-CP-45-09. July 2. Available at www.ceiops.org

CEIOPS. 2009d08. Draft L2 Advice Own Funds. Article 97 and 99. Classification and eligibility. CEIOPS-CP-46-09. July 2. Available at www.ceiops.org

CEIOPS. 2009d09. Draft L2 Advice SCR Standard Formula. Article 109. Structure and design of Market Risk Module. CEIOPS-CP-47-09. July 2. Available at www.ceiops.org

CEIOPS. 2009d10. Draft L2 Advice SCR Standard Formula. Non-Life Underwriting Risk. CEIOPS-CP-48-09. July 2. Available at www.ceiops.org

CEIOPS. 2009d11. Draft L2 Advice SCR Standard Formula. Article 109c. Life Underwriting Risk. CEIOPS-CP-49-09. July 2. Available at www.ceiops.org

CEIOPS. 2009d12. Draft L2 Advice SCR Standard Formula. Health Underwriting Risk Module. CEIOPS-CP-50-09. July 2. Available at www.ceiops.org

CEIOPS. 2009d13. Draft L2 Advice SCR Standard Formula. Further advice on the counterparty default risk module (Complementary to CEIOPS' CP no. 28). CEIOPS-CP-51-09. July 2. Available at www.ceiops.org

CEIOPS. 2009d14. Draft L2 Advice SCR Standard Formula. Article 109e and f. Allowance of Reinsurance Mitigation Techniques. CEIOPS-CP-52-09. July 2. Available at www.ceiops.org

CEIOPS. 2009d15. Draft L2 Advice SCR Standard Formula. Article 109 1(g). Operational Risk. CEIOPS-CP-53-09. July 2. Available at www.ceiops.org

CEIOPS. 2009d16. Draft L2 Advice SCR Standard Formula. Loss-absorbing capacity of technical provisions and deferred taxes. CEIOPS-CP-54-09. July 2. Available at www.ceiops.org

CEIOPS. 2009d17. Draft L2 Advice Article 128. Calculation of the MCR. CEIOPS-CP-55-09. July 2. Available at www.ceiops.org

CEIOPS. 2009d18. Draft L2 Advice Articles 118 and 124. Tests and Standards for Internal Model Approval. CEIOPS-CP-56-09. July 2. Available at www.ceiops.org

CEIOPS. 2009d19. Draft L2 Advice Capital Add-On. CEIOPS-CP-57-09. July 2. Available at www.ceiops.org

CEIOPS. 2009d20. Draft L2 Advice Supervisory Reporting and Public Disclosure Requirements. CEIOPS-CP-58-09. July 2. Available at www.ceiops.org

CEIOPS. 2009d21. Draft L2 Advice Assessment of Group Solvency. CEIOPS-CP-60-09. July 2. Available at www.ceiops.org

CEIOPS. 2009d22. Draft L2 Advice Supervision of Risk Concentration and Intra-Group Transactions. CEIOPS-CP-61-09. July 2. Available at www.ceiops.org

CEIOPS. 2009d23. Draft L2 Advice Cooperation and Colleges of Supervisors. CEIOPS-CP-62-09. July 2. Available at www.ceiops.org

CEIOPS. 2009e01. Draft L2 Advice on Repackaged Loans Investment. CEIOPS-CP-63-09. November 2. Available at www.ceiops.org

CEIOPS. 2009e02. Draft L2 Advice on the Extension of the Recovery Period—Pillar II Dampener. CEIOPS-CP-64-09. November 2. Available at www.ceiops.org

CEIOPS. 2009e03. Draft L2 Advice on Partial Internal Models. CEIOPS-CP-65-09. November 2. Available at www.ceiops.org

CEIOPS. 2009e04. Draft L2 Advice on the Group Solvency for Groups with Centralised Risk Management. CEIOPS-CP-66-09. November 2. Available at www.ceiops.org

CEIOPS. 2009e05. Draft L2 Advice on SCR Standard Formula and Own Funds—Treatment of Participations. CEIOPS-CP-67-09. November 2. Available at www.ceiops.org

CEIOPS. 2009e06. Draft L2 Advice on SCR Standard Formula and Own Funds—Treatment of Ring Fenced Funds. CEIOPS-CP-68-09. November 2. Available at www.ceiops.org

CEIOPS. 2009e07. Draft L2 Advice on SCR Standard Formula—Design of the Equity Risk Sub-Module. CEIOPS-CP-69-09. November 2. Available at www.ceiops.org

CEIOPS. 2009e08. Draft L2 Advice on SCR Standard Formula—Calibration of Market Risk. CEIOPS-CP-70-09. November 2. Available at www.ceiops.org

CEIOPS. 2009e09. Draft L2 Advice on SCR Standard Formula—Calibration of Non-Life Underwriting Risk. CEIOPS-CP-71-09. November 2. Available at www.ceiops.org

CEIOPS. 2009e10. L2 Advice on SCR Standard Formula—Calibration of Health Underwriting Risk. CEIOPS-CP-72-09. November 2. Available at www.ceiops.org

CEIOPS. 2009e11. Draft L2 Advice on MCR—Calibration. CEIOPS-CP-73-09. November 2. Available at www.ceiops.org

CEIOPS. 2009e12. Draft L2 Advice on SCR Standard Formula—Correlation Parameters. CEIOPS-CP-74-09. November 2. Available at www.ceiops.org

CEIOPS. 2009e13. Draft L2 Advice on SCR Standard Formula—Undertaking Specific Parameters. CEIOPS-CP-75-09. November 2. Available at www.ceiops.org

CEIOPS. 2009e14. Draft L2 Advice on Technical Provisions—Simplifications. CEIOPS-CP-76-09. November 2. Available at www.ceiops.org

CEIOPS. 2009e15. Draft L2 Advice on SCR Standard Formula—Simplifications. CEIOPS-CP-77-09. November 2. Available at www.ceiops.org

CEIOPS. 2009e16. Draft L2 Advice on simplifications for captives. CEIOPS-CP-79-09. November 2. Available at www.ceiops.org

CEIOPS. 2009f01. CEIOPS' Advice for L2 Implementing Measures on SII: Elements of Actuarial and Statistical Methodologies for the Calculation of the Best Estimate (former Consultation Paper no. 26). CEIOPS-DOC-21-09. November 10. Available at www.ceiops.org

CEIOPS. 2009f02. CEIOPS' Advice for L2 Implementing Measures on SII: Segmentation for the Calculations of Technical Provisions (former Consultation Paper no. 27). CEIOPS-DOC-22-09. November 10. Available at www.ceiops.org

CEIOPS. 2009f03. CEIOPS' Advice for L2 Implementing Measures on SII: SCR-Standard Formula—Counterparty Default Risk (former Consultation Paper no. 28 and Consultation Paper no. 51). CEIOPS-DOC-23-09. November 10. Available at www.ceiops.org

CEIOPS. 2009f04. CEIOPS' Advice for L2 Implementing Measures on SII: Supervisory Approval of Ancillary Own Funds (former Consultation Paper no. 29). CEIOPS-DOC-24-09. November 10. Available at www.ceiops.org

CEIOPS. 2009f05. CEIOPS' Advice for L2 Implementing Measures on SII: Technical Provisions—Treatment of future premiums (former Consultation Paper no. 30). CEIOPS-DOC-25-09. November 10. Available at www.ceiops.org

CEIOPS. 2009f06. CEIOPS' Advice for L2 Implementing Measures on SII: SCR-Standard Formula—Allowance of Risk Mitigation Techniques (former Consultation Paper no. 31). CEIOPS-DOC-26-09. November 10. Available at www.ceiops.org

CEIOPS. 2009f07. CEIOPS' Advice for L2 Implementing Measures on SII: Technical Provisions—Assumptions About Future Management Actions (former Consultation Paper no. 32). CEIOPS-DOC-27-09. November 10. Available at www.ceiops.org

CEIOPS. 2009f08. CEIOPS' Advice for L2 Implementing Measures on SII: Procedure to be Followed for the Approval of an Internal Model, Including the Addendum on Group Specificities (former Consultation Paper no. 37). CEIOPS-DOC-28-09. November 10. Available at www.ceiops.org

CEIOPS. 2009f09. CEIOPS' Advice for L2 Implementing Measures on SII: System of Governance (former Consultation Paper no. 33). CEIOPS-DOC-29-09. November 10. Available at www.ceiops.org

CEIOPS. 2009f10. CEIOPS' Advice for L2 Implementing Measures on SII: Transparency and Accountability (former Consultation Paper no. 34). CEIOPS-DOC-30-09. November 10. Available at www.ceiops.org

CEIOPS. 2009f11. CEIOPS' Advice for L2 Implementing Measures on SII: Valuation of Assets and "Other Liabilities" (former Consultation Paper no. 35). CEIOPS-DOC-31-09. November 10. Available at www.ceiops.org

CEIOPS. 2009f12. CEIOPS' Advice for L2 Implementing Measures on SII: Special Purpose Vehicles (former Consultation Paper no. 36). CEIOPS-DOC-32-09. November 10. Available at www.ceiops.org

CEIOPS. 2009f13. CEIOPS' Advice for L2 Implementing Measures on SII: Technical Provisions—Actuarial and Statistical Methodologies to Calculate the Best Estimate (former Consultation Paper no. 39). CEIOPS-DOC-33-09. November 10. Available at www.ceiops.org

CEIOPS. 2009f14. CEIOPS' Advice for L2 Implementing Measures on SII: Technical Provisions—Risk-Free Interest Rate (former Consultation Paper no. 40). CEIOPS-DOC-34-09. November 10. Available at www.ceiops.org

CEIOPS. 2009f15. CEIOPS' Advice for L2 Implementing Measures on SII: Calculation of Technical Provisions as a Whole (former Consultation Paper no. 41). CEIOPS-DOC-35-09. November 10. Available at www.ceiops.org

CEIOPS. 2009f16. CEIOPS' Advice for L2 Implementing Measures on SII: Risk Margin (former Consultation Paper no. 42). CEIOPS-DOC-36-09. November 10. Available at www.ceiops.org

CEIOPS. 2009f17. CEIOPS' Advice for L2 Implementing Measures on SII: Technical Provisions—Standard for data quality (former Consultation Paper no. 43). CEIOPS-DOC-37-09. November 10. Available at www.ceiops.org

CEIOPS. 2009f18. CEIOPS' Advice for L2 Implementing Measures on SII: Technical Provisions—Counterparty Default Adjustment (former Consultation Paper no. 44). CEIOPS-DOC-38-09. November 10. Available at www.ceiops.org

CEIOPS. 2009f19. CEIOPS' Advice for L2 Implementing Measures on SII: Classification and Eligibility of Own Funds (former Consultation Paper no. 46). CEIOPS-DOC-39-09. November 10. Available at www.ceiops.org

CEIOPS. 2009f20. CEIOPS' Advice for L2 Implementing Measures on SII: SCR-Standard Formula—Market Risk (former Consultation Paper no. 47). CEIOPS-DOC-40-09. November 10. Available at www.ceiops.org

CEIOPS. 2009f21. CEIOPS' Advice for L2 Implementing Measures on SII: SCR-Standard formula—Non-Life Underwriting Risk (former Consultation Paper no. 48). CEIOPS-DOC-41-09. November 10. Available at www.ceiops.org

CEIOPS. 2009f22. CEIOPS' Advice for L2 Implementing Measures on SII: SCR-Standard formula—Life Underwriting Risk (former Consultation Paper no. 49). CEIOPS-DOC-42-09. November 10. Available at www.ceiops.org

CEIOPS. 2009f23. CEIOPS' Advice for L2 Implementing Measures on SII: SCR-Standard formula—Health Underwriting Risk (former Consultation Paper no. 50). CEIOPS-DOC-43-09. November 10. Available at www.ceiops.org

CEIOPS. 2009f24. CEIOPS' Advice for L2 Implementing Measures on SII: SCR-Standard Formula—Reinsurance mitigation (former Consultation Paper no. 52). CEIOPS-DOC-44-09. November 10. Available at www.ceiops.org

CEIOPS. 2009f25. CEIOPS' Advice for L2 Implementing Measures on SII: SCR-Standard Formula—Operational Risk (former Consultation Paper no. 53). CEIOPS-DOC-45-09. November 10. Available at www.ceiops.org

CEIOPS. 2009f26. CEIOPS' Advice for L2 Implementing Measures on SII: SCR-Standard formula—Loss Absorbing Capacity of Technical Provisions (former Consultation Paper no. 54). CEIOPS-DOC-46-09. November 10. Available at www.ceiops.org

CEIOPS. 2009f27. CEIOPS' Advice for L2 Implementing Measures on SII: MCR Calculation (former Consultation Paper no. 55). CEIOPS-DOC-47-09. November 10. Available at www.ceiops.org

CEIOPS. 2009f28. CEIOPS' Advice for L2 Implementing Measures on SII: Tests and Standards for Internal Model Approval (former Consultation Paper no. 56). CEIOPS-DOC-48-09. November 10. Available at www.ceiops.org

CEIOPS. 2009f29. CEIOPS' Advice for L2 Implementing Measures on SII: Capital Add-ons (former Consultation Paper no. 57). CEIOPS-DOC-49-09. November 10. Available at www.ceiops.org

CEIOPS. 2009f30. CEIOPS' Advice for L2 Implementing Measures on SII: Supervisory Reporting and Disclosure (former Consultation Paper no. 58). CEIOPS-DOC-50-09. November 10. Available at www.ceiops.org

CEIOPS. 2009f31. CEIOPS' Advice for L2 Implementing Measures on SII: Remuneration (former Consultation Paper no. 59). CEIOPS-DOC-51-09. November 10. Available at www.ceiops.org

CEIOPS. 2009f32. CEIOPS' Advice for L2 Implementing Measures on SII: Group Solvency Assessment (former Consultation Paper no. 60). CEIOPS-DOC-52-09. November 10. Available at www.ceiops.org

CEIOPS. 2009f33. CEIOPS' Advice for L2 Implementing Measures on SII: Intra-Group Transactions and Risk Concentration (former Consultation Paper no. 61). CEIOPS-DOC-53-09. November 10. Available at www.ceiops.org

CEIOPS. 2009f34. CEIOPS' Advice for L2 Implementing Measures on SII: Cooperation and Colleges of Supervisors (former Consultation Paper no. 62). CEIOPS-DOC-54-09. November 10. Available at www.ceiops.org

CEIOPS. 2009f35. CEIOPS Feedback Statement on the 1st and 2nd Set of Solvency II Final L2 Advice to EC. CEIOPS-DOC-57-09. November 10. Available at www.ceiops.org

CEIOPS. 2009g. Draft L2 Advice on Technical Criteria for Assessing 3rd Country Equivalence in Relation to Articles 172, 227 and 260. CEIOPS-CP-78. November 30. Available at www.ceiops.org

CEIOPS. 2010a01. Repackagd Loans Investment (former Consultation Paper no. 63). CEIOPS-DOC-59-10. January 29. Available at www.ceiops.org

CEIOPS. 2010a02. CEIOPS' Advice for L2 Implementing Measures on SII: Extension of the Recovery Period (former Consultation Paper no. 64). CEIOPS-DOC-60-10. January 29. Available at www.ceiops.org

CEIOPS. 2010a03. CEIOPS' Advice for L2 Implementing Measures on SII: Partial Internal Models (former Consultation Paper no. 65). CEIOPS-DOC-61-10. January 29. Available at www.ceiops.org

CEIOPS. 2010a04. CEIOPS' Advice for L2 Implementing Measures on SII: Supervision of Group Solvency for Groups with Centralised Risk Management (former Consultation Paper no. 66). CEIOPS-DOC-62-10. January 29. Available at www.ceiops.org

CEIOPS. 2010a05. CEIOPS' Advice for L2 Implementing Measures on SII: Treatment of Participations (former Consultation Paper no. 67). CEIOPS-DOC-63-10. January 29. Available at www.ceiops.org

CEIOPS. 2010a06. CEIOPS' Advice for L2 Implementing Measures on SII: Ring-Fenced Funds (former Consultation Paper no. 68). CEIOPS-DOC-64-10. January 29. Available at www.ceiops.org

CEIOPS. 2010a07. CEIOPS' Advice for L2 Implementing Measures on SII: Equity Risk Sub-Module (former Consultation Paper no. 69). CEIOPS-DOC-65-10. January 29. Available at www.ceiops.org

CEIOPS. 2010a08. CEIOPS' Advice for L2 Implementing Measures on SII: SCR Standard Formula—Calibration of Market Risk Module (former Consultation Paper no. 70). CEIOPS-DOC-66-10. January 29. Available at www.ceiops.org

CEIOPS. 2010a09. CEIOPS' Advice for L2 Implementing Measures on SII: SCR Standard Formula—Correlations (former Consultation Paper no. 74). CEIOPS-DOC-70-10. January 29. Available at www.ceiops.org

CEIOPS. 2010a10. CEIOPS' Advice for L2 Implementing Measures on SII: SCR standard Formula—Undertaking-Specific Parameters (former Consultation Paper no. 75). CEIOPS-DOC-71-10. January 29. Available at www.ceiops.org

CEIOPS. 2010a11. CEIOPS' Advice for L2 Implementing Measures on SII: Technical Provisions—Simplified Methods and Techniques to Calculate Technical Provisions (former Consultation Paper no. 76). CEIOPS-DOC-72-10. January 29. Available at www.ceiops.org

CEIOPS. 2010a12. CEIOPS' Advice for L2 Implementing Measures on SII: SCR Standard Formula—Simplified Calculations in the Standard Formula (former Consultation Paper no. 77). CEIOPS-DOC-73-10. January 29. Available at www.ceiops.org

CEIOPS. 2010a13. CEIOPS' Advice for L2 Implementing Measures on SII: SCR Standard Formula—Simplifications/Specifications for Captives (former Consultation Paper no. 79). CEIOPS-DOC-74-10. January 29. Available at www.ceiops.org

CEIOPS-GC. 2008. Report on Proxies. CEIOPS-Groupe Consultatif Coordination Group. CEIOPS-DOC-27/08. July. Available at www.ceiops.org

Centeno, Lourdes. 1989. The Bühlmann-Straub Model with the premium calculated according to the variance principle. *Insurance: Mathematics and Economics*, Vol. 8, No. 1, pp. 3–10.

CFO. 2004. European Embedded Value Principles. CFO Forum. May 5. Available at http://www.cfoforum.nl

Charpentier, Arthur. 2003. Tail distribution and dependence measures. 34th ASTIN Colloquium, Berlin.

Chavez-Demoulin, Valérie, Paul Embrechts, and Johanna Nešlehová. 2006. Quantitative models for operational risk: Extremes, dependence and aggregation. *Journal of Banking and Finance*, Vol. 30, No. 10, pp. 2635–2658.

Chen, Re-Raw. 2003. *Credit Risk Modeling: A General Framework*. Rutgers University, New Brunswick, NJ. January.

Christoffersen, Peter and Sílva Goncalves. 2005. Estimation risk in financial risk management. *Journal of Risk*, Vol. 7, pp. 1–28.

CIA. 1996. *Dynamic Capital Adequacy Testing*. Educational Note. Committee of Solvency Standards for Financial Institutions, Canadian Institute of Actuaries, Ottawa, Canada. January.

Collard, Fabrice and Harris Dellas. 2004. *The Great Iinflation of the 1970s*. International Finance Discussion Papers, No. 799, April. Board of Governors of the Federal Reserve System. Available at www.federalreserve.gov/pubs/ifdp.

COM. 1998. Directive 98/78/EC of the European Parliament and of the council of 27 October 1998 on the supplementary supervision of insurance undertakings in an insurance group. *Official Journal* L 330, 12.5.98.

COM. 2002a. Regulation (EC) No 1606/2002 of the European Parliament and of the Council of 19 July 2002 on the application of international accounting standards. *Official Journal* L 243, 9.11.2002, pp. 0001–0004.

COM. 2002b. Directive 2002/87/EC of the European Parliament and of the Council of 16 December 2002 on the supplementary supervision of credit institutions, insurance undertakings and investment firms in a financial conglomerate and amending Council Directives 73/239/EEC, 79/267/EEC, 92/49/EEC, 92/96/EEC, 93/6/EEC and 93/22/EEC, and Directives 98/78/EC and 2000/12/EC of the European Parliament and of the Council. *Official Journal* L 35, 2.11. 2003.

COM. 2003. Directive 2003/41/EC of the European Parliament and of the Council of 3 June 2003 on the activities and supervision of institutions for occupational retirement provision. *Official Journal* L 235, 9.23.2003, pp. 0010–0021.

COM. 2006. Amended Framework for Consultation on Solvency II, MARKT/2515/06, April. European Commission, Internal Market and Services DG, Financial Institutions, Insurance and Pensions, Brussels.

COM. 2007a. Proposal for a Directive of the European Parliament and of the Council on the taking-up and pursuit of the business of Insurance and Reinsurance. SOLVENCY II.COM (2007) 361 final, 2007/0143 (COD), Brussels, July 19.

COM. 2007b. Letter from Jörgen Holmquist, European Commission, to the chairman of CEIOPS, Thomas Steffen, regarding further work on Solvency II, Markt/H2/BC/sv D(2007) 10830, Brussels, July 19. Available at http://ec.europa.eu/internal_market/insurance/solvency/index_en.htm

COM, 2008. Amended Proposal for a Directive of the European Parliament and of the Council on the taking up and pursuit of the business of Insurance and Reinsurance ("Solvency II"), COM (2008)119, Brussels. February 26.

Cont, R. 2006. Model uncertainty and its impact on the pricing of derivative instruments. *Mathematical Finance*, Vol. 16, No. 3 (July), 519–547.

Cornish, Edmond Alfred and Ronald A. Fisher. 1937. Moments and cumulants in the specification of distributions. *Revue de l'Institut International de Statistique*, Vol. 4. pp. 307–320.

COSO. 2004a. Enterprise Risk Management—Integrated Framework. Committee of Sponsoring Organizations of the Treadway Commission, September. Available at http://www.coso.org

COSO. 2004b. Enterprise Risk Management—Integrated Framework, Executive Summary, Committee of Sponsoring Organizations of the Treadway Commission, September. Available at http://www.coso.org

Coutts, Stewart M. and Russell Devitt. 1989. The assessment of the financial strength of insurance companies by a generalized cash flow model. In Cummins, J. David and Richard A. Derrig. 1989. *Financial Models of Insurance Solvency, Huebner International Series on Risk, Insurance and Economic Security*. Kluwer Academic Publishers, Boston, ISBN: 0-7923-9018-0; Volume II of the Proceedings of the First International Conference on Insurance Solvency 1986.

CRD. 2009. Capital Requirements Directive as adopted in May. European Parliament legislative resolution of 6 May 2009 on the proposal for a directive of the European Parliament and of the Council amending Directives 2006/48/EC and 2006/49/EC as regards banks affiliated to central institutions, certain own funds items, large exposures, supervisory arrangements, and crisis management [COM(2008)0602-C6–0339/2008-2008/0191(COD)]. Available at http://www.europarl.europa.eu

CRO. 2006. *A Market Cost of Capital Approach to Market Value Margins*. Discussion Paper. The Chief Risk Officer Forum, March 17. Available at http://www.croforum.org

CRO. 2008. *Liquidity Risk Management. Best Risk Management Practices*. The Chief Risk Officer Forum, October. Available at http://www.croforum.org

CRO. 2009a. *Operational Risk Management*. The Chief Risk Officer Forum, May. Available at http://www.croforum.org

CRO. 2009b. *Position on Liquidity Premium*. The Chief Risk Officer Forum and CFO Forum, October 15. Available at http://www.croforum.org

Crosbie, Peter and Jeff Bohn. 2003. *Modeling Default Risk*. December 18. Moody's KMV. Available at http://www.moodyskmv.com

Crouhy, Michel, Dan Galai, and Robert Mark. 2000. A comparative analysis of current credit risk models. *Journal of Banking & Finance*, Vol. 24, pp. 59–117.

Cruz, Marcelo, Editor. 2009. *The Solvency II Handbook. Developing ERM Frameworks in Insurance and Reinsurance Companies*. Risk Books, London, ISBN: 978-1-906348-19-9.

Cummins, J. David and Richard A. Derrig. 1988. *Classical Insurance Solvency Theory, Huebner International Series on Risk, Insurance and Economic Security*. Kluwer Academic Publishers, Boston, ISBN: 0-89838-272-6; Volume I of the Proceedings of the first International Conference on Insurance Solvency 1986.

Cummins, J. David and Richard A. Derrig. 1989. *Financial Models of Insurance Solvency, Huebner International Series on Risk, Insurance and Economic Security*. Kluwer Academic Publishers, Boston, ISBN: 0-7923-9018-0; Volume II of the Proceedings of the first International Conference on Insurance Solvency 1986.

Cummins, J. David and Richard A. Derrig. 1991. *Managing the Insolvency Risk of Insurance Companies, Huebner International Series on Risk, Insurance and Economic Security*. Kluwer Academic Publishers, Boston, ISBN: 0-7923-9152-7; Proceedings of the second International Conference on Insurance Solvency 1988.

Cummins, J. David and Richard D. Phillips. 2005. Estimating the cost of equity capital for property-liability insurers. *Journal of Risk and Insurance*, Vol. 72, No. 3, pp. 441–478.

Cummins, J. David, Martin F. Grace, and Richard D. Phillips. 1999. Regulatory solvency prediction in property-liability insurance: Risk-based capital, audit ratios and cash flow simulation. *Actuarial Research Clearing House*, Vol. 1, pp. 15–68.

Cuthbertson, Keith. 2004. Derivative securities. In Teugels, Jozef L. and Bjørn Sundt, Editors. 2004. *Encyclopaedia of Actuarial Science*. John Wiley & Sons, Chichester, ISBN: 0-470-84676-3.

Dahms, René. 2008. *A Loss Reserving Method for Incomplete Data*. Paper presented at the 38th ASTIN Colloquium, Manchester. July. Available at www.actuaries.org. To appear in *Journal of the Swiss Association of Actuaries*.

Dahms, René, Michael Merz, and Mario V. Wüthrich. 2008. Claims development result for combined claims incurred and claims paid data. Submitted to *Scandinavian Actuarial Journal*.

D'Arcy, Stephen P. 1988. *Financial Economics*. Paper presented to Seminar at the Institute of Actuaries, London, March.

D'Arcy, Stephen P. 1989. On becoming an actuary of the third kind. *Proceedings of the Casualty Actuarial Society*, Arlington, VA, pp. 45–76.

D'Arcy, Stephen P. 2009. Risk appetite. *Risk Management*, No. 15, March 2009. Society of Actuaries. Available at http://www.soa.org

D'Arcy, Stephen P. and Neil A. Doherty. 1988. *The Financial Theory of Pricing Property-Liability Insurance Contracts*. Huebner Foundation Monograph 15. Wharton School, University of Pennsylvania, Philadelphia, ISBN: 0-256-06124-6.

Das, Sanjiv R. and Paul Hanouna. 2009. Hedging credit: Equity liquidity matters. *Journal of Financial Intermediation*, Vol. 18, pp. 112–123.

Daul, Stéphane, Enrico De Giorgi, Filip Lindskog, and Alexande McNeil. 2003. The grouped t-copula with an application to credit risk. *Risk*, Vol. 16, No. 11, pp. 73–76.

Daykin, Chris D. 1984. *The Development of Concepts of Adequacy and Solvency in Non-Life Insurance in the EEC*. 22nd International Congress of Actuaries, Sydney, pp. 299–309.

Daykin, Chris D. and G.B. Hey. 1990. Managing uncertainty in a general insurance company. *Journal of the Institute of Actuaries*, Vol. 117, pp. 173–277.

Daykin, Chris D., G.D. Bernstein, S.M. Coutts, E.R.F. Devitt, G.B. Hey, D.I.W. Reynolds, and P.D. Smith. 1987. Assessing the solvency and financial strength of a general insurance company. *Journal of the Institute of Actuaries*, Vol. 114, pp. 227–310.

Daykin, Chris D., G.D. Bernstein, S.M. Coutts, E.R.F. Devitt, G.B. Hey, D.I.W. Reynolds, and P.D. Smith. 1989. The solvency of a general insurance company in terms of emerging costs. In Cummins, J. David and Richard A. Derrig. 1989. *Financial Models of Insurance Solvency, Huebner International Series on Risk, Insurance and Economic Security*. Kluwer Academic Publishers, Boston, ISBN: 0-7923-9018-0; Volume II of the Proceedings of the first International Conference on Insurance Solvency 1986.

Daykin, Chris D., E.R. Devitt, M.R. Kahn, and J.P. McCaughan. 1984. The solvency of general insurance companies. *Journal of the Institute of Actuaries*, Vol. 111, pp. 279–336.

Daykin, Chris D., Teivo Pentikäinen, and Martti Pesonen. 1994. *Practical Risk Theory for Actuaries*. Chapman and Hall, Monographs on Statistics and Applied Probability 53, London, ISBN: 0-412-42850-4.

DB. 2006. *Concentration in Credit Portfolios*. Deutsche Bundesbank. Monthly report. June. Available at http://www.bundesbank.de

De Felice, Massimo and Franco Moriconi. 2004. Market consistent valuation in life insurance measuring fair value and embedded options. *Giornale dell'Istituto Italiano degli Attuari*, Vol. LXVII, Roma, pp. 95–117.

De Felice, Massimo and Franco Moriconi. 2005. Market-based tools for managing the life insurance company. *ASTIN Bulletin*, Vol. 35, No. 1, pp. 95–111.

De Long, J. Bradford. 1995. *America's Only Peacetime Inflation: The 1970s.* University of California at Berkeley and National Bureau of Economic Research. December 19. Available at econ161.berkeley.edu/pdf_files/Peacetime_Inflation.pdf.

De Mori, Bruno. 1965. Possibilite d'etablir des bases techniques acceptables pour le calcul d'une marge minimum de solvabilite des enterprises d'assurances contre les dommages, *ASTIN Bulletin*, Vol. III, pp. 286–313.

Degen, Matthias, Paul Embrechts, and Dominik D. Lambrigger. 2007. The quantitative modeling of operational risk: Between g-and-h and EVT. *Astin Bulletin*, Vol. 37, No. 2, pp. 265–291.

Demarta, Stefano and Alexander McNeil. 2005. The t-copula and related copulas. *International Statistical Review*, Vol. 73. No. 1, pp. 111–129.

Dembo, Ron and Dan Rosen. 2000. The practice of portfolio replication. *Algo Research Quarterly*, Vol. 3, No. 2, September, pp. 11–22. Available at http://www.algorithmics.com. (A versio of the paper has also been published in the Annals of Operational Research, Vol. 85, 1999, pp. 267–284.)

Denuit, Michel and Jan Dhaene. 2003. *Simple Characterizations of Comonotonicity and Countermonotonicity by Extremal Correlations.* Report, Catholic University of Leuven, Belgium, March 19. *Belgian Actuarial Bulletin*, Vol. 3, pp. 22–27.

Denuit, Michel, Jan Dhaene, and Carmen Ribas. 1999. *Some Positive Dependence Notions, With Applications in Actuarial Sciences.* Research report 9942, ETEW KU Leuven. Available at www.kuleuven.be/insurance.

Denuit, Michel, Jan Dhaene, and Carmen Ribas. 2001. Does positive dependence between individual risks increase stop-loss premiums? *Insurance: Mathematics & Economics*, Vol. 28, No. 3, pp. 305–308.

Denuit, Michel, Jan Dhaene, Marc Goovaerts, and Rob Kaas. 2005. *Actuarial Theory for Dependent Risks: Measures, Orders and Models.* John Wiley & Sons, Chichester, ISBN: 0-470-01492-X.

Denuit, Michel, Jan Dhaene, Marc Goovaerts, Rob Kaas, and Roger Laeven. 2006. Risk measurement with the equivalent utility principles. In *Risk Measures: General Aspects and Applications*, Editor Ludger Rüschendorf, *Statistics and Decisions*, Vol. 24, No. 1, pp. 1–26.

Derrig, Richard A. and Krzysztof M. Ostaszeewski. 2004. Fuzzy set theory. In Teugels, Jozef L. and Bjørn Sundt, Editors. 2004. *Encyclopaedia of Actuarial Science.* John Wiley & Sons, Chichester, ISBN: 0-470-84676-3.

Desmedt, Stijn and Jean-Francois Walhin. 2008. On the subadditivity of tail value at risk: An investigation with copulas. *Variance*, Vol. 2, No. 2, pp. 231–252.

Dhaene, Jan, Andreas Tsanakas, Emiliano A. Valdez, and Steven Vanduffel. 2009. *Optimal Capital Allocation Principles.* Working Paper. Available at www.kuleuven.be/insurance (submitted).

Dhaene, Jan, Marc Goovaerts, M. Lundin, and Steven Vanduffel. 2005. Aggregating economic capital. *Belgian Actuarial Bulletin*, Vol. 5, pp. 14–25.

Dhaene, Jan, Michel Denuit, Marc J. Goovaerts, Rob Kaas, and David Vyncke. 2002a. The concept of comonotonicity in actuarial science and finance: Theory. *Insurance: Mathematics & Economics*, Vol. 31, pp. 3–33.

Dhaene, Jan, Michel Denuit, Marc J. Goovaerts, Rob Kaas, and David Vyncke. 2002b. The concept of comonotonicity in actuarial science and finance: Applications. *Insurance: Mathematics & Economics*, Vol. 31, pp. 133–161.

Dhaene, Jan, Steven Vanduffel, and Marc Goovaerts. 2007. *Comonotonicity.* Report, Catholic University of Leuven, Belgium, March 19.

Dhaene, Jan, Steven Vanduffel, Marc J. Goovaerts, Rob Kaas, Q. Tang, and David Vyncke. 2006. Risk measures and comonotonicity: A review. *Stochastic Models*, Vol. 22, pp. 573–606.

Dhaene, Jan, Roger J.A. Laeven, Steven Vanduffel, Grzegorz Darkiewicz, and Marc J. Goovaerts. 2008a. Can a coherent risk measure be too subadditive? The *Journal of Risk and Insurance*, Vol. 75, No. 2, pp. 365–386.

Dhaene, Jan, L. Henrard, Z. Landsman, A. Vandendorpe, and Steven Vanduffel. 2008c. Some results on the CTE-based capital allocation rule. *Insurance: Mathematics and Economics*, Vol. 42, pp. 855–863.

Dhaene, Jan, Michel Denuit, and Steven Vanduffel. 2009. Correlation order, merging and diversification. Catholic University of Leuven, Belgium, June 25. *Insurance: Mathematics and Economics*. Vol. 25, No. 3, pp. 325–332.

Djehiche, Boualem and Per Hörfelt. 2005. *Standard Approaches to Asset and Liability Risk. Scandinavian Actuarial Journal*, No. 5, pp. 377–400.

Doff, René. 2007. *Risk Management for Insurers. Risk Control, Economic Capital and Solvency II*. Risk Books, London, ISBN: 978-1-904339-79-3.

Donnelly, Catherine and Paul Embrechts. 2009. *The Devil is in the Tails: Actuarial Mathematics and the Subprime Mortgage Crisis*. Report, RiskLab, ETH Zürich, November 6.

Dowd, Kevin. 2004. Value-at-risk. In Teugels, Jozef L. and Bjørn Sundt, Editors. 2004. *Encyclopaedia of Actuarial Science*. John Wiley & Sons, Chichester, ISBN: 0-470-84676-3.

Dowd, Kevin and David Blake. 2006. *After VaR: The Theory, Estimation, and Insurance Applications of Quantile-Based Risk Measures*. CRIS Discussion Papers—2006.II, Centre for Risk & Insurance Studies, The University of Nottingham. *Journal of Risk and Insurance*, Vol. 73, No. 2, pp. 193–229.

Düllmann, Klaus and Nancy Masschelein. 2006. *A Tractable Model to Measure Sector Concentration Risk in Credit Portfolios*. Working Paper, Deutsche Bundesbank and National Bank of Belgium. October. Available at http://papers.ssrn.com/sol3/papers.cfm?abstract_id=944778

DUS. 2007. *Försäkrade i Sverige—dödlighet och livslängder. Prognoser 2007–2050*. The Research Council for Actuarial Science, Swedish Insurance Federation, Svenska Försäkringsföreningen, ISBN: 91-974960-4-9 (in Swedish, with an English summary).

Dvorak, Brian. 2008. *A Brief History of Active Credit Portfolio Management*. March 25. Moody's KMV. Available at http://www.moodyskmv.com

Dwyer, Douglas and Shisheng Qu. 2007. *EDF^{TM} 8.0 Model Enhancements*. January 27. Moody's KMV. Available at http://www.moodyskmv.com

EC. 1999. Solvency margin review. Commission Services' working document. Annex to DIV 9049 (06/99). XV/2025/99/Rev.1.

EC. 2000. *ART Market Study*. Final Report, Study Contract ETD/99/B5-3000/C/51, October 2. European Commission, Internal Market DG. Report by Tillinghast Towers-Perrin.

EEC. 1973. First Council Directive of 24 July 1973 on the coordination of laws, regulations and administrative provisions relating to the taking-up and pursuit of the business of direct insurance other than life assurance. *Official Journal of the European Communities*, No. L 228/3 (73/239/EEC).

EEC. 1979. First Council Directive of 5 March 1979 on the coordination of laws, regulations and administrative provisions relating to the taking-up and pursuit of the business of direct life assurance. *Official Journal of the European Communities*, No. L 63/1 (79/267/EEC).

EEC. 1983. Seventh Council Directive 83/349/EEC of 13 June 1983 based on Article 54 (3) (g) of the Treaty on consolidated accounts. *Official Journal of the European Communities*, No. L 193, 7.18.1983, pp. 0001–0017.

EEC. 1991. Council Directive 91/674/EEC of 19 December 1991 on the annual accounts and consolidated accounts of insurance undertaking. *Official Journal of the European Communities*, L 374, 12.31.91, pp. 0007–0031.

EEC. 1992. Council Directive 92/49/EEC of 18 June 1992 on the coordination of laws, regulations and administrative provisions relating to direct insurance other than life assurance and amending

Directives 73/239/EEC and 88/357/EEC (third non-life insurance Directive). *Official Journal of the European Communities*, L 228, 11.8.1992, p. 1–23.

EIOPC. 2006. Choice of a risk measure for supervisory purposes: Possible amendments to the Framework for Consultation. European Insurance and Occupational Pensions Committee. MARKT/2534/06-EN, November 2006. The European Commission, Available at http://ec.europa.eu/internal_market

Eling, Martin and Denis Toplek. 2009. Modeling and management of non-linear dependencies-copulas in dynamic financial analysis. The *Journal of Risk and Insurance*, Vol. 76, No. 3, pp. 651–681.

Elliott, Robert J. and P. Ekkehard Kopp. 2005. *Mathematics of Financial Markets*, 2nd edition. Springer Finance Textbook, New York, ISBN: 0-387-21292-2.

Embrechts, Paul. 2007. *Copulas: A personal View*. Department of Mathmatics, ETH, Zürich. Published in *Journal of Risk and Insurance*, 2009, Vol. 76, No. 3, pp. 639–650.

Embrechts, Paul, Alexander McNeil, and Daniel Straumann. 2002. Correlations and dependence in risk management: Properties and pitfalls. In *Risk management: Value at risk and beyond*, Editor M. Dempster, Cambridge University Press, Cambridge, pp. 176–223. (A shorter version of this paper was presented at the XXX ASTIN meeting in Tokyo 1999.)

Embrechts, Paul, Filip Lindskog, and Alexander McNeil. 2003. Modelling dependence with copulas and applications to risk management. In *Handbook of Heavy Tailed Distributions in Finance*, edited by S. Racher, Elsevier, Amsterdam. Chapter 8, pp. 329–384.

Embrechts, Paul, Hansjörg Furrer, and Roger Kaufmann. 2008. Different kinds of risk. In *Handbook of Financial Time Series*, Editors T.G. Andersen, R.A. Davies, J.-P. Kreiss, and T. Mikosch, Springer Verlag, Berlin/Heidelberg, ISBN: 978-3-540-71296-1.

Embrechts, Paul, Roger Kaufmann, and Gennady Samorodnitsky. 2004. *Ruin Theory Revisited: Stochastic Models for Operational Risk*. Risk Management for Central Bank Foreign Reserves, ECB, Frankfurt, pp. 243–261. Web-book: http://www.ecb.int/pub/pdf/other/riskmanagecbreserves2004en.pdf

Embrechts, P. and Giovanni Puccetti. 2006. Aggregating risk capital, with an application to operational risk. *The Geneva Risk and Insurance Review*, Vol. 31, No. 2, pp. 71–90.

Engel, Robert. 2001. GARCH 101: The use of ARCH/GARCH models in applied econometrics. *Journal of Economic Perspective*, Vol. 15, No. 4, pp. 157–168.

Eom, Young Ho, Jean Helwege, and Jing-zhi Huang. 2002. *Structural Models of Corporate Bond Pricing: An Empirical Analysis*. Working Paper no. 2000-16, Pennsylvania State University, University Park, PA. First draft March 1, 2000. Last draft February 8, 2002.

EP. 2008. Report on the amended proposal for a directive of the European parliament and of the Council on the taking-up and pursuit of the business of Insurance and Reinsurance (recast). [COM(2008)0119-C6-0231/2007-2007/0143(COD)]. Committee on Economic and Monetary Affairs. European Parliament. A6-9999/2008. 10.16.2008.

Erling, Martin, and Thomas Parnitzke. 2007. Dynamic financial analysis: Classification, conception, and implementation. *Risk Management and Insurance Review*, Vol. 10, No. 1, pp. 33–50.

Ernst and Young. 2008. *The Meaning of Market Consistency in Europe*. Ernst & Young report. Available at www.ey.com/solvencyii

Esscher and Fredrik. 1932. On the probability function in the collective theory of risk. *Skandinavisk Aktuarietidskrift*, Vol. XV, pp. 175–195.

EU Commission. 2006. Amended Framework for Consultation on Solvency II, MARKT/2515/06, April. European Commission, Internal Market and Services DG, Financial Institutions, Insurance and Pensions, Brussels.

EU Commission. 2007. Proposal for a Directive of the European Parliament and of the Council on the Taking-Up and pursuit of the Business of Insurance and Reinsurance. Solvency II. COM(2008) 361 final, 2007/0143 (COD), {SEC(2007) 870}, {SEC(2007) 871}, Brussels, 7.10.2007.

Faltinsen, Stig and Ellinor Samuelsson. 2009a. Position Paper on Level Two Rules for the Directive on Solvency II—A Method Based on Macroeconomic Principles for the Valuation of Technical Reserves When No True Market Price Exists. Swedish Insurance Federation & Finansnæringens Hovedorganisasjon. June 5. Available at www.forsakringsforbundet.se.

Faltinsen, Stig and Ellinor Samuelsson. 2009b. Macroeconomic extrapolation. Swedish Insurance Federation & Finansnæringens Hovedorganisasjon. September 9. Available at www.forsakringsforbundet.se.

Feldblum, Sholom. 1992. *European Approaches to Insurance Solvency*. Casualty of Actuarial Society Discussion Paper Program, Arlington, VA, pp. 511–552.

Ferguson, Ronald E. 1983. Duration. *Proceeding of the Casualty of Actuaries*, Vol. LXX, pp. 265–288.

FFSA. 2003. *Standard Approach Drafted by the Chairman Jukka Rantala: Extension Proposal for Asset Liability Risk*. Working Paper CEA, Féderation Francaise des Sociétés d'Assurance, Paris. 7.30.03.

Field, Peter, Introducer. 2003. *Modern Risk Management: A History*. Risk Books, London, ISBN: 1-904-339-05-0.

Finger, Christopher C. 1999. Conditional approaches for CreditMetrics portfolio distributions. *CreditMetrics Monitor*, April, pp. 14–26.

Finger, Christopher C. 2001. The one-factor CreditMetrics model in the New Basel Capital Accord. *RiskMetrics Journal*, Vol. 2, No. 1, pp. 9–18.

Finkelstein, Gary. 2005. *Exotic Options. Through 100 years to Current State of the Art Risk Management of Investment Guarantees*. Presentation at the XV AFIR Colloquium in Zürich, September 7–9, Milliman Consultants and Actuaries.

Fisher, Lawrence. 2005. *Yield Elasticity: A New, Objective Measure of Interest-Rate Risk*. Working Paper presented at the Financial Management Association International, Chicago, October 14.

Fisher, Ronald A. and E.A. Cornish. 1960. The percentile points of distributions having known cumulants. *Technometrics*, Vol. 2, No. 2, pp. 209–225.

Fisher, Jeff and David Geltner. 2000. *Quarterly Unsmoothing of the NCREIF Index Without Assuming an Efficient Market: A Transactions-Based Version of the NCREIF Index*. Working Paper, January. Published in Real Estate Finance, Bloomington, IN, and Cincinnati, OH. 2000. Available at http://www.ncreif.com/download.aspx?file=Quartely_Unsmoothing_of

Folpmers, Marco. 2008a. A practical guide to measuring operational risk using subjective data through copulas and scenario analysis. *The Journal of Operational Risk*, Vol. 3, No. 3, Fall, pp. 63–74.

Folpmers, Marco. 2008b. Operational risk according to solvency II: SCR, Pillar 2 and copula's. *AENORM*, Magazine 61, November. Available at http://www.aenorm.eu

Frees, Edward W. and Emiliano A. Valdez. 1998. Understanding relationships using copulas. *North American Actuarial Journal (NAAJ)*, Vol. 2, No. 1, pp. 1–25. Discussions by Christian Genest, Kilani Ghoudi, and Louis-Paul Rivest, pp. 143–149, in *NAAJ*, Vol. 2, No. 3, and by Shaun S. Wang, pp. 137–141, in *NAAJ*, Vol. 3, No. 1.

Frey, Rüdiger, Alexander J. McNeil, and Mark A. Nyfeler. 2001. Copulas and credit models. *Risk*, October, pp. 111–114. Also published in *Theory and Practice of Credit Risk Modelling*. Editor Alexander Lipton. Risk Books, 2008, pp. 103–114, ISBN: 978-1-904339-64-9 and in *Credit Risk Modelling*, Technical Papers published in *Risk* 1999–2003. Editor Michael B. Gordy. Risk Books, 2004, pp. 109–114, ISBN: 978-1-904339-08-5.

Furrer, H. 2004. *On the Calculation of the Risk Margin Within the SST*. Working Paper, Swiss Life, Zûrich, August.

GC. 2006. Solvency II: Risk Margin Comparison, a letter from Groupe Consultatif to CEIOPS, February 13.

GC. 2008. *Valuation of Best Estimate under Solvency II for Non-life Insurance*. Interim Report, Groupe Consultatif, November 11. Available at http://www.gcactuaries.org

GC. 2009. *Valuation of Best Estimate for Non-life Insurance. Current Practice—A Country Comparison*. Groupe Consultatif, February 4. Available at http://www.gcactuaries.org

GDV. 2005. *Discussion Paper for a Solvency II Compatible Standard Approach (Pillar I) Model Description*. Version 1.0, December 1. GDV, Berlin. Available at http://www.gdv.de/Downloads/English/Documentation_Sol_II.pdf

GDV. 2009. *Position Paper on the Determination of the Risk-Free Interest Rate Term Structure Under Solvency II*. November, GDV, Berlin. Available at http://www.gdv.de

Genest, Christian and Anne-Catherine Favre. 2007. Everything you always wanted to know about copula modeling but were afraid to ask. *Journal of Hydrologic Engineering*, July/August, pp. 347–368.

Genest, Christian and Johanna Nešlehova. 2007. A primer on copulas for count data. *ASTIN Bulletin*, Vol. 37, No. 2, pp. 475–515.

Genest, Christian and Louis-Paul Rivest. 1993. Statistical inference procedures for bivariate archimedean copulas. *JASA*, Vol. 88, No. 423, pp. 1034–1043.

Genest, Christian, Bruno Rémillard, and David Beaudoin. 2009. Goodness-of-fit test for copulas: A review and a power study. *Insurance: Mathematics and Economics*, Vol. 44, No. 2, pp. 199–213.

Genest, Christian, Jean-Francois Quessy, and Bruno Rémillard. 2006. Goodness-of-fit procedures for copula models based on the probability integral transformation. *Scandinavian Journal of Statistics*, Vol. 33, pp. 337–366.

Giamouridis, Daniel. 2006. Estimation risk in financial risk management: A correction. *Journal of Risk*, Vol. 8, pp. 121–125.

Gielens, Geert. 2004. Credit risk. In Teugels, Jozef L. and Bjørn Sundt, Editors. 2004. *Encyclopaedia of Actuarial Science*. John Wiley & Sons, Chichester, ISBN: 0-470-84676-3.

GISG. 1987. *Report on Solvency Working Party*. Institute of Actuaries, General Insurance Study Group, Torquay, October 29–30.

Gisler, Alois. 2009. *The Insurance Risk in the SST and in Solvency II: Modelling and Parameter Estimation*. Paper presented at the 39th ASTIN Colloquium, Helsinki, June. Available at www.actuaries.org

Gordy, Michael B. 2000. A comparative anatomy of credit risk models. *Journal of Banking & Finance*, Vol. 24, No. 1, pp. 119–149.

Gordy, Michael B. 2003. A risk-factor model foundation for ratings-based bank capital rules. *Journal of Financial Intermediation*, Vol. 12, pp. 199–232.

Gordy, Michael B. 2004. Model foundations for the supervisory formula approach, In *Structured Credit Products: Pricing, Rating, Risk Management and Basel II*, Editor William Perraudin. Risk Books, London.

Groupe Consultatif. 2005. *Diversification*. Technical Paper, Version 3.1, Oxford, October 17 (written by Henk van Broekhoven).

Gupton, Greg M., Christopher C. Finger, and Mickey Bhatia. 1997. *CreditMetrics*. Technical Document, J.P. Morgan & Co. Incorporated. Available at http://www.riskmetrics.com

Haaf, Hermann and Dirk Tasche. 2002. *Calculating Value-at-Risk Contributions in CreditRisk+*. Report. February 28. Available at http://arxiv.org/abs/cond-mat/0112045

Hamerle, Alfred, Thilo Liebig, and Daniel Rösch. 2003. *Credit Risk Factor Modeling and the Basel II IRB Approach*. Deutsche Bundesbank. Discussion Paper, Series 2: Banking and Financial Supervision, Frankfurt am Main. No. 02/2003.

Hardie, A. C., Limb, A. P., Loades, D. H., Lumsden, I. C., Mason, D. C., Pollock, G., Robertson, E. S., Scott, W. F., and Wilkie, A. D. 1984. *The Solvency of Life Assurance Companies*. Paper presented to the Faculty of Actuaries, Working Party on Solvency, Edinburgh. October.

Hardy, Mary R. 2004. Dynamic financial modeling of an insurance enterprise. In Teugels, Jozef L. and Bjørn Sundt, Editors. 2004. *Encyclopaedia of Actuarial Science*. John Wiley & Sons, Chichester, ISBN: 0-470-84676-3.

Haynes, A.T. and R.J. Kirton. 1952. The financial structure of a life office. *Transactions of the Faculty of Actuaries*, Vol. 21, pp. 141–218.

Henderson, Vicky. 2004. Black-Scholes model. In Teugels, Jozef L. and Bjørn Sundt, Editors. 2004. *Encyclopaedia of Actuarial Science*. John Wiley & Sons, Chichester, ISBN: 0-470-84676-3.

Hershbarger, Robert, and Ran BarNiv. 1991. Classifying financial distress in the life insurance industry, Chapter 3. In Cummins, J. David, and Richard A. Derrig. Editors 1991. *Managing the Insolvency Risk of Insurance Companies, Huebner International Series on Risk, Insurance and Economic Security*. Kluwer Academic Publishers, Boston, ISBN: 0-7923-9152-7; Proceedings of the second International Conference on Insurance Solvency 1988.

Hibbert, John. 2009. Liquidity premium: Myth or reality? Insight. Barrie and Hibbert, September. Available at www.barrhibb.com

Hibbert, John, Axel Kirchner, Gavin Kretzschmar, Ruosha Li, and Alexander McNeil. 2009. *Liquidity Premium—Literature Review of Theoretical and Empirical Evidence*. Research Report. Barrie and Hibbert, September, Version 1.1. Available at www.barrhibb.com

Hovinen, Esa. 1969. Procedures and basic statistics to be used in magnitude control of equalization reserves in Finland. *ASTIN Bulletin*, Vol. 5, pp. 227–238.

HS Treasury. 2006. *Supervising Insurance Groups Under Solvency II*. Discussion Paper, November, HM Treasury and FSA, UK.

Hughston, Lane, Editor. 1999. *Options: Classic Approaches to Pricing and Modelling*. Risk Books, London, ISBN: 1-899332-66-9.

Hull, John, Izzy Nelken, and Alan White. 2004. *Merton's Model, Credit Risk, and Volatility Skews*. University of Toronto and Super Computing Consulting group, Chicago, January. Available at http://papers.ssrn.com/sol3/papers.cfm?abstract_id=707603

Hürlimann, Werner. 2004. Distortion risk measures and economic capital. *North American Actuarial Journal*, Vol. 8, No. 1, pp. 86–95.

Hürlimann, Werner. 2008a. *On the Non-Life Solvency II Model*. Paper presented at the 38th ASTIN Colloquium, Manchester, July. Available at www.actuaries.org

Hürlimann, Werner. 2008b. Solvency II: Reinsurance counterparty default risk. *Life & Pensions*, December, pp. 39–44. Available at http://www.life-pensions.com

Hürlimann, Werner. 2009. *Optimization of the Non-Life Insurance Risk Diversification in Solvency II*. Paper presented at the 39th ASTIN Colloquium in Helsinki, June. Available at www.actuaries.org

IAA. 2004. *A Global Framework for Insurer Solvency Assessment*. International Actuarial Association, Ontario, ISBN: 0-9733449-0-3.

IAA. 2005. *Current Estimates under International Financial Reporting Standards IFRS*. International Actuarial Standard of Practice IASP 5, June 16, International Actuarial Association, Ontario.

IAA. 2006. A Comment on the Draft second set of IAIS Observations, Issues arising as a result of the IASB's Insurance Contracts Project—Phase II, Released by the International Association of Insurance Supervisors: April 27, International Actuarial Association, Ontario.

IAA. 2009a. *Note on Enterprise Risk Management for Capital and Solvency Purposes in the Insurance Industry*. International Actuarial Association, Ontario, March 31. An International Actuarial Research Paper, ISBN: 978-0-9812787-4-2.

IAA. 2009b. *Measurement of Liabilities for Insurance Contracts: Current Estimates and Risk Margins*. Prepared by the ad hoc Risk Margin Working Group, International Actuarial Association, Ontario, April 15. An International Actuarial Research Paper, ISBN: 978-0-9812787-0-4.

IAIS. 2004. *Guidance Paper on Investment Risk Management*. IAIS, Basel. October

IAIS. 2005. *A New Framework for Insurance Supervision: Towards a Common Structure and Common Standards for the Assessment of Insurer Solvency*. International Association of Insurance Supervisors, Basel. October.

IAIS. 2006. *Issues arising as a result of the IASB's Insurance Contracts Project—Phase II*. Second Set of IAIS Observations, May. International Association of Insurance Supervisors, Basel.

IAIS. 2007a. *The IAIS Common Structure for the Assessment of Insurer Solvency*. International Association of Insurance Supervisors, Basel. February 14.

IAIS. 2007b. *Guidance Paper on Enterprise Risk Management for Capital Adequacy and Solvency Purposes*. International Association of Insurance Supervisors, Basel. Guidance Paper No. 2.2.6, October.

IAIS. 2007c. *Guidance Paper on the Use of Internal Models for Risk and Capital Management Purposes by Insurers*. International Association of Insurance Supervisors, Basel. Guidance Paper No. 2.2.7, October.

IAIS. 2007d. *Guidance Paper on the Structure of Regulatory Capital Requirements*. International Association of Insurance Supervisors, Basel. Guidance Paper No. 2.2.1, October.

IAIS. 2007e. *The IAIS Common Structure for the Assessment of Insurer Solvency*. International Association of Insurance Supervisors, Basel. February 14.

IAIS. 2007f. *Summary of IAIS Positions on the Valuation of Technical Provisions*. International Association of Insurance Supervisors, Basel. October.

IAIS. 2008a. *Standard on Enterprise Risk Management for Capital Adequacy and Solvency Purposes*. International Association of Insurance Supervisors, Basel. Approved October, Standard No. 2.2.6.

IAIS. 2008b. *Guidance Paper on Enterprise Risk Management for Capital Adequacy and Solvency Purposes*. International Association of Insurance Supervisors, Basel. Approved October, Guidance Paper No. 2.2.5.

IAIS. 2008c. *Standard on the Structure of Regulatory Capital Requirements*. International Association of Insurance Supervisors, Basel, Approved October, Standard No. 2.2.1.

IAIS. 2008d. *Guidance Paper on the Structure of Regulatory Capital Requirements*. International Association of Insurance Supervisors, Basel, Approved October, Guidance Paper No. 2.2.1.

IASB. 2001. *Draft Statement of Principles (DSOP) on Insurance Contracts*. IASB, November 16. Available at www.iasb.org

IASB. 2004. *IFRS 4 Insurance Contracts*. IASB, ISBN: 1-904230-49-0 (including (i) IFRS 4 Insurance Contracts, ISBN: 1-904230-50-4, (ii) Basis for Conclusions, ISBN: 1-904230-51-2, (iii) Implementation Guidance, ISBN 1-904230-52-0). Available at www.iasb.org

IASB. 2006. *Fair Value Measurement. Part 1: Invitation to Comment and Relevant IFRS Guidance, Part 2: SFAS 157 Fair Value Measurement*. Discussion Paper. International Accounting Standard Board, ISBN: 1-905590-21-0. Available at www.iasb.org

IASB. 2007a. *Preliminary Views on Insurance Contracts. Part 1: Invitation to Comment and Main Text*. International Accounting Standard Board, ISBN: 978-1-905590-33-9. Available at www.iasb.org

IASB. 2007b. *Preliminary Views on Insurance Contracts. Part 2: Appendices*. International Accounting Standard Board, ISBN: 978-1-905590-34-6.

IASB. 2008a. *Fulfillment Value* (Agenda paper 14A). Board meeting September 18, 2008, London. Insurance Contracts, International Accounting Standard Board. Available at www.iasb.org

IASB. 2008b. *Comparison Between Current Exit Value, Current Fulfillment Value and Value in Use* (Agenda paper 14B). Board meeting September 18, 2008, London. Insurance Contracts, International Accounting Standard Board. Available at www.iasb.org

IASB. 2008c. *Education Session: Candidate Measurement Approaches* (Agenda paper 3B). Board meeting October 2008, London. Insurance Contracts, International Accounting Standard Board. Available at www.iasb.org

IASB. 2009a. *Candidate Measurement Approaches—Tabular Comparison* (Agenda paper 10E). Board meeting February 2009, London. Insurance Contracts, International Accounting Standard Board. Available at www.iasb.org

IASB. 2009b. *Margins* (Agenda reference 5A). Board meeting April 2009, London. Insurance Contracts, International Accounting Standard Board. Available at www.iasb.org

IASB. 2009c. *Candidate Measurement Approaches—Tabular Comparison* (Agenda reference 6A). Insurance Contracts WG June 2009, London. Insurance Contracts, International Accounting Standard Board. Available at www.iasb.org

IASB. 2009d. Financial Instruments: Classification and Measurement. Exposure Draft ED/2009/7. International Accounting Standard Board. Available at www.iasb.org

IASB. 2009e. *Measurement Approach for Insurance Contracts* (Agenda reference 11A). IASB/FASB Meeting July 2009, London. Insurance Contracts, International Accounting Standard Board. Available at www.iasb.org

IASB. 2009f. *Candidate Measurement Approaches—Tabular Comparison* (Agenda reference 11C). IASB/FASB Meeting July 2009, London. Insurance Contracts, International Accounting Standard Board. Available at www.iasb.org

IASB. 2009g. IASB Update. September. International Accounting Standard Board. Available at www.iasb.org

IASB. 2009h. IFRS 9 Financial Instruments. Near final draft. November. International Accounting Standard Board. Available at www.iasb.org

IASB. 2009i. Financial Instruments: Amortised Cost and Impairment. Exposure Draft ED/2009/12 and Basis for Conclusions. November. International Accounting Standard Board. Available at www.iasb.org

Ingram, David. 2006. Standard and Poor's enterprise risk management evaluation of insurers. *Risk Management*, No. 7, March, CAS/SOA/CIA Risk Management Section, Available at http://www.soa.org

Jarvis, S., F. Southall, and E. Varnell. 2001. *Modern Valuation Techniques*. Paper presented to the Staple Inn Actuarial Society, February 6, Available at http://www.sias.org.uk

Johnson, Norman L. and Samuel Kotz. 1970. *Distributions in Statistics: Continuous Univariate Distributions—1*. John Wiley & Sons, New York, ISBN: 0-471-44626-2.

Joint Forum. 2001. Risk Management Practices and Regulatory Capital, Cross-Sectoral Comparison. November. Bank for International Settlements. The Joint Forum. Available at www.bis.org/publ/joint04.htm.

Joint Forum. 2006. The Management of Liquidity Risk in Financial Groups. May. Bank for International Settlements. The Joint Forum. Available at http://www.bis.org/publ/joint16.htm

Jones, Bruce L. and Rièardas Zitikis. 2003. Empirical estimation of risk measures and related quantities. *North American Actuarial Journal*, Vol. 7. No. 4. pp. 44–54.

Jonk, Carlo. 2008. Comparing approximations for sums of dependent random variables. *AENORM*, Magazine 60, July. Available at http://www.aenorm.eu

JPMorgan. 2006. JPMorgan Liability-Driven Investment (LDI) Survey. JPMorgan Asset Management. Available at http://www.jpmorgan.com/pages/jpmorgan/am

JTF. 2006. *Draft Report on the Comparison of the Sectoral Rules for the Eligibility of Capital Instruments*. Joint Task Force on Capital, Basel. JTF/06/06, October 13, CEBS and CEIOPS.

JRMS. 2008. *Risk Management: The Current Financial Crisis, Lessons Learned and Future Implications*. Essays presented by the Joint Risk Management Section of the Society of Actuaries, Casualty Actuarial Society and Canadian Institute of Actuaries. Available at http://www.soa.org/library/essays/rm-essay-2008-toc.aspx

Kahane, Yehuda. 1979. Solidity, leverage and the regulation on insurance companies. *The Geneva Papers on Risk and Insurance*, No. 14, December, pp. 3–19.

Kahane, Yehuda, Charles S. Tapiro, and Laurent Jacques. 1989. Concepts and trends in the study of insurer's solvency. In Cummins, J. David and Richard A. Derrig. Editors. 1989. *Financial Models of Insurance Solvency, Huebner International Series on Risk, Insurance and Economic Security*. Kluwer Academic Publishers, Boston, ISBN: 0-7923-9018-0; Volume II of the Proceedings of the first International Conference on Insurance Solvency 1986.

Kastelijn, W.M. and J.C.M. Remmerswaal. 1986. *Solvency, Surveys of Actuarial Studies 3*. Nationale-Nederlanden N.V., Rotterdam, The Netherlands.

Kaufmann, Roger and Pierre Patie. 2003. *Strategic Long-Term Financial Risks: The One-Dimensional Case*. Research Report, RiskLab, ETH Zürich, December 2.

Kendall, Sir Maurice and Alan Stuart. 1977. The Advanced Theory of Statistics. Distribution Theory. Volume 1, Fourth Edition. Charles Griffin & Company Ltd, London & High Wycombe.

Klugman, Stuart A. and Rahul Parsa. 1999. Fitting bivariate loss distributions with copulas. *Insurance: Mathematics and Economics*, Vol. 24, pp. 139–148.

Koopmans, T.C. 1942. *The Risk of Interest Fluctuations in Life Insurance Companies.* Penn Mutual Life Insurance Company, Philadelphia.

Kotz, S., N.L. Johnson, and C.B. Read. 1985. *Encyclopedia of Statistical Sciences.* John Wiley & Sons, New York, ISBN: 0-471-05558-2 (v.5).

KPMG. 2002. Study into the methodologies to assess the overall financial position of an insurance undertaking from the perspective of prudential supervision. Contract no: ETD/2000/BS-3001/C/45, May.

Kriele, Marcus and Jochen Wolf. 2007. On market value margin and cost of capital. *Blätter der Deutschen Gesellschaft für Versicherungsmathematik*, Vol. 28, No. 2, pp. 195–219.

Kuritzkes, Andrew, Til Schuermann, and Scott M. Weiner. 2003. *Risk Measurement, Risk Management and Capital Adequacy in Financial Conglomerates.* Wharton Working Paper, April 22. Printed in Herring, R. and R. Litan. Editors. Brookings-Wharton Papers in Financial Services.

Kwok, Yue-Kuen. 1998. *Mathematical Models of Financial Derivatives.* Springer Finance, Springer-Verlag, Singapore, ISBN: 981-3083-255.

Lambrigger, Dominik D., Pavel V. Shevchenko, and Mario V. Wüthrich. 2007. The quantification of operational risk using internal data, relevant external data and expert opinion. *The Journal of Operational Risk*, Vol. 2, No. 3, pp. 3–27.

Lambrigger, Dominik D., Pavel V. Shevchenko, and Mario V. Wüthrich. 2008. Give credit where credit is due: Operational risk goes Bayesian. *Bulletin Français d'Actuariat*, Vol. 8, No. 16, pp. 4–13.

Landsman, Z. and E.A. Valdez. 2003. Tail conditional expectations for elliptical distributions. *North American Actuarial Journal*, Vol. 7, No. 4, October, pp. 55–71.

Large, Andrew. 2004. Why we should worry about liquidity. *Financial Times*, November 11, p. 15.

de Larosière, Jacques. 2009. *The High-Level Group on Financial Supervision in the EU.* Report, February 25. Available at http://ec.europa.eu

Le Saout, Erwan. 2002. *Incorporating Liquidity Risk in VaR Models.* Université de Rennes, June. Available at www.gloriamundi.org/.

Lechkar, Mohamed and Dennis van Welie. 2008. Putting a dampener on Solvency II. *Life & Pensions*, November, pp. 36–40. Available at http://www.life-pensions.com

Leibowitz, Martin L. 1983. How financial theory evolves into the real world—Or not: The case of duration and immunization. *The Financial Review*, Vol. 18, No. 4, pp. 271–280.

Leigh, Julian. 2004. *Fair Value Accounting: Implications for General Insurance.* Paper presented to the 2003 GIRO Convention, The Staple Inn Actuarial Society. Available at www.sias.org.uk/data/papers/FairValueAccounting/DownloadPDF

Lelyveld, Iman van, Editor. 2006. *Economic Capital Modelling: Concepts, Measurement and Implementation.* Risk Books, London, ISBN: 1-904339-39-5.

Lowe, Stephen P. and James N. Standard. 1997. An integrated dynamic financial analysis and decision support system for a property catastrophe reinsurer. *ASTIN Bulletin*, Vol. 27, No. 2, pp. 339–371.

Luder, Thomas. 2005. *Swiss Solvency Test in Non-Life Insurance.* Working Paper, Federal Office of Private Insurance, Bern, July 28.

Luenberger, David G. 1998. *Investment Science.* Oxford University Press, Oxford, New York, ISBN: 0-19-510809-4.

Lütkebohmert, Eva. 2008. *Concentration Risk in Credit Portfolio.* Springer Verlag, Heidelberg, ISBN: 978-3-540-70869-8, EAA Lecture Notes.

Macaulay, Frederick R. 1938. *Some Theoretical Problems Suggested by the Movement of Interest Rates, Bond Yield, and Stock Prices in the United States Since 1856.* National Bureau of Economic Research, New York.

Macdonald, Angus. 1994. A note on life office models. *Transactions of the Faculty of Actuaries*, Vol. 44, pp. 64–72.

Macdonald, Angus. 1997. Current actuarial modeling practice and related issues and questions. *North American Actuarial Journal*, Vol. 1, No. 3, July, pp. 24–37.

Macdonald, Angus. 2004. Model office. In Teugels, Jozef L. and Bjørn Sundt, Editors. 2004. *Encyclopaedia of Actuarial Science*. John Wiley & Sons, Chichester, ISBN: 0-470-84676-3.

Madrid-report. 2003. *Internal Control for Insurance Undertakings*. CEIOPS, December. Available at www.ceiops.org

Malamud, Semyon, Eugene Trubowitz, and Mario V. Wüthrich. 2007. *Market Consistent Pricing of Insurance Products*. Research Report, ETH Zürich, Department of Mathematics, Zürich, Switzerland.

Manganelli, Simone and Robert F. Engle. 2001. *Value at Risk Models in Finance*. Working Paper No. 75, European Central Bank, Frankfurt am Main. August.

Manghetti-report. 2001. *Technical Provisions in Non-Life Insurance*. Conference of Insurance Supervisory Authorities of the Member States of the European Union (now CEIOPS). DT/I/223/00.Rev 2 (The WG chaired by Giovanni Manghetti, Italy). Available at www.ceiops.org

Manistre, B. John and Geoffrey H. Hancock. 2005. Variance of the CTE estimator. *North American Actuarial Journal*, Vol. 9, No. 2. pp. 129–156.

Manistre, B. John and Geoffrey H. Hancock. 2008. Summary of "Variance of the CTE Estimator." *Risk Management*, No. 13, August, CAS/SOA/CIA Risk Management Section. Available at http://www.soa.org

Manley, Henry William. 1869. A comparison of the values of policies as found by means of the various tables of mortality and the different methods of valuation in use among actuaries. *Journal of the Institute of Actuaries*, Vol. 14, pp. 249–305.

MARKT. 1999. The review of the overall financial position of an insurance undertaking (Solvency II Review). MARKT/2095/99, EC Internal Market DG.

MARKT. 2001a. Note to the Solvency Subcommittee of the Insurance Committee, Solvency II: Presentation of the proposed work. MARKT/2027/01, EC DG Internal Market.

MARKT. 2001b. Note to the Solvency Subcommittee, Banking rules: Relevance for the insurance sector? MARKT/2056/01, EC DG Internal Market.

MARKT. 2001c. Note to the Solvency Subcommittee: Risk-based capital system. MARKT/2085/01, EC DG Internal Market.

MARKT. 2002a. Note to the Solvency Subcommittee, Risk models of insurance companies or groups. MARKT/2515/02, EC DG Internal Market.

MARKT. 2002b. Solvency II—Review of the work. MARKT/2518/02, EC DG Internal Market.

MARKT. 2002c. Discussion note to the members of the IC Solvency Subcommittee, Current and future solvency work in the IAIS and within the actuarial profession from a Solvency II point of view. MARKT/2520/02, EC DG Internal Market.

MARKT. 2002d. Note to the members of the IC Solvency Subcommittee, Considerations on the links between the Solvency 2 Project and the extension of the "Lamfalussy" approach to insurance regulation. MARKT/2519/02, EC DG Internal Market.

MARKT. 2002e. Report of the working group on life assurance to the IC Solvency Subcommittee, MARKT/2528/02, EC DG Internal Market.

MARKT. 2002f. Report of the working group on non-life insurance to the IC Solvency Subcommittee. MARKT/2529/02, EC DG Internal Market.

MARKT. 2002g. Solvency II: Review of work (November 2002). MARKT/2536/02, EC DG Internal Market.

MARKT. 2002h. Paper for the Solvency Subcommittee: Considerations on the design of a future prudential supervisory system. MARKT/2535/02, EC DG Internal Market.

MARKT. 2003a. Design of a future prudential supervisory system in the EU—Recommendations by the Commissions Services. MARKT/2509/03, EC DG Internal Market.

MARKT. 2003b. Reflections on the general outline of a framework directive and mandates for further technical work. MARKT/2539/03, EC DG Internal Market.

MARKT. 2003c. Note to the Solvency Subcommittee. Questionnaire concerning the paper MARKT/2535/02, January 9. MARKT/2500/03, EC DG Internal Market.

MARKT. 2003d. Solvency II: Orientation debate—Design of a future prudential supervisory system in the EU (Recommendations by the Commission Services). MARKT/2503/03 (Insurance Committee).

MARKT. 2003e. Draft Decision Points and Conclusions of the 32nd meeting of the Insurance Committee. Brussels, April 9. MARKT/2511/03 (Insurance Committee).

MARKT. 2003f. Solvency II. Update on developments and issues for consideration by the Insurance Committee. MARKT/2530/03 (Insurance Committee).

MARKT. 2004a. Organisation of work, discussion on pillar 1 work areas and suggestions of further work on pillar 2 for CEIOPS—Issues paper for the meeting of the IC Solvency Subcommittee on March 12, 2004. MARKT/2543/03, EC DG Internal Market.

MARKT. 2004a1. Framework for consultation on Solvency II. 7.14.2004. European Commission, Internal Market DG.

MARKT. 2004a2. Solvency II: Road Map for the Development of Future Work: Proposed Framework for Consultation and Proposed First Wave of Specific Calls for Advice from CEIOPS. MARKT/2506/04, EC Insurance Committee.

MARKT. 2004b. Note to the Members of the IC Solvency Subcommittee. The draft second wave Calls for Advice from CEIOPS and stakeholders consultation on Solvency II. MARKT/2515/04, October 11.

MARKT. 2004c. Solvency II: Proposed second wave of specific Calls for Advice from CEIOPS. MARKT/2519/04, November 2004 (Annex to MARKT/2506/04, Specific Calls for Advice from CEIOPS (Second Wave), November 15).

MARKT. 2005a. Solvency II: Consultative Document. Draft Specific Calls for Advice from CEIOPS (Third Wave). MARKT/2501/05, February.

MARKT. 2005b. Amended Framework for consultation on Solvency II. July. European Commission, Internal Market and Services DG.

MARKT. 2006a. Draft Minutes of the 2nd meeting of the European Insurance and Occupational Pensions Committee. Brussels, December 2, 2005. MARKT/2503/06. March 7. Available at http://ec.europa.eu/internal_market/insurance/committee_en.htm

MARKT. 2006b. Amended Framework for consultation on Solvency II. April. MARKT/2515/06, European Commission, Internal Market and Services DG.

MARKT. 2006c. Draft Minutes of the 3rd meeting of the European Insurance and Occupational Pensions Committee. Brussels, April 5. MARKT/2516/06, June. Available at http://ec.europa.eu/internal_market/insurance/committee_en.htm

MARKT. 2006d. Draft Minutes of the 4th meeting of the European Insurance and Occupational Pensions Committee. Brussels, July 5. MARKT/2526/06, August. Available at http://ec.europa.eu/internal_market/insurance/committee_en.htm

MARKT. 2007a. Draft Minutes of the 5th meeting of the European Insurance and Occupational Pensions Committee. Brussels, November 29, 2006. MARKT/2508/07, June. Available at http://ec.europa.eu/internal_market/insurance/committee_en.htm

MARKT. 2007b. Draft Minutes of the 7th meeting of the European Insurance and Occupational Pensions Committee. Brussels, July 4, 2007. MARKT/2514/07, November. Available at http://ec.europa.eu/internal_market/insurance/committee_en.htm

MARKT. 2008a. Draft Minutes of the 8th meeting of the European Insurance and Occupational Pensions Committee. Brussels, November 29, 2007. MARKT/2502/08, January. Available at http://ec.europa.eu/internal_market/insurance/committee_en.htm

MARKT. 2008b. Draft Minutes of the 10th meeting of the European Insurance and Occupational Pensions Committee. Brussels, June 27, 2008. Ares 34079, December. Available at http://ec.europa.eu/internal_market/insurance/committee_en.htm

Martellini, Lionel. 2006. *Managing Pension Assets: From Surplus Optimization to Liability-Driven Investment*. Research Report, EDHEC, EDHEC Risk and Asset Management Research Centre, March. Available at www.edhec-risk.com

Martin, Richard. 2004. *Credit Portfolio Modeling Handbook*. Credit Suisse First Boston, October 29. Available at http://www.csfb.com

McCune, E.D. and H.L. Gray. 1982. Cornish-Fisher and Edgeworth expansions. In *Encyclopedia of Statistical Sciences*, Vol. 2, Samuel Kotz and Norman L. Johnson (Editors-in-chief), John Wiley & Sons, New York, ISBN 0-471-05547-6.

McLeish, Don L. 2005. *Monte Carlo Simulation and Finance*. John Wiley & Sons, Hoboken, ISBN 978-0-471-67778-9 (0-471-67778-7).

McNeil, Alexander, Rüdiger Frey, and Paul Embrechts. 2005. *Quantitative Risk Management. Princeton University Press*. Princeton series in Finance, Princeton, ISBN: 0-691-12255-5.

Merton, Robert C. 1974. On the pricing of corporate debt: the risk structure of interest rates. *Journal of Finance*, Vol. 29, pp. 449–470. Also reprinted in Hughston (1999).

Merz, Michael and Mario V. Wüthrich. 2007. Prediction error of the expected claims development result in the chain ladder method. *Journal of the Swiss Association of Actuaries*, Hefte 1, pp. 117–136.

Merz, Michael and Mario V. Wüthrich. 2008. *Modeling the Claims Development Result for Solvency Purposes*. Presented at the 38th ASTIN Colloquium, Manchester, July (www.actuaries.org) and also at Casualty Actuarial Society E-Forum, Fall 2008. Available at http://www.casact.org/pubs/forum/08fforum

Meyers, Glenn. 2000. Coherent Measures of Risk. An Exposition for the Lay Actuary. CAS Forum, Dynamic Financial Analysis Call Papers, Arlington, VA. Summer 2000.

Meyers, Glenn. 2003. *The Economics of Capital Allocation*. Paper presented to the Thomas J. Bowles Symposium, April 10. Available at http://www.casact.org/pubs/forum/03fforum/03ff391.pdf

Meyers, Glenn. 2009. *Proxies*. Paper presented at the 39th ASTIN Colloquium, Helsinki, June. Available at www.actuaries.org

Møller, Thomas and Mogens Steffensen. 2007. *Market-Valuation Methods in Life and Pension Insurance*. Cambridge University Press, Cambridge, ISBN 978-0-521-86877-8.

Mourik, Teus. 2005. *Market Value Margin Versus Economic Capital*. Working Paper, KPMG, The Netherlands.

Müller Report. 1997. Report of the Working group "Solvency of Insurance Undertakings" set up by the Conference of the European Union Member States, DT/D/209/97. Available at www.ceiops.org

Münzner, Hans. 1963. Probleme der Konzentrationsmessung. *Allgemeines Statistisches Archiv*, Bd. 47, pp. 1–9.

NAIC. 2006. Life Risk-Based Capital (E) Working Group, Agenda items, Saturday, March 4, 2006, Disney Dolphin.

Nelsen, Roger B. 1999. *An Introduction to Copulas*. Lecture Notes in Statistics No. 139. Springer Verlag, New York, ISBN: 0-387-98623-5.

Neuhaus, Walther. 2004. Alternative risk transfer. In Teugels, Jozef L. and Bjørn Sundt, Editors. 2004. *Encyclopaedia of Actuarial Science*. John Wiley & Sons, Chichester, ISBN: 0-470-84676-3.

Norberg, Ragnar. 1986. A contribution to modelling of IBNR claims. *Scanidnavian Actuarial Journal*, 1986, No. 3–4, pp. 155–203.

Norberg, Ragnar. 1993. *A Solvency Study in Life Insurance.* III AFIR Colloquium, Rome, Italy, March 30–April 3.

Norberg, Ragnar and Bjørn Sundt. 1985. Draft of a system for solvency control in non-life insurance. *ASTIN Bulletin*, Vol. 15, No. 2, pp. 149–169.

Nygård, Fredrik and Arne Sandström. 1981. *Measuring Income Inequality.* Almqvist & Wiksell International, Stockholm, ISBN 91-22-00439-4.

Odeh, Robert E., Donald B. Owen, Z.W. Birnbaum, and Lloyd Fisher, 1977. *Pocket Book of Statistical Tables.* Marcel Dekker, Inc., New York and Basel, Statistics: Textbooks and Monographs, Vol. 22, ISBN 0-8247-6515-X.

Ogrodzki, Jerome. 2007. *Replicating Portfolios. Insurance and Financial Services Review, Europe.* November, Watson Wyatt. Available at https://www.watsonwyatt.com/europe/pubs/insurancefinancial/

Ohlsson, Esbjörn and Jan Lauzeningks. 2008. *The One-Year Non-Life Insurance Risk.* Paper presented at the 38th ASTIN Colloquium, Manchester, July. Available at www.actuaries.org

OJ. 2006. Directive 2006/48/EC of the European parliament and of the council of 14 June 2006 relating to the taking up and pursuit of the business of credit institutions (recast). *Official Journal of the European Union*, 6.30.2006, L/177. Available at http://eur-lex.europa.eu

OJ. 2009. Directive 2009/138/EC of the European parliament and of the council of 25 November 2009 on the taking-up and pursuit of the business of Insurance and Reinsurance (Solvency II) (recast). *Official Journal of the European Union*, 12.17.2009, L/335. Available at http://eur-lex.europa.eu

Owen, Gary and Jonathan Law, Editors, 2005. *Oxford Dictionary of Accounting*, 3rd edition. Oxford University Press, Oxford, ISBN: 0-19-280627-0.

Pallister, John and Alan Isaacs, Editors. 2003. *A Dictionary of Business*, 3rd edition. Oxford University Press, Oxford, ISBN: 0-19-860755-5.

Panjer, Harry H. Editor. 2001. *Financial Economics: With Applications to Investments, Insurance and Pensions.* The Actuarial Foundation, The Society of Actuaries, Schaumberg, IL, ISBN: 0-938959-48-4 (Authors: P.P. Boyle, S.H. Cox, D. Dufresne, H.U. Gerber, H.H. Mueller, H.W. Pedersen, S.R. Pliska, M. Sherris, E.S. Shiu, and K.S. Tan).

Panjer, Harry H. 2002. *Measurement of Risk, Solvency Requirements and Allocation of Capital Within Financial Conglomerates.* Department of Statistics and Actuarial Science, University of Waterloo, Canada. The paper was presented to the ICA/AFIR in March 2002, Cancun, Mexico.

Panjer, Harry. 2006. *Operational Risk, Modeling Analytics.* John Wiley & Sons, Hoboken, NJ, ISBN: 13-978-0-471-76089-4.

Parker, Gary. 1997. Stochastic analysis of the interaction between investment and insurance risks. *North American Actuarial Journal*, Vol. 1, No. 2, pp. 55–84 (incl. discussion).

Pedersen, Christian S. and Stephen E. Satchell. 1998. An extended family of financial risk measures. *Geneva Papers on Risk and Insurance Theory*, December, Vol. 23, pp. 89–117.

Pentikäinen, Teivo. 1952. On the net retention and solvency of insurance companies. *Skandinavisk Aktuarietidskrift*, Vol. 35, pp. 71–92.

Pentikäinen, Teivo. 1962. Reserves of motor-vehicle insurance in Finland, *ASTIN Bulletin*, Vol. 2, pp. 161–173.

Pentikäinen, Teivo. 1967. On the solvency of insurance companies. *ASTIN Bulletin*, Vol. IV, Part III, pp. 236–247.

Pentikäinen, Teivo, Editor. 1982. *Solvency of Insurers and Equalization Reserves. Volume I General Aspects.* Insurance Publ. Comp. Ltd, Helsinki, ISBN: 951-9174-20-6.

Pentikäinen, Teivo. 1984. *Aspects on the Solvency of Insurers.* 22nd International Congress of Actuaries, Sydney, pp. 61–73.

Pentikäinen, Teivo. 2004. Solvency. In Teugels, Jozef L. and Bjørn Sundt, Editors. Encyclopaedia of Actuarial Science. John Wiley & Sons, Chichester, ISBN: 0-470-84676-3.

Pentikäinen, Teivo, and Jukka Rantala. 1986. Run-off risk as a part of claims fluctuation. *ASTIN Bulletin*, Vol. 16, pp. 113–147.

Pentikäinen, Teivo, Heikki Bonsdorff, Martti Pesonen, Jukka Rantala, and Martti Ruohonen. 1989. *Insurance Solvency and Financial Strength*. Finnish Insurance Training and Publishing Company Ltd, Helsinki, ISBN: 951-9174-75-3.

Pesonen, Erkki. 1967. Magnitude control of technical reserves in Finland. *ASTIN Bulletin*, Vol. 4, pp. 248–251.

Pfeifer, Dietmar and Doreen Strassburger. 2008. Solvency II: Stability problems with the SCR aggregation formula. *Scandinavian Actuarial Journal*, No. 1, pp. 61–77.

Piesch, Walter. 1975. *Statistische Konzentrationsmasse*. J.C.B. Mohr (Paul Siebeck), Tübingen, ISBN: 3-16-337452-2.

Pool, Bill. 1990. *The Creation of the Internal Market in Insurance*. Commission of the European Communities, Brussels.

Pörn, Kurt. 1968. A study in risk theory and its application to the computation of the fluctuation reserve used in Finland. *Skandinavisk Aktuarietidskrift*, Vol. 51, pp. 1–25.

QIS3. 2007. QIS 3 Technical Specification. CEIOPS-DOC-01/07. Available at http://www.ceiops.eu/content/view/118/124/; see also CEIOPS (2007a).

QIS4. 2008. QIS 4 Technical Specifications. European Commission, MARKT/2505/08, March 31. Available at http://www.ceiops.eu/content/view/118/124

Ramlau-Hansen, Henrik, 1982. An application of credibility theory to solvency margins: Some comments on a paper by G.W. De Wit and W.M. Kastelijn. *ASTIN Bulletin*, Vol. 13, pp. 37–45.

Rantala, Jukka, Editor. 1982. *Solvency of Insurers and Equalization Reserves. Volume II Risk Theoretical Model*. Insurance Publ. Comp. Ltd, Helsinki, ISBN: 951-9174-20-6.

Redington, Frank M. 1952. Review of the principles of life-office valuations. *Journal of the Institute of Actuaries*, Vol. 78, pp. 286–340.

Redington, Frank M. 1982. The origin of immunization. *The Actuary, The Newsletter of the Society of Actuaries*, Vol. 16, No. 1, January.

Resti, Andrea. 2002. *The New Basel Capital Accord. Structure, Possible Changes and Micro- and Macroeconomic Effects*. CEPS Research Report No. 30, September.

Riksbank. 2006. Praktiska hänsyn vid beräkningen av diskonteringsräntor, Avdelningen för finansiell stabilitet. Sveriges Riksbank, 2006-11-15 (in Swedish). Available at www.riksbank.com

Robert Eric Beard. 1964. Some thoughts on solvency of insurance companies. Quarterly letter from Algemeene Reinsurance Companies, Jubilee Number Volume 1, July, pp. 79–108.

Robert Eric Beard, Teivo Pentikäinen, and Erkki Pesonen. 1984. *Risk Theory: The Stochastic Basis of Insurance*, 3rd edition. Chapman & Hall, Monographs on Statistics and Applied Probability 20, London/New York, ISBN: 0-412-24260-5.

Rockafellar, R.T, Stanislav Uryasev, and Michael Zabarankin. 2003. *Deviation Measures in Risk Analysis and Optimization*. Research Report 2002-7. Risk Management and Financial Engineering Lab/Center for Applied Optimization, University of Florida, Gainsville, June 8.

Rockafellar, R.T., Stanislav Uryasev, and Michael Zabarankin. 2006a. Generalized deviations in risk analysis. *Finance and Stochastics*, Vol. 10, pp. 51–74.

Rockafellar, R.T, Stanislav Uryasev, and Michael Zabarankin. 2006b. Optimality conditions in portfolio analysis with general deviation measures. *Mathematical Programming, Series B*, Vol. 108, pp. 515–540.

Rosenberg, Joshua V. and Til Schuermann. 2004. *A General Approach to Integrated Risk Management With Skewed, Fat-Tailed Risks*. Federal Reserve Bank of New York Staff Report No. 185, May.

RSAS. 2003. *Time-Series Econometrics: Cointegration and Autoregressive Conditional Heteroskedasticity*. The Royal Swedish Academy of Science, Advanced information on the Bank of Sweden Prize in Economic Sciences in Memory of Alfred Nobel, October 8. Available at www.kva.se.

Ruygt, Francis. 2006. *Cost of Capital Approach for Setting Risk Margins for Insurance Contract Liabilities.* Background Paper, Insurance Working Group, IASB, London, January (Agenda paper 5A).

Ryan, John P. 1984. *Application of Simulation Techniques to Solvency Testing for a Non-Life Office.* 22nd International Congress of Actuaries, Sydney, pp. 269–277.

Sachs, Rainer. 2007. *Reinsurance Credit Risk.* Report, Münchener Rück, October (a first version was available in January 29. Available at http://www.munichre.com. Also in Solvency II and Reinsurance. GDV, Berlin. 2007, pp. 53–67. Available at http://www.gdv.de/Downloads/Themen/sII_reinsurance.pdf).

Salzmann, Robert and Mario V. Wüthrich. 2009. *Cost-of-Capital Margin for a General Insurance Liability Runoff.* Working Paper, Department of Mathematics, ETH Zurich.

Sandström, Arne. 2005. *Solvency: Models, Assessment and Regulation.* Chapman and Hall, Boca Raton, ISBN: 1-58488-554-8.

Sandström, Arne. 2007a. Solvency II: Calibration for skewness. *Scandinavian Actuarial Journal*, No. 2, pp. 126–134.

Sandström, Arne. 2007b. Solvency—A historical review and some pragmatic solutions. *Journal of the Swiss Association of Actuaries*, Hefte 1, pp. 11–34.

Sandström, Arne. 2008. Solvency. In *Encyclopedia of Quantitative Risk Assessment*, Editors Edward Melnick and Brian Everitt, John Wiley & Sons, Chichester, ISBN: 978-0-470-03549-8.

Savelli, Nino and Gian Paolo Clemente. 2009. *Hierarchical Structures in the Aggregation of Premium Risk for Insurance Underwriting.* Paper presented at the 39th ASTIN Colloquium, Helsinki, June. Available at www.actuaries.org

Schlude, Herfried. 1979. The EEC solvency rules for non-life insurance companies. *Nordisk FörsäkringsTidskrift*, Årgång 59, No. 1, pp. 24–30.

Schrager, David. 2008. Replicating portfolios for insurance liabilities. *AENORM*, Magazine 59, April. Available at http://www.aenorm.eu

Schuermann, Til. 2004. What do we know about loss given default. In *Credit Risk Models and Management*, Editor David Shimko, Risk Books, London, ISBN: 1-904339-21-2.

Schuermann, Til. 2008. Credit migration matrices, In *Encyclopedia of Quantitative Risk Assessment*, Editors Edward Melnick and Brian Everitt, John Wiley & Sons, New York, ISBN: 978-0-470-03549-8.

Sharma, Paul. 2002. Prudential Supervision of Insurance Undertakings. Conference of Insurance Supervisory Services of the Member States of the European Union (now CEIOPS), December (The WG chaired by Paul Sharma, U.K.). Available at www.ceiops.org

Shaw, Richard A. 2007. *The Modelling of Reinsurance Credit Risk.* GIRO 2007 Working Party, Institute of Actuaries. Available at http://www.actuaries.org.uk

Sherris, Michael. 2004. Asset management. In Teugels, Jozef L. and Bjørn Sundt, Editors. 2004. *Encyclopaedia of Actuarial Science.* John Wiley & Sons, Chichester, ISBN: 0-470-84676-3.

Shimpi, Prakash. 2009. Financial Crisis: Time to Improve ERM. Towers Perrin. *Emphasis* 2009/1, pp. 10–13.

Shimpi, Prakash A. and Stephen P. Lowe. 2006. Risk management—ERM for insurers—From compliance to value. *Risk Management*, No. 8, July. Society of Actuaries. Available at http://www.soa.org

Shiu, Elias S.W. 1986. *A Generalization of Redington's Theory of Immunization.* Proceedings of the Twentieth Annual Actuarial Research Conference, University of Texas, Austin, November 21–23, 1985. *Actuarial Research Clearing House*, 1986.2 Issue.

Shiu, Elias S.W. 2004. Matching. In Teugels, Jozef L. and Bjørn Sundt, Editors. 2004. *Encyclopaedia of Actuarial Science.* John Wiley & Sons, Chichester, ISBN: 0-470-84676-3.

Shiu, Yung-Ming. 2006. *Dynamic Financial Analysis in Insurance.* Working Paper, Department of Finance, Tunghai University, Taiwan.

Selvaggi, Mariano. 2009. *Analysing Operational Losses in Insurance*. ABI Research Paper 16, Association of British Insurers. Available at http://www.abi.org.uk

Siegel, Sidney. 1956. *Nonparametric Statistics for the Behavioral Sciences*. McGraw-Hill Kogakusha, Ltd., Tokyo, Library of Congress Catalog Card Number: 56-8185.

Sigma. 1999. Alternative Risk Transfer (ART) for corporations: A passing fashion or risk management for the 21st century? Sigma. No. 2/1999, Swiss Re. Available at http://www.swissre.com

Sigma. 2000. Asset-liability management for insurers. Sigma. No. 6, Swiss Re. Available at http://www.swissre.com

Sigma. 2006. Solvency II: An integrated risk approach for European insurers. Sigma No. 4, Swiss Re. Available at http://www.swissre.com

Simester, Thomas T., Moderator. 1982. Immunized and dedicated portfolios for pension funds. *Record of Society of Actuaries*, Vol. 8, No. 4, pp. 1277–1297.

Skerman, R.S. 1984. *The Responsibility of the Actuary for the Adequacy of Life Insurance Reserves*. 22nd International Congress of Actuaries, Sydney, pp. 49–59.

Skinner, Peter. 2008. Draft Report on the amended proposal for a directive of the European Parliament and of the Council on the taking-up and pursuit of the business of Insurance and Reinsurance (recast). [COM(2008)0119-C6-0231/2007-2007/0143(COD)], Committee on Economic and Monetary Affairs. 2007/0143(COD). 3.13.2008, European Parliament.

Slee, D.J. 1984. *Solvency and Adequacy of Reserves for a Direct Writer of Worker's Compensation Insurance in Australia*. 22nd International Congress of Actuaries, Sydney, pp. 237–254.

Smith, Andrew D. 2009. Dependency and copulas. In Cruz, Marcelo, Editor. 2009. *The Solvency II Handbook. Developing ERM Frameworks in Insurance and Reinsurance Companies*. Risk Books, London, pp. 541–578. ISBN: 978-1-906348-19-9.

Smith, Andrew D. and T.J. Sheldon. 2004. Market consistent valuation of life assurance business. *British Actuarial Journal*, Vol. 10, No. 3, pp. 543–605.

SOA. 1996. *Dynamic Financial Condition Analysis Handbook*. Task Force, Record of the Society of Actuaries, Schaumburg, IL, Vol. 22, No. 3.

SOA. 1998. *Asset-Liability Management*. Professional Actuarial Specialty Guide, Society of Actuaries, August. Available at http://www.soa.org

SOA. 2006. *Enterprise Risk Management Specialty Guide*. Society of Actuaries, May. Available at http://www.soa.org

SOA. 2008. Financial Reporting for Insurance Contracts under Possible Future International Accounting Standards. Society of Actuaries Research Project. January 29. Prepared by PricewaterhouseCoopers. Available at http://www.soa.org

SST. 2004. The Risk Margin for the Swiss Solvency Test. Federal Office for Private Insurance, Bern, September 17. Available at http://www.finma.ch/archiv/bpv/e/index.html

SST. 2006a. The Swiss Experience with Market Consistent Technical Provisions—The Cost of Capital Approach. Federal Office for Private Insurance, Bern, February 24. Available at http://www.finma.ch/archiv/bpv/e/index.html

SST. 2006b. Technical document on the Swiss Solvency Test. Federal Office of Private Insurance. Version of October 2. Available at http://www.finma.ch/archiv/bpv/e/index.html

Standard & Poor's. 2006. *Annual 2005 Global Corporate Default Study and Rating Transitions*. Global Fixed Income Research, January. Available at http://www.standardandpoors.com

Stone, Steve. 2007. Understanding the Black-Scholes Equation. *Risks and Rewards. Investment Section*, No. 50, August. Available at www.soa.org

Strassburger, Doreen. 2006. *Risk management and solvency—Mathematical methods in theory and practice*. PhD Thesis, Carl von Ossietky Universität, Oldenburg.

Strommen, Stephen J. 2006. Setting the level of margins in a principle-based valuation using a cost-of-capital approach with exponential utility. *The Financial Reporter*, No. 65, June.

Strommen, Stephen J. 2008. *Risk Margins and their Role in an Accounting Framework*. Discussion Paper, May 5.

Sundt, Bjørn and Raluca Venic. 2009. *Recursions for Convolutions and Compound Distributions with Insurance Applications*. Springer Verlag, Heidelberg, ISBN: 978-3-540-92899-7, EAA Lecture Notes.

SwissRe. 1998. Proportional and non-proportional reinsurance. Swiss Re, 01/98, Available at http://www.swissre.com. Research and Publications.

Tang, Andrew and Emiliano A. Valdez. 2006. *Economic Capital and the Aggregation of Risks using Copulas*. Presented at the 28th International Congress of Actuaries, Paris. Available at www.actuaries.org

Teugels, Jozef L. and Bjørn Sundt, Editors. 2004. *Encyclopaedia of Actuarial Science*. John Wiley & Sons, Chichester, ISBN: 0-470-84676-3.

Tilley, J.A. 1980. The matching of assets and liabilities. *Transactions of the Society of Actuaries*, Vol. XXXII, pp. 263–304.

Tilley, J.A. 1988. The application of modern techniques to the investment of insurance and pension funds. *Transactions of the 23rd International Congress of Actuaries*, Helsinki, Vol. R, pp. 301–326.

Tobin, Patrick and Alan Brown. 2004. Estimation of liquidity risk in banking. *ANZIAM Journal*, Vol. 45, pp. 519–533. Available at http://anziamj.austms.org.au/V45/CTAC2003/Tobi/Tobi.pdf

Trehan, Bharat. 2005. Oil Price Shocks an Inflation, Economic Research and Data, FRBSF Economic Letter, 2005–28, October 28, The Federal Reserve Bank of San Francisco.

Trowbridge Report. 1979. Discussion of the preliminary report of the committee on valuation and related problems. *Record of Society of Actuaries*, Vol. 5, No. 1, pp. 241–284.

Tsanakas, Andreas. 2007. *Capital Allocation With Risk Measures*. 5th Actuarial and Financial Mathematics Day, February 9. Koninklijke Vlaamse Academie Van Belgie voor Wetenschappen en Kunsten, Brussels.

Valentine, James. 1875. A comparison of reserves brought out by the use of different data in the valuation of the liabilities of a life office. *Journal of the Institute of Actuaries*, Vol. 18, pp. 229–242.

Vanderhoof, Irvin T. 1972. The interest rate assumption and the maturity structure of the assets of a life insurance company. *Transactions of the Society of Actuaries*, Vol. XXIV, pp. 157–205.

Vanderhoof, Irvin T. 1973. Choice and justification of an interest rate. *Transactions of the Society of Actuaries*, Vol. XXV, pp. 417–458.

Vasicek, Oldrich. 1987. *Probability of Loss on Loan Portfolio*. Technical Paper, KMV Corporation. Available at http://www.moodyskmv.com

Vasicek, Oldrich. 1991. *Limiting Loan Loss Probability Distribution*. Technical Paper, KMV Corporation. Available at http://www.moodyskmv.com

Vasicek, Oldrich. 2002. *The Distribution of Loan Portfolio Value*. Technical Paper. Available at http://www.moodyskmv.com. Published in *Risk*, Vol. 15, No. 12, December, pp. 160–162. Also published in *Theory and Practice of Credit Risk Modelling*. Editor Alexander Lipton, Risk Books, 2008, pp. 115–124, ISBN: 978-1-904339-64-9.

Venter, Gary G. 2002. Tail of copulas. *Proceedings of the CAS*, Vol. LXXXIX, pp. 68–113.

Venter, Gary G. 2003. Quantifying Correlated Reinsurance Exposures with Copulas. CAS Forum, Spring, pp. 215–229.

Venter, Gary G. 2007. Strategic planning models. *Risk Management*, No. 10, March. Society of Actuaries. Available at http://www.soa.org

Venter, Gary, John A. Major, and Rodney E. Kreps. 2006. Marginal decomposition of risk measures. *ASTIN Bulletin*, Vol. 36, No. 2, pp. 375–413.

Venter, Gary, Jack Barnett, Rodney Kreps, and John Major. 2007. Multivariate copulas for financial modeling. *Variance*, Vol. 1, No. 1, pp. 103–119.

de Vylder, F. 1978. A practical solution to the problem of ultimate ruin. *Scandinavian Actuarial Journal*, pp. 114–119.

Wang, Shuan. 1998. Aggregation of correlated risk portfolios: Models and Algorithms. *Proceeding Casualty Actuarial Society*, Vol. LXXXV, pp. 848–939.

Wang, Shuan. 2000. A class of distortion operators for pricing financial and insurance risks. *Journal of Risk and Insurance*, Vol. 67, pp. 15–36.

Wang, Shuan. 2002a. A Set of New Methods and Tools for Enterprise Risk Capital Management and Portfolio Optimization, CAS Risk Capital Management Call Paper Program, CPP-26, Summer 2002, pp. 43–78.

Wang, Shuan. 2002b. A universal framework for pricing financial and insurance risks. *ASTIN Bulletin*, Vol. 32, pp. 213–234.

Wang, Shuan. 2002c. *A Risk Measure that goes Beyond Coherence*. Working Paper, SCOR Reinsurance Co. Available at www.gloriamundi.org/.

Wang, Shuan. 2004. Where is ERM heading? *Risk Management*, No. 1, March. Society of Actuaries. Available at http://www.soa.org

Wang, Shuan. 2006. Latest research enterprise risk management for property/casualty insurance companies. *Risk Management*, No. 9, December, CAS/SOA/CIA Risk Management Section. Available at http://www.soa.org (Reprint from Actuarial Review, Vol. 33, No. 4, 2006).

Wang, Shuan and Jan Dhaene. 1998. Comonotonicity, correlation order and premium principles. *Insurance: Mathematics and Economics*, Vol. 22, No. 3, pp. 235–242.

Wang, Shuan and Robert Faber. 2006. *Enterprise Risk management for Property-Casualty Insurance Companies*. ERM-II, CAS and CAS/SOA Risk management Section Joint report, August 1. Available at http://www.ermii.org

Wang, Shuan, Virginia R. Young, and Harry Panjer. 1997. Axiomatic characterization of insurance prices. *Insurance: Mathematics and Economics*, Vol. 21, pp. 173–183.

Ward, G.C. 1968. The use of model offices in Australia. *Transactions of the 18th International Congress of Actuaries*, Munich, Vol. 2, pp. 1065–1080.

Ward, Lisa S. and David H. Lee. 2002. Practical Application of the Risk-Adjusted Return on Capital Framework. CAS Forum Summer 2002, Dynamic Financial Analysis Discussion Paper.

Watson Wyatt. 2004a. *Calibration of the Enhanced Capital Requirement for With-Profit Life Insurers*. Report to FSA. June. Available at http://www.fsa.gov.uk

Watson Wyatt. 2004b. Untitled report to FSA regarding survey results. Available at http://www.fsa.gov.uk

Widder, David V. 1965. *Advanced Calculus*, 2nd edition. Prentice-Hall Mathematics Series, Englewood Cliffs, NJ, fifth printing.

Wilde, Tom. 1997. *CreditRisk+: A Credit Risk Management Framework*. October. Credit Suisse First Boston. Available at http://www.csfb.com

Wilkie, A. David. 1984. *Steps Towards a Comprehensive Stochastic Investment Model*. Occasional Actuarial Research Discussion Paper, No. 36, Institute of Actuaries, London.

Wilkie, A. David. 1986a. A stochastic investment model for actuarial use. *Transactions of the Faculty of Actuaries*, Vol. 39, pp. 341–403.

Wilkie, A. David. 1986b. Some applications of stochastic investment models. *Journal of the Institute of Actuaries Students' Society*, Vol. 29, pp. 25–51.

Wilkie, A. David. 1995. More on a stochastic investment model for actuarial use. *British Actuarial Journal*, Vol. 1, pp. 777–964.

Willemse, W.J. and H. Wolthuis. 2005. *Risk-Based Solvency Norms and their Validity*. Report (September 1, 2005). Presented at the 28th International Congress of Actuaries, Paris, 2006.

Willets, R.C., A.P. Gallop, P.A. Leandro, J.L.C. Lu, A.S. Macdonald, K.A. Miller, S.J. Richards, N. Robjohns, J.P. Ryan, and H.R. Waters. 2004. Longevity in the 21st century. *British Actuarial Journal*, Vol. 10, Part IV, pp. 685–898.

Wirch, J.L. 1997. *Value-at-Risk for Risk Portfolios*. Actuarial Research Conference, Calgary, Alberta, Canada, August.

Wirch, J.L. and M.R. Hardy. 1999. A synthesis of risk measures for capital adequacy. *Insurance: Mathematics and Economics*, Vol. 25, pp. 337–347.

de Wit, G.W. and W.M. Kastelijn. 1980. The solvency margin in non-life insurance companies. *ASTIN Bulletin*, Vol. 11, pp. 136–144.

Witcraft, Susan. 2004. Risk statistics. In Teugels, Jozef L. and Bjørn Sundt, Editors. 2004. *Encyclopaedia of Actuarial Science*. John Wiley & Sons, Chichester, ISBN: 0-470-84676-3.

Wolthuis, Henk and Mark J. Goovaerts, Editors. 1997. *Reserving and Solvency in Insurance in the EC*, 2nd edition. Institute of Actuarial Science and Econometrics (IAE), Universiteit van Amsterdam, Amsterdam, ISBN: 90-802117-6-1.

Wright, Peter. 2006. *A View of the IASB's Work on Accounting for Insurance Contracts and Financial Instruments*. Report presented at the 28th International Congress of Actuaries, Paris, 2006.

Wüthrich, Mario V. and Hans Bühlmann. 2009. *The One-Year Runoff Uncertainty for Discounted Claims Reserves*. Research Report, Department of Mathematics, ETH Zurich, Version from May 27. To appear in Giornale dell Istituto Italiano degli Attuari.

Wüthrich, Mario V., Hans Bühlmann, and Hansjörg Furrer. 2007. *Market-Consistent Actuarial Valuation*. Springer Verlag, Heidelberg, ISBN: 978-3-540-73642-4, EAA Lecture Notes.

Wüthrich, Mario V. and Michael Merz. 2008. *Stochastic Claims Reserving methods in Insurance*. Wiley, Chichester, ISBN: 978-0-470-72346-3.

Wüthrich, Mario V., Michael Merz, and Natalia Lysenko. 2009. Uncertainty of the claims development result in the chain ladder method. *Scandinavian Actuarial Journal*, No. 1, pp. 63–84.

Zeng, Bin and Jing Zhang. 2001. *An Empirical Assessment of Asset Correlation Models*, November 4, Moody's KMV. Available at http://www.moodyskmv.com

Index

M

P

Q

R

S

T

U